The Illustrated Encyclopedia of
Astronomy and Space
Revised Edition

Editor Ian Ridpath

Thomas Y. Crowell, Publishers
New York
Established 1834

PHOTOCREDITS: Aerofilms; Aldus Books; Anglo-
Australian Observatory; Australian News and Information
Bureau; Big Bear Solar Observatory; British Insulated
Callender's Cables Ltd; Camera Press; A. C. Cooper Ltd;
Heather Couper; Cornell University; Dominion
Astrophysical Observatory; European Southern Observatory;
Hale Observatories; Jet Propulsion Laboratory; Kitt Peak
National Observatory; Lick Observatory; D. F. Malin; Mcdonnell
Douglas; Mount Wilson and Palomar Observatories;
Museum of the History of Science, Oxford; NASA;
National Film Archive; National Radio Astronomy
Observatory; National Research Council; Novosti Press
Agency; P. Popper Ltd; Rainforth Collection; I. Ridpath;
Rockwell International; Royal Astronomical Society; Royal
Observatory, Edinburgh; R. Scagell; The Science Museum,
London; Science Research Council; U. K. Schmidt Telescope
Project; Westerbork Radio Observatory; Yerkes
Observatory.

THE ILLUSTRATED ENCYCLOPEDIA OF
ASTRONOMY AND SPACE.

LC no. 79–7098

ISBN 0–690–01838–X

   80 81 82 83 10 9 8 7 6 5 4 3 2

Printed in the United States of America

The Illustrated Encyclopedia of Astronomy and Space

Editor **IAN RIDPATH**

Major contributors

HEATHER A. COUPER
Greenwich Planetarium

Dr. PAUL C. W. DAVIES
Department of Mathematics, King's College,
University of London

Dr. JOHN GRIBBIN
Science Policy Research Unit, University of Sussex;
Physics consultant, *New Scientist*

NIGEL HENBEST
Science writer

Dr. KEITH HINDLEY
Director, Meteor Section,
British Astronomical Association

JAMES MUIRDEN
Manufacturer of astronomical instruments
and science writer

Dr. PAUL MURDIN
Royal Greenwich Observatory;
Anglo-Australian Observatory

IAN RIDPATH
Science author and editor, astronomy and
astronautics contributor to *New Scientist, Astronomy,
The Observer*

COLIN A. RONAN
Editor, *Journal of the British
Astronomical Association*

Dr. JOHN ROSS
Astrophysicist

Professor IAN W. ROXBURGH
Department of Mathematics, Queen Mary College,
University of London

ROBIN S. SCAGELL
Science writer

Foreword

Astronomy, the oldest of the sciences, has changed dramatically during the last few decades. At the turn of the century an encyclopedia of astronomy would have been mainly concerned with the dynamics of the solar system together with a description of the sky as seen by the naked eye. Telescopes had very little auxiliary instrumentation: they were used mainly for compiling catalogues from direct visual observation. Spectroscopy was in its infancy, and the only observations outside the visible part of the electromagnetic spectrum were rudimentary demonstrations of infrared radiation from the Sun.

A modern encyclopedia must cover a very different field. Observations are now made over more than 50 octaves of the electromagnetic spectrum, from long radio waves to high energy gamma rays. In contrast, the visible spectrum, although it remains vitally important, covers only one octave. Other means of studying the Universe — cosmic rays, neutrinos, gravitational waves — all play their part. But the most dramatic, and the most readily comprehensible revolution in technique has been the direct exploration of the solar system through satellites and space probes, the supreme achievement being man's visit to the Moon.

Most astronomers are not, however, much concerned about the solar system. Our horizons have been widened to comprehend the Galaxy, of which our Sun is a member along with a hundred billion other stars, and other galaxies stretching away to the limits of measurable distances of time and space. This is the real revolution in astronomy: we are now able to study the Universe as a whole. Astrophysics now merges with cosmology in describing not only what stars and galaxies are made of but also, in some detail, how they were formed and are still forming, and what will be their ultimate fate.

This new astronomy burst into life with the development of radio astronomy in the early 1950s. The radio spectrum, from wavelengths of 10 meters down to 1 centimeter, is the only part of the electromagnetic spectrum, apart from the visible and near-visible, that reaches the surface of the Earth almost unaffected by the atmosphere. The discovery that radio waves carried new and exciting information about the Universe has led to the construction of radio telescopes of many different types, some of immediately impressive appearance such as the famous 250-foot telescope at Jodrell Bank, others of extraordinary technical ingenuity such as the Very Long Baseline Interferometers (VLBI) that now combine receiving stations spread over several different continents for the precise and delicate measurements of the sizes and shapes of quasars. The discoveries of radio astronomy have changed our view of the Universe, and added new words to our language: quasars, pulsars, big bang, background radiation. X-ray astronomy followed on with its own revelations of new and unforeseen celestial objects: X-ray binaries, slow pulsars, possibly — but not certainly — black holes. The techniques of X-ray astronomy are, however, entirely different and very demanding. The atmosphere is a complete block to X-rays, and observations can be made only from high-altitude balloons, rockets, or satellites. The same applies to gamma-ray astronomy, and to the far infrared and ultraviolet.

A wide choice of techniques is now presented to astronomers. Preliminary survey of the previously unexplored parts of the spectrum is now complete and the choice must be based on the returns of information that can be expected from new and improved telescopes. Most modern astronomers now favor putting their resources of money and effort once again into large optical telescopes, for two main reasons. First and foremost, astronomy is now seen as a whole, in which the same objects must be studied by all possible different techniques. In this the techniques of optical astronomy have fallen behind, and new telescopes and instruments are badly needed. Second, it turns out that the visible and near-visible spectrum contains by far the greatest concentration of information, particularly in the wealth of spectral lines from atoms and molecules which are the basic constituents of the Universe.

The new astronomies, radio, X-ray, gamma ray, ultraviolet and infrared, have thus paradoxically brought about a resurgence in optical astronomy. New major observatories have recently been constructed at exceptionally good sites such as Kitt Peak in Arizona, Cerro Tololo and La Silla in Chile, and Siding Spring in Australia, and a major European observatory is under construction on the island of La Palma in the Canary Islands. New techniques of photon-counting detectors and television systems give the new telescopes an unprecedented sensitivity. Nevertheless, the atmosphere is still a limitation. Light traveling through the atmosphere suffers a random refraction, which spoils star images. Just beyond the visible spectrum, and particularly in the ultraviolet, there are regions rich in spectral lines but which are totally absorbed by the atmosphere. This barrier can only be overcome by the Space Telescope.

The excitement of the new discoveries, such as pulsars and quasars, and the glamor of the new techniques such as telescopes in space and radio interferometers, should not mislead a would-be astronomer into a desperate anxiety to join a major project lest he be condemned to mediocrity. Astronomy is now an important part of science education, and research groups are flourishing at observatories and universities throughout the world. Most of the major telescopes and instruments are available for use by any serious research worker, and international scientific exchange of all kinds is commonplace in astronomy. It hardly needs to be said that there are other simpler ways of enjoying astronomy: it is a never-failing pleasure just to look at the sky on any clear night, tracing out the constellations and identifying the planets. At whatever level one enjoys astronomy, this encyclopedia will provide a true guide to the old, familiar Universe, and to the new Universe which has so recently been revealed to us.

Professor F. Graham Smith, FRS
Director of the Royal Greenwich
Observatory

A note on units and symbols used in this Encyclopedia

The U.S. system of designating large numbers is used. Thus the U.S. billion equals the British thousand million, and the U.S. trillion equals the British billion. To save writing endless zeros in large numbers, scientists often use a power of ten notation. In this the figure 10 is given with a superscript which shows the total number of zeros to be written. Thus, 10^2 is 100, 10^6 a million (1,000,000), 10^9 a billion (1,000,000,000), and 10^{12} a trillion (one million million). This notation is sometimes used in the book. All units are given in standard and metric measure; the tons are metric (1,000 kg or 2,205 pounds). To the accuracy with which most astronomical figures are known, metric tonnes can be considered equivalent to American long tons or British tons.

When labeling bright stars the tradition is to use Greek letters. For convenience, we have spelled out the names of these letters. The complete Greek alphabet is given below.

α, A	alpha	ι, I	iota	ρ, P	rho
β, B	beta	κ, K	kappa	σ, Σ	sigma
γ, Γ	gamma	λ, Λ	lambda	τ, T	tau
δ, Δ	delta	μ, M	mu	υ, Y	upsilon
ε, E	epsilon	ν, N	nu	ϕ, Φ	phi
ζ, Z	zeta	ξ, Ξ	xi	χ, X	chi
η, H	eta	o, O	omicron	ψ, Ψ	psi
$\vartheta, \theta, \Theta$	theta	π, Π	pi	ω, Ω	omega

The 150-foot diameter radio telescope of the Algonquin Radio Observatory has a paraboloidal surface accurate to a few millimeters, and can therefore focus radio waves as short as 2.8 centimeters on to the receiver mounted in front. It is used extensively to monitor the changing emission from quasars.

A

α

The symbol for the astronomical coordinate known as
RIGHT ASCENSION.

Å

The symbol for the unit of measure of the wavelength
of light called an ANGSTROM.

aberration

Any of a number of faults that mar the image in an
optical instrument. In SPHERICAL ABERRATION, light
from different parts of a lens or mirror is brought to
different foci, causing a blurred image. Lenses suffer
also from CHROMATIC ABERRATION, in which light of
different colors is bent by differing amounts, as in a
prism, producing color fringes around the object. In
COMA, the images are progressively elongated toward
the edge of the telescope field. ASTIGMATISM is the
failure of an optical system to bring light rays from
different planes, such as horizontal and vertical, to a
common focus, causing elliptical images. A focal plane
that is not flat produces so-called *curvature of field* and
arises when different parts of the image are formed at
different distances from the lens or mirror. Where the
telescope's magnification varies across the field, the
image is distorted so that straight lines look curved.
So-called *barrel distortion* occurs where the telescope's
magnification decreases toward the edge of the field,
bowing straight lines outwards. Where magnification
increases toward the edge of field, straight lines are
bowed inwards in what is termed *pincushion distortion*.

aberration of starlight

An effect caused by the Earth's motion in orbit which
produces slight changes in the apparent positions of
stars. Aberration of starlight was discovered by James
BRADLEY in 1728, after careful observations showed a
displacement in position of the star Gamma Draconis
during the year. Bradley realized that the effect, which
amounted to only about 20 seconds of arc, was caused
by the motion of the Earth across the path of light rays
from the star. This made the light rays appear to be
coming in at a slant, in the same way that vertically
falling raindrops seem to be moving at an angle as seen
from a speeding vehicle. This effect gives a slightly false
position for all stars, which must be corrected for in
observations. The aberration of starlight was the first
direct observational proof of the Earth's movement
around the Sun.

absolute magnitude

A measurement used in comparing the total light
output of stars. The absolute magnitude of a star is the
brightness it would exhibit if it were at a standard
distance of 10 parsecs (32.6 light-years). Absolute
magnitude can usually be calculated from knowledge
of the star's nature. By comparing its calculated
absolute magnitude with the brightness with which it
actually appears, astronomers can determine how far
away it must be. (See APPARENT MAGNITUDE; DISTANCE
MODULUS; MAGNITUDE.)

absolute zero

The coldest temperature possible. At absolute zero all
thermal motion of molecules ceases because all heat

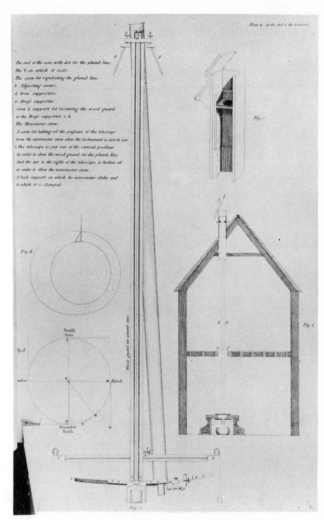

The zenith telescope with which James Bradley discovered
the aberration of light. A weight held the telescope
firmly against a micrometer screw on the scale at the
bottom of the telescope, enabling accurate position
measurements of Gamma Draconis and other stars to be
made.

has been removed; the object is therefore totally cold.
Absolute zero is equal to −273.16°C or −459.67°F.

absorption lines

Wavelengths in an object's spectrum which have been
absorbed by cooler gas. The wavelengths of the
absorption lines depend on the elements that cause
them (see SPECTROSCOPY). The FRAUNHOFER LINES in
the Sun's visual spectrum are caused by absorption of
sunlight by the cooler gases in the Sun's outer layers.
All stars show such dark absorption lines in their
spectra, which allow astronomers to analyze their
composition. Absorption lines are also caused by
clouds of gas between the stars. Absorption can occur
at any wavelength, including long wavelengths which
can be observed only by radio astronomers.

achondrite

A stony meteorite lacking the inclusions called
chondrules common in most other stony meteorites
(see CHONDRITE). Achondrites are similar in
composition to terrestrial basalt rocks.

achromatic

Term given to lenses that are specially corrected to prevent the appearance of color fringes around the image; this fault, common in cheap optics, is called CHROMATIC ABERRATION. Achromatic lenses are made from more than one piece of glass, each piece canceling out the chromatic effect of the other. The first achromatic lenses were made in 1733 by the English amateur optician Chester Moor Hall (1703–1771), after a study of the human eye had convinced him that achromatic lenses were possible in spite of the well-known prism experiments of Isaac Newton, which had seemed to show that refraction of light must inevitably cause rainbow-colored images. However, Hall never published his discovery, and the achromatic lens is usually credited to John DOLLOND.

Adams, John Couch (1819–1892)

British mathematician and astronomer, best known for his calculations that predicted the existence of the planet Neptune. Adams graduated with highest honors from Cambridge University in 1843, having already begun to investigate irregularities that were being observed in the motion of Uranus, the outermost planet of the solar system known at that time. Adams calculated that these irregularities could be caused by the gravitational effect of a hitherto unknown planet, and predicted where it was likely to be found. In 1845 he presented his calculations to the astronomer royal, Sir George Airy, who took no action. Only after the French mathematician Urbain LEVERRIER had published similar results was a search ordered. By then, however, the German astronomer Johann Galle was hot on the trail at the Berlin Observatory, and on September 23, 1846, he found Neptune close to the position calculated by Adams and Leverrier. Although the honors at first went to the Frenchman, Adams eventually won recognition for having been first to arrive at the solution.

Adams, Walter Sydney (1876–1956)

American astronomer, from 1923 to 1946 director of the Mount Wilson Observatory, known for his important spectroscopic studies of stars and planets. In 1914 he and Arnold Kohlschütter found that the nature of the dark lines in a star's spectrum shows whether it is a dwarf or giant star, thus allowing its ABSOLUTE MAGNITUDE to be deduced. Comparing a star's absolute magnitude with its observed brightness gives its distance (see DISTANCE MODULUS). This valuable distance-measuring technique developed by Adams is termed the *method of spectroscopic parallaxes*. In 1915 Adams examined the small, faint companion star of Sirius and found from its spectrum that its surface is white-hot; this was the first so-called WHITE DWARF to be recognized. In 1925 Adams found that this star's light shows a slight red shift, known as the Einstein shift, caused by its high gravity. Later, in 1932, he showed that the atmosphere of Venus consists of carbon dioxide; he worked with Theodore Dunham, Jr. (b. 1897), who himself made important spectroscopic studies of planetary atmospheres.

aerolite

Technical name for a stony meteorite, the most common type of meteorite that falls to Earth (see CHONDRITE).

Agena

An upper stage used on American launch vehicles, introduced in 1959. The Agena's engine, of 16,000-lb. (7,260-kg) thrust, can be stopped and started at will. Agenas were used with Atlas and Thor first stages to launch probes toward the Moon and planets, as well as to put into orbit numerous Earth satellites. The Agena is 5 feet (1.5 m) in diameter and approximately 23 feet (7 m) long. Agena stages were used as orbiting rendezvous targets during the Gemini series of manned space flights.

Airy, Sir George Biddell (1801–1892)

English astronomer royal from 1835 to 1881; previously director of the Cambridge University Observatory, where he set up the 12-inch (30-cm) Great Northumberland refractor later used in the search for Neptune. Airy totally reorganized the Royal Greenwich Observatory, installing a new series of instruments he himself had designed. Most famous of these was the transit telescope used to determine time from the passage of stars across the meridian. The position of this instrument defines 0° longitude, the Greenwich meridian, established by international agreement in 1884. Airy supervised the analysis and publication of all lunar and planetary observations made at Greenwich between 1750 and 1830, which astronomers still use, and which allowed John Couch ADAMS to predict the existence of Neptune. Airy failed, however, to act promptly on Adams' calculations. The British government used Airy as a consultant on numerous scientific and technical matters.

Aitken, Robert Grant (1864–1951)

American astronomer, discoverer of over 3,000 double stars. In 1932 he published a major listing of double stars, the *New General Catalogue of Double Stars within 120° of the North Pole*, containing measurements of 17,180 double stars made by him and William Joseph Hussey (1862–1926) with the 36-inch (91-cm) telescope at Lick Observatory. This superseded the *General Catalogue of Double Stars*, published in 1906 by the American double-star observer Sherburne Wesley Burnham (1838–1921), who had discovered a total of 1,274 pairs. From 1930 to 1935 Aitken was director of the Lick Observatory in California.

Albategnius (c.858–929)

The greatest Arab astronomer, also known as al-Battani. Using the most accurate astronomical instruments of his day, Albategnius redetermined the length of the year, timed the occurrence of the spring equinox to within a few hours, improved the value for precession, and accurately measured the tilt of the Earth's axis relative to its orbit. His tables of lunar, solar, and planetary positions were better than those in Ptolemy's *Almagest*. Albategnius also studied the changing apparent diameter of the Sun during the year, caused by the eccentricity of the Earth's orbit. He found that the point where the Sun's apparent diameter is smallest (and thus the Earth–Sun distance greatest) had moved from its position at the time of Ptolemy. Modern astronomy has since confirmed that the Earth's farthest point from the Sun does indeed move.

albedo

A measure of the proportion of light reflected by a

non-shiny surface. Dark objects have a lower albedo than light-colored ones. An object with an albedo of 1 would reflect all the light that hits it, making it brilliant white; an object with zero albedo would be totally black. The albedo of a planet or satellite is therefore a guide to the nature of its surface.

Aldebaran
The brightest star in the constellation Taurus, the bull, of magnitude 0.86. Aldebaran, also called Alpha Tauri, is a red giant star, held to represent the bull's glinting eye. It is 64 light-years away, and about 36 times the diameter of the Sun, giving out as much light as over 100 Suns.

Aldrin, Edwin Eugene (b. 1930)
Lunar module pilot of the Apollo 11 crew, which made the first lunar landing on July 20, 1969. Aldrin, a qualified engineer, was selected as an astronaut in 1963. He first flew in space on the Gemini 12 mission in November 1966, during which he made a record space walk of over two hours. On the Apollo 11 flight, Aldrin joined crew commander Neil Armstrong on the lunar surface for a walk of some two hours. Aldrin practiced various ways of moving about under the low lunar gravity, and helped set up experiments and collect rocks. Like the other crew members of Apollo 11, Aldrin left the astronaut corps after his return to Earth.

Alfvén, Hannes Olof Gösta (b. 1908)
Swedish physicist who received the Nobel Prize for physics in 1970 for his work on ionized gases and their interaction with magnetic fields. This research has been vital in attempts to produce controlled nuclear fusion, since powerful magnetic fields must be used to bottle up the hot gases in which the fusion reactions take place. It has also contributed to our understanding of the origin of the solar system, by showing that the fast rotation of the young Sun could have been transferred by magnetic fields to the tenuous gas cloud around it from which the planets formed, explaining why most of the spin, or angular momentum, of the solar system is present today in the planets rather than in the Sun. This work has also aided understanding of the outer regions of the Earth's magnetic field called the MAGNETOSPHERE; in 1939 Alfvén published a theory that linked solar storms with the activity of particles in the Earth's magnetosphere that cause aurorae. Alfvén's work ranges over a wide field, from studies of asteroid groupings which suggest that the asteroid belt is a failed attempt at forming a small planet, to the field of cosmology, in which he has strongly supported the view that equal amounts of matter and antimatter must exist in the Universe (see ANTIMATTER COSMOLOGY).

Algol
Famous star in the constellation Perseus, also called Beta Persei. Algol appears to vary in brightness between magnitudes 2.2 and 3.5 every 2.87 days. The variations are in fact caused by a fainter companion that periodically eclipses the main star of Algol, and the system was the first such ECLIPSING BINARY to be discovered. The English amateur John Goodricke suggested the correct explanation for Algol's behavior in 1782. Later astronomers discovered that Algol is actually a triple star, although the third companion does not take part in the eclipses. In 1971 Algol was identified as a radio source; some of the radio emission is believed to be caused by transfer of gas between the two main stars. Algol is 82 light-years away.

Algonquin Radio Observatory
Site of the largest radio telescope in Canada, 150 feet (46 m) in diameter, opened in 1966. The observatory, at Algonquin Park, Ontario, is operated by the National Research Council of Canada. The 150-foot telescope works at wavelengths down to 3 centimeters. It has been used to pioneer the technique known as very long baseline interferometry (VLBI), by comparing signals it receives with those received by an 84-foot (26-m) telescope at the associated observatory at Penticton, British Columbia. The Algonquin dish has also been linked with the radio telescope at Parkes, Australia, giving a baseline almost equal to the diameter of the Earth, for distinguishing small details in objects such as quasars.

Alouette satellites
Two Canadian scientific satellites, launched by the United States, designed to study the Earth's ionosphere. Successor to Alouette is the ISIS (International Satellites for Ionospheric Studies) series, a joint Canadian–U.S. program.

Satellite	Launch date
Alouette 1	September 28, 1962
Alouette 2	November 28, 1965 (launched with Explorer 31, which made supporting studies)
ISIS 1	January 30, 1969
ISIS 2	April 1, 1971

Alpha Centauri
A triple-star system (also called Rigil Kent), containing the nearest stars to our Sun, 4.3 light-years away. To the naked eye, Alpha Centauri appears as one star, the third-brightest in the entire sky, of magnitude −0.27. A small telescope shows two stars, of magnitudes 0 and 1.4; it is their combined light that makes Alpha Centauri appear so bright. These stars orbit each other once every 80 years. The main star is very similar to our Sun; its mass is 1.1 times the Sun's, and its radius 1.23 times that of the Sun. The second star is slightly smaller and cooler. Bigger telescopes reveal a magnitude 10.7 red dwarf star in the Alpha Centauri system. This third star is called PROXIMA CENTAURI, because in part of its orbit around the other two it comes closer to us than either of the main stars. The name Alpha Centauri when used without qualification usually refers to all three stars. The Alpha Centauri system is approaching us at about 15.5 miles (25 km) per second. In about 28,000 years it will be 3.1 light-years away, and 0.7 magnitudes brighter.

Altair
Brightest star in the constellation Aquila, also called Alpha Aquilae. It appears of magnitude 0.77. Altair is far hotter than the Sun, and is several times more massive. It is 16 light-years away, and 1.65 times the Sun's diameter.

altazimuth
The simplest form of mounting for telescopes. As the name implies the instrument is free to move independently in altitude (up and down) and in azimuth (left to right). Continual adjustments must thus be made in both axes to follow a star across the sky. EQUATORIAL MOUNTING allows a telescope to follow an object in one movement.

altitude
The angle between a celestial object and the horizon, measured at right angles to the horizon.

Ambartsumian, Viktor Amazaspovich (b. 1908)
Soviet (Armenian) astrophysicist who in 1947 discovered the existence of STELLAR ASSOCIATIONS, vast fields of hot, young stars in our Galaxy, which have apparently been born together. Their existence showed that large-scale star formation still continues in our Galaxy's spiral arms. In 1955 Ambartsumian suggested that the radio-emitting galaxies then being discovered had suffered explosions at their centers, and he proposed that galaxies grow by ejecting material; subsequent observations apparently showing just such a process in operation have helped substantiate his view (see RADIO GALAXY). In 1946 Ambartsumian founded the BYURAKAN ASTROPHYSICAL OBSERVATORY in Soviet Armenia.

Ames Research Center
NASA scientific and engineering establishment, adjoining the U.S. Naval Air Station at Moffett Field, California. It conducts studies on the possibility of life elsewhere in space, as well as on the effects of spaceflight on humans. Aeronautical studies at Ames include research on reentry into the Earth's atmosphere of space capsules and the winged SPACE SHUTTLE. The Ames Research Center has directed the PIONEER series of space probes. The Center was set up in 1940 by NASA's forerunner, the National Advisory Committee for Aeronautics (NACA), and is named for former NACA president Joseph Sweetman Ames (1864–1943).

Anaxagoras (c.500–c.428 B.C.)
Greek philosopher, who maintained that the Earth and the objects in the sky were made of essentially the same substance. He believed that the Sun was a large, red-hot stone, and that the Moon was a dark, solid body like the Earth, illuminated only by light from the Sun. This led him to the true explanation of eclipses: that they are caused by the blocking off of light from the Sun.

Anaximander (c.610–c.546 B.C.)
Greek philosopher and astronomer, a pupil of THALES and sometimes called the father of astronomy. Anaximander put forward the first known cosmological theory, in which he described the Earth as a cold, hard body surrounded by luminous heavenly objects like wheels of fire in space. Realizing that the sky rotates around the pole star, he visualized the heavens as a spinning sphere—thus setting an unfortunate precedent, for astronomers until the time of Kepler assumed that there were real spheres in the sky. Anaximander could see that the Earth's surface was curved, but imagined our planet as a squat cylinder, floating in space at the center of the Universe.

Anders, William Alan (b. 1933)
American astronaut who flew in Apollo 8, the first manned spacecraft to orbit the Moon. Anders photographed the lunar surface during Apollo 8's 10 orbits, showing that the selected landing sites would be safe for later Apollo missions to touch down on, and improving astronomers' knowledge of lunar features. Anders, a qualified nuclear engineer, was selected as an astronaut in 1963.

Andromeda
Constellation in the northern hemisphere of the sky, named for a princess of Greek mythology; it is best placed for viewing during the northern hemisphere autumn. Andromeda can be easily located, lying next to the famous square of Pegasus; one corner of the square is actually the star Alpha Andromedae. The planetary nebula NGC 7662 in Andromeda is 5,000 light-years distant. At the heart of the constellation lies its most famous feature—the ANDROMEDA GALAXY.

Andromeda galaxy
The most distant object in space visible to the naked eye, also known to astronomers by its catalog numbers of M31 and NGC 224. It is a separate spiral galaxy in space, like our own Milky Way but apparently twice the diameter. It is sometimes incorrectly called the Andromeda nebula. The galaxy appears as a fuzzy oval patch to the naked eye, in the center of the constellation Andromeda, but large telescopes reveal that it is composed of individual stars; one estimate puts their number at 300 billion. Observations of CEPHEID VARIABLE stars in the Andromeda galaxy have shown that it is about 2.2 million light-years away, making it the nearest major galaxy to our own. It shows, like our own Galaxy, areas of dark and light gas, clusters of stars, and exploding stars. The Andromeda galaxy has two small bright companions, similar to the MAGELLANIC CLOUDS that accompany our own Galaxy. Recent investigations have found at least three other midget companions to the Andromeda spiral. The Andromeda galaxy, along with our own Galaxy and several others, is part of what is termed the LOCAL GROUP.

Anglo-Australian Observatory
An astronomical observatory at Siding Spring, New South Wales, at which the UK and Australian governments jointly operate the 153-inch (3.9-m) Anglo-Australian telescope. At the same site, the UK operates a 42/78-inch (122/183-cm) Schmidt telescope. See also SIDING SPRING OBSERVATORY.

angstrom
Unit of measure of the wavelength of light, equal to one ten-billionth (10^{-10}) of a meter, symbol Å. It is named for the Swedish physicist Anders Jonas Ångström (1814–1874), a pioneer of spectroscopy who made studies of the solar spectrum and in 1862 discovered hydrogen in the Sun. In 1868 he mapped the Sun's spectrum, using what became known as the angstrom unit; it was officially named for him in 1905. Ten thousand angstroms equal 1 micron (μ).

angular diameter
The apparent diameter of an object measured in

degrees, or fractions of a degree. As seen from the Earth, for instance, the angular diameters of the Sun and Moon are very similar—about half a degree each. Their actual diameters, of course, are very different.

angular distance
The apparent separation of two points or objects in the sky, measured in degrees or parts of a degree.

Anik satellites
Satellites for communications within Canada; also called Telesat, after the Canadian company that owns and operates them. Anik 1 (the name is the Eskimo for "brother") was launched by the United States on November 9, 1972, followed by Anik 2 on April 20, 1973, and Anik 3 on May 7, 1975. The satellites are in SYNCHRONOUS orbit, so that they appear to hang stationary in the sky above the Earth's equator; they make up the world's first domestic synchronous satellite system. A satellite of improved design was launched in December 1978 to replace the first two Aniks. A series of third-generation satellites, Anik-C, is due for launch beginning in 1981, to provide considerably increased capacity.

Antares
The star alpha in the constellation Scorpius. Its name, meaning "rival of Mars," was given it because of its distinctive red tint, much like the color of the planet. Antares is a red supergiant star of magnitude 1.08, with a diameter about 285 times that of the Sun. It has a peculiar hot companion star of magnitude 6.8. The luminosity of Antares is equal to about 5,000 Suns, which allows it to appear the brightest star in Scorpius even though it is 430 light-years away. In 1971 its companion star was discovered to give out radio emission, due to gas streaming between the two stars or to matter being ejected from the companion.

antimatter
Matter consisting of so-called antiparticles, which are the same as the elementary particles of ordinary matter, such as protons and electrons, but have the opposite electric charge. For example, an antielectron, or *positron*, has a positive, not a negative charge, and the antiproton has a negative, not positive charge. If matter and antimatter come into contact, they annihilate each other, releasing vast amounts of energy.

antimatter cosmology
Theory which maintains that equal amounts of matter and antimatter exist, and always have existed, in the Universe. According to standard cosmology, on the other hand, the Universe began with slightly more matter than antimatter; all the antimatter was destroyed by annihilation together with most of the matter, leaving behind our present Universe, made up entirely of matter. Since matter and antimatter annihilate, they could coexist only if the Universe were to contain separate regions of matter and antimatter, with annihilation taking place at the boundary between them. Antimatter cosmologists believe this separation occurred in the BIG BANG, and as the Universe

expanded, regions with ordinary galaxies and regions with antimatter galaxies developed. The energy released by annihilation between the regions should produce gamma rays that could be observed as a gamma-ray background radiation. At present we cannot measure such radiation accurately enough to know if the Universe really is symmetrical.

Antlia (the air pump)

A faint constellation in the southern hemisphere of the sky, south of Hydra. It was named by Nicolas Louis de Lacaille in the 1750s.

Antoniadi, Eugène M. (1870–1944)

French astronomer born of Greek parents, one of the greatest visual observers of the planets. He is best known for his painstaking and expertly drawn maps of the planets Mars and Mercury. Many of his finest observations were made with the 32-inch (81-cm) refractor of the Meudon Observatory near Paris. Antoniadi originated what is termed the *Antoniadi scale*, a measure of the quality of the observing conditions, or *seeing*, recorded in numbers from I (perfect) to V (atrocious).

apastron

The point in the mutual orbits of a double star at which the two components are farthest apart, or in the orbit of a planet at which it is farthest from its parent star.

aperture synthesis

Technique in radio astronomy which uses a number of small radio dishes to build up the same view of the sky as would be seen by one enormous dish. It overcomes the impossibility of building fully steerable radio telescopes with diameters much greater than a few hundred yards or meters. In aperture synthesis, a line of interconnected radio dishes tracks a source for 12 hours as it crosses the sky. During this period, and as a result of the Earth's rotation, the line traces out a ring in space. As seen from the pole this ring is a perfect circle; but in practice, because radio sources are usually far from the pole, the line of dishes will be foreshortened, giving an elliptical aperture. Some of the dishes may then be moved and the observations repeated the next day. All the readings are stored in a computer and eventually combined to make a complete map. The largest aperture synthesis instrument is the 3-mile (5-km)-long radio telescope at Cambridge, England. It is able to see the radio sky in more detail than any other radio telescope.

aphelion

The point in an object's orbit that is farthest from the Sun. It is on the opposite side of the orbit from the *perihelion*, the nearest point to the Sun. For objects such as planets, with nearly circular orbits, the difference between aphelion and perihelion distances is relatively slight. But for comets, which have very elongated orbits, the difference can be considerable.

apochromat

Telescope object glass that has been highly corrected for color error (CHROMATIC ABERRATION). The typical ACHROMATIC lens brings two colors (usually red and blue) to a common focus by using two different types

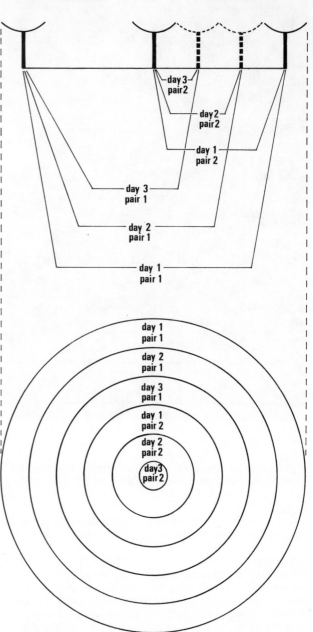

The dishes of an aperture synthesis telescope are connected together into pairs. Any number of dishes can be used. The greater the number, the quicker the observation is completed, but the greater the construction cost. By making one or more of the dishes movable, a series of pairs with different spacings can be built up on successive days. As the Earth spins, the pairs trace out the area of an imaginary single dish whose diameter equals the maximum length of the line.

of glass in combination. In an apochromat, three lenses are used, thus bringing a third color, usually violet, to the same focus. This greatly reduces the bluish halo of so-called secondary spectrum that a refracting telescope shows around bright stars. Such lenses are often called *photovisual*, because the violet rays to which early photographic emulsions were primarily sensitive focus near the same plane as the visual (yellow) rays. A near-approach to apochromatic quality can be obtained by using special varieties of glass in a two-lens objective.

apogee
The greatest distance from Earth of an orbiting object such as a satellite. The opposite point in its orbit is termed the *perigee*.

Apollo program
The space project that landed a total of 12 American astronauts on the Moon. On May 25, 1961, President John F. Kennedy set the goal of landing a man on the Moon and returning him safely to Earth before 1970. At that time there had only been one American manned spaceflight—the short suborbital mission of Alan Shepard. One early idea for reaching the Moon was to build an enormous rocket, to be called Nova, which would launch a single spacecraft to land on the lunar surface and then take off again; this method of reaching the Moon was called *direct ascent*. A rival idea was called *Earth-orbit rendezvous*, in which the spacecraft would be assembled from parts launched into Earth orbit by several smaller rockets. By the end of 1962 NASA had decided on a third method, called *lunar orbit rendezvous*. Only one rocket would be needed to launch both the Apollo spacecraft and a smaller vehicle called the lunar module, which would separate to make the eventual lunar landing. The idea had first been proposed by the Russian rocket theorist Yuri Vasilievich Kondratyuk (1897–1942) about 1916.

Apollo hardware. The three-man Apollo command module is conical in shape, with a maximum width of 12 feet 10 inches (3.9 m) and total height of 10 feet 7 inches (3.2 m). It has five windows: one in the hatch, two at the sides, and two facing forward for observing rendezvous operations. At the command module's apex is a tunnel and removable hatch, through which astronauts can crawl when docked with another craft. Around this tunnel are stored the parachutes for return to Earth and inflatable bags to right the spacecraft after splashdown.

The command module's heat-insulated walls vary in thickness from 0.7 inch (1.8 cm) at the apex to 2.7 inches (6.9 cm) at the base, which takes the full friction of reentry. The walls contain small gas jets for aligning the craft during reentry.

Inside the command module are three couches for the astronauts, an equipment bay for navigation, and stowage lockers for food and spacesuits (they wear the suits during launch, reentry, and other critical maneuvers such as docking). The couch armrests contain hand controls for guiding the spacecraft. Two astronauts can sleep in hammocks slung below the couches.

Above the spacecraft at launch is the launch escape tower, with small rockets powerful enough to pull the command module away to safety in the event of the launch vehicle's malfunctioning. The escape tower is jettisoned once the launcher's second stage has successfully ignited.

Behind the command module is the cylindrical service module, which supplies oxygen, water, and electric power to the command module. It houses the antennae used for high-quality communication with Earth. The service module is the same width as the command module, but over twice the length—24 feet 5 inches (7.4 m). Its main feature is the large service propulsion system (SPS) engine, used to make major speed changes. Around the walls of the service module are four sets of rockets, called the reaction control

system (RCS). These each contain four small thrusters for various maneuvers and for orienting the combined command and service modules (called CSM for short) in space. On the later Apollo missions, previously empty bays of the service module were fitted with equipment to photograph and study the Moon from orbit. Tape and film from these experiments had to be retrieved by spacewalk, because the service module is jettisoned before reentry.

The overall weight of the CSM is about 65,000 lb. (29,500 kg). When bound for the Moon the CSM was launched by the Saturn V rocket; but the CSM can be put into Earth orbit by the smaller Saturn 1B (see SATURN ROCKETS), as was done in the Apollo 7 flight. For lunar landings, the LUNAR MODULE was stored at the top of the Saturn V's third stage, below the CSM, and was extracted by a docking maneuver once underway.

The missions. The spacecraft and its rockets were checked initially in experimental unmanned flights. The first manned flight of the CSM was to have been on February 21, 1967, with astronauts Virgil I. Grissom, Edward H. White, and Roger B. Chaffee. But during a simulated countdown on January 27 a fire in the capsule killed all three men. The Apollo program was delayed for 18 months while the accident was investigated and the Apollo craft redesigned to eliminate fire hazards and to introduce a quick-opening hatch; the rocket that was to have launched the ill-fated crew was instead used for the Apollo 5 test flight.

When it finally flew, the first manned Apollo, numbered Apollo 7, was described as "101 percent successful," accomplishing more tests in orbit than the original flight plan had envisaged. The next Apollo was the first manned launching by a Saturn 5; it put the first three men in orbit around the Moon. Apollo 9, which stayed in Earth orbit, was a trial flight of the spidery lunar module; and Apollo 10 was a complete dress rehearsal for the first lunar landing, including a trip in the lunar module to within 9 miles (14.4 km) of the lunar surface.

In the event, Apollo 11 met Kennedy's deadline with five months to spare; Apollo 12 repeated the achievement before the decade of the sixties was out. Although Apollo 11 had landed a few miles off target, Apollo 12 came down within a few hundred yards of the long-dead Surveyor 3 pathfinder probe, parts of which the astronauts brought back for examination on Earth. Apollo 13 was a near disaster; an explosion in an oxygen tank in the service module cut out nearly all the spacecraft's electrical supply, and the astronauts had to rely on the lunar module's electrical power and rockets to return them safely to Earth.

Apollo 14 marked a trend toward longer Moon walks and more scientific experiments—the astronauts set up explosive charges and operated a thumper device for seismic sounding of the Moon's outer layer. Apollo 15 saw the introduction of an electrically powered LUNAR ROVER vehicle, which allowed the astronauts to drive around and collect samples, while from orbit a mapping camera in the service module photographed the Moon in detail, and a small automatic satellite was released to send back data on the Moon's gravitational field.

The Apollo 16 mission made the only landing in the lunar highlands, to collect material for comparison with the lowland plains samples gathered by the other

Apollo flights. All records were broken by the final lunar mission, Apollo 17, which included the longest Moon walk and the greatest amount of Moon samples ever returned. One discovery was orange soil, at first thought to be signs of volcanism but later identified as glass beads produced by melting from the heat of a meteorite impact.

The Apollo program ceased with Apollo 17, at least in part because of cuts in the NASA budget and the lack of challenge from the Soviet Union, which had run into difficulties with its own manned space activities. Probably no further Americans will visit the Moon until the 1980s at the soonest. However, some remaining Apollo hardware has been used in exploits such as SKYLAB and the joint APOLLO–SOYUZ TEST PROJECT.

Spectacular lunar scenery photographed on the last Apollo Moon mission, Apollo 17, in December 1972. Astronaut Harrison Schmitt is seen collecting samples near a massive boulder that has cracked in two. The boulder is covered by thin dust, from erosion by micrometeorites. Small fragments broken from the boulder are scattered around. In the background, the plain of Mare Serenitatis stretches away to the distant Taurus mountains.

Unmanned Apollo test flights

Mission	Launch date	Results
Apollo 1	February 26, 1966	Suborbital test launch with Saturn 1B; CSM not sent into orbit
Apollo 2	July 5, 1966	Orbital test of Saturn 1B; second stage sent into orbit, but no spacecraft carried
Apollo 3	August 25, 1966	Suborbital test with Saturn 1B; CSM survived high-speed reentry
Apollo 4	November 9, 1967	First launch of a Saturn 5 rocket, and first test of Apollo CSM in orbit. The unmanned command module was blasted back into the atmosphere at the same speed as reentry from the Moon, to test its heat shield
Apollo 5	January 22, 1968	Unmanned test flight of the lunar module on its own; launched into Earth orbit by Saturn 1B
Apollo 6	April 4, 1968	Second test flight of Saturn 5, launching CSM into Earth orbit

Manned Apollo missions

Mission/Crew	Launch date/ Splashdown	Results
Apollo 7 Walter M. Schirra Donn F. Eisele R. Walter Cunningham	October 11/ October 22, 1968	Earth-orbital test flight of three-man CSM; launch by Saturn 1B
Apollo 8 Frank Borman James A. Lovell William A. Anders	December 21/ December 27, 1968	First manned Saturn 5 launch; 10 orbits of Moon in Apollo CSM
Apollo 9 James A. McDivitt David R. Scott Russell L. Schweickart	March 3/ March 13, 1969	Earth orbital test of CSM and lunar module; launched by Saturn V
Apollo 10 Thomas P. Stafford John W. Young Eugene A. Cernan	May 18/ May 26, 1969	Full dress rehearsal of Moon landing, in lunar orbit; 2½ days spent orbiting Moon
Apollo 11 Neil A. Armstrong Michael Collins Edwin E. Aldrin	July 16/ July 24, 1969	Armstrong and Aldrin make first manned lunar landing, on July 20 in Sea of Tranquillity
Apollo 12 Charles Conrad Richard F. Gordon Alan L. Bean	November 14/ November 24, 1969	Conrad and Bean land on November 19 in Ocean of Storms
Apollo 13 James A. Lovell John L. Swigert Fred W. Haise	April 11/ April 17, 1970	Landing attempt canceled after explosion in oxygen tank damages spacecraft
Apollo 14 Alan B. Shepard Stuart A. Roosa Edgar D. Mitchell	January 31/ February 9, 1971	Shepard and Mitchell land on February 5 in Frau Mauro region of Moon
Apollo 15 David R. Scott Alfred M. Worden James B. Irwin	July 26/ August 7, 1971	Scott and Irwin land on July 30 at Hadley rill. First use of lunar roving vehicle
Apollo 16 John W. Young Thomas K. Mattingly Charles M. Duke	April 16/ April 27, 1972	Young and Duke land in Descartes highlands on April 21
Apollo 17 Eugene A. Cernan Ronald E. Evans Harrison H. Schmitt	December 7/ December 19, 1972	Cernan and Schmitt land on December 11 at the edge of the Sea of Serenity, near the crater Littrow

Apollonius of Perga (c.262–c.190 B.C.)

Greek mathematician credited with introducing the idea of EPICYCLES and eccentrics (off-center circles) to help explain the erratic motion of the planets in the sky. Such geometrical devices were subsequently prominent in the cosmological system of PTOLEMY. In his work on geometry, Apollonius studied and named the ellipse, the parabola, and the hyperbola. Although he did not realize it, the ellipse is in fact the shape that describes the actual orbits of the planets.

Apollo–Soyuz Test Project (ASTP)

Joint mission in Earth orbit made by an American Apollo craft and a Soviet Soyuz spaceship in July 1975. The two were launched into similar orbits, and the Apollo caught up and docked with Soyuz. Apollo carried a special docking tunnel to serve as an airlock, enabling astronauts to move between the two spacecraft. One purpose of the mission was to test the docking tunnel for possible rescue of stranded spacemen, although the main importance of ASTP was political.

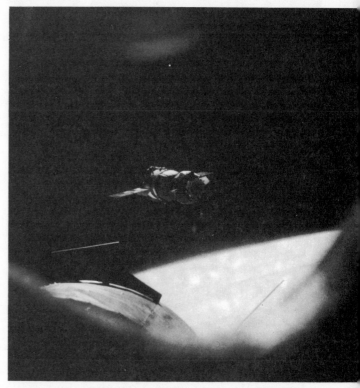

The Apollo-Soyuz flight in July 1975 offered the opportunity for the U.S. and Soviet Union to make their spacecraft mutually compatible. Here, the 6.6 ton Soviet Soyuz craft, with solar panels extended, is seen from the Apollo Command Module. Soyuz is 24 feet (7 m) in length.

apparent magnitude

The brightness with which a star appears as seen from Earth. It differs from intrinsic brightness, or ABSOLUTE MAGNITUDE.

apsides

The points in an orbit at which the bodies are either at their closest or at their farthest, such as perihelion and aphelion for objects in the solar system, or periastron and apastron for objects in orbit around another star. The line joining the two points is called the *line of apsides*.

Apus (the bird of paradise)

A faint constellation in the south polar region of the sky. It was named by Johann Bayer in 1603.

Aquarius (the water carrier)

One of the constellations of the zodiac, lying in the
equatorial region of the sky. The Sun passes through
Aquarius in the last half of February and first half of
March. Aquarius is prominent in the evening sky
during the northern hemisphere autumn. Although
there are no particularly brilliant stars in Aquarius, it
contains the globular cluster M2 as well as other
objects of interest to astronomers, including the
planetary nebula NGC 7009, called the Saturn nebula
because of its resemblance to that planet.

Aquila (the eagle)

A prominent constellation lying on the celestial equator
in part of the Milky Way. It is visible in the evening
sky during the northern hemisphere summer. The
brightest star in Aquila is ALTAIR. With the stars Vega
in Lyra and Deneb in Cygnus, Altair forms the so-
called summer triangle. There are several interesting
variable stars in Aquila, and a number of novae have
occurred within its boundaries.

Ara (the altar)

A southern hemisphere constellation that lies in the
Milky Way south of Scorpius. It once formed part of
the zodiac. There are no particularly bright stars.

arc minute, arc second

Small units of angular measure, 1/60 and 1/360
of a degree, respectively, used to record angular
diameter or angular separation. They are not to be
confused with minutes and seconds of time.

Arcturus

Red giant star in the constellation Boötes, the
herdsman. Arcturus, also called Alpha Boötis, is the
fourth-brightest star in the sky, of magnitude − 0.06.
Its radius is 23 times that of the Sun, and its
luminosity is equal to about 100 Suns. Arcturus is
36 light-years away.

Aerial view of the Arecibo 1,000-foot fixed radio telescope.
The receiver is mounted in the box under the curved
horizontal truss, 870 feet above the wire mesh surface. It
is reached from the control room, **foreground**, by a walkway
stretching from the base of the nearest tower.

Arecibo Observatory

Site of the world's largest single radio astronomy dish,
1,000 feet (305 m) in diameter. The dish is slung like
a hammock in a natural hollow between hills in
northern Puerto Rico near the city of Arecibo. The
instrument, completed in 1963 and resurfaced in 1974,
is used for ionospheric studies and radar mapping of
the Moon and planets as well as for radio astronomy.
The telescope cannot be steered, but the sky from $43°$
north to $6°$ south can be covered by moving feed
aerials, which are supported above the dish from three
towers around its rim.

Argelander, Friedrich Wilhelm August (1799–1875)

German astronomer responsible for the compilation
of the BONNER DURCHMUSTERUNG, one of the world's
greatest star catalogs, which lists positions and
magnitudes for stars down to magnitude 9.5. The
catalog sacrificed accuracy for quantity, however, and
in 1867 Argelander proposed a new and greater
project to the Astronomische Gesellschaft (German
Astronomical Society). This resulted in the highly
accurate AGK series of catalogs. In 1837 Argelander
had published a study of the observed motions of stars
(their PROPER MOTIONS), which confirmed the 1783
suggestion of William HERSCHEL that the Sun is moving
through space toward a point in the constellation
Hercules; this point is termed the SOLAR APEX.
Argelander also introduced the so-called *step method*
of visually estimating the brightness of a star by
comparing it with stars of known magnitude. This
technique in effect founded the serious study of
variable stars, only a few of which were known at the

Above Saturn, photographed by Stephen Larson with the 61-inch (155-cm) Catalina telescope in Arizona. The different colors of the outer rings, A and B, show up well; ring C, inside them, is faint and can only be seen here where it hides the planet's surface.

Below These superimposed images of the Sun were taken by astronauts on a Skylab mission. The colors, produced in a laboratory reconstruction, are false; the brightness of the image at any point corresponds to its X-ray temperature. The spike at the top of the view extends for a million miles into space.

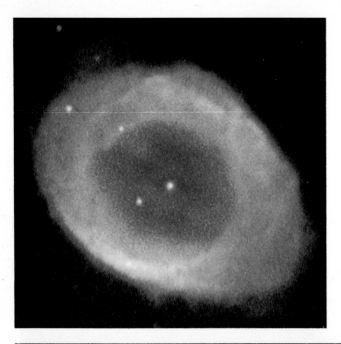

Two famous planetary nebulae. **Left** The Ring nebula in Lyra. The "smoke-ring" effect is believed to have been caused by the expulsion of the outer layers of a former red giant star. At the center of the nebula is a white dwarf star, the hot core of the former red giant. This photograph was taken with the 4-m reflector of Kitt Peak National Observatory in Arizona.

Below The Dumbbell nebula in Vulpecula, photographed with the 200-inch Mount Palomar reflector. Ultraviolet light from the central star in a planetary nebula excites the surrounding gases, making them glow.

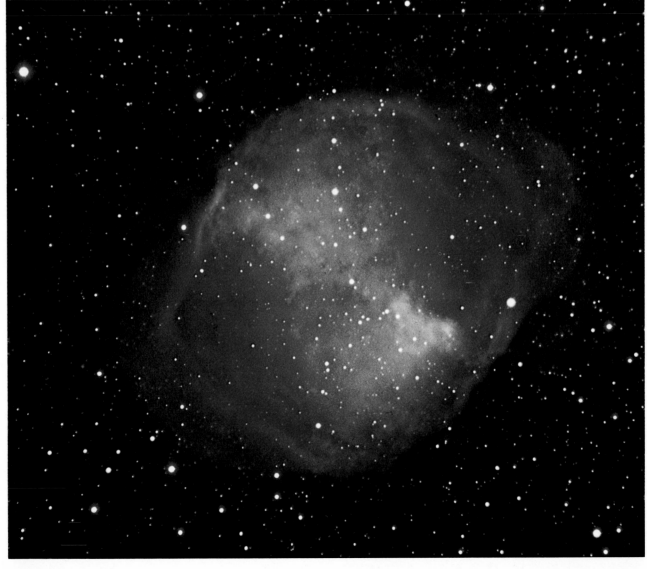

time. The step method is still used by amateur variable-star observers today.

Ariane
A three-stage rocket developed by the European Space Agency, capable of launching up to 2,200 lb. (1,000 kg) into geostationary orbit. Ariane will be launched from a site at Kouru in French Guiana. Its first test flight was scheduled for November 1979, and it is due to enter service in late 1980.

Ariel
A satellite of Uranus, discovered in 1851 by the English amateur astronomer William Lassell. Ariel orbits Uranus once every $2\frac{1}{2}$ days at a distance of 119,000 miles (192,000 km). Its diameter is about 900 miles (1,500 km).

Ariel satellites
Series of British scientific satellites, launched by the United States. Ariel 1 was the first international satellite; like Ariel 2 it was built in the U.S., with only the experiment packages supplied by Britain. From Ariel 3 onward the entire spacecraft has been British-built.

Satellite	Launch date	Remarks
Ariel 1	April 26, 1962	Studies of the ionosphere and solar radiation
Ariel 2	March 27, 1964	Atmospheric and radio-astronomy studies
Ariel 3	May 5, 1967	Ionospheric and radio-astronomy studies
Ariel 4	December 11, 1971	Radio-astronomy experiments
Ariel 5	October 15, 1974	X-ray studies; mapped the sky at X-ray wavelengths and examined specific sources in detail
Ariel 6	June 2, 1979	Cosmic rays and X-ray studies

Aries (the ram)
A constellation of the zodiac, visible in the night sky during the northern autumn, and lying between Taurus and Pisces. Aries is not a particularly bright group; its importance has been in its location. About 2,000 years ago the Sun was in Aries as it passed from south to north of the celestial equator, an event called the vernal, or spring, EQUINOX. This point became known as the first point of Aries. However, through a slight wobble of the Earth's axis called PRECESSION, the location of the vernal equinox no longer lies in Aries; nevertheless, the term "first point of Aries" is still occasionally used.

Aristarchus of Samos (flourished 280–264 B.C.)
Greek astronomer who calculated the relative sizes of the Earth, Moon, and Sun. He attempted first to measure the relative distances from Earth of the Moon and Sun. With his results he compared the sizes of the Sun and Moon with that of the Earth. (The actual size of the Earth was later measured by ERATOSTHENES.) Although Aristarchus' value for the Sun's size was too small, he was still able to show it is far larger than the Earth. He therefore proposed that the objects in the sky should revolve around the Sun, rather than the Earth as was generally supposed. His contemporaries did not agree with his logic, and the concept of a heliocentric Universe was rejected until the time of COPERNICUS.

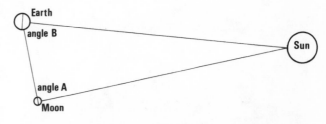

2   Both the Sun and Moon appear the same size in the sky; but since the Sun is more distant than the Moon, it must be correspondingly larger in size. The Sun and Moon can then be put on a scale drawing, as shown here, using Aristarchus' value for their relative distances. Observations of lunar eclipses showed the Greeks that the size of the Earth's shadow at the distance of the Moon was $2\frac{2}{3}$ the Moon's diameter. Putting this shadow onto the drawing allowed Aristarchus to add the Earth to scale; he then read off the sizes and distances of the Sun and Moon in terms of the Earth's size. He made the Moon's diameter $\frac{1}{3}$ that of the Earth, which is close to the true value of about $\frac{1}{4}$. He thought the Sun was seven times the Earth's diameter, which is too small because of the error in his estimate of the Sun's distance.

1   How Aristarchus sought to measure the distance of the Sun and Moon. At exactly half Moon, the angle between the Sun, Moon and Earth (angle A) is a right angle (90°). To complete the triangle, Aristarchus measured the angle between the Sun, Earth and Moon (angle B) at the same instant; he could then calculate the relative lengths of the sides Earth–Moon and Earth–Sun. He found that the Sun is about 19 times farther away than the Moon. The true figure is actually about 400 times; the error arose because angle B is difficult to measure with accuracy.

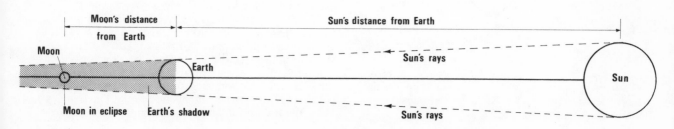

Aristotle (384–322 B.C.)

One of the major Greek philosophers, whose authority was cited by astronomers and physicists for almost 2,000 years. Aristotle took the system of heavenly spheres that EUDOXUS had originated to explain the motions of celestial bodies, and developed it by adding another 22 spheres, whose purpose was to counteract the effect of one planet's spheres on another's, so that neighboring planets did not affect each other's motion. Although Eudoxus probably regarded his spheres as only a mathematical device, Aristotle believed that they were real, solid orbs of crystal. Though Aristotle's scheme of the heavens was superseded by that of PTOLEMY, the notion of solid crystalline spheres persisted until the time of Tycho BRAHE.

Aristotle rejected the idea that the Earth spins on its axis or that it orbits the Sun, and his authority was used to dismiss all such notions until the time of GALILEO. Yet Aristotle's astronomical views were not all misleading: he argued convincingly for the spherical nature of the Earth, pointing out the way in which stars rose above or dipped below the horizon as one traveled north or south, and also noting the curved edge of the Earth's shadow at lunar eclipses. His extensive writings contained both his own investigations and summaries of existing Greek thought; thus we find from Aristotle that the Greeks knew the Moon was spherical, and also understood the true reason for the Moon's phases.

This armillary sphere, made in 1554 by Volpaja, enabled the Earth to be set to any position beneath the celestial sphere. The wide band is the Zodiac, which runs between the rings marking the two tropics.

armillary sphere

An obsolete instrument once used to convert star positions directly from the observed AZIMUTH and ALTITUDE to the celestial equivalents (RIGHT ASCENSION and DECLINATION). It consisted of a number of interlocking circles representing the celestial equator, the ecliptic, and other GREAT CIRCLES such as the horizon, so that a star, once sighted, could have its celestial coordinates read off on the scales. Early armillaries, as used by Hipparchus, Ptolemy, and Arab astronomers, worked in a vertical position; they achieved their highest development when Tycho BRAHE installed several, up to $9\frac{1}{2}$ feet (3 m) in diameter, at his observatory (1576–1601). They were mounted on polar axes and in some cases were graduated in units of 10 arc seconds. Armillary spheres were never, however, as accurate as quadrants, which obtained direct readings of altitude and azimuth and could be constructed more rigidly, and they were little used after Tycho's time.

Armstrong, Neil Alden (b. 1930)

The first man to walk on the Moon, he was commander of the historic Apollo 11 mission of July 1969. Armstrong, a test pilot, was selected as an astronaut in 1962. He was command pilot of the Gemini 8 mission in March 1966, which performed the world's first space docking. However, a jammed thruster caused the spacecraft to spin dangerously, and Armstrong and copilot David Scott had to undock and make an emergency landing.

As commander of Apollo 11, Armstrong piloted the lunar module *Eagle* to a safe landing on the Moon's Sea of Tranquillity. As he stepped on to the Moon's surface on July 21, 1969, Armstrong said: "That's one small step for a man, one giant leap for mankind." Armstrong spent a total of $2\frac{1}{4}$ hours on the lunar surface. He collected the first lunar rock, and helped Edwin Aldrin fill boxes with a total of 50 lb. (22 kg) of samples. He also put out scientific experiments and took photographs of the surroundings. After the Apollo 11 mission Armstrong left the astronaut corps to become professor of aerospace engineering at the University of Cincinnati.

ascending node

See NODES.

asteroid

A small rocky body moving in an elliptical orbit around the Sun; asteroids are also called *minor planets*. Thousands exist, ranging from about 600 miles (1,000 km) in diameter down to small boulders. They are too small to have atmospheres and their Sun-scorched and meteorite-pounded dusty surfaces would be hostile to all life-forms. Asteroids frequently collide among themselves, producing debris that occasionally falls to Earth as METEORITES.

Discovery of asteroids.   Most asteroids orbit the Sun in the gap between Mars and Jupiter, where the German astronomer Johann BODE predicted in 1772 that an undiscovered planet would lie. A group calling themselves the "celestial police" were organized by the German astronomer Baron Franz Xaver von Zach (1754–1832) to search for this new body. On January 1, 1801, the Italian astronomer Giuseppe PIAZZI, not one of the group, discovered the first asteroid, CERES. But the celestial police soon discovered other bodies moving in similar orbits. The German astronomer Wilhelm OLBERS found the asteroids PALLAS and VESTA in 1802 and 1807; JUNO was discovered by the German astronomer Karl Ludwig Harding (1765–1834) in 1804. These four asteroids are considerably brighter than the rest of the swarm. A fifth, Astraea, was not discovered until 1845; it was found by the German Karl Ludwig Hencke (1793–1866), who two years later found Hebe. Discoveries then followed rapidly, with 100 asteroids found by 1868, 200 by 1879, and 300 by 1890. This early hunting involved laborious visual searching; it was revolutionized in December 1891 when Max Wolf discovered an asteroid by the use of photography. The large numbers of faint asteroids discovered with this technique presented serious problems of identification until modern computers allowed all known asteroids to be readily followed. Minor-planet centers have now been established at Cincinnati and at Leningrad.

About 2,500 asteroids now have precisely known orbits, of which about 1,850 have been officially numbered. Some 2,000 more were sighted once but were lost through lack of further observations. The full number of minor planets is immense. Surveys suggest that there are about 50,000 asteroids which become brighter than, or equal to, magnitude 21. The total mass of the asteroid belt is roughly 0.0004 the Earth's mass, or about $2\frac{1}{2}$ million million million tons; this is twice the mass of the largest asteroid, Ceres.

Asteroid groups. The asteroids can be conveniently classed in three groups according to their orbits: the main belt, the Apollo and Amor group, and the TROJAN asteroids. The main belt comprises over 95 percent of known asteroids, moving in slightly elliptical paths between the orbits of Mars and Jupiter. Those on the well-defined inner edge of the belt lie firmly in the plane of the solar system, but moving outward the orbits become more scattered. The Apollo and Amor group have highly elliptical, more sharply inclined orbits; at their farthest points (aphelion) they are in the main belt, but their closest points to the Sun (perihelion) lie among the inner planets. Thus, bodies such as 433 Eros, 1566 Icarus, 1620 Geographos, and 1685 Toro, can come quite close to the Earth. The Trojan asteroids move along in the same orbit as the planet Jupiter.

The asteroids of the main belt orbit the Sun with periods between about 2 and 6 years, compared with 11.86 years for Jupiter. The distribution is not smooth, and there are distinct groups and gaps (the so-called KIRKWOOD GAPS). The gaps occur at precise fractions of Jupiter's period, where orbits would be repeatedly perturbed by Jupiter's gravitational pull. The principal breaks occur at periods of 4.0, 4.8, and 5.9 years, corresponding to ratios of 3:1, 5:2, and 2:1 of Jupiter's period, and there are many minor gaps.

In the 1920s, the Japanese astronomer Hirayama Seiji (1874–1943) found certain orbital groupings among minor planets. About 25 percent of asteroids can be placed in 10 such Hirayama families, each caused when two major asteroids collided in the past. Some families are compact, while others are very dispersed, suggesting that collisions have occurred at widely different times in solar system history.

Nature of asteroids. The name *asteroid* (star-like) was coined 150 years ago by William Herschel; even with today's giant telescopes few asteroids appear

Characteristics of Representative Asteroids

Asteroid	Period years	Orbit Perihelion distance a.u.	inclination degrees	diameter km	rotation period hours	Albedo %	Surface Type	Comment
1566 Icarus	1.12	0.187	23.0	1.4	2.3	17.8	Silicaceous	Apollo group
1620 Geographos	1.40	0.83	13.0	$0.8 \times 1.0 \times 4.0$	5.2	20.9	Silicaceous	Apollo group
433 Eros	1.76	1.13	10.8	$7 \times 16 \times 35$	5.3	14.2	Silicaceous	Amor group
8 Flora	3.27	1.86	5.9	150	13.6	16.8	Silicaceous	main belt
18 Melpomene	3.48	1.80	10.2	141	14.0	14.0	Silicaceous	main belt
4 Vesta	3.63	2.55	7.1	503	10.7	26.4	Basaltic?	main belt
192 Nausikaa	3.72	1.82	6.9	92	—	20.0	Silicaceous	main belt
887 Alinda	4.00	1.16	9.0	4.4	—	14.6	Silicaceous	Amor group
5 Astraea	4.13	2.10	5.3	116	16.8	17.7	Silicaceous	main belt
15 Eunomia	4.30	2.15	11.8	270	6.1	15.5	Silicaceous	main belt
3 Juno	4.36	1.99	13.0	226	7.2	19.0	Silicaceous	main belt
324 Bamberga	4.39	1.78	11.3	230	8.0	3.6	Carbonaceous	main belt
1 Ceres	4.60	2.55	10.6	955	9.1	7.2	Carbonaceous	main belt
2 Pallas	4.61	2.11	34.8	538	10.5	10.3	Carbonaceous	main belt
624 Hektor	11.55	5.02	18.0	$50 \times 50 \times 210$	6.9	2.8	Unusual—not identified	Trojan group
944 Hidalgo	13.70	1.98	43.1	—	—	—	—	Unique body

as anything other than star-like points, making diameter measurement very difficult. In 1971, however, two new methods of determining asteroid diameters were developed. The radiometric method involves the careful measurement of a minor planet's brightness at visual and infrared wavelengths; these figures indicate whether an object of given brightness is small and reflective, or larger but darker, from which can be calculated both visual ALBEDO (reflectivity) and asteroid diameter. The polarimetric method involves a study of the polarization of sunlight reflected by an asteroid to yield the asteroid's albedo which, with the visual brightness, reveals the diameter. This new work gives diameters which are considerably larger than early estimates, and which indicate that asteroids are very dark bodies. Albedos vary from the exceptionally low 2.8 percent for 624 Hektor and 747 Winchester, to as much as 26.4 percent for 4 Vesta, with the majority in the range 7 to 18 percent. (In comparison, carbon black has an albedo of about 3 percent, the Moon—a very dark body—6.7 percent, and the planet Venus 76 percent.)

The amount of light reflected by asteroids generally varies regularly by up to 1.5 magnitudes. This suggests they are rotating, irregular-shaped bodies, probably fragments from asteroid collisions. A typical example is the Martian moon PHOBOS which, like its companion DEIMOS, is believed to be a captured asteroid. In some cases, the deduced shapes are extreme: 433 Eros, for instance, is a spindle-shaped body, while the Trojan asteroid 624 Hektor is either a tumbling cylinder made of iron—a stone cylinder would disintegrate—or a twin system with two close stony bodies revolving around their common center of gravity. The three largest asteroids and a handful of smaller ones show little light variation and are probably near-spherical in shape.

Light reflected from asteroids at different wavelengths can be compared with powdered minerals in the laboratory to give broad indications of asteroid surface compositions. About 90 percent of asteroids so far studied have surfaces resembling either carbonaceous meteorites or normal stony or stony-iron meteorites (see METEORITE). Most stony asteroids have diameters of about 60 to 125 miles (100–200 km), while carbonaceous bodies are common in larger and smaller sizes and tend to predominate in the outer parts of the belt. 16 Psyche and a group of smaller asteroids have featureless spectra suggesting nickel-iron bodies, and 4 Vesta is unique both in its high albedo and in a spectral curve which resembles that of ACHONDRITE meteorites.

Origin of the asteroids.   Wilhelm Olbers first suggested that the asteroids might be the fragments of a disrupted planet. But the general opinion now is that, at the origin of the solar system, a cloud between two and four times the Earth's distance from the Sun formed an original family of perhaps 15 to 30 asteroids with diameters from 60 to 600 miles (100–1,000 km). Collisions among these led to the highly fragmented system now visible, with only Ceres, Pallas, Vesta, and possibly a handful of others of the original bodies surviving intact. Asteroids of the Apollo and Amor group have such eccentric and highly tilted orbits that they may not have been formed in the same fashion at all, and may instead be the remains of extinct short-period comets.

astigmatism

The image fault produced by a lens or mirror in which different diameters focus rays into different planes. *Axial astigmatism* (the formation of an astigmatic image when the instrument is pointed directly at the object) is caused by a deformed optical component, and the image of a star at the best focus appears as a cross or lozenge. Off-axis or *abaxial astigmatism* arises when rays strike an uncorrected optical system at an angle, so limiting the diameter of the usable field. Off-axis astigmatism increases with decreasing focal ratio of the mirror or object glass. It is rarely important visually because of the restricted field of view, but is particularly objectionable in wide-field photographic instruments. The SCHMIDT TELESCOPE was designed primarily to solve the problem.

astrolabe

An ancient astronomical instrument, used for observing the altitudes of stars. In its simplest form it consisted of a gradual disk hanging vertically, with a sight or *alidade* which was turned to point to the object. The Arabs adopted the instrument, graduating the *tablet* or face to show the altitudes of various bright stars at different times, so that the local time could be determined. By substituting different tablets, the instrument could be used in different latitudes. In this form it was widely employed as a navigational aid until the more accurate sextant supplanted it in the 18th century.

A Persian astrolabe dating from AD 1221, without the alidade or sighting rule. Gears are used on this version to operate a calendar movement on the reverse side.

Its modern form, the *prismatic astrolabe,* is a very accurate instrument designed to measure the instant at which a given star comes to an altitude of 60°. This enables the latitude of the observing site to be determined. The instrument works by comparing the direct position of the star in the sky with that observed reflected from a mercury bath. Both images are fed, via a 60° prism, into a horizontal telescope. The observer then sees two stars apparently moving toward each other. When they merge, or come alongside, the star's altitude is exactly 60°. If the latitude of the site is already known, the prismatic astrolabe can be used to determine local time.

astrometry

The measurement of positions of objects in the sky; it is thus also termed *positional astronomy.* Astrometry is the basic work of national institutions such as the U.S. Naval Observatory and the Royal Greenwich Observatory. The accurate star positions they provide are needed by navigators and surveyors as well as by astronomers. The observatories also monitor the Earth's rate of spin, by reference to the passage of stars across the sky, to provide a basis for our time system. Although regular time signals are now derived from atomic clocks (see ATOMIC TIME), star checks are still needed to ensure that the signals do not get out of step with the Earth's slightly varying rate of rotation.

Foundations.   Modern astrometry grew from the foundations laid by star catalogers such as HIPPARCHUS, Tycho BRAHE, James BRADLEY, Friedrich BESSEL, Friedrich ARGELANDER, and Sir George AIRY. Their work is distinct from that of observational astronomers such as Galileo and his successors, who were concerned with the visible features of celestial objects, or that of astrophysicists, who study the actual nature of celestial objects.

The first astronomical observations were entirely astrometrical; they led to the establishment of the first crude calendars, and eventually to a true understanding of the movement of the planets around the Sun. Since the invention of the telescope, astrometric observations have become more accurate; they have revealed that stars are not truly still in the sky as the ancient astronomers had believed. Therefore, even if all star positions were accurately tabulated, they would need to be resurveyed after a decade or so to take account of changes. Actually, the resurveying is continuous, and it is marked by an ever-increasing standard of accuracy.

Star movements.   The celestial coordinates of an object drift continually because of the wobbles in the Earth's motion called PRECESSION and NUTATION. In addition, stars orbit the Galaxy at various speeds; they thus show varying shifts, known as their PROPER MOTION, the study of which aids astronomers in understanding the Galaxy's structure.

Nearby stars show a shift in position, called PARALLAX, as we view them from different places in the Earth's orbit. The parallax shift is greater the nearer the star is to us, and provides an important direct measurement of star distances; the study of parallax is therefore a major concern of astrometrists. Eventually, far better parallax measures will be possible by placing two telescopes wide apart in the solar system, giving a far longer baseline for observations

than the diameter of the Earth's orbit to which we are currently restricted.

Our most reliable information on the masses of stars is deduced from the orbits of visual binary stars, which are another important astrometric target. Astrometrists can also detect companion objects, too faint to be seen directly, from the slight wobble they produce in the proper motions of some stars. This technique has revealed the existence of a planetary system round at least one nearby star (see BARNARD'S STAR).

New developments.   Modern radio astronomy instruments such as APERTURE SYNTHESIS telescopes and giant INTERFEROMETERS produce positional results far more precise than optical observations. To correlate the two systems, astronomers must now carefully measure the positions of known optical counterparts of radio sources, such as quasars, pulsars, and radio galaxies. In the 1980s, astrometry will be vastly improved by telescopes in orbit, such as the LARGE SPACE TELESCOPE. From space, the blurring effect of the Earth's turbulent atmosphere is avoided, which will allow positional measurements far more accurate than any obtainable from the ground.

Astronomer Royal

Honorary post in British astronomy, originally associated with the directorship of the Royal Greenwich Observatory. When King Charles II founded the Royal Observatory in 1675 he appointed John Flamsteed his first "astronomical observator." In 1971 the post of astronomer royal was separated from the Royal Observatory's directorship; the first man to be appointed astronomer royal under this new arrangement was the Cambridge radio astronomer Professor Sir Martin Ryle.

A separate post of Astronomer Royal for Scotland, created in 1834, is attached to the directorship of the Royal Observatory, Edinburgh. From 1791 to 1921 there was a post of Royal Astronomer for Ireland.

astronomical unit (a.u.)

The average distance between the Earth and Sun, equivalent to 92,955,832 miles (149,597,910 km). Distances within the solar system are frequently expressed in astronomical units. There are several ways of establishing its precise value, the modern method being to measure by radar the distance of another body in the solar system, usually the planet Venus. The time taken for a radar echo to return from Venus reveals its distance from Earth with great accuracy. The relative distances of the planets from the Sun are known from their orbital periods (see KEPLER'S LAWS), as if on a scale map. Measuring the separation between two planets reveals the scale, and allows the actual distances of the planets from the Sun to be worked out. The distance between the Earth and a space probe in orbit around the Sun, as revealed by tracking data, can provide a useful check on the scale of the solar system. An earlier means of calculating the distance between two objects in the solar system involved measuring the parallax of minor planets that pass close to the Earth. This method was used in 1931 when the minor planet Eros passed about 0.15 a.u. (14 million miles, $22\frac{1}{2}$ million km) from Earth. But the new radar techniques give far better results, accurate to within a few miles.

astronomy, history of

Astronomy is probably man's oldest natural science. The passing year was marked by the progress of the seasons, and the Sun's altitude above the horizon indicated the time of day. From these simple calculations of time and position astronomy grew.

The earliest written astronomical records come from the fertile zone at low northerly latitudes that includes the flood-plains of the Nile in Egypt, the Tigris and Euphrates in Mesopotamia, and similar regions in India and China, and later Central America. In these areas, large-scale agriculture and trade developed as early as 4000 B.C.

Time-keeping was essential, in order to determine the correct season for planting and harvesting, and the appropriate days for religious festivals. The construction of a good working calendar was thus the major task facing the earliest astronomers. The length of the Moon's cycle of phases and the length of the year bear no simple relationship to one another: the Moon's cycle is just over $29\frac{1}{2}$ days, while the solar year lasts approximately 365 days (about 11 days more than 12 lunar months). The easiest solution seems to have been adopted—an extra month when the calendar grew too far out of step with the crops. From more detailed observations of the Moon, the Babylonians discovered by about 300 B.C. that almost a complete number of lunar months fitted into both 8 and 19 solar years; this latter relationship was rediscovered by the Greeks, and is known as the METONIC CYCLE.

Practical requirements fostered the development of astronomy. More accurate observations revealed that both the Sun and Moon moved across specific constellations during a year, so that the stars visible at a particular time of night are characteristic of the season. This resulted in the regular observation of stars and star groups and enabled a further check to be made on the calendar. For example, the Egyptians noted the correlation between the Nile floods and the *heliacal rising* of the star Sirius at Memphis (the star's first appearance in the morning sky before the Sun). Though the early astronomers gained considerable knowledge of the movements of celestial bodies, they gave little thought to the physical explanation of these motions. Not until the age of the Greek philosophers were theories sought to explain the phenomena.

Greek astronomy. While the Babylonians and Egyptians laid the foundations, it was left to the Greeks to consolidate early astronomy. Greek writers claimed to have learned from the Babylonians, but the difference between Greek and Babylonian astronomy was that while Babylonia was a single school of thought the Greeks lived in smaller city states, each producing its own scholars. Greek writings reveal a wide variety of notions, not all of them eventually accepted. The greatest names are those of ARISTOTLE, HIPPARCHUS, and PTOLEMY, and it was their views which eventually held sway.

Aristotle taught that the Earth is a sphere—indeed its diameter was later measured by ERATOSTHENES. His ideas about the structure of the Universe were less accurate, however: he fixed the Sun, Moon, and stars to crystal spheres which rotated about the Earth. In order to account for the RETROGRADE MOTIONS of some of the planets (an apparent backward movement

actually caused by the Earth overtaking them), a total of 55 spheres were necessary.

The greatest contributions to ancient astronomy came from one man—Hipparchus. His accurate measurements of the planets and his stellar catalog were unrivaled. Some three centuries later, Ptolemy took Hipparchus' work as the basis of his "Greater Collection," which eventually became known as the *Almagest.* Although Ptolemy's own observations have been questioned, it was through his work that astronomy was carried through to the Middle Ages. His geocentric theory of the Universe, with its numerous EPICYCLES and DEFERENTS, held sway for over 1,000 years.

After the decline of Greco-Roman civilization, the world's scientific knowledge was preserved in the hands of the Arabs. To them we owe many of our present star names. They refined the science of measurement, but few real theoretical advances can be attributed to them.

The revolution. When COPERNICUS published in 1543 details of his new *heliocentric* theory of the solar system, with the Sun, rather than the Earth, at the center of the planets' orbits, it did not immediately revolutionize astronomy. The classical Greek notions were too well established for it to be regarded as anything other than an aid to the calculation of planetary positions—and it seemed only slightly better than the accepted Ptolemaic *geocentric* system, in which the planets moved around the Earth. For 50 years, the situation simmered. But slowly the new ideas of Copernicus began to gain popularity, and the Church eventually had to take steps to suppress theories contrary to the classical view of the Universe.

To establish the new view of the Universe, it took the courage of GALILEO to report on the evidence of his own eyes that Aristotle was wrong, and the genius of KEPLER to refine the crude Copernican theory. The change began with the work of the Danish astronomer Tycho BRAHE, who realized the importance of good observations if any theory was to be confirmed. His observations of the Sun, Moon, and planets over many years were the most accurate that could be made without a telescope. It was fortunate that Tycho chose Kepler as his assistant, for on Tycho's death, Kepler brought his mathematical skill to bear on the new data. Eventually, he concluded that the Earth and planets *must* move around the Sun, in ellipses rather than in circles; and he went on to derive his famous laws of planetary motion.

With the invention of the telescope, Galileo in 1609 attacked Aristotle's and Ptolemy's ideas from another direction. He saw spots on the Sun, innumerable stars in the Milky Way—and moons moving around Jupiter, showing that there were bodies in the solar system of which Aristotle had not dreamed. His stand against Aristotelian dogma made him a natural champion of the Copernican theory.

While in Italy the Church was trying to suppress Galileo and his followers, the scientists of Protestant northern Europe were able to say and teach more or less what they thought. Thus it was men such as Christiaan HUYGENS, Edmond HALLEY, and above all Isaac NEWTON, who were eventually to break through the barriers which had existed for 2,000 years.

Newton and telescopes. Newton was not an astronomer. But in his theories of gravitation and of motion, and in his observations of the nature of light, he did more for astronomy than many an observer. Newton's laws, combined with those of Kepler, gave a virtually complete description of the movements of the planets, a description still valid today.

For a hundred years after Newton's invention of the reflecting telescope, little use was made of its advantages. Great observers such as CASSINI, FLAMSTEED, and Halley, continued to use simple refracting telescopes for their charting and measuring of heavenly bodies. In 1758 John DOLLOND introduced the achromatic lens, which gave improved, color-free images, and telescopic astronomy could begin in earnest. Yet the man who more than anyone else begun the study of the stars themselves, William HERSCHEL, used reflecting telescopes.

His discovery of Uranus and his subsequent studies of stars and nebulae were entirely due to the excellent telescopes which he constructed himself. Other astronomers still preferred to use refractors, but the high quality of objective lenses by men such as FRAUNHOFER could not really offset the light-gathering power of large reflectors.

During the 19th century, ASTROPHYSICS began to emerge in its own right. The studies of the spectrum began by Fraunhofer mark the first attempts at an understanding of the nature of the stars, work which was refined and extended by such men as William HUGGINS and Norman LOCKYER.

The many advances that have taken place during this century are due not so much to the work of individuals as to the improvement in observational techniques, such as the combination of large reflectors and photography, and to the introduction of powerful computers. The trend is toward larger and more advanced instrumentation used by teams of scientists rather than by individuals.

astrophotography

The use of photography has almost completely supplanted visual observation in modern astronomical research. Early emulsions (c.1840) were so slow that several minutes' exposure was required to record the Moon, a feat first performed by John W. DRAPER. The first star to be photographed was Vega, by W. C. BOND in 1850, with an exposure of 100 seconds using the 15-inch (38-cm) refractor of Harvard College Observatory. The British amateur astronomer Warren de la Rue (1815–1889) invented the spectroheliograph in 1858 and began a daily series of photographs of the Sun. Since the time of these pioneers, photographic sensitivity has increased by about 50,000 times, and practically all new telescopes are designed as photographic instruments.

Techniques. Except for small wide-angle patrol cameras, pure refracting systems are rarely used in astrophotography, because no combination of lenses can bring all colors to exactly the same focus. Mirrors are almost universally employed, often in combination with weak correcting lenses to give a flat focal plane or one of wider angle. Most new large telescopes are CASSEGRAIN TELESCOPES, modified to a form known as Ritchey-Chrétien, which is free from COMA, an aberration that reduces the useful field of view. The SCHMIDT TELESCOPE is used where fields of several degrees have to be covered on one plate.

Research has produced greatly improved photographic materials. Silver bromide, the main light-sensitive agent, is affected most strongly by violet light, but treating the emulsion with suitable dyes has extended its sensitivity across the visible spectrum and into the infrared. Special sensitizing has produced emulsions suitable for the long-exposure photographs needed for faint astronomical objects. Glass plates are preferred to film because of their permanence and suitability for accurate measurement.

Applications. Astrophotography can supply many kinds of information. The recording of faint stars and galaxies requires large aperture and long exposure times (up to perhaps 5 hours). Positional work for stars, minor planets, or comets, needs a large plate scale and a short exposure time. The determination of star brightnesses (PHOTOMETRY) is done on plates covering a wide field of view, to provide comparison stars of known brightness. Multicolor photometry requires filters of different color.

A number of specialized instruments have been developed for solar photography. In general, fixed telescopes of very long focal length fed by a moving mirror system (see COELOSTAT) are used, giving a direct solar image several inches across; whole-disk records now date back over a century. The invention of the CORONAGRAPH makes it possible to photograph the brighter regions of the solar atmosphere in the absence of an eclipse.

Photography has had little, or only very recent impact, in the observation of close double stars and the resolution of fine planetary detail. Both are highly dependent upon atmospheric steadiness, and the best planetary photographs have been obtained from high-altitude observatories such as the PIC DU MIDI OBSERVATORY in France.

The application of photography to astronomy has increased the size of the usefully-observable Universe a hundredfold, and the vital field of spectrum analysis would be practically impossible without it because of the faintness of the images. In general, the photographic plate can record stars up to three magnitudes fainter than the eye could detect using the same instrument; without it we should know little about the nature of our Galaxy, stellar evolution, or the construction of the Universe.

astrophysics

The application of physics to the study of the Universe, as distinct from the measurement of accurate celestial positions (ASTROMETRY). The Universe acts as a giant natural laboratory in which we can observe matter under conditions far more extreme than exist on Earth. Astrophysics developed from the new science of SPECTROSCOPY—the analysis of starlight—which 100 years ago showed that the stars are balls of incandescent gas, of similar physical nature to our Sun. Under the guidance of pioneers such as George Ellery HALE, astrophysics rapidly became a major part of observational astronomy. More recently, the study of matter outside the Earth has been increasingly important to our general understanding of physics, especially processes involving strong gravitational fields and nuclear energy.

Modern solar studies. Astrophysicists are still baffled by many mysteries involving our own Sun. The broad outline of a typical star's structure was

explained 50 years ago by Sir Arthur EDDINGTON, and in 1939 Hans BETHE outlined the nuclear processes that make stars glow. But details of the Sun's behavior do not fit the pattern. One problem is that, although theories predict that particles called neutrinos should be produced in great quantities by the nuclear reactions that power the Sun, no neutrinos have yet been detected. Either the theories are wrong, or our Sun is not a typical star.

Even the origin and behavior of major features such as the Sun's magnetic field and SUNSPOTS are poorly understood. Other phenomena, such as the approximately 11-year variation in the Sun's activity (the SOLAR CYCLE) have yet to be explained by astrophysics. Research into these and other problems of solar structure is now aided by observations from satellites and space probes that study the Sun at wavelengths blocked by the Earth's atmosphere from astronomers on the ground, and which measure the so-called SOLAR WIND of particles streaming from the Sun as it gusts with variations in solar activity.

Other stars. The great wealth of detail we can observe on the Sun is not visible on other stars because of their vast distances. They are therefore treated en masse rather than individually, by combining measurements from many different stars to give average values of characteristics such as brightness, temperature, and size. Astrophysicists study how stars are born, evolve, and die; their composition and its variation between different generations (see POPULATIONS, STELLAR); why they occur in galaxies; and why galaxies themselves cluster together.

Spectroscopy not only reveals details of the temperature and composition of stars; sometimes,

The Southern Cross rises to the left of the moonlit dome of the Anglo-Australian Telescope in this time-exposure showing star trails circling the South Celestial Pole— the simplest type of astrophotography.

Spectra are essential in interpreting the light from stars. A thin objective prism mounted in front of a telescope elongates all the star images into spectra, and in this star field in Cygnus normal stars (showing vertical dark absorption lines on a bright background) can be readily distinguished from the very hot Wolf-Rayet stars (with bright emission lines).

where direct observation shows one star, the spectroscope reveals light from two stars (see SPECTROSCOPIC BINARY). The motions of such twin stars in orbit around each other produce changes in the spectral lines (caused by the DOPPLER EFFECT), from which astronomers can deduce the period of the orbits and the masses of the stars.

Information on the masses, diameters, and temperatures of stars provides the raw data for creating mathematical models of stars in computers. The computer calculates how temperature, pressure, and density vary from the center of a given star to its surface. The model can even be allowed to evolve in the computer, revealing how stars age. The predictions of the star's surface temperature and composition can then be checked with spectroscopic observations of real stars.

Between the stars. Astrophysics is now probing the tenuous matter between the stars. Radio and infrared astronomy reveal many complex molecules in the cool dust and gas clouds of our Galaxy (see INTERSTELLAR MOLECULES). These include so-called organic molecules which are based on carbon, the backbone of life. If such complicated molecules exist in space, they must have been present on Earth from the time it formed, providing building blocks for the assembly of the first life forms (see LIFE IN THE UNIVERSE). Also between the stars are heavy, high-speed particles called COSMIC RAYS, ejected by the violent processes that astronomers observe in the Universe.

Deep space. Astrophysics provides the basic observations to help determine the origin of the Universe. In the 1920s, the American astronomer Edwin HUBBLE proved that there are galaxies outside our own, and that their motions show the Universe is expanding. This led to the idea that the Universe began in a BIG-BANG explosion and has been evolving ever since. Astrophysicists looking deep into space (and therefore far back into time) have found confirming evidence that the Universe looked different in the past: long ago there were objects called QUASARS, emitting vast amounts of radiation at all wavelengths. The advent of RADIO ASTRONOMY has given astrophysicists a new window on the Universe, revealing what appear to be vast explosions in distant galaxies (see RADIO GALAXY). Such powerful events strain the limits of known physical processes. Radio observations have also revealed the so-called

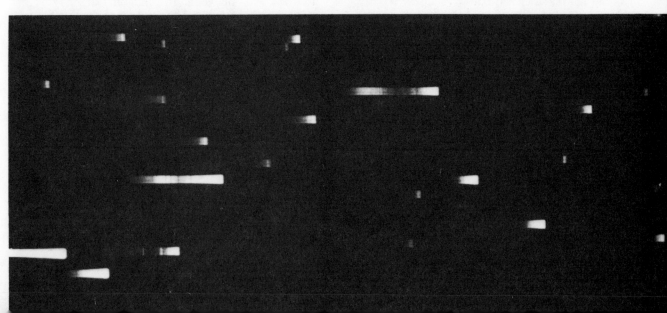

Atlas rocket

BACKGROUND RADIATION, believed to be a relic of the big-bang fireball in which the Universe formed (see also COSMOLOGY).

The new astrophysics. Artificial satellites are opening up new windows on the Universe, particularly for ultraviolet, X-ray, and gamma-ray emissions, which come from intensely hot, energetic objects—quasars in deepest space and small, compact stars in our own Galaxy, possibly including black holes. The objects involved have high gravitational fields; and this provides new information about the way gravity can be converted into energy. The observations help test Einstein's general theory of RELATIVITY, which is concerned with the nature of gravity and provides the best overall theory for describing the Universe. The theory also predicts that gravity waves should ripple through the Universe; one aim of astrophysics in the next decade will be to detect these waves, using sensitive laboratory detectors which oscillate as the wave passes, and observations from space platforms monitored by laser beams.

Back to earth. Astrophysics embraces both old knowledge and new discoveries. The laws of CELESTIAL MECHANICS, which describe the motions of planets in gravitational fields, are today also applied to the motion of spacecraft and the orbits of double stars that include X-ray sources (see X-RAY ASTRONOMY).

But to most people, developments in understanding the Earth and its immediate environment are likely to be the outstanding product of astrophysics. Space-probe studies of other planets in our solar system, and of the relations between the Sun and Earth, are building a new and better picture of our planet. The discovery of the Earth's MAGNETOSPHERE (the extension of the magnetic field into space), and how it interacts with the SOLAR WIND, may explain changes in the atmosphere that affect weather patterns and climate. Studies of the whole solar system's movement through space, and its interaction with interstellar material, could explain such dramatic changes in the Earth's history as the ice ages.

An Atlas rocket with Centaur second stage, launching the Surveyor 4 Moon probe in 1967. Atlas by itself could only be used for orbital launches; with Centaur or Agena second stages, payloads could be sent to the Moon.

Atlas rocket

Major American space launcher, modified from an intercontinental military missile which was first successfully test flown on December 17, 1957. The Atlas was used to launch America's first men into orbit in the MERCURY PROJECT. With upper stages such as AGENA and CENTAUR, Atlas rockets are still used to launch U.S. satellites and space probes. The Atlas has three engines which ignite at lift-off. The two outer engines, each developing 185,000-lb. (84,000-kg) thrust, fall away after $2\frac{1}{2}$ minutes, while the central sustainer engine of 60,000-lb. (27,300-kg) thrust continues to burn; because of this arrangement, the Atlas is often termed a $1\frac{1}{2}$-stage vehicle. The main body of Atlas has a diameter of 10 feet (3 m). Its height and weight vary according to mission, but average around $82\frac{1}{2}$ feet (25 m) and 300,000 lb. (140,000 kg).

atomic time

A means of timekeeping based on oscillations of an atom of cesium, a soft, silver-white metal. The cesium atom oscillates over 9 billion times a second. Counting these oscillations gives the basis of an extremely accurate method of timekeeping, which has now been adopted worldwide. The first such clock was built in 1955; modern versions are accurate to one second in many tens of thousands of years, and accuracies of one second in a million years may eventually be possible. The rotation of the Earth was the previous basis of time signals. But since the start of 1972, the world's legal time system has been based on the readings of an atomic clock. Atomic clocks give far greater accuracy than any previous method of timekeeping, and have revealed that the rotation of the Earth is not constant. In addition to occasional fluctuations, there is an overall tendency for the Earth's rotation to slow down. Extra seconds, called leap seconds, must occasionally be introduced into atomic time to keep it roughly in step with the Earth's rotation. One or two leap seconds are introduced each year.

ATS satellites

Series of satellites testing satellite applications of new scientific and technical developments in communications, observation, weather and

Earth-resource monitoring, and navigation. The initials ATS stand for Applications Technology Satellite. The original ATS program grew out of experience with the SYNCOM communications satellites, and the emphasis has been on communication, including the relaying of television pictures, and experiments in educational broadcasts and medical consultation.

A new generation of ATS satellites started with ATS-6. Using a 30-foot (9.1-m) antenna and far more powerful transmitters than any previous communications satellite, it relays education and health instruction broadcasts to small ground stations; it was first located over the United States to provide links with remote areas in the Rockies, Appalachians, and Alaska. ATS-6 was later moved to a point over Kenya, where it broadcast to several thousand stations in India. ATS-6 also acts as a link between other satellites and the Earth, as for example in relaying data and pictures from the Apollo-Soyuz joint flight in 1975; this is possible because of the much higher orbit of ATS.

a.u.
Abbreviation for ASTRONOMICAL UNIT, a yardstick for expressing distances within the solar system.

Auriga (the charioteer)
Prominent constellation in the northern hemisphere of the sky. It is visible in the night sky during the northern winter, lying on the edge of the Milky Way between Gemini and Taurus. The brightest star in Auriga is the yellow-colored CAPELLA. Auriga contains two star clusters visible in binoculars, and several double stars of interest to astronomers. One of these is the yellow supergiant Epsilon Aurigae, 3,300 light-years distant, which is orbited by a dark companion that eclipses it, causing the star to vary from magnitude 3.4 to 4.5 every 27 years. This companion is believed to be a star surrounded by a cloud of dark dust or gas—possibly a forming planetary system. A spectacular double star in Auriga, Zeta Aurigae, consists of a red giant orbited by a bright blue star.

aurora
An atmospheric phenomenon caused by the impact on the upper atmosphere of atomic particles from the Sun that become trapped in the Earth's magnetic field. Aurorae are most commonly seen near the magnetic poles. In the northern hemisphere they are called the *aurora borealis,* in the southern hemisphere the *aurora australis.* Their popular name is the northern lights or southern lights.

Aurorae appear as diffuse light areas in the sky, often red, yellow, or green in color. Auroral displays can be shaped like arches or folds of drapery, and aurorae often shimmer and move to produce a truly astounding spectacle.

Aurorae occur at heights above 60 miles (100 km). They are more frequent during maximum periods of the Sun's 11-year cycle of activity, when particles are sprayed out from solar flares. Norwegian scientists were foremost in studying aurorae in the early part of this century. They found that the colors of aurorae were mostly produced by molecules of oxygen and nitrogen in the upper atmosphere as they were

bombarded by particles from the Sun, chiefly electrons. One recent discovery has been that aurorae in the north and south hemispheres are linked, with atomic particles oscillating back and forth between the hemispheres along the Earth's magnetic lines of force. Numerous satellite and rocket experiments have been made to learn more about the aurorae, but not all the processes occurring are fully understood yet.

azimuth
The bearing of an object around the observer's horizon; it is the coordinate at right angles to the object's ALTITUDE. Azimuth is measured in degrees from north to a point on the horizon directly beneath the object.

B

Baade, Walter (1893–1960)
German-born American astronomer who discovered there are two essentially different populations of stars, and whose determination of a new distance for the Andromeda galaxy led to a doubling in the scale of intergalactic distances. Using the 100-inch (254-cm) Mount Wilson telescope during the early 1940s, Baade made the first photographs of individual stars in the central region of the Andromeda galaxy. He found that they were reddish in color, in contrast to the blue-white stars found by Edwin HUBBLE in the spiral arms. Baade termed these population II and I respectively (see POPULATIONS, STELLAR). After World War II, Baade used the new 200-inch (508-cm) telescope to study CEPHEID VARIABLE stars in both populations, and in 1952 found that the population I (spiral-arm) stars were brighter—and, thus, farther away—than had previously been supposed. Technically, each population of Cepheids thus had its own PERIOD-LUMINOSITY RELATION. The distance to the Andromeda galaxy had been calculated from observations of population I stars, but using a period-luminosity law that was in fact applicable only to population II stars. The calculated distance was therefore wrong. Baade's new value showed that the Andromeda galaxy was roughly 2 million light-years away, over twice as far as had been thought. Baade's discovery of stellar populations advanced the study of stellar evolution, and his recalibration of intergalactic distances substantially altered concepts about the size and age of the Universe.

Babcock, Harold Delos (1882–1968)
American astrophysicist who, with his son Horace Welcome Babcock (b. 1912), in 1951 developed an instrument for recording magnetic fields on the Sun. This instrument, the solar magnetograph, operates by measuring the effect of magnetic fields on spectral lines (the ZEEMAN EFFECT). The solar magnetograph has revealed regular reversals of the Sun's magnetic polarity. In 1964, Horace Babcock was appointed director of the Mount Wilson and Palomar Observatories (now Hale Observatories). Horace Babcock is also noted for his discovery in 1946 and subsequent study of MAGNETIC STARS.

background radiation
Radiation coming from space at radio and far

infrared wavelengths, believed to be energy left over from the BIG-BANG origin of the Universe. Such radiation was independently predicted by the American physicists George GAMOW in 1948 and Robert DICKE in 1964, and first measured by the American physicists Arno Penzias (b. 1933) and Robert Woodrow Wilson (b. 1936) in 1965. (They received the Nobel prize in physics for this discovery in 1978). Measurements at various radio and infrared wavelengths show that the radiation is consistent with that emitted by a source at a temperature of 2.7°K (−454.8°F), the actual temperature of the Universe. Theory predicts that the Universe would have reached its present temperature as it expanded after the initial big-bang explosion. The radiation was actually emitted, at a temperature of several thousand degrees, but cooled as the density of matter in space thinned out and could not absorb all the energy from the big-bang explosion. This probably happened when the Universe was between 100,000 and 1 million years old (one hundred-thousandth to one ten-thousandth its present age). The exact figure is uncertain because the present density of the Universe is unknown.

Baikonur

Official name for the Soviet space launching site at TYURATAM, near the Aral Sea about 1,300 miles (2,100 km) southeast of Moscow.

Baily, Francis (1774–1844)

English astronomer, best known for describing the phenomena known as BAILY'S BEADS, which appear at a solar eclipse. Baily was a founder of the Royal Astronomical Society. He became widely known for his painstaking analysis and editing of many classic star catalogs, notably those of Ptolemy, Tycho Brahe, Hevelius, and Flamsteed, and for improvements in the

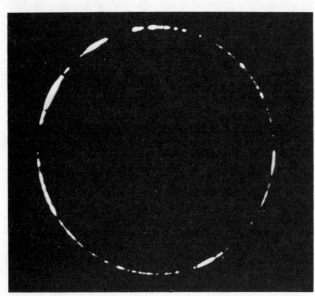

A complete ring of Baily's beads, photographed during the near-total solar eclipse of May 1966. These "beads" are seen during eclipses which occur when the apparent size of the Moon is very slightly less than that of the Sun, allowing the edge of the brilliant solar disk to shine through irregularities in the Moon's rugged limb.

Nautical Almanac. In 1842 he produced a figure for the density of the Earth—5.66 times that of water—which is within 3 percent of the modern value.

Baily's beads

Beads of light at a solar eclipse caused by the Sun's light shining between mountain peaks along the edge, or limb, of the Moon. As the Moon moves across the Sun's face just before totality, the last sliver of sunlight is broken into shining drops by the jagged lunar mountains. As the eclipse ends the same effect occurs at the opposite edge of the Moon. The phenomenon lasts for only a few seconds. Often one bead shines much brighter than others to give what is termed the "diamond ring" effect. At an annular eclipse, the Moon's disk does not quite cover all the Sun, and a complete ring of Baily's beads is seen. The phenomenon is named after the English astronomer Francis BAILY who described it at an annular eclipse in 1836.

Barnard, Edward Emerson (1857–1923)

American astronomer, the leading visual observer of his day and a pioneer of astrophotography. His most famous discovery was of a rapidly moving star close to the Sun, now called BARNARD'S STAR. In 1892 he discovered Jupiter's fifth moon, the first to be found since Galileo's time. In 1889 he began photographing the Milky Way in detail, and showed that its dark regions are not true gaps but are caused by areas of obscuring gas; his *Photographic Atlas of Selected Areas of the Milky Way* was published posthumously in 1927. Barnard also discovered 16 comets. Observing with the Lick 36-inch (91-cm) refractor in 1892–1893, he discovered craters on Mars, though his observations were not published at the time.

Barnard's star

The second-closest star to the Sun, 5.9 light-years away. Barnard's star, lying in the constellation Ophiuchus, is a faint red dwarf of visual magnitude 9.5. The star is named for E. E. BARNARD, who discovered in 1916 that it has the fastest PROPER MOTION of any star across the sky, amounting to 10.3 arc seconds a year. In about 180 years, therefore, it moves across an area of sky equal to the apparent diameter of the Moon. According to Peter VAN DE KAMP, Barnard's star is orbited by at least two planets about the size of Jupiter. Barnard's star is approaching us at 67 miles (108 km) per second. In about 10,000 years' time it will be the closest star of all, 3.8 light-years distant, of magnitude 8.6 and with a proper motion of over 25 arc seconds a year.

Bayer, Johann (1572–1625)

German astronomer who in 1603 published *Uranometria,* the first atlas covering the entire sky. It charts over 2,000 stars, 1,000 of them taken from the observations of Tycho Brahe. Observations of the southern heavens came from the work of Dutch navigators. Bayer created 12 new constellations—Apus, Chamaeleon, Dorado, Grus, Hydrus, Indus, Musca, Pavo, Phoenix, Triangulum Australe, Tucana, and Volans—to accommodate many of the southern stars. He assigned Greek and Latin letters to stars in approximate order of their brightness, a system still used today.

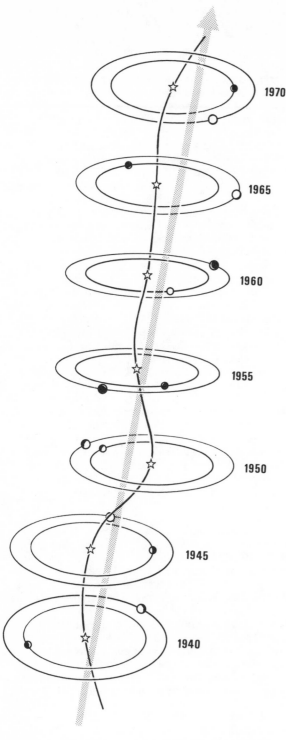

The orbits of two planets around Barnard's star are deduced from the wobble of the star's observed motion. After Peter van de Kamp.

B.D.
Abbreviation for the BONNER DURCHMUSTERUNG star catalog.

Bean, Alan Lavern (b. 1932)
American astronaut who landed on the Moon with the Apollo 12 mission, and commanded the second

crew to visit the Skylab space station. Bean made his first space flight in November 1969 as lunar module pilot on Apollo 12. With his commander Charles Conrad, Bean made two Moon walks, spending a total of seven hours on the lunar surface. In 1973 he led the second three-man Skylab crew on what was then a record-breaking stay in space of 59 days.

Belyaev, Pavel Ivanovich (1925–1970)
Soviet cosmonaut, commander of the Voskhod 2 mission in March 1965 during which Alexei Leonov crawled from the craft to perform the first walk in space. Belyaev had to control the spacecraft's return to Earth manually when the automatic landing system failed. Belyaev died from peritonitis following an operation for stomach ulcers.

Bessel, Friedrich Wilhelm (1784–1846)
German astronomer who made the first measurement of a star's distance. Bessel became director of Prussia's Königsberg Observatory in 1813. There he measured the exact positions of over 50,000 stars. The PROPER MOTIONS of a number of these stars indicated they were relatively close to the Sun. Bessel chose a star with large proper motion, 61 Cygni, and from careful observation found in 1838 that it showed a slight yearly shift in position as observed from different parts of the Earth's orbit (see PARALLAX). The amount of shift revealed the star's distance, which Bessel calculated for 61 Cygni at 10.3 light-years; the modern value is 11.2. In 1844 Bessel announced the presence of unseen companions to Sirius and Procyon, based on an analysis of their irregular proper motions. The companion stars later were discovered visually and are now known to be white dwarfs. Bessel's method is still used to reveal faint companions of stars.

Betelgeuse
A red giant star which marks the right shoulder of Orion; it is also known as Alpha Orionis. Betelgeuse is so large that the orbit of Mars would fit inside it. Because of its huge size it is slightly unstable, and varies irregularly in size and light output. Its magnitude changes from about 0.4 to 1.3 and its size from 300 to 420 times that of the Sun. Betelgeuse is roughly 650 light-years from the Sun.

Bethe, Hans Albrecht (b. 1906)
German-born American physicist who showed in 1938 how stars could generate energy in their interiors by the fusion of hydrogen atoms into helium, through what is called the CARBON-NITROGEN-OXYGEN CYCLE. With the American physicist Charles Louis Critchfield (b. 1910), Bethe also showed that a direct reaction between hydrogen nuclei to produce helium, now called the PROTON-PROTON CHAIN, could account for energy generation in cooler stars. The carbon-nitrogen-oxygen cycle is now believed to operate inside stars hotter than the Sun, while the proton-proton chain predominates in the Sun and cooler stars. Bethe's was the first detailed explanation of stellar energy processes, and it earned him the 1967 Nobel Prize in physics.

Big bang
The supposed event which started the Universe's expansion. Measurements of the current expansion of

the Universe (see HUBBLE'S CONSTANT) suggest that the big bang occurred about 18 billion years ago. At the time of the big bang, all matter in the Universe must have been squeezed into a hot, dense, primeval fireball. Conditions in the fireball were so extreme that not even individual atoms could have existed there. The cosmological expansion was much faster immediately after the big bang explosion than it is now. In this rapid expansion, many complicated processes occurred in a very short time. After about a millionth of a second from the beginning of the expansion, the fireball temperature was on the order of a trillion degrees—hotter than the interiors of the hottest stars, and not until after a few minutes was the temperature low enough for atomic nuclei to form. Calculations show that only helium would have formed in abundance; the rest of the matter remained as uncombined protons, later to form hydrogen atoms. Observations reveal that about 25 percent of the present Universe is helium, presumably mostly of primeval origin (see also ELEMENTS, ABUNDANCE OF).

Within a few hundred seconds the temperature had dropped too low for further nuclear processes to take place, but not until about a million years later did the fireball cool sufficiently for atoms to form. When this happened the material became transparent, allowing the fading glow of the cooling primeval fireball to remain visible ever since. It is now detected as a feeble BACKGROUND RADIATION at a mere 2.7° above ABSOLUTE ZERO.

Much interest attaches to the very earliest stages of the big bang. Before a millionth of a second, little is known of the physical condition of the fireball, but very general considerations of gravity, which governs the explosive motion of the early Universe, lead us to predict that the density of matter would have been limitless, forming a so-called *singularity* at the beginning of the expansion. If so, this moment represents the real "creation" of the Universe. In the future, physics may allow us to probe further back into the earliest stages of the expansion, even to the moment when the structure of space-time was breaking up; possibly this will change our present simple picture of the big bang.

BL Lac objects

A class of elliptical galaxies with variable bright centers, related to quasars; they are also known as Lacertids. Their prototype, BL Lacertae, was originally thought to be a strange variable star.

black hole

A theoretical object whose gravitational pull is so strong that nothing can escape, not even light. Black holes were first suggested in 1795 by the French mathematician Pierre LAPLACE, on the basis of Newton's theory of gravity. In 1939 the American physicist J. Robert Oppenheimer (1904–1967) showed that similar objects are also predicted by Einstein's theory of gravity (the general theory of RELATIVITY).

If a star is greatly compressed—to a few miles or kilometers across in the case of our Sun—its surface gravity would then be so great that nothing could prevent it from shrinking without limit. Such an event may happen in a SUPERNOVA explosion, when the core of the exploding star is tightly compressed. Inside a critical radius, known as the SCHWARZSCHILD RADIUS,

space and time become highly distorted. The effect is to imprison anything which falls through this radius. Therefore black holes can grow in size as they sweep up more material, but no amount of energy can enable a body inside the Schwarzschild radius to reach the outside again. Even light cannot escape; the collapsing star appears totally black.

The fate of the compressed object strains our comprehension. Theory predicts that in a few thousandths of a second the collapsing matter collides in the center of the star at a single mathematical point. The density of the star would then rise to infinity; the very fabric of space-time itself would become so violently curved and distorted that it could no longer exist within the black hole. No one knows what actually happens at the center, except that the star (or any observer who has fallen in after it) cannot reemerge through the Schwarzschild radius into the same Universe again. For during the fleeting destruction of the star, an infinite span of time will have elapsed in the outside world. The star has traveled, quite literally, out of our Universe, leaving only a "hole" in space—a black hole. Recently, it has been discovered that even this black hole slowly disappears. Subtle effects cause the hole to glow faintly. As time goes on, the temperature slowly rises, until eventually the hole simply disappears completely in a flash of radiation.

Although black holes cannot be observed directly, they may be detectable in orbit around other stars. In such a situation, matter from the visible star would be pulled into the black hole, heating up strongly and emitting X-rays as it falls in. Recently, satellites have discovered a number of X-ray sources in the direction of double-star systems. The X-ray source in Cygnus called Cyg X-1 may be just such a system containing a black hole (see X-RAY ASTRONOMY).

blue giant

A large, hot, and bright star. A typical blue giant has a temperature of 20,000° K (36,000° F), a radius 10 times that of the Sun and a brightness of 20,000 Suns. Blue giants can arise in two ways. They may be massive young stars that burn out very quickly, some or all exploding as SUPERNOVAE. The brightest stars of the PLEIADES cluster are blue giants of this type. Or they may be old stars at a late phase of their evolution, passing between RED GIANT and WHITE DWARF stages. This type of blue giant is found in globular clusters. Both types lie at the top right of the MAIN SEQUENCE of stars in the HERTZSPRUNG-RUSSELL DIAGRAM, a plot of star brightnesses against their color.

Bode, Johann Elert (1747–1826)

German astronomer, popularizer of the so-called BODE'S LAW, a formula that roughly describes the distances of the planets from the Sun out to Uranus. Bode was director of the Berlin Observatory, where in 1801 he produced *Uranographia,* the first successful attempt to chart all stars visible to the naked eye, and in which he also included the first systematic delineation of constellation boundaries.

Bode's law

A number series that roughly coincides with the average distance of the planets from the Sun outward to Uranus. It is named for the German astronomer

Johann Bode, who pointed out the relationship in the 1772 edition of his book *Introduction to the Study of the Starry Sky*. According to Bode, if the distance from the Sun to Saturn (then the outermost planet known) is taken as 100, then Mercury is separated from the Sun by 4 such parts. The distance to Venus is $4 + 3 = 7$, to Earth $4 + 6 = 10$, and Mars $4 + 12 = 16$. Then at $4 + 24 = 28$ comes a gap. Jupiter is $4 + 48 = 52$ parts away, and Saturn $4 + 96 = 100$.

When the planet Uranus was discovered in 1781, it seemed to fit this scheme, though not with perfect accuracy. Bode's law led to a search for the missing planet between Mars and Jupiter, resulting in the discovery of the asteroids. The law was also used as a starting point by astronomers in search of planets beyond Uranus. However, it clearly breaks down toward the edge of the solar system. According to modern views, the planets would have come naturally to take up station in orbits where they least perturbed each other; such an arrangement would spontaneously produce some kind of Bode's law progression of distances in any planetary system.

Bode's "law" is thus not really a law at all. What is more, it was first put forward in 1772 by the German mathematician Johann Daniel Titius (1729–1796), and is now often referred to as the Titius-Bode law.

bolide
An exploding or fragmenting bright meteor (see FIREBALL).

bolometer
A device for measuring radiation emitted by an object at all wavelengths, invented in 1878 by the American astronomer Samuel Pierpont Langley (1834–1906). Radiation falling on the bolometer's detector causes a rise in temperature, which changes the electrical resistance of a circuit; the amount of resistance change reveals the intensity of the incident radiation. The so-called *bolometric magnitude* of a star is its total brightness over all wavelengths; *bolometric correction* is the difference between the visual and bolometric magnitudes of a star.

Bond, William Cranch (1789–1859)
American astronomer, pioneer of the use of photography in astronomy, who in 1850 took the first photograph of a star, Vega. Bond was the founder and first director of the Harvard Observatory, and was succeeded by his son George Phillips Bond (1825–1865). Together the two mapped the sky photographically, studied comets, and discovered the eighth satellite of Saturn, named Hyperion, and Saturn's transparent inner ring, called the crepe ring. George P. Bond explored the potential of photography for measuring star brightnesses and recording positions. In 1857, he took the first photograph of a double star, Mizar.

Bondi, Sir Hermann (b. 1919)
Austrian-born British mathematician, who with Thomas GOLD proposed the STEADY STATE theory of cosmology in 1948, maintaining that the appearance of the Universe has been constant throughout time. Bondi has done major work on the theory of gravitational radiation and other areas of astrophysics, notably the internal structure of normal stars like the Sun. From 1967 to 1971 he was director-general of the European Space Research Organization, ESRO; he then became chief scientific adviser to the British Ministry of Defence.

Bonner Durchmusterung
One of the world's major star catalogs, prepared by the German astronomer Friedrich ARGELANDER over a period of 25 years. The catalog was first published in 1859–62 and contained 324,198 stars down to about magnitude 9.5, from the north pole to $-2°$ in declination. The stars were plotted on charts published in 1863 the best of their time. Although not as accurate as later catalogs, the *Bonner Durchmusterung* retains its importance because of the large number of stars listed. Argelander's colleague Eduard Schönfeld (1828–1891) extended the catalog to $-23°$ in 1886, plotting 133,659 stars.

booster
Term often used to describe the first stage of a launch rocket, or sometimes the whole rocket. *Strap-on boosters* are subsidiary rockets that augment the thrust of the first stage at lift-off. They fall away as the rocket climbs. VOSTOK and TITAN III are examples of rockets that use strap-on boosters.

Boötes (the herdsman)
A large constellation in the northern hemisphere of the sky, best placed for observation in the spring. The brightest star in Boötes is the red giant ARCTURUS. Boötes is found at the end of the handle of the Plow (Ursa Major); it is shaped like a kite, with Arcturus at its tail. Boötes contains a number of interesting double stars, including Epsilon Boötis, also called Izar or Pulcherrima. Of visual magnitude 2.59, it is 230 light-years away; the main star is also a SPECTROSCOPIC BINARY.

Borman, Frank (b. 1928)
American astronaut, commander of the Apollo 8 mission, which made 10 orbits of the Moon on December 24–25, 1968, before returning to Earth. Borman, a qualified aeronautical engineer, became an astronaut in 1962. His first space flight was with copilot James Lovell in Gemini 7 during December 1965, when Gemini 6 joined them in orbit for the first rendezvous between two manned spacecraft. Borman and Lovell went on to break the duration record for space flight, becoming the first men to spend two weeks in space.

Boss, Lewis (1846–1912)
American astronomer, responsible for one of the best-known and most accurate modern star catalogs, the *Boss General Catalogue*. Lewis Boss died after publishing only a preliminary catalog containing 6,188 stars. But his son Benjamin Boss (1880–1970) completed the catalog, finally published in 1937, with 33,342 stars. From his work on star positions, Lewis Boss found in 1908 that the stars in the HYADES appear to be moving away from the Sun en masse. This common movement provides an important trigonometrical method of finding the distance of the cluster (see MOVING CLUSTER METHOD).

Bowen, Ira Sprague (1898–1973)

American astrophysicist, who made extensive and
important studies of the spectra of nebulae, showing
that mysterious green-colored emission lines which
had been attributed to an unknown element called
"nebulium" were actually caused by ionized oxygen
and nitrogen atoms under the near-vacuum conditions
of space (see FORBIDDEN LINES). Bowen became
director of the Mount Wilson and Palomar
Observatories (now Hale Observatories) in 1948.
He was responsible for the installation of the 200-inch
(508-cm) reflector, and designed many subsidiary
pieces of equipment for use with large telescopes.

Bradley, James (1693–1762)

English astronomer, discoverer of the ABERRATION OF
STARLIGHT and NUTATION. Bradley's uncle was James
Pound (1669–1724), a clergyman and amateur
astronomer who introduced him to the study of the
stars. Bradley became professor of astronomy at
Oxford in 1721 and devoted the rest of his life to
science.

At the private observatory in Kew, near London,
of Samuel Molyneux (1689–1728), a politician and
amateur astronomer, Bradley began in 1725 to look
for the effect of PARALLAX in star positions. Nearby
stars should show a slight motion during the year
because of the Earth's changing position in its orbit.
Bradley concentrated on the star Gamma Draconis,
which appeared nearly overhead. He soon found a
shift, but it was too large and in the wrong direction to
be parallax. Bradley realized the shift was due to the
Earth's motion across the path of the incoming
starlight, which slightly distorted the true position of
the star. This unexpected effect, which he announced in
1728, is called the aberration of starlight. It was not
what Bradley was looking for, but it was nonetheless
the first direct observational proof of the Earth's
motion around the Sun. From the amount of
aberration Bradley calculated the speed of light at
183,000 miles (295,000 km) per second—very close to
the modern value.

Bradley was made astronomer royal in 1742. At
Greenwich his accurate observations confirmed
another shift in the position of several stars, which
could not be explained by aberration. In this case the
entire Earth was nodding slightly in space; this effect
is called nutation. Bradley believed it occurred
because the Moon's orbit moves once around the
Earth every 18.6 years. He extended his observations
over two decades before he finally announced the
effect and its cause in 1748.

Bradley continued to log accurate star positions
throughout his time at Greenwich. His observations
were later assembled into an important star catalog
by the German astronomer Friedrich BESSEL.

Brahe, Tycho (1546–1601)

Danish astronomer, the greatest observer of the
pretelescopic era. His painstaking records of planetary
positions provided the German mathematician
Johannes KEPLER with the raw data for his laws of
planetary motion, which for the first time gave an
accurate description of planetary orbits and finally
established the modern view of the solar system.

Tycho, as he is usually called, was born into a noble
family and had an adventurous upbringing. Although

The greatest observer of the pre-telescopic era, Tycho
Brahe. His discoveries brought him wealth and fame
during his lifetime, yet today we remember him for the
observations which he bequeathed to his assistant—
Johannes Kepler.

sent to study law, Tycho was so impressed by a solar
eclipse in 1560 that his interest was turned to the
study of the sky. All his time was spent reading about
and observing the heavens, and he became dismayed
when he realized from his observations that existing
tables of planetary motion were seriously in error. He
therefore resolved to prepare the most accurate
observational records possible, and built instruments
larger and more accurate than any before.

The quality of his work became clear in 1572, when
an exploding star, or supernova, flared up in the sky.
Tycho's observations showed that the supernova must
lie far beyond the atmosphere and in the realm of the
stars, which had always been considered perfect and
unchanging. This was a major blow to astronomical
tradition. Tycho's book on the star, De nova stella,
made his reputation, which was confirmed in 1577
when he was able to show that a bright new comet
also lay far beyond the Moon. Tycho proved that the
comet of 1577 moved in an orbit that took it among
the planets. This refuted the age-old view that the
planets orbited the Earth on solid crystal spheres;
the comet would have shattered those spheres had
they existed. But Tycho could not bring himself to
accept the alternative view of COPERNICUS that all
objects orbited the Sun. Instead, he proposed his own
system, in which the planets (and comets) orbited the
Sun, which in turn went around the stationary Earth.

In 1576 the king of Denmark granted Tycho the
island of Ven between Denmark and Sweden, where
he set up a castle called Uraniborg and an observatory
called Stjerneborg. This became the astronomical
center of the world, over which Tycho reigned like a

Eta Carinae nebula is a beautiful and complex cloud of luminous gas in the southern Milky Way, photographed here with the UK Schmidt telescope in Australia. The dark patch in the center is known as the Keyhole, because of its shape.

Above The famous naked-eye star cluster known as the Pleiades, or Seven Sisters, is made up of young, hot stars. Fuzzy wisps around the brightest stars are the remains of the cloud of matter from which the Pleiades formed.

Below left NGC 3324 is an open star cluster, seen in a photograph from the Anglo-Australian Telescope. The cluster's biggest and brightest star appears different in color from the blue stars of the rest of the cluster because it has aged faster and has already evolved to become a red supergiant.

Below right The Coalsack is a dark cloud of dust in the constellation of Crux, the southern cross. It absorbs light from the stars behind it, thereby creating the appearance of a hole in the Milky Way.

true king. At Ven he produced series of observations more accurate than any before, taking positions of stars and planets to the limit of naked-eye accuracy and measuring the length of the year to within one second. In Tycho's lifetime nearly every astronomical constant was revalued, and the calendar drastically reformed.

Tycho, however, was an exceptionally arrogant and argumentative man. In a student duel in 1565 he had lost the bridge of his nose, which he replaced with a bizarre metallic substitute. In 1588, the king of Denmark died; the new king was not prepared to tolerate him, and so after 20 years on Ven, Tycho left. Eventually, in 1599, he settled at Benatky Castle near Prague, where he had been appointed imperial mathematician to the Emperor Rudolf II. There he embarked on an analysis of the lunar and planetary observations he had collected at Ven. In 1600 he was joined by the young Johannes Kepler. The following year Tycho died, leaving his observations to Kepler in the hope that he would be able to prove the Tychonic theory of the heavens. Ironically, Kepler used Tycho's remarkable work to establish instead the Copernican theory of the solar system.

Brans-Dicke cosmology
A variant of Einstein's theory of gravity (general RELATIVITY) developed in the early 1960s by the American physicists Carl Henry Brans (b. 1935) and Robert H. DICKE, in which a new type of field produced by matter throughout the Universe plays an important role. The Brans-Dicke theory proposes that this field changes the geometry of space and time, thus affecting the motion of bodies in a gravitational field. One consequence is that the force of gravity would appear to grow progressively weaker as the Universe expands. Although the predicted weakening would be very slight (less than one part in ten billion each year), the accumulated effect over astronomical time scales would have a drastic effect on the solar system, making the Sun, for example, much brighter in the distant past. Observations of lunar and planetary motions have so far failed to provide convincing evidence for the weakening of gravity.

bremsstrahlung
Radiation emitted by high-speed particles, particularly electrons, as they are suddenly slowed down or scattered by atoms. The word *bremsstrahlung* is German for "braking radiation." As the particles are retarded, part of their energy is converted into electromagnetic radiation; the intensity and wavelength of the radiation depend on the rate of retardation. Bremsstrahlung is usually associated with the emission of radiation in stars and galaxies.

Bruno, Giordano (1548–1600)
Italian philosopher, a Dominican monk, who was an early supporter of COPERNICUS' theory that the Earth and other planets orbit the Sun. Bruno envisaged an infinite Universe, with endless suns each having their own retinue of planets. He believed that these planets might be populated, and pointed out that the inhabitants would think they were at the center of the Universe. He was thus among the first to show that our view of the Universe is purely relative. Bruno was burned at the stake by the Inquisition for his

opposition to the idea that any absolute truth could exist. Fear of the revival of similar heretical philosophies was also behind the Inquisition's persecution of GALILEO.

Burbidge, Geoffrey (b. 1925) and Eleanor Margaret (b. 1922)
British husband-and-wife team of astrophysicists, who with William Fowler and Fred HOYLE showed in 1956 that heavy elements are constantly being built up from light ones inside stars, and were therefore not all produced in the supposed big-bang explosion at the start of the Universe. Margaret Burbidge's observations of the rotation of galaxies led to the first accurate estimates of galactic masses. The Burbidges also discovered that QUASARS show several different red shifts in their spectral lines, indicating material is being ejected at high speed. From 1972 to 1973 Margaret Burbidge was director of the Royal Greenwich Observatory, the first woman to hold the post.

Bykovsky, Valery Fyodorovich (b. 1934)
Soviet cosmonaut who made the longest solo space flight in history. He spent almost five full days in the Vostok 5 spacecraft during June 1963. He was joined in orbit for three days by female cosmonaut Valentina Tereshkova in Vostok 6. Bykovsky was selected as a cosmonaut in 1960.

Byurakan Astrophysical Observatory
Major Soviet observatory, located near the city of Yerevan in Armenia. The observatory was founded in 1946 by Viktor AMBARTSUMIAN, its current director. Among other instruments are a Schmidt telescope with a 39-inch (100-cm) corrector lens and a 52-inch (132-cm) mirror. Its largest telescope is a 102-inch (260-cm) reflector.

C

Caelum (the chisel)
A small and faint constellation in the southern hemisphere of the sky near the foot of ERIDANUS. Its name was given by de Lacaille in the 1750s.

calendar
A timetable for reckoning days and months of the year. The first calendars go back to primitive times, when the Moon's $29\frac{1}{2}$-day cycle of phases was the guide. As man developed an agricultural way of life, he required a calendar linked to the seasons, which entailed using the Sun, not the Moon. The calendar now adopted throughout the world is solar. Some lunar calendars still exist for determining religious festivals, but since the year contains 365 days, and no $29\frac{1}{2}$-day period can be exactly divided into it, lunar and solar calendars remain separate. The basis of our present calendar is Roman, although the Romans originally used a lunar, and not a solar system of reckoning. In the eighth or seventh century B.C. a ten-month lunar calendar had been drawn up with months of 30 days and 31 days, giving a total of 304 days; the balance of 61 days was left as a gap. In due course extra days were added until the total reached 355. Yet this "Roman Republican Calendar" still did not keep in step with the seasons, and it was

modified to a solar one by inserting an extra or
"intercalary" month. This was added between
February 23 and 24 once every two years; it contained
either 27 or 28 days and, when it appeared, the
remaining five days of February were omitted. The
intercalary month thus gave an additional 22 or 23 days
every two years which, added to the 355-day year,
gave a solar calendar of 366¼ days.

The intercalary month was inserted on the order of
the priesthood, but by the middle of the first century
B.C. the calendar had reached a most unsatisfactory
state. In consequence, Julius Caesar consulted the
Greek astronomer Sosigenes (fl. 1st century B.C.), and
decided to abandon a lunar calendar altogether. The
year was taken as 365¼ days in length, a total of
90 days were intercalated, and the year beginning
March 1, 45 B.C. became January 1 of the Julian
calendar. To obtain the odd quarter day, one
intercalary day was inserted every fourth year
between February 23 and 24.

Various changes in the lengths of months were
introduced into the Julian calendar; our present
system arose after July was named for Julius Caesar,
August after Augustus, and the arrangement of 30
and 31 days was adopted, with 28 days allotted for
February. The Julian calendar contained no weeks,
merely business and nonbusiness days. The seven-day
week did not arrive until the fourth century A.D.

The adoption of 365.25 days for the length of the
year is only an approximation; the precise figure is
365.242199 days. This small difference of only
11 minutes 14 seconds per annum mounts up to
1½ days in two centuries, and 7 days in 1,000 years. By
1545 the date of the vernal equinox was 10 days out,
and the determination of Easter was affected. By
1572, the year of the accession of Pope Gregory XIII,
a new and more correct calendar was prepared with
the help of the Bavarian Jesuit astronomer
Christopher Clavius (1537–1612). The year was taken
as 365.2422 days; and it was agreed that 3 out of 4
century years should not be leap years; in practice
a century year is not a leap year unless exactly
divisible by 4. Thus 1700, 1800, and 1900 were not
leap years, but 2000 will be. Most Roman Catholic
countries adopted the Gregorian calendar in 1582;
other countries followed much later, with Britain
and America changing in 1752.

Callisto
Second-largest satellite of Jupiter, 3,100 miles
(5,000 km) in diameter, discovered by Galileo in 1610.
It orbits Jupiter every 16 days 16 hours 32 minutes at
an average distance of 1,169,800 miles (1,882,600 km).
Callisto's mass is about 1.5 times that of our Moon,
but its density is only 1.65 times that of water; to
account for this low density, one theoretical model
suggests that the satellite has a muddy core
surrounded by a deep, slushy mantle topped with a
thin icy crust. Some dark markings have been charted
on Callisto, so that at least part of its surface may
be covered with dust.

Cambridge Radio Observatory
See MULLARD RADIO ASTRONOMY OBSERVATORY.

Camelopardalis (the giraffe)
A large but faint constellation near the north pole of

the sky, introduced in 1624 by the German
mathematician Jakob Bartsch (1600–1633), a
son-in-law of Johannes Kepler. Its name is sometimes
also written as Camelopardus.

Cancer (the crab)
A constellation of the zodiac, best visible in the
northern winter. The Sun passes through Cancer
from late July to mid-August. Cancer is the faintest
of the zodiacal constellations, having no particularly
bright stars. Its most famous feature is the star cluster
called Praesepe, also known as the beehive, bearing
the catalog numbers M44 and NGC 2632. Praesepe
contains several hundred stars 572 light-years away.

Cancer, Tropic of
The farthest latitude north of the equator at which the
Sun appears overhead during the year. The latitude
of the Tropic of Cancer corresponds to the angle of the
Earth's axial tilt—23° 26′ 32″ in 1976 (the angle
changes slightly with time because of the perturbing
effects of the Sun, Moon, and planets). The Sun is
overhead at noon on the Tropic of Cancer at the
summer SOLSTICE, around June 21 each year. About
2,000 years ago, the Sun lay in the constellation of
Cancer at the summer solstice. But the drifting effect
of the Earth's axis called PRECESSION has carried the
summer solstice into Gemini, and before the end of
the century it will have moved into Taurus. Its
southern hemisphere equivalent is the Tropic of
CAPRICORN.

Canes Venatici (the hunting dogs)
A constellation of the northern hemisphere of the sky
lying between Boötes and Ursa Major, best seen in
the northern hemisphere spring. It was introduced
in 1536 by the German cartographer Petrus Apianus
(Peter Bienewitz or Bennewitz; 1495–1552), but
reached its present form on the star map of Johannes
HEVELIUS. The brightest star, Alpha, was named Cor
Caroli (Charles' heart) by Edmond Halley; it is a
double star of total magnitude 2.8. Canes Venatici
contains the famous Whirlpool galaxy, M51 (NGC
5194), the first galaxy in which spiral structure was
recognized, by Lord ROSSE in 1845. The constellation
also contains a bright globular cluster, M3 (NGC
5272).

Canis Major (the greater dog)
A prominent constellation in the southern hemisphere
of the sky, containing the brightest star in the sky,
SIRIUS. From the northern hemisphere, Canis Major
is best seen during winter. The star Epsilon, also known
as Adhara, is 18 times the Sun's diameter, of magnitude
1.50; it has a companion of magnitude 8.1.

Canis Minor (the lesser dog)
A small constellation lying below Gemini in the
equatorial region of the sky, best seen during the
northern hemisphere winter; with Canis Major it
represents the dogs of Orion, near whose feet it lies.
The major star of Canis Minor is PROCYON.

Cannon, Annie Jump (1863–1941)
American astronomer, responsible for the Harvard
system of classifying star spectra, now universally
used by astronomers. Miss Cannon's work at the

Harvard College Observatory showed that stars can be grouped into a small number of classes related to their color. These types are assigned the letters O, B, A, F, G, K, M—ranging from hottest to coolest; each letter is also subdivided from 0 to 9 (see SPECTRAL TYPE). Miss Cannon was the major author of the massive *Henry Draper Catalogue,* issued in nine volumes between 1918 and 1924, which contains the spectral classification of 225,300 stars; later supplements brought the total to over 350,000 (see also DRAPER, HENRY). Miss Cannon also published two volumes of variable stars, in 1903 and 1907, the second containing about 2,000.

Canopus
The second-brightest star in the sky, of magnitude −0.73, also called Alpha Carinae. Canopus is a yellow supergiant, 25 times the diameter of the Sun, and 110 light-years away. Canopus, named for the mythical Greek helmsman, is often used as a guide star for spacecraft navigation.

Cape Canaveral
A sandy promontory on the Atlantic coast of Florida, the main launching site for U.S. space missions. Cape Canaveral was first used as a missile test site in 1950, with the launch of modified V-2 rockets. Rows of permanent launchpads were later constructed close to the shoreline for missile tests and space launchings; each type of rocket requires its own launch facilities. The Cape is now shared by NASA's KENNEDY SPACE CENTER, which controls civilian space programs, and the Cape Canaveral Air Force Base, which handles military launches. NASA built the Vehicle Assembly Building (V.A.B.) farther inland on Merritt Island; in this enormous building the giant Saturn family of rockets for the Apollo lunar landing program and the Skylab space station missions were assembled before being driven to the launchpad. The V.A.B. has been modified to handle the SPACE SHUTTLE. Other rockets are assembled on the launchpad. Rockets launched from the Cape fly southeastward over the Atlantic, where they are tracked by the Air Force's Eastern Test Range facilities based on islands, ships, and aircraft. The Eastern Test Range was originally established for tracking missile tests and has its headquarters at Patrick Air Force Base, 15 miles (24 km) south of the Cape. Cape Canaveral is used for launches by the Army, Navy, and Air Force as well as by NASA. In 1963 it was renamed Cape Kennedy, but in 1973 reverted to its original name.

Cape Observatory
Familiar name for the Royal Observatory, Cape of Good Hope, which was founded in 1820. The Scottish astronomer Sir David Gill (1843–1914) was Royal Astronomer at the Cape from 1879 to 1907, where he took numerous photographs of the southern skies. These were analyzed in the Netherlands by Jacobus KAPTEYN to produce the *Cape Photographic Durchmusterung,* a listing of 454,875 stars down to magnitude 9.5 between declination −19° and the south pole, published between 1896 and 1900. At the start of 1972, the Cape Observatory merged with the Republic Observatory in Johannesburg to form the SOUTH AFRICAN ASTRONOMICAL OBSERVATORY, which has its headquarters at the Cape.

Capella
The seventh-brightest star in the sky, also known as Alpha Aurigae. It is a double-star system consisting of two yellow stars each about three times the Sun's mass, orbiting every 104 days. Their light combines to give Capella an apparent magnitude of 0.09; the system is 45 light-years away.

Capricorn, Tropic of
The southernmost latitude on Earth at which the Sun appears directly overhead at noon. This occurs on the winter SOLSTICE (about December 22). The latitude corresponds to the inclination of the Sun's path or ECLIPTIC to the celestial equator (23° 26′ 32″ in 1976). Although the Sun used to lie in the constellation of Capricorn at the solstice, the drifting or PRECESSION of the Earth's axis has now carried the position into Sagittarius (see also CANCER, TROPIC OF).

Capricornus (the sea goat)
An inconspicuous constellation of the zodiac, best seen during the northern hemisphere late summer; the Sun passes through the constellation from late January to mid-February. Its brightest star, Alpha, appears to be double; in fact, the two stars are unconnected, one being 116 light-years away and the other 1,100.

carbon-nitrogen-oxygen cycle (CNO cycle)
A chain of nuclear reactions by which energy is released in stars twice or more as massive as the Sun. The principal result of the reactions is to turn hydrogen into helium. The nucleus of a helium atom is a little

A representation of the carbon cycle (more fully, the carbon-nitrogen-oxygen cycle). Beginning at the bottom, a carbon 12 nucleus and a hydrogen nucleus (a proton) together form nitrogen plus a photon at gamma ray wavelengths (shown by γ). The nitrogen decays to carbon 13, giving off an electron and a neutrino (ν) The reactions continue round the cycle with the emission of more photons, electrons and neutrinos until carbon 12 is reached again, with the overall consumption of four protons and the production of one helium nucleus.

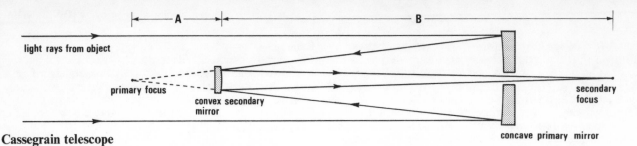

light rays from object

primary focus

convex secondary mirror

secondary focus

concave primary mirror

Cassegrain telescope

less heavy than the four hydrogen nuclei (protons) which fuse to make it; this small amount of missing mass is turned into energy, which is released to power the star. The reaction begins when protons are added one by one to a carbon nucleus, turning it into nitrogen and then oxygen nuclei, before it decays to its original form, ejecting a helium nucleus as it does so. The net result of the cycle is to convert hydrogen to helium, leaving the carbon unchanged to begin another cycle.

The excess mass is released as energy in the form of neutrinos, elusive subatomic particles, and gamma rays, energetic photons. The neutrinos speed directly out of the center of the star, where the nuclear reactions occur; but the gamma rays diffuse out only slowly, becoming degraded along the way to a much larger number of lower energy photons, which leave the star's surface in the form of visible light, and ultraviolet and infrared radiation. During the CNO cycle, three gamma rays and two neutrinos are created from the mass which is destroyed. In stars of about the Sun's mass or less, the CNO cycle still occurs, but it produces less of the star's energy than the PROTON-PROTON CHAIN.

carbonaceous chrondrite
A fragile type of stony meteorite, containing carbon. See CHONDRITE.

Carina (the keel)
A constellation in the southern hemisphere of the sky, formerly part of the ancient constellation of Argo Navis but made separate by Nicolas Louis de Lacaille. The main star in Carina is the brilliant CANOPUS, but the constellation's most famous feature is Eta Carinae, a spectacularly variable star about 6,800 light-years away, embedded in a bright, diffuse nebula called NGC 3372. After varying between fourth and second magnitude for two centuries, Eta Carinae surged up to become the second-brightest star in the sky in 1843. It has faded since to the limit of naked-eye visibility, at which it remains. The star is surrounded by an expanding gas shell that seems to have been thrown off in its outburst; some astronomers speculate that Eta Carinae may be a peculiar kind of supernova.

Carpenter, Malcolm Scott (b. 1925)
Second American astronaut to orbit the Earth. Carpenter, who had been backup to John Glenn for the first orbital flight, circled the Earth three times on May 24, 1962. He used up more fuel for maneuvering the spacecraft while in orbit than planned, and landed 250 miles (400 km) off target.

Cassegrain telescope
A type of reflecting telescope in which the image is observed through a hole in the main mirror. Light is collected, as in a normal reflecting telescope, by a

large concave primary mirror, and is reflected to a smaller convex mirror facing it. This secondary mirror, positioned inside the main mirror's focus, reflects the light back to a hole in the center of the main mirror. The secondary mirror thereby increases the FOCAL LENGTH of the telescope without lengthening the tube. The compactness of the Cassegrain design makes it favored for large reflecting telescopes. The primary mirror usually has a FOCAL RATIO between 3 and 5; the secondary mirror can increase this to an effective focal ratio of between 12 and 30.

The design was invented in 1672 by the French physicist N. Cassegrain. In the classical Cassegrain telescope, the primary mirror is a PARABOLOID and the secondary a HYPERBOLOID. Other combinations of curvatures may also be used. The Dall-Kirkham modification employs a spherical secondary mirror and a primary that is between a sphere and a paraboloid. The Ritchey-Chrétien type, named for the American G. W. RITCHEY and the French astronomer and optician Henri Chrétien (1879-1956), has hyperbolic curves on both mirrors and is free from COMA; it therefore has a wider usable field than other types. For this reason it is employed in most large telescopes.

Cassini, Giovanni Domenico (1625–1712)
Italian-French astronomer whose observations added significantly to knowledge of the solar system. As professor of astronomy at the University of Bologna, Cassini in 1665 measured the rotation period of Jupiter to within a few minutes, and the next year did the same for Mars. In 1668 he made tables of the motion of Jupiter's four bright satellites, which were later used by Olaus ROEMER in his measurement of the speed of light. In 1669 Cassini was invited to Paris, becoming director of the new Paris Observatory on its opening in 1671; he took French nationality in 1673 and changed his first names to Jean Dominique. At Paris, Cassini discovered four satellites of Saturn —Iapetus, Rhea, Tethys, and Dione—between 1671 and 1684, and in 1675 drew attention to the dark gap in the rings of Saturn now called Cassini's division. He also realized that the rings are not solid, but consist instead of countless tiny particles orbiting like little moons. In 1672, with the help of observations by Jean Richer (1630–1696), Cassini measured the distance of Mars, thereby producing by far the most accurate estimate of the scale of the solar system.

On his death, his son, Jacques Cassini (1677–1756), succeeded him as director of the Paris Observatory. Jacques compiled the first tables of the motion of Saturn's satellites, and determined the proper motion of the star Arcturus. César François Cassini de Thury (1714–1784), son of Jacques, became director of the Paris Observatory in 1756, and began a major map of France. This was completed by his own son, Jacques Dominique, comte de Cassini (1748–1845), who succeeded him as director.

Cassini's division

Gap about 1,700 miles (2,700 km) wide in Saturn's rings, separating ring A and ring B and named for its discoverer Giovanni Domenico CASSINI. It is not totally empty, but contains far fewer particles than do the adjacent rings. The division is caused by the gravitational effects of Saturn's satellites, notably Mimas. Particles in Cassini's division orbit Saturn with half the period of Mimas, and are thus strongly perturbed and tend to move into a different orbit. This explanation was advanced in 1867 by Daniel Kirkwood (see also KIRKWOOD GAPS).

Cassiopeia

A prominent constellation in the north polar region of the sky, lying on the edge of the Milky Way and named for a queen of Greek mythology. Its brightest stars form a shape like a letter W, the apex of which points toward the pole star. The apex star, Gamma, is an irregular variable star, which seems to throw off shells of material. Near the star Kappa was the supernova observed by Tycho Brahe in 1572; this is now a radio source, 11,400 light-years away. One of the most powerful radio sources in the sky, Cassiopeia A, lies near the star cluster labeled M52; this source is also the remains of a supernova.

Castor

Second-brightest star in the constellation Gemini, also known as Alpha Geminorum. It is a system of six different stars, with a total visual magnitude of 1.58. The two main stars, magnitudes 1.96 and 2.89, are in mutual orbit every 420 years; they are each SPECTROSCOPIC BINARIES, of periods 9.2 and 2.9 days respectively. The system is completed by a far-off pair of red dwarf stars, eclipsing every 19 hours 33 minutes and varying in magnitude between 9.1 and 9.6. Castor is 45 light-years away.

catadioptric system

An optical system that uses both lenses and mirrors to form an image. Combining reflection and refraction helps to improve telescopic performance (particularly in suppressing of COMA). One of the first such attempts was made by the Swiss optician Emile Schaer (1862–1931), who between 1913 and 1922 used a lens near the focus of a 39-inch (100-cm) reflecting telescope to improve its field of view for photography. Later, the American astronomer Frank E. Ross designed coma-correcting lenses for use with the large American reflectors. The first truly catadioptric system, however, was the wide-field telescope designed by Bernhard SCHMIDT. A similar idea, with a different type of lens, was employed in the MAKSUTOV telescope (1944). Another important catadioptric instrument is the extremely fast super-Schmidt telescope designed by the American optical engineer James Gilbert Baker (b. 1914). By introducing extra lenses into the system, catadioptric patterns allow the designer to correct simultaneously aberrations including coma, ASTIGMATISM, and field curvature.

Catalina Observatory

The observing station of the University of Arizona's Lunar and Planetary Laboratory, situated at 8,235 feet (2,510 m) in the Catalina Mountains. The LPL operates a 61-inch (155-cm) reflector for lunar and planetary photography, inaugurated in 1965. Nearby is the Mount Lemmon Infrared Observatory.

celestial mechanics

The study of the motion and gravitational interactions of bodies in space. Celestial mechanics is used to calculate the orbits of bodies, and to predict their positions for constructing tables of their motions. Similar techniques are now used to calculate the orbits of satellites and space probes; this branch is often termed *astrodynamics*. Celestial mechanics was born in 1687, when Isaac NEWTON set down his three laws of motion, and his discovery that all bodies attract each other by the force of gravity. The Swiss mathematician Leonhard Euler (1707–1783) then established classical methods for determining the motions of the Moon and comets. Careful measurements indicated that the long-range gravitational effects of the planets produced perturbations on solar system bodies. These effects were investigated by a series of brilliant French mathematicians in the mid-18th century. Alexis Claude Clairaut (1713–1765) computed the perturbations of HALLEY'S COMET by the major planets; Joseph Louis LAGRANGE discovered secular (time-dependent) effects in the solar system; and the marquis de LAPLACE proposed his nebular hypothesis for the formation of the solar system after considering the stability of rotating fluids. The greatest achievement of this period was the prediction of the planet Neptune from its perturbations on Uranus by John Couch ADAMS and independently by Urbain LEVERRIER. The start of the 20th century saw the introduction of further refinements by such workers as Simon NEWCOMB and the French mathematician Jules-Henri Poincaré (1854–1912). Shortly afterward, Albert EINSTEIN proposed his theory of RELATIVITY, which helped explain details in the motion of some bodies, particularly Mercury. The computational tradition of Newcomb and his colleagues was continued and extended by Dirk Brouwer (1902–1966), Gerald Maurice Clemence (1908–1974), and Wallace John Eckert (1902–1971); the latter introduced computer techniques which have revolutionized celestial mechanics. In 1951, the three published the monumental *Coordinates of the Five Outer Planets, 1653–2060,* which serves as the basis of all research involving the motions of the planets from Jupiter to Pluto.

celestial sphere

The imaginary sphere of the heavens, with the Earth at its center, which appears to rotate once every day. All astronomical objects appear to lie on the surface of this sphere. In ancient times the stars were believed to be points on a real sphere revolving around the Earth; the concept is retained as a useful device in establishing a coordinate system to specify the position of an object in the sky.

The Earth's daily west-to-east rotation makes the celestial sphere appear to rotate from east to west every 23 hours 56 minutes 4 seconds (a SIDEREAL DAY). The celestial sphere rotates about the north and south celestial poles, which are in line with the Earth's own axis; an observer at one of the terrestrial poles would therefore find the corresponding celestial pole directly overhead, while an equatorial observer would have them at his north and south horizons. Near the north celestial pole lies the bright POLE STAR. The altitude of the celestial pole above the observer's horizon is equal to his latitude. The celestial equator is the projection of the Earth's equator on the celestial sphere. Its maximum altitude on the celestial sphere is equal to 90° minus the latitude.

The position of an object on the celestial sphere is defined by the coordinate called *declination* (equivalent to latitude on Earth) and *right ascension* (the equivalent of longitude). Declination is measured in degrees north (+) or south (−) of the equator; the poles lie at 90° declination. Right ascension is measured in hours, from 0 to 24, corresponding to the celestial sphere's daily rotation; the hour number increases from west to east. The zero point of right ascension lies where the Sun crosses the celestial equator on its way north at the beginning of northern spring, the VERNAL EQUINOX. The Sun's apparent yearly path around the celestial sphere is termed the ECLIPTIC; it is actually a projection of the plane of the Earth's orbit. The ecliptic cuts the celestial equator at about $23\frac{1}{2}°$, the angle of the Earth's axial tilt.

Because the reference points on the celestial sphere are reflections of the Earth's own rotation, the wobble of the Earth's axis, called PRECESSION, causes the celestial poles to move, so that the lines of right ascension and declination are slowly changing. The annual drift averages about 3 seconds of right ascension and 15 arc seconds of declination (varying considerably in different parts of the sky), and for precise positional work the EPOCH (reference date) of observation must be stated.

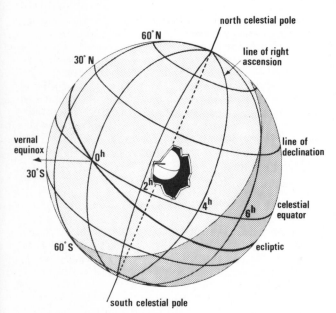

Centaur
American upper-stage rocket burning liquid hydrogen. It is 30 feet (9.1 m) long and 10 feet (3 m) wide, and has two engines that together produce 30,000 lb. (13,600 kg) of thrust, almost twice that of the AGENA second-stage rocket. The Centaur's first mission was on May 30, 1966, when, with an ATLAS first-stage booster, it launched Surveyor 1 to the Moon. The 117-foot (35.7-m) high Atlas-Centaur combination has since launched numerous planetary probes. The Centaur now serves as the top stage on a modified Titan III booster, to launch still heavier probes, such as the Vikings sent to Mars.

Centaurus (the centaur)
A prominent constellation in the southern hemisphere of the sky, lying in a rich part of the Milky Way. Its brightest star, ALPHA CENTAURI, is actually a triple system whose three components are the closest stars to the Sun. At the heart of the constellation is the globular cluster Omega Centauri, cataloged as NGC 5139. This is one of the brightest and richest of all globular clusters, containing hundreds of thousands of stars, 17,000 light-years away. The galaxy NGC 5128 is a famous radio source, Centaurus A; it seems to have ejected clouds of gas in explosions.

cepheid variable
A type of star which swells and contracts in size like a beating heart, varying in brightness as it does so. The time taken for a cepheid variable to complete one cycle of light changes is directly related to its average brightness: the longer the period, the brighter the star. Cepheid variables go through their pulsation cycle over periods of between 2 and 40 days, varying in brightness by up to about one magnitude. They are named for the prototype, Delta Cephei, discovered in 1784 by the English amateur astronomer John Goodricke. The cepheids are yellow supergiant stars, similar in color to the Sun but with masses 5 to 10 times greater. They are also much brighter, which makes them visible over long distances. Since their period of pulsation and their average brightness are linked, (the so-called PERIOD-LUMINOSITY RELATION), by measuring the period astronomers can calculate the cepheid's intrinsic brightness, its ABSOLUTE MAGNITUDE. Comparing this with how bright it appears in the sky its APPARENT MAGNITUDE reveals its distance. Cepheid variables therefore act as a standard for measuring distances, and have been of great importance to astronomers. In our Galaxy, cepheids lie along the spiral arms and occur in GALACTIC CLUSTERS; they are young and belong to the so-called population I of stars. To distinguish them from *W Virginis stars*, a similar but older and

The celestial sphere, as seen from the outside. The Earth rotates counterclockwise as seen from above the north pole, so if one imagines the Earth to be fixed in space, the celestial sphere rotates clockwise—in this illustration, from left to right—once a day. The celestial equator, poles and lines of declination (latitude) are above their Earth counterparts. Where the ecliptic (the Sun's track) crosses the equator is the vernal equinox, where the Sun is located on March 21. This is the 0ʰ point of right ascension, and as the sphere apparently rotates all other points in the sky follow it after a given time interval. This gives rise to the lines of right ascension.

fainter class of stars with which they were formerly confused, population I cepheids are sometimes called *classical cepheids* (see VARIABLE STARS). In the 1971 *General Catalogue of Variable Stars,* 696 cepheid variables were listed.

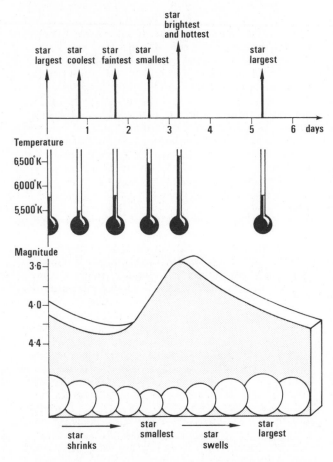

As a cepheid star pulsates in size, so its temperature and surface area change. The maximum brightness does not occur when the star is at its largest and comparatively cool, but shortly after it passes through its smallest stage.

Cepheus

A constellation near the north pole of the sky named for a king in Greek mythology. Its most famous feature is Delta Cephei, the prototype of the class of stars known as CEPHEID VARIABLES. Near Delta Cephei lies the famous double star Krüger 60, composed of 10th- and 11th-magnitude red dwarfs which orbit each other every 44.6 years; they are 12.8 light-years away. The star is named for the German astronomer Adalbert Krüger (1832–1896), who mistook two stars in the same line of sight for a real double. In 1890 the American astronomer S. W. Burnham discovered that one of the stars seen by Krüger did in fact have a genuine companion.

Ceres

Largest of the asteroids, and the first to be discovered, by Giuseppe PIAZZI in 1801. Its diameter is about 593 miles (955 km), and it orbits the Sun every 1,682 days (4.6 years) at an average distance of

257,120,000 miles (413,800,000 km). At its brightest, Ceres is just visible to the naked eye, and is easily followed in binoculars. The surface temperature of Ceres has been measured as around 160° K. Its mass is about 6.7×10^{-10} that of the Sun, and it has an average density about that of Mars or the Moon.

Cernan, Eugene Andrew (b. 1934)

American astronaut, commander of the final Apollo mission to the Moon. Cernan, a qualified aeronautical engineer, was selected as an astronaut in 1963. On his first space flight, Gemini 9 in June 1966, he performed a space walk lasting over two hours. In May 1969 Cernan flew on the Apollo 10 rehearsal for the first Moon landing, descending in the lunar module with Tom Stafford to within 10 miles (16 km) of the Moon. Cernan landed on the Moon on December 11, 1972, in the Apollo 17 mission. With geologist Harrison Hagan ("Jack") Schmitt (b. 1935) he roamed the lunar surface for a record total of 22 hours during a three-day stay.

Cerro-Tololo Inter-American Observatory

Astronomical observatory on Cerro Tololo Mountain in the foothills of the Chilean Andes, run by the Association of Universities for Research in Astronomy (AURA). The observatory, at an altitude of 7,100 feet (2,160 m), has headquarters 35 miles (55 km) northwest in La Serena on the Pacific coast. The observatory was officially opened in November 1967 and is used by astronomers from throughout the world. Its largest telescope is a 158-inch (400-cm) reflector, a twin of the reflector at KITT PEAK NATIONAL OBSERVATORY and the largest telescope in the southern hemisphere; it was installed in late 1974. Also at the observatory are a 60-inch (152-cm), a 36-inch (91-cm), and two 16-inch (40.6-cm) reflectors, and a 24-inch (61-cm) Schmidt telescope.

Cetus (the whale)

A large and straggling constellation of the equatorial zone of the sky, best seen during the northern hemisphere autumn. Its most famous object is the star Omicron, also called MIRA. This is the prototype of a class of stars which vary erratically in brightness over long periods of time. The star Tau Ceti is very much like our Sun; it is 11 light-years away, and thus one of the closest stars. Nearby in the sky is UV Ceti, a pair of twin red dwarfs among the smallest stars known; one of them is a special kind of FLARE STAR, which has given its name to a small group known as UV CETI STARS.

Chamaeleon (the chameleon)

A faint and insignificant constellation lying near the south pole of the sky. It was named by Johann Bayer in 1603.

Chandrasekhar, Subrahmanyan (b. 1910)

Indian-born American astrophysicist, responsible for important studies of stellar structure and evolution, including the nature of WHITE DWARF stars. During the early 1930s Chandrasekhar calculated that a white dwarf could not have a mass more than 1.44 times the Sun's mass; above this, it would compress itself into something denser still. This upper limit for a white dwarf's mass is know known as the

Chandrasekhar limit. Stars that are more massive must lose matter before they can become white dwarfs; this is believed to happen at the end of a star's life when its outer layers expand to form a PLANETARY NEBULA. However, if the star's mass is still above the Chandrasekhar limit, it will collapse into a BLACK HOLE. Chandrasekhar has also studied how stars transfer energy by radiation in their atmospheres; his classic book on the subject is *Radiative Transfer* (1950).

chondrite

The commonest stony meteorite, containing tiny round inclusions called *chondrules,* measuring about 1 millimeter across. Stony meteorites make up more than 92 percent of meteorite falls, and of these over 91 percent are chondrites. Chondrites have a mean density 3.6 times that of water, and consist of tiny fragments of various minerals and flecks of nickel-iron, with the chondrules packed among them. They are quite different in nature from terrestrial rocks, which suggests they have never been part of any large planetary body. The chondrules are normally the minerals olivine or pyroxene, melted and recooled. These inclusions are thought to represent primordial material, particles which condensed from the gas and dust cloud around the Sun.

Carbonaceous chondrites. An exceptionally fragile group of rare meteorites with mean densities about twice that of water. They contain up to 5 percent by weight of a black, tarlike, carbon-rich material. They also have considerable amounts of water, and so can never have been heated to too great a temperature. Many asteroids seem to have surfaces similar in composition to carbonaceous chondrites. Carbonaceous chondrites are probably the commonest type of meteorite in space, but they seldom reach Earth intact because of their fragility.

The material of which meteorites are made formed within a short period of about 100 million years some 4.6 billion years ago. The majority of chondrites originated in about six asteroids with diameters from 120 to 400 miles (200–650 km), orbiting 1.9 to 1.8 a.u. from the Sun. The chondrites were apparently broken off from these during collisions some 10 million years ago. The carbonaceous meteorites came from the surfaces of these asteroids, while the progressively more reheated chondrites originated at greater depths in the parent bodies. Some carbonaceous meteorites may also have originated in periodic comet nuclei.

Christie, Sir William Henry Mahoney (1845–1922)
Eighth astronomer royal. He joined the Royal Greenwich Observatory in 1870 as chief assistant to Sir George AIRY. With Edward Walter Maunder (1851–1928) Christie began the famous Greenwich series of solar photographs which provide valuable records of daily solar activity. Christie succeeded Airy as astronomer royal in 1881. His main contribution was in reequipping the observatory for astrophotography and spectroscopic studies, which continue today; he introduced some of the most famous Greenwich telescopes, such as a 28-inch (71-cm) refractor, and the 26-inch (66-cm) photographic refractor and 30-inch (76-cm) reflector jointly named for their donor, the surgeon and amateur astronomer Sir Henry Thompson (1820–1904).

chromatic aberration

The failure of a lens to bring light of all wavelengths to the same focal point, thereby producing color fringes around an image. Chromatic aberration occurs because each lens tends to act like a prism, splitting white light into a spectrum of colors. Glass refracts blue light more sharply than red, with the result that the "blue" image is formed slightly closer to the lens than the "red" image; the other colors of the spectrum are distributed in between. A single lens therefore produces a short line of colored images of the original object. The amount of chromatic aberration depends on the dispersive (color-spreading) power of the glass. Flint glass has a higher dispersion than crown glass, and a lens made of flint glass thus suffers from more serious chromatic aberration than a similar one made of crown glass. By combining two lenses of different dispersive power, the effects of chromatic aberration can be largely canceled out to produce a near color-free image (see OBJECT GLASS).

chromosphere

The layer of gas about 10,000 miles (16,000 km) thick above the Sun's visible surface (the PHOTOSPHERE). The chromosphere is less than one-thousandth the density of the photosphere; consequently it emits only a relatively weak light which is usually lost in the photosphere's brilliance. The chromosphere is visible only at eclipses or through special instruments such as the CORONAGRAPH. The name, meaning "sphere of color," arises because of the layer's distinct pinkish-red tone, caused by light emitted from hydrogen atoms at a specific wavelength termed Hα. The chromosphere is homogeneous only to heights of about 1,000 to 2,000 miles (1,600–3,200 km). Its upper region looks like a flaming forest because of the jets of hot gas, called SPICULES, which surge up from the photosphere to heights of as much as 10,000 miles (16,000 km), injecting material into the Sun's thin outer atmosphere, the CORONA. The temperature of the chromosphere rises from about 5,000° K at the top of the photosphere to over 1 million degrees where it gives way to the corona.

Circinus (the compass)
A faint and insignificant constellation in the southern hemisphere of the sky, adjacent to Centaurus. It was given its name by Nicolas Louis de Lacaille in the 1750s.

circumpolar

Term describing a celestial object that does not set when seen from a given latitude on Earth; as the term implies, the object appears to circle around the pole. For a star to be circumpolar, its angular distance from the pole must be less than the observer's latitude. From latitude 45° only objects within 45° of the pole are circumpolar; all other objects rise and set during the night. At the pole (90°) all celestial objects visible are circumpolar. But at the equator none are circumpolar; all objects seem to rise and set.

Clark, Alvan (1804–1887)
American instrument-maker whose company five times set the world record for manufacturing the largest telescope lens. His sons George Bassett Clark (1827–1891) and Alvan Graham Clark (1832–1897)

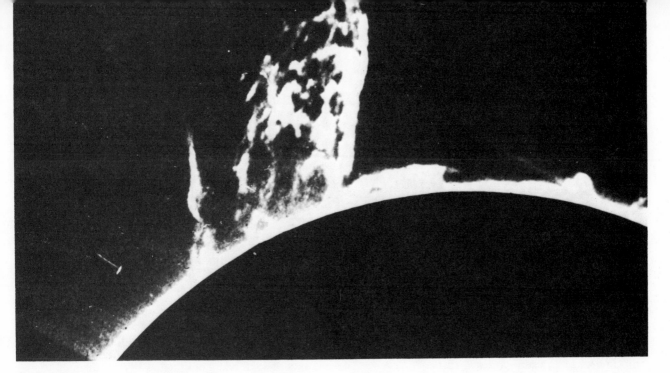

A shining arc of the Sun's chromosphere, seen here in a picture taken with a coronagraph, in which the brilliant photosphere is artificially eclipsed with a diaphragm. The great surge of gases, **left center**, is an eruptive prominence, some 190,000 miles (300,000 km) high, comprising chromospheric material suspended in the corona.

helped make the firm at Cambridge, Massachusetts, the world's leader in astronomical optics. The company's reputation was established in 1862, when Alvan Graham Clark, testing a newly completed 18½-inch (47-cm) refractor, discovered the white dwarf companion to the star SIRIUS. The Clark firm made the 26-inch (66-cm) refractor of the U.S. Naval Observatory in 1871, a 30-inch (76-cm) refractor for Pulkovo Observatory in 1884, and the 36-inch (91-cm) refractor of Lick Observatory in 1888. They completed the 40-inch (101-cm) telescope for Yerkes Observatory in 1897; this remains the largest refracting telescope ever made.

coalsack
A large dark cloud of dust and gas, lying in the Milky Way near the Southern Cross. The minute dust particles of the cloud scatter light from the stars behind it, giving the appearance of a jagged hole in the bright background of stars. The coalsack is the most prominent of all the dark NEBULAE, and lies 400 light-years from us. It is 40 light-years across and probably over 100 times more massive than the Sun. It is no different in composition from the bright nebulae, like that in Orion; it is dark only because no stars are embedded within to illuminate it. Such nebulae may be the birthplaces of stars; the coalsack is apparently starting to fragment into smaller, darker globules which will eventually form a star cluster. In time, the nebula will begin to glow from the light of protostars within. Similar clouds can be seen in the northern Milky Way. The so-called *northern coalsack* in Cygnus is almost as notable as its southern counterpart.

coelostat
A mirror system used to reflect the light from a celestial body into a fixed telescope; the name coelostat means "stationary sky." The method is frequently employed with long-focus solar tower telescopes. A plane mirror is mounted facing the celestial equator on an axis pointing to the celestial pole; if it is driven around this axis as the Earth rotates, the reflected beam not only remains stationary but does not alter its orientation during the object's passage across the sky. Normally, a second plane mirror is used to reflect the light from the coelostat into the telescope. For general astronomical use the simpler *siderostat* ("stationary star"), which gives a rotating field, is used. The siderostat consists of a single adjustable mirror, turning equatorially once in 24 hours. It reflects the beam toward the north or south celestial pole and into the objective of a suitably inclined telescope; such instruments are often known as *polar telescopes* since the tube is parallel to the Earth's axis.

Collins, Michael (b. 1930)
Command-module pilot on the Apollo 11 flight to the Moon in July 1969. Collins remained in orbit around the Moon while astronauts Neil Armstrong and Edwin Aldrin made the first manned lunar landing. Collins was selected as an astronaut in 1963. He first flew in space in July 1966 on the Gemini 10 mission, during which he made three space walks, on one of them moving across to a rocket the Gemini spacecraft had docked with. After Apollo 11, Collins left the astronaut corps to become director of the Smithsonian Air and Space Museum in Washington.

color index
A measure of the apparent color of a star, and hence of its temperature. Cooler stars emit comparatively more light at longer wavelengths and thus appear redder than hot stars. The color index is expressed as the difference in a star's brightness when measured at two selected wavelengths. The *international color*

index, defined by E. C. PICKERING about 1890, is the difference between the PHOTOGRAPHIC MAGNITUDE (blue light) and the PHOTOVISUAL MAGNITUDE (yellow light). It is zero for white stars (SPECTRAL TYPE A0), positive for red stars, and negative for blue stars. Magnitudes are now seldom measured photographically, and the use of color filters with PHOTOELECTRIC CELLS allows the color index between any two wavelengths to be found. The widely used UBV SYSTEM utilizes the ultraviolet, blue, and yellow (visual) magnitudes.

color-magnitude diagram
A plot of the apparent brightness of stars against their COLOR INDEX. It is similar to the HERTZSPRUNG-RUSSELL DIAGRAM, which plots absolute magnitude against color. The color-magnitude diagram is often used for star clusters, since all members of the cluster are at the same distance, and the correction between apparent and absolute magnitude is constant for each star. In fact, the distance of the cluster can be accurately found from the amount of adjustment needed to bring its MAIN SEQUENCE, based on apparent magnitudes, into line with the main sequence of the Hertzsprung-Russell diagram, which is based on absolute magnitudes. The difference between the apparent and absolute magnitudes is the DISTANCE MODULUS for the cluster.

Columba (the dove)
An insignificant constellation in the southern hemisphere of the sky near Canis Major; it was introduced during the 17th century.

coma
An optical defect in which a star appears distorted into a comet- or pear-like shape toward the edge of the field of view. It is caused by zones toward the edge of a lens or mirror which focus rays arriving obliquely into a short line of images of progressively larger diameter. Coma is one of the two principal off-axis aberrations (the other is ASTIGMATISM) limiting the field of view of a camera or telescope. Its seriousness increases toward the edge of the field, but it is eventually swamped by astigmatism. Coma is more awkward than astigmatism in measuring star positions because comatic star images are not symmetrical. Achromatic objective lenses for refracting telescopes can be designed free from coma (if they are also free from SPHERICAL ABERRATION they are said to be *aplanatic*), but the Newtonian or classical Cassegrain reflecting telescope needs special correcting lenses.

Coma Berenices
A faint constellation between Leo and Boötes in the northern hemisphere of the sky, representing the hair of the Egyptian queen Berenice. Its main feature is an enormous cluster of many thousands of galaxies, several hundred million light-years away. Much nearer is the spiral galaxy M67, known as the black-eye galaxy because it contains a large area of dark dust. The north galactic pole lies in Coma Berenices, between the stars Beta and Gamma.

comet
A small icy body embedded in a cloud of gas and dust moving in a highly elliptical orbit around the Sun.

Comets spend most of their lives in frozen reaches far from the Sun, but periodically their orbits bring them close enough to be heated up and to release gas and dust clouds to form a hazy head, developing tails which always point away from the Sun. These have given comets their name: from the Greek meaning "long-haired one."

Comets were thought to be atmospheric phenomena until Tycho Brahe demonstrated that the comet he discovered in 1577 showed no shift in position through parallax, proving it was far beyond the Moon. Isaac Newton demonstrated that comet Kirch, discovered in 1680, moved in an orbit around the Sun in accordance with his theory of gravity. Edmond HALLEY, and later Wilhelm OLBERS, greatly improved methods of determining the orbits of comets. The study of their physical structure and behavior began in the mid-18th century and is now an increasingly important field of research.

Discovery and naming. Comets are named for their discoverer or discoverers, with up to three names permitted. Lower-case letters are added as preliminary designations to record the order of discovery, the first comet of 1976 being designated 1976 a, the second 1976 b, and so on. Permanent designations are eventually allotted in the order of perihelion passage (closest approach to the Sun), the first comet to pass in 1976 being designated 1976 I, the second 1976 II, and so on. Amateur astronomers systematically scan the sky with telescopes and large binoculars in search of new comets; professional astronomers find faint comets accidentally on their photographic plates. The annual discovery rate has slowly increased; in 1750 one comet was found, in 1800 two, 1850 three, 1900 five, 1950 eight, and in 1975 eleven.

Orbits. By 1975, astronomers had observed 964 comets well enough to compute their orbits with accuracy. This total consists of 523 so-called *new* comets, moving in very long-period near-parabolic orbits, and 441 appearances of some 102 objects known as *periodic* comets, moving in short-period elliptical orbits (periods of less than 200 years). New comets move in orbits so long they will not return for thousands of years. Thus comet Bennett 1970 II, with an aphelion at 288 a.u., will not return for more than 1,000 years, while the famous comet Kohoutek 1973 f will take 75,000 years to complete one orbit. All comets are clearly members of the solar system moving in closed orbits. There is not one known case of a comet approaching from deep space. However, some comet orbits have been so perturbed by the planets' gravity they have left the solar system forever.

Comet perihelion distances vary from the record of comet van den Bergh 1974 g, 75 million miles (120 million km) beyond the orbit of Jupiter, to only 0.0048 a.u. for comet Thome 1887 I, within 14,000 miles (22,500 km) of the Sun's surface. But the majority cluster around 1 a.u., the Earth's own distance from the Sun. New comets have near-parabolic orbits which can be inclined at any angle to the planetary system. By contrast, short-period comets move in ellipses close to the plane of the planets. The short-period comets have their aphelions (farthest points from Sun) close to the paths of the giant planets, principally Jupiter. The so-called *comet families* were probably caused by the gravitational attractions of the planets.

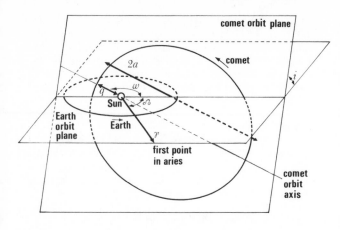

This diagram shows how a comet's orbit is defined in space. The longitude of its ascending node (Ω) from the First Point of Aries is measured in the plane of the ecliptic, while the longitude of perihelion (ω) is measured in the comet's plane. The inclination (i) is the angle between the two planes, while the perihelion distance (q) and the eccentricity define the size of the orbit. In any particular case, the moment of perihelion (T) must be given.

Anatomy of a comet. Despite the popular picture of a comet as a brilliant object with a fine tail, most comets are actually faint, diffuse, tail-less objects visible only with large telescopes. Comets consist of a nucleus, a head or *coma,* and a tail. The nucleus is the only solid body, with a diameter of from 1 to 30 miles (1.6–48 km), a low overall density, and a mass from about 10 billion to 100 trillion tons. A comet's nucleus is made of dust particles loosely compacted with water ice, together with frozen carbon monoxide and methane. The Sun's heat melts the nucleus, releasing huge volumes of gas which carry away dust and ice particles. This gas produces the head or *coma* of the comet consisting of water vapor, carbon monoxide, and OH (hydroxyl), with minor amounts of other molecules containing carbon, hydrogen, and nitrogen. These molecules are then broken up into smaller fragments to produce a PLASMA, or ionized gas. The coma is normally about 10,000 to 100,000 miles (16,000–160,000 km) across; the record is 1.1 million miles (1.8 million km) for comet Flaugergues 1811 I. Hydrogen, carbon, and oxygen atoms form an invisible enveloping cloud some 10 to 20 million miles (16 to 32 million km) across. If a comet becomes very active, the SOLAR WIND carries away dust and gas to form its tail. Prominent tails develop only in large

Characteristics of some Periodic Comets and their Orbits

Comet	Designation	Period (years)	Perihelion (astronomical units)	Aphelion	Orbital inclination (degrees)	Number of observed returns	Nucleus diameter (miles)	Current absolute magnitude*
Encke	1786 I	3.30	0.339	4.10	12.0	52	2	11.4
Tempel (2)	1873 II	5.26	1.364	4.68	12.5	15	1.5	10.2
Finlay	1886 VII	6.90	1.080	6.17	3.6	9	0.5	13.5
Faye	1843 III	7.41	1.616	5.98	9.1	17	1.0	11.5
Kearns-Kwee	1963 VIII	9.01	2.229	6.43	9.0	2	8	9.5
Tempel-Tuttle	1866 I	32.91	0.982	19.56	162.7	4	10	9.0
Halley	1682 I	76.09	0.587	35.33	162.2	27	25	4.6

*Absolute magnitude is the brightness of the comet when 1 a.u. from both Sun and Earth

Characteristics of some New Comets and their Orbits

Comet	Designation	Perihelion (astronomical units)	Orbital inclination (degrees)	Orbital eccentricity	Current absolute magnitude
Ikeya-Seki	1965 VIII	0.0078	141.86	0.99915	6.0
Seki-Lines	1962 III	0.0314	65.01	1.00000	6.5
Daido-Fujikawa	1970 I	0.0657	100.17	1.00000	10.5
Kohoutek	1973 f	0.1424	14.30	1.00000	6.5
Arend-Roland	1957 III	0.3160	119.94	1.00021	4.5
Bennett	1970 II	0.5376	90.05	0.99627	5.0
Suzuki-Saigusa-Mori	1975 k	0.8375	118.51	1.00000	9.5
Kohoutek	1970 III	1.7189	86.31	0.99911	10.0
Humason	1962 VIII	2.1334	153.28	0.98957	1.5
Wild	1968 III	2.6091	135.17	1.00000	8.5
van den Bergh	1974 g	6.0197	60.77	1.00000	9.0

active comets close to the Sun, and always point away from the Sun. The plasma in the coma consists mainly of ionized molecules of carbon monoxide, carbon dioxide, methylidine, and nitrogen, which are accelerated away from the head at speeds up to 400 miles (640 km) per second, forming a straight-rayed gas tail pointing directly away from the Sun. The inner coma is rich in dust grains about one micron (millionth of a meter) across. Being larger than the ions of the gas tail, these move more slowly away from the head, producing a fan-like dust-tail. These are always curved, but can appear straight when seen as the Earth passes through the plane of the comet's orbit. At such times, the dust particles may appear to stretch in front of the comet to produce an anomalous spike, or *anti-tail*. Comet Arend-Roland 1957 III was very dusty, releasing 70 tons per second of fine particles about 1 micron across, while the famous comet Kohoutek 1973 f released over 1,000 tons of mainly large particles every second. Comets also release small amounts of much heavier particles, of millimeter-, centimeter-, and even meter-size. These spread out slowly ahead of and behind the comet in its orbit to form a meteor stream. The tails of bright comets can be extremely long, usually about 5 to 25 million miles (8–40 million km), the record being 200 million miles (320 million km) for the great comet 1843 I.

Brightness. Comets brighten rapidly as they move toward the Sun, because sunlight excites their gases to fluorescence. New comets are dust-rich and do not brighten as much as periodic comets, which are gas-rich. Comets also show unexpected brightenings and fadings caused by changes in solar activity. They are very sensitive to the solar wind and provide excellent probes for the study of solar activity.

Periodic comets. Comets lose from 0.1 to 1 percent of their mass at each perihelion passage. Short-period comets have periods generally from five to seven years and thus come to perihelion frequently; they fade by 2 magnitudes or more per century as they decay. Periodic comets tend to be gas-rich because they have lost the dusty outer layers of their nucleus, revealing a more compacted center, which contains larger, meteor-sized particles. The force of gas escaping from a periodic comet's nucleus slightly changes its path over long periods of time. In some comets this effect increases as the comet ages, while in others the effect decreases. This suggests there are two types of nuclei: one a uniform icy-conglomerate which eventually dissipates completely, the other a nucleus with a central core of heavier compacted stony material which ultimately loses all its gas, leaving a rocky body like an asteroid. Some of the Apollo and Amor ASTEROIDS might well be such "dead" comets.

Numbers and origin. The total number of comets is enormous—about 10 million comets must have their perihelion points within the orbit of Neptune alone. Most new comets move sunward from a great distance, suggesting they originate in a vast cloud at a distance of 20,000 to 60,000 a.u. This is known as Oort's cloud, after the Dutch astronomer Jan OORT, and contains an estimated 100 billion comets. It is believed to have formed at the very edge of the solar system as the Sun and planets were being born. The comets in the cloud move in circular orbits inclined at all angles to the planetary system, until they are perturbed by passing stars into orbits that eventually carry them toward the inner solar system. Comets are clearly composed of material that has been deep-frozen for millions of years; and their investigation by spacecraft, currently being planned, will yield much valuable information on the early history of the solar system.

command module

The compartment in a manned spacecraft in which the crew sit during launch and landing. The command module contains equipment for communication, navigation, and life support, and controls for guiding the spacecraft. The idea of building spacecraft in sections, or modules, was introduced for the GEMINI series of manned flights; in Gemini, the crew compartment, called the reentry module, was attached to an adapter module, which contained bulky equipment such as fuel cells and oxygen tanks. It was in the APOLLO PROGRAM that the names command module and service module were introduced for these parts of the spacecraft. The single-man American MERCURY craft was entirely self-contained.

commensurability

An exact ratio between the orbital periods of two or more orbiting bodies. Orbital periods whose ratio is any whole number, or whole fraction, are said to be commensurable. If, for example, one body takes two or three times as long to complete an orbit as another, or a half or a third of the time, then the orbits of the two bodies are commensurable. The bodies will return to the same relative positions after two or three orbits; in consequence, they will have strong gravitational effects on each other. Some of the KIRKWOOD GAPS in the asteroid belt are believed to be due at least in part to commensurability with the orbit of Jupiter; and CASSINI'S DIVISION in Saturn's rings may be due to commensurability with Saturn's satellites.

communications satellites

Relays in space for sending telephone, radio, and television signals around the world. They provide many communications circuits far more cheaply than long-distance cables, and they can link any places in the world that have suitable ground equipment. Most countries are now in satellite communication via the INTELSAT network. Sets of three satellites are used, stationed at points above the Atlantic, Indian, and Pacific Oceans, because each satellite can only see about a third of the Earth below it.

Communications satellites use very short radio wavelengths called *microwaves,* which pass straight through the Earth's ionosphere without being reflected, unlike longer waves used for conventional radio transmission. Television transmission also uses very short wavelengths that cannot be bounced off the ionosphere. Without communications satellites it is difficult to send television pictures over long distances without using cables or chains of relay stations. Using communications satellites is much more reliable than bouncing radio waves off the ionosphere, where solar storms can cause sudden radio "blackouts."

A communications satellite ground station has an aerial like a radio telescope, usually 80 to 100 feet (25–30 m) in diameter. It sends radio signals out to

The large satellite is an Intelsat IVA communications satellite, workhorse of international 'phone and video circuits, with a capacity for 11,000 simultaneous two-way voice-quality channels. Compared with it is Early Bird, which in 1965 became the world's first synchronous orbit communications satellite.

the satellite, and receives others. Messages in each direction are carried on waves of different frequency; each of these so-called *carrier* waves can have hundreds of telephone calls superimposed on it. Depending on the number of carrier waves, the most powerful satellites can deal with several thousand telephone conversations at a time. Television pictures take up a great deal more space on the carrier, leaving room for fewer telephone messages.

History. The first attempt at communication using an artificial satellite was the U.S. government's Project SCORE (Signal Communication by Orbiting Relay Equipment), in December 1958. Score was simply an Atlas rocket in low orbit carrying a tape recorder and radio equipment. It broadcast a prerecorded Christmas message from President Eisenhower, and also recorded messages from ground stations and retransmitted them. Following this success, the U.S. Army Signal Corps in 1960 experimented with a satellite called Courier. This was a 500-lb. (227-kg), 52-inch (132-cm) diameter sphere studded with photocells to draw power from sunlight, that relayed teletype, voice, and facsimile data between military ground stations, before breaking down after 17 days.

Score and Courier were *active-repeater* satellites: they accepted messages and actively rebroadcast them. But at the same time, NASA was experimenting with the so-called *passive reflector* type of communications satellite. These were giant balloons that simply reflected signals back to Earth (see ECHO SATELLITES).

Telstar and Relay. Balloon satellites such as Echo have the advantage that anyone can bounce signals from them, at any frequency, and there are no electronic parts to go wrong. But the signal returned from a simple balloon reflector is as weak as a reflection from the Moon. Much better reception is obtained from satellites that amplify the signal before rebroadcasting it. This was spectacularly demonstrated in 1962 by TELSTAR, which carried the first transatlantic television signals and effectively opened the era of commercial satellite communications. Telstar was followed by a more powerful NASA satellite called RELAY.

Both Telstar and Relay were put into comparatively low orbits around the Earth. One possible communications system is to have a succession of such low-altitude satellites passing continuously across the sky; but rapidly moving satellites provide a tracking problem for ground stations, and satellites in low orbit can be damaged by radiation from the Earth's Van Allen belts.

Syncom and Intelsat. Satellites in higher orbits can cover a much greater area of the Earth. In the synchronous or geostationary type of orbit first envisaged by the English author Arthur Charles Clarke (b. 1917), they appear to hang stationary over one point on the equator. The first experimental satellite of this type was SYNCOM, introduced in 1963 by NASA. Geostationary satellites soon became standard for international communications. In 1964 an international organization named Intelsat was created to arrange a network of communications satellites for commercial use. The organization has continued to build, launch, and operate the Intelsat series of satellites.

Military communications. Because of their reliability, both the United States and the Soviet Union use satellites for military communications. In 1966 the first U.S. military satellites were launched in the Initial Defense Satellite Communication System (IDSCS); the system of 26 satellites was completed in 1968. Placed just below synchronous altitude, they drifted slowly around the Earth so that if one malfunctioned, another would soon appear to take its place. They were followed in 1971 by the first launches of the Defense Satellite Communication System (DSCS), which are larger and more powerful satellites in synchronous orbit, capable of communicating with small, mobile ground stations. Other tactical satellites being developed, such as the U.S. Army's Tacsat, allow communication with individual ships, aircraft, tanks, and jeeps.

Other satellites. Synchronous satellites cannot cover areas at latitudes higher than about 70°. The Soviet Union, because of its large land area in high northerly latitudes, instead uses a system of satellites inclined at 65° to the equator in elliptical 12-hour orbits. There is always a satellite in view of a ground station to ensure round-the-clock satellite coverage (see MOLNIYA SATELLITES). Countries with similar remote communities are turning to domestic satellites for ease of communication, such as with the ANIK SATELLITES, WESTAR SATELLITES, and ATS SATELLITES.

Other nations with special communications requirements are now producing their own satellites. The Franco-German Symphonie satellite, launched on December 18, 1974, is placed in synchronous orbit at 11.5° west longitude for communications between Europe, Africa, and North and South America. On May 11, 1978, the European Space Agency launched OTS (Orbital Test Satellite) for European communications.

Future satellites will need to work at higher frequencies, in order to carry more messages, and to have the information packed onto their carrier waves by more efficient techniques. Across the busy North Atlantic satellites will need special antennae to focus on the most important ground stations. Eventually, satellites may have on-board switching, so that they act like automatic telephone exchanges in the sky.

Comsat

The Communications Satellite Corporation, an American corporation set up in 1963 to own and operate commercial communications satellites. With other countries the Comsat Corporation operates the INTELSAT series; it also manages the Intelsat system on behalf of the other participants. The word comsat, without a capital letter, is often used as an abbreviation for COMMUNICATIONS SATELLITES.

Congreve, Sir William (1772–1828)

British rocket pioneer who developed a series of gunpowder-powered projectiles for military use; these were the first real improvements since the rocket's invention. Congreve's rockets carried explosive warheads and were used in several campaigns during the Napoleonic Wars; artillery companies and even

ships were equipped to fire Congreve rockets. The phrase "the rockets' red glare" in the "Star-Spangled Banner" is a reference to the bombardment of Fort McHenry by Congreve rockets during the War of 1812. Congreve rockets were later modified for firing life lines and signal flares.

conjunction

An alignment of astronomical bodies. A planet closer to the Sun than the Earth is said to be at *inferior conjunction* when it is between Earth and Sun, and at *superior conjunction* when it is on the far side of the Sun from Earth; the exact moment of conjunction is when the centers of the Sun and the planet have the same celestial longitude. A planet farther from the Sun than Earth is said to be in conjunction when it is behind the Sun as seen from Earth; when such a planet is opposite in the sky to the Sun, it is said to be at opposition. Two planets can be in conjunction with each other if their celestial longitudes are the same. By extension, the term is often used to mean any close approach between two planets.

Conrad, Charles (b. 1930)

American astronaut, commander of the second Apollo Moon-landing mission, and of the first Skylab crew. Conrad, a qualified aeronautical engineer, became an astronaut in 1962. His first flights were on Gemini 5 in August 1965, and Gemini 11 in September 1966. On November 19, 1969, he piloted the Apollo 12 lunar module to a precision landing near an old automatic Moon probe, Surveyor 3. During a $31\frac{1}{2}$-hour stay on the Moon with Alan BEAN, he spent a total of 8 hours roaming the lunar surface. Conrad's fourth space mission was in May-June 1973, when he and his two-man crew spent a month in the Skylab space station. In early 1974 Conrad resigned from the astronaut corps to enter private business.

Constellations

Constellation	Genitive case	English name	Abbreviation	Approximate position		Area (square degrees)	Order of size
				α h	δ °		
Andromeda	Andromedae	Andromeda	And	1	+40	722	19
Antlia	Antliae	Air pump	Ant	10	−35	239	62
Apus	Apodis	Bird of Paradise	Aps	16	−75	206	67
Aquarius	Aquarii	Water carrier	Aqr	23	−15	980	10
Aquila	Aquilae	Eagle	Aql	20	+5	652	22
Ara	Arae	Altar	Ara	17	−55	237	63
Aries	Arietis	Ram	Ari	3	+20	441	39
Auriga	Aurigae	Charioteer	Aur	6	+40	657	21
Boötes	Boötis	Herdsman	Boo	15	+30	907	13
Caelum	Caeli	Chisel	Cae	5	−40	125	81
Camelopardalis	Camelopardalis	Giraffe	Cam	6	+70	757	18
Cancer	Cancri	Crab	Cnc	9	+20	506	31
Canes Venatici	Canum Venaticorum	Hunting dogs	CVn	13	+40	465	38
Canis Major	Canis Majoris	Greater dog	CMa	7	−20	380	43

Constellation	Genitive case	English name	Abbreviation	Approximate position α h	δ o	Area (square degrees)	Order of size
Canis Minor	Canis Minoris	Lesser dog	CMi	8	+5	183	71
Capricornus	Capricorni	Goat	Cap	21	−20	414	40
†Carina	Carinae	Keel	Car	9	−60	494	34
Cassiopeia	Cassiopeiae	Cassiopeia	Cas	1	+60	598	25
Centaurus	Centauri	Centaur	Cen	13	−50	1,060	9
Cepheus	Cephei	Cepheus	Cep	22	+70	588	27
Cetus	Ceti	Whale	Cet	2	−10	1,231	4
Chamaeleon	Chamaeleontis	Chameleon	Cha	11	−80	132	79
Circinus	Circini	Compasses	Cir	15	−60	93	85
Columba	Columbae	Dove	Col	6	−35	270	54
Coma Berenices	Comae Berenicis	Berenice's hair	Com	13	+20	386	42
Corona Australis	Coronae Australis	Southern crown	CrA	19	−40	128	80
Corona Borealis	Coronae Borealis	Northern crown	CrB	16	+30	179	73
Corvus	Corvi	Crow	Crv	12	−20	184	70
Crater	Crateris	Cup	Crt	11	−15	282	53
Crux	Crucis	Southern cross	Cru	12	−60	68	88
Cygnus	Cygni	Swan	Cyg	21	+40	804	16
Delphinus	Delphini	Dolphin	Del	21	+10	189	69
Dorado	Doradus	Swordfish	Dor	5	−65	179	72
Draco	Draconis	Dragon	Dra	17	+65	1,083	8
Equuleus	Equulei	Little horse	Equ	21	+10	72	87
Eridanus	Eridani	River	Eri	3	−20	1,138	6
Fornax	Fornacis	Furnace	For	3	−30	398	41
Gemini	Geminorum	Twins	Gem	7	+20	514	30
Grus	Gruis	Crane	Gru	22	−45	366	45
Hercules	Herculis	Hercules	Her	17	+30	1,225	5
Horologium	Horologii	Pendulum clock	Hor	3	−60	249	58
Hydra	Hydrae	Water snake	Hya	10	−20	1,303	1
Hydrus	Hydri	Lesser water snake	Hyi	2	−75	243	61
Indus	Indi	Indian	Ind	21	−55	294	49
Lacerta	Lacertae	Lizard	Lac	22	+45	201	68
Leo	Leonis	Lion	Leo	11	+15	947	12
Leo Minor	Leonis Minoris	Lesser lion	LMi	10	+35	232	64
Lepus	Leporis	Hare	Lep	6	−20	290	51
Libra	Librae	Scales	Lib	15	−15	538	29
Lupus	Lupi	Wolf	Lup	15	−45	334	46
Lynx	Lyncis	Lynx	Lyn	8	+45	545	28
Lyra	Lyrae	Lyre	Lyr	19	+40	286	52
Mensa	Mensae	Table Mountain	Men	5	−80	153	75

Constellations (continued from previous page)

Constellation	Genitive case	English name	Abbreviation	Approximate position		Area (square degrees)	Order of size
				α h	δ o		
Microscopium	Microscopii	Microscope	Mic	21	−35	210	66
Monoceros	Monocerotis	Unicorn	Mon	7	−5	482	35
Musca	Muscae	Fly	Mus	12	−70	138	77
Norma	Normae	Level	Nor	16	−50	165	74
Octans	Octantis	Octant	Oct	22	−85	291	50
Ophiuchus	Ophiuchi	Serpent holder	Oph	17	0	948	11
Orion	Orionis	Orion	Ori	5	+5	594	26
Pavo	Pavonis	Peacock	Pav	20	65	378	44
Pegasus	Pegasi	Pegasus	Peg	22	+20	1,121	7
Perseus	Persei	Perseus	Per	3	+45	615	24
Phoenix	Phoenicis	Phoenix	Phe	1	−50	469	37
Pictor	Pictoris	Easel	Pic	6	−55	247	59
Pisces	Piscium	Fishes	Psc	1	+15	889	14
Piscis Austrinus	Piscis Austrini	Southern fish	PsA	22	−30	245	60
†Puppis	Puppis	Stern	Pup	8	−40	673	20
†Pyxis (= Malus)	Pyxidis	Compass	Pyx	9	−30	221	65
Reticulum	Reticuli	Net	Ret	4	−60	114	82
Sagitta	Sagittae	Arrow	Sge	20	+10	80	86
Sagittarius	Sagittarii	Archer	Sgr	19	−25	867	15
Scorpius	Scorpii	Scorpion	Sco	17	−40	497	33
Sculptor	Sculptoris	Sculptor	Scl	0	−30	475	36
Scutum	Scuti	Shield	Sct	19	−10	109	84
Serpens	Serpentis	Serpent	Ser	17	0	637	23
Sextans	Sextantis	Sextant	Sex	10	0	314	47
Taurus	Tauri	Bull	Tau	4	+15	797	17
Telescopium	Telescopii	Telescope	Tel	19	−50	252	57
Triangulum	Trianguli	Triangle	Tri	2	+30	132	78
Triangulum Australe	Trianguli Australis	Southern triangle	TrA	16	−65	110	83
Tucana	Tucanae	Toucan	Tuc	0	−65	295	48
Ursa Major	Ursae Majoris	Great bear	UMa	11	+50	1,280	3
Ursa Minor	Ursae Minoris	Little bear	UMi	15	+70	256	56
†Vela	Velorum	Sail	Vel	9	−50	500	32
Virgo	Virginis	Virgin	Vir	13	0	1,294	2
Volans	Volantis	Flying fish	Vol	8	−70	141	76
Vulpecula	Vulpeculae	Fox	Vul	20	+25	268	55

†The four constellations Carina, Puppis, Pyxis, and Vela originally formed the single constellation, Argo Navis, the Argonauts' Ship

constellations

Star patterns as seen from Earth, which provide a set of references for the recognition and identification of objects in the sky. Most ancient civilizations recognized similar patterns among the stars; from these our modern constellations have grown. The constellations in worldwide use today stem from those of the Greeks, listed by PTOLEMY in 150 A.D.; many of them probably originated with the Babylonians. Ptolemy listed 48 constellations. Twelve of them were the constellations of the ZODIAC, of much earlier origin; other constellations represented Greek mythological figures. New constellations were added in the 17th and 19th centuries, notably by star mappers such as Johann BAYER, Johannes HEVELIUS, and Nicolas Louis de LACAILLE; many of the more insignificant patterns eventually fell into disuse, leaving a total of 88 constellations. There were, however, no generally agreed constellation boundaries until 1930, when the International Astronomical Union decided upon regular boundaries following lines of right ascension and declination for 1875. These were drawn up in a two-volume atlas by the Belgian astronomer Eugene Joseph Delporte (1882–1955). The boundaries themselves are fixed with respect to the stars; however, because of PRECESSION they have now moved away somewhat from their original lines of right ascension and declination. Constellation names are in Latin, with Latin case endings. Thus the constellation name *Capricornus* becomes *Capricorni* when referring to a star in Capricornus (the genitive case): alpha Capricorni means "alpha of Capricornus." Modern astronomers retain the ancient constellations as convenient guides to the location of objects.

continuous creation

The continuous production of matter from nothing, postulated by the STEADY-STATE THEORY of the Universe. The steady-state theory starts from the assumption that the Universe looks the same at all times, as well as from all points in space. But since the Universe is observed to be expanding, the matter in it would become more spread out with time; and the proponents of the steady-state theory had to assume that matter is being continuously created to fill the space caused by the expansion.

The rate of creation was calculated at one hydrogen atom per liter of volume (about one quart) every 500 billion years, which is far too small to check in the laboratory. The matter created was thought to condense into galaxies in a continuous process. Fred HOYLE proposed that the created matter appeared from a *C-field,* a form of negative energy which he supposed to fill the Universe.

Strong doubt has now been cast on the steady-state theory by the discovery of the weak BACKGROUND RADIATION in space, which apparently originated in a completely different state of the Universe in the past. The steady-state theory now has few adherents, and the BIG-BANG cosmology currently in favor does not require the continuous creation of matter.

Cooke, Thomas (1807–1868)

One of the foremost telescope makers of the 19th century. He became well known in 1856, when Charles Piazzi Smyth (1819–1900), astronomer royal for Scotland from 1845 to 1888, took a 7-inch (18-cm) Cooke telescope to test seeing conditions on the island of Tenerife. The American firm of Alvan CLARK had just made an 18½-inch (47-cm) lens, the world's largest, and Cooke was commissioned in 1863 to build a 25-inch (64-cm) refractor by Robert Stirling Newall (1812–1889), a wealthy telegraph cable manufacturer and amateur astronomer. The lens was

completed in 1868, but Cooke, exhausted, died soon after. The Newall telescope was given to the University of Cambridge in 1889, and in 1959 was installed at Mount Pendeli Observatory, 12 miles (19 km) northeast of Athens. In 1922 the Cook firm merged with an instrument company set up by Edward Troughton (1753–1835) and William Simms (1793–1860).

Cooper, Leroy Gordon (b. 1927)

Member of the first group of astronauts, selected in 1959, who made the sixth and last flight in the MERCURY series in May 1963. He orbited the Earth 22 times, more than all previous Mercury flights combined. In August 1965 he flew on the Gemini 5 mission, spending eight days in space and making 120 revolutions of the Earth, at that time a record. In mid-1970 he resigned from the astronaut corps to enter private business.

Copernicus, Nicolaus (1473–1543)

Polish astronomer whose heliocentric theory of the Universe demoted the Earth to the status of an ordinary planet, revolutionizing science and profoundly altering Man's conception of his world.

To Copernicus, a canon at a cathedral on the shores of the Baltic, astronomy was little more than a hobby, although his scientific knowledge was well known and he was consulted by the papacy on possible calendar reform. As a student of wide learning, Copernicus knew that certain Greek philosophers had suggested schemes of the Universe in which the Earth was not centrally placed. In the light of these alternative Greek ideas, he decided to take a fresh look at the contemporary view of the Universe, which he found unsatisfactory. According to the teaching he had received, Earth was stationary at the center of the Universe, surrounded by a sphere on which the stars were fixed. The movements of the planets were explained by a system of rotating circles known as deferents centered on the Earth and smaller ones, epicycles, whose centers lay on the circumference of the deferents. The movements of the circles around the central Earth should be regular; but to account properly for the observed motions, the deferents' centers had to be displaced from the center of the Earth, and thus from the center of the Universe. Copernicus believed he could overcome this inconsistency by placing the Sun at the center of the Universe.

Copernicus worked for many years to perfect his theory. A summary of his ideas was circulated, but he was reluctant to publish anything in detail in case so revolutionary an idea might meet with ridicule. In 1539 he was visited by the German mathematician Rheticus (Georg Joachim von Lauchen, 1514–1576). Rheticus stayed for two years and persuaded Copernicus to allow him to publish a small treatise on the theory and then to prepare a full text, which was published at Nuremberg in 1543. Called *De Revolutionibus Orbium Coelestium* (On the Revolution of the Celestial Spheres) it was marred by an unsigned preface by the German theologian Andreas Osiander (1498–1552), describing the theory as no more than a mathematical convenience. A copy reached Copernicus as he lay dying.

There is no doubt that Copernicus believed his

heliocentric theory to be a true description of the Universe as it actually was and, considering the storm of controversy it raised, so did others. It led eventually to new laws of physics, and a vastly bigger Universe than previously imagined; while the dethronement of Man as the center of all creation had the most profound moral and theological implications.

Coriolis force

A fictional force which appears to deflect the motion of bodies over a rotating surface, such as the surface of a spinning planet. On Earth, an object shot or thrown due north from the equator will land a little to the right (the east) of the target. This is because the eastward velocity of a point on the equator, which the projectile shares when it is fired, is greater than that of a point nearer the pole. Similarly, an object propelled toward the equator from the northern hemisphere would seem to veer to the right (in this case the west) of the intended course, as it dropped behind the rotational rate of the Earth nearer the equator. In the southern hemisphere, the projectile would seem to veer left.

This so-called *Coriolis effect*, named for the French physicist Gaspard Gustave de Coriolis (1792–1843), who drew attention to it in 1835, is responsible for the circular movements of weather features such as depressions and cyclones. It must be corrected for when plotting the path of a missile.

corona

The outermost layer of the Sun's atmosphere, beginning about 10,000 miles (16,000 km) above the visible surface. The name "corona" is the Latin for crown. The corona is observable at a total eclipse as a pearly light, comparable in brightness to the full

Moon, streaming outward in fans and rays. It has no upper boundary but thins out gradually into interplanetary space. The outermost region of the corona is detected streaming past the Earth as the SOLAR WIND. The appearance of the corona varies through the SOLAR CYCLE. At solar minimum, when activity on the Sun's surface is concentrated toward the equator, the corona contains bright streamers that extend mostly from the equatorial region. At cycle maximum, there are streamers all around the Sun and the corona becomes more circular in appearance. The bright inner regions of the corona are caused by sunlight scattered from electrons; this is termed the K corona, from the German *kontinuum*. Farther out, the corona consists of sunlight scattered by dust particles; this is the F, or Fraunhofer, corona, which shows the dark FRAUNHOFER LINES of the Sun's spectrum. The smallest part of the corona's light is emitted from hot atoms. Mysterious emission lines in the corona's spectrum were once attributed to an unknown element named "coronium." In 1942 the Swedish physicist Bengt Edlén (b. 1906) showed that these lines were due to highly ionized atoms, which meant that the gases of the corona were exceptionally hot—modern investigations suggest a temperature of 2 million degrees K. The corona is probably heated by shock waves rising through the chromosphere from the photosphere.

The solar corona, during the total eclipse of February 25, 1952. Seen here at sunspot minimum, the corona is less dense than at other times and is concentrated toward the equator. Streamers of coronal gas, associated with active regions on the surface, surround lower latitudes, while radial plumes project from the poles.

Corona Australis (the southern crown)
A small but attractive constellation lying below
Sagittarius in the southern hemisphere, a counterpart
to the northern crown (CORONA BOREALIS). It lies in an
interesting region of the Milky Way, and contains
the globular cluster NGC 6541 about 14,000
light-years away.

Corona Borealis (the northern crown)
A constellation of the northern hemisphere of the sky
between Hercules and Boötes, best seen during spring.
It includes a famous irregular variable star,
R Coronae Borealis, which is the prototype of a class
of eruptive variables; its brightness changes from
5.8 to 12.5 erratically, and it can stay at maximum
for a year or two before it rapidly fades. Another
remarkable star, T Coronae Borealis, is a recurrent
nova; it erupted in 1866 and again in 1946.

coronagraph
A device invented by Bernard LYOT in 1930 for
observing the Sun's atmosphere in the absence of a
total eclipse. The inner brightest part of the corona
is about a million times fainter than the solar surface,
and is swamped by the glare of sunlight scattered in
the Earth's atmosphere. Lyot circumvented this
difficulty by establishing his coronagraph at the
high-altitude PIC DU MIDI OBSERVATORY, so reducing
atmospheric glare. To overcome the equally serious
glare caused by dust and diffraction in the telescope,
he designed the highly ingenious coronagraph. A
perfectly polished lens forms an image of the Sun
that is intercepted by an occulting disk, producing a
miniature eclipse, and then focused by a second lens
either for visual observation with an eyepiece, or on
a photographic plate. A series of diaphragms blocks
off the bright ring of diffracted light formed around
the objective. With these and other refinements,
observation of the corona in full daylight thus became
possible, and Lyot was able to observe PROMINENCES
visually and to photograph the corona out to about
7 arc minutes from the edge of the Sun. Other
coronagraphs have since been established at
high-altitude solar observatories throughout the
world.

Corvus (the crow)
A small and insignificant constellation in the
southern hemisphere of the sky, lying below Virgo.

cosmic microwave background
See BACKGROUND RADIATION.

cosmic rays
Nuclei of atoms, stripped of all their electrons,
shooting through space at speeds close to that of
light. Ninety percent of cosmic rays are hydrogen
nuclei (protons), and nine percent are helium nuclei
(alpha particles); the nuclei of all other elements
contribute only one percent of the total. High-speed
electrons, present in space in smaller numbers, are
also classified as cosmic rays. (The term cosmic "ray"
dates from early this century, when these fast particles
were thought to be high-energy X rays.)

The energy of a cosmic ray particle is a combined
measure of its mass and its speed. It is usually
expressed in electron volts (eV); 1 eV is the energy
gained by an electron when accelerated through an
electric potential of one volt. The energy of an air
molecule at room temperature is 0.1 eV, and the
energy of particles in an X-ray machine is about
10,000 eV. The fastest particles in the accelerators
used by nuclear physicists reach about 10^{11} (a
hundred billion) eV. Some cosmic rays have energies
as high as 10^{19} (ten million million million) eV.
Collisions of these very energetic particles with other
matter allow nuclear physicists to study reactions
impossible to achieve in man-made accelerators.

The so-called primary cosmic ray particles from
space do not reach the Earth's surface. Instead, they
collide with the molecules of the atmosphere,
smashing them into fragments, which in turn shatter
other nuclei in the atmosphere. This produces an *air
shower* of particles (secondary cosmic rays), which can
cover several thousand square yards or meters at
ground level.

The relatively rare cosmic rays with very high
energy (greater than 10^{14}, or 100 trillion, eV) can be
studied only by the air showers they produce.
Instruments scattered over several square miles on
the ground detect the shower, the size of which is
related to the energy of the particle which caused it.
The difference in arrival time at the various detectors
reveals from what direction the particle has come.
But lower-energy cosmic rays are abundant enough to
be studied directly, by instruments carried above the
atmosphere in balloons or satellites. Many of these
low-energy cosmic rays originate in the Sun during
solar flares and radio bursts, and their numbers
consequently fluctuate with the 11-year solar cycle.
True cosmic rays from deep space are thought to be
accelerated to their high speeds in exploding stars
(SUPERNOVAE) and their remnants (the fast-rotating
PULSARS). Some of the most energetic cosmic rays
may be produced in QUASARS.

Because cosmic ray particles are electrically
charged, their motion is influenced by magnetic
fields in space. The Earth's own magnetic field focuses
cosmic rays toward the poles, which makes it
difficult to determine their true direction or origin.
The degree to which a path is bent by a magnetic
field depends on the particle's energy. The most
energetic cosmic rays probably escape from the
magnetic fields of galaxies into space. Those that are
trapped in a galaxy produce electromagnetic
SYNCHROTRON RADIATION as they spiral in the
magnetic field. This process is most efficient for
electrons, which are several thousand times lighter
than atomic nuclei. Cosmic-ray electrons produce the
radio noise observed from our own and other
galaxies.

Cosmic rays can have important biological effects:
brain cells of men in space can be damaged by
primary cosmic rays, while on Earth, air-shower
particles are energetic enough to pass through the
body and alter the basic genetic material, DNA.
Mutations affecting the process of evolution may
therefore be caused by cosmic rays. Even major
changes may be possible once every several hundred
million years, when a supernova explodes within a
hundred light-years of the Sun. The resulting influx of
cosmic rays could wipe out entire species, and it has
even been suggested that the extinction of the
dinosaurs was due to such an event.

cosmogony

The study of the origin and evolution of individual objects in the Universe, such as stars and galaxies. Most of cosmogony is usually treated under ASTROPHYSICS and COSMOLOGY. The term cosmogony is now often restricted to the origin of the solar system (see PLANETS).

cosmology

The study of the origin and evolution of the Universe. Cosmology began as what we would now call physical geography, the study of the Earth, and became in Greek times the theory of planetary motions. After the invention of the telescope, cosmologists started to determine the structure of our Galaxy, the Milky Way, and then to study the motion of nearby galaxies. Present-day cosmology is largely concerned with the theory of space and time under the gravitational influence of matter.

Modern theories of cosmology. The light from galaxies shows a so-called RED SHIFT, in which all the SPECTRAL LINES appear at slightly longer wavelengths than normal. This is usually interpreted to be a result of the DOPPLER EFFECT, implying therefore that all galaxies are receding as seen from the Earth. The more distant the galaxy, the more it is red-shifted (HUBBLE'S LAW), meaning that the distances between all galaxies are increasing. An observer in another galaxy would thus see all other galaxies receding from him, and would conclude that he lives in an expanding Universe. Cosmologists generally believe that the Universe is expanding from an explosion, the so-called BIG BANG, which occurred some 18 billion years ago. The major competitor to the big-bang theory was the STEADY-STATE THEORY, which claimed that the CONTINUOUS CREATION of matter causes the Universe to expand. The steady-state theory has been under observational attack on two related fronts. First, the weak BACKGROUND RADIATION in space observed by radio telescopes is most easily explained as the left-over radiation from the big bang itself, much cooled by the expansion of the Universe. Second, distant galaxies appear to be packed together more closely than those nearer, showing that the Universe was more compact in the past; it thus cannot be unchanging. The steady-state idea has now lost favor, although it survives in mathematical variants by Fred HOYLE and Jayant Vishnu Narlikar (b. 1938).

The ultimate fate of the Universe may be either to expand for ever, or else to coast to a halt and collapse again, perhaps to re-explode. This is the theory of the OSCILLATING UNIVERSE, in which the big bang is sometimes called the *big bounce*. The Universe will collapse if it contains sufficient matter for gravitational attraction to overcome the outward momentum of the galaxies. To calculate the amount of matter in the Universe, we must know both the number of galaxies, which is relatively easy to calculate, and the average mass of a galaxy. Estimates of galaxy masses give answers differing by a factor of 100; moreover, there are no accurate estimates of the amount of matter (if any) between galaxies. Thus, the total mass of the Universe is not well enough known for its fate to be predicted by this test. An alternative approach is to examine the velocities of very distant galaxies. The light we see left these galaxies billions of years ago when, according to the oscillatory theory, the Universe would have been expanding faster than it is now. The most recent results show little slow down, and suggest that the Universe will expand indefinitely (see HUBBLE'S CONSTANT).

At the present stage, cosmological theory is also experimenting with various kinds of modifications to Einstein's General Theory of RELATIVITY. This theory sees the Universe as a skeleton of space and time on which the galaxies are studded, enabling us to see the space-time structure, just as the arrangement of atoms in a crystal enables us to see the crystal lattice. One such alternative is the BRANS-DICKE COSMOLOGY, which predicts a change in the force of gravity with time; the rate of change would affect the structure of the Earth and stars over their lifetimes. There is every possibility that astronomical observations will in the near future distinguish which space-time cosmology is correct (see also UNIVERSE).

Cosmos satellites

Continuing series of Russian Earth satellites, successors to the SPUTNIK series, introduced by Cosmos 1 on March 16, 1962. Among the stated intentions of the Cosmos program is the study of the Earth's upper atmosphere, radiation from space and from the Sun, radio propagation in the ionosphere, the Earth's magnetic field and radiation belts and meteoric matter, as well as the testing of new spacecraft; but the name Cosmos has also been used to conceal military satellites and probes that have failed to work.

The Russians introduced a new rocket, which they call the Cosmos launcher, for the start of the Cosmos program. It is a two-stage rocket, based on an intermediate-range ballistic missile code-named Sandal by NATO; the first stage has a thrust of 161,000 lb. (73,000 kg) and the second stage 24,250 lb. (11,000 kg). The first Cosmos satellite was launched from a former missile site called KAPUSTIN YAR, about 60 miles (100 km) southeast of Volgograd; all previous Soviet orbital launchings had been from TYURATAM. The Cosmos launcher, known in the West as B1, is capable of orbiting small satellites, up to about 1,000 lb. (450 kg). On August 18, 1964, the Russians introduced a more powerful launcher when they put three satellites, Cosmos 38, 39, and 40, into orbit together. This launcher, not named by the Russians but known in the West as C1, is believed to be based on the Skean intermediate-range missile, using a new restartable upper stage: it can orbit payloads of up to 2,200 lb. (1,000 kg). Heavier satellites are launched by the VOSTOK rocket also used for manned missions.

The launch of Cosmos 112 on March 17, 1966, saw the introduction of a third launch site, called PLESETSK, near Archangel in northern Russia. It can handle a wider range of rockets than Kapustin Yar, and has since become the major Cosmos launch site. Plesetsk is often used for military satellite launches; about half the Cosmos satellites are believed to have military purposes. Many of them are reconnaissance satellites, which eject a package of film for recovery after about two weeks in orbit, although some have been tests of a system for intercepting and destroying other satellites in orbit. Other military Cosmos are communications and navigation systems.

The 500th Cosmos satellite was launched on July 10, 1972. Among the achievements of the series up until that date had been the automatic link-up in space between Cosmos 186 and 188 on October 30, 1967 — this was the world's first unmanned rendezvous, and was a test flight of two prototype SOYUZ spacecraft. The Cosmos series was also used to develop an operational WEATHER SATELLITE system. Cosmos 122, launched on June 25, 1966, was an acknowledged forerunner of the system, carrying Earth sensors; on February 28, 1967, Cosmos 144 was launched as a prototype of the Meteor series of weather satellites.

A typical small Cosmos satellite is a cylinder about $5\frac{1}{2}$ feet (1.8 m) long and $3\frac{1}{2}$ feet (1.2 m) wide with hemispherical ends, weighing roughly 900 lb. (400 kg). However, the recoverable military reconnaissance satellites, and satellites used for biological studies, are modified Vostok capsules.

Cooperation with other Soviet bloc countries led to multinational experiments on board Cosmos 261, launched on December 20, 1968. This was followed by the Intercosmos series, the first of which, Intercosmos 1, was launched on October 14, 1969. On March 31, 1978, the Soviet Union launched the 1,000th satellite of the Cosmos series.

coudé focus

A focal point in a CASSEGRAIN TELESCOPE, produced by an arrangement of plane mirrors which diverts the light through the hollow polar axis to a stationary observing position. A coudé focus is particularly desirable for mounting heavy equipment such as a spectrograph, and the long light path gives a large-scale image valuable for detailed studies of spectra. Most large Cassegrain telescopes have a coudé focus. In the 200-inch (508-cm) telescope on Mount Palomar, for example, the coudé focus is formed in a constant-temperature room below ground level at an effective FOCAL LENGTH (mirror-image distance) of 500 feet (150 m). In refracting telescopes, plane mirrors may be placed behind the object glass to reflect the light along the polar axis to a coudé focus.

Crab Nebula

The expanding cloud of gas ejected by a star seen to explode in July 1054 A.D. This exploded star, or *supernova*, was observed by Oriental and Arab astronomers to reach magnitude −5 (brighter than Venus), remaining visible in daylight over 23 days. The resulting nebula was first noted telescopically in 1731 by the English amateur astronomer John Bevis (1693–1771), and later became the first object (M1) in the list of nebulae compiled by the French astronomer Charles MESSIER. It is easily visible in small telescopes as a hazy oval, lying in the constellation Taurus. Larger telescopes show that its total extent is 7 by 4 minutes of arc (12 by 7 light-years), although it is very faint toward the edges.

The Crab Nebula is about 6,300 light-years from the Sun. It emits powerful radiation at all wavelengths, from radio to X rays and gamma rays. This vast output, 25,000 times the luminosity of the Sun, is maintained by a rotating PULSAR at the nebula's center. The pulsar, only a few miles in diameter but as massive as the Sun, was originally the dense central core of the exploding star. It now rotates 30 times a second, flashing each time at radio, optical, and X-ray wavelengths. It is the fastest-rotating pulsar known, and the only one visible optically. Its radio pulses were discovered in 1968, and its optical flashes in 1969.

The pulsar is linked to the nebula by magnetic fields, which are braking its spin by one part in a million per day. The rotational energy lost is emitted as radiation by electrons spiraling in the magnetic field (SYNCHROTRON EMISSION). In this respect the optical emission from the Crab differs from that of all other known nebulae, which shine by the light emitted when electrons recombine with ionized atoms (see NEBULA). The Crab Nebula is also unique among

The Crab Nebula photographed in red light, which emphasizes the long gaseous filaments emitting the hydrogen red spectral line. The diffuse light from the center is synchrotron radition, caused by fast electrons from the pulsar, **lower of the two central stars**, moving through the nebula's magnetic field.

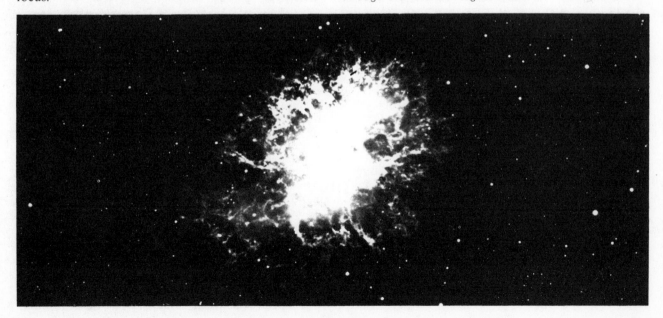

supernova remnants, which are generally ring-shaped and optically faint. This difference is probably again due to the pulsar. The expansion of the nebula was discovered in 1921 by the American astronomer John Charles Duncan (1882–1967), who compared photographs taken in 1909 and 1921; he took a confirming plate in 1938.

The Crab Nebula contains long thin filaments, whose light originates from recombining atoms. Each filament is surrounded by a magnetic field, which carries current of about a thousand million million amperes. These filaments were first observed by Lord ROSSE in 1844. His drawing vaguely resembled the pincers of a crab, which gave the nebula its name. The pulsar accelerates the filaments outward at about 700 miles (1,000 km) per second. The proportion of helium to hydrogen in the filaments is seven times greater than normal at the surfaces of stars. This additional helium was produced by fusion of hydrogen inside the star while it was shining steadily; the helium-rich interior became mixed with the surface layers after the explosion. The mass of luminous gas in the filaments is about equal to that of the Sun. There may also be several times more mass within the filaments which cannot be seen. The original star must thus have been considerably more massive than the Sun to have produced both the nebula and the central neutron star.

Crater (the cup)
A small and insignificant constellation in the southern hemisphere of the sky, lying between Leo and Hydra.

craters
Bowl-shaped depressions with raised rims formed by the fall of METEORITES. Shortly after the formation of the solar system, interplanetary space was dense with debris of all sizes; this has steadily been swept up by the planets, leaving visible effects on their surfaces. The unweathered faces of the Moon and Mercury abound with craters, while on Mars erosion has obliterated older features. On Earth, erosion is severe; even giant craters weather away rapidly, to leave only a few small recently formed craters clearly visible.

The Earth's atmosphere melts away the outer layers of any falling object from space and strongly decelerates bodies below 100 tons mass. The Earth's atmosphere cannot slow larger meteorites completely; their tremendous speed is converted into heat, causing a violent explosion. A 1,000-ton body would produce a 300-ton meteorite, striking the ground at 3 miles (5 km) per second. The ensuing explosion would shatter the meteorite, scattering fragments around a 500-foot (150-m) crater. Many such meteorite craters have been found. The Haviland crater in Kansas is 56 feet (17 m) across and was surrounded by more than 10 tons of pallasite meteorites. The Sikhote-Alin group of 24 craters in the U.S.S.R. was formed on February 12, 1947, by the fall of a giant fragmenting iron meteorite; the largest crater, 85 feet (26 m) across and 20 feet (6 m) deep, was caused by a 3-ton mass. The Henbury group of 13 craters in Australia, the largest 722 feet (220 m) across, are surrounded by many tons of iron meteorites.

An even larger example is the great meteor crater near Flagstaff, Arizona, one of the earliest meteorite impact features identified. It is a $\frac{3}{4}$-mile (1.2-km)-diameter, 600-foot(180-m)-deep bowl with a 150-foot (45-m) raised rim. A 250,000-ton, 240-foot(73-m)-diameter iron body impacting at 10 miles (16 km) per second produced the crater. Its rim is composed of folded and inverted layers of material pushed out from the center, while the rock under the floor of the crater is badly shattered and jumbled. Beneath this, however, the local sandstone is undisturbed. This impact explosion was so violent that it almost vaporized the meteorite; only the rear 20 feet (6 m) survived to scatter hundreds of tons of iron meteorites for 30 square miles (80 km^2) around the crater. The vaporized iron is now present as billions of tiny iron spherules in the surrounding desert soil.

Larger meteorites would be completely destroyed in the explosion. A $\frac{1}{2}$-mile (0.8-km)-diameter asteroid would penetrate the surface rock for about 2 miles

(3 km). The resulting explosion would have a peak temperature of 20,000°K at more than 5 million atmospheres pressure, vaporizing everything within a mile (1.6 km) of the impact and melting everything out to $1\frac{1}{2}$ miles (2.4 km). The resultant crater would be 14 miles (22.5 km) across and $1\frac{1}{2}$ miles (2.4 km) deep. The Ries basin in Germany is the eroded and weathered remains of just such a 14-mile crater. Even larger fossil craters (also known as *astroblemes*—literally, "star-wounds") have now been identified. The Vredefort ring in South Africa is the remains of a 25-mile (40-km) crater, an impact by a $1\frac{1}{2}$-mile (2.4-km)-diameter asteroid, while the Sudbury structure in Canada is a similar crater which initiated widespread volcanic activity.

Several even larger craters are suspected, including the basin 217 miles (350 km) in diameter containing Lake Tengiz in Siberia, while the Nastapoka Island arc in Canada's Hudson Bay may be the only surviving portion of a 275-mile (440-km) crater. A steady stream of new probable fossil meteorite craters has been found in recent years, bringing the total number to about 200.

Crimean Astrophysical Observatory

Russian observatory near Simferopol in the Crimean Peninsula, at an altitude of 1,700 feet (560 m). It was originally sited at Simeis on the Black Sea shore, but was relocated after damage in World War II. Its main telescope is a 102-inch (260-cm) reflector completed in 1960; there is also a 49-inch (125-cm) reflector and other smaller telescopes. Located at the same site is the southern observing station of Moscow's Sternberg Astronomical Institute, which shares the observatory's facilities.

Crux (the southern cross)

The smallest, but one of the most famous of constellations. It lies in the southern hemisphere of the sky, next to Centaurus, of which it was a part until made a constellation of its own in 1673 by the French sailor Augustin Royer. Binoculars show that Alpha Crucis, also known as Acrux, is a double star, both components being brilliant white; their combined magnitude is 0.79. Beta Crucis, magnitude 1.24, is also a brilliant white star; next to it is the cluster NGC 4755, about 1,000 light-years away, centred on the star Kappa Crucis. Crux also contains the COALSACK nebula.

Culgoora Radioheliograph

A unique instrument at Culgoora, Australia, designed for radio observations of the Sun, particularly of the solar flares. It consists of 96 antennae, each 45 feet (13.7 m) in diameter, mounted in a circle 1.9 miles (3 km) in diameter; the aerials produce instantaneous radio pictures of the active areas around the Sun on cathode ray screens in the central control room. The radioheliograph, devised by the Australian radio astronomer John Paul Wild (b. 1923), began operation in 1967 at a frequency of 80 MHz; it has since been modified to work at double and half this frequency as well. The instrument forms part of the Culgoora Solar Observatory, which includes telescopes for optical observation of the Sun. The observatory is operated by the Radiophysics Laboratory of the Commonwealth Scientific and Industrial Research Organization (CSIRO).

culmination

The maximum altitude above the horizon reached by a celestial body on a given date. Unless the body moves significantly during the night, culmination is the moment that it crosses the *meridian,* the north–south line in the sky.

curvature of space

A distortion of space caused by the presence of matter. Such a distortion can be thought of as analogous to the distortion familiar on maps of the Earth using the Mercator projection. On the map the distortion occurs because the surface of the Earth is spherical, but the map's surface is flat. Polar regions such as Greenland consequently appear "stretched out." High-school geometry is "flat" (Euclidean) geometry, and is not correct on a spherical surface where, for example, initially "parallel" lines may meet. Until Einstein, it was believed that the geometry of space was Euclidean. However, Einstein's theory of RELATIVITY proposed that gravity "curves" space in a fashion analogous to the curvature of the Earth. Therefore the geometry of the Universe, or of the space surrounding any object, curves the path of matter and of light. Such a space distortion around the Sun was first confirmed from observations by Sir Arthur EDDINGTON during a solar eclipse in 1919. Radio telescopes have recently been used to measure the deflection by the Sun of radio waves from a distant QUASAR, and these very accurate measurements confirm the optical results.

Although difficult to visualize, the effects of space curvature are readily calculated. A curved Universe can be finite in size, yet have no boundary; it would thus be analogous to the Earth's surface, also finite but without any edge. If a person could travel in an apparently straight line far enough in any direction, he would find himself returning to the point from which he began.

Cygnus (the swan)

A prominent constellation of the northern hemisphere of the sky, best seen during the northern summer; because of its shape it is often called the Northern Cross. Cygnus lies in a dense part of the Milky Way and contains many objects of interest. Its major star is DENEB, one of the brightest in the sky. Beta Cygni, called Albireo, is one of the most beautiful double stars; its component stars are yellow and blue in color, of magnitudes 3.2 and 5.4 respectively. The first star other than the Sun to have its distance calculated was 61 Cygni; this was done by Friedrich BESSEL in 1838. One remarkable star in the constellation is P Cygni, a nova-like variable that was seen to flare up in the 17th century. Cygnus also contains the North American nebula (NGC 7000), an illuminated mass of gas so named because of its shape, and the Veil nebula (NGC 6992), which is part of the so-called Cygnus loop, the remnants of a supernova which exploded 60,000–70,000 years ago. One of the most powerful radio sources in the sky is Cygnus A, which lies near the star Gamma Cygni; it is thought to be an exploding galaxy (see RADIO GALAXY). X rays are detected coming from a double star near Eta Cygni; this source, called Cyg X-1, is believed to give the first direct observational evidence of a BLACK HOLE in space (see X-RAY ASTRONOMY).

D

δ

The symbol for the astronomical coordinate known as DECLINATION.

David Dunlap Observatory

Canadian observatory situated at an altitude of 800 feet (244 m) at Richmond Hill, Ontario, operated by the University of Toronto. Its main telescope is a 74-inch (188-cm) reflector, the largest optical telescope in Canada, which began operation in 1935. Since 1971 the University of Toronto has also operated a 24-inch (61-cm) reflector at the Las Campanas observatory in Chile.

Dawes, William Rutter (1799–1868)

English astronomer, originator of the famous *Dawes limit* formula for determining telescopic RESOLUTION. Dawes was a pioneer double-star observer, publishing a total of almost 2,800 measurements. From 1839 to 1844 he was in charge of the private observatory of George Bishop (1785–1861) in Regent's Park, London, where he made numerous observations. He was succeeded by John Russell Hind (1823–1895) who discovered 10 asteroids while there. Dawes calculated from his observations of double stars the so-called Dawes limit: the resolving power of a telescope measured in seconds of arc is 4.56 divided by the aperture measured in inches. This holds true assuming average seeing; under exceptional conditions the Dawes limit can be exceeded, though in many cases it will not be reached.

day

The time taken for the Earth to spin once on its axis; by extension, the rotation period of any planet. The rotation of the Earth can be measured relative to the stars (a SIDEREAL DAY) or to the Sun (a SOLAR DAY). Astronomers employ the sidereal day, but for other purposes the Earth's rotation with respect to the Sun is used. The two kinds of days differ slightly in length because the Sun's position against the star background changes as the Earth goes around its orbit. Actual solar time is, however, slightly irregular (see EQUATION OF TIME), and for normal timekeeping we use the *mean solar day*, which measures the Earth's rotation relative to an imaginary MEAN SUN, moving across the sky at a uniform rate. The mean solar day is divided into 24 hours of mean solar time. In terms of mean solar time, a sidereal day lasts 23 hours 56 minutes 4 seconds. The mean solar day is 24 hours 3 minutes 56.55 seconds of sidereal time.

declination

The celestial equivalent of latitude on Earth. The declination of an object north or south of the celestial equator is given in degrees and fractions of a degree, marked positive (+) for northerly declinations and negative (−) for southerly ones (see CELESTIAL SPHERE).

deferent

In early systems of planetary motion, the hypothetical circular path along which moved the center of a smaller circle, the EPICYCLE.

Deimos

Smaller and more distant of the two moons of Mars, discovered in 1877 by Asaph HALL. It orbits Mars every 30 hours 18 minutes at a distance of 14,590 miles (23,490 km) from the planet's center. Mariner 9 photographs show that Deimos is irregular in form and dotted with small craters. Its general shape resembles that of a squashed potato, with dimensions of 9 by 7½ by 7 miles (15 × 12 × 11 km).

Delphinus (the dolphin)

A small constellation in the northern hemisphere of the sky between Pegasus and Aquila. Four stars form a rectangle called Job's coffin. If the names of its two brightest stars, Sualocin and Rotanev, are spelled backwards, they give *Nicolaus Venator* (in Italian, Niccoló Cacciatore, 1780–1841), the assistant to Giuseppe PIAZZI, who cleverly named them.

Delta

A medium-sized rocket used to launch numerous scientific and communications satellites, notably the INTELSAT series. Delta is based on the Thor intermediate-range missile, with a liquid-fuel second stage and an optional solid-fuel third stage. There have been over 100 successful Delta launches since its introduction with Echo 1 in 1960. Improvements in design have steadily increased the size of the rocket and its payload capacity. In 1964, three small solid-fuel strap-on boosters were added to the first stage; this vehicle was named the Thrust Augmented Delta (TAD). In 1967 an improved version of the third stage produced the Thrust Augmented Improved Delta (TAID). The following year the Thor first stage was lengthened to give the Long-Tank Delta. The addition of strap-ons turns it into the Long-Tank Thrust Augmented Delta (LTTAD). Six or even nine strap-on solid rocket motors (SRMs) can now be added to the first stage; they are each 23 feet (7 m) long and 31 inches (0.8 m) in diameter, and have a thrust of 52,000 lb. (23,500 kg). More advanced strap-ons each give a thrust of 84,000 lb. (38,100 kg). In 1974 a new variant was introduced called the Straight-Eight Delta because of its constant 8-foot (2.4-m) diameter. This has a first-stage thrust of 205,000 lb. (93,000 kg), second-stage thrust of 9,800 lb. (4,450 kg), and optional third stages of 10,000-lb. (4,540-kg) or 15,000 lb. (6,800-kg) thrust. The complete rocket, including a protective shroud covering the satellite, is 116 feet (35.4 m) high. Delta is capable of placing a satellite of 1,500 lb. (680 kg) into synchronous orbit, or one of 4,000 lb. (1,800 kg) into low Earth orbit.

Deneb

The brightest star in the constellation Cygnus, also called Alpha Cygni. It is a brilliant white star about 50,000 times more luminous that the Sun, of apparent magnitude 1.26, and about 1,500 light-years away. The name Deneb means "tail."

descending node

See NODES.

Dicke, Robert Henry (b. 1916)

American physicist who played an important part in the development of lasers, and later, in the 1960s, performed a delicate experiment which confirmed to one part in one hundred million that all bodies fall equally fast under gravity. This is called the principle of equivalence, and provides the physical basis of the general theory of RELATIVITY. Dicke developed a detailed extension of general relativity in which gravity grows progressively weaker with time (see BRANS-DICKE COSMOLOGY), and in 1964 suggested a search for the now-famous cosmic BACKGROUND RADIATION from the big bang.

diffraction
The apparent bending of light (or other radiation) around obstacles. It is because of diffraction that the edges of shadows are not sharp. A light beam passing the edge of an object is diffracted through a small angle, which varies with the wavelength: the longer the wavelength, the greater the bending. Scientists make use of this effect by means of a DIFFRACTION GRATING, which spreads white light into a spectrum of colors. Light from a star passing through a telescope is diffracted to form an image which is not a point but a disk, called the *Airy disk,* surrounded by successively fainter concentric circles or fringes. The smaller the telescope, the larger the Airy disk and hence the poorer its resolution, its ability to discriminate fine detail. In a reflecting telescope, the struts supporting the secondary mirror at the prime focus also cause diffraction, visible on photographs as the spikes on the images of bright stars. Diffraction was first noted by the Italian physicist Francesco Maria Grimaldi (1618–1663).

diffraction grating
A device for diffracting light into a spectrum, usually by employing a series of fine parallel lines ruled on a material such as glass. Light passing an obstacle is partly spread into a spectrum; by arranging a large number of narrowly spaced obstacles, light can be split into a series of spectra. Some of the light will pass straight through, but some will be partly diffracted into beams on either side of the undeviated light. These are called first-order beams; they will be dispersed into colors, with the longest wavelengths deviated the most. Farther away from the undeviated beam are fainter, more dispersed spectra; these are termed second-order, third-order, and so on. Thus, a grating can be used in a SPECTROSCOPE. Gratings are made by ruling machines that produce several thousand lines per centimeter. This is costly, and it is common to make transfer-like replicas using the original as a mold, in a material such as collodion or plastic. If the grating is in front of the telescope rather than at the focus, a coarser spacing is used, known as an *objective grating.* Reflection gratings are formed by ruling lines of an aluminized mirror. By varying the angle the grating can be *blazed* to throw most of its light into one order spectrum. The *Rowland grating,* invented in 1885 by the American physicist Henry Augustus Rowland (1848–1901), has lines ruled on a concave mirror, thus bringing the spectrum to a focus without the need for further optics.

Dione
Satellite of Saturn, fifth in order of distance from the planet, discovered by G. D. CASSINI in 1684. Its diameter is about 500 miles (800 km), and it is probably a rocky body, with a density of 3.2, similar to that of our Moon. Dione orbits Saturn every 2 days 17 hours 41 minutes at a distance of 234,600 miles (377,500 km).

direct motion
The movement of an object in its orbit from west to east. This is the normal direction of movement in the solar system; it is also called *prograde motion.* The opposite direction of movement, from east to west, is known as RETROGRADE MOTION.

Discoverer satellites
Series of U.S. Air Force satellites for scientific and military research. The Discoverer satellites were put into orbits over the Earth's poles. They were launched from the Western Test Range in California by Thor rockets, using an early version of the AGENA as an upper stage. In most Discoverers, an instrument package was ejected from the nose of the Agena after a specified number of orbits; it was either snared by an aircraft as it floated back to Earth under parachutes, or retrieved from the ocean after splashdown. The reentry practice gained with Discoverer was valuable for later manned programs; the capsule of Discoverer 13 was in fact the first object to be recovered from orbit. The first 15 Discoverers were engineering test satellites, but later capsules contained biological specimens to study radiation hazards in space and prototype equipment for military surveillance systems. After Discoverer 38, military surveillance and capsule recovery became classified subjects and subsequent satellites were not described.

Discoverer series

Serial number	Launch date	Results
1	February 28, 1959	No reentry capsule. World's first polar-orbiting satellite
2	April 13, 1959	Capsule ejected on orbit 17, but lost in Arctic
3	June 3, 1959	Failed to orbit
4	June 25, 1959	Failed to orbit
5	August 13, 1959	Capsule ejected, but into another orbit
6	August 19, 1959	Capsule ejected on orbit 17, but recovery failed
7	November 7, 1959	Capsule not ejected
8	November 20, 1959	Launched into wrong orbit; capsule overshot recovery area on orbit 15
9	February 4, 1960	Failed to orbit
10	February 19, 1960	Failed to orbit
11	April 15, 1960	Capsule ejected on orbit 17, but recovery failed
12	June 29, 1960	Failed to orbit
13	August 10, 1960	Capsule recovered from sea after 17 orbits; first recovery from orbit

(continued p. 58)

Discoverer Series

Serial number	Launch date	Results
14	August 18, 1960	First midair recovery, on orbit 17
15	September 13, 1960	Capsule ejected after 17 orbits, but lost in storm
16	October 26, 1960	Failed to orbit
17	November 12, 1960	Midair recovery after 31 orbits
18	December 7, 1960	Midair recovery after 48 orbits
19	December 20, 1960	No reentry capsule; carried infrared experiments
20	February 17, 1961	Capsule not ejected; equipment failure
21	February 18, 1961	No reentry capsule; infrared experiments
22	March 30, 1961	Failed to orbit
23	April 8, 1961	Capsule shot into different orbit; not recovered
24	June 8, 1961	Failed to orbit
25	June 16, 1961	Ocean recovery after 33 orbits
26	July 7, 1961	Midair recovery after 32 orbits
27	July 21, 1961	Failed to orbit
28	August 3, 1961	Failed to orbit
29	August 30, 1961	Ocean recovery after 33 orbits
30	September 12, 1961	Midair recovery after 33 orbits
31	September 17, 1961	Capsule failed to eject
32	October 13, 1961	Midair recovery after 18 orbits
33	October 23, 1961	Failed to orbit
34	November 5, 1961	Capsule not ejected
35	November 15, 1961	Midair recovery after 18 orbits
36	December 12, 1961	Ocean recovery after 64 orbits. Same launch rocket as Oscar 1 amateur radio satellite
37	January 13, 1962	Failed to orbit
38	February 27, 1962	Midair recovery after 65 orbits

dispersion

The spreading out of light into its constituent wavelengths to form a spectrum, occurring, for example, when light passes through a prism.

distance modulus

Technique for determining distances in astronomy by comparing an object's intrinsic brightness with its brightness as observed on Earth. An object's distance modulus therefore expresses the difference between the ABSOLUTE MAGNITUDE (actual light output) and the observed magnitude (apparent magnitude) of an object. One unit of magnitude corresponds to a difference in brightness of 2.512 times (see MAGNITUDE). Thus an object with a distance modulus of 15 magnitudes appears with only a millionth (2.512^{-15}) the brightness it would exhibit if it were at a distance of 10 parsecs (about 32.6 light-years), the standard distance used for calibrating absolute magnitude. Since the intensity of light diminishes according to the inverse square of distance, each magnitude of distance modulus corresponds to an increase in distance by a factor of $\sqrt{2.512}$, or 1.585. Thus a distance modulus of 15 magnitudes means a distance one thousand times (1.585^{15}) greater than the standard distance of 10 parsecs—in other words, 10,000 parsecs. In practice, brightness is also reduced by light-absorbing thin gas and dust in space, an effect which must be taken into account. The distances of objects such as galaxies are obtained by comparing the brightness of supergiant stars with similar stars of known distance in our own Galaxy. This produces an average distance for the entire galaxy.

Dollfus, Audouin Charles (b. 1924)

French astronomer and leading planetary expert, discoverer of Saturn's 10th moon, Janus. The son of an aviation pioneer, Dollfus made balloon ascents to obtain better views of the planets. He pioneered observations of the polarization of light from planetary surfaces, and concluded that the surface of Mars is covered by particles of an iron oxide, probably limonite (Fe_2O_3). His discovery of Janus was made on photographs taken at the Pic du Midi Observatory when the Earth passed through the plane of Saturn's rings in December 1966.

Dollond, John (1706–1761)

English lens maker who discovered by experiments in 1757 that an ACHROMATIC lens could be made by using two different types of glass, crown glass and flint glass. In 1752 John Dollond had joined the optical firm set up in 1750 by his son Peter Dollond (1730–1820); the lenses they made became world famous. In 1754 John Dollond made the first successful HELIOMETER. Peter Dollond in 1765 invented the triple achromatic lens, by placing two convex lenses of crown glass either side of the biconcave flint glass lens. In 1805 the firm was joined by Peter's nephew, George Dollond (1774–1852). John Dollond's daughter Sarah married Jesse Ramsden (1735–1800), who became a famous instrument maker in his own right, and in 1782 invented the eyepiece design that bears his name.

Dominion Astrophysical Observatory

Canadian observatory situated near Victoria, British Columbia, at an altitude of 750 feet (229 m). Its main telescope, a 73-inch (185-cm) reflector, began operation in 1918, and a 48-inch (122-cm) reflector was added in 1961. The associated Dominion Radio Astrophysical Observatory at Penticton, British Columbia, has an 84-foot (26-m) radio telescope and two large MILLS CROSS arrays. Both establishments are operated by the National Research Council of Canada.

Doppler effect

The change in frequency of waves emitted by an object as it moves toward or away from an observer. A familiar example, involving sound waves, is the change in pitch of sirens on emergency vehicles: the sound becomes higher pitched as the vehicle approaches but then lowers in pitch after the vehicle passes and moves away. The Austrian physicist Christian Johann Doppler (1803–1853) first drew attention to this effect in 1842. It holds true for any type of wave motion, including that of light.

Doppler used the analogy of a ship moving through equally spaced ocean waves (i.e., the distance from crest to crest is constant). If the ship is stationary, the waves will break on the bow, say, once every second. If, however, it is moving into the waves, they will break more frequently. To the observer on the ship, it would seem that the frequency of the waves was higher, or that the space between them, their wavelength, was shorter. The wavelength would grow longer, and the frequency drop, if the ship were moving away from the source of the waves. If, instead, the wave source is moving and the observer is stationary, the wavelength will appear longer as the source moves away, or shorter as the source approaches. For light, the longer wavelengths are redder; for sound, longer wavelengths are lower in pitch.

By measuring the shift in an object's spectral lines—toward the red end of the spectrum if it is receding, or toward the violet end if it is approaching—astronomers can establish the direction and speed of the object.

The RADIAL VELOCITIES of stars, the motions of DOUBLE STARS about each other, the rotation of stars, and the motions of galaxies (usually seen as a RED SHIFT) can all be determined in this way.

Dorado (the swordfish)
A constellation in the southern hemisphere of the sky, named by Johann Bayer in 1603. None of its stars is of particular interest, but Dorado contains the larger MAGELLANIC CLOUD. This has within it the so-called Tarantula nebula, NGC 2070, around the star 30 Doradus.

double star
Two stars linked by gravity; an alternative name for such a pair is a *binary star.* The first genuine double star to be discovered was MIZAR, observed by the Italian astronomer Giovanni Riccioli in 1650. This is a physical double star, in contrast to an apparent or optical double, a relatively rare occurrence in which two stars appear close together in the sky, but in reality are widely separated along the line of sight. It is estimated that only about 15 percent of all stars are single bodies like the Sun, while 46 percent are in double systems, in which one star is gravitationally attracted to the other and orbits it. The remaining 39 percent occur in multiple systems of three members or more. The separation between double stars ranges from the so-called *contact binaries* like W Ursae Majoris, in which the stars actually touch each other and share a common atmosphere, to stars so widely separated that they can be distinguished in a telescope. Astronomers suggest that the closest double stars—those with orbital periods less than about 100 years—were formed when a gas cloud condensing into a star split into two or more parts. Double stars with the longest periods may have been formed separately.

The closest binaries are revealed as double when their light is analyzed with a spectroscope (SPECTROSCOPIC BINARIES). Close binaries can also be detected if the Moon happens to pass in front of them, for their light will fade in two steps. The new technique of SPECKLE INTERFEROMETRY, in which a high-speed photograph of the star is de-blurred by illuminating it with a laser, can also produce images of two stars close together. The orbits of binary stars lie at completely random angles; they do not, for example, line up with the plane of the Galaxy. If the orbit is seen edge-on (an inclination angle of $90°$) the two stars may periodically pass in front of each other as an ECLIPSING BINARY.

Periods. In accordance with Kepler's third law (see KEPLER'S LAWS), the closer double stars tend to have shorter periods. The longest well-established period, 480 years, is of the binary Eta Cassiopeiae, easily seen to be double with binoculars. There are, however, stars whose periods are so long that they have nowhere near completed a single orbit since observations began—the period of PROXIMA CENTAURI around ALPHA CENTAURI is about a million years. The twin nature of some double stars is inferred simply because the stars have the same distance and share the same motion (PROPER MOTION) through space. The shortest periods are found among the binary stars having a small star such as a WHITE DWARF as one component. These can approach quite close, like AM Canum Venaticorum, with the shortest-known period—$17\frac{1}{2}$ minutes—of any ordinary double star. However, the X-ray novae having periods as short as 120 seconds discovered by the Ariel 5 satellite in 1975 may be binary stars in which both components are white dwarfs or NEUTRON STARS.

Evolution. The brighter star in a binary system is called the *primary,* the fainter the *secondary.* During the course of its evolution the primary expands, and its atmosphere may be gravitationally attracted to the secondary, the gas streaming from one star to the other. When this happens, the primary is said to fill its *Roche lobe,* and the binary star is said to be semidetached. So much gas may flow from the primary to the secondary that the secondary eventually becomes the new primary. Sirius, which has a companion that has evolved to a white dwarf, has now reached this state. The new primary also eventually expands, fills its Roche lobe, and the binary star again becomes semidetached, with the gas stream pouring from the new primary to the new secondary. Algol is such a star. The secondary may have become a white dwarf, neutron star, or black hole. When the gas stream falls onto such a small star it creates X rays; this is the cause of many of the X-ray stars discovered by the Uhuru (SAS-A) satellite in 1971. If so much material falls onto a white dwarf that its mass exceeds the maximum possible for such a star (the Chandrasekhar limit), it explodes as a NOVA.

Draco (the dragon)
A large and straggling constellation near the north pole of the sky. Its brightest star, Gamma Draconis, was observed by James Bradley in his discovery of the ABERRATION OF STARLIGHT. Thuban, Alpha Draconis,

was the north pole star about 2832 B.C.; the pole has since moved because of PRECESSION. Draco contains the planetary nebula NGC 6543, 1,700 light-years away.

Drake, Frank Donald (b. 1930)

American radio astronomer who made the first attempt to detect signals from other civilizations, called Project Ozma. At the National Radio Astronomy Observatory in West Virginia, Drake used an 85-foot (26-m) radio telescope to "listen" to two nearby stars resembling the Sun, Tau Ceti and Epsilon Eridani, for a total of 150 hours in 1960. Although no signals were heard, Project Ozma turned attention to the possibilities of interstellar contact by radio. In 1959, Drake's radio observations revealed the Van Allen radiation belts around Jupiter. Drake became director in 1971 of the National Astronomy and Ionosphere Center, which runs the 1,000-foot (305-m) Arecibo radio telescope. With this instrument Drake began to listen in 1975 for possible radio messages from super-civilizations in nearby galaxies (see LIFE IN THE UNIVERSE).

Draper, Henry (1837–1882)

American physician and amateur astronomer, a pioneer of astrophotography. His father, John William Draper (1811–1882), took the first photographs of the Moon in 1840, and of the Sun's spectrum in 1843. Henry Draper took the first successful photograph of a star's spectrum, that of Vega, in 1872; in 1880 he photographed the Orion nebula, and later also recorded its spectrum.

Henry Draper's widow endowed the Harvard College Observatory with funds to produce a classification of stellar spectra. A preliminary catalog containing 10,351 stars, the work of Williamina Paton Fleming (1857–1911), appeared in 1890. The main work, the *Henry Draper Catalogue,* containing 225,300 stars down to eighth magnitude, was published between 1918 and 1924. It was compiled by Annie J. CANNON, and it introduced the now-standard Harvard classification system (see SPECTRAL TYPE). She also compiled the *Henry Draper Extension* (1925–1936), listing 47,000 additional spectra of fainter stars. A further volume, published posthumously in 1949, contained another 86,000 stars. An entirely new edition of the catalog is now being prepared.

Dreyer, Johan Ludwig Emil (1852–1926)

Danish astronomer, compiler of the *New General Catalogue of Nebulae and Clusters of Stars,* containing 7,840 objects and published in 1888. Dreyer was appointed astronomer at the observatory of Lord ROSSE in 1874, and his interest in nebulae was stimulated by observations with Rosse's large telescope. Dreyer compiled his famous catalog, often simply called the NGC, while at Armagh Observatory where he was director from 1882 to 1916; it superseded John HERSCHEL's previous catalog. In 1895 and 1908 Dreyer published supplements, called the *Index Catalogues* (IC), of 1,529 and 3,857 newly found nebulae and clusters. Objects are often referred to by the NGC and IC numbers that Dreyer gave them. Dreyer was also a prominent historian of astronomy, editing works of Tycho Brahe (1908) and William Herschel (1912), and in 1906 publishing the standard *History of Astronomy from Thales to Kepler.*

dwarf stars

Name applied to ordinary stars like the Sun that are in the prime of their lives, shining by converting hydrogen into helium in their centers. The brighter, hotter dwarfs are as much as 65 times as massive as the Sun. Beyond this size, known as the *Eddington limit,* radiation pressure in their interiors blows them apart. The dimmer, cooler dwarfs (RED DWARFS) can be as small as 1 percent of the solar mass. Below this limit, the object would never become hot enough to make fusion occur and would be classified as a planet. In the HERTZSPRUNG-RUSSELL DIAGRAM the dwarfs lie on the MAIN SEQUENCE. Stars which are substantially larger and brighter than dwarfs are called GIANTS and SUPERGIANTS. There are a few stars lying just below the main sequence, called subdwarfs. These probably resemble ordinary dwarfs, but being older are of slightly different composition.

Dyson, Sir Frank Watson (1868–1939)

English astronomer royal from 1910 to 1933 (and astronomer royal for Scotland from 1906 to 1910). Dyson organized expeditions to observe the total eclipse of 1919; these showed that light from stars is bent by gravity as it passes close to the Sun, thus confirming a key prediction of Einstein's theory of general relativity. Dyson's own observations at numerous eclipses produced important information on the Sun's faint outer regions, the CHROMOSPHERE and CORONA. At Greenwich Dyson inaugurated the radio transmission of time signals in 1924, extending it to a worldwide service in 1927. In 1924 he introduced the free-pendulum type clock, designed by the engineer William Hamilton Shortt (1882–1971), which was used to maintain Greenwich time until the introduction of the QUARTZ-CRYSTAL CLOCK.

E

Early Bird

The first communications satellite launched by the INTELSAT organization. It was placed into synchronous orbit on April 6, 1965, over the Atlantic at longitude 35° west. Early Bird could carry 240 two-way voice circuits, or one television channel. It operated for $3\frac{1}{2}$ years, providing the first commercial satellite link between the United States and Europe; it was reactivated during July 1969 because of a temporary failure with an Intelsat 3 satellite. Early Bird was developed from the SYNCOM design.

Earth

The third planet from the Sun, and the only one known to support life. As seen from space, Earth would be termed the blue planet, because two-thirds of its surface is covered by seas. It is this predominance of water in liquid form that makes Earth hospitable to life as we know it. Often, however, much of the surface is obscured by white clouds caused by evaporating water vapor. By watching the Earth from space, another civilization would soon deduce the major features of the Earth's atmospheric circulation and of its weather patterns. We get such a view by the use of WEATHER SATELLITES. The Earth spins once on its axis every DAY and completes one orbit around the Sun roughly every $365\frac{1}{4}$ days (the YEAR). The Earth is roughly spherical in shape, with an average diameter of 7,918 miles (12,742 km); strictly, it has a slight bulge at the equator, giving an equatorial diameter 26.6 miles (42.8 km) greater than that from pole to pole. Its mass is 5.976×10^{21} tons, and its average density is 5.52 times that of water.

The Earth is only one of nine major planets in our SOLAR SYSTEM, orbiting the Sun. The Sun itself is merely one of about one hundred billion stars in the MILKY WAY Galaxy, many of which probably have planets of their own. Current astronomical opinion is that planet Earth is far from unique.

Formation of the Earth. The Earth and the rest of the solar system formed together with the Sun about 4.6 billion years ago when a cloud of cool gas and dust in space began to condense under the effect of its own gravity (see PLANETS). In the large central portion of the collapsing cloud, heat generated by the condensation process was sufficient to trigger nuclear reactions, forming a star—our Sun. But in the outer regions of the cloud, smaller, cooler objects formed— the planets, of which the Earth is a small, rocky example. Heat generated by the decay of radioactive elements in its interior, together with heating by the young Sun, would have driven away the lightest gases, leaving only the heaviest materials, such as the metal and rock that make up most of the Earth today. In its molten state, the elements making up young Earth would have become separated, with the heaviest (such as iron) sinking to the center to make a core, while the lighter rocks floated above, forming a mantle and crust. The continents have been subsequently built up by volcanic activity, and modified by erosion. Continuing radioactive decay still heats the Earth's interior, so that parts of it remain liquid. Slow convection in the warm mantle is believed to be the force that moves parts of the crust in the phenomenon known as continental drift.

Present structure of the interior. Our knowledge of the interior comes from studies of seismic waves from earthquakes and, in recent years, from nuclear explosions underground. Vibrations travel through the different layers of the Earth at different speeds, and are reflected and refracted like light waves traveling through glasses. Analysis of seismic vibrations recorded at many stations around the world allows seismologists to determine the nature and density of the underlying rocks.

The solid crust is only about 3 miles (5 km) thick under the oceans, but under the continents has an average thickness of 20 miles (32 km). Beneath this thin skin is a solid area of the mantle, extending down to about 60 miles (100 km), to which the crust is attached. This crust-mantle slab floats on a slush of upper mantle that extends down to about 150 miles (250 km). Under this slush lies the mantle proper, divided into two layers of different density and extending to 1,800 miles (2,900 km) depth. The outer, liquid core extends down from the bottom of the mantle to a transition layer at about 3,000 miles (5,000 km). The inner 800 miles (1,300 km) of our planet is made up of the iron-rich solid core.

The boundary between the crust and mantle is named the Mohorovičić discontinuity, or Moho, after the Croatian geophysicist Andrija Mohorovičić (1857–1936), who postulated it in 1909. The density of the material inside the Earth ranges from about 13.5 times the density of water at the center, down to 10 at the edge of the outer core, from 5.5 to 3.5 in the

The interior of the Earth, deduced from studies of the way earthquake waves are reflected and refracted as they pass through the Earth.

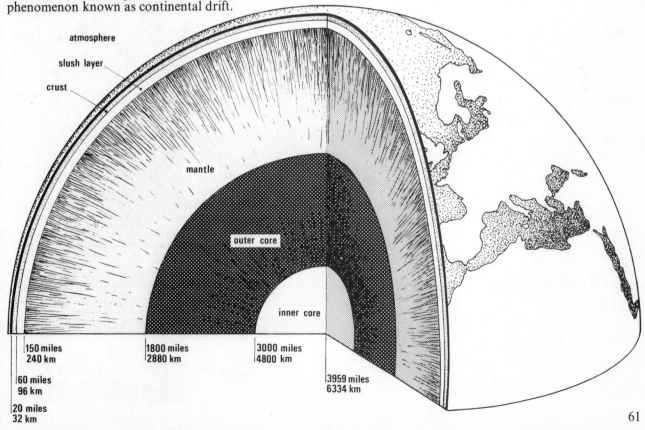

atmosphere
slush layer
crust
mantle
outer core
inner core

150 miles
240 km

1800 miles
2880 km

3000 miles
4800 km

60 miles
96 km

3959 miles
6334 km

20 miles
32 km

This view of Earth, taken by the Soviet spacecraft Zond 5 shows Africa and parts of Europe and Asia. It was taken on 21st September 1968, and thus at an equinox—the terminator line lies exactly along a line of longitude and passes through the poles. A belt of cloud lies along the equator, over the tropical rain forests, while a low pressure system covers Europe.

A profile of the Earth's atmosphere, showing the altitudes at which various phenomena appear. Even at the top of the mountains, men would find it difficult to breathe without additional oxygen.

mantle, and from 3.3 to 2.8 in the crust. The composition of the core is probably much the same as that of iron-nickel meteorites; the upper mantle is thought to be chiefly a silicate rock rich in the mineral olivine.

Surface features. The Earth's outer crust-and-mantle laminate is cracked like an eggshell into segments, termed plates, which are moved by the convection currents in the mantle below. Where two plates are separating, as along the center of the Atlantic Ocean, material wells up from inside the Earth to create new seafloor. Where two plates meet, old seafloor dips down into the Earth's interior, forming a deep trench, as at the Pacific edge of South America. Continents can be split apart as new oceans form, or they may collide to form mountain ranges. There is no evidence that similar processes occur on the other rocky planets of the solar system.

The Earth's surface is constantly being changed by erosion, caused by wind, ice, and water. Because of this, it is difficult to pick out on the surface of the Earth features corresponding to the craters of the Moon, Mars, and Mercury. A few large craters are known, and it seems very likely that all the inner planets of the solar system were subjected to the same intense bombardment by meteorites at some early stage in the evolution of the solar system (see CRATERS).

EARTH'S ATMOSPHERE

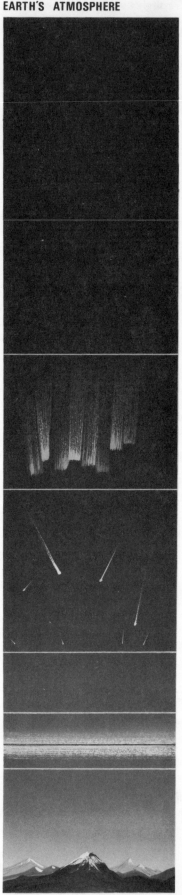

EXOSPHERE
250 mi
400 km

200 mi
320 km

IONOSPHERE

150 mi
240 km

90 mi
144 km

CHEMOSPHERE

45 mi
72 km

STRATOSPHERE

TROPOSPHERE

The atmosphere. The Earth's present atmosphere is the creation of volcanic activity, which released new gases, including large amounts of steam, to form a new atmosphere after the original mantle of light gases was lost to space. Most of the steam condensed to water, filling the oceans. The remainder of this early atmosphere seems to have been chiefly carbon dioxide, carbon monoxide, and nitrogen. The breakdown of these gases by sunlight and early forms of plant life produced an atmosphere that now contains about 78 percent nitrogen, 21 percent oxygen, a little less than 1 percent argon, together with traces of carbon dioxide, neon, helium, methane, krypton, and still rarer elements, as well as a variable amount of water vapor.

Like the Earth's interior, the atmosphere can be divided into several layers. The *troposphere* extends from the surface to an altitude of about 10 miles (16 km); this is the layer in which weather occurs, and at the top of the troposphere (the tropopause) the temperature is about $-60°C$. Above the troposphere is the *stratosphere,* a region in which temperature increases with altitude to a maximum of about $0°C$ at 30 miles (50 km). The increase in temperature is caused by absorption of ultraviolet radiation from the Sun as oxygen (O_2) is converted into ozone (O_3). The ozone layer protects life on Earth by absorbing harmful ultraviolet radiation. Because the stratosphere becomes warmer with increasing altitude it is a very stable layer, preventing upward convection and acting as a "lid" on the weather systems of the troposphere below.

Above the stratosphere, temperatures decrease again in the *mesosphere,* down to a minimum of about $-100°C$ at an altitude of 55 miles (90 km), and then increase again in the *thermosphere* until reaching equilibrium with space conditions. In this layer heating occurs because solar radiation splits oxygen molecules (O_2) into individual atoms (O). Speaking of the "temperature" of the tenuous outer layers is somewhat misleading, because although the molecules in the upper atmosphere may move at speeds corresponding to particular temperatures at sea level, the air is too thin to contain very much heat. It is in this region that the electrically-charged layers known collectively as the IONOSPHERE lie. AURORAE also occur in the thermosphere. Above about 300 miles (500 km) the thermosphere is also called the *exosphere,* because it is from this region that molecules from the atmosphere can leak away into space. Very high in the atmosphere charged particles interact with the Earth's magnetic field producing the radiation zones known as the VAN ALLEN BELTS, and the region known as the MAGNETOSPHERE, which for practical purposes marks the boundary of our home in space.

eccentricity

A measure of how elliptical an orbit is, or how far it deviates from being a true circle. The nearer the eccentricity is to zero, the closer the orbit is to being a circle, the nearer to one, the nearer the orbit is to a parabola.

Echo satellites

Two giant plastic balloons, inflated in space and used as *passive reflector* communications satellites. They were originally planned solely to study the effects of atmospheric drag, but they had a reflective aluminum coating added for communications purposes at the suggestion of the American communications engineer John Robinson Pierce (b. 1910). To the naked eye the Echo satellites appeared like bright stars drifting slowly across the sky. The Echo satellites were used for voice, teletype, and facsimile transmission; Echo 2 was used to exchange data with the Soviet Union, the first such joint East–West program.

Name	Launch date	Initial orbital elements			Notes
		perigee	*apogee*	*period*	
Echo 1	August 12, 1960	941 miles (1,514 km)	1,052 miles (1,693 km)	118 min.	Weight 166 lb. (75 kg), diameter 100 feet (30 m). Reentered May 24, 1968
Echo 2	January 25, 1964	642 miles (1,033 km)	816 miles (1,313 km)	109 min.	Weight 547 lb. (248 kg), diameter 135 feet (41 m). Reentered June 7, 1969

eclipse

Normally, the passage of one astronomical body into the shadow of another; but the term is also applied to the passage of the Moon in front of the Sun (a solar eclipse), though the event is more correctly termed an OCCULTATION of the Sun by the Moon.

Solar eclipses can occur only at new Moon, when the Moon comes between the Sun and Earth. The Sun is not eclipsed at each new Moon because the Moon's orbit is inclined by about $5°$ to the plane of the Earth's orbit. A solar eclipse can therefore happen only when the new Moon occurs near the two points where the lunar and terrestrial orbits cross (called the NODES). When the Moon is near its most distant from Earth (apogee), the tip of its conical umbra (dark central shadow) falls short of the Earth's surface by up to 20,300 miles (32,700 km). This produces a so-called annular eclipse, in which the dark disk of the Moon is surrounded by a bright ring (annulus) of sunlight. When the Moon is at its closest (perigee), its umbra covers a strip along the Earth's surface about 167 miles (269 km) wide. Outside this track, in the outer partial shade called the penumbra, observers over several thousand miles see a partial eclipse. Total solar eclipses last longest when the Sun is at its greatest distance (aphelion) and therefore appears smallest (early July), and the Moon is near perigee; this circumstance produced the longest eclipse of modern times, 7 minutes 14 seconds, on June 30, 1973.

Lunar eclipses occur when the Moon passes into the Earth's shadow, which it can do only at full Moon, when it is opposite the Sun in the sky. At the mean distance of the Moon, the Earth's shadow is about 5,700 miles (9,180 km) wide. A total lunar eclipse can last for up to 100 minutes, plus a further two hours during the Moon's passage in and out of the shadow.

Around this central umbra is the penumbra (diameter 10,200 miles or 16,400 km), in which the sunlight is only partly cut off by the Earth; taking this into account, a lunar eclipse as a whole can last up to nearly six hours. Because the atmosphere refracts sunlight into the Earth's shadow, the totally eclipsed Moon usually appears a dull copper color; variations are caused by changing atmospheric transparency. Lunar eclipses are rarer than solar eclipses, but, since they can be seen wherever the Moon is above the horizon, they are about equally frequent to any given observer. There are seldom more than two lunar eclipses in a year (1928 is an exception, with three), while there may be up to five solar eclipses.

Because the motions of the Earth, Moon, and Sun have been precisely calculated, eclipses can be accurately computed far into the past or future. A major work of this kind is the *Canon of Eclipses,* published in 1887 by the Austrian astronomer Theodor von Oppolzer (1841–1886), which contains tables of eclipses from 1208 B.C. to 2162 A.D.

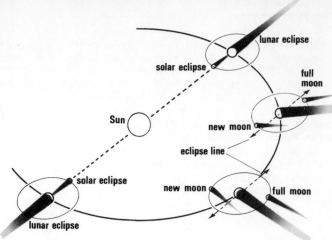

Eclipses can only happen when the plane of the Moon's orbit around the Earth and the Earth's orbit around the Sun coincide. At intermediate points, the new or full Moons are either above or below the Earth–Sun line and no eclipses appear.

Major Solar Eclipses until the year 2000

Date	Type	Duration of Totality		Central track
1979 February 26	Total	2	48	Pacific Ocean, U.S.A. Canada, Greenland
1980 February 16	Total	4	08	Atlantic Ocean, Congo, Kenya, Indian Ocean, India, China
1980 August 10	Annular			S. Pacific Ocean, Bolivia, Brazil
1981 February 4	Annular			Australia, New Zealand, Pacific Ocean
1981 July 31	Total	2	03	U.S.S.R., N. Pacific Ocean
1983 June 11	Total	5	11	Indian Ocean, East Indies, Pacific Ocean
1983 December 4	Annular			Atlantic Ocean, Equatorial Africa, Somalia
1984 May 30	Annular			Pacific Ocean, Mexico, U.S.A., Atlantic Ocean, Algeria
1984 November 22/23	Total	1	59	East Indies, S. Pacific Ocean
1985 November 12	Total	1	55	S. Pacific Ocean, Antarctic
1986 October 3	Annular/ Total	0	01	N. Atlantic Ocean
1987 March 29	Annular/ Total	0	56	Argentina, Atlantic Ocean, Congo, Indian Ocean
1987 September 23	Annular			U.S.S.R., China, Pacific Ocean
1988 March 18	Total	3	46	Indian Ocean, East Indies, Pacific Ocean

Date	Type	Duration of Totality		Central track
1988 September 11	Annular			Indian Ocean, south of Australia, Antarctic
1990 July 22	Total	2	33	Finland, U.S.S.R., Pacific Ocean
1991 January 15/16	Annular			Australia, New Zealand, Pacific Ocean
1991 July 11	Total	6	54	Pacific Ocean, Central America, Brazil
1992 June 30	Total	5	20	S. Atlantic Ocean
1994 May 10	Annular			Pacific Ocean, Mexico, U.S.A., Canada, Atlantic Ocean
1994 November 3	Total	4	23	Peru, Brazil, S. Atlantic Ocean
1995 April 29	Annular			S. Pacific Ocean, Peru, Brazil, S. Atlantic Ocean
1995 October 24	Total	2	05	Iran, India, East Indies, Pacific Ocean
1997 March 9	Total	2	50	U.S.S.R., Arctic Ocean
1998 February 26	Total	3	56	Pacific Ocean, Central America, Atlantic Ocean
1998 August 22	Annular			Indian Ocean, East Indies, Pacific Ocean
1999 February 16	Annular			Indian Ocean, Australia, Pacific Ocean
1999 August 11	Total	2	23	Atlantic Ocean, England, France, Central Europe, Turkey, India

Lunar Eclipses until the year 2000

	Date	Type	Duration of Totality		Moon overhead
1979	September 6	Total	0	52	Polynesia
1981	July 17	Partial			N. Chile
1982	January 9	Total	1	24	Arabian Sea
1982	July 6	Total	1	42	Easter Island
1982	December 30	Total	1	06	Midway Island
1983	June 25	Partial			Pitcairn Island
1985	May 4	Total	1	10	Malagasy
1985	October 28	Total	0	42	Bay of Bengal
1986	April 24	Total	1	08	New Hebrides
1986	October 17	Total	1	14	Arabian Sea
1987	October 7	Partial (very small)			Venezuela
1988	August 27	Partial			Samoa
1989	February 20	Total	1	16	Philippines
1989	August 17	Total	1	38	Brasilia
1990	February 9	Total	0	46	Bangalore

	Date	Type	Duration of Totality		Moon overhead
1990	August 6	Partial			W. Coral Sea
1991	December 21	Partial			Honolulu
1992	June 15	Partial			N. Chile
1992	December 10	Total	1	14	Sahara
1993	June 4	Total	1	38	New Hebrides
1993	November 29	Total	0	50	Mexico City
1994	May 25	Partial			S. Brazil
1995	April 15	Partial			Fiji
1996	April 4	Total	1	24	Gulf of Guinea
1996	September 27	Total	1	12	N.E. Brazil
1997	March 24	Partial			S. Colombia
1997	September 16	Total	1	06	Indian Ocean
1999	July 28	Partial			Tonga
2000	January 21	Total	1	24	West Indies
2000	July 16	Total	1	42	Coral Sea

The Moon blocks the Sun's brilliant disk from view during a total solar eclipse, revealing intricate structures in the Sun's chromosphere and lower corona. The symmetrical shape of the corona indicates that this photograph was taken near sunspot maximum, as do the large numbers of bright prominences.

eclipsing binary

A system of two stars circling one another, with each periodically passing in front of the other and blocking off its light. The total light output of the star system therefore appears to vary periodically (see also VARIABLE STARS). ALGOL was the first such eclipsing binary to be noticed. There are two eclipses of Algol at each revolution; the deepest, so-called primary eclipse, occurs when the fainter yellow giant companion passes over the brighter blue dwarf star. Half an orbit later the blue star, which though brighter is smaller, blots out part of the yellow giant producing a shallow eclipse, the secondary minimum. The time taken for one star to move in front of another during eclipse reveals the size of the stars. Analysis of the light-curve of an eclipsing binary also reveals the angle at which the orbit of the stars is inclined to Earth. Varying X-ray sources discovered by recent scientific satellites (see X-RAY ASTRONOMY) are believed to be eclipsing binaries in which one of the components is a NEUTRON STAR or BLACK HOLE.

ecliptic

The plane of the Earth's orbit around the Sun. The path traced out by the Sun against the star background during the year corresponds to the plane of the ecliptic. Because of the tilt or inclination of the Earth's axis, the ecliptic is inclined to the celestial equator (the projection of the Earth's equator on the celestial sphere) at an angle known as the *obliquity of the ecliptic*. This angle slowly changes, due to the gradually changing inclination of the Earth's axis caused by the gravitational pulls of the Sun, Moon, and planets. At the beginning of 1976 it was 23° 26′ 32.66″, decreasing by about 0.45″ annually. The ecliptic takes its name from the fact that all lunar and solar eclipses occur when the Moon is on or near this plane.

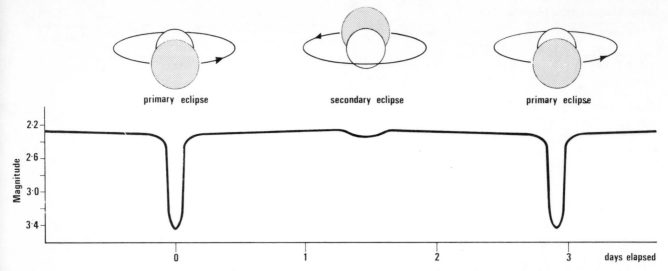

primary eclipse secondary eclipse primary eclipse

Algol is a typical eclipsing binary star. Its main drop in light (primary eclipse) occurs when the larger but cooler star passes in front of the smaller but hotter one. Secondary eclipse occurs when the larger star is behind the smaller; since the surface brightness of the larger star is so much lower than that of the smaller star, the drop in the total light at secondary eclipse is scarcely noticeable.

Eddington, Sir Arthur Stanley (1882–1944)

British astrophysicist, noted for his pioneering investigations of the internal structure of stars, and his contributions to the understanding of relativity and cosmology. Eddington's early studies were of stellar distribution and motion, including an analysis of STAR STREAMING. In *Stellar Movements and the Structure of the Universe* (1914), he proposed that, contrary to general opinion, our Galaxy was only a small part of the total Universe. In 1917 Eddington established the basic theory of CEPHEID VARIABLE pulsations and in the early 1920s showed that the temperature of stars must rise to several million degrees at their centers to keep them from collapsing, and that the material inside them would be in the form of an ionized gas throughout. He developed a simple model for the interior of stars, showing how their gravity, gas pressure, and radiation pressure are interrelated. He found that radiation pressure increased so rapidly with mass that stars above about 50 solar masses could not exist because they would blow themselves apart. Eddington also championed the view that an atomic process of some kind was responsible for energy generation inside stars. From his theoretical work he calculated diameters for red giant stars which were confirmed by the measures of Albert MICHELSON. In 1924 Eddington announced the MASS-LUMINOSITY RELATION, which demonstrates the way the brightnesses of stars depend on their mass. Also in 1924 he predicted the ultra-high density of the WHITE DWARF companion of Sirius, which was confirmed when Walter S. ADAMS observed the gravitational red shift in its light. He summarized his work in *The Internal Constitution of the Stars* (1926).

Eddington is also noted for introducing the theory of relativity to the English-speaking public. He was early in grasping the significance of Einstein's work, and in 1919 led an expedition to Principe, an island off the west coast of Africa, to make the measurements at a total eclipse that first confirmed the prediction that starlight should bend as it passes through the Sun's gravitational field.

Effelsberg Radio Observatory

Location of the world's largest fully steerable radio telescope, 328 feet (100 m) in diameter. Effelsberg lies in the Eifel Mountains of West Germany, about 25 miles (40 km) west of Bonn. and is operated by the Max Planck Institute for Radio Astronomy in Bonn. The dish as a whole can operate at wavelengths down to 5 centimeters; the central part, 262 feet (80 m) wide, is surfaced to operate at wavelengths down to 1.2 centimeters. The telescope was put into operation in 1971.

Einstein, Albert (1879–1955)

German-born physicist, regarded as the greatest theoretical physicist of the century, and responsible for fundamental advances in a wide variety of fields. Einstein is most noted for his theory of RELATIVITY—a description of the structure of space and time. In his so-called Special Theory, published in 1905, Einstein replaced Newton's concepts of space, time, and the motion of bodies, which had gone unchallenged for more than two centuries, with a unified picture of *space-time*. The General Theory, which followed in 1915, was an even more radical departure, describing gravity in terms of such new concepts as the CURVATURE OF SPACE. Entirely new cosmological models have emerged from these ideas. Einstein's predictions were confirmed in 1919, when measurements at a total solar eclipse proved that starlight is bent on passing the Sun. In 1921 he was awarded the Nobel Prize in physics; this was in specific recognition of his explanation of the *photoelectric effect,* in which electrons are emitted from a surface struck by a beam of light. Einstein had demonstrated in 1905 that this could be explained by assuming that light acted as a stream of particles.

ELDO

The European Launcher Development Organization, set up in February 1964 to develop a space launcher based on the British Blue Streak strategic missile as a first stage. Other participants were Belgium, France,

West Germany, Italy, and the Netherlands, with Australia providing the Woomera launch facilities. Although the consortium built two rocket designs, Europa I and Europa II, all launch attempts failed. The Europa program was subsequently canceled, and all European space activities were pooled in the EUROPEAN SPACE AGENCY. French desire for an independent launcher led to the development of the ARIANE rocket.

electromagnetic radiation

The range of radiation, from gamma rays through the spectrum of visible light, to radio waves. Virtually all our information about the Universe, outside those parts of the solar system that can be sampled directly, reaches us in the form of electromagnetic radiation. Its EMISSION is the result of a change in speed or direction of charged particles, whether in the atom (producing SPECTRAL LINES) or on their own (as often for radio waves).

It can be thought of either as radiating waves, or as particles. Every charged particle has an electric and magnetic field associated with it, and as the particle is moved, the change in its position can be thought of as creating ripples in these fields. These ripples move at 186,000 miles (300,000 km) per second (see LIGHT, VELOCITY OF), the same for all forms of electromagnetic radiation. These ripples can also behave as particles, usually called *photons,* a term originally referring to light but now including all wavelengths.

The different forms of radiation are characterized by their WAVELENGTH or FREQUENCY. At the shorter wavelengths the particles are highly energetic, as demonstrated by the destructive nature of gamma rays and ultraviolet light. Longer wavelengths are increasingly less energetic.

Elektron satellites

Soviet satellites placed in orbit to study the Earth's Van Allen radiation belts. Two pairs were launched; one of each pair was sent to examine the outer section of the belt, the other the inner section. Elektrons 1 and 2 were launched in January 1964; Elektrons 3 and 4 followed in July of that year.

elements, abundance of

The chemical composition of the Sun and bright nebulae is revealed by analysis of their light (see SPECTROSCOPY). Taken as a whole, they have the so-called "present cosmic abundances" shown below. Although the Earth, Moon, and meteorites lack the light gases hydrogen and helium predominant elsewhere in space, the abundance of their other, heavier elements is remarkably similar to that in the Sun and other young stars. The oldest stars in our Galaxy are, however, deficient in elements heavier than helium. These stars often have a thousandth the heavy elements that the Sun has, and also slightly less helium. Since the surface layers of a star retain the cosmic abundance of the time at which it formed, it seems that the Milky Way originally condensed from a cloud of only hydrogen and helium. All the heavier elements (and some additional helium) have been produced by nuclear fusion in stars, and have been scattered throughout interstellar space by NOVA and SUPERNOVA explosions. Subsequent generations of

stars condensing from this gas are therefore progressively more enriched in the HEAVY ELEMENTS.

The helium present when the Milky Way formed (about 80 percent of the amount now observed) was produced from hydrogen in the very hot and dense early Universe, only 100 seconds after the BIG BANG in which the Universe originated. Deuterium (heavy hydrogen) can only have been formed in the big bang, since it reacts to form helium during the formation of stars. Lithium, beryllium, and boron are also destroyed in stars, and so are very rare in space. These three elements seem to be formed when carbon, nitrogen, and oxygen nuclei are broken up by the very fast COSMIC RAY particles (a process called SPALLATION).

Present Cosmic Abundances of the Elements

Element	Symbol	Weight relative to hydrogen	Number of atoms to one million hydrogen atoms
Hydrogen	H	1	1,000,000
Deuterium (H_2)	D	2	100
Helium	He	4	73,000
Lithium	Li	7	0.002
Beryllium	Be	9	0.00003
Boron	B	11	0.001
Carbon	C	12	370
Nitrogen	N	14	120
Oxygen	O	16	680
Neon	Ne	20	110
Magnesium	Mg	24	33
Silicon	Si	28	31
Sulfur	S	32	16
Iron	Fe	56	26
All other elements			15

ellipse

A shape somewhat like a squashed circle, produced by cutting a cone at an angle. An ellipse, being symmetrical, has two *foci*; the farther apart these are, the more flattened, or eccentric, the ellipse (see ECCENTRICITY).

ellipsoid

A solid surface of which any cross section is an ELLIPSE. So-called ellipsoidal mirrors for reflecting telescopes are in fact parts of prolate SPHEROIDS, since the cross section in the plane of the mirror is a circle and not an ellipse.

elliptical galaxy

A galaxy that is elliptical in cross-section. Some elliptical galaxies are virtually spherical in shape, like enlarged globular clusters, while others are so flattened as to be lens-shaped, like spiral galaxies without arms. Elliptical galaxies contain mostly very old stars, and little or no gas or dust. Therefore star formation in elliptical galaxies has ceased, whereas it continues in spiral galaxies like the Milky Way. About one-quarter of all known galaxies are ellipticals. The largest and brightest galaxies in the Universe are supergiant ellipticals, containing the mass of up to 10 million million stars, and with diameters of several hundred thousand light years. Dwarf ellipticals, containing the mass of perhaps only a few million stars and with diameters down to a few thousand light years, are the most abundant type of elliptical galaxy. They may in fact be the most numerous of all types of galaxy in the Universe, but their extreme faintness makes them invisible over large distances. Most of the galaxies in our LOCAL GROUP are dwarf ellipticals.

elongation
The angle between the Sun and a planet (generally Venus or Mercury), or between a planet and its satellite, as seen from Earth. *Greatest elongation* is the maximum angular separation.

emission
The production of ELECTROMAGNETIC RADIATION, including visible light and radio waves. Radiation is emitted when atomic particles (in practice, usually electrons) are accelerated or decelerated. Electrons can jump between one orbit and another in atoms when excited by the addition of energy. This produces the bright SPECTRAL LINES characteristic of nebulae. By a change in the direction of spin of the atomic particles, the 21-centimeter wavelength radio line of cold hydrogen gas is produced. Electrons moving freely in space may be decelerated or accelerated by close approaches to other particles. The radiation produced in this way is called BREMSSTRAHLUNG. Where electrons move at many different speeds, as in a hot object, the same process gives rise to the more familiar black body radiation, as emitted by stars or any hot bodies. Electrons may also be forced to move in curved paths by magnetic fields. This gives rise to SYNCHROTRON RADIATION.

emission nebula
A nebula that glows by the emission of light from its constituent gas (see NEBULA).

Enceladus
Third satellite of Saturn in order of distance from the planet, discovered in 1789 by William Herschel. It orbits Saturn every 32 hours 53 minutes at a distance of 147,950 miles (238,100 km). Enceladus has a diameter of about 375 miles (600 km).

Encke, Johann Franz (1791–1865)
German astronomer, who calculated in 1819 that a comet observed the previous year had the shortest period of any comet known; it is now called ENCKE'S COMET. He later deduced the masses of Mercury and Jupiter from the effects of their gravity on the comet's orbit. In 1825 Encke was appointed director of Berlin Observatory and began its reconstruction; the new observatory was opened in 1835. There he began the creation of a new star chart, which made possible the discovery by J. G. GALLE of the planet Neptune. Encke also analyzed in detail the observations of the transits of Venus in 1761 and 1769, deriving a distance for the Sun from Earth of just over 95 million miles.

In 1837 Encke discovered a partial gap in Saturn's outer ring, now called *Encke's division*.

Encke's comet
The comet with the shortest known period—3.3 years. It is named for J. F. ENCKE, who in 1819 computed the orbit of a comet discovered the previous year by the French comet-hunter Jean Louis Pons (1761–1831), and identified it with comets observed in 1786, 1795, and 1805. In its orbit of the Sun, Encke's comet moves from inside the orbit of Mercury to about three-fourths the distance of Jupiter; it is a member of Jupiter's family of comets (see COMET). Encke's comet can be seen throughout its entire orbit, and it makes its next return to perihelion in 1977. The comet has a nucleus some 2 miles (3.2 km) across, with a mass of 100 billion tons. Since 1800 it has been slowly fading by 2 magnitudes a century; the gases it produces at each return have declined from 10 million tons to less than 1 million tons. Now decaying rapidly as an active object, it is expected to degas completely to form a dark, asteroid-like body within 60 to 100 years. Encke's comet is responsible for the TAURID meteor shower.

ephemeris
Collection of tables of the positions of a planet, satellite, or comet in the sky (plural: *ephemerides*). Planetary ephemerides can be calculated with confidence for many years ahead, as can the positions of most of their satellites, but the orbits of most comets are less certain because they can be observed over only a small part of their total path. Ephemerides for newly-discovered comets thus often require considerable revision as more observations are made available. The principal publications dealing with planetary and satellite ephemerides are the annual *American Ephemeris* and *Astronomical Ephemeris,* published in the United States and Britain respectively, with identical contents and produced in collaboration. Predictions for minor planets (asteroids) are issued by the Institute of Theoretical Astronomy at Leningrad.

epicycle
Device once used by astronomers who explained the movement of the planets by a combination of circles. An epicycle was a small circle on the rim of a bigger circle called the *deferent*. By adjusting the rate of movement of a planet around its epicycle, and the epicycle's rate of movement around the deferent, any required final motion could be reproduced. Such complex and clumsy systems were rendered unnecessary when Johannes KEPLER showed in 1609 that the planets' orbits are actually ellipses.

epoch
A standard date to which star positions are referred. The effects of PRECESSION and NUTATION change the celestial coordinates of an object over time. The epoch gives a reference date on which to base the positions in a star catalog or atlas. Dates used currently are the years 1950 or 2000.

equation of time
The correction needed to account for the difference between mean solar time, as shown by a regularly

running watch, and apparent solar time, as recorded by a sundial. The difference arises because the real Sun progresses across the sky at an irregular rate during the year, since the Earth's orbit is elliptical and its axis is tilted so that the Sun does not move along the celestial equator. Astronomers have invented a fictional MEAN SUN, which moves around the celestial equator at a constant rate, to serve as a basis for regular timekeeping. The greatest difference between time by the mean Sun and time by the real Sun occurs at the beginning of November, when the real Sun is due south about $16\frac{1}{2}$ minutes before noon, mean time. In February it is nearly 15 minutes late. The equation of time is zero about April 15, June 13, September 1, and December 25.

equator

The line around a rotating celestial body that is equidistant from the poles, defined as 0° latitude. The *celestial equator* is the projection of the Earth's equator on the CELESTIAL SPHERE, and from it the celestial equivalent of latitude, called DECLINATION, is measured.

equatorial mounting

A mounting for astronomical telescopes that has one axis (the *polar axis*) aligned with the axis of the Earth. The daily rotation of the Earth can be counteracted simply by driving the telescope about its polar axis at a rate of one revolution per day. Adjustment of the other axis (the *declination axis*) is made only during acquisition of the object and to compensate for guidance errors. Once the telescope is pointed at an object, the driving motor keeps the target in view as it moves across the sky. In contrast, the simple ALTAZIMUTH mounting, with axes in the vertical and horizontal planes, must be constantly adjusted in both planes to follow an object's movement across the sky.

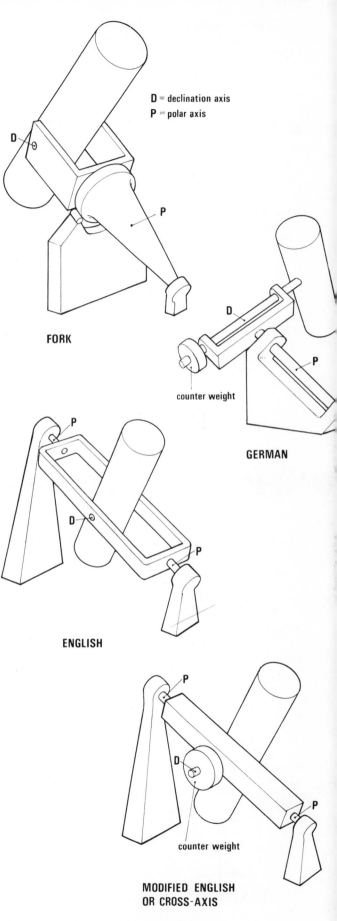

D = declination axis
P = polar axis

FORK

counter weight

GERMAN

ENGLISH

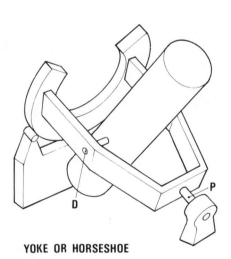

YOKE OR HORSESHOE

counter weight

**MODIFIED ENGLISH
OR CROSS-AXIS**

In these diagrams of the various types of equatorial mounting, the polar axis in each case is shown by P, the declination axis by D. All are set to the same latitude—the angle which the polar axis makes with the horizontal is always the same.

The arrangement of the polar and declination axes can take various forms. The *German mounting* is the most compact type, and is used for almost all refracting telescopes because of the ease with which it can be set on a relatively tall pier; however, it is not very well balanced. Large reflecting telescopes are usually mounted on variations of the *English mounting,* in which the tube is situated between the two polar bearings, giving improved stability at the cost of much greater bulk. The celestial polar regions are inaccessible with the classical English mounting; they can be reached with the *modified English mounting,* or the U-shaped *Yoke mounting* like that used on the Mount Palomar 200-inch (508-cm). The *Fork mounting* is widely used for cameras, because it is reasonably compact, requires no counterweight, and can command the entire sky; but its use is limited to very large instruments.

equinox

The moment when the Sun crosses the celestial equator, and day and night are of equal length anywhere in the world (equinox means "equal night"). There are two equinoxes in each year. The *spring* or *vernal equinox* occurs when the Sun passes north of the celestial equator, on or near March 21; it marks the beginning of spring in the northern hemisphere. At the *autumnal equinox,* around September 23, the Sun passes south of the celestial equator and northern autumn begins. (The seasons are reversed for the southern hemisphere.)

Equuleus (the little horse)

A small and insignificant constellation in the northern hemisphere of the sky adjacent to Pegasus. Its brightest star is only of the fourth magnitude.

Eratosthenes of Cyrene (c.276–c.194 B.C.)

Greek astronomer who about 240 B.C. made the first accurate measurement of the Earth's size. He did so by measuring the difference in the altitude of the Sun as seen from two different places on Earth at the same moment. However, his work was later ignored in favor of a smaller value obtained by the Greek philosopher Poseidonius (c.135–c.50 B.C.) using a similar method. This underestimate inspired Columbus in his belief that only a few weeks' sailing separated Europe from the Indies.

Eridanus (the river)

A long constellation extending from the equator far into the southern sky. Its brightest star, Alpha Eridani, known as Achernar, is one of the brightest in the sky, of magnitude 0.47. Epsilon Eridani is a nearby star quite like the Sun, to which astronomers have listened for possible radio messages from other civilizations (see LIFE IN THE UNIVERSE). The star Omicron Eridani (also known as 40 Eridani), 15.9 light-years away, is a remarkable triple system; the main star resembles the Sun, the second star is a faint white dwarf, and the third an even fainter red dwarf. NGC 1300 is a well-known spiral galaxy in Eridanus.

Eros

An asteroid of the Amor group, also known as minor planet 433, discovered on photographs taken by Gustav Witt in Berlin and Auguste Charlois in Nice on August 13, 1898. It moves in a 1.81-year orbit between 1.13 and 1.8 a.u. from the Sun, inclined 11° to the ecliptic. Eros is a cigar-shaped body about 22 by 10 by 4 miles (35 × 16 × 6 km), rotating about its smallest axis every $5\frac{1}{4}$ hours. The irregular shape suggests it has been broken off a larger body. Eros has a reflectivity (albedo) of 14.2 percent and probably possesses a stony or stony-iron surface. In exceptional circumstances, Eros can pass quite close to the Earth, as in 1894, 1931, and 1975. In January 1975 the planet passed within 14 million miles (22.5 million km), appearing as a swiftly moving seventh-magnitude object. Until the advent of radar astronomy, observations of such close passages were used to define the exact value of the ASTRONOMICAL UNIT.

ERTS

Abbreviation for Earth Resources Technology Satellite, the original name for NASA's LANDSAT program. The series was renamed prior to the launch of Landsat 2 on January 22, 1975.

escape velocity

The speed an object must attain to escape from a gravitational field. Escape velocity depends on the mass of the body responsible for the gravitational field; the Earth, for instance, has an escape velocity at its surface of 6.95 miles (11.2 km) per second, while the less massive Moon has an escape velocity

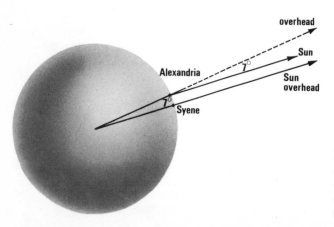

Eratosthenes knew that at noon on the summer solstice, the Sun appeared exactly overhead at a place in Egypt called Syene, near the modern Aswan. Yet on the same date at Alexandria, some way north, the Sun appeared about 7° from the vertical—roughly one fiftieth of a circle. Therefore the distance round the entire globe should be about fifty times the distance between Syene and Alexandria, which, in Greek units, was 5,000 *stadia*; hence the circumference of the Earth, according to Eratosthenes, was about 250,000 stadia. Modern research shows that the Greek *stadium* was equal to about one tenth of a mile, giving a figure for the Earth's circumference in modern units of 25,000 miles—almost exactly right.

of 1.5 miles (2.4 km) per second. A body with a small mass cannot retain an atmosphere, because gas particles of the atmosphere will move faster than the escape velocity. For an object already in circular orbit at a given distance, the escape velocity is $\sqrt{2}$ times the velocity required to maintain it in that orbit.

ESRO
The European Space Research Organization, an association of 10 countries—Belgium, Denmark, France, West Germany, Italy, Netherlands, Spain, Sweden, Switzerland, and the United Kingdom— set up in 1964 for cooperation in space research and satellite development. ESRO launched 184 sounding rockets for upper-atmosphere research, and built seven scientific satellites for launch by the United States. In 1975 ESRO merged with the European Launcher Development Organization ELDO to form the EUROPEAN SPACE AGENCY (ESA).

ESSA satellites
Series of meteorological satellites launched by the Environmental Science Services Administration. They superseded the earlier Tiros series, and were also known as the Tiros Operational Satellite (TOS) system. ESSAs 2, 6, and 8 used television cameras with automatic picture transmission (APT) to broadcast continuous cloud-cover photographs to any ground station. The other ESSA satellites stored their photographs for transmission on command to specific ground stations. The ESSA satellites were placed into near-polar orbits so that they covered the entire Earth each day; new satellites were orbited to replace previous ones as picture quality declined with age. In October 1970 the functions of the Environmental Science Services Administration were assumed by the newly created National Oceanic and Atmospheric Administration, which launched the subsequent NOAA series of WEATHER SATELLITES.

ether
The imaginary medium in which light was assumed to travel. The concept arose in the 19th century, when it was discovered that light travels in waves. By analogy with sound waves, which need air through which to travel, scientists supposed that light waves were transmitted through a medium called the ether. It was assigned curious properties: it had to fill all of space, be transparent and weightless, and allow bodies such as the Earth to move without any friction. By the turn of the century, the theory was abandoned because the MICHELSON-MORLEY EXPERIMENT had failed to confirm it, and because the behavior of light could be explained adequately without its use.

Eudoxus of Cnidus (c.400–c.350 B.C.)
Greek astronomer who produced the first detailed, if fanciful, system to account for the observed motions of the planets. Eudoxus was a pupil of the Greek philosopher Plato (c.427–c.347 B.C.) who, like PYTHAGORAS before him, believed that the heavens were perfect, and that celestial objects could thus move only in circles, which were considered the "perfect" shape. Eudoxus realized from his own observations that the planets did not move across the sky on simple circular paths. To explain the observed motions he introduced for each celestial object a nest

of spheres, with the axis of each sphere set into the surface of the others. Eudoxus' scheme required a total of 27 spheres: one for the stars, three each for the Sun and Moon, and four each for the planets Mercury, Venus, Mars, Jupiter, and Saturn. The motions of the celestial objects could thus be accounted for by combinations of "perfect" motion, and the concept of heavenly spheres, particularly as later developed by ARISTOTLE, became firmly entrenched in astronomical thought. Eudoxus' scheme was in fact not totally successful in reproducing the motions of the celestial bodies, and his follower Callippus (c.370–c.300 B.C.) added another seven spheres, bringing the total to 34.

Europa
Smallest of the four main satellites of Jupiter, discovered in 1610 by Galileo. It orbits Jupiter every 3 days, 13 hours, 14 minutes, at a distance of 416,970 miles (671,050 km). Europa's diameter is about 1,900 miles (3,100 km), its density is three times that of water, and its mass two-thirds that of our Moon. Part of Europa's surface is covered with water frost, and detailed observations suggest that it possesses polar caps extending to about $30°$ from its equator. Europa is believed to be a rocky body, like the Moon.

European Southern Observatory
Astronomical observatory on Cerro La Silla, a mountain in northern Chile about 55 miles (90 km) northeast of the city of La Serena. The observatory, at an altitude of 8,000 feet (2,500 m), has as its major telescope a 142-inch (360-cm) reflector; other important instruments include a 39-inch (100-cm) Schmidt telescope and a 59-inch (150-cm) spectrographic telescope. Formally opened in 1969, the observatory is run by a consortium of Belgium, Denmark, France, West Germany, the Netherlands, and Sweden.

European Space Agency (ESA)
An organization of 11 European countries—Belgium, Denmark, Eire, France, West Germany, Italy, Netherlands, Spain, Sweden, Switzerland, and the United Kingdom—for space research, satellite applications, and rocket development. ESA began formal operation on May 31, 1975, as successor to the joint European Space Research Organization (ESRO) and the European Launcher Development Organization (ELDO).

The first ESA satellite was COS B, launched for gamma-ray studies on August 9, 1975. Other ESA satellites have included the GEOS geophysical satellite, and the Meteosat weather satellite. Among ESA satellites under development are the European Communications Satellite (ECS), and Exosat, for X-ray astronomy. Many of these ESA satellites will be launched by the European rocket ARIANE. ESA is also involved in collaborative projects with NASA, such as the SPACE TELESCOPE and the International Solar Polar Mission. The most adventurous of all ESA's projects is SPACELAB, a manned scientific laboratory to be flown in the cargo bay of the American Space Shuttle. Spacelab will be manned by American and European crews, each sharing its facilities to conduct experiments in orbit.

evening star

Popular term for the planet Venus when it appears shining brilliantly in the evening sky shortly after sunset. When Venus is rising in the morning sky before the Sun, it appears as the brilliant morning star.

exobiology

The study of life beyond the Earth (see LIFE IN THE UNIVERSE).

expanding Universe

The apparent recession of distant galaxies with speeds which increase proportionally to their distance. Because light from a receding object is shifted by the DOPPLER EFFECT toward the red end of the spectrum, the observational evidence for the expansion of the Universe is often referred to as the RED SHIFT. The relationship between a galaxy's red shift and its distance is called HUBBLE'S LAW. At very large distances, as the recession speed approaches the speed of light, the relationship breaks down. Different cosmological theories predict different amounts of breakdown (different values of the *deceleration parameter*), so that they could in principle be tested by observations, but in practice observations of the faint galaxies distant enough to show this breakdown are still too difficult to make with accuracy. Although we see all galaxies receding from us, we are not a center of repulsion. Actually, the expansion of the Universe resembles a spotted balloon being blown up: the distance between all spots is increasing, at a rate proportional to the distance between them (see also UNIVERSE).

Explorer satellites

Series of American scientific satellites, begun by the Advanced Research Projects Agency (ARPA) and transferred to NASA on its formation. Explorer 1 was the first U.S. satellite in orbit and discovered the Earth's Van Allen radiation belts. Explorer 6, nicknamed the paddlewheel satellite because of its outspread solar panels, transmitted the first crude television pictures of the Earth from orbit. A series of Interplanetary Monitoring Platform (IMP) spacecraft was later started within the Explorer program, as were the Small Astronomy Satellites (SAS). Members of the Navy Solrad (solar radiation) and Injun satellite series also came within the Explorer program. Other branches of the Explorer program are concerned with the structure of the atmosphere and with radio astronomy. The overall Explorer title has now been dropped in favour of individual satellite names, and international cooperation is being encouraged, as in the International Ultraviolet Explorer and International Sun-Earth Explorer satellites.

Explorer Satellites

Satellite	Launch date	Remarks
Explorer 1	January 31, 1958	First U.S. satellite in orbit. Discovered Van Allen belt. Reentered March 31, 1970
Explorer 2	March 5, 1958	Launch failure
Explorer 3	March 26, 1958	Radiation and micrometeoroid measurements

Satellite	Launch date	Remarks
Explorer 4	July 26, 1958	Radiation data
Explorer 5	August 24, 1958	Launch failure. (After this, launch failures were not given Explorer numbers)
Explorer 6	August 7, 1959	First television pictures of Earth
Explorer 7	October 13, 1959	Magnetic field and solar flare measurements
Explorer 8	November 3, 1960	Ionospheric research
Explorer 9	February 16, 1961	12-foot balloon to study atmospheric density
Explorer 10	March 25, 1961	Magnetic field measurements from near Earth to halfway to Moon
Explorer 11	April 27, 1961	Gamma ray measurements
Explorer 12	August 16, 1961	Radiation and solar wind measurements
Explorer 13	August 25, 1961	Micrometeoroid satellite
Explorer 14	October 2, 1962	Studied Earth's magnetosphere
Explorer 15	October 27, 1962	Radiation monitor
Explorer 16	December 16, 1962	Micrometeoroid measurements
Explorer 17 (AE-A)	April 3, 1963	Atmospheric Explorer
Explorer 18 (IMP A)	November 27, 1963	First Interplanetary Monitoring Platform; measured radiation between Earth and halfway to Moon
Explorer 19	December 19, 1963	12-foot balloon; identical to Explorer 9
Explorer 20	August 25, 1964	Ionospheric research
Explorer 21 (IMP B)	October 4, 1964	Magnetic field, cosmic ray, and solar wind studies
Explorer 22	October 10, 1964	Ionospheric and geodetic research
Explorer 23	November 6, 1964	Micrometeoroid measurements
Explorer 24	November 21, 1964	12-foot balloon to study atmospheric density
Explorer 25	November 21, 1964	Radiation data; same launch rocket as Explorer 24
Explorer 26	December 21, 1964	Radiation monitor
Explorer 27	April 29, 1965	Geodetic and ionospheric research
Explorer 28 (IMP C)	May 29, 1965	Studied magnetic field and radiation between Earth and Moon
Explorer 29 (GEOS-1)	November 6, 1965	First Geodetic Earth-Orbiting Satellite (GEOS). Carried flashing lights and laser reflectors for geodetic measurements
Explorer 30	November 19, 1965	Solar radiation satellite

Satellite	Launch date	Remarks
Explorer 31	November 29, 1965	Ionospheric studies; complemented work of Canadian Alouette satellite launched with same rocket
Explorer 32 (AE-B)	May 25, 1966	Atmospheric Explorer
Explorer 33 (IMP D)	July 1, 1966	Magnetic field studies from Earth to beyond Moon. Lunar orbit intended, but failed
Explorer 34 (IMP F)	May 24, 1967	Measured radiation and magnetic field between Earth and halfway to Moon
Explorer 35 (IMP E)	July 19, 1967	In orbit around Moon; measured Earth's magnetic "tail"
Explorer 36 (GEOS-2)	January 11, 1968	Geodetic measurements
Explorer 37	March 5, 1968	Solar radiation satellite
Explorer 38 (RAE-A)	July 4, 1968	First Radio Astronomy Explorer, monitoring radio emissions from celestial and Earth sources
Explorer 39	August 8, 1968	Air Density Explorer; 12-foot balloon
Explorer 40	August 8, 1968	Injun Explorer, measured radiation; launched by same rocket as Explorer 39
Explorer 41 (IMP G)	June 21, 1969	Magnetic fields and radiation measurements between Earth and halfway to Moon
Explorer 42 (SAS-A Uhuru)	December 12, 1970	First Small Astronomy Satellite, also known as Uhuru. Measured X rays from space
Explorer 43 (IMP I)	March 13, 1971	Studied Earth's magnetosphere out halfway to Moon
Explorer 44	July 8, 1971	Solar radiation satellite
Explorer 45	November 15, 1971	Magnetospheric research
Explorer 46	August 13, 1972	Meteoroid technology satellite
Explorer 47 (IMP H)	September 23, 1972	Measured radiation and magnetism at a distance about halfway to the Moon
Explorer 48 (SAS-B)	November 16, 1972	Measured gamma rays from space
Explorer 49 (RAE-B)	June 10, 1973	Radio Astronomy Explorer, in orbit around Moon, recording radio noise from Sun and Galaxy
Explorer 50 (IMP J)	October 26, 1973	Monitored solar flares and radiation from high orbit. Final IMP
Explorer 51 (AE-C)	December 15, 1973	Atmospheric Explorer
Explorer 52	June 3, 1974	Hawkeye scientific satellite, studied interaction between solar wind and Earth's magnetic field
Explorer 53 (SAS-C)	May 7, 1975	Third Small Astronomy Satellite, studying X-ray sources

Satellite	Launch date	Remarks
Explorer 54 (AE-D)	October 6, 1975	Atmospheric Explorer
Explorer 55 (AE-E)	November 20, 1975	Atmospheric Explorer, including upper-atmosphere ozone monitor
ISEE 1/2 (International Sun-Earth Explorer)	October 22, 1977	Two satellites launched by same rocket. Joint NASA/ESA project to study effect of Sun on near-Earth space
IUE (International Ultraviolet Explorer)	January 26, 1978	Joint NASA/ESA/UK satellite for ultraviolet astronomy
ISEE 3 (International Sun-Earth Explorer)	August 12, 1978	Monitors solar wind between Earth and Sun

eyepiece

A lens or system of lenses for examining the image formed by a telescope's object glass or mirror. An eyepiece allows the eye to view the image from very close range; the magnification increases as the distance is reduced. The magnification can be expressed by dividing the FOCAL LENGTH of the telescope by the focal length of the eyepiece. Eyepiece focal lengths are normally quoted in millimeters, and values of between 4 and 40 mm are common. Most eyepieces consist of two spaced lenses or systems of lenses; that nearest the eye is called the *eye lens,* the other the *field lens,* because its purpose is to increase the field of view. Different eyepieces designs give different performance, and there are a number of well-known types, generally named for their designers. The main aspects to be considered are apparent field of view (the diameter, in degrees, of the circle of light seen by the eye); correction for SPHERICAL and CHROMATIC ABERRATION; eye relief, or the clearance between the face of the eye lens and the observer's eye; and freedom from scattered light and ghost images of bright objects caused by internal reflections.

Among simpler eyepieces are the *Huygenian* and *Ramsden* types, consisting of two separated plano-convex lenses. The *achromatic Ramsden,* with a double eye lens, is a common improvement. Among highly-corrected designs are the *orthoscopic* (a family of eyepieces of wide variation, but consisting generally of a single eye lens and a triple cemented field lens); the *monocentric* (a triple cemented lens), and the *Erfle* (a complicated system with three separate elements). The *Tolles* is an unusual but effective eyepiece design, consisting of a single short cylinder of glass with convex faces.

F

facula

A bright patch seen near the edge of the Sun. Faculae usually occur near sunspots, but appear before them and also tend to outlive them. They are brighter and hotter than their surroundings high up in the solar atmosphere, so that the contrast is greatest at the

Sun's edge or *limb*. Faculae seem to correspond to regions of high magnetic field, which are presumably involved in the heating of the facular regions.

fireball

A meteor of magnitude −5 or brighter. Occasionally fireballs can be as bright as magnitude −20 to −25 (the full Moon is magnitude −12.7). They are produced by bodies as large as several feet across. Exploding or flaring fireballs, known as *bolides,* are common; they occur when the parent meteoroid disintegrates shortly after entering the atmosphere. Bodies producing fireballs of magnitude −10 or brighter are massive enough to lead to meteorite falls, but fewer than 1 percent reach the ground without disintegrating into dust.

Flamsteed, John (1646–1719)

First astronomer royal of England, appointed on March 4, 1675, by King Charles II. He was set the task of producing tables of the Moon's motion and star positions to enable seamen to determine their longitude, then the major problem in navigation. Finding longitude meant comparing the time locally, obtained by observing the Sun or stars, with a standard time reference. One suggestion for such a "standard clock" was the position of the Moon among the stars as it moved round its orbit. But the positions of objects in the sky were not then known precisely enough to provide the required accuracy, and so Flamsteed, a promising young astronomer, was given the task.

In 1676 the Royal Observatory was built for him in Greenwich Park, but he was given no instruments or assistants to accomplish the job. He borrowed and bought instruments and clocks, and began to teach private pupils to supplement his income. Flamsteed's ambition was to produce a major catalog of unprecedented accuracy. He was reluctant to part with his observations until they were complete, but Isaac Newton nevertheless arranged for some of the work to be published. The result was an inaccurate and incomplete catalog, many copies of which Flamsteed later destroyed.

The catalog on which Flamsteed's fame securely rests, the three-volume *Historia Coelestis Britannica* containing the positions of 2,935 stars, was published only posthumously, in 1725; it was the first major star catalog produced with the aid of a telescope, and is one of the foundation stones of modern astronomy. Flamsteed's catalog superseded that of the great Tycho BRAHE; it was far more extensive, and also six times as accurate. In 1729 a set of star maps based on the catalog was published, called *Atlas Coelestis.*

Flamsteed cataloged the stars in each constellation in order of increasing right ascension; other astronomers later numbered the stars to give the so-called Flamsteed numbers still used today. In 1694 he found a regular shift in position of the pole star which he thought was evidence of its PARALLAX; but its actual explanation was the ABERRATION OF STARLIGHT, which was later discovered by James BRADLEY. Flamsteed's planetary observations showed that the motions of Jupiter and Saturn are affected by each other's gravitational pull. And he deduced from observations of sunspots that the Sun's rotation period is about $25\frac{1}{4}$ days. The original building of the observatory at Greenwich is now called Flamsteed House.

flare

A brilliant burst of light occuring near a SUNSPOT. Flares are short-lived phenomena, lasting no more than a few hours, and are more common near SOLAR CYCLE maximum. They send out streams of particles which produce effects on Earth such as aurorae and radio blackouts. Flares appear to be produced by bursts of magnetic energy. They occur in the chromosphere and, being composed mainly of hydrogen, are best seen at the hydrogen wavelength known as the Hα spectral line.

An intensely bright solar flare of August 7, 1972, photographed in the red spectral line of hydrogen. This flare was one of many which occurred during that month, all associated with a particularly complex sunspot group. Several of these gave rise to extremely large disturbances in the Earth's magnetosphere.

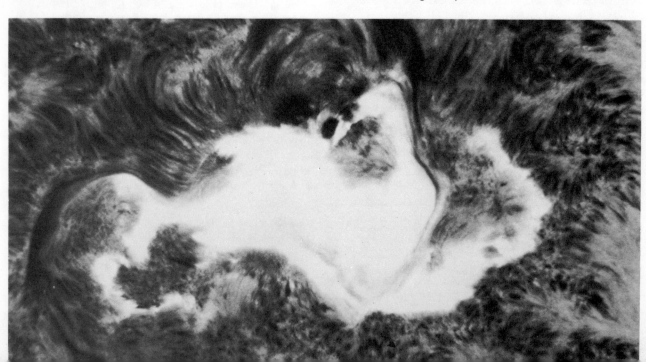

flare star

Any star which suddenly increases its brightness and then fades to its former level. Some 500 are known. Flares typically last a few minutes and can be so intense as to multiply the brightness of the star a hundredfold. A flare star is usually a young star just beginning to burn hydrogen, and the flare activity represents its attempts to adjust its structure in order to achieve stability. The flares probably die away in intensity and frequency as the star becomes a stable dwarf, until they are as infrequent and as small as those occurring on the Sun (see also UV CETI STARS).

floccule

A striking bright or dark feature of the Sun's CHROMOSPHERE. Observations of the chromosphere at specific wavelengths emitted by hydrogen or calcium reveal a network of chains, roughly 20,000 miles (32,000 km) in diameter, which outline giant convection cells in the chromosphere called *supergranules*. In hydrogen light, the dark floccules are more prominent; in calcium light bright floccules outline the network. The floccules are themselves composed of smaller bright and dark mottles. The dark floccules are, in fact, not completely dark but show a "rosette" structure of bright and dark mottles. The flow of gas in the convection cells is thought to concentrate magnetic fields at the edge of the supergranules, thus forming floccules. Floccules are actually the same phenomena as SPICULES, only seen head-on directly in front of the Sun's disk.

flying saucers

See UFO.

focal length

The distance between the lens or mirror of an optical instrument and the image it forms of an object which is at infinite distance (as, in optics, all astronomical objects are assumed to be). The longer the focal length, the larger the scale of the image formed, but the smaller the field of view.

focal plane

The plane in which an optical instrument forms an image. Many systems, however, form their images in a curved field, rather than a plane, including the SCHMIDT TELESCOPE. To record such an image, a photographic plate must also be curved.

focal ratio

The focal length of a lens or mirror divided by its aperture; also termed *f* ratio. For an instrument of given aperture, a small focal ratio gives a shorter tube length, a wider field of view, and greater image brightness (it is said to be "faster," like a camera lens). Long-focus instruments give a larger but fainter image and are unwieldy. Typical focal ratios for visual instruments are $f/5$ to 8 for reflecting telescopes, and $f/12$ to 20 for refracting telescopes; compound and photographic instruments range from about $f/2$ to 30.

Fomalhaut

One of the brightest stars in the sky, of magnitude 1.16. It is a white star, 1.56 times the Sun's diameter, and 23 light-years away. Fomalhaut is the brightest star in the constellation of Piscis Austrinus.

forbidden lines

Bright lines which can appear in the spectra of certain objects such as gaseous nebulae, but which are not found in laboratory spectra. They are emission lines, caused when an atom's electron jumps from one orbit to another. In the denser gases familiar on Earth, the electron is normally knocked out of such an orbit by a collision before it has a chance to jump. Therefore, the probability of observing such an energy jump under terrestrial conditions is very low—it is said to be forbidden. In thin gases, however, such as in a nebula, the chances of a collision are much reduced and the electron can remain in its orbit until ready to jump. Before the development of the concept of forbidden lines, two mysterious "elements" had been postulated to explain mysterious emission lines. "Nebulium" had unique green lines; it is in fact due to oxygen with two electrons missing. "Coronium," observed in the Sun's corona, produced emission lines now known to be forbidden lines principally of iron and nickel, heavily ionized.

Fornax (the furnace)

A small and faint constellation in the southern hemisphere of the sky adjacent to Eridanus, introduced by Nicolas Louis de Lacaille in the 1750s; he originally called it Fornax Chemica, the chemical furnace. Its main significance to astronomers is that it contains a dwarf elliptical galaxy, a member of our LOCAL GROUP, about 800,000 light-years away.

Fraunhofer, Joseph (1787–1826)

German optician and physicist, best known for charting the dark lines in the solar spectrum that are named after him (see FRAUNHOFER LINES). In 1814, while testing the optical properties of glass with the aim of improving the design and construction of ACHROMATIC (color-free) lenses, Fraunhofer noted a number of dark lines crossing the spectrum of the Sun. These had been discovered in 1802 by the English physicist William Hyde Wollaston (1766–1828), who had supposed they marked the boundaries between different colors. Fraunhofer systematically charted the lines, assigning the most prominent letters of the alphabet that are still used today. His instruments were sensitive enough to determine that the pattern of lines was the same for light from the planets (because they were reflecting sunlight), but was different for light from Sirius and other bright stars.

In the early 1820s Fraunhofer went on to study the phenomenon of diffraction, comparing the effects of prisms and diffraction gratings and calculating the wavelengths of certain spectral lines.

His unique theoretical knowledge and great practical skill made Fraunhofer the leading lens maker of his day, producing objectives remarkably free of spherical and chromatic aberration and coma. His most famous instruments were the $9\frac{1}{2}$-inch (24-cm) refractor at Dorpat and the $6\frac{1}{4}$-inch (16-cm) heliometer at Königsberg with which Friedrich BESSEL measured the parallax of 61 Cygni. A successor to Fraunhofer was George Merz (1793–1867), who in 1859 supplied the $12\frac{3}{4}$-inch (32.4-cm) Great Equatorial refractor to the Royal Greenwich Observatory.

Fraunhofer lines

The dark lines seen crossing the Sun's spectrum at certain wavelengths, caused by the absorption of light by the gases in its atmosphere. Over 500 were cataloged by Joseph FRAUNHOFER in 1815. All the Fraunhofer lines are ABSORPTION LINES; the Sun's atmosphere, up to a few hundred miles from its visible surface, is cooler than the photosphere itself, and atoms of various elements absorb light at various specific wavelengths. In this way, astronomers can analyze the chemical composition of other stars besides the Sun, because they all show dark Fraunhofer lines in their spectra. About 25,000 Fraunhofer lines have now been identified in the spectrum of the Sun.

frequency

The number of waves that pass a given point in a given period of time. In astronomy, the waves are of ELECTROMAGNETIC RADIATION (such as light or radio waves). Frequency is given in cycles per second, or *hertz* (Hz), after the German discoverer of radio waves, Heinrich Rudolf Hertz (1857–1894). The velocity of the waves divided by their WAVELENGTH will give the frequency.

G

Gagarin, Yuri Alekseyevich (1934–1968)

Soviet cosmonaut, the first man to fly in space. On April 12, 1961, he made one orbit of the Earth in the spaceship Vostok 1; his orbit had a maximum distance from Earth of 203 miles (344 km) and a minimum distance of 112 miles (190 km). The total flight lasted 108 minutes from lift-off to touchdown. Gagarin ate and drank during his flight and practiced writing under weightless conditions. He reported no problem in adjusting to weightlessness. For reentry the craft was aligned and the retro-rockets fired automatically. Despite some Soviet reports to the contrary, Gagarin ejected from the capsule at an altitude of 23,000 feet (7,000 m) and landed separately by parachute. Gagarin died on March 27, 1968, in a plane crash while training for another space mission.

galactic clusters

Shapeless star clusters found in the spiral arms of the Galaxy; they are also known as OPEN CLUSTERS, because their individual stars are scattered thinly within them, unlike the densely packed GLOBULAR CLUSTERS. Galactic clusters contain young, population I stars (see POPULATIONS, STELLAR). About 1,000 galactic clusters are known, though there may be a total of 18,000 in the Galaxy. The nearest 150, lying within 10,000 light-years, are strongly concentrated in the plane of our Galaxy, and outline three spiral arms, with the Sun located on the inside edge of one. The HYADES and PLEIADES are examples of galactic clusters. Typical galactic clusters contain some 100 stars and measure 10 light-years across. The stars in them will probably disperse over periods of billions of years. The giant star clouds known as STELLAR associations are sometimes also classified as galactic clusters.

galaxies

Systems of billions of stars bound together by their own gravity. Our own Milky Way is a typical galaxy. Other galaxies appear as fuzzy patches of light in small telescopes because of their great distances; as a result, they were originally called extragalactic nebulae and were cataloged in the same lists as star clusters and gaseous nebulae lying within the Milky Way. The three catalogs most commonly encountered are those of Charles MESSIER, published in 1784; the *New General Catalogue* (NGC) compiled by J. L. E. DREYER in 1888; and its supplement, the *Index Catalogue* (1895 and 1908). Galaxies listed in more than one can have different numbers: for example, the ANDROMEDA GALAXY is both M31 and NGC 224. Only three external galaxies are visible to the naked eye: M31 and the two MAGELLANIC CLOUDS.

Structure. Three different types of galaxies are found in the Universe. The Milky Way is an example of a SPIRAL GALAXY: these have a bright central region (nucleus) and a large spherical halo composed of old (population II) stars, with a disk of young stars, gas, and dust (population I) surrounding the nucleus. The disk is typically 100,000 light-years in diameter but only 2,000 light-years thick; its material is concentrated into two arms which spiral outwards from opposite sides of the nucleus. Some galaxies have a bar-like nucleus, and the spiral arms start at the ends of the bar. The individual stars and gas clouds in the disk rotate around the center of the galaxy in almost circular orbits, like the planets around the Sun. From the stars' velocities, the total mass of the galaxy can be calculated. A typical spiral is a hundred billion times as massive as the Sun. Since the Sun is a typical star, these galaxies must contain some hundred billion stars.

ELLIPTICAL GALAXIES lack spiral arms and appear as oval star systems consisting of old (population II) stars with no gas or dust clouds. They range in shape from circular (EO) systems, to flattened ovals three times as long as they are broad (E6). Even flatter galaxies are generally lens-shaped, and are classified SO because they seem to be intermediate in form between an elliptical and a spiral seen edge-on.

The smallest elliptical galaxies are no larger than the GLOBULAR CLUSTERS in the Milky Way, about a million times the Sun's mass. The most massive ellipticals (classified cD) are supergiants a hundred times heavier than our Galaxy; they are spheres about 300,000 light-years in diameter. Such elliptical galaxies are the largest and most massive single bodies known in the Universe.

The third class, *irregular galaxies,* are generally smaller than spirals. They contain both young and old stars, but lack any regular structure. The Magellanic Clouds, visible to the naked eye in the southern hemisphere, are intermediate in form between irregulars and small barred spirals, and orbit about our own Galaxy. Some irregulars appear to have been disrupted by explosions.

A few spiral and elliptical galaxies have small bright nuclei, and are known as SEYFERT or N-GALAXIES. These nuclei are often bright at infrared and radio wavelengths, and are apparently the sites of very violent explosions. Many very massive elliptical galaxies also have extended regions of radio emission on either side and are classified as RADIO GALAXIES.

Elliptical galaxy NGC 205, one of the companions of the Andromeda galaxy. Unlike spiral galaxies, ellipticals contain little dust or gas, and their smooth light distribution arises from millions of old, red stars. The bright stars visible on this galaxy photograph (and all others) are foreground stars belonging to our own Milky Way.

The "Whirlpool" galaxy (M51) was the first in which spiral structure was detected, by Lord Rosse in 1860. It is comparable in size and appearance to our Milky Way Galaxy, but is joined by a thin arm to a massive irregular companion galaxy, bottom. Similar luminous bridges are found to link other pairs of galaxies.

The apparently largest galaxy, left, of **Stephan's quintet** is not associated with the other members, but is merely seen superimposed on a more distant group of four galaxies. The galaxy at the lower left is an elliptical, while the other three seem to be distorted spiral galaxies, two of which are interacting.

Formation and evolution. Galaxies were once thought to evolve from the spherical EO type, becoming progressively more elongated until they formed slowly unwinding spiral arms. The most favored theory today, however, is that the structure of a galaxy is determined at its initial formation; there is probably no evolutionary sequence from elliptical to spiral, or vice versa. Theories of galaxy formation generally assume that the stars formed during the gravitational collapse of a *protogalaxy,* a huge cloud of hydrogen gas. A slowly rotating protogalaxy collapses to a nearly spherical shape. The resulting high density of gas causes stars to form rapidly, creating an elliptical galaxy, all the stars of which are now old, and in which no gas remains. In the collapse of a faster-rotating protogalaxy, only some of the gas falls to the center to condense into stars immediately. The rest forms a rotating disk condensing more slowly, so that a significant amount of gas still remains, along with young stars only recently formed. This disk is unstable and forms arms which give the galaxy a spiral appearance. Irregular galaxies probably also originate in a fast-rotating gas cloud.

Clusters. Most galaxies are found in groups or clusters, gravitationally bound together. The Milky Way is in a small LOCAL GROUP which contains about 30 members, including the Andromeda galaxy. Larger clusters can have thousands of members. Many of these clusters appear spherical in form, and often surround a supergiant (cD) elliptical galaxy. The intrinsic brightness of cD galaxies is very similar from one cluster to another, and thus a cluster's apparent brightness is a good indicator of its relative distance.

Although the galaxies in a cluster orbit a common center of gravity, their actual motions across the sky are far too small to be seen directly. However, the DOPPLER EFFECT in their light can be measured, and reveals their orbital velocities. From these astronomers have calculated how much mass the average galaxy must have for the cluster to be gravitationally bound. The result is 10 to 100 times larger than is usually measured for galaxies. This indicates that there must be a large amount of invisible mass present.

Distances and velocities. Galaxies provide markers for cosmologists studying the expansion of the Universe. Individual stars can be seen in such nearby galaxies as M31. Measuring the brightness of these stars, particularly CEPHEID VARIABLES, allows astronomers to determine the distance of the galaxies. The Andromeda galaxy has been found to be about 2.2 million light-years away, yet on the scale of the Universe it is a near neighbor.

The nearest cluster of galaxies to the Local Group lies in the constellation Virgo. The brightest individual stars in this cluster can just be resolved, indicating a distance of 65 million light-years. The distance of more remote galaxies can be derived from the apparent magnitude of the very bright Type I SUPERNOVAE. These exploding stars always have the same intrinsic maximum brightness, and they can outshine their parent galaxy for a few days.

Astronomers find that the distance of a galaxy is directly proportional to the RED SHIFT in its spectrum; this shift is believed to be the result of the Doppler effect caused by the recession of the galaxies as the Universe expands. The red shift-distance relation is known as HUBBLE'S LAW.

Galileo lived in an age when to challenge the ideas which had been accepted since Greek times was regarded as heretical. He could use an acid tongue and sarcastic wit to great effect in arguments—which earned him many enemies. But his perseverance led to a revolution in thought.

Galileo Galilei (1564–1642)
Italian mathematician, astronomer, and physicist, responsible for a great series of contributions to modern scientific thought. While still a young man studying medicine at Pisa, Galileo discovered that a pendulum of a given length always takes the same time to complete one swing, whether that swing is large or small. Later in life he suggested applying the principle to clocks. Galileo was attracted to physics and was soon lecturing to the Florentine Academy on the hydrostatic balance and the center of gravity of solids; he returned to Pisa in 1589 as professor of mathematics. Here Galileo determined that all bodies fall at the same rate, although it is uncertain that he publicly demonstrated the fact from the Leaning Tower. In 1592 he moved to Padua, where he designed and made a calculating device known as a "military and geometrical compass" and wrote a short treatise on mechanics. Late in 1609 he heard of the invention of the telescope, built one himself, and was probably the first to survey the heavens telescopically. His discoveries of the phases of the planet Venus, of mountains on the Moon, and of the four largest satellites of Jupiter, helped convince him that the views of COPERNICUS were right. He published his results early in 1610 in *Sidereus Nuncius* (The Starry Messenger).

Later in 1610 Galileo returned to Florence as the Grand Duke's mathematician and philosopher, and published a book about his observations of sunspots. All the evidence obtained with the telescope and from his investigations into physics convinced Galileo that the age-old doctrines about the nature of the Universe were in error, and he set about writing his famous *Dialogue Concerning the Two Chief World Systems—Ptolemaic and Copernican,* published in

Florence in 1632. A witty, penetrating discussion of ancient and modern views, it raised a storm of controversy. Unfortunately Galileo, who had a fiery temper and a sarcastic wit, had already made many enemies, and his opponents had the book banned. Galileo was brought before the Inquisition and forced to recant his belief in the Copernican theory.

Galileo was treated leniently, but had to live the rest of his life under virtual house arrest, moving to Arcetri on the outskirts of Florence in 1633, the year after his trial. Although he kept clear of controversy, he continued his research, and in 1638 a new book of his was published at Leiden in the Netherlands from a manuscript smuggled out of Italy.

This last work, *Discourses Concerning Two New Sciences,* discussed the principles of mechanics. Together with the physical principles published in the *Dialogue,* it demolished old and outdated ideas and laid the foundations of the new mathematical physics. At the same time, Galileo's telescopic observations of the heavens opened up a new dimension in astronomy. By observation, experiment, and the use of mathematics, Galileo was a founder of modern scientific method in the physical sciences.

Galle, Johann Gottfried (1812–1910)
German astronomer, discoverer of the planet Neptune. In 1835 Galle became assistant to Johann ENCKE at the Berlin Observatory. On September 23, 1846, the French mathematician Urbain LEVERRIER asked Galle to look for a new planet whose position he had just calculated. That same night Galle and Heinrich Louis d'Arrest (1822–1875), a discoverer of comets and asteroids, set to work using the new star chart being compiled by Encke, and spotted the planet almost immediately; it was subsequently named Neptune. Galle was interested in comets (he discovered three) and asteroids, and spent much of his career computing their orbits. His studies helped prove the connection between the orbits of comets and certain meteor showers. In 1872 he suggested measuring the distances of certain asteroids during their close approaches to establish an accurate scale of the solar system; this method has been used successfully to yield a precise value for the ASTRONOMICAL UNIT.

gamma-ray astronomy
The study of radiation emitted by celestial objects at wavelengths shorter than X rays (i.e. less than 0.1 angstrom). These rays are absorbed by the Earth's atmosphere, and can be studied only by rockets and satellites. The first gamma-ray satellites were the Vela satellites, launched by the United States and intended to monitor nuclear explosions. They began operating in 1967, and immediately detected bursts of gamma rays from space. There are four or five bursts a year, which seem to come from random directions.

This discovery was completely unexpected, and the origin of the bursts is still disputed. They usually last only a few seconds, and must come from an object only a few light-seconds across, about the size of the Sun. Their random distribution means that the gamma-ray sources are either within a few hundred light-years of the Sun, or else are far outside our Galaxy. Objects of intermediate distance would be concentrated toward the plane of the Milky Way, like the fainter stars in the sky.

Gamma-ray sources lying relatively near the Sun could be produced by collisions of comets with compact NEUTRON STARS, or by giant stellar flares a billion times more powerful than solar flares. Stellar explosions (SUPERNOVAE) in distant galaxies could also produce these short bursts.

Sources producing a continuous stream of gamma rays have also been located. The first such source was the CRAB NEBULA, the remains of a supernova within the Milky Way. The gamma rays originate in the very small dense neutron star (or PULSAR) which formed in the explosion. They are found to "flash" 30 times a second as the rapidly rotating star sweeps a beam of radiation past the Earth.

The center of our Galaxy is another gamma-ray source, and satellites have also found a general "background" of gamma rays which arrive continuously from all directions. This background may be due to very many weak sources scattered randomly, like the background observed by X-RAY ASTRONOMY.

Large satellites for gamma-ray research, such as the American SAS B and the European COS B, have now been launched and are leading to major advances in gamma-ray astronomy.

Gamow, George (1904–1968)
Russian-born American astrophysicist, best known for his association with the BIG-BANG theory of cosmology. Gamow proposed that the matter of the Universe originally existed in a primordial state called the "ylem," and that helium and perhaps other elements were formed from the ylem shortly after the big bang that started the Universe's expansion. The theoretical work was performed with the American physicist Ralph Asher Alpher (b. 1921); together with Hans BETHE, they published the famous Alpher-Bethe-Gamow paper (α, β, γ) in 1948. They predicted the existence of a BACKGROUND RADIATION in the Universe, left over from the big bang. Gamow also made pioneering studies of energy-generation inside stars, and the evolution of normal stars to the red giant stage.

Ganymede
The main moon of Jupiter, 3,275 miles (5,270 km) in diameter, and one of the largest moons in the solar system (Titan and Triton may be larger); it was discovered by Galileo in 1610. Ganymede orbits Jupiter once every 7 days 3 hours 42.5 minutes at a distance of 665,120 miles (1,070,400 km). Its mass is about twice that of our Moon, and its density roughly twice that of water; one model suggests that Ganymede may be a muddy ball coated with a layer of frozen water about 300 miles (480 km) deep. Observations show that about half its visible surface is covered by water frost. Visual mapping and space probe photographs have revealed dark markings, possibly due to dust or craters on its surface.

Gauss, Carl Friedrich (1777–1855)
German mathematician and astronomer who made fundamental contributions to celestial mechanics (the study of orbital motion and gravitational interaction). His most striking achievement came in 1801, when he "rediscovered" on paper the asteroid

Ceres. It had recently been located by Giuseppe PIAZZI, but it then disappeared behind the Sun, and the combined efforts of astronomers were unable to relocate it. Gauss had devised a method for working out an object's orbit from only three good positional measurements; he had also invented the mathematical technique known as "least squares," which can be used to fit a smooth curve to observations of uneven quality. With these techniques, Gauss calculated the orbit of Ceres and told observers where to find the asteroid. In 1807 Gauss became director of the observatory at Göttingen, where he spent the rest of his life. His *Theoria motus* of 1809 contained his work on orbital motion and gravitational perturbation, which were used by later mathematicians in tracking down Neptune from its effects on the motion of Uranus. Gauss also turned his attention to geodesy and terrestrial magnetism.

gegenschein

A faint, diffuse patch of hazy light close to the ecliptic, directly opposite the Sun in the sky; the word is German for "counterglow." The gegenschein was first noted by the Danish astronomer Theodor Brorsen in 1854. The glow is about 20° across at maximum, produced by sunlight reflected from dust particles, like the ZODIACAL LIGHT. It is much fainter than the Milky Way.

Gemini (the twins)

Constellation of the zodiac, in the northern hemisphere of the sky, best seen in the northern winter. The Sun is in Gemini at the SUMMER SOLSTICE, passing through the constellation from late June to late July. It is best known for its two brightest stars, CASTOR and POLLUX. It contains a bright star cluster, M35 (also known as NGC 2168), containing about 120 stars, and 2,600 light-years away.

Gemini project

American space program to practice rendezvous and docking techniques and to gain experience of long-duration space missions. The Gemini missions showed that men could live safely in space for periods long enough to reach the Moon and return, and that the docking operations required for Moon missions could be successfully accomplished. The achievements of the Gemini program established American leadership in the space race.

Gemini hardware.   The Gemini project was born late in 1961, while the first Mercury flights were underway. The capsule had to be bigger than Mercury and it had to be maneuverable, with considerable on-board control. To assist with rendezvous and docking, and to make pinpoint landings possible, a microcomputer was installed, the forerunner of the Apollo on-board computers without which the Moon landings could never have been made.

The crew capsule in Gemini, called the reentry module, was conical, with an extended nose called the rendezvous and recovery section, which contained rendezvous radar and landing parachutes. Also in the nose were 16 thrusters for attitude control during reentry; these added considerably to the accuracy of Gemini splashdowns. The base diameter of the reentry module was 7 feet 6 inches (2.3 m), and the diameter of the extended nose

The first close rendezvous in space was between Geminis 6 and 7 in December 1965. This view, taken through the window of Gemini 6, shows the whole of Gemini 7 in its orbital mode, including the white-painted equipment section. This part carried the attitude control rockets, and was jettisoned before reentry.

39 inches (1 m). The conical crew section was 5 feet 10½ inches (1.8 m) long; including the rendezvous and recovery section, it was 11 feet (3.35 m) long. Behind the reentry module was the adapter module, in two parts: the retrograde section, containing retro-rockets; and the equipment section, containing fuel cells for electricity generation, and attitude control rockets for in-flight maneuvering. Before reentry, the equipment section was jettisoned to reveal the retro-rockets, which later were also jettisoned to leave the heat shield at the base of the reentry module. The base diameter of the adapter module was 10 feet (3 m), and its total length 7 feet 6 inches (2.3 m). The complete Gemini weighed about 8,000 lb. (3,600 kg).

Astronaut access to the capsule was through two hatches, one above each crew seat. Each hatch had a crescent-shaped forward-facing window. Gemini spacecraft were launched by a two-stage Titan II rocket. In the event of a launch failure, the astronauts would be catapulted to safety by ejector seats. For spacewalks, both astronauts donned suits, the craft was depressurized, and one of the hatches was opened. An astronaut either left the capsule entirely, or performed a "stand-up" EVA by looking out the hatch.

The missions.   Gemini was tested in orbit before manned flights began. The complex nature of the missions demanded a new control center; this was the Manned Spacecraft Center (now Johnson Space Center) in Houston. Gemini 4 was the first mission not controlled from the Cape.

The Gemini series accomplished the first orbit change by a manned spacecraft (Gemini 3), the first space docking (Gemini 8), and various duration records for spaceflight and extravehicular activity. Considerable photography of the Earth's surface and weather patterns was also accomplished during the Gemini missions, paving the way for later Earth resource surveys.

Mission	Launch date	Results
Gemini 1	April 8, 1964	Unmanned orbital test flight
Gemini 2	January 19, 1965	Unmanned suborbital flight to test reentry heat shield
Gemini 3	March 23, 1965	Virgil I. Grissom and John W. Young made 3 orbits of Earth. First manned spacecraft to change orbit
Gemini 4	June 3, 1965	James A. McDivitt and Edward H. White; White became first American to walk in space, maneuvering with a hand-held jet gun for 21 minutes. 62 orbits
Gemini 5	August 21, 1965	Leroy G. Cooper and Charles Conrad made 8-day, 120-orbit flight
Gemini 7	December 4, 1965	Frank Borman and James A. Lovell made record-breaking 14-day, 206-orbit flight
Gemini 6	December 15, 1965	Walter M. Schirra and Thomas P. Stafford made first space rendezvous, maneuvering with Gemini 7. 15 orbits
Gemini 8	March 16, 1966	Neil A. Armstrong and David R. Scott made first space docking, with an Agena target vehicle. A stuck thruster caused the spacecraft to roll dangerously, and Gemini 8 undocked for an emergency splashdown. 7 orbits
Gemini 9	June 3, 1966	Thomas P. Stafford and Eugene A. Cernan; intended docking with Agena target vehicle frustrated by a shroud that failed to jettison. Cernan performed a total of 2 hours 7 minutes of EVA. 45 orbits
Gemini 10	July 18, 1966	John W. Young and Michael Collins rendezvoused and docked with Agena target vehicle, and used its engine to boost themselves into a new orbit of apogee 476 miles (766 km). Disengaged from first Agena and then docked with Agena vehicle used in Gemini 8, which had been parked in a new orbit. Collins retrieved a micrometeoroid detector from the side of the Agena during a 30-minute EVA. 43 orbits
Gemini 11	September 12, 1966	Charles Conrad and Richard F. Gordon docked with Agena target vehicle and used its propulsion system to boost themselves into a new orbit with a record-breaking apogee of 850 miles (1,368 km). Gordon attached a tether to the Agena during a spacewalk; Gemini undocked and kept station with the tethered Agena. 44 orbits
Gemini 12	November 11, 1966	James A. Lovell and Edwin E. Aldrin docked with Agena target vehicle. Aldrin performed a total of 200 minutes of stand-up EVA, photographing a solar eclipse. On the third day Aldrin worked for 129 minutes on Agena. 59 orbits

Geminid meteors

One of the principal annual meteor streams, first detectable about December 7 each year. The Geminids reach a maximum rate of about 55 meteors an hour on December 13–14, then rapidly decline within a day. The Geminid radiant is 1° west of Castor in Gemini at maximum, but moves eastward at 1° a day; the radiant reaches its greatest altitude at 2 A.M. The stream follows an elliptical 1.63-year orbit, ranging from 0.129 a.u. to 2.63 a.u. from the Sun. The Geminids are the strongest of the meteor showers, but they have no ascertainable parent comet.

geocentric

Term meaning Earth-centered. A *geocentric orbit* is an orbit about the Earth. *Geocentric coordinates* are positions of celestial objects as measured from the center of the Earth. The *geocentric system* was the ancient belief that the Earth was the center of the Universe.

geostationary

Term describing an orbit in which a satellite appears to hang stationary over a point on the Earth's equator, frequently used for COMMUNICATIONS SATELLITES. It is also termed a synchronous orbit, because the satellite's orbital period is synchronized with the Earth's rotation. A geostationary orbit is circular, 22,300 miles (35,900 km) above the equator. In practice, slight errors in positioning the satellite, together with the perturbing effects of the Sun's and the Moon's gravity and the irregular gravitational field of the Earth, all tend to pull the satellite off station. It is therefore not absolutely stationary, but describes a slight figure-of-eight motion in the sky. Ground stations must therefore make slight tracking movements to follow the path of the satellite, and its position must be corrected occasionally by small gas thrusters on board.

giant star

A large bright star; the very biggest and brightest stars are called SUPERGIANTS, but there is no firm dividing line. The brightness of a star depends on its surface temperature and size. A giant star is brighter than a dwarf star of the same temperature; a brightness difference of five magnitudes corresponds to a diameter difference of ten times. Stars enter a giant stage toward the end of their evolution; they pass through it relatively quickly, so that giants are a hundred times rarer than main sequence stars. Yet their brightness allows them to be easily spotted: three-fourths of the hundred brightest stars are giants, including a sprinkling of supergiants (see also BLUE GIANT; RED GIANT).

gibbous

Term describing the phase of the Moon or a planet between half and full illumination.

Glenn, John Herschel (b. 1921)

First American to orbit the Earth. He circled the Earth three times in his craft *Friendship 7* on February 20, 1962, landing after a total flight time of 4 hours 55 minutes 23 seconds. His altitude varied between 100 miles (161 km) and 163 miles (262 km). During the second orbit Glenn saw what he termed

globular cluster

"fireflies," bright specks around the spacecraft which were later realized to be flakes of paint illuminated by the Sun. Glenn controlled the orientation of his capsule by "flying" it with a joystick. During reentry the retro-rockets were not jettisoned for fear that the heat shield had come loose; it was a false alarm caused by a faulty warning light. Glenn, who was selected as an astronaut in 1959, retired from the space program in 1964. He later entered politics and was elected to the U.S. Senate from Ohio in 1974.

globular cluster

A spherical-shaped cluster of old stars. Globular clusters are distributed in a halo around the center of our Galaxy. Some 125 globular clusters are known in our Galaxy, with between 100,000 and 10 million stars in each. The average diameter of a globular cluster is about 100 light-years. The two brightest clusters are easily visible to the naked eye and were once mistakenly cataloged as stars. They are Omega Centauri and 47 Tucanae, both too far south to be seen from mid-northern latitudes. The brightest globular cluster which can be distinguished further north is M13, also known as the Great Cluster in Hercules. It is just visible to the naked eye.

Globular clusters swarm about the center of our Galaxy, with half the known number lying in the constellations Scorpio and Sagittarius. From this distribution, Harlow SHAPLEY in 1917 deduced the true size of the Milky Way system and the Sun's position toward its edge. Similar globular clusters are found distributed about the center of other galaxies including the Andromeda Galaxy and M 104 in Virgo, known as the Sombrero Hat. Most globular clusters lie well away from the plane of the Galaxy, unlike the much younger GALACTIC CLUSTERS which are concentrated in the spiral arms.

Globular clusters are among the most ancient objects in the Galaxy; the oldest are estimated to be 13 billion years of age, and they all contain population II stars (see POPULATIONS, STELLAR), such as RR LYRAE VARIABLES. Globular clusters are believed to be the first parts of a galaxy to form as a giant cloud of gas collapses into a disk. They are believed to orbit the center of the Galaxy on elliptical paths like comets around the Sun. Some "tramp" globulars not bound to any galaxy exist in our LOCAL GROUP.

globule

A circular or oval dark nebula made of dust, seen silhouetted against a background of stars or a bright nebula. Large globules are several light-years in diameter, and contain as much mass as several dozen stars. The smallest globules, however, are only about the size of the solar system, and contain about as much mass as the Sun. These are sometimes termed *Bok globules,* after the Dutch-born American astronomer Bart Jan Bok (b. 1906), who first drew attention to them in 1947. Approximately 100 globules are known, but there must be many more that remain unseen because they are projected against dark areas of sky or are hidden behind larger nebulae. Globules may be clouds collapsing to create star clusters or individual stars.

Goddard, Robert Hutchings (1882–1945)

American rocket pioneer, who built and launched the

A blazing ball of stars, 47 Tucanae, is one of the brightest and finest globular clusters. A typical globular cluster contains about 100,000 stars, in a volume about 150 light years across. Globular clusters form a halo around our Galaxy, and contain some of the Galaxy's most ancient stars.

world's first liquid-fueled rocket. Like the other two visionaries of spaceflight, Konstantin TSIOLKOVSKY and Hermann OBERTH, Goddard worked out the theory of rocket propulsion very much on his own and could see that rockets provided the only way of traveling in space. Goddard was a professor of physics at Clark University at Worcester, Massachusetts. He took out his first rocket patents in 1914; they included designs for propulsion systems and for a form of multistage rocket. He eventually registered over 200 patents, and in 1960 the federal government finally paid $1 million for use of these patents to his widow and the Guggenheim Foundation, which had supported him.

During World War I Goddard developed a tube-launched rocket that was to become the bazooka. But his real interest since boyhood was spaceflight. Goddard had written a paper in 1916 for the Smithsonian Institution, which awarded him a small grant; the paper was published in 1919 under the title *A Method of Reaching Extreme Altitudes*. In it, Goddard outlined the value of rockets for upper atmosphere research, and even dwelt on the possibility of firing rockets to the Moon.

This notion brought him considerable publicity, some of it hostile; Goddard thereafter continued his work in seclusion. On March 16, 1926, he made history by flying the world's first liquid-propellant rocket. Using liquid oxygen and gasoline, it flew for 2.5 seconds, rising 41 feet (12.5 m). In 1929 the aviator Charles A. Lindbergh visited Goddard, and later arranged for grants from the Daniel Guggenheim Fund. With these, Goddard moved to a ranch near Roswell, New Mexico, in 1931, where he continued his experiments on a larger scale.

A 1936 publication, *Liquid-Propellant Rocket Development,* summarized his work, which included the development of new stabilization and combustion techniques that helped propel his rockets more than a mile in height at near-supersonic speeds. He then turned to the development of propellant pumps to force the liquids into the combustion chamber, launching rockets 22 feet (6.6 m) in length, 18 inches (45 cm) in diameter, and weighing 450 lb. (200 kg) or more.

With the help of only four assistants, Goddard had developed most of the basic components needed for successful long-range rocketry. There was little official interest, except during World War II when he supervised development of jet-assisted takeoff for aircraft. He died before he could return to his peacetime activities—by then, the potential of the rocket as a long-range weapon had been demonstrated by the German V-2. Perhaps the greatest tribute was paid to him by Wernher von Braun when questioned shortly after the war. "Don't you know about your own rocket pioneer?" he asked. "Dr. Goddard was ahead of us all."

Goddard Space Flight Center
Facility of the National Aeronautics and Space Administration (NASA) for space science research and satellite tracking, established in 1959 at Greenbelt, Maryland. The Center has been responsible for over half NASA's Earth satellites, including members of the Explorer series, the Orbiting Astronomical Observatory, Orbiting Geophysical Observatory, and Orbiting Solar Observatory programs, Landsat, and various weather and communications satellites. Goddard scientists have also sent experiments aboard satellites of other agencies. Goddard Space Flight Center is the headquarters of STADAN, the worldwide Space Tracking and Data Acquisition Network for tracking unmanned satellites, and it is also the switching center for the manned and deep-space tracking networks operated by the Johnson

Robert Goddard with his first liquid-fueled rocket near Auburn, Mass, in 1926. The Goddard Space Flight Center at Greenbelt, Md, is named after him.

Space Center and Jet Propulsion Laboratory. The Center is named for the American rocket pioneer Robert H. Goddard.

Gold, Thomas (b. 1920)
Austrian-born American astronomer, who with Hermann BONDI proposed the famous STEADY-STATE THEORY of the Universe in 1948. In a later cosmological theory he proposed that if the Universe ceased expanding and began to contract, time might also run in reverse. More recently, he has provided the standard theoretical model of a PULSAR, describing it as a rotating neutron star with a strong magnetic field, emitting lighthouse-beams of energy. He is also known for his prediction that the surface of the Moon is covered in dust.

Gould, Benjamin Apthorp (1824–1896)
American astronomer, the first director (1870) of Cordoba Observatory in Argentina, where he initiated production of the *Cordoba Durchmusterung,* a southern extension of the BONNER DURCHMUSTERUNG star catalog. The *Cordoba Durchmusterung* was eventually published in 1930 under the editorship of Gould's successor, the American astronomer Charles Dillon Perrine (1867–1951), who discovered two moons of Jupiter and 13 comets. The catalog

covers the sky from −23° to the south pole, listing 613,953 stars to the 10th magnitude. Gould established the existence of a band of hot, young stars around the sky inclined at an angle of 20° to the Milky Way; this band, apparently first noted by John Herschel, is now called *Gould's belt*. It is a flattened system of stars within about 1,000 light-years of the Sun which has been expanding for about 40 million years; the stars of the Pleiades cluster are part of Gould's belt. In 1849 Gould founded the *Astronomical Journal*.

granulation

A mottling effect on the Sun's visible surface (the PHOTOSPHERE) caused by columns of gas rising from the Sun's hot interior like water boiling in a pan. This mottling is also termed the "rice-grain" effect. Granules are between about 200 and 1,000 miles (300–1,500 km) in diameter, with bright centers where the hot gas rises and darker rims where cooler gas descends. Granules are in constant motion and exist for only a few minutes. This continual seething is believed to transmit energy into the upper layers of the Sun's atmosphere.

Supergranules are similar but independent large-scale convection phenomena in the chromosphere.

gravity

A fundamental property of matter, which produces a mutual attraction between all bodies. Although gravity is by far the weakest of the known forces of nature, on the astronomical scale it usually overwhelms all other forces, determining the motions of planets, stars, galaxies, and even the Universe. Two theories of gravity have been widely accepted. The first was proposed by Isaac NEWTON, and described gravity as a force acting across the space between two bodies, reducing in intensity with the square of the distance (the inverse-square law of force). With this theory

A closeup view of the Sun's photosphere, showing the fine network of granules, each only 600 miles (1,000 km) across. The bright centers of these roughly hexagonal cells are sites where hot material carried from the interior reaches the surface; the dark boundaries are regions of cooling, descending gas.

Newton was able to account for the orbits of the planets in our solar system. The other theory, proposed by Albert Einstein, is called general RELATIVITY, and treats gravity not as a force, but as a distortion in the geometry of space and time. This remarkable identification of gravity with geometry is founded upon the *equivalence principle* first discovered by Galileo in the 17th century. Anyone who accelerates rapidly in an automobile experiences a pushing force that feels exactly like a gravitational force. This equivalence of the forces of acceleration and gravity is also responsible for the weightlessness which astronauts experience in orbit. Although they are still within the Earth's gravity, its force is neutralized by the acceleration in free fall as they orbit the Earth. The apparent weightlessness arises because all the surrounding objects are falling equally fast through space together, a phenomenon which suggests that gravity is better regarded as a property of space and time, rather than of matter itself.

The extreme feebleness of gravity means that it is difficult to perform convincing laboratory experiments to understand its true nature. One example of this concerns *gravity waves*, which are predicted from theory by analogy with electromagnetic waves. Gravity waves might be produced as "ripples" from supernova explosions, or by the swallowing of matter into black holes, and would travel outward at the speed of light. In 1969 the American physicist Joseph Weber (b. 1919) claimed to have detected gravity waves coming from the center of our Galaxy. To detect these waves, Weber had set up two large aluminum cylinders that would be slightly distorted whenever a wave passed through them. The expected effect would be so weak it would be necessary to measure changes in the length of the bar of less than one atomic nucleus. Other researchers have been unable to confirm Weber's claims, although experiments to detect gravity waves continue.

Great Bear

See URSA MAJOR.

great circle

A line on the surface of a sphere which divides it into two equal hemispheres; on the Earth, lines of

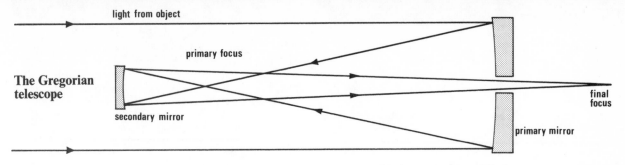

The Gregorian telescope

light from object

primary focus

secondary mirror

primary mirror

final focus

longitude are great circles, but lines of latitude (except for the equator) are not. The horizon produces an approximate great circle when projected onto the CELESTIAL SPHERE (the imaginary sphere carrying the celestial objects and rotating around the Earth once a day).

Green Bank Observatory
See NATIONAL RADIO ASTRONOMY OBSERVATORY.

greenhouse effect
The warming of a planet by its atmosphere. Short-wavelength radiation such as visible light reaches the surface of the planet, which warms up and radiates heat. Some of this heat is absorbed by the atmosphere, which acts as an insulating blanket around the planet. The effect is the same as in a greenhouse. Without an atmosphere, the temperature of the Earth would be about $-30°C$, rather than the average $16°C$ observed. The atmosphere of Venus is mostly carbon dioxide, a strong heat absorber. This, along with some traces of water vapour, produces unusually high temperatures on Venus, in what is sometimes termed the runaway greenhouse effect.

Greenwich Mean Time (GMT)
Local time as calculated at the longitude of the Royal Greenwich Observatory (the *Greenwich meridian*). Greenwich Mean Time became the international time reference in 1884, but in 1972 it was superseded by ATOMIC TIME. GMT is often called Universal Time (UT) by astronomers.

Greenwich Observatory
See ROYAL GREENWICH OBSERVATORY.

Gregorian telescope
The earliest proposed form of reflecting telescope, suggested by the Scottish astronomer-mathematician James GREGORY. The image formed by a concave primary mirror is refocused by a much smaller concave secondary mirror, which reflects the light back through a central hole in the primary. The original design, published in 1663, called for a PARABOLOID figure on the primary mirror and a so-called prolate ELLIPSOID secondary, although different combinations are possible. The Gregorian telescope did not become a practical proposition until the 18th century, when instrument makers were able to make the mirror shapes required. The Gregorian design gives an upright image, but the tube length of a given system is about 50 percent longer than that of the equivalent CASSEGRAIN TELESCOPE, and the design has fallen into disuse. Its former popularity was due to the fact that the concave secondary mirror, which can be made to form a real image for testing purposes, was easier to make than the convex

secondary of a Cassegrain, which cannot be tested directly. Improved techniques have removed this objection to the Cassegrain system.

Gregory, James (1638–1675)
Scottish mathematician and astronomer, who in 1663 published the first design for a reflecting telescope, now known as the GREGORIAN TELESCOPE, which, however, he was unable to construct. In 1668 Gregory proposed that the distances of the stars could be estimated by comparing their brightness to that of the Sun. Using this method he calculated Sirius to be 83,190 times the distance of the Sun. Newton amended his method in 1685 and arrived at a more accurate value of 950,000 solar distances; the true figure is 550,000. James' nephew, David Gregory (1659–1708), was an early champion of Newton's theory of gravity. In 1695, David Gregory suggested that chromatic aberration could be eliminated by combining lenses of different composition; this was the first proposal for an ACHROMATIC lens.

Grissom, Virgil Ivan "Gus" (1926–1967)
American astronaut, first man to make two space flights. His first flight, a suborbital mission on July 21, 1961, was the second space launching in the Mercury series. On March 23, 1965, he and John Young flew the first manned spacecraft in the Gemini series. Grissom was also scheduled to command the first of the three-man Apollo flights. But during a simulated countdown at the launchpad on January 27, 1967, a fire in the spacecraft killed him and fellow astronauts Roger Bruce Chaffee (1935–1967) and Edward WHITE.

Grus (the crane)
A constellation of the southern hemisphere of the sky, introduced by Johann Bayer in 1603. Its brightest star, Alpha, known as Alnair, is a white star of magnitude 2.16.

H

H I, H II regions
Areas of interstellar space containing hydrogen gas. In H I regions the hydrogen is in its so-called neutral form—cool and un-ionized. In H II regions it is hot and ionized.

H I clouds emit no visible light, although they can produce dark absorption lines in light passing through them, but the cool (about $100°K$) hydrogen atoms emit radio waves at a wavelength of 21 centimeters, as first predicted by H. C. van de Hulst in 1944. This 21-cm line is produced when the electron in the hydrogen atom changes its direction of spin. Hydrogen is the principal constituent of interstellar matter, and 21-cm observations of H I regions provide considerable information about the structure of our MILKY WAY Galaxy. The motions of the gas can be studied from the slight Doppler shift in the 21-cm line.

H II regions (gaseous NEBULAE) contain hotter material, at a temperature of about 10,000°K, in which hydrogen atoms exist as IONS, with the electrons no longer bound to the protons. Interactions between free electrons and protons produce radio noise over a broad range of wavelengths. It was this radiation that was observed coming from the Milky Way by the first radio astronomers. Certain optical spectral lines are also emitted, produced by atoms in which the electron has not gained enough energy for IONIZATION. The ionization is caused by ultraviolet radiation from very hot stars; the ORION NEBULA, for example, is ionized by four central stars (the Trapezium). These stars have condensed from the H II region which they now ionize. H II nebulae are typically a thousand times denser than the interstellar H I clouds, and are 10 to 100 light-years across.

Hale, George Ellery (1868–1938)
American astronomer, a major pioneer of the experimental branch of astronomy called astrophysics. He founded the Yerkes, Mount Wilson, and Mount Palomar observatories, and was responsible for the construction of the world's largest optical telescopes. As a young man in 1889, he invented the spectroheliograph, a device for photographing the Sun's prominences and other solar phenomena; this success turned his interest permanently toward solar physics. Hale's father, a wealthy manufacturer, supported his scientific interests and built him a private observatory. A 12-inch (30-cm) refractor was installed there in 1891, and Hale began regular observations with his spectroheliograph. In 1892 he moved to the University of Chicago, where he used his organizational talent to set up the Yerkes Observatory, opened in 1897, and containing the world's largest telescope, a 40-inch (102-cm) refractor with which he continued his work on the Sun and stellar spectroscopy. In 1895 Hale began the *Astrophysical Journal,* still one of the major outlets for research papers.

In 1904, with a grant from the Carnegie Foundation, Hale founded the Mount Wilson Solar Observatory, on a peak near Pasadena, California. There in 1905 he proved that sunspots were cooler areas on the Sun, and in 1908, with a tower telescope 60 feet (18.3 m) high feeding underground instruments, found that intense magnetic fields are associated with sunspots. Hale's advances produced the first understanding of these strange solar blemishes, which had puzzled astronomers for centuries. Hale then made the even more important discovery that the magnetic polarity of sunspots is reversed in each successive 11-year solar cycle. In 1912 a 150-foot (45.7-m) tower telescope was completed on Mount Wilson. Observations with this instrument led Hale to predict an overall solar magnetic field, eventually confirmed in the 1950s by H. D. and H. W. BABCOCK.

In 1908 a 60-inch (152-cm) reflecting telescope was set up on Mount Wilson, using a glass disk donated by Hale's father. This allowed stellar spectra to be studied in considerably greater detail than ever before.

But Hale soon persuaded the businessman John D. Hooker to provide funds for a 100-inch (254-cm) mirror, which was installed on the mountain in 1917 and was responsible for a major step forward in knowledge of the Universe through the work of Edwin HUBBLE.

Hale then began thinking of a still larger project. Once the existence of other galaxies had been proved with the 100-inch, Hale wanted an instrument of greater power to study them in detail: a 200-inch (508-cm) reflector. The Rockefeller Foundation provided $6 million in funds, and Mount Palomar was picked as the site for the new observatory. Hale had died by the time the 200-inch opened in 1948; in 1970 the Mount Wilson and Palomar Observatories were renamed the HALE OBSERVATORIES in his honor.

Hale Observatories
Name since 1970 of the Mount Wilson and Palomar Observatories, founded by George Ellery HALE and operated by the Carnegie Foundation and the California Institute of Technology. Mount Wilson, where Hale began observations in 1904, is 20 miles (32 km) northeast of Los Angeles. The observatory, at an elevation of 5,705 feet (1,742 m), has tower telescopes 60 feet (18.3 m) and 150 feet (45.7 m) high for solar studies, a 60-inch (152-cm) reflector opened in 1908, and the famous 100-inch (254-cm) Hooker telescope opened in 1917, named after the Los Angeles businessman John D. Hooker who in 1906 provided $45,000 to purchase the mirror. The observatory on Mount Palomar, 50 miles (80 km) northeast of San Diego, is at an elevation of 5,597 feet (1,706 m). Its main telescope is the 200-inch (508-cm) Hale reflector, opened in 1948. Another major instrument is a Schmidt telescope with a 72-inch (183-cm) mirror and 48-inch (122-cm) corrector plate, which took the famous Sky Survey photographic atlas. In 1970 a 60-inch (152-cm) photometric telescope was installed on Mount Palomar.

In 1969 the Big Bear Lake Solar Observatory, operated by Hale Observatories, opened 100 miles (160 km) east of Los Angeles, at an elevation of 6,700 feet (2,042 m). The Hale observatories also operate the LAS CAMPANAS OBSERVATORY, a southern outstation in Chile.

Hall, Asaph (1829–1907)
American astronomer, discoverer of the two moons of Mars during the planet's close approach to the Earth in 1877. On August 11 of that year Hall first glimpsed Deimos, the smaller and more distant moon; he confirmed his sighting on August 17, and the same night found Phobos. Hall made his discoveries with the 26-inch (66-cm) refractor of the U.S. Naval Observatory in Washington, where he was chief observer. He went on to become an authority on the satellites of the solar system. Hall was also an assiduous observer of double stars, and in 1892 showed that the two stars of 61 Cygni were in orbit around each other.

Halley, Edmond or Edmund (1656–1742)
Second English astronomer royal, computer of the orbit of the famous comet that now bears his name. Halley was inspired by the star-cataloging activities of John FLAMSTEED at Greenwich to perform similar

work for the southern stars. In 1676 he traveled to St. Helena, an island in the South Atlantic, and on his return two years later published a catalog of 341 stars, the first southern catalog made with the aid of a telescope, which established his scientific reputation. Halley became a friend and admirer of Isaac Newton, and in 1684 encouraged him to publish his work on gravitation and motion. The outcome was the famous *Principia,* which Halley paid for and saw through the press.

Halley contributed to several branches of science, but his most famous work came in 1705, with the publication of *Synopsis of Cometary Astronomy,* in which he computed the orbits of 24 comets. Three of them—the comets of 1531, 1607, and 1682—were so similar that he concluded they must all be the same body, moving around the Sun with a 76-year period. He predicted a return in 1758, though he did not live to see it confirmed. It is now named HALLEY'S COMET in his honor.

In 1679 Halley had first suggested that observations of the transits of Venus across the face of the Sun could be used to measure the scale of the solar system. In 1716 he made specific suggestions for the predicted transits of 1761 and 1769, which were observed (long after his death) throughout the world. In 1718 Halley found that the stars Sirius, Aldebaran, and Arcturus had moved perceptibly from their positions as recorded in Ptolemy's star catalog, thus revealing their PROPER MOTIONS.

Halley's major work had therefore already been completed when in 1720, at the age of 64, he was appointed astronomer royal in succession to Flamsteed. However, he had long been interested in the motion of the Moon, and as early as 1684 had observed regular deviations from its predicted motion. Despite his age, Halley spent the remainder of his life assiduously observing the Moon through one entire SAROS cycle of 18 years. These observations were eventually published in 1749 as additions to tables of the Moon and planets he had prepared in 1719, but had kept from publication. He died before he could analyze these observations, but they proved of enormous value to later astronomers in calculating the complex nature of the Moon's motion.

Halley's comet
The first periodic comet to be identified, and the most prominent of its type. When Edmond HALLEY calculated the orbit of the comet of 1682, he noted that it was very similar to those of the comets of 1531 and 1607. He concluded that all three were the same object, moving in an elliptical 76-year orbit, and he predicted a return in about 1758. More accurate figures were later computed by the French mathematician Alexis Claude Clairaut (1713–1765). The comet was first seen on its return by the German amateur astronomer Johann Georg Palitzsch (1723–1788) on Christmas Day, 1758. The comet reached perihelion on March 13, 1759, returned again in 1835, and was particularly well placed for observations in 1910. It will return to perihelion in February 1986, but is expected to be a disappointing object since it will be badly placed for observation. Halley's comet moves between 0.59 and 35.3 a.u. of the Sun (from between the orbits of Mercury and Venus to beyond the orbit of Neptune). The comet has

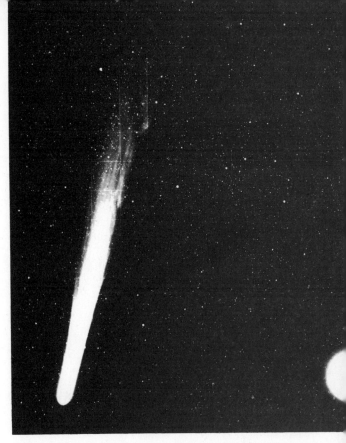

Halley's comet, close in the sky to Venus, **right**, on its last approach to the Sun in 1910. The comet is due to return in 1986, but will not be as spectacular as in previous appearances.

now been identified at 21 returns prior to 1531, the earliest being in 87 B.C. It has a mass of about 50 trillion tons and develops a 20-million-mile (32-million-km) tail at perihelion. The Earth passed through the tail in 1910. The comet is thought to be the parent source of meteors in the Eta Aquarid and Orionid meteor streams.

Harvard College Observatory
Astronomical facility of Harvard University, founded in 1839, which in 1973 joined with the SMITHSONIAN ASTROPHYSICAL OBSERVATORY to form the Harvard-Smithsonian Center for Astrophysics. At its Agassiz Station it operates a 61-inch (155-cm) reflecting telescope and a 60-ft (18.3-m) radio telescope. It also has a radio astronomy station at Fort Davis, Texas, and shares in the operation of the 4,550-foot (1,387-m) Boyden Observatory, 15 miles (27 km) east of Bloemfontein, South Africa, founded in 1927 as the southern station of Harvard. This contains a 60-inch (152-cm) reflector and 32/36-inch (81/91-cm) Schmidt-Baker camera.

HEAO satellites
The High Energy Astronomy Observatory series of three American satellites, designed to observe the sky at X-ray and gamma-ray wavelengths, superseding the previous SMALL ASTRONOMY SATELLITES. HEAO-1 was launched on August 12, 1977 to make an all-sky X-ray survey. HEAO-2, launched on November 13, 1978, studies individual X-ray sources. HEAO-3, for launch in 1979, will survey the sky at gamma-ray wavelengths.

heavy elements

In astronomy, all the chemical elements except HYDROGEN and HELIUM. Astronomers often refer to heavy elements as "metals," even though they include such nonmetallic elements as carbon and oxygen. All have been built up by nuclear fusion reactions inside stars. They are scattered through space by nova and supernova explosions, and stars subsequently formed from the debris are richer in heavy elements than were preceding generations. The variation in the abundance of elements among stars thus provides a clue to the evolution of stars and galaxies. The oldest stars in our Galaxy, which occur in the globular clusters and in the extended "halo" of the Galaxy, have only a thousandth the proportion of heavy elements found in younger stars like the Sun (see ELEMENTS, ABUNDANCE OF).

heliacal rising and setting

The closest observable rising or setting of a star to its conjunction with the Sun. The heliacal risings of bright stars such as Sirius were used by ancient astronomers as the basis of their calendar; the stars acted as time-markers of the Sun's yearly journey around the sky. The term is now sometimes also used to denote the simultaneous rising or setting of a star with the Sun.

heliocentric

Term meaning Sun-centered. A *heliocentric orbit* is an orbit about the Sun. The *heliocentric system* was the belief that the Sun is the center of the Universe. Copernicus' heliocentric theory superseded the geocentric (Earth-centered) system of the Greeks, although modern astronomy has shown that the Sun is only the center of the solar system, not of the Universe.

heliometer

A device for measuring the angular distance between objects in the sky. The first successful heliometer was made in 1754 by John DOLLOND. The heliometer's OBJECT GLASS is divided into two equal D-shaped halves that can slide along the join, producing a split image. When the two halves form a circular lens, a normal image is produced; but as they slide past each other two sets of images are produced, one by each semicircular half of the lens. The angular separation of two objects is found by measuring the amount by which the lenses have to be moved in order to superimpose their two images. The heliometer derives its name from the original purpose—to measure the diameter of the Sun. An advantage of the heliometer over ordinary micrometers is that it can easily measure angular distances larger than the field of view of the eyepiece, and for many purposes it can produce more accurate results. Although the measurement of large angular distances on photographs has made the heliometer itself obsolete, double-image micrometers using the heliometer principle are often used with ordinary telescopes. The German astronomer Friedrich BESSEL used a $6\frac{1}{4}$-inch (15.8-cm) heliometer to obtain the first reliable PARALLAX measurement of a star in 1838.

Helios probes

Joint space program of Germany and the United States to study the Sun and interplanetary space. The first Helios probe was launched from Cape Canaveral on December 10, 1974, into an elliptical orbit that takes it to within 28 million miles (45 million km) of the Sun every 192 days, closer than any previous probe; the first such approach was on March 15, 1975. An identical craft was launched on January 15, 1976, for a close approach to the Sun of within 27 million miles (43 million km) on April 17.

heliostat

An arrangement of two plane mirrors which feeds the image of a celestial object into a stationary telescope. One mirror is equatorially mounted (on an axis parallel to that of the Earth), and reflects light onto the other mirror which directs the beam into the telescope. The advantage of this system over the COELOSTAT is the ease with which different objects can be brought into view in quick succession. On the other hand, the field of view slowly rotates as the Earth turns, while the field of the coelostat is fixed in orientation. Another form of heliostat, more correctly called a *siderostat,* uses a single equatorially-mounted plane mirror to reflect the light toward one of the celestial poles and directly into the telescope, which is set up parallel to the Earth's axis (see COUDÉ FOCUS).

helium in space

Helium is the second-lightest and second most abundant element in space; about one helium atom exists for every ten hydrogen atoms in the Universe. Helium is several hundred times more abundant than any of the other elements (the so-called HEAVY ELEMENTS). It is formed from hydrogen by nuclear reactions in the center of normal stars like the Sun, and may be ejected into space by NOVA or SUPERNOVA explosions. But only about 15 percent of interstellar helium has been created in this way, the largest part was probably produced in the BIG-BANG explosion from which the Universe is believed to have originated. The oldest (population II) stars in the Milky Way Galaxy, for example, contain almost the same proportion of helium as do the younger stars; but they lack the heavier elements created since the big bang in stars and stellar explosions (see also ELEMENTS, ABUNDANCE OF; HYDROGEN IN SPACE).

Heracleides Ponticus (c.388–c.315 B.C.)

Greek astronomer, who made the first known suggestion that the Earth rotates on its axis. Heracleides also proposed that since Mercury and Venus never appear to stray far from the Sun, they could actually be in orbit around it. This was the earliest suggestion that not all objects revolved around the Earth. The ideas of Heracleides, though ignored at the time, had an influence on Renaissance astronomers such as Copernicus and Tycho Brahe.

Hercules

A major constellation of the northern hemisphere of the sky, best seen in summer. Alpha Herculis, also known as Ras Algethi, is a double star, the primary of which is a red supergiant 500 to 600 times the Sun's diameter, and one of the largest stars known; it varies irregularly between third and fourth magnitude. The fainter star, a yellow giant of magnitude 5.4, is also a SPECTROSCOPIC BINARY of period 51.6 years. Hercules contains two globular clusters. One is M13 (NGC 6205), one of the brightest globular clusters visible, about 22,500 light-years away and containing some 300,000 stars. The other globular, M92 (NGC 6341), is roughly 36,000 light-years away.

Hermes

An asteroid of the Apollo group, discovered by Karl Reinmuth at Heidelberg in November 1937. Hermes is about $\frac{1}{2}$ mile (0.8 km) in diameter. The asteroid passed within 500,000 miles (800,000 km) of the Earth's sunlit side (only twice the distance of the Moon), and observations were only possible for five days. The 2.1-year orbit which carries Hermes between 0.62 and 2.7 a.u. of the Sun is therefore poorly determined and the asteroid has been lost. It retains its preliminary designation of Minor Planet 1937 UB and has never been formally numbered. Hermes passes closer to the Earth than any other known asteroid and should be relocated during a future encounter.

Herschel family

A family of British astronomers of German origin, who made pioneering contributions to astronomy.

Sir William Herschel (1738–1822) was trained as a musician, coming to England in 1757, and becoming an organist at the resort of Bath. He was an avid amateur astronomer, and being dissatisfied with commercially available telescopes he began to construct his own reflectors. Herschel had a flair for grinding and polishing the metal mirrors then in use; when he wanted larger ones than those readily available, he cast his own metal blanks at home.

William Herschel was an indefatigable observer who regularly surveyed the night sky. In 1781, during his third survey of the heavens, he found an object that turned out to be Uranus, the first planet to be discovered since prehistoric times. Herschel became famous; he was granted a royal pension as astronomer to King George III, and forsook music for astronomy. He moved from Bath to be near Windsor and the king, finally settling at Slough where he spent the rest of his life. Unlike his contemporaries, who concentrated their attention on the Sun, Moon, and planets, Herschel occupied himself with the stars, which required large mirrors to gather increasing amounts of light as he probed ever-further into space. In 1783 he built a reflector with an aperture of 18 inches (46 cm); this proved to be of fine optical quality and became Herschel's favorite instrument. In 1785 the king financed the construction of a gargantuan telescope with a mirror 48 inches (122 cm) in diameter. Completed in 1789, it was the largest telescope in the world and a sight important visitors to England were taken to see. However it proved cumbersome, and although Herschel made a number of discoveries with it, he still preferred the 18-inch telescope.

Herschel tried to determine stellar parallax by observing the changing separation between two stars close together in the sky. Although he did not measure any parallaxes in this way, he did discover that many such pairs were actually two stars in orbit about each other. In 1782, 1785, and 1821, he published positions

Sir William Herschel, **top**, the "father of stellar astronomy"— at 56. His fame was widespread by this time, though he was not knighted until 1816, when he was 78. The discovery of Uranus, which made him famous, was made from his back garden in Bath in 1781.

Caroline Herschel, **below left**, at the age of 97, a year before her death in 1848. Her work in discovering comets earned her recognition in her own right. From her notes and diary, we have a detailed record of her brother William's life.

The famous photograph of Sir John Herschel, at the age of 75, by family friend Mrs Julia Cameron. Herschel had a keen interest in photography: he had invented his own method, essentially the modern process, in 1839 within days of hearing of the early Daguerrotype system. He made the first glass photographic negative, of his father's great "Forty Foot" reflector, just before it was dismantled in that year.

and other details of a total of 848 double stars. Herschel also devised methods to determine stellar brightness and found that stars differ widely in their intrinsic luminosity.

Herschel spent most of his time cataloging all celestial objects visible in his telescopes. He made a particular study of hazy objects or nebulae, recording a total of 2,500 of them. When his 48-inch telescope resolved some into separate stars, he speculated on whether an even larger telescope would show them all to be merely clusters of stars. Further observations convinced him, correctly, that this was not so; some nebulae were patches of what he called a "shining fluid."

In 1783 Herschel determined that the Sun was moving through space toward a point in the constellation Hercules (see SOLAR APEX). Perhaps his most significant contribution was an analysis of more than 3,000 selected areas of the sky, which suggested that the stars are arranged into a circular, lens-shaped slab—our Galaxy.

Caroline Herschel. In 1772 William brought his sister Caroline Lucretia Herschel (1750–1848) over from Germany, and in due course she became her brother's devoted and untiring assistant. Herschel observed every night if the sky was suitable, employing watchers on overcast nights to notify him as soon as the clouds cleared; Caroline was always at his side, helping him and writing up the results by day. She herself discovered no less than eight comets and revised the star catalog of John FLAMSTEED. She was awarded many honors for her own astronomical research, which would certainly have been more extensive had she not spent so much of her time helping William, with whom she remained until his death.

Sir John Herschel. William Herschel's son John Frederick William Herschel (1792–1871) refurbished the 18-inch and took it to South Africa where, between 1834 and 1838, he made extensive surveys of the southern skies. He also observed at Slough, cataloging stars, measuring their brightnesses with a new device of his own, and discovering 525 nebulae. He devised new ways of determining binary star orbits, and cooperated in making accurate observations of the difference in longitude between London and Paris.

John's interests were extremely broad. He experimented in photography, coining the words "positive" and "negative," and helped the photographer William Henry Fox Talbot (1800–1877) in his research. He also made useful contributions in the fields of biology, botany, and meteorology. An able administrator as well as research scientist, John Herschel was closely connected with the Royal Society and was instrumental in forming and running what is now the Royal Astronomical Society. He undertook a host of public duties, and was a great popular writer on science, particularly astronomy. Although his second son, Alexander Stewart Herschel (1836–1907), made important meteor observations, establishing the radiant of several streams, after John's death the great scientific contributions of the Herschel family effectively came to an end.

Hertzsprung, Ejnar (1873–1967)
Danish astronomer, who showed that the color and luminosity of stars are related, and discovered the division of stars into giants and dwarfs. These discoveries were first outlined in papers published in 1905 and 1907, and were confirmed in 1913 by the work of the American astronomer Henry Norris RUSSELL. The Hertzsprung-Russell diagram, a plot of stars' luminosity against their spectral types, is now named for the two men. Hertzsprung discovered on his graphs that most stars fell into a broad band, now called the MAIN SEQUENCE, which indicated the relationship between color and brightness for the majority of stars. The group of giant stars, however, lay above this main sequence in an area separated by what is called the *Hertzsprung gap*.

Hertzsprung made a study of the stars of the Pleiades, separating members of the cluster from foreground stars by their different motions. In 1911 he discovered that the star Polaris was a CEPHEID VARIABLE, and in 1913 determined the distance of the Small Magellanic Cloud by comparing the brightness of the Cepheid variable stars in it with the brightness of nearby Cepheid stars in our own Galaxy.

Hertzsprung-Russell diagram
A graph on which the color or temperature of stars is plotted against their brightness. It is named for the astronomers who independently developed it in the early part of this century: the Dane Ejnar HERTZSPRUNG and the American Henry Norris RUSSELL. The position of a star on the H-R diagram reveals its physical nature and the stage it has reached in its evolution. The diagram is useful for reading off the intrinsic brightness (ABSOLUTE MAGNITUDE) of a star when only its spectrum is known; by comparing the star's absolute magnitude with its observed apparent magnitude, its distance can be readily computed.

Features of the H-R diagram. Most stars lie in a band stretching diagonally across the diagram and termed the *main sequence*. Stars on the main sequence are ordinary stars in the prime of their lives, like the Sun. A star's position on the main sequence depends on its mass: the more massive stars lie toward the upper end, while the smallest stars lie near the bottom. The Sun, being average, is placed about halfway along. Above and to the right of the main sequence are stars which are much brighter than the main-sequence stars of the same temperature; these are giants, and represent a later stage in a star's evolution, when it has exhausted its hydrogen fuel and begun to swell in size. Eventually the Sun, like all main-sequence stars (except, perhaps, the smallest and coolest) will evolve off the main sequence to become a giant star. Above the giants is a scattering of supergiants, formed when stars much larger than the Sun evolved off the main sequence. The more massive stars evolve faster, and thus move off the main sequence first; astronomers can estimate the age of a star cluster (whose stars all formed together) by noting the point at which stars have evolved off the main sequence to become giants. Certain types of inherently variable stars, such as the CEPHEID VARIABLES, are found in specific regions among the giants and supergiants on the H-R diagram. Eventually, stars like the Sun are believed to puff off their outer layers into space, leaving the small but intensely hot core which forms a WHITE DWARF; these are found at the bottom left of the H-R diagram.

History of the H-R diagram. Russell, in his 1913

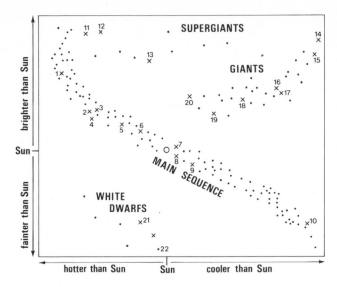

Schematic representation of the Hertzsprung-Russell diagram: the dots represent schematically placed stars, crosses represent actual stars. Brightnesses and temperatures of stars are shown relative to the Sun (circle). Main sequence stars are: 1.Spica; 2.Vega; 3.Castor; 4.Sirius; 5.Altair; 6.Procyon; 7.Alpha Centauri; 8.Tau Ceti; 9.Epsilon Eridani; 10.Barnard's star. Giants and supergiants: 11.Rigel; 12.Deneb; 13.Polaris; 14.Betelgeuse; 15.Antares; 16.Mira; 17.Aldebaran; 18.Arcturus; 19.Pollux; 20.Capella. White dwarfs: 21.Sirius B; 22.Procyon B.

graph, plotted the absolute magnitude against spectral class for nearby stars whose distances he measured from their PARALLAX; he was plotting a star's brightness against its temperature. Hertzsprung, working on the Hyades and Pleiades star clusters in 1911, plotted the apparent magnitude of the stars in each cluster against their COLOR INDEX. (This plot is sometimes distinguished from Russell's version by the name *color-magnitude diagram*.) Since the stars in each cluster are at virtually the same distance, their apparent magnitudes are in direct proportion to their absolute magnitudes; and, like spectral class, color index measures temperature. Hertzsprung and Russell's diagrams were thus essentially the same. The measurement of stellar brightnesses and colors has now become far more refined, and the H-R diagram is established as a major analytical tool of astrophysics.

Hevelius, Johannes (1611–1687)
Polish astronomer, best known for his famous atlases of the Moon and stars. Hevelius' volume of 1647, *Selenographia*, contained his Moon map, the first major attempt of its kind, notable for showing the Moon's libration (rocking motion). Hevelius' map also introduced the first lunar nomenclature, though few of his names for lunar features survive today, those of Riccioli having been adopted in preference. Hevelius observed the Moon and planets with telescopes of his own construction, up to 150 feet (46 m) long and supported from masts by a complex system of pulleys and stays. To make star observations he built copies of the instruments used by Tycho

Brahe, which did not involve optics. His best-known work is his catalog of 1,564 stars, the most extensive of its day, published posthumously in 1690. He engraved accompanying maps which introduced seven now-familiar constellations—Canes Venatici, Lacerta, Leo Minor, Lynx, Scutum, Sextans, and Vulpecula.

Hewish, Antony (b. 1924)
British radio astronomer, discoverer of the rapidly flashing radio sources known as PULSARS. Hewish and his colleagues at the Mullard Radio Astronomy Observatory in Cambridge, England, first detected rapidly pulsating radio signals in mid-1967, with an instrument intended to measure how radio signals from quasars flickered as they passed through gusts in the stream of atomic particles flowing from the Sun. In early 1968 Hewish's team announced that pulsars were probably tiny rotating NEUTRON STARS, predicted by physicists but never previously detected, in which protons and electrons have been crushed together to form neutrons. Hewish was corecipient of the 1974 Nobel Prize in physics for his discovery.

Hipparchus of Nicaea (fl. 146–127 B.C.)
Greek astronomer, considered the greatest figure of ancient astronomy. His most notable discovery was that the Earth wobbles like a spinning top in space over a period of about 26,000 years, an effect which he termed PRECESSION. He deduced this effect from a regular shift between the positions of stars that he cataloged and those recorded by earlier observers; the shift meant that the Sun preceded the stars in its yearly motion around the sky, hence the name.

Hipparchus measured the length of the year from the Sun's motion to an accuracy of within six minutes. He abandoned the complex system of multiple spheres proposed by ARISTOTLE to explain the movement of heavenly bodies, and turned instead to the geometrical devices involving circles that the Greek APOLLONIUS had suggested. Hipparchus found that the Sun's motion could be explained by assuming it followed an off-center circle, known as an *eccentric*, around the Earth. Hipparchus also investigated the motion of the Moon, a complex problem that has occupied astronomers ever since, and was unable to account for it with total success. His work did, however, lead to major advances in the prediction of eclipses of the Sun and Moon.

Hipparchus also left an important legacy in his catalog of 850 stars, completed in 129 B.C.; it was during its compilation that he discovered precession. Hipparchus divided the stars in his catalog into six classes of brightness, from first magnitude (the brightest) to sixth magnitude (faintest visible to the naked eye), a system still used today. He collated observations of the planets to aid later astronomers in calculating their motions; the observations existing in his time were too sketchy for him to work out their orbits accurately. The work of Hipparchus had a great effect on PTOLEMY, much of whose great book the *Almagest* is actually based on Hipparchus.

Horologium (the pendulum clock)
A faint and unremarkable constellation adjacent to Eridanus in the southern hemisphere of the sky, introduced in the 1750s by Nicolas Louis de Lacaille.

Hoyle, Sir Fred (b. 1915)

British astronomer, best known for his support of the STEADY-STATE THEORY of cosmology, which holds that the Universe has appeared essentially the same at all times, and that matter is being continuously created as the Universe expands. This idea was proposed in 1948 by Thomas GOLD and Hermann BONDI; Hoyle then developed the theory in mathematical terms, modifying Einstein's theory of general relativity. Hoyle has also made fundamental contributions to our understanding of the structure and evolution of stars, notably the evolution of Sun-type stars into red giants; much of this work was presented in his famous 1955 book *Frontiers of Astronomy*. Collaborating with Geoffrey and Margaret BURBIDGE and the American astrophysicist William Alfred Fowler (b. 1911), he described how chemical elements can be built up by nuclear reactions inside stars, eventually being scattered into space by supernova explosions (see ELEMENTS, ABUNDANCE OF).

H-R diagram

Abbreviation for HERTZSPRUNG-RUSSELL DIAGRAM, on which stars are plotted according to spectral type and magnitude.

Hubble, Edwin Powell (1889–1953)

American astronomer who proved that galaxies of stars exist outside our own Milky Way, and discovered that the Universe is expanding. These advances have been compared in fundamental importance to the establishment of the heliocentric theory of the solar system three centuries earlier. Hubble's first observations, at Yerkes Observatory, were of the fuzzy patches called nebulae. In 1917 he came to the conclusion that some nebulae were patches of gas within our Galaxy, while the spiral-shaped nebulae were probably far more distant. In 1919 Hubble joined George Ellery HALE at Mount Wilson Observatory, and with the 60-inch (152-cm) telescope there showed that the gaseous nebulae in our own Galaxy were made luminous by stars embedded in them.

Then he turned to other nebulae with the new 100-inch (254-cm) telescope. In 1923 he found a CEPHEID VARIABLE star in the outer regions of the Andromeda nebula, M31; this and other stars showed conclusively for the first time that M31 was in fact a separate star system, far beyond our own.

Hubble went on to study numerous other galaxies, introducing in 1925 the now-standard classification system of spiral galaxies, barred spirals. ellipticals, and irregulars (see GALAXIES). In 1929, when comparing the distances of galaxies with their radial velocities (speed of movement deduced from spectroscopic observations), he found that the galaxies seemed to be receding with speeds that increased with distance. This relationship, termed HUBBLE'S LAW, revealed that the Universe is apparently expanding. Hubble's law was confirmed over even greater distances by the American astronomer Milton LaSalle Humason (1891–1972).

Hubble found that galaxies were distributed evenly over the sky except for a "zone of avoidance" along the plane of the Milky Way, caused by obscuring clouds of dust. Hubble helped plan the 200-inch (508-cm) telescope at Mount Palomar.

Hubble's constant

A measure of the rate at which the Universe is expanding. HUBBLE'S LAW shows that a galaxy's speed of recession depends on its distance; Hubble's constant is the figure that relates velocity to distance, and is determined by observation. Current figures put Hubble's constant at about 10 miles per second per million light-years (55 km/sec/megaparsec). Galaxies, in other words, recede at a rate of 10 miles per second for every million light-years of their distance. Some theories of cosmology predict that the rate of expansion of the Universe may change over time. A rough measure of any such change can be found by comparing the local expansion rate with the rate at very distant regions of the Universe. Because of the time the light from these distant regions has taken to reach us, we see them as they appeared many billions of years ago. Current results suggest scarcely any measurable change, which seems to indicate that the expansion of the Universe will continue indefinitely. If the expansion of the Universe has been constant over time, then the inverse of Hubble's constant yields the age of the Universe—that is, the time since expansion began from the BIG BANG. This period is 18 billion years.

Hubble's law

The relationship between a galaxy's distance and its speed of recession, announced in 1929 by Edwin HUBBLE. Hubble found that distant galaxies were receding at speeds which increased in proportion to their distance: the farther the galaxy, the faster its recession. This result, termed Hubble's law, shows that the Universe is apparently expanding. Like spots on an inflating balloon, every galaxy moves away from every other, and no galaxy is at the center of expansion; therefore, the expansion of the Universe appears everywhere the same. It follows that the rate at which two galaxies move apart is proportional to their separation—galaxies twice as far apart recede at twice the speed. Hubble confirmed this simple model by demonstrating that the RED SHIFT in a galaxy's light (which measures the recessional velocity) was proportional to the apparent brightness of the galaxy (which is dependent on its distance). Hubble's law received an immediate explanation in the theory of relativity, which regards the expansion of the Universe as a uniform expansion of space itself. The actual rate of expansion is given by HUBBLE'S CONSTANT.

Huggins, Sir William (1824–1910)

English astronomer, a pioneer of stellar spectroscopy. His first observations, published in 1863, showed that the stars were made of incandescent gas like the Sun, and that although stars differed in composition they still contained elements that were familiar on Earth. In 1864 he found that, although some of the fuzzy patches in the sky called nebulae were groups of distant stars, others were actually clouds of glowing gas. He was unable to identify the bright spectral lines emitted by the gaseous nebulae and attributed them to an unknown substance he called "nebulium;" it was not until 1927 that the American astronomer Ira S. BOWEN finally proved that these lines were caused by known elements under the rarefied conditions of space (see FORBIDDEN LINES). The

Italian comet observer Giovanni Battista Donati (1826–1873) had taken the first spectrum of a comet in 1864, showing it to be made of gas; Huggins went on to find evidence of carbon compounds in comets. He also observed the clouds of hot gas at the Sun's edge called prominences, and discovered a cloud of hot gas that had been expelled from a nova that flared up in the constellation Corona Borealis in 1866. In 1868 he measured the first RADIAL VELOCITY for a star, Sirius, by observing the shift in position of its spectral lines caused by its speed of recession.

Huygens, Christiaan (1629–1695)
Dutch scientist, best known in astronomy for his description of the rings of Saturn as a swarm of particles orbiting the planet like tiny moons. He gave this explanation in 1655, the same year that he discovered Saturn's largest moon, Titan. Although Saturn's rings had been seen before, no one had fully understood their true nature. Huygens built his own telescopes, the best of their time. In 1659 he discovered the first markings on Mars, notably the dark area called Syrtis Major. His optical studies led him in the 1650s to invent the two-lens Huygenian EYEPIECE, still popular today. He is also credited with first suggesting that light travels as a wave, which allowed him to give explanations of refraction and reflection far better than any provided by the theory that visualized light purely as a stream of particles. Among his many other scientific advances, Huygens invented the first successful pendulum clock in 1656, thus aiding astronomers in their problem of accurate timekeeping.

Hyades
A V-shaped cluster of stars in the constellation Taurus. The Hyades are about 148 light-years away; they are the nearest cluster to the Sun. They are moving through space at a speed of about 27 miles (43 km) per second. The distance to the Hyades is found by observing the motions of the stars (see MOVING-CLUSTER METHOD). The result is extremely important, because it is used as the first step for determining distances to most other stars and galaxies. The Hyades itself contains about 200 known stars spread over a diameter of about 12 light-years. Accompanying the cluster is a stream of about 200 other scattered stars together with the Praesepe cluster ("the Beehive"). The age of these stars is estimated at 600 million years.

Hydra (the water snake)
The largest constellation in the sky. It extends from just north of the celestial equator, in its head region, into the southern hemisphere. Its brightest star, Alpha Hydri, also called Alphard, is an orange giant star of magnitude 2.16.

hydrogen in space
Hydrogen, the lightest element, is the most abundant substance in the Universe. It occurs in three forms in interstellar space: neutral atoms (H I), ions (H II), and molecules (H_2). Hydrogen atoms are distributed between the stars in our Milky Way Galaxy. Although the average density is about one atom to 5 cubic centimeters, individual areas of cold hydrogen (H I REGIONS) vary considerably. Cold hydrogen gas emits the 21-centimeter radiation observed in RADIO

ASTRONOMY. Areas of luminous hydrogen gas (H II regions) have densities some thousand times greater; their electrons have been stripped from the atoms by the ultraviolet radiation from the hot stars condensing within them. Hydrogen molecules (H_2) are rare in interstellar space, because they are broken into individual atoms by ultraviolet radiation. However, satellite observations at ultraviolet wavelengths have shown that some do occur, and the study of interstellar molecules suggests that H_2 is common in very dense interstellar clouds where there is enough dust to shield the molecules from ultraviolet radiation. Most of the hydrogen in dense clouds may be in this form, with a density of more than a million molecules per cubic centimeter.

Ordinary hydrogen atoms consist of an electron orbiting a proton. The heavier isotope, deuterium (symbol D), is also found in space. It possesses a neutron attached to the central proton, and is twice as heavy as ordinary hydrogen. Deuterium is easily destroyed by nuclear reactions in stars, and the deuterium in space cannot have been made in stars by nuclear fusion reactions, like the heavy elements. It was probably formed very soon after the BIG-BANG explosion in which our Universe originated, when the whole Universe was at a temperature of 1 billion degrees K. The observed abundance of deuterium seems to indicate that the Universe contains a relatively low density of matter, and that there is insufficient gravitational attraction to prevent the expansion of the Universe from continuing indefinitely.

Hydrus (the lesser water snake)
A constellation near the south pole of the sky, introduced in 1603 by Johann Bayer. Beta Hydri is a star similar to the Sun.

hyperbola
A symmetrical curve which never closes in on itself. A hyperbola is formed by cutting a cone at an angle steeper than either of the cone's sides. A hyperbola looks like a PARABOLA whose arms diverge, instead of becoming parallel at infinity. The orbit of an object passing through the solar system, but never captured by the Sun, would be a hyperbola.

hyperboloid
A solid shape formed by rotating a HYPERBOLA around its central axis (the line about which it is symmetrical). The mirrors of certain types of reflecting telescopes are occasionally hyperboloid.

hypergolic propellants
Rocket fuels that are self-igniting; they burn spontaneously when mixed, thus eliminating the need for an ignition system. Hypergolic fuels were used on the main engine in the Apollo service module, and in the lunar module.

Hyperion
A satellite of Saturn, the eighth in order of distance from the planet; it was discovered in 1848 by W. C. and G. P. BOND. Hyperion is about 300 miles (480 km) in diameter and orbits Saturn every 21 days 6 hours 38 minutes at an average distance of 921,350 miles (1,483,000 km).

I

Iapetus

Second most distant moon of Saturn, discovered in 1671 by Giovanni Domenico CASSINI. Iapetus orbits Saturn every 79 days 7 hours 56 minutes at a distance of 2,212,200 miles (3,560,200 km). It has a density three times that of water, indicating it is probably a rocky body like the Moon; the diameter of Iapetus is about 1,000 miles (1,600 km). The satellite probably presents the same face to Saturn at all times. Its trailing side appears six times brighter than its other side; it is believed that Iapetus is mostly snow-covered, with the dark side representing a large patch of bare rock.

IAU

Abbreviation for the INTERNATIONAL ASTRONOMICAL UNION.

Icarus

An asteroid of the Apollo group, discovered by Walter BAADE at Mount Palomar in June 1949. Icarus, also known as minor planet 1566, is 0.87 miles (1.4 km) in diameter. It moves in an exceptionally eccentric ellipse passing from within 0.19 a.u. of the Sun (where it must be heated to a dull red), out to as far as 2.0 a.u. Inclined at 23° to the ecliptic, the 1.1-year orbit brings Icarus close to the Earth every 19 years. In June 1968 it passed within 4 million miles (6.4 million km) of the Earth, appearing as a fast moving 12th-magnitude object. It will pass close to the Earth again in 1987. Icarus is a near-spherical body with a $2\frac{1}{4}$-hour rotation period, the shortest of any asteroid. It has a reflectivity (albedo) of 17.8 percent and a stony or stony-iron surface composition. Icarus is probably the rocky central core of an extinct and degassed periodic comet nucleus rather than a true asteroid (see ASTEROIDS).

image intensifier

An electronic device which increases the brightness of a faint image. Image intensifiers are now becoming widely used in professional astronomy; they reduce photographic exposure times and reveal faint detail never before visible. Low levels of light are amplified by using a *photoemitter,* a layer of material which emits electrons when struck by light. The emitted electrons, which represent the image, are accelerated by electric fields through an evacuated tube, where they strike a phosphor screen, like a television screen. The electrons cause the screen to glow, reproducing the original image. Electromagnets or charged plates are placed around the tube to focus the electrons on the phosphor screen. A variant is the *electronographic tube,* or electronic camera. In this case, no phosphor is used, and the electrons strike the photographic emulsion directly. In other forms of image tubes, the image on the photoemitter is allowed to build up and is then scanned as in a television camera. This allows a remote display on a television screen and permits "time exposures" to increase light collection.

inclination

The angle between the plane of a particular orbit and a reference plane, which for objects in the solar system is usually the plane of the Earth's orbit. For objects orbiting the Earth, the reference plane is more usually the plane of the Earth's equator.

Indus (the Indian)

An insignificant constellation near the south pole of the sky, named in 1603 by Johann Bayer. The star Epsilon Indi, 11.2 light-years away, is one of the closest stars to us, and slightly cooler than the Sun.

inferior conjunction

The instant at which Mercury or Venus is in a direct line between the Earth and Sun (see CONJUNCTION).

inferior planet

A planet with an orbit closer to the Sun than Earth, i.e. Venus and Mercury.

infrared astronomy

The study of radiation from space between the wavelengths of red light (7800 Å) and about 1 millimeter (where microwave radio astronomy begins). This includes the range of wavelengths we feel as heat. An object which releases most of its energy in the infrared is cooler than our Sun, which has a surface temperature of 6,000°K and radiates most strongly in the visible part of the spectrum.

The infrared sky. Because the brightness of a star at a particular wavelength depends on its temperature, eyes sensitive to infrared radiation would see a very different pattern of constellations from those we are familiar with. The pattern of bright stars "seen" by an infrared detector depends on exactly which wavelengths it is most sensitive to.

The first survey of the sky in the infrared was made at Mount Wilson in the mid-1960s, and used detectors most sensitive at 2.2 microns, about four times the wavelength of yellow light. The survey covered 75 percent of the sky, and revealed 20,000 sources, most of them stars with surface temperatures of 1,000 to 2,000°K. The brightest 5,500 of these sources made up the first catalog of infrared stars.

Very few of these bright infrared stars are visible at optical wavelengths, and very few of the brightest visible stars emit strongly in the infrared. The 300 brightest infrared stars are spread at random across the sky, as are the brightest stars at visual wavelengths. But the faintest 300 infrared stars in the Mount Wilson catalog were concentrated strongly in the plane of the Milky Way, particularly in the direction of the galactic center.

This distribution indicates that these sources are more than 1,000 light-years distant, and must therefore be relatively powerful. They are probably bright, hot stars, whose light has passed through interstellar dust clouds on the way to our telescopes. The dust scatters the shorter wavelengths, but the longer wavelengths can penetrate through the dust, so that the infrared light from the star is detected. This is similar to the way dust in the atmosphere scatters blue light from the Sun at sunrise and sunset, but allows red light to penetrate.

Dust in our Galaxy is concentrated in the plane of the Milky Way and obscures the galactic center at optical wavelengths. Although ordinary light is 99.9999999 percent absorbed, the absorption at 2 microns wavelength is only 90 percent. Infrared studies thus provide a means of penetrating the obscuring dust, and are of great importance in understanding the structure of our Galaxy.

The galactic center.   Many galaxies have a bright central nucleus, visible with optical telescopes, but such a nucleus in our Galaxy would be obscured by dust. Radio and infrared observations, however, show an extended source of energy at the galactic center. The ANDROMEDA GALAXY also shows a peak of infrared emission at its nucleus, and a comparison indicates that the center of our Galaxy contains millions of ordinary stars packed into a space only a few light-years across. At the core, stars are 200 times closer together than they are near the Sun, so the night sky on a planet orbiting one of those stars would be 40,000 times brighter than ours.

There appears to be a point source of very intense infrared radiation at the galactic center itself, only $\frac{1}{3}$ light-year in diameter but radiating as much energy as 300,000 Suns. Its nature is not known; possibly it is a giant BLACK HOLE.

In some galaxies with very bright optical nuclei (SEYFERT GALAXIES) the infrared emission from the nucleus comprises most of the total radiation from the galaxy. Some QUASARS are also strong infrared sources. In both cases the radiation is strongest at long wavelengths (about 70 microns). Further sky surveys at these wavelengths will probably reveal many similar sources.

Very cool objects.   Much of the early excitement of infrared astronomy centered on the prospect of detecting PROTOSTARS, clouds of gas and dust, at only a few hundred degrees K, early in the process of collapsing to form stars. These hopes have not been realized, but the search for very cool objects has revealed some interesting phenomena.

Most of the cool stars detected so far (with temperatures below 1,700°K) are long-period variables, perhaps related to the cool variables visible optically (such as MIRA); but a few of the sources are very cool but unvarying, quite unlike anything observed before. An early discovery was a bright infrared star in Cygnus, as bright at 2 microns as Vega (the fourth-brightest star in the visible sky), and brighter at 20 microns than any object except the Sun. The spectrum of this source corresponds to a temperature of 1,000°K. Although it may simply be a cool young star, there is a possibility that it is a very bright supergiant star, of a kind never before observed, surrounded by obscuring dust.

Infrared detectors.   Infrared radiation is absorbed by water vapor in the atmosphere, and so it is important to observe from high altitudes. Almost any optical reflecting telescope can be adapted for infrared studies, using detectors sensitive at infrared wavelengths (such as lead sulfide PHOTOELECTRIC CELLS) mounted at the focus, and usually cooled by liquid nitrogen to improve sensitivity. Using the best optical telescopes is wasteful since observations at the longer infrared wavelengths do not require the precision of optical mirrors. Infrared observations can be made at twilight, however, when the telescope cannot be used optically. Most large telescopes have been used for infrared work, but new infrared "flux collectors" of large aperture are now being planned. The biggest such instrument is the 152-inch (390-cm) U.K. infrared telescope on Mauna Kea in Hawaii.

Intelsat

The International Telecommunications Satellite Corporation, an organization set up by a number of nations in 1964 to produce and operate satellites for

Satellite	Launch date	Remarks
Intelsat I (Early Bird)	April 6, 1965	Stationed over Atlantic
Intelsat II-A (Lani Bird)	October 26, 1966	Failed to achieve synchronous orbit
Intelsat II-B (Pacific 1)	January 11, 1967	Stationed over Pacific
Intelsat II-C (Atlantic 2)	March 22, 1967	Stationed over Atlantic
Intelsat II-D (Pacific 2)	September 27, 1967	Stationed over Pacific
Intelsat III-A	September 18, 1968	Launch failure
Intelsat III-B	December 18, 1968	Stationed over Atlantic
Intelsat III-C	February 5, 1969	Initially located over Pacific, but moved to Indian Ocean when malfunction reduced capacity
Intelsat III-D	May 21, 1969	Stationed over Pacific; completed first commercial global comsat system
Intelsat III-E	July 26, 1969	Intended to be stationed over Atlantic, but entered improper orbit due to launch-vehicle malfunction
Intelsat III-F	January 15, 1970	Stationed over Atlantic
Intelsat III-G	April 23, 1970	Stationed over Atlantic
Intelsat III-H	July 23, 1970	Intended to be stationed over Pacific, but failed to enter synchronous orbit
Intelsat IV-A	January 26, 1971	Stationed over Atlantic
Intelsat IV-B	December 20, 1971	Stationed over Atlantic
Intelsat IV-C	January 23, 1972	Stationed over Pacific
Intelsat IV-D	June 13, 1972	Stationed over Indian Ocean
Intelsat IV-E	August 23, 1973	Stationed over Atlantic
Intelsat IV-F	November 21, 1974	Stationed over Pacific
Intelsat IV-G	May 22, 1975	Stationed over Indian Ocean
Intelsat IVA-F1	September 25, 1975	Stationed over Atlantic
Intelsat IVA-F2	January 29, 1976	Stationed over Atlantic
Intelsat IVA-F4	May 26, 1977	Stationed over Atlantic
Intelsat IVA-F5	September 29, 1977	Launch failure
Intelsat IVA-F3	January 7, 1978	Stationed over Indian Ocean
Intelsat IVA-F6	March 31, 1978	Stationed over Indian Ocean

international communications; the countries contribute to the cost and share the profits of operating a global satellite system. The Corporation's satellites are named Intelsat, and there have been several different designs since the launching of the first, Early Bird, in 1965. Early Bird, based on the experimental SYNCOM satellites, was the only Intelsat I type of satellite. Stationed in synchronous orbit above the Atlantic, it provided the first commercial satellite link between North America and Europe, carrying 240 two-way telephone conversations, or one television channel. The 85-lb. (38.5-kg) satellite covered only the North Atlantic with its antenna. The succeeding 190-lb. (86-kg) Intelsat II series, of which there were four satellites, each covered a wider area, although they carried the same number of circuits.

The first global system arrived with satellites of the Intelsat III series, which were put into orbit above the Atlantic, Pacific, and Indian Oceans. Each Intelsat III could handle up to 1,200 simultaneous telephone conversations, or four television channels, and their antennae covered all the Earth visible to them. They weighed 332 lb. (150.5 kg) each. Their successors, of the advanced Intelsat IV design, weigh 1,585 lb. (719 kg), and include spot-beam antennae to cover small areas of high traffic, as well as antennae to cover the entire globe. Their capacity is 5,000 telephone circuits or 12 color-television channels. Intelsat IV-A, a modified design, incorporates more spot-beam antennae to give a capacity of up to 9,000 telephone circuits. As each new generation of satellites has been orbited, their predecessors have been retired or put on standby. With the growth in international communications traffic, particularly across the North Atlantic, satellites of an entirely new generation— Intelsat V—will be needed by the early 1980s.

interference

An effect which occurs when two waves of the same wavelength are combined. If the waves coincide, peak for peak and trough for trough (*in phase*), they produce a wave of twice the height, or *amplitude*; but if the peaks of one coincide with the troughs of the other (*out of phase*) they cancel out each other completely. These conditions are known respectively as *constructive* and *destructive* interference. All ELECTROMAGNETIC RADIATION, including visible light, consists of waves, and can undergo interference.

An optical interference filter consists of several thin layers of transparent material coated on glass. Light reflected within the layers interferes with itself. The thickness of the layers is designed so that only a narrow range of wavelengths passes through without destructive interference. The rest of the light is reflected back. These filters are often used in narrow band PHOTOMETRY, to measure the magnitudes of stars at different wavelengths.

Two waves not exactly in or out of phase combine to give a wave whose amplitude is intermediate between constructive and destructive interference. This amplitude can thus be used to measure the phase difference between the waves. The INTERFEROMETER is based on this principle.

interferometer

A pair of receiving devices, such as radio or optical telescopes, linked to make the equivalent of a much larger receiver. Interferometers are most important in radio astronomy, where the long wavelengths mean that a dish several miles across would be needed to give a resolution equal to that of an optical telescope (about 1 arc second). In practice, two radio telescopes observing the same source are linked and their outputs combined electronically. Unless the source is directly overhead, the radio waves must travel further to one telescope than to the other. The waves received are thus to some extent out of step with each other. As the Earth rotates this *phase difference* changes, and the combined signal varies in amplitude (see INTERFERENCE). The rate of change depends on the position of the source, which can thus be measured very accurately. For a radio source that is extended over a part of the sky, different parts of the source will be in or out of phase at the same moment. The combined signal is an average of all these, and so its total variation is less than that from a point source, giving clues to the radio source's structure.

A shorter wavelength, or a longer baseline, reveals finer details. Radio telescopes separated by tens of miles can be linked by telemetry. In *very long baseline interferometry* (VLBI), which uses telescopes in different continents, the outputs of each are recorded on magnetic tape and combined later. A resolution of a few millionths of an arc-second can be achieved by VLBI. By using many fairly short interferometers and combining the results in a computer it is possible to make detailed maps of radio sources. This technique of APERTURE SYNTHESIS effectively creates a telescope as big as the separation of the longest interferometer used. A common technique is to use the Earth's rotation to simulate the effect of a large dish.

Optical interferometers can help resolve details too small for ordinary optical telescopes to distinguish. In 1920 the American physicist Albert MICHELSON used a stellar interferometer attached to the Mount Wilson 100-inch (254-cm) telescope to measure the diameters of some large nearby stars by interferometry. At Narrabri Observatory, about 250 miles (400 km) northwest of Sydney, Australia, is a stellar interferometer consisting of two reflectors 21 feet (6.5 m) in diameter, each made of 251 smaller mirrors, moving on a track 600 feet (200 m) in diameter. This optical interferometer has measured the diameters of many nearby bright stars.

International Astronomical Union (IAU)

An organization containing several thousand leading astronomers, founded in 1919 for international cooperation in astronomical research; it grew out of an international solar union founded by George Ellery Hale. The IAU holds a general assembly every three years, at which recent findings are discussed, and which decides on such matters as the standardization of constellation boundaries and the nomenclature of astronomical objects. The IAU has a telegrams bureau at the Smithsonian Astrophysical Observatory, Cambridge, Massachusetts, which notifies observatories of objects such as novae and newly discovered comets that require immediate additional observations.

interstellar absorption

The dimming of light by dust particles in space. These

dust grains, about .00001 centimeter in size, scatter some of the light from stars and galaxies that would otherwise reach the Earth. Red light is less affected than blue, so that distant objects appear redder as well as fainter. From the degree reddening, the total amount of light absorbed can be determined, and the apparent magnitude which the object would have in the absence of absorption can then be calculated. This revised magnitude must be used if the object's distance is to be found from the DISTANCE MODULUS.

Interstellar dust is concentrated in the plane of the Milky Way. The long "rift" down the middle of the Milky Way, seen best in the constellation Cygnus, is due to the greater depth of dust in that direction, which absorbs more of the light from distant stars. Dark nebulae, such as the COALSACK in the Southern Cross, which were once thought to be holes through the Milky Way, are in fact denser clouds of dust which block off distant stars in patches.

Interstellar absorption lines, seen in the spectra of distant stars, are caused by atoms of elements such as calcium or sodium in the interstellar gas.

interstellar molecules
Many simple molecules are found in clouds of gas and dust in the Milky Way, and are revealed by their characteristic radio emission lines. A few molecules can also be detected optically; methylidyne and cyanogen were found in the late 1930s by the absorption lines they produced in the spectra of starlight. Since the late 1960s, however, the development of microwave radio techniques has revealed over 50 molecules, some of them fairly complex and containing up to 11 atoms, in the interstellar material of our Galaxy. A few compounds, including formaldehyde, have also been detected in other galaxies.

Detection. The radio emission from these molecules is produced by changes in their spin and oscillation. These motions produce a characteristic set of radio wavelengths by which each compound can be identified. It is not always straightforward to assign an observed radio line to a particular molecule (there are several molecules producing emission at 3.4 millimeters, for example), and so it is important that molecules are identified by two or more lines whenever possible. Some hydroxyl, water, and methyl alcohol sources are small and very powerful. They appear to be caused by natural MASERS.

Distribution. About half the interstellar molecules so far detected have been found in only one or two sources, such as the ORION NEBULA. Only hydroxyl radical, formaldehyde, and carbon monoxide have been found distributed widely around our Galaxy. Suitable conditions for their formation in space seem to occur in dark dust clouds and GLOBULES, H II REGIONS, emission nebulae, and certain infrared stars similar to MIRA-type variable stars. The reasons why different molecules are found in different sources are far from fully understood.

Interstellar chemistry. The existence of fairly complex molecules in detectable quantities indicates that some interstellar clouds have densities greater than previously suspected. Atoms can only form molecules if they come into contact; to produce the observed molecular abundances, clouds must have densities of at least 10 million atoms per cubic

centimeter. The density of these clouds also prevents the rapid destruction of molecules, since the dust particles shield out the interstellar ultraviolet radiation which would otherwise break the molecules up.

Leaving aside the chemically unreactive substance helium, most of the molecules are made up of the four most common elements: hydrogen, nitrogen, carbon, and oxygen. The next most common atoms, sulfur and silicon, are also found in some molecules. But the molecules observed are not those which would be produced by random sticking together of the common atoms; the formation of organic (carbon-containing) compounds seems to be particularly favored by the conditions in interstellar space. The relative abundances of molecules also vary from source to source. Astronomers believe that molecules build up on the surface of dust grains in the cloud, rather than by simple collisions between the gas atoms. Interstellar molecules provide good evidence that the material of the solar system is typical of our whole Galaxy. The presence of organic molecules in dense interstellar dust clouds may also be significant for the origin of LIFE IN THE UNIVERSE.

invariable plane
The plane of average spin, both orbital and axial, of the solar system. It is obtained by adding up the total angular momentum of each planet. The invariable plane is inclined by 1° 43' to the plane of the Earth's orbit.

Io
Second-closest satellite of Jupiter, discovered in 1610 by Galileo. Io orbits Jupiter every 42 hours $27\frac{1}{2}$ minutes at a distance of 262,070 miles (421,760 km). Io's diameter is 2,273 miles (3,658 km) and its density 3.48 times that of water. Io has orange and brown markings, believed to be caused by salt and sulphur deposits on its surface, and it is surrounded by a tenuous atmosphere of sodium. An ionosphere has also been detected. Voyager 1 in 1979 discovered that Io is volcanically active. Radio astronomers discovered in 1964 that sudden radio bursts from Jupiter are linked with the position of Io in its orbit, and it is probable that the satellite interacts with Jupiter's magnetic field to cause occasional discharges of particles from Jupiter's radiation belts.

ions
Atoms which have either lost or gained electrons, and thus have acquired a positive or a negative charge, respectively. Stars and much of the interstellar gas are composed of positive ions and the electrons they have lost, a mixture known as a PLASMA.

The process by which ions are formed is termed *ionization*. It usually refers to the loss of electrons by atoms in high-speed collisions in a hot gas. The voltage required to ionize an atom is its *ionization potential*; for hydrogen this is 13.6 volts. Removal of successive electrons from multi-electron atoms becomes progressively more difficult; the second ionization potential is thus greater than the first, and so on.

ionosphere
The region of the Earth's atmosphere containing atoms and molecules that have been *ionized* (had electrons removed) by solar radiation, making them electrically

charged. The ionosphere actually consists of several layers at altitudes extending from about 30 miles (50 km) to about 300 miles (480 km) where it merges with the MAGNETOSPHERE. It extends through the regions of the atmosphere known as the mesosphere and thermosphere (see EARTH). Because it contains electrically charged particles, the ionosphere can reflect radio waves of certain wavelengths.

The ionosphere is divided into three main regions: the D, E, and F layers. The D layer, below about 55 miles (90 km), is weakly ionized and plays little part in reflecting radio waves; it largely vanishes at night. The E layer, between 55 and 100 miles (90 and 160 km), contains more strongly ionized molecules, and the F layer above has the greatest concentration of ionized atoms. Variations in the ionization of the E and F layers are closely linked to changes in solar activity, including flares, the 27-day rotation of the Sun, and the roughly 11-year SOLAR CYCLE. These two layers are also known as the Heaviside-Kennelly and Appleton layers, respectively. The F (Appleton) layer is sometimes further subdivided into F1 and F2 layers centered at about 110 miles (180 km) and 160 miles (250 km) in altitude (the heights at which many spacecraft orbit). Probing the ionosphere by sounding rockets and satellites has been of major importance in understanding the ionosphere.

ISIS satellites
International Satellites for Ionospheric Studies, two joint Canadian-U.S. satellites for ionospheric research, continuing the Canadian ALOUETTE satellite program.

J

Jansky, Karl Guthe (1905–1950)
American communications engineer, the founder of radio astronomy. Working at the Bell Telephone Laboratories, Jansky began to investigate the causes of interference in long-distance telephony in 1931. He tracked some of them down to natural sources such as thunderstorms, but he also found another source of radio noise that seemed to move with the stars. In 1932 he identified the source as lying in the direction of the constellation Sagittarius, which is toward the center of our Galaxy. Listening at a wavelength of 15 meters, Jansky was the first to detect the radio emission of the gas lying between the stars of the Milky Way. Jansky never followed up these first radio astronomy observations, and it was left to the radio amateur Grote REBER to pursue them. The unit of strength of cosmic radio waves has been named the jansky in his honor; it is equal to 10^{-26} watt per square meter per hertz.

Janssen, Pierre Jules César (1824–1907)
French pioneer of solar physics, inventor of the spectrohelioscope, an instrument for examining the Sun in the bright lines of the spectrum emitted by certain atoms. At a total solar eclipse in 1868 Janssen took spectra of the bright flames called prominences seen at the Sun's edge, and found that they emitted bright lines due to the gas hydrogen. Janssen realized that these bright lines could be observed when the Sun was not eclipsed, and developed an instrument (the spectrohelioscope) that allowed him to observe the

Sun and its prominences at a particular hydrogen wavelength. Janssen's observations showed that the bright envelope seen around the Sun at eclipses and named the CHROMOSPHERE is in fact a cloud of hot hydrogen. Also at the 1868 eclipse, Janssen observed puzzling spectral lines which the English astronomer Norman LOCKYER interpreted as being due to a previously unknown element he named helium; only in 1895 was this element detected on Earth. In 1862 Janssen had discovered that absorption of light by the Earth's atmosphere produced dark lines in the solar spectrum, and he suggested (correctly) that the same effect in light reflected from planets would allow study of the composition of their atmospheres. He attempted to make such analyses, but with little success. In 1876 the French government granted him an observatory site at Meudon near Paris, where he set up a 33-inch (83-cm) refractor and used it to take a pioneering atlas of solar photographs.

Janus
Tenth satellite of Saturn, discovered in December 1966 by the French astronomer Audouin DOLLFUS. The satellite was found when the rings of Saturn were seen edge-on, so that it was not hidden by glare. Janus orbits Saturn every 17 hours $58\frac{1}{2}$ minutes at a distance of 98,500 miles (159,000 km), which is just at the edge of Saturn's rings; the satellite cannot normally be seen because of the rings' brilliance. Janus has a diameter of about 200 miles (320 km).

JD
Abbreviation for JULIAN DATE, a system of day numbering commonly used in astronomy.

Jeans, Sir James Hopwood (1877–1946)
British astrophysicist, best known for his support of the theory that the solar system originated from the Sun's encounter with a passing star. From his study of rotating bodies, Jeans concluded that a fast-spinning star would either split in two, forming a binary system, or spin off a stream of gas into space (he thought that something similar might be occurring, on a large scale, to account for the spiral shapes of galaxies). He could find no intermediate case in which a cloud of gas around the star would form into planets, and he concluded that the so-called nebular hypothesis of the solar system's origin, as advanced by KANT and LAPLACE, was untenable. Instead, in 1919, he developed the idea put forward by the Americans Thomas Chrowder Chamberlin (1843–1928) and Forest Ray Moulton (1872–1952) that a passing star drew out a cigar-shaped filament of material from the Sun that condensed into planets. This famous theory, however, is no longer accepted (see PLANETS). One remarkably foresighted suggestion of Jeans, made in 1928, was that matter is continually created in the Universe, an idea which is now of great prominence in cosmology (see CONTINUOUS CREATION; STEADY-STATE THEORY). From 1928 onward Jeans devoted himself to the popularization of astronomy, producing such well-known books as *The Universe Around Us, The Mysterious Universe,* and *The Stars in their Courses.*

Jet Propulson Laboratory (JPL)
A division of the National Aeronautics and Space

Administration (NASA), situated in Pasadena, California. JPL was set up in November 1944 for missile development, and became part of NASA at the end of 1958. It is operated for NASA by the California Institute of Technology. JPL supervises the Deep-Space Network of NASA tracking stations, which include its own 210-foot (64-m) antenna at Goldstone. JPL scientists are also engaged in unmanned space projects to explore the solar system. America's first satellite, Explorer 1, was developed at JPL, and the laboratory has since been responsible for the Ranger, Surveyor, and Mariner series of probes. The Orbiter section of the Viking spacecraft was also the work of JPL. NASA's Space Flight Operations Facility, the mission control center for unmanned Moon and space probes, is situated at JPL. The laboratory operates an astronomical observatory at Table Mountain, California.

Jodrell Bank

British radio-astronomy observatory near Macclesfield, Cheshire, operated by the University of Manchester; it is also known as the Nuffield Radio Astronomy Laboratories. It was founded after World

The 250-foot diameter Mark I radio telescope at Jodrell Bank was for many years the largest fully steerable dish in the world. Bearings in the housings at either side allow the dish to tip, while the whole structure rotates on a circular rail track not visible here.

War II by Bernard LOVELL, and began by using radar equipment to study cosmic rays and meteors. In 1952 construction began on the famous 250-foot (76-m) dish, which was the world's largest fully steerable radio telescope until the opening of the 328-foot (100-m) EFFELSBERG dish in Germany in 1971. The great Jodrell Bank dish was completed in 1957 in time to track Sputnik 1, which put it in the world's headlines. Jodrell Bank has also received signals from other space probes, but its main work is in radio astronomy, in such areas as the detection of radio emission from flare stars, the measurement of radio-source diameters, and the location of large numbers of pulsars. In 1964 an elliptical-shaped dish, 125 by 83 feet (38 × 25 m), was installed at Jodrell Bank for observations at shorter wavelengths.

Johnson Space Center
A NASA installation near Clear Lake, southeast of Houston, Texas. Opened in 1963, the facility was originally known as the Manned Spacecraft Center: it was renamed in 1973 in honor of former President Lyndon Johnson. The center houses Mission Control, which directs manned space missions after lift-off. The first mission to be controlled from Houston was Gemini 4 in 1965; previous manned flights had been controlled from Cape Canaveral. In addition to the planning and direction of space missions, Johnson Space Center is also responsible for the design and management of spacecraft, and for astronaut selection and training: astronauts practice maneuvers such as spacecraft docking in full-scale simulators at JSC. Also housed at JSC is the Lunar Receiving Laboratory, where Moon rock samples are stored and studied. Rice University operates the Lunar Scientific Institute at JSC under a NASA grant.

Jones, Sir Harold Spencer (1890–1960)
Tenth British astronomer royal, who made precise calculations of the distance between the Earth and the Sun (the ASTRONOMICAL UNIT), thus establishing the accurate scale of the solar system. The calculations were based on the close approach of the asteroid EROS in 1930–1931, of which Spencer Jones organized worldwide observations. Finding the distance between two objects orbiting the Sun provides a baseline for determining the scale of the entire solar system; before the advent of radar, close approaches of asteroids were the most accurate way of achieving this. After 10 years' work, Spencer Jones in 1941 announced that the Sun's distance is almost exactly 93 million miles, a figure that has been scarcely altered since. Spencer Jones was appointed astronomer royal in 1933. He showed that the Earth's speed of rotation is not truly constant, and in 1938 introduced quartz-crystal clocks to supplant purely astronomical observations as an accurate basis of timekeeping; in 1936 he introduced a system of time distribution by telephone. After World War II, Spencer Jones supervised the removal of the Royal Observatory from its centuries-old site at Greenwich, London, to a new and better site at Herstmonceux, Sussex.

Julian date
Chronological system which acts as a standard reference for logging historical events recorded on the different calendars of various civilizations. The Julian date is particularly useful to astronomers for coordinating ancient records of eclipses and other astronomical events. The system was devised in 1582 by the French scholar Joseph Justus Scaliger (1540–1609), and named for his father, Julius Caesar Scaliger; it has no connection with the Julian calendar. Scaliger chose January 1, 4713 B.C. as a starting date, from which Julian days are numbered consecutively. Thus, January 1, 1975, is JD 2,442,414 and January 1, 1980 is JD 2,444,240. Julian days begin at noon. The system is still used in astronomical computing for such purposes as predicting eclipses or plotting the light-curve of a variable-brightness star.

Juno
The third asteroid to be discovered, also known as minor planet 3. Juno was found by the German astronomer Karl Ludwig Harding (1765–1834) on March 29, 1807. It orbits the Sun every 4.36 years, moving from the inner portion of the asteroid belt at 1.98 a.u. out to its fringes at 3.36 a.u.; the orbit is inclined at 13° to the ecliptic. Juno is an irregular object some 140 miles (226 km) across, with a mass of about 3×10^{16} tons. It rotates in $7\frac{1}{4}$ hours with a pole near Gamma Camelopardalis. Juno is a reddish asteroid with a reflectivity (albedo) of 19 percent. Its spectrum suggests a stony-iron surface composition akin to the SIDEROLITE meteorites.

Juno rockets
American space launchers used for early satellites and probes. The Juno I was a Jupiter-C rocket modified by adding a solid-fuel top stage with satellite attached. This combination launched the Explorers 1, 3, and 4 satellites. The Juno II was based on the JUPITER ballistic missile (not to be confused with the smaller Jupiter-C, which was a development of the Redstone missile). Juno II had the same solid-fuel upper stages as Juno I, contained within a jettisonable nose cone. Juno II launched the Pioneer 3 and 4 probes, as well as Explorers 7, 8, and 11.

Jupiter
The fifth planet in order of distance from the Sun, and the largest planet in the solar system, with $2\frac{1}{2}$ times the mass of all the other planets combined. The planet's dimensions are enormous. Its equatorial diameter is 88,730 miles (142,800 km) or 11 Earth diameters. Its volume is equal to 1,319 Earths, but with a mass of 1.9×10^{24} tons, its density is only 1.33 the density of water, a quarter that of the Earth. Jupiter moves at 8 miles (13 km) per second in an elliptical path averaging 483,631,000 miles (778,329,000 km) from the Sun, taking 11.9 years to complete one orbit.

Jupiter spins on its axis faster than any other planet. Its rotational period of 9 hours 50 minutes 30 seconds is particularly remarkable for a body of such size; a point on Jupiter's equator spins around at 22,000 miles (35,000 km) per hour. Its rate of rotation is so rapid that the planet is slightly flattened at the poles and bulging at the equator; its equatorial diameter is nearly 5,500 miles (8,800 km) greater than its polar diameter. The rapid rotation, flattening at the poles, and unusually low density, all suggest that Jupiter is not a solid planet like the Earth, but a largely fluid body.

Surface. Small telescopes reveal a regular pattern of light zones and dark belts running parallel to Jupiter's equator, usually yellow to reddish-brown in color. Larger instruments show a wealth of fine detail which can change in a few hours. The relative prominence of the various belts varies over a period of years, while the region around the south equatorial belt can undergo sudden and rapid upheavals, darkening and developing many dark and light patches and spots. Such outbreaks occur at intervals of about 6 to 10 years, and particularly active disturbances, such as that of 1971, spread right around the planet before slowly subsiding. Detailed photographs taken by Pioneer spacecraft in 1974 and 1975 show that the equatorial regions are remarkably devoid of fine structure, but consist instead of large-scale swirls and plumes of cloud, formed because the planet's rapid

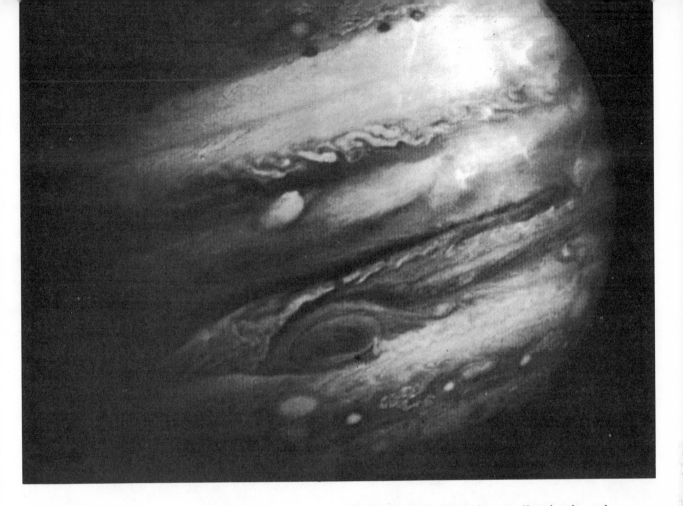

Jupiter's swirling cloud bands, photographed from 20 million miles (32 million km) by Voyager 1 in February 1979, showing details down to 375 miles (600 km) across. Note various dark and light ovals, as well as clouds circulating around the great red spot (lower center).

rotation has "smeared" the individual convection cells of hot gas rising from below the clouds. Away from the equator, the alternating cloud bands and darker belts become progressively more ragged and full of fine detail. Close to the poles there are no organized belts at all, and the convection cells are visible as a mass of dark and light spots, swirls, eddies, and disorganized mottlings.

The most remarkable feature on Jupiter, and its only permanent surface marking, is the RED SPOT, discovered in 1666. Smaller, temporary, red spots have been seen at high latitudes. The only other features to rival the longevity of the great red spot are three small white ovals close to the south temperate belt, first glimpsed in the 1920s and recorded intermittently ever since.

Atmosphere. Jupiter is remarkable in that it emits almost twice as much heat as it receives from the Sun. This may be heat left over from its formation. There is a strong GREENHOUSE EFFECT which keeps the temperature of the upper atmosphere at about 120° to 138°K at both equator and pole, even on the dark side of the planet. The deeper parts of the atmosphere are probably stagnant, with turbulence and "weather" occurring only in the top 40 to 50 miles (64 to 80 km). Water condenses to form clouds at low levels,

ammonia condenses at intermediate levels, and methane forms the highest clouds. The light zones on Jupiter are dense cloud layers at about 130°K, formed from upwelling currents of gas. Dark objects, especially the belts, are actually the denser, warmer (230° to 270°K) and lower regions of the atmosphere. Occasionally, very dark small features have shown temperatures as high as 310°K, representing transient windows into even lower regions.

The main colors on Jupiter are probably due to various combinations of ammonia and hydrogen sulfide. These can produce a wide variety of green, yellow, orange, red, and brown colorations carried into the visible levels by updrafts. The brightest red areas, such as the red spot, may be due to the presence of red phosphorus. When the atmosphere is disturbed, large amounts of colored material rise to the cloud tops where they are blown around and produce a general darkening. The great equatorial disturbances at 6 to 10-year intervals are probably caused by the release of heat which has slowly built up below a temperature inversion in the low weather levels.

The rotation periods of Jupiter's features vary for different latitudes. At high latitudes, the major white cloud zones rotate with a period of 9 hours 55 minutes, similar to that of 20 MHz radio noise detected from Jupiter. The radio noise is thought to be caused by activity below the weather zone, so that its rotation also represents the rotation period of the dense main body of Jupiter. Materials in the region of the dark belts, however, show rotation periods which can vary by as much as six minutes. The 20,000-mile-(32,000-km)-wide equatorial region of Jupiter's clouds

101

rotates five minutes faster than the dense core, demonstrating the fluid nature of the visible surface.

Jupiter is rich in the building blocks of living organisms—all the molecules thought to have been involved in the beginnings of life on Earth. At the bottom level of the weather zone, the temperature reaches 300° to 350°K, at which there would be a plentiful supply of liquid water. It is conceivable that forms of life have evolved here, possibly simple forms of stress-adapted bacteria.

Below the clouds. Jupiter consists mainly of hydrogen, with a much smaller amount of helium. Heavier elements are present in combination with hydrogen. The volatile products form the weather zone, while at greater depths the planet is believed to consist of a hydrogen/helium mix with a rocky-metallic core some 6,000 miles (10,000 km) across at the center. The central temperature must be about 30,000°K, with a pressure of 100 million Earth atmospheres. The hydrogen mix changes from a dense fluid into a metallic state about 15,000 miles (25,000 km) below the surface of the planet. The slow change of fluid to metal, or perhaps the slow gravitational contraction of the entire central core, could be the source of Jupiter's excess heat flow. The depth of Jupiter's atmosphere is about 600 miles (1,000 km). Unlike the planets nearer the Sun, Jupiter has retained a large proportion of volatile materials and is closer to the Sun in total composition.

Magnetosphere. Jupiter possesses a strong magnetic field and intense radiation belts, stretching out to 90 planetary radii. A dipole magnetic field about 10 times as strong as the Earth's field, with north and south poles reversed, relative to the Earth, is found close to the planet. Beyond this is a non-dipole field about three times stronger. Jupiter's radiation belts, containing trapped high-energy protons and electrons, are some 10,000 times stronger than the Earth's Van Allen belts. They emit strong bursts of radio waves partly influenced by Io, the innermost of the large moons of Jupiter. Io orbits within the radiation belts, and material eroded from its surface forms a huge cloud of hydrogen and sodium which encircles Jupiter.

Moons. Jupiter has at least 14 moons. The four brightest—Io, EUROPA, GANYMEDE, and CALLISTO— are termed the Galilean satellites, after their discoverer. The only other man to discover four Jovian moons was the American astronomer Seth Barnes Nicholson (1891–1963). Other discoverers of Jovian moons were E. E. BARNARD, the American astronomer Charles Dillon Perrine (1867–1951), and the British astronomer Philibert Jacques Melotte (1880–1961). Most recently, Charles T. Kowal of Hale Observatories has conducted a fruitful search for faint extra satellites. Jupiter's moons fall into three main groups.

The innermost five have periods from 12 hours to 17 days and move in near-circular orbits in the equatorial plane of the planet. The next group of four have periods of from 239 to 260 days, moving in elliptical orbits inclined at about 28° to the planet's equator. The final group of four are tiny bodies with periods of 631 to 758 days, moving in retrograde ellipses at great distances from Jupiter. The two outer groups are probably captured asteroids, and must contain other as yet undiscovered members.

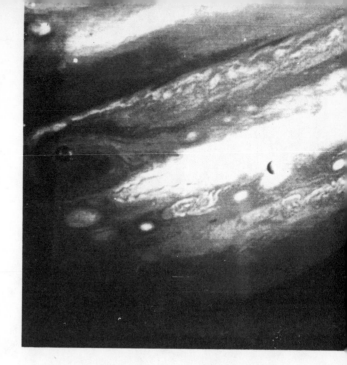

Jupiter's moons Io (left) and Europa seen against the backdrop of the planet. Io appears dark, almost merging with the red spot behind, because of orange-colored salt and sulfer deposits which cover its surface. Europa, a rocky body with icy patches, is much brighter.

In 1979 the Voyager 1 spacecraft found that Jupiter has a faint ring of particles at a distance of about 130,000 km from the planet's center, orbiting every 7 hours.

Moons of Jupiter

Moon		Mean distance from Jupiter (km)	Orbital period (days)	Diameter (km)	Discoverer and date
V	Amalthea	181,000	0.498	240	Barnard, 1892
I	Io	421,760	1.769	3,659	Galileo, 1610
II	Europa	671,050	3.551	3,100	Galileo, 1610
III	Ganymede	1,070,400	7.155	5,270	Galileo, 1610
IV	Callisto	1,882,600	16.689	5,000	Galileo, 1610
XIII	Leda	11,100,000	239	15	Kowal, 1974
VI	Himalia	11,477,600	250.566	100	Perrine, 1904
X	Lysithea	11,720,250	259.219	20	Nicholson, 1938
VII	Elara	11,736,700	259.653	30	Perrine, 1905
XII	Ananke	21,200,000	631*	20	Nicholson, 1951
XI	Carme	22,600,000	692*	20	Nicholson, 1938
VIII	Pasiphae	23,500,000	744*	20	Melotte, 1908
IX	Sinope	23,600,000	758*	20	Nicholson, 1914

(The characteristics of Jupiter's 14th moon, discovered by Kowal in 1975, are not yet known.)
*Retrograde

Jupiter rocket

American medium-range ballistic missile, which in 1958 was installed in Italy and Turkey. The missile was 58 feet (17.7 m) high, 105 inches (2.7 m) wide, and weighed 110,000 lb. (50,000 kg). It was successfully test fired on March 1, 1957, making it the first operational U.S. missile of its range. A Jupiter missile blasted the monkeys Able and Baker on a suborbital ride from Cape Canaveral on May 28, 1959, in preparation for later manned launches. To aid in testing components for the Jupiter, the smaller REDSTONE rocket was modified to produce the so-called Jupiter-C design. The Redstone was lengthened and had a revolving drum attached at the top. The drum contained two concentric rings of solid-fuel rockets, which comprised the second and third stages of the vehicle; the drum was spun by small jets to even out the thrust of the solid-propellant rings. Nose cones for reentry into the atmosphere were tested on Jupiter-C firings. A small solid-fuel fourth stage with a satellite attached turned Jupiter-C into the launcher named Juno I. It was this rocket that launched the first American satellite, Explorer 1 (see JUNO ROCKETS). Confusingly, the Jupiter military missile also had extra stages added to it to act as a satellite launcher. In this configuration it was called Juno II.

K

°K

The symbol for degrees Kelvin, the temperature scale measured from ABSOLUTE ZERO, the coldest temperature possible. A Kelvin degree is equivalent to a centigrade degree.

Kant, Immanuel (1724–1804)

German philosopher who, though not known as a scientist, did contribute to modern theories of the origin of the solar system. In his 1755 book *General History of Nature and Theory of the Heavens* (subtitled *Concerning the Structure and the Mechanical Origin of the Whole Universe, According to Newtonian Principles*) he described our solar system as part of a lens-shaped system of stars, the Galaxy, and suggested that the fuzzy patches called nebulae seen by astronomers were in fact other galaxies. Kant proposed that the Sun and planets arose from the condensation of diffuse material in the Universe, as a result of gravitational attraction between particles. This is now called the nebular hypothesis, and was further developed by the French mathematician the Marquis de LAPLACE.

Kapteyn, Jacobus Cornelius (1851–1922)

Dutch astronomer who measured positions and brightnesses of 454,875 stars on photographs taken at the CAPE OBSERVATORY, to produce the *Cape Photographic Durchmusterung,* a major catalog of southern-hemisphere stars and the first large-scale photographic survey in astronomy. Kapteyn began work in 1886, and his results appeared in three volumes between 1896 and 1900. He studied the distribution of stars in space, finding that giant stars are a thousand times less common than fainter, Sun-type stars. Kapteyn concluded that the Sun was near-centrally placed in a rotating, lens-shaped Galaxy of stars;

however, the obscuring effect of gas prevented him from seeing the true extent of our Galaxy, and we now know that the Sun is far from the center. Kapteyn found that the motions of stars with respect to the Sun seem to cluster in two opposite directions, a phenomenon known as STAR STREAMING. It is now known that this effect is due to the Galaxy's rotation.

Kapustin Yar

Soviet rocket-launch site about 60 miles (100 km) southeast of Volgograd and northwest of the Caspian Sea. It was originally an intercontinental missile test site, and was later modified for space use. Early Soviet sounding rockets were launched from Kapustin Yar, including modified V-2s that carried animals. The first orbital launch was Cosmos 1 in 1962. Subsequent space launches from Kapustin Yar have also been of small Cosmos scientific satellites. Although the second space launch site to be used, it is far less important than either TYURATAM or PLESETSK.

Kennedy Space Center (KSC)

NASA installation at Cape Canaveral, Florida, responsible for the preparation and launching of manned and unmanned space rockets. Most U.S. space missions use the KSC facilities at Cape Canaveral, although some are launched from facilities operated by KSC at the Western Test Range in Lompoc, California. Headquarters of KSC are on Merritt Island, which is the site of the Vehicle Assembly Building (VAB) 525 feet (160 m) tall, inside which the giant Saturn rockets were assembled for Apollo and Skylab missions, and which is now modified for Space Shuttle assembly. Rockets assembled in the VAB are fired from one of two pads at nearby launch complex 39. Kennedy Space Centre covers 84,000 acres. (*See picture p. 104.*)

Kepler, Johannes (1571–1630)

German mathematician and astronomer, who discovered that the orbits of the planets are ellipses, not circles, and derived the three fundamental laws of planetary motion known as KEPLER'S LAWS. Kepler was destined for a career as a Lutheran minister, but after being appointed a teacher of mathematics at Graz, he pursued the subject of mathematical astronomy for the rest of his life. Kepler accepted the novel theory of Copernicus, which placed the Sun at the center of the Universe. Kepler tried to see if a series of regular geometric solids would fit between the spheres which were thought to carry the planets around the sky. This ingenious attempt, published as the *Mysterium Cosmographicum* (The Cosmographic Mystery), in 1596, brought him into contact with the famous Danish astronomer Tycho BRAHE. In 1600 Kepler moved to Prague to work with Tycho, and on the latter's death in 1601 succeeded him as Imperial Mathematician. From an analysis of Tycho's observations, Kepler found that Mars moves in an elliptical orbit at varying speeds depending on its distance from the Sun. These momentous findings, which firmly established the truth of the Copernican scheme and banished forever the ancient notion that planetary motions must be based on circles, were published in his book of 1609, *Astronomia Nova* (New Astronomy).

Kepler had an underlying belief that God had

The Vehicle Assembly Building dominates the Kennedy Space Center, Fla. Here the Saturn 5 rocket which launched Apollo 11, the historic first manned Moon flight, is moving along the 3½ mile Crawlerway to Launch Complex 39A.

constructed the Universe according to a divine plan. The idea in his *Mysterium Cosmographicum* was one reflection of this, but he now returned to the question in the light of new studies of Tycho's observations which showed that the motion of all planets resembled that of Mars. Kepler found a relationship between the size of a planet's orbit and the time it takes to complete a circuit of the Sun. He also discovered relationships between the orbital speeds of the planets and the notes in a musical scale, developing a scheme of divine musical harmony which he published as *Harmonices Mundi* (Harmonies of the World) in 1619. At the same time he devised a physical theory to account for elliptical motion and published all his planetary results in *Epitome Astronomiae Copernicanae* (Epitome of Copernican Astronomy) between 1618 and 1621. In 1627 he also published the *Rudolphine Tables,* which gave planetary positions based on Tycho's observations.

Kepler also studied and wrote on the subject of optics. After he had used one of Galileo's telescopes, he published a design of his own that later became more widely adopted. He observed and wrote about the nova of 1606, another proof of how mistaken were the old views about the changelessness of the heavens.

Kepler's laws
Three laws governing the motions of the planets in their orbits around the Sun, derived by Johannes KEPLER from the observations of the Danish astronomer Tycho Brahe. Kepler found the premise that the planets orbit in circles at a uniform speed untrue whatever scheme of planetary motion he tried. He therefore boldly broke with tradition and tried

various oval orbits, finally discovering that only elliptical orbits along which the planets traveled at varying speeds would fit the facts. His three laws state:

  (i) that every planet orbits the Sun in an ellipse, with the Sun itself at one focus of that ellipse;
 (ii) that the radius vector (line from the Sun to the planet) sweeps out equal areas within the ellipse in equal times;
(iii) that a fixed ratio exists between the time taken to complete an orbit (the orbital period) and the size of the orbit, and that this ratio is the same for every planet. The ratio may be expressed as P^2/a^3, where P is the orbital period and a the semi-major axis of the ellipse (the distance from the center of the ellipse to a point on the ellipse nearest to one focus). If P and a are known for one planet, then a may be found for any other planet whose period is already known. From this the scale of the solar system can thus be calculated.

Kirkwood gaps
Areas in the asteroid belt in which few or no asteroids orbit. They are named for the American astronomer Daniel Kirkwood (1814–1895), who first noted them in 1857. He pointed out that they lie at exact fractions of the orbital period of Jupiter (the orbital periods are thus said to be COMMENSURABLE); asteroids in these regions would therefore be swept into another orbit by Jupiter's regular gravitational pull. Kirkwood also explained the Cassini division in Saturn's rings by commensurability of the ring particles with the orbital periods of Saturn's satellites.

Kitt Peak National Observatory
Observatory located in the Quinlan Mountains 56 miles (90 km) southwest of Tucson, Arizona, at an elevation of 6,770 feet (2,064 m), opened in 1960 and operated by the Association of Universities for

Research in Astronomy (AURA). Kitt Peak hosts 16 telescopes, the largest single concentration of astronomical instruments in the world. Its main telescope is a 158-inch (400-cm) reflector, opened in 1973. The McMath Solar Telescope is the world's largest solar telescope, with a 60-inch (150-cm) main mirror and a 300-foot (91.4-m) focal length. Among the observatory's other telescopes is an 84-inch (210-cm) reflector used for optical and infrared observations; a 50-inch (130-cm) reflector for infrared studies; two 36-inch (91-cm) reflectors; two 16-inch (40-cm) reflectors; and a 24-inch (61-cm) solar telescope. In addition, the University of Arizona's STEWARD OBSERVATORY operates a 90-inch (230-cm) and 36-inch (91-cm) reflector on the same site. Also on the mountain is a 52-inch (140-cm) reflector operated jointly by the University of Michigan, the Massachusetts Institute of Technology, and Dartmouth College; and a 36-foot (10.75-m) radio telescope of the National Radio Astronomy Observatory.

Komarov, Vladimir Mikhailovich (1927–1967)
First Soviet cosmonaut to make two space flights, and the first man to be killed during a space mission. Komarov commanded the three-man Voskhod 1 flight launched on October 12, 1964. The capsule, a Vostok modified to hold three men, remained in space for a day. On April 23, 1967, he was launched alone in Soyuz 1, the first of a new generation of Soviet manned spacecraft. But during this test flight the capsule became difficult to control, and Komarov had to make an emergency reentry after 18 orbits. According to official reports, the parachute became twisted, apparently due to the spin of the spacecraft, and the Soyuz crashed to Earth, killing Komarov.

Korolev, Sergei Pavlovich (1907–1966)
The anonymous Russian "Chief Designer of Carrier Rockets and Spacecraft." As such, he was responsible for the development of the first Soviet intercontinental missile, which in modified form became the basic Soviet space launcher; he also supervised the design of the first Soviet artificial satellites and space probes, as well as the Vostok, Voskhod, and Soyuz manned capsules. In 1931 Korolev was a founder member of the Moscow Group for the Study of Rocket Propulsion called GIRD, where he headed work by pioneers such as Friedrich TSANDER and Mikhail TIKHONRAVOV on the first Soviet liquid-fueled rockets. In 1933 this organization was merged with the work of rocketeers in Leningrad to form the Rocket-Science Research Institute (RNII), where work progressed on simple missiles. Imprisoned by Stalin in 1938, Korolev was set to work during World War II to develop jet-assisted takeoff for aircraft. Freed after the war, he began to develop improved versions of the V-2, using them for atmospheric sounding and animal flights. In 1954 his group embarked on the construction of the first Soviet intercontinental missile, which in 1957 was turned to space use with Sputnik 1. Korolev masterminded all Soviet space activities from that date until his death.

Kuiper, Gerard Peter (1905–1973)
Dutch-born American solar-system expert. In 1944 he discovered the atmosphere of Saturn's satellite Titan, and in 1948 found that the Martian atmosphere is made of carbon dioxide. In 1948 he discovered the fifth moon of Uranus, Miranda, following it in 1949 with the discovery of Nereid, the second moon of Neptune. Kuiper measured the diameter of Pluto in 1950 and found it to be similar to that of Mars, which was much smaller than had previously been imagined. He proposed that the solar system formed from a turbulent cloud of dust and gas around the Sun, in which small regions, called protoplanets, grew dense enough to contract into planets. This idea was inspired by his work on double stars, and he viewed the solar system as a "failed" twin-star system. This led him to suggest that planetary systems exist around at least one star in a hundred, making them far more common than supposed; this view is now widely supported. Kuiper was also interested in the surface features of the Moon and compiled several atlases of lunar photographs. In 1960 he founded the Lunar and Planetary Laboratory at the University of Arizona.

L

Lacaille, Nicolas Louis de (1713–1762)
French astronomer, called the "father of southern astronomy" for his pioneer mapping of the southern skies. From 1751 to 1753 Lacaille made observations from the Cape of Good Hope, mapping nearly 10,000 stars and introducing 14 new constellations: Antlia, Caelum, Circinus, Fornax, Horologium, Mensa, Microscopium, Norma, Octans, Pictor, Pyxis, Reticulum, Sculptor, and Telescopium. He measured the positions of the Moon, Mars, and Venus, simultaneously with similar observations in Europe, in order to determine more accurately the scale of the solar system. From his observations he arrived at a value for the Sun's distance that was only 10 percent too small.

Lacerta (the lizard)
An inconspicuous constellation sandwiched between Andromeda and Cygnus, introduced by the Polish astronomer Johannes HEVELIUS.

Lagrange, Joseph Louis (1736–1813)
French mathematician, who in 1772 discovered what are now known as the LAGRANGIAN POINTS. Lagrange made studies of the overall stability of the solar system, showing that no long-term changes could be expected in the planets' orbits. Together with LAPLACE, he dominated the field of mathematical astronomy for a generation. Lagrange later headed the commission that established the metric system of weights and measures.

Lagrangian points
Five points in space at which a very small body can remain in a stable orbit with two very massive bodies. The points were first recognized by Joseph Louis LAGRANGE and are rare cases in which the relative motions of three bodies can be computed exactly. In the case of a massive planet orbiting the Sun in a circular path, the first stable point (L_1) lies on that orbit diametrically opposite the large planet itself. The points L_2 and L_3 are both on the Sun-planet line, one closer to the Sun than the planet and the other farther away. The remaining two points, L_4 and L_5,

are positions on the planet's orbit that each form an equilateral triangle with the planet and the Sun. Clouds of gas and dust are believed to collect at the Lagrangian points in the Moon's orbit around the Earth, and the TROJAN asteroids are found at the Lagrangian points in the orbit of Jupiter.

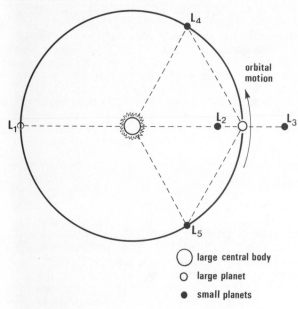

○ large central body

○ large planet

● small planets

The five positions to which a small body can move in fixed relative position with a massive planet orbiting a large central body.

Landsat

Three NASA satellites for surveying the Earth's resources, originally called Earth Resources Technology Satellites (ERTS). The Landsats were launched into orbits approx. 565 miles (910 km) above the Earth, circling from pole to pole 14 times a day. As the satellite orbits, its sensors scan a strip 115 miles (185 km) wide. After 252 orbits (18 days) the satellite has completed one survey of the entire Earth and is back over its starting point. The satellite passes over the Earth's equator at about 9.30 A.M. local time on each orbit.

The Landsats were developed from the Nimbus weather satellite. They possess two main sensor systems: a so-called multispectral scanner (MSS), which scans the Earth at four wavelengths, two in the visible region and two in the infrared; and three television cameras, which take pictures simultaneously in three wavelengths, from which color images can be assembled. Different features of the Earth are prominent in different wavelengths, enabling experts to examine large areas of forest and crops, study water and air pollution, marine resources and biology. Several countries, led by Brazil and Canada, have built ground stations to receive pictures direct from Landsat as it passes overhead. Landsats 1 and 2 were launched in 1972 and 1975. Landsat 3, containing improved sensors, was launched on March 5, 1978.

Langley Research Center

A NASA installation at Hampton, Virginia, for research and development in aeronautical and astronautical technology. The installation is named for the American aviation pioneer Samuel Pierpont Langley (1834–1906); it was established in 1917 as Langley Field, a center for aircraft development. Langley came under NASA when that body was set up in 1958, and provided experienced personnel for many of NASA's other branches, notably the Manned Spacecraft Center (now JOHNSON SPACE CENTER) in Houston, Texas. Langley scientists helped develop the high-flying rocket planes that were the precursors to manned spaceflight. The Mercury, Gemini, and Apollo spacecraft were all developed at Langley, where wind-tunnel tests solved the problem of a capsule's reentry into the atmosphere. Facilities also allow structural testing of rocket models to study their behavior in flight. Simulators at Langley permitted astronauts to practice docking and lunar landings in full-scale mock-ups. The Langley Research Center was responsible for the Lunar Orbiter series of unmanned spacecraft and the Viking program.

Laplace, Pierre-Simon, marquis de (1749–1827)

French mathematical astronomer, formulator of the so-called nebular hypothesis to explain the solar system's origin. In his *Exposition du système du monde* (1796), he described the solar system as having formed from a giant cloud or nebula of material. As this cloud shrank, it threw off rings of material that condensed to form the planets; at the center was formed the Sun. Laplace suggested that the many spiral-shaped nebulae observed in space were examples of such a process in action, although we now know the spiral nebulae to be enormous galaxies of stars. Laplace's hypothesis was similar to an idea put forward some time earlier by the German philosopher Immanuel KANT. In outline it is similar to modern ideas of the solar system's origin.

Laplace published a massive five-volume *Traité de mécanique céleste* between 1799 and 1825, summarizing work on gravitational astronomy since Newton. With his contemporary LAGRANGE, he made many important investigations in the field of celestial mechanics, including a major analysis of the Moon's motion which led to the construction of new and more accurate tables. Laplace showed that there were limits to the possible variations of inclination, eccentricity, and size of the planets' orbits, demonstrating that no major changes could be expected from interactions among the planets.

Las Campanas Observatory

An astronomical observatory on Cerro Las Campanas, a mountain in the Chilean Andes, 120 miles (193 km) north-east of La Serena. The observatory, part of HALE OBSERVATORIES and operated by the Carnegie Institution of Washington, is built on a ridge at 7,480 ft (2,280 m) altitude. Its main telescope is a 101-inch (256-cm) reflector, inaugurated in 1976. There is also a 40-inch (1-m) reflector, in operation since 1971, and a 24-inch (61-cm) reflector operated by the University of Toronto.

laser

A device for producing an intense beam of light; the name is an acronym for *Light Amplification by*

*S*timulated *E*mission of *R*adiation. Invented in 1960, lasers are optical forms of MASERS, and work on similar principles. A laser tube contains a helium-neon mixture (gas laser) or cylindrical ruby crystal (solid-state laser), whose atoms are excited. Light of a particular wavelength stimulates these excited atoms to emit light at that same wavelength, producing a precisely aimed beam of very intense light by reflections from mirrors at each end of the tube.

One notable astronomical application of the laser is its use in plate-scanning machines to provide small, brilliant spots of light for measuring photographic density. More spectacularly, the later Apollo missions carried a laser altimeter in the command module, which worked by timing the return of a laser beam reflected from the lunar surface. Intense laser beams have even been directed from major observatories on the Earth to reflectors placed on the lunar surface by Apollo astronauts and carried on the Soviet LUNOKHOD automatic rovers. As a result, the Earth–Moon distance is now known to within a few centimeters.

Lassell, William (1799–1880)
British astronomer and instrument maker, discoverer of two satellites of Uranus and one of Neptune. Lassell improved construction techniques for large reflector telescopes, which then used mirrors of speculum metal. In 1846 he built a 24-inch (61-cm) metal-mirror reflector, using it that year to discover Triton, the largest satellite of Neptune. In 1851 he discovered the satellites Ariel and Umbriel of Uranus. Lassell also independently discovered the satellite Hyperion of Saturn, two days after W. C. BOND in 1848. He built a 48-inch (122-cm) reflector in Malta in 1860, with which he located 600 new nebulae.

latitude
A coordinate for determining positions on Earth north or south of the equator. An object's latitude is its angle north or south of the equator; all lines of latitude are parallel. The equivalent of latitude on the CELESTIAL SPHERE is DECLINATION; and it should not be confused with *celestial latitude,* which measures instead the angle north or south of the ecliptic (the Sun's apparent path around the sky). The ecliptic is inclined at $23\frac{1}{2}°$ to the celestial equator because the Earth's axis is tilted relative to the plane of its orbit.

launch window
The period during which conditions are right for the successful launch of a satellite or space probe. The launch window is affected by factors such as the local time at the launch site and the position in orbit of another craft or planet for an intended rendezvous. For instance, launch windows to Mars occur every 26 months and to Venus every 19 months.

Leavitt, Henrietta Swan (1868–1921)
American astronomer, who discovered that the period and brightness of CEPHEID VARIABLE stars is related (see PERIOD-LUMINOSITY RELATION). She made her discovery from a study of variable stars in the Magellanic Clouds, which are satellite galaxies of our own; because all the stars in each cloud are at roughly the same distance from us, their relative brightnesses can be directly compared. The period-luminosity

relationship, published by Miss Leavitt in 1912, was used by Harlow SHAPLEY to determine the size of our Galaxy. Miss Leavitt discovered a total of 2,400 variable stars. She also measured photographic magnitudes of stars to produce standard lists of star brightnesses, such as the famous *North Polar Sequence* published in 1917.

Lemaître, Georges Édouard (1894–1966)
Belgian astronomer who in 1927 formulated the BIG-BANG theory of cosmology, which holds that the Universe began in a giant explosion. Lemaître studied at the Massachusetts Institute of Technology in the mid-1920s, where he learned of Edwin HUBBLE's work showing that the Universe seems to be expanding. Lemaître proposed that all the matter in the Universe was once concentrated into what he termed the primeval atom, whose explosion scattered material into space to form galaxies, which have been flying outward ever since. Lemaître's idea has been developed by other astronomers, notably George GAMOW, and is now the most widely held cosmological theory.

lens
An optical glass component which either converges or diverges a beam of light passing through it. A converging lens is said to be *positive*; a diverging one is *negative*. A positive lens is capable of forming a real image of an object on a screen; a negative lens is not, and is used only for special purposes.

The usual task of a lens (or a system of lenses) is to form an image of an object. If the object is at infinity (as, in effect, all astronomical objects are), its image is formed at a distance from the lens called the FOCAL LENGTH. The focal length of a lens determines the size of the image it produces. If the focal length is 100 units, then the image of an object 1° across in the sky will measure 1.75 units. The scale is entirely independent of the aperture of the lens, which affects only the brightness of the image. If the lens is to be used for a powerful telescope, a long focal length, to give a large image-scale, is desirable, while a large aperture, giving a brighter image, will allow fainter objects to be seen.

The performance of a lens, by which is generally meant the sharpness of its images, is governed by the curvature of its faces and the type of glass employed. The image formed by any single lens will be colored, because of its tendency to disperse the light passing through it into a spectrum; this deformity is known as CHROMATIC ABERRATION. It can be minimized by combining two lenses made of different types of glass. Such a combination is said to be *achromatic*. The other principal fault to which a lens may be subject is SPHERICAL ABERRATION, brought about because different concentric regions have different focal lengths, and thus produce no single sharp focal point. Spherical aberration is strongly affected by the choice of curvature on the faces; by a careful choice of curves, and the correct combination of two or more lenses into a unit, spherical aberration can be practically eliminated. Astronomical OBJECT GLASSES rarely consist of more than two lenses, but camera lenses, which must work at relatively wide aperture and cover a large field of view in sharp focus, must contain a large number of individual lenses to give good results (see also REFRACTING TELESCOPE).

Leo (the lion)

A major constellation of the zodiac, best seen during the northern hemisphere spring; the Sun passes through Leo from mid-August to mid-September. The head of the lion is formed by a characteristic sickle-shaped line of stars, at the foot of which is the constellation's brightest star, REGULUS (Alpha Leonis). The second-brightest star is Denebola, magnitude 2.23 and 42 light-years distant. The LEONID METEORS appear to radiate from the constellation each November.

Leo Minor (the lesser lion)

A small and faint constellation in the northern hemisphere of the sky, north of Leo itself, introduced on the 1690 star map of Johannes HEVELIUS.

Leonid meteors

A regular meteor shower which appears for about two days around November 17 each year. The stream has a radiant 3° northwest of Gamma Leonis and reaches its greatest altitude at 6.30 A.M. Although activity from the stream is usually a modest 5 to 20 meteors each hour, major meteor storms occur at infrequent intervals. The last took place on the morning of November 17, 1966, when astonished observers in the central United States witnessed meteor rates peaking for 15 to 20 minutes at 100,000 meteors an hour. A similar storm on November 12, 1833, attracted the attention of the American astronomers Denison Olmsted (1791–1859) and Alexander Catlin Twining (1801–1844). They demonstrated that the storm phenomenon was caused by particles moving together in a solar system orbit, thereby founding the scientific study of meteors. An earlier storm had been recorded in 1799, and another took place in 1866. The stream had produced only mediocre displays in 1899 and 1933, and the great storm of 1966 came as a surprise. The stream is debris from comet Tempel-Tuttle 1866 I and moves in the same orbit. Storms occur when the Earth passes through the stream close to the position of the comet, and the next is timed to occur on November 18, 1999.

Leonov, Alexei Arkhipovich (b. 1934)

Soviet cosmonaut, the first man to walk in space. Leonov crawled through an airlock in space on March 18, 1965, during the flight of the two-man Voskhod 2 capsule, remaining outside for 10 minutes. He had a portable life-support system strapped to his back during the walk, but was tethered to the craft by a lifeline. Leonov was commander of the Soviet crew for the APOLLO-SOYUZ TEST PROJECT in 1975.

Lepus (the hare)

A constellation of the southern hemisphere of the sky at the foot of Orion, best seen during the northern hemisphere winter. The name of its brightest star, the 2.69-magnitude Arneb, is Arabic for hare. R Leporis, which lies near the borders of the constellation with Eridanus, is one of the reddest stars known. It is a variable star of long period, like MIRA, ranging between magnitudes 5.9 and 10.5 every $432\frac{1}{2}$ days.

Leverrier, Urbain Jean Joseph (1811–1877)

French mathematician and astronomer, who predicted the existence of the planet Neptune. In 1845 he began to study the motion of the planet Uranus, which was not following its predicted path in the sky. Leverrier calculated that it was being pulled out of place by an unknown body. In 1846 he told astronomers at the Berlin Observatory where he believed the new planet lay; it was immediately discovered by Johann GALLE, and eventually named Neptune. After the discovery came news that similar calculations had been made by the Englishman John Couch ADAMS. From 1847 until his death Leverrier was engaged on the monumental task of calculating accurate tables for all the planets, producing a set of standard references for astronomers into the 20th century. During this work, in 1859, Leverrier predicted the existence of an asteroid belt close to the Sun in order to explain the observed motion of the planet Mercury. The apparent acceleration of Mercury is now known to be an effect of RELATIVITY.

Libra (the scales)

A constellation of the zodiac, lying between Virgo and Scorpius, best seen during the northern hemisphere spring; the Sun passes through Libra during November. Alpha Librae is a double star, just resolvable to keen eyesight. Beta Librae, a SPECTROSCOPIC BINARY, is one of the few bright stars that appear green in color. Delta Librae is an ECLIPSING BINARY, similar in type to ALGOL, ranging from magnitudes 4.8 to 5.9 every 2.33 days.

libration

The slow east–west and north–south rocking motions of the visible face of the Moon about its mean position. The Moon's axial rotation period precisely equals its orbital revolution period, and so it should always in theory present the same face to the

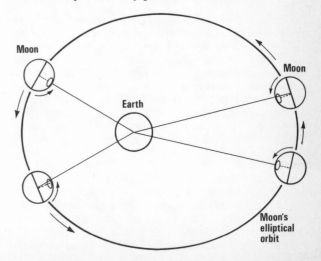

The movement of the Moon in its orbit when close to the Earth, **left**, and at its greatest distance, **right**. The rapid orbital revolution at left is faster than the steady rotation and the usual visible hemisphere drifts to one side, moving sideways normally central features such as the crater shown here. On the right, the steady rotation is faster than the slow orbital revolution, causing the visible hemisphere to drift in the opposite direction. The ellipticity of the lunar orbit and magnitude of the effect are greatly exaggerated here.

Earth. However, the Moon's orbit is elliptical, and it thus moves more quickly when closest to the Earth and more slowly when farthest away. The Moon's speed in orbit therefore changes rhythmically while its axial rotation remains uniform. When closest to the Earth, the Moon's axial rotation falls behind its orbital revolution so that it turns its face to the east; at its greatest distance, its axial rotation moves ahead of its orbital revolution, so that it turns its face to the west. This rocking effect is east–west libration. At the same time, the Moon's axis of rotation in inclined not at right angles but 83° to its orbital plane, producing a less pronounced north–south libration. The principal effect of libration is to allow features which would otherwise just be hidden to rock into view at the edge of the Moon's disk. In this way it is possible for Earth-bound observers to see 59 percent of the lunar surface.

Lick Observatory

The astronomical observatory of the University of California, situated at an elevation of 4,209 feet (1,283 m) on Mount Hamilton, 13 miles (21 km) east of San Jose, California. The observatory was established by a bequest from landowner James Lick (1796–1876). Its main telescopes are a 120-inch (304-cm) reflector opened in 1959, and a 36-inch (91-cm) refractor which, when the observatory was opened in 1888, was the largest in the world. Headquarters of the observatory moved in 1966 to Santa Cruz. In 1968 Lick astronomers selected the 5,862-foot (1,787-m) Junipero Serra Peak in Monterey County as a dark-sky outstation.

life in the Universe

Many scientists now accept that life of some sort may exist on other bodies in space. Their view stems from recent advances in our understanding of the origin and early evolution of life, as well as evidence from astronomers that planetary systems may be very common in space (see PLANETS).

Attempts have been made to estimate the number of other civilizations that might exist in our own Galaxy, the Milky Way. These take into account many factors: the rate at which stars are formed; the number of stars that probably have planets; the number of planets on which life is likely to have arisen; the probability that such life might rise to a technologically advanced stage; and the length of time for which such civilizations might exist. These estimates—they are little more than educated guesses—suggest there may be a million advanced civilizations in the Galaxy—that is, one for every 100,000 stars. The average distance between two such civilizations (such as ourselves and another) would be a few hundred light-years.

Origin of life. Remains of algae, similar to those living today, are found in rocks 3,200,000,000 years old, showing that life has existed on Earth for at least three-quarters of our planet's existence. Traces of life may exist in even older rocks, and this suggests that simple forms of life arose rapidly once conditions on Earth became suitable.

The very first chemicals of life were probably created from the gases that surrounded the Earth at its birth—ammonia, methane, and water vapor. These same gases are found today in the atmosphere

James Lick, who founded the Lick Observatory on Mt Hamilton, Cal., lies buried beneath the pedestal of the 36-inch (91-cm) refractor, shown here. This telescope has been in use since 1888, and is still the second largest refractor in the world. The focal length of the lens is 57.8 feet (17.6 m). The weights on the right of the picture, on the end of the declination axis of the German-type mounting, counterbalance the telescope.

around Jupiter, which is thought to have changed little since the planets first formed. Experiments with mixtures of such gases have shown that they can combine to form the simple basic building blocks of life called amino acids. Amino acids form chains to make protein, the structural material of life. The gases of the Earth's early atmosphere would have been stuck together into amino acids with the energy from the Sun's ultraviolet radiation, or from lightning or shock waves. Raining into the oceans below, the amino acids and other chemicals would have formed an organic soup from which life could start to form. Although the origin of even the simplest organism from inert chemicals remains a mystery, biochemists are beginning to understand some of the pathways by which the very complex molecules of life may have built up on the early Earth, or any other planet, from simpler precursors.

Molecules between the stars. One of the most exciting developments in understanding the origin of life has been the discovery of organic molecules in space in regions such as the Orion nebula where stars are being born. In the dense gas-and-dust clouds of our Galaxy, radio astronomers have detected many molecules of varying complexity (see INTERSTELLAR MOLECULES). Similar molecules are also found in the gases of comets, which are believed to represent the left-over portions of the cloud around the ancient Sun from which the planets formed. The compounds now known in space include formic acid and methylamine, which can combine to give the amino acid called glycine. The predominance of carbon in all these compounds encourages biochemists in their belief that other life-forms will most likely be based on carbon, as life on Earth is.

Meteorites, lumps of rock from space that crash to Earth, have brought further evidence that complex organic chemistry once occurred in the cloud surrounding our growing Sun. Researchers have found a crop of amino acids in two stony meteorites that fell in recent years at Murchison, Australia, and in Kentucky. These meteorites have also revealed evidence of other important organic molecules, evidently formed nonbiologically in space.

Stars and life. Life could not form in the vicinity of all stars. Some stars flare up irregularly, emitting dangerous radiation; others change in brightness more regularly (see VARIABLE STARS). The very biggest stars burn out too quickly for life to begin around them. And the longest-lived stars are dim dwarfs that send out insufficient light and heat to warm up any planets they may have. The best sort of stars for life to form around are those which are average in age and brightness. Our Sun is just such a star, and there are many stars like it in the Galaxy.

For life like ours to form, there must be a planet in a region around the star that is neither so hot that all water evaporates, nor so cold that water freezes. Water is vital to life on Earth, and it so happens that the Earth lies in the very center of the region around our Sun where liquid water can exist. However, some biochemists think that life might form using liquid ammonia as a solvent instead of water, and this means that life could form in colder regions, perhaps among the clouds of a planet like Jupiter.

Mars is one planet in our solar system where Earth-type organisms might have arisen. The two American Viking spacecraft of 1976 were designed to search for life on Mars.

Communicating with the stars. No advanced creatures like ourselves exist on any of the planets in our solar system. They may, however, exist on planets of other stars, and our best chance of finding out is by radio contact. The first attempt to listen for radio messages from other civilizations was made in 1960 by Frank DRAKE in Project Ozma. He used the 85-foot (26-m) diameter radio telescope of the National Radio Astronomy Observatory at Green Bank, West Virginia, to listen to the stars Tau Ceti and Epsilon Eridani. These are both stars like the Sun, and are 12 and 11 light-years away from us, respectively.

Drake listened at a wavelength of 21 centimeters, at which hydrogen in the Galaxy naturally emits

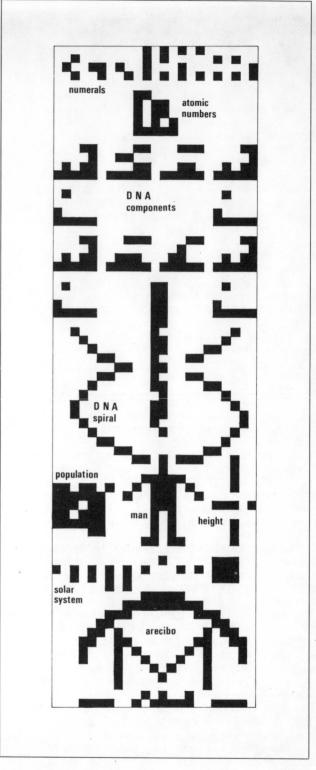

The test message sent from the Arecibo radio telescope on November 16, 1974, to the 300,000 stars of the globular cluster M13 in Hercules, 24,000 light-years distant. The message, sent at 12.6 centimeters wavelength, contained 1,679 on-off pulses which could be arranged into a pictogram, showing, first, the numbers 1 to 10 in the binary code used by computers, **top line, reading right to left**; then the atomic numbers of the basic elements that make up living cells; the formulae for the molecules of DNA, the genetic material of cells; a representation of the familiar double-helix spiral of DNA, with a "backbone" showing the number of DNA components; a figure of a human, flanked by numbers showing the approximate population of the Earth and the human's height; a sketch of the solar system indicating that planet three (Earth) is associated with the human; and a representation of the Arecibo dish transmitting the message.

radio waves. Anyone trying to attract the attention of other civilizations might choose this wavelength to signal on, because it is the one at which radio astronomers most frequently listen. Despite a total of 150 hours of listening to both stars, no signs of a message were detected. However, since only one civilization like our own would be expected for every 100,000 stars, the negative result was not at all unexpected.

Astronomers in the United States, the Soviet Union, and Canada have since extended the Project Ozma work to a total of nearly 1,000 stars. In 1975 Frank Drake and Carl SAGAN began to listen to nearby galaxies with the 1,000-foot (305-m) radio telescope at Arecibo, in search of signals from possible super-civilizations. All these searches have so far proved fruitless.

Stars farther and farther from the Sun will be searched, at various wavelengths, until success is achieved, or the possibilities are exhausted.

light

The part of the spectrum of electromagnetic radiation to which the human eye is sensitive; by extension, the term is often applied to all wavelengths of electromagnetic radiation.

Electromagnetic waves are regular periodic fluctuations in space of electric and magnetic fields. The distance between successive waves or troughs (the WAVELENGTH) determines the color of light. Blue light has a wavelength of about .0004 millimeter and red light about .0007 millimeter. Waves outside the visible region are of the same nature and travel with the same speed (see LIGHT, VELOCITY OF). The types of radiation, in order of increasing wavelength, are gamma rays, X rays, ultraviolet, visible and infrared light, microwaves, and radio waves.

Color, and other properties of light such as refraction, polarization, diffraction, and interference, can be explained by regarding light as a wave motion. But other properties can be described only if we assume that light is a stream of particles (photons), each a packet of energy of a definite size. Since as long ago as the time of Newton, there has been a conflict between the wave and particle theories of light. This has been resolved by the modern *quantum theory,* developed about 1900 by the German physicist Max Karl Ernst Ludwig Planck (1858–1947), which shows that the two views are compatible. The large-scale properties of the particles or photons approximate to those of waves, and light is therefore regarded as having a dual nature.

Early astronomers could exploit only visible light. Using modern techniques, we can observe the Universe at almost all wavelengths. Radio telescopes, in particular, have extended our perception of the Universe. Any limitations that are imposed are mainly those of the Earth's atmosphere. Infrared radiation is absorbed by water vapor in the atmosphere and ultraviolet light is cut off by the ozone layer. These problems are now being overcome by observing from rockets and satellites.

light, velocity of

The speed at which light travels in a vacuum; in denser mediums, such as glass or water, light travels more slowly. The speed of light was first estimated by the Danish astronomer Ole ROEMER in 1676. He found that when Jupiter was distant from the Earth, the eclipses of its satellites took place later than predicted, due to the greater traveling time of the light (called *light time*). He could not, however, obtain a good estimate for the velocity of light because the size of the solar system was not then accurately known.

The theoretical basis for assessing the velocity of light was established by the Scottish physicist James Clerk Maxwell (1831–1879). In Maxwell's equations of ELECTROMAGNETIC RADIATION, published in 1864, he showed that light always has the same velocity, regardless of the velocity of the object emitting it relative to the observer. This result was the starting point of Einstein's 1905 Special Theory of RELATIVITY.

The velocity of light (c) can be measured by determining independently the wavelength (λ) and frequency (v) of a particular SPECTRAL LINE. These are linked by the equation $c = \lambda v$. The modern value of the speed of light is 186,282.397 miles (299,792.458 km) per second. The speed of light in a vacuum is the same at all wavelengths.

light-year

The distance traveled by light in a vacuum during one year (strictly, a tropical year, see YEAR). It is equivalent to 5.8786 trillion miles (9.4607 trillion km), which represents 63,240 astronomical units, or 0.3066 parsec.

limb

The edge of a celestial body as seen from Earth.

Lindblad, Bertil (1895–1965)

Swedish astronomer, who in 1926 suggested that our Galaxy is rotating, to account for the phenomenon of star motion called STAR STREAMING, discovered by Jacobus KAPTEYN. With his son Per Olöf Lindblad (b. 1927), he made studies of the rotation of galaxies, suggesting that their spiral arms are caused by waves of density sweeping around galaxies, a concept developed in more detail by the Chinese-born American mathematician Chia-Chiao Lin (b. 1916). Lindblad developed a method of classifying spectra to determine the absolute magnitudes of stars. He also made surveys of faint stars in selected regions of the sky, improving knowledge of stellar statistics.

Local Group

A cluster of over 30 known galaxies including the Milky Way. The Local Group is believed to be a permanent cluster, held together by the gravity of its members. The Milky Way is near one edge of the group, and the ANDROMEDA GALAXY (M31), 2 million light-years away, marks the other edge. M31 is the largest and most massive galaxy in the Local Group, 1½ times as large as the Milky Way. The third-largest galaxy, M33 in the constellation Triangulum, is only a tenth as massive as the Andromeda galaxy.

Apart from these three spiral galaxies, all the members of the Local Group are dwarf galaxies, with masses less than a hundredth that of the Milky Way. They contain between a million and a billion stars each, and some of the smallest are probably globular clusters that have escaped from the Milky Way. Some of these dwarf galaxies are in orbit around the spirals; the two MAGELLANIC CLOUDS are satellites of the

Milky Way, while the Andromeda galaxy has several small companions.

A large nearby galaxy on the edge of the Local Group was discovered in 1968. It cannot be seen optically because the view from Earth is obscured by dust clouds in the Milky Way, but it can be detected by infrared astronomy. This galaxy, Maffei 1, seems to be about as massive as the Milky Way and is of the elliptical type, containing no interstellar gas. Its distance is about 3 million light-years.

Lockyer, Sir Joseph Norman (1836–1920)

English astronomer, who in 1868 discovered the element helium, from lines observed by the French astronomer Pierre JANSSEN in the spectrum of the Sun's atmosphere during a total solar eclipse. Lockyer's name for the element came from the Greek *helios,* meaning Sun. At the Solar Physics Observatory in South Kensington, London, Lockyer made spectroscopic observations of sunspots in 1866, finding Doppler shifts in their spectral lines that indicated strong convective currents of gas in the Sun's outer layers. In the same year he devised a scheme to observe the bright prominences at the Sun's edge, normally visible only at an eclipse, by spreading out the Sun's light in a spectroscope, and observing the prominences at the wavelengths they emit most strongly. But he was unable to make his first observations until October 1868, by which time Pierre Janssen had already put the technique to work. Lockyer and Janssen found from their observations that prominences were eruptions from the outer layer of the Sun, which Lockyer called the chromosphere. Lockyer also pioneered the study of the astronomical alignments of stone circles and ancient temples. Perhaps his greatest contribution to science came in 1869, when he founded and became first editor of *Nature,* the world's leading scientific journal.

longitude

A coordinate for determining the position of an object east or west of the prime meridian of Greenwich, which is designated 0° longitude. Unlike lines of latitude, lines of longitude are not parallel; they originate in a common point at each pole, and are widest apart on the Earth at the equator. The Earth spins through 15° of longitude in an hour, and so the longitude of a location thus indicates its local time relative to Greenwich. The equivalent of longitude on the CELESTIAL SPHERE is RIGHT ASCENSION, usually measured in hours, minutes, and seconds. It is not to be confused with *celestial longitude,* a separate coordinate measured in degrees east of the vernal equinox along the ecliptic.

Lovell, Sir Alfred Charles Bernard (b. 1913)

British radio astronomy pioneer, founder of the famous Jodrell Bank Radio Observatory with its 250-foot (76.2-m) dish, for many years the world's largest fully steerable radio telescope. After World War II, Lovell used his wartime experience with radar to study radio echoes from cosmic ray showers. Many echoes turned out in fact to come from the trains of meteors, as Lovell's team conclusively proved in 1946. Lovell's radio studies located many daytime meteor showers that were previously unknown. In 1949 he began plans for the 250-foot

dish, which finally came into operation in 1957, in time to track the first Sputnik. The Jodrell Bank telescope was occasionally called upon for communications with space probes, but mostly Lovell pursued astronomical work, including the detection of radio emissions from outbursts on FLARE STARS.

Lovell, James Arthur (b. 1928)

American astronaut, commander of the Apollo 13 flight of April 1970, whose projected Moon landing was canceled because of an explosion in the service module. Lovell and his crew used the engine and electrical power of the attached lunar module to make an emergency return to Earth. Lovell's first flight was in the record-breaking Gemini 7 mission of December 1965, in which he and Frank BORMAN spent a total of 14 days in space. Lovell commanded the Gemini 12 flight in November 1966, during which Edwin ALDRIN made an extensive space walk. Lovell was also a crew member of Apollo 8 in December 1968, the first manned spacecraft to reach the vicinity of the Moon.

Lowell, Percival (1855–1916)

American astronomer, who instituted the search that led to the discovery of the planet Pluto. In 1894 Lowell, a wealthy amateur, founded the Lowell Observatory at Flagstaff, Arizona, the best observing conditions he could find, to observe Mars. Inspired by reports of "canals" on Mars by the Italian astronomer Giovanni SCHIAPARELLI, Lowell believed he detected an entire network of canals on the planet, and wrote works such as *Mars and Its Canals* (1906) and *Mars as the Abode of Life* (1908), in which he described a dying and arid planet whose inhabitants were forced to channel water from the poles to their crops nearer the equator. Despite its immense popular appeal, few astronomers agreed with Lowell's theory, nor could they confirm his canal observations. At the turn of the century Lowell began analyzing the motion of Uranus to determine if it revealed the gravitational effects of an unknown planet beyond Neptune. In 1905 he produced the first of a series of predictions of where a new planet might lie, which culminated in his famous 1915 memoir on the subject. A major search was instituted at Lowell Observatory, but only in 1930 was Pluto discovered. It is such a small body, however, that it could not have had any observable effects on Uranus, and most astronomers now assign Pluto's discovery to chance.

Lowell Observatory

An astronomical observatory founded by Percival LOWELL in 1894 at Flagstaff, Arizona, at an elevation of 7,250 feet (2,210 m). The observatory contains a 24-inch (61-cm) refractor installed by Lowell in 1896, together with 30-inch (76-cm), 24-inch, and 21-inch (53-cm) reflectors used mostly for photoelectric observations. The observatory also possesses the famous Pluto telescope, a 13-inch (33-cm) astrographic telescope used to discover Pluto. This is now situated at the dark-sky observing site opened in 1961 at Anderson Mesa, 12 miles (19 km) southeast of Flagstaff. The 72-inch (183-cm) Perkins reflector of the Ohio State and Ohio Wesleyan universities was moved to this site in 1961, and in 1968 a 42-inch

(107-cm) reflector was also installed. In 1965 a NASA-sponsored Planetary Research Center was opened in the grounds of Lowell Observatory; this is the headquarters of an international photographic survey of the planets, and contains the world's most extensive collection of planetary photographs.

luminosity

A measure of the total amount of radiation emitted by a star or other glowing object per second, usually expressed in relation to the luminosity of the Sun. A star's luminosity depends on its temperature and size. One way of expressing the luminosity of a star is its ABSOLUTE MAGNITUDE. Astronomers divide stars into six *luminosity classes*: Ia, Brightest supergiants; Ib, Less luminous supergiants; II, Bright giants; III, Giants; IV, Subgiants; V, Main-sequence stars (often termed dwarfs).

Lunar Module (LM)

The two-stage craft in which Apollo astronauts landed on the Moon. The Lunar Module was stored under the Apollo command and service modules at launch, and extracted by a docking maneuver once Apollo was on its way to the Moon. The Lunar Module had a total height with its four legs extended of 22 feet 11 inches (7 m), divided between the descent (lower) stage of 10 feet 7 inches (3.2 m) and the ascent (upper) stage of 12 feet 4 inches (3.8 m); its overall width was 14 feet 1 inch (4.3 m), and launch weight 32,000 lb. (14,500 kg). The two-man LM crew occupied the ascent stage, which was pressurized; the unpressurized descent stage contained the powerful braking engine, which could thrust from 1,050 lb. (476 kg) to 9,870 lb. (4,480 kg) for controlling the Moon landing. The LM had a roof hatch for docking with the command module, and a front hatch for exit onto the lunar surface; the astronauts, standing in the ascent stage, watched their descent through two triangular-shaped forward windows. The ascent stage contained room for lunar space suits and rock samples. To lift off from the Moon's surface the ascent stage explosively separated from the descent stage, and the 3,500-lb. (1,600-kg) thrust ascent stage engine boosted the astronauts into Moon orbit to meet the command module. Sixteen small reaction control system (RCS) thrusters, arranged in groups of four, were used for delicate maneuvers and attitude control of the LM.

Apollo 14 Lunar Module *Antares* stands deserted in the Moon's Fra Mauro crater in February 1971, while astronauts Mitchell and Shepard carry out EVA. On the foremost leg is the ladder which the astronauts used for access to the module. In this view the top of the module partly shields the Sun's glare.

A technician checks a Lunar Orbiter spacecraft before launch. In the center of the framework are the lenses of the two cameras; behind them is the self-contained darkroom and picture readout mechanism. Above the propellant tanks on the upper part of the craft can be seen a thruster nozzle; to the left is the directional antenna used to transmit the data to Earth.

Lunar Orbiter

A series of five American Moon-orbiting craft which photographed the entire lunar surface, front and back, revealing its features in detail and surveying possible landing sites for astronauts. Each Lunar Orbiter carried two cameras, one for detail shots and the other for wide-angle views. The photographs were recorded on film in the spacecraft and electronically transmitted to Earth. The first three probes entered orbits around the Moon's equatorial region, while Lunar Orbiters 4 and 5 were put into polar orbit and thus surveyed the whole Moon. Tracking the Lunar Orbiters revealed the existence of MASCONS, areas of dense rock on the Moon that cause a higher gravitational pull than normal. After their flights the Lunar Orbiters were crashed onto the Moon, so as not to interfere with later missions.

Probe	Launch date	Remarks
Lunar Orbiter 1	August 10, 1966	Entered lunar orbit August 14; impacted Moon October 29
Lunar Orbiter 2	November 6, 1966	Entered lunar orbit November 10; impacted Moon October 11, 1967
Lunar Orbiter 3	February 4, 1967	Entered lunar orbit February 8; impacted Moon October 9
Lunar Orbiter 4	May 4, 1967	Entered lunar orbit May 8; impacted Moon October 6
Lunar Orbiter 5	August 1, 1967	Entered lunar orbit August 5; impacted Moon January 31, 1968

Lunar Roving Vehicle (LRV)

The electrically powered Moon car used by Apollo astronauts for exploring the lunar surface. Lunar rovers were used on the flights of Apollo 15, 16, and 17. The wire-wheeled vehicle, 122 inches (310 cm) long and 45 inches (114 cm) high, allowed astronauts to drive at speeds up to 8 miles (13 km) per hour in search of rock and soil samples. Instruments on the control panel told the astronauts how far they had progressed from the Lunar Module, and in what direction. An umbrella-shaped antenna on the vehicle kept the astronauts in direct touch with Earth, and allowed mission controllers to point the on-board television camera by radio commands. The 480-lb. (218-kg) lunar rover, capable of negotiating foot-high obstacles, two-foot crevices, and slopes of up to 20°, was stored folded in the Lunar Module's descent stage; it was deployed by the astronauts once on the Moon, and left there after use.

Luna spacecraft

A series of Soviet Moon probes, originally called Lunik. Luna 1 missed its target but was the first object to leave Earth and enter an orbit around the Sun. Luna 2, in September 1959, became the first probe to hit the Moon and Luna 3 the following month flew behind the Moon, sending the first photographs of its far side. In 1966 Luna 9 became the first probe to make a successful soft landing, and Luna 10 became the first probe to orbit the Moon. The Luna program broadened in 1970 when Luna 16 automatically returned a small sample of Moon soil to Earth, and Luna 17 landed an automatic roving Moon car called LUNOKHOD.

Probe	Launch date	Remarks
Luna 1	January 2, 1959	Missed Moon by 3,728 miles (6,000 km); in solar orbit
Luna 2	September 12, 1959	Hit Moon September 13
Luna 3	October 4, 1959	Sent back first photographs of lunar far side
Luna 4	April 2, 1963	Missed Moon by 5,282 miles (8,500 km); possible soft-landing failure
Luna 5	May 9, 1965	Impacted Moon May 12; failed soft-lander
Luna 6	June 8, 1965	Missed Moon by 100,000 miles (160,000 km); failed soft-lander
Luna 7	October 4, 1965	Impacted Moon October 7; failed soft-lander
Luna 8	December 3, 1965	Impacted Moon December 6; failed soft-lander
Luna 9	January 31, 1966	Soft-landed on Moon February 3, in western Oceanus Procellarum. Returned photos for three days
Luna 10	March 31, 1966	Entered lunar orbit April 3; measured magnetic field, meteoroids
Luna 11	August 24, 1966	Entered lunar orbit August 28; successor to Luna 10

Probe	Launch date	Remarks
Luna 12	October 22, 1966	Entered lunar orbit October 25; took photographs, made measurements
Luna 13	December 21, 1966	Soft-landed on Moon December 24; returned photos, tested soil
Luna 14	April 7, 1968	Entered lunar orbit April 10; measured near-Moon conditions such as magnetic and gravitational field, solar wind particles
Luna 15	July 13, 1969	Impacted Moon July 21 in Mare Crisium; failed sample-return attempt
Luna 16	September 12, 1970	Landed in Mare Fecunditatis September 20; returned to Earth September 24 with 0.2 lb. (100 g) of Moon soil
Luna 17	November 10, 1970	Landed on Moon November 17, carrying Lunokhod 1 automatic Moon rover
Luna 18	September 2, 1971	Impacted Moon September 11; probably failed landing attempt
Luna 19	September 28, 1971	Entered lunar orbit October 3; studied lunar surface and near-lunar space
Luna 20	February 14, 1972	Landed near Mare Fecunditatis February 21; small soil sample returned on February 25
Luna 21	January 8, 1973	Landed in Mare Serenitatis on January 15, carrying Lunokhod 2 lunar rover
Luna 22	May 29, 1974	Entered lunar orbit June 2; studied Moon and near-lunar space
Luna 23	October 28, 1974	Landed on Moon November 6. Damaged drill prevented sample return
Luna 24	August 9, 1976	Landed on Moon August 18. Soil samples returned on August 22

lunation

A complete cycle of lunar phases, the period from one new Moon, for example, to another. A lunation is therefore equivalent to a synodic MONTH, and lasts 29.53059 days.

Lunokhod

A Soviet automatic Moon car, driven by radio command from Earth. Lunokhod 1 was carried to the Moon aboard the Luna 17 soft-lander, which touched down on November 17, 1970. The eight-wheeled vehicle rolled off the lander, transmitting television pictures of its surroundings. Instruments detected cosmic rays and measured the chemical composition and physical nature of lunar rock; a small reflector panel bounced laser beams back to Earth, allowing scientists to determine the precise distance of the Moon. The 1,667-lb. (756-kg) craft had a length along its four wheels of 87 inches (222 cm) and a width of 63 inches (160 cm). By the end of its active life on October 4, 1971, Lunokhod 1 had covered a total distance of 34,588 feet (10,542 m), running in and out of craters and examining rocks as it moved over the lunar surface. An improved version, the 1,848-lb. (838-kg) Lunokhod 2, was delivered to the Moon's surface by Luna 21 on January 15, 1973. Lunokhod 2, traveling at twice the speed of its predecessor, completed its mission on June 4, having covered 23 miles (37 km) inside the crater Le Monnier. Among other features, it examined a large slab of Moon rock believed to have been ejected from a crater, and ran alongside a deep fissure.

Lupus (the wolf)

A constellation of the southern hemisphere of the sky, lying on the edge of the Milky Way between Scorpius and Centaurus. It contains no objects of particular importance.

Lynx (the lynx)

A faint constellation in the northern hemisphere of the sky near Ursa Major, introduced by Johannes HEVELIUS. Its brightest star is a red giant appearing of magnitude 3; there are no objects of major interest in the constellation.

Lyot, Bernard Ferdinand (1897–1952)

French astronomer, inventor of the CORONAGRAPH, an instrument that makes the Sun's corona visible without a solar eclipse. Lyot pioneered the study of the surface nature of planets by measuring the polarization of light reflected from them; he showed the dusty nature of the lunar topsoil and the existence of sandstorms on Mars in this way. To study the polarization of the Sun's corona, Lyot mounted his coronagraph at the high-altitude Pic du Midi Observatory in the French Pyrenees. He made the first spectrogram of the inner corona in 1930 and took the first corona photograph through his instrument in

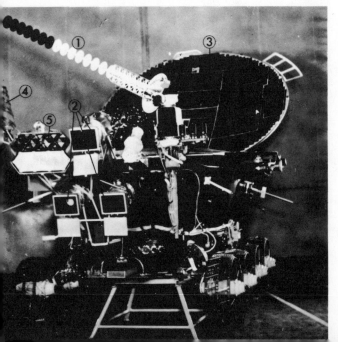

Lunokhod 2, which landed on the Moon in January 1973 in the crater Le Monnier, at the edge of the Mare Serenitatis. Visible here are the directional antenna 1, TV cameras 2, solar cell array 3, omnidirectional antenna 4 and laser reflector array 5.

Lyra

1931. With a special monochromatic filter developed in 1933 to pass only narrow bands of light, Lyot in 1935 began to take the first motion picture films of prominences at the Sun's edge. The instruments he developed have revolutionized solar observations.

Lyra (the lyre)
A small but prominent constellation of the northern hemisphere of the sky, seen overhead in middle latitudes during summer. Lyra is particularly distinctive because its main star, VEGA, is one of the brightest in the sky: Beta Lyrae, also called Sheliak, is a remarkable ECLIPSING BINARY, varying between magnitudes 3.4 and 4.3 every 12.9 days. Matter is believed to be flowing from one star to the other, each of which is distorted in shape by the other's gravitational pull; hot gas is probably also spiraling away from them into space. Epsilon Lyrae is a famous multiple star called the "double double." Keen eyesight, or a pair of binoculars, shows two stars, of magnitude 4.68 and 4.5; in a telescope, each of these is also found to be double. Delta Lyrae and Zeta Lyrae are also double stars. Between the stars Gamma and Beta is the famous ring nebula in Lyra, cataloged as M57 or NGC 6720; it is actually a PLANETARY NEBULA, 4,100 light-years distant. Near the border of the constellation with Cygnus lies RR Lyrae, the prototype of a famous class of variable stars (see RR LYRAE VARIABLES).

Lyrid meteors
A regular meteor shower reaching a brief peak of about 8 meteors an hour on April 21 each year. The stream radiant, which moves eastward 1° a day, lies 8° southwest of Vega in Lyra at maximum, and reaches its greatest altitude at 4 A.M. Splendid displays about 2,000 years ago are recorded, but the stream is now very much in decline. It follows an orbit of very long period and is derived from periodic comet Thatcher 1861 I.

M

Magellanic Clouds
Two small galaxies that orbit our Milky Way Galaxy. They are easily visible to the naked eye from latitudes south of 20°N, and were first described by the Portuguese navigator Ferdinand Magellan (c.1480–1521) in 1521 during his voyage around the world. Both Clouds have the same fuzzy appearance as the Milky Way. The Large Magellanic Cloud (LMC), lying in the constellations Dorado and Mensa, is about 6° across. About 20° away in Tucana is the Small Magellanic Cloud (SMC), 2° in size. Long-exposure photographs reveal fainter outer regions of each extending to three times the naked-eye size.

Both Magellanic Clouds are 160,000 light-years distant, and are the nearest external galaxies to our own (twelve times closer than the Andromeda galaxy). They are comparatively small, with masses about 1/30 and 1/200 that of the Milky Way, containing both young and old stars. The LMC contains the largest-known gaseous nebula, 30 Doradus, called the Tarantula nebula, which has a mass of over a million Suns, and may be condensing to a globular cluster of

The Large Magellanic Cloud, **extending off the top of this photograph,** has a bright central bar, like some spiral galaxies. The huge bright nebula 30 Doradus is just above the left end of the bar. The superimposed squares are 1 degree across.

Many gaseous nebulae are visible in the Small Magellanic Cloud as small diffuse patches, and the stars can also be resolved by large telescopes. The brightest stars in this photograph and the two globular clusters, **right and top**, are foreground objects in our Milky Way Galaxy.

stars. Young globular clusters are comparatively common in the LMC, although unknown in the Milky Way.

Recent radio-astronomy observations have shown a streamer of hydrogen connecting the Clouds to the edge of the Milky Way. This hydrogen was apparently torn away from the Clouds as they passed through the disk of our Galaxy.

magnetic stars
Stars remarkable for their very strong magnetic fields. The nearest star to us, the Sun, has a generally weak magnetic field like that of the Earth, except locally in sunspots, where the fields can be thousands of times greater. In 1946 the American astronomer H. W. BABCOCK showed that some stars have general magnetic fields far stronger than the Earth's, the most powerful being over 50,000 times as strong. Both the magnetic field and the magnitude of many magnetic stars vary with periods generally between $\frac{1}{2}$ and 20 days. These variations represent the time the stars take to rotate on their axes. Magnetic stars may have enormous "starspots" (like sunspots) that appear and disappear as the star spins, causing the variations in light output. The magnetic field in the starspot is thought to be much stronger than elsewhere on the star's surface. Magnetic stars have a peculiar excess of certain rare metallic elements in their outer layers, presumably produced by nuclear reactions in giant flares, which are associated with the spots and magnetic fields.

magnetosphere
The outermost region of the Earth's atmosphere, extending beyond the IONOSPHERE, from an altitude of 300 miles (500 km) upward. In this region, also called the exosphere, ionized particles are controlled by the Earth's magnetic field. The magnetosphere forms the boundary region between the Earth's atmosphere and the SOLAR WIND in space. At the magnetosphere's edge, particles from the Earth's atmosphere escape into space. The outer edge of the magnetosphere (the *magnetopause*) extends about 10 Earth radii toward the Sun; beyond this is an unstable region called the *magnetosheath,* produced by shock waves as the solar wind encounters the magnetosphere. Away from the Sun, in the lee of the solar wind, the magnetosphere forms a long *magnetotail*.

Two zones of high density are centered above the equator; these are the VAN ALLEN radiation belts. The Van Allen belts were discovered in the late 1950s by early artificial satellites. In shielding the Earth from the solar wind, the magnetosphere deflects charged particles into the Van Allen belts, but some spill over into polar regions of the upper atmosphere where they excite atmospheric atoms to produce the colorful AURORAE.

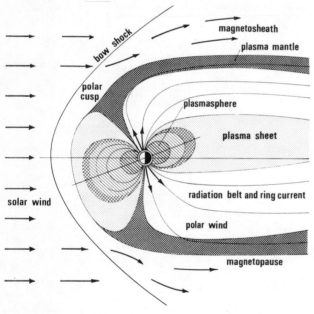

The Earth's invisible surroundings of magnetic zones. These regions have been detected by satellites and probes bearing magnetometers and particle counters; the arrows show the directions in which charged particles would move.

magnitude
A measure of an object's brightness. The Greek astronomer HIPPARCHUS first classified the naked-eye stars into six magnitude steps, with the brightest being first magnitude and the faintest he could see sixth magnitude. When stars were first measured with a PHOTOMETER, it was found that first-magnitude stars were roughly 100 times brighter than sixth-magnitude stars. In 1856 the English astronomer Norman Robert

Pogson (1829–1891) proposed the current classification system, in which a difference of five magnitudes is taken as representing exactly a factor of 100 in brightness. This means that one magnitude difference corresponds to a difference in actual brightness of 2.512 times (2.512 is the fifth root of 100). Objects more than 100 times brighter than sixth-magnitude stars are given negative (minus) magnitudes; the faintest objects are given increasingly large positive magnitudes. The term "magnitude" when used without further qualification usually means APPARENT MAGNITUDE, the brightness with which a star appears in the sky. Astronomers also use ABSOLUTE MAGNITUDE, which is a calibration of the actual light output of a star. (See also DISTANCE MODULUS; PHOTOGRAPHIC MAGNITUDE; PHOTOVISUAL MAGNITUDE.)

main sequence
The diagonal band ranging from bright blue stars to faint red stars on the HERTZSPRUNG-RUSSELL DIAGRAM, a plot of star temperatures against brightness. Main-sequence stars are termed DWARFS; consequently, the main sequence is also known as the *dwarf sequence*. The position of a star on the main sequence depends on its mass, the heaviest stars being the brightest. The majority of stars lie on the main sequence because the dwarf stage of a star's evolution, when it is in the prime of its life and producing energy by burning hydrogen in its core, lasts longer than any other. In theory, at the time they begin burning hydrogen in their centers, all dwarfs lie on a line in the H-R diagram called the *zero-age main sequence* (ZAMS). But in the actual observation of a young star cluster the line is broadened into a band, and the ZAMS is taken to be the lower edge of the observed main sequence.

major axis
The longest diameter of an ELLIPSE, one which passes through its two foci.

Maksutov telescope
A telescope system using both a lens and mirror, named for the Russian Dmitri Dmitrievich Maksutov (1896–1964), who published his design in 1944, although it was discovered independently by the Dutch optical manufacturer Albert Bouwers (1894–1972). The earlier SCHMIDT TELESCOPE corrects the blurred image, or SPHERICAL ABERRATION, of a spherical concave mirror by using a thin glass plate placed some distance in front of it. The cross-sectional shape or *figure* of this plate is difficult to achieve by mass-production methods. Maksutov substituted a thick, steeply-curved lens or *shell*, offering fewer production problems. Used as a wide-field camera, the performance of the Maksutov matches that of the Schmidt, with the added advantage that the tube is only about three-quarters the length. However, few large Maksutov-type cameras have been constructed because large shells are much thicker and heavier than the equivalent Schmidt plates. The Maksutov system has been exploited most widely in the production of relatively small-aperture telescopes for amateur use, and in telephoto *mirror-lens* systems for photography. In both cases, the central region of the convex face of the shell is coated with aluminum,

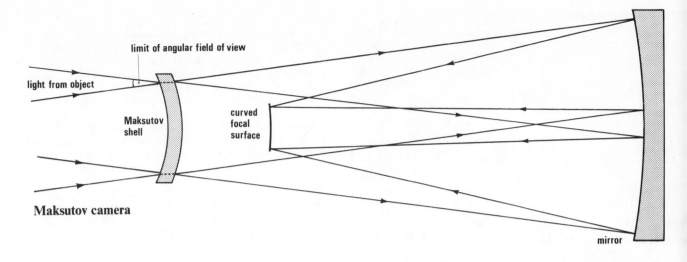

limit of angular field of view

light from object

Maksutov
shell

curved
focal
surface

mirror

Maksutov camera

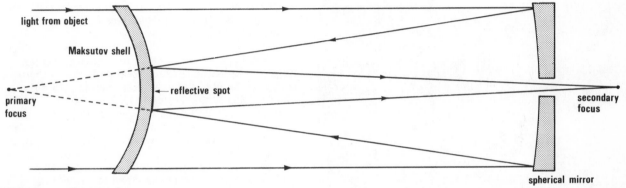

light from object

Maksutov shell

primary
focus

reflective spot

secondary
focus

spherical mirror

Maksutov—Cassegrain

the mirror so formed acting as a Cassegrain-type secondary (see CASSEGRAIN TELESCOPE). Such instruments are highly portable, and the coated reflecting surfaces, unlike those of ordinary reflecting systems, are protected from dust and corrosion in a sealed tube. Maksutov systems are relatively expensive, due to the care required in matching the curves and thickness of the shell to prevent false color in the image.

Mariner spacecraft

A series of American planetary probes. Mariner 2 was the first probe to reach another planet—Venus—successfully; Mariner 4 was the first probe to send results from Mars, revealing craters on the Martian surface. Mariner 9 became the first object to go into orbit around another planet when it reached Mars for a year-long photographic reconnaissance. Mariner 10 made history as the first double-planet mission, flying past Venus and then moving on to give astronomers their first detailed look at the surface of Mercury. Mariner 10 went into a close orbit around the Sun, re-encountering Mercury in September 1974 and March 1975. In 1977, two Mariner-type spacecraft, named VOYAGER, were launched to survey in detail the giant planets Jupiter and Saturn, arriving in 1979 and 1981.

Probe	Launch date	Remarks
Mariner 1	July 22, 1962	Launch failure; intended Venus probe
Mariner 2	August 26, 1962	Flew past Venus December 14 at a distance of 21,594 miles (34,752 km)
Mariner 3	November 5, 1964	Intended Mars probe; contact lost because spacecraft shroud failed to jettison
Mariner 4	November 28, 1964	Flew past Mars July 14, 1965, at a distance of 6,118 miles (9,846 km)
Mariner 5	June 14, 1967	Flew past Venus October 19, 1967, at a distance of 2,480 miles (3,990 km)
Mariner 6	February 25, 1969	Flew past Mars July 31 at a distance of 2,120 miles (3,412 km)
Mariner 7	March 27, 1969	Flew past Mars August 5 at a distance of 2,190 miles (3,534 km)
Mariner 8	May 8, 1971	Launch failure
Mariner 9	May 30, 1971	Went into Mars orbit November 13
Mariner 10	November 3, 1973	Flew past Venus February 5, 1974, at a distance of 3,585 miles (5,769 km); passed Mercury on March 29 at a distance of 431 miles (694 km)

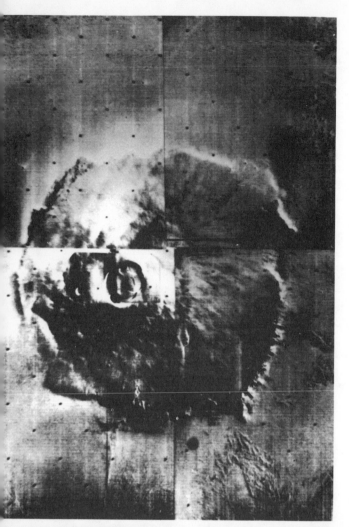

Mars

The fourth planet in order of distance from the Sun, commonly called the red planet because of its color. Mars is the planet that most resembles the Earth: it spins on its axis every 24 hours 37 minutes 23 seconds, and shows permanent surface markings, polar ice caps, and atmospheric clouds. Space-probe photographs have revealed canyons, shield volcanoes, ancient water courses, sand dunes, and heavily cratered landscapes resembling the highlands of the Moon. The climate is extremely cold and dry with violent dust storms. The VIKING probes in 1976 failed to detect any definite signs of life on the planet.

Mars moves at 15 miles (24 km) per second in an elliptical path an average of 141,636,000 miles (227,941,000 km) from the Sun, taking 687 days to complete one orbit. The planet's equatorial diameter is 4,217 miles (6,787 km), about half that of the Earth, with a mass of 6.4×10^{20} tons. Mars has a mean density of 3.94 times that of water, compared with the Earth's 5.52. Its rotational axis is inclined at 25° (the Earth's tilt is $23\frac{1}{2}°$), and its surface gravity is 40 percent of Earth's.

Mars shows a wealth of fine detail—bright orange areas contrasting with dark bluish-gray markings, topped with brilliant white polar caps. Early theories held that the light areas were continents, the dark areas oceans, and the polar caps deep polar snow drifts. Eventually, however, it was realized that the presence of only minor cloud activity and thin polar caps indicated that water was very scarce. Scientists then postulated that Mars was dry and dusty, and that the dark areas were lowlands covered with primitive vegetation. This seemed to explain the darkening of the dusky areas in spring as the polar cap began to melt. As moisture was released the plants would burst into growth, breaking through the layers of dust deposited by the winter storms and darkening

Top: The largest volcano in the solar system, Olympus Mons, photographed on Mars by Mariner 9. Olympus Mons measures over 300 miles (500 km) across and is 17 miles (27 km) high. At its summit is a complex crater 43 miles (70 km) wide. Volcanic lava appears to have flowed from Olympus Mons over the surrounding plain.

Above: Part of the complex Martian rift valley, called Valles Marineris (Mariner Valleys), photographed by the Viking 1 orbiter. The valley system, a total of 3,000 miles (4,800 km) long and 45 miles (70 km) wide, is apparently the result of crustal faulting, with subsequent erosion by wind and landslides.

the lowlands. But we now know that Martian conditions are apparently too hostile for life.

Atmosphere. The Viking probes in 1976 showed that the Martian atmosphere is 95 per cent carbon dioxide, with about two per cent nitrogen and traces of carbon monoxide, oxygen, and argon. The surface pressure at the Viking lander sites was about 7.5 millibars, less than one per cent that of the Earth's atmospheric pressure. The atmospheric pressure of Mars varies seasonally, because a large percentage of the atmosphere becomes frozen into the polar caps during the winter. The first Viking lander registered a maximum mid-afternoon air temperature in the summer of $-29°C$, falling to $-85°C$ at dawn. At the second Viking lander site farther north, winter air temperatures fell to a minimum of $-123°C$. At neither Viking site does the ground temperature rise above freezing point at any time of the year, and there is no liquid water on the planet today. Scientists believe that the atmosphere of Mars was denser in the past, and the climate more clement.

Clouds appear on Mars, but they cover only about 5 percent of the surface even when most active. White clouds forming downwind from highland areas may be stationary, like those forming in the lee of mountains on Earth. Blue clouds, which form at an altitude of about 8 miles (13 km), appear to be similar to terrestrial NOCTILUCENT CLOUDS, formed by ice crystals around meteoritic dust high in the atmosphere.

Weather. Martian weather varies greatly with season and time of day. In winter, a massive temperature difference between equator and pole produces brisk westerly gales and creates intense low-pressure areas. The strong frontal activity lifts dust particles of all sizes to considerable heights. Great dust storms develop from hurricane-like disturbances over the central highlands, throwing up vast quantities of dust to form a layer at 12 to 18 miles (19–29 km) altitude. This can envelop the entire equatorial and temperate regions of the planet, as in 1909, 1911, 1956, and 1971. In winter, therefore, dust is being spread around the planet, deposited at ground level as winds abate in spring. As the polar cap begins to retreat, the atmosphere clears to reveal the Martian surface blanketed with dust and with indistinct light and dark areas. As temperatures rise, the gradient between the pole and equator is reduced, and the westerly winds drop. They are replaced by breezes toward the equator around dawn and dusk, with general easterly winds during the main part of the day.

Mars is covered with red-colored soil, the result of amounts of iron oxide in the planet's crust. As seen from the surface, the sky is pink, caused by dust particles suspended in the atmosphere. Since fine particles will always appear lighter in shade than larger dust particles of the same material, a meteorological explanation for the springtime wave of darkening is possible. The spring easterly winds can have little effect on heavy dust, but will scour exposed highland areas free of fine dust, carrying and depositing it on adjacent bright areas, increasing the contrast between dark highlands and light lowlands. Thus the dark areas on Mars are probably the exposed eastern slopes of highland regions.

Martian topography. The highlands of Mars, which rise to heights of 7 miles (11 km), are concentrated mainly in the southern hemisphere. In contrast, the northern half of the planet consists of dust-filled lowland areas with extensive fields of recent volcanic activity. Photographs returned by space probes in orbit around Mars provide a detailed picture of the surface. The southern highlands have been heavily cratered by meteorites, indicating a great age—perhaps as much as 3.5 billion years. The largest craters are two immense impact basins, Hellas and Argyre. Hellas is a double-rimmed 3-mile-(5-km)-deep bowl some 1,000 miles (1,600 km) across resembling the Mare Imbrium basin on the Moon. The uplands also contain a number of degraded, weathered volcanic areas of great age; the whole area is similar to the highlands of the Moon. Part of the equatorial region is occupied by an extension of the southern uplands, dissected in spectacular fashion by a great 3,000-mile (4,800-km) rift zone. The major fault valley reaches a width of 45 miles (70 km) and a depth of 4 miles (6.4 km) dwarfing the Earth's Grand Canyon. There are many subsidiary faults of varying ages, some heavily eroded, others clearly younger.

The northern hemisphere is strikingly different. It contains low-lying smooth basins and plains, with young, sparsely scattered impact craters. The lowlands form a wide belt around 65° north latitude and have many relatively young basalt flows. Close to the pole is an extensive volcanic plateau, 250 miles (400 km) across and 2 miles (3 km) high, with central calderas and extensive fault systems. But the most remarkable features are four huge shield volcanoes close to the equator. Reaching an astounding height of 18 miles (29 km), they consist of complex summit calderas atop cone-shaped mountains. The largest, Olympus Mons, visible from Earth as a dusky spot, is some 320 miles (515 km) across at the base; all four volcanoes are quite clearly young features. Dust has collected in several sites at high temperate latitudes, forming fields of dunes resembling the Earth's sandy deserts. North of the equatorial plateau is a vast area of chaotic landscape, hundreds of square miles of jumbled surface blocks resembling the site of a major earthquake. Throughout the equatorial regions, and especially emerging from the chaotic zones, are numerous meandering valleys, many of them with extensive tributary systems. Hundreds of small channels complete a picture clearly indicating that once water flowed on Mars. Water in a liquid state must have existed at some distant epoch, when the atmosphere was much denser. The water courses appear of different ages, which suggests that Mars has experienced several short periods of warmth separated by lengthy glacial eras. The warm spells would periodically release water from the permafrost that must exist underground close to the poles. Remarkably, the close up photographs of Mars do not show any indication of the Martian "canals" drawn by such astronomers as Percival LOWELL. These observers apparently joined up quite disconnected scattered dusky markings and interpreted them as hard, dark lines.

The polar caps of Mars consist of a mixture of frozen carbon dioxide and frozen water. In fact, the entire planet is believed to be locked in a permafrost shell just below the visible crust. Collapse or melting of this shell in places is believed to be the cause of several of the depressions and valleys on Mars. The caps re-form each winter on top of layered sediments as much as 4 miles (6 km) thick. Surrounding them is a pitted landscape where evaporating ice and wind-erosion have scoured depressions in which ice lingers as the cap melts. Beyond these are extensive dune fields, which become scarcer toward the equator. There seems to be a slow transport of dust from the windblown equatorial regions to build up the dune-fields and layered polar deposits.

The present geological structure of Mars, with ancient highlands in one hemisphere and youthful lowlands in the other, bears a certain resemblance to the Earth before continental drift began to break up the original giant continent Pangea. Internally, Mars does not seem to possess a liquid core; instead, it consists simply of mantle and crustal regions. Its internal composition may be uniform, with iron bound up with the bulk of Martian oxygen as oxides instead of being separated into a central iron core.

The Viking lander probes in 1976 failed to find any sign of organic molecules in the soil of Mars, and their photographs showed no living things on the surface. At these two landing sites, at least, there is apparently no life on Mars.

Moons. Mars has two tiny moons, Phobos and Deimos, which revolve in circular orbits in the planet's equatorial plane. These irregularly shaped, impact-scarred blocks of rock, may be captured asteroids (see DEIMOS; PHOBOS).

Mars probes

A series of Russian space probes to investigate the planet Mars (for American exploration of Mars see MARINER SPACECRAFT and VIKING). The first Soviet probes to reach their target successfully were Mars 2 and Mars 3 in 1971, both of which ejected lander capsules and then entered orbit around the planet. The Mars 2 lander, which crashed, was the first man-made object to reach the surface of Mars. The Mars 3 lander descended successfully; its transmissions failed 20 seconds after it had begun to send a television picture, which showed nothing. In 1974, Mars 4's braking engine failed and it swept past the planet; Mars 5 successfully entered orbit around Mars and sent back useful photographs. These orbiter craft were meant to act as communications relays for lander probes ejected from Mars 6 and 7. But contact was lost with the Mars 6 lander as it neared the surface, and the Mars 7 lander missed the planet entirely.

Probe	Launch date	Remarks
Mars 1	November 1, 1962	Radio contact lost after 66 million miles (106 million km) on March 21, 1963
Mars 2	May 19, 1971	Entered Mars orbit November 27; surveyed surface and atmosphere of planet. Lander capsule ejected but crashed
Mars 3	May 28, 1971	Entered Mars orbit December 2 and surveyed planet. Ejected lander, but transmissions ceased after 20 seconds
Mars 4	July 21, 1973	Passed Mars at a distance of 1,367 miles (2,200 km) on February 10, 1974, due to braking rocket failure

Probe	Launch date	Remarks
Mars 5	July 25, 1973	Entered Mars orbit February 12, 1974
Mars 6	August 5, 1973	Flew past Mars March 12, 1974, and ejected lander capsule which crashed
Mars 7	August 9, 1973	Flew past Mars March 9, 1974, and ejected lander capsule which missed planet

Marshall Space Flight Center (MSFC)

A NASA facility at Huntsville, Alabama, for the development of launch vehicles and spacecraft, named for General George C. Marshall. The center, established in 1960, grew out of missile work at the U.S. Army's Redstone Arsenal at Huntsville, which produced the Redstone, Jupiter, and Juno rockets. Under the directorship of Wernher VON BRAUN, the Marshall Space Flight Center developed the SATURN family of launch vehicles, in addition to the Skylab space station. MSFC provided the lunar roving vehicle used on the Apollo Moon missions, and it manages the High Energy Astronomical Observatory (HEAO) satellite series (see HEAO SATELLITES). MSFC is developing the engines to be used for the SPACE SHUTTLE, and is also working on the SPACE TELESCOPE and Space Tug assemblies to be flown in the Shuttle. MSFC has test stands for experimental firings of rockets under development. Rocket stages are manufactured at the Michoud Assembly Facility in New Orleans, and checked at the National Space Technology Laboratories in Bay St. Louis, which is now a separate NASA facility but until 1974 was part of MSFC, under the name of the Mississippi Test Facility.

mascon

An area where the Moon's gravitational field is increased; the name is a contraction of *mass concentration. Mascons were first noted in 1968 from tracking the paths of space probes orbiting the Moon; minor irregularities in the space probes' orbits revealed the areas of increased gravity. A major mascon is associated with the Mare Imbrium lowland plain on the Moon, and further mascons are associated with a number of other lunar seas. Mascons are probably aggregations of denser rock under the Moon's surface where molten lava has collected in one area and solidified.

maser

A source of very intense microwave radiation (a few centimeters in wavelength); the name is an acronym for *M*icrowave *A*mplification by *S*timulated *E*mission of *R*adiation. The first maser device, invented in 1953 by the American physicist Charles Hard Townes (b. 1915), contained excited molecules of ammonia gas; these were stimulated to emit a very intense beam of microwaves when triggered by radiation of the same wavelength. Such maser amplifiers can be used in radio telescopes as very sensitive receivers.

In 1963, radio astronomers discovered an extremely powerful source of 6-centimeter wavelength radiation associated with clouds of dust and gas near our galactic center. It was identified as the hydroxyl radical (OH), but at an improbably high temperature. With the discovery of similar sources, including water at 1.35 centimeters, and weak emission from more complex molecules like methyl alcohol, astronomers concluded that they were observing cosmic masers. These small objects (comparable in size with the solar system) occur in larger dust clouds which radiate at infrared wavelengths. They are thought to be PROTOSTARS condensing from the interstellar medium. At the other end of the stellar lifespan, red giant stars show maser action from a number of molecules, including hydroxyl, water, and silicon monoxide (SiO). This is believed to arise in a shell of dust and gas ejected from the star as it dies.

Maskelyne, Nevil (1732–1811)

Fifth British astronomer royal, who in 1766 began publication of the *Nautical Almanac,* a set of tables containing astronomical information for navigators. Maskelyne was sent in 1761 to the island of St. Helena to observe the transit of Venus across the Sun. Cloud prevented observations, but during the voyage Maskelyne tested a means of finding longitude at sea called the *method of lunar distances.* This involved measuring the Moon's position against the background of stars (the lunar distances were the stars' distances from the Moon). The Moon's position in its orbit acted as a standard clock in the sky, to give a universal time scale (like modern Greenwich time) that could be read off from tables. Comparing this standard time with the local time revealed the observer's longitude. But the success of the method depended on the accuracy of lunar tables. Maskelyne's trials showed the suitability of tables prepared by the German astronomer Johann Tobias Mayer (1723–1762), and in 1763 Maskelyne wrote *The British Mariner's Guide* to describe the technique. Maskelyne also helped test a rival "standard clock"—the chronometers of the English instrument maker John Harrison (1693–1776). Despite their greater accuracy the cost of chronometers meant that the method of lunar distances was long preferred, and the *Nautical Almanac* contained tables of lunar positions.

Eleven years later, Maskelyne began publication of the regular series of Greenwich Observations—previous astronomers royal had kept their observations to themselves. His catalog of 1790 showed the PROPER MOTIONS of 36 stars, from which William HERSCHEL determined the Sun's motion in space. Maskelyne successfully observed the 1769 transit of Venus, and from it calculated the Sun's distance to an accuracy of within one percent. In 1774 he performed a celebrated experiment to measure the density of the Earth. By measuring the deflection of a plumbline hung near the Schiehallion, a mountain in Perthshire, Scotland, he found a value for the Earth's density of 4.5 times that of water—lower than the true value of 5.5, but the best result of the time.

mass-luminosity relation

The relationship between a star's mass and its brightness, discovered by Sir Arthur EDDINGTON in 1924. The masses of stars are known reliably only from observations of the orbits of double-star systems. These masses bear a direct relation to the stars' intrinsic brightness or absolute magnitude. The relationship indicates that the biggest stars are also

the hottest; a doubling of mass corresponds to an increase in brightness of 2.5 magnitudes. This mass-luminosity relation applies to most ordinary stars. The highly condensed white dwarfs, however, do not conform because of their exceptional nature; nor do the faint companions in certain types of double-star systems (known as the W Ursae Majoris stars). These latter may be the central cores of red giants, stripped of their outer layers by the gravitational attraction of their companion stars.

Mauna Kea Observatory

The world's highest astronomical observatory, operated by the University of Hawaii at an elevation of 13,800 feet (4,205 m) on Mauna Kea, a dormant volcano on the island of Hawaii. Opened in 1970, it contains an 88-inch (224-cm) reflector; there are also two 24-inch (61-cm) reflectors. Canada, France, and Hawaii have jointly constructed a 142-inch (360-cm) reflector on the mountain, scheduled for opening in 1979. Mauna Kea Observatory is an ideal spot for infrared observations; its altitude places it above most of the water vapour in the atmosphere that absorbs infrared radiation. NASA has constructed a 125-inch (320-cm) infrared telescope on Mauna Kea, opened in 1979. A UK 152-inch (386-cm) infrared telescope began operation on the same site in 1978; this is the world's largest infrared telescope.

McDonald Observatory

An astronomical observatory jointly operated by the University of Chicago and the University of Texas at an altitude of 6,828 ft (2,081 m) on Mount Locke, 17 miles (27 km) from Fort Davis, Texas. McDonald Observatory is named for financier William Johnson McDonald (1844–1926) whose will donated funds for an observatory. It was opened in 1939 with an 82-inch (208-cm) reflector. In 1968 a 107-inch (270-cm) reflector was installed. The observatory also has a 36-inch .91-cm) reflector built in 1957.

mean Sun

A fictitious body representing the annual motion the true Sun would have around the CELESTIAL SPHERE if the Earth's orbit were circular instead of elliptical, and if its axis were not tilted but upright. The mean Sun therefore progresses along the celestial equator at a uniform rate; the true Sun appears to move along the ecliptic at a varying rate (see EQUATION OF TIME).

Mensa (the table mountain)

A faint constellation near the south pole of the sky, containing part of the Large Magellanic Cloud. It was introduced by Nicolas Louis de Lacaille.

Mercury

The closest planet to the Sun, Mercury is an arid, airless world with a meteorite-battered surface baked to above the melting point of lead at midday, and cooled to bitter temperatures at night. It can never have supported life and more closely resembles the Moon than any of the other planets.

Physical nature. Mercury is 3,032 miles (4,880 km) in diameter, intermediate in size between the Moon and Mars. Mercury is remarkably dense, with a mass of 3.30×10^{20} tons and a density of 5.45 times that of water, very nearly the density of the Earth. It rotates

in 58.65 days about an axis perpendicular to the
ecliptic. Mercury's orbit is unusually elliptical, taking
it between 28.6 and 43.4 million miles (46 and 69.8
million km) from the Sun. The orbit is inclined at 7° to
the ecliptic, and Mercury completes one circuit in
87.97 days. The position of Mercury's perihelion, the
axis of the orbit, advances slowly around the Sun
some 43 seconds of arc more per century than can be
accounted for by planetary perturbations alone. This
additional movement has been successfully accounted
for by the theory of relativity, as a consequence of the
CURVATURE OF SPACE near the Sun.

Visibility. Mercury shows a series of phases as it
orbits the Sun. On the far side of the Sun, the tiny
disk is seen nearly fully illuminated; as Mercury moves
around into the evening sky it shows a decreasing
gibbous phase until it reaches half phase at greatest
elongation. As the planet starts to move between us
and the Sun, its apparent size increases as it approaches
us and its phase becomes an increasingly slender
crescent. After passing between the Earth and Sun, the
reverse sequence of phases follows as the planet
emerges into the morning sky.

Occasionally, Mercury can be seen in TRANSIT
across the solar disk, appearing as a slow-moving dark
spot. This happens relatively rarely, because of the
relative inclinations of the orbits of Mercury and the
Earth. In transit Mercury is at its closest to the Earth,
as little as 48 million miles (77 million km) distant.
The last transit of Mercury took place on November
11, 1973; the next will occur on November 12, 1986,
and November 14, 1999.

Mercury never strays far from the Sun in the sky.
Its greatest apparent angle from the Sun varies from
28°, when the planet is at aphelion, to only 18° at
perihelion. This makes the planet very difficult to
observe, despite its maximum brightness of magnitude
−1. Mercury can only be viewed against a bright,
daytime sky, or low down after sunset or before
sunrise.

Surface markings and conditions. Experienced
observers detected faint dusky shadings on Mercury's
disk, and careful study of these suggested that Mercury
rotated on its axis every 88 days, with the same face
permanently pointing toward the Sun—one
hemisphere in perpetual sunlight, the other in
perpetual night. This belief persisted for more than
80 years, until radar studies from 1965 onward
showed that the correct rotation period was 58.65 days.
This is exactly two-thirds of the 87.97-day orbital
period and leads to a Mercurian day (from sunrise to
sunrise) lasting 176 Earthdays, or Mercury's orbital
periods. Another unusual effect is that at perihelion,
the planet's rotation is slower than its motion around
the Sun, so that the Sun ceases its slow westward
movement, slipping back eastward for some hours
before resuming its normal motion.

When Mercury is at perihelion, the Sun appears
50 percent larger than at aphelion, and shines with
more than twice the intensity. The equatorial surface
temperature reaches 415°C at perihelion, enough to
melt lead and tin; even at aphelion it is 285°C. In
sharp contrast, the dark-side temperature is as low as
−175°C. This great temperature range between day
and night is partly the result of an extremely thin
atmosphere, a million millionth the density of the
Earth's at sea level, or only ten times denser than the

This mosaic of Mariner 10 photographs of Mercury, taken
from 124,000 miles, shows a heavily cratered terrain very
similar to the Moon's highlands. The largest craters
are about 120 miles in diameter. The brightest crater on
Mercury, **center**, has been named after the late Gerard
P. Kuiper, an eminent planetary scientist.

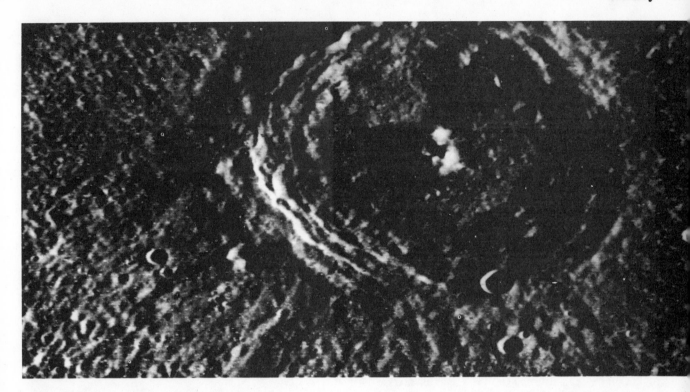

Moon's. The surface composition is very similar to that of the Moon, with a 7 percent reflectivity (ALBEDO), identical polarization, and a radar reflectivity like the Moon's. The faint dusky shadings and bright areas are the equivalents of the lunar maria and highlands.

Surface features. Previously sketchy information about Mercury was dramatically augmented by the three close passages of Mariner 10 to the planet in 1974 and 1975. Over 2,800 photographs were taken, including some with a resolution of only a hundred yards.

The photographs reveal a heavily cratered surface more closely resembling the Moon than Mars. The lack of erosion suggests that Mercury has been devoid of an appreciable atmosphere for practically its whole existence. Craters abound everywhere on the surface, from giant basin structures down to small craters only a few hundred yards across at the limit of the photographs. A study of about 25 percent of the surface has revealed no less than 17 impact basins more than 125 miles (200 km) across, and some larger than the mighty Mare Imbrium, the 420-mile (675-km) basin on the Moon. The largest, some 800 miles (1,300 km) across, has been named the Caloris Basin and contains a heavily fractured, lava-flooded floor.

Like the Moon, Mercury has two distinct hemispheres. One is covered with heavily cratered highlands, while the other contains smooth, lightly cratered plains similar to the lunar maria. The craters on Mercury and the Moon are similar, varying from sharply defined recent craters surrounded by bright halos and ray systems, to very ancient eroded craters distorted by later impacts. However, Mercurian craters tend to be shallower than their lunar counterparts. Central peaks caused by upward movement of underlying layers, or terracing of the crater walls caused by slumping, are present in most craters larger than about 9 miles (14 km) across,

Giant impact crater on Mercury, 85 miles (140 km) in diameter, photographed by the Mariner 10 space probe. Note the complex central peak, and terracing of the walls. Small secondary craters and ridges surround the formation, caused by material splashed out from the explosion which produced the original crater.

while they are not common in lunar craters less than 35 miles (56 km) across. The ejecta blanket and secondary craters surrounding large craters extend outward only half the distance normal on the Moon. All these effects are undoubtedly caused by the higher gravity on Mercury, more than twice that on the Moon.

The mare areas are very similar to those on the Moon, with similar amounts of cratering. A unique feature on Mercury, however, are the irregular scarps up to 2 miles (3.2 km) high, which can extend for several hundred miles. These are quite unlike anything on the Moon and cut through whatever lies in their path—basins, craters, or maria. Another highly unusual feature, a weird terrain of finely dissected hills, large rough valleys, heavily degraded craters, and great landslides, is found opposite the great Caloris Basin. Similar terrain on a much smaller scale has also been found opposite the Imbrium and Orientale basins on the Moon. Evidently, the basin impacts produced intense shock waves in the body of the planet which met at the antipodal point, causing great upheavals.

The detailed nature of the surface is probably identical with what we know about the Moon—a surface soil layer (*regolith*) perhaps 50 to 100 feet (15–30 m) deep comprising rock debris of all sizes embedded in a glassy soil, with an underlying rock mass shattered by impacts. The high overall density of Mercury suggests a central iron-nickel core extending some 80 percent of the distance from the center toward the surface, with an overlying crust of silicate

rocks. Mariner 10 detected a magnetic field around
Mercury about one percent the intensity of the
Earth's field. It is probably caused by the permanent
magnetization of the surface rocks of the planet.

Evolution. The sequence of events in Mercurian
history parallels the evolution of the Moon. After the
planet formed, the surface was melted by the heat of
intense meteoric bombardment. As this died away,
the highland regions solidified. Further impacts
scattered debris over the entire surface, leaving the
craters and basins. Subsurface radioactive melting then
filled some of the great basins with lavas to produce
the mare areas. The surface was then further modified
by smaller meteorite impacts. Internally, Mercury still
possesses a partially molten core. The gradual cooling
of the liquid core must have led to considerable
contraction, and this may be the cause of the great
scarps—the wrinkles that lace the planet's surface.

Mercury has no moon. Mariner 10 failed to
detect any satellites as small as 3 miles (5 km) across.

Mercury project

The American program to launch a man into space.
Its origins go back to experiments during the 1950s
with high-altitude rocket planes, the X-1 and X-2
series that took men toward the edge of the Earth's
atmosphere. From these developed the famous X-15
rocket plane, which during the 1960s established
records for altitude—67 miles (108 km)—and speed—
4,520 miles (7,274 km) per hour. Although proposals
were made for boosting a winged vehicle like X-15 into
orbit—similar to the concept of the SPACE SHUTTLE—
it was decided in 1958 to extend the flight program to
higher altitudes and greater speeds by launching an
astronaut in a wingless capsule carried aloft by a
rocket. By contrast, the Soviet manned space program
was an ad hoc adaptation of large unmanned satellites,
already well-advanced in design.

The Mercury project set itself the target of putting
a man in orbit, testing his ability to function in
spaceflight, and returning him safely to Earth. At the
time the scheme was proposed, no one knew how well
humans could withstand the rigors of launch,
followed by weightlessness and the deceleration of
reentry. Therefore the Mercury project was preceded
by test firings carrying primates, whose successful
flights on so-called suborbital paths, (dropping just
short of orbit) and then into orbit itself, paved the
way for the men who were to follow.

Mercury hardware. The Mercury capsule
introduced the conical shape followed by Gemini,
Apollo, and even the Soviet Soyuz spacecraft (the
first Russian design, Vostok, was a sphere). Mercury
was designed to hold one man. Its total length was
9 feet 7 inches (2.9 m) and its maximum diameter
6 feet 2 inches (1.9 m); it weighed nearly 3,000 lb.
(1,360 kg). At its blunt end was the heat shield, to
which was strapped a pack of retro-rockets which
were jettisoned before reentry. Once the craft had
entered the atmosphere and the parachutes began to
deploy from the nose, the heat shield dropped down to
extend an air bag; this cushioned the impact of
splashdown, and then slowly filled with water to
stabilize the Mercury capsule in the sea.

The astronaut sat in a contoured couch. Above him
was an observation window. The astronaut wore a
simple space suit, developed from high-altitude

The launch of a Mercury capsule by an Atlas booster.
The capsule is at the top, just below the escape tower
which could pull it clear of the main booster during the
first few minutes of the flight in case of danger. The
controlling vernier rockets can be clearly seen maintaining the rocket attitude.

aviation suits, to be inflated only in an emergency. He had a hand controller which fired gas jets to adjust the spacecraft's attitude in space, although orbital changes were not possible as they were with the later Gemini. An escape tower above the spacecraft could pull the capsule free in the event of a launch rocket malfunction; the escape tower was jettisoned once the capsule was safely on its intended course.

The first two manned Mercury flights were suborbital lobs into the Atlantic Ocean. These flights followed a long, looping path into space and back again. The astronauts were able to control the orientation of the spacecraft, but they became weightless for only about five minutes. The two suborbital flights were launched by REDSTONE rockets, modified from their original role as intermediate-range missiles. Four other manned Mercury capsules were sent into orbit atop ATLAS ROCKETS, America's first long-range ballistic missiles.

The missions. Alan B. Shepard became the first American astronaut on May 5, 1961, when he was launched on a 304-mile (489-km) flight lasting 15 minutes 22 seconds and reaching a maximum altitude of 116.5 miles (187.5 km). On July 21, 1961, Virgil I. Grissom made a similar flight, lasting 15 minutes 37 seconds, reaching a maximum altitude of 118 miles (190 km). On splashdown, the capsule's hatch was ejected by accident, and the capsule itself sank, although Grissom was safely recovered. The first American in orbit was John H. Glenn, who circled the Earth three times on February 20, 1962. The flight was repeated three months later by M. Scott Carpenter. In October 1962 Walter M. Schirra made six orbits of the Earth, and in May 1963 L. Gordon Cooper made the last and by far the longest Mercury flight, 22 orbits (34.3 hours) in length.

Mercury flights

Mission	Launch date	Results
Mercury-Redstone 3 (*Freedom 7*)	May 5, 1961	Alan Shepard made suborbital flight
Mercury-Redstone 4 (*Liberty Bell 7*)	July 21, 1961	Virgil Grissom made suborbital flight
Mercury-Atlas 6 (*Friendship 7*)	February 20, 1962	John Glenn made 3-orbit flight
Mercury-Atlas 7 (*Aurora 7*)	May 24, 1962	Scott Carpenter made 3-orbit flight
Mercury-Atlas 8 (*Sigma 7*)	October 3, 1962	Walter Schirra made 6-orbit flight
Mercury-Atlas 9 (*Faith 7*)	May 15, 1963	Gordon Cooper made 22-orbit flight

meridian

The great circle passing overhead and through the observer's north and south horizon. Celestial objects attain their greatest altitude (*culmination*) on the meridian. The moment at which a celestial object crosses the meridian depends on the observer's longitude. One degree of longitude amounts to 4 minutes' difference in time. The timing of objects crossing the meridian, or in TRANSIT, therefore provides a basis for time measurement or determination of longitude.

Messier, Charles (1730–1817)

French astronomer best known for compiling a famous list of nebulae and star clusters. Messier became assistant to the French astronomer Joseph-Nicolas Delisle (1688–1768) in 1751; his interest was turned to comet hunting in 1758 with the predicted return of Halley's comet, which Messier was one of the first to see. In all, Messier discovered 15 new comets and claimed another six that were actually found to be first seen by others. In 1758 Messier was deceived by the comet-like appearance in his telescope of the Crab nebula, and made a special note to avoid it. From that point onward he began cataloging fuzzy-looking objects in the sky that might be mistaken for comets. Messier produced his first list, containing 45 objects, in 1771; a second list in 1780 added another 23 objects; and his final list of 1781 (published in 1784) had 103 objects. Not all were Messier's own discoveries; several had been seen first by other observers, notably the Swiss comet-hunter Jean Philippe Loys de Chéseaux (1718–1751). Another six objects were added to Messier's list by the French astronomer Pierre François André Méchain (1744–1804). Astronomers still refer to objects by their Messier, or M, numbers: for example, M1 (the Crab nebula) and M31 (the Andromeda galaxy).

meteor

The streak of light produced when a solid particle from space, known as a METEOROID, enters the Earth's atmosphere at high speed and burns up to produce fine dust. Meteors as bright as naked-eye stars are known popularly as *shooting stars*, while brighter meteors are termed FIREBALLS or bolides. Meteors occur between heights of about 55 and 65 miles (88 and 105 km). The majority of them have velocities of about 20 or 35 miles (32 or 56 km) per second, depending on whether they overtake the Earth or the Earth overtakes them. Significantly, no meteors have velocities clearly in excess of the escape velocity for the solar system, which indicates that meteors are solar system members.

Many meteors leave behind them a glowing train, which may persist for several seconds or even minutes. These occur at 55 to 60 miles (88–97 km) and are caused by the recombination of ionized atoms and molecules. Very bright daytime meteors or fireballs that penetrate to low altitudes can leave dark persistent trains behind them. These occur at 20 to 25 miles (32–40 km) and are suspended dust particles.

Sporadics and shower meteors. Most naked-eye meteors are lone travelers, termed *sporadic* meteors. The number of sporadics visible per hour rises from 6 after sunset to a peak of about 14 just before dawn, when the observer is facing in the direction of the Earth's motion. On several occasions each year, large numbers of meteors are seen moving outward from a RADIANT point in a meteor shower. These meteors orbit the Sun in parallel paths at the same velocity, forming a meteor stream. There are about 10 annual meteor streams, plus many minor streams of low

Characteristics of the Major Meteor Streams and their Orbits

Stream	Period of Visibility	Date of display peak	Radiant R.A. h m	Point Dec. o	Average peak rate meteors/ hour	Velocity miles/sec.	Perihelion distance a. u.	Stream orbit Inclination o	Period years	Associated comet
Quadrantids	January 2–5	January 4	15 28	+50	110	24.1	0.98	72.9	6.3	none
Lyrids	April 20–24	April 22	18 08	+32	8	29.2	0.92	79.5	200	Thatcher 1861 I
eta Aquarids	May 2–7	May 5	22 24	00	18	41.3	0.70	158.0	80	Halley
delta Aquarids	July 22–Aug. 10	July 31	22 36	−08	30	25.2	0.085	22.2	4.2	none
Perseids	July 27–Aug. 16	August 12	03 04	+58	65	37.0	0.95	113.0	140	Swift-Tuttle 1862 III
Orionids	October 17–25	October 21	06 24	+15	25	40.3	0.58	164.3	80	Halley
Taurids	Oct. 25–Nov. 25	November 8	03 44	+18	10	18.3	0.31	4.5	3.1	Encke 1786 I
Leonids	November 16–19	November 17	10 08	+22	15	43.6	0.98	162.1	33.3	Tempel 1866 I
Geminids	December 7–15	December 14	07 28	+32	55	21.7	0.129	24.9	1.63	none

activity almost indistinguishable from the background of sporadic meteors.

Meteors and comets. A meteor stream is formed by particles scattered by a periodic comet; the stream and comet occupy the same orbit. The stream is initially a compact, dense swarm of meteoroids, but slowly spreads out, eventually disintegrating completely to add to the stock of sporadic meteors. Very young streams give intense displays of up to 1,000 meteors per minute lasting for only a few hours at the longest. Such encounters, although rare, occurred in the Leonid meteor storms of 1799, 1833, 1866, and 1966, the Andromedid storms of 1872 and 1885, and the Giacobinid storms of 1933 and 1946. Meteor streams are named for the constellation in which their radiant lies, or for their parent comet. Meteor showers are best observed when their radiant is high in the sky. The ZENITHAL HOURLY RATE of a shower is the predicted rate with the radiant overhead; but because the radiant seldom is directly overhead, the actual observed rate is somewhat lower.

Observing meteors. The basis of modern meteor astronomy was laid down a century ago by naked-eye observers who recorded meteor paths, velocity, and brightness. An outstanding pioneer was the English amateur William Frederick Denning (1848–1931). Far more accurate photographic methods were later applied, culminating in the super-Schmidt cameras designed specifically for meteor work. With focal ratios of $f/0.65$ and a field of 56°, these can photograph meteors as faint as magnitude +4 at the rate of two every hour. Radio waves are reflected from the ionization path left by a meteor, and meteors as faint as magnitude +10 can be recorded by radar. This method has detected several meteor streams whose radiants are in the daylight sky and can never be observed visually.

By using all these techniques, meteors from magnitude −20 down to +10 can be observed, corresponding to bodies with masses from 10 tons down to 0.0001 gram. Most meteors brighter than magnitude −12 are sporadics, and a few of these are sufficiently massive and strong to withstand entry into the Earth's atmosphere to drop METEORITES. Meteor streams account for many fainter meteors—about 40 percent of meteors recorded photographically (mean magnitude −2) and 35 percent of naked-eye visual meteors (mean magnitude +2). However, shower meteors make up only a few percent of the faintest objects, as detected by radar and telescope (mean magnitude +8). Faint meteors are, of course, more numerous than bright ones. For ordinary naked-eye meteors, the numbers increase by 3.5 times for each magnitude step downward, while very faint meteors increase by a factor of 2.5.

Over a year, the Earth encounters about 10 trillion meteors brighter than magnitude +10, gaining some 2,000 tons of material. This is insignificant, however, when compared to the 200,000 tons added each year by micrometeorites and meteors fainter than magnitude +10.

The spectra of meteors reveal that their composition is generally similar to either stony or iron meteorites. Although for many years meteors were thought to be fragile dust-balls, many scientists now believe that most are composed of a very fragile form of gas-rich carbonaceous CHONDRITE, a type of stony meteorite. Ordinary stream meteors are clearly derived from the outer layers of young periodic comet nuclei, while fireballs may be scattered boulders from the disrupted nuclei of ancient periodic comets or debris from asteroid collisions.

meteorite

A lump of rock or metal from space that crashes to Earth. The study of such material is known as *meteoritics.* Accounts of pieces of stone or iron falling from the sky occur in the records of many early civilizations, but the earliest witnessed fall from which material still survives is the stone that fell near Ensisheim, Alsace, on November 16, 1492. The true nature of meteorites was first recognized by the German physicist Ernst Florens Friedrich Chladni (1756–1827) in 1794, but his views were only accepted after the fall of meteorites at L'Aigle, France, on April 26, 1803, documented by the French physicist Jean Baptiste Biot (1774–1862).

Frequency of falls and finds. A meteorite actually seen to hit the ground is known as a *fall;* a meteorite which is come across accidentally, having fallen at some undetermined time in the past, is known as a *find*. About 2,000 falls and finds have now been recovered, with roughly six falls and 10 finds being added each year. An estimated 500 meteorites fall each year over the whole surface of the Earth. Of these, 300 drop into the oceans and seas, while the bulk of the remainder fall either in uninhabited areas, or in places where they are not noticed. An increasing number have damaged buildings and other man-made structures, but there have been very few cases of injuries recorded.

Meteorites are broadly divided into three types: stones or aerolites, which are subdivided into CHONDRITES and achondrites; irons or SIDERITES, subdivided into hexahedrites, octahedrites, and ataxites; and stony-irons or SIDEROLITES, subdivided into pallasites and mesosiderites. The relative proportions of meteorites is shown below:

Distribution of Meteorite Types

Meteorite type	Finds %	Falls %
chondrites	38.6	84.8
achondrites	0.9	7.9
total aerolites (stones)	39.7	92.7
siderolites (stony-irons)	5.8	1.7
siderites (irons)	54.5	5.6

Although aerolites are the most common meteorites that fall to Earth, they weather rapidly and are difficult to trace long after a fall. By contrast, the rarer iron meteorites resist weathering, and can thus be found long after they have fallen.

Size of meteorites. Small meteorites tend to be stones, but the proportion of irons increases with size, presumably because large irons more easily withstand collisions in space and entry into the Earth's atmosphere. The largest stone meteorite weighs just over 1 ton and was part of the Norton County, Kansas, achondrite fall of February 18, 1948. There are many irons much larger; the biggest is the ataxite Hoba, found in 1920 near Grootfontein in South West Africa (Namibia). This is the largest meteorite in the world, weighing over 60 tons, and has never been moved.

Fall phenomena. Meteorite-producing bodies are heated by friction as they enter the Earth's atmosphere. The meteor produced is extremely bright, and fireballs with a peak brilliance of more than magnitude −22 have been recorded. During entry, the surface of the meteorite is heated to several thousand degrees; the stone or iron is vaporized and melted away more quickly than the heat penetrates into the meteorite, and the interior is unaffected. The meteorite slows to free fall at a height of about 10 to 15 miles (16–24 km) when the fireball fades; the meteorite then drops at about 150 to 200 miles (240–320 km) per hour, taking about 1½ to 2 minutes to reach the ground. The meteorite arrives at ground level quite cold to the touch, and coated with a dull black fusion crust where the once-molten outer layer has cooled. Most meteorites break up into a number of large pieces during the fireball phase of descent. These fragments fall to the ground in an elliptical area, the heaviest pieces carrying the farthest; this is known as a

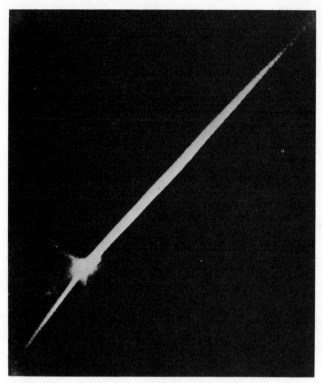

A meteor exploding as it ploughs through the Earth's atmosphere at a speed of about 30 miles a second (50 km/sec). These stony or metallic particles of interplanetary debris are usually only millimeters in size; this extremely bright example may have been caused by a chunk several centimeters across, weighing a few grams.

dispersion ellipse. Though most meteorites produce only a few fragments, in rare cases hundreds of individual meteorites are produced. The Allende meteorite fall of February 8, 1969, in Chihuahua, Mexico, scattered an estimated 5 tons of chondrite meteorite specimens up to 230 lb. (105 kg) in weight over an area some 30 by 4 miles (48 × 6.4 km).

Despite recent intensive efforts using networks of cameras to follow the passage of the bright fireballs that lead to meteorite falls, only two have so far been recorded and their orbits calculated. These are the chondrite falls of April 7, 1959, near Pribram, Czechoslovakia, and of January 3, 1970, near Lost City, Oklahoma.

The origins of meteorites. Meteorite specimens recovered on Earth and preserved in museum collections are not representative of interplanetary debris. These meteorites represent only the 5 percent of debris which is strong enough to survive entry into the Earth's atmosphere. The other 95 percent disintegrate to produce flaring fireballs. These are thought to be composed of two different types of fragile material possibly similar to the carbonaceous chondrites. This material may come from comets, while most aerolites, siderolites, and siderites are believed to be fragments of asteroids. Meteorites contain a number of minerals not found on Earth, but these are always composed of known elements. Organic molecules have been found in some meteorites, presumably produced by chemical processes in space (see also LIFE IN THE UNIVERSE).

meteoroid

A solid object moving in interplanetary space, larger than a single molecule but smaller than an asteroid. Meteoroids smaller than about 5 microns (5 millionths of a meter) are decelerated in the Earth's atmosphere to fall as MICROMETEORITES. Meteoroids between 5 microns and about 6 inches (15 cm) in diameter are destroyed by ablation (frictional vaporization and melting) in the Earth's atmosphere, producing METEORS. Meteoroids larger than 6 inches, if they possess sufficient structural strength, can fall to earth as METEORITES.

meteorological satellites

See WEATHER SATELLITES

Metonic cycle

The period after which the Moon's phases recur on the same day of the year. The cycle was discovered in 432 B.C. by the Greek astronomer Meton (c.460 B.C.—?). The period lasts just over 6,939 days (exactly 19 years), during which there are 235 LUNATIONS. The Metonic cycle was used as a basis for lunar calendars.

Michelson, Albert Abraham (1852–1931)

Polish-born American physicist, who made a series of remarkably accurate measurements of the speed of light, and was the first to determine the diameter of a star other than the Sun. Michelson made his first measurement of the speed of light in 1878, and continued his experiments for the rest of his life. He obtained his best value, 186,285 miles (299,798 km) per second in 1926, over a 22-mile (35-km) path between Mount Wilson and Mount San Antonio in California. In 1881 Michelson began a series of tests culminating in the famous MICHELSON-MORLEY EXPERIMENT of 1887, which proved that light travels at a constant speed in space and refuted the ether hypothesis. In 1893 Michelson measured the length of the standard meter in terms of the wavelength of red light from heated cadmium, thus providing a precise standard of length that could be reproduced exactly. At Mount Wilson Observatory, Michelson set up an INTERFEROMETER to measure the diameters of several nearby giant stars in 1920. He used mirrors to feed two beams of light from a star onto the main mirror of the 100-inch (254-cm) telescope, which focused the light; the interference fringes produced by the two beams revealed the diameters of the star, the first time such a direct measurement had been made. In 1907 he was awarded the Nobel Prize in physics for his painstaking optical measurements. He was the first American scientist to receive the honor.

Michelson-Morley experiment

An attempt to measure the Earth's motion through the ETHER, carried out in 1887 by the American physicists Albert MICHELSON and Edward Williams Morley (1838–1923). According to the theory prevailing at that time, light moved at constant speed in an invisible substance called the ether. The Earth's motion through the ether would produce a kind of wind, so that the speed of light measured perpendicular to the Earth's direction of motion would be different from that measured in line with the Earth's direction of motion. Michelson and Morley compared the travel time of two light beams sent along optical paths at right angles to each other. Because of the predicted "ether wind" it was expected that the beam directed along the Earth's line of motion would return later than the perpendicular beam. In fact, no detectable difference was found, a result of stunning unexpectedness, which virtually invalidated the ether theory. The experiment has been repeated many times with increasing precision, but no difference in travel time has ever been detected. The result is consistent with the theory of RELATIVITY, which abandons the idea of any "fixed" medium in space and instead holds that the speed of light is constant relative to any observer.

micrometeorite

A particle from space which is small enough to be decelerated in the Earth's atmosphere without being vaporized. Smaller than 5 microns (5 millionths of a meter) across, micrometeorites fall gently to Earth at a daily rate of about 50 particles per square yard. Micrometeorites have been detected in ocean sediments, polar ice caps, and on clean exposed surfaces at ocean sites. They have also been collected by rockets and satellites above the Earth's atmosphere. Micrometeorites are probably derived from the periodic comets which release about 100,000 tons of fine dust a year into the solar system.

Microscopium (the microscope)

A small and faint constellation lying next to Sagittarius in the southern hemisphere of the sky, introduced by Nicolas Louis de Lacaille.

microwave background

See BACKGROUND RADIATION.

Milky Way

A spiral galaxy of some 100 billion stars, of which the Sun is one. The Milky Way is a thin disk 100,000 light-years across and 2,000 light-years thick; the disk widens near its center into a flattened nuclear bulge. The Sun lies in the disk, about 30,000 light-years from the center. Surrounding the disk is a sparsely populated, spheroidal halo, up to 500,000 light-years in diameter. The three regions (halo, nucleus, and disk) contain very different types of objects.

Halo. The halo represents the original extent of the gas cloud from which the Milky Way condensed 10 billion years ago. Rotation made it collapse into a disk, which now forms the most densely-populated part of the Galaxy; but some stars condensed out of the gas before the collapse occurred. Halo stars are therefore the oldest in the Galaxy. Many of them are found in the GLOBULAR CLUSTERS that are distributed symmetrically about the Galaxy, with a marked concentration toward the galactic center in Sagittarius. This distribution is shared by the RR LYRAE VARIABLE stars, found both in and out of globular clusters. Their short periods of variability (less than one day) and similar intrinsic brightness (ABSOLUTE MAGNITUDE +0.5) make them valuable probes for determining the structure of the halo. Non-cluster stars make up the greatest part of the halo's mass. Some areas of neutral hydrogen gas (H I REGIONS) still remain in the halo, distributed in

clouds which appear to be falling onto the disk at high velocities (up to 110 miles, or 177 km, per second); the reason for this is not known.

Nucleus. Apart from its outer margins, the nucleus can only be observed at radio or infrared wavelengths, which penetrate the dust clouds that obscure it. The nucleus measures 20,000 by 10,000 light-years, and contains most of the mass of the Galaxy, largely in the form of old stars (although younger than the halo population). Toward the galactic center these stars crowd together many thousands of times more densely than those near the Sun.

A thin, rotating disk of hydrogen extends 2,500 light-years from the center, containing a mass equivalent to 5 million Suns in gas, and probably 1,000 times more in stars. Many millions more stars lie within this zone, which also contains giant clouds of ionized hydrogen gas (H II REGIONS) nearer the center. The exact center of the Galaxy is marked by clouds of hot hydrogen and a compact radio source with a diameter of about 200 astronomical units. This is possibly a supermassive object, 10,000 times heavier and 300,000 times brighter than the Sun; or it may be a BLACK HOLE.

Disk. The youngest stars in the Galaxy are concentrated in a thin plane surrounding the nucleus. The most recently-formed objects occupy the thinnest zone, while the older disk stars gradually merge with those of the halo, marking the progress of the Galaxy's collapse from a spherical gas cloud. About 10 per cent of the mass of the Galaxy is still in the form of cool hydrogen gas lying in the disk and not yet condensed into stars. Neutral hydrogen emits strong radio signals at a wavelength of 21 centimeters, enabling radio astronomers to map both its location and velocity. The gas forms a layer of uniform thickness close to the galactic plane in the inner regions, but becomes thicker and distorted at the rim, possibly due to tidal effects from the MAGELLANIC

Our Milky Way Galaxy, seen on a chart prepared by plotting the 7,000 brightest stars in the sky, and mapping in the distribution of nebulae from photographs. Gas and star clouds concentrate toward its nucleus, **center**, crossed by dark obscuring dust lanes, **left center**. The Magellanic Clouds, our companion galaxies, look like detached portions, **below right**.

CLOUDS. Within the uniform layer, the hydrogen concentrates into subclouds along two spiral arms winding outward from the nucleus. Two portions of these arms can be seen: the inner Sagittarius arm; and the Perseus arm, 6,000 light-years beyond the Sun. The Orion and Cygnus region in which our Sun lies may be a spur on the inside of the Perseus arm. We see the stars of these arms superimposed on each other in the sky, forming the diffuse band of the Milky Way.

The Galaxy's spiral arms are rich in the gas from which stars form, and contain many hot, bright young stars. The clumpiness of the gas produces STELLAR ASSOCIATIONS, some of which are only a few million years old. These stars heat the hydrogen to temperatures above 10,000°K, forming regions like the ORION NEBULA. Interstellar clouds also contain quantities of dust particles, only a few millionths of a centimeter across. These consist of silica and graphite coated with ice, and are believed to provide surfaces for the formation of INTERSTELLAR MOLECULES. Dust grains are distributed throughout the disk, as well as in dark clouds like the COALSACK, absorbing and scattering starlight. As a result, optical telescopes can only penetrate about 6,000 light-years in the galactic plane, so that much of our information about the spiral arms comes from radio studies. A general background of galactic radio emission is also observed, arising from the acceleration of COSMIC RAY electrons in a large-scale magnetic field (SYNCHROTRON RADIATION) which follows the spiral arms.

The entire Galaxy rotates, with the inner parts moving faster than those farther out. This differential rotation indicates that the Galaxy's mass is mainly concentrated toward its center. The Sun, traveling at 170 miles (274 km) per second, takes between 200 and 250 million years to make one circuit.

Mills Cross
A type of radio telescope invented in 1953 by the Australian astronomer Bernard Yarnton Mills (b. 1920). The Mills Cross design uses two rows of aerials arranged at right angles to give a large cross-shaped array. The pattern of vision of each line of aerials is a narrow fan, but automatically switching between the two crossed aerial systems produces a narrow beam where the two fans overlap. The Mills Cross was an important forerunner of the APERTURE SYNTHESIS telescopes used to map fine details of radio sources.

Mimas
Saturn's second satellite in order of distance from the planet, with a diameter of about 300 miles (500 km). Mimas orbits Saturn every 22 hours 37 minutes at a distance of 115,320 miles (185,590 km). It was discovered in 1789 by William Herschel.

minor axis
The shortest diameter of an ELLIPSE, perpendicular to the MAJOR AXIS.

minor planet
See ASTEROID.

Mira
The first variable star to be discovered; its name means "wonderful." Mira is a red giant star in Cetus, 820 light-years away, and is the prototype of long-period variables. Mira was first noted by David Fabricius in 1596, but soon faded. It was bright again in 1603 when Johann BAYER labeled it Omicron Ceti on his star map. Mira has a period of roughly 332 days, expanding and contracting between 2.8 and 3.8 astronomical units in diameter. At its brightest, Mira is of the second magnitude, but it can fade to tenth magnitude. It has a blue dwarf companion orbiting it

every 14 years at a distance of some 40 a.u. During its expansion, Mira puffs off a cloud of gas which falls onto this companion, changing its brightness.

Miranda
The nearest satellite of Uranus, discovered in 1948 by Gerard Kuiper. It orbits Uranus every 33 hours $55\frac{1}{2}$ minutes, at a distance of 81,060 miles (130,450 km). Miranda is about 350 miles (550 km) in diameter.

mirror
A reflecting component in an astronomical instrument. Unlike an ordinary mirror, those used for precise optical purposes have their reflective coating on the front surface rather than the back. This avoids the optical imperfections that would be caused if the light actually passed through the glass. The reflective coating is therefore relatively unprotected. The modern technique is to evaporate a very thin film of aluminum onto the mirror, and to coat it with a very thin layer of transparent silica; this can produce a surface lasting several years before it becomes tarnished and corroded. Mirrors have the advantage over lenses that they do not produce false color, or CHROMATIC ABERRATION. With the advent of photography this became particularly important, because most photographic emulsions were primarily sensitive to one special color, and a lens system could not be corrected to perform properly for all colors of the spectrum.

Furthermore, aluminum reflects very well in the ultraviolet part of the spectrum—an important region photographically—whereas normal glass transmits poorly at these wavelengths. Physically, mirrors have further advantages as light-collectors. The glass does not have to be perfectly clear, since only the front surface is used to form the image, and the disk can be supported across the back to reduce distortion. The difficulty of making large lenses and mounting them well has limited the refracting telescope to an aperture of 40 inches (102 cm); in contrast, there is no reason to suppose that the size limit for mirror disks has been reached even with the Soviet 236-inch (600-cm) reflector. Another advantage is that the FOCAL LENGTH (mirror-image distance) of a mirror can be much shorter than that of a lens, a very important consideration when a large-aperture instrument is being designed.

Although the material used for a mirror has no direct optical effect on its performance, glass is usually used because it can be polished to a smooth, accurate finish. Stainless steel, aluminum, and other materials have also been tried. Astronomical mirrors usually have a thickness about one-sixth their diameter for rigidity, making large disks extremely heavy. Mirrors were once made of plate glass cast into disks or cut from thick sheets, but the development of low-expansion, heat-resistant glass for ovenware led to its use in telescopes—the 200-inch (504-cm) Mount Palomar telescope mirror is made of Pyrex glass. Further development has produced glass with virtually zero expansion, the best-known being *Zerodur,* made by the West German firm of Schott and used in many very large modern instruments. The advantage of low-expansion glass is that temperature changes hardly affect the mirror's surface shape (see also REFLECTING TELESCOPE).

Mizar
The central star in the handle of the Big Dipper, also known as Zeta Ursae Majoris and 88 light-years from Earth. Mizar was the first double star to be discovered by telescopic observation, by G. B. Riccioli in 1650. It was also the first to be photographed, by G. P. BOND in 1857. The orbital period of the two stars is at least 20,000 years. The brighter star was the first SPECTROSCOPIC BINARY discovered, by E. C. PICKERING in 1889; its orbital period is 20.5386 days. The fainter is also a spectroscopic binary, with a period of 175.55 days. All four stars of Mizar are orbited every 10 million years by Alcor, which is visible to the naked eye nearby. The radial velocity of Alcor is variable, which suggests that it, too, may be a spectroscopic binary.

Molniya satellites
A series of Soviet communications satellites, the first of which was launched on April 23, 1965. They have orbits very different from those of the geostationary communications satellites situated over the equator, such as INTELSAT, which most other countries use. Satellites over the equator cannot adequately cover regions in high latitudes, which include much of the Soviet Union, and so the Molniya satellites have been put into orbits inclined at 65° to the equator. Their paths are so arranged that they spend as much time as possible over the Soviet Union. This is done by making them move in highly elliptical orbits, which rise to a maximum height of about 25,000 miles (40,000 km) over the northern hemisphere, but dip to 300 miles (500 km) on the opposite side of the Earth. This orbit, which it takes about 11 hours 50 minutes to complete, ensures that each satellite is visible from the Orbita network of Soviet ground stations for 8 to 10 hours at a time. A system of three Molniyas can therefore provide 24-hour coverage. The main body of each Molniya is a cylinder, roughly 11.3 feet (3.4 m) long by 5.2 feet (1.6 m) wide. The satellites gain power from six solar-cell panels like the petals of a flower. The satellite, weighing about 1,800 lb. (800 kg) communicates with Earth via two 3-foot (0.9-m) dish antennae. Several Molniyas have also carried television cameras to photograph Earth's cloud patterns. On November 24, 1971, the first of an improved and heavier generation of satellites, called Molniya 2, was launched. These use frequencies similar to those of the Intelsat series; the Molniya 1 satellites use lower frequencies. On July 29, 1974, a Molniya satellite was launched into a geostationary orbit over the equator, but others have reverted to the usual highly elliptical orbit. On November 21, 1974, the first of a third generation of satellites called Molniya 3 was launched, apparently carrying improved transmission equipment.

Monoceros (the unicorn)
A constellation in the equatorial region of the sky, between Canis Major and Orion, best seen during winter, introduced in 1624 by Jakob Bartsch (1600–1633), German mathematician and son-in-law of Johannes Kepler. Monoceros lies in the Milky Way, and contains several interesting objects, including the famous Rosette nebula, NGC 2237–9. In front of the large cluster NGC 2264 is the irregular variable star S Mon (also known as 15 Mon), which varies between magnitudes 4.2 and 4.6 every few days. Nearby is a sixth-magnitude star named Plaskett's star, after the Canadian astronomer John Stanley Plaskett (1845–1941) who discovered in 1922 that it was a SPECTROSCOPIC BINARY with masses 76 and 63 times that of the Sun, the most massive pair known. The stars orbit each other every 14.4 days. Also nearby is NGC 2261 (otherwise known as R Mon), Hubble's variable nebula.

month
A unit of time based on the Moon's motion around the Earth. Astronomers refer to several types of months. The *synodic month* is the time taken for the Moon to go through a complete cycle of phases (a *lunation*); it lasts 29.530588 solar days (29d 12h 44m 2.9s). A *sidereal month* is the time taken for the Moon to return to the same position against the star background; it lasts 27.321661 days (27d 7h 43m 11.5s). The difference between the sidereal and synodic months is due to the Earth's motion in orbit around the Sun. The *draconic month* is the time between two passages of the Moon through the ascending node (from south to north across the plane of the Earth's orbit); it lasts 27.21222 days (27d 5h 5m 35.8s). The *tropical month* measures the passage of the Moon across the longitude of the equinox and back again; it lasts 27.321582 days (27d 7h 43m 4.7s). The *anomalistic month* is the time between successive close points (perigees) in the Moon's orbit; it takes 27.554551 days (27d 13h 18m 33.2s). The *solar month* is an artificial unit devised to fit into our calendar; it lasts one-twelfth of a solar year (30.43685 days).

Moon
The Earth's only natural satellite. The Moon is a stark, lifeless world whose surface, unprotected by substantial atmosphere, is alternately scorched by the Sun at lunar midday and frozen to bitter

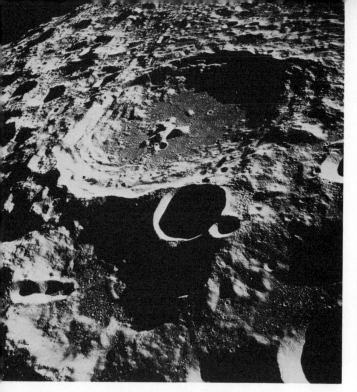

A view of the rugged lunar far side, taken in 1969 from the orbiting Apollo 11 spacecraft. The large crater, center, with terraced slopes and central peaks is Daedalus, about 50 miles (80 km) in diameter.

temperatures at night. The shattered surface is very ancient, showing the scars of more than 3 billion years of meteorite impacts. Having changed so little, the Moon can provide a vast amount of information on the early history of the solar system.

Physical characteristics. The Moon is 2,160 miles (3,476 km) across, less than one-third the diameter of the Earth, and has a mass of 7.34×10^{19} tons. Its mean density is only 3.34 times that of water, in contrast to the Earth's 5.52, indicating a quite different internal constitution. The Moon moves in an elliptical path inclined at 5° to the ecliptic, taking 27.3 days to complete one orbit. The Moon also rotates on its axis in 27.3 days; it thus keeps the same face pointing Earthward. As the Moon orbits the Earth, we see different amounts of its sunlit side, leading to the familiar lunar phases, a complete set of which takes 29.5 days.

The Moon can come within 218,000 miles (351,000 km) of the Earth's surface; its average distance is 238,855 miles (384,400 km). It reaches magnitude −12.7 at full Moon, even though its dark surface reflects only some 7 percent of the incoming light. The Moon's proximity produces reactions in the Earth's surface, atmosphere, and seas, which most clearly show as the familiar ocean TIDES.

Other planets have satellites with no more than a few thousandths their own mass. By contrast, the Moon's mass is about 1/81 that of the Earth, and the Earth-Moon system can almost be regarded as a double planet.

Lunar craters. The Moon's surface is divided into sharply contrasting bright rugged highland areas and dark smooth lowlands. The craters, scattered across the entire surface, are 16 times more abundant on the bright highlands than on the lowland plains.

Smaller craters are everywhere more numerous—halving the size leads to a fourfold increase in numbers. This simple rule holds roughly true for craters down to only a few yards across.

Most lunar craters are now believed to be the result of impacts by large meteorites. Eroded impact craters have been found on the Earth's surface (see CRATERS) and are also present on Mars and Mercury. However, some small craters—particularly those in chains—may be the result of eruptions, gas explosions, or collapse along lines of weakness. The lunar craters are almost perfectly circular, a feature not shared by volcanic craters. The volume of the central bowl is equal to that of the crater's raised rim, which spreads out to about $1\frac{1}{2}$ times the diameter of the central bowl. The ratio of crater depth to diameter decreases for larger craters, being 1:5 for 1-mile (1.6-km) craters, 1:10 at 6 miles (9.7 km), and only 1:50 for the largest lunar craters. These figures belie the impression given by photographs—that lunar craters are deep cavities with steep slopes. In reality, the craters are very shallow, with gentle 10° to 12° slopes. In the case of the 146-mile (235-km) crater Clavius, an observer standing at the center of the floor would see little of the rim because it would be below the horizon.

Craters are surrounded by an ejecta blanket, consisting of crushed rock blasted out of the crater by the meteorite impact. The ejecta appears bright around recent craters, but darkens with age. In some cases, the ejecta forms long rays.

Lunar craters show a wide range of ages. Some are clear and sharply defined, while others are weathered and partially filled in, often being obscured and distorted by later impacts. The youngest-appearing craters are usually quite small, suggesting that few large impacts have occurred in recent lunar history.

The maria. The largest impact features are the Moon's dark circular plains called the maria. The Mare Imbrium, for example, is a giant crater 420 miles (676 km) in diameter, rimmed by mountain ranges. This giant basin was produced by the impact of an asteroid 80 miles (130 km) in diameter, an event whose violence is still evident in the form of gouges, valleys, and secondary craters visible for more than 1,000 miles (1,600 km). The basins were subsequently covered by later, smaller impact craters and eventually filled with lava when the Moon's interior had heated up through radioactive decay. Each mare was formed by many lava flows over a considerable period of time. The far side of the Moon contains several basins up to $3\frac{1}{2}$ miles (5.5 km) deep, but these have not been subsequently filled in with lava flows.

Mountains and valleys. Instead of Earth-type mountain ranges, other, less spectacular features are present on the Moon. The dark lunar plains show wrinkle ridges a few hundred feet high formed by contraction of the centers of the maria. An unusual type of valley, known as a rille and akin to a terrestrial rift valley, has a flat floor and steep parallel walls. Rilles can be hundreds of miles long and are found close to the edges of the maria. Other rilles—the sinuous rilles—may be lava tubes under the maria surfaces that have collapsed. Finally, faults up to 100 miles (160 km) long occur, forming rock faces up to 1,200 feet (365 m) high.

The surface structure. Lunar landings have

Apollo 15's dramatic view of the lunar craters Aristarchus (left) and Herodotus. Aristarchus, 25 miles (40 km) in diameter, the brightest large crater on the Moon, is the center of a major system of rays.

allowed detailed photography of the lunar surface, laboratory analysis of lunar rocks, and active investigation of the lunar surface and interior. The surface is clearly very ancient, a jumble of debris from more than 3 billion years of meteorite impacts. Large and small boulders and smaller rocks are interspersed with huge amounts of fine dust. Many of the fine particles are glass spheres caused by melting of the surface in the heat of meteorite impacts. This jumble of debris (regolith) is from 35 to 60 feet (11–18 m) deep and consists of 1 to $2\frac{1}{2}$ percent of meteoritic material, the shattered remains of the original impacting meteorites. Below the regolith is a region of intensely shattered but more solid rock, which probably extends to depths of more than 10 miles (16 km). The continual impact of meteorites, which still continues, is constantly adding to and turning over the regolith. In addition, the exposed surface of rocks is being eroded by some 1 millimeter every million years through micrometeorite impacts. Ancient boulders would by now have lost some 6 feet (1.8 m) of material since the lunar surface was formed, thereby adding more dust to the lunar regolith.

The Moon's atmosphere is ten trillion times as rare as the Earth's atmosphere at sea level. It is formed from gases released by heating, impact melting, radioactive decay, or possibly volcanic venting during the lunar night, and is partially dispersed by the solar wind each lunar day. The lack of an appreciable atmosphere leads to enormous temperature ranges on the lunar surface, from 105°C at midday on the equator down to −155°C during the long lunar night.

The lunar rocks are unlike those of the Earth, being $6\frac{1}{2}$ times richer in aluminum, calcium, and titanium, but four times poorer in sodium, magnesium, and iron. The highland rocks also differ in many ways from the mare rocks, which accounts for their lighter color.

The interior. The network of seismometers left on the surface by the Apollo landings indicates that the Moon is seismically very quiet. Moonquakes are only a thousandth as strong as earthquakes, and instead of producing intense short-duration signals, as on Earth, they are dampened by the shattered surface to produce long-enduring ringing signals. About 40 major quake sources have now been identified at depths from 370 to 750 miles (595–1,200 km), in striking contrast to the Earth, where earthquakes only occur in the outer layers.

The interior of the Moon, like that of the Earth, consists of several distinct regions. The 40-mile(65-km)-thick outer crust has a density three times that of water; below this is the 740-mile(1,190-km)-deep lunar mantle, only slightly denser. Finally, there is a central core some 600 miles (1,000 km) across. Although crust and mantle are solid, the temperature of the core is perhaps 1,500°C, enough to melt its center. The Moon lacks a large nickel-iron core like the Earth's which accounts for its low density.

Lunar history. Dating of lunar samples has allowed scientists to piece together the history of the Moon. The Moon formed along with the rest of the objects in the solar system about 4.6 billion years ago. Almost immediately, its surface was melted by the heat of a heavy meteorite bombardment, which eventually died away. The lunar highlands formed about 3.95 billion years ago, and the declining meteorite bombardment scattered a thick blanket of impact debris across the lunar surface. Liquid magma then flowed out from under the Moon's nearside surface on various occasions from 3.9 to 3.3 billion years ago, forming the lunar maria. The surface has remained cold ever since.

The interior of the Moon has had a complex history.

Mare Imbrium, a vast, lava-flooded plain some 600 miles (960 km) across, was formed during a final burst of internal lunar activity over three billion years ago. It is encircled by the 18,000 ft (5,400 m) high Lunar Alps, **top right**, and Apennines, **bottom right**; the dark-floored crater, **top**, is Plato, and the bright ray crater Copernicus is at bottom of picture.

Initially the Moon was formed with a molten core that solidified with a strong magnetic field frozen into it. The molten crustal materials then solidified in this magnetic field, to become partially magnetized. Later, the core melted again from radioactive heating, losing its magnetism to produce the present low lunar magnetic field.

Lunar origins. A vast amount of information is now available, but it is still far from certain how the Moon originated. The Moon is currently slowly receding from the Earth; at one time it was as close as 40,000 miles (64,000 km). Theories of lunar origin therefore center on whether the Moon was formed in that position or was somehow captured. One theory suggests that the Moon was formed from material thrown out from the Earth's equator, while according to another, the Moon and Earth formed as a double planet from the same dust cloud. The strikingly different composition of the Earth and Moon is, however, a serious obstacle to these proposals. Yet another theory proposes that the Moon was formed as a small independent planet and was captured by the Earth. Although possible, this would involve an extremely unlikely series of events. At present, the most plausible theory traces the Moon's formation to material captured into Earth orbit. Planetesimals, low-density bodies like asteroids several tens of miles across, would have been plentiful in the early solar system; on passing very close to the Earth these

would have been disrupted, their lighter surface materials being captured and their heavier core materials continuing in solar orbit. This captured material in orbit around the Earth would eventually coalesce to form the Moon. This suggested origin satisfactorily explains the Moon's light density and its nonterrestrial composition.

Morgan, William Wilson (b. 1906)

American astronomer, who found the first evidence for the spiral structure of our Galaxy; previously, the existence of spiral arms in our Galaxy had only been assumed. Studying the distribution of blue giant stars in the Milky Way, Morgan in 1951 showed that they traced the outline of at least two spiral arms: one, toward the edge of the Galaxy, is called the Perseus arm, because it contains the stars of the double cluster in Perseus; the other, called the Orion arm, includes the Orion nebula and our Sun. Morgan also found evidence of another arm in the Sagittarius region, toward the center of the Galaxy. The spiral arms have since been mapped in detail by the radio-wavelength radiation given off by the hydrogen gas they contain. At Yerkes Observatory Morgan, together with Philip Childs Keenan (b. 1908) and Edith Kellman, developed a new classification system for stellar spectra, based on earlier work by Walter S. ADAMS and Bertil LINDBLAD. This was published in the 1943 Yale *Atlas of Stellar Spectra*; the system is usually referred to as the MORGAN-KEENAN CLASSIFICATION. In 1953, Morgan established a now-standard system for measuring the COLOR INDEX of a star, called the UBV SYSTEM. Morgan has also worked on the classification of galaxies, and in 1963 drew attention to supergiant elliptical galaxies that lie at the center of clusters of galaxies; these are the largest galaxies known.

Morgan-Keenan classification

A shorthand notation invented by W. W. MORGAN and Philip Childs Keenan (b. 1908) to describe the appearance of the spectrum of a star. The temperature is coded with one of a series of letters, called a spectral class, which, in order of decreasing temperature, run: OBAFGKM (mnemonic: "Oh, be a fine girl, kiss me"). Up to 10 subdivisions may be added as numbers after the letter: thus, B5 is four-tenths of a class cooler than B1. Roman numerals I, II, III, IV, and V, called the luminosity class, are used to distinguish supergiants, giants, and dwarfs (the smaller the numerals the brighter the star). The Sun's spectral classification, G2V, shows that it is a coolish dwarf, very similar to Alpha Centauri, also G2V. Betelgeuse, M2I, is a cool red supergiant, cooler and brighter than the yellow giant Aldebaran, K5III. Sometimes lower case letters are added to give extra detail in the spectral classification, such as B1Vpnne! meaning a hot (B1) dwarf (V) with some peculiarities (p) in its spectrum, very rapidly rotating (nn) and with strong emission (e!) lines. Some peculiar-looking spectra have special spectral classes.

morning star

Not a true star, but rather the planet Venus, seen shining brilliantly in the morning sky as it rises a few hours before the Sun. When it sets after the Sun in the evening sky, it becomes the brilliant evening star.

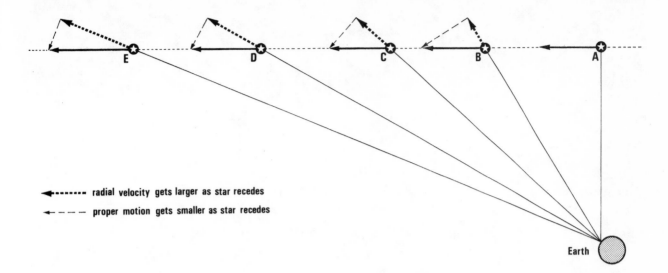

◀━━━━ ······· radial velocity gets larger as star recedes

◀━ ─ ─ ─ proper motion gets smaller as star recedes

Mount Palomar
The site of the 200-inch (508-cm) reflecting telescope of HALE OBSERVATORIES, more correctly called Palomar Mountain.

Mount Stromlo Observatory
An astronomical observatory operated by the Australian National University at an altitude of 2,520 feet (768 m) on Mount Stromlo, 7 miles (11 km) west of Canberra; it was founded in 1930. Its main telescope is a 74-inch (188-cm) reflector, completed in 1955; there are also 50-inch (127-cm) and 30-inch (76-cm) reflectors. A 26-inch (66-cm) refractor operated jointly by Yale and Columbia universities is situated on Mount Stromlo, as is the 20/26-inch (50/66-cm) Schmidt camera of the University of Uppsala, Sweden. Mount Stromlo Observatory operates a field station near Coonabarabran, New South Wales; this is called SIDING SPRING OBSERVATORY.

Mount Wilson
The location in California of the 100-inch (254-cm) reflecting telescope of the HALE OBSERVATORIES.

moving cluster method
A technique for finding the distance of a nearby star cluster whose stars are moving through space along parallel paths. If the cluster is receding, its stars appear to stream toward a single point in space, the *convergent point,* just as parallel railroad tracks seem to converge at a point on the horizon. The nearer a star to the convergent point, the more its speed is directed away from us (RADIAL VELOCITY); the closer a star is to the Earth, the greater its motion across the line of sight (PROPER MOTION). The relative amounts of radial velocity and proper motion for a star depend on its angular distance from the convergent point and its true distance from Earth. Once the convergent point has been identified, the actual distance of the star can be found.
The distance of the HYADES star cluster, determined in this way, forms the foundation for all distance-finding in astronomy (apart from the distances of a few nearby stars, which can be deduced from their parallaxes).

Five stars in a moving cluster belong to a star stream, with same space velocity. But the star A, nearest to Earth, has a zero radial velocity (speed along line of sight) and maximum proper motion (speed perpendicular to line of sight), while the more distant stars like E have a large radial velocity and small proper motion. The exact trade-off between proper motion and radial velocity depends on the angle between the convergent point and the line of sight. The measure of this angle, combined with the radial velocity and proper motion, gives the star's distance.

Mullard Radio Astronomy Observatory (MRAO)
The radio astronomy observatory of the University of Cambridge, situated at Lord's Bridge, near Cambridge, and founded in 1951. Cambridge radio astronomers, led by Sir Martin RYLE, have developed the APERTURE SYNTHESIS method of combining results from several small dishes to produce the view that would be obtained by one much larger dish. Their major instruments of this type are the One-Mile Telescope, opened in 1964, which consists of three 60-foot (18.3-m) dishes (two fixed, one movable) arranged in a line 1 mile (1.6 km) long; and the Five Kilometer Telescope opened in 1972, which has eight dishes (four fixed, four movable) of 42-foot (12.8-m) diameter arranged in a line 5 km (3 miles) long. The Half-Mile Telescope, using two 30-foot (9.1-m) dishes, is mounted on the ½-mile (0.8-km) length·of railway track used by the movable dish of the One-Mile Telescope. Astronomers at MRAO have made several famous surveys of the radio sky, and many radio sources are referred to by their numbers in the third, fourth, or fifth Cambridge catalogs (3C, 4C, 5C). A fixed aerial array, intended to study scintillation of signals from quasars, produced the information that led to the 1967 discovery of PULSARS at MRAO by Antony HEWISH. (See picture p. 138.)

Musca (the fly)
A small constellation in the southern hemisphere of the sky, at the edge of the Coalsack nebula near Crux. It was introduced by Johann Bayer in 1603.

The 200-inch (5-m) Hale telescope at Mt Palomar, Cal. In this view, the telescope's open framework tube points north, along the polar axis. The mirror itself is hidden by the petals of a cover at the lower end of the tube. The scale of the instrument can be judged from the tables at ground level.

Aerial view of five of the eight dishes comprising the Five Kilometer radio telescope at the Mullard radio astronomy observatory, Cambridge. The furthest four can each be moved to eight positions on a ¾ mile long rail track. The other three dishes lie to the left, fixed at intervals of ¾ mile.

N

nadir
The point directly beneath an observer's feet, 90° below the true horizon and directly opposite the ZENITH. The nadir is the direction indicated by a freely suspended plumb bob.

NASA
The National Aeronautics and Space Administration, a federal agency established October 1, 1958, to administer all nonmilitary aeronautical research and space programs. It was successor to the National Advisory Committee for Aeronautics (NACA). NASA headquarters at Washington, D.C., coordinates the activities of the various NASA field stations: AMES

RESEARCH CENTER; Hugh L. Dryden Flight Center (formerly Edwards Flight Center), Edwards, California, concerned with manned flight in aircraft and spacecraft; GODDARD SPACE FLIGHT CENTER; JET PROPULSION LABORATORY; JOHNSON SPACE CENTER; KENNEDY SPACE CENTER; LANGLEY RESEARCH CENTER; Lewis Research Center, Cleveland, Ohio, concerned with aircraft and rocket propulsion and responsible for development of the Agena and Centaur rockets; MARSHALL SPACE FLIGHT CENTER; and WALLOPS FLIGHT CENTER.

National Radio Astronomy Observatory (NRAO)
The largest radio astronomy observatory in the United States, founded in 1957 at Green Bank, West Virginia. The first major telescope, 85 feet (26 m) in diameter, was completed in 1959. Two similar dishes, both movable, have since joined it to make an INTERFEROMETER with a maximum length of 1.65 miles (2.7 km). The interferometer is used in conjunction with a 45-foot (13.7-m) dish located 22 miles (35 km) away, which can receive wavelengths down to 3.7 centimeters. The observatory's largest telescope is a 300-foot (91.4-m) dish erected in 1962 and resurfaced in 1970; initially intended for observations at the 21-centimeter hydrogen line, it is now used down to a 6-centimeter wavelength. In 1965 an equatorially mounted 140-foot (42.7-m) telescope came into operation, working at wavelengths as low as 1.3 centimeters. Two years later a 36-foot (11-m) telescope designed for wavelengths of a few millimeters was installed by the observatory at Kitt Peak, Arizona; it has subsequently made important discoveries of INTERSTELLAR MOLECULES. The National Radio

The major instruments at the National radio astronomy observatory are the 140-foot fully steerable dish, **far left, in the distance**, the 300-foot transit telescope which tilts only north–south, **near left**, and three 85-foot dishes connected electronically to form an interferometer, **right**.

Astronomy Observatory contains a replica of the aerial with which Karl JANSKY detected the first radio waves from space, and it preserves the 31-foot (9.5-m) dish used by Grote REBER to continue Jansky's work. The NRAO is constructing the world's largest radio-astronomy instrument, the VERY LARGE ARRAY, near Socorro, New Mexico.

navigation satellites
Satellites that aid aircraft and ships in pinpointing their position. Initially satellite navigation was used mostly for naval vessels, but its use in merchant shipping and civil aviation is now increasing. A navigation satellite transmits information on its position and orbit. This can be picked up by small receivers, and the Doppler shift in the signals reveals the receiving station's location relative to the satellite, from which an extremely accurate position can be computed. The first navigation satellite was Transit 1B, launched on April 13, 1960 (see TRANSIT SATELLITES).

A series of satellites called Timation was begun in 1967. These carry atomic clocks and provide both navigational data and accurate time signals. The signals have been used to compare atomic clocks in the United States and Britain. The Timation system, initiated by the U.S. Navy, has been merged with an Air Force project to produce a coordinated navsat network known as the Global Positioning System, which involves a series of satellites called Navstar.

neap tide
See TIDES.

nebula
A region of gas and dust in the Galaxy, usually fuzzy in appearance. The name is derived from the Latin word for cloud. The obsolete terms "spiral nebulae" and "extragalactic nebulae" actually refer to GALAXIES, and they have nothing to do with genuine nebulae. Some nebulae, such as the famous COALSACK,

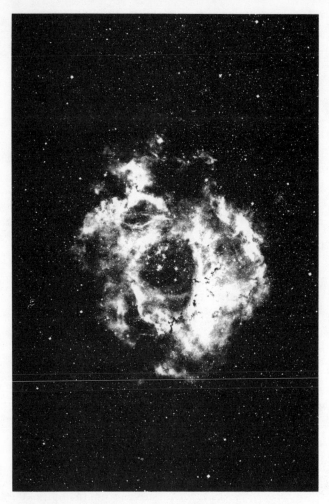

Glowing gas cloud, the Rosette nebula, in the constellation Monoceros. At the nebula's center is a cluster of young stars, whose radiation lights up the hydrogen gas around them. Darker material festooned across the lower right portion of the nebula may mark the formation sites of new stars.

are cold clouds of dust and gas that absorb light from stars behind them, thereby appearing as dark patches (see INTERSTELLAR ABSORPTION). Such *dark nebulae* look like "holes" in the Milky Way. *Bright nebulae* are similar clouds illuminated by stars.

Reflection nebulae are bright nebulae that shine by reflecting light from a nearby star; the light therefore exhibits the same spectrum as that of the star. The dust in the nebula tends to scatter the starlight, however, and because blue light is scattered more than red light, the reflection nebula appears bluer than the star.

The nebula around the star Merope in the PLEIADES cluster was the first reflection nebula to be recognized as such, by V. M. SLIPHER in 1913. A temporary reflection nebula was seen near Nova Persei, the year after its eruption in 1901. The burst of light produced by this NOVA spread into the surrounding space, illuminating a formerly invisible dark nebula that eventually disappeared as the nova faded back into insignificance. A curious reflection nebula discovered in 1852 by the English amateur astronomer John Russell Hind (1823–1895) lies near the star T Tauri. Known now as Hind's variable nebula, it has fluctuated in brightness ever since. Changes in its brightness are not correlated with the brightness of T Tauri itself; instead, its variability is thought to be caused by clouds of dark dust sweeping between the nebula and T Tauri, thus temporarily blocking the star's light.

Emission nebulae are nebulae whose gas is excited to fluorescence by the ultraviolet light from a nearby hot star, thus producing emission of light at certain specific wavelengths. Such emission lines were first seen by William HUGGINS in 1864, when he examined the spectrum of the ORION NEBULA, the brightest of all emission nebulae. Some of the wavelengths of light given off by emission nebulae were immediately recognized as coming from such elements as hydrogen. Some however were unknown, and ascribed to a hypothetical element, "nebulium." This eventually proved to be oxygen which, in the rarefied conditions of space, emitted so-called FORBIDDEN LINES unknown in terrestrial laboratories. Such wavelengths of oxygen make emission nebulae appear green to the naked eye, although in photographs they often appear red because the photographic emulsion is more sensitive to the red wavelength of hydrogen called $H\alpha$. Emission nebulae are of three main kinds: PLANETARY NEBULAE, so called because they resemble the disk of a faint planet; *loops,* due to various causes including supernovae; and *diffuse nebulae* of irregular shapes. Inside diffuse nebulae, such as the Orion nebula, new stars are forming which will make the nebula shine. Diffuse nebulae and hot stars together are called H II REGIONS, areas composed principally of ionized hydrogen.

Neptune

The eighth planet in average distance from the Sun. The eccentric orbit of the outermost planet Pluto, however, brings it within the path of Neptune from January 1979 to March 1999. The existence of Neptune was calculated by mathematicians before the planet was discovered telescopically. Early in the 19th century, astronomers found that the newly discovered planet Uranus was not moving as expected. It appeared to be perturbed in its orbit by the gravitational pull of an unseen body. The perturbations of Uranus were analyzed by the English mathematician John Couch ADAMS, who in 1845 predicted the position of the unseen body. He sent his results to the Astronomer Royal, Sir George AIRY, but no search was begun until after similar results were announced the following year by the Frenchman U. J. J. LEVERRIER. Neptune was first recognized by J. G. GALLE at Berlin Observatory on September 23, 1846, in a search requested by Leverrier. Like Uranus, Neptune had been seen by previous astronomers, but was mistaken for a star.

Physical properties. Neptune's equatorial diameter is 30,750 miles (49,500 km), slightly smaller than Uranus. Neptune's axis is inclined at 28° 48′ to the vertical. Its rotation period is not certain but is believed to be around $18\frac{1}{2}$ hours. The planet is slightly oblate (squashed) in shape, with a polar diameter about 600 miles (1,000 km) smaller than its equatorial diameter. Neptune orbits the Sun every 164.8 years at a mean distance of 30.06 astronomical units (2,794,100,000 miles, 4,496,700,000 km). The near-circular orbit, with an eccentricity of only 0.0086, is inclined at 1° 46′ to the ecliptic. Neptune's mass is 17·2 times the Earth's, giving it an average density of 1.7 times that of water. Neptune appears blue-green in color, and is visible as scarcely more than a small disk in the largest telescopes.

Structure of Neptune. In size and structure, Uranus and Neptune are the most similar of all the planets in the solar system. Neptune's extensive atmosphere contains large amounts of methane, which are responsible for its greenish color (methane absorbs yellow and red light), although the major atmospheric constituents are believed to be hydrogen and helium. As on Uranus, ammonia probably forms clouds below the visible layers of Neptune's atmosphere. Observations of an occultation of a star by Neptune in 1968 suggests that Neptune has a warm stratosphere, with a temperature of 140°K, which is nearly 100° higher than what would be expected at the top of the atmosphere. The warming of the upper layers of Neptune's atmosphere is probably caused by absorption of solar heat by methane. The interior structure of Neptune is believed to be virtually identical to that of Uranus, with a central rocky core 10,000 miles (16,000 km) in diameter covered by a layer of ice 5,000 miles (8,000 km) thick; over three-fourths of the planet's mass is believed to be contained in this ice-coated core.

Satellites. Neptune has two known moons, TRITON and NEREID. Triton may be the largest satellite in the solar system. It moves around Neptune in a circular retrograde (east-to-west) orbit inclined 20° to the planet's equator. It has been suggested that Pluto was once a moon of Neptune, and that a close encounter of Pluto with Triton ejected Pluto into a separate orbit around the Sun and threw Triton into its retrograde orbit. Further evidence of some highly unusual event in Neptune's past is the orbital eccentricity of Nereid, which is the largest for any known satellite; the distance of Nereid from Neptune varies by over 5¼ million miles (8.5 million km).

Satellites of Neptune

	Discoverer	Diameter (km)	Orbit radius (10^3km)	Period (days)	Eccentricity
Triton	Lassell (1846)	6,000	355	5.877	0.00
Nereid	Kuiper (1949)	500	5562	359.881	0.75

Nereid
The more distant of Neptune's two satellites, discovered in 1949 by Gerard Kuiper. Nereid, with a diameter of about 300 miles (500 km), moves around Neptune every 359 days 21 hours 9 minutes in an eccentric orbit that takes it between 826,000 miles (1,330,000 km) and 6,100,000 miles (9,760,000 km) from the planet.

neutron star
A very small, dense star, so tightly packed that its protons and electrons have been compressed together to form the particles called neutrons. A neutron star's mass is roughly equal to that of the Sun, but its diameter is only a few miles. The resulting density is such that a pinhead of the star's material would weigh a million tons. In 1934 the astronomers Walter BAADE and Fritz ZWICKY first proposed that when a SUPERNOVA explosion blows off the outer layers of a star, its remaining dense core is compressed into a neutron star. The basic structure of neutron stars was calculated by the American physicist J. Robert Oppenheimer (1904–1967) in 1939, and although the details are still poorly understood, it is thought that the interior of a neutron star is liquid, with a relatively thin solid crust surrounded by an atmosphere of iron atoms less than a centimeter thick.

If the collapsing star is more than three times as massive as the Sun, it cannot become a neutron star because its gravity would be strong enough to crush even the neutrons, and it would continue to shrink until even light could not escape, thus becoming a BLACK HOLE.

The first neutron stars were discovered in 1967, as the regularly flashing objects called PULSARS. As expected, pulsars seem to be the remains of supernova explosions, as for instance in the CRAB NEBULA. Neutron stars also account for some of the X-ray sources in our Galaxy. Many X-ray sources are double stars, with the two components very different in size. As gas from a giant star's atmosphere falls onto the compact star, it is heated to a temperature of hundreds of millions of degrees and emits X rays.

In many sources the X-ray flux changes rapidly, indicating that the small star is only a few miles across, and must therefore be a neutron star.

new Moon
The Moon at the instant it lines up between the Earth and Sun. The illuminated side is therefore turned away from us, and the Moon is invisible. A solar eclipse would occur at each new Moon, were it not that the Moon's orbit around the Earth is inclined to the path of the Earth around the Sun.

Newcomb, Simon (1835–1909)
Canadian-born American astronomer, one of the foremost of all mathematical astronomers. He first made his mark in 1860, with a paper showing the orbits of the asteroids did not diverge from one point, a conclusion which undermined the existing belief that they were fragments of a single larger body; modern astronomers support Newcomb's view. In 1857 Newcomb had joined the American *Nautical Almanac* office, and at the U.S. Naval Observatory in Washington he began a series of observations to prepare more accurate tables of lunar and planetary motions. In 1877 he became head of the office, and began preparation of the most accurate tables ever made of the movements of objects in the solar system. The results, published in the *Nautical Almanac,* were used throughout the first half of the 20th century. Newcomb computed position tables for the Sun, Mercury, Venus, Mars, Uranus, and Neptune. His colleague George William Hill (1838–1914) undertook the calculations for Jupiter and Saturn. Newcomb studied in detail the complex problem of the Moon's motion, finding errors in the previously accepted lunar tables of the Danish astronomer Peter Andreas Hansen (1795–1874). From the work of Newcomb and Hill

on the Moon's motion, the English mathematician Ernest William Brown (1866–1938) produced in 1919 new lunar tables for publication in the *Nautical Almanac*.

Newcomb also produced a new and more accurate set of astronomical constants, and in 1880–1882 he remeasured the velocity of light. Newcomb called an international conference in Paris in 1896, at which astronomers from Europe and the United States agreed to adopt a new, standardized set of astronomical constants, based on Newcomb's figures.

Newton, Sir Isaac (1642–1727)

British mathematical physicist who developed the concept of universal gravitation and made important discoveries in optics. Newton's major discoveries began in 1665 at his home in Woolsthorpe, Lincolnshire, where he had retreated from Cambridge University during the Great Plague. His work continued when he returned to Cambridge and culminated in 1687 with the publication of his *Philosophiae Naturalis Principia Mathematica* (*Mathematical Principles of Natural Philosophy*), usually known simply as the *Principia*.

Newton's great contribution to science was to explain the physical Universe in mathematical terms, thus bringing to completion the work of both Kepler and Galileo. In the *Principia* he expounded in mathematical form the laws of movement and reaction of physical bodies on Earth and in space, enshrining his results in his three now-famous laws of motion. He also laid down the principle of universal gravitation and gave this principle mathematical precision by stating that every body attracts every other body with a force that depends on how massive the bodies are and the distance between them.

In the *Principia* Newton examined the actions of bodies under gravitation, and his results solved the problems of planetary motion that had faced astronomers since ancient times. He showed how KEPLER'S LAWS of planetary motion were a natural consequence of universal gravitation, and he demonstrated how the planets interacted with one another in their orbits. Newton developed a new mathematics—the calculus, later to be of great use in science—so that he could obtain his results. Newton's work was later pursued by his friend Edmond Halley, who used Newton's principles to compute the orbits of comets. In the late-18th century William Herschel discovered pairs of orbiting stars that obeyed Newton's law of gravity, while in the 19th century the astronomers J. C. ADAMS and Urbain LEVERRIER discovered the planet Neptune by applying Newtonian theory to the perturbed

Sir Isaac Newton in 1702, when he was 60. His most important work was done when he was in his early 20s, though much of it was not published until later. In his 60s, however, Newton turned from academic work to become an administrator at the Royal Mint—felt to be a more suitable occupation for a great man.

orbital movements of the planet Uranus. During the 18th and 19th centuries, mathematicians worked out other consequences of the Newtonian theory of gravity as applied to planetary orbits, and investigated the difficult problems of applying it to the Moon's orbit.

Observational astronomy also owes an important debt to Newton. His experiments on the dispersion of white light into a colored spectrum showed that white light was compounded of light of every color. Newton believed that lenses would always form images with colored fringes (CHROMATIC ABERRATION). He therefore designed a reflecting telescope, using mirrors instead of lenses, and in 1668 built such an instrument himself. Over the following centuries the reflecting telescope developed into one of the most powerful of all astronomy's tools.

Newton's pioneering research in optics was developed in the 19th century into the science of spectroscopy which, in turn, gave rise to ASTROPHYSICS—the study of the physical and chemical nature of the stars. His theory of the nature of light, eventually published in his *Opticks* of 1704, proved invaluable to still later research.

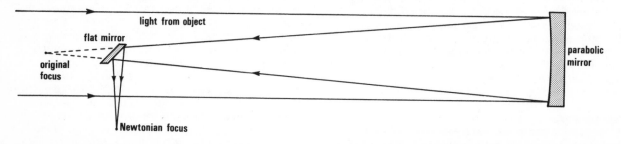

Newtonian Telescope

N-galaxy

A type of distant galaxy with a very small but bright central nucleus. N-galaxies were first discovered as the optical counterparts of some strong RADIO GALAXIES, and the name was devised by W. W. MORGAN to indicate the prominence of the nuclei (N) of these galaxies. They are probably enormously powerful SEYFERT GALAXIES, seen at larger distances than would otherwise be possible. Their nuclei vary in brightness over the years and can sometimes become so luminous that the rest of the galaxy cannot be seen. The star-like nuclei then resemble QUASARS, which are generally thought to be extremely powerful N-galaxy nuclei so distant that the surrounding galaxy is invisible. The first quasar to be identified, 3C 48, has in fact been reclassified as an N-galaxy because long-exposure photographs show the faint fuzz of a galaxy around it.

NGC

Abbreviation for *New General Catalogue of Nebulae and Clusters of Stars,* a now-standard list of deep-sky objects compiled by the Danish astronomer Johan Ludwig Emil DREYER. The NGC, published in 1888, contained nearly 8,000 objects, with another 5,000 added in two supplements, called the *Index Catalogues,* published in 1895 and 1908. Many nebulae and star clusters are referred to by their NGC numbers, although the M numbers given to the most prominent objects by the French astronomer Charles MESSIER also remain in use.

Nikolayev, Andrian Grigoryevich (b. 1929)

Soviet cosmonaut who in August 1962 piloted the Vostok 3 capsule during the joint flight with Pavel Popovich in Vostok 4. The following year Nikolayev married the Soviet female cosmonaut Valentina TERESHKOVA. He made a second flight in June 1970 on the Soyuz 9 mission.

Nimbus satellites

A series of American weather satellites from which the Earth-sensing LANDSAT design was developed. Nimbus satellites tested new sensing devices which have been incorporated into other weather satellites, such as the NOAA series (see WEATHER SATELLITES). Nimbus was a butterfly-shaped satellite, 10 feet (3 m) long and 11 feet (3.4 m) wide across its two winglike solar panels. A circular tray 5 feet (1.5 m) wide carried the sensing devices, acting at wavelengths from infrared to ultraviolet. In addition to returning photographs of Earth's cloud cover, ice caps, and landforms, satellites of the Nimbus series also made temperature and humidity measurements at various depths in the atmosphere. The Nimbus satellites were launched into polar orbits so that they could scan the entire Earth. Nimbus I was sent into orbit in August 1964; Nimbus 7, a pollution monitoring satellite, was launched in October 1978.

noctilucent clouds

Clouds similar in appearance to cirrus or cirrostratus clouds, formed of ice crystals frozen around micrometeorites at heights of 50 to 55 miles (80–90 km) in the Earth's atmosphere. They appear as bright blue-white patches against the twilight sky, being visible only by reflected sunlight when the Sun is 10° to 15° below the horizon. Particles collected from the clouds by high-altitude rockets are found to be either true micrometeorites or dust derived from burnt-up meteors.

nodes

The points at which the orbit of a body intersects a reference plane, such as the plane of the Earth's orbit. Where the object is moving from south to north, the intersection is termed the *ascending node*; the *descending node* is the point at which the object moves across the reference plane from north to south. REGRESSION OF NODES is movement of the nodes caused by gravitational influence of other bodies.

Norma (the level)

A small, faint constellation in the northern hemisphere of the sky, lying on the edge of the Milky Way, introduced by Nicolas Louis de Lacaille.

northern lights

The popular name for the aurora borealis (see AURORA).

nova

A faint star that suddenly erupts in brightness, becoming visible where no star had been seen before; the term comes from the Latin word for "new." Two or three novae are discovered every year, often by amateur astronomers. Novae increase in brightness by thousands or tens of thousands of times, and can often be distinguished by the naked eye. They climb to their maximum magnitude very rapidly— sometimes in as little as a day or two—before fading again over several days, months, or even years. Novae are named after their constellation and the year of their maximum brightness. Nova Persei 1901, discovered by the Scottish amateur astronomer Thomas David Anderson (1853–1933), brightened by a factor of at least 10,000 in one day, temporarily becoming one of the brightest stars in the sky. Its spectrum showed that a shell of gas had been thrown off at a speed of 1,250 miles (2,000 km) per second; it is this rapid expansion of the surface area of a nova that is responsible for the dramatic increase in brightness. The total mass thrown off, however, is only .00001 of the star's original mass, so that a nova outburst does not greatly disrupt it. A SUPERNOVA explosion on the other hand virtually destroys a star. At maximum, the intrinsic brightness of a nova (its ABSOLUTE MAGNITUDE) reaches −7 or −8, about 10,000 times (10 magnitudes) less than that of a supernova.

Most novae, possibly all, are actually binary stars, as shown in 1962 by the American astronomer Robert Paul Kraft (b. 1927). One of the stars in each binary system is a WHITE DWARF, and the nova outburst is believed to be caused by gas flowing from the companion star onto the white dwarf, where it ignites and is ejected in a small nuclear explosion. Several novae have been seen to erupt more than once; Nova Pyxis holds the records for recurrent outbursts with four, in 1890, 1902, 1920, and 1944. Satellites have recently revealed the existence of X-ray novae (see X-RAY ASTRONOMY). These are probably also binary stars, with one of the components being a NEUTRON STAR or a BLACK HOLE.

nutation

A slight nodding of the Earth's axis in space, the result of a change in the angle of the Earth's axial tilt. Nutation is superimposed on the general motion of PRECESSION, which is caused by the gravitational attractions of the Sun and Moon on the Earth's equatorial bulge. Since the pulls of the Sun and Moon are not identical, the S-shaped wobble of nutation results. Each nutation cycle takes 18.6 years, during which the Moon's orbit shifts once around the Earth. This orbital shift has a similar cause to that of precession: the Moon's orbit is inclined at 5° to the Earth's own orbit in space, and the Sun exerts a force that tends to pull the two into the same plane, with the result that the whole orbit of the Moon precesses around the sky in 18.6 years. There are tiny extra cycles of a fortnight (half a lunar orbit) and six months (half the Earth's orbit) within the main nutational movement. The maximum amount of nutation is about 9 seconds of arc in each direction. Nutation was discovered in 1748 by James BRADLEY.

OAO satellites

The Orbiting Astronomical Observatory series of American satellites, which observed the Universe at ultraviolet and X-ray wavelengths. Since hot objects emit most of their radiation in the short-wavelength region of the spectrum, which is blocked by the Earth's atmosphere, they can only be studied accurately by orbiting observatories. Using clusters of telescopes, spectrometers, and X-ray and gamma ray detectors, the OAO satellites observed young, massive stars, nebulae believed to be the sites of star formation, and galaxies and quasars. They found that the most massive stars are even hotter than previously believed, and that galaxies emit considerably more energy than expected at ultraviolet wavelengths.

Satellite	Launch date	Remarks
OAO 1	April 8, 1966	Carried one 16-inch (41-cm) telescope and four 8-inch (20-cm) telescopes for ultraviolet studies, plus X-ray and gamma-ray detectors. Ceased operation after 2 days owing to power failure
OAO 2	December 7, 1968	Carried Celescope, a group of four 12½-inch (32-cm) telescopes for ultraviolet observations of hot main-sequence stars; a 16-inch (41-cm) reflector for studies of nebulae; four 8-inch (20-cm) telescopes for stellar photometry; and two ultraviolet stellar spectrometers with 8-inch (20-cm) mirrors
OAO B	November 30, 1970	Failed to reach orbit because spacecraft fairing failed to jettison. Main telescope was 36-inch (91-cm) reflector
OAO 3 (Copernicus)	August 21, 1972	Carried 32-inch (81-cm) telescope for ultraviolet spectroscopy of hot stars, plus three X-ray detectors operated by British experimenters. Final OAO

Oberon

The most distant satellite of Uranus. Oberon's diameter is about 1,000 miles (1,600 km); it orbits Uranus at a distance of 364,270 miles (586,230 km) every 13 days 11 hours 7 minutes. It was discovered in 1787 by William Herschel.

Oberth, Hermann Julius (b. 1894)

German rocket pioneer, with the Russian Konstantin TSIOLKOVSKY and the American Robert GODDARD one of the three founding fathers of astronautics. Oberth was primarily a theorist and not an inventor. His writings, however, inspired the German interest in rocket research that led directly to the space age. As early as 1917 Oberth had drawn up designs for a long-range liquid-fueled rocket for military use. This design work was rejected for a Ph.D. thesis in 1922, but the following year he turned his research into a now-classic book, *The Rocket into Interplanetary Space*, which included the first detailed discussion of orbiting space stations. In 1929 Oberth published his major work, *The Road to Space Travel*, in which he foresaw the development of electric rockets and ion propulsion. The book won an award established by the French rocket pioneer Robert Esnault-Pelterie, the prize money supported Oberth's experiments with rocket motors in the German Society for Space Travel, one of his assistants being Wernher VON BRAUN. Oberth worked briefly at Peenemünde during World War II, and in 1955 spent three years with von Braun in the United States before returning to retirement in Germany.

object glass

The lens used in a REFRACTING TELESCOPE to form an image. A single lens, because of the defect known as CHROMATIC ABERRATION, forms a colored image; all precision object glasses therefore consist of two separate lenses to minimize this effect. The front lens is of crown glass (light in weight, consisting mainly of silica), and the rear lens is of flint glass, which is heavy because it contains lead oxide. Even with this modification, residual color still degrades the image if the FOCAL RATIO (lens-image distance divided by lens aperture) is less than about 12. Most object glasses therefore have focal ratios of between 15 and 20. In small sizes, up to about 6 inches (15 cm), the lenses are usually mounted together in a single cell, separated by three pieces of foil located around the edge; larger lenses may have separate cells with mutual adjustment.

observatories, history of

The earliest known observatories were constructed in Egypt about 2600 B.C., and in Babylonia some six centuries later. These buildings were as much religious as astronomical in purpose. In Britain and in Brittany (northwest France), astronomically aligned stone circles were being built around the same time as the Babylonian ziggurats. The earliest observatories were constructed to help make accurate calendars; stone circles such as STONEHENGE probably also produced eclipse predictions. Astronomers were mainly concerned with the stars, the Sun, and the Moon. Other astronomical observations were made, primarily of the five naked-eye planets, but there appear to have been few if any formal observatories as such. Ancient astronomers like HIPPARCHUS and PTOLEMY used

Orion nebula is the most famous gas cloud in space, easily visible to the naked eye. It is a star factory, inside which new stars are still forming. The Orion nebula is illuminated by new-born stars within it. Photograph taken by UKSTU.

Above The Veil nebula is the remains of an old exploded star or supernova. Wisps of gently glowing gas loop across the background stars of the Milky Way in Cygnus, as seen in this photograph taken with the Schmidt telescope of Hale Observatories.

Below left Face of the red giant star Betelgeuse, as resolved by the technique of speckle interferometry, using the 4-m reflector at Kitt Peak in Arizona. Darker patches are believed to be cooler spots on the star's surface. The colors are not real, but are added electronically for clarity.

Below right Messier 83, a spiral galaxy in the constellation of Hydra. Note the difference in color between the old, red stars of the galaxy's center and the younger, bluer stars of the spiral arms. This photograph was taken by the 3.9-m Anglo-Australian Telescope.

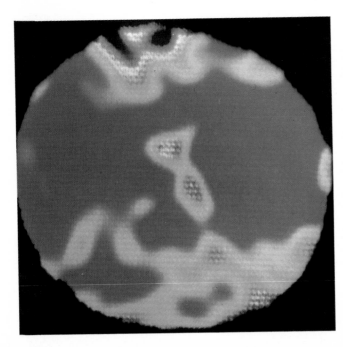

observing instruments, but there is no indication they used permanent observing stations.

The first genuine astronomical observatories, in the sense of permanent places set aside and equipped with instruments, were privately owned. The Moslem astronomer Ulugh Beg (1394–1449) had an observatory with large stone instruments at Samarkand in the 15th century, where he produced a catalog of 994 stars; at almost the same time the German astronomer Regiomontanus (Johann Müller, 1436–1476) used an observatory in Nuremberg, at which he observed the comet of 1472, later identified as HALLEY'S COMET. But probably the most notable observatory of the pre-telescope era was established in 1576 by Tycho BRAHE on the island of Hven (modern Ven) in the sound between Copenhagen and Landskrona. Tycho constructed "Uraniborg," a large observing building with living accommodation, and "Stjerneborg," in which the instruments were housed in buildings largely underground and acting as windbreaks. Tycho made positional measurements using small instruments of the highest precision. This contrasted with Ulugh Beg, who used large stone instruments, including a quadrant with a vertical wall 180 feet (55 m) high, in the hope that large size would lessen the effect of errors in marking the scales; the same principle was followed in the 1720s by the Indian astronomer Jai Singh at Delhi. But Tycho had demonstrated that an observatory with small, precise instruments gave the best results.

The advent of the telescope in 1609 began a new era in astronomy, and in the 1670s national observatories were founded at Paris (see PARIS OBSERVATORY) and at Greenwich near London (see ROYAL GREENWICH OBSERVATORY). These were permanent institutions financed by the government and staffed by professional astronomers. Although amateur observatories proliferated, Paris and Greenwich set the pattern; other countries took many years, however, to establish their own national observatories.

Subsequent developments followed advances in observational techniques and in the teaching of science in the universities. In Britain, for example, Oxford and Cambridge had small observatories in the 17th century, yet not until 1778 was a full-sized university observatory—the Radcliffe at Oxford—in use. In the United States college observatories were not established until the following century, the HARVARD COLLEGE OBSERVATORY being the first, in 1839. Large telescopes were introduced by private astronomers; they included the 48-inch (122-cm) reflector of William HERSCHEL (1789) and the 72-inch (182-cm) reflector of Lord ROSSE (1845). Institutional observatories initially favored large refractors, notably the 36-inch (91-cm) refractor of LICK OBSERVATORY (1888) and the world's largest refractor, the 40-inch (102-cm) of YERKES OBSERVATORY (1897).

Leading astronomers such as George Ellery Hale realized early this century that the large telescopes of the future would be reflectors, and that large-aperture telescopes had to be sited in the best possible observing conditions. Southern California proved particularly suitable and in 1904 Mount Wilson Observatory was established near Los Angeles. It installed its famous 100-inch (254-cm) reflector in 1917, and it set the pattern for many later

observatories. The giant 200-inch (508-cm) reflector was erected on nearby Palomar Mountain after World War II (see HALE OBSERVATORIES). The famous French PIC DU MIDI OBSERVATORY is situated in the clear air of the Pyrenees, and a mountain-top site was chosen for the KITT PEAK NATIONAL OBSERVATORY near Tucson, Arizona. The need for clearer skies has forced even old-established observatories like the Royal Greenwich Observatory and the Paris Observatory to move to better sites.

In the southern hemisphere major observatories were relatively slow to be established. Not until 1813 was a permanent observatory established for southern hemisphere research, the CAPE OBSERVATORY in South Africa. Later in the century other southern observatories followed in South Africa, Australia, New Zealand, and in South America. Only since World War II, however, have really large observatories run by universities and research institutes spread widely across both hemispheres.

The observatory has changed markedly in appearance over the years. Originally it consisted of buildings housing measuring instruments; telescopes when they came were of very long focal length and had to be kept outside. Later, with improved technology, they were mounted in domes. Now, with the arrival of radio astronomy, astronomers have reverted to the outdoor observatory, for radio telescopes are far too large to keep in a dome (particularly APERTURE SYNTHESIS arrays). This also applies to the intensity INTERFEROMETER, for which two or more optical telescopes, mounted on rails, are used in the open. The outdoor observatory has returned as a feature of 20th-century astronomy.

The latest development of the observatory is the orbiting observatory or space station. In 1962 the satellite ARIEL 1 began radio-astronomy studies, and soon a series of radio-controlled Orbiting Solar Observatory (OSO) and Orbiting Astronomical Observatory (OAO) satellites were in space, observing in wavelengths that the Earth's atmosphere blocks from the ground. The SPACE TELESCOPE, which will be put into orbit in the 1980s, will achieve a level of performance far higher than that of the finest instruments on Earth. The SKYLAB space station has already presaged a totally new kind of astronomical observatory, which will come to fruition with the Spacelab module to be launched by the SPACE SHUTTLE, in which astronomers will operate their own instruments in orbit.

occultation

The obscuring of one astronomical body by another. The Moon, for example, occults numerous stars as it moves across the sky. A star disappears behind the Moon almost instantaneously; by timing the moment of occultation precisely, astronomers can compute the Moon's exact position in its orbit. Such observations have improved knowledge of the Moon's motion considerably. When a star appears to slide along the Moon's limb near the poles (a grazing occultation), its light flashes as it passes behind lunar mountains and reappears in valleys. The occultation of planetary satellites, asteroids, and radio sources by the Moon gives precise information about their sizes and positions. The slow fading of stars as they pass behind planets with dense atmospheres, such as

Venus or Jupiter, reveals much about the nature of these planetary atmospheres. Objects of uncertain size, such as Pluto and certain asteroids, have had their diameters more accurately determined by observing their occultations of stars.

Octans (the octant)

A faint constellation at the south pole of the sky, introduced by Nicolas Louis de Lacaille. Its brightest stars are only of the fourth magnitude.

o.g.

Abbreviation for OBJECT GLASS, the front lens of a refracting telescope.

OGO satellites

The Orbiting Geophysical Observatory series of American satellites, which studied the Earth and its surroundings in space. OGO satellites were box-shaped with extendable booms carrying sensors. The satellites measured cosmic ray particles and solar emissions, the atomic processes in the Earth's upper atmosphere that cause the AURORAE, and the Earth's magnetic shell in space called the MAGNETOSPHERE. OGO results have included better mapping of the Earth's magnetic field and radiation belts. OGO satellites were launched into near-polar or very eccentric orbits that took them a third of the way to the Moon at their farthest. OGO 1 was launched in September 1964; the final satellite, OGO 6, went into orbit in June 1969.

Olbers, Heinrich Wilhelm Matthäus (1758–1840)

German astronomer, discoverer of two asteroids and five comets, and author of a famous cosmological paradox. In 1796 Olbers invented a new and simplified way of calculating comet orbits called *Olbers' method,* which was soon adopted by many astronomers. On January 1, 1802, Olbers relocated the first asteroid, Ceres, in the position calculated by Carl Friedrich GAUSS, after it had been lost by its initial discoverer, Giuseppe PIAZZI. While following Ceres, Olbers on March 28, 1802, discovered another asteroid, Pallas, the second to be discovered. Olbers became convinced that the asteroids were the shattered remains of a larger body; he searched for more fragments, and on March 29, 1807, discovered Vesta. He proposed in 1811 that comet tails always point away from the Sun because material is blown away from the head by solar radiation; this was before the discovery of RADIATION PRESSURE. The comet he found in 1815 is now called Olbers comet. Olbers is possibly best known for the so-called OLBERS PARADOX, stated in 1826. According to this paradox, light from all the stars in space should make the night sky bright. Olbers himself believed that the sky appeared dark at night because dust clouds blocked off light from the most distant stars. Astronomers now believe that the expansion of the Universe weakens the light from other galaxies that reaches Earth.

Olbers' paradox

The darkness of the night sky has long been thought to be a cosmological paradox, because in an unchanging Universe filled with an infinite number of stars the whole sky should shine with the brightness of a typical star like the Sun. In 1826 Heinrich Wilhelm OLBERS suggested that wherever one looks in the sky there must be a star, even if it is at a very great distance. Every tiny patch of the sky should thus be shining like the surface of a typical star, and the whole sky should be bright.

The discovery that all stars are concentrated into galaxies does not alter the paradox, but it is now known that the Universe is neither infinite in extent nor unchanging with time. The general expansion of the Universe was discovered a century after Olbers had formulated his paradox. Light from the stars in distant galaxies is shifted to longer wavelengths by this cosmological expansion, and in the process its energy is reduced. The most distant stars are thus progressively dimmer than they would have been in a nonexpanding Universe, and Olbers' argument breaks down.

Oort, Jan Hendrik (b. 1900)

Dutch astronomer, who has made major studies of the size and structure of our Galaxy. Oort's mentor was Jacobus KAPTEYN, who had proposed that the Sun was near the center of a lens-shaped Galaxy of stars. But in 1917 the American astronomer Harlow SHAPLEY showed that the Galaxy was much larger than Kapteyn had envisaged, and that the Sun lay toward the edge. Oort analyzed the systematic star motions discovered by Kapteyn and termed STAR STREAMING. Like the Swede Bertil LINDBLAD, Oort realized the streaming was due to the Galaxy's rotation about a center far from the Sun, with stars at different distances orbiting at different rates. From a detailed analysis of star motions and distribution Oort in 1927 accurately calculated the mass of our Galaxy (about 200 billion times the Sun's mass), its dimensions (100,000 light-years in diameter), and the Sun's distance from the center (about 30,000 light-years), thereby confirming and extending Shapley's discovery. During World War II Oort realized that radio astronomy, as pioneered by Grote REBER, provided a new technique for observing the Galaxy's structure. Oort's student Hendrik van de Hulst calculated in 1944 that hydrogen would radiate at a wavelength of 21 centimeters, and in 1954 Oort's team published a famous radio map of the spiral arms of our Galaxy, as revealed by hydrogen emission. Motions in this gas confirmed Oort's calculation that the Galaxy rotates once every 225 million years at the Sun's distance. Other studies by Oort showed that clouds of gas are being ejected from the nucleus of our Galaxy as if from an explosion; he also drew attention to gas clouds from space that the Galaxy appears to be sweeping up. Observations of the Crab nebula in 1956 showed that radio waves from it are strongly polarized, and are thus emitted by so-called SYNCHROTRON RADIATION (electrons spiraling in magnetic fields). In 1950 Oort proposed that comets exist in a vast cloud at the edge of the solar system, about 1 light-year from the Sun; they are perturbed toward the Sun by the gravitational pulls of nearby stars. This "comet cloud" hypothesis is now widely accepted.

open cluster

A shapeless, loosely packed cluster of stars. Open clusters occur in the spiral arms of the Galaxy, and are thus also called GALACTIC CLUSTERS. The HYADES and PLEIADES are typical open star clusters, showing a loose concentration of bright, blue stars and up to 400 known fainter members, in a volume typically less than 100 cubic parsecs. Stars in some open clusters are so sparsely scattered that it is difficult to decide where the cluster ends. Star-formation in a cluster occurs in a short burst until the birth of massive stars, which radiate so much energy they prevent new PROTOSTARS from condensing. Because the stars in open clusters formed virtually simultaneously, astronomers can see the different rates at which stars of different masses evolve. Open clusters have thus provided valuable data on the evolution of stars. The distances to open clusters can be found by comparing the brightness of their MAIN SEQUENCE stars with similar stars of known distance (*main-sequence fitting*), or by the MOVING CLUSTER METHOD.

Ophiuchus (the serpent holder)
A constellation in the equatorial region of the sky, visible in the northern summer; it is depicted with a serpent (represented by the constellation SERPENS) twined around it. Its brightest star, called Ras Alhague, is a hot white star of magnitude 2.04. The star 70 Ophiuchi, 16.7 light-years away, is a pair of yellow dwarf stars cooler than the Sun, orbiting each other every 87.8 years; their magnitudes are 4.27 and 6.0. The brighter of the two is also a SPECTROSCOPIC BINARY, with a period of 18 years. Another nearby star is 36 Ophiuchi, 18.3 light-years distant, also composed of two yellow dwarfs, both of magnitude 5.3, orbiting every 549 years. The star RS Ophiuchi is a well-known recurrent NOVA. Near the star 66 Ophiuchi is BARNARD'S STAR, the second-closest star to the Sun. Ophiuchus lies in the Milky Way, and includes several interesting star clusters. Although Ophiuchus is not a constellation of the zodiac, the Sun spends 20 days within its boundaries, from November 27 to December 17.

opposition
The instant at which a planet farther from the Sun than Earth appears opposite the Sun in the sky. (Planets between the Earth and Sun cannot come to opposition.) At opposition, a planet appears on the meridian (north–south line in the sky) around midnight. Opposition is the best time to observe a planet, for it is then at its closest to Earth.

orbit
The path followed by one astronomical body as it moves around another. In reality, both bodies move around their common center of gravity. For example, the center of gravity of the Earth-Moon system is 2,900 miles (4,700 km) from the Earth's center, roughly 1,000 miles (1,600 km) under the Earth's surface. The Moon follows an elliptical orbit about this point while the Earth wobbles slightly each month. For a very tiny body orbiting around a massive one, however, the movement of the large body is insignificant.

The shape of an orbit is governed by the laws of gravity and motion formulated by Isaac NEWTON and Johannes KEPLER. An orbit is precisely defined by a set of factors known as the *elements of the orbit*. These can be shown by considering a comet

following an elliptical path around the Sun (see diagram accompanying the entry COMET). The orientation of the orbital plane is defined by three angles: i is the inclination of the orbital plane; Ω defines the point where the orbit plane crosses the Earth's orbit; and ω fixes the comet orbit axis. The semimajor axis a defines the length of the ellipse, and q defines the perihelion distance, the closest point on the orbit to the Sun. The position of the comet at a precise moment is also required, normally, the time of perihelion passage, T, is given. There are thus six elements defining a comet's orbit.

The eccentricity of the ellipse, e, is often quoted and equals $1 - q/a$. For a circle, e is exactly zero; for a parabola e is exactly 1. Circular and parabolic orbits are ideal cases and are never found in reality because the slightest perturbation turns them into either ellipses or hyperbolas. An elliptical orbit is closed; an object in a hyperbolic orbit makes only one close passage to the central body and then escapes along an open orbit.

Orbits are computed by the careful analysis of observed positions, taking into account the gravitational perturbations caused by other bodies. The study of orbits is not confined to the solar system. Astronomers also calculate the relative motions of binary stars and can follow the orbit of star clusters and individual stars around our Galaxy.

Orbiting Astronomical Observatory, Orbiting Geophysical Observatory, Orbiting Solar Observatory
See OAO, OGO, OSO.

Orion
One of the major constellations of the sky, representing the hunter of Greek legend. Orion lies astride the celestial equator; it is visible in the south during the northern hemisphere winter. Its main stars are BETELGEUSE and RIGEL. Gamma Orionis, called Bellatrix, is the constellation's third-brightest star, of magnitude 1.64 and 360 light-years distant; it is 8.2 times the Sun's diameter. The central star in Orion's belt, Epsilon Orionis or Alnilam, is a blue-white star of magnitude 1.70, 1,800 light-years away and 54 times the Sun's diameter. Zeta Orionis, called Alnitak, is a double star of total magnitude 1.74, 1,200 light-years away; just south of it lies the famous *horsehead nebula* of dark gas. The third star in the belt, Delta Orionis or Mintaka, is a triple star of combined magnitude 2.46. The two visual components have individual magnitudes of 2.48 and 6.87; the brighter component is in fact a slightly variable eclipsing binary of period 5.7 days. Probably the constellation's most famous feature is the ORION NEBULA, which makes up the sword of Orion hanging from his belt; at its center is the quadruple star Theta Orionis. The entire constellation is bathed in a tenuous nebulosity; it is the center of a large field of new-formed stars in our own spiral arm of the Galaxy.

Orion nebula
A giant cloud of gas and dust in the constellation Orion, also called M42 and NGC 1976. The nebula, which makes up the sword of Orion, is about 1,500 light-years away and 15 light-years in diameter. New stars are being formed inside it. Among them are four which make up the *Trapezium,* collectively

Visible to the unaided eye as a hazy spot under Orion's belt, the Orion nebula is revealed here as a delicate fan of glowing gas, penetrated by lanes of obscuring dust. Its central regions contain luminous, young blue stars, some of which are known to have been born in the last fifty years.

termed Theta Orionis. The brightest of these is chiefly responsible for making the nebula glow (see NEBULA). The Orion nebula is the brightest nebula in the sky, visible easily in binoculars. It was discovered in 1610 by the French astronomer Nicolas Claude Fabri de Peiresc (1580–1637). Behind the bright part of the nebula is a much larger nonluminous cloud in which INTERSTELLAR MOLECULES are found. The Orion nebula is the core of a much larger STELLAR ASSOCIATION. Encircling the entire constellation is Barnard's Loop nebula, a "bubble" being blown in the interstellar gas and dust by the combined output of so many bright stars in one region.

OSCAR satellites

A series of amateur-built satellites for use by radio hams, carried into space in the spare payload capacity of American rockets. The name OSCAR stands for *O*rbiting *S*atellite *C*arrying *A*mateur *R*adio. The first two Oscar satellites carried simple beacon transmitters in the 2-meter amateur band for tracking. Oscar 3 had a repeater in the 2-meter band, allowing the first satellite contacts between radio hams; Oscar 4 worked at both 2 meters and 70 centimeters. Oscar 5, the first to be built by the Amsat amateur radio satellite corporation, broadcast telemetry at 2 meters and 10 meters, wavelengths used on Oscar 6 for amateur communications. Oscar 7 relayed communications at 70 centimeters, 2 meters, and 10 meters. The Oscar satellites have been used for educational demonstrations of space techniques. Future Oscars may include geostationary satellites. The first Oscar was sent in orbit on December 12, 1961. Oscar 8 was launched on March 5, 1978.

oscillating Universe

A theory of cosmology which holds that the Universe expands and contracts in cycles. If its density of matter is sufficiently great, the present Universe will start to contract, slowly at first, and then catastrophically, to a final fireball similar to the BIG BANG. In this holocaust, space, time, and perhaps all laws of nature must break down completely. It is therefore impossible to know what, if anything, comes after the end of the collapse. Some astronomers, however, have suggested that the Universe "bounces" out on another cycle of expansion and contraction. This process could continue indefinitely. Such a Universe would then have no determinable beginning or end. The fireball phase might destroy all physical structure, so that each cycle would begin with new matter, and possibly even new laws of physics. The time direction of all physical processes might even be reversed, so that the next cycle would run "backwards." Unless some such phenomenon occurs, the oscillating Universe would become progressively hotter with each cycle. Current observations suggest, however, that the expansion of the Universe will not be halted by gravity, but will continue indefinitely (see HUBBLE'S CONSTANT).

OSO satellites

Orbiting Solar Observatory series of American satellites, which studied the Sun and its terrestrial effects. Experiments measured radiation from the Sun and space at ultraviolet, X-ray, and gamma-ray wavelengths, and studied interplanetary dust and emissions from the Earth's upper atmosphere. Numerous solar flares, which emit most of their intense energy at short wavelengths, were monitored by the OSOs. Mapping the Sun at different wavelengths also produced pictures of the Sun's atmosphere at different levels. Each OSO consisted of a spinning platform for all-sky scanning, surmounted by a sail containing solar cells and pointed directly at the Sun. The OSO series monitored solar activity through a complete 11-year cycle.

Satellite	Launch date	Remarks
OSO 1	March 7, 1962	Carried 13 experiments
OSO 2	February 3, 1965	Carried 8 experiments
OSO C	August 25, 1965	Launch failure
OSO 3	March 8, 1967	Carried 9 experiments
OSO 4	October 18, 1967	Carried 9 experiments
OSO 5	January 22, 1969	Carried 8 experiments
OSO 6	August 9, 1969	Carried 7 experiments
OSO 7	September 29, 1971	Carried 6 experiments
OSO 8	June 21, 1975	Carried 8 experiments. Last OSO

P

Pallas

The second asteroid to be discovered, also known as minor planet 2, found by Wilhelm OLBERS on March 28, 1802. Pallas orbits the Sun every 4.61 years in an elliptical path between 2.11 a.u. and 3.43 a.u.; its orbit is inclined at 43°. At its brightest Pallas reaches magnitude 6, at which it is just visible to the naked eye in clear dark skies. It has a diameter of 334 miles (538 km) and a mass of about 4×10^{17} tons. Pallas has a reflectivity of about 10 per cent; its dark surface is probably similar to the carbonaceous meteorites in composition.

pallasite
A type of stony-iron meteorite (see SIDEROLITE).

parabola
The curve produced by cutting a cone parallel to one of its sides, identical to the trajectory of a ball thrown in the air. A parabola can be thought of as an ELLIPSE so elongated that it has an infinite distance between its foci, and an ECCENTRICITY of 1. It thus has one focus. The path of an object falling from an infinite distance toward the Sun would be a parabola.

paraboloid
The solid surface formed by rotating a PARABOLA about its axis of symmetry. A concave paraboloidal telescope mirror reflects light rays traveling in a line parallel to the axis to one single point, the focus of the parabola, thus forming an image.

parallax
The change in position of an object when viewed from two different positions. A finger held at arm's length jumps from right to left when viewed alternately with each eye. The jump from side to side is termed a *parallactic shift*. In astronomy, the nearer stars show a parallactic shift against the backdrop of more distant stars when viewed from opposite sides of the Earth's orbit. The change from the star's mean position is called its parallax, and the amount of parallax is a measure of the star's distance: the nearest stars show the largest parallax. The reciprocal (inverse) of a star's parallax gives its distance in PARSECS. The first star to have its parallax measured was 61 Cygni, by Friedrich BESSEL in 1838. Almost simultaneously, the parallax of Alpha Centauri was measured at the Cape of Good Hope by the Scottish astronomer Thomas Henderson (1798–1844), and that of Vega was measured by Friedrich STRUVE. The parallax of the nearest star, PROXIMA CENTAURI, is 0.762 arc seconds. The measurement of these tiny parallax shifts is a prime concern of ASTROMETRY.

The Moon, Sun, and planets also reveal a parallactic shift when seen from different points on Earth. This was the method first used to determine their distance. The parallactic shift of the Moon is almost 1° when seen from opposite sides of the Earth. The Earth's rotation carries an observer from one side to the other each day; the daily shift shown by nearby objects is termed a *diurnal parallax*.

Paris Observatory
Observatory founded in 1667 at the suggestion of the French astronomer Jean Picard (1620–1682), noted for his accurate determination in 1671 of the size of the Earth. The observatory's first directors were the distinguished CASSINI family; they were succeeded by Dominique Francois Jean Arago (1786–1853), and then by Urbain LEVERRIER, whom Arago had

encouraged to analyze planetary motions, leading to the discovery of Neptune. In 1876 an observatory for solar-system studies was established at Meudon, $2\frac{1}{2}$ miles (4 km) south of Paris; it was taken over by the main Paris Observatory in 1926. The Paris Observatory is now headquarters of the International Time Bureau, which standardizes the time systems of world observatories. The main telescope at Paris is a 12-inch (30-cm) refractor. Meudon has a 39-inch (100-cm) and a 24-inch (61-cm) reflector, and a 33-inch (83-cm) refractor. The Paris Observatory operates a major radio-astronomy station at Nançay, 100 miles (160 km) south of Paris; this has a MILLS CROSS radio telescope with arms 1 mile (1.6 km) and $\frac{1}{2}$ mile (0.8 km) long, and a meridian radio telescope with a fixed aerial 1,000 feet (305 m) long and 110 feet (33.5 m) high, and a movable antenna 700 by 130 feet (213 × 40 m).

Parkes Observatory
The National Radio Astronomy Observatory of Australia, situated 15 miles (24 km) north of Parkes, New South Wales, at an altitude of 1,230 feet (375 m), run by the Commonwealth Scientific and Industrial Research Organization (CSIRO). Its main telescope is a 210-foot (64-m) dish, opened in 1961 to work at wavelengths of 75, 21, and 10 centimeters; it has since been resurfaced to work at wavelengths down to 1.35 centimeters. This telescope was responsible for several important discoveries of INTERSTELLAR MOLECULES. A 60-foot (18-m) dish runs on rails nearby to provide interferometer spacings with the main radio telescope.

The 210-foot diameter radio telescope at Parkes, soon after completion. The height of the supporting building, which contains the control room, allows the dish to be tipped into an almost vertical position.

parsec

A measure of astronomical distance. A parsec is the distance at which the Earth and Sun would appear to be 1 second of arc apart. A star at this distance would therefore show a shift in position (PARALLAX) of 1 arc second in the sky as observed from opposite sides of the Earth's orbit. (Actually, no star is quite this close.) The origin of the term parsec—a contraction of *par*allax of one *sec*ond—is attributed to the English astronomer Herbert Hall Turner (1861–1930), an expert on measuring star positions. A thousand parsecs is termed a *kiloparsec*; a million parsecs is a *megaparsec*. The distance of a star in parsecs is the reciprocal (inverse) of its parallax in seconds of arc. One parsec is 3.2616 light-years, 206,265 astronomical units, 19.174 trillion miles, or 30.857 trillion km.

Pavo (the peacock)

A constellation of the southern hemisphere of the sky, introduced on the 1603 star map of Johann BAYER. It contains the bright globular star cluster NGC 6752, about 20,500 light-years away and 100 light-years in diameter. None of the stars of Pavo is of particular note.

Pegasus

A prominent constellation of the northern sky, best seen during the northern hemisphere fall; it represents the winged horse of Greek mythology. Its most famous feature is the so-called square of Pegasus, one corner of which is actually marked by the star Sirrah of the neighboring constellation Andromeda. The star Alpha Pegasi, of magnitude 2.57, is called Markab, an Arabic name meaning saddle. Beta Pegasi, called Scheat, is a red supergiant about 90 times the Sun's diameter; it varies irregularly between magnitudes 2.1 and 3.0 about every month.

Pegasus satellites

Three American satellites designed to monitor micrometeorites around the Earth. Each Pegasus satellite had long wings covered with aluminum foil that registered micrometeorite impacts. Their results showed that micrometeorites were not a significant hazard to men in space. The Pegasus satellites were launched on test flights of the SATURN 1 rocket: Pegasus 1 on February 16, 1965; Pegasus 2 on May 25, 1965; and Pegasus 3 on July 30, 1965. They were switched off on August 29, 1968.

penumbra

The partially shaded area in an eclipse. Areas in the Moon's penumbra at a solar eclipse see the Sun only partially covered. The term penumbra is also used to denote the outer, light-shaded portion of a SUNSPOT.

perfect cosmological principle

The principle that the Universe is always basically the same, wherever and whenever it is observed. Cosmologists have always assumed that the appearance of the Universe is independent of the observer's position in space (the cosmological principle), but that it must change with time as the stars age, and as the galaxies move apart in what is now believed to be an expanding Universe. In 1948 Hermann BONDI and Thomas GOLD suggested that the Universe as a whole

is the same at all times. For this to be so, new galaxies must be forming continuously to replace dying galaxies and to preserve the average "density" of the Universe as the older galaxies move away (see CONTINUOUS CREATION). This principle formed the foundation of the STEADY-STATE THEORY of the Universe.

periastron

Point at which an object in orbit around a star comes closest to that star. The term is usually used to describe the point at which the two stars of a binary system are closest together, although it can also indicate a planet's closest approach to a star.

perigee

Closest approach to Earth by an object in the solar system. When the Earth is at PERIHELION, the Sun is at perigee. The farthest point from Earth is termed the *apogee*.

perihelion

Closest point to the Sun of a body in orbit around it. The Earth is at perihelion on about January 2 each year at a distance of 91,397,000 miles (147,090,000 km). The farthest point from the Sun is termed the *aphelion*.

period-luminosity relation

The very precise way in which CEPHEID VARIABLE stars of longer period are systematically more luminous than shorter period Cepheids. Henrietta LEAVITT discovered this relation in 1912, when investigating Cepheid variables in the Small MAGELLANIC CLOUD. All these Cepheids are at effectively the same distance from the Sun, and so the relation between period and luminosity appeared as a correlation between period and apparent magnitude.

Harlow SHAPLEY first determined the ABSOLUTE MAGNITUDE (intrinsic luminosity) of Cepheids from a statistical study of these stars in our Galaxy. By using the results of Leavitt's and Shapley's work, a Cepheid's absolute magnitude can be deduced from its period, and a comparison with its apparent

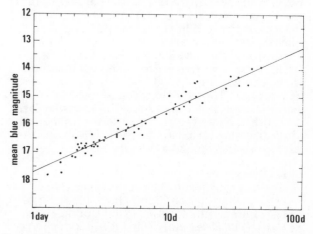

The relation between the apparent magnitude of the cepheids in the Small Magellanic Cloud and their period, plotted logarithmically.

magnitude allows its distance to be determined
(see DISTANCE MODULUS). In 1929 Edwin HUBBLE
used the apparent magnitudes of the Cepheid variables
in the ANDROMEDA GALAXY to prove that it is outside
our own Galaxy, the Milky Way. The Cepheid
period-luminosity relation is still the most accurate
method for determining the distances of nearby
galaxies.

Perseid meteors

A major annual meteor stream, first detectable about
July 25, and reaching a peak of about 65 meteors an
hour on August 12 to 13, then declining over a period
of five days. The stream radiant, which moves eastward
$1\frac{1}{4}°$ a day, is 8° north of Alpha Persei at maximum
and is at its greatest altitude at 6 A.M. Perseid meteors
enter the Earth's atmosphere at $37\frac{1}{2}$ miles (60 km) per
second, and are noted for their brilliant terminal
flares and persistent trains. The stream follows an
elliptical 140-year orbit, moving out beyond the orbit
of Uranus, and is associated with periodic comet
Swift-Tuttle 1862 III.

Perseus

A prominent constellation of the northern hemisphere
of the sky, lying in a rich part of the Milky Way;
it is best seen during the northern hemisphere fall.
The constellation represents the figure Perseus of
Greek mythology. Alpha Persei, called Mirfak, is a
supergiant yellow-white star of magnitude 1.90.
The most famous star in Perseus is the variable-
brightness ALGOL, or Beta Persei. The star Rho
Persei is a red giant that varies irregularly between
magnitudes 3.3 and 4.0 every one to two months.
In 1901, the brilliant Nova Persei flared up at the
center of the constellation, throwing off a shell
of gas still visible today. A prominent feature of
the constellation is the so-called *double cluster*, a
twin star group cataloged as NGC 869 and 884
but better known as *h* and Chi Persei. Both clusters
have 300 to 400 stars and are 7,350 light-years away,
with diameters of 75 to 80 light-years. M34
(NGC 1039) is a loose cluster of about 100 stars,
1,400 light-years away. The strong radio source
Perseus A is associated with galaxy NGC 1275
near the center of Perseus. Perseus is the radiant
of the PERSEID METEORS.

perturbations

Slight disturbances in the motion of a body caused
by the gravitational pull of another object.

phase

The proportion of an illuminated body that is
visible to an observer. Objects such as the Moon,
Venus, and Mercury go through a complete cycle
of phases from new (illuminated side not visible)
through a crescent, half phase (dichotomy), gibbous
phase, to full, and back again. Planets farther from
the Sun than Earth do not show such a cycle of
phases.

Phobos

The larger of the two satellites of Mars, discovered in
1877 by Asaph HALL. Phobos orbits a mere
3,700 miles (5,955 km) above the surface of Mars
(5,810 miles, 9,350 km from the planet's center). Its

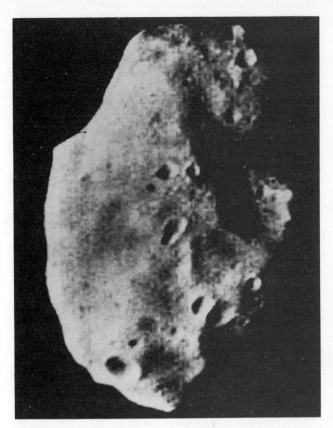

The larger of Mars' two satellites, Phobos, is shown to be
heavily cratered in this photograph taken by Mariner
9 from a distance of 3,400 miles. The indentation in the
outline at top left is a large crater seen in profile; another
large crater is visible at center right.

orbital period is 7 hours 39 minutes 14 seconds, and
it could thus be seen rising, moving across the sky,
and setting at least twice a day. Phobos rotates on its
own axis in the same time as it takes to orbit, thereby
keeping one face permanently turned toward Mars.
As photographed by the Mariner 9 space probe in
1971, Phobos appears a rocky, potato-shaped body of
dimensions 17 by $13\frac{1}{2}$ by 12 miles ($27 \times 21.5 \times 19$ km).
Like its companion moon DEIMOS, Phobos is believed
to be a captured asteroid.

Phoebe

The most distant satellite of Saturn, discovered in
1898 by W. H. PICKERING. Phoebe orbits Saturn every
550 days 8 hours 5 minutes at an average distance of
8,047,650 miles (12,951,440 km). Its diameter is about
120 miles (200 km). Phoebe orbits Saturn in a
retrograde direction (east to west).

Phoenix (the phoenix)

An inconspicuous constellation in the southern
hemisphere of the sky near the foot of Eridanus,
introduced on the 1603 star map of Johann BAYER.
There are no objects of particular interest.

photoelectric cell

Device which produces an electrical output
corresponding to the amount of light striking a
sensitive surface (the photoelectric effect). Simple
photocells, such as those used in light meters or

cameras, are rarely sensitive enough for use in astronomy. Instead, the most widely used device is the *photomultiplier tube,* which amplifies the energy of electrons liberated by the impact of light on the sensitive surface.

The tube is made of glass, with the sensitive layer at one end. Metal rings along the interior of the tube carry increasing electrical potentials which accelerate the electrons given off. The electron flow constitutes an electric current proportional to the light's intensity, and can be measured on a suitable meter. Such devices are known as photoelectric photometers.

photographic magnitude

A star's apparent brightness as measured on a photographic plate sensitive to the blue end of the spectrum. Until the 1920s, black-and-white photographic plates used for astronomy were sensitive only to blue and ultraviolet light, and brightnesses measured from them did not compare with eye estimates. A blue star, for example, would appear bright on a photograph while a red star, which might appear equally bright to the eye, would show up only faintly on the photograph. Photographic magnitudes (symbol: m_{pg}) are still occasionally used, but they have been largely replaced by the B magnitude of the UBV SYSTEM.

photometer

A device for measuring the brightness of objects. Early photometers required a human observer to make visual comparisons between the object under study and a reference of known brightness, such as a standard star. Today, electronic devices incorporating PHOTOELECTRIC CELLS (or, more usually, *photomultiplier tubes*) are used; these are known as photoelectric photometers. Filters can also help measure the brightness of the object in different wavelengths (see UBV SYSTEM).

photometry

The measurement of an object's brightness; the devices used are called *photometers.* Astronomers employ the MAGNITUDE scale for recording brightnesses. The simplest photometry is based on visual estimates of star magnitudes, as first used by the Greek HIPPARCHUS. Many amateur astronomers still make accurate visual estimates of star brightnesses to chart the fluctuations of VARIABLE STARS. However, most modern photometry is carried out either by measuring the brightnesses of individual stars with a telescope and photometer, or by photographing a field of stars and measuring the sizes of the images on the developed plate.

Photoelectric photometry—using photoelectric cells—can record brightness to within a hundredth of a magnitude and gives immediate results, but it can use too much valuable telescope time if a large number of stars are to be measured. Photographic photometry requires simpler apparatus and less telescope time, but is generally not as accurate.

The light from a bright star will spread out more in the photographic emulsion than would light from a faint one; it "burns" a larger image, and the diameter of a star's photographic image thus reveals its brightness. Because of the nature of photographic emulsions, however, a doubling in star brightness does not produce a doubling in image area—the relationship varies with brightness. Consequently, stars of known magnitude must be used as references. A photoelectric photometer, by contrast, has more regular characteristics, and fewer reference stars over the entire range are needed.

In either system, filters may be placed in the light path in order to match the color sensitivity of the detector to the UBV SYSTEM. This is known as *multicolor photometry.*

photosphere

The visible surface of the Sun. The photosphere is not an actual solid surface, but, instead, a layer of relatively dense gas about 200 miles (300 km) thick, whose temperature ranges from 9,000° at the bottom to 4,300° where it merges with the CHROMOSPHERE. Almost all the light we receive from the Sun comes from the photosphere, although the energy source lies far deeper inside. It is believed that energy travels by radiation through the Sun's core and is transported to the surface by great convection currents, the tops of which form GRANULATION cells making up the fine structure of the photosphere. Each hexagonal granule is only about 600 miles (1,000 km) across, and so the network is extremely difficult to see through the Earth's turbulent atmosphere. The centers of granules consist of rising gas about 100° hotter than the surroundings, while cooler gas flows downward at their dark boundaries; their lifetime is only 10 minutes. In addition to the granules, the photosphere also shows larger-scale turbulent motions. Other features, such as SUNSPOTS and FACULAE, are visible on the surface; but these only appear when strong magnetic fields prevent the normal convective motions.

The spectrum of the photosphere is that of a hot body at 6,000° (continuous spectrum) crossed by dark absorption lines (the FRAUNHOFER LINES). These lines reveal the relative abundances of chemical elements making up the Sun's outer layers, found to be 90 percent hydrogen and 8 percent helium, with small amounts of oxygen, carbon, nitrogen, magnesium, silicon, and iron. The gases of the photosphere are thought to be representative of the original material from which the Sun and planets formed; only the innermost parts of the Sun have been altered by nuclear reactions.

photovisual magnitude

The magnitude of brightness of a star measured by using a photographic system sensitive to the same range of colors as the eye. The color sensitivity of photographic emulsions does not quite match that of the eye; some colors are slightly emphasized and some diminished on a photograph. Filters used in combination with panchromatic emulsions do, however, provide an overall color sensitivity very similar to that of the eye. Magnitudes measured from a photograph taken this way are photovisual (symbol: m_{pv}).

Piazzi, Giuseppe (1746–1826)

Italian astronomer, who on January 1, 1801, discovered the first asteroid, Ceres, while making a star catalog at the Observatory of Palermo, of which he was founder and was appointed first director in 1790. Although Ceres looked like an ordinary star, Piazzi was

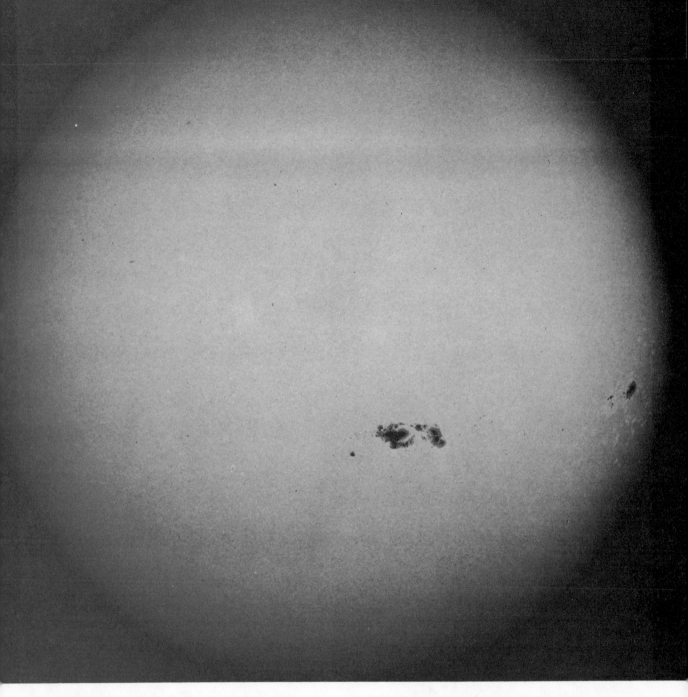

The Sun's photosphere, showing typical surface features; a large sunspot group, **center**, and a smaller foreshortened group, **right**, which is about to be carried around the limb by the Sun's rotation. The pronounced limb-darkening allows the bright facules surrounding the right-hand group to be clearly seen.

suspicious because it was not marked on the star list of Nicolas Louis de LACAILLE. He found that it moved from night to night like a small planet or comet; he lost it on February 11, 1801, as it moved into the Sun's glare. Calculations by the German mathematician Carl Friedrich GAUSS showed that the object was moving in the large gap between the orbits of Mars and Jupiter, where a group of German astronomers were searching for an unknown planet;

one of these astronomers, Heinrich OLBERS, relocated Ceres in 1802 from Gauss' predictions. In 1814 Piazzi published his catalog, containing 7,646 stars and showing the proper motions of many of them. One faint star, 61 Cygni, showed a particularly large proper motion, indicating its closeness to the Sun, a fact verified by the German Friedrich BESSEL. The thousandth asteroid, discovered in 1923, was named Piazzia in his honor.

Pic du Midi Observatory
An astronomical observatory operated by the University of Toulouse, at an altitude of 9,390 feet (2,862 m) in the French Pyrenees. Founded in 1882, it was one of the world's first high-altitude observatories, designed to exploit the advantages of improved seeing high above the turbulent layers of the Earth's

atmosphere. It was here that Bernard LYOT installed his coronagraph in 1930. A 24-inch (61-cm) refractor, installed in 1943, was used for renowned observations of the Moon and planets; it was removed in 1972. In 1964 a 43-inch (110-cm) reflector came into operation, and a 78-inch (200-cm) reflector is under construction.

Pickering, Edward Charles (1846–1919)

American astrophysicist, renowned for his major catalogs of star brightness and spectra. Trained as a physicist, Pickering became director of Harvard College Observatory in 1877 and inaugurated astrophysical studies there. He developed the meridian photometer, a device for visually comparing the brightness of a given star with Polaris, chosen as a magnitude standard. The first great star-brightness catalog, containing 4,260 stars, was published in 1884; it is known as the *Harvard Photometry*. The *Revised Harvard Photometry* later extended the work to a total of over 50,000 stars. (Unfortunately, Polaris has since been found to be slightly variable in brightness and therefore provides an unreliable comparison.) Another major project of the observatory was the *Henry Draper Catalogue,* a classification of stellar spectra; most of this work was performed by Annie J. CANNON. Pickering believed in the value of astrophotography, and set about producing a PHOTOGRAPHIC MAGNITUDE scale for stars. In 1903 he issued the first-ever *Photographic Map of the Entire Sky,* consisting of 55 plates down to magnitude 12 taken at Harvard and its southern station at Arequipa, Peru. Additionally, Pickering photographed large areas of sky on each clear night, building up a 300,000-plate Harvard photographic library, invaluable to astronomers searching for changes in the brightness and position of objects. In 1889 Pickering found the first SPECTROSCOPIC BINARY star, Mizar; he worked with Antonia Caetana de Paiva Pereira Maury (1866–1952), a niece of Henry Draper, who also developed a classification system for stellar spectra. Pickering, with his international contacts, made Harvard the center for dissemination of astronomical news and information, a role it has maintained in conjunction with the SMITHSONIAN ASTROPHYSICAL OBSERVATORY.

Pickering, William Henry (1858–1938)

American astronomer, brother of E. C. PICKERING. In 1891 he helped found Harvard's southern station at Arequipa, Peru, where in 1898 he discovered Saturn's ninth satellite, Phoebe. Pickering also reported a tenth satellite, which he called Themis, in 1905; it has never been seen again. Probably he mistook a faint star, passing asteroid, or blemish on the photographic plate for a new satellite. In 1903 Pickering published a famous photographic atlas of the Moon, which he had taken at Harvard's station in Jamaica; this became his private observatory after retirement in 1924. Pickering claimed he saw evidence of vegetation and frost on the Moon, but in fact he misinterpreted small spots on the surface whose contrast changes under varying angles of illumination. Pickering helped set up Percival LOWELL's observatory at Flagstaff, Arizona. Pickering, like Lowell, was interested in Mars, on which he discovered the dark spots termed "oases;" also like Lowell, he tried to predict the position of a planet beyond Neptune. Photographic plates taken

for Pickering at Mount Wilson Observatory actually contained images of Pluto, but one was masked by a blemish and the other was superimposed on a star; consequently they were not noticed until after Pluto had been officially discovered by Clyde TOMBAUGH in 1930.

Pictor (the painter)

A faint constellation of the southern hemisphere of the sky, introduced by Nicolas Louis de Lacaille; it contains no important objects.

Pioneer spacecraft

A continuing series of American space probes to explore the solar system. The first Pioneers were intended as Moon probes, though none succeeded. Pioneers 5 through 9 were interplanetary monitors, measuring radiation from solar storms and changes in the interplanetary magnetic field. Their results helped evaluate the danger to astronauts of solar flares, and by monitoring the side of the Sun turned away from Earth gave notice of solar outbursts that the Sun's rotation would bring into view. Pioneers 10 and 11 were the first space probes to reach Jupiter. Two Pioneer-type probes, called Pioneer-Venus 1 and 2, were sent to Venus in 1978. Pioneer-Venus 1 orbited the planet, while Pioneer-Venus 2 plunged through its atmosphere to study conditions.

Craft	Launch date	Remarks
Pioneer 1	October 11, 1958	Intended lunar orbiter, to send back TV pictures. Fell short due to insufficient thrust, but reached 70,717 miles (113,800 km) from Earth, mapping extent of Van Allen radiation belts
Pioneer 2	November 8, 1958	As Pioneer 1. Launch failure
Pioneer 3	December 6, 1958	Intended lunar flyby; insufficient launch thrust. Reached 63,580 miles (102,300 km) from Earth, mapping intensity variations of Van Allen radiation belts
Pioneer 4	March 3, 1959	Passed 37,300 miles (60,000 km) from Moon
Pioneer 5	March 11, 1960	Interplanetary probe, orbiting Sun between Earth and Venus; sent data on solar flares and particles until June 26, 1960
Pioneer 6	December 16, 1965	Interplanetary probe, orbiting Sun between Earth and Venus
Pioneer 7	August 17, 1966	Interplanetary probe, orbiting Sun between Earth and Mars; with Pioneer 6 monitored solar activity
Pioneer 8	December 13, 1967	Interplanetary probe; orbiting Sun slightly farther than Earth
Pioneer 9	November 8, 1968	Interplanetary probe; orbiting Sun between Earth and Venus
Pioneer 10	March 3, 1972	Bypassed Jupiter at 81,000 miles (130,000 km) on December 3, 1973. Now on a trajectory that will eventually take it out of the solar system

Craft	Launch date	Remarks
Pioneer 11	April 5, 1973	Bypassed Jupiter at 26,725 miles (43,000 km) on December 3, 1974. Due to reach Saturn in September 1979
Pioneer-Venus 1	May 20, 1978	Went into orbit around Venus December 4, 1978. Photographed clouds and made radar map of the surface
Pioneer-Venus 2	August 8, 1978	Entered Venus atmosphere December 9, 1978, ejecting four sub-probes to examine atmospheric conditions

Pisces (the fishes)

A constellation of the zodiac, best seen during the northern hemisphere autumn. The Sun passes through Pisces from mid-March to mid-April. None of its stars is particularly prominent; it is best located by its proximity to the famous square of Pegasus. Alpha Piscium, or Alrisha, is a multiple star of combined magnitude 3.94; the components are of magnitudes 4.33 and 5.23, revolving with a period of 720 years. Each star is also a spectroscopic binary. The most important feature of Pisces is that it contains the VERNAL EQUINOX—the point where the Sun's path around the sky cuts the celestial equator. This is the point from which the celestial coordinate of RIGHT ASCENSION is measured.

Piscis Austrinus (the southern fish)

A constellation of the southern hemisphere of the sky below Cetus, best seen during the northern hemisphere summer; its name is also written Piscis Australis. Its brightest star is FOMALHAUT.

plage

A brighter, denser region found at all levels of the solar atmosphere. The term is often reserved for a bright, dense region in the CHROMOSPHERE, where it can be studied with a spectrohelioscope. A similar region in the upper PHOTOSPHERE is often called a FACULA. X-ray and radio plages can be observed in the CORONA.

planet

A nonluminous body that shines by reflecting sunlight. Planets can be made of rock and metal, like our Earth, or of gas, like Jupiter. Any such object orbiting a star can be termed a planet; some planets may also drift dark and unknown in space, free from any star. In our own solar system there are nine major planets orbiting a glowing star, the Sun. Smaller bodies, which range from about the size of bricks to several hundred miles across, are termed minor planets or ASTEROIDS; sometimes the name *planetesimal* or *planetoid* is used. However, no real distinction can be drawn between a major planet and what is termed an asteroid. The upper limit to the size of a planet is about 1 percent the mass of our Sun (roughly 10 times the mass of Jupiter); above this the temperature and pressure at the object's center are enough for nuclear reactions to begin, so that it becomes a small star in its own right.

Formation of planetary systems. Planetary systems are believed to form as natural byproducts of the origin of stars. A star is born when a cloud of dust and gas in the Galaxy begins to callapse under its own gravitational pull (see STAR). The embryo star forms at the cloud's center, surrounded by a rotating disk of left-over gas and dust. The dust grains stick together to form a carpet of solid particles in the plane of the doughnut-shaped disk, growing bigger by collisions until their gravity is enough to pull more material toward them, eventually forming a planet. The heat of the young star evaporates the most volatile substances near to it, leaving an inner band of rocky planets. Farther away, the forming planets can acquire a dense cloak of gas from the cloud around the young star to produce gaseous giants like Jupiter; while in the farthest reaches of the planetary system the gases freeze to ice, producing a cloud of comets.

The remaining gas is driven away from the planetary system forever by the SOLAR WIND of atomic particles from the star, and left-over rock and dust particles are gradually swept up by the fully-formed planets, producing a bombardment such as scarred the surfaces of the Moon, Mars, and Mercury. Our Sun and planets are believed to have formed in this way 4.6 billion years ago. Computer models of planetary formation show that other planetary systems would look very similar.

Planets of other stars. About half the stars in the sky have one or more visible companions, forming double or multiple star systems. Given enough material in the disk around a young star, a second star may be formed, producing a close binary system. Twin systems of wider separation, with orbital periods over 100 years, may arise when the initial collapsing cloud of gas and dust fragments into several parts. The remaining stars, which seem to have no companions, may instead have planets. If each planetary system contains as many planets as our own solar system, there may even be more planets than stars in the Galaxy. Planets of other stars are too faint to be visible with Earth-based telescopes, although sensitive telescopes in space may eventually detect them. Instead, astronomers must look for the wobble in a star's PROPER MOTION across the sky, caused as a star and its planetary system rotate around their common center of gravity, like a dumbbell. Such a wobble has been detected by Peter VAN DE KAMP for BARNARD'S STAR, a red dwarf that is the second-closest star to the Sun (the closest star is actually the triple system of ALPHA CENTAURI). According to Van De Kamp, two giant planets similar to Jupiter may orbit Barnard's star; smaller planets the size of Earth might also exist, but their effects would be too small to notice. Other nearby stars suspected of having planetary companions are Epsilon Eridani, Lalande 21185 (in the constellation Ursa Major), and possibly Tau Ceti (see also LIFE IN THE UNIVERSE).

planetary nebula

A gaseous shell surrounding a hot central star. Their resemblance to planets (especially URANUS) when viewed through a small telescope led Sir William HERSCHEL to give them their name in 1785. They exhibit a variety of forms, as shown by such names as

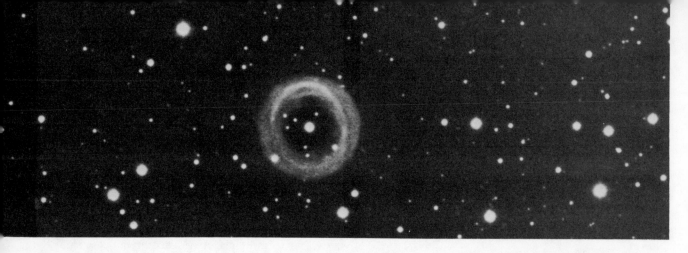

The planetary nebula Shapley I, showing the faint central star surrounded by its huge ejected shell of gas. Although the gas cloud is made to glow brightly by the radiation from the star, it is very tenuous and has barely one-tenth of the Sun's mass.

the "Dumb-bell" (M27), the "Ring" (M57), and the "Owl" (M97), although others have a more formless structure. Some 1,000 are known (it is estimated that perhaps 60,000 exist), concentrated toward the center of our Galaxy. This location demonstrates that they are old, POPULATION II objects.

The central star of a planetary nebula is very hot, with a surface temperature between 30,000°C and 150,000°C. At these temperatures, most of the radiation is emitted at ultraviolet wavelengths, so that the star appears faint optically; in some cases it is completely invisible. Although most of these stars are about the same mass as the Sun, a few are more massive WOLF-RAYET STARS. The nebula itself is a thin shell of gas that glows by absorbing ultraviolet radiation from the central star and reemitting it as visible light. Most of the gas is hot hydrogen, only 1,000 times denser than the hydrogen in space. The nebulae have diameters of the order of 40,000 a.u.; yet the amount of mass in this volume is estimated to be only 10 percent that of the Sun. Spectroscopic observations show planetary nebulae to be in a state of turbulent internal motion due to intense heating from the central star; they are also expanding at 12 to 30 miles (20–50 km) per second. This indicates they have been ejected from the central star, and that they disperse into space over a period of only about 10,000 years.

It has been suggested that planetary nebulae explosions are one way in which a massive star can throw off enough material to come within the Chandrasekhar limit (1.5 times the mass of the Sun), below which a star must fall in order to become a WHITE DWARF. Rough calculations involving the numbers of white dwarfs and planetary nebulae in the Galaxy tentatively support this theory, indicating that the planetary nebula phase may be a rapid one in the evolution of many stars.

plasma

A very high-temperature gas consisting of negatively charged electrons and positively charged atomic nuclei, or IONS. The atoms of the gas are broken up into these constituent parts either by collisions occurring between the atoms in a hot gas, or as a result of being struck by high-energy ultraviolet radiation from a nearby hot star. As a whole, a plasma cloud has no electric charge, but unlike an ordinary gas it is an electrical conductor and can thus be strongly affected by magnetic fields in space. Most of the matter in the Universe is in a plasma state (at temperatures from 10,000 to 100 million degrees), including the interiors of stars, interplanetary gas (like the SOLAR WIND), much of the interstellar gas (H II REGIONS), and the gas that fills giant clusters of galaxies.

Pleiades

A cluster of stars in the constellation Taurus, also known as M45. The Pleiades contain six or seven naked-eye stars, together with some 200 fainter members within a radius of 1°. The Pleiades are young stars, born within about the last 50 million years, and are still surrounded by remnants of the nebula from which they formed. The densest part of the nebula surrounds the magnitude-4.25 star Merope. The Pleiades are about 415 light-years away, three

This long-exposure photograph reveals clouds of dusty gas which reflect the light from the young, bright stars of the Pleiades cluster. Only six or seven stars are seen by the unaided eye, but the telescope reveals about 200 members, formed only about 60 million years ago. *The Kitt Peak National Observatory*

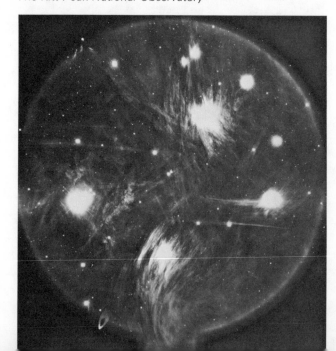

times farther than the HYADES cluster which lies nearby in the sky. The brightest Pleiades stars are BLUE GIANTS, about one-twelfth the age of the fainter and less massive stars of the Hyades. One of the bright stars, Pleione, is a so-called *shell star,* which is rotating rapidly and occasionally throws off a shell of gas in a small NOVA-like explosion. The brightest of the Pleiades is Alcyone (Eta Tauri), of magnitude 2.96.

Plesetsk
Soviet space launch site 105 miles (170 km) south of Arkhangel in northern Russia. The first space launch from Plesetsk, a former missile site, was Cosmos 112, on March 17, 1966; it has since become the most important Soviet launch site for unmanned satellites. Satellites from Plesetsk are sent into polar or high-inclination orbits, for communications, weather monitoring, and military reconnaissance.

Plow
See URSA MAJOR.

Pluto
The planet of greatest average distance from the Sun. Pluto follows a highly unusual elliptical path inclined at more than 17° to the solar system plane, moving between 2.750 and 4.582 billion miles (4.425 and 7.375 billion km) from the Sun; its average distance is 3.666 billion miles (5.90 billion km). The planet takes 247.7 years to complete one orbit and will not return to its discovery position until the year 2177. Because of its highly elliptical orbit Pluto is closer to the Sun than is Neptune between January 1979 and March 1999. Pluto rotates on its axis in 6.39 days. Its diameter is not known precisely, but it seems to be no more than about 2,000 miles (3,000 km), making it the smallest known planet. Its surface is believed to be largely covered by frozen methane, at a temperature of about −230°C. In 1978 a satellite of Pluto was found orbiting at a center-to-center distance of 10,500 miles (17,000 km). The moon's orbital period, 6.39 days, is the same as the rotation period of Pluto. The diameter of this moon is about 0·4 that of Pluto. Its existence allows astronomers to make their first accurate calculation of the mass of Pluto, which turns out to be only about 0.002 the mass of the Earth. The planet's density is similar to that of water. Pluto is evidently like a snowball of frozen gases.

Discovery. Pluto was the second planet to be discovered as the result of a deliberate search, and the first to be found by photography. After the discovery of Neptune in 1846 through calculations of its predicted gravitational effect on the motion of Uranus, Uranus still did not seem to be moving as expected. Astronomers therefore suspected the perturbing effect of yet another planet. Various astronomers, notably Percival LOWELL and William H. PICKERING, issued predictions of the unknown planet's position. Nothing was discovered, however, and so in 1929 the Lowell Observatory in Arizona began a concerted search using a 13-inch (33-cm) photographic telescope. The planet Pluto was discovered on February 18, 1930, by Clyde TOMBAUGH, from plates taken on January 21 and 29.

Pluto's nature. Unfortunately, the newcomer was smaller and fainter than expected, for it showed no rounded disk like Uranus or Neptune.

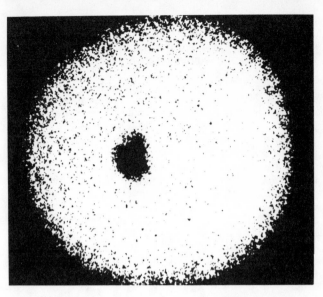

Pluto and its satellite, provisionally named Charon, merge to form an elongated blob in this discovery photograph taken with the 155-cm reflector of the U.S. Naval Observatory. The moon, about 40 per cent Pluto's diameter, orbits at the same rate as the planet spins, therefore remaining stationary above one spot on the planet.

Estimates of Pluto's diameter in 1950 by Gerard KUIPER, using the 200-inch (505-cm) Hale telescope, gave a figure of under 4,000 miles (6,400 km). Latest studies show Pluto is smaller still, and is in fact the smallest and lightest planet of the solar system. Pluto is therefore far too small to have had a significant effect on the paths of Uranus and Neptune; it was in fact found quite by accident. The apparent perturbations in the motion of the two outer giant planets must have been caused by inaccurate early observations, together with a degree of uncertainty about their true orbits.

It is conceivable that Pluto is not a true planet at all, but an escaped satellite of Neptune. Another possibility is that Neptune is the most distant of the family of giant planets, and that Pluto is the brightest member of a swarm of small planets moving in inclined, elliptical orbits in the outer regions of the solar system.

polar distance
The angle between an object and the celestial pole.

Polaris
The north pole star, lying in the constellation of Ursa Minor, the lesser bear, and also known as Alpha Ursae Minoris; an alternative name is Cynosura. Polaris is a double star. The main star, a yellow supergiant, is actually a Cepheid variable, of the so-called W Virginis type, varying between magnitudes 2.1 and 2.2 every 3.97 days (the average value is 2.12). It has a spectroscopic companion orbiting it every 29.6 years. Polaris also has an optical companion (not physically connected) of magnitude 9. Polaris does not lie exactly at the celestial pole, but about 1° from it. Because of PRECESSION, the celestial pole will pass closest to Polaris about the year 2012, and then move away again.

polarization

Radiation whose waves oscillate only in certain planes along the direction of travel is said to be polarized. Ordinary unpolarized light can be thought of as consisting of waves which oscillate at right angles to the direction of travel—not only up and down but in all planes. Polarization occurs if the oscillations are restricted so that only those waves in a particular plane are transmitted; this is known as *linear polarization. Circular polarization* occurs when the plane of polarization rotates with time. Radiation can be polarized when it is reflected from interstellar dust or from the dust in a nebula such as that around the Pleiades (see NEBULA). Radiation emitted by electrons in a magnetic field (SYNCHROTRON RADIATION) is also polarized. The polarization of light waves can be analyzed by a polarizing filter. Polarized radio waves are detected by a dipole antenna (a simple rod shape). The rod will receive only those waves that are polarized at right angles to it. Turning the dipole at different angles will vary the signal strength if the source is polarized.

pole

The end of an axis of rotation. The Earth's geographical poles are at the north and south ends of its rotation axis. Projected onto the CELESTIAL SPHERE, these points form the north and south celestial poles, about which the sky seems to rotate (actually an effect of the Earth's spin). The *galactic pole* is at 90° to the plane of the Galaxy, and the *ecliptic pole* lies at 90° to the plane of the Earth's orbit round the Sun (the ecliptic).

pole star

The bright star nearest the celestial pole. The north pole star is currently POLARIS. Because of the drifting effect of the Earth's axis called PRECESSION, the position of the celestial pole against the stars changes with time, so that other stars can become the pole star. In about 5,000 years, for example, the star Alpha Cephei will be the pole star, and in 12,000 years Vega. There is currently no bright star near the south celestial pole.

Pollux

The brightest star in the constellation Gemini, the twins; it is also known as Beta Geminorum. Pollux is an orange giant star, 35 light-years away, and of magnitude 1.15. It is the nearest giant star to the Sun.

Pond, John (1767–1836)

Sixth British astronomer royal, appointed in 1811. Pond's private observations, made before he had become astronomer royal, had revealed errors in star positions measured at Greenwich, caused by warping of the observatory's quadrant instrument with old age. As astronomer royal Pond therefore began to reequip the Greenwich Observatory for modern astronomy; he also established new observing methods and increased the staff from one to six. The result was an 1833 catalog listing 1,113 stars to hitherto unattained standards of accuracy. In the same year Pond introduced the first daily public time signal at Greenwich, by dropping a time ball at one o'clock down a long pole. In 1820 he recommended the founding of the CAPE OBSERVATORY in South Africa. Pond's predecessor, Nevil MASKELYNE, had begun publication of tables for seamen, called the *Nautical Almanac.* During Pond's time a separate *Nautical Almanac* Office was set up to supervise the publication; it continues today. Pond retired in 1835.

populations, stellar

The division of stars and star systems into two categories on the basis of age, population I being young stars, and population II old stars. Walter BAADE was the first to make this distinction, after observing that the bright stars in the center of the ANDROMEDA GALAXY are red, while those in its spiral arms are blue. He called the former region population II, and noted the similarity with GLOBULAR CLUSTERS in our own Galaxy. The population I spiral arms resemble our younger OPEN CLUSTERS. Baade's classification is still useful, although it is now recognized that there is a continuous range of intermediate types.

Population II stars formed some 10 billion years ago, when the galaxies first condensed. Since that time, all the massive, hot, blue stars have evolved to become red giants, and only red stars (both giants and dwarfs) remain. These stars condensed before SPIRAL GALAXIES like our own flattened to a disk, and they consequently still form a spherical halo around the Galaxy. (In elliptical galaxies all the gas condensed at the same time, and they contain only population II stars.)

Population II stars near the Sun are often found to have high velocities, because they are in fact relatively stationary while the Sun orbits the galactic center at 150 miles (250 km) per second. Because the galaxies condensed from gas containing only hydrogen and helium, population II stars contain very little of the so-called HEAVY ELEMENTS.

Many of the stars in the Galaxy's disk are only slightly younger than the population II stars, and are known as *intermediate population stars.*

Concentrated in a thin layer in the disk (the galactic plane) are the young population I stars, the youngest of which (extreme population I) are still forming from the interstellar gas and dust clouds. These nebulae and their associated young, hot, blue stars are also observed in small irregular galaxies (like the MAGELLANIC CLOUDS), and in the disks of other spiral galaxies.

Population I stars near the Sun share its motion around the Galaxy, and therefore have low measured velocities. Their composition is similar to the Sun's (a population I star). They contain heavy elements, which were created within previous generations of stars and returned to the nebulae by SUPERNOVA explosions.

position angle

Relative positions of two objects, such as the two components of a double star, measured as an angle from north through east.

Poynting-Robertson effect

The slowing in the motion of small meteor particles through their collision with solar radiation. This causes their orbits to become smaller, so that they spiral into the Sun. The effect was first predicted in

1903 by the British physicist John Henry Poynting (1852–1914), and calculated in detail in 1937 by the American physicist Howard Percy Robertson (1903–1961). Very large bodies, such as the planets, are unaffected by the collisions, but millimeter-sized particles in the asteroid belt are forced into the Sun in about 60 million years. However, the very smallest specks of dust and gas molecules in the solar system are actually blown away from the Sun by RADIATION PRESSURE.

precession

A slow wobbling of the Earth on its axis, like the wobble of a spinning top whose axis is not upright, but far slower. The net effect is to change the part of the sky at which the Earth's axis points. The Greek astronomer HIPPARCHUS discovered this effect 2,000 years ago, although its explanation had to wait for Isaac Newton's work on gravitation.

Precession is caused by the gravitational pulls of the Moon and Sun on the Earth's slight equatorial bulge. This bulge arises because of the Earth's rotation; the planet's equatorial diameter is about 26 miles (42 km) greater than its polar diameter. The Earth's equator, and hence also the bulge, is inclined at about $23\frac{1}{2}°$ to the plane of the Earth's orbit (the *ecliptic*); the Sun and Moon pull on the bulge, as if to tilt our planet back to the vertical. But instead of tilting upright, the Earth's axis swings in a cone-shaped motion, still at $23\frac{1}{2}°$ to the vertical; neither the tilt of the Earth's axis, nor the position of the poles on the globe, is changed by precession. The Earth's axis takes about 26,000 years to swing around once; this is called a *cycle of precession*.

During each cycle of precession the Earth's poles trace out a circle on the sky. Precession therefore slowly changes the position of the celestial poles. Although POLARIS is the pole star today, in 12,000 years the pole will have drifted near to Vega. The changing orientation of the Earth with respect to the stars affects the positioning against the star background of the *equinoxes,* the points at which the Earth's equator intersects the ecliptic (the plane of the Earth's orbit). The equinoxes slide once around the sky every 26,000 years, in what is termed the *precession of the equinoxes*; the equinoxes move about 50 seconds of arc westward against the star background each year. Precession also shifts the star coordinates known as right ascension and declination; this is why star positions are always given for a certain *epoch,* or reference date (currently 1950 or 2000). A winter constellation such as Orion will be seen in the summer skies after half a cycle of precession. The seasons will not be affected, however, because our calendar is based on the movement of the Sun, and the first day of northern spring will always fall around March 21 (the spring equinox). The spring equinox lay in the constellation of Aries 2,000 years ago, and is still referred to as the First Point of Aries. However, precession has now moved it into the constellation of Pisces, and it will reach the constellation of Aquarius in about 600 years. So the much-heralded age of Aquarius will not be with us for some time yet.

prime focus

The point at which the primary mirror in a reflecting telescope focuses an image. The advantages of using the prime focus for photographic work include greater efficiency, since only one reflection occurs; smaller image scale because of the shorter focal length (mirror-image distance), and hence faster registration of extended objects such as nebulae; and a wider field of view. Very large instruments, such as the 158-inch (400-cm) reflector at Kitt Peak, Arizona, and the 200-inch (508-cm) reflector at Mount Palomar, California, have a "cage" inside the tube so that the observer can sit and guide from the prime focus while taking a photograph.

Procyon

The eighth-brightest star in the sky, and the brightest star in the constellation Canis Minor; it is also called Alpha Canis Minoris. Procyon is a brilliant yellow star of magnitude 0.34, 11.4 light-years away. Its diameter is 2.17 times that of the Sun, and its mass 1.74 Suns. Procyon is actually a double star, with a faint white dwarf companion of magnitude 10.8, mass 0.63 times the Sun, and 0.01 the Sun's diameter, orbiting every 41 years.

Prognoz satellites

A series of Soviet scientific satellites to monitor solar activity and the interaction of solar atomic particles with the Earth's surroundings. The name Prognoz means "forecast." The Prognoz satellites travel in elliptical orbits, 124,000 miles (200,000 km) from Earth at their farthest, almost halfway to the Moon. They are therefore similar in nature to the American IMP (Interplanetary Monitoring Platform) series. Prognoz 1 was launched in April 1972; the most recent (October 1978) was Prognoz 7.

prograde motion

Movement of an object from west to east, the usual direction in the solar system. An alternative name is *direct motion*. Movement in the opposite direction (east to west) is called RETROGRADE MOTION.

prominence

A hot, bright cloud of gas projecting from the Sun's CHROMOSPHERE into the CORONA. There are two main types: quiescent prominences, which may last for months; and eruptive prominences, which have lifetimes of a few hours and are often associated with FLARES.

Quiescent prominences are also called "filaments," because they appear as dark, ribbonlike structures projected against the Sun's disk in spectroheliograms. They can be thousands of miles long. About a third are associated with sunspot groups, in which the prominence material, mainly hydrogen gas at a temperature of 10,000°, is observed to stream downward into the spots. Such filaments may suddenly become activated and their material blown away from the Sun at speeds of hundreds of miles per second. Most, however, are long-lived and unchanging; they are supported in the corona by the magnetic field in a manner still not understood.

Eruptive prominences may be composed either of material ejected from the Sun after a flare (surge prominences), or of condensing material falling back to the surface long after a flare has taken place (loop prominences). Their trajectories also seem to be

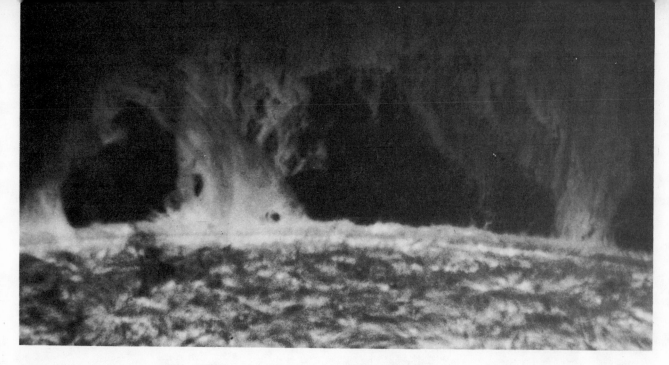

A quiescent prominence arching 38,000 miles (60,000 km) above the solar limb, following the loops of magnetic field between two sunspots on the photosphere below. The material flows downward, toward the points where the field comes out of the Sun; but if activated, it may suddenly blow off into space.

determined by the local magnetic field, although their velocities are higher than those of quiescent prominences—up to 600 miles (1,000 km) per second. The gas of which they are composed is also at a higher temperature, and may reach 30,000°.

Both types of prominences owe their origin to loops of magnetic field breaking through the solar photosphere, a phenomenon which also gives rise to SUNSPOTS, PLAGES, and flares. Like these, the frequency of prominences is governed by the SOLAR CYCLE.

proper motion

The motion of a star across the sky, expressed in arc seconds per year. The absolute proper motion of bright stars can be determined from precise position measurements over a period of years. The relative proper motion of a fainter star is found by comparing its position relative to other stars on photographs taken with the same telescope, some 5 to 50 years apart. Its absolute proper motion can then be calculated from the known proper motions of the brighter comparison stars. The average proper motion of all naked-eye stars is about 0.1 arc second per year; nearby stars generally show larger proper motions. The largest known proper motion, 10.27 arc seconds a year, belongs to BARNARD'S STAR.

Proton launcher

A Soviet rocket first used in 1965 to launch the series of PROTON SATELLITES, and subsequently employed for Soviet lunar and planetary probes and the SALYUT space station. In its performance it is midway between the American Saturn 1B and Saturn V rockets. The Proton launcher has a first stage made of a central core surrounded by six strap-on boosters, with upper stages added as necessary. It can place about 50,000 lb. (23,000 kg) in Earth orbit, send 12,000 lb. (5,500 kg) to the Moon, or 10,000 lb. (4,500 kg) to the planets. The diameter of its central body is about 13 feet (4 m), and its overall length is up to 225 feet (68.5 m), including the Salyut space station. Its first stage has a thrust of about 3,300,000 lb. (1,500,000 kg). The Proton launcher is sometimes known in the West as the D type.

proton-proton chain

A nuclear process, similar to that which occurs in the hydrogen bomb, by which energy is produced in stars like the Sun. At the beginning of the proton-proton chain, two protons (hydrogen nuclei) combine to form a deuterium nucleus or *deuteron*. A positron (positively charged electron) and a neutrino are also produced in this reaction. Because the neutrino is electrically neutral, its passage through the star is unimpeded. The deuteron is short-lived; within about five seconds, it interacts with further protons to form helium, releasing electromagnetic radiation in the form of gamma rays. This initial form of helium, helium-3, is unstable, and the final step in the chain is the reaction between two colliding helium-3 nuclei. A stable helium-4 nucleus is formed with the release of two protons. The net effect of the chain is to transform four hydrogen atoms into one helium atom. In the Sun, about 600 million tons of hydrogen are converted to helium each second; in the course of this process, roughly 4 million tons of matter are turned into energy. The proton-proton reaction predominates at temperatures below about 15 million degrees K. At the hotter temperatures inside large stars, the CARBON-NITROGEN-OXYGEN CYCLE prevails.

Proton satellites

A series of Soviet Earth satellites whose stated intention was to study high-energy cosmic rays. However, their great weight—12 tons for the first three and 17 tons for the last—together with the fact that they were launched with a new high-performance rocket (the PROTON LAUNCHER) suggested that they

Above Lagoon nebula, also known as M8, is a glowing cloud of gas in the constellation Sagittarius. The bright nebula is bisected by a dark lane of dust. This photograph was taken by the Anglo-Australian Observatory.

Below Centaurus A is a famous radio galaxy. In this picture from the Anglo-Australian Telescope it appears as a giant elliptical galaxy crossed by a dark lane of dust. Radio emission comes from two invisible lobes on either side of the galaxy.

Left The Earth, photographed by Apollo astronauts between the Earth and Moon. The outlines of Africa and Arabia are easily recognized. Fine cirrus cloud, too faint to show up, cuts down the visibility of the Sahara region. Clouds rim many coastlines, while hurricanes rage in the southern oceans. The south polar cap is also cloud-covered.

Below left An Apollo Moon mission takes off using the giant Saturn V booster. Nine Apollo craft went to the Moon; three of them (8, 10 and 13) did not land, however. The total cost of the Apollo program was 25 billion dollars, of which one quarter was spent on developing Saturn V.

Below The Soviet Soyuz 9 flight being prepared for launch from the Tyuratam Cosmodrome in 1970 — the first Soviet launch to be televized. This same pad on the barren steppes of Kazakhstan was used for the launches of Sputnik 1 in 1957 and Yuri Gagarin — the world's first spaceman — in 1961.

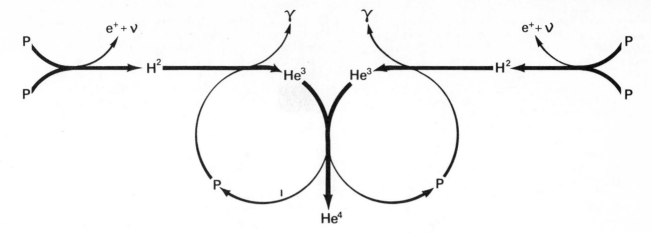

Four protons (p) are converted to a helium nucleus with release of radiation (r) neutrinos (ν) and protons (e^+) in a chain involving deuterium (H^2) and the lighter isotope of helium (He^3).

were actually associated with development of future space projects such as the SALYUT manned space station. Proton 1 was launched on July 16, 1965; Proton 2 on November 2, 1965; Proton 3 on July 6, 1966; and Proton 4 on November 16, 1968.

protoplanet
A cloud of gas, dust particles, and rocks in the process of forming into a planet.

protostar
The early stage in a star's formation, between the collapse of a gas cloud and the onset of nuclear burning at the star's core. Some astronomers think that small, spherical interstellar clouds called GLOBULES are such contracting clouds. As the cloud contracts it becomes hot, radiating energy at infrared wavelengths. Such radiation has been detected from protostars in the Orion nebula. T TAURI STARS are believed to be protostars at a late stage, but still surrounded by remnants of the cloud from which they formed.

Proxima Centauri
The nearest star to the Sun, 4.3 light-years away. It is a faint red dwarf, of magnitude 10.7, the third member of the ALPHA CENTAURI star system. It is called Proxima because it is about 0.1 light-year nearer to us than the other two stars of Alpha Centauri. Proxima Centauri was discovered in 1915 at Johannesburg by Robert Thorburn Aytoun Innes (1861–1933), Union Astronomer of South Africa and a double-star specialist who in 1927 published the *Southern Double Star Catalogue*. Proxima Centauri appears over 2° away from the other two stars of Alpha Centauri; it must take about 1 million years to orbit them. Proxima is a FLARE STAR.

Ptolemy (Claudius Ptolemaeus; c.100–c.178)
Alexandrian astronomer, geographer, and mathematician, the last great astronomer of the ancient world. Ptolemy wrote many works: the *Tetrabiblos* on astrology, treatises on planetary

hypotheses, on mechanics, on music, on map projections, and one on optics in which he discussed the effect of refraction in the Earth's atmosphere on the apparent positions of celestial bodies. His most important works were his *Guide to Geography* and *Almagest*.

The *Guide to Geography* included details about mapmaking, and lists of places with their longitudes and latitudes. Although Ptolemy placed the equator in too high a latitude and took a value for the Earth's size nearly one third less than that determined by ERATOSTHENES, the book was regarded as authoritative until the great age of discovery began.

Ptolemy's greatest work was *The Mathematical Collection*, later known to the Arabs (through whom the book came to us) by the Greek superlative "megistee" (the greatest), corrupted into the name *Almagest*. In this encyclopedic work Ptolemy discussed the motion of the planets and the layout of the Universe. He extended the star catalog of the great Greek astronomer HIPPARCHUS with his own observations; and he discussed in detail whether or not the Earth is fixed in space, concluding that all evidence indicated it must be stationary. Ptolemy explained the movements of the planets using the basic concept of uniform motion in a circle, together with mathematical devices proposed in the third century B.C. by APOLLONIUS OF PERGA. These made use of large and small circles (*deferents* and *epicycles*) and a movable eccentric (a large eccentrically mounted circle), and with them Ptolemy was able to describe all planetary motions in detail.

A persuasive synthesis of all Greek astronomical knowledge, Ptolemy's *Almagest* exerted a profound and lasting influence on all subsequent generations of astronomers. It would not be seriously challenged for 1,300 years.

Pulkovo Observatory
The Central Observatory of the U.S.S.R. Academy of Sciences, located at an elevation of 250 feet (75 m) near Leningrad. It was founded in 1839 by F. G. W. STRUVE, its first director, with a 15-inch (38-cm) refractor that was then the largest in the world. In 1885 a 30-inch (76-cm) refractor by Alvan CLARK was installed, the world's largest until the opening of the LICK OBSERVATORY in 1888. Destroyed during World War II, the observatory was reopened in 1954 and now has a $27\frac{1}{2}$-inch (70-cm) reflector, $25\frac{1}{2}$-inch

(65-cm) refractor, and a 20/27$\frac{1}{2}$-inch (50/70-cm) Maksutov camera. The observatory also operates a transit radio telescope measuring 345 by 10 feet (105 × 3 m).

pulsar

A radio source that emits short pulses of radiation at very regular intervals, typically of about a second. The radiation is produced by a very small and dense NEUTRON STAR, which rotates and "flashes" a beam of radio waves like the beam from a lighthouse.

Pulsars were discovered accidentally in 1967 by Antony HEWISH, who was investigating the scintillation (or "twinkling") of distant radio sources. The regularly pulsing sources seemed so artificial they were first dubbed "Little Green Men" signals; but Thomas GOLD soon showed that a rotating neutron star could in theory produce the pulses. This hypothesis was confirmed in 1968, when a pulsar was discovered in the center of the CRAB NEBULA, the debris of a star which was seen to explode in the year 1054. Walter BAADE and Fritz ZWICKY had predicted that neutron stars are formed in such SUPERNOVA explosions, and the radio pulsar coincided with a star long thought to be the remains of the supernova. In 1969 the light from this star was found to be flashing in time with the radio pulsar, which has a period of only 1/30 second. In the same year the pulsar was found to emit X rays with the same periodicity. The radiation at all wavelengths is produced by electrons moving in the neutron star's magnetic field, which is a trillion times as strong as the Earth's field. This SYNCHROTRON RADIATION travels outward in a beam, causing the pulsar to flash as it rotates.

Over 300 pulsars have now been discovered, and their periods range from the 1/30-second of the Crab pulsar to 4.3 seconds. Apart from the Crab and Vela pulsars, they can only be detected at fairly long radio wavelengths of about 1 meter.

The electrons in space between the pulsar and the Earth slow down the radiation, so that different wavelengths from each pulse arrive at different times. The measured delay between wavelengths reveals the distance of the pulsar. Most pulsars lie in the disk of our Galaxy, where the stars and gas are concentrated, and are at distances of 300 to 30,000 light-years from the Sun.

Although their timekeeping is almost as good as that of atomic clocks, all pulsars are slowing down at a rate of about a billionth of a second a day as they radiate away their energy. This provides a key to a pulsar's age. The Crab pulsar is found to be about 1,000 years old, in good agreement with the date of the supernova outburst. Most pulsars, however, are millions of years old, and the comparative youth of the Crab pulsar probably accounts for its easy visibility at optical and X-ray, as well as at radio wavelengths. The second-fastest pulsar is about 10,000 years old; it lies in the constellation Vela amid a filamentary nebula, which is the gaseous remnant of the supernova that produced it. Occasionally, both the Crab and Vela pulsars suddenly speed up during their otherwise steady slow-down in spin. This is probably the result of an internal change in the neutron star (a "starquake").

After 10 million years a pulsar's radio emission fades away. Older neutron stars can only be detected if they are orbiting another star, so that gas falls onto them, heating them up and emitting X rays (see X-RAY ASTRONOMY). Some sources emit X rays in pulses, with a period of a few seconds; in one star (Hercules X-1) optical pulses are also seen. In these X-ray pulsars the radiation is produced by hot gas, and not by the synchrotron process which operates in radio pulsars.

Puppis (the stern)

A constellation of the southern hemisphere of the sky next to Canis Major; it originally formed part of the larger constellation Argo Navis, the Argonauts' ship, which was broken into smaller parts by Nicolas Louis de LACAILLE. Puppis lies in the Milky Way and contains several star clusters; the most prominent is M47 (NGC 2422), 3,700 light-years away and about 27 light-years in diameter. L_1 and L_2 Puppis are two stars that appear close to each other in the sky but are not physically connected (they are an *optical double*). L_1 is of magnitude 5.04 and 540 light-years away. L_2 is a red giant, 180 light-years away, which varies irregularly between magnitudes 2.6 and 6.0 about every 140 days.

Pythagoras (c.580 B.C.–c.500 B.C.)

Greek philosopher and mathematician, who believed that the Earth was spherical in shape and that it lay at the center of a spherical Universe. He realized that the Sun, Moon, and planets follow their own paths around the sky. He introduced the idea that each is carried around the Earth on a crystalline sphere; each sphere produced a musical note (the music of the spheres) as it turned. The concept of the heavenly spheres was developed into an extremely intricate system by EUDOXUS and others.

Pyxis (the compass)

A faint and insignificant constellation of the southern hemisphere of the sky, below Hydra, once part of the larger constellation of Argo Navis; it was made a separate constellation by Nicolas Louis de LACAILLE. Pyxis contains no objects of interest.

Q

QSO

Abbreviation for quasi-stellar object, popularly known as a QUASAR.

Quadrantid meteors

One of the principal meteor streams, reaching a sharp peak of from 45 to 250 (average about 110) meteors an hour on January 3 to 4 each year; the display lasts only a few hours. The stream is named for the obsolete constellation of Quadrans Muralis. At maximum the radiant is 25° east of Eta Ursae Majoris. Currently the strongest annual shower, the Quadrantids are bluish meteors with fine persistent trains; they have no known associated comet.

quartz-crystal clock

A highly accurate clock based on the rapid oscillations of a crystal of the mineral quartz. When a quartz crystal is compressed, a small voltage forms across it.

Similarly, a voltage applied across a quartz crystal distorts its shape. (Technically, this is termed the *piezoelectric effect*.) Feeding back the voltages formed by the distortion of the quartz crystal will set it oscillating about 100,000 times a second. These oscillations act like a very fast but highly accurate pendulum, which can be used to control an electric circuit. The oscillations are reduced by frequency dividers to drive an electric clock. Quartz-crystal clocks superseded pendulum clocks for keeping Greenwich Mean Time in 1942, and they have since been widely adopted as accurate timekeepers throughout the world. At their best they are accurate to about .000001 second a day. The quartz crystal is kept in an evacuated bulb at constant temperature to insulate it from outside effects. However, the frequency of a quartz crystal will slowly change with age, and to maintain its accuracy a quartz crystal clock must therefore be regularly adjusted. This is done by calibrating it against the standard frequency produced by an ATOMIC CLOCK.

quasar

An object which appears as a star-like point of light, but which emits far more energy than an entire galaxy. The name quasar is a contraction of quasi-stellar object (QSO). Many quasars also emit radio waves, and these are sometimes called quasi-stellar radio sources (QSS).

Quasars were discovered in 1963, when the first accurate measurements of radio-source positions showed that some coincided with star-like objects with very unusual optical spectra. Maarten SCHMIDT interpreted these spectra as having very large RED SHIFTS. Over 650 quasars are now known, showing red shifts up to 3.53. This means that, in the case of the latter quasar, OQ 172, the wavelength of its radiation has been lengthened 4.53 times, so that the visible light received from the quasar was actually emitted at ultraviolet wavelengths, 4.53 times shorter.

All distant galaxies also have red-shifted spectra, caused by their recession in the expanding Universe; the red shift, and thus the velocity, increases with distance (HUBBLE'S LAW). Most astronomers accept that the large red shifts of quasars are caused by their fast rates of recession, and therefore that they are very remote—hundreds of millions or billions of light-years from our Galaxy. They appear comparatively bright, however, and for them to do so at such an enormous distance, many of them must be a hundred times more luminous than even the most luminous galaxies we know. Moreover, many quasars are variable in light output, halving or doubling their brightness in a few months. This indicates that the light-emitting region can be only about a light-year across, 1/100,000 the size of a galaxy. A quasar is probably, in fact, the very active central region (nucleus) of a galaxy too distant to be seen.

Quasar spectra are very similar to those of the bright nuclei of the much nearer N-GALAXIES, and some quasars show a surrounding "fuzz" on long exposure photographs which appears to be an enveloping galaxy. How such a large amount of energy is produced in a volume only a light-year in diameter remains a mystery; it could be due to a disk of intensely hot gas around a supermassive black hole of a few billion solar masses; the tidal force of the black hole would tear apart stars to form the disk, and subsequently suck in the gas.

Some astronomers have speculated that quasars are much nearer, and therefore much less luminous; but the explanation of their red shifts then becomes a problem. One suggestion is that quasars have been ejected from galaxies relatively nearby, and that their red shifts are due to some undiscovered physical law.

Many quasar spectra show absorption lines produced by gas clouds lying between the quasar and the Earth. The number of clouds observed can be as many as 17, always with a lower red shift than the quasar emission lines (which indicate its distance). The clouds may be in intergalactic space along the line of sight to the quasar, or else they may surround the quasar, having been ejected from it some time ago.

A new class of objects, probably related to quasars, was discovered about 1970. They are named BL LAC OBJECTS after their prototype BL Lacertae, which was once thought to be a variable star. These objects are galactic nuclei which vary even more rapidly than quasars and have no lines in their spectra. They are probably about as far away as the nearest quasars.

R

radar astronomy

The investigation of bodies in the solar system by reflecting radio waves off them. Usually the same radio telescope is used both to transmit a powerful pulse of radio waves and to detect the very faint echo returning moments later. Because the relative amount of energy that returns drops off sharply as increasingly distant objects are investigated, radar astronomy is limited to distances within the solar system.

Radar astronomy began in the 1950s, when "contact" was made with the Moon. Because the speed of the radio waves (the velocity of light) is very accurately known, the time taken for the echo to return allowed a precise determination of the Earth-Moon distance to be made.

Contact was made with Venus in the 1960s. Since the relative sizes of the planets' orbits are known from KEPLER'S LAWS, an accurate distance for Venus led to an extremely precise absolute scale of size for the whole solar system. As a result, the distance from the Earth to the Sun (the ASTRONOMICAL UNIT) is now established with an accuracy of well within a mile.

The way in which radio waves are reflected depends both on the nature of the surface structure and on the presence of any large irregularities like hills or craters. Mercury and Mars, for example, must have loose surface layers, because their radar properties are similar to the Moon's. Radar echoes have also revealed craters on Venus, whose surface is permanently hidden from optical astronomers by clouds.

The rotation of a planet slightly alters the wavelengths reflected from its edges as a result of the DOPPLER EFFECT. In 1962 the rotation period of Venus was first determined by radar techniques; it was found to be 243 days in a retrograde direction (opposite to that of the other planets). Another surprise followed in 1965, when radar astronomers determined that Mercury rotates in 58.6 days, not the 88 days which had long been accepted on the basis of visual observations.

radial velocity

The velocity of a star along the line of sight, expressed as a negative figure if the star is moving toward us, and positive if moving away. It is measured by observing the shift in wavelength of a moving star's spectral lines arising from the DOPPLER EFFECT. Corrections for the Earth's motion are applied to give a radial velocity relative to the Sun. The average radial velocity for stars in the solar neighborhood is about ±12 miles (20 km) per second.

radiant

The direction from which a METEOR appears to radiate as it enters the atmosphere. The radiant of a sporadic meteor can be determined by triangulation of its path from two sites about 100 miles (160 km) apart. If the meteor's radiant and velocity are known, its orbit around the Sun can be calculated. Meteor streams contain many meteoroids moving in parallel paths with almost identical radiants. Thus stream meteors, which can occur anywhere in the sky, have paths which all appear to radiate from one point. The radiant is actually caused by a perspective effect, similar to that which makes parallel railroad tracks appear to radiate from a distant point.

radiation

See ELECTROMAGNETIC RADIATION.

radiation belt

See VAN ALLEN BELT.

radiation pressure

The tiny force exerted on bodies by a beam of light. The pressure of sunlight on a giant sail could conceivably be used to propel craft through the solar system (this idea is termed "solar sailing"). Radiation pressure helps push the dust in comet tails away from the Sun (the main effect actually comes from the atomic particles of the SOLAR WIND). Larger bodies are not pushed away by radiation pressure, but instead lose momentum and spiral into the Sun (POYNTING-ROBERTSON EFFECT). Radiation pressure from the hot insides of a star is thought to play a part in supporting the star's structure.

radio astronomy

The observation of the Universe at radio wavelengths. In 1931 Karl JANSKY, a physicist working at the Bell Telephone Laboratories, first discovered that some of the "static" that interferes with radio communications comes from space. Jansky and Grote REBER showed that this noise comes from the whole of the MILKY WAY. But radio astronomy developed rapidly only after World War II had pushed forward the development of radio techniques and equipment.

Nature of radio noise from space. Radio waves are a part of the electromagnetic spectrum, with wavelengths roughly a million times longer than those of visible light. The Earth's atmosphere absorbs most wavelengths of electromagnetic radiation; only visible light, some infrared radiation, and some radio waves actually reach the ground from space. Radio waves shorter than about 1 centimeter are absorbed by atmospheric water vapor, and extra-terrestrial radio waves longer than 30 meters are reflected back into space by the Earth's ionosphere. But the wavelengths between 1 centimeter and 30 meters constitute the radio "window" through which a vast amount of astronomical information has been gathered since the 1950s.

All hot objects, such as the Sun, emit radiation over a wide range of wavelengths; such *thermal* radiation accounts for part of the radio radiation received from space. But most of the powerful radio sources are due to the interaction of very fast electrons with magnetic fields. The electrons spiral along the magnetic field, emitting SYNCHROTRON RADIATION. Some gases emit radio waves at certain specific wavelengths ("lines"). The most important is hydrogen, which emits at 21 centimeters wavelength; but fairly complex interstellar molecules have recently been identified in space from their characteristic radio wavelengths (see INTERSTELLAR MOLECULES).

Solar system radio astronomy. An early application of radio techniques in astronomy was to track the trail of hot, electrically conducting gas (IONS) which meteors leave in the atmosphere. This revealed many previously unknown daytime meteor showers.

The Sun and its features such as sunspots and flares can be studied at radio wavelengths. The Moon and most of the planets, however, are quiet at radio wavelengths, except for their weak thermal radiation. Much more can be learned about them by "bouncing" radio waves off their surfaces (RADAR ASTRONOMY). The exception is JUPITER, which has a strong magnetic field and trapped radiation belts of charged particles (like the Earth's VAN ALLEN BELTS), which emit synchrotron radiation, and occasional radio bursts.

The radio sky. The brightest object in the radio sky is the Milky Way. Its radio emission comes not from stars, but from synchrotron emission by the COSMIC RAY electrons moving in the interstellar magnetic fields. Very few stars emit radiation enough to be detected by even the most sensitive radio telescopes. However, the sky is full of individual sources, ranging in apparent size from mere points to disks the size of the full Moon and larger. Many of these sources are within our own Galaxy, and appear close to the line of the Milky Way in the sky. Extragalactic sources, on the other hand, are evenly scattered over the sky.

Radio sources were originally named for the constellations in which they occur, with a letter indicating their order of brightness: for example, the CRAB NEBULA, the brightest radio source in Taurus, was designated Taurus A. Surveys of the radio sky have now revealed so many sources that this system has been dropped in favor of more logical schemes. For the northern hemisphere of the sky, the most widely used catalog is that from the third Cambridge survey (3C), produced at the MULLARD RADIO ASTRONOMY OBSERVATORY; the most extensive surveys of the southern sky have been made at PARKES OBSERVATORY in Australia.

The Parkes survey introduced the now widespread convention of identifying each source by a six-digit number which gives its coordinates in the sky. Thus the source 1954–55 is at right ascension 19 hours 54 minutes and declination −55°. This number is prefixed by letters identifying the observatory which discovered the source—PKS for Parkes, B for Bologna, and so on. (The prefix PSR is commonly used for pulsars, so that the Crab nebula pulsar, for example, is PSR 0531 + 21.) Many early surveys, including the 3C catalog, do not use this logical approach, but simply allocate numbers to the sources in order of right ascension. The 3C catalog runs from 3C 1 to 3C 470.

Galactic radio astronomy. Most strong radio sources within our Galaxy are either the expanding shells of hot gas from SUPERNOVAE (exploded stars), such as the Crab nebula; or else they are the hot gaseous nebulae (H II REGIONS) in which stars are forming. Such sources can also be seen optically, but one class of galactic objects that can be studied virtually only by radio observation are PULSARS, the dense, rapidly rotating remnants of supernova explosions. Since their accidental discovery in 1967 over 100 have been detected, but only one or two are visible optically.

The structure of the Milky Way as a whole can be readily studied by radio astronomy, because radio waves are not absorbed by the interstellar dust that limits the view of optical astronomers. The large-scale distribution of neutral hydrogen gas (H I REGIONS) is revealed by 21-centimeter line observations, which show the hydrogen concentrated into spiral arms. The magnetic field threading the Galaxy can be investigated both by the general synchrotron emission of the Milky Way, and by its effect on the polarization of radio waves from sources outside the Milky Way. The arrival times of pulsar pulses also give a measure of the number of free electrons in interstellar space.

Extragalactic radio sources. Nearby spiral galaxies, like the Andromeda galaxy, appear similar at radio wavelengths to the Milky Way. They show both the general synchrotron radiation and the 21-centimeter line from hydrogen clouds in the spiral arms. Study of the latter are improving our understanding of how stars form from hydrogen clouds, and of the mechanism which produces the spiral arms in galaxies.

The exact wavelength received from a hydrogen cloud depends on its velocity toward or away from us (the DOPPLER EFFECT). Careful measurements reveal the rotation of the gas in a spiral galaxy, enabling its mass to be calculated. The Doppler shift of wavelength also gives the speed of the galaxy along the line of sight (*radial velocity*). This technique can only be applied to galaxies less than about 50 million light-years away because of the limited sensitivity of present-day radio receivers. The Doppler shift of more distant galaxies is found from the optical spectral lines; the distance of the galaxies can then be derived using HUBBLE'S LAW.

One early surprise in radio astronomy was the discovery that one of the brightest sources in the sky (Cygnus A) is associated with a very distant galaxy, 1 trillion light-years away. Nine years later, in 1963, Maarten SCHMIDT found even more remarkably that some radio sources which appear as star-like objects in optical telescopes have very large Doppler shifts; according to Hubble's law they must be as far away as the RADIO GALAXIES like Cygnus A. These QUASARS (or quasi-stellar objects) are now thought to be very powerful explosions at the centers of distant galaxies.

Radio galaxies and quasars emit up to a million times more radio energy than a normal galaxy like the Milky Way, often from vast clouds on either side.

However, in many quasars, the radio emission comes from a central region only a few light-years across, where the explosion is taking place. Some radio galaxies and quasars have both a small central component and extended clouds. The radio emission from the central source usually varies slowly over time, and the rate of change at different wavelengths should help explain how vast amounts of energy are converted into radiation in these sources.

Radio astronomy and cosmology. Radio astronomy has contributed to COSMOLOGY, the study of the whole Universe, in a number of important ways. Radio sources act as probes of the expanding Universe. Even traveling at the velocity of light, the radio waves from the most distant quasars have taken billions of years to reach us, and these distant sources thus give a view of conditions early in the development of the Universe. By counting the numbers of sources of different apparent intensities, cosmologists can determine how the Universe is evolving, and it is found that the number of radio sources in the Universe was greater in the past. This evidence was the first to indicate that the STEADY-STATE THEORY (of an unchanging Universe) is incorrect.

One of radio astronomy's main contributions to cosmology was the 1965 discovery of the BACKGROUND RADIATION, a faint, uniform "hiss" of radio waves from all directions in space, apparently a result of the BIG BANG from which the Universe began.

radio galaxies

Distant galaxies that are very powerful sources of radio waves. Their radio output can be up to a million times that of our own Galaxy. This radiation originates not in the radio galaxy itself, but in two huge clouds, one on either side of the galaxy. In the largest radio source yet discovered, known as 3C 236, these clouds lie along a line 20 million light-years in total length.

The first radio galaxy to be identified was Cygnus A, the brightest radio source in the constellation Cygnus and the second-brightest in the whole sky. In 1954 it was found to coincide with a faint object resembling two galaxies close together. At first, astronomers thought that two galaxies were colliding, and that the radiation originated in the collisions of gas clouds. As more radio galaxies were identified, it became clear that most are single galaxies, not involved in any collision. They are in fact giant elliptical galaxies, containing 10 trillion stars in a roughly spherical volume 300,000 light-years across. The Cygnus A galaxy seems to have a dark dust lane across its center, giving it the divided appearance originally suggesting two close galaxies.

At the center of some radio galaxies (the N-GALAXIES) is a small, highly luminous nucleus, which often produces as much light as all the stars in the rest of the galaxy. When such a small nucleus far outshines the rest of the galaxy, only it will be seen on photographic plates. This apparently star-like object is known as a "quasi-stellar object," or QUASAR.

All types of radio galaxies generally have the two radio-emitting clouds outside the optical galaxy, although the galactic nucleus itself also emits shorter radio wavelengths. The energy supply to the radio clouds must come from the nucleus, where a vast amount of energy is being produced by processes not yet understood. The clouds themselves emit radio waves through a process known as SYNCHROTRON RADIATION, produced when high-speed electrons move in a magnetic field. How the energy reaches the clouds is still a subject of dispute. One theory envisages two continuous streams of electrons and magnetic fields leaving the nucleus, and spreading into clouds where they strike the thin gas between the galaxies. Alternatively, the "slingshot" theory suggests that gravitational interactions in the nucleus can fling out BLACK HOLES in opposite directions, each accompanied by magnetic fields and electrons.

The radio emission from the largest radio galaxy known, 3C 236, is shown to extend over 20 million light-years on this "photograph" of the radio brightness. The galaxy powering the source lies at the center of the cross, and its optical size is only 1/10 the width of the cross.

The second largest radio galaxy known, DA 240, comprises two radio-emitting clouds, totaling 7 million light-years in length. On this "photograph" showing the radio brightness, the bright spot in the center is radiation from the galaxy's nucleus. Other bright spots are distant radio sources, while the arcs are artifacts produced by the telescope.

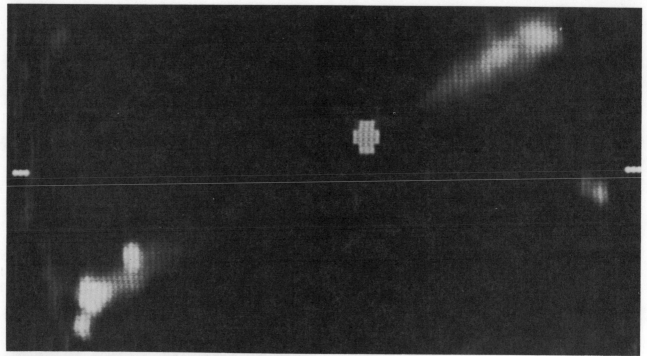

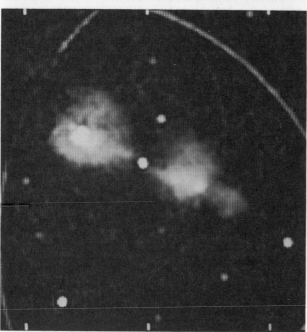

radio telescopes

Instruments for receiving radio waves from space. The telescope includes an antenna, an amplifying system, and some form of recorder to preserve the observation.

An antenna (or dipole) is basically a short metal rod, in which electric current is produced by incoming radio waves. In radio astronomy, radio waves must be collected over a large area, or aperture, because the signals are so weak, and because a large aperture gives a better ability to resolve small details (the telescope is said to have a smaller *beamwidth*). A large number of dipoles spread over a field and connected together forms a very sensitive aerial, and in the MILLS CROSS arrangement the aperture can be as large as 1 mile (1.6 km). Rapid-intensity changes (scintillation) in distant radio sources were first investigated with a 4-acre (1.6-ha) field of dipoles at the Cambridge radio observatory. The high sensitivity of this telescope led to the accidental discovery of the faint, regularly "ticking" PULSARS.

Dish antennae. Most radio telescopes have in fact only one dipole, and the radio waves are collected and reflected onto it by a large metal dish, analogous to the mirror of an optical reflecting telescope.

Because radio waves are about a million times longer than light waves, a radio dish must be much larger than an optical telescope to have satisfactory resolving power. But for the same reason, the mirror surface need not be as perfectly smooth as that of an optical instrument, and by using a steel frame covered with wire mesh relatively large dishes can be built without great difficulty.

Parabolic dishes are the most versatile antennae for radio-telescope systems, useful over a wide range of wavelengths. Radio astronomers often prefer to observe at short wavelengths, however, at which the resolution obtainable with a given aperture is better. The smoothness of the dish sets a limit on the shortest wavelength that can be used. Most dish instruments have the added advantage of being fully steerable (they can be pointed to any part of the sky).

As radio astronomy has developed, technological advances have progressively improved the angular resolution of dish antennae, and some of the largest can now achieve a beamwidth of no more than 1 arc minute (1/30 the apparent diameter of the Moon). At the same time, the shortest wavelengths that can be amplified electronically have been reduced from about 10 centimeters in the early 1960s to a few millimeters today.

As a result, parabolic reflectors can now be divided into two groups. Smaller instruments with very smooth surfaces are used for millimeter wavelength studies, such as the detection of INTERSTELLAR MOLECULES. They were pioneered by a 36-foot (11-m) instrument at the KITT PEAK OBSERVATORY. On the other hand, the giant antennae familiar from the 1950s and 1960s are now used chiefly at centimeter wavelengths. The largest instruments are the 330-foot (100-m) dish at the EFFELSBERG RADIO OBSERVATORY in West Germany, which has a resolving power of 0.8 arc minute at 2 centimeters wavelength; the 250-foot (75-m) JODRELL BANK dish (5 arc minutes at 10 centimeters); and the 210-foot (65-m) dish at PARKES OBSERVATORY in Australia (3 arc minutes at 5 centimeters).

In a class of its own, the giant 1,000-foot (305-m) dish of the ARECIBO RADIO OBSERVATORY was built simply by bulldozing smooth a natural depression in the mountains of northern Puerto Rico and covering it with fine wire mesh. This sacrifices steerability for size, giving a resolution of 0.3 arc minute at 3 centimeters over the strip of sky scanned by the telescope as the Earth rotates. Similar fixed antennae on a smaller scale have been used for Moon studies and for observing the Crab nebula. As an inexpensive means of undertaking particular limited tasks, such instruments are very useful.

Interferometers. The outputs from two dish antennae can be combined electronically to form an INTERFEROMETER system. By moving the antennae relative to one another a radio "picture" of a source can be built up, equivalent to that which would be obtained from a much larger dish. This technique of APERTURE SYNTHESIS was developed by Sir Martin RYLE at the MULLARD RADIO ASTRONOMY OBSERVATORY at Cambridge, England, where the world's largest aperture-synthesis telescope is located. It is equivalent to a single dish 3 miles (5 km) in diameter, and produces radio maps with an angular resolution of 0.01 arc minute, equal to photographs

Each of the identical 90-foot dishes at the Owens Valley radio observatory, California, reflects radio waves on to a receiver in the box supported in front of the dish. Their outputs are combined electronically to produce the resolution of a dish up to 1,600 feet across, the maximum distance between the movable telescopes.

obtained with the best optical telescopes. (The resolving power is comparable to that needed to read newsprint at a distance of over 300 yards.) There are "one-mile" aperture-synthesis telescopes at Cambridge and at the WESTERBORK RADIO OBSERVATORY in the Netherlands. A much larger synthesis instrument, the VERY LARGE ARRAY, is now under construction.

Amplifiers and recorders. The electronic receiving and recording components of a radio telescope must be highly sensitive and capable of amplifying the small voltage fluctuations from the antenna. In practice, all amplifiers introduce a certain amount of receiver noise and prevent very faint radio sources from being detected. Transistors have now completely replaced triode tubes in radio telescope receivers, and new techniques involving parametric amplifiers and MASERS are leading to receivers 10 times more sensitive than those now in use.

The final component of the receiver system is a computer to record the output from the amplifier, to control the movements of the antenna, and to produce a contour map of the radio source as the telescope beam scans across it. The computer can also be set any number of repetitive tasks. In the search for pulsars made at Jodrell Bank using the 250-foot dish, for example, most of the northern hemisphere sky was surveyed in successive 0.5° strips. In each 10 minutes of the search, the computer recorded 16,000 samples of the signal, and while doing so analyzed the previous 10 minutes' data, testing the samples for any of 10,000 periodic variations and printing out any signals suspected of coming from a pulsar.

Ranger probes

A series of American Moon probes. The first Ranger probes were intended to send television pictures as they approached the Moon, and to eject a balsa-insulated instrument package which, after slowing by retro-rockets, would hard-land on the Moon and send back seismometer readings of Moonquakes. All these attempts failed, although the Soviet Union successfully used the same technique in 1966 to land their Luna 9 probe. The Ranger program was reorganized into purely photographic missions to return increasingly detailed television pictures as the probe sped toward the Moon, destroying itself on impact. The final members of the Ranger series successfully produced the first close-ups of the Moon's surface, showing detail far too small to be seen from Earth. The Ranger pictures revealed that even the apparently flattest parts of the Moon (the *mare* areas) are actually pockmarked with tiny craters down to only a few feet in diameter. The detailed views of the Moon returned by the Ranger cameras aided designers of future soft-landing spacecraft.

Probe	Launch date	Remarks
Ranger 1	August 23, 1961	Test launch into Earth orbit
Ranger 2	November 18, 1961	Test launch into Earth orbit
Ranger 3	January 26, 1962	Missed Moon on January 28 by 22,862 miles (36,793 km)
Ranger 4	April 23, 1962	Impacted Moon's far side on April 26; on-board command system failed
Ranger 5	October 18, 1962	Missed Moon on October 21 by 450 miles (724 km)
Ranger 6	January 30, 1964	Impacted Moon February 2; television system failed
Ranger 7	July 28, 1964	Impacted Moon July 31; returned 4,308 photographs
Ranger 8	February 17, 1965	Impacted Moon February 20; returned 7,137 photographs
Ranger 9	March 21, 1965	Impacted Moon March 24; returned 5,814 photographs

Reber, Grote (b. 1911)

American pioneer radio astronomer, the first individual to follow up Karl JANSKY's initial discovery of radio waves from space. Reber built a 31-foot-(9.4-m)-diameter dish aerial in his back garden in 1937, and used it to map radiation from the sky at around a 1-meter wavelength. His first radio maps were published in 1940 and 1942, revealing a radio sky very different from the sky at optical wavelengths. He found a major source of radio noise in Sagittarius, the direction of our Galaxy's center, and he also located several areas of strong radio emission unrelated to visible objects; these have since been identified as sources such as the Crab nebula, a supernova remnant in Cassiopeia, and a radio galaxy in Cygnus. During the war years, Reber was the world's only radio astronomer; his dish aerial was also the world's first specially built radio telescope. His published results attracted the interest of other astronomers, including Jan OORT, who after the war also began to investigate the sky at radio wavelengths. Reber set up a new instrument in Hawaii to investigate longer-wave emissions in 1951. Three years later he moved to the Commonwealth Scientific and Industrial Research Organization in Tasmania, where he continues his radio-astronomy studies.

red dwarf

A star of low surface temperature (2,000–3,000°C) and a diameter about half that of the Sun. Such stars are very faint, with less than 1 percent of the Sun's luminosity, but they are the longest-lived stars in our Galaxy, because of their small energy output. All red dwarfs are still in their MAIN SEQUENCE phase of stellar evolution, shining by converting hydrogen to helium in their interiors; their expected life spans are in excess of 10 billion years. Some 80 percent of the stars in the vicinity of the Sun are known to be red dwarfs, and although they are so faint they cannot be seen at distances much greater than 100 light-years, it is believed that red dwarfs are by far the most numerous stars in the Galaxy. Some (for example, PROXIMA CENTAURI, the Sun's nearest neighbor) are FLARE STARS.

red giant

A star with a low surface temperature (2,000–3,000°C) and a diameter between 10 and 100 times that of the Sun. All stars spend a part of their lives as red giants, once they have converted the hydrogen in their cores to helium, although the period is considerably less than that spent on the MAIN SEQUENCE. Red giants shine by converting helium to carbon (and heavier elements) in a very small, dense core, which is surrounded by an extended, tenuous envelope. Since the gravity at the surface of this envelope is small, matter can easily escape into space in the form of stellar winds and prominences. This process is important in returning matter to the interstellar medium. Red giants are typically 100 times brighter than the Sun, and many of them are long-period VARIABLE STARS of the MIRA type.

red shift

The amount by which the wavelengths of light and other forms of ELECTROMAGNETIC RADIATION from distant galaxies and quasars are increased because of the expansion of the Universe. It is often given the symbol z. This cosmological red shift is one example of the DOPPLER EFFECT, the wavelength change caused by any motion of a light source along the line of sight. The farther the light must travel to reach us, the more the waves are stretched. The wavelengths from distant galaxies are therefore consistently longer than those from nearby galaxies (HUBBLE'S LAW). The shift of SPECTRAL LINES of known wavelength toward the red (longer wavelength) end of the spectrum can be converted to a distance in light-years by using HUBBLE'S CONSTANT.

The fractional change in wavelength is the same for all the wavelengths in an object's spectrum. Thus a galaxy with a red shift of 0.1 has its hydrogen spectral line at 4861 Ångstrom units (Å) increased by 486 Å to 5347 Å, while that at 6563 Å is increased by 656 Å to 7219 Å. The most distant galaxies have red shifts of about 0.7 (corresponding to a distance of 10 billion light-years), but the very bright QUASARS can be seen at much greater distances. The farthest of all quasars has a red shift of 3.53. We see this object, OQ 172, by its very short ultraviolet radiation stretched to visible wavelengths by the expanding Universe.

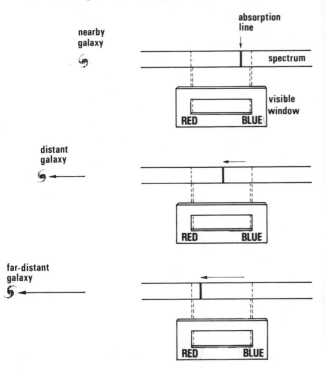

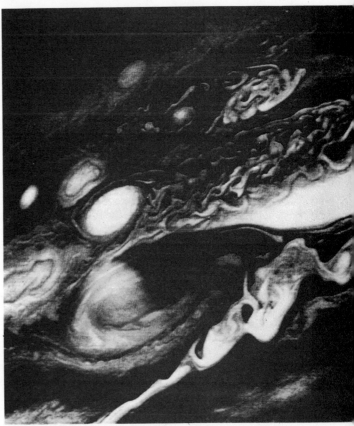

Jupiter's celebrated red spot (lower center), photographed in close-up by the Voyager 1 space probe, is a gigantic whirlpool in Jupiter's clouds. Smaller white ovals are also seen in this photograph, which shows details as small as 100 miles (160 km) in diameter.

The red shift, or lengthening of the wavelength of light from a receding body, is shown here with reference to the visible "window" on the spectrum to which the eye is sensitive. As galaxies recede at ever increasing speeds, light from them is red-shifted by a corresponding amount, as revealed by the differing position of a specific wavelength marked by a dark absorption line. The red shift causes the entire spectrum from the object to move across the visible window, taking some wavelengths out of sight beyond the red end, and bringing previously invisible wavelengths into view at the blue end. When an object approaches, a similar effect occurs, but in the opposite direction (blue shift).

red spot
The only apparently permanent feature among the swirling cloud bands of Jupiter, first observed by Giovanni CASSINI in 1666. The red spot is elliptical in shape, measuring 20,000 by 8,000 miles (32,000 × 13,000 km) and nestles in a bay on the southern edge of Jupiter's south equatorial belt. Its intensity varies from a striking orange-red to a pallid almost invisible cream, but its shape is constant. The spot was particularly prominent in the late 1870s and again in the mid-1970s. Photographs from the Voyager space probes in 1979 show the red spot to be a spinning whirlpool in Jupiter's clouds that circulates counter-clockwise like an anti-cyclone. Clouds in the neighboring belts swirl around the outside of the spot, setting up a turbulent wake behind it. Similar effects, on a smaller scale, seem to operate around other, smaller spots of various color in Jupiter's clouds. There is no sign of convection within the spot itself, invalidating a previous theory that the spot marked the top of a

storm cloud like that over a hurricane. The spot's color must be due to complex chemicals in Jupiter's clouds. One theory attributes the spot's color to red phosphorus. Similar reddish spots have been observed at higher latitudes on Jupiter, but these are less than half the size and are transient in nature.

Redstone rocket
An American rocket for space launches, developed from a short-range ballistic missile. The Redstone rocket was built by Wernher von Braun's team at the U.S. Army Redstone Arsenal in Huntsville, Alabama (later the MARSHALL SPACE FLIGHT CENTER). The Redstone launched America's first two spacemen, Alan Shepard and Virgil Grissom, on short suborbital flights at the start of the MERCURY PROJECT. In a modified form, known as JUNO 1, the Redstone launched America's first satellite, Explorer 1. The basic Redstone was a single-stage vehicle, 70 inches (178 cm) in diameter and 69 feet (21 m) long, with a single engine of 78,000 lb. (35,380 kg) thrust. Because of its low payload capacity, the Redstone was soon superseded by more powerful launchers, such as the ATLAS ROCKET.

reflecting telescope
A telescope in which a concave mirror collects and focuses light from a celestial object. Most modern telescopes are reflectors because they are less

expensive, more compact, and can be made in larger apertures than the refracting telescope, which uses lenses. The heart of any reflecting telescope is the primary mirror, which is concave and reflects light rays to form an image at the primary focus. The distance between the mirror and its prime focus is called the FOCAL LENGTH, and it is affected by the mirror's amount of concavity. For example, the depth of the curved surface of a 12-inch(30-cm)-diameter mirror of focal length 48 inches (120 cm) will be 0.18 inch (4.7 mm). Such a mirror is said to have a FOCAL RATIO (focal length/aperture) of $f/4$. The primary mirror is normally made of glass, coated on its curved front surface with aluminum (see MIRROR). Smaller secondary mirrors are used to divert the light to various observing positions.

In the telescope system known as the *Newtonian,* after its inventor Isaac Newton, the primary image is diverted by a small flat mirror out through the side of the tube. This is the system used in many amateur-owned reflecting telescopes. However, most modern professional reflectors are of the CASSEGRAIN design, in which a convex secondary mirror about a quarter the diameter of the primary mirror reflects the light back through a hole in the primary's center. This convex secondary mirror multiplies the focal length several times, giving the benefit of a long effective focal length in a relatively short tube. The longer the focal length the larger the image. Many telescopes also have a COUDÉ focus, particularly useful in spectroscopy. By swinging the secondary mirror out of the way, the PRIME FOCUS can be used for photography; a photographic plate takes the place of the secondary mirror. The prime focus

The World's Largest Reflecting Telescopes

	Aperture in.	m	Mounting	Observatory	Date
1	236	6.0	Altazimuth	Zelenchukskaya, Caucasus, U.S.S.R.	1976
2	200	5.0	Horseshoe	Mount Palomar, California	1948
3	158	4.0	Horseshoe	Kitt Peak, Arizona	1973
4	158	4.0	Horseshoe	Cerro Tololo, Chile	1975
5	153	3.9	Horseshoe	Siding Spring, Australia	1974
6	142	3.6	Horseshoe	Mauna Kea, Hawaii	1979
7	142	3.6	Horseshoe	European Southern Observatory, La Silla, Chile	1976
8	120	3.0	Fork	Lick Observatory, California	1959
9	107	2.7	Cross-axis	McDonald Observatory, Texas	1968
10	102	2.6	Fork	Crimean Astrophysical Observatory, U.S.S.R.	1960
11	102	2.6	Fork	Byurakan, Armenia, U.S.S.R.	1975
12	101	2.6	Fork	Las Campanas, Chile	1975
13	100	2.5	English	Mount Wilson, California	1917

provides wide-angle views; correcting lenses are often used at the prime focus to improve the definition over a wide field. An obsolete reflector design, which uses a concave secondary instead of a convex one, is called the GREGORIAN TELESCOPE.

reflection nebula
See NEBULA.

refracting telescope
A telescope in which a large lens or OBJECT GLASS collects and focuses the light from the object being observed. Refractors have lost much of their long-established popularity with the introduction of photography in celestial observation. For photographic work the reflector, in its many varied forms, is on the whole superior because it produces color-free (*achromatic*) images. The image produced by a refracting telescope can never be made absolutely colorless, because no combination of lenses in the object glass can bring light of all colors to precisely the same focus. In principle, the flint glass lens in an achromatic object glass should correct the red-to-violet band of images formed by the crown glass lens, but in practice some of the violet and red light is not correctly focused. A refractor's image therefore exhibits a purplish halo when corrected for normal visual use. The so-called *photographic* correction, in which blue and green light is brought to a sharp focus, is no longer relevant now that photographic emulsions are sensitive to all colors of light, not merely the blue, as was the case in the early days of photography (see ASTROPHOTOGRAPHY).

Refracting telescopes are very much more expensive than reflectors of the same aperture. First, optical (transparent, flaw-free) glass must be used, which is about 10 times as costly as ordinary glass. Furthermore, since the tubes of refracting telescopes are much longer than those of reflectors, the observatory will be large and costly. However, the relatively long FOCAL LENGTH (objective-image distance), and the fact that the tube is completely enclosed, with the light passing along it only once, produce steady images and good definition, while the absence of reflecting surface reduces scattered light and so increases the image contrast.

The main use of the surviving large refracting telescopes today is in ASTROMETRY, a field covering the measurement of close DOUBLE STARS and of stellar positions for determining PROPER MOTIONS and PARALLAX; many visual observers also prefer its use in planetary observation. The current desire for large-aperture telescopes to probe deep into space means that no large refractors have been built for many years, and most of the instruments currently used were built in the 19th century, including the world's largest, the 40-inch (102-cm) at Yerkes Observatory, and the 36-inch (91-cm) at Lick Observatory. The third-largest refracting telescope, the 32.6-inch (83-cm) at the Meudon branch of Paris Observatory, is used regularly for the visual measurement of double stars, as was the 28-inch (71-cm) at the Royal Greenwich Observatory until very recently. Two famous refractors used for both visual and photographic observation, both 24 inches (61 cm) in aperture, were set up at Lowell Observatory, Arizona, and at the Pic du Midi

Observatory in the Pyrenees. The Flagstaff instrument, originally used by Percival LOWELL in his Martian "canal" observations, is now employed photographically in the international planetary patrol project.

regression of nodes
The westward precession of an orbit caused by the gravitational influence of other bodies. The orbit's movement is noted by reference to the *nodes,* the points where the orbit crosses a reference plane such as the ecliptic or celestial equator.

Regulus
The brightest star in the constellation Leo; also known as Alpha Leonis. Regulus, of magnitude 1.36, is a blue-white star 3.8 times the Sun's diameter, 84 light-years away. It has companions of magnitude 7.64 and 13, making it a triple star.

relativity
A theory of the structure of space and time, and its relation to motion and gravity, which was proposed by Albert Einstein.

The special theory (1905) is founded on the principle that the velocity of light is the same for all observers, no matter how they move relative to each other. This apparent paradox was required both theoretically and to explain the unexpected result of the MICHELSON-MORLEY EXPERIMENT, and Einstein showed that our familiar notions of space and time must be abandoned as a result. Among other things, the special theory predicts that time runs at different rates for observers who move at different velocities, and that two events that appear simultaneous to one observer may not do so to another.

In addition, the theory predicts new phenomena in mechanics: among them, the possibility of changing matter into energy, the inability of a material body to reach the speed of light, and the famous "twins paradox" in which a fast-moving space traveler returns to Earth younger than his twin who remains behind. In spite of these apparently bizarre possibilities, the special theory is now an established cornerstone of modern physics. Its predictions have been confirmed in many ways by experiment.

The general theory (1915) is an extension of relativity to situations in which gravity is an important factor. It treats gravity not as a force, but as a consequence of the CURVATURE OF SPACE. In special relativity, Einstein had treated time as an extra dimension, very like the familiar three dimensions of space. Now he showed that the four-dimensional Universe (extending over both space and time) becomes "curved" in the presence of matter. Any body "falling" under gravity is actually following a straight line, at constant speed, relative to the curved background of "space-time."

If space-time is, in fact, curved, then even light should be affected by gravity. Observations have confirmed this, showing that starlight passing close to the Sun is slightly deflected. Moreover, light that leaves the surface of a massive body is shifted to a longer wavelength, an indication that time passes more quickly in an intense gravitational field.

When the force of gravity is weak, Einstein's theory differs little from the conventional Newtonian view of the Universe. The motion of the planets, for example, is hardly affected by relativity. However, a slight disturbance in the predicted motion of Mercury, caused by the effects of relativity, is actually detectable, and the explanation of Mercury's behavior is one of the greatest successes of the theory.

When the force of gravity becomes very strong, strange new effects are predicted. The immense force of gravity exerted by the entire Universe is capable of bending the whole of space. In BLACK HOLES, the space-time around a collapsing star shrinks catastrophically, until even light is trapped within it.

Because gravity is generally such a relatively weak force, general relativity has been experimentally checked in only a very few cases. Although other theories of space-time have been proposed (for example the BRANS-DICKE COSMOLOGY), general relativity is still the most widely accepted.

Relay satellites
Two NASA experimental communications satellites, launched into medium-altitude orbits. The satellites were 172-lb. (78-kg) tapered octagons, studded with solar cells. Receiving on 1725 MHz and retransmitting on 4170 MHz, they could carry 300 one-way voice transmissions, or one black-and-white television channel. The Relay satellites were used to send messages and pictures across the Atlantic and Pacific Oceans, and together with the privately-built TELSTAR satellites proved the value of active-repeater communications satellites.

Relay 1 (launched December 13, 1962) orbited every 186 minutes; Relay 2 (launched January 21, 1964) orbited every 195 minutes.

resolution
The ability of a telescope to distinguish fine detail. The larger the telescope's aperture, the better its resolving power. A telescope's resolution measured in arc seconds is related to its aperture by the famous Dawes' limit (see DAWES, WILLIAM).

If, for example, the components of a double star lie closer together than this limit, they cannot be separated no matter how high a magnifying power is applied. The Dawes' limit does not apply to the resolution of fine extended detail, such as that on planetary disks, where contrast rather than angular width sets the limit.

Reticulum (the net)
A small and faint constellation of the southern hemisphere of the sky, introduced on the star chart of Nicolas Louis de LACAILLE. Zeta Reticuli is a widely separated pair of stars similar to the Sun, of magnitudes 5.2 and 5.5, about 30 light-years from Earth.

retrograde motion
Motion from east to west, opposite to the normal west-to-east direction of motion in the solar system. The four outer satellites of Jupiter, the outermost moon of Saturn, and Neptune's moon Triton, all have retrograde orbits; the planet Venus has a retrograde axial rotation. Planets farther from the Sun than Earth seem to move retrograde (westward against the star background) for a short time as the Earth catches up and overtakes them in its orbit around the Sun.

Rhea

The second-largest satellite of Saturn, sixth in order of distance from the planet. Rhea, discovered in 1672 by Giovanni CASSINI, orbits Saturn every 4 days 12 hours 25 minutes at a distance of 327,590 miles (527,200 km). Its diameter is about 980 miles (1,600 km).

Rigel

The brightest star in the constellation Orion, and the sixth-brightest in the sky; it is also known as Beta Orionis. Rigel is a blue-white supergiant of magnitude 0.08, 78 times the Sun's diameter, and about 850 light-years away. Rigel is a multiple star; it has twin companions of magnitude 7.7, one of which is a SPECTROSCOPIC BINARY of period 9.9 days.

right ascension

The celestial equivalent of longitude on Earth. Right ascension is measured in hours, minutes, and seconds, from 0h to 24h, eastward along the celestial equator from the VERNAL EQUINOX. The symbol for right ascension is α.

Rigil Kent

See ALPHA CENTAURI.

Ritchey, George Willis (1864–1945)

American astronomer and optical instrument maker, coinventor of a modified type of CASSEGRAIN TELESCOPE known as the *Ritchey-Chrétien* design. In 1896 Ritchey joined George Ellery HALE at the Yerkes Observatory, where he made a correcting plate for the 40-inch (101-cm) refractor, thereby turning it into a camera with which he took a famous series of lunar photographs. Ritchey built a 24-inch (61-cm) reflector at Yerkes, and used it to photograph nebulae with a fellow optician and astronomer, Francis Gladheim Pease (1881–1938). Ritchey also ground the mirror for the Mount Wilson 60-inch (152-cm) reflector, with which he photographed faint nebulae; his discovery in 1917 of a nova in one such object, NGC 6946, helped confirm that these were actually distant galaxies, separate from our own Milky Way system. Ritchey later supervised the grinding of the 100-inch (254-cm) Mount Wilson mirror, being joined in 1911 on the telescope's construction by Pease. At Mount Wilson Ritchey began to develop his new optical designs, but his ideas were not adopted. In 1919 he left Mount Wilson and later moved to the optical institute at Paris Observatory. In 1931 he returned to the U.S. Naval Observatory, where he designed and built a 40-inch reflector.

Roche's limit

The distance from a planet inside which a satellite will be torn apart by tidal forces. The disrupting forces occur because the portion of the satellite nearest the planet is tugged more strongly by gravity than the portion farthest away. Small meteorite-sized lumps of rock and man-made satellites can hold together well inside Roche's limit, but where the planet and its satellite are made of similar material, the limit is $2\frac{1}{4}$ times the planet's radius. For satellites of lesser density, Roche's limit is correspondingly farther away. The tiny particles of Saturn's rings all lie within Roche's limit; if the Moon came too close to our Earth, it too would end up as a set of rings. Roche's limit is named for the French astronomer Edouard Roche (1820–1883), who calculated its existence in 1848.

rocket

An engine used for propulsion which works on the reaction principle, forward propulsion being achieved by reaction to a jet of gases streaming backward out of a nozzle. It differs from the jet engine, which also works by reaction, in that it carries an oxygen supply (usually liquid oxygen) to burn its fuel, unlike the jet engine, which obtains oxygen from the air. Rockets can therefore function in airless space better than in the atmosphere, where air pressure and drag impede progress. The fuel and the oxidizer are carried in separate tanks, and they are pumped into the combusion chamber and ignited by a spark. (Some small rocket motors use other propellants, such as HYPERGOLIC FUELS, which ignite spontaneously on contact and need no ignition system.)

Origin and development. The rocket was probably invented about the start of the 13th century by the Chinese; it is known that rockets were used at the Battle of K'ai-fung-fu in 1232. These were made with gunpowder, and were similar to today's firework rockets. News of the invention spread to Europe, but rockets were superseded for military purposes by the development of guns. However, Indian soldiers used rockets so effectively against the British at the Battles of Seringapatam in 1792 and 1799 that a British artillery officer, William CONGREVE, decided to investigate their possibilities. Congreve greatly improved the range, weight, and accuracy of rockets, which became a standard part of British artillery in the early 19th century. Soon, rocket battalions were also used by other European nations.

A Congreve rocket was stabilized by a long stick, like a firework. In 1843 the British inventor William Hale (1797–1870) drilled angled vents around the base of his rockets, so that the escaping exhaust gases caused the projectile to spin; this spin stabilization did away with the need for sticks. By 1865, Hale had perfected a design in which the rocket was spun by curved vanes inserted into its exhaust stream. Hale rockets superseded those of Congreve by 1867. However, by the 20th century the increasing range and accuracy of guns again overtook rocket development, and it was not until World War II that rockets returned to the field of battle.

Pioneers of astronautics. Toward the end of the 19th century, visionaries began to glimpse the value of rockets for space travel. In 1881 the Russian Nikolai Ivanovich Kibalchich (1853–1881), in prison awaiting execution for the assassination of Tsar Alexander II, conceived a man-carrying rocket, driven by the explosion of a series of gunpowder cartridges. In about 1890 the German Hermann Ganswindt (1856–1934) proposed a manned spaceship powered by the thrust of artillery-type shells. However, the true founder of astronautics was the Russian Konstantin TSIOLKOVSKY, who by 1898 had worked out the mathematical formulae governing a rocket's operation. In subsequent papers he established the need for liquid fuels, which are more controllable and have greater power than solid fuels, and analyzed the theory of the multistage, or step, rocket, which allows a much higher final speed. But

Tsiolkovsky was only a theoretician. The first man to experiment with liquid-fueled rockets was the American Robert H. GODDARD, who flew the world's first liquid-fueled rocket in 1926. At the same time in Germany, the writings of Hermann OBERTH inspired rocketeers. Europe's first liquid-fueled rocket was flown in 1931 by the German Johannes Winkler (1897–1947). The German Society for Space Travel, founded in 1927, also flew liquid-propellant rockets in 1931, one of the experimenters being the young Wernher VON BRAUN. The following year this work came under German army sponsorship, which led in 1937 to the setting up of the Peenemünde experimental

The design of a multistage rocket is shown well by the basic launcher used by the USSR for Soyuz and other manned spacecraft. The first stage, at the bottom, has four strap-on motors and an additional central motor. Above this, the four motors of the second stage can be seen through a separating framework. Above the dark colored section is the third stage, with the spaceship above that. At the top is the escape tower, which can pull the manned spaceship clear of danger in case of a launch disaster.

station, where the V-2 was developed. This advanced liquid-fueled rocket, a 1-ton warhead, was a forerunner of today's ballistic missiles. It both changed the pattern of warfare for the future, and brought the world to the threshold of the Space Age.

Rockets in the space age. At the end of World War II, Wernher von Braun and his team surrendered to the U.S. Army, with whom they continued their rocket experiments on a small scale, while the U.S. Air Force developed the more powerful ATLAS, TITAN, and THOR missiles. During this time the Soviet Union began to develop a missile that could cross the Atlantic. Following Tsiolkovsky's inspiration before the war, Soviet engineers such as Friedrich TSANDER had made considerable progress in rocket development. Tsander's coworkers Sergei KOROLEV, Valentin Glushko, and Mikhail TIKHONRAVOV, headed Soviet missile development after the war. Their first intercontinental missile flew in August 1957; in October that year it was used to launch the first Earth satellite, Sputnik 1. With upper stages added, the same booster launched early Soviet Moon probes, and the Vostok manned spacecraft. Not until 1967 was this booster publicly unveiled, in the version that launched the first Soviet cosmonauts; consequently, it is usually referred to as the VOSTOK LAUNCHER (an alternative name is Standard launcher). In improved form, with new upper stages, it launches manned Soyuz spacecraft.

After the development of the REDSTONE and JUNO missiles for early American space launches, von Braun's team began to design the SATURN rocket family specifically for manned space applications, principally the Apollo program. This led to the giant Saturn V, the world's most powerful rocket, first successfully fired in 1967. Two years previously the Soviet Union had introduced its PROTON rocket, of under half Saturn V's power, which was intended to send a manned Soyuz spacecraft looping around the Moon. Trial shots of unmanned ZOND craft were launched, but the Soviet lunar goal was deferred after Apollo 8 made the first manned lunar orbit in 1968. Because of technical difficulties the Proton launcher has never been considered safe for manned launchings, and its uses have been confined to launching unmanned space stations and lunar and planetary probes.

The Soviet Union has also attempted to build a giant rocket, known in the West as the G type, more powerful than the Saturn V. This would be used to launch large space stations into Earth orbit, or to send substantial payloads to the Moon for the first Russian lunar landing. The G-type booster is believed to be a three-stage vehicle with a total height of about 340 feet (105 m), first-stage diameter 52 feet (16 m), and with first-stage thrust about 11,000,000 lb. (5,000,000 kg). It could put over 100 tons into Earth orbit, send 60 tons to the Moon, or 25 tons to the planets. Reports suggest that one rocket exploded on the launchpad in mid-1969; another broke up during flight in mid-1971; and a third vehicle was destroyed by ground command after two minutes in late 1972. Further development of the booster, initially for a manned lunar landing, is expected.

Rockets in the future. In 1980 the standard disposable type of rocket will be largely displaced by the reusable SPACE SHUTTLE, a vehicle boosted away from the Earth by conventional rockets, but which will glide back to Earth like an aircraft. The Shuttle's power is provided by rockets similar to those in current use, because only chemical fuels produce the sudden surge of thrust necessary to break away from the Earth's gravity. But rockets of the future, designed to perform in space alone, will use forms of propulsion more efficient than bulky and heavy chemical fuels. One idea is the so-called ion rocket, or electric rocket, in which atoms of an element such as the metal cesium are heated to the point that their atoms are stripped of electrons; these electrically charged particles (*ions*) can then be accelerated out of the rocket by electric fields to produce a high-speed exhaust. Although the thrust of such an engine is very low, it can operate for long periods of time, thus building up a very high final speed. Small electric rockets have been tested in space.

Another suggestion is for a fission rocket, in which atomic power from a reactor like that in a submarine is used to heat a liquid such as liquid hydrogen or even water, turning it into a gas which is expelled from the rocket. Eventually, rockets may be able to tap nuclear fusion, duplicating the atomic processes that power stars. A dream for the distant future is the nuclear ramjet, which would scoop up hydrogen atoms from gas clouds between the stars to feed an on-board fusion reactor. The most exotic propulsion system of all is the photon rocket, which would use a beam of light particles—photons—to produce thrust. The exhaust speed of such a rocket would be the fastest speed possible—the speed of light. But such a propulsion system is completely beyond the capabilities of our current or foreseeable technology.

Roemer, Ole or Olaus (1644–1710)

Danish astronomer who in 1676 attempted the first measurement of the speed of light. In 1672 Roemer was appointed to the new Paris Observatory. While there, he noted that the times between successive eclipses of Jupiter's moons varied; as the Earth was approaching Jupiter the time intervals diminished, and as the Earth and Jupiter drew apart the intervals increased. He surmised that the difference was due to the time taken for light to cross the space between Earth and Jupiter, and from his observations announced that the speed of light was 140,000 miles (225,000 km) per second. Although about 25 percent too small, this was the first demonstration that light in fact had a finite speed. In 1681 Roemer returned to Denmark, becoming director of Copenhagen Observatory, where he is credited with inventing the transit instrument.

Rosse, Lord (1800–1867)

Irish astronomer (full name, William Parsons, third earl of Rosse), who built a famous 72-inch (183-cm) reflecting telescope, the world's largest until the opening of the Mount Wilson 100-inch (254-cm) in 1917. Rosse at his family seat of Birr Castle, Parsonstown, began in 1827 to improve the design and construction of large reflecting telescopes, which were still at an experimental stage. He developed a new alloy of copper and tin that would take the maximum polish (telescope mirrors were then made exclusively of metal), and in 1828 invented a steam-driven engine for automatic mirror polishing. In 1839 he cast a trial 36-inch (91-cm) mirror and used it for observing, before embarking on his intended giant telescope. In 1842 and 1843 he successfully made two 72-inch (182-cm) mirrors of 54-foot (16.5-m) focal length, weighing 4 tons each; they were used in rotation, one being repolished while the other was mounted in the telescope. The giant telescope, called the Leviathan of Parsonstown, had a tube 7 feet (2.1 m) wide and 58 feet (17.7 m) long. It began operation in 1845, mounted between two walls 56 feet (17 m) high which acted as windbreaks, but which restricted its field of movement to 10° either side of the meridian. With the telescope, Rosse discovered the spiral shape of certain nebulae, beginning in 1845 with the so-called Whirlpool nebula, M51; 75 years later these spiral nebulae were recognized to be distant galaxies. Rosse noted the ring-like structure of planetary nebulae, and in 1848 gave the CRAB NEBULA its name, because he drew it as a pincer-like shape. Rosse employed distinguished observers, such as Ralph Copeland (1837–1905) who, from 1889 to 1905, was astronomer royal for Scotland, and J. L. E. DREYER. After Rosse's death the telescope was little used; it was dismantled in 1908, although the walls and tube remain.

Royal Greenwich Observatory

Britain's national astronomical observatory, located at Herstmonceux Castle near Hailsham, Sussex. The observatory was founded at Greenwich, London, by King Charles II in 1675. Its original purpose was to make accurate observations of the Moon and stars in order to aid navigators in establishing longitude. The first director was John FLAMSTEED; he and his successors bore the title ASTRONOMER ROYAL until 1972, when the post of astronomer royal was separated from the observatory's directorship. The first observatory building, designed by Sir Christopher Wren (1632–1723), himself an amateur astronomer, was opened in 1676 and is now called Flamsteed House. In addition to basic positional astronomy, the work of the observatory progressed to include the accurate measurement of time; in 1880 GREENWICH MEAN TIME was adopted as legal time in Britain, and in 1884 the meridian passing through Greenwich was chosen as the world's prime meridian (0° longitude). Both the *Nautical Almanac* Office, producing catalogs for astronomers, navigators, and surveyors, and the Time Department, are important sections of the observatory. Because of deteriorating observing conditions in London, the observatory was moved from Greenwich to Herstmonceux after World War II, although it retains the name Royal Greenwich Observatory (RGO); the move was completed in 1958. The observatory, however, no longer lies exactly on the Greenwich meridian. Its main telescope is the 98-inch (250-cm) Isaac Newton reflector, installed in 1967. The observatory also contains a 38-inch (96.5-cm) and a 36-inch (91-cm) reflector, two 30-inch (76-cm) reflectors, and a 26-inch (66-cm) refractor. The Royal Greenwich Observatory plans to establish a new observing station, the Northern Hemisphere Observatory, on La Palma in the Canary Islands, with a 177-inch (450-cm) reflector as its main telescope. Also at the observatory will be a 100-inch (254-cm) reflector (the Isaac Newton telescope with a new mirror), and a 39-inch (100-cm) reflector. The Northern Hemisphere Observatory should be in operation by 1980.

RR Lyrae variables

Pulsating stars (see VARIABLE STARS) whose average luminosity is exactly the same (absolute magnitude + 0.5) and which can thus be used to calibrate distances throughout the Galaxy. Belonging to stellar population II (older stars), they are found in the galactic nucleus and halo and also in GLOBULAR CLUSTERS (they are often called "cluster variables"). RR Lyrae stars belong to SPECTRAL TYPES A to F and have periods of between 0.3 and 0.9 days varying by about a magnitude. They are named for their seventh-magnitude prototype, RR Lyrae. In 1920, Harlow SHAPLEY used the apparent magnitudes of RR Lyrae stars in globular clusters to determine the clusters' distances, showing that the Galaxy was much larger than had previously been thought. RR Lyrae stars in the galactic nucleus have also been used to determine the distance to the center of the Galaxy. It is believed that RR Lyrae stars represent a short phase (lasting about 80 million years) in the final evolution of population II RED GIANT stars.

Russell, Henry Norris (1877–1957)

American astronomer, codiscoverer with Ejnar HERTZSPRUNG of the relation between a star's brightness and color, now graphically displayed on the HERTZSPRUNG-RUSSELL DIAGRAM, which Russell first produced in 1913. From this diagram, Russell proposed a scheme of stellar evolution in which stars began as red giants, shrinking and increasing in temperature, ending their lives as cool red dwarfs. However, by the 1930s it was realized that stars derive their energy from nuclear reactions in their interiors, not from simple contraction. Russell was a leading pioneer in the study of eclipsing variable stars. In 1912 he published a major paper on the analysis of eclipsing-binary light curves, in which he showed how to calculate the sizes of the stars and their orbits. From this, Russell was able to compute their distance from Earth. His work on star sizes and distances led in 1913 to the discovery, independent of Hertzsprung, that stars fall into two brightness classes, which Hertzsprung had termed giants and dwarfs. Russell's plot of brightness against spectrum type showed the band of dwarf stars now known as the MAIN SEQUENCE. In 1929 Russell published an analysis of the Sun's spectrum, noting the presence of 56 elements; he measured the relative abundances of the elements, showing that hydrogen was the major constituent. Termed the Russell mixture, this was the first detailed breakdown of a typical star's true composition.

Ryle, Sir Martin (b. 1918)

English radio astronomer who developed the technique of APERTURE SYNTHESIS, by which signals from several small radio dishes are combined to synthesize the performance of a much larger dish. This was pioneered by Ryle at the MULLARD RADIO ASTRONOMY OBSERVATORY, Cambridge, which he founded and became director of in 1957. With a succession of increasingly powerful radio telescopes, Ryle has been able to catalog progressively fainter radio sources, showing that the number of sources increases with distance before falling away abruptly at the observable edge of the Universe. This suggests that the Universe has been evolving from a definite origin, which supports the BIG-BANG theory of cosmology. Ryle became Britain's twelfth astronomer royal in 1972, in recognition of the importance of his work in radio astronomy. In 1974 he shared the Nobel Prize in physics with his colleague Antony HEWISH.

S

Sagan, Carl Edward (b. 1934)

American astronomer and biologist, who has done pioneering work on the possibility of life elsewhere in the Universe. He demonstrated processes leading to the origin of life by bombarding mixtures of gas with ultraviolet light in order to simulate the effect of the Sun's radiation on the atmosphere of the early Earth. The experiments produced amino acids, the building blocks of protein. Sagan suggested in 1960 that water vapor in the clouds of Venus reinforced the GREENHOUSE EFFECT of its carbon-dioxide atmosphere, preventing heat from escaping and building up the planet's high observed temperature. He proposed a scheme for making Venus habitable by using algae to break down the carbon dioxide and release oxygen, and has also suggested a similar scheme for Mars. In 1966 Sagan discovered the existence of high elevations on Mars by analysis of radar echoes from the planet. He also suggested that wind-blown dust causes the seasonal changes in the dark areas on Mars, rather than growing vegetation as was previously supposed; photographs from the Mariner 9 probe confirmed this view. With the radio astronomer Frank DRAKE, Sagan has investigated the problem of interstellar communication, devising the message plaque fixed to the Pioneer 10 and 11 spacecraft, and the radio message briefly transmitted from the Arecibo radio telescope. Drake and Sagan have also listened with the Arecibo dish for radio messages from nearby galaxies (see LIFE IN THE UNIVERSE).

Sagitta (the arrow)

A small constellation lying in the Milky Way near Aquila, visible during the northern hemisphere summer. Despite its faintness, it is one of the 48 constellations that were listed by PTOLEMY almost 2,000 years ago. The constellation's most interesting feature is the recurrent nova WZ Sagittae, seen to flare up in 1913 and again in 1946.

Sagittarius (the archer)

A constellation of the zodiac, lying in the south celestial hemisphere and best seen from the northern hemisphere during summer. The center of our Galaxy lies in the direction of Sagittarius, giving rise to the dense Milky Way star fields that are a feature of the constellation. The actual nucleus of the Galaxy is believed to be marked by the radio source Sagittarius A. Famous bright nebulae in Sagittarius include the Lagoon nebula (M8, NGC 6523), the Omega nebula (M17, NGC 6618), and the Trifid nebula (M20, NGC 6514). The constellation also includes numerous open and globular star clusters. The Sun passes through Sagittarius from mid-December to mid-January, reaching its farthest point south (the winter SOLSTICE) on about December 22.

Salyut

Soviet space station, measuring 39 feet (12 m) in length in the shape of three connected cylinders $6\frac{1}{2}$ feet (2 m), 10 feet (3 m), and $13\frac{1}{2}$ feet (4 m) in diameter. Salyut has a volume of about 3,500 cubic feet (100 m³) roughly a quarter that of the American SKYLAB, and weighs $18\frac{1}{2}$ tons. It is launched by the Soviet PROTON rocket. The first Salyut, launched in April 1971, drew power from four wing-like solar cells, two at each end of the station. Subsequent Salyuts were of modified design, using three rotable solar panels mounted amidship. The Salyuts contain equipment for astronomical and Earth observation, and for biological studies. Salyuts 1, 4 and 6 were believed to be mainly scientific in purpose; and Salyuts 2, 3 and 5 were principally military surveillance stations. Salyuts 3 and 5 operated semi-automatically, telemetering data and ejecting a film package. Crews are ferried up to the space station by Soyuz craft; a simplified Soyuz design, omitting solar panels, has been introduced for Salyut missions. Starting in January 1978, during the Salyut 6 mission, automatic ferry craft called Progress (a modified Soyuz) have been used to refuel the space stations and carry other supplies, enabling the Salyut crews to set new space endurance records. Cosmonauts from various Communist countries have worked aboard Salyut stations.

Mission	Launch date	Results
Salyut 1	April 19, 1971	Soyuz 10 crew docked on April 24 but did not enter, probably because of hatch failure. Soyuz 11 crew spent 23 days aboard in June, but died due to spacecraft pressure failure during reentry. Salyut 1 reentered October 11
Salyut 2	April 3, 1973	Disintegrated in orbit; reentered May 28
Salyut 3	June 24, 1974	Soyuz 14 docked and transferred crew for 14-day mission. Rendezvous attempt by Soyuz 15 failed. Salyut 3 reentered January 24, 1975
Salyut 4	December 26, 1974	Soyuz 17 docked and transferred crew for 29-day mission. Subsequent Soyuz 18 crew completed 63-day mission in Space station. Unmanned Soyuz 20 docked automatically. Salyut 4 re-entered February 3, 1977
Salyut 5	June 22, 1976	Soyuz 21 docked and transferred crew for 48-day mission. Soyuz 23 failed to dock. Soyuz 24 docked and transferred crew for 17-day mission. Salyut 5 re-entered on August 8, 1977
Salyut 6	September 29, 1977	Soyuz 25 failed to dock. Soyuz 26 docked and transferred crew for record 96-day flight, during which Soyuz 27 and Soyuz 28 paid week-long visits. Subsequently Soyuz 29 docked and transferred crew for new record flight of 140 days, with visits from Soyuz 30 and 31

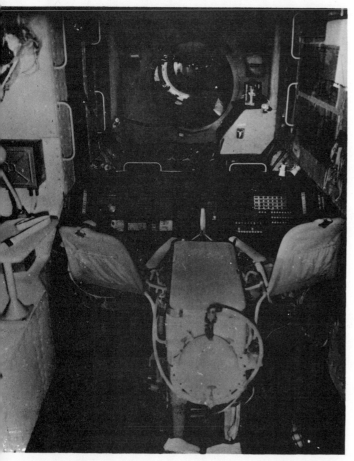

The interior of the Soviet Salyut 1 space station, the first space laboratory. Numerous hand holds make maneuvering in free fall easier. In the background is the hatchway for access to the Soyuz spaceship.

Sänger, Eugen (1905–1964)

Austrian pioneer of astronautics, who developed the concept of a reusable space transporter, now known as the SPACE SHUTTLE. Sänger's proposals for a space transporter were born in the late 1920s; he published his first description of a rocket plane in 1933. In the 1930s at the University of Vienna he developed rocket motors using diesel fuel with burning times of up to 30 minutes. From 1936 to 1945 he directed a rocket research institute for the Luftwaffe, and while there Sänger produced detailed plans for a long-range winged bomber, boosted into the air by a rocket-powered sled and accelerated by its own engines to the edge of the atmosphere; it would skip in and out of the atmosphere to extend its range before finally dropping toward its target. This work provided a basis for many postwar studies on reusable rocket systems, and shortly before his death he drew up designs for such a two-staged winged space transporter. Sänger foresaw the use of such vehicles to ferry men and materials to orbiting space stations, the role now outlined for the space shuttle.

San Marco

An Italian rocket launching site based on two converted oil rigs in Ngwana Bay off the Kenyan coast, operated by Rome University's Aerospace Research Center. Being 2° south of the equator it is excellently placed for launches into equatorial orbit. The San Marco launch platform is supported by

20 legs resting on the seabed; it measures 295 by 88½ feet (90 × 27 m). Nearby is the triangular Santa Rica platform, which houses the launch control center and launch personnel. The San Marco platform is used for launching American Scout rockets and sounding rockets. It became operational in 1966, and made its first orbital launch with the second of Italy's San Marco series of satellites for atmospheric studies. The platform has also been used to launch America's series of Small Astronomy Satellites (the first U.S. satellites to be launched by another country), and Britain's Ariel 5 X-ray satellite.

saros

The period of 6,585.3 days (18 years 11⅓ days) after which the Earth, Moon, and Sun return to their same positions relative to each other and eclipses repeat themselves. During the saros there are 223 lunations.

satellite

Any small object orbiting a larger body. The world's first artificial satellite was SPUTNIK 1, launched on October 4, 1957. The Sputniks were superseded in 1962 by the COSMOS SATELLITES. America's first Earth satellite was EXPLORER 1, launched on January 31, 1958. By the end of 1974, 1,606 Earth satellites had been launched, 674 of which were still in orbit. See also COMMUNICATIONS SATELLITE; NAVIGATION SATELLITE; WEATHER SATELLITE; and the names of individual satellite programs.

satellites of the planets

See under separate planets, and under the names of the individual satellites.

Saturn

The sixth planet from the Sun, renowned for its bright ring system. Its equatorial diameter is 119,300 km (74,100 miles); of all the planets in the solar system Saturn is surpassed in size only by Jupiter. Like its huge neighbor it is composed mostly of gas and wreathed by dense clouds. Saturn's shape is flattened by its rapid rotation, which varies from as little as 10 hours 14 minutes at the equator to 10 hours 40 minutes at latitude 60°; the axis is inclined at 26° 44'. Saturn moves at 6 miles (9.6 km) per second in an elliptical path between 9.01 and 10.07 astronomical units from the Sun (average distance 886,703,000 miles; 1,427,010,000 km), taking 29½ years to complete one orbit.

Structure. Viewed through a medium-sized telescope, Saturn's disk shows bands and belts like those on Jupiter. On Saturn, however, they are much more diffuse, fewer in number, and rarely showing much detail. Colors are also much less pronounced, with a golden yellow predominating. The equatorial band is yellow-white, with an orange-yellow equatorial belt and yellow tropical and temperate zones where orange belts are sometimes faintly glimpsed. The polar regions which appear greenish, are the darkest part of the surface. The cloud tops are a bitter −170°C, some 50° colder than on Jupiter, because of the planet's greater distance from the Sun. As a result of the lower temperature, there are fewer outbursts of activity in the clouds, and there are no distinct markings like the great red spot or white ovals on Jupiter. The most celebrated outburst was the great

Saturn and its rings photographed with the 100-inch Mount Wilson reflector. The bright outer rings, separated by Cassini's division, are easily visible; but the two inner rings are too faint to be seen on this photograph. Light and dark bands, similar to Jupiter's, can be seen on Saturn's disk.

white equatorial spot which appeared in August 1933, and which slowly spread until the whole equatorial zone brightened. High-latitude white and dark spots are less conspicuous but more common, appearing every few months.

Saturn's atmosphere is so cold that most of the ammonia freezes into clouds at lower levels. The outer atmosphere therefore consists mainly of hydrogen, helium, and methane, the latter forming particularly strong absorption bands in the planet's spectrum. Saturn's overall composition resembles that of Jupiter, but with a somewhat smaller proportion of hydrogen. Saturn is believed to have a rocky core about 12,000 miles (20,000 km) in diameter, about half the size of Jupiter's core, covered with a 3,000-mile (5,000-km) coating of ice and surrounded by a 5,000-mile- (8,000-km)-thick layer of metallic hydrogen. The rest is hydrogen gas, which gives the planet its low average density of 0.7 that of water. The rocky core accounts for only about 15 percent of Saturn's mass. Because Saturn is less than one-third the mass of Jupiter, its center is not compressed as greatly. But pressure at the center is still some 50 million atmospheres, and the temperature is more than 15,000°C. Like Jupiter, Saturn is radiating slightly more heat than it receives from the Sun. Either the planet is still slowly contracting, or the heat is left over from its formation.

The rings. Saturn's rings are perhaps the most beautiful in the solar system. They form a complete unbroken band around the planet's equator and have a maximum diameter of about 170,000 miles (275,000 km) from rim to rim. There are three main rings (A, B, and C), together with two fainter ones. Ring B is the wide and brilliant ring, flanked on the inside by the elusive crepe ring C, and bordered on the outside by ring A. Rings A and B are separated by the dark CASSINI DIVISION. In 1969 a much fainter ring D was discovered inside ring C, almost touching the planet's surface, while a ring of very scattered debris seems to exist outside ring A.

The rings were recorded almost as soon as the

telescope was invented, but early observers were unable to interpret the poor images produced by their primitive telescopes. Christiaan HUYGENS first realized the nature of the rings in 1655 and also explained their changes in appearance. During one 29½-year orbit of Saturn around the Sun, there are two periods when the rings are tilted toward us edge-on. When they are presented edge-on, the rings disappear from view, which implies that they must be very remarkably thin—measuring no more than about a kilometer in thickness.

The true nature of the rings was shown in 1895 by the American astronomer James Keeler, who found by the Doppler shift in light from the rings that they rotate like a swarm of particles on independent orbits, rather than as a solid disk. Careful observation of the rings has revealed several bands in which the density of ring particles is lower than elsewhere. These are believed to be due to the gravitational effects of Saturn's nearest moons, particularly Mimas, which perturb ring particles into different orbits. A ring particle moving in an orbit with a period which is an exact fraction of a satellite's period will suffer repeated gravitational tugs by the satellite at the same point in its orbit. These perturbations will produce a thinning of particle density. The gaps in the particle distribution are direct analogies of the KIRKWOOD GAPS in the asteroid belt, which are caused by perturbations from Jupiter.

The planet's globe is clearly visible through the crepe ring C, which confirms that the ring is made of particles. In addition, when the rings pass in front of a bright star, the star is visible in the Cassini division, visible slightly dimmed through ring A, and even seen faintly through ring B. Also, moons which pass into the shadow of the rings are still faintly visible even when shaded by the densest part of ring B. The ring

particles reflect about 70 percent of the light hitting them, and the rings can actually be brighter than the rest of Saturn.

The true nature of the ring particles is still something of a puzzle. They are probably blocks about the size of bricks, coated with frozen water. The rings lie inside ROCHE'S LIMIT for Saturn, within which a satellite would be pulled apart by Saturn's gravitational force. The rings may be the shattered remains of a moon that strayed too close, or they may be the building blocks of a satellite that never formed.

Satellites. Saturn has ten moons, eight of which move in near-circular orbits close to the planet in the plane of the rings. Beyond these eight, Iapetus moves in an orbit inclined at 15°, and Phoebe, much more distant, moves in a retrograde elliptical orbit.

Saturn rockets

A family of large space launchers developed for manned applications by a team under Wernher VON BRAUN at the MARSHALL SPACE FLIGHT CENTER, Huntsville, Alabama. Planning began in 1958 on a booster originally called Juno V, which had eight propellant tanks from REDSTONE rockets clustered around a central large JUPITER tank, and used eight engines developed from Thor and Jupiter rockets. In 1959 the name Saturn was adopted, and the project was officially taken up by NASA; by the following year its role as a launcher for Apollo had been defined, an S-IV second-stage design was chosen, and there were plans for an S-V third stage which never materialized. On October 27, 1961, the first Saturn 1 was test-flown from Cape Canaveral. By then, the Apollo Moon-landing goal had been announced and plans were laid for more powerful successors to Saturn 1. Largest of these was the so-called Nova class of launchers, with a lift-off thrust of 12,000,000 lb. (5,500,000 kg), eight times that of the Saturn 1. Eventually, however, the designers settled on the three-stage Saturn V. In 1964 the Saturn 1 test flights put dummy Apollo capsules into orbit, concluding in 1965 with the launch of three PEGASUS micrometeorite-detection satellites.

Saturn 1B. For actual manned Apollo flights an uprated Saturn 1, called the Saturn 1B, had begun development in 1962, with a modified upper stage called the S-IVB. This improved booster was first flown on February 26, 1966, and on October 11, 1968, launched Apollo 7, the first manned Apollo. Saturn 1B was subsequently used for launching Skylab crews and the American half of the Apollo-Soyuz mission. For these later launches, the Saturn 1B was erected on a pedestal to bring its total height to that of the larger Saturn V, so that it could use the same launch facilities. Saturn 1B's first stage was 80 feet 4 inches (24.5 m) long and 21 feet 5 inches (6.5 m) in diameter, producing 1,640,000 lb. (744,000 kg) thrust from eight H-1 engines. The S-IVB second stage, 58 feet 5 inches (17.8 m) long and 21 feet 8 inches (6.6 m) in diameter, produced up to 225,000 lb. (102,000 kg) thrust from a single engine called the J-2. It was topped by a ring-shaped instrument unit containing the electronic guidance equipment to control the rocket's operation. Saturn 1B's overall height with payload was 224 feet (68 m), and it weighed 650 tons loaded.

Satellites of Saturn

	Discoverer	Diameter (km)	Orbit radius (10^3 km)	Inclination	Eccentricity
Janus	Dollfus (1966)	300	159	0°	0
Mimas	Herschel (1789)	500	186	1.5°	0.02
Enceladus	Herschel (1789)	600	238	0°	0.0045
Tethys	Cassini (1684)	1,000	295	1.1°	0
Dione	Cassini (1684)	800	378	0°	0.002
Rhea	Cassini (1672)	1,600	527	0.3°	0.001
Titan	Huygens (1655)	5,800	1,222	0.3°	0.029
Hyperion	Bond (1848)	500	1,483	0.6°	0.104
Iapetus	Cassini (1671)	1,600	3,560	14.7°	0.028
Phoebe	Pickering (1898)	200	12,951	150° (retrograde)	0.163

A Saturn V rocket lifts the Apollo 11 spacecraft off the pad at the Kennedy Space Center. This is the full Saturn configuration weighing a total of 2,850 tons (with fuel) and capable of lifting a 150-ton payload into orbit or sending a 50-ton payload to the Moon.

Saturn V. Saturn 1B allowed the Apollo program to progress until the larger Saturn V Moon rocket was available. The design for this three-stage vehicle was outlined in 1961, and construction began the next year. Its first stage was based on the F-1 engine developed by the Rocketdyne Corporation, with a thrust of 1,500,000 lb. (680,000 kg); this is equal to the entire first-stage thrust of the original Saturn. The Saturn V's first stage (called S-1C) had five such engines. Its S-II second stage contained five J-2 engines, and the S-IVB third stage had one J-2 engine. The first Saturn V was launched on a test flight on November 9, 1967, sending a dummy Apollo command module into orbit; it was the most powerful rocket ever launched. This flight also initiated the new launch complex 39 at Kennedy Space Center, from which the Moon missions were to leave. After one more test flight the Saturn V was used for a manned launch, the Apollo 8 mission in December 1968. Subsequent Moon flights, and the Skylab space station, were also carried by Saturn V.

The rocket's overall height, including spacecraft, was 363 feet (111 m). Its first stage was 138 feet (42 m) long and 33 feet (10 m) wide; its total thrust was about 7,600,000 lb. (3,450,000 kg). The S-II second stage was 81 feet 7 inches (24.9 m) high, 33 feet (10 m) in diameter, with a thrust of up to 1,160,000 lb. (526,000 kg). The S-IVB top stage was 58 feet 7 inches (17.9 m) long, 21 feet 8 inches (6.6 m) in diameter, and produced up to 230,000 lb. (104,000 kg) thrust. Total loaded weight of the Saturn V at lift-off was about 2,850 tons. Because of improving technology and different mission requirements, engine performance and total weight varied slightly at each launch. Saturn V's capacity was about 150 tons in Earth orbit, or 50 tons to the Moon.

scattering

The deflection of light and other forms of radiation. The most familiar form of scattering is that produced by molecules and dust particles, as in the Earth's atmosphere. This is named *Rayleigh scattering*, for the English physicist Lord Rayleigh (1842–1919), who in 1871 first explained its operation. Shorter-wavelength (blue) light is scattered more than red, which is the reason the sky is blue. Another form of scattering is the *Compton effect*, named for the American physicist Arthur Holly Compton (1892–1962), who described it in 1923. In this process, particles of light (photons) collide with electrons, thereby transferring some of their energy to the electrons. The photons then have less energy, and therefore a longer wavelength, and they are scattered in random directions. The *inverse Compton effect*, which is important in astronomy, occurs when the electrons themselves have very high energies and are moving close to the speed of light—around quasars or supernovae, for example. In this case, energy is transferred from the electrons to the photons, which gain energy. This can account for the production of X rays from quasars and supernovae.

Schiaparelli, Giovanni Virginio (1835–1910)

Italian astronomer who first reported "canals" on Mars. Schiaparelli noted these straight markings at the near approach of Mars in 1877 (the same occasion on which Asaph HALL discovered the moons of Mars). He named them *canali*, meaning channels; but the word was mistranslated as canals, implying they were artificial. This aroused the interest of Percival LOWELL, who founded the once-popular view of life on Mars. Schiaparelli did not subscribe to Lowell's theories. He continued mapping Mars, and originated the current nomenclature of Martian surface features. Schiaparelli also made extensive observations of Mercury and Venus, concluding that they both kept the same face turned to the Sun, which has since been disproved.

Schiaparelli is perhaps best known for his discovery that meteor showers follow the same orbits as comets, and he proposed that meteors were produced by the disintegration of comets. He showed that the Perseid meteors were associated with a comet seen in 1862 (1862 III), and the Leonids with comet 1866 I; his ideas were confirmed in 1872 when Biela's comet, which had disintegrated on a previous approach to the Sun, was replaced by the Andromedid meteor shower.

Schirra, Walter Marty (b. 1923)

The only astronaut to fly in all three types of American spacecraft: Mercury, Gemini, and Apollo. Schirra was one of the original seven astronauts selected in 1959. He flew in *Sigma 7*, the fifth Mercury mission, on October 3, 1962, completing six orbits of the Earth in 9.2 hours. In December 1965 he commanded the Gemini 6 mission, which, with Gemini 7, made the world's first space rendezvous. In October 1968 he commanded Apollo 7, the Earth-orbital maiden flight of the Apollo capsule. Schirra left the astronaut corps in 1969 to enter private business.

Schmidt, Maarten (b. 1929)

Dutch-born American astronomer, the first to interpret the spectrum of a QUASAR, establishing the enormous speed at which these objects are receding. In the early 1960s, Jesse Greenstein, Allan Sandage, and Schmidt used the Hale 200-inch (508-cm) telescope to study the optical counterparts of some compact radio sources. These faint, star-like objects with strange spectra were called "quasi-stellar sources" (later shortened to *quasars*). In 1963, Schmidt showed that the enormous RED SHIFT in the spectrum of the brightest quasar, 3C 273, meant it was moving away from us at over 25,000 miles (40,000 km) per second. Other quasars were subsequently found to have even greater red shifts. Schmidt has spent the last decade making detailed studies of these exciting but poorly-understood objects.

Schmidt telescope

A wide-angle photographic telescope first constructed in 1930 by the Estonian optician Bernhard Voldemar Schmidt (1879–1935). Normal reflecting telescopes have a field of view of not more than about $\frac{1}{2}°$ (equivalent to the apparent diameter of the Moon), while wide-field camera lenses suffer from CHROMATIC ABERRATION and distortion. Schmidt's wide-field reflector eradicated the optical defects of ordinary reflectors by using a spherical mirror with a correcting lens in front.

Although a mirror with a spherical curve is afflicted with SPHERICAL ABERRATION, which destroys the sharpness of its images, by placing in front of it a thin lens or *plate*, polished with a double curve, Schmidt was able to correct the aberration and obtain star images of superb quality across a previously unheard-of field of view. His first camera, using a 10.2-inch (36-cm) correcting plate in conjunction with a 17.3-inch (44-cm) mirror, covered a field of view of 16° with excellent definition.

The difficulty of making the correcting plate delayed the development of the Schmidt camera until after its inventor's death, but its eventual acceptance was rapid. One of the largest Schmidt cameras in the world is the 48 × 72 inch (122 × 183 cm) at Mount Palomar, installed as a scouting instrument for the 200-inch (508-cm) reflector; in 1973, a similar instrument was installed at the Anglo-Australian Observatory in New South Wales. The largest Schmidt, with a 79-inch (200-cm) mirror, was installed at the Karl Schwarzschild Observatory, Tautenberg, East Germany, in 1960. These instruments have between them now mapped the entire sky, covering stars down to the 20th magnitude and lower. Another large Schmidt, the 39-inch (100-cm) at the European Southern Observatory, La Silla, Chile, also took part in the work.

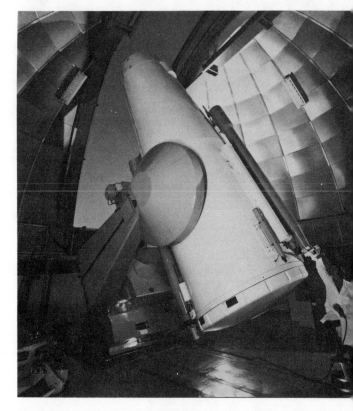

The 48-inch (1.22-m) Schmidt telescope at Siding Spring, Australia, is carrying out a survey of the southern skies to match that of the northern heavens made with the famous Mt Palomar Schmidt of the same size in the early 1950s. Like the Palomar instrument, it photographs areas of the sky 6° square on plates 14 inches (36 cm) square.

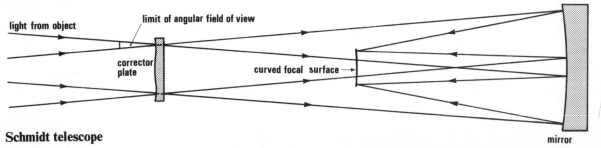

Schmidt telescope

Schwarzschild, Karl (1873–1916)

German astronomer, noted for his pioneering work on the theory of stellar structure, particularly the transport of heat by radiation in the outer layers of stars. Arthur EDDINGTON extended this work to the interior of stars, producing the first modern view of a star's structure. Schwarzschild developed the technique of measuring star brightnesses by photography, comparing their visual and photographic magnitudes to obtain their COLOR INDEX. He worked on the development of relativity theory, predicting the existence of the SCHWARZSCHILD RADIUS around high-density objects. His son, the German-born American astronomer Martin Schwarzschild (b. 1912), helped create the modern understanding of a star's life-history, as outlined in his classic 1958 book *The Structure and Evolution of the Stars*. Martin Schwarzschild originated and directed Project Stratoscope, a balloon-borne camera-telescope for celestial observations, which paved the way for current astronomical satellites.

Schwarzschild radius

The distance from an extremely dense or massive body at which the ESCAPE VELOCITY equals the velocity of light. The theory of RELATIVITY forbids any greater speeds than the velocity of light, and Karl SCHWARZSCHILD first showed that neither matter nor radiation from within such a radius can escape to the rest of the Universe. The surface of the imaginary sphere around a body at the Schwarzschild radius is called an *event horizon* or BLACK HOLE.

The Schwarzschild radius for a mass equal to the Sun's is only 2 miles (3 km); a star's core must be compressed to about this size by a SUPERNOVA explosion before a black hole is formed. By comparison, a body of the Earth's mass would have a Schwarzschild radius of only 0.4 inch (1 cm); no known process can reduce a planet to this size and density.

scintillation

The "twinkling" of a celestial object. Rapid variations in brightness of a star or other point source are caused by the Earth's atmosphere—generally the lowest 5 miles (8 km), which is rarely completely steady. Changes in air temperature produce turbulence, which varies the amount by which the atmosphere bends light passing through it. The resultant scintillation is visible through a telescope as the bad "seeing," which mars the steadiness of an image. To overcome scintillation, observatories must be sited high on mountains in areas of particularly stable air. Radio sources also scintillate, because the radiation from them undergoes similar effects when passing through clouds of charged particles in space.

Scorpius (the scorpion)

A constellation of the zodiac lying in the southern hemisphere of the sky, visible in the northern hemisphere during summer. Its brightest star is ANTARES. Beta Scorpii is a multiple star of magnitudes 2.9, 5.06, and 9.7; the brightest of the three is also a spectroscopic binary of period 6.8 days. Mu^1 Mu^2 Scorpii form an optical (unrelated) pair, as do $Omega^1$ $Omega^2$ Scorpii and several other apparent doubles. Other objects of note are the bright globular cluster M4

NASA's smallest complete launcher, Scout, used for orbiting small payloads (up to 300 lb.) into Earth orbit. Its four stages are all solid fueled, unlike other launchers which use liquid fuel This adds to its versatility, since it can be fired from small launch sites.

(NGC 6121) and the open cluster M6 (NGC 6405), called the jewel-box cluster. The brightest X-ray source in the sky, Sco X-1, is identified with a faint star in the constellation. The Sun passes through Scorpius briefly during the last week of November.

Scott, David Randolph (b. 1932)

Commander of the Apollo 15 mission, which landed on the Moon on July 30, 1971. With fellow astronaut James Benson Irwin (b. 1930) he explored the lunar surface for over 18 hours during a $2\frac{3}{4}$-day stay, traveling 17 miles (27 km) in the Lunar Roving Vehicle. Scott, a qualified aeronautical engineer, flew with Neil ARMSTRONG on the Gemini 8 mission in March 1966, which accomplished the world's first space docking. In 1969 he was command module pilot on the Apollo 9 mission, which tested the lunar module in Earth orbit. Scott was selected as an astronaut in 1963. In 1975 he left the astronaut corps to become director of NASA's Flight Research Center at Edwards, California.

Scout rocket

An American solid-fuel space launcher for orbiting small satellites. It is NASA's smallest launch rocket, and the only one employing solid fuels on all stages. Scout is a four-stage rocket, 75 feet (22.9 m) tall with payload attached, weighing 47,300 lb. (21,450 kg). It was introduced in 1960, and has since launched numerous scientific satellites, including members of the Explorer series, from Wallops Island, the Western Test Range, and the San Marco platform. Modern Scout rockets can place payloads up to 390 lb. (175 kg) in Earth orbit. Scout's Algol II first stage is 30 feet (9 m) long and 40 inches (1 m) in diameter, with a thrust of 105,000 lb. (47,600 kg). An alternative first stage, Algol III, has an increased width of 45 inches (1.14 m), giving a thrust of around 140,000 lb. (63,500 kg) for heavier payloads. Its Castor II second stage, 20 feet (6 m) long and 30 inches (76 cm) wide, has a thrust of 61,000 lb. (27,500 kg). Antares II, the third stage, is 10 feet (3 m) long and 30 inches (76 cm) wide, with a thrust of 21,000 lb. (9,500 kg). The fourth stage, Altair III, is 6 feet (1.8 m) long and 20 inches (50 cm) wide, with a thrust of 6,000 lb. (2,700 kg). A fifth stage is now also available for high-orbit missions.

Sculptor (the sculptor)
A faint constellation in the southern hemisphere of the sky, below Cetus, introduced by Nicolas Louis de LACAILLE. The south galactic pole lies in Sculptor. The constellation also contains a nearby dwarf galaxy in our local group.

Scutum (the shield)
A faint constellation of the southern hemisphere of the sky introduced by the Polish astronomer Johannes HEVELIUS, who originally called it Scutum Sobieskii, or Sobieski's shield, in honor of a patron. It is visible between Aquila and Serpens during the northern hemisphere summer. Delta Scuti varies from magnitudes 4.9 to 5.2 every 0.194 days, and is the prototype of a rare class of pulsating variable stars. The cluster M11 (NGC 6705) contains about 200 stars.

seasons
Divisions of the year caused by the varying presentation of the Earth's northern and southern hemispheres to the Sun. If the Earth's axis were upright with respect to the plane of its orbit, the Sun would always seem to pass directly overhead to an observer on the equator. Since the axis is in fact inclined from the upright at an angle of about $23\frac{1}{2}°$, the north and south poles are alternately tilted toward the Sun. The Sun's altitude therefore increases and decreases during the year, producing the seasons. Spring in the northern hemisphere starts at the spring EQUINOX, about March 21. Summer begins at the summer SOLSTICE (around June 21), autumn at the autumnal equinox (roughly September 23), and winter at the winter solstice (about December 22).

selenocentric
Term meaning centered on the Moon. A *selenocentric orbit* is an orbit around the Moon.

selenology
Study of the lunar surface, equivalent to geology on Earth. A *selenologist* is a lunar geologist.

Serpens (the serpent)
A large constellation straddling the celestial equator, representing a serpent wound around OPHIUCHUS. The constellation is split into two parts: Serpens Caput, the head, and Serpens Cauda, the tail; it is best seen during the northern hemisphere spring and summer. Delta Serpentis is a double star, of magnitudes 5.16 and 4.23. M5 (NGC 5904) is a bright globular cluster, 27,000 light-years away. M16 (NGC 6611) is a famous star cluster and nebula of gas which contains dark GLOBULES.

service module
Unmanned section of a spacecraft containing engines for course corrections and reentry, air and water supplies, and electrical power for the spacecraft's instruments. America's first manned spacecraft, Mercury, was self-contained, but an embryonic service module, called the equipment module, was introduced on Gemini flights. In the Soviet Vostok and Voskhod craft the last stage of the launch rocket went into orbit attached to the capsule and acted as a service module. The true Service Module (SM), with a large propulsion engine, was introduced on the Apollo spacecraft. A similar rear section, though with a less powerful engine, is attached to the Soviet Soyuz craft.

Sextans (the sextant)
A faint and insignificant constellation of the equatorial region of the sky, lying between Leo and Hydra, visible in the northern hemisphere spring. It was introduced on the 1690 star map of Johannes HEVELIUS.

Seyfert galaxy
A galaxy with a very small, bright nucleus, containing high-speed clouds of gas indicating recent explosions. Such galaxies were first studied by Carl K. Seyfert (1911–1960) in 1943, and about 100 are now known. One percent of SPIRAL GALAXIES seem to have these bright nuclei, but there are also a few Seyfert galaxies which are not spirals.

A Seyfert galaxy nucleus contains a very small core, less than a light-year across, which emits predominantly blue light as well as ultraviolet, infrared, and sometimes radio waves. This core seems to be a "mini-quasar," over 100 times fainter than normal QUASARS. Seyfert galaxies may therefore provide a link between normal galaxies and quasars. Around the core are clouds of hot gas ionized by the ultraviolet light from the core, and moving at hundreds of miles per second. Filling the rest of the nucleus (100 to 1,000 light-years across) is much more tenuous gas that emits optical FORBIDDEN LINES.

The light from the core of Seyfert galaxies varies in brightness over a period of months, and is probably produced by electrons moving in strong magnetic fields (SYNCHROTRON RADIATION). The infrared radiation from some Seyferts also varies, and may be produced by the same synchrotron process. In other Seyferts the infrared radiation comes from dust grains in the nucleus that have been heated by the light from the core. The amount of radiation emitted at infrared wavelengths from the nucleus is often considerably greater than the luminosity of the entire galaxy at optical wavelengths.

Shapley, Harlow (1885–1972)
American astronomer, who discovered that our Galaxy is much larger than was previously supposed, and that

This short exposure photograph shows only the bright central regions of the Seyfert galaxy NGC 4151; the faint outer parts would cover the entire photograph. Even so, the galaxy's center is overexposed, and the very small bright nucleus appears larger than its true size. The barred spiral, **top left**, is a very distant background galaxy.

the Sun is not centrally placed within it. Shapley made his discovery from a study of CEPHEID VARIABLE stars in globular clusters around our Galaxy. In 1912, Henrietta LEAVITT had discovered the relationship between period of variation and average brightness of Cepheid variables. In 1916–1917 at Mount Wilson Observatory, Shapley photographed similar variable stars in globular clusters, and carefully calibrated the period-luminosity law to determine the clusters' distances. Shapley also noted that most globular clusters are concentrated in one part of the sky, toward the direction of Sagittarius. If globular clusters are in fact scattered symmetrically around our Milky Way system, then this apparent direction of concentration must mean that the Sun is not centrally placed in the Milky Way. Shapley's bold conclusion overthrew all previous thinking about the Sun's place in the Universe—a revelation in many ways comparable to the theory of Copernicus, which ejected the Earth from the center of the heavens. From his calculations of the clusters' distances, Shapley deduced that the Galaxy was about 10 times larger than previously estimated, and that the Sun lay about 50,000 light-years from the center. However, the effect of interstellar gas and dust, dimming stars and making them appear too faint, made his estimates too high. By 1930 he had reduced his figures to a diameter of 100,000 light-years, with the Sun about 30,000 light-years from the center. This is in accord with modern values. These results were independently confirmed by the Dutch astronomer Jan OORT. Shapley's conclusions led in 1920 to the so-called Great Debate, a public discussion with the American astronomer Heber Doust Curtis (1872–1942) about the nature of faint nebulae. Although Shapley believed they were either part of our own enlarged Milky Way system or near

neighbors of it, Curtis suggested they were separate star systems, or galaxies, far distant from the Milky Way. On this point, Shapley was defeated.

From 1921 until his retirement in 1952, Shapley was director of the Harvard College Observatory. As a Harvard student in 1914, Shapley had shown that Cepheid variables were giant stars whose light changes were caused by actual pulsations, a theory later developed by Arthur EDDINGTON. Shapley continued his studies of variable stars; and in the 1930s he discovered the first two dwarf galaxies, both members of our LOCAL GROUP, which lie in the constellations of Sculptor and Fornax; he also showed that galaxies occur in clusters, which he called *metagalaxies*.

Shatalov, Vladimir Alexandrovich (b. 1927)
Soviet cosmonaut, pilot of the Soyuz 4 spacecraft, which docked with Soyuz 5, in January 1969. While the craft were docked, two space-suited cosmonauts walked in space from Soyuz 5 to Soyuz 4, returning to Earth with Shatalov in the Soyuz 4 command module. In October 1969 Shatalov was overall commander of the first joint flight of three manned spacecraft; he flew in Soyuz 8, which maneuvered in space with Soyuz 6 and Soyuz 7. Shatalov's third flight came in April 1971 as commander of Soyuz 10, which docked with the Salyut 1 space station. Despite expectations, the cosmonauts did not enter the space station but undocked after $5\frac{1}{2}$ hours and returned to Earth; some reports spoke of a hatch failure that could have proved fatal. Shatalov, one of the Soviet Union's most experienced spacemen, subsequently became director of cosmonaut training.

Shepard, Alan Bartlett (b. 1923)
The first American to be launched into space, a naval pilot selected as an astronaut in 1959.

Shepard was boosted in the *Freedom 7* capsule on a suborbital flight lasting 15 minutes 22 seconds on May 5, 1961. He reached a maximum altitude of 116.5 miles (187.5 km), covering a distance of 303.8 miles (488.9 km). Shepard was temporarily grounded because of an ear disorder, and became chief of the astronaut office. After an operation in

1969 he was fit to fly again, and commanded the Apollo 14 mission, which landed on the Moon on February 5, 1971. He and Edgar Dean Mitchell (b. 1930) made two Moon walks, lasting a total of $9\frac{1}{4}$ hours, during which they ascended to the rim of Cone crater to take samples. Shepard left the astronaut corps in 1974.

shooting star
The popular term for a METEOR.

sidereal day
The time the Earth takes to rotate on its axis with respect to a fixed point in space, such as is indicated by a star. Unlike the apparent SOLAR DAY, which is subject to considerable fluctuations, the length of the sidereal day was long thought to be constant. The rotation of the Earth, however, is now known to be slowing down by about 0.003 second a year, which means that the apparent sidereal day is gradually lengthening. The length of the *mean* sidereal day is 23 hours 56 minutes 4 seconds (see DAY).

sidereal period
The time taken by a planet or satellite to go once around its orbit as measured against the star background. The sidereal period of a planet can be said to be its "year." It differs from the period observed from the Earth, because the Earth itself is moving. For example, the Moon's sidereal period is 27.32 days, but its cycle of phases appears to last 29.53 days. The latter, the *apparent* period, is known as the SYNODIC PERIOD.

sidereal time
Time as determined by the stars, used for finding the position of a celestial object in the sky. A sidereal day is divided into 24 hours of sidereal time, which therefore represents the apparent rotation of the CELESTIAL SPHERE (the imaginary stellar vault) around the Earth.

sidereal year
The time taken for the Earth to revolve once around the Sun with respect to the star background. This period is equivalent to 365.26 SOLAR DAYS, or 366.26 SIDEREAL DAYS.

siderite
An iron meteorite; its name is derived from the Greek *sideros*, meaning "iron." Siderites are composed of an iron-nickel mixture with small amounts of cobalt, copper, phosphorus, carbon, iron sulfide, and traces of other elements and minerals. Siderites are solid masses with a density about 7.8 times that of water. Their variable nickel content divides them into three distinct types. Siderites with 5 to 6 percent nickel are named *hexahedrites*; those with 6 to 20 percent nickel are called *octahedrites*; and siderites with more than 20 percent nickel form *ataxites*. All iron meteorites were formed deep inside a parent body where an iron-nickel core slowly cooled (at a rate of less than 250° per million years) and crystallized. Siderites must have originated in about 6 to 10 asteroid-like parent bodies 130 to 320 miles (210–515 km) across, and were released in collisions between 1,000,000 and 100,000 years ago.

siderolite
A stony-iron meteorite, containing about equal amounts of nickel-iron metal and stony minerals. There are two main types of siderolites, although the name is often also applied to meteorites that are difficult to classify. The *pallasites* consist of either complete or shattered crystals of the mineral olivine, averaging about 5 millimeters in diameter, embedded in nickel-iron. The olivine solidified while the metal was still molten, and it must have formed near the edge of the parent body's molten core. The *mesosiderites* contain a near-equal mixture of nickel-iron and the minerals pyroxene and plagioclase. The metal may be interlaced between the stone fragments or may appear as separate grains. Siderolites seem to have cooled extremely slowly, at less than 1° per million years, and must have originated deep in their asteroid-like parent planets.

Siding Spring Observatory
Astronomical observatory located at an elevation of 3,822 feet (1,165 m), 18 miles (29 km) from Coonabarabran, New South Wales. It was founded in 1965 as a field station of MOUNT STROMLO OBSERVATORY, and has reflectors of 40-inch (104-cm), 26-inch (66-cm), and 24-inch (61-cm) aperture. In 1973 the United Kingdom's Science Research Council opened a 48/72-inch (122/183-cm) Schmidt telescope at Siding Spring. At the same site, the British and Australian governments jointly operate the 153-inch (390-cm) Anglo-Australian Telescope (AAT), inaugurated in 1974.

Sirius
The brightest star in the sky, of magnitude -1.47; it is 8.6 light-years away, the fifth-closest star to the Sun. Sirius lies in the constellation Canis Major, and is also known as Alpha Canis Majoris. It is a white star, 1.68 times the Sun's diameter and 23 times as luminous; its mass is 2.14 times that of the Sun. Sirius is actually double, with a magnitude 8.4 white dwarf companion orbiting it every 50 years; the white dwarf is 1.05 times the Sun's mass, 0.007 times its diameter, and 0.002 times its luminosity. Sirius' HELIACAL RISING was a familiar sign to the ancient Egyptians that the Nile floods were about to begin; the Egyptians used Sirius observations to measure the length of the year.

Skylab
First American space station, modified from the third stage of a Saturn V rocket. Skylab, weighing 75 tons (about four times that of the Soviet SALYUT), was launched into a near-circular orbit about 270 miles (435 km) high on May 14, 1973. About a minute into the launch the space station's micrometeoroid shield deployed prematurely and was ripped away, destroying one solar panel and jamming another. The meteoroid shield had been painted white to reflect the Sun; with it gone, the space station began to overheat. The Skylab mission was saved by the first crew, who erected a sunshield, and during a space walk cut free the jammed solar panel. Skylab contained four sections. Largest was the orbital workshop 10,644 cubic feet (301 m^3) in volume, 48.1 feet (14.7 m) long and 21.6 feet (6.6 m) in diameter; this was divided into living quarters and a work section.

Skylab. America's space laboratory, seen from an Apollo transit craft in June 1973. One solar paddle is missing, having been torn off accidentally, the makeshift sunshield which was used to prevent the spaceship overheating can be seen. The cross-shaped solar paddles generated power for the Apollo Telescope Mount.

The airlock module, 17.6 feet (5.4 m) long and varying in diameter from 5.5 feet (1.7 m) to 10 feet (3 m), contained equipment for controlling the station's operation, and a hatch for space walks. The 622-cubic-foot (17.6-m³) airlock module allowed astronauts to leave Skylab without depressurizing the entire station. The multiple docking adapter, 17.3 feet (5.3-m) long and 10 feet (3 m) in diameter, contained an Apollo docking port at its forward end and a reserve or rescue port in its side. This 1,140-cubic-foot (32.3-m³) section contained controls for the Earth-resources scanners, a furnace and vacuum chamber for experiments in processing materials, and controls and display console for the Apollo telescope mount. The telescope mount contained six telescopes for solar observation, and was powered by a windmill-shaped array of four solar panels. At launch it was lined up with the rest of Skylab, but swung to one side when in orbit. The three Skylab crews showed that men can successfully work for long periods in weightless conditions, and demonstrated the value of manned space stations for observations of Earth and sky and for developing new industrial techniques. Total returns from the missions were 45 miles (72 km) of magnetic tape, 2,500 square feet (232 m²) of Earth-resources photographs, and over 200,000 images of the Sun. Skylab is expected to remain in orbit until the 1980s.

Mission	Launch date	Results
Skylab 1	May 14, 1973	World's largest payload; orbited by two-stage Saturn V
Skylab 2	May 25, 1973	Crew of Charles Conrad, Joseph P. Kerwin, and Paul J. Weitz. Deployed parasol sunshade to cool space station, and freed solar wing during EVA. Crew returned June 22, setting new space duration record of 28 days, 49 minutes
Skylab 3	July 28, 1973	Crew of Alan L. Bean, Owen K. Garriott, and Jack R. Lousma. Installed new Sun shield during space walk, made extensive Earth surveys and solar observations that provided new data on formation of solar flares. Returned September 25, setting new duration record of 59½ days
Skylab 4	November 16, 1973	Crew of Gerald P. Carr, Edward G. Gibson, and William R. Pogue. 7-hour EVA on December 25 by Pogue and Gibson to change film in the telescope mount and observe comet Kohoutek. Returned February 8, 1974, after record 84-day mission

Slayton, Donald Kent (b. 1924)
One of America's original seven astronauts, selected in 1959. Slayton was named to fly the second Mercury Earth-orbital mission, but was grounded in 1962 because of a heart murmur. Instead, he became director of flight crew operations until 1972 when he was eventually passed as fit to fly. Slayton was docking module pilot on the Apollo-Soyuz mission.

Slipher, Vesto Melvin (1875–1969)

American pioneer of spectroscopy, who discovered that many galaxies are receding at high velocities. This led to Edwin HUBBLE's conclusion that the Universe is expanding. Slipher began to measure the radial velocities (motions in the line of sight) of galaxies in 1913. These motions were revealed by shifts in spectral lines, and Slipher was the first man to detect them. Slipher's observations helped show that galaxies were indeed separate star systems from our own, but their full meaning became clear only in the 1920s, when Hubble found that a galaxy's speed of recession corresponds to its distance. This has provided the foundation for all current cosmological studies.

Slipher spent his working life at the Lowell Observatory, joining it in 1901 and serving as director from 1916 to 1952. He made spectroscopic studies of the rotation of planets, and discovered dark bands in the spectra of the major planets that revealed the components of their atmospheres, notably ammonia and methane. He showed that calcium and sodium are widely distributed between the stars, since they appear superimposed on stellar spectra. In 1913 he found that the nebulosity around the Pleiades shines by reflecting starlight; this was the first discovery of a reflection nebula, as distinct from a bright-line emission nebula like that in Orion (see NEBULA), and it proved that dust as well as gas exists between the stars. Slipher detected the rotation of a spiral galaxy in 1913; his later studies showed that galaxies rotate with their arms trailing. After Percival Lowell's death, Slipher directed the photographic search that led to the discovery of the planet Pluto by Clyde TOMBAUGH in 1930.

Slipher's brother, Earl Carl Slipher (1883–1964), also worked at Lowell Observatory, and was a noted observer of Mars. He produced two famous photographic atlases, one of Mars and one of the planets.

Small Astronomy Satellites (SAS)

Three American satellites in the EXPLORER series, designed to observe the sky at short wavelengths blocked from the ground by the Earth's atmosphere. They were launched from the SAN MARCO platform off the coast of Kenya. The first Small Astronomy Satellite, also called Explorer 42, was launched on Independence Day in Kenya and was named Uhuru, the Swahili word for Freedom. The 315-lb. (143-kg) satellite made the first systematic scans of the sky at X-ray wavelengths, work continued by the OAO-3 and ARIEL-V satellites. The 410-lb. (186-kg) second SAS, otherwise known as Explorer 48, carried a spark chamber to study gamma rays. The 430-lb. (195-kg) SAS-C, also called Explorer 53, was launched for a more detailed investigation of individual X-ray sources in our Galaxy.

Smithsonian Astrophysical Observatory

An astrophysical research institute of the Smithsonian Institution, founded in 1890 by Samuel Pierpont Langley (1834–1906). In 1955 Smithsonian Astrophysical Observatory (SAO) headquarters moved to the grounds of the HARVARD COLLEGE OBSERVATORY, and in 1973 the two institutions established a joint Center for Astrophysics. The SAO has a worldwide network of satellite-tracking cameras, and operates the Mount Hopkins Observatory, 35 miles (56 km) south of Tucson, Arizona. On a ridge 7,600 feet (2,316 m) high at Hopkins, a telescope 39-feet (10-m) wide, made of 248 hexagonal mirrors, was installed in 1968 to observe faint light emitted by gamma-ray impacts with the atmosphere; in 1970 a 60-inch (152-cm) reflector came into operation. A multiple-mirror telescope, consisting of six 72-inch (183-cm) mirrors whose light-collecting area equals that of a single 176-inch (450-cm) mirror, was opened on the mountains 8,585-foot (2,617-m) summit in 1979 by the SAO in conjunction with the University of Arizona.

solar apex

The point on the CELESTIAL SPHERE toward which the Sun appears to be moving; the antapex is the point directly opposite in the sky. Systematic motion of the Sun relative to other stars is revealed by analyzing their PROPER MOTIONS and RADIAL VELOCITIES. In 1783, Sir William HERSCHEL used this method to locate the position of the apex in the constellation Hercules. It was later shown, however, that the position of the apex, and the velocity of the Sun toward it, depend on the group of stars chosen as a reference frame. Modern measurements indicate that the Sun travels toward the constellation Lyra with a velocity of 12 miles (20 km) per second, relative to stars within 60 light-years.

solar constant

The amount of energy received from the Sun on a given area on the edge of the Earth's atmosphere, when the Earth is at its average distance from the Sun. The accepted value of the solar constant is 2 calories per minute per square centimeter, equal to about 1.3 kilowatts per square meter. The solar "constant" varies by one or two percent because of changing solar activity. The solar constant was first measured by the American astrophysicist Charles Greeley Abbot (1872–1973).

solar cycle

The period of about 11 years over which changes in the SUN's surface activity appear to go through a cycle. The periodic behavior of SUNSPOTS was established in 1843 by Heinrich Schwabe, who found a period of 10 years between the successive times at which the number of sunspots reached a maximum. It was later established that the average period was closer to 11 years, but the length of an individual cycle can be anything from 7 to 17 years.

F. G. W. Sporer discovered that at the beginning of a new cycle (after a minimum), the first spots always appear between latitudes 30° and 45° on either side of the solar equator. Later, spots appear progressively closer to the equator, increasing in number for $4\frac{1}{2}$ years, and then declining for the remaining $6\frac{1}{2}$ years of the cycle. Before the last spots of the old cycle have disappeared, the first spots of the new cycle can be seen at higher latitudes. Successive cycles of sunspots have opposite magnetic field directions, so that from a magnetic standpoint the solar cycle repeats itself only after 22 years.

The number of sunspots is related to the number and intensity of solar active regions; thus variations in other types of solar activity go through the same cycle. These include the number of PROMINENCES,

FLARES, and PLAGES in the PHOTOSPHERE and CHROMOSPHERE, and the intensity of streamers in the CORONA. The latter are related to the SOLAR WIND of particles blowing into space, which is in turn responsible for magnetic storms and aurorae on the Earth. An 11-year cycle for these geomagnetic effects has been known for a century. More recent research suggests that even the Earth's weather may follow the solar cycle, though the reasons are still unknown. Occasionally, the solar cycle seems to stop temporarily, as apparently happened between 1645 and 1715.

solar day

The time the Earth takes to rotate on its axis with respect to the Sun. On average, this is equal to 1.002738 mean SIDEREAL DAYS. The difference between the two periods occurs because the Earth is also revolving around the Sun as it rotates on its axis. The Earth must spin approximately an extra 1/365th of a revolution each day to bring the Sun facing the same hemisphere again.

solar system

The group of planets, comets, and asteroids orbiting the Sun, whose gravitational pull dominates space in all directions out to a distance of 2.4 light-years. The Sun makes up more than 99.95 percent of the mass of the solar system and is its only significant source of light and heat.

The planets can be divided into two groups: the terrestrial planets and the giant planets. The first group stretches from 0.3 to 1.7 ASTRONOMICAL UNITS from the Sun (an astronomical unit, or a.u., is the average distance from the Earth to the Sun). Five bodies make up the terrestrial planets: Mercury, Venus, the Earth-Moon system and Mars. All are similar in nature, being made predominantly of rock, some with central metallic cores. The presence or absence of atmosphere is dictated by the planet's surface gravity and temperature.

From 1.7 to 4.9 a.u. there is a gap in the system, beyond which comes the giant planet group stretching from 4.9 to 30.3 a.u. from the Sun. This consists of four essentially similar bodies, Jupiter, Saturn, Uranus, and Neptune, all globes of light-weight gases, probably with rocky cores. Jupiter is the dominant planet, making up 75 percent of the mass of the entire planetary system. Beyond Neptune is Pluto, a small frozen terrestrial-type planet which may be either an escaped satellite of Neptune or perhaps the largest of a group of small planets that occur beyond 35 a.u. It is unlikely that further giant planets exist beyond Neptune.

The gap between Mars and Jupiter contains a vast swarm of small rocky bodies known as the asteroids, with diameters from 600 miles (1,000 km) down to a few inches. These have orbits which scatter from 0.2 to 15 a.u., with the main swarm at 2 to 4 a.u. The inner solar system also contains the periodic comets, comets which have been perturbed by the planets' gravity into short-period elliptical orbits. The vast majority have aphelia close to Jupiter's orbit, but some are associated with the other giant planets. Finally, beyond all the planets, a great swarm of tiny, frozen comet nuclei is thought to exist at distances from 20,000 to 60,000 a.u. A small proportion of these are perturbed into elongated orbits so that they move

toward the Sun. The comet cloud is so far from the Sun that it is only loosely bound to the solar system and many comets must be perturbed into interstellar space by the gravitational pull of stars passing within a few light-years of the Sun.

The space between the planets is not empty, but packed with areas of fine dust particles and gas molecules. The asteroids constantly suffer collision and form debris, while the comets steadily shed both gas and dust as they are heated by the Sun. Fine dust and gas are ejected from the system by the SOLAR WIND, a stream of atomic particles flowing from the Sun. Dust too heavy to be eliminated in this way slowly spirals into the Sun over a period of several million years.

solar time

Time used for all ordinary purposes, measured in terms of the Earth's rotation with respect to the Sun rather than to the stars (SIDEREAL DAY). *Apparent* solar time, as indicated by a sundial, does not run at a constant rate because of the Earth's slightly eccentric orbit and the inclination of its axis (see EQUATION OF TIME). *Mean* solar time, in which these variations are smoothed out, is therefore always used.

solar wind

A continuous stream of protons (hydrogen nuclei) and electrons, together with a few helium nuclei, which is constantly emitted by the Sun in all directions. It can be regarded as the uppermost part of the Sun's CORONA, forced away into interplanetary space by energy traveling up from the solar surface (the PHOTOSPHERE). The particles travel away from the Sun at speeds between 220 and 500 miles (350 and 800 km) per second; the solar wind has an average density of 5 protons and 5 electrons per cubic centimeter as it passes the Earth.

Although a solar wind had long been suspected for a number of reasons, among them the link between solar activity and changes in the Earth's magnetic field, direct measurements were not possible until the advent of probes exploring interplanetary space. The first extensive measurements were made by MARINER 2 on its voyage to Venus in 1962, and the latest results extend from only 0.3 a.u. from the Sun, as recorded by the HELIOS PROBE, to the 8 a.u. plumbed by the Jupiter probe PIONEER 10. The wind probably extends out to about 100 a.u., where it becomes so weak it is stopped by interstellar gas.

Many effects of the solar wind can be observed indirectly. Radio waves from distant sources are made to scintillate ("twinkle") as they pass through irregularities in the wind. Solar wind particles striking the nucleus of a COMET ionize molecules, and propel them into a straight tail pointing away from the Sun. "Knots" in the tail are seen to accelerate under the continuous pressure of the wind.

The Earth's magnetic field traps solar wind particles into the MAGNETOSPHERE, especially into the region of the VAN ALLEN BELTS. The particles in turn affect geomagnetic activity, such as magnetic storms, AURORAE, and radio fadeouts. These follow a 27-day cycle, as the more powerful wind from above the Sun's most active regions is swept past the Earth by the 27-day rotation of the Sun. Magnetic effects also vary with the 11-year SOLAR CYCLE, which governs changes in solar wind strength.

solstice

The moment when the Earth's axis is inclined at its maximum ($23\frac{1}{2}°$) toward the Sun. The north pole is tilted sunward at its maximum about June 21 (producing the beginning of the northern summer and the southern winter) while the maximum sunward inclination of the south pole, giving the opposite conditions, is about December 22. At these times the Sun is at its greatest angular distance, $23\frac{1}{2}°$, north and south of the *celestial equator*.

South African Astronomical Observatory

An observatory opened in 1973 at an altitude of 6,004 feet (1,830 m), 9 miles (14 km) from Sutherland in Cape Province, South Africa. The SAAO is an amalgamation of the CAPE OBSERVATORY, where its headquarters lie, and the former Republic Observatory, Johannesburg; it is operated jointly by the British and South African governments. It contains a 39-inch (100-cm) reflector, transferred from the Cape Observatory, together with a 20-inch (51-cm) reflector from the Republic Observatory. In 1975, the 74-inch (188-cm) reflector of the Radcliffe Observatory, Pretoria, was moved to the SAAO.

Southern Cross

See CRUX.

Soyuz

Soviet manned spacecraft for long-duration flights and rendezvous and docking missions. Soyuz is built in three parts. At the front is a near-spherical orbital compartment 8.7 feet long by 7.3 feet wide (2.65 m × 2.25 m) used for working in space. The central section is a bell-shaped command module 7.2 feet long by 7.1 feet wide (2.2 m × 2.15 m), in which cosmonauts sit during take-off and reentry. At the rear is a cylindrical service module 7.5 feet long by 7.2 feet wide (2.3 m × 2.2 m) which contains supplies, maneuvering engines, and retro-rockets. Soyuz has a habitable volume of 360 cubic feet (10.2 m^3) and weighs 14,750 lb. (6,690 kg). The Soyuz service module has two wing-like solar panels attached to generate electricity, but these are removed in the simplified Soyuz design that ferries crews to and from the SALYUT space station. The Soyuz docking tunnel is located at the forward end of the orbital compartment, through which cosmonauts must crawl to transfer between docked craft. The orbital compartment can be used as additional space for resting or for scientific equipment. Both the orbital compartment and the service module are jettisoned before reentry, burning up in the atmosphere.

Soyuz was originally intended to carry three men without space suits. But this was changed after the Soyuz 11 mission, when a pressure loss prior to reentry killed the three crewmen. Soyuz crews now wear space suits during launch and reentry; but the weight and space that this takes up means that Soyuz can now hold only two cosmonauts. During the first Soyuz flight, the craft went out of control in orbit and crashed to Earth after reentry when its parachute lines became twisted, killing its test pilot. Soyuz may have been intended for a manned round-the-Moon flight; for such missions it would have had the orbital compartment removed. Unmanned test flights to the Moon of the Soyuz command and service modules were made under the name of Zond, and for lunar missions Soyuz would have been launched by the PROTON booster. However, for all flights into Earth-orbit Soyuz has been launched by a rocket using the same lower stages as the VOSTOK LAUNCHER, but with a more powerful upper stage.

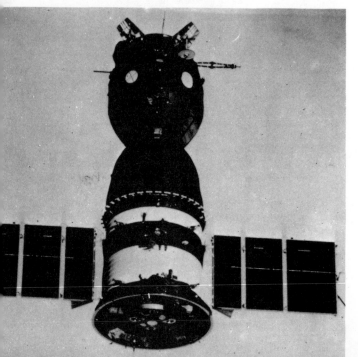

A Soyuz spaceship, as modified for use in the the Apollo-Soyuz Test Project. The smaller compartment is the orbital module; on its end can be seen part of the docking mechanism. At the other end is the instrument module, with its solar panels; in the middle is the descent module, the only part which returns to Earth.

Mission	Launch date	Results
Soyuz 1	April 23, 1967	Vladimir Komarov killed during reentry on April 24 after 18 orbits
Soyuz 2	October 25, 1968	Unmanned target for Soyuz 3
Soyuz 3	October 26, 1968	Georgi Beregovoi maneuvered close to Soyuz 2 but did not dock with it
Soyuz 4	January 14, 1969	Vladimir Shatalov docked with Soyuz 5; Yevgeny Khrunov and
Soyuz 5	January 15, 1969	Alexei Yeliseyev transferred into Soyuz 4 by a space walk, leaving Boris Volynov to return to Earth alone in Soyuz 5
Soyuz 6	October 11, 1969	Georgi Shonin and Valeri Kubasov made joint maneuvers with Soyuz 7 and 8, and conducted welding experiments
Soyuz 7	October 12, 1969	Anatoli Filipchenko, Vladislav Volkov, and Viktor Gorbatko carried out joint maneuvers with Soyuz 6 and 8
Soyuz 8	October 13, 1969	Vladimir Shatalov and Alexei Yeliseyev commanded group flight of Soyuz 6, 7, and 8; first flight involving three craft, seven cosmonauts

Mission	Launch date	Results
Soyuz 9	June 2, 1970	Andrian Nikolayev and Vitaly Sevastyanov made record 17½-day flight
Soyuz 10	April 23, 1971	Vladimir Shatalov, Alexei Yeliseyev, and Nikolai Rukavishnikov docked with space station Salyut 1 but did not enter, possibly due to hatch problem
Soyuz 11	June 6, 1971	Georgi Dobrovolsky, Viktor Patsayev, and Vladislav Volkov docked with Salyut 1 and transferred for record 23 days; crew members were killed during reentry because of capsule pressure loss
Soyuz 12	September 27, 1973	Vasily Lazarev and Oleg Makarov made two-day test flight of simplified Soyuz for space-station ferry missions
Soyuz 13	December 18, 1973	Pyotr Klimuk and Valentin Lebedev made week-long scientific flight
Soyuz 14	July 3, 1974	Pavel Popovich and Yuri Artyukhin docked with Salyut 3 space station for 16-day mission
Soyuz 15	August 26, 1974	Gennady Sarafanov and Lev Demin failed in attempts to rendezvous automatically and dock with Salyut 3
Soyuz 16	December 2, 1974	Anatoly Filipchenko and Nikolai Rukavishnikov made 6-day rehearsal for Apollo-Soyuz mission
Soyuz 17	January 11, 1975	Alexei Gubarev and Georgi Grechko docked with Salyut 4 space station for 29-day mission
Soyuz	April 5, 1975	Vasily Lazarev and Oleg Makarov failed to reach orbit because launch rocket upper stage failed to separate. Soyuz capsule returned to Earth safely, but was not numbered
Soyuz 18	May 24, 1975	Pyotr Klimuk and Vitaly Sevastyanov completed 64-day mission aboard Salyut 4 space station

Mission	Launch date	Results
Soyuz 19	July 15, 1975	Alexei Leonov and Valeri Kubasov performed joint docking with American Apollo in Apollo-Soyuz Test Project
Soyuz 20	November 17, 1975	Unmanned test of automatic shuttle craft; docked with Salyut 4 on November 19 and transferred fuel into space station. Landed automatically February 16, 1976
Soyuz 21	July 6, 1976	Boris Volynov and Vitaly Zholobov docked with Salyut 5 for 48-day mission
Soyuz 22	September 15, 1976	Valery Bykovsky and Vladimir Aksenov made eight-day scientific flight
Soyuz 23	October 14, 1976	Vyacheslav Zudov and Valery Rozhdestvensky failed to dock with Salyut 5
Soyuz 24	February 7, 1977	Viktor Gorbatko and Yuri Glazkov docked with Salyut 5 for 17-day flight
Soyuz 25	October 9, 1977	Vladimir Kovalyonok and Valery Ryumin failed to dock with Salyut 6
Soyuz 26	December 10, 1977	Yuri Romanenko and Georgi Grechko docked with Salyut 6 record-breaking 96-day flight
Soyuz 27	January 10, 1978	Vladimir Dzanibekov and Oleg Makarov paid week-long visit to Salyut 6
Soyuz 28	March 2, 1978	Alexei Gubarev and Czechoslovak Vladimir Remek paid week-long visit to Salyut 6
Soyuz 29	June 15, 1978	Vladimir Kovalyonok and Alexander Ivanchenkov docked with Salyut 6 for record-breaking 140-day mission
Soyuz 30	June 27, 1978	Pyotr Klimuk and Pole Miroslaw Hermaszewski paid week-long visit to Salyut 6
Soyuz 31	August 26, 1978	Valery Bykovsky and East German Sigmund Jahn paid week-long visit to Salyut 6

Spacelab

A space station built by the European Space Agency, to be carried into orbit in the cargo bay of the SPACE SHUTTLE, from which it will make observations of Earth and sky, and allow scientists to experiment with conditions of zero gravity. Spacelab is 49 ft (15 m) long and 14 ft (4.3 m) diameter. When not working in Spacelab, the crew of up to four will rest in the Shuttle Orbiter. Spacelab comes in two main versions. It can consist of a pressurized module from which astronauts control the operation of instruments mounted outside on unpressurized pallets, open to the vacuum of space. Or it can consist entirely of unpressurized pallets, the equipment on which is controlled from within the Orbiter. Each orbital flight of Spacelab will last from 7 to 30 days; after each flight it will return in the Shuttle Orbiter, and thus be used up to 50 times during a 10-year life. The payload can be modified for each flight, and will include manufacturing tests and space medicine experiments. Spacelab will be used jointly by NASA and the European Space Agency, and will fly with international crews. It will be the only orbiting workshop available in the West during the early years of the Shuttle's operation.

space probe

Any instrument-carrying device sent from the Earth to examine another celestial body, or to take measurements of conditions in space. The first space probe was LUNA 1 in 1959, which flew past the Moon. Starting in February 1961 the Soviet Union began its series of VENUS PROBES, and in November 1962 launched the first of its MARS PROBES. However, the first probes successfully to send results from other planets were those in the American MARINER series.

The Space Shuttle, the winged reusable space transporter of the 1980s, will carry various payloads into orbit in its cargo bay, including the European-built Spacelab space station, here shown in cutaway.

For other American planetary exploration series see PIONEER and VIKING. For American Moon probes see RANGER, SURVEYOR, and LUNAR ORBITER.

Space Shuttle

Reusable winged space transporter, which is launched like a conventional rocket but which glides back to Earth like an aircraft. The concept originated in the 1930s with the rocket-plane design of the Austrian spaceflight pioneer Eugen SANGER, and was developed during World War II at Peenemünde by Wernher VON BRAUN and his group, who drew up plans for a multistage winged rocket. Following the war, several designs for winged, recoverable launch systems were outlined, and trial flights began during the 1960s with simple aerodynamic shapes known as lifting bodies. However, non-reusable spacecraft were developed for the first manned flights because of their relative simplicity and lower initial cost.

Modern Space Shuttle. NASA officially began investigating Space Shuttle designs in 1968. Although initially envisaged with a winged reusable booster as well as a winged reusable orbiter, development costs meant that by 1972 the winged, fly-back booster was discarded in favor of conventional rockets. As finally defined, the Orbiter is 122 feet (37.2 m) long and 57 feet (17.4 m) high, with a wingspan of 78 feet (23.8 m). It has three main engines of 470,000 lb. (213,200 kg) thrust each, fed by propellants from an external tank 154 feet (46.9 m) long and 27.5 feet (8.4 m) in diameter. This tank is jettisoned just before reaching orbit and reenters the atmosphere; it is the only part of the system not planned for reuse. Two solid-fuel strap-on boosters are also ignited at lift-off; these are 149 feet (45.4 m) long and 12 feet 2 inches (3.7 m) in diameter, and each produces a thrust of 2,650,000 lb. (1,202,000 kg). They drop away at a height of 27 miles (43 km) and parachute back for recovery and reuse. Total weight of the Shuttle at launch is approximately 4,400,000 lb. (2,000,000 kg). The Orbiter, which is the size of a modern jetliner, is designed to carry up to four crew members and three passengers into orbit. Its cargo bay measures 60 feet (18.3 m) long by 15 feet (4.6 m) wide, capable of carrying as much as 65,000 lb. (29,500 kg). It will be able to bring back up to 32,000 lb. (14,500 kg) from orbit; one of the advantages of the Shuttle is that it can be used to retrieve satellites for repair.

Shuttle missions. The Shuttle's large capacity will be used to take many satellites into space simultaneously, to launch major scientific payloads such as the SPACE TELESCOPE, and to ferry men and material into orbit for purposes such as space station construction. A space station known as SPACELAB is being developed by the European Space Agency, to be carried in the Shuttle's payload bay for international scientific and engineering experiments in orbit.

The Shuttle can spend from one to four weeks in orbit before returning to Earth for a landing on a 15,000-foot (4.6-km) runway at Kennedy Space Center. After about two weeks the Orbiter will be ready to fly again with a new payload. The Shuttle will be assembled and launched from Cape Canaveral, using facilities modified from the Saturn-Apollo program. Each Orbiter is expected to make 100 or more missions. Reuse will slash the cost of space travel by about 90 percent, and the introduction of the

190

Shuttle is expected largely to replace conventional rockets. The first Orbiter made a series of test landings in 1977, being released from the back of a Boeing 747 Jumbo jet. The first manned launch of the Shuttle is planned for late 1979.

space station
A large orbiting spacecraft for long-term human habitation. Space stations allow detailed astronomical observations to be made above the blurring effect of the Earth's atmosphere; they also act as platforms for Earth surveys, and permit experiments to take place in a total vacuum and at zero gravity, conditions unattainable on Earth. The idea was current as early as 1869, when the American author Edward Everett Hale (1822–1909) wrote a story entitled "The Brick Moon," in which he proposed the launching of an artificial satellite (made of brick) into a 4,000-mile (6,400-km) polar orbit to aid navigation. The moon was to be hurled aloft by rolling against a rapidly-spinning flywheel. Unfortunately, the brick moon slips too early, and is propelled into orbit along with 37 construction workers and their families, who continue to live there.

Scientific proposals for a space station were made in 1923 by the German astronautical pioneer Hermann OBERTH, who envisaged space stations as refueling posts for space rockets. Grandiose schemes since have envisioned large space stations being constructed from parts ferried up from Earth, perhaps built in a wheel shape and spinning to provide artificial gravity.

In the mid-1960s the U.S. Air Force began development of the Manned Orbiting Laboratory, a two-man space station for military reconnaissance, intended to be launched atop a Titan 3 rocket with the astronauts in a modified Gemini capsule. The project was canceled in 1969. About that time, NASA was studying designs for a 12-man scientific space station to be launched during the late 1970s as a follow-on to the Apollo program. Plans envisaged eventual extension of the station to hold 50 men. The Soviet Union is also believed to have similar advanced designs under consideration. However, NASA's plans were curtailed because of budgetary cuts.

The first space station to be launched was the Soviet Union's SALYUT, which went into orbit on April 19, 1971, and was boarded by its first crew in June of that year. Several subsequent Salyuts were launched. America's larger SKYLAB station was launched on May 14, 1973; its third and final crew returned to Earth on February 8, 1974. Both Salyut and Skylab were somewhat rudimentary space stations, being converted from the upper stages of rockets. No further American space-station missions are planned until the European-built SPACELAB becomes available in the early 1980s. This module will travel in the cargo bay of the SPACE SHUTTLE. Eventually, the Shuttle may be used to ferry into orbit components for a larger permanent space station.

space suit
A device worn by astronauts to protect them from the airless conditions of space. A space suit is a kind of personal spacecraft for the astronaut. It supplies air for him to breathe, insulates him from the extremes of hot and cold, and provides a vital pressurized container. The basic components of a space suit are

Apollo 11 astronauts practice their lunar surface activities in a Houston laboratory while wearing full space suits and back-packs.

an inner pressure garment made of coated nylon, covered with insulating layers of aluminum-coated plastic film and glass-fiber cloth. A space suit becomes very rigid when inflated, and careful design is needed to allow a space-suited astronaut some mobility.

The first Mercury space suits were modified from high-altitude aviation suits, and were only intended to be inflated in an emergency, if the spacecraft lost pressure. Soviet Vostok space suits were concealed by a loose-fitting cloth garment, but were probably similar in design. For Gemini missions a more advanced suit was designed, pressurized by an umbilical cord from the spacecraft. Wearing such space suits, American astronauts made their first space walks (EVAs). For the first Soviet EVA, in the Voskhod 2 craft, Alexei Leonov wore a suit with a back-pack similar to those later used in the Soyuz 4 and 5 missions for crew transfers. A back-pack allows the astronaut to move independently of the spacecraft, as was required for the Apollo Moon-landing missions. For Moon walks a special Moon suit was devised with a back pack called the Portable Life-Support System (PLSS).

Apollo suits were made more flexible by introducing bellows-like joints. They also had additional layers to withstand the more extreme conditions of walking and working on the Moon. Next to the astronaut's skin was a liquid cooling garment that prevented him from overheating. Around this was the pressure garment, a restraint layer to prevent the suit from ballooning, layers of insulation, and a micrometeorite-protection layer, topped with an abrasion-proof outer covering. Helmet and gloves were attached by air-tight rings.

The plastic helmet, with a movable visor to shield against solar glare, was fixed in position, and the astronaut was free to move his head inside; he could take sips from a water bag mounted in the neck of the suit. The gloves had special insulation against heat and abrasion. For Moon walks rubber-soled overshoes were slipped on and the suit contained a urine-collection bag. The Portable Life-Support System provided air-conditioning for the space suit and supplied water for the cooling garment; it also embodied a radio transmitter. The PLSS allowed astronauts to spend up to eight hours on EVAs. The Apollo Moon suit's total weight was 180 lb.; similar suits were used on Skylab spacewalks.

Space Telescope

An astronomical telescope planned to be orbited by the SPACE SHUTTLE in late 1983. The Space Telescope is a remote-controlled observatory in space, receiving commands from Earth and returning data by radio link. The telescope has a main mirror 94 inches (240-cm) aperture. Orbiting above the blurring effects of the atmosphere, the Space Telescope will achieve a level of performance far higher than that of the best Earth-based telescopes; it will be able to discriminate detail 10 times as fine, and objects 100 times as faint. The telescope will also be able to function at wavelengths that the Earth's atmosphere prevents from reaching the ground. Astronauts will visit the Space Telescope to replace equipment and make repairs; this will give it an operating life of at least 10 years, and possibly as long as that of a ground-based observatory.

spallation

The erosion of surfaces by the impact of small particles and high-energy radiation.

spec.

Abbreviation for the Latin SPECULUM, meaning mirror.

specific impulse

A measure of a rocket engine's performance. Specific impulse is calculated by dividing the rocket's thrust by the weight of fuel used per second. The result gives the time in seconds for which a unit of fuel produces a unit of thrust. The higher a rocket's specific impulse, the more efficient it is.

speckle interferometry

A technique for reconstructing star images that have been distorted by atmospheric turbulence, or poor seeing. Light from a star is distorted by constantly moving "cells" in the lower atmosphere, usually 4 to 12 inches (10–30 cm) in size. The image of a star seen through a large telescope thus consists at any instant of a group of "speckles," each a poor-quality image of the star, like the multiple images produced by a fly's eye. To reconstruct the star image the speckle pattern is first photographed in a limited wavelength band, using an IMAGE INTENSIFIER to allow a short exposure time. In one technique, the speckle pattern is then illuminated by a LASER, so that the combined interference of the individual poor images produces one single image. Several photographs of such single images are combined to reconstruct the final high-quality image. Alternatively, individual speckles

are combined using a computer to produce a final good image. By such methods, resolutions better than 0.01 arc second can be achieved, thus revealing the disks of large stars, or separating close doubles.

spectral lines

The narrow lines observed when an object's light is dispersed into a spread of wavelengths (a SPECTRUM). Each line represents light of one particular wavelength. Spectral lines may either be bright (emission lines), or occur as dark ABSORPTION LINES against the bright background of a continuous spectrum of all wavelengths. Gaseous nebulae show emission-line spectra, while ordinary stars have absorption-line spectra. The lines in the Sun's spectrum were discovered in 1814 by Joseph Fraunhofer, and are known as FRAUNHOFER LINES. The spectral lines of other astronomical bodies were first detected by William HUGGINS and Pietro Secchi in the 1860s.

Each type of atom has its own unique set of spectral lines. These reveal the elements present in the star's surface or in the nebula, like a spectral "fingerprint." The positions of the lines in the spectrum are the same whether the lines are absorption or emission.

The spectrum of hydrogen contains only a few lines, while a metal like iron has thousands. For any one element the number of lines, their wavelengths, and their relative strengths all depend on properties of the atom that can be calculated or measured in the laboratory. The relative strengths of a hydrogen line and an iron line, for example, depend on three factors: the atomic parameters, the proportions of each element in the star, and the star's surface temperature. Since the majority of stars have nearly the same relative abundance of elements, the effects of temperature allow stars to be readily classified by their spectral appearance into a sequence of SPECTRAL TYPES, governed by the star's surface temperature.

A comparison of the strengths of the various lines of any particular element ("curve of growth analysis") reveals the exact abundance of that element, when the atomic parameters and the surface temperature are known.

A detailed study of the way a particular spectral line decreases in strength from its center out to the continuous spectrum on either side (the *profile*) produces more information about the physical conditions at the star's surface. Spectral lines are made wider by higher temperatures or higher pressures, and both these quantities can be measured from a good profile. In particular, a giant star can be distinguished by its very narrow spectral lines, due to the low pressure at the surface of such a large object, where the force of gravity is relatively small. The nature of a star's magnetic field can also be determined, since it produces a splitting of the spectral lines through the ZEEMAN EFFECT.

spectral type

The category into which a star can be classified on the basis of the SPECTRAL LINES that occur in its spectrum.

From the 1860s, when William HUGGINS and Pietro Secchi first observed the stars spectroscopically, it had been noticed that stellar spectra differ considerably, both in the number of lines present and in the darkness of the ABSORPTION LINES relative to the bright background (the *continuum*). Spectra were first classified systematically by E. C. PICKERING at the Harvard College Observatory at the beginning of this century, and this work culminated in the monumental *Henry Draper Catalogue* of 225,300 stars published in 1924. Originally the letters of the alphabet had been used to classify spectra in order of increasing complexity. But it was soon shown that the most meaningful order is that representing the stars' surface temperatures, which can be deduced from the stars' colors. The consequent rearrangement of the letters produced the sequence as used today, which runs (from hottest to coolest) O B A F G K M. The peculiar, extremely hot WOLF-RAYET stars are often included at the beginning of the list as class W. There are also the classes R, N, and S, which are about the same temperature as the K and M stars, but differ from the standard sequence in having an unusually high concentration of certain HEAVY ELEMENTS. R and N stars have an excess of carbon, and are often called "carbon stars" (class C), while S stars contain zirconium and yttrium.

Stars at the beginning of the sequence, up to class G, are referred to as "early-type," while those after G are "late-type." These names are purely historical in origin, and do not imply that stars evolve along the sequence.

The spectra change from one class to the next. For example, between classes F and G the spectral lines of hydrogen become less pronounced, while those of calcium become stronger. This enables each class to be subdivided into ten *types*. The Sun is thus type G2, one-fifth the way between a G0 and a K0 spectrum.

Classification of spectra is done by eye alone, and involves the study of certain easily recognizable lines. The most important are those of hydrogen, the H and K FRAUNHOFER LINES of calcium, and the numerous lines due to such heavy elements as carbon and titanium oxide.

When the spectra of giant and dwarf stars of the same type were compared in detail, slight differences were found. This prompted W. MORGAN to add a LUMINOSITY classification to the Harvard star classification. Luminosity class V corresponds to a normal dwarf (MAIN SEQUENCE) star like the Sun, III to a giant, and I to a supergiant. (A few stars fall into the intermediate categories IV and II.) As an example, the Sun is G2 V, Aldebaran is K5 III, and Rigel is B8 I. This MORGAN-KEENAN CLASSIFICATION (MK system), published in 1943, is now used universally to categorize stars by their spectral lines.

spectroscope

A device for observing the SPECTRUM. Strictly speaking, a spectroscope is an instrument used by the naked eye, but in practice the term can be loosely employed for all devices which record the spectrum.

A simple prism or diffraction grating used to observe a white light-source will split the image into the rainbow colors of the spectrum. This is in effect a continuous series of overlapping images of the source, spread out by wavelength. If a small band of wavelengths are deficient, as in an ABSORPTION LINE, they will not be seen unless they cover a greater width of the spectrum than a single image of the source—in other words, they will not be resolved. To improve the resolving power, the light to be analyzed

A spectrograph attached to the 73-inch (1.88-m) reflector at Victoria, British Columbia. By attaching the plate holder shown on the lower branch to either of the other two, a higher dispersion spectrum may be obtained.

is passed through a slit which is imaged by a lens system. The width of the slit compared to the dispersion of the spectrum determines the narrowest feature that can be seen. Thus the resolving power depends on the spectroscope, rather than on the size of the telescope. Visual spectroscopes are of this basic type. Substituting a camera for the eye gives a permanent record of the spectrum, and the instrument becomes a *spectrograph,* the resulting photograph being called a *spectrogram.*

If the spectroscope has a scale for reading off the wavelengths, it is strictly speaking a *spectrometer.* This name is becoming widely used for devices in which a PHOTOELECTRIC CELL either scans the spectrum or views it through filters with limited transmission bandwidths, thus giving an electrical read-out corresponding to the light intensity. Such an instrument should correctly be termed a *spectrophotometer.*

When observing the Sun, it can be inconvenient to see only the spectrum of a single strip, as represented by the slit. Consequently, the slit is arranged to scan the solar disk, and another slit, moving in conjunction isolates the spectral line of interest. An image of the Sun in one wavelength only can be built up. Where a photograph is taken, the device is called a *spectroheliograph.* If an observer wishes to view the disk in the light of one line only, the slits must scan rapidly so that persistence of vision enables the whole disk to be seen at once; the device is known as a *spectrohelioscope.* An alternative to this system now widely used is to view the Sun through a narrow band-pass interference filter in the light beam of a telescope; no other apparatus is necessary. With bandwidths of the order of an angstrom, the disk is seen in one wavelength only.

spectroscopic binary

A DOUBLE STAR in which the motion of one star
around the other is detected by a spectroscope.
Although the two stars in a spectroscopic binary are
too close to be seen individually, the movement of
one star around the other produces a DOPPLER
EFFECT in its spectral lines (provided the orbit is not
exactly at right angles to the line of sight). As the star
alternately approaches the Earth and then recedes,
its spectral lines shift first one way and then the other,
revealing its orbital period. The first spectroscopic
binary discovered was MIZAR, by E. C. Pickering in
1889. If the orbit is edge-on to us, one star will
periodically obscure the other, causing an
ECLIPSING BINARY, such as ALGOL.

If the two stars in a spectroscopic binary are of
nearly equal brightness, two spectra can be seen (the
pair are known as a *two-line spectroscopic binary*).
In this case, astronomers can deduce the mass of both
stars. (If the plane of the star's orbit is at right angles
to the line of sight so that no Doppler effect is visible,
the two spectra may still be individually detectable.
This is termed a *spectrum binary*.) Usually, however,
the light from the brighter star swamps the light from
the fainter, and only one spectrum can be seen (the
pair are a *single-line spectroscopic binary*). In this case
the astronomer can only deduce the minimum
possible mass for the unseen secondary star. In the
case of the single-line spectroscopic binary Cygnus X-1,
the minimum mass of the unseen companion is six
solar masses. It is therefore too massive to be a white
dwarf or neutron star (which could be no larger than
about 1.5 solar masses, the Chandrasekhar limit).
Consequently, many astronomers think that the unseen
companion of Cygnus X-1 is a BLACK HOLE.

spectroscopy

The study of the spectra of heavenly bodies. By using a
SPECTROSCOPE white light can be analyzed into its
component colors, producing the visible SPECTRUM
of the rainbow, from dark red through violet. There
are three general types of spectra: *continuum, emission,*
and *absorption*. A continuum is the complete band of
color, with no other features; an absorption spectrum
is a continuum crossed by certain dark lines; and an
emission spectrum consists of bright colored lines only.

The German scientists Gustave Robert Kirchhoff
(1824–1887) and Robert Wilhelm Bunsen (1811–1899),
put forward the basic principles governing spectra in
1859. What has become known as Kirchhoff's law of
radiation states that, for a given temperature and
wavelength, the ratio between energy radiated and
energy absorbed by a body is fixed. Thus, if a body
emits all wavelengths, it must also absorb all
wavelengths. This leads to the somewhat surprising
result that an object which radiates perfectly also
absorbs perfectly, and will therefore be perfectly
black—a so-called *black body*. Furthermore, a
body which emits light at a certain wavelength
only, also absorbs light at the same wavelength
only.

The other basic principle of spectroscopy is that an
incandescent body—either solid, liquid or gaseous—
will emit at all wavelengths; that is, a continuum is
emitted. A gas at comparatively low temperature,
however, will emit only a line spectrum.
Putting these two principles together shows that the

hot surface of a star will emit a continuum, just as will
the glowing filament of a light bulb or molten metal.
The atmosphere of the star, however, is cooler, and it
will absorb certain wavelengths characteristic of the
gases present. If these gases could be viewed by
themselves, they would be seen to be emitting light at
the same wavelengths.

Only certain lines are permitted because the electrons
of atoms can only exist in specified energy levels. If a
certain quantity of energy is added to the atom,
causing it to be in an "excited" state, the electrons can
increase their energy only by fixed amounts, between
energy levels. If the atom loses the energy again, it can
only do so by radiating at the same energy levels—that
is, at particular wavelengths of radiation.

Each atom or molecule has its own distinct set of
energy levels. Taking the hydrogen atom, the simplest
example because it has only one electron, the lowest
energy level, in which the atom is completely neutral,
is called the ground state. An electron jumping between
this and the next energy level—in either direction—
will produce a line of wavelength 1,216 angstroms, in
the ultraviolet.

Jumps (or *transitions*) up to higher energy levels
produce lines of greater energy, further into the
ultraviolet. The difference between levels soon
becomes small, however, and the lines become closer
together until they reach a series limit. Transitions
between the next energy level up and higher ones
produce a similar set of lines in the visible part of the
spectrum called the Balmer series, which are easily
recognized in stellar spectra (see SPECTRAL TYPE). If
the jumps are upward, then energy has been added to,
or absorbed by, the atom, and so absorption lines
result. If the jumps are downward, then the atom is
releasing its energy and produces emission lines.

spectrum

The entire range of ELECTROMAGNETIC RADIATION,
from gamma rays to radio waves. More usually, the
visible wavelengths only—the colors of the
rainbow from violet to red—are known as the
spectrum. Astronomers often also use the word to
refer to a photograph of a star's visible-light spectrum,
more properly called a *spectrogram*.

speculum

Latin word for mirror. The first mirrors for reflecting
telescopes were made from a special alloy, two-thirds
copper and one-third tin, called speculum metal
because it was designed to take a high polish. However,
metal mirrors do not keep their figure accurately, and
soon tarnish so that they must frequently be repolished.
About the turn of the century metal mirrors were
superseded by glass mirrors with a thin reflective
coating of silver or aluminum (see MIRROR).

spherical aberration

Loss of sharpness in an image when different parts of
a lens or mirror do not bring light rays to the same
focus. A spherical concave mirror produces a blurred
image of a distant object, such as a star, because the
central part of the mirror has a longer focal length
than the outer zone. The same effect is produced by a
simple lens, although the focal point is harder to
define because of the added effect of CHROMATIC
ABERRATION. For astronomical use, the spherical

aberration of a mirror is eliminated by deepening the center slightly, and thus shortening the focal length; the curve of a mirror so deepened is a PARABOLA. The effect can also be counteracted in special optical systems such as the SCHMIDT TELESCOPE. The usual method of curing the spherical aberration of a lens is to combine it with another lens, which is also necessary to remove chromatic aberration (see OBJECT GLASS).

spheroid
A regular, rounded object, shaped like a slightly squashed sphere. A spheroid can be thought of as the solid surface formed by rotating an ellipse about either its longer or its shorter axis. If about its shorter axis, the result is a *prolate spheroid*; if about its longer axis, the result is an *oblate spheroid*. The Earth is an example of an oblate spheroid—it is slightly flattened at the poles.

Spica
The brightest star in the constellation Virgo, and one of the brightest in the entire sky; it is also called Alpha Virginis. Spica, a blue-white star eight times the Sun's diameter, has a magnitude of 0.96, and is 260 light-years away. It is a spectroscopic eclipsing binary with a period of four days; the masses of the two stars are 10.9 and 6.8 times that of the Sun.

spicules
Vertical, jetlike features of the solar CHROMOSPHERE, giving it a fine, hairlike appearance where it merges with the CORONA. These jets of hot gas, at temperatures of 10,000°, shoot 6,000 miles (10,000 km) up into the corona and subside again in less than five minutes. There are perhaps 500,000 spicules on the Sun at any given time, aligned with the magnetic fields around the edges of supergranulation cells in the chromosphere. These cells, each 20,000 miles (35,000 km) across, are part of a network of convective motions in the chromosphere, and it has been suggested that spicules are involved in carrying energy from the lower to the upper chromosphere.

Although spicules appear to be governed by magnetic fields, they are absent over active regions where the field is particularly strong. Recently, observations at ultraviolet wavelengths from Skylab have revealed the presence of "macrospicules," giant spicules 20,000 miles (35,000 km) long, which last for up to 40 minutes.

spiral galaxy
A type of galaxy in which many of the stars and nebulae lie along spiral arms that appear to wind out from the center. Lord ROSSE discovered the spiral pattern in some fuzzy "nebulae" with his 72-inch (180-cm) reflecting telescope in 1845, but not until the 1920s were these proved to be external galaxies comparable to our MILKY WAY. Photographs taken with large telescopes show that about three-fourths of the galaxies in the Universe are spirals.

A spiral galaxy has a central, rounded nucleus (noticeably elongated in a *barred spiral*) composed of old stars, surrounded by a spherical halo also made up of old stars, many of which are concentrated in GLOBULAR CLUSTERS. Young stars, and gas which has not yet formed stars, lie in a disk, about 100,000 light-years in diameter but only 2,000 light-years thick. This material orbits the galactic center just as the planets move around the Sun, with the more distant stars traveling more slowly (*differential rotation*). The rate of the disk's rotation reveals the galaxy's total mass, which is typically a hundred billion times that of the Sun.

The spiral arms prominent on photographs are regions of the disk where stars, gaseous nebulae, and dust clouds are particularly concentrated. The spiral pattern is thought to rotate around the galactic nucleus as a whole, so that the outer parts must travel

A closeup of the chromosphere in the red spectral line of hydrogen showing spicules aligned with the magnetic field. These short-lived jets of hot gas, each 6,000 miles (10,000 km) long, can be seen to form a pattern, following the borders of the supergranulation cells.

faster than the inner. Since the opposite is true of individual stars, gas and dust, whose rotation speed decreases with distance from the center, the material comprising a spiral arm is constantly changing. It seems that the disk material becomes bunched together as it orbits through that part of the disk where the spiral arm is. The bunching is probably caused by the gravitational attraction of the material already comprising the spiral arm; although this material soon orbits out of the arm, the newly bunched material continues the process. The result is a self-perpetuating arm, of a type known as a "density wave."

Spiral galaxies are classified by the prominence of their spiral structure. A galaxy with a nearly "smooth" disk and a bright nucleus is termed Sa (or SBa if it is a barred spiral); a galaxy nucleus is Sc, or SBc. Intermediate types are Sb or SBb. The Milky Way is a typical spiral galaxy, in type between Sb and Sc.

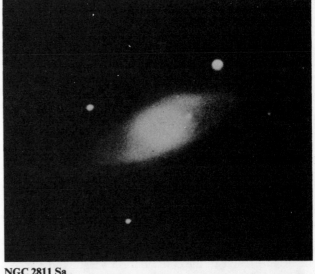

NGC 2811 Sa

NGC 3031 Sb
NGC 1300 SBb

NGC 628 Sc

Four spiral galaxies, showing the variety of structure among some of the types. NGC 2811 (Type Sa) has a large nucleus and tightly-wound arms; NGC 3031 (Type Sb) has more open arms; while NGC 628 (Type Sc) has loose, clumpy arms surrounding a small nucleus. NGC 1300 (Type SBb) is a moderately open-armed barred spiral, in which the arms spring from a central bar.

sporadic
A meteor which does not belong to a recognized meteor stream, but orbits the Sun in its individual path. Sporadic meteors can be seen on any clear dark night, their numbers rising from 6 per hour at 6 P.M. to 14 per hour at 6 A.M. The numbers increase because the Earth's leading edge (its morning hemisphere) sweeps up more particles from space than does its evening side. Sporadic meteors tend to be caused by meteors; they are remains of ancient meteor streams or debris from collisions between larger meteoroids.

spring tide
See TIDES.

Sputnik

A series of Soviet satellites; Sputniks 1 and 2 were the Earth's first two artificial satellites. Sputnik 1 was a 184.3-lb. (83.6-kg) aluminum sphere 23 inches (58 cm) in diameter, filled with nitrogen gas to maintain an even temperature, and with four aerials of length 8 feet to 9½ feet (2.4–2.9 m). The Sputnik transmitted "beep-beep" signals as it orbited Earth every 96 minutes between 142 and 588 miles (228 and 947 km); the frequency of the signals indicated the on-board temperature. Contrary to many reports, Sputnik 1 carried no other on-board instrumentation. Propagation of the radio signals from the Sputnik revealed characteristics of the Earth's ionosphere, and tracking of the satellite and the rocket stage that carried it into orbit gave information about the density of the upper atmosphere. Sputnik 2, weighing an impressive 1,121 lb. (508 kg), carried the dog Laika into space, the first living creature to orbit the Earth. Laika lived in a cylindrical compartment for 10 days, dying in orbit before the craft burned up in the atmosphere. The dog's flight showed that living organisms could survive in space. Sputnik 3, weighing 2,926 lb. (1,327 kg), was the first extensively instrumented Soviet satellite. It confirmed the existence of the Van Allen radiation belts discovered by the American Explorer 1. Sputnik 4 was the first of a series of unmanned tests of the Vostok spacecraft (the so-called Korabl Sputniks), and Sputniks 7 and 8 were associated with the first Soviet Venus probes. After Sputnik 10, Soviet Earth satellites were renamed COSMOS SATELLITES. The Sputnik series were launched by the so-called VOSTOK LAUNCHER, without an upper stage for the first three satellites but with upper stages for the remaining Sputniks.

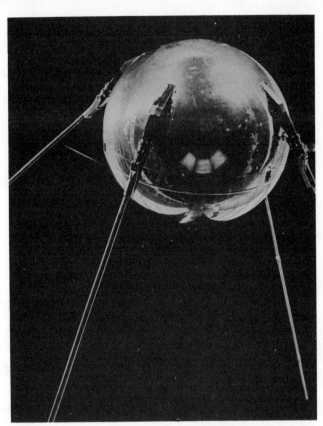

Sputnik 1, the world's first artificial satellite, launched on October 4, 1957. The 184-lb. (83.6-kg) satellite orbited for just three months transmitting data on its internal temperatures and pressures.

Mission	Launch date	Results
Sputnik 1	October 4, 1957	First artificial satellite. Reentered January 4, 1958
Sputnik 2	November 3, 1957	Carried dog Laika. Reentered April 14, 1958
Sputnik 3	May 15, 1958	Scientific satellite. Reentered April 6, 1960
Sputnik 4	May 15, 1960	Vostok test flight. Intended recovery failed when retro-rockets sent capsule into higher orbit
Sputnik 5	August 19, 1960	Vostok test; dogs Belka and Strelka recovered in capsule on August 20 after 18 orbits
Sputnik 6	December 1, 1960	Vostok test flight. Dogs Ptsyolka and Mushka perished when spacecraft burned up after 17 orbits due to incorrect angle of reentry
Sputnik 7	February 4, 1961	Failed Venus probe
Sputnik 8	February 12, 1961	Launched Venus 1 from Earth-orbit
Sputnik 9	March 9, 1961	Vostok test flight. Dog Chernushka recovered after one orbit
Sputnik 10	March 25, 1961	Vostok test flight. Dog Zvezdochka recovered after one orbit

Stafford, Thomas Patten (b. 1930)

One of the few spacemen to fly on four missions, commander of the Apollo spacecraft in the Apollo-Soyuz joint docking mission in July 1975. Stafford was selected as an astronaut in 1962, and in December 1965 flew in the Gemini 6 craft that rendezvoused with Gemini 7. In June 1966 he commanded the Gemini 9 mission, with Eugene CERNAN. In May 1969 Stafford commanded the Apollo 10 mission with Cernan and John YOUNG; this was a rehearsal for the first lunar landing attempt. After the Apollo-Soyuz flight, Stafford left the astronaut corps to command the Air Force's Flight Test Center at Edwards Air Force Base.

star

A self-luminous ball of gas; the Sun is a typical star. The night-sky stars appear as points of light because they are much farther away than the Sun; light reaches us in 8.3 minutes from the Sun, but takes 4.3 years from the nearest star. Under clear skies, about 3,000 stars are visible to the naked eye at any given time; normally, however, the effects of atmospheric pollution and street lighting mean that far fewer stars can be seen. Astronomers have cataloged over one million stars, but little is known about most of them except their approximate brightness and position. In the Palomar Sky Survey, the 48-inch (122-cm) Schmidt telescope at Mount Palomar photographed the sky from the north

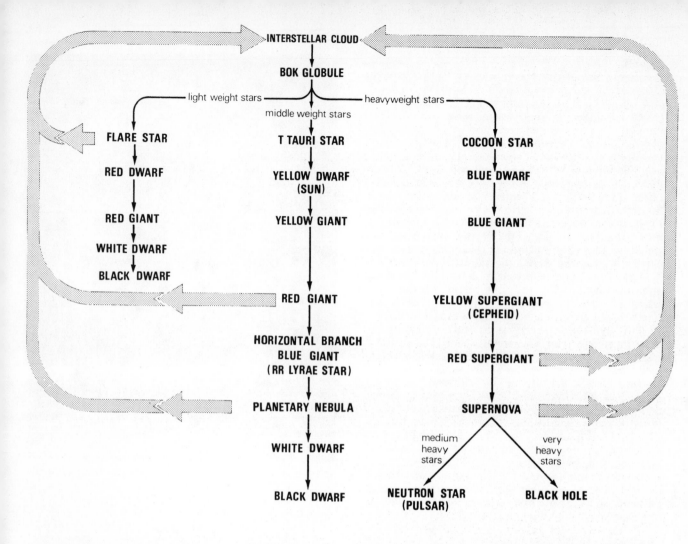

Simplified diagram showing the main stages in the evolution of stars from interstellar clouds. Mass loss back to the interstellar medium is indicated with grey arrows. Protostars are shown following the Bok Globule, and the long-lived main sequence stages after the protostars. The planetary nebula and the supernova represent the principal explosive stages, while the stellar deaths appear at the very last level.

pole to 33° south of the equator; about 800 million stars are estimated to be shown on these photographs. However, our Galaxy (the Milky Way) is estimated to contain at least 100 billion stars.

Nature of stars. Stars vary widely in size and temperature. The largest are the red SUPERGIANTS, up to 1,000 times the size of the Sun. The smallest known stars are NEUTRON STARS like the one in the Crab nebula, which can be as small as 10 miles (16 km) in diameter. The surface temperatures of stars range from up to 100,000°K down to less than 2,000°K. Variations in temperature produce corresponding differences in colors, from the hot blue of VEGA, through the warm yellow of CAPELLA, to the cool red of BETELGEUSE. Astronomers learn about the nature and composition of stars by studying their light (SPECTROSCOPY). Atoms in the star's atmosphere absorb some light from the star at specific wavelengths; the relative strengths of these absorption lines in a spectrum reveal the density, temperature, and chemical composition of a star's outer layers. Most stars consist largely of hydrogen, with some helium and a sprinkling of certain other elements.

Formation of stars. Stars are formed by the collapse of the great clouds of gas and dust like the ORION NEBULA, in our Galaxy. Often the collapse is triggered by compression, as clouds bump into each other, or by changes in the heating of the cloud by nearby stars. The cloud then collapses under the effect of its own gravity, breaking up into smaller subcondensations, which continue to collapse until their internal temperature rises above about 10 million degrees. The pressure of the gas then increases rapidly and halts the contraction; a star has been born. A fragmenting cloud will give rise to a star cluster like the PLEIADES. The smaller subcondensations are believed to look like the small, dark GLOBULES seen in various parts of the Galaxy. Often, the rotation of a globule will make it split into two or more parts, perhaps forming a DOUBLE STAR. Other globules may give rise to a star and PLANETS.

Inside a star. The energy of the stars comes from nuclear reactions in their very hot central regions. Atoms are stripped down to their nuclei and crushed together, releasing energy. In stars like the Sun, hydrogen combines to form helium directly by the collision of hydrogen nuclei in the PROTON-PROTON CHAIN; in hotter stars, with central temperatures above about 15 million degrees, the main process is the CARBON-NITROGEN-OXYGEN CYCLE, in which hydrogen is added to carbon nuclei to form nitrogen, which then breaks up into carbon and helium. At temperatures around 100 million degrees helium is turned into carbon and oxygen, and then at still higher temperatures the matter inside the star is successively converted into neon, magnesium, silicon, and iron, each reaction releasing energy to heat the star.

Evolution of stars like the Sun. When a star like our Sun first forms out of an interstellar gas cloud, it is about 50 times the size of the present Sun and 500 times as luminous. The central regions are too cool for nuclear reactions, so that the star contracts, drawing on its gravitational energy to make good the loss by radiation into space. As the star contracts, its luminosity decreases, and the central regions heat up. After 30 million years the center is hot enough for nuclear reactions to take place, and the contraction begins.

The star remains much the same for about 10 billion years, the major part of its lifetime, with a luminosity and size similar to that of our Sun. This phase of hydrogen burning is the period when the star is on the MAIN SEQUENCE; it lasts for so long because there is an enormous energy reserve in the hydrogen of which the star is primarily made.

Eventually the star will use up its hydrogen fuel in the hot central regions, producing an inert helium core. Since there is plenty of fuel left in the rest of the star, the nuclear burning advances outward until the hydrogen is burning in a shell surrounding the core. When this happens, the star begins to increase in size and luminosity, slowly at first but then at an increasing pace until the star becomes a vast RED GIANT producing 1,000 times the light output of the Sun. Not only does the light output increase, but the star expands to become as much as 100 times the size of the Sun. When our Sun reaches this size, it will have engulfed the Earth.

As the star evolves, the helium in the center increases in temperature until it becomes hot enough to burn. The onset of this is very sudden, and the whole star must readjust to the new energy source. In a matter of days it contracts, decreasing in light output to about 100 times that of the Sun. But as the helium burns, the star again begins to grow in size and brightness, burning up its helium to form a carbon core and then burning helium and hydrogen in shells around the core. The star expands to even greater size than in its previous giant phase, up to 400 times the size of the Sun with 10,000 times its light output. It becomes so large its own gravity can hardly hold it together, and eventually the surface layers are blown off into space. This ejected matter forms a shell expanding away from the hot remnant central core, as we see in a PLANETARY NEBULA. The central core remains as a hot WHITE DWARF, which slowly cools to a cinder-like black dwarf star as it radiates the rest of its heat into space.

Evolution of massive stars. Stars more massive than the Sun undergo the same early evolution. The centers of these stars are so hot that most energy comes from the carbon-nitrogen-oxygen cycle. The star changes little while it is burning the hydrogen in the core; it is also a main sequence star and spends most of its life in this stage. But it is considerably brighter than the Sun: a star 10 times the Sun's mass has a luminosity about 10,000 times

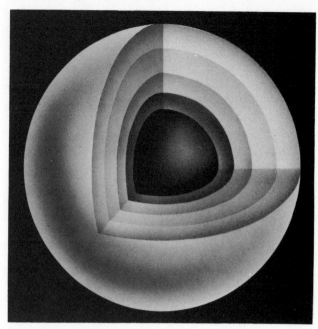

The interior of a large mass star before it becomes a supernova. The star has shells of different material. In the center neon is being converted into iron. When the star has an iron core it collapses, the outer shells heat up and the star explodes.

that of the Sun. Such large stars evolve much more rapidly since they burn up fuel at a greater rate; their main-sequence stage lasts only 10 million years.

When the central store of hydrogen is exhausted, hydrogen burning moves out into a shell surrounding the burnt-out helium core. However, this core contracts, heating up to a temperature of 100 million degrees, at which it starts to be converted into carbon and oxygen. The star swells up to become a giant or supergiant, remaining like this for the rest of its life. When the helium in the center is all converted into carbon and oxygen, the core again contracts until it is hot enough to burn carbon into magnesium. When this is burnt out, the magnesium burns into neon which itself burns into iron. The evolution of the center of the star consists of burning up one fuel, contracting and heating up to ignite the next fuel, while the lower-temperature reactions are taking place in successive shells away from the center.

When an iron core has been formed, the star is near the end of its life. Iron is the most stable element; to convert iron into heavier elements consumes energy, instead of releasing it. The required energy supply can only come from gravitational contraction; therefore, the core collapses. As the central core does so, the outside falls in on top of it, rapidly increasing its temperature and accelerating the nuclear burning in the regions where there is still some fuel. The sudden release of energy makes the star explode violently, producing a SUPERNOVA. In the explosion numerous elements are produced and scattered into space. In some cases the star may blow itself entirely to bits; in others, the star's compressed core may be left as a neutron star. In still other cases the remnant core may be so massive that it continues to contract, disappearing from sight as a BLACK HOLE.

Kappa Crucis, a young galactic (or "open") star cluster in the Southern Cross. It comprises about a hundred stars which show a great variety of colors, and have earned it the name of the "Jewel Box." Fifteen light-years across, this cluster is believed to be only sixteen million years old.

star cluster

An aggregation of stars, held together by their own gravity. Two distinct types of star clusters are observed, the huge, round GLOBULAR CLUSTERS of about 1 million stars, and the much smaller and sparser OPEN CLUSTERS.

Globular clusters formed early in the history of our Galaxy (10 billion years ago), before it settled down to the flat disk-like arrangement of stars which we now see. As a result, the globular clusters, about 125 in number, occur all around the Milky Way, and not merely in the disk. They orbit around the center of our Galaxy in highly elongated ellipses, with a period of about 100 million years. The number of globular clusters increases toward the galactic center, and most therefore are observed in the direction of the constellation Sagittarius, where the Milky Way's nucleus lies.

The brightest globular clusters, Omega Centauri and 47 Tucanae, are easily visible to the naked eye in the southern hemisphere. Large telescopes can readily resolve the outer parts of most globulars into individual stars. Their distances (10,000 to 150,000 light-years) can be determined from the apparent magnitudes of stars known as RR LYRAE VARIABLES, which always have the same intrinsic luminosity. They are usually about 100 light-years in diameter, and they contain only stars composed almost entirely of hydrogen and helium (POPULATION II stars).

Open clusters occur in the flat disk of our Galaxy, and are far more numerous than the globulars. About 1,000 are known, and it is estimated that the Milky Way may contain a total of 18,000. Open clusters are difficult to detect beyond 5,000 light-years, both because they contain far fewer stars than the globulars, and because the dust in the Milky Way's disk dims the light from them (INTERSTELLAR ABSORPTION). The nearest is the HYADES (in the constellation Taurus) at 130 light-years, while the PLEIADES (also in Taurus) and Praesepe (in Cancer) are 400 light-years distant. Open clusters average 10 light-years in diameter, and they contain only a few

hundred stars, similar in chemical composition to the Sun (POPULATION I stars).

Open clusters have formed continuously since the galactic disk appeared. The oldest are of almost the same age as the globular clusters, while the youngest are only a few million years old. Open clusters are being created in places where stars condense out of gas clouds, for example in the ORION NEBULA. Most of the stars in these newly formed STELLAR ASSOCIATIONS will disperse among the other stars of the disk, but those near the center of the nebula may remain bound by their gravity to form clusters that will remain together for billions of years.

star streaming

The tendency of stars that occur over a comparatively large area of the sky to be traveling in the same direction in space. The term was first used by J. C. KAPTEYN in 1904, when an investigation into the PROPER MOTIONS of nearby stars showed that their directions of motion are not random, and that two particular directions in space are favored by a large number of them. If a correction is made for the Sun's motion, these two directions point toward and away from the center of our own Galaxy. Kapteyn's two star streams are thus reducible to one, with stars moving in both directions along it. When astronomers found that the Sun is orbiting the center of the Milky Way, Karl SCHWARZSCHILD and Bertil LINDBLAD showed that this star stream is merely a reflection of the fact that the nearby stars travel in elliptical rather than circular orbits around the galaxy.

Today the term "star stream" is used in a different sense. It refers to small, sparsely scattered groups of stars moving in the same direction. Five of the stars in the Plow, together with Sirius, are part of one stream, while others surround OPEN CLUSTERS of stars like the Hyades.

steady-state theory

A theory of COSMOLOGY, now credited by very few astronomers, which regards the Universe as essentially unchanging over time. In the BIG-BANG cosmological model, now generally accepted, the Universe has a birth, growth, and perhaps death. The steady-state theory, as proposed by Hermann BONDI and Thomas GOLD (1948) took a very different standpoint, beginning with the PERFECT COSMOLOGICAL PRINCIPLE of an unchanging Universe. Although individual stars and galaxies pass through their life-cycles, it was suggested that new galaxies are constantly forming out of freshly created matter. As the Universe expands, this new matter appears to fill in the "gaps" left by the receding galaxies in the EXPANDING UNIVERSE, and thereby maintains a constant overall density.

The CONTINUOUS CREATION rate is very low (about one atom per cubic mile of space per hour), and Fred HOYLE, one of the theory's main protagonists, showed that Einstein's theory of relativity could be modified to explain the creation of matter from a reservoir of negative energy (the C-field) which fills the Universe.

The steady-state Universe has no beginning and no end, and this feature made it a very popular choice among cosmologists. By 1965, however, support had dwindled. The theory was unable to account for either the radio astronomers' "source counts" data (which showed many more radio sources in the past), or the newly discovered microwave BACKGROUND RADIATION. The latter can be simply explained according to the big-bang theory as the "echo" of the big bang itself at the beginning of an evolving Universe. Moreover, the strongest reason for the original introduction of the perfect cosmological principle, the fact that the Universe appeared to be younger than the Earth, had been removed in 1952, when Walter BAADE showed that the size and age of the Universe had been grossly underestimated.

Hoyle experimented with variations on the original steady-state theory, postulating that we live in a temporary expanding "bubble" in a Universe which is constant on a much larger scale. The background radiation he attributed to dust grains in space. Hoyle eventually abandoned the theory altogether, though his coworker J. V. Narlikar still supports it.

stellar association
An aggregation of 10 to 1,000 stars which are close together merely because they have not moved far from their common place of origin. Unlike the stars in STAR CLUSTERS, those in associations are moving too fast for their gravity to hold them together permanently.

Although a few of the stars may in the end stay together as multiple stars or OPEN CLUSTERS, most members of an association disperse into the general background of stars in our Galaxy in about 10 million years, only 1 percent of our Galaxy's age. As a result associations can be recognized by the concentration of young stars, the very hot and highly luminous O and B SPECTRAL TYPES, or by the occurrence of the even younger T TAURI STARS.

The OB associations and T associations were first investigated in 1947 by V. A. AMBARTSUMIAN, who found that they often occur together. For example, many of the stars forming the constellation of Orion are part of an OB association of 1,000 stars that surrounds a T association containing a few hundred members. At the center, more stars are still condensing out of the gas of the ORION NEBULA.

step rocket
See ROCKET.

Steward Observatory
An astronomical observatory operated by the University of Arizona at an elevation of 6,811 feet (2,076 m) on Kitt Peak (see KITT PEAK NATIONAL OBSERVATORY). The Steward Observatory was founded at Tucson in 1922, with a 36-inch (91-cm) reflector, and moved to Kitt Peak in 1963. A 90-inch (229-cm) reflector was opened in 1969.

Stonehenge
A prehistoric circle of giant stones near Salisbury, southern England, once used as an astronomical observatory of remarkable sophistication. Stonehenge, which has been carefully excavated and studied, appears to have been built at four different periods. Stonehenge I, consisting of an outer circle of 56 stones and bank together with an outlying "heel" stone, was constructed about 1800 B.C. In Stonehenge II, built sometime during the 17th century B.C., the avenue containing the heel stone was

continued some 2 miles (3 km) eastward to the River Avon, and large igneous stone pillars—"bluestones"—were brought from over 150 miles (240 km) to form two circles. Stonehenge III, set up soon after 1600 B.C. and Stonehenge IIIb, entailed some changes and the addition of a circle of stone uprights with a continuous ring of stone lintels above them. These stones, some of which weigh 50 tons, came from a site 20 miles (30 km) away. Final refinements—Stonehenge IIIc—were made about 1400 B.C.

The heel stone and central uprights were aligned with the sunrise at the summer solstice, and the outer 56 ring holes—the "Aubrey" holes—could be used as a "computer" for calculating, with the help of observations, calendar dates and, possibly, eclipses. Recent research has shown that Stonehenge was part of a network of megalithic observatories spread over Britain and parts of France.

Strömgren, Bengt Georg Daniel (b. 1908)
Swedish astronomer, who showed that bright nebulae are made to glow by radiation from hot, young stars embedded within them, which ionizes the hydrogen gas of the cloud to form a so-called H II REGION. During the 1930s he suggested the existence of what became known as the *Strömgren sphere,* an area in the gas cloud made luminous by the star's radiation, sharply bounded by a cool, dark area of unionized gas (an H I REGION). The size of the luminous Strömgren sphere could be calculated from the density of the gas and the temperature of the embedded star. This theory accounted for the observed shapes of glowing H II regions in our Galaxy, and allowed astronomers to estimate the true sizes of H II regions in other galaxies, vital knowledge in attempts to determine galactic distances. Strömgren also showed that the scatter of stars about the MAIN SEQUENCE was due to their differing chemical composition. Strömgren succeeded his father Svante Elis Strömgren (1870–1947) as director of Copenhagen Observatory in 1940. In 1951 he became joint director of Yerkes and McDonald observatories. There he pioneered the classification of stellar spectra by carefully measuring the star's brightness at specific selected wavelengths.

Struve family
A succession of four generations of astronomers, founded by Friedrich Georg Wilhelm von Struve (1793–1864), a German who in 1817 became director of Dorpat Observatory (now Tartu in Estonia). He equipped the observatory with a $9\frac{1}{2}$-inch (24-cm) refractor by Joseph FRAUNHOFER, which was then the largest instrument of its kind in the world. Struve then began a series of measurements of binary stars, surveying the heavens from 15° south of the celestial equator to the north celestial pole. He examined 120,000 stars and increased the number of known binary systems from 700 to 3,112, accurately measuring them all. In 1835 Struve was invited by Tsar Nicholas I to supervise the building of a new observatory at PULKOVO near St. Petersburg (Leningrad), and here, using a new refractor 15 inches (38 cm) in aperture, he continued his binary-star work, also measuring the PARALLAX of Vega.

In 1864 F. G. W. Struve was succeeded at Pulkovo

by his son Otto Wilhelm Struve (1819–1905), who himself measured some 500 other binaries and, in the mid-1880s, equipped the observatory with a new 30-inch (76-cm) refractor. O. W. Struve had two sons, Hermann and Ludwig. Hermann Struve (1854–1920) became director of the Berlin Observatory, and made his name with observations of Saturn's rings and binary stars. His son, Georg Struve (1886–1933), studied planetary satellites and completely remodeled the observatory at Neubabelsberg, of which he became director. Ludwig Struve (1858–1920) was appointed to the chair of astronomy at Kharkov, and his son, Otto Struve (1897–1963), also became a noted astronomer. Otto left Russia in 1921 for the United States. At Yerkes Observatory, he discovered the existence of interstellar calcium from the absorption lines it made in stellar spectra. In 1932 he became joint director of Yerkes and McDonald observatories. He discovered the fast rotation rates of many hot, large stars, and was an influential champion of the view that many stars like the Sun may have planetary systems.

Sun

The star that is the central body of the SOLAR SYSTEM. The Sun contains 99.9 percent of the mass of this system, and it is by far the most important member, controlling the motions of all the other bodies.

The Sun is, however, only one of some 100 billion stars which make up our Galaxy, the MILKY WAY, and is located in the galactic disk at a distance of about 30,000 light-years from the center. It is a typical POPULATION I (comparatively young) star, situated on the inner edge of a spiral arm. Even with a velocity of 150 miles (250 km) per second, the Sun takes 225 million years to orbit the Galaxy.

Overall properties. At a distance of only 93,000,000 miles from the Earth, the Sun is some 270,000 times closer than the next nearest star (PROXIMA CENTAURI). It presents a disk measuring just over half a degree of angular diameter; it is therefore the only star whose surface features can be studied. The Sun is a very average star; a yellow dwarf of SPECTRAL TYPE G2, which corresponds to a surface temperature of 6,000°. Although its apparent magnitude in our skies is −26.5, the Sun's ABSOLUTE MAGNITUDE—the brightness it would appear to have if it were situated at a standard distance of 10 PARSECS (33 light-years)—is only +4.8. The absolute magnitudes of other stars range from −8 to +19; the Sun falls in about the middle of this range. It is, however, quite a small star, with a diameter of 865,000 miles (1,392,000 km), only 109 times that of the Earth; although its volume could contain a million Earth-sized bodies. The Sun is 330,000 times heavier than the Earth, with a mass of 1.99×10^{27} tons; its average density is only about a quarter that of the Earth's, indicating a fundamental difference in structure and composition.

The Sun's rotation is not uniform like that of a solid planet, but becomes slower toward the poles. The period of rotation is 24.7 days at the equator, 28.2 days at 45° latitude, and about 34 days near the poles. The adopted mean value, that which occurs at latitude 15° is 25.38 days, called the sidereal period of rotation. The mean period observed from Earth is 27.27 days, because the Earth revolves around the

Sun in the same direction as the rotation. This is known as the synodic period of rotation. The solar equator is inclined to the ECLIPTIC at an angle of 7° 15′.

Astronomers are now coming to realize that the Sun may be slightly variable is energy output, not just during the 11-year SOLAR CYCLE but also over longer periods of time. In particular, in a 70-year interval between 1645 and 1715 scarcely any sunspots appeared, a time which coincided with the so-called Little Ice Age on Earth when winters in Europe were unusually severe. This period is known as the Maunder minimum after the British astronomer E. Walter Maunder who drew attention to it in 1890. The amount of carbon-14 in the growth rings of trees, which varies with solar activity, confirmed the Maunder minimum and also showed the existence of several previous solar minima, such as from 1460 to 1540, again coinciding with cold conditions on Earth. Evidently solar activity does vary markedly over long periods, and affects the climate of the Earth.

Surface layers and atmosphere. The temperature at the center of the Sun is estimated to be 15 to 20 million degrees, but the "surface," from which we receive most of its light, is only about 6,000°. This PHOTOSPHERE is not a true solid surface, but a layer of relatively dense gas a few hundred miles thick, whose temperature falls from 9,000° at the bottom to 4,300° at the top. The upper level marks the transition to the CHROMOSPHERE, a far more rarefied and nonhomogeneous layer 10,000 miles (16,000 km) in thickness, in which temperature rises rapidly with height to 1 million degrees. At this point, the CORONA is reached, a very low-density region of completely ionized gas (PLASMA). Some parts of the corona may reach 4 million degrees. The outermost layers of the corona are in fact streaming away from the Sun into space and constitute the SOLAR WIND, which has been detected even beyond the orbit of Jupiter.

The spectrum of the Sun is crossed by thousands of dark ABSORPTION LINES and bands (FRAUNHOFER LINES), from which the chemical composition of the solar photosphere can be determined. The relative abundances of elements are a good guide to the composition of the gas from which the Sun formed, although they will differ from abundances in the interior, which have been altered by nuclear reactions. The photosphere consists of 90 percent hydrogen and 8 percent helium. It also contains small amounts of all the other elements (the HEAVY ELEMENTS), whose proportions are remarkably close to those observed on Earth.

It was previously possible to obtain the spectrum of the chromosphere and corona only during a total solar eclipse, when the dazzling light of the photosphere is blocked off. Bernard LYOT's development of the coronagraph, which artificially blocks off the photosphere, now allows observations to be made at any time. More recently, results at X-ray and ultraviolet wavelengths from rockets and satellites have yielded new information about the coronal spectrum. It contains many emission lines from elements such as iron and calcium, which have been stripped of nearly all their electrons by the extremely high temperatures. Many of these are so-called FORBIDDEN LINES, indicating a very low density for the corona. The cooler, denser

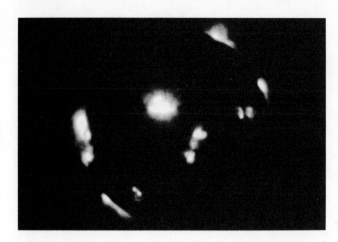

A photograph of the Sun taken at X-ray wavelengths from a rocket above the Earth's atmosphere. The bright regions are hot spots in the corona where the low-density gas is at temperatures of several million degrees. The surface we see at optical wavelengths (photosphere) is too cool to emit X rays, and appears black in this picture.

chromosphere also shows several emission lines, which generally correspond to the absorption lines in the photospheric spectrum. Its pinky-red color (visible during a total eclipse) is due to emission from a strong hydrogen line (Hα) at 6563 Å. The element helium was discovered through its emission lines in the solar chromosphere.

Solar activity. The Sun is by no means an undisturbed or unchanging body. The photosphere is threaded with a weak magnetic field measuring about 1 gauss. For reasons that are still not certain, this field can be compressed and intensified so that loops of magnetic field are forced up through the surface, giving rise to an "active region." The field strengths may become as high as 4,000 gauss, and are thought to prevent the usual motions of photospheric gas over areas several hundreds or thousands of miles across. These SUNSPOT regions are cooler and therefore darker than the undisturbed photosphere; they may last for many weeks. An active region also reveals its presence in the chromosphere and corona. In the lower levels of the chromosphere, PLAGES and PROMINENCES (both quiescent and eruptive) are associated with active regions, even though sunspots are not always present on the photosphere beneath. When a particularly complex spot group is visible, highly energetic FLARES occur in the upper chromosphere and lower corona, resulting in the emission of radiation and charged particles which interact with the Earth's MAGNETOSPHERE. It is believed that flares are a result of discharges between intense magnetic fields of opposite polarity.

The level of solar activity varies over a period of about 11 years, the SOLAR CYCLE. Similar activity has also been detected on several cooler stars, which are thought to have sunspots and flares on a much larger scale than the Sun.

Energy generation. The energy output from the Sun is 3.8×10^{33} ergs per second; each square centimeter of the Sun's surface is continuously radiating energy at the rate of a 9-horsepower engine.

Since the Earth is small and far away, it receives only a small proportion of this energy, yet this tiny fraction amounts to 4,690,000 horsepower per square mile (see SOLAR CONSTANT). The source of the Sun's enormous energy output was a mystery for many years. If the energy were derived solely from heat stored internally, the Sun's brightness would last only a few thousand years. Although many theories had been proposed, it became apparent in the 1930s that the only form of energy generation which would persist for billions of years was nuclear energy. Two mechanisms have been suggested—the PROTON-PROTON CHAIN and the CARBON-NITROGEN-OXYGEN-CYCLE. Both are thermonuclear processes occurring at extremely high temperatures, but the proton-proton chain is thought to be by far the more common process occurring in the Sun. In both processes, a mass of hydrogen is transformed into a slightly smaller mass of helium. The residual mass is transformed into energy according to Einstein's formula $E = mc^2$ (where c is the velocity of light). One gram of material therefore produces 22 trillion calories of energy. This means the Sun is losing 4 million tons of mass each second, but this amount will correspond to only 7 percent of its total mass in 1 trillion years. The process of thermonuclear fusion as we understand it is compatible with an age for the Sun of about 5 billion years. The Sun is at present in a stable state; the outflowing of energy tends to make it expand, but this is balanced by gravitation which tends to make it collapse. The Sun's supply of hydrogen will last at least another 5 billion years. At the end of that period it is believed the Sun will expand to become a RED GIANT star. This description of solar energy production predicts that the thermonuclear reactions should produce neutrinos as a byproduct. Being small and electrically neutral, these particles would pass unimpeded through the Sun and eventually reach the Earth. Neutrinos should therefore provide an insight into the interior of the Sun that is not given by electromagnetic radiation. The search for neutrinos has, however, been so far unsuccessful, even though the predicted flux should easily have been detected. Since the basic theory of hydrogen fusion is unlikely to be seriously wrong, astronomers are currently investigating modifications which would yield a lower neutrino flux.

sunspot

A relatively cool, dark area on the solar PHOTOSPHERE. Individual spots range in size from small "pores" about 900 miles (1,500 km) across, to huge, complex spots 100 times as large. Generally, sunspots form in groups, which may cover hundreds of millions of square miles. Each spot has a dark central area, the umbra, surrounded by a lighter penumbra, which has a fine, radial structure; both regions are cooler than the photosphere, the umbra by about 1,600°, the penumbra by 500°. The lower temperatures permit the formation of molecules, which would not otherwise exist in the photosphere. These can be studied with a spectroscope, which also reveals that the sunspot gases are in motion, streaming from the umbra into the penumbra with velocities of about 1.2 miles (2 km) per second (the Evershed effect).

It is believed that spots are regions where the normal

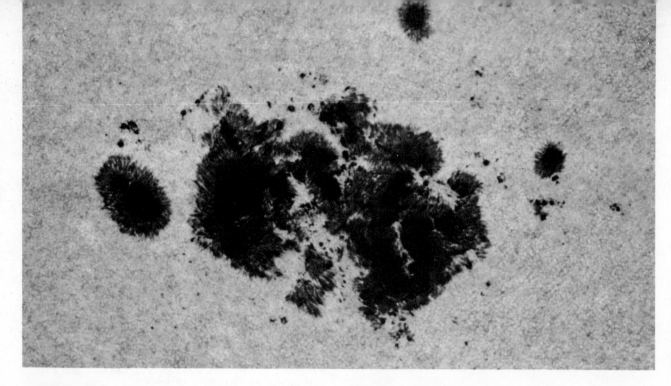

The giant sunspot group of May 17, 1951, 125,000 miles (200,000 km) in length. This photograph, taken with the 60-foot Mt. Wilson tower telescope, shows small-scale features such as pores, and fine radial structure in the penumbra. The larger leading spot, **right**, has a very complex structure with bright bridges crossing its umbra.

convective motions of the photosphere have been inhibited by strong magnetic fields. Splitting of spectral lines (the ZEEMAN EFFECT) reveals fields of up to 4,000 gauss at the centers of sunspots, with the lines of force directed upward in the umbra, but becoming more nearly horizontal toward the penumbra. Like all other magnetic phenomena, the frequency of spots is governed by the SOLAR CYCLE; there are 100 times more spots on the disk at sunspot maximum than at minimum. The earliest spots of a cycle are found at latitudes 40° north and south of the solar equator, but as the cycle proceeds, they form progressively closer to the equator. It is rare, however, to find spots within 10° of the equator, or more than 40° away.

The first indication of the birth of a sunspot is the appearance of a bright PLAGE region in the lower CHROMOSPHERE. Pores develop on the photosphere below, which grow rapidly as the magnetic field increases. After a week, the spot group is fully developed, and the "leading" spot (which "leads" in the direction of the Sun's rotation) is usually the largest. The two main spots in a group always have opposite magnetic polarity, and all leading spots in one hemisphere are of the same polarity. In a large spot group, the lines of force are very complex, and the energy released by the reconnection of field lines to produce a more simple configuration gives rise to FLARES. Most spot groups die away slowly after 10 days, taking about a month to decrease in area and disappear. The magnetic field remains until the spot has almost completely vanished.

sunspot cycle
See SOLAR CYCLE.

supergiant stars
The very brightest, largest stars, formed when stars heavier than the Sun evolve into old age. The red supergiant ANTARES is so large that if it were at the center of the solar system it would engulf the orbit of Mars and reach almost to the asteroid belt. Supergiants are rare, because they burn out very quickly. But they are so bright that most of the patchy features seen in photographs of the spiral arms of external galaxies are in fact associations of supergiants. Sixteen of the hundred brightest stars are supergiants, five in the Orion region (Rigel, Alnilam, Alnitak, Saiph, and Betelgeuse).

superior conjunction
The instant at which Mercury or Venus is directly behind the Sun as seen from Earth.

superior planet
A planet whose orbit is farther from the Sun than the orbit of the Earth. Mars, Jupiter, Saturn, Uranus, Neptune, and Pluto are the superior planets.

supernova
A star that explodes and ejects most of its mass at very high velocities. The energy released is a million times that of a NOVA, an explosion in a star's atmosphere that leaves the star fundamentally unchanged. After a supernova explosion the star is either totally destroyed, or its central core collapses into a very dense NEUTRON star, or possibly even a BLACK HOLE.

In a supernova explosion, a star beomes as bright as a small galaxy for a few days. About 500 supernovae have been discovered in other galaxies since systematic searches were begun by Fritz ZWICKY in 1937. Most fall into two main categories.

Type I supernovae brighten to an absolute magnitude of −19.4, decrease by 3 magnitudes in a month, and then fade at a constant rate of 1/70 magnitude per day. Since all Type I supernovae are of similar intrinsic brightness, their apparent magnitude

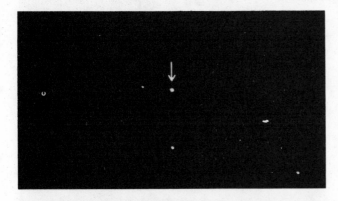

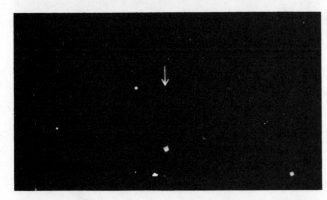

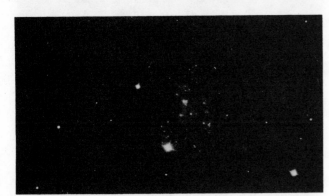

Three views of a type I supernova, **arrowed**, in the galaxy IC 4182, showing its decline in brightness from maximum in 1939. The galaxy and the foreground stars appear more prominently in the later pictures (taken in 1940 and 1942) as progressively longer exposures are required to show the fading supernova.

reveals the distance of their parent galaxy. The spectrum of a Type 1 supernova shows that about one solar mass is ejected, at a speed of about 7,500 miles (12,000 km) per second, and that this material contains only about 10 per cent hydrogen. The pre-supernova is probably the core of an old red giant that has lost its hydrogen-rich envelope.

Type II supernovae have more varied light curves. Their maximum absolute magnitude is about −17.6, and most of them drop to an almost constant magnitude for several weeks before fading further. Their spectra show that up to 10 times the Sun's mass is ejected, at a velocity of about 5,000 miles (8,000 km) per second. The pre-supernova must thus have at least 10 times the Sun's mass; theory shows

that such very massive stars evolve over less than 100 million years, becoming unstable and exploding (see STAR).

Supernovae in our Galaxy were recorded in the years 1604, 1572, 1181, 1054, 1006, and 185. The expanding debris from the supernova of 1054 is visible as the CRAB NEBULA. Other supernovae have left expanding remnants that are strong radio emitters but generally faint optically. After about 10,000 years, however, the gas in these remnants begins to form into bright filamentary nebulae, like the Veil Nebula in Cygnus. A supernova is expected in the Milky Way on average every 30 to 50 years. Most will appear faint, due to the INTERSTELLAR ABSORPTION of light by dust grains, but a third of them should be visible to the naked eye.

Surveyor probes

A series of American lunar soft-landers that paved the way for the first Apollo manned landings. Surveyor was a triangular-shaped craft, 10 feet (3 m) tall with three footpads of crushable aluminum honeycomb. Surveyor carried a solar panel and flat high-grain antenna at the top of a mast. As the spacecraft approached the Moon, it was decelerated by a 10,000-lb. (4,536-kg) thrust retro-rocket, which was then jettisoned. The Surveyor was braked onto the Moon by three smaller jets that switched off at an altitude of 14 feet (4.3 m), allowing the craft to

Probe	Launch date	Remarks
Surveyor 1	May 30, 1966	Landed in Oceanus Procellarum near crater Flamsteed on June 2. Returned 11,150 photographs until July 13
Surveyor 2	September 20, 1966	Impacted Moon September 23 southeast of crater Copernicus after control system failed
Surveyor 3	April 17, 1967	Landed in Oceanus Procellarum on April 20. Surface sampler dug in lunar soil. Returned 6,315 pictures until May 3. Visited by Apollo 12 astronauts in November 1969
Surveyor 4	July 14, 1967	Landed in Sinus Medii on July 17. Radio contact lost prior to touchdown
Surveyor 5	September 8, 1967	Landed by remote control from Earth in southern Mare Tranquillitatis on September 11. Carried box to analyze soil by bombardment with alpha particles. Returned 18,006 photographs until September 24
Surveyor 6	November 7, 1967	Landed in Sinus Medii on November 10. Analyzed soil with alpha-scattering device. Landing rockets refired on November 17, causing Surveyor to lift off and resettle 8 feet (2.5 m) away. Returned 30,000 photographs
Surveyor 7	January 7, 1968	Landed near crater Tycho on January 10. First highland landing. Carried sampling scoop and chemical analysis device. Returned 21,000 photographs

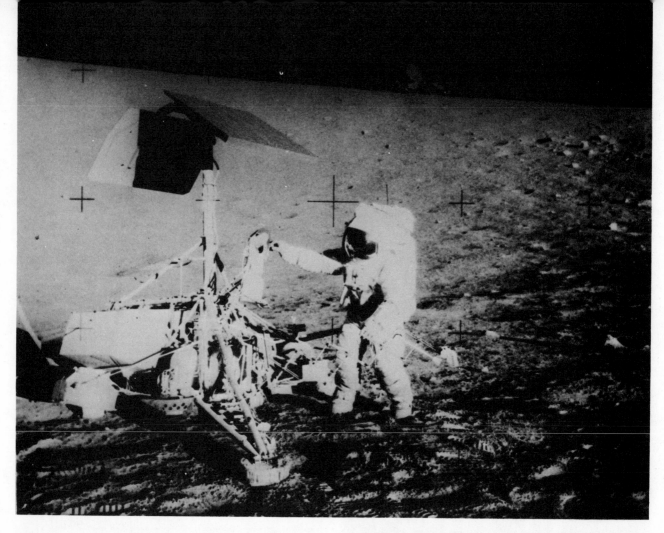

The U.S. Surveyor 3 Moon probe landed in the Oceanus Procellarum in 1967, where it tested the soil structure with a mechanical digger and photographed the interior of the craterlet it landed in. 2½ years later, Apollo 12 astronauts landed close by, examined the spacecraft and removed its TV camera for study back on Earth. It was found that some Earth bacteria on it had survived the hostile lunar environment.

drop to the surface at about 8 miles (5 km) per hour. Its weight on the surface was 620 lb. (280 kg). Each Surveyor carried a television camera of variable focal length which pointed at a movable mirror; tilting the mirror changed the television camera's field of view. Surveyors 3 and 7 carried mechanical scoops to dig in the lunar soil, and Surveyors 5, 6, and 7 carried small chemical-laboratory boxes that were lowered to the lunar surface to give rough soil analyses. These analyses indicated the basalt-like composition of lunar rocks, confirmed in greater detail by the Apollo missions. The Surveyor program showed that the lunar surface was safe for Apollo landings.

synchronous orbit
An orbit in which a satellite moves around a body at the same rate as the body spins on its own axis; this is also termed a GEOSTATIONARY ORBIT. The term is also applied to an object that takes as long to rotate once about its axis as it takes to complete one orbit. The Moon has such a synchronous rotation, always turning one face toward Earth; several other moons of the solar system, such as Phobos and Deimos of Mars, also have synchronous rotations. Such a synchronous spin arises because of the braking effect of the planet's gravity, and is also termed a *captured rotation*.

synchrotron radiation
Electromagnetic radiation emitted by charged particles moving near the velocity of light in a magnetic field. This emission was first observed in particle accelerators on Earth, called synchrotrons, from which the name arose. The wavelength of synchrotron radiation depends on the velocity of the particle and the strength of the magnetic field; and it is polarized (see POLARIZATION) at right angles to the lines of magnetic force.

Electrons produce synchrotron radiation more efficiently than do the heavier protons. Electrons moving in a magnetic field produce the radio emission from solar flares, from the remnants of SUPERNOVA explosions, and from RADIO GALAXIES and QUASARS. X rays produced by the synchrotron process account for some sources observed in X-RAY ASTRONOMY. The visible light from some astronomical objects, such as the CRAB NEBULA (a supernova remnant) is also synchrotron radiation.

Syncom
A series of NASA experimental communications satellites in 24-hour synchronous orbit. Their purpose was to test the suitability of synchronous orbits for

communications satellites, and to enable the difficult technique of inserting a satellite into such an orbit to be practiced. The advantage of a synchronous, or geostationary, orbit, is that the satellite hangs near-motionless over the equator, unlike the earliest forms of communications satellite which moved rapidly across the sky. However, if the satellite's orbit is inclined slightly to the equator it will move alternately north and south of the equator each day, tracing out a figure-of-eight pattern in the sky. The 86-lb. (39-kg) first Syncom was scheduled for an orbit inclined about 33° to the equator. However, the satellite went dead shortly before it reached its station in orbit, and was unusable. The identical Syncom 2 became the first successful communications satellite in synchronous orbit, although it too was inclined at 33° to the equator. Syncom 2 was stationed over Brazil and relayed telephone conversations from North and South America to Africa. Syncom 3 became the first truly geostationary satellite in August 1964, when it was maneuvered into equatorial orbit above the Pacific Ocean. It relayed television coverage of the Olympics from Japan to the United States. The Syncom satellites could carry one two-way telephone conversation, 16 teletype channels, or one black-and-white television channel. The satellites received at 7.4 GHz and transmitted at 1.8 GHz. Their success led directly to the INTELSAT series of commercial communications satellites. Syncom 1 was launched in February 1963; Syncom 2 in July 1963; and Syncom 3 in August 1964.

synodic period

The time taken for an object to return to the same position in the sky as seen from Earth. Examples of synodic periods are the intervals between two oppositions of Mars or between two full Moons. The synodic period differs from the orbital period relative to the stars (*sidereal period*), because of the Earth's motion in its own orbit around the Sun.

T

Taurid meteors

One of the regular annual meteor streams. They are first detectable on October 20 each year and reach a broad peak of 12 meteors per hour on November 8, declining over three weeks. The radiant, which is double, moves eastward at less than 1° a day; it lies 15° southwest of the Pleiades at maximum and reaches its greatest altitude at 1 A.M. A millennium ago, the Taurids were the strongest annual stream, but they are now very much in decline. They are closely associated with ENCKE'S COMET.

Taurus (the bull)

A major constellation of the zodiac, in the northern hemisphere of the sky. Taurus is best seen during the northern hemisphere winter. The constellation is famous for its two large star clusters, the HYADES and PLEIADES. Its brightest star is the red giant ALDEBARAN, which lies in front of the Hyades cluster but is not associated with it. Between the horns of Taurus is the famous CRAB NEBULA, the remains of an exploded star. The TAURID METEORS appear to radiate

from a point near Epsilon Tauri. Near the same point is the faint patch known as Hind's variable nebula (NGC 1554–5), discovered by the English astronomer John Russell Hind (1823–1895). This contains at its heart the star T Tauri, the prototype of a class of irregular variable stars (see T TAURI STARS). The Sun passes through Taurus from mid-May to late June.

tektites

Small, glassy objects, measuring about an inch across, found scattered in a number of localities in the world. The drop-like and button shapes of tektites and their flow structures indicate they were formed from the rapid cooling of molten material, and their extraordinarily low water content distinguishes them from the volcanic glass formed by eruptions on Earth. Tektites are, however, similar to impact glass found at meteorite crater sites, and some tektites contain metallic blobs resembling meteoritic iron.

Tektites could have been formed during the impact of a meteorite, which splashed out molten rock. Some of the melted globules were projected into space, where they cooled before being further melted on reentry into the Earth's atmosphere. Tektites show no traces of encounter with cosmic rays, so that they cannot have traveled for long in space. The impact that caused them would therefore have occurred either on the Earth or the Moon. An alternative theory says they were ejected from volcanoes on the Moon.

Certain smaller fields of tektites are linked to known terrestrial meteorite craters. Impact glasses found at the 14-mile (22.5 km) diameter Ries crater in Germany have been dated at 14.8 million years, exactly the age of the Czechoslovakian tektites (called moldavites); both were melted from rocks about 300 million years old. Impact glass from the $6\frac{1}{2}$-mile (10.5-km) diameter Lake Bosumtwi crater in Ghana has been dated at 1.3 million years, the same age as the nearby Ivory Coast tektites; both these were formed from rocks 2 billion years old. The North American (Texas and Georgia) and Far Eastern tektite fields have no known associated terrestrial meteorite crater. However, the Far Eastern field could have come from the impact which produced the crater Tycho on the Moon. Debris ejected along Tycho's major bright ray would follow a $3\frac{1}{2}$-day path to Earth and fall in exactly the same pattern as the Far Eastern tektite field.

telescope

A device for collecting and magnifying light, which has made possible the development of modern astronomy. The first telescope was made in 1608 by Hans Lippershey, a Dutch lens maker; telescopes were first turned to the sky by the Italian physicist GALILEO. In 1609 Galileo constructed an instrument magnifying 30 times, with which he discovered the four bright satellites of Jupiter, the phases of Venus, and the mountains of the Moon.

A telescope is superior to the naked eye for two principal reasons: it collects more light, thus revealing fainter objects; and it can be made to magnify the images it receives. Some work in astronomy, involving the discovery of fine planetary detail or the separation of two close stars, requires high magnification. On the other hand, the detection of very faint and nebulous objects, such as remote galaxies, is aided much more by a telescope's light-gathering power.

Telescopium

The earliest instruments were REFRACTING TELESCOPES, using a large lens or OBJECT GLASS to collect and focus the light. The drawback of a simple lens system is that it acts like a prism, causing light to spread out and form a colored image (CHROMATIC ABERRATION). The only way then known of combating this was to make the FOCAL LENGTH of the lens very long in relation to its aperture. In the later years of the 17th century, telescopes 150 feet (46 m) or more in length were being built, but with apertures of only a few inches. These *aerial telescopes,* though very cumbersome, were used to good effect by such observers as Christiaan HUYGENS and Giovanni CASSINI to discover objects that included the rings of Saturn and several of its satellites. However, a great advance was achieved in 1721, when the first truly workable REFLECTING TELESCOPE was constructed by the English instrument maker John Hadley (1682–1744); although reflectors had been made previously, they were of poor optical quality and unsuitable for astronomy. Using a concave mirror instead of a lens to form the image, Hadley's reflector was less then 6 feet (1.8 m) long, yet had an aperture of 6 inches (15 cm). This was possible because a mirror produces no chromatic aberration. Reflecting telescopes, in their various forms, have been used for astronomical research ever since. Most have been either of the *Newtonian* type introduced by Hadley, with a small plane mirror diverting the light through the side of the tube, or of the CASSEGRAIN configuration, which effectively folds a telescope of long focal length into a short tube.

The refracting telescope did not regain its position until the middle of the 18th century, when the London lens maker John DOLLOND marketed an achromatic object glass, consisting of two lenses, each of which compensated for the other's chromatic aberration. Achromatic refractors soon became popular because of their highly efficient light-transmission capabilities. After Dollond, the great makers of refracting telescopes included Joseph FRAUNHOFER, and the firm of Alvan CLARK, which made the two largest refracting telescopes in the world, the 36-inch (91-cm) LICK OBSERVATORY instrument, and the 40-inch (102-cm) at YERKES OBSERVATORY. However, mirrors can now be made much larger than lenses, which gives reflectors a major advantage in light-gathering. The age of huge reflectors began with William HERSCHEL, who built a 48-inch (122-cm) reflector in 1789 at Slough, England. This was followed by Lord ROSSE's 72-inch (183-cm) instrument of 1845.

All early reflectors used a shiny alloy called *speculum metal* for their mirrors. This did not reflect light very efficiently, and tarnished quickly. The most important development in the reflecting telescope came in the late 19th century with the introduction of mirrors made from glass, coated with a highly reflective layer of silver or (much more commonly) aluminum. The first of the modern giants was the 60-inch (152-cm) at Mount Wilson Observatory, California, in 1908. A 100-inch (254-cm) was installed at the same observatory in 1917, and the giant 200-inch (508-cm) at Mount Palomar, California, was completed in 1948. At present several instruments approaching this aperture are in existence or being completed (see REFLECTING TELESCOPE). In the few remaining professional fields where visual work,

The first reflecting telescope, made by Isaac Newton in 1671. Its mirror, shown beside it, is 2 inches (5 cm) in diameter and 8 inches (20 cm) focal length. Like all early mirrors, it is made of metal. The telescope tube is made of oiled parchment. A trial eyepiece hole, covered over and waxed by Newton, is still visible.

rather than photographic recording, is still undertaken, refracting telescopes are generally preferred because of their more critical definition and the relative permanence of their adjustments. The same instrument may be used continuously for years or even decades. (See also CATADIOPTRIC TELESCOPE; MIRROR; SCHMIDT TELESCOPE.)

Telescopium (the telescope)
A faint constellation in the southern hemisphere of the sky, introduced by Nicolas Louis de Lacaille. Telescopium lies below Sagittarius. Its brightest star is of only the fourth magnitude.

Telstar
Two American communications satellites, built by the American Telephone and Telegraph Company for experiments in long-distance television, telephone, and telegraph transmissions. Telstar 1 carried the first live transatlantic television broadcast, beamed from a ground station at Andover, Maine, to Europe. The Telstar satellites were 32-inch (81-cm) spheres, studded with solar cells. They received transmissions at frequencies of 6.39 GHz and retransmitted at 4.18 GHz; they could carry 600 telephone circuits or one television channel. The Telstars orbited at medium heights, so that they moved across the sky and had to be continuously tracked by ground-station antennae. The first Telstar was damaged by radiation

from the Earth's Van Allen belts, and the second Telstar was inserted into a slightly higher orbit. Their immediate success established the potential of commercial communications satellites.

Satellite	Launch date	Remarks
Telstar 1	July 10, 1962	Orbited between 593 and 3,503 miles (954 and 5,638 km) every 158 minutes. Weight 170 lb. (77 kg)
Telstar 2	May 7, 1963	Orbited between 604 and 6,713 miles (972 and 10,803 km) every 225 minutes. Weight 175 lb. (79 kg)

Tereshkova, Valentina Vladimirovna (b. 1937)

Soviet cosmonaut, the first woman to fly in space. She piloted the Vostok 6 spacecraft in June 1963, making 48 orbits of the Earth over a period of 70.8 hours, while Valery Bykovsky was orbiting separately in Vostok 5. Although not a pilot, Tereshkova was an avid parachutist, and learned to fly during her cosmonaut training. She is said to have volunteered for spaceflight in a letter following the flight of Vostok 2. Some reports suggest she was actually the back-up pilot for Vostok 6, and was substituted at the last moment when the woman chosen to make the flight became indisposed. Tereshkova married cosmonaut Andrian NIKOLAYEV on November 3, 1963.

terminator

The sunrise–sunset line dividing the lit and unlit sides of a planet, moon, or asteroid.

Tethys

The fourth satellite of Saturn in order of distance from the planet, probably only about 600 miles (1,000 km) in diameter. Tethys orbits Saturn every 45 hours 18½ minutes in a circular path 183,150 miles (294,750 km) above the planet's surface. It has a density about 1.1 times that of water, indicating that it is made of loosely-compacted frozen gas. Tethys was discovered in 1684 by G. D. CASSINI.

Thales of Miletus (c.624 B.C.–c.546 B.C.)

Greek philosopher, regarded as one of the founders of physical science. Thales believed that water is the primary substance of the Universe, and he visualized the Earth as a flat floating disk. Thales plotted the course of the Sun around the sky, and predicted the eclipse of May 28, 585 B.C. The eclipse was said to have stopped a battle between the Lydians and the Medes. Thales is known to have written on the movements of the Sun, and to have made suggestions for celestial navigation. One of his pupils was ANAXIMANDER.

Thor rocket

An American space launcher, based on the Thor intermediate-range ballistic missile. Thor was the first American intermediate-range missile, test fired in 1957. The basic Thor was 65 feet (19.8 m) tall and 8 feet (2.4 m) in diameter, tapering toward the top. It had a single engine similar to the two booster engines used in the ATLAS ROCKET, of 150,000 lb. (70,000 kg) thrust, burning liquid oxygen and kerosene.

By the time Thor was withdrawn from military use in 1963, it had already been modified into a reliable space launcher. With upper stages such as the Able and AbleStar (modified from the VANGUARD rocket) and the AGENA, Thors were used to launch many early U.S. satellites and probes. These included several Pioneers and Explorers, together with the Discoverer series. In 1963 the Thrust-Augmented Thor (TAT) was introduced, with lift-off power augmented to 330,000 lb. (150,000 kg) by three solid-fuel strap-on motors. In 1966 a long-tank Thor, known as Thorad, was developed; this was 71 feet (21.5 m) long and gave Thor a 20 percent increase in payload capacity. With strap-ons, Thorad became known as the Long-Tank Thrust Augmented Thor (LTTAT). Thor is also used as part of the DELTA launch rocket.

tides

The rhythmic rise and fall of the surface of the sea that occurs twice each day and is due to the gravitational attraction of the Moon and Sun. Because the Moon is so much nearer, it has about twice the effect on the tides as the Sun. The Moon's gravity pulls the Earth's water surface into two bulges, one on the side facing the Moon and the other on the opposite side. Similar effects are caused by the Sun. Twice a month, at new and full Moon, the Sun and Moon are pulling in line, and their tidal effects combine to produce tides higher than normal (called "spring" tides because they spring up). When the Moon is at first or last quarter, it is pulling at right angles to the Sun, and the tidal effects work against each other. This produces tides of a small range (called "neap" tides, meaning scanty). As the Earth rotates under these two tidal bulges, points on its surface experience a cycle of low tides twice each day. The actual height of the tide is partly determined by the shape of the coastline and the depth of water.

Tikhonravov, Mikhail Klavdiyevich (1900–1974)

Soviet rocket engineer, designer of early Russian rockets and later the anonymous "Chief Theoretician of Cosmonautics." During the 1920s Tikhonravov began research into aerodynamics, and in about 1931 joined the Moscow Group for the Study of Reactive Propulsion (MosGRID). He and Sergei KOROLEV tested their first successful rocket, the GIRD 09, on August 17, 1933; this used a solidified form of gasoline as fuel. Later, engines built by Valentin Glushko were used to power the rockets Tikhonravov had designed; one reached an altitude of over 6 miles (9.5 km), and was apparently being developed as an antiaircraft missile. In 1939 and 1940 Tikhonravov worked on a long-range ballistic missile, a forerunner of the later Soviet space launchers. Tikhonravov eventually became responsible for calculating trajectories and flight plans for Russian space rockets, satellites, and probes.

time

The measurement of time is one of the basic functions of astronomy. Early communities measured time by the obvious alternation of day and night, the Moon's cycle of phases, and the progress of the seasons. The day, the month, and the year are the fundamental astronomical divisions of time on which the CALENDAR is based. Hours, minutes, and seconds are only arbitrary units of convenience.

Primitive clocks, which worked by measuring sand running through an orifice, or water dripping into a receptacle, had to be checked and regulated by observations of the rotation of the Earth. These observations were made by noting the moment at which the Sun or a star returned to its original position (usually due south) in the sky. Such crude methods could produce huge errors. Serious attempts at more precise timekeeping came when long sea voyages were first attempted in the 16th and 17th centuries. Only through having an accurate knowledge of the time could a navigator calculate his longitude from observations of the Sun and stars. Eventually, observatory clocks and marine chronometers became essential equipment for any maritime nation. As ordinary life became more regulated, domestic timepieces also came into demand.

Until the 1920s, it was assumed that the rotation of the Earth must be the prime standard for timekeeping; no clock could be as steady and unvarying. The introduction of the Synchronome-Shortt free pendulum clock at the Royal Greenwich Observatory in 1925 destroyed this assumption. The clock was so accurate that the tiny effects on the Earth's rotation caused by NUTATION were revealed. Nutation can produce a change of up to 0.003 second per day. Even more accurate clocks soon followed, devices measuring the rate of vibration of quartz crystals (see QUARTZ-CRYSTAL CLOCK) and, more recently, cesium atoms (see ATOMIC CLOCK). By comparing the time derived from such equipment with direct observations of the rotation of the Earth, not only nutation but a gradual slowing-down of the Earth's rotation, amounting to about one second per year, can be demonstrated. The most accurate measurements of the Earth's rotation are now made with a photographic zenith tube (PZT), a telescopic camera aimed directly at the zenith. This photographs stars as they pass overhead, from which the Earth's rotation relative to the stars can be obtained to an accuracy of a few thousandths of a second. Since nutation is a cyclic phenomenon, with a main period of 18.6 years and a range of ± 1.2 seconds, it is ignored for all civil purposes. But the slowing of the Earth's spin, which involves a cumulative error, is allowed for by the occasional inclusion of an extra leap second in time signals. Astronomical computing practice, which demands a constant time interval for its basis, ignores these effects and uses ephemeris time, which assumes a constant rotational speed for the Earth.

Tiros satellites

A series of American weather satellites; the name TIROS stands for Television and Infra-Red Observation Satellite. The Tiros series returned photographs of the Earth's cloud cover and monitored the flow of heat from the Earth into space; the "heat balance" of the Earth had never before been measured. The Tiros series returned a total of over 500,000 photographs, establishing the value of satellites for weather monitoring and forecasting. They were succeeded by the Tiros Operational System (TOS) satellites, which were designated ESSA SATELLITES once in orbit. An Improved Tiros Operational System (ITOS) later came into operation; these satellites were given a NOAA

designation in orbit (see WEATHER SATELLITES). In October 1978 the first of a new series of eight improved Tiros satellites, the Tiros-N series, was launched into near-polar orbit to replace the NOAA satellites. In addition to scanning cloud and snow cover, the Tiros-N series makes temperature and moisture readings of the atmosphere and ocean, and relays data from remote platforms around the globe. Tiros-N pictures and data can be picked up by ground stations worldwide, to improve both short-range and long-range weather forecasting.

Satellite	Launch date	Remarks
Tiros 1	April 1, 1960	First weather satellite. Sent 22,952 photographs until June 17
Tiros 2	November 23, 1960	Sent 36,156 photographs until December 4, 1961
Tiros 3	July 12, 1961	Returned 35,033 photographs until February 27, 1962
Tiros 4	February 8, 1962	Returned 32,593 photographs until June 10
Tiros 5	June 19, 1962	Returned 58,226 photographs until May 4, 1963
Tiros 6	September 18, 1962	Returned 66,674 photographs until October 11, 1963
Tiros 7	June 19, 1963	Over 125,000 photographs returned until February 3, 1966
Tiros 8	December 21, 1963	Over 100,000 photographs returned until July 1, 1967
Tiros 9	January 22, 1965	In near-polar orbit; gave first photographs of Earth's entire cloud cover until February 15, 1967
Tiros 10	July 2, 1965	In polar orbit; abandoned July 3, 1967

Titan

The largest satellite of Saturn, and the only moon in the solar system known to have a substantial atmosphere. Titan was discovered in 1655 by Christiaan HUYGENS. It orbits Saturn every 15 days 22 hours $41\frac{1}{2}$ minutes at an average distance of 759,080 miles (1,221,620 km). In 1944 the American planetary astronomer Gerard KUIPER detected an atmosphere of methane gas around Titan. More recent observations suggest that Titan may have a cloudy atmosphere with between one and ten percent the Earth's atmospheric pressure, composed not only of methane but also of hydrogen. These gases seem to produce a mild GREENHOUSE EFFECT on Titan; however, its surface temperature is probably only about 115°K. Titan's clouds are reddish in color, like the clouds of Jupiter and Saturn. Accurate measurements place the visible diameter of Titan at 3,600 miles (5,800 km). But since the clouds are probably 100 miles (160 km) deep, the actual diameter of Titan must be about 3,400 miles (5,500 km). Titan therefore is larger than Mercury and has a denser atmosphere than Mars. It is bigger than any of Jupiter's moons, but Neptune's satellite TRITON may be larger still. Titan has a low density of about 1.6 times that of water. It is probably made mostly of frozen gases; these would be slushy near the surface, releasing clouds of vapor to enrich the atmosphere. Some of the lightest gases probably escape from the atmosphere of Titan, spreading out into a gas ring around Saturn.

Titan rocket

A family of American space launchers, developed from the Titan intercontinental ballistic missile. The two-stage Titan I missile entered service in 1962. Titan I was fueled by liquid oxygen and kerosene, which had to be pumped in at the last minute, and this could delay launches in an emergency. The improved Titan II used storable HYPERGOLIC propellants. Titan II was modified to launch the Gemini series of manned spacecraft. The first stage of this modified version was 63 feet (19.2 m) long and its second stage 27 feet (8.2 m) long; first-stage thrust was 430,000 lb. (195,000 kg), and second-stage thrust was 100,000 lb. (45,000 kg). The Air Force, which developed Titan, produced the more powerful Titan III specifically for space launches. Titan III has several versions. Titan IIIA was a three-stage extension of the Titan II; Titan IIIB is a Titan II with an AGENA upper stage. Titan IIIC is based on the Titan IIIA, with the addition of two solid-fuel strap-on boosters. Each of these 10-foot(3-m)-diameter boosters is built in segments. The five-segment version, 85 feet (25.9 m) long, produces a thrust of 1,200,000 lb. (540,000 kg). An advanced vehicle, the Titan IIIM, using seven-segment strap-on boosters each of 1,400,000 lb. (640,000 kg) thrust, was planned to launch the Air Force's Manned Orbiting Laboratory space station, but the project was canceled in 1969.

Titan III operations. Titan III rockets are assembled vertically in special buildings and rolled out to the launchpad like a Saturn V. The Air Force operates two Titan III launchpads, complexes 40 and 41 at Cape Canaveral. Similar facilities exist at Vandenberg Air Force Base, California (see WESTERN TEST RANGE). Titan IIIC orbits military satellites, particularly communications satellites which it carries eight at a time. A modified version, Titan IIID, launches the Air Force's Big Bird reconnaissance satellites. The most advanced member of the series is the Titan IIIE, which has a CENTAUR upper stage. This combination launched the VIKING probes to Mars, and the HELIOS Sun probes. The Titan IIIE is boosted at launch by two solid-fuel strap-ons, like those on the Titan IIIC, which drop away after two minutes. The central core's 520,000-lb. (236,000-kg) thrust first stage, 72.9 feet (22.2 m) long, then takes over and burns for 2½ minutes. After this the 23.3-foot (7.1-m) second stage of 101,000 lb. (46,000 kg) thrust ignites. Finally, the Centaur top stage fires to place the payload in orbit. It is later restarted to carry the probe onto the planets. Overall height of Titan IIIE on the launchpad, including payload and its large shroud, is 160 feet (48.8 m); lift-off weight is over 1,400,000 lb. (640,000 kg). Titan IIIE can place up to 38,000 lb. (17,000 kg) in low Earth orbit, or send 8,000 lb. (3,600 kg) to the planets.

Titania

The fourth satellite of Uranus in order of distance from the planet, discovered in 1787 by William Herschel. Titania orbits Uranus every 8 days 16 hours 56½ minutes at an average distance of 272,390 miles (438,370 km). Its diameter is about 1,100 miles (1,800 km).

Titius-Bode law

See BODE'S LAW.

Titov, Gherman Stepanovich (b. 1935)

Soviet cosmonaut, the first man to spend a full day in space. Titov made a 17-orbit flight lasting 25.3 hours in Vostok 2 in August 1961. He was the second Russian in space; he had already been backup to Yuri Gagarin, who made the first manned spaceflight. Early in his flight, Titov reported feelings of nausea, or space sickness, and he suffered from an inner-ear disorder for some time after his return to Earth. Titov was made commander of the cosmonaut corps in 1964, and he is believed to be involved in the development of a number of important future space projects.

Tombaugh, Clyde William (b. 1906)

American astronomer, who discovered the planet Pluto on February 18, 1930, from photographs taken at the Lowell Observatory, Arizona. Tombaugh, not a professionally trained astronomer, had been hired by the observatory to photograph the sky in search of a new planet beyond Neptune. Tombaugh compared two plates in a device called a BLINK MICROSCOPE; Pluto revealed itself because it had moved in position during the few days between each exposure. Tombaugh's studies also revealed many previously unknown asteroids and variable stars. After Pluto had been discovered, the planet search continued, and showed that no other bodies of any size existed at the edge of the solar system. Tombaugh later made a similar search for possible small natural satellites of the Earth, finding that no bodies bright enough to be photographed were orbiting Earth except our one Moon. Tombaugh has since become professor of astronomy at New Mexico State University, where he continues planetary studies.

total eclipse

See ECLIPSE.

transit

The moment when a celestial body crosses the MERIDIAN, the north–south line passing overhead in the sky. The meridian is the normal reference line for observing stars either to check their position or, if this is already known, to measure the rotation of the Earth. The precise observation of transits has been vital for accurate timekeeping.

Transits are observed by a *transit instrument,* a refracting telescope mounted on a horizontal shaft so that it can be pointed only along the meridian. Cross-wires run vertically through the field of view, sharply in focus with the star, and the passage across the meridian can be timed to an accuracy of ±.05 second or less. Currently, however, the most precise timing observations are made with a prismatic astrolabe (see ASTROLABE) or with a photographic zenith tube (see TIME). The word *transit* also applies to the passage of a planet (such as Mercury or Venus) across the face of the Sun, or of a satellite or a surface feature across the disk of a planet.

Transit satellites

A series of U.S. Navy NAVIGATION SATELLITES, initially intended to provide accurate position-fixing for Polaris submarines, but later used by merchant shipping and airliners. Transit 4A was the first satellite to have a nuclear power source, the small radioisotope generator known as SNAP-3 (Systems for Nuclear Auxiliary Power). The Transit series has been superseded by the polar-orbiting Defense Navigation Satellite System (DNSS), also nuclear powered. Transit 4A was launched on June 29, 1961.

Triangulum (the triangle)

A small constellation in the northern hemisphere of the sky between Andromeda and Perseus, known since ancient times. None of its stars is of particular interest, but it contains the spiral galaxy M33 (NGC 598), a prominent member of our LOCAL GROUP about 2 million light-years away.

Triangulum Australe (the southern triangle)

A small but prominent constellation of the southern hemisphere of the sky, introduced on the 1603 star map of Johann BAYER. It lies on the edge of the Milky Way. Its brightest star, Alpha Trianguli Australis, 99 light-years away, is of magnitude 1.88. Beta Trianguli is of magnitude 3.04, 38 light-years away. The triangle is completed by Gamma, magnitude 3.06.

Triton

The largest satellite of Neptune, and possibly the largest satellite in the solar system; one recent estimate gives it a diameter of roughly 3,700 miles (6,000 km). Triton is also remarkable because it orbits Neptune from east to west, a motion termed RETROGRADE; this is opposite to the general direction of motion in the solar system. A possible explanation suggests that Triton's orbit was disturbed by undergoing a near-collision with another Neptunian moon, which was ejected to become the planet PLUTO. Triton moves around Neptune every 5 days 21 hours 2 minutes in a circular orbit at 220,740 miles (355,250 km). Triton may have an atmosphere.

Trojans

Two groups of asteroids moving in the same orbit as the planet Jupiter. They occupy two of the LAGRANGIAN POINTS, and form equilateral triangles with Jupiter and the Sun. The orbit of Jupiter is, however, elliptical, and the Trojans are perturbed by other planets, so that they wander considerably either side of the Lagrangian points known as L_4 and L_5. The first Trojan, minor planet 588 Achilles, was found in 1906. The Trojans are named for the legendary heroes of the Trojan Wars. They are large, dark, and slightly reddish bodies; their generally elongated shapes suggest they are the debris from collisions of larger asteroids. One Trojan particularly unusual in shape is 624 Hektor (see ASTEROID). There are only 15 numbered Trojans, but a recent survey suggests there may be a total of as many as 700 brighter than magnitude $+21$ at the two Lagrangian points.

Tsander, Friedrich Arturovich (1887–1933)

Russian rocket pioneer, who developed early Soviet liquid-fueled rocket engines. Tsander, a disciple of Konstantin TSIOLKOVSKY, published in 1924 a design for a hybrid vehicle that would fly like an airplane to the edge of the atmosphere, then blast into space with its rocket motors. The plane's wings would be taken off and melted down to reduce weight and provide additional fuel. During the 1920s Tsander worked on designs for liquid-fuel rocket engines, and in 1930 built his first rocket motor, the OR-1, which ran on gasoline and compressed air. The following year he became head of engine development in the Moscow Group for the Study of Rocket Propulsion (MosGIRD), where with younger colleagues such as Sergei KOROLEV he produced the more powerful OR-2, which used liquid oxygen instead of compressed air. This engine was test fired in 1933, shortly before Tsander's premature death. His followers later used a modified OR-2 to launch a Tsander-designed rocket, the GIRD-10. This was the first fully liquid-fueled Soviet rocket; it flew on November 25, 1933, reaching an altitude of over 3 miles (5 km).

Tsiolkovsky, Konstantin Eduardovich (1857–1935)

Russian spaceflight pioneer, called the Father of Astronautics. Tsiolkovsky became partly deaf at the age of 10; he was introspective and studious by nature. He qualified to become a teacher, but he dreamed of flight, and in his spare time he developed plans for a gas-filled metal dirigible, and built the first Russian wind tunnel to carry out research in aerodynamics. His studies in aeronautics led him to even bolder speculations, long before the Wright Brothers had made even their first flights. His earliest article on spaceflight, "Free Space" (1883), accurately described the weightless conditions of space; he later wrote a science-fiction story about a trip to the Moon. In his book *Dreams of Earth and Sky* (1895), he wrote of an artificial Earth satellite, orbiting at a height of 200 miles (320 km).

By 1898 he had worked out the simple theory of rocket propulsion, relating the rocket's final speed to its exhaust velocity and showing the amount of propellant needed for a rocket of given mass. In 1903 he published an article titled "Exploration of Space by Reactive Devices;" this was the first theoretical demonstration that space travel by rocket was possible. It also contained a design for a rocket that would work on liquid hydrogen and liquid oxygen—the fuels of many modern rockets. Tsiolkovsky realized that single-stage rockets were not powerful enough to escape from Earth on their own, and that a multistage rocket would be needed. He termed this a *rocket train*. His book *Cosmic Rocket Trains* (1924) described a multistage rocket with the various sections stacked on top of each other—what is termed series staging. But in 1935, shortly before his death, he wrote of a rocket squadron, consisting of boosters arranged side-by-side, similar to the parallel staging technique actually adopted by Russian space boosters. Tsiolkovsky established the theoretical basis of spaceflight, and his achievement acted as an inspiration to rocketeers such as Friedrich TSANDER, culminating in the first Soviet Earth satellite.

T Tauri stars

Irregular variable stars, often found in groups (*T-associations*) together with large amounts of dust and gas. They are named after the prototype star in the constellation Taurus. Many T Tauri stars show high outputs of infrared radiation thought to come from warm dust clouds that surround them. The variability of these stars is possibly due to the obscuring effect of the cloud as it swirls around the star. T Tauri stars are PROTOSTARS, settling down into a stable existence as dwarfs. As they contract from a large cloud of gas and dust they spin progressively faster, throwing off material to form a disk from which planets may form, as happened early in the history of our own solar system (see PLANETS). The class of T Tauri variables was first recognized in 1945 by the American astronomer Alfred Harrison Joy (1882–1973).

Tucana (the toucan)
A constellation near the south pole of the sky, containing the smaller of the two MAGELLANIC CLOUDS. The constellation also contains one of the brightest globular star clusters, 47 Tucanae (NGC 104), of magnitude 3, about 19,000 light-years away. Beta Tucanae is a multiple star; Beta[1] Beta[2] are a binary pair, of magnitudes 4.52 and 4.48, 148 light-years away; however, Beta[1] itself has a magnitude 14 companion, and Beta[2] is also binary, made up of components of magnitudes 4.9 and 5.7 orbiting each other every 43 years. Beta[3] is an optical companion, only 93 light-years away; however, it too is double, with components of magnitudes 5.7 and 6.1.

Tycho
See BRAHE, TYCHO.

Tyuratam
The main Russian space launch site, near the town of the same name northeast of the Aral Sea. The Tyuratam cosmodrome is the Soviet equivalent of Cape Canaveral; confusingly, the Russians refer to it as Baikonur, which is actually a place 173 miles (278 km) to the northeast. A new city, Leninsk, has grown up 15 miles (24 km) north of Tyuratam to service the space center; it is not marked on official maps, but has a population of some 50,000. The cosmodrome has an area five times that of Cape Canaveral. It contains an estimated 80 to 85 launchpads for firing military test missiles, Earth satellites, manned rockets such as Soyuz, and unmanned PROTON rockets, together with two additional pads for a giant superrocket that will be larger than the American Saturn V. The first Sputniks, and all Soviet manned launches and space probes, have been launched from Tyuratam. It was first used as a launch site in the 1950s, during development of missiles and sounding rockets. Tyuratam remains the site from which the most important space missions are launched, but it is now less busy a place than PLESETSK, used mainly for military purposes.

U

UBV system
The measurement of the apparent brightness of a star at three specific wavelengths in its spectrum. In a photoelectric PHOTOMETER, the required wavelength region is selected by a colored glass filter, and the APPARENT MAGNITUDE is measured by a PHOTOELECTRIC CELL. In 1951 Harold Lester Johnson (b. 1921) and W. W. MORGAN devised the most commonly used set of three filters: U (ultraviolet), B (blue) and V (visual, or yellow). More recently other filters, passing red and infrared bands have been added to the system.

The difference in magnitude between two wavelengths is known as a COLOR INDEX. The color index B-V, the ratio of blue to yellow light, is determined by the temperature of the star. The second color index, U-B, the ratio of ultraviolet to blue, is determined by the surface gravity of a star, and distinguishes dwarf, giant, and supergiant stars.

All color indices can be affected by INTERSTELLAR ABSORPTION, which must be taken into account.

UFO
Unidentified Flying Objects, or UFOs, are regularly reported by both the general public and by trained observers. Many of these sightings are found on investigation to be misidentifications of aircraft, bright stars or planets, meteors, balloons and satellites, birds, or atmospheric and electrical phenomena. Only when a sighting cannot be readily explained in known terms does it become a genuine UFO. UFOs are often described as cigar-shaped or disk-shaped. Some appear as glowing objects at night, others as opaque disks in the daytime sky, and some have been tracked on radar. Occasional reports speak of apparent "landings" by these objects, and even encounters with supposed "occupants." Some or all of these cases may be hoaxes or hallucinations. Some scientists believe that with sufficient investigation, all UFO cases could be explained in known terms. Others believe that among the most puzzling UFO reports may be genuine phenomena new to science. Certain individuals, however, will always regard the classification of an object as a UFO as a sufficient explanation in itself. These people equate UFOs with the mythical "flying saucers," supposed flying craft reputedly under the control of other beings. Although many scientists believe we may not be alone in space (see LIFE IN THE UNIVERSE), there is no scientifically accepted evidence that we have been visited by alien beings, either in the distant past or more recently.

ultraviolet astronomy
The study of the Universe in the ultraviolet part of the electromagnetic spectrum, which lies between visible light and X rays, from a wavelength of 3000 Å down to 300 Å. Ultraviolet radiation is absorbed by the Earth's atmosphere, and detectors must be taken above it by rocket or satellite.
Ultraviolet satellites. The Orbiting Astronomical Observatory OAO-II, launched in 1968, surveyed the ultraviolet sky for the first time. Its successor, the $2\frac{1}{2}$-ton Copernicus satellite (OAO-III), launched in 1972, carried the largest reflecting telescope yet sent into orbit, a 32-inch (81-cm) mirror with associated sensors, to detect and measure the ultraviolet spectra of stars and interstellar molecules.

One of the most important Copernicus research projects is the study of hydrogen molecules in space, by the ultraviolet absorption lines they produce in the spectra of background stars. Possibly half the interstellar gas in our Galaxy is in this form, and it can only be detected at ultraviolet wavelengths. Hydrogen atoms are also easily detected at short ultraviolet wavelengths, and their distribution near the Sun will be determined by future satellites.

In 1978 the International Ultraviolet Explorer (IUE) satellite was launched, carrying a 17.7-inch (45-cm) telescope. The satellite, a cooperative project between NASA, ESA, and the UK, can be controlled from the ground like a telescope in an observatory. IUE has studied ultraviolet light from planets, stars of all types including X-ray sources, planetary nebulae, supernova remnants, galaxies and quasars. Long exposures allowed detection of an 11th-magnitude galaxy.

The Sun. Long before the advent of such long-lived, accurately pointing space platforms, many ultraviolet observations of the SUN were made using sounding rockets. The first of these marked the beginning of space-age astronomy: in October 1946 a U.S. Naval Research Laboratory experiment photographed the solar spectrum down to a wavelength of 2200 Å from a V-2 ROCKET at an altitude of 50 miles (80 km). As well as many other sounding rocket experiments, a series of Orbiting Solar Observatory (OSO) satellites was launched by NASA from 1962 onward, each carrying about 70 lb. (21.5 kg) of instruments. These could be pointed with an accuracy of about 1 arc minute, and scanned about 1/30 of the Sun's diameter at a time.

Many spectral lines of highly ionized silicon, oxygen, iron, and other elements have been studied at ultraviolet wavelengths in the solar spectrum, providing valuable information about the region of rapidly changing temperature that forms the boundary between the CHROMOSPHERE and the CORONA. Ultraviolet observations are also providing new insights into solar FLARE activity. Observations from the OAO satellites have now shown that, as expected, there are lines of similar ionized elements in the spectra of other stars. This implies that a high-temperature corona is a common feature among stars, and could provide a means of monitoring "starspot" activity. The SOLAR CYCLE of sunspot activity is roughly 11 years long, and it affects the intensity of the ultraviolet lines with the same 11-year period. Although it is not possible to observe equivalent spots on other stars directly, a similar regular variation in the strength of the ultraviolet lines would show that they, too, go through similar cycles of activity. A period of 11 years would probably not be expected, but if such variations were found after several years of monitoring, they could provide a valuable insight into the workings of all stars, including the Sun.

Galaxies and quasars. Ultraviolet observations of galaxies indicate what proportion of their stars are young blue stars, which radiate strongly in the ultraviolet. The first ultraviolet observations of the ANDROMEDA GALAXY showed an unexpected amount of radiation at wavelengths less than 2500 Å. If the radiation from distant galaxies is similar to that from Andromeda, this ultraviolet "peak" in their spectra might be moved into the visible region by the cosmological RED SHIFT. Galaxies farther from us would then appear brighter and bluer than they really are, and this effect could have important implications for the entire field of cosmology, which uses observations of galaxy luminosities.

The same red shift moves ultraviolet lines in the spectra of QUASARS into the visible part of the spectrum, and in many cases it is the identification of these lines that makes measurement of the red shift possible. In the mid-1960s, when the first quasar red shifts were being measured, observers drew on calculations of strong ultraviolet emission lines made originally to help in the study of gaseous nebulae within our Galaxy. The ionization of these H II REGIONS is caused primarily by ultraviolet radiation from very hot stars, and the successful application of this work to quasar observations extends "ultraviolet astronomy" to include optical observations under these special conditions of high red shift.

umbra

The darkest part of a shadow. In the solar system, the umbra is that part of an object's shadow in which light from the Sun is totally cut off; therefore a body entering another's umbra is totally eclipsed. The umbra is surrounded by a much larger, partially shaded area termed the penumbra. The dark central portion of a sunspot is also referred to as the umbra.

Umbriel

The third satellite in order of distance from Uranus, discovered in 1851 by William LASSELL. It orbits Uranus every 4 days 3 hours 27½ minutes at an average distance of 166,020 miles (267,180 km). Umbriel is about 600 miles (1,000 km) in diameter.

Universe

The Universe contains everything that exists—all of space, time, and matter. Most of the Universe is near-empty space. The matter that we see is clustered into vast systems called galaxies, which may be a billion billion miles across. Much of this matter is in the form of luminous stars. Our own Sun is an undistinguished star belonging to a typical galaxy called the Milky Way, which contains about 100 billion other stars, together with some gas and dust. Many millions of other galaxies are visible through large telescopes. Some of these are so distant that the light we see left them before the Earth was born. By looking deep into space we therefore see the Universe as it appeared in the remote past. The study of the structure and evolution of the Universe as a whole is the subject of COSMOLOGY.

Early ideas about the Universe. To ancient peoples, the Earth seemed the center of a small Universe, consisting of spheres on which the celestial bodies were fixed (see ASTRONOMY, HISTORY OF). In the 16th and 17th centuries the Earth was removed from the center of the Universe, to be replaced by the Sun. But scientists' ideas about the total extent of the Universe were little changed until they slowly began to realize that the stars are separate Suns, much more distant than the planets; the astronomer James GREGORY is credited with one estimate of their distance. Only toward the end of the 19th century, with the development of ASTROPHYSICS, was the gaseous nature of stars fully verified. In 1917 the American astronomer Harlow SHAPLEY transformed the known scale of the Universe by showing that our Galaxy was

Clusters of galaxies are visible deep into the Universe
to the limits of our telescopes, like this irregular grouping
called Abell 1060, photographed by the Anglo-Australian
telescope. The cluster contains galaxies of all kinds, from
spirals to ellipticals.

much larger than had previously been supposed. But
most astronomers still assumed that the limits of our
own Galaxy were the limits of the entire Universe, until
in the 1920s Edwin HUBBLE proved the existence of
other galaxies of stars, stretching far into space. Since
then, progressively more powerful telescopes reaching
ever-deeper into the Universe at optical and radio
wavelengths have attempted to trace the structure of
the Universe to its visible edge.

Modern views of the Universe. Hubble laid the
foundation for modern theories of cosmology with
his discovery that the Universe is apparently
expanding. The concept of an expanding Universe
suggested that all the galaxies were once compressed
close together, and that the Universe originated in a
BIG-BANG explosion; this theory was given support
by the discovery in 1965 of a weak BACKGROUND
RADIATION in the Universe, apparently caused by heat
left over from the big bang itself. According to

current estimates, the Universe originated about
18 billion years ago.

As astronomers probe deep into the Universe,
ordinary galaxies become increasingly faint and
difficult to see. At optical and radio wavelengths the
super-bright RADIO GALAXIES and QUASARS dominate
the picture. Observations suggest that the appearance
of objects in the Universe has changed over time, as
the Universe has evolved. By observing the rate of
expansion at great distances, we can determine whether
the expansion will continue for ever, or whether it will
slow down and possibly reverse (see OSCILLATING
UNIVERSE).

An alternative view of the Universe, called the
STEADY-STATE THEORY, held that the Universe has no
origin and has always appeared the same.
Unfortunately, there seems to be no way of fitting
either background radiation or the apparent evolution
of the Universe into the steady-state theory, and most
astronomers have therefore abandoned it.

But it is conceivable that we see only a small
section of a much larger Universe, and that our
restricted view prevents us from perceiving the true
nature of the cosmos. Our present models of the
Universe may seem as naive to future astronomers as
do the ideas of the Greeks to us today.

Uranus

The seventh planet from the Sun, and the first to be discovered telescopically. Uranus is a large planet, about four times the diameter of the Earth, with a thick atmosphere. Although at its brightest it is just visible to the naked eye, Uranus was not discovered until March 13, 1781, when it was found by Sir William Herschel during his systematic search of the sky. Uranus had in fact been sighted but not recognized as a planet by earlier observers; listings of it as a star in a number of catalogs stretched back nearly a century. These prediscovery observations allowed the orbital motion of Uranus to be calculated, and revealed irregularities which later led to the discovery of Neptune.

Physical properties. Uranus has a mean distance from the Sun of 19.18 astronomical units (1,783,000,000 miles; 2,869,600,000 km). The orbit is inclined by only 0° 46′ to the ecliptic and has an eccentricity of 0.047; the actual distance from Uranus to the Sun varies by 167,000,000 miles (269,000,000 km). Uranus orbits the Sun every 84.01 years. As a result of its varying distance from the Earth, Uranus changes in brightness by 26 percent between conjunction and opposition. The most remarkable feature of Uranus is that its axis is nearly in the plane of its orbit; the north pole is inclined at 98° to the vertical, so that the planet appears to rotate about its axis in a retrograde direction. The orbital motion is, of course, direct. A peculiar result of its highly tilted axis is that Uranus has extreme seasons, each pole experiencing a 42-year "summer" and a 42-year "winter." The planet's rotation period is not known accurately, but it is believed to be between 15 and 24 hours. In 1977, thin rings of rocky debris were discovered around Uranus. Nine such rings are now known, 42,000 to 51,000 km from the planet's center.

Uranus appears through a telescope as a greenish disk, on which some observers have reported seeing faint bands. However, high-quality photographs taken from balloon-borne telescopes have detected no markings of any kind on the planet. These photographs reveal that the equatorial diameter of Uranus is 32,200 miles (51,800 km). The planet is slightly oblate (squashed) in shape; its polar diameter is about 2,000 miles (3,000 km) smaller. The mass of Uranus, determined from the motions of its satellites and the perturbations that it produces on the motion of Saturn, is 14.6 times the Earth's. Uranus has a density of 1.2 times that of water.

Structure of Uranus. Uranus and Neptune are smaller and colder than their companion giant planets Jupiter and Saturn, and they form a separate family of planets. They are composed of a smaller proportion of hydrogen and helium than are Jupiter and Saturn, and a larger proportion of heavy elements. While Jupiter and Saturn are similar to the Sun in composition, Uranus and Neptune are more like the comets.

The atmosphere of Uranus shows a strong presence of methane, which gives the planet its greenish tinge. Ammonia probably also exists, but because of the low temperature (about −210°C) at the top of the atmosphere, it has condensed into clouds at a lower level. Hydrogen is the major constituent of the atmosphere, and helium is also thought to be present. Cloud layers of methane and ammonia shield the deeper levels from sunlight.

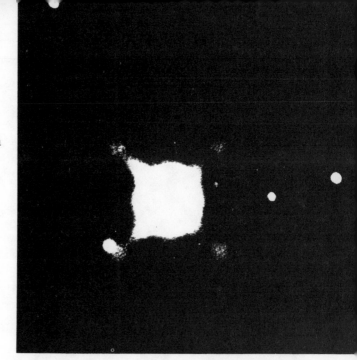

All five of Uranus' satellites are visible on this long exposure photograph taken with the McDonald 82-inch reflector; from left to right they are Oberon, Ariel, Miranda, Umbriel, and Titania. The planet itself is highly over-exposed, and a combination of photographic and telescopic effects causes the apparent cross and ring around it.

Beneath its atmosphere Uranus is believed to have a rocky core about 10,000 miles (16,000 km) in diameter, coated with a layer of ice 5,000 miles (8,000 km) thick. The pressure at the center is over 2 million atmospheres and the temperature about 4,000°C. The rocky core and the layer of ice together account for about four-fifths of the planet's mass, but the overall low density of Uranus is due to its extensive gaseous atmosphere.

Satellites. Uranus has five known moons: Miranda, Ariel, Umbriel, Titania, and Oberon. The moons orbit in the plane of the planet's highly tilted equator, moving in the same direction as the planet's rotation. It is not known how the satellites could have so exactly taken up the planet's extreme axial tilt. One suggestion is that the formation of the satellites was somehow part of the same event that tilted the planet's axis. The system of moons presents itself edge-on every 42 years, as, for example, in 1882, 1924, and 1966, while in 1861, 1903, and 1945, the orbital paths appeared to observers as circles.

Satellites of Uranus

	Discoverer	Diameter (km)	Orbit radius (10³ km)	Period (days)	Eccentricity
Miranda	Kuiper (1948)	550	130	1.4135	0.017
Ariel	Lassell (1851)	1,500	192	2.5204	0.0028
Umbriel	Lassell (1851)	1,000	267	4.1442	0.0035
Titania	Herschel (1787)	1,800	438	8.7059	0.0024
Oberon	Herschel (1787)	1,600	586	13.463	0.0007

Urey, Harold Clayton (b. 1893)
American chemist and Nobel Prize winner, notable in
the field of astronomy for computing the table of the
relative cosmic abundances of the elements, and for
championing the view that the planets were formed by
the coalescence of asteroid-sized bodies. He proposed
that the early atmosphere of the Earth was rich in
hydrogen, methane, and ammonia, and suggested it
was from this mixture that the first building blocks of
life formed. A student of his, Stanley Lloyd Miller
(b. 1930), confirmed this in 1952 when he passed an
electric spark through such a mixture of gases,
simulating the effect of lightning in the atmosphere
of the early Earth, and produced a host of amino
acids, the basic components of protein. This work
suggested that there may be many other planets in
space on which life has evolved (see also LIFE IN THE
UNIVERSE).

Ursa Major (the great bear)
A major constellation of the northern hemisphere of
the sky, circumpolar from mid-northern latitudes but
best seen during the northern hemisphere winter. Its
most conspicuous part is a pot-shaped group of seven
bright stars, forming what is commonly termed the
Big Dipper or Plow. The star Beta, also called Merak,
of magnitude 2.44, and Dubhe, Alpha Ursae Majoris,
magnitude 1.95, form the Pointers, so called because
a line drawn through them leads to POLARIS, the
Pole Star. Dubhe is actually double, with components
of magnitude 2.0 and 4.8 orbiting every 44 years.
Alioth, Epsilon Ursae Majoris, is a spectroscopic
binary varying between magnitudes 1.68 and 1.83 every
five days. Among other double and multiple stars in
Ursa Major is MIZAR, Zeta Ursae Majoris. Gamma
Ursae Majoris, Phekda, is of magnitude 2.54, and Eta
Ursae Majoris, Benetnash, of magnitude 1.91. The
Plow's faintest star is Megrez, Delta Ursae Majoris, of
magnitude 3.44. The stars of the Dipper, except Dubhe
and Benetnash, form an associated group, moving
through space together. There are numerous galaxies
in the region of Ursa Major, including the beautiful
spiral M81 (NGC 3031), 12 million light-years away,
and M101 (NGC 5457), 23.5 million light-years

distant. M82 (NGC 3034) is a well-known exploding
galaxy, and M97 (NGC 3587) is a planetary nebula
known as the Owl, 2,600 light-years distant. Lalande
21185, a magnitude 7.48 red dwarf, lies near
coordinates 11^h, $+36°$. It is the fourth-closest star to
the Sun, 8.1 light-years away, and may possess a
planetary system. Xi Ursae Majoris, 26 light-years
away, was the first double star to have its orbit
computed. It is composed of stars with masses 1.28
and 0.99 that of the Sun, orbiting each other every
59.84 years; they are of magnitudes 4.41 and 4.87.
Both components are also spectroscopic binaries, of
periods 669 and 3.98 days.

Ursa Minor (the little bear)
A constellation at the north pole of the sky, containing
the north pole star, POLARIS, Alpha Ursae Minoris. The
constellation's second-brightest star is Beta, called
Kochab, of magnitude 2.24; this was the north pole
star between about 1500 B.C. and 300 A.D. The
constellation is also known as the little dipper.

U.S. Naval Observatory
Astronomical facility of the U.S. government, with
headquarters in Washington D.C. It originated in
1830 as a part of the Navy's department of charts and
instruments, and was established under its current
name in 1844. It moved to its present site in northwest
Washington in 1893, when it also absorbed the U.S.
Navy's *Nautical Almanac* Office. The Naval
Observatory is responsible for the nation's time service,
and for the production of almanacs for astronomers,
navigators, and surveyors; these publications are
prepared in collaboration with the *Nautical Almanac*
Office of the Royal Greenwich Observatory in England.
The Naval Observatory operates a time service
substation in Richmond, Florida, near Miami. Among
the observatory's instruments at Washington is the
famous 26-inch (66-cm) refractor, installed in 1873,
with which Asaph HALL discovered the two moons of

A flare on UV Ceti observed simultaneously by Russian
astronomers at Odessa and by the Jodrell Bank 250-foot
radio telescope.

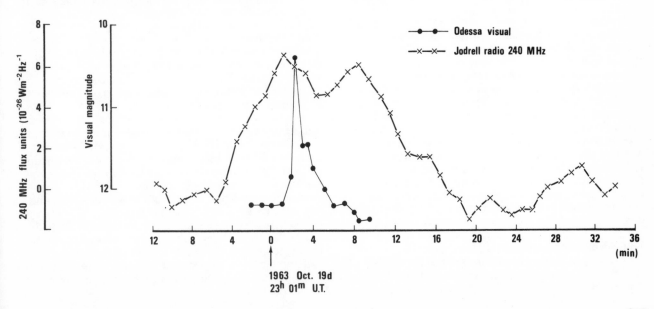

217

Mars. This telescope was remounted when the observatory moved in 1893, and was completely modernized in 1960. The observatory also operates a 24-inch (61-cm) reflector and 15-inch (38-cm) astrographic telescope at its Washington headquarters. In 1955 the U.S. Naval Observatory established an observing station near Flagstaff, Arizona, at an elevation of 7,579 feet (2,310 m). This houses a 61-inch (155-cm) astrometric reflector with a 50-foot (15.2 m) focus, installed in 1963 for accurate measurements of star positions. Another telescope at the Flagstaff station, a 40-inch (102-cm) reflector, is used for observations of comets and asteroids.

UT

Abbreviation for Universal Time, the standard time reference of the world. UT is equivalent to Greenwich Mean Time (GMT), and the name Universal Time was introduced at the 1928 meeting of the International Astronomical Union; GMT has been the world's standard time system since 1884. Universal Time is based on the rotation of the Earth, and is obtained by observations of star positions as the Earth spins; because the rate of the Earth's rotation is not quite constant, UT differs from the more regular scale of ATOMIC TIME.

UV Ceti stars

Variable stars which suddenly flare in brightness by up to 250 times (six magnitudes). The flare-up takes only a few seconds, and is followed by a steady decline to normal brightness over a few minutes. UV Ceti stars are small, cool dwarfs, intrinsically very faint. Only about 30 are known, because they are normally so faint, but they may actually be the commonest type of variable star in the Galaxy. Sir Bernard LOVELL has observed bursts of radio waves from UV Ceti—the prototype star—and others of its type, occurring simultaneously with the optical flares. UV Ceti stars are apparently rapidly rotating protostars, not yet having reached the mature stable state of a dwarf (see also FLARE STAR).

V

V-2 rocket

The forerunner of modern ballistic missiles and space rockets, developed for the German army during World War II by Wernher VON BRAUN. The V-2, first successfully fired on October 3, 1942, was the first rocket to exceed the speed of sound. The V-2 was 46 feet (14 m) long and 65 inches (1.65 m) in diameter; it weighed about 28,000 lb. (12,700 kg), had a thrust of 56,000 lb. (25,400 kg), and could carry a 1-ton payload 200 miles (320 km). The first V-2 was launched against an enemy target on September 6, 1944. (The earlier V-1 was a subsonic pulsejet, not developed by the von Braun group.) After the war, remaining V-2 rockets were used by the American government for upper-atmosphere research, and from them developed the American REDSTONE rocket.

Van Allen, James Alfred (b. 1914)

American physicist, who supervised the use of V-2 rockets for upper-atmosphere research after World War II, and developed the Aerobee sounding rocket.

The ancestor of the giant rockets of the U.S. space program, the V2 rocket was developed in Germany during World War 2 and used to attack London. The firing shown here was made by Allied scientists after the capture of the rockets and launch sites.

Van Allen was among those who proposed the 1957–1958 International Geophysical Year and devised the experiments sent aboard the first U.S. satellite, Explorer 1, launched as part of the IGY program. Explorer 1 carried a geiger counter to measure cosmic ray particles in the upper atmosphere. Its results, confirmed by those of its successors, revealed two doughnut-shaped belts of charged particles around the Earth, known as the VAN ALLEN BELTS. The discovery provided new knowledge about the Earth's magnetic shell, called the MAGNETOSPHERE.

Van Allen belts

Two zones surrounding the Earth in which charged particles are concentrated and trapped by the Earth's magnetic field. James VAN ALLEN deduced their existence in 1958 from measurements made by the early Explorer satellites.

The two belts are both toroidal (doughnut-shaped), and lie at heights of 1,900 and 14,000 miles (3,000 and 22,000 km) above the Earth's equator. High-energy electrons fill both belts, but the principal constituents of the inner belt are the heavier protons (hydrogen nuclei). Both types of particles have probably been ejected from the Sun as part of the SOLAR WIND, and captured by the Earth's magnetic field.

Space probes have shown that Venus and Mars do not possess Van Allen belts, and therefore cannot have magnetic fields comparable to that of the Earth. But Jupiter has a very strong magnetic field, and with it, associated Van Allen belts 10,000 times as intense as the Earth's.

Van De Kamp, Peter (b. 1901)

Dutch-born American astronomer who discovered the nearest planetary system to our own, orbiting the red dwarf known as BARNARD'S STAR, the second-closest star to the Sun. At Sproul Observatory in Pennsylvania, Van De Kamp began in 1937 to measure the positions of nearby stars in the hope of finding evidence for planets around them. A planetary system would cause a slight wobble in the PROPER MOTION of a star across the sky. Van De Kamp first noted such a wobble in the motion of Barnard's star in 1956, and in 1963 he proposed that a planet 1.6 times the mass of Jupiter orbited the star. From further observations he concluded in 1969 that two planets orbited Barnard's star. According to an analysis published in 1975, these planets have masses 1.0 and 0.4 times that of Jupiter, and orbit every 11.5 and 22 years at a distance of 2.71 and 4.17 astronomical units. There may be other, smaller planets whose effects are undetectable. In 1974 Van De Kamp announced the presence of a possible planetary companion or degenerate star orbiting Epsilon Eridani.

Vandenberg Air Force Base

Headquarters at Lompoc, California, of the WESTERN TEST RANGE for rockets and missiles.

Vanguard project

A U.S. Navy project begun in 1955, intended to launch the first American satellite. The Vanguard launcher was based on a sounding rocket called Viking, with a modified Aerobee sounding rocket as a second stage, and a new solid-propellant third stage. The Vanguard rocket stood 72 feet (22 m) tall overall, and had a maximum diameter of 45 inches (114 cm); first-stage thrust was 27,000 lb. (12,250 kg), second stage thrust 7,500 lb. (3,400 kg), and third-stage thrust 3,100 lb. (1,400 kg). Vanguard was beaten into orbit by Sputnik; the earliest Vanguard launch attempts failed, and the first American satellite was Explorer 1, launched by a JUNO ROCKET.

variable stars

Stars whose light reaching the Earth varies in brightness. The star's actual light output may change, or part of its light may be temporarily blocked by another star or a dust cloud. In 1975 the *Moscow General Star Catalog* listed 25,140 variable stars, and they are being discovered at an ever-increasing rate. Variability down to about a hundredth of a magnitude can be measured with a photoelectric photometer, which converts the light received from a star into an easily measured electric current. This technique is time-consuming, and can only be applied to one star at a time. A more usual method is to take two photographs of a star field at different times and examine them in a blink microscope. The operator can spot stars which vary from one plate to another by as little as 0.2 magnitude.

Except for those stars already named when their variability was discovered, such as Delta Cephei, variable stars are noted with a letter or pair of letters plus the constellation (RR Lyrae, for example). Where even these possible permutations are insufficient, variable stars are numbered with a prefix V.

Variable stars are normally classified into three

Vanguard project

Satellite	Launch date	Remarks
Vanguard TV3 (TV = test vehicle)	December 6, 1957	Lost thrust after 2 seconds; launch failure
Vanguard TV3 backup	February 5, 1958	Broke up in flight due to control system malfunction
Vanguard 1 (Vanguard TV4)	March 17, 1958	Second U.S. satellite; 3¼-lb. (1.5-kg) sphere orbited between 405 and 2,462 miles (652 and 3,962 km) every 134.3 minutes; tracking revealed Earth slightly pear-shaped
Vanguard TV5	April 28, 1958	Third-stage failure
Vanguard SLV1 (SLV = satellite launch vehicle)	May 27, 1958	Second-stage malfunction
Vanguard SLV2	June 26, 1958	Premature second-stage cutoff
Vanguard SLV3	September 26, 1958	Insufficient second-stage thrust
Vanguard 2	February 17, 1959	22-lb. (10-kg) satellite orbited between 347 and 2,064 miles (558 and 3,322 km) every 125.9 minutes; attempted TV pictures of cloud cover spoiled by satellite wobble
Vanguard SLV5	April 13, 1959	Second-stage failure
Vanguard SLV6	June 22, 1959	Second-stage malfunction
Vanguard 3	September 18, 1959	100-lb. (45.4 kg) satellite orbited between 317 and 2,329 miles (510 and 3,748 km) every 130.2 minutes; mapped Earth's magnetic field, recorded micrometeorites, and monitored solar radiation. Final Vanguard

groups, although there are some which do not fit well into any of them. The groups are pulsating, eruptive, and eclipsing stars.

Pulsating variable stars throb like a heart beat, some regularly (like CEPHEIDS) and some irregularly (like many red giants). Pulsating stars change in size, and also in surface temperature, both of which affect the star's brightness. The temperature change is usually most important, so that the star is brightest when hottest; however, maximum temperature does not always coincide with maximum (or minimum) size. The surface of the star rises and falls as it pulsates, but this motion does not penetrate deeply, and the central core containing 95 percent of the star's mass remains unaffected. The pulsation is caused by the ionization of helium and hydrogen.

Pulsating variable stars include CEPHEIDS and RR LYRAE stars. These, together with the very short period stars called Delta Scuti stars (periods less than 5 hours, range of variation about 0.1 magnitude), occupy only a small area of the HERTZSPRUNG-RUSSELL DIAGRAM, called the *instability strip*. Nearly all stars pass through this area sometime in their evolution; presumably all are pulsators at that time.

Other pulsators include the *Beta Cephei stars* (also called *Beta Canis Majoris stars,* after another well-

Name	prototype	Pulsating variable stars period (days)	variation (magnitudes)	Kind of star	Where found
cepheids	Delta Cephei	1 to 70	0.1 to 2	yellow supergiants	spiral arms, galactic clusters
Type II cepheids	W Virginis	1 to 70	0.1 to 2	yellow supergiants	halo, galactic center
long-period variables	Mira	80 to 1000	2.5 to 5	red giants	everywhere
semi-regular variables	Z Aquarii	about 100	0.5 to 2.5	red giants	everywhere
irregular variables	Betelgeuse	irregular	about 1	red supergiants	spiral arms
RV Tauri stars	RV Tauri	30 to 150	1 to 3	yellow supergiants	halo, galactic center
RR Lyrae stars	RR Lyrae	0.05 to 1.2	0.1 to 2	blue giants	halo, globular clusters
Beta Cephei stars	Beta Cephei	0.1 to 0.6	0.1	blue giants	spiral arms
Delta Scuti stars	Delta Scuti	less than 0.2	0.1	blue/yellow giants	disk, galactic clusters

studied example), which are periodically varying blue giants. Red giants also vary more or less regularly and are known as *Mira-type, semi-regular variables,* and *irregular variables,* in decreasing order of regularity. Being large stars with low densities, their periods are long (typically 100 days).

Eruptive variable stars are stars on whose surfaces occur explosions (flares) or which themselves explode (like *novae*). These unpredictable stars are the most spectacular variables, increasing in brightness up to 100 million times (SUPERNOVAE). Eruptive variables include FLARE STARS, like T TAURI STARS and UV CETI STARS, in which the flares are finished in a few minutes, as well as U Geminorum and Z Camelopardis variables, which show slow flares lasting a day or two. These latter are close double stars which contain a WHITE DWARF star, surrounded by a disk of gas. Gas streams from the companion star to the white dwarf, colliding with the disk to form a hot spot which comes and goes as the binary system rotates. These variable stars show a fascinating range of light variations.

In their gross light output, the *R Coronae Borealis stars* are the reverse of flare stars. These enigmatic stars from time to time suddenly *drop* in brightness, staying faint for weeks or months, and then gradually recover their former brightness. These stars have an excess of carbon in their atmosphere, and possibly they periodically puff a cloud of atmosphere into space. Away from the star the carbon cools to form graphite flakes, which temporarily obscure the star. It recovers its former brightness as the puff of graphite disperses into space.

Eclipsing variable stars are DOUBLE STAR systems in which one star periodically passes in front of the other, blocking some of the light from Earth (see ECLIPSING BINARY). Even when the orbit is angled slightly, so the two stars do not eclipse, the total light may still vary during each revolution because the gravitational pull of each may be strong enough to distort the other into a non-spherical shape. When seen side-on, an ellipsoid, a shape like a football, presents a larger surface area than when seen end-on (along its long axis); for this reason ellipsoidal stars vary as they rotate. The closer together two stars are in the double star system, the more pronounced is their distortion and the greater the variation. The star Beta Lyrae is a typical close eclipsing variable star,

having a period of 12.9 days and a range of brightness of 0.7 magnitude.

Although not eclipsing variables, MAGNETIC STARS also vary during their rotation, as large spots are carried across their face. They are known after their typical example as *Alpha^2 Canum Venaticorum variables.*

Vega
The brightest star in the constellation Lyra, and the fifth-brightest in the sky. Vega, of magnitude 0.04, is a white star 26 light-years from Earth. Its diameter is three times that of the Sun. Vega, also known as Alpha Lyrae, will be the pole star in about 14000 A.D.

Vela (the sails)
A constellation of the southern skies, once part of the larger constellation Argo Navis until made separate by Nicolas Louis de LACAILLE. It lies in a bright part of the Milky Way next to Centaurus, and contains several interesting star clusters. Its brightest star, Gamma, is a visual double of magnitudes 2.22 and 4.79; the brighter component is the brightest WOLF-RAYET STAR known. Both components are also spectroscopic binaries, making Gamma Velorum a quadruple star. Vela also contains a famous PULSAR, which was seen flashing optically in 1977. The supernova explosion which caused this pulsar may have been recorded by the Sumerians 6,000 years ago.

Venus
The second planet from the Sun. Venus is a rocky planet with an extremely hot, dry surface enveloped by a dense, smog-ridden atmosphere. Although the Earth's twin in size, Venus has evolved in a quite different way, producing a planet where there is little chance of finding even the simplest forms of life.

Physical nature. Venus is 7,521 miles (12,104 km) in diameter, slightly smaller than the Earth, with a mass of 4.87×10^{21} tons and a mean density of 5.25 times that of water, compared with the Earth's 5.52. The planet takes 243 days to rotate on its axis, and it does so in a retrograde direction, opposite to that of all the other planets except Uranus. Its axis is inclined at only 3°. Venus moves at about 22 miles (35 km) per second in an almost circular orbit some 67,238,000 miles (108,210,000 km) from the Sun. The orbit is inclined at 3° 24′ to the ecliptic, and Venus completes

Clouds of Venus photographed in ultraviolet light by Mariner 10. Clouds spiral away from the planet's equator towards the poles.

one circuit every 224.7 days, showing phases in the same way as Mercury.

Venus is the morning and evening star, the Phosphorus and Hesperus of antiquity, which were long thought to be different bodies. It is the most brilliant object in the sky after the Sun and Moon, reaching magnitude −4.3. It is often visible in daylight and can cast shadows at night. Venus passes closer to the Earth than does any other major planet, but its position inside the Earth's orbit means that the planet's dark side then faces us.

Venus stays close to the Sun in the sky, reaching a maximum separation (elongation) of 47°. At maximum evening elongation, the planet appears as a half-illuminated disk. It then rapidly moves sunward, increasing in size with the phase changing into a slender crescent. On rare occasions Venus crosses the solar disk, passing directly between the Earth and Sun, and it is then visible as a slow-moving dark spot. The last such transits occurred in 1874 and 1882, and the next will occur on June 7, 2004 and June 5, 2012.

Markings and rotation. Venus is unrewarding to the telescopic observer because its surface is always obscured by an unbroken layer of opaque yellowish clouds, showing only occasional dusky markings. When Venus is close to the Sun in the sky at inferior conjunction, it appears as a fine crescent. Frequently the whole outline of the planet is then visible by light scattered in the planet's dense cloudy atmosphere. The first clear glimpse of markings on Venus came in 1928, when Frank E. Ross took photographs of the planet in ultraviolet light. These showed dark patches and streaks near the equatorial and temperate regions, while the poles always appeared bright. The markings changed rapidly from day to day and were clearly cloud formations, as confirmed by photographs from the Mariner 10 space probe in 1974, which showed clouds spiraling outward from the equator to the poles. As first found by Earth-based observers in the 1960s, the clouds of Venus rotate in a retrograde direction (from east to west) around the planet every four days.

The rotation period of the solid body of Venus remained a mystery until the study of radar echoes from the planet in the early 1960s showed that Venus has a very slow east-to-west rotation of 243.0 days. This may have been caused by the impact on Venus of a former moon. The planet has no satellites today.

Surface conditions. In the mid-1950s, radio astronomers detected radiation from Venus, which suggested that the planet's surface was very hot. The high surface temperature has since been confirmed by space probes, which have also measured temperatures throughout the atmosphere. The temperature is about −40°C at the cloud-top level of 55 miles (99 km), but rises sharply deeper into the atmosphere. It reaches 70°C at 30 miles (50 km) in altitude, 200°C at 20 miles (32 km), 350°C at 10 miles (16 km), and a blistering 475°C on the planet's surface. Heat and light absorbed by the planet are re-radiated as infrared radiation, which is trapped by the carbon dioxide and water vapor in the atmosphere to build up the high observed temperatures (see GREENHOUSE EFFECT). The surface of Venus reflects radio waves much as do the surfaces of the Moon and Mercury, although its higher radio reflectivity of 12 per cent suggests that it is smoother than either of these. Computer analysis of complex radar echoes has enabled scientists to map Venus showing mountainous areas and craters up to several hundred miles across.

The first direct information about the nature of the planet's surface came in October 1975, when the Soviet spacecraft Venus 9 and 10 each made a soft landing and sent back a panoramic photograph. The photographs were surprising, showing clear landscapes with large and small boulders that cast clear shadows. The rocks are only slightly rounded by wind erosion during dust storms, and the landscapes seem to be relatively young. The deep layer of dust and debris, or regolith, that characterizes the ancient surfaces of the Moon and Mercury, is absent on Venus. This suggests that geological forces may still be active on Venus, as on Earth. About 2 percent of the Sun's light penetrates the planet's thick atmosphere. Internally, Venus is believed to be much like the Earth, with a liquid core probably smaller than the Earth's, overlain by a mantle and a granite-like crust.

The atmosphere. The dense blanketing atmosphere of Venus keeps the surface temperature constant at about 475°C from the equator to the poles, both midday and midnight. The atmospheric pressure is 91 times that at the Earth's surface. The atmosphere is over 90 percent carbon dioxide, with some hydrogen, oxygen, helium, atomic carbon, carbon monoxide, water, hydrogen chloride, hydrogen fluoride and sulfuric acid. The Pioneer-Venus entry probes in 1978 also found argon in the planet's atmosphere. The so-called clouds of Venus are more like smogs than the clouds of Earth. The main cloud region on Venus occurs at about 30 to 40 miles (50 to 65 km) in altitude with less conspicuous haze extending above this. This is in striking contrast to the Earth, where normal clouds rarely extend to heights above 8 miles (13 km). Initially, the clouds were thought to be water clouds topped with ice crystals, but the water content of the atmosphere is now known to be very slight. It has now been found

that the clouds are composed of droplets of sulfuric acid with some hydrochloric acid and hydrofluoric acid, together with other minor constituents forming individual haze layers.

The weather in the bottom few miles of the atmosphere is fairly quiet, with only gentle breezes of a few miles an hour most of the time. At greater altitudes, however, the atmosphere becomes much more turbulent with quite violent winds. The Mariner 10 photographs of the planet show that the atmosphere is heated strongly by the Sun and jet streams blow away toward the poles. Great belts of cloud cover the equatorial and temperate latitudes, while the poles lie beneath permanent smooth cloud caps. Above this activity is a fierce 220 mile (350 km) per hour gale moving westward, which seems to be a permanent feature of the visible cloud layer. This motion causes the four-day cloud rotation period clearly visible in ultraviolet photographs taken from Earth.

The small amount of hydrogen present in the atmosphere is carried to Venus by the solar wind. Unlike the Earth, which has a strong magnetic field and is protected by an extensive magnetosphere, Venus apparently lacks such a protective field. The planet's slow rotation can hardly generate a significant field through dynamo action even though Venus probably has an iron core. The planet's only protection is its ionosphere.

The evolution of Venus and Earth. Although the Earth and Venus must have begun their evolution as almost identical bodies, they have clearly evolved in quite different ways. On the Earth, the temperature was low enough for water eventually to condense to form oceans, dissolving carbon dioxide and enabling it to be incorporated into carbonate rocks. This left an atmosphere of nitrogen, oxygen, and lesser trace gases. On Venus, however, the temperature never permitted water to condense but led to a permanent cloud layer which promoted the greenhouse effect, kept carbon dioxide as an atmospheric constituent, and allowed water molecules to be broken up and lost in the high atmosphere.

Venus has no moons. Scrutiny of spacecraft photographs of the planet that would have revealed any bodies down to a size of a few hundred yards has been fruitless.

The surface of Venus photographed by the Russian craft Venus 9. The light curved area in the foreground is part of the spacecraft; the vertical stripes result from breaks in the transmission. In this distorted view, the horizon is seen crossing the top right corner, and rocks of all shapes and sizes are evident.

Venus probes

A series of Soviet space probes (also called Venera) to the planet Venus; for American Venus probes see MARINER and PIONEER. Venus 1 was the world's first planetary probe, though it failed in its mission, as did its two successors. Venus 4 began a new series of probes that ejected lander capsules as they sped past the planet. The ball-shaped capsule, 3.3 feet (1 m) in diameter, descended under parachutes, sending back data on temperature, pressure, and atmospheric composition. The first three capsules were all crushed by the enormous atmospheric pressure on Venus before they reached the surface. The improved descent capsule of Venus 7 was the first object to land successfully on the planet. All these capsules had entered the night side of Venus, the side which is turned toward us when Venus is at its closest. The Venus 8 lander touched down on the daylight side of the planet, and showed that conditions there are very similar. The heavier Venus 9 and 10 capsules each returned one photographic panorama of the surface

Probe	Launch date	Remarks
Venus 1	February 12, 1961	Contact lost at 4.7 million miles (7.5 million km). Bypassed planet at 60,000 miles (100,000 km)
Venus 2	November 12, 1965	Passed Venus at 15,000 miles (24,000 km) on February 27, 1966, but failed to return data
Venus 3	November 16, 1965	Impacted Venus March 1, 1966, but failed to return data
Venus 4	June 12, 1967	Ejected capsule into Venus atmosphere on October 18, transmitted for 94 minutes during descent
Venus 5	January 5, 1969	Ejected capsule into Venus atmosphere on May 16, transmitted data for 53 minutes during descent
Venus 6	January 10, 1969	Ejected capsule into Venus atmosphere on May 17, data returned for 51 minutes during descent
Venus 7	August 17, 1970	Ejected capsule into Venus atmosphere December 15, which transmitted data from surface for 23 minutes
Venus 8	March 27, 1972	Ejected capsule into Venus atmosphere on July 22, which soft-landed and returned data from the surface for 50 minutes
Venus 9	June 8, 1975	Lander capsule descended on October 22, returning panoramic photograph and other data from surface for 53 minutes. Orbiter section continued in orbit around Venus
Venus 10	June 14, 1975	Lander capsule descended on October 25, returning panoramic photograph and other data for 65 minutes. Orbiter section continued around Venus
Venus 11	September 9, 1978	Landed on Venus December 25
Venus 12	September 14, 1978	Landed on Venus December 21

after soft landing. Venus 11 and 12 landed on the planet, but did not return pictures.

vernal equinox

The moment when the Sun, moving north, lies on the celestial equator, it is also termed the spring equinox. The vernal equinox indicates the end of winter in the northern hemisphere, and the end of summer in the southern, being the commencement of the six-month period when the Sun lies north of the celestial equator. The vernal equinox is the zero point for the reckoning of the celestial coordinate called RIGHT ASCENSION, the equivalent of longitude on Earth. The vernal point is moving slowly westward around the sky because of the slow wobbling or PRECESSION of the Earth's axis; when first established by the Greek astronomers it lay in the constellation Aries and was referred to as the *First Point of Aries,* a name it still bears even though precession has now carried it into Pisces.

Very Large Array (VLA)

The world's largest and most sensitive radio telescope, located about 40 miles (64 km) west of Socorro, New Mexico, and operated by the National Radio Astronomy Observatory. Construction began in 1973, and is planned for completion in 1981; limited observations started in 1976. When finished, the VLA will consist of 27 movable antennae, each 82 feet (25 m) in diameter, arranged in a Y-shaped pattern along arms 13 miles (21 km) long. The instrument will work at wavelengths down to 1 centimeter, and will use the principle of APERTURE SYNTHESIS to give the resolution of a single dish 17 miles (27 km) in diameter.

Vesta

The fourth asteroid to be discovered, also known as minor planet 4, found by Wilhelm OLBERS on March 29, 1807. Vesta moves in a slightly elliptical orbit between 2.15 and 2.57 a.u. from the Sun, in the inner regions of the asteroid belt. The orbit has a 3.63-year period and is inclined at only 7° to the planetary system. Vesta is roughly spherical and rotates in $10\frac{1}{2}$ hours. It has a diameter of $312\frac{1}{2}$ miles (503 km) and a mass of about 3×10^{17} tons. Vesta's basalt-like surface gives it a relatively high reflectivity of 26.4 percent. This, combined with its position on the inner edge of the asteroid belt, means that Vesta can become brighter than magnitude 6, making it the brightest of the main-belt asteroids, and just visible to the naked eye in clear dark skies.

Viking

Two American space probes which reached Mars in 1976. Each Viking consisted of two halves: an orbiter craft which surveyed Mars from orbit, and a lander, which touched down on Mars to sample its soil in search of life. Each lander also carried two cameras to take panoramic photographs in colour. Instruments on board the landers analyzed the Martian atmosphere and sensed ground tremors. Viking lander 1 touched down in a lowland basin called Chryse (22.27°N. 47.94°W); the second lander came down in the area called Utopia (47.67°N, 225.71°W). Photographs from the landers showed the surface of Mars to be a red, stony desert with no visible signs of life. The

Top: Sand dunes and rocks are seen in this panorama of Mars from Viking lander 1. Viking's meteorology boom crosses the picture. **Above:** Porous rocks on the surface of Mars photographed by Viking lander 2. Part of the lander's footpad is at bottom right.

landers carried miniature biological laboratories which tested for the presence of microorganisms in the soil in three different ways; another experiment analyzed the soil for organic (carbon-containing) molecules. Reactions occurred in the biological experiments, but no organic molecules were detected in the soil; therefore the results of the biology experiments have been attributed to chemical reactions in the soil, not to the existence of Martian micro-life. Follow-up probes to Viking may roam the surface in search of the most favourable spots for life on Mars.

Probe	Launch date	Remarks
Viking 1	August 20, 1975	Entered Mars orbit June 19, 1976. Lander descended July 20, 1976
Viking 2	September 9, 1975	Entered Mars orbit August 7, 1976. Lander descended September 3, 1976

Virgo (the virgin)

A constellation of the equatorial region of the sky and one of the constellations of the zodiac, best seen during the northern hemisphere spring. The Sun passes through Virgo from mid-September to early November; the autumnal equinox lies in Virgo. The constellation's brightest star is SPICA. Gamma Virginis, called Porrima, is a double star, of magnitudes 3.63 and 3.6, orbiting every 171.76 years. Virgo is well known for the thousands of distant galaxies that lie within its boundaries. A nearer galaxy, M104 (NGC 4594), is called the Sombrero galaxy because of its distinctive shape.

Volans (the flying fish)

A constellation of the southern hemisphere of the sky, introduced on the 1603 star map of Johann BAYER. None of its stars is brighter than magnitude 3.5.

von Braun, Wernher (1912–1977)

German-born American rocket engineer, designer of the world's largest rocket, the Saturn V Moon launcher. While still in his teens von Braun assisted the German rocket pioneer Hermann OBERTH with engineering experiments and test flights in the German Society for Space Travel. Other experimenters were Klaus Riedel (1910–1944), Rudolf Nebel (b. 1897), and Willy Ley (1906–1969). The German army soon took an interest, and von Braun's Ph.D. thesis on rocket engine development was completed under the army's auspices. This work laid the foundation for the V-2— and, ultimately, the space age itself.

In 1937 the German army moved its rocket development center from Kummersdorf, near Berlin, to Peenemünde on the Baltic coast. There, under the direction of the army general Walter Robert Dornberger (b. 1895), the world's first missiles for long-range bombardment took shape, as well as radically new shorter-range missiles and rocket-propelled aircraft.

Foremost of von Braun's developments was the V-2 ROCKET, the world's first successful large missile.

The V-2 was first successfully fired on October 3, 1942, reaching a maximum altitude of 53 miles (85 km); by the end of the war over 5,000 V-2s had been built. In 1944 von Braun was arrested by the Gestapo and charged with being more interested in space flight than in military rockets. In part, this was true, for among designs by the von Braun team at Peenemünde were three-stage rockets for putting men and spacecraft into orbit. However, Dornberger pointed out that without von Braun there would be no V-2, and he was released.

At the end of the war von Braun and most of his Peenemünde colleagues fled south to surrender to the American army, which allowed them to continue V-2 development at White Sands Proving Ground in New Mexico. In 1950 von Braun was moved to missile development in Huntsville, Alabama, where he designed the REDSTONE ROCKET, which was eventually used for the first U.S. manned suborbital launches, and which provided the basis for the Jupiter and Juno rockets that launched early American satellites.

After the setting up of NASA, von Braun in 1960 was placed in charge of the George C. Marshall Space Flight Center at Huntsville. Von Braun had already begun to design powerful rockets made up from clusters of existing engines, and this approach led to the family of SATURN ROCKETS used in the Apollo program to put men on the Moon. With the run-down of the American space program, von Braun retired from NASA in 1972 to enter private industry.

Voskhod

A modified Soviet VOSTOK spacecraft for carrying two or three cosmonauts. Voskhod contained permanent couches in place of the Vostok ejector seat; escape in an emergency was therefore impossible. The size of the Voskhod capsule was the same as that of Vostok. Voskhod 1 carried the first three-man crew into space; because of weight problems and the lack of space, the cosmonauts wore no space suits. On Voskhod 2 the third couch was removed and replaced by a concertina-like airlock for a space walk; both cosmonauts wore space suits on this mission. When extended, the airlock measured 6 feet by 3 feet (1.8 × 0.9 m) and was jettisoned before reentry. Voskhod had small braking rockets added to cushion touchdown; the cosmonauts remained inside the craft as it landed. The Voskhod rocket was the VOSTOK LAUNCHER with a more powerful upper stage added. Because of this extra power, Voskhod entered a higher orbit than

Mission	Launch date	Results
Voskhod 1	October 12, 1964	Cosmonauts Vladimir Komarov, Konstantin Feoktistov, and Boris Yegorov made day-long, 16-orbit flight in 11,728-lb. (5,320-kg) first multi-man craft
Voskhod 2	March 18, 1965	Alexei Leonov made first space walk from craft piloted by Pavel Belyaev during day-long mission. Manual entry on 18th orbit after automatic control system failed on previous orbit brought 12,527-lb. (5,682-kg) craft down over 1,000 miles (1,600 km) off course

Vostok, and thus could not rely on atmospheric drag to pull it down if the retro-rockets failed; therefore a reserve retro-rocket was added at the front of the spherical crew compartment.

Vostok

The first Soviet manned spacecraft, designed to carry a cosmonaut in orbit for up to 10 days. Vostok was basically a hermetically sealed sphere, $7\frac{1}{2}$ feet (2.3 m) in diameter weighing about 5,290 lb. (2,400 kg). It contained normal air at atmospheric pressure. The sphere was coated with a honeycomb ablative heat shield, surfaced with metal foil strips to reflect sunlight in orbit. The capsule had no steering rockets, but relied on an offset center of gravity to align itself automatically as it encountered air resistance during reentry. Vostok went into orbit attached to the cylindrical last stage of its carrier rocket; during launch it was covered by a conical fairing. Joining the rocket stage to the capsule was a cylindrical instrument section, which contained air bottles, communications equipment, and a retro-rocket; this brought the craft's total weight up to about 10,400 lb. (4,717 kg). Before reentry, the retro-rocket was fired and jettisoned. However, Vostok was put into a sufficiently low orbit that atmospheric drag would

Mission	Launch date	Results
Vostok 1	April 12, 1961	First manned spaceflight; Yuri Gagarin made one orbit of Earth
Vostok 2	August 6, 1961	Gherman Titov made day-long flight
Vostok 3	August 11, 1962	Andrian Nikolayev made 64 orbits, landing on August 15
Vostok 4	August 12, 1962	Pavel Popovich made 48 orbits simultaneous with Vostok 3, landing on August 15
Vostok 5	June 14, 1963	Valery Bykovsky made 81 orbits, longest-ever individual flight, landing on June 19
Vostok 6	June 16, 1963	Valentina Tereshkova became first spacewoman, making 48 orbits simultaneous with Vostok 5, landing June 19

pull the capsule back to Earth after 10 days even if the retro-rocket failed. Inside the capsule, the cosmonaut sat in an ejector seat, with a porthole in front of him. After reentry, the hatch cover was blown off and the cosmonaut ejected to parachute to Earth independently. He could also eject in case of a launch emergency. On-board control was minimal and most functions occurred automatically. Despite its greater size and weight as compared to the American MERCURY, the Vostok was little more than an automatic satellite with a passenger. Vostok capsules were tested with animal passengers as part of the SPUTNIK series. Unmanned Vostok-type capsules have continued to be used for biological payloads, and for reconnaissance satellites in the COSMOS series. Vostok was launched by a modification of the rockets that launched Sputnik, called the VOSTOK LAUNCHER.

Vostok launcher

Soviet space rocket, developed from the first Soviet intercontinental military missile. This rocket was first shown to the West in 1967 in the version that launched the VOSTOK manned spacecraft. However, the same first stage was used to launch the first Sputniks, and with different upper stages it also launched the first Soviet lunar and planetary probes, and the VOSKHOD and SOYUZ manned craft. In its basic configuration, as used for Sputnik, the launcher is known in the West as the A type. It contains a central core, 92 feet (28 m) long and 9.7 feet (2.95 m) in diameter, to which are attached four tapered strap-on boosters 62 feet (19 m) long. Maximum diameter across the base is 33.8 feet (10.3 m). All engines ignite at launch, and the strap-on boosters fall away at altitude while the central core (which the Russians term the second stage) continues burning. The central core and four strap-ons each contain similar engines, producing a total first-stage thrust of 1,111,000 lb. (504,000 kg). Vostok launches were made with the addition of the so-called Luna top stage, introduced for the first three Luna flights, which gave a thrust of around 11,000 lb. (5,000 kg); this version of the launcher is termed the A-1 type. With a longer and more powerful top stage—the Venus stage, first used for launching Venus probes—the rocket became the A-2 type, as used for VOSKHOD and SOYUZ manned craft. This top stage, 25 feet (7.5 m) long, produces about 66,000 lb. (30,000 kg) of thrust. For Vostok launches, the overall height of the rocket was 125 feet (38 m); for Soyuz launches (including escape tower) the height is 167 feet (51 m).

Voyager

Two American probes to the outer planets. Both passed Jupiter, taking photographs of its clouds and moons before flying on to examine Saturn and its rings. The second was scheduled to be sent to Uranus and possibly also Neptune, reaching these remote planets in January 1986 and September 1989 respectively. Like Pioneer 10 and 11 before them, the Voyagers will eventually leave the solar system.

Probe	Launch date	Remarks
Voyager 1	September 5, 1977	Jupiter fly-by on March 5, 1979, and Saturn encounter on November 12, 1980
Voyager 2	August 20, 1977	Jupiter fly-by on July 10, 1979, and Saturn encounter on August 27, 1981. May also fly on to Uranus and Neptune

Vulpecula (the fox)

A faint constellation of the northern hemisphere of the sky next to Cygnus, best seen during the northern summer. It was introduced by Johannes HEVELIUS. Its brightest star, Alpha, is only of magnitude 4.63. However, the constellation contains the famous planetary nebula M27 (NGC 6853) known as the Dumbbell, 1,250 light-years distant. The first PULSAR, CP 1919, was discovered in Vulpecula.

W

Wallops Flight Center

A NASA installation at Wallops Island, Virginia, for aeronautical and astronautical research. It was founded in 1945 as a field station of the LANGLEY RESEARCH CENTER, but is now a separate facility. The center's headquarters and experimental research airport are located on the mainland, along with tracking instruments and the Range Control Center. On the island itself are rocket assembly buildings and launchpads. Spacecraft components have been tested on short flights from Wallops Island, but the center's main responsibility is for atmospheric and space environment research, as well as ecological sensing. Rockets fired from Wallops Island range from small sounding rockets to the SCOUT satellite launcher. The first satellite to be launched was Explorer 9, on February 16, 1961. Scout rockets have since launched several more small scientific satellites from the island.

wavelength

A term applied to the measurement of wave motions, particularly of ELECTROMAGNETIC RADIATION (such as light waves or radio waves). If the waves are visualized like sea waves, the wavelength is the distance from crest to crest. Optical wavelengths are often measured in ANGSTROMS (10^{-10} m); the wavelength of yellow light is about 5,000 angstroms. Wavelength is the wave motion's velocity divided by its FREQUENCY.

weather satellites

Earth satellites which monitor the Earth's atmosphere and surface, assisting meteorologists in their understanding and prediction of weather patterns. Weather satellites send back pictures of cloud, snow, and ice cover. They also have sensors which record the temperature of the atmosphere and oceans, thus measuring the overall heat balance of the Earth that determines weather systems. Weather satellites give advance warning of impending natural disasters, such as hurricanes forming at sea and floods from the melting of heavy snows, producing savings in lives and property that are said to have paid back the cost of the entire space program. The first weather satellites were the TIROS series; Tiros 1 was launched on April 1, 1960. The Tiros series was superseded by the ESSA satellites of the Environmental Science Services Administration, while new equipment was tested in the NIMBUS series. An advanced series, initially called the Improved Tiros Operational System (ITOS) was introduced early in 1970; the series designation was changed to NOAA on the formation of the National Oceanic and Atmospheric Administration in late 1970. NOAA 1 was launched on December 11, 1970; NOAA 2 in October 1972; NOAA 3 in November 1973; and NOAA 4 in November 1974; and NOAA5 in July 1976. These satellites move in near-polar orbis for worldwide coverage, orbiting every 115 minutes at an altitude of around 900 miles (1,500 km).

In May 1974 and February 1975 a series of synchronous meteorological satellites (SMS) went into service in stationary orbit above the Earth's equator. These scan both the Earth's cloud cover and the Sun's activity, and relay data from remote automatic weather stations. From their high orbit, the satellites can cover a large area of the Earth in each picture, which they transmit at 30-minute intervals. The change of the system's name to GOES (Geostationary Operational Environmental Satellite) with a new launch in October 1975 indicates that the system became operational. These satellites provide the American contribution to the International Global Atmospheric Research Program (GARP). The European Space Agency and Japan are also cooperating in this program.

The Soviet Union has its own weather satellite system, called Meteor. Meteor satellites are cylindrical in shape, 16 feet (5 m) long by 5 feet (1.5 m) in diameter, with wing-like solar panels. They move in similar orbits to the NOAA satellites, and return equivalent data. By international agreement, all nations exchange weather satellite results. Eventually, the Meteor system will be supplemented by low-altitude observations from Salyut space stations, and high-altitude monitoring from synchronous weather satellites. The Meteor system officially began on March 26, 1969, with the launch of Meteor 1, although there has been a series of experimental weather satellites launched previously in the Cosmos series. Several Meteor satellites are launched each year.

Weizsäcker, Carl Friedrich von (b. 1912)
German astrophysicist, who in 1945 proposed a theory of the origin of the solar system which suggested that the planets formed by the aggregation of dust particles from a disk orbiting the primeval Sun. Weizsäcker's view was derived ultimately from the ideas of Immanuel KANT and the marquis de LAPLACE, but provided a plausible mechanism for the formation of planets, which the Kant-Laplace theory lacked. It was the first of the modern theories of the solar

system's origin, from which the currently accepted views were developed (see PLANETS). In 1938 Weizsäcker proposed the carbon-nitrogen-oxygen cycle as the origin of stellar energy, independently of Hans BETHE.

Westar satellites
American domestic communications satellites, similar to the Canadian ANIK series, owned and operated by Western Union. The drum-shaped satellites, 6.3 feet (1.9 m) in diameter and 11.8 feet (3.6 m) tall to the top of their 5-foot (1.5-m) mesh antenna, weigh 1,265 lb. (573 kg) and can carry 7,200 two-way telephone circuits or 12 color television channels. The Westar satellites are placed in geostationary orbit, and are the first U.S. domestic communications satellites. They were launched in 1974. The system is planned to be extended as traffic grows.

Westerbork Radio Observatory
A radio astronomy observatory near Groningen, the Netherlands, operated by the Netherlands Foundation for Radio Astronomy, which also operates the 82-foot (25-m) dish at Dwingeloo. Westerbork contains a 1-mile (1.6-km) APERTURE SYNTHESIS telescope consisting of a line of ten fixed and two movable dishes of 82 feet diameter. The telescope, which began observations in 1970, works at wavelengths of 49, 21, and 6 centimeters. It is one of the world's most sensitive radio telescopes. Two more dishes are planned to be added about 1978 to extend the telescope's total aperture to 2 miles (3 km).

Nine of the twelve dishes of the Westerbork Synthesis Radio Telescope; the other three continue the line to the right. The two most distant dishes can be moved on a 300-yard-long track. Each dish is surfaced with fine wire mesh, and is counterbalanced by the large cylindrical weight.

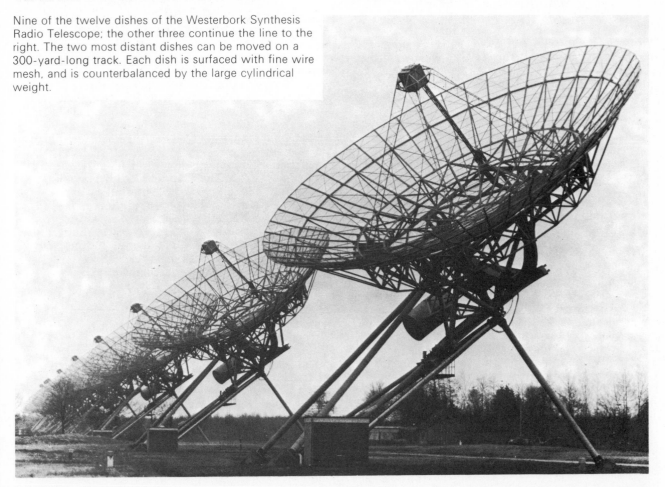

Western Test Range

A missile firing range based at Vandenberg Air Force Base, Lompoc, California, previously known as the Pacific Missile Range. The range stretches southward for more than 5,000 miles (8,000 km) into the Pacific, and is used for test firing Minuteman and Titan II ballistic missiles. It is also used to launch sounding rockets, and to put satellites into polar orbit. The first satellite launch from the Western Test Range was Discoverer 1, on February 28, 1959. Facilities are now being installed for polar launches and landings of the SPACE SHUTTLE.

Whipple, Fred Lawrence (b. 1906)

American astronomer, who in 1949 suggested that a comet's nucleus can be compared to a dirty snowball, in which dusty and rocky materials are cemented into a ball a few miles across by frozen gases such as methane and ammonia; this is now the accepted view. Whipple discovered a total of six comets, the last in 1942. He devised a system of special cameras with rotating shutters to photograph meteors, and found from their orbits that they were debris from comets. From 1955 to 1973 Whipple was director of the Smithsonian Astrophysical Observatory, where he organized the optical tracking system for Earth satellites, setting up the worldwide system of Baker-Nunn cameras and the amateur Moonwatch teams.

White, Edward Higgins II (1930–1967)

First American to walk in space. White made a 21-minute spacewalk, maneuvering at the end of a tether with a hand-held gas gun, on June 3, 1965, during the flight of Gemini 4, commanded by James McDivitt. White was chosen as a crew member of the first manned Apollo flight; but all three crewmen were killed by a fire in the spacecraft during a simulated countdown.

white dwarf

A star whose diameter is only 1 percent that of the Sun, and whose luminosity is 10,000 times less. Although called "white" dwarfs, they can in fact be any color, depending on their surface temperature; the hotter white stars (temperature around 10,000°C) were generally the first to be observed. Best known is the companion to SIRIUS (Sirius B).

They represent the final stage of stellar evolution—dying stars, taking billions of years to cool to black globes. Their light originates in a thin hydrogen atmosphere made to shine by heat leaking away from the interior, and unlike other stars, they have no nuclear energy source. Consequently, there has been nothing to prevent them from collapsing until they are of planetary size. Their matter is in what is known as a degenerate state, in which atoms are stripped of their electrons and the electrons packed tightly together. White dwarfs are degenerate in their central regions, and become increasingly degenerate as they cool. Their density, as a result, is very high, about 1 million times that of water.

The gravity at the surface of a white dwarf star is tens of thousands of times greater than the Earth's, a consequence of its small size and relatively high mass. Even light loses energy and suffers a RED SHIFT when leaving these stars (see RELATIVITY). There is a limiting mass, called the Chandrasekhar limit (about 1.5 times that of the Sun), above which a white dwarf cannot exist. It is still not fully known how heavier stars lose enough mass to enable them to become white dwarfs.

White Sands Proving Ground

A U.S. Army rocket testing range at White Sands, New Mexico, used for launches of sounding rockets and missiles. White Sands was activated in 1945 as a testing ground for the American Wac Corporal sounding rocket; later, it was used for launching captured V-2 rockets by Wernher VON BRAUN, who was then at Fort Bliss, Texas. In 1950 the von Braun group moved to what is now the MARSHALL SPACE FLIGHT CENTER at Huntsville and their test launches switched to CAPE CANAVERAL, although White Sands remained in use for launches of smaller missiles and sounding rockets.

Wolf, Rudolf (1816–1893)

Swiss astronomer who confirmed the discovery by Heinrich Schwabe of the sunspot cycle, and more correctly determined its period as an average of about 11 years. Wolf showed that the sunspot cycle was correlated with aurorae and disturbances in the Earth's magnetic field, thus initiating the study of solar-terrestrial effects. Around 1850 he devised the system of using sunspot counts as a measure of solar activity; Wolf's system of sunspot numbers is continued to this day.

Wolf-Rayet stars

A rare group of small, hot stars surrounded by luminous clouds of ejected material. The first three such stars were discovered in 1867 by the French astronomers Charles Joseph Etienne Wolf (1827–1918) and Georges Antoine Pons Rayet (1839–1906). The spectra of Wolf-Rayet stars show broad emission lines caused by atoms of helium, carbon, and nitrogen in a hot atmosphere surrounding the star. A typical Wolf-Rayet star has a temperature in excess of 30,000°K and emits an abundance of ultraviolet energy which ionizes the carbon and nitrogen in its atmosphere. The star's energy also makes its atmosphere turbulent, and it streams off the star and into space at thousands of miles per second, thereby blurring the emission into broad bands. Wolf-Rayet stars may be the central cores of massive red giants that have been exposed to view by the stripping off of their outer layers. About 200 Wolf-Rayet stars are known. They are also called WC and WN stars, depending on whether they show carbon or nitrogen in their spectra.

Woolley, Sir Richard van der Riet (b. 1906)

Eleventh English astronomer royal (1956–1971). Woolley made fundamental analyses of the movements of stars around our Galaxy from their observed radial velocities and proper motions; this clarified knowledge of the Galaxy's structure. His studies helped distinguish several groups of stars which formed at different times during the Galaxy's evolution. These investigations led to the modern view that our Galaxy originated from a large collapsing cloud, forming first a widespread "halo" of old stars and finally flattening out to produce a disk of young stars, including the spiral arms. In 1972 Woolley became director of the SOUTH AFRICAN ASTRONOMICAL OBSERVATORY.

Woomera
A rocket range in South Australia, 280 miles (450 km) northwest of Adelaide, set up in 1947 by Britain and Australia for the development of missiles such as Blue Streak. Woomera was also used to launch upper-atmosphere sounding rockets, and later became the launch site for the ELDO Europa 1 launcher, based on Blue Streak, which unsuccessfully attempted to orbit a European satellite. Two satellites have been orbited from Woomera: WRESAT 1, of Australia's Weapons Research Establishment, launched by a modified REDSTONE rocket on November 29, 1967, to study the upper atmosphere and solar radiation; and the British technology satellite Prospero on October 28, 1971, by a British Black Arrow rocket.

X-ray astronomy
The study of the Universe at X-ray wavelengths, comprising ELECTROMAGNETIC RADIATION between 0.1 and 300Å in wavelength. It began on June 18, 1962, when an Aerobee sounding rocket was launched to study the X rays that were expected to arise from the interaction of solar radiation with the Moon's surface. No such X rays were found, but to the surprise of all astronomers a powerful source emitting X rays of a few angstroms in wavelength was detected in the direction of the constellation Scorpius. There was also evidence for a background of diffuse X-radiation coming from all parts of the sky, which has not yet been completely explained.

Throughout the 1960s other X-ray sources were discovered by rocket and balloon flights, but the real growth of this branch of astronomy occurred only in the early 1970s, with the launching of a succession of satellites carrying X-ray telescopes and detectors: Uhuru, Copernicus, ANS-1 and Ariel V. The Sun has also been extensively studied by rockets, and by the Orbiting Solar Observatory satellites (OSO). Solar X rays arise primarily in the very hot gas of the CORONA over active sunspot regions, but very small bright patches are also observed near solar flares.

X-ray surveys. Uhuru, launched in December 1970, provided the first X-ray sky surveys. These allowed many new sources to be identified, and showed that some are not long-lived. Before it ceased to function effectively, Uhuru completed three surveys, the most complete being the 3U catalog of 160 sources.

The British Ariel V satellite, launched four years after Uhuru, has shown that there are many extremely short-lived X-ray sources occurring all the time, flaring up briefly and dying away after a few days or weeks. These "X-ray novae" occur at the rate of at least one a month in our Galaxy.

In 1977 a more sensitive survey of the X-ray sky was begun by HEAO 1, the first of a series of three High Energy Astronomy Observatories (see HEAO). HEAO 1 increased the number of known X-ray sources from 350 to 1,500, and detected X-ray emissions coming from numerous distant galaxies and quasars. HEAO 2, launched in 1978, was designed to examine specific sources in more detail.

Scorpius X-1. The first X-ray source discovered, named Sco X-1, was the most intensively studied and best-known source for a decade, because it was the brightest source known, and because it was also identified at optical and radio wavelengths. The optical counterpart emits strongly in ultraviolet wavelengths, shows "flickering" changes in brightness and occasional large flares, and its spectrum has many emission lines. These characteristics are in some ways similar to those of old NOVAE, and since these occur in binary-star systems it was suggested that Sco X-1 is also a double star.

The energy generated in Sco X-1 as X rays has been roughly constant for more than 10 years, and is estimated at 100,000 times the *total* luminosity of the Sun. Such a massive amount of energy can be liberated in a binary system when matter from the extended atmosphere of a giant star flows onto a small companion such as a NEUTRON STAR. According to this model, as material falls onto the companion, its rotation forms it into a disk (the *accretion disk*). The gas in the disk spirals inward and heats up to tens of millions of degrees; X rays are naturally emitted from gas at this temperature by the BREMSSTRAHLUNG process.

X-ray binaries. Two-thirds of the identified sources in the 3U survey are stars in our Galaxy, and in many cases their optical spectra show them to be binaries. Optically, we usually see the giant star of the pair, while the X rays comes from the compact companion, too faint to be seen in visible light. (Sco X-1 is one of the rare exceptions where the compact star outshines the giant.)

The X-ray intensities from the best-studied sources vary periodically, and the periods fall into two main groups: hours to days, and seconds to minutes. The longer periods represent the time taken by the compact star to orbit its companion (like the terrestrial year); while the short periods represent the time taken for the compact object to rotate on its axis (analogous to the terrestrial day).

The rotational periods are considerably longer than those of the radio PULSARS, which are probably younger neutron stars.

An X-ray binary evolves from a normal close DOUBLE STAR by what is known as mass exchange. The originally more massive star (the primary) evolves more rapidly at first, and expands into a giant star. The gravity of its lighter and smaller companion (the secondary) pulls matter out of the primary, and the secondary eventually grows to become the more massive of the pair. The original primary then explodes as a SUPERNOVA, whose compact remnant, a neutron star, is left orbiting the original secondary. When matter from this star flows back onto the neutron star, X rays may be produced. At this stage, theory predicts that the visible star must be a giant or supergiant, and this is confirmed by observation in most cases. Emission of X rays should thus be a short-lived phase in the evolution of many binaries.

Transient sources. Of the transient sources detected by X-ray satellites, the first to be identified optically was discovered by Ariel V in August 1975, in the constellation Monoceros. This quickly became an

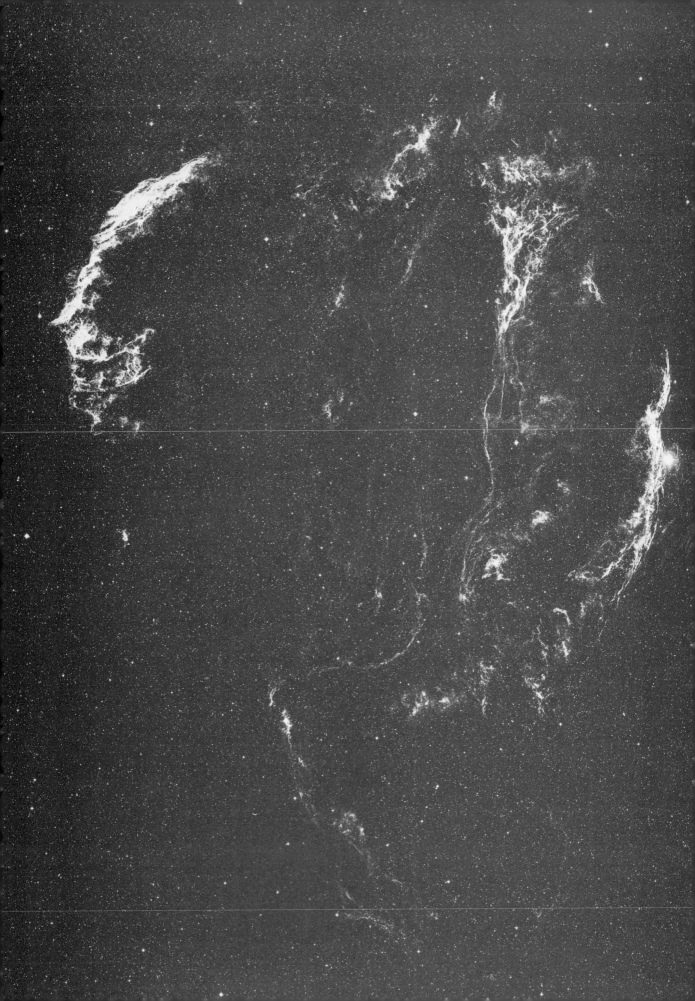

X-ray source more intense than even Sco X-1. Optical astronomers identified it with a star of apparent magnitude 11, which had brightened from magnitude 20. Old photographs showed a previous brightening in 1917, and the star may thus belong to the class of VARIABLE STARS known as recurrent novae, which are in fact binary stars. Since all the X-ray binaries vary erratically to some extent, the transient sources may be no more than extreme examples in which mass transfer occurs spasmodically.

Black holes? Much attention has recently been focused on X-ray astronomy by the claim that it has revealed a BLACK HOLE in space, an object so dense that even light cannot escape from its gravity. Observations of binary stars can provide an indication of the masses of the two components, and in the case of the star seen at the position of the source Cygnus X-1, this technique suggests that the mass of the compact star is at least six times that of the Sun. Although an ordinary star can maintain its structure through the pressure produced by its nuclear reactions, theory predicts that any compact object more than three times the Sun's mass must collapse through its own gravity into a black hole. The observations of Cygnus X-1 are widely taken as showing a black hole orbiting a normal star, with X rays being emitted by gas falling toward the black hole, but not yet inside it. There is, however, an alternative explanation. Cygnus X-1 could be a triple system, in which the bright, optically visible star is orbited by a dim binary pair. The latter could be a conventional X-ray source, with the two individual stars each less massive than three Suns.

Other X-ray sources. The 3U catalog also contains supernova remnants, the expanding shells of gas from exploded stars. These move outward so fast that they form a shock wave in the interstellar gas, heating it to enormous temperatures and causing the bremsstrahlung emission of X rays. In a category of its own is the CRAB NEBULA, formed in a supernova observed to explode in the year 1054. This remnant contains a powerful radio and optical pulsar, and X rays are generated both in the pulsar and in the nebula by the interaction of very fast electrons from the pulsar with a magnetic field (SYNCHROTRON RADIATION).

A few extragalactic sources of X rays are known. Some powerful binary-type sources have been detected in the MAGELLANIC CLOUDS, the nearest external GALAXIES to our own Galaxy, the Milky Way. Much stronger X-radiation, however, is produced in the centers of galaxies with disturbed, energetic nuclei, such as SEYFERT GALAXIES, RADIO GALAXIES, and QUASARS. Extremely powerful extended regions of X-ray emission occur between galaxies at the centers of giant clusters of galaxies. This emission comes from gas heated to tens of millions of degrees, either because it has fallen into the strong gravitational field of the cluster, or because it is "stirred up" by the galaxies moving through it.

The Cygnus Loop, or Veil Nebula, photographed here in the red light of hydrogen, is a relatively powerful emitter of X rays. X-ray telescopes show that the radiation comes from the same circular arc as the visible light, a shell of gas ejected by a stellar explosion (supernova) 20,000 years ago.

Y

year

The time taken for the Earth to make one orbit of the Sun. The year on which we base our calendar lasts 365.2422 days; it is termed the *tropical* year. It is shorter than the orbit of the Earth with respect to the star background (the *sidereal year,* lasting 365.2564 days) because of PRECESSION, which effectively brings the Sun back to its starting point slightly earlier each year. A third type of year is the interval between two successive perihelion passages of the Earth. This, the so-called *anomalistic year,* lasts 365.2596 days, and arises because gravitational perturbations by the planets advance the perihelion of the Earth's orbit.

Yegorov, Boris Borisovich (b. 1937)

Soviet physician who became the first doctor in space when he flew on the Voskhod 1 mission in October 1964. Yegorov, an expert on the balancing mechanism of the inner ear, joined the team of Soviet space doctors in 1961, and joined the cosmonaut corps in the summer of 1964 purely for the Voskhod 1 flight.

Yeliseyev, Alexei Stanislovovich (b. 1934)

Soviet cosmonaut who flew on the Soyuz 5 flight in January 1969, which made the first docking between two manned spacecraft; Yeliseyev then walked in space to Soyuz 4. In October 1969 he flew with Vladimir SHATALOV on the Soyuz 8 mission, which performed joint maneuvers in space with Soyuz 6 and 7. On April 23, 1971, he made his third flight, on the Soyuz 10 mission, which docked briefly with the Salyut 1 space station. Yeliseyev, one of the Soviet Union's most experienced cosmonauts, was flight director for the Apollo-Soyuz mission.

Yerkes Observatory

The astronomical observatory of the University of Chicago, situated at Williams Bay, Wisconsin, at an altitude of 1,096 feet (334 m) on the shores of Lake Geneva. Yerkes Observatory was founded by George Ellery HALE with a grant from the businessman Charles Tyson Yerkes (1837–1905). It opened in 1897 with the famous 40-inch (102-cm) refractor, which remains the world's largest refracting telescope. In 1901 a 24-inch (61-cm) reflector was added, made by G. W. RITCHEY; in 1968 this was removed to make way for a new 41-inch (104-cm) reflector. The University of Chicago also shares in the operations of the McDONALD OBSERVATORY.

Young, John Watts (b. 1930)

Commander of the Apollo 16 mission, which landed on April 21, 1972, in the Descartes highland region of the Moon. Young and fellow astronaut Charles M. Duke (b. 1935) made three lunar-surface excursions, totaling 20 hours 15 minutes, during a stay on the Moon of 71 hours 2 minutes. Young, a former Navy test pilot, made his space debut on March 23, 1965, in Gemini 3, the first American two-man mission. He commanded the Gemini 10 flight in July 1966. Young was also command-module pilot of Apollo 10 in May 1969, the dress rehearsal for the first lunar landing mission.

Z

Zeeman effect
The splitting of SPECTRAL LINES by a magnetic field, named for the Dutch physicist Pieter Zeeman (1865–1943), who discovered the effect in 1896. The amount of splitting varies with the strength of the field. The lines split, usually into three, because the magnetic field changes the energy levels of electrons in an atom (see SPECTROSCOPY). The outer two lines are polarized in a different direction from the inner one (see POLARIZATION); this distinguishes the Zeeman effect from other line-splitting or broadening effects. A device called a magnetograph, developed by H. D. and H. W. BABCOCK at Mount Wilson in the early 1950s, compares the intensity and polarization on alternate sides of suitable lines in the Sun's spectrum. This produces an accurate value of the magnetic field at a single point on the solar surface. To build up a picture of the entire solar disk, the magnetoheliograph is used. This employs the same principle as the spectroheliograph (see SPECTROSCOPE). Magnetic fields have also been detected on some stars by noting the Zeeman effect (see MAGNETIC STARS).

Zelenchukskaya Astrophysical Observatory
An astronomical observatory at an altitude of 6,800 feet (2,070 m) in the Caucasus Mountains of Russia, containing the world's largest optical telescope, the 236-inch (600-cm) reflector, which began operation in early 1976. At the same site is the RATAN 600 (Radio Astronomical Telescope of the Academy of Sciences), consisting of 900 aluminum reflectors mounted in a circle 1,890 feet (576 m) in diameter, capable of working at wavelengths from 0.4 to 21 centimeters. Each of the individual panels can be pointed up and down, and the instrument can work as a unit or as four separate sections.

zenith
The point in the sky directly above an observer and opposite the NADIR. *Zenith distance* is the angle between an object and the zenith.

zenithal hourly rate
The rate of a meteor shower that would be recorded by an experienced observer with a clear dark sky and the meteor radiant in the zenith (abbreviation ZHR). The altitude of the radiant in the sky has a major effect on the observed meteor rate. To derive the ZHR multiply by 1.1, 1.3, 1.5, 1.7, and 2.0 for radiant altitudes of 66°, 52°, 43°, 35°, and 27° respectively.

zodiac
A belt of constellations about 9° wide over which the Sun, Moon, and planets (except Pluto) appear to move. The constellations are Aries (The Ram) ♈, Taurus (The Bull) ♉, Gemini (The Twins) ♊, Cancer (The Crab) ♋, Leo (The Lion) ♌, Virgo (The Virgin) ♍, Libra (The Balance) ♎, Scorpius (The Scorpion) ♏, Sagittarius (The Archer) ♐, Capricornus (The Goat) ♑, Aquarius (The Water Bearer) ♒, and Pisces (The Fish) ♓. Of the twelve constellations, seven were said to represent animals, and the Greeks called the zone "zodiakos kyklos" or circle of animals; from this the word zodiac was derived. The zodiacal signs given above appear in manuscrips of the late Middle Ages, and their origin is obscure.

The twelve signs of the zodiac are taken by astrologers to cover a belt of the sky 30° wide, and the Sun is supposed to pass through them at specific, equally spaced dates. However, the constellations do not, in fact, occupy equal spaces, and PRECESSION (a movement of the Earth's axis) has now altered the dates at which the Sun is in a particular constellation. The Sun also now passes through Ophiuchus (The Serpent Bearer), which is not a zodiacal constellation at all.

zodiacal light
A diffuse pointed cone of faint light rising obliquely into the sky above either the last trace of twilight in the evening sky or the first trace of twilight in the morning sky. The zodiacal light lies along the ecliptic or zodiac, and is caused by sunlight reflected from a disk-shaped cloud of dust particles in the solar system. The cloud has a total mass of about 30 million million tons and contains particles down to about 10^{-12} gram in mass. These particles can enter the Earth's atmosphere to produce MICROMETEORITES, eventually falling to ground level. The solar wind sweeps about 100,000 tons of fine dust a year out of the solar system; larger particles spiral in toward the Sun (see POYNTING-ROBERTSON EFFECT). However, fresh dust is continually scattered by the periodic comets. The zodiacal light is not difficult to observe, but it does require a dark, haze-free sky. For observers in northern temperate latitudes, it is best seen after sunset in spring or before sunrise in autumn.

Zond spacecraft
A series of Soviet space probes that act as tests for future missions. Zonds 1 and 2 were launched toward Venus and Mars, respectively, after communications failures with the first Soviet Venus and Mars probes; however, radio contact was also lost with the Zond probes. Zond 3 took lunar farside pictures which it transmitted as it receded into space, to test picture transmission systems for future planetary probes. Subsequent Zonds were unmanned test flights of Soyuz craft (minus the spherical orbital module), in preparation for an intended manned circumlunar mission. However, this was postponed indefinitely after the manned circumlunar flight by Apollo 8 in December 1968.

Zond spacecraft

Probe	Launch date	Remarks
Zond 1	April 2, 1964	Launched toward Venus; communications failed
Zond 2	November 30, 1964	Launched toward Mars; communications failed
Zond 3	July 18, 1965	Flew behind Moon, photographing area not covered by Luna 3. Headed toward Mars, retransmitting lunar photographs in communications test
Zond 4	March 2, 1968	Unsuccessful test of circumlunar Soyuz

Probe	Launch date	Remarks
Zond 5	September 15, 1968	First flight to vicinity of Moon and back. Carried biological specimens to assess radiation hazard and tape recording to test voice transmission between capsule and Earth. Splashed down in Indian Ocean September 21
Zond 6	November 10, 1968	Photographed lunar farside. Skip-glide reentry using aerodynamic lift; landed in Soviet Union November 17
Zond 7	August 8, 1969	Repeat of Zond 6 mission. Landed August 14
Zond 8	October 20, 1970	Similar to previous Zonds, with modified reentry trajectory; splashed down in Indian Ocean October 27

Zwicky, Fritz (1898–1974)
Swiss astrophysicist, who in 1934 first clearly distinguished with Walter BAADE that supernovae are different in nature from the less bright novae, and proposed that a star is largely destroyed in a supernova explosion but may leave its compressed core as a NEUTRON STAR. Zwicky began an international program to search for supernovae in distant galaxies, personally finding many new examples that helped clarify the nature of these enormous outbursts. His observational studies led to the discovery of many galaxies and clusters of galaxies, which he studied in detail. Zwicky discovered that some nearby galaxies seem to be linked by bridges of stars, as if the galaxies were interacting. From 1943 to 1949 Zwicky worked with the Aerojet Engineering Corporation on development of jet engines and jet-assisted takeoff for aircraft.

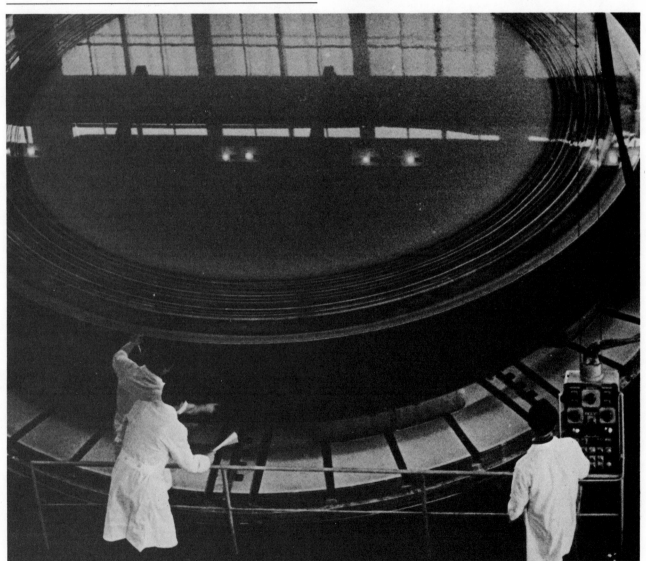

First stages in the grinding of the mirror of the Zelenchuk 236-inch (6-m) telescope, the world's largest. The 42-ton borosilicate glass disk is here being milled to a rough concave shape with a diamond wheel before polishing in a Leningrad optical workshop.

Index

《Index》

U.S. Price Level				Selected U.S. Monetary Data			
Year	GNP Price Deflator (1972 = 100)	Consumer Price Index (CPI), (1967 = 100)	Percent Change in CPI from Previous Year	M-1 Money Supply (billions of dollars)	Percent Change in M-1 from Previous Year	Interest Rate on Three-month U.S. Treasury Bills (annual percent)	Prime Interest Rate (annual percent)
1929	32.76	51.3	0				5.50–6.00
1933	25.13	38.8	−5.1			.52	1.50–4.00
1939	28.43	41.6	−1.4			.02	1.50
1940	29.06	42.0	1.0			.01	1.50
1941	31.23	44.1	9.7			.10	1.50
1942	34.32	48.8	9.3			.33	1.50
1943	36.14	51.8	3.2			.37	1.50
1944	37.01	52.7	2.1			.38	1.50
1945	37.91	53.9	2.3			.38	1.50
1946	43.88	58.5	18.2			.38	1.50
1947	49.55	66.9	9.0			.59	1.50–1.75
1948	52.98	72.1	2.7			1.04	1.75–2.00
1949	52.49	71.4	−1.8			1.10	2.00
1950	53.56	72.1	5.8			1.22	2.07
1951	57.09	77.8	5.9			1.55	2.56
1952	57.92	79.5	.9			1.77	3.00
1953	58.82	80.1	.6			1.93	3.17
1954	59.55	80.5	−.5			.95	3.05
1955	60.84	80.2	.4			1.75	3.16
1956	62.79	81.4	2.9			2.66	3.77
1957	64.93	84.3	3.0			3.27	4.20
1958	66.04	86.6	1.8			1.84	3.83
1959	67.60	87.3	1.5	141.0		3.41	4.48
1960	68.70	88.7	1.5	141.8	0.5	2.93	4.82
1961	69.33	89.6	.7	146.5	3.3	2.38	4.50
1962	70.61	90.6	1.2	149.2	1.8	2.78	4.50
1963	71.67	91.7	1.6	154.7	3.6	3.16	4.50
1964	72.77	92.9	1.2	161.9	4.6	3.16	4.50
1965	74.36	94.5	1.9	169.5	4.7	3.55	4.54
1966	76.76	97.2	3.4	173.7	2.5	3.95	5.63
1967	79.06	100.0	3.0	185.1	6.6	4.88	5.61
1968	82.54	104.2	4.7	199.4	7.7	4.32	6.30
1969	86.79	109.8	6.1	205.8	3.2	6.68	7.96
1970	91.45	116.3	5.5	216.6	5.2	6.46	7.91
1971	96.01	121.3	3.4	230.8	7.0	4.39	5.72
1972	100.0	125.3	3.4	252.0	9.2	4.07	5.25
1973	105.75	133.1	8.8	265.9	5.5	7.04	8.03
1974	115.08	147.7	12.2	277.6	4.4	7.89	10.81
1975	125.79	161.2	7.0	291.2	4.9	5.84	7.86
1976	132.34	170.0	4.8	310.4	6.6	4.99	6.84
1977	140.05	181.5	6.8	335.4	8.1	5.27	6.83
1978	150.42	195.4	9.0	363.1	8.3	7.22	9.06
1979	163.42	217.4	13.3	389.1	7.2	10.04	12.67
1980	178.42	246.8	12.4	414.9	6.6	11.51	15.27
1981	195.60	272.4	8.9	441.9	6.5	14.03	18.87
1982	207.38	289.1	3.9	480.5	8.7	10.69	14.86
1983	215.34	298.4	3.8	525.4	9.3	8.63	10.79
1984[a]	223.38	311.1	4.0	554.5	5.5	9.58	12.04

[a]preliminary estimate

Source: Economic Report of the President, 1985, Tables B–3, B–52, B–56, B–61, B–66.

Source: Economic Report of the President, 1985, Tables B–3, B–52, B–56, B–61, B–66.

Economics

Economics

(Robert Burton)

Robert B. Ekelund, Jr.
Auburn University

Robert D. Tollison
George Mason University

LITTLE, BROWN and COMPANY
Boston   Toronto

Library of Congress Cataloging in Publication Data

Ekelund, Robert B. (Robert Burton), 1940–
  Economics.

  Includes index.
  1. Economics.   I. Tollison, Robert D.
II. Title.
HB171.5.E47   1985   330   85-10256
ISBN 0-316-23123-1

Library of Congress Catalog Card No. 85-10256

ISBN 0-316-23123-1

9 8 7 6 5 4 3 2 1
MU

Published simultaneously in Canada
by Little, Brown & Company (Canada) Limited

Printed in the United States of America

CREDITS

Photos: Page 11: Sharon A. Bazarian/The Picture Cube; Page 53: Courtesy of the Trustees of the Boston
Public Library; Page 65: Jerry Gordon/Archive; Page 95: *left* and *right,* The Granger Collection, New
York; Page 152: AP/Wide World Photos; Page 204: Historical Pictures Service, Chicago; Page 240: Russell
French; Page 257: Wide World Photos; Page 259: *left,* Historical Pictures Service, Chicago; *right,* Peter
Lofts Photography, Cambridge, England; Page 320: Courtesy of the Public Services Department, Federal
Reserve Bank of Boston; Page 340: *left,* Courtesy of Apple Computer, Inc.; *right,* Courtesy of Atari
Industries; Page 370: *left,* The Granger Collection, New York; *right,* Courtesy of Gary Becker; Page 399:
Courtesy of GTE SPRINT Communications Corp; Page 419: Photo courtesy of Chicago Transit Author-
ity; Page 453: The Granger Collection, New York; Page 459: *left,* Photo by V. H. Mottram; National
Portrait Gallery, London, England; *right,* Courtesy of Ronald Coase; Page 502: Historical Pictures Service,
Chicago; Page 508: *left,* The Granger Collection, New York; *right,* Wide World Photos; Page 509: The
Granger Collection, New York; Page 552: Reprinted by permission of United Feature Syndicate, Inc.;
Page 589: Shoemaker's Shop, Greek 520–510 B.C., B/F amphora. H: .361 m. H. L. Pierce Fund 01.8035,
Courtesy Museum of Fine Arts, Boston; Page 590: *left* and *right,* The Bettmann Archive; Page 612: Neg.
#A53972, American Red Cross; Page 627: Courtesy Shawmut Banks, Advertising Department; Page 632:
Courtesy Irish Tourist Board; Page 635: Courtesy Federal Reserve Bank of Minneapolis, Office of Public
Information; Page 652: Wide World Photos; Page 673: *left,* Courtesy of Milton Friedman; *right,* T. Charles
Erickson, Yale University, Office of Public Information; Page 692: Reprinted by permission, Tribune
Media Services, Inc.; Page 734: The Bettmann Archive; Page 761: *left,* Courtesy of Paul Samuelson; *right,*
Courtesy of Robert Lucas; Page 819: *left,* Courtesy of Bettina Bien Greaves, Foundation For Economic
Education, Irvington-on-Hudson, NY; *right,* The Bettmann Archive; Page 826: Nicholas Daniloff, *U.S.
News and World Report;* Page 837. Kirschenbaum/Stock Boston; Page 841: Wide World Photos. Page 843:
left, The Granger Collection, New York; *right,* The Bettmann Archive.

(continued on page 861)

For My Mother
and in Memory of My Father, Bob, and My Friend, Terry

RBE

For Anna, April, and My Parents

RDT

Preface

In this introduction to economics, we pursue two goals: (1) to build a solid foundation in basic and modern economic theory; and (2) to connect theory with policy and issues in today's world. Our approach stresses the interrelationship of these goals. In each chapter we rely on applications to test the insight of theory, and we rely on theory to provide the context for understanding real-world events. We think this is the best way to prepare students to appreciate the power of economic thinking, to see the vital connection between the logic of economics and the issues confronting them on television, in newspapers, in the voting booth, and in their careers.

To build a solid foundation in theory, we present micro- and macroeconomics step-by-step, illustrating and reinforcing key concepts at every opportunity. We focus patiently on critical concepts such as supply and demand to make sure students are secure in the basics. We also move strongly into current theories such as public choice, aggregate-demand/aggregate-supply analysis, modern monetarism, and rational expectations. Although the range and diversity of topics is wide, we have tried not to be encyclopaedic. Students gain little from exposure to a smattering of topics strung together with little organizing force. We have tried instead to integrate modern perspectives into the book by building upon the fundamentals set forth carefully and consistently throughout.

To connect theory with the world around us, we have woven in hundreds of applications throughout the book, ranging from the effects of the federal deficit on inflation to the effects of competition in the market for Halloween pumpkins. The diversity of these applications demonstrates the versatility of economic reasoning. (We have a bit more to say about applications under "Special Features" below.)

Since our approach stresses the relationship between theory and real-world issues, the technical presentations in this book are concise and straightforward. We rely heavily on examples to illustrate abstract theory. We describe graphs and graphical relationships in patient detail. And we use an abundance of visual aids to motivate and clarify.

In areas where disagreements over economic theory have arisen—and contemporary economics has its share of disputes—we seek to give all views a fair and proper hearing. Although this equal time rule is largely unnecessary in the more settled areas of microeconomics, it is invoked fairly often in the macroeconomic theory and policy sections of the book.

In sum, we feel that the approach of *Economics* reflects our sense that students too often merely receive the facts and theories of economics in an introductory textbook and too often grow uncomfortable with its abstractness. We want our readers to discover economics' intuitive appeal, to share our enthusiasm for a discipline rich in insight.

Special Features

- Many of the applications in *Economics* are set off in "Economics in Action" boxes that appear at the ends of chapters. Here is a sample of some of the topics explored: "The U.S. Farm Problem," "The Economics of Law

Firms," "All That Glitters: The De Beers Diamond Cartel," "Is Urban Mass Transit Worth the Cost?" "Should Congress Be Forced to Balance the Budget?" "International Capital Movements and the Dollar." For a full listing of "Economics in Action" boxes, see the table of contents.

- Within each chapter, we have set off brief, interesting perspectives on theory and institutions in "Focus" boxes. Some of the "Focus" topics include "Corporate Takeovers and 'Greenmail,' " "The Pros and Cons of Advertising," "Money and Computers: Banking at Home," "Inflation, Bracket Creep, and Indexation," "The Soviet Underground Economy." Again, see the table of contents for a complete listing.

- A special historical feature titled "Point-Counterpoint" concludes each Part of the book. Here we offer side-by-side biographies of some of the most important thinkers in economics and compare and contrast their theories. Part VI, "Money: Its Creation and Management," for example, concludes with a "Point-Counterpoint" on Milton Friedman and James Tobin, presenting their different views on the importance of setting targets for the nation's money supply. All "Point-Counterpoint" sections are listed in the table of contents.

Design and Pedagogy

- Hundreds of two-color graphs are integral to our presentation. The graphs are large and easy to read. Captions carefully summarize the major points in each graph.

- Large, flow-chart diagrams and other illustrations and photographs enhance the visual appeal of the book.

- Key terms are printed in the margins and listed at the end of each chapter so that students can review material systematically.

- All of the key terms are also gathered together in a complete glossary at the end of the book.

- Each chapter concludes with a concise chapter "Summary" and a useful selection of "Questions for Review and Discussion."

Organization

Economics is divided into eight Parts. In the general sequence, microeconomics precedes macroeconomics. We believe this is the most effective ordering for introductory courses, since many of the issues in macroeconomic theory have their roots in microeconomics. We realize, of course, that many instructors prefer to teach macroeconomics first. The text is written so that either macro- or microeconomics can be taught first. In other words, a student can move directly from Chapter 4, "Markets and Prices: The Laws of Demand and Supply," to Chapter 21, "Macroeconomics: Contemporary Problems and Issues," without any loss in continuity.

Introduction

Part I of *Economics* is a general introduction to the basic tools of economic analysis, including opportunity cost, marginal analysis, comparative advantage, supply and demand, and market equilibrium. We have packed these chapters with lively, interesting issues and applications. Chapter 2, for instance, looks briefly at the effects of trade barriers between the United States and Japan, and in Chapter 4 we show how usury laws affect the supply and demand for loanable funds.

Microeconomics

Part II begins the formal presentation of microeconomic theory. It covers elasticity, consumer choice, the firm, the firm's costs of production, and output market structures. Although the sequence and coverage of these chapters is relatively standard (and effective), we do include a number of interesting innovations. Chapter 7, "The Firm," for example, introduces students not only to the various firm structures but also to the broader economic questions of why firms exist, what roles they serve, and why they take the forms they do. Chapter 9, "The Competitive Firm and Industry," introduces students to the concept of rivalrous behavior and to the dynamic process of competition as well as to the more static model of "pure" competition. Chapter 11, "Monopolistic Competition, Oligopoly, and Cartels," explores recent theory about cartel behavior, relating this theory to problems now experienced by OPEC.

Part III covers demand and supply in factor markets for labor, land, and capital. Perhaps the most unusual chapter in this section is Chapter 15, "Rents, Profits, and Entrepreneurship," which, among other things, investigates the modern theory of rent seeking behavior and the special role entrepreneurial ability plays in our economy. At the end of this chapter we profile the remarkable careers of two well-known computer entrepreneurs, Stephen Jobs of Apple Computer and Jack Tramiel of Atari. Part III closes with a detailed look at the issue of income distribution, including a careful examination of economic discrimination against women and minorities.

In Part IV we integrate microeconomic theory and public policy. We cover four broad areas: industry concentration, market failure, taxation, and public choice. Chapter 17, "Market Structure and Public Policy," traces the evolution of antitrust policy from the days of the Standard Oil trust to the present era of decontrol. Chapter 18, "Market Failure and Public Policy," applies the modern theory of externalities to such diverse problems as endangered species, international defense alliances, shrinking oil reserves, as well as the familiar smoke-belching factory. Chapter 19, "Taxation," lays the groundwork for understanding the current plans for tax reform in the United States. Chapter 20, "The Theory of Public Choice," shows how economic analysis yields startling new insights into the political decision-making process. For example, students will discover one of the reasons why lobbyists are such a major part of our political machinery.

Macroeconomics

Part V begins with the basic tools of macroeconomics: concepts such as real and nominal GNP, two-flow models of national income, price indexes, and aggregate demand and aggregate supply. In Chapter 23, "From Classical to Keynesian Theory," we start presenting formal macroeconomic models of the economy. Chapter 23 offers a fuller than usual treatment of Classical theory as a prelude both to Keynes and modern theories of aggregate supply. The basic Keynesian approach to private consumption and investment begins in Chapter 23 and concludes in Chapter 24, "Output Fluctuations and the Public Sector." After we have built the Keynesian model for income determination, we conclude Chapter 24 with a pertinent look at recent proposals for a balanced budget and their possible effects under Keynesian theory. Chapter 25, "Aggregate Demand, Aggregate Supply, and Demand Management," introduces the variable price model for income determination. At the conclusion of this chapter, we consider the effects of fiscal policies on aggregate supply, and introduce the idea of "supply-side" economics and the emergence of the rational expectations hypothesis. Chapter 26, "Aggregate Supply" investigates how market conditions for labor and labor expectations help determine supply-side shifts. For instructors who prefer not to cover these concepts in depth, Chapter 25 can stand alone as a complete introduction to aggregate supply/aggregate demand analysis.

Part VI is a complete introduction to money, monetary institutions, monetary policy, and monetary theory. We devote particular attention to the enormous changes occurring within our financial institutions, including the effects of the Monetary Decontrol Act of 1980 and the changing role of commercial banks within the financial system. Unlike most introductory books, we devote a complete chapter, Chapter 29, to the monetary tools of the Federal Reserve. In Chapter 30, we present the elements of the modern quantity theory of money and the monetarist approach to inflation. The "Economics in Action" for Chapter 30 highlights the current debate over monetary policy.

The four chapters of Part VII extend macroeconomic theory into current policy debates. Chapter 31 reviews for students three separate models of the macroeconomy—the Keynesian theory, the monetarist theory, and the rational expectations theory—and shows how each model leads to various fiscal and monetary policies. In Chapter 32 we examine public policy on the problems of inflation and unemployment, and in Chapter 33 we discuss policy in relation to business cycles. Part VII ends by focusing on the issue of long-term economic growth and the possible reasons for the slow-down of U.S. growth rates in the 1960s and 1970s.

International Trade and Economic Development

In Part VIII we conclude the book with three chapters on international trade, international finance, and comparative economic systems and development. Chapter 35 applies the basic theory of comparative advantage to an analysis of tariffs and quotas. The "Economics in Action" for this chapter examines how protectionist sentiments extend even to a small industry, domestic flower growers. Chapter 36 covers many of the most pressing international financial problems, including the recent debt problems of several Third World countries. Finally, Chapter 37 analyzes the relationship between a country's economic system of incentives and its economic growth. We take a close look at the Soviet Union's economy in light of its weak performance. We also examine the various problems of economically underdeveloped countries.

The Complete *Economics* Package

- *Economics*
- *Microeconomics* and *Macroeconomics*, paperback split volumes of the hardcover text. International trade and development chapters are included in both volumes.
- The *Test Bank* to accompany *Economics* includes over 2,000 multiple-choice items. The broad assortment of items test students in recall, inference, calculation, and graph interpretation.
- All test items are available on microcomputer software.
- The student *Study Guide* to accompany *Economics* helps students review and relearn key concepts and provides a broad assortment of exercises and applications to test and reinforce student understanding. Each chapter in the guide is divided into six parts: "Chapter in Perspective," "Learning

Objectives," "Review of Key Concepts," "Helpful Hints," "Self-test," and "Something to Think About."
- The *Instructor's Manual* to accompany *Economics* includes brief introductions to the rationale and scope of each chapter, a complete list of suggested readings for each chapter, and additional teaching resources.
- Two-color overhead transparencies.

Acknowledgments

No book of this scope can be written without a great deal of help and advice. Our book is certainly no exception. A large number of "official" reviewers—listed at the end of this section—were invaluable in improving the quality of our work, and we express our deep gratitude to them. In the category "friendly unofficial critics," we gratefully acknowledge the advice of the following: Richard Ault, Don Ballante, Andy Barnett, Raymond Battalio, Steve Caudill, Charles DeLorme, Roger Garrison, Randy Holcombe, George Horton, John Jackson, Mark Jackson, Charles Maurice, François Melese, Steve Morrell, Richard Saba, David Saurman, and David Whitten. For special help and assistance we wish to thank Gary Anderson, Bob Hebert, Bill Shughart, and Mark Thornton. To our friend Keith Watson, who authored the Study Guide and the Instructor's Manual to this book and who provided expert assistance throughout the entire project, we owe a very special debt. In this same category we wish to thank Greg Tobin, our developmental editor at Little, Brown, whose impact appears on every page of this book. His sound determination to make improvements from concept to the final product is the most valuable kind of assistance that authors could obtain. We also wish to thank Al Hockwalt and Will Ethridge, former and present economics editors at Little, Brown, for their faith and support. We extend our gratitude to development editor Shelley Roth, production editor Sally Stickney, designer George McLean, copyeditor Barbara Flanagan, and editorial assistant Max Cavitch. We also want to thank graduate students Don Boudreaux, Brian Goff, Ladd Jones, Ivan Kelly, Karen Palasek, Kendall Somppi, Deborah Walker, and Biff Woodruff for their aid at various stages of the project. Secretaries Cynthia Spinks and, most especially, Pat Watson, who typed the bulk of the manuscript several times, were very able help. Our official reviewers were:

Richard K. Anderson, Texas A & M University
Ian Bain, University of Wisconsin—Milwaukee
Robert Barry, College of William and Mary
W. Carl Biven, Georgia Institute of Technology
Ronald G. Brandolini, Valencia Community College
Jacquelene M. Browning, Texas A & M University
Bobby N. Corcoran, Middle Tennessee State University
Judith Cox, University of Washington
Larry Daellenbach, University of Wisconsin—La Crosse
Harold W. Elder, University of Alabama
Donald Ellickson, University of Wisconsin—Eau Claire
Keith D. Evans, California State University—Northridge
Susan Feiner, Virginia Commonwealth University
David Gay, Brigham Young University
Kathie Gilbert, Mississippi State University
Otis Gilley, University of Texas—Austin
William R. Hart, Miami University (Oxford)

J. Paul Jewell, Kansas City, Kansas Community College
Ki Hoon Kim, Central Connecticut State University
Patrick M. Lenihan, Eastern Illinois University
Jim McKinsey, Northeastern University
Herbert Milikien, American River College
Norman Obst, Michigan State University
Samuel Parigi, Lamar University
Glenn Perrone, Pace University
John Pisciotta, Baylor University
E. O. Price III, Oklahoma State University
John Price, San Francisco State University
Robert Pulsinelli, Western Kentucky University
Mark Rush, University of Florida
Don Tailby, University of New Mexico
Allan J. Taub, Cleveland State University
Chris Thomas, University of South Florida
Abdul M. Turay, Mississippi State University
Michael Watts, Purdue University
Donald A. Wells, University of Arizona
Walter J. Wessels, North Carolina State University
George Zodrow, Rice University
Armand J. Zottola, Central Connecticut State University

RBE and RDT

Suggested One-Semester and One-Quarter Outlines for *Economics*

Chapters	Micro Emphasis	Macro Emphasis	Contemporary Problems and Policy Emphasis	Combined Micro-Macro Emphasis
1 Economics in Perspective	●	●	●	●
2 Economic Principles	●	●	●	●
3 Markets and the U.S. Economy	●	●	●	●
4 Markets and Prices: The Laws of Demand and Supply	●	●	●	●
5 Elasticity	●			
6 The Logic of Consumer Choice	●		●	●
7 The Firm	■			■
8 Production Principles and Costs to the Firm	●		●	●
9 The Competitive Firm and Industry	●		●	●
10 Monopoly: The Firm as Industry	●		●	●
11 Monopolistic Competition, Oligopoly, and Cartels	■			
12 Marginal Productivity Theory and Wages	●			●
13 Labor Unions	■			
14 Capital and Interest	■			
15 Rents, Profits, and Entrepreneurship	●			●
16 The Distribution of Income	●			●
17 Market Structure and Public Policy	●		●	●
18 Market Failure and Public Policy	●		●	■
19 Taxation	●		■	
20 The Theory of Public Choice	●			
21 Macroeconomics: Contemporary Problems and Issues		●	●	●
22 Measuring the Macroeconomy		●		
23 From Classical to Keynesian Theory		●	●	●
24 Output Fluctuations and the Public Sector		●	●	●
25 Aggregate Demand, Aggregate Supply, and Demand Management		●	●	●
26 Aggregate Supply		●		
27 An Introduction to Money and the Banking System		●	●	●
28 The Creation of Money		●		■
29 The Federal Reserve System		●		■
30 Money Demand, Inflation, and Monetarism		●	■	●
31 From Macroeconomic Theory to Policy: An Overview		●	●	●
32 Inflation and Unemployment		●	●	●
33 Business Cycles		■	■	
34 Economic Growth and Productivity		■		
35 International Trade	●	●		■
36 The International Monetary System		■		
37 Economic Systems and Economic Development		■	●	

● Recommended for both one-semester and one-quarter courses
■ Recommended for one-semester course but optional for one-quarter course

A Note to Students

We have designed special features in *Economics* that make it easier for you to preview, read, and review the contents of each chapter. Become acquainted with these features and you will give yourself a better chance to succeed on quizzes and exams.

Before you read each chapter, take a few minutes to preview its topics. Begin with the brief chapter overview in the first few paragraphs. Then glance over the section headings and subheadings and read over the chapter summary at the end. After reading the chapter through slowly for the first time, check your comprehension by using the review questions and the list of key terms at the end of each chapter. Reading the "Focus" boxes and the "Economics in Action" box is another good way to test your understanding of concepts. (For each chapter, the *Study Guide* to accompany *Economics* includes a section of additional "Helpful Hints.")

During your second reading, be sure to go over each graph and table in detail, reading the caption and rereading the text description slowly. Try to ask yourself questions as you read and review each graph, and try drawing the graph yourself. After reading the chapter a second time, test your comprehension by going over the end-of-chapter questions and working through the problems there. (The *Study Guide* contains additional short-answer problems.)

Before each quiz, use the key-term definitions printed in the margins to help you review chapter concepts. Also reread the chapter summary. (The *Study Guide* includes a sample selection of multiple-choice test items for each chapter.) With all of these tools at hand, you should find the study of economics a bit less of a chore!

Brief Contents

Contents

V Private-Sector and Public-Sector Macroeconomics   461

VI Money: Its Creation and Management   593

Economics

I

The Power
of Economic
Thinking

1

Economics in Perspective

D oes economics matter? More to the point, why should you spend precious time and money learning economics when there are so many other activities, products, and services—not to mention other college courses—competing for your attention? The answer: Economics touches all facets of our lives as consumers of hamburgers and home computers, as voters for political candidates, and as workers and employers. Economics analyzes why we are poor or rich as individuals and extends its scope to government policies about inflation, unemployment, economic growth, and international trade. Close study of economics thus gives an entirely new perspective on a wide variety of human activities and institutions.

What Economics Is (and What It Isn't)

Most people would say that economics deals with the stock market or with how to make money by buying and selling gold, land, or some other commodity. This common view contains a grain of truth but does not touch on the richness, depth, and breadth of the matter. Economics is a social science—the oldest and best developed of the social sciences. As such, it studies human behavior in relation to three basic questions: What **goods** and **services** are produced? How are goods and services produced? For whom are goods and services produced?

Goods: All tangible things that humans desire.

Services: All forms of work done for others—such as medical care and car washing—that do not result in production of tangible goods.

All societies and all individuals have faced these three questions. Since goods and the wherewithal to produce them have never existed in limitless amounts, the insistent questions—How? What? and For whom?—must be asked; for at least two hundred years, economists have tried to analyze how individuals and societies answer them. Consider some famous economists' definitions of economics:

Adam Smith (1776): Economics or political economy is "an inquiry into the nature and causes of the wealth of nations."[1]

Nassau William Senior (1836): Political economy is the science that treats the nature, the production, and the distribution of wealth.[2]

Karl Marx (1848): Economics is the science of production. Production is a social force insofar as it channels human activity into useful ends.[3]

Alfred Marshall (1890): "Political Economy or Economics is a study of mankind in the ordinary business of life . . . it is on the one side a study of wealth; and on the other, and more important side, a part of the study of man."[4]

Lionel Robbins (1935): "Economics is the science which studies human behavior as a relationship between ends and some means which have alternative uses."[5]

Ludwig von Mises (1949): Economics "is a science of the means to be applied for the attainment of ends chosen, not . . . a science of the choosing of ends."[6]

Jacob Viner (1958): "Economics is what economists do."[7]

Milton Friedman (1962): "Economics is the science of how a particular society solves its economic problems."[8]

Economics: A Working Definition

There is merit in each of the preceding definitions; economists like them all. Economics *is* the study of how nations produce and increase wealth. Economics also studies the activities of people in producing, distributing, and consuming wealth. It analyzes how people and particular societies choose among competing goals or alternatives, but not how or why the goals are chosen. All this is "what economists do."

A common thread runs through all definitions of economics. However narrow or broad, each definition emphasizes the inescapable fact that **resources**—the wherewithal to produce goods and services—are not available in limitless quantities and that people and societies, with unlimited desires for goods and services, must make some hard choices about what to do with the resources that are available. Our working definition of economics includes these elements and may be expressed as follows:

> Economics is the study of how individuals and societies, experiencing limitless wants, choose to allocate scarce resources to satisfy their wants.

The Economic Condition: Scarcity

What, exactly, is **scarcity?** More to the point, what are scarce resources? Dorothy Parker, an American humorist, once said, "If you can't get what

Resources: Inputs necessary to supply goods and services. Such inputs include land, minerals, machines, energy, and human labor and ingenuity (called the factors of production).

---

[1]Adam Smith, *An Inquiry into the Nature and Causes of the Wealth of Nations,* ed. Edwin Cannan (1776; reprint, New York: Modern Library, 1937).

[2]N. W. Senior, *An Outline of the Science of Political Economy* (1836; reprint, New York: A. M. Kelley, 1938).

[3]Karl Marx, *Capital,* trans. Ernest Untermann, ed. F. Engels, 3 vols. (Chicago: Charles Kerr, 1906–1909).

[4]Alfred Marshall, *Principles of Economics* (London: Macmillan, 1920), p. 1.

[5]Lionel Robbins, *The Nature and Significance of Economic Science,* 2nd ed. (London: Macmillan, 1935), p. 16.

[6]Ludwig von Mises, *Human Action* (New York: Regnery, 1966), p. 18.

[7]Jacob Viner in Kenneth Boulding, *The Skills of the Economist* (Cleveland: Howard Allen, 1958), p. 1.

[8]Milton Friedman, *Price Theory* (Chicago: Aldine, 1962), p. 1.

Scarcity: Limitation of the amount of resources available to individuals and societies relative to their desires for the products that resources produce.

you want, you'd better damn well settle for what you can get." The entire study of economics amplifies and expands on Parker's proposition. As individuals and as a society, we cannot get all of what we want because the amount of available resources is limited. Our ancestor *homo erectus* faced this problem, as do the bush people of the African Kalahari desert today; King Louis XIV of France and Howard Hughes could not get all of what they wanted; Americans and the people of underdeveloped Chad face limits. The role of the economist and of economics in general is to explain how we can make the most of this problem of scarcity—in Parker's terms, how to get as much as we can of what we want.

The most important problem in economics is that while the wants of individuals and societies must be satisfied by limited resources, the wants themselves are not limited; rather, they are endless. We are never satisfied with what we have. Individuals are forever lured by more tempting foods, more cleverly engineered computers, more up-to-date fashions. Societies continually desire safer highways, more accurate missiles, greater Social Security benefits, or more cancer research.

Scarcity is of course relative to time and fortune. Our generation has many more services and goods to choose from than our parents and grandparents did. The quantity and quality of goods and services may have grown from primitive to modern times, but the supply of resources needed to produce them is limited, and human wants are not.

Free goods: Things that are available in sufficient quantity to fill all desires.

You may feel that some things are not scarce and that the best things in life—such as love, sunshine, and water—are free. In economic terms, **free goods** are goods that are available in sufficient supply to satisfy all possible demands. But are many things truly free? Surface water is usually unfit for drinking except in areas far from human habitation. Water suitable for drinking must be raised to the surface from deep wells or piped from reservoirs and treatment plants, operations involving resources that are scarce even when water itself is not. Scarcity of winter sun in the North results in costly winter vacations in warm states. And if you think love is free . . . .

Economic goods: Scarce goods.

Costs: The value of opportunities forgone in making choices among scarce goods.

All scarce goods—from television to chlorinated water—are called **economic goods.** Their scarcity leads to **costs.** While it is customary to associate cost with the money price of goods, economists define *cost* as the value of what individuals have to forgo to acquire a scarce good. Since all unlimited wants cannot be met with scarce resources, individuals have to make choices—between, for instance, more steaks and more computer games; societies may have to choose between safer highways and more accurate missiles. Cost is therefore the direct result of scarcity of resources. Scarcity of resources means that both individuals and societies must endure the costs of acquiring more of any good or service. That cost is the value of the good given up in place of the good chosen.

Scarce Resources and Economic Problems

Human resources: All forms of labor used to produce goods and services.

Nonhuman resources: Inputs other than human labor involved in producing goods and services.

There are basically two categories of scarce resources: human resources and nonhuman resources (see Table 1–1). **Human resources** encompass all types of labor, including specialized forms of labor such as management or entrepreneurship. **Nonhuman resources** include the rest of the bounty: land, natural resources such as minerals and water, capital, and still other resources such as technology and time.

Examples of human resources abound. By definition, all human resources apply talent and energy to produce goods and services. The cook at the Chicken Shack, the hair stylist at the Mad Hacker, the chief executive of a

TABLE 1–1   Economic Resources

Economic resources include all human and nonhuman resources that are scarce in supply. Technology and time are scarce and can be categorized as resources.

Human Resources	Nonhuman and Other Resources
Labor, including entrepreneurship and management	Land
	Natural resources, including minerals and water
	Capital
	Technology
	Time

computer firm, and the assembly-line worker at a General Motors plant all represent human resources. Obviously, labor includes a huge variety of skills, both general and precise. Economists are interested not only in the scarcity of labor but in its equally scarce quality. The quality of human resources can be enhanced through investments in education and training.

Economists view *entrepreneurship* as a special form of labor. An entrepreneur is a person who perceives profitable opportunities and who combines resources to produce salable goods or services. Entrepreneurs attempt to move resources from lower- to higher-valued uses in the economy and take the risk that, by so doing, they can make profits. Lemonade-stand entrepreneurs, for example, see an opportunity to make a profit by combining lemons, ice, and cups and by selling the final product. *Management*, a second special form of labor, guides and oversees the process by which separate resources are turned into goods or services. The successful lemonade entrepreneur could hire a manager to oversee the opening of new stands around the neighborhood.

Human resources utilize nonhuman resources such as land, minerals, and natural resources to produce goods and services. A plot in Manhattan, an acre in Iowa, a coal deposit in Pennsylvania, a uranium mine in South Dakota, and a timber stand in Oregon are all scarce nonhuman resources. New deposits of minerals can be discovered, forests can be replanted, and agricultural land can be reclaimed from swamps. But at any one time, the available supply of nonhuman resources is limited.

Capital, a second category of nonhuman resources, comprises all machines, implements, and buildings used to produce goods and services either directly or indirectly. A surgeon's scalpel, a factory, an electric generator, and an artist's brush are all used to produce goods and services and thus are considered capital.

Many different forms of capital may be needed to produce a single economic good. With a wheat harvesting machine, a South Dakota farmer can reap a huge crop. But the wheat must also be milled into flour and transported from South Dakota to bakeries in California. Once the wheat has arrived, bakeries must utilize brick or convection ovens to produce bread. The harvesting and milling machines, the railroad, and the baker's ovens are all capital goods, created to increase the amount of final production.

Capital—and the resources used to produce it—is scarce. To create capital, we must sacrifice consumer goods and services because the production of capital takes time away from the production of goods that can be consumed in the present. Societies and individuals must therefore choose between immediate consumption and future consumption. That choice is crucial to growth and ultimate economic well-being. We return to this important issue in Chapter 2.

Technology and *time* can also be regarded as resources. Technology, in general, is a resource composed of all "know-how," inventions, and innovations that help us get more from scarce resources. Finer distinctions can be made. Technology is knowledge of production methods. An improvement in technology implies that we produce more with a certain amount of inputs. Existing technology is the outcome of many inventions, some of which were the invention of new resources—such as aluminum, radium, hybrid plants. All inventions that increase the productivity of labor and capital can be thought of as improvements in technology. Innovation is the application of technology to the production of goods and services.

Information is a scarce and costly resource. The acquisition of information on which to base purchasing or managerial decisions has never been free. In the nineteenth century and well into the twentieth, businesses hired armies of bookkeepers to provide sufficient information for managers to use to make decisions. The digital computer, developed in the mid-twentieth century, made information storage and retrieval far less costly. In fact, technology has progressed so rapidly that the quantity of information that could be stored in a warehouse-sized computer in 1950 can now be placed on a chip the size of a fingernail with a lot of room to spare! Technology, then, is a resource that helps make other resources less scarce.

Time is another scarce resource because while its quantity is fixed, the things we could do with it proliferate. We must therefore constantly make choices. We may choose to spend an evening listening to a New Wave rock band instead of studying accounting or hearing a Bach organ recital. We may choose a career in acting or dance instead of in law or computer programming. Retirees may choose to return to the classroom rather than spend time fishing or playing bridge. Like all human and nonhuman resources, time is scarce and is a cost that we inevitably incur when we make a choice.

Scarcity of resources is at the core of all economic problems. Resources can be augmented over time; indeed, we are much better off materially than our grandparents, and our grandparents were better off than their grandparents. At any one time, however, individuals and societies cannot get all of what they want. Given scarcity, individuals and societies must make choices, and a primary role of economists is to analyze scarcity and the process of choosing.

The Power of Economic Thinking

All economists of all political stripes have common fundamental ways of thinking. These perspectives, at the core of the science of economics, appear many times throughout this book. When properly and consistently used, they brand a person as adept in the economic way of thinking. A look at these economic perspectives in simple, commonsense language should convince you that economics and economic reasoning are closely related to decisions you make every day.

Resources Cost More Than You Think

What does it cost you to take a skiing weekend in the mountains or to make a trip to the beach during spring break? Your instant reply might include the costs of gasoline, auto depreciation, air fare, lift tickets, food, drink, enter-

Accounting costs: Direct costs of an activity measured in dollar terms; out-of-pocket costs.

tainment, and a motel room. These money expenditures are called **accounting costs.**

Economists do not look upon costs in the traditional manner of the accountant. Rather, economists look at what must be given up—in the case of the ski trip, what must be given up to make the trip. Economists ask what opportunities must be forgone when we choose anything, and they call the forgone opportunities **opportunity costs.**

Opportunity costs: The value placed on opportunities forgone in choosing to produce or consume scarce goods.

Mary makes a beach trip for three days. She is a college student and has a part-time job at a local restaurant earning $30 per night. Mary could have worked and earned $90 rather than take the trip. Economists would say $90 is one of the opportunity costs of making the trip.

Such opportunity costs also exist for society. Use of government lands in Wyoming and Montana as recreation areas or national forests entails a cost. Through lease or purchase, these lands could be used as a source of oil, minerals, and timber. Such use would contribute to society's well-being, but park land serves the recreational needs of society as well. Whatever choice society makes for the use of the land will incur a cost in economic terms. An opportunity must be forgone when the land is used in either manner.

A favorite saying of economists is "There is no such thing as a free lunch." The first fundamental principle of economics is that most things in life come at the opportunity cost of something forgone. They are never free.

Economic Behavior Is Rational

Rational self-interest: The view of human behavior espoused by economists. People will act to maximize the difference between benefits and costs as determined by their circumstances and their personal preferences.

The second fundamental principle of economics is that people behave according to **rational self-interest.** Economists selected from a number of alternative views of human behavior and settled on a very simple one—that of *homo economicus* (economic man or woman). Rather than viewing humans as inconsistent, incompetent, selfish, or altruistic, economists argue that human behavior is predictably based on a person's weighing the costs and benefits of decisions. This choice making is influenced by constraints such as the availability of options, personal tastes, values, and social philosophy. A student will choose to eat lunch at the local health food restaurant rather than the fast-food cafeteria on campus if the personal benefits of doing so, say eating nutritious food in a pleasant atmosphere, outweigh the costs, such as longer lines and greater distance to be walked.

When costs or benefits change, behavior may change. Consider Jack, a poor but honest man. When he finds $100 cash in a phone booth, he turns it in to authorities. Had Jack found $1000, he might have acted similarly, but discovery of $100,000 might have caused Jack to pause. It may be more costly to be honest under some circumstances.

The view of the individual espoused by economists has always been subject to misinterpretation. When economists say that humans are self-interested, they do not mean that other views of human behavior and motivation are unimportant or irrelevant. Altruistic or charitable behavior, for example, is perfectly compatible with *homo economicus.* If a person values altruism, then an act that benefits others will carry emotional benefits and will therefore be in that person's rational self-interest. Economists simply maintain that people calculate the costs and benefits of their decisions in acting in their own self-interest.

If the IRS were to quadruple the income tax deduction for charitable contributions, economists would predict an increase in charity. No economist would deny that love, charity, and justice are important aspects of human behavior. Indeed, an economist has argued that "that scarce resource Love" is in fact "the most precious thing in the world" and that humans are therefore compelled to conserve it in ordering the affairs of the world.[9] Economists have merely advanced the simple but powerful proposition that, given personal tastes, values, and social philosophy, rational self-interest is a better guide to predicting behavior than any other assumption about why people act as they do.

The economists' view of the rational self-interested individual applies not only to what is usually termed "economic behavior" but also to other realms of behavior. Animals other than humans, such as birds or rats, can be shown to be self-interested in that they act to maximize their own well-being given their constraints (see Focus, "Grubbing the Rational Way"). Politicians maximize their self-interest by wooing voters to elect and reelect them. Since politicians face periodic elections, political behavior such as garnering support from particular voters can be predicted using the economic self-interest assumption. Even matters such as dating, marriage, and divorce have been subjected to analysis using the self-interest assumption! After more than two hundred years of economic theorizing, rational self-interest remains the most powerful predictor of most behavior.

Choices Are Made at the Margin

Margin: The difference in costs or benefits between the existing situation and a proposed change.

Economists do not ordinarily analyze all-or-nothing decisions but are concerned with decisions made at the **margin**—the additional costs or benefits of a specific change in the current situation. An individual consumer, for example, does not decide to spend his or her entire budget solely on food, cassette tapes, or weekends at the beach. Consumers purchase hundreds of goods and services. Their choice to purchase or not purchase additional units of any one good is based on the additional (or *marginal*) satisfaction that that single unit would bring to them.

Marginal analysis: Study of the difference in costs and benefits between the status quo and the production or consumption of an additional unit of a specific good or service. This, not the average cost of all goods produced or consumed, is the actual basis for rational economic choices.

Marginal analysis is a method of finding the optimal, or most desirable, level of any activity—how much coffee to drink, how much bread to produce, how many store detectives to hire, and so on. In an economic sense, every activity we undertake involves *both* benefits and costs, so an optimal level of an activity is the point at which the activity's benefits outweigh its costs by the greatest amount. For example, the optimal level of coffee drinking is reached when coffee's total benefits (its taste and stimulating effects, perhaps) exceed its costs (the expense, the health risks, the acid indigestion) by the greatest degree.

You might find it easy to decide how much coffee to drink on a particular evening, but what of other, more complicated decisions? How, for example, would a baker decide how much bread to produce? He or she might try to add up the total costs and benefits of baking each loaf, but this could be a costly if not impossible task. Through marginal analysis, the baker would look at his activity in small steps. If the net benefit (total benefit − total costs) increases every time a loaf goes in the oven, then the baking will continue. If the net benefit decreases, then the baking will decrease or stop as well.

Or suppose you are the manager of a large department store and want to reduce the amount of shoplifting that takes place. By hiring additional store

[9]D. M. Robertson, "What Does the Economist Economize?" in *Economic Commentaries* (London: Staples, 1956), p. 154.

FOCUS   **Grubbing the Rational Way**

The economists' view of rational behavior—that humans react, predictably, to their perceptions of costs and benefits—has been recognized in many forms for some time. Rational behavior under scarcity was certainly recognized by Charles Darwin and other early natural scientists. Is the economists' view time-bound and applicable only to twentieth-century industrialized countries and only to humans, as some critics argue, or does it apply in all times and cultures and among all other species as well?

The coal tit is a common forest-dwelling English bird that resembles the American chickadee. This small bird evidently hasn't paid any attention to claims that economic behavior is not predictable. Surprisingly, the behavior of the coal tit provides an excellent example of economic principles in operation. This bird responds "rationally" to "price" changes and even searches for the "cheapest market" to shop in!

The coal tit survives through the winter by eating the larvae of the eucosmid moth, which it finds lying dormant in small cavities just under the surface of pine cones. The bird finds the moths by tapping on the outside of the pine cones. Naturalists can distinguish in the spring between moth larvae that were eaten by the coal tit, were destroyed by some other cause, or survived to become moths. An ecologist and naturalist named Gibb determined the number of moth larvae in pine cones in different areas of the forest in the fall. The next spring, Gibb returned to see how many had fallen prey to the coal tit over the winter.

Using the data collected by the naturalist, economist Gordon Tullock[a] derived the graph shown in Figure 1–1. (The appendix to this chapter contains a general explanation of graphs.) The horizontal axis represents the number of larvae in different localities. The vertical axis signifies the number of larvae per 100 pine cones in each locality. Curve A plots the number of larvae found in the fall, before coal tit predation began; Curve B shows the number of larvae that survived predation, counted in the following spring. Hence the space between curve A and curve B repre-

[a]Gordon Tullock, "The Coal Tit as a Careful Shopper," *American Naturalist* (November 1969), pp. 77–80.

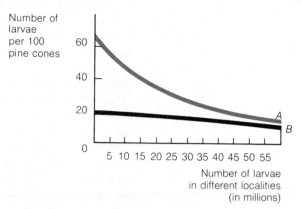

FIGURE 1–1   **Larvae Density Before and After the Coal Tits' Predation**

sents the number of larvae consumed during the winter by the birds in different localities.

An interesting relation is immediately apparent: The coal tits have obviously concentrated their predation in localities where the density of larvae is over 60 per 100 cones. As density of available larvae decreases, so does predation. Tullock explains that the amount of energy spent in seeking out each larva is the "price" the bird must pay to consume the larva. Where the density of larvae is high, this price is low, and vice versa. In other words, the coal tit consumes more larvae when the price is low and less larvae when the price is high.

But this is not all. The fact that the coal tits systematically seek out those areas of the forest with higher density of larvae and avoid the low-density areas indicates that the birds are behaving much like we would expect a careful shopper to behave. They are searching for the cheapest market and then shopping there.

This apparently rational behavior of the coal tit does not imply that the tiny bird engages in conscious economic calculation. It is more likely that searching for the higher-density areas is an instinctive response that developed in the birds as the result of natural selection. Its value in increasing the coal tits' chances of survival over a long, cold winter when food is scarce is obvious. But the economic efficiency of the behavior remains striking regardless of the coal tits' inspiration.

detectives, you will benefit from fewer thefts. But the detectives must be paid, so the net benefit is the amount you save minus the costs of the detectives. If store detectives cost $200 per week, you can perform an experiment. Hire an additional detective and see whether losses from shoplifting exceed $200 per week. If additional losses are more than $200, hire additional detectives, if less, let detectives go. Stop hiring when losses per week are exactly equal to the weekly wage of detectives. *(Note that the optimal level of shoplifting you will permit is not zero!)*

For another example, suppose a shipbuilding company that produces four ships at a time consistently receives more orders than it can fill. The land on which the company is located allows room for expansion. In deciding whether to enlarge the operation, the rational manager must weigh the marginal costs and benefits of producing more ships. Will it be profitable to produce one more ship? That is, will the net benefits increase? To accommodate work on five ships at one time, the company would have to enlarge its boat yard, construct storage space for more shipbuilding materials, and pay its current workers overtime or train new workers. Will the extra costs of building one more ship be warranted, given the expected benefit, the current selling price of a ship? Can the company better maximize profits by building ten ships at a time? By making no change unless ship prices rise? Rational self-interest requires isolating and considering the specific effects of specific changes—of choices made at the margin, in the area of change. (Economics in Action at the end of this chapter provides further illustrations of marginal analysis.)

Prices Are the Signals to Produce More

Prices: The opportunity costs established in markets for scarce goods, services, and resources.

What products and services are produced in our economic system and how are they produced? In a market economy **prices** are the essential signals that tell producers and resource suppliers what and how to produce. Take a common product such as apple jelly. Apple jelly is not free at the supermarket. Since it bears a price, you can be sure that scarce resources—such as apples, sugar, and jars—are being used to produce it and that someone finds it profitable to do so.

Buyers and sellers in action at a commodity exchange

What happens if consumers suddenly develop a craving for apple jelly, preferring it to all other kinds? Consumers express this craving by buying as much apple jelly as they can, leaving such products as grape jam and apricot preserves gathering dust on the shelves. Meanwhile, the relative unavailability of apple jelly causes the price to rise because there are insufficient resources devoted to apple jelly production to meet the new demand for it. Manufacturers, noting the increased sales and rising prices, recognize that additional production of apple jelly will be profitable. Their reaction is to produce additional jelly by ordering more jars, buying more apples from apple farmers, and hiring more labor and machinery. Again, higher prices of apples, jars, and other resources are the signal to the apple farmers and the other resource suppliers to sell more of their inputs and services. Why do they do it? Because higher prices mean greater profitability to them.

All goods, including apple jelly, are produced in this manner. Consumers are sovereign in that they decide what is to be produced—TVs, automobiles, beets—with the scarce resources in our society. Economists say that consumers transmit their desires through **markets,** which are simply arrangements whereby buyers and sellers exchange goods or services for money or some other medium. Buyers and sellers can be physically together in a market, as in a livestock market, or they can be separated by geographic distance or by wholesalers and other intermediaries, as when the buying and selling of a company's stock is carried out through stockbrokers. A market may even be an abstraction, as is the labor market, in which the supply of and demand for jobs and workers are juggled on a broad scale. The prices established in these markets (through mechanics to be explored in Chapter 4) are the key to what gets produced and how much is produced in our economic system. Economists therefore give a great deal of attention to the role and function of prices and markets in explaining how things get done.

Market: An arrangement that brings together buyers and sellers of products and resources.

Economists Won't Say Who Should Get What

Economists study poverty and wealth but they will not, as economists, say how rich people should be or how much poverty is tolerable. Economists do not decide whether a college football coach deserves $16 million on a five-year contract or whether a Nobel Prize–winning physicist deserves more. Economists simply explain that the value of all individuals—expressed as their income—is determined by the relative desires for their services on the one hand and by the relative scarcity of their talents and abilities on the other. Perhaps good football coaches are relatively more scarce and more highly desired than good physicists.

Most products and services—such as food, vacations, and domestic help—are allocated to those with the greatest desire and ability to pay for them. Our ability to pay is determined by the value of our own particular services or resources in the marketplace. This value depends on our education, on-the-job training, health, luck, inheritance, and a host of other factors.

The distribution of "public" goods and services—such as highways and bridges, MX missiles, and "free" movies on campus—is conducted by national, local, and student governments through the filter of politics and voting. A decision whether to add seating capacity to the football stadium or to double the size of the English department at a public university may be made by a university committee appointed by the president and board of trustees, who may be appointed by the governor, who is elected by the citizens of the state. Ultimately, therefore, decisions on how resources are to be distributed are made by voter choices. If voters do not like the decisions made in the political process, they have a periodic right to change the decision makers. Chosen by voters, politicians make economic decisions based on their perceptions of costs and benefits to the public—as well as to their own chances of reelection. In government as in the workings of the market, the chief role of economists is not to say who should get what but to describe and analyze the process—the costs, benefits, and incentives—through which the distribution of all products and services takes place.

We All Need Money, But Not Too Much

Barter: Direct exchange of goods or services.

Economists also place a special significance on the role of money in our lives. The trade of goods for goods, called **barter,** will not work in any economy that completes several million transactions every day. Barter may have been sufficient in primitive societies in which few things were produced and traded, but as individuals became specialized, using more and more of their talents and skills, the number and kind of traded commodities and services grew. As the number of commodities grew, the cost of transacting by barter grew enormously. A common denominator, acceptable to all, developed within economic societies to reduce the costs and inconveniences of barter. That common denominator is **money,** which serves as a medium of exchange.

Money: A generally accepted medium of exchange.

Money is not limited to coins and paper dollars. Almost anything—shells, feathers, paper, gold, and cattle, for instance—can serve as money. The important point to economists is that money, whatever it is, reduces the costs of barter and increases specialization and trade. At the same time, economists are concerned that money should not be available in such quantities as to become unacceptable and valueless to people who produce, buy, and sell. For society, at least, huge increases in the money supply may be too much of a good thing.

Inflation: A sustained increase in prices; a reduction in the purchasing power of money.

When the quantity of money increases beyond its use as a means of making transactions, confidence in the medium of exchange and its value (in terms of the quantity of goods and services a unit of it can purchase) deteriorates. Economists call this phenomenon **inflation.** Inflation was an especially sad fact of American life during the 1970s, and it is a persistent element in less-developed nations around the world. Inflation, if left untreated, can result in sheer collapse of entire economies with reversion to primitive barter conditions. Economists therefore are very concerned with the relation between the production and trade of goods and the quantity of money available to facilitate these crucial activities. Breakdowns must be prevented, and economists have valuable insights on preventive measures.

Voters Choose the Role of Government—
The Economist Only Criticizes

In a functioning democratic system, society at large chooses an economic role for government by electing politicians. In general, that role has included such activities as taxation for and provision of collective goods such as national defense, highways, and education; the regulation of monopoly; pollution control; control of the money supply; and alleviation of poverty through welfare programs. In its economic role of taxing and spending, moreover, government can affect economic factors such as inflation, the degree of unemployment, economic growth, and international trade.

Economists have always been concerned with the effects of government activity in these areas. In the United States, economists are in the thick of government. Since 1946 an official Council of Economic Advisers has been appointed by the president to aid in the formulation and implementation of economic policy. Almost every agency of government employs teams of economists to give advice in their areas of expertise.

In their role as advisers, economists primarily evaluate, from the perspective of economic theory or analysis, the effects of proposals or, more correctly, of marginal changes in economic policies. Economists do not argue, for example, that additional funds should be allocated to Social Security and taken away from water reclamation projects. They simply evaluate the costs and benefits of such proposals.

Economists are not confined to giving advice on specific issues such as the Social Security program or the effects of advertising regulation by the Federal Trade Commission. Larger issues are within economists' purview as well. Predicting the effects of government taxing and spending policies on employment, interest rates, and inflation is a very large part of economists' role, one that receives our attention in almost half of this book. Economists might predict, for example, the impact of a tax cut on private spending, interest rates, the private production of goods and services, and inflation. As such, economists are concerned with what is termed *economic stabilization*.

Despite economists' influence, however, the economic role of government is ultimately decided by voters in periodic elections. Specific economic functions are not decided in elections, but politicians run on platforms pledging to "increase social welfare spending," "increase defense spending," "reduce the debt," "balance the budget," "increase environmental regulation," and so on. In a loose sense, then, voters decide economic policies. Economists' role, although critical, is simply to criticize and evaluate the implications of government's economic policies. Economists advise and implement. They do not make ultimate decisions about what the government should do.

The Role of Theory in Economics

Model: An abstraction from real-world phenomena that approximates reality and makes it easier to deal with; a theory.

Economists must be able to discern fundamental regularities about human behavior. Only if behavior is regular and consistent, on average, among individuals or from generation to generation will economists be able to predict the results of behavior. The means through which economists organize their thoughts about human behavior and its results are called **models.** Economists, like all scientists, construct models to isolate specific phenomena for study. Constructing models requires assumptions and abstractions from the real world.

The Need for Abstraction

The economist, the biochemist, or the physicist could never handle all the details surrounding any event. Thousands of details are involved in even the seemingly simple act of buying a loaf of bread. The friendliness of the clerk, the way the buyer is dressed, and the freshness of the bread are all details surrounding the purchase, and these details may be different each time a buyer makes a purchase. Likewise many details surround a particular chemical reaction or an astronomical event like the appearance of a comet. Economists, however, like chemists or astronomers, are interested in particular details about an event, such as how much the bread cost or how many loaves the buyer purchased.

Economists must *abstract* from extraneous factors to isolate and understand some other factor because, as one economist put it, people's minds are limited and nature's riddles are complex. Economists must assume that the extraneous factors are constant—that "other things are equal"[10]—or that, if altered, they would have no effect on the relations or model under consideration. Humanity has never progressed very far in understanding anything—be it chemistry, astronomy, or economics—without abstracting from the multiplicity of irrelevant facts. *Thus, all economic models are of necessity abstractions. Good models, those that perform and predict well, use relevant factors; poor models do not.*

An economic model (or, essentially the same thing, a theory) can be expressed in verbal, graphical, or mathematical form. Sometimes verbal explanation is sufficient, but graphs or algebra often serve as convenient shorthand means of expressing models. Economics is like other sciences in this regard. A complicated chemical or physical process can be described in words, but it is much more convenient to use mathematics. All methods of expressing models will be used in this book, but we rely primarily on words and graphs. The appendix to this chapter explains how to read and interpret graphs.

The Usefulness and Limitations of Theory

In recognizing the regularities of human behavior and in committing them to a model, economists or scientists do not have to rethink every event as it occurs. Theory saves time and permits extremely useful predictions about future events based upon regularities of human behavior.

Economic theory has its limits, though. Since a complex world is the economist's workshop, economic theory cannot be tested in the way chemical or physical theories can. Economists can seldom find naturally existing conditions that provide a good test of a model, let alone a sequence of such conditions, and it is difficult to conduct controlled laboratory experiments on human beings to determine their economic behavior. Because of the difficulties of testing, economists are further from certainty than, say, geneticists or biologists. The accuracy of economics is somewhat more like that of meteorology. Total accuracy is not within the meteorologist's abilities. At least, total accuracy is not worth what it would cost to obtain in either weather forecasting or economics. The very bizarre weather of 1983—including a great drought in Australia, coastal storms in California, and floods along the lower Mississippi River—were caused by a complicated host of factors related to the jet steam, volcanic eruptions, and equatorial currents. It took meteorologists a good deal of time to sort out the relevant causes by testing and applying their models.

In economics, the problem of imperfect conditions for testing is a limitation to theory, but theory is essential nonetheless. Nothing can substitute for the usefulness of theory in organizing our thoughts about the real world and in describing the regularities of economic behavior.

In constructing models, moreover, economists and all other scientists must be careful to follow the rules of logic and to avoid common fallacies. Two of these are the **fallacy of composition** and the *post hoc* **fallacy.** The fallacy of composition involves generalizing based only on a particular experience. Suppose that you had never seen a fox, and then the first one you saw was white. Should you conclude that all foxes are white? If you did, you would

Fallacy of composition: Generalization that what is true for a part is also true for the whole.

---

[10]The Latin phrase *ceteris paribus*, meaning "other things being equal," is common economic shorthand for an assumption about human behavior.

Post hoc fallacy: From *post hoc, ergo propter hoc*, "after this, therefore because of this." The inaccurate linking of unrelated events as causes and effects.

commit the fallacy of composition. The assumption that if everyone stands up at a crowded football game each will get a better view founders on the same fallacy.

The *post hoc* fallacy (from *post hoc, ergo propter hoc*, "after this, therefore because of this") is the false assumption of cause-and-effect relations between events. The ancient Aztec Indians of Mexico believed that the spring ritual of sacrificing children and virgins appeased the gods of agriculture, bringing a bountiful harvest. When crops appeared in the fall, high priests insisted that the good crop was caused by the sacrifice. If the crop was insufficient, more children and virgins had to be wary the following spring. In economic analysis, the *post hoc* fallacy may be involved in statements such as "high interest rates cause inflation" or "after hoarding sugar, the price of sugar rises" or "the government subsidy to Chrysler Motor Company saved the business." Economists and all who are interested in sound thinking must constantly guard against such fallacies.

Positive and Normative Economics

Positive economics: Observations or predictions of the facts of economic life.

Normative economics: Value judgments about how economics should operate, based on certain moral principles or preferences.

Economics is usually a positive rather than a normative science. Positive statements describe what is or predict what will be under certain circumstances, whereas normative prescriptions entail value judgments about what should be. For the most part, economists confine themselves to positive statements. Individual values, concepts of justice, or tastes are normative matters and generally do not enter economists' analyses of issues. These matters are not testable in any accepted scientific sense and are therefore excluded from economics.

Consider U.S. welfare programs. To say that additional tax dollars should be spent on welfare would involve a normative judgment on the part of economists. Once society has decided to spend more on welfare, however, economists can make positive statements about the effects of such spending. Or economists may observe the welfare system as it stands and analyze how the system could work more efficiently in achieving its goals. In other words, economists must show, once society has decided on a welfare system and on some dollar amount to be devoted to welfare, how alternative systems of welfare distribution would alter the effectiveness of the program. Economists make positive statements in presenting alternatives and their effects so that society might get the most for the dollars spent.

On a smaller scale, imagine a student facing an afternoon choice between watching soap operas and studying for an economics quiz. In positive terms, economists can only describe the alternatives open to the student. An observer would be on normative grounds if he or she insisted that the student should avoid TV all afternoon and spend the time preparing for the quiz. Free choice is, in this as in all cases, the prerogative of the economic actor. Economists can only array the alternatives. On occasion, normative statements slip into economic discussion, as they undoubtedly do in this book. Beware of them and learn to distinguish them from positive economics.

Microeconomics and Macroeconomics

Microeconomics: Analysis of the behavior of individual decision-making units within an economic system, from specific households to specific business firms.

The first half of this book concerns what economists call microeconomics, and the second part deals with macroeconomics. **Microeconomics,** like microbiology, concerns the components of a system. Just as a frog is made up of individual cells of various kinds, individual sales and purchases of commodities from potatoes to health care are the stuff of which the whole economy is composed. Microeconomists are thus concerned with individual markets and with the determination of relative prices within those markets—the price of potatoes versus the price of hot dogs. Supply and demand for all goods, services, and factors of production are the subject matter of microeconomics. Microeconomists address questions such as the following: Will the use of larger quantities of solar energy reduce the total energy bill of Americans? Will an increased tax on cigarettes increase or decrease federal revenues from the tax? Will the quality and quantity of nursing services in the state of Wyoming be changed by the licensing of nursing in that state?

Macroeconomics is the study of the economy as a whole. The overall price level, inflation rate, international exchange rate, unemployment rate, economic growth rate, and interest rate are some issues of concern to macroeconomists. Macroeconomists analyze how these crucial quantities are determined and how and why they change.

Some issues contain elements of both microeconomics and macroeconomics; for instance, a proposed U.S. tariff on Japanese auto imports. Microeconomists would be interested in the effects of the tarrift on auto prices, on the American auto industry, and on the quantity of cars bought by U.S. consumers. Macroeconomists would be primarily interested in the effects of the tariff on international trade and on total spending. They might address such questions as: Would the tariff increase or decrease economic growth in the United States and in Japan? Would total employment in the United States or Japan change because of the tax?

Daily television reports and newspapers are filled with discussions of both macroeconomic and microeconomic issues: Do tax cuts to business spur investment and economic growth? Will an OPEC price decrease or an alteration in supply affect oil and gas prices in America? Is the economy headed for more inflation, greater unemployment, or both? Will a tightening of the money supply cause an increase or decrease in interest rates? Does foreign competition mean fewer jobs for domestic autoworkers? Will price competition cause some airlines or trucking firms to shut down? Should the federal government balance the budget or not? A mastery of the fundamental and general principles of economics will give you a good foundation for formulating intelligent and reasoned answers to important questions such as these.

Macroeconomics: Analysis of aspects of the economy as a whole.

Why Do Economists Disagree?

Government or academic economists are forever predicting economic doom or economic prosperity for the same future time period. If economics contains fundamental principles agreed on by all, why do economists always seem to be fighting over predictions or over causes and effects? Why do economists disagree? As we noted earlier, one of the limitations of economic science is its inability to test theory. This imperfection means that alternative theories may be offered to explain the same events, such as the effects of deficits or of the Great Depression of the 1930s. Alternative theories mean divergent policy prescriptions and different perspectives on the effects of policy prescriptions.

Even when economists agree on theoretical apparatus, they may differ on the magnitude of the effects suggested by the theory. And secondary effects—effects not considered in a theory—can occur after the initial impact of a policy change has taken place. Economists often disagree on the nature and importance of these secondary effects. Would the addition of new monetary incentives for an all-volunteer army increase or decrease our military effectiveness? Economists of good faith and common fundamentals might disagree. We do not wish to overemphasize economists' disagreements or the

"All the experts say the market is going to turn around, but I say maybe."
Drawing by Saxon; © 1984 The New Yorker Magazine, Inc.

"iffy" or unfinished nature of economics. Meteorologists of equal training and ability might well disagree over whether it will rain tomorrow.

Economics, like meteorology, is an inexact science, and it is likely to remain so. But economic theory and prediction are accurate and testable enough to provide extremely important insights into numerous issues that touch our lives. And the science of choice and scarcity is (like meteorology) constantly being improved from both theoretical and empirical perspectives. The well-established power of economic thinking, as revealed in the following chapters, is the result of two hundred years of such continuing improvements. Sound understanding of the principles of economic thinking will improve your understanding of the world.

Summary

1. Economics is basically concerned with three questions: What goods and services are produced? How are goods and services produced? For whom are goods and services produced?

2. Economics is the study of how individuals and societies choose to allocate scarce resources given unlimited wants.

3. Scarce resources include human and nonhuman resources as well as time and technology. Human resources consist of labor, including entrepreneurship, and management. Nonhuman resources include land, all natural resources such as minerals, and capital.

4. All scarce resources bear an opportunity cost. This means that individuals or societies forgo opportunities whenever scarce resources are used to produce anything.

5. Economic behavior is rational: Human beings are assumed to behave predictably by weighing the costs and benefits of their decisions and their potential actions.

6. Economic choices are made at the margin, that is, the additional costs or benefits of a change in the current situation. Prices are the signal that indicates individual and collective choices for goods and services as well as the relative scarcity of the resources necessary to produce the goods and services.

7. Economic thinking also concerns the role of money, inflation, unemployment, and stabilization of the overall economy. The study of prices of individual products and inputs is called microeconomics, and the study of inflation, unemployment, and related problems is called macroeconomics.

8. Economics is a more inexact and imprecise science than chemistry or physics. It is accurate enough to predict much economic behavior, however, and to provide insight into a large number of important problems.

Key Terms

goods	costs	margin	money	microeconomics
services	human resources	marginal analysis	model	macroeconomics
resources	nonhuman resources	prices	fallacy of composition	
scarcity	accounting costs	market	*post hoc* fallacy	
free goods	opportunity costs	barter	positive economics	
economic goods	rational self-interest	inflation	normative economics	

Questions for Review and Discussion

1. What problem creates the foundation of economic analysis? Is this problem restricted to the poor? Is it relevant to the animal kingdom?
2. Can wants be satisfied with existing resources?
3. What are resources? How is the resource capital different from the resources land and labor?
4. What is the difference between behaving in one's self-interest and behaving selfishly?
5. What functions do prices have in the economy? Why were things rationed during World War II?
6. What is the basic function of money? What happens when there is too much money?
7. Do economists run the economy?
8. Why do economists abstract from reality when formulating theories? Does this imply that their theories have no relevance in the real world?
9. "The stock market crash of 1929 preceded the Great Depression. Therefore, the fall in stock prices caused the Great Depression." "Periods of inflation are frequently followed by increases in the stock of money. Therefore, inflation causes increases in the money supply." "A occurs after B. Therefore, B causes A." These statements violate which fallacy?
10. "The distribution of income in the United States is not fair." What type of statement is this? Can "fair" be used to describe something without being normative?

ECONOMICS IN ACTION

Marginal Analysis in Everyday Decisions

Far from being an abstract, theoretical idea, marginal analysis—making decisions at the margin, or letting "bygones be bygones"—is central to everyday life. Individuals, governments, and businesses are always making decisions based on considerations of marginal benefits and marginal costs. Here are some examples.

You, as a student, may face the following types of decisions on any given day: (1) whether to change majors; (2) whether to take an optional quiz in your economics course. You will make such decisions at the margin—that is, on the basis of marginal (not total) costs and benefits. Contemplating a change in majors would involve both costs and benefits, some monetary, others nonmonetary. How many more credit hours, for example, are required before your current degree program is finished? How does this figure compare to the number of hours required in the alternative degree program? If the latter number is greater, how much income would you forgo by spending more time in school? Clearly, your decision hinges on all of the additional (or marginal) costs and benefits from changing majors. The same principle holds in even less important decisions. Should you take an optional economics quiz to

improve your grade? The marginal costs of the quiz include your time spent studying (you could be studying for some other exam), and the marginal benefits depend on whether and how much an excellent quiz grade will improve your average in economics.

Government, through the political process, also makes decisions at the margin. The completion of the MX missile program or the decision to finish a nuclear power plant are examples. These decisions will not be made on the basis of how much has already been spent on the projects but on all of the additional costs and benefits attached to completion. Political opposition, for example, will increase the marginal costs to those who vote to complete the MX project, whereas political (voter) support will increase their marginal benefits. If the costs of completing a nuclear power plant no longer outweigh the benefits of ample electrical power, then it is likely the plant will sit half-finished.

Businesses also make decisions at the margin. Suppose you are the manager of a large grocery store located close to campus. On what basis would you decide to stay open all night rather than close at midnight and open at 6 A.M.? Clearly, the additional costs would include the additional electricity required plus the additional labor cost of cashiers, stock persons, and so on. The additional benefit is the sales revenue the store would make from staying open six more hours each day. In other words, the decision would not be based on the costs of building the grocery store, installing shelves, and so on. The decision, like all decisions, will be based on marginal considerations.

Questions

1. On what will you base your decision whether to eat another piece of pizza at a pizza parlor that advertises "all you can eat" for $5?
2. Apply marginal analysis to a decision you are facing right now. What are the marginal costs and benefits? How will you make your decision?

APPENDIX
Working with Graphs

Economists frequently use graphs to demonstrate economic theories or models. This appendix explains how graphs are constructed and how they can illustrate economic relations in simplified form. By understanding the mechanics and usefulness of graphs, you will find it much easier to grasp the economic concepts presented in this book.

The Purpose of Graphs

Most graphs in this book are simply pictures showing the relation between economic variables, such as the price of a good and the quantity of the good that people are willing to purchase. There are many such pairs of variables in economics: the costs of production and the level of output, the rate of inflation and the level of unemployment, the interest rate and the supply of capital goods, for example. Graphs are the most concise means of expressing the variety of relations that exist between such variables.

Figure 1–2 illustrates these ideas. It is a bar graph that shows the relation between the U.S. gross national product (GNP), a measure of the value of all goods and services produced annually, and time, in this case the period between 1960 and 1980. The GNP variable is plotted along the vertical line on the left side of the graph. The time variable is plotted along the horizontal line at the bottom of the graph. From this graph you can roughly estimate the level of GNP for each year between 1960 and 1980. For instance, GNP in 1967 was somewhere around $800 billion. You can also see that GNP tended to increase over the years 1960–1980. By working with the graph, you could estimate the rate of increase of GNP from year to year or the percentage increase over a particular period, such as 1965–1970.

This one concise picture—the graph—contains a great deal of information. Aside from economy of expression, graphs have a great deal of cognitive appeal: People can more easily grasp concepts demonstrated through pictures than concepts demonstrated through words.

Some of the graphs in this book are bar graphs; many others use the Cartesian coordinate system, which consists of points plotted on a grid formed by the intersection of two perpendicular lines. See Figure 1–3. The horizontal line is the **x-axis,** and the vertical line is the **y-axis.** The intersec-

x-axis, y-axis: Perpendicular lines in a coordinate grid system for mapping variables on a two-dimensional graph. The x-axis is the horizontal line; the y-axis is the vertical line. The intersection of the x- and y- axes is the origin.

FIGURE 1–2

A Bar Graph

This graph shows the levels of U.S. GNP between the years 1960 and 1980 at five-year intervals.

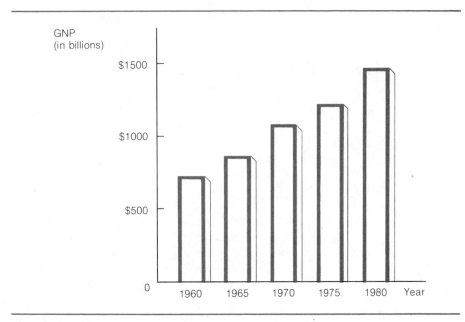

FIGURE 1–3

A Linear Graph

Any pair of numerical values can be plotted on this grid. The upper-right quadrant, shaded here, is the portion of the graph most often used in economics.

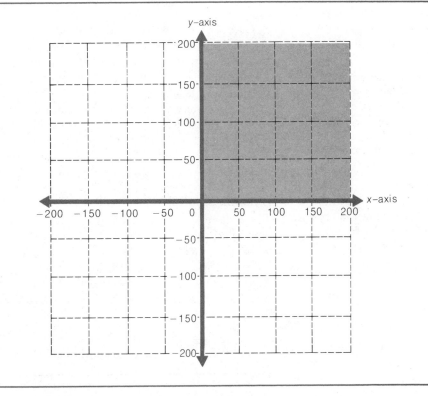

tion of the two lines is the point 0, called the *origin*. Above the origin on the vertical axis, all values are positive. Below the origin on the vertical axis, all values are negative. Values to the right of the origin are positive; values to the left are negative. The economic variables we use in this book are usually positive; that is, to plot economic relations we will usually use the upper-right portion, or quadrant, of a graph such as the one in Figure 1–3.

How to Draw a Graph

Suppose you wish to graph the relation between two variables. Variable y is the number of words memorized, and variable x is the number of minutes spent memorizing words. Table 1–2 presents a set, or schedule, of hypothetical data for these two variables.

We can plot the data from Table 1–2 on a graph using points in the upper-right quadrant formed by the x-axis and the y-axis. See Figure 1–4. The variable number of words memorized is measured along the y-axis and is considered the y value. The variable number of minutes spent memorizing is measured along the x-axis and is the x value. From the data in Table 1–2 we see that each increase of one minute of memorizing resulted in four more memorized words. The graph in Figure 1–4 illustrates this relation.

The points marked as large dots in Figure 1–4 represent pairs of variables. Point A represents the pair on the first line of Table 1–2: 4 words and 1 minute. Point B represents 8 words and 2 minutes, and so on. When we connect the points we have a straight line running upward and to the right of the origin. Lines showing the intersection of x- and y-values on a graph are referred to as **curves** in this book, whether they are straight or curved lines.

Curve: Any line, straight or curved, showing the correlation between two variables on a graph.

Curves show two types of relations between variables: positive and negative. The relation between the x and y values in Figure 1–4 is positive: as the x value increases, so does the y value. (Or, as the x value decreases, the y value decreases. Either way, the relation is positive.) On a graph, a **positive relation** is shown by a curve that slopes upward and to the right of the origin. In a **negative, or inverse relation,** the two variables change in opposite directions. An increase in y is paired with a decrease in x. Or a decrease in x is paired with an increase in y. Figure 1–5 shows negative, or inverse, relations on a graph. The curve for a negative relation slopes downward from left to right—the opposite of a positive relation.

Positive relation: A direct relation between variables in which the variables change in the same direction. A positive relation has an upward-sloping curve.

Negative, or inverse, relation: A relation between variables in which the variables change in opposite directions. A negative relation has a downward-sloping curve.

General Relations

Throughout this book, some graphs display general relations rather than specific relations. A general relation does not depend on particular numerical values of variables, as was the case in Figure 1–4. For example, the relation between the number of calories a person ingests per week and that person's weight (other things being equal) is positive. This suggests an upward-sloping line, as shown in Figure 1–6a, with calories on the x-axis and weight on the

TABLE 1–2   Schedule of Hypothetical Data

Number of Words Memorized (y)	Number of Minutes Spent Memorizing (x)
4	1
8	2
12	3
16	4
20	5

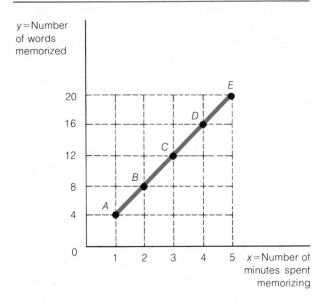

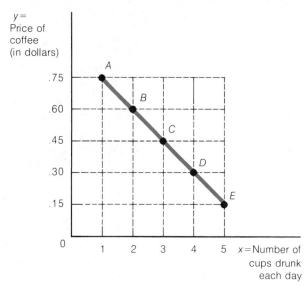

FIGURE 1–4   A Simple Line Graph

Curve *AE* shows the essential relation between two sets of variables: the number of words memorized and the number of minutes spent memorizing. The relation is positive; as the variable on the *x*-axis increases or decreases, the variable on the *y*-axis increases or decreases, respectively. A positive relation is shown by an upward-sloping line tracing the intersection of each pair of variables.

FIGURE 1–5   A Negative Relation Between Two Variables

As the price of a cup of coffee increases, the number of cups drunk each day decreases. When variables move in opposite directions, their relation is negative, or inverse. Negative relations are shown by a curve that slopes downward and to the right of the origin.

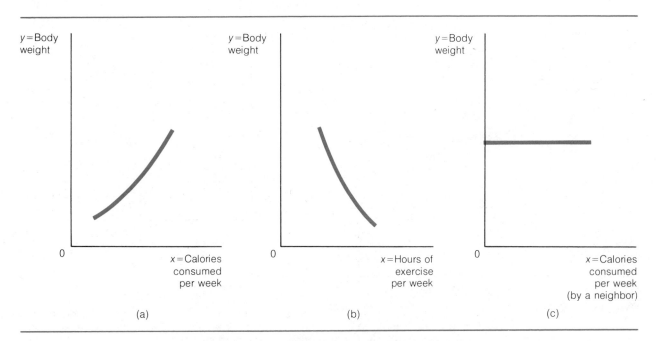

FIGURE 1–6   General Graphical Relations

General relations are (a) positive, (b) inverse, or (c) independent. General relations are not measured numerically on the *y*- or *x*-axis.

y-axis. Figure 1–6 represents a general relation, so it does not need numerical values on either axis. It does not specify the number of calories required to maintain a particular body weight; it simply shows that an increase in the number of calories consumed increases body weight. Figure 1–6b shows an inverse relation between body weight and the amount of exercise per week, again a general relation. Sometimes two variables are not related. Figure 1–6c shows that a person's weight is independent of a neighbor's caloric intake. Thankfully, no matter how much your neighbor eats, it has no effect on your weight.

At this point you should be able to construct a graph showing simple relations. For example, graph the relation between the weight of a car and the miles per gallon it achieves or between the length of the line at the school cafeteria and the time of day.

Complex Relations

Occasionally relations are more complex and do not fall into the simple positive or negative category. Figure 1–7 shows that for some relatively low x values, the variables on the graph are positively related, but as the x value increases, a point is reached where the two variables become inversely related.

From the origin to x^* the curve slopes positively. But for values greater than x^*, the curve slopes negatively. To illustrate, we let the x-axis show the number of pieces of pizza that a friend consumes in an evening and the y-axis show the total amount of pleasure she receives from each additional

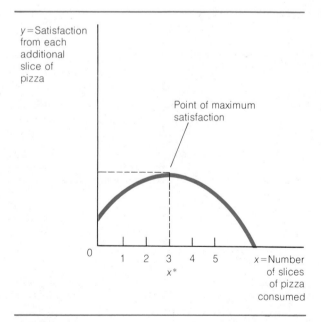

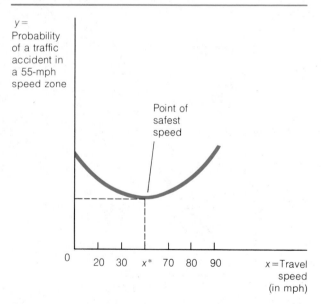

FIGURE 1–7   Variables Both Positively and Negatively Related

For lower values of x—in this example, the number of pizza slices eaten—the y value—the amount of satisfaction—increases. After some point, say 3 slices, the y value decreases with increasing values of x.

FIGURE 1–8   A U-shaped Curve

Some relations yield a U-shape. Variables along the x- and y-axes are inversely related at first, but after some critical value, here shown as x^*, they become positively related.

slice. At first, the more she eats the happier she becomes, but eventually a critical point is reached, after which the more she eats the less pleasure she receives from each additional slice.

Many other relations can exist. A U-shaped curve, as graphed in Figure 1–8, shows that the *x* value is at first inversely related to the *y* value and then positively related. After some critical value, x^*, the *y* value begins to rise as the *x* value increases. The *y* axis in this example might represent the probability of having an accident on an interstate highway and the *x*-axis might show the miles per hour at which a car travels. To avoid accidents, x^* is the optimal speed. Driving at speeds lower than x^* increases the probability of an accident, but rates faster than x^* mph also increase the chances of an accident.

Slope of the Curve

Slope: The ratio of the change in (Δ) the *x* value to the change in (Δ) the *y* value; $\Delta y/\Delta x$.

In graphical analysis the amount by which the *y* value increases or decreases as the result of an increase or decrease in the *x* value is the **slope** of the curve. The slope of the curve is an important concept in economics. Much economic analysis studies the margin of change in a variable or in a relation between variables, and the slope of a curve measures the marginal rate of change. In Figure 1–4, for example, every change in the *x* value—every increase of 1 minute—is associated with an increase of 4 words memorized. The slope of the curve in Figure 1–4 is the rate of change. The concept "change in" is expressed by the symbol Δ. So in Figure 1–4 the slope of the line *AE* is $\Delta y/\Delta x$, or $4/1 = 4$.

For the straight-line curves in Figures 1–4 and 1–5, the slope is constant. Along most curves, however, the slope is not constant. In Figure 1–8, an increase in speed from 55 mph to 60 mph may increase the probability of an accident only slightly, but an increase from 60 mph to 65 mph may increase the probability by a greater amount. Not only does the probability increase for each mile-per-hour increase but it increases by a greater amount. This indicates that the slope not only is positive but also is increasing.

The slope of a curved line is different at every point along the line. To find the slope of a curved line at a particular point, draw a straight-line tangent to the curve. (A tangent touches the curve at one point without crossing the curve.) Consider point A in Figure 1–9. The slope at A on the curved line is equal to the slope of the straight-line tangent to the curved line. Dividing the change in the *y* variable (Δy) by the change in the *x* variable (Δx) yields the slope at A. In Figure 1–9, every change in *x* from 0 to 1 and 1 to 2, and so on, results in a change in *y* along the curve. At point A, the change is $y = -4$. So $\Delta y/\Delta x = -4/1 = -4$. At lower points along the curve, such as B, the curve is flatter, that is, less steep. The line tangent at B has a slope of -2. As we move from point A to point B, equal increases in *x* result in smaller and smaller decreases in *y*. In other words, the *rate* at which *y* falls decreases as *x* increases along this particular curve.

Other relations of course lead to different slopes and different changes in the slope along the curve. Along the curve in Figure 1–10, not only does the *y* value increase as the *x* value increases but the slope increases as the *x* value increases. For equal increases in the *x* value, the incremental changes in the *y* value become larger.

Interpreting the essential concepts illustrated in graphs is a necessary part

FIGURE 1–9

Slope Along a Curve

The slope of a curve at a particular point is the slope of a straight-line tangent to the curve at that point.

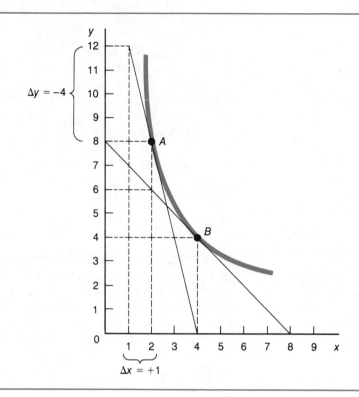

FIGURE 1–10

A Curve with Increasing Slope

As the x variable increases, the change in the y variable increases.

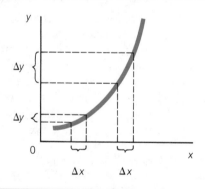

of learning economic principles. Most of the graphs used in this book are really quite simple and require only a few minutes of examination to grasp the relation between the two variables. With a certain amount of practice you can easily understand any graph in the book.

Summary

1. Graphs are a concise expression of economic models, or theories. Graphs usually show the relation between two variables, such as price and quantity.
2. Linear graphs are drawn with two perpendicular

lines, called axes. The x-axis is a horizontal line. Variables measured along the x-axis are called x values. The y-axis is a vertical line. Variables along it are called y values.

3. Lines showing the correlation between x and y values are called curves. An upward-sloping curve indicates a positive relation between variables: As the x value increases or decreases, the y value does likewise. A downward-sloping curve indicates a negative, or inverse, relation between variables: As the x value increases, the y value decreases and vice versa.
4. Relations between variables may be specific or general. Specific relations are based on numerical quantities measured on either the x- or y-axis or both.

General relations are not based on specific numerical quantities.
5. Variables can be both positively and negatively related. In such complex relations, the curve is either bow-shaped or U-shaped.
6. The slope of a curve measures the ratio of change in the x-value to change in the y value, expressed as $\Delta y/\Delta x$. Along a straight-line curve, the slope is constant. Along a curved line, the slope is different at every point.

Key Terms

x-axis	curve	negative, or inverse, relation
y-axis	positive relation	slope

Questions for Review and Discussion

1. Explain how the slope of a line tells you whether the relation between the two variables illustrated is direct or inverse. If you drew a line illustrating your score on a history test as you study more hours, what would its slope be?
2. Plot the following points on a graph with X on the horizontal axis and Y on the vertical. Connect the dots to show the relation between X and Y.

	X	Y
A	0	2
B	2	4
C	4	8
D	6	14
E	8	22

What is the slope of the line between the points A and B? What is the slope between C and D? Is this a straight line?

3. Curves can be represented in algebraic form by an equation. For example, the points along a certain line can be demonstrated by the following equation:

$$Y = 10 - 2X$$

The value of Y can be found by inserting different values of X and solving the equation for Y. Some of the values of Y for different values of X are shown in the following table. Fill in the blanks by solving the equation for the values of Y for each value of X given:

X	Y
0	10
2	6
4	—
6	—
8	—
10	—
12	—

Draw this set of points on a graph. Does this show a direct or inverse relation between X and Y? What is the slope of this line between the first and second points? Between the second and third? Is this a straight line?

4. If you measured body weight on the horizontal axis and the probability of dying of heart disease on the vertical axis, what would be the slope of a curve that showed the relation between these two variables?

Economic Principles

The essence of the economic problem is how to get as much value as we can from limited and costly resources. To do so, we constantly make choices. A college student may choose to spend a weekend at the beach rather than buy a new sweater and skirt. Another may choose to sleep late rather than attend an eight o'clock chemistry class. Every choice must be made at some cost.

Economists emphasize that decisions to use scarce resources bear an opportunity cost. The **opportunity cost** of a decision is the next most preferred or next-best alternative to a good or activity (the new clothes or the class instruction) that one chose to forgo to obtain some other good (a beach weekend or a late sleep). Economists have developed tools for expressing the choice and opportunity cost that are central to all decisions.

Opportunity cost: The next-best alternative that is lost when undertaking any activity.

Opportunity Cost: The Individual Must Choose

Suppose that a student is trying to decide between typing an overdue paper and studying for an economics quiz and that these two activities are the most desired of all activities on a particular afternoon. The student chooses among five combinations of hours spent typing and numerical grades on the economics quiz (see Table 2–1). Choice (1), no typing, results in a perfect grade on the quiz. Choice (2), 1 hour of typing, gives the student 4 hours of studying and a grade of 90 on the quiz. The third, fourth, and fifth alternatives mean still more typing finished but lower grades on the quiz. In fact, in the fifth and final choice, the student does not study for the quiz at all—yielding a failing grade—but does manage to type the entire paper.

These alternatives are graphed in Figure 2–1. The vertical axis plots the possible hours of typing, and the horizontal axis presents the possible quiz

TABLE 2–1   The Individual Chooses: Typing or Grades

Individuals must make choices from many alternatives. Most frequently these choices are based on evaluations of the combined costs and benefits of the alternatives. The benefit of typing the entire paper would be chosen at the cost of earning a failing grade on the quiz. The choice is not just all or nothing, however. The individual has intermediate options involving varying combinations of some studying and some typing.

Choice	Pages Typed	Grade Earned
(1) 0 hours typing, 4 hours studying	0	100
(2) 1 hour typing, 3 hours studying	4	90
(3) 2 hours typing, 2 hours studying	8	75
(4) 3 hours typing, 1 hour studying	12	55
(5) 4 hours typing, 0 hours studying	16	0

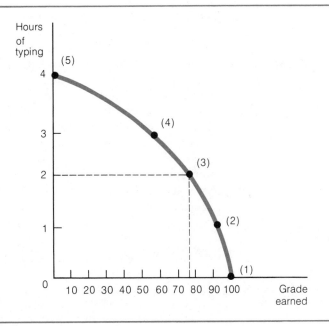

FIGURE 2–1   The Individual Chooses: Typing or Grades

As the hours of typing increase, the score on the economics quiz falls.

grades. The points representing the combinations of grade and pages typed, as given in Table 2–1, can be connected as a curve. Choice (3), 2 hours of typing and 2 hours of studying, is shown as point (3). With that choice, the student types 8 pages and gets a 75 on the quiz.

Several aspects of Figure 2–1 are important. *All possible combinations* of typing time and study time are represented by the curved line, meaning that choices between typing and studying are *continuous*. Since they are continuous, the student could choose any combination of hours of typing and quiz grade—say a half-hour of typing and a grade of 95. If the choices were *discrete*, the student would be limited to the five choices labeled (1) to (5). For convenience, economists ordinarily assume that economic relations are continuous.

A second crucial point is that Figure 2–1 and Table 2–1 constitute what economists call a *model*. Economists construct models to isolate specific phenomena for study. The model summarized in Figure 2–1 and Table 2–1 assumes a number of things about the student in the example. First, the choices are continuous rather than discrete. Second, we assume as given the amount of time the student puts into the course before the afternoon in question. Also assumed is the student's IQ. Does the student have the flu, a factor that would detract from performance on the quiz? Is the overdue paper more important to the student's academic standing than the economics quiz? Are there other alternatives such as washing a car or spending time in language lab? Is the fact that one alternative has a future payoff, such as a high-paying economics-related job, relevant to the choice? Certainly matters such as these are significant, but we assume them as given with respect to the choice at hand.

Economists also emphasize that the decision to study or to type is made at the margin. Practically, making decisions at the margin means that we are faced not only with all-or-nothing choices but also with degrees of balancing units of one choice against another. Each choice would be made in terms of the costs of an additional hour of studying or typing forgone. Satisfactions from both activities are thus balanced **at the margin.**

In making decisions at the margin we assume that all other factors are equal, or fixed. Each time we make a decision it is based on some fixed or given factors resulting from previous decisions. In our example, the student would weigh, for instance, the amount of previous study, a fixed factor in choosing typing or studying on the afternoon in question.

Previous decisions cannot be changed and are gone forever, but that does not mean that they have no impact on current choices. It simply means that we make choices in the present with unchangeable past choices as given and unalterable. This is another aspect of making choices at the margin.

Choices at the margin: Decisions made by examining the benefits and costs of small, or one-unit, changes in a particular activity.

Opportunity Cost and Production Possibilities: Society Must Choose

So far we have focused on individual choices, but the same principles apply to society's choices. Just as an individual is limited by the scarcity of time in choosing among options for an afternoon, society is constrained by scarce resources in producing and consuming alternative goods. Let us examine an abstract example. Suppose a society produces only two goods—oranges and peanut butter—and that all its resources and technology are fixed and constant in amount and level.

Given the state of technology and assuming that the society uses all its resources, the society could produce either 100 million jars of peanut butter or 16 million tons of oranges (see Table 2–2). Or the society may choose to devote some of its resources to producing peanut butter and some to producing oranges. These choices, called society's production possibilities, are numbered A–E in Table 2–2. From this information a production possibilities

TABLE 2–2   Society Decides: Oranges or Peanut Butter

Society must choose among many alternative goods to produce. The cost of one good is the lost production of other goods.

Choice	Peanut Butter (millions of jars)	Oranges (millions of tons)
A	100	0
B	90	4
C	75	8
D	50	12
E	0	16

Production possibilities frontier: The situation represented by a curve that shows all of the possible combinations of two goods that a country or an economic entity can produce when all resources and technology are fully utilized and fixed in supply.

curve can be constructed (see Figure 2–2). A production possibilities curve depicts the alternatives open to society for the production of two goods, given all existing resources, human and nonhuman, and an existing state of technology. When resources are fully employed, moreover, society is said to be located on its **production possibilities frontier.**

At choice A in Figure 2–2, society produces 100 million tons of peanut butter and no oranges, while at choice E society uses all of its resources in orange production. Choices in between, with some of both commodities produced, are shown at points B, C, and D. Continuous choices of all other possible combinations are shown along the curve; resources are assumed to

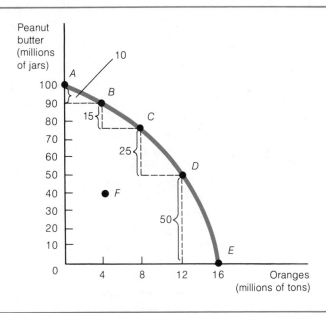

FIGURE 2–2   Production Possibilities Curve

As more and more oranges are produced, the opportunity cost per orange increases. The opportunity cost of producing 4 million tons of oranges is 10 million jars of peanut butter. The opportunity cost of producing 8 million tons of oranges is an additional 15 million jars of peanut butter, and so on.

be fully employed at all points on the curve. The curve therefore shows all of the possibilities for the production of oranges and peanut butter.

The Law of Increasing Costs

Just as there was an opportunity cost to the student in choosing to study for the quiz or to type the paper, there is an opportunity cost for society in any choice between producing oranges and producing peanut butter.

Consider a society that chooses nothing but peanut butter, choice A in Figure 2–2. The opportunity cost is that no oranges can be produced. A move to choice B, representing a shift of some resources from peanut butter production to orange production, means a sacrifice of 10 million jars of peanut butter, but society gains 4 million tons of oranges. Thus the opportunity cost of the first four million tons of oranges is 10 million jars of peanut butter. To get the second four million tons of oranges, however, a larger quantity of peanut butter must be given up—15 million jars. Thus as society becomes more and more specialized in production of either oranges or peanut butter, the opportunity cost per unit rises. This effect is called the **law of increasing costs.**

> **Law of increasing costs:** As more scarce resources are devoted to producing one good, the opportunity costs per unit of the good tend to rise.

Why does the law of increasing costs hold? As we know from experience, resources, training, and talents are not all alike and are not perfectly adaptable to alternative uses. Land suited to orange growing is not equally suited to peanut farming and vice versa. Orange pickers are not equally adept at the manufacture of metal lids for peanut butter jars or even peanut farming without additional training. The most-suited human and nonhuman resources are moved into production first, but as production of a good increases and becomes more specialized, less-adaptable resources must be used to produce it. The costs of the good produced increasingly rise because greater quantities of another good must be given up to get additional quantities of the good produced. A society of peanut butter lovers might be growing peanuts in greenhouses in North Dakota, and concert pianists and nuclear technicians might be operating the machinery in peanut butter processing plants.

Unemployment of Resources

The law of increasing costs always applies when resources are fully employed. When there is a degree of **unemployment of resources,** however—such as a certain amount of land lying idle—society finds itself at a point *within* the production possibilities curve. Point F in Figure 2–2 represents such a situation. From point F to the curve, additional oranges or peanut butter can be produced without any increase in opportunity costs until resources become fully employed. The production possibilities curve is thus referred to as a *frontier* when it represents full employment of resources. Full employment means that society is realizing its maximum output potential. Many economic problems are the result of unemployment, a subject that gets a good deal of attention in this book.

> **Unemployment of resources:** A situation in which human or nonhuman resources that can be used in production are not so used.

Choices Are Made at the Margin

As with individuals, society's choices are made at the margin. Suppose, for example, that a society has for years been one of peanut butter lovers and orange haters. Suppose further that long-time devotion of society's resources to peanut butter production "warped" resources—made them more adaptable over time—to peanut butter production. Past decisions to warp the resources would make current additional orange production more costly. A sudden decision by society to produce more oranges would have to be made at the margin. Thus, past decisions obviously affect current costs, but past decisions cannot be changed.

How the Production Possibilities Frontier Shifts

If we keep in mind the meaning of the production possibilities frontier, we will also understand that an increase or decrease in society's human or non-human resources can shift society's output potential. An understanding of such changes is central to the understanding of economic growth—or the lack of it.

Factors Causing a Shift

The production possibility frontier shifts inward or outward in response to any change in the quantity of human resources, any change in the quantity of nonhuman resources, or any change in technology. It makes sense that changes in these factors shift the production possibilities frontier because we previously assumed they were constant.

Suppose a technological improvement occurs in peanut butter production that increases the productivity of peanut harvesters—perhaps a fertilizer or new crop techniques. In such a case, society's production possibilities relating to peanut butter are increased without a similar decrease in orange production. The result is a shift outward and upward on the peanut butter end of the production possibilities curve. Such a change is shown in Figure 2–3a as a shift from the PP_1 curve to the PP_2 curve.

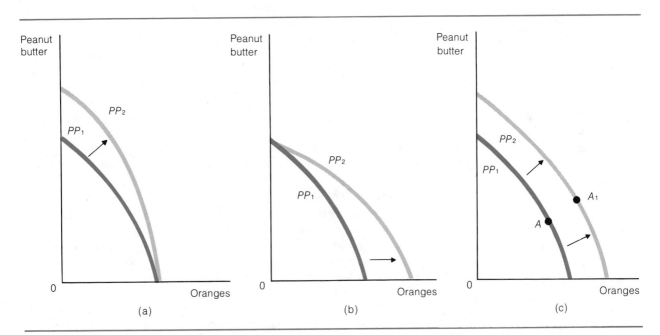

FIGURE 2–3   Shifting the Production Possibilities Frontier

(a) A technological improvement in peanut butter production shifts the production possibilities curve outward and upward at the peanut butter end. (b) A technological improvement in orange production shifts the production possibilities curve outward and rightward at the orange end. (c) A general increase in technology or resource supply shifts the entire production possibilities curve outward.

In Figure 2–3b, a productivity change brought about by technological improvement occurs in orange production. Here the production possibilities curve shifts outward and rightward on the orange end; if society prefers, additional oranges can be produced without any sacrifice of peanut butter.

In Figure 2–3c, a general change in productivity, technology, or resource supply is assumed. An increase in the quantity of labor (through a simple growth in population), for example, shifts the entire production possibilities frontier outward from PP_1 to PP_2. Suppose society now chooses the peanut butter/orange combination A_1. More of both goods are produced and consumed as a result of the technological improvement. But, society faces new trade-offs once it is producing on its new frontier. Opportunity costs arise again in resource utilization, just as they did at combination A on production possibility curve PP_1. Society cannot escape scarcity, yet it can improve its production possibilities through growth in technology, quantity of human or nonhuman resources, or resource productivity.

Economic Growth: How Economies Progress

We can understand the nature of economic growth with the aid of the production possibilities curve model. The key to economic growth in society is related to the growth or change in capital stock and other crucial resources.

Before considering the concept of capital stock, think about how life must have been for primitive peoples. Humans, having just barely earned the designation *homo sapiens* (wise man), initially had to use most of their time and resources to survive. Later they began to use tools to hunt and kill wild animals. It is not enough to say that primitive peoples found better ways of hunting. In addition to developing the technology of spear-making, primitives had to refrain from some consumption of wild game to have more in their future.

Capital stock: Supply of items used in the production of goods and services; these items include tools, machinery, plant and equipment, and so on.

The creation of such **capital stock** (a supply of items used to produce other items) takes time and bears an opportunity cost in present consumption. The benefits of creating capital, however, are in increased amounts of future consumption. In other words, forming new capital stock requires that people save—that is, abstain from consuming in the present. When investment in capital stock results from this saving, capital formation and growth occur. The formation of capital goods—spears, lathes, tractors, and so on—requires a redirection of resources from consumption of goods to production of capital goods. The opportunity cost of acquiring capital is thus consumption goods, but the reward for society is an increase in future productive capacity in consumption goods, capital goods, or both.

We can illustrate this phenomenon with the production possibilities curve. Figure 2–4 shows two production possibilities frontiers that contrast capital goods production (the vertical axis) with consumption goods production (the horizontal axis) for two hypothetical societies with initially similar production possibilities curves. Consider society A's choice of consumption and capital goods in 1985, represented by point K_A in Figure 2–4a. Clearly, society A devotes a larger proportion of its resources to capital goods than to consumption goods. Saving (abstaining from current consumption), investing, and capital formation are all high in that society. The payoff for society A will come later, say in 1995, with vastly enlarged productive capacity, designated by the new production possibility curve, PP_{1995}, in Figure 2–4a.

Society B chooses to use almost all of its resources in 1985 for consumption goods, with only a small quantity of resources devoted to capital goods, a combination represented by point K_B in Figure 2–4b. The consequence of

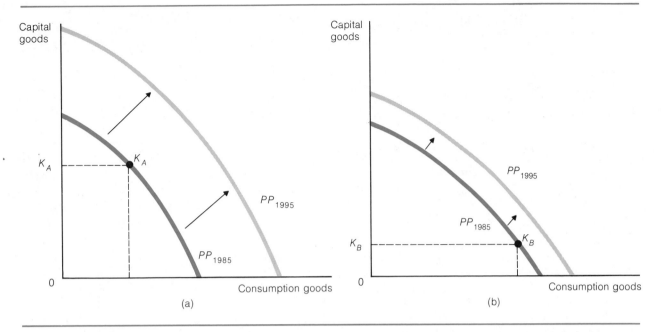

FIGURE 2–4 Growth Choices on the Production Possibilities Curve

(a) By producing more capital goods, K_A, in 1985, society A can produce more consumption goods in the future than (b) society B, which produces only K_B capital goods in 1985.

Economic growth: An increase in the sustainable productive capacity of a society.

Specialization: Performance of a single task in the production of a good or service to increase productivity.

this choice is low economic growth over the following decade. By 1995, the production possibilities of that society have grown by only a small amount, as curve PP_{1995} in Figure 2–4b indicates. Thus **economic growth** in the future is largely determined by current decisions about production. Technological changes are important factors in the progress of society, but to take advantage of technology, society must sacrifice some present consumption. (See Focus, "How Societies Can Regress: Armageddon Economics," and Focus, "Less-Developed Countries, Recent U.S. Growth, and Production Possibilities.")

In addition to choosing to put some resources into the creation of capital stock, societies can increase their output by specializing in and trading, or exchanging, goods and services. **Specialization** simply means that the tasks associated with the production of a product or service are divided—and sometimes subdivided—and performed by many different individuals to increase the total production of the good or service. The principles of specialization are therefore related to how all societies and economic organizations try to overcome the basic problem of scarcity.

Specialization and Trade: A Feature of All Societies

Virtually all known peoples have engaged in specialization and trade, and theories and principles related to specialization and trade have been part of economists' tool kits for more than two hundred years. Adam Smith (1723–1790), the Scottish philosopher and recognized founder of economics, pub-

FOCUS   How Societies Can Regress: Armageddon Economics

As we know from the ancient Greek and Roman experiences, civilization and its production possibilities can decline or be lost. An example from recent history also illustrates this point. In 1939 and 1940 the productive capabilities of Nazi Germany were huge and were growing larger. Massive quantities of resources were feeding a war machine for conquering the world.

Just as in the case of orange and peanut butter production, there was a trade-off, here between military goods and private goods (sometimes called a choice between guns and butter). The opportunity cost of the huge arms buildup in Germany was the resources that could have been devoted to other production. We can conceptualize this position as point A_{1940} in Figure 2–5.

The result of Germany's militarism is well known. By 1945, the country lay in ruins. More than half of its prewar population and 85 percent of its prewar productive capacity were destroyed. We might depict the production possibilities curve as having shrunk to the one labeled 1945 in the figure. On this curve, at choice B_{1945}, society must devote all of its resources to private consumption goods (such as food) just to survive.

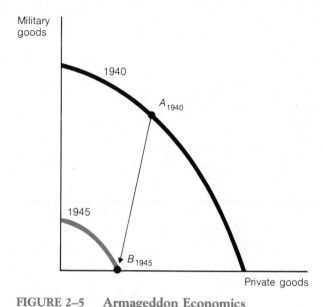

FIGURE 2–5   Armageddon Economics

The loss of resources shifts the production possibilities curve toward the origin, indicating that society must use all of its resources to produce private consumption goods just to survive.

lished his great work *An Inquiry into the Nature and Causes of the Wealth of Nations* in 1776 (a year of declarations). It formally established the science of what we now call economics.

At the very core of economics, according to Smith, is the ability of individuals and societies to deal with the facts of scarcity through specialization and trade. In the *Wealth of Nations*, Smith used the example of a pin factory, a seemingly trivial manufacturing activity he had observed directly, to evaluate specialization:

> . . . a workman not educated to this business . . . nor acquainted with the use of the machinery employed in it . . . could scarce, perhaps, with his utmost industry, make one pin in a day, and certainly could not make twenty. But in the way in which this business is now carried on, not only the whole work is a peculiar trade, but it is divided into a number of branches, of which the greater part are likewise peculiar trades. One man draws the wire, another straights it, a third cuts it, a fourth points it, a fifth grinds it at the top for receiving the head; to make the head requires two or three distinct operations; to put it on, is a peculiar business, to whiten the pins is another; it is even a trade by itself to put them into the paper; and the important business of making a pin is, in this manner, divided into about eighteen distinct operations, which, in some manufactories, are all performed by distinct hands, though in others the same man will sometimes perform two or three of them.[1]

[1]Adam Smith, *An Inquiry into the Nature and Causes of the Wealth of Nations*, ed. Edwin Cannan (1776; reprint, New York: Modern Library, 1937), pp. 4–5.

FOCUS   **Less-Developed Countries,
Recent U.S. Growth,
and Production Possibilities**

The production possibilities curve offers insights into many important matters. The economic growth rates of many less-developed countries, for example, are quite low compared with those of the United States and other, more-developed countries. Yet many poor developing countries teem with resources, and modern technology is available to them. Because they must devote a huge quantity of resources to present consumption to meet the demands of high population growth, their future economic growth is severely limited. Some economists point not only to high population growth rates but also to low levels of education as probable causes of the failure of some countries to grow economically.

The production possibilities curve can also help explain the slowed rate of U.S. economic growth in the 1970s and early 1980s. The problems have been a lagging growth in labor productivity, sharply curtailed access to vital natural resources such as oil and other essential inputs, and a slower growth rate of saving and investment, with consequent reductions in the rate of production of capital goods. The U.S. production possibilities frontier expanded more slowly over this period than it did, on average, over the past one hundred years. We discuss these issues in more detail in Chapter 34.

Division of labor: Individual specialization in separate tasks involved in production of a good or service; increases overall productivity and economic efficiency.

Smith's point in describing the pin factory is that when tasks are divided, permitting each individual to concentrate on a single element in the production of a good, output increases over what it would be if each individual produced the entire good. Specialization and the **division of labor** led to increased output. As a result, people became more dependent on one another for all goods. We may help produce autos, for example, but we still want eggs for breakfast. Smith recognized that mutual dependence could be a problem but thought that the potential for increased output with given resources was well worth the cost.

If Smith's pin factory does not stir your imagination, think of any organization of which you have ever been a part and try to list the number of distinct divisions of labor in it. Think of a larger factory, a steel mill, a football team, a fraternity, a church. Specialization exists in almost everything we do.

In the modern world, nations, states, firms, and individuals specialize and trade according to two principles—absolute advantage and comparative advantage.

Absolute Advantage

Absolute advantage: Ability of a nation or an economic entity to produce a good with fewer resources than some other entity requires.

Absolute advantage simply means that one production unit (individual, firm, state, or nation), given an equal expenditure of resources, is more efficient than another at producing some good or service. In other words, that production unit can produce more from the same resources.

For example, Sally, the best physician in town, meets Sam, the best auto mechanic in town. Given equal expenditures of resources, Sally is more efficient than Sam at health care, while Sam is more efficient than Sally at auto repair. That is, with an equal expenditure of all resources, including time, Sally can produce more medical services than Sam and Sam more auto repair services than Sally. Therefore, it is more efficient for Sally to specialize in medicine and to trade her services for Sam's.

Similarly, states and nations may find mutual benefit in specializing and trading when each has an absolute advantage in producing different goods. Assume that the states of Texas and Idaho are isolated and that both are

TABLE 2–3   Absolute Advantage: Output Before Specialization

Texas has an absolute advantage in the production of beef: Texas can produce one pound of beef with fewer resources than Idaho needs. Idaho has an absolute advantage in the production of potatoes.

| | Texas | | Idaho | | |
	Yearly Production (millions of pounds)	Resources Spent (percent)	Yearly Production (millions of pounds)	Resources Spent (percent)	Total Output
Beef	40	25	25	25	65
Potatoes	100	75	300	75	400

TABLE 2–4   Absolute Advantage: Output After Specialization

Through specialization, Idaho's output of potatoes increases and Texas's output of beef increases. The increase in total output allows both states to gain from trade.

| | Texas | | Idaho | | |
	Yearly Production (millions of pounds)	Resources Spent (percent)	Yearly Production (millions of pounds)	Resources Spent (percent)	Total Output
Beef	160	100	0	0	160
Potatoes	0	0	400	100	400

able, with the same resources, to produce the quantities of beef and potatoes reported in Table 2–3. Further assume that the quantities are actually demanded and consumed within the two states. From Table 2–3, it is clear that Texas is more efficient at producing beef than is Idaho. It is 1.6 times more efficient (40/25), to be exact. Idaho, on the other hand, is absolutely more efficient at producing potatoes—in fact, 3 times (300/100) more efficient. The reasons for these absolute advantages, though somewhat irrelevant for the working of the principle, may include climatic differences, different endowments or qualities of resources including labor or land, or differing levels of technology. Whatever the underlying reasons, Table 2–3 shows the respective outputs and consumptions before specialization.

Suppose that Texas and Idaho agree to specialize: Idaho will produce only potatoes, Texas will produce only beef, and they will trade to help meet each other's needs. If we assume that output proportions remain the same—that Idaho, for example, will be able to produce 400 million pounds of potatoes with 100 percent of its resources since it could produce 300 million pounds with 75 percent—the respective outputs after specialization appear as in Table 2–4. Texas now produces 160 million pounds of beef and no potatoes, while Idaho produces 400 million pounds of potatoes and no beef. Note the result: There is a clear gain from specialization—an increase of 95 million pounds of beef.

On what terms will the parties trade? How will they determine a fair trade ratio between beef and potatoes? Though we can discern the limits to the trading terms, it is impossible to determine what the exact trade will be without additional information. The actual **terms of trade** would be determined by the relative bargaining strength of Texans and Idahoans, which is itself determined by the demands for the two goods within the two states. But we do know the limits. Idaho would be willing to trade 100 million pounds of potatoes for a quantity of beef greater than 25 million pounds (the quantity of beef Idaho produced and consumed before specialization). If Idaho accepted any less beef, the state would be worse off than before and would not have agreed to specialize. At the same time, for 100 pounds of potatoes (Texas's prespecialization production and consumption), Texas would not be willing to part with more than 120 million pounds of beef—a quantity that would leave Texas only as well off as before specialization.

If the two states split the gain, 100 million pounds of potatoes would trade for 60 million pounds of beef. Texas would then be able to consume 100 million pounds of both potatoes and beef, and Idaho would have 300 million pounds of potatoes and 60 million of beef. Both states would be better off.

To summarize, our example shows that Texas and Idaho have specialized, traded, and gained according to the principle of absolute advantage. Specialization permitted Texas and Idaho to devote all of their respective resources to the production to which they were best suited. Trade ensued, after which both states are better off.

Comparative Advantage

Adam Smith explained the principle of absolute advantage in his *Wealth of Nations* in 1776. It was left to his English followers, economists David Ricardo (1772–1823) and Robert Torrens (1780–1864) to develop a somewhat less obvious but critically important principle of specialization—the theory of **comparative advantage.** In a situation of comparative advantage, nations or economic entities can benefit from trade of two goods even though one has an absolute advantage in the production of both. This is because of a difference in opportunity costs for each nation or entity, giving both the less productive and the more productive entity a relative advantage in specializing in production of one of the goods. Comparative, rather than absolute, advantage is the modern principle most closely associated with the basis for trade between any two parties.

Ricardo developed the principle of comparative advantage by observing and analyzing Britain's international trade position early in the nineteenth century, especially during the Napoleonic Wars with France. As in Ricardo's original examples, international specialization can illustrate the principle in modern terms. Look at Table 2–5, where the hypothetical outputs of oil and cosmetics are given for the United States and West Germany. While the United States is able to produce more (in absolute terms) of both goods, the United States is 1.5 times (90/60) more efficient at producing oil than cosmetics, whereas West Germany is twice as efficient (20/10) at producing cosmetics than oil. If these production rates are also the rates at which the two commodities can be exchanged within the two countries before specialization, oil is cheaper relative to cosmetics in the United States, and cosmetics are cheaper than oil in West Germany (half as cheap, to be exact).

After specialization (see Table 2–6), the same total quantity of cosmetics is produced, but a net increase of 20 million barrels of oil is obtained. The 20 million barrels of oil is the net gain from specialization.

TABLE 2–5   Comparative Advantage: Output Before Specialization

Oil has a lower opportunity cost in the United States than in West Germany. West Germany has a comparative advantage in cosmetics, for it is relatively more efficient at producing cosmetics than at producing oil.

	United States		West Germany		
	Yearly Production (millions of barrels)	Resources Spent (percent)	Yearly Production (millions of barrels)	Resources Spent (percent)	Total Output
Oil	90	75	10	75	100
Cosmetics	60	25	20	25	80

TABLE 2–6   Comparative Advantage: Output After Specialization

Specialization according to comparative advantage results in a net increase of 20 million barrels of oil.

	United States		West Germany		
	Yearly Production (millions of barrels)	Resources Spent (percent)	Yearly Production (millions of barrels)	Resources Spent (percent)	Total Output
Oil	120	100	0	0	120
Cosmetics	0	0	80	100	80

How can this gain be divided between the United States and West Germany? As stated earlier, we cannot determine that division without knowing the relative bargaining strength of the two nations. But the limits to the bargain can be determined.

In exchange for 60 units of cosmetics (the United States' prespecialization consumption), West Germany will not be willing to take less than 10 million barrels of oil. If West Germany took less than 10 million barrels of oil, it would be worse off than before specialization; in that case West Germany would not specialize. The United States would not give more than 30 million barrels of oil for 60 million barrels of cosmetics. To do so would make the United States worse off than before. Thus the trade must be something greater than 10 million barrels of oil for 60 million barrels of cosmetics but something less than 30 million barrels of oil for 60 million barrels of cosmetics. If they split the net gain, the United States gives up 20 million barrels of oil, retaining 100 million barrels of oil and 60 million barrels of cosmetics, and West Germany retains 20 million barrels of cosmetics and acquires 20 million barrels of oil through trade. Both nations are better off after specialization and trade, although the United States is absolutely more efficient in producing both of the traded goods. For both nations, specialization and trade, according to the principle of comparative advantage, expanded the consumption possibilities in oil.

Exchange Costs

The models of absolute and comparative advantage—along with all other economic models—are simplifications of reality. A number of important assumptions hide behind our simple discussions, some of which we have already mentioned. But there are also costs to the process of the exchange that must be accounted for when calculating the benefits of specialization and trade. We classify these **exchange costs** as transaction costs, transportation costs, and artificial barriers to trade.

Transaction Costs. Transaction costs are all the resource costs (including time-associated costs) that are incurred because of exchange. Transaction costs occur every time goods and services are traded, whether exchanges are simple (purchase of a pack of gum) or complex (a long-term negotiated contract with many contingencies).

Here's a simple example: Gwen goes to the supermarket to purchase a pound of coffee. What are the costs of the transaction? Gasoline and auto depreciation must of course be considered as resource costs, but the principal cost is the opportunity cost of Gwen's time. Gwen might have spent this time working or playing tennis instead of grocery shopping. Gwen's wage rate might then serve as her opportunity cost. If the shopping trip takes 30 minutes and if Gwen's wage rate is $15 an hour, the time part of her transaction costs is $7.50.

In this simple case the contracting and negotiating are instantaneous—Gwen simply gives the money to the checkout person, takes her coffee, and the transaction is complete. In other, more complex exchanges—such as the purchase of a house, a car, or a major appliance or the negotiation of a long-term labor contract—contracting and negotiating costs can be substantial. Think of the time and other resource costs associated with long-term supply contracts such as U.S. arms deals with allies or negotiations for ammonia plants in China. Every detail must be studied, formulated, and then spelled out.

The important point is that transaction costs include resources used by each party to the exchange. The higher the resource costs, the lower the benefit that comes from specialization and trade. Institutions—new legal arrangements, new marketing techniques, new methods of selling—have emerged and are continuously emerging, to reduce all forms of transaction costs. Gwen could have purchased coffee at a convenience store nearer to her house than the supermarket is. The price of the coffee might have been somewhat higher, but Gwen's time costs, and therefore her total transaction costs, would have been lower. In a broader sense, laws about contracts, and law itself, are means by which transaction costs are reduced. The invention of money and the development of various forms of money and financial instruments are responses to transaction costs associated with barter and more primitive means of exchange. We will return to these issues in Chapter 3 and later.

In addition to money, intermediaries such as wholesalers and advertisers developed over the ages to facilitate exchange. The creative marketing of goods and services from bazaars to discount stores to media advertising has increased consumer information and thus reduced the costs of making transactions. Middlemen have lowered transaction costs by decreasing the risk of exchange. The production of some goods entails some risk and uncertainty on the part of buyers and sellers. Planting wheat in the spring for sale in the

Exchange costs: The value of resources used to make a trade; includes transportation costs, transaction costs, and artificial barriers to trade.

Transaction costs: The value of resources used to make a purchase, including time, broker's fees, contract fees, and so on.

fall obviously entails some uncertainty about what prices will be at harvest. Middlemen-speculators who deal in futures provide sellers and buyers assurance of prices in the future. This type of middleman makes profit on the miscalculations of buyers and sellers but performs the service of reducing uncertainty and thereby increasing trade and specialization.

Transportation costs: The value of resources used in the transportation of goods.

Transportation Costs. A second impediment to trade is transportation costs, resource costs associated with the physical transport of products from place to place. The higher these resource costs, the lower the benefits from specialization and trade.

We did not include transportation costs in our initial examples of absolute and comparative advantage. Suppose that transport of beef to Idaho and potatoes to Texas was costly because of the rugged terrain between the two states. High transportation costs would reduce the gains that Texans and Idahoans might enjoy from specialization and trade. If transportation costs were high enough, possible advantages to specialization and trade could be wiped out completely. Cheaper transportation costs, such as those created by new trade routes in the Renaissance or by the invention and spread of the railroad in nineteenth-century America, permit more trade and open up opportunities for new forms of and increases in specialization. The Pony Express, the invention of the automobile and truck, and the dawning of air freight transport were all boons to specialization and increased output.

Artificial barriers to trade: Any restrictions created by government that inhibit trade, including quotas and tariffs.

Artificial Barriers to Trade. The final impediments to specialization and trade are government-imposed restrictions such as tariffs, quotas, and outright prohibitions on the import or export of goods. More localized restrictions include minimum-wage laws and specific restrictions on an industry or in an area. Such impositions either reduce or eliminate the benefits of specialization and trade. Governments always have reasons for these restrictions, but the reasons must be closely scrutinized because the benefits from international and domestic specialization and trade are potentially huge for consumers in all nations. Artificial barriers have the power to reduce economic welfare by reducing or eliminating the benefits of specialization according to the law of comparative advantage (see Economics in Action, "Comparative Advantage: The Case of Bio Bio," at the end of this chapter).

The possible effects of restrictions on trade can be seen in the large and growing volume of trade between Japan and the United States in commodities such as TVs, automobiles, stereo equipment, and musical instruments. Special-interest groups—such as American autoworkers and manufacturers—have lobbied for import tariffs and other trade restrictions to protect their own interests, which include increased demand for American-made products and therefore increased domestic production. Government-enforced tariffs make Japanese goods more expensive, however, causing a reduction in the general well-being of Americans. Artificial restrictions on trade, whatever their purpose, reduce the advantages of specialization and trade. As a matter of economic principle, therefore, most economists are generally advocates of the advantages of free trade over any type of trade restriction. Further, any institution or mechanism that reduces the costs of exchange usually gets the support of economists because greater specialization permits better utilization of scarce resources. Specialization, in other words, helps us get more of what we want, given a limited amount of resources.

Summary

This chapter described several principles that provide a foundation to economic thinking:

1. Economic choices, both for the individual and for society, always involve an opportunity cost.
2. Opportunity cost includes all costs or opportunities forgone in the decision to engage in a particular activity.
3. The law of increasing costs means that as more of one good is consumed by a society, the opportunity costs of obtaining additional units of that commodity rise. The cost increase results because resources become less adaptable as production becomes more specialized.
4. Choices are ordinarily made not in all-or-nothing fashion but at the margin. Both individuals and societies, therefore, calculate the cost of consuming additional units of some good or service.
5. The production possibilities curve shows the possible quantities that could be produced of any two goods given the state of technology and society's scarce resources.
6. Changes in technology or increases (or decreases) in the amount of resources cause outward or inward movements in the production possibilities frontier.
7. Greater quantities of output can be obtained with society's scarce resources when people specialize and trade. Trade takes place according to one of two principles: absolute advantage or comparative advantage.
8. Trade can take place between two individuals or economic entities even if one of the entities is more efficient at producing all goods. All that is required is that each entity be relatively more efficient than the other in some production.
9. Transaction costs, transportation costs, and artificial trade barriers such as tariffs and quotas reduce the benefits obtainable from specialization and trade.

Key Terms

opportunity cost
choices at the margin
production possibilities frontier
law of increasing costs
unemployment of resources
capital stock

economic growth
specialization
division of labor
absolute advantage
terms of trade
comparative advantage

exchange costs
transaction costs
transportation costs
artificial barriers to trade

Questions for Review and Discussion

1. What did reading this chapter cost you? Did you include the price of the book? What will reading the next chapter cost? Does that include the price of the book?
2. Do government-sponsored financial aid programs for college students influence the amount of education produced? Do these programs shift the production possibilities curve?
3. What does a movement along the production possibilities frontier suggest? What does a point inside the curve suggest?
4. A subsidy to farmers who purchase tractors and combines increases the production of this farm machinery. Does this cause an increase in the production possibilities curve or just a movement along the curve? Can subsidies cause economic growth?
5. Why would a country with an absolute advantage in the production of all goods be willing to trade with other countries?
6. Alpha can produce 60 bottles of wine or 40 pounds of cheese. Beta can produce 90 bottles of wine or 30 pounds of cheese. Both have constant costs of production. Draw their production possibilities curves. What is Alpha's cost of one bottle of wine? What is Beta's cost of one pound of cheese? If they trade, who should specialize in cheese?
7. What are the costs of going to college? Does the marginal benefit outweigh the marginal cost?
8. Is the lost present consumption associated with the production of capital goods worth the benefit of the new capital?
9. Does Japan have an absolute advantage over the United States in the production of televisions and stereo equipment or is it just a comparative advantage?
10. Who is hurt by and who benefits from an import quota on foreign beef?
11. How does the cost of purchasing a loaf of bread at a supermarket compare with the cost of purchasing a loaf of bread at a convenience store?

ECONOMICS IN ACTION   Comparative Advantage: The Case of Bio Bio

There are terrible economic problems in Bio Bio, a coastal region of western Chile with a land mass about the size of Massachusetts and Connecticut combined. In 1983, 33 percent of its 1.4 million inhabitants were unemployed. Bio Bio lawyers, business people, and even economists blame the region's current economic problems on the theory of comparative advantage. Is the situation in Bio Bio a case of comparative advantages gone wrong?

Bio Bio's comparative advantage, prior to the opening of the Panama Canal in 1914, was in shipping. The region's coastal cities enjoyed prosperity by offering refueling and resupplying services to ships that had to pass through the Strait of Magellan. After the Panama Canal opened, the Chilean government sought to prop up its fragile economy by enforcing protective tariffs on textiles and by offering direct subsidies to coal miners. The government poured money into state-owned enterprises such as steel mills and petrochemical plants. For years, these actions created jobs and forestalled economic chaos.

The region's present economic problems emerged, it is argued, after the overthrow of the Marxist government of Salvador Allende by the Chilean military in 1973. Economic reforms, at the suggestion of University of Chicago-educated free-market economic advisers, included the elimination of all protective subsidies and tariffs to industries across Chile and, in general, the institution of free-market production on the basis of comparative advantage.

Free-market economic advisers of the new Pinochet government argued that Bio Bio had a comparative advantage in fish and timber production. Other industries that were rapidly developing new markets included ceramics and the production of rose hips, an ingredient in natural foods and vitamins. These industries have grown substantially since free-market policies were instituted but not enough to take up the slack in employment. The region suffered bitterly during the worldwide recession of the early 1980s. Are the government's free-trade policies a failure? Is comparative advantage a cruel hoax to these people?

The answer is no: The Bio Bio problem is essentially an "adjustment problem." Comparative advantage in ship service and related trade worked in favor of resi-

Source: Everett G. Martin and Fernando Paulson, "Victims of a Theory: Chilean Region, Competitive in Few Products, Hits Hard Times Under Rule of 'Chicago Boys,'" *Wall Street Journal,* May 5, 1983, p. 60.

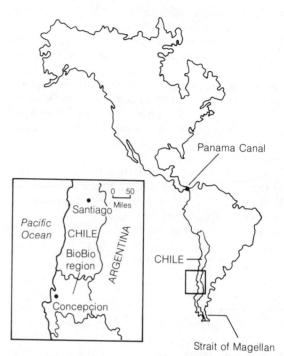

Prior to 1914 and the opening of the Panama Canal, the Bio Bio region of Chile was a popular refueling stop for ships en route between the Atlantic and Pacific oceans. Since 1914, the region has suffered severe unemployment problems.

dents until the Panama Canal was opened. Afterward the government located industries in Bio Bio through subsidies and tariffs. Subsidized workers did well in every industry. But when subsidies and tariffs ended, Bio Bio found its real comparative advantage in fish, lumber, and rose hips. Of course, capital and labor resources of the area were concentrated in the previously subsidized industries.

The cost of retraining textile workers to be fishermen is not zero. The quality and skills of a large part of the labor force cannot be changed quickly. Hence, problems arise when the government encourages people, for extended periods of time, to invest capital and skills in activities where no comparative advantage exists.

It is not clear, in other words, that the previous Bio Bio "prosperity" was genuine. Consumers in Bio Bio paid higher prices for both types of goods produced—those that were subsidized through the tariff protection and those for which a comparative advantage existed but which were not produced. Taxpayers in all regions of Chile subsidized the owners and workers in govern-ment-favored industries. The many consumers and tax-payers paid for the prosperity of the few.

Artificial barriers to trade are costly to any society because resources are not directed to their most valued uses by comparative advantage. Comparative advantage will work in Bio Bio, as elsewhere, once resources adapt to new, more suitable production.

Question

Assume that the United States decides to become self-sufficient in energy production and reduces imported energy sources to zero by means of a prohibitive tariff. What effects would the oil tariff have on the benefits due to specialization and exchange in the short term and over a longer period of time?

3

Markets and the U.S. Economy

*T*he vehicles for economic growth, as we have seen, are technology and resource development, specialization, and trade. But what makes these vehicles possible? Why are they more apparent in some societies than in others? To better understand the answers to these questions and the workings of the American market economy—the system most closely analyzed in this book—we present an overview of how the American economy works, and we contrast it with the ways in which other societies have chosen to deal with production and distribution in the face of scarcity.

The U.S. Market System in Perspective

Rice paddies in China, nut-and-fruit-gathering societies in Africa, government enterprise in Russia, free-trade zones in Hong Kong, computer hardware development in California's Silicon Valley—all these arrangements represent different answers to the "what," "how," and "for whom" questions under different economic systems. An **economic system** is the particular form of social arrangements through which the three fundamental questions are answered.

Economic system: A means of determining what, how, and for whom goods and services are produced.

Comparative Systems

Three basic types of economic systems have emerged in response to scarcity of resources: traditional, command, and market societies. But all societies, including the United States, actually contain some elements of each type. It is very difficult to identify a pure type of economic organization in contemporary societies.

Traditional Societies. A **traditional society** is characterized by subsistence food gathering, primitive agriculture, or nomadic herding. Such societies

Traditional society: An economic system in which the "what," "how," and "for whom" questions are determined by customs and habits handed down from generation to generation.

typically suffer acute scarcity. Specialization and trade, crucial factors in expanded production, are minimal. The answer to "what is produced" is limited to fulfillment of the basic needs of the community (food, shelter, clothing). Most activities are geared to basic survival. Little or no time is left for innovation or for the development of technology after basic needs have been met.

The "how" of production—means such as hunting tools, agricultural implements, and the construction of shelter or storage—change very little through time as skills and roles are passed on from father to son, mother to daughter. Specialization exists in the traditional society, but it is narrowly restricted by the social arrangements surrounding production.

Distribution—the answer to the "for whom" question—is based solely on social arrangements rooted in what the society regards as "superior qualities" of its members—the strongest and most successful hunters, best firekeepers, most talented rugmakers, the oldest or wisest people. Members of society with such superior qualities receive the best of the traditional society's output.

Command society: An economic system in which the questions of "what," "how," and "for whom" are determined by a central authority.

Command Societies. In a **command society,** "what," "how," and "for whom" questions are answered by some central authority—a single individual (king, dictator, pharaoh) or a group of individuals (Communist party, military junta). The central authority determines what is going to be produced. In ancient Egypt, pharaohs, the personifications of the gods, were able to direct massive quantities of resources into the construction of pyramids and other monuments through slave labor. These monuments were constructed at the opportunity cost of consumption goods and services for the Egyptian population.

Expressions of command are found in contemporary societies as well. Perhaps the most familiar contemporary example of a command society is the Soviet Union. Over the last fifty years, the members of the ruling Politburo in the Soviet Union have ordered that a vast quantity of resources be poured into military goods at the expense of consumer goods and capital goods used to produce consumer goods. Other advanced nations, including the United States, invest large quantities of resources in national defense, although decisions to do so are ultimately directed through a democratic voting process in many of the nations.

Most production in a command society takes place in government-owned or -sponsored enterprises. Invention and innovation can be prominent in command societies, but their development and the use of their results are ordinarily controlled by authority. Free thought, the handmaiden of innovation, is not encouraged, and particular sciences and arts (for example, space technology and weaponry in the Soviet Union) tend to thrive while others (in the Soviet Union, those necessary for production of consumer goods, such as TVs and refrigerators) languish.

A large degree of discretionary power exists in a command society, and this power leads to a selective distribution of wealth. Those highly placed or those in the favor of the highly placed have first choice of products and services produced. In the Soviet Union rewards are based on membership and power within the Communist party bureaucracy. Opulent resort homes in the Soviet Union, for example, belong mostly to well-placed government and military bureaucrats. (We discuss the economics of the command society in more detail in Chapter 37.)

Historically, traditional and command societies have been most common.

Egyptian, Greek, Roman, and medieval societies (dictatorships, monarchies, and aristocracies) were composed of both traditional and command characteristics. In the seventeenth and eighteenth centuries, however, England and other European countries underwent a significant decline of centrally controlled and regulated economic life. These developments ushered in the market society as a new form of economic organization.

Market society: An economic system in which individuals acting in their self-interest determine what, how, and for whom goods and services are produced, with little or no government intervention.

Market Societies. In a **market society,** impersonal forces lead consumers and producers to answer the three fundamental questions of production. Consumers answer the "what" question. Production of goods and services such as personal computers and auto repair is determined by "dollar votes." Just as political votes elect a president, dollar votes—or money spent on products and services—express consumers' demands. Suppliers of products and services, interested in making profits, respond to these votes by providing the products and services in just the right quantities. (Chapter 4 explains exactly how this occurs, through the laws of supply and demand.)

The "how" question in the market society is, at any one time, answered by available technology and by suppliers' profit-motivated desire to produce goods most cheaply given the price of resources. Most goods can be produced by a variety of methods. Avocadoes, for example, can be picked by hand or by harvesting machines. The avocado grower will choose whatever method minimizes the costs of production. In a market economy prices are the signals not only for what to produce and how to produce in the present but also for how new technologies and new resources can be brought into production over time.

When one highly demanded resource becomes scarce and therefore high-priced, producers will attempt to substitute other resources for it. If possible substitutions are limited, alternative technologies and new types of resources may be developed. In mid-nineteenth-century America, for example, whale oil—used primarily for lighting—became scarce and high-priced as the supply of whales depleted. The high price, however, provided the incentive to develop alternative lighting fuel such as fossil fuels. Some historians believe that the depletion of the whales encouraged the discovery of petroleum and the use of oil derived from it.

Prices also answer the "for whom" question in a market society. Just as the prices and quantities of all goods and services available in a market system are determined by the demands of buyers and sellers, so are the "prices" (such as wage rates, rental rates, and interest rates) of resources used to produce them. The value of particular resources in general depends on the demand for the product that the resource helps produce. For example, the owner of property bordering Central Park in New York City may expect to receive a higher rental than one who owns desert property in Nevada. Luciano Pavarotti, a world-renowned operatic tenor, may be expected to receive a higher wage than a college president. The money rewards to these resources are determined not only by the demand for them but also by their relative scarcity. First-class tenors are scarcer than college presidents.

Contemporary Economic Organizations

There are no pure economic systems—all societies contain mixed elements of traditional, command, and market systems. In most contemporary societies, however, one type of economic organization predominates. Within some Third World countries, especially in tribal cultures, tradition predominates. But more advanced countries also include some elements of tradition. For instance, generations of families of doctors, lawyers, and crafts people are common in American life. Commandlike elements also exist in the American economy. Taxes in general require societal consent, but the power to print money is a command function given to government through the Constitution until changed by the will of the people. Indeed, we may think of all of the functions of local, state, and federal governments in the United States as temporary command functions that may be changed periodically through the elective process.

The United States is in general a market economy with elements of command and tradition, while the Soviet Union and China are primarily command societies with some traditional and market characteristics. The Russians have at times used prices to allocate resources, but within an overall command scheme. Chinese society, since the death of Mao Tse-tung in 1976, has begun using market incentive in the production and sale of goods, especially agricultural products. These economies are primarily command-oriented, however. Basic decisions to allocate most available resources are left to central authority in the Soviet Union and China. Individual citizens cannot decide, either through dollar votes or through the political process, to change the direction of major resource allocations. In the United States, on the other hand, consumers and voters are sovereign decision makers with regard to how resources are allocated and to how the three fundamental questions are answered.

The important question is which of these economic systems expands or retards specialization, production, and trade. Of market economies and command economies—the two leading systems in the modern world—which is better at furthering economic growth and well-being? The question is loaded because the answer in part depends on the value one places on political and economic freedom. But most Western economists see the market system as the greatest force for economic growth within the context of political and economic freedom. The U.S. market economy—its functioning and its problems—therefore receives the lion's share of attention in this book. We show that the American economy is a one-of-a-kind economy but that it shares some important features with other, similar market systems. One shared feature is the use of money and a circular flow between money, goods, and resources.

Money in Modern Economies

Barter: The trading of goods for goods with no medium of exchange such as money.

Despite their differences, a feature of all developed economic systems is the use of money. **Barter**—the exchange of goods for goods—is cumbersome and costly. For a system to take advantage of specialization, output must grow steadily. For output to expand, trade must take place at a low cost. But the costs of bartering goods are numerous, for the value of each good must be recorded in terms of each other good traded. As the number of goods traded increases, calculations become overwhelming. Money was invented as a medium of exchange and as a unit of accounting. It is a good in terms of which the value of all other goods can be calculated. Money lowers the cost of transacting and is therefore essential to any modern society.

We can think of money as a substitute for real things traded. Wages, for example, represent a trade of the output of a worker for the goods and services consumed by the worker. We may earn wages as a bricklayer and purchase fried chicken at a fast-food establishment. In a real sense, we are bartering bricklayer labor for fried chicken. Money simply substitutes for real trades.

The Circular Flow in a Market Economy

The exchange of money for goods and services takes place throughout the United States and other developed economies. One can conceptualize this exchange as a **circular flow of income** (Figure 3–1), a cyclical pattern of money payments and services and goods produced.

Circular flow of income: The flow of real goods and services, payments, and receipts between producers and suppliers.

In Figure 3–1, the arrows on the outer circle represent the flow of real goods and services between business firms and households, while the arrows on the inner circle represent the reciprocal flow of money payments. In the circular flow of products, business firms produce products and services with resources (land, labor, and capital) from resource suppliers. To complete the circle, households that purchase products and services are virtually all resource suppliers—of labor, land, or capital. The flow of products has a corresponding and counterbalancing money flow in the opposite direction. For goods and services, firms receive money payments (business receipts) from households, and for resources (labor, land, and capital) resource owners receive money income in the form of rents, wages, and interest from firms. The quantities and the mix of resources used and products produced are determined by price signals. These signals, as we will show more clearly in Chapter 4, are the result of the impersonal forces of supply and demand.

Figure 3–1 is not meant to depict a functioning economy in detail but to serve as a model for real and money flows. Missing are the complexities of flows by which producers supply goods to other producers and of the tax and expenditure flows by which government provides public goods and services. These complexities are discussed later in the book.

Institutions of American Capitalism

While money and real goods circulate through all modern economic systems like blood through the veins and arteries of the body, societies vary in the institutions they create for ownership and use of goods and money. In a **socialist economy,** goods and services and resources are owned by the society as a whole and distributed, in command fashion, by the government. A **capitalist economy** relies primarily on market forces and the profit motive for production, distribution, and consumption of goods and services. Capitalism is characterized by individual ownership of property, free enterprise, open competition, and a minimal role for government.

Socialist economy: An economic system in which the means of production are owned and controlled by the government.

Capitalist economy: An economic system in which the means of production are privately owned.

Property and the Law

In the American economy the individual's right to own and dispose of property is regarded as basic. Property includes both physical property (such as houses and automobiles) and intellectual or intangible property. For example, the ownership of poems, songs, and books by their authors is protected by copyright laws; inventions are protected by patents. The legal apparatus set up to protect such rights may even make a distinction between property and property rights. Rental of a carpet cleaner from U-Rent-Um gives the renter certain property rights over the cleaner but not ownership of the property itself. In all societies, rights to use property are limited. For example, the Environmental Protection Agency has used its legislatively derived power to limit the rights of businesses to pollute air and water and has established

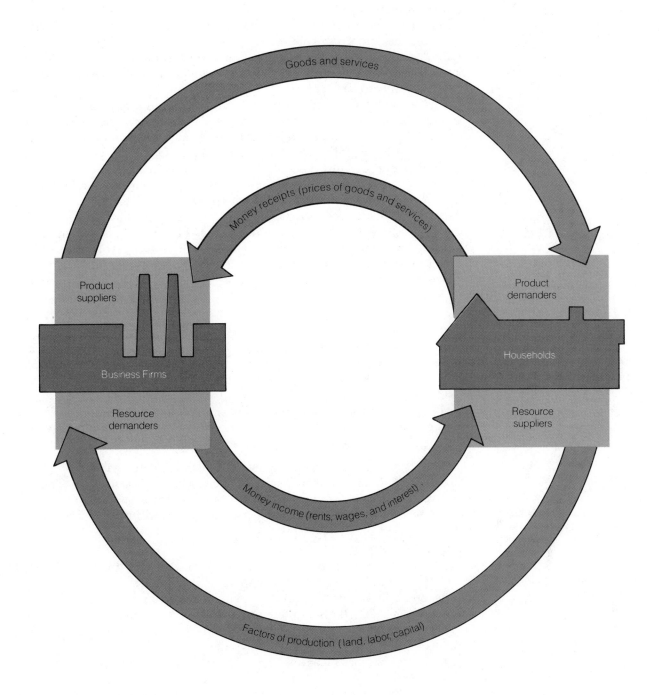

FIGURE 3–1   The Circular Flow of Income

Economic activity between business firms (producers) and households (consumers) takes place in a cyclical pattern. The arrows on the outer circle represent the flow of real goods and services (consumer products and production factors such as labor); the arrows on the inner circle represent the reciprocal flow of money payments for real goods and services.

worker protection standards (such as those limiting the use of asbestos) that restrict businesspeople's free use of property.

In a capitalist society, the law also protects property rights. Without some guarantee that property rights will be protected, there would be little incentive to accumulate capital stock and therefore to grow economically. Without state guarantees of rights to property, individuals would have to protect their own property at high cost.

Legal protection of property rights often emerges as a response to market activities. Consider an example from the American frontier west. In the early days of silver and gold prospecting, individuals were forced to protect their own mining claims. As their claims grew, individual protection entailed higher opportunity costs; that is, higher costs in time spent away from mining or prospecting and in potential loss of equipment or mined ore. Thus, the collective benefits of enforcement through police, courts, and prisons outweighed the costs of such enforcement to individual prospectors—paying taxes, serving on posses, and the like. Property rights and the laws that protect them emerge when benefits to the parties involved begin to exceed the costs of acquiring and enforcing the rights. Such rights to private property, established by law, have been a part of American capitalism since the foundation of the nation.

Free Enterprise

Free enterprise: Economic freedom to produce and sell or purchase and consume goods without government intervention.

Free enterprise, the freedom to pursue one's economic self-interest, is an intrinsic part of the capitalist system. Men and women are free to choose their line of work with few or no governmental restraints or subsidies, and businesspeople are free to combine any resources at their command to produce products and services for profit. Laborer-consumers are free to produce, purchase, and exchange any good or service so long as their activity does not infringe on others' rights. Free enterprise, in sum, means that

1. laborers are free to work at any job for which they are qualified;
2. business firms and entrepreneurs can freely combine resources, at competitive market prices, to take advantage of profit opportunities; and
3. consumers can decide what products and services will be produced.

Competitive Economic Markets

Competition: A market situation satisfying two conditions—a large number of buyers and sellers and free entry and exit in the market—and resulting in prices equal to the costs of production plus a normal profit for the sellers.

The American economic system is also characterized by free competitive markets. **Competition** entails two important conditions: a large number of buyers and sellers and free entry and exit in the market. When these two conditions are met, the self-interested actions of buyers and sellers tend to keep prices of goods and services at a reasonable level, usually the costs of production plus a normal profit for the sellers.

When the number of buyers and sellers is large, no individual buyer or seller can affect the market price of a product or service. Many millions of individuals purchase canned soup, for example, but no one buyer purchases enough to affect the market price of the soup. Likewise, the existence of competing suppliers means that no individual seller can acquire enough power to alter the market for his or her gain. Sellers of canned soup are numerous enough so that no one seller can affect the price of soup by increasing or decreasing output.

Crucial to the effects of large numbers is the condition that firms be free to enter and leave markets in response to profit opportunities or actual losses. New firms entering particular lines of business, bankruptcies, and business

FOCUS   Adam Smith's "Invisible Hand"

Economists tend to praise a free and unfettered competitive market system because the rational and self-interested forces that characterize economic behavior lead not to a permanent state of chaos but to a harmony of interests. Adam Smith had great insight into the matter more than two hundred years ago in his "invisible hand" passage in the *Wealth of Nations:*

> Every individual necessarily labours to render the annual revenue of the society as great as he can. He generally, indeed, neither intends to promote the public interest, nor knows how much he is promoting it. By preferring the support of domestic to that of foreign industry, he intends only his own security; and by directing that industry in such a manner as its produce may be of the greatest value, he intends only his own gain, and he is in this, as in many other cases, led by an invisible hand to promote an end which was not part of his intention. Nor is it always the worse for the society that it was no part of it. By pursuing his own interest he frequently promotes that of the society more effectively than when he really intends to promote it. I have never known much good done by those who affected to trade for the public good. It is an affectation, indeed, not very common among merchants, and very few words need be employed in dissuading them from it.[a]

Smith felt that individuals' tendency to act in their own self-interest is a natural law and a natural right that precedes the existence of government. The exertion of these individual rights in a competitive market setting, furthermore, creates the greatest good for the

[a]Adam Smith, *An Inquiry into the Nature and Causes of the Wealth of Nations,* ed. Edwin Cannan (1776; reprint, New York: Modern Library, 1937), p. 423.

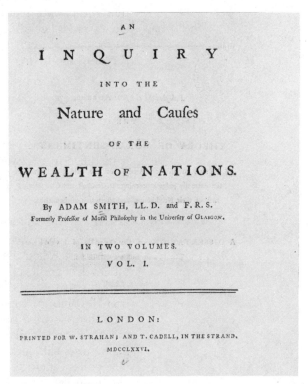

Title page from the original Wealth of Nations, *1776*

greatest number in society. Smith's view, although a mainstream perspective in American capitalism, has been amended to accommodate government provisions of goods when the market fails to provide them in sufficient quantities.

failures are expected consequences of a competitive system. New fast-food restaurants open every day in anticipation of profits. Airlines declare bankruptcy and leave the industry—a sure sign of losses. Competition requires that entry and exit into business be free and unregulated. The Focus, "Adam Smith's 'Invisible Hand,' " discusses some aspects of the competitive system.

Coordinating the billions of individual decisions involved in competition is an interconnected system of prices for inputs and outputs that is so complex that no individual or computer can fully comprehend it. We begin to study the intricacies of the price system in Chapter 4.

The Limited Role of Government

American capitalism as an economic system requires an attitude of laissez-faire (from the French, meaning roughly "to let do"). Laissez-faire has come to mean minimum government interference and regulation in private and

Laissez-faire economy: A market economy that is allowed to operate according to competitive forces with little or no government intervention.

economic lives. In a pure **laissez-faire economy,** government has a role limited to setting the rules—a system of law establishing and defining contract and property rights, ensuring national defense, and providing certain goods that the private sector cannot or would not provide. The last category includes roads, canals, and a banking system. Another important feature of laissez-faire society is the government's legally limited or sharply curtailed access to the taxing power.

The Mixed System of American Market Capitalism

In truth, no society ever conforms totally to the laissez-faire ideal. The ideal is modified in two ways: in an altered notion of the competitive process and in an expanded role for government. Such a modification is called **mixed capitalism.**

Mixed capitalism: An economy in which both market forces and government forces determine the allocation of resources.

An Evolving Competitive Process

Was any country ever composed of so many competing buyers and sellers to become, in Adam Smith's phrase, a "nation of shopkeepers"? Although historical data are less than perfect, we are fairly certain that purely competitive market structures did not exist even in Adam Smith's time. With the Industrial Revolution, capital requirements of firms were such that the most efficient firms—those producing goods at lowest cost—became larger. Certain industries and markets no longer had large numbers of competing sellers. Economists call such markets imperfectly competitive. They are also called oligopolies (characterized by a few competitors) or monopolies (having a single producer).

Some economists have argued that the decline in the number of competitors in some markets has led to concentrations of economic power in the hands of a few and to the demise of the laissez-faire competitive system. Modern economic research into the competitive process disputes this position, however. In the new view, competition is not to be described by a given number of sellers and buyers but rather by a rivalry for profits—that is, a process. Such rivalry—or even the potential for it, as long as individuals and businesses are free to enter and exit the market—produces results similar to competition among many buyers and sellers. One or two sellers in an industry can be competitive as long as entry and exit in the market is possible.[1]

The Expanded Role of Government

The most important modification in the traditional conception of laissez-faire capitalism is an expanded social and economic role of government. Since the turn of the century, and especially since the 1930s, the relative size of government in the United States has grown dramatically in both social and economic spheres. In the 1960s and 1970s we saw large increases in government payments to individuals through Social Security, Aid to Families with Dependent Children, Medicare and Medicaid, unemployment compensation, and other welfare programs. The direct economic activity of government has grown apace.

[1]See Isreal M. Kirzner, *Competition and Entrepreneurship* (Chicago: University of Chicago Press, 1973), for more details on rivalrous competition.

Public Goods and Externalities. Theoretically, underlying the government's role in economic life is the failure of a free-market society to satisfy all of its members' needs. The market society can fail in its ability to provide **public goods** such as national defense. Since national defense protects all citizens regardless of whether they pay for it, no one is likely to contribute to defense voluntarily. The private market fails in the sense that public goods such as defense would not be provided (or provided in sufficient quantity) unless government assumed responsibility.

Another cause for government intervention in a free-market economy is what economists call an *externality*. An externality is an unintended by-product of some activity, and it often involves environmental protection. A beautiful garden creates a **positive externality** in that it confers benefits to neighbors for which they do not pay. A **negative externality** might arise from a factory belching smoke or a firm dumping chemical wastes into a stream. In such a case, costs are imposed on members or segments of society rather than limited to the perpetrators of the externality. Negative externalities have led to various government interventions when the market has failed to limit the cost to the perpetrator. Taxes, subsidies, quotas, prohibitions, and assignment of legal liability are examples of government intervention (see Chapter 18).

Antitrust and Monopoly Regulation. Another broad area of increased government participation in the U.S. free-market economy is **industry regulation.** In the early decades of this century, antitrust laws prohibiting price discrimination, collusion among producers, and deceptive advertising practices were passed in an attempt to restore competition where it no longer existed. Such laws continue to be enforced today.

Even earlier, however, economists and politicians believed that government had a role whenever competition could not exist, perhaps because of economies of large-scale production, or what economists call *natural monopolies*. Such monopolies are created when each seller can produce more and more output at lower and lower costs. Eventually it becomes profitable for only one seller to supply the *total* quantity demanded of a good, thereby creating a monopoly. Federal, state, and local governments undertook the regulation—not the ownership or operation—of transportation, communications, energy, and many other industries that were regarded as natural monopolies. Government regulation, in this view, was regarded as a substitute for competition where viable competition could not exist because of industry production and cost conditions.

Some economists (see Chapter 17) have strongly disputed this view and question the existence of large-scale economies (natural monopoly) in many of these regulated industries. Some contemporary economists believe that regulation of prices and profits must fail, either because regulation has been ineffective or because these industries are more competitive than previously thought. Questions have been raised about the self-interested supply of regulation by politicians combined with the self-interested demand for regulation by firms and industries. Do industries and other interest groups use the government regulatory apparatus for their own benefit? Should broad areas of regulation of industry, such as regulation of transportation and trucking, be eliminated? These and many other issues concerning the expanded role of government are in hot debate. A firm foundation in economic theory is required to answer these important questions.

Public goods: Goods that no individual can be excluded from consuming, once it has been provided to another.

Positive externality: A benefit of producing or consuming a good that does not accrue to the sellers or buyers but can be realized by a larger segment of society.

Negative externality: A cost of producing or consuming a good that is not paid entirely by the sellers or buyers but is imposed on a larger segment of society.

Industry regulation: Government rules to control the behavior of firms, particularly regarding prices and production techniques.

Economic stabilization: A situation in which aggregate economic variables such as inflation rate, unemployment rate, interest rate, and growth rate are fairly constant over time.

Economic Stabilization. A final, but crucially important, part of the expanded role of government is in macroeconomic **stabilization** of the economy—that is, the government's efforts to promote full employment of resources without creating increases in the price level (or inflation). Taxation, expenditure policies, and the money supply can be intentionally changed by the federal government to help maintain full employment and promote economic growth at noninflationary levels.

Whether the government is or is not capable of achieving macroeconomic goals of full employment without inflation is a subject of debate among economists. Indeed, fully half of this book is devoted to alternative views on the role of government in economic stabilization. For the present, however, we should note that the mission to stabilize business cycles of inflation and unemployment has been a generally accepted role of government since the 1930s—the years of massive unemployment of resources known as the Great Depression. Much of the impetus to assign this new function to government came from the writings of the British economist John Maynard Keynes (1883–1946). Keynes believed that government could actually influence employment and inflation through spending and taxing policies and could thereby prevent depressions or severe reductions in economic activity. Parts V, VI, and VII of this book studies Keynesian macroeconomics in depth along with its more modern extensions and alternatives.

The role of government in American economic life is much larger than it was fifty years ago. The American market system is modified laissez-faire. Government has provided a large number of social and economic goods, regulated markets and externalities such as pollution, and attempted to establish a high rate of economic growth through full employment without inflation. These microeconomic and macroeconomic functions of government receive a great deal of attention later in this book. A brief look now at the relative size of government's role in the private economy will provide some perspective on the modified system of laissez-faire that constitutes the American market system.

Growth of Government

Private sector: All parts of the economy and activities that are not part of government.

How big is government in our mixed American economy? Should the economic role of government be larger or smaller in relation to that of the **private sector**—all nongovernment activities? This critical matter is the subject of an ongoing debate among political candidates, members of Congress, journalists, local, state, and federal voters, intellectuals, academics, and private citizens. Rather than developing value judgments about how big government should be, economists analyze the probable effects of political decisions about the role of government in the present and in the past. To provide background information for such an economic analysis, we examine the size of the U.S. government's economic role by looking at its expenditures and the taxes it levies to support them, contrast the level of U.S. government taxation and spending with that of some other developed nations, and then relate this information to patterns of economic growth.

The Size of the Federal Budget

We can get an overview of the role of the federal government in the U.S. economy by looking at a gross measure of government expenditures and receipts over the past three and a half decades. Figure 3–2 depicts the federal budget (expenditures and tax receipts) for the years 1950–1984. After a slow

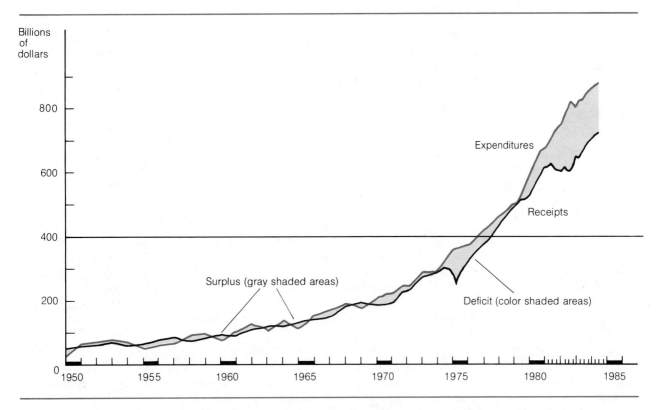

FIGURE 3–2   Growth of the Federal Budget, 1950–1984

Since 1950 the federal budget (the total of expenditures and receipts) has grown both absolutely, in total dollars, and as a percentage of the gross national product. Federal deficits have also grown dramatically in recent years. In 1984, the federal deficit was over $175 billion.

Source: Board of Governors of the Federal Reserve System, *Historical Chart Book* (Washington, D.C.: Government Printing Office, 1984), p. 51.

rate of growth through the mid-1960s, expenditures and tax receipts began to increase at a faster pace.

The Great Society welfare programs begun by President Lyndon B. Johnson in the mid-1960s and the defense expenditures of the Vietnam War were partly responsible for the absolute increases in government spending and taxation. Especially during the 1970s, deficits appeared when expenditures exceeded tax revenues. (The economic effects of these deficits are explored in later chapters.) The government's economic growth is not only of recent origin, however. The Franklin D. Roosevelt administration's social and economic programs in the 1930s—a response to the largest worldwide depression in modern history—were an initial and important force for the expansion of government into a mixed economy. The participation of government at all levels—federal, state, and local—has increased dramatically since 1930.

Since prices of goods and services have risen considerably during the twentieth century, the absolute dollar increase in government expenditures does not necessarily indicate whether government has grown bigger relative to the private sector. For this information, economists often look at government expenditures as a percentage of the **gross national product,** or GNP. The GNP is the aggregate value of all goods and services produced in the country over some period, usually a year. Using this measure, econo-

Gross national product:
The dollar value in terms of market prices of all final goods and services produced in an economy in one year.

mists have found that while the government accounted for less than 10 percent of all purchases of goods and services as a percent of gross national product in 1929, by 1984 government purchases of goods and services were responsible for more than 20 percent of GNP. Since 1960, government's percentage of purchases of goods and services has remained fairly constant at 20 percent, but this constancy understates the growth of the government's role in the economy. The government's tax receipts at all levels in 1984 accounted for more than 30 percent of GNP. We can understand the discrepancy between expenditures and receipts and the expanding role of government by examining the kinds or distributions of expenditures at the various levels of government and then at the ways the government collects revenues.

Government Expenditures

Direct government purchases: Real goods and services such as equipment, buildings, and consulting services purchased by the government.

Government transfer payments: Money transferred by government through taxes from one group to another, either directly or indirectly; also called income security transfers.

There are two kinds of government expenditures: direct purchases of goods and services and transfer payments. **Direct purchases** of newly produced goods and services include such items as missiles, highway construction, police and fire stations, consulting services, and the like. In other words, the government purchases real goods and services. **Transfer payments** are the transfers of income from some citizens (via taxation) to other citizens; these are sometimes called *income security transfers* or payments. Examples of transfer payments are Social Security contributions and payments, Aid to Families with Dependent Children, food stamp programs, and other welfare payments. These transfers do not represent direct purchase by the government of new goods and services, but they influence purchases of goods and services in the private sector. They are a growing part of government's role in the mixed economy.

The Distribution of Federal Expenditures. Out of the thousands of items in the federal budget, we can use six major categories to compare expenditures as a percentage of the total federal budget in 1960 and 1984 (see Figure 3–3). Since providing national defense is one of the major functions of the federal government, we would expect defense to account for a large proportion of federal outlays, and it does. National defense expenditures represented about one-quarter of all federal outlays in 1984. The largest single item in the 1984 federal budget, however, was not defense expenditures but income security transfers, which made up about 33 percent of total outlays. Expenditures on interest service of the federal debt, education, and natural resources ranged from over 12 percent to about 1 percent, respectively. The remainder, accounting for only 17 percent of the federal budget, went to such activities and projects as the administration of justice, science and technology, transportation, agriculture, international affairs, energy, the environment, revenue sharing, and the running of general government.

A mere recital of the proportions of outlays in the 1984 budget is not as interesting as a more dynamic picture of how these outlays changed over the two and one-half decades preceding 1984. Using Figure 3–3 we can compare the distribution of expenditures in 1960 and 1984. In 1960, fully 50 percent of federal expenditures was for national defense, while only 19 percent went to income security. Health-related expenditures have also grown from less than 1 percent in 1960 to more than 10 percent in 1984.

Changes in defense spending and in transfer payments between 1960 and 1984 are part of a clear trend over the period, shown in Figure 3–4. Over most of the 1960s, transfer payments grew at a faster rate than defense pur-

FIGURE 3–3
Federal Expenditures, 1960 and 1984

These two illustrations show percentage shares of federal spending. Expenditures at the federal level between 1960 and 1984 underwent dramatic changes. The percentage spent on defense was almost cut in half and the percentage of the budget spent for income security transfers increased by 14 percent. Education and health also received increased shares of expenditures. (Data for 1983 and 1984 are estimated from budget revisions in April 1983.)

Source: Facts and Figures on Government Finance (Washington, D.C.: The Tax Foundation, 1983), p. 108.

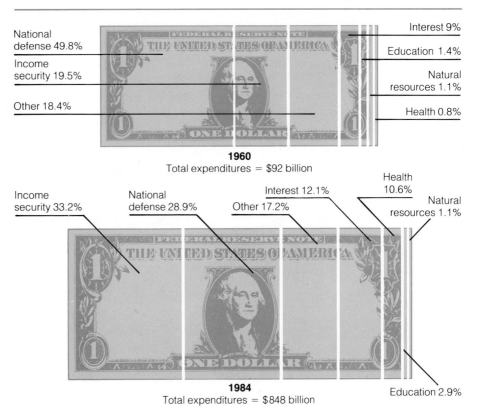

National defense 49.8%
Income security 19.5%
Other 18.4%
Interest 9%
Education 1.4%
Natural resources 1.1%
Health 0.8%

1960
Total expenditures = $92 billion

Income security 33.2%
National defense 28.9%
Other 17.2%
Interest 12.1%
Health 10.6%
Natural resources 1.1%
Education 2.9%

1984
Total expenditures = $848 billion

FIGURE 3–4
Growth in Government Expenditures, 1950–1984

The dramatic growth rate in transfer payments over the 1960s and 1970s is shown in the figure. Absolute amounts spent on transfer payments overtook defense expenditures in about 1970.

Source: Board of Governors of the Federal Reserve System, *Historical Chart Book* (Washington, D.C.: Government Printing Office, 1984), p. 53.

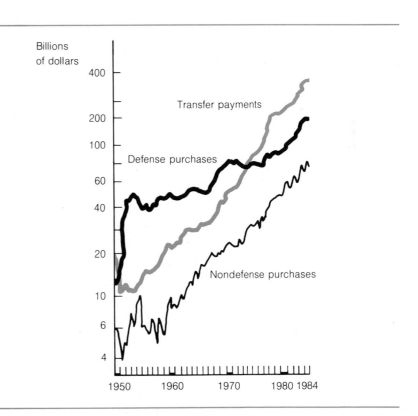

Billions of dollars

Transfer payments
Defense purchases
Nondefense purchases

1950   1960   1970   1980 1984

chases, reflecting the decisions of President Johnson and Congress to attack poverty and social imbalance. In spite of the fiscal pressures of the Vietnam conflict, transfer payments overtook defense purchases in absolute amounts—that is, in actual billions of dollars spent—in about 1970 and have exceeded them every year since.

Will this trend continue? Ronald Reagan campaigned for and won the presidency in 1980 and 1984 partly on this issue, promising to slow the growth rate in income security expenditures and to raise it on defense expenditures. The final outcome is unclear.

State and Local Expenditures. Figure 3–5 shows that the primary public goods provided by state and local government are education, highways, public welfare, hospitals and police, fire, and correctional institutions. Economists, voters, and other observers tend to view the federal government as the principal economic agent in the mixed economy. The truth is, however, that state and local governments combined are larger purchasers of goods and services (as a percentage of GNP) than the federal government. The big difference between the economic impact of the federal government and state and local governments is the huge federal redistribution of funds through the tax system from some citizens to other citizens. When income security transfers are included, the economic impact of the federal budget is larger than that of state and local governments.

Government Receipts: The U.S. Tax System in Brief

Goods and social transfers provided at all levels of government are paid for out of taxation. The type of taxes levied at federal, state, and local levels varies a great deal.

Federal Taxation. The principal source of federal revenues, as shown in Figure 3–6, is the individual income tax. In 1984, the income tax accounted for 45 percent of total federal receipts.

Second in order of importance at the federal level, representing 36.7 per-

FIGURE 3–5

State and Local Expenditures by Category

Principal expenditures at the state and local levels of government are on education, highways, public welfare, and health and hospitals.

Source: Facts and Figures on Government Finance (Washington, D.C.: The Tax Foundation, 1983), p. 178.

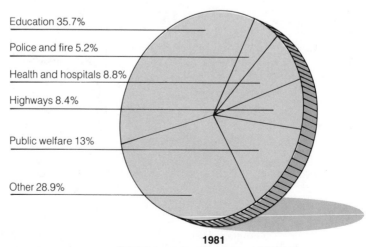

Education 35.7%

Police and fire 5.2%

Health and hospitals 8.8%

Highways 8.4%

Public welfare 13%

Other 28.9%

1981
Total state and local outlays = $407 billion

FIGURE 3–6

Distribution of Federal Tax Receipts, 1959 and 1984

Social insurance tax receipts have more than doubled as a percentage of total receipts between 1959 and 1984, while the relative contribution of the individual income tax has remained almost constant. (Data for 1983 and 1984 are estimated from budget revisions in April 1983.)

Source: Facts and Figures on Government Finance (Washington, D.C.: The Tax Foundation, 1983) pp. 26, 101.

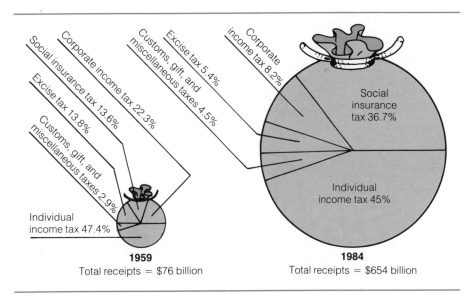

1959
Total receipts = $76 billion

1984
Total receipts = $654 billion

cent of receipts in 1984, are receipts from social insurance taxes and other contributions. These taxes are, principally, the payroll taxes paid jointly by employees and employers that are used to finance Social Security, disability compensation, and other payments. Receipts from these taxes have grown, dramatically, from less than 10 billion in 1950 to more than 200 billion in 1984.

Taxes on corporate income accounted for about 8 percent of revenues in 1984 and have generally declined since 1970 as a percentage of federal revenue. Other sources of federal revenues include federal excise taxes on goods such as liquor, tobacco, and gasoline, customs deposits paid on imports and exports, and estate and gift taxes.

State and Local Receipts. State and local governments rely primarily on property taxes and sales taxes for revenue. An additional revenue source, of varying importance from state to state, is the state income tax. Only about 10 percent of state receipts were from state income tax in 1984. Transfers of revenue from the federal to state and local governments, called grants-in-aid, have assumed increasing importance over the past twenty years. In 1984, for example, federal grants accounted for the highest single percentage of revenue for states and municipalities.

The United States and Other Mixed Economies

The preceding discussions give some indication of the kinds of activities pursued by government in a mixed economy as well as the kinds of taxes the government relies on for revenue. The relative size of government is only hinted at by the breakdowns of outlays and receipts, however. To understand how mixed the American economy is, we can consider some international comparisons.

Figure 3–7 shows the growth of the government's public expenditures as a percentage of gross domestic product for five Western industrialized nations between 1965 and 1977. **Gross domestic product** (GDP), like GNP, is a measure of a country's production of goods and services, but GDP measures final goods and services produced with resources located within the country.

Gross domestic product: A measure of the final goods and services produced by a country with resources located within that country.

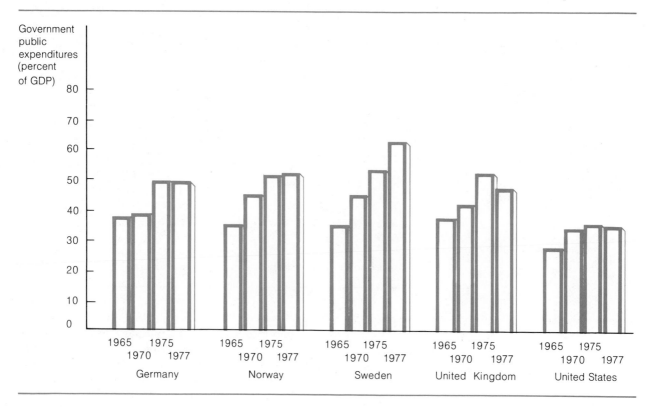

FIGURE 3–7   Relative Growth Rates of Government Public Expenditures in Five Western Democracies, 1965–1977

The bars represent government expenditures as a percentage of gross domestic product (GDP). By this measure, the U.S. economy is the least mixed of five Western democracies.

Source: Organization for Economic Cooperation and Development, *Main Economic Indicators,* various issues.

(GDP is determined by subtracting income earned by foreign investments from the GNP.)

Public expenditures as a percentage of GDP grew between 1965 and 1977 in all of the countries shown in Figure 3–7. The percentage of government expenditures in Sweden almost doubled between 1965 and 1977. By 1977, more than 60 percent of spending in Sweden was directed by the government. Germany, Norway, and the United Kingdom were about evenly divided in their mix between public and private participation in the economy by 1977. The bottom line: Of the leading Western industrialized nations, the U.S. economy is the least mixed in that the government directs only one-third of the country's spending. The trend over the past fifty years, however, has been toward more government participation in Western democracies.

The Economic Effects of Increased Government

Do economists care about how mixed economies are? As noted in Chapter 1, most economists believe that rational self-interested forces and private enterprise are the key factors leading to maximum output, growth, and efficiency. In this view, governmental-directed enterprises—or self-perpetuating bureaucracies—do not normally provide maximum incentives for work, creativity, technology, and economic progress.

High income taxation needed to finance growing government budgets tends to reduce private savings and work effort, forcing taxpayers to engage in more leisure activity and other nonmarket activity. High corporate taxation discourages business investment in capital goods, technology, and innovation, resulting in low rates of economic growth. Most economists would predict that countries with low rates of government participation will have higher rates of capital formation over time, and vice versa. Japan, for example, has only about one-third of its spending directed by the government, but it has one of the highest rates of private investment in the world. In 1980, investment as a percentage of GDP was over 31 percent in Japan; that is, 31 percent of the country's resources were devoted to the production of capital goods for *future* production. The United Kingdom, on the other hand, where government spending accounted for more than 43 percent of GDP in 1978, had the lowest rate of capital formation of Western democracies (17.8 percent of GDP in 1980).

While these relations are merely suggestive, many economists are concerned about the economic impact of larger and larger governments on growth and economic progress (see Economics in Action, "The Mixed Economy of Sweden," at the end of this chapter). Government, as we have seen, has important and legitimate functions. Few economists, moreover, would question some redistribution of income based on a concept of justice since a market system does not automatically produce a just distribution of income or wealth. The question plaguing economists is, as usual, a marginal one: Will added government control over a given amount of private resources increase or decrease satisfaction and economic incentives to work, produce, and invest? Is the relative size of government a tonic to economic society or a sedative? These are extraordinarily difficult questions to answer. Much of this book is devoted to both microeconomic and macroeconomic analyses of these critical questions since American capitalism is, increasingly, a blend of larger government combined with free-market forces.

Summary

This chapter discussed alternative economic systems and their characteristics, with particular emphasis on the American economy.

1. Specialization and trade, which are responses to the problem of scarcity, take place within economic systems. The three main types of economic systems are traditional, command, and market systems.
2. The primitive, traditional society handles the fundamental "what," "how," and "for whom" questions by employing skills and customs passed down from generation to generation. Such societies are economically static because the level of output is strictly limited.
3. Command societies answer the three fundamental questions through a central authority that directs the allocation and distribution of scarce resources. Such societies can be very specialized and technologically advanced.
4. Market societies answer the fundamental questions through the interplay between consumers' dollar votes and the free and unregulated decisions of producers and resource suppliers.
5. All real-world systems are combinations of traditional, command, and market characteristics. Further, modern economic systems are mixed economies because they contain elements of both free-market forces and government provision of goods, services, and income security transfers.
6. Some of the important features of American capitalism include the individual's right to own and dispose of property, a legal system protecting property and contracts, free enterprise, competitive economic markets, and a traditionally limited economic role for government.
7. An expanded economic role for government characterizes contemporary American capitalism. The basis for this role is the provision of goods such as

national defense that would not be produced in sufficient quantity by the private sector. Government also intervenes in the market to block the effects of negative externalities, to regulate industry competition, and to help achieve economic stabilization.

8. Local, state, and federal governments directed about one-third of America's resources in 1984 through purchases of goods and services and through the redistribution of income. Income security transfers are, in absolute terms and as a percentage of total spending, the largest item in the federal budget, whereas expenditures on defense are half the percentage today that they were in 1960.

9. Economists study mixed economies and the role of government to determine whether an enlarged public sector increases economic well-being or reduces incentives to work and to invest capital.

Key Terms

economic system	socialist economy	public goods	gross national product
traditional society	capitalist economy	positive externality	direct government purchases
command society	free enterprise	negative externality	government transfer payments
market society	competition	industry regulation	gross domestic product
barter	laissez-faire economy	economic stabilization	
circular flow of income	mixed capitalism	private sector	

Questions for Review and Discussion

1. How are the "what," "how," and "for whom" questions determined in a market society and in a command society?

2. "They don't build cars like they used to. These days cars wear out before they are paid for." This type of statement is heard frequently. Who determines the quality and durability of products?

3. What is capitalism and what does it have to do with property rights and economic freedom?

4. What are some of the roles of government in the U.S. economy? Has the role of government increased in size and scope? Does this hinder or help the rate of economic growth?

5. What are the primary activities of state and local governments?

6. Can free enterprise exist in a country where a dictator or king has absolute power?

7. "The local cable TV company provides slow services, and it doesn't have many channels." If this cable company has a monopoly granted by the government, is Adam Smith's invisible hand at work?

8. The Oakland city government was disappointed when the Raiders, a professional football team, moved to Los Angeles. Is a professional football team a public good?

9. Does individual self-interest hinder economic growth and well-being if there is competition?

10. What can government do to improve the general economic welfare? Has it done such things?

⌁ ECONOMICS IN ACTION   Exchange in the Bazaar Economy

The main features of a market economy—a well-defined division of labor, the determination of production and distribution by impersonal forces, and specialized institutions serving to minimize transaction costs—are ordinarily associated by economists with industrialized, highly technological societies such as the United States. But this association is apparently not a necessary relation. Complex institutions of market exchange can emerge in a setting that, at least in technological terms, seems quite backward.

An interesting case in point is the bazaar economy of Morocco, which has been studied in detail by Clifford Geertz.[a] Bazaars exist in a society that in most respects has changed little in hundreds of years; by far the most important economic activity is subsistence agriculture.

Bazaars (in Arabic, *suqs*) are quite literally market-

[a]Clifford Geertz, "Suq: The Bazaar Economy in Sefrou," in *Meaning and Order in Moroccan Society: Three Essays in Cultural Analysis,* ed. L. Rosen (Cambridge: Cambridge University Press, 1979); "The Bazaar Economy: Information and Search in Peasant Marketing," *American Economic Review* 28 (May 1978), pp. 28–32.

Scene from a Moroccan bazaar

places. They represent either permanent or in some cases periodic centers in which almost all trade occurs. Their principal features have remained virtually the same for hundreds of years. To the eyes of the typical tourist, a bazaar may appear to be chaos incarnate—a blur of shouting, wildly gesticulating men and women in colorful costumes hawking a confusing array of goods, from bolts of cloth to dates, from cattle to utensils. Goods do not have fixed prices—every trade is assumed to be entirely negotiable. To make matters seem even more baffling, the bazaar is only partially "monetized" in its exchange; much trade occurs as barter.

However, the bazaar is anything but chaotic. In fact, the *suq* represents a complex market structure that resembles the market economies of modern industrialized countries to a remarkable degree.

Despite the apparent technological backwardness of Moroccan society, participants in the bazaar economy invariably occupy specialized niches in a well-defined division of labor. For example, virtually no artisans market their own wares. Shopkeepers and traders in general tend to specialize in one good or in a narrow range of goods. Geertz identified at least 110 different specialized occupational categories in the bazaars around Segrou, a town at the foot of the Middle Atlas Mountains, adding that many more occupations exist but are hard to categorize. This degree of specialization alone suggests a complex and highly organized economic system. Moreover, movement from one occupation to another is common; according to Geertz, "no one in the bazaar can afford to remain immobile; it's a scrambler's life" (1979, p. 185).

Specialized institutions have emerged that minimize the costs of exchange. There is a group of professional auctioneers; another of specialized brokers who act as agents for bazaar traders; and another of *arbitrateurs*, who travel from one bazaar where goods are relatively low-priced to sell them at another bazaar where they can be relatively high-priced. As Geertz explains, in a bazaar—as in any modern market economy—"search is the paramount economic activity, the one upon which virtually everything else turns, and much of the apparatus . . . is concerned with rendering it practicable" (1978, p. 216).

But perhaps the most striking feature of the bazaar is its impersonality. Prices are determined by a process of impersonal competitive bidding in which any buyers and sellers can participate. This is not to imply that bazaar participants (*bazaaris*) are total strangers to one another—usually they are not—but rather that in the bazaar all participants are equals, regardless of their cultural background. Groups who might never associate with each other under any other circumstances (and who may even dislike each other) trade with one another in the bazaar. Jews and Arabs trade freely, although outside of the bazaar they have little or nothing to do with one another. The bazaar does not depend on any particular group for its characteristics but is an impersonal mechanism of exchange. As Geertz notes, "A suq is a suq, in Fez or in the Atlas, in cloth or in camels. The players differ (and the stakes), but not the shape of the game" (1979, p. 175).

Finally, the bazaar economy is completely unregulated. There is no government intervention of any kind. There is not even a government court system—disputes in the bazaar are settled by private arbitrators (specialists in the business of settling disputes). But disputes are relatively uncommon because in the highly competitive bazaar economy, one of the most valuable capital assets a trader can have is a reputation for honesty. Overall, the system seems to function smoothly—if noisily—without government regulation.

Question

Would one expect to find much race or sex discrimination in highly organized markets—such as in the U.S. stock or commodity exchanges—characterized by large numbers of buyers and sellers? Why or why not? Do you think that the possibilities for discrimination are greater when the numbers of buyers and/or sellers are small? Why?

ECONOMICS IN ACTION

The Mixed Economy of Sweden: How Much Government Is Too Much?

Economists have long defended the important economic roles of government in Western democracies. There is little debate about government's role in providing public goods, handling externalities such as pollution, and dealing with economic stabilization. But can the size of government get out of control relative to the size of the private sector? Some economists think so, and many point to Sweden as an example.

In 1977, nearly two-thirds of the gross domestic product of Sweden passed through the hands of either the national or local governments. In Sweden, the public sector, as in other Western democratic governments, has assumed responsibility for many services, such as education, employment, care of the sick and aged, and protection of the environment. But the Swedish government's role as redistributor of income, when coupled with its function as purchaser of goods and services, has made it a far more pervasive economic force than the country's private sector. What have been the economic consequences of the growth of government in Sweden?

Public goods and income transfers must be paid for by citizens, and Sweden, a nation of some 8.3 million people, has the highest tax rates of all Western democracies. Consider the national income tax and the local income tax rates for 1977 presented in Table 3–1. Taxes are translated to U.S. dollars in the table.

For single people earning $40,000 per year in Sweden, the total tax payment is more than 60 percent of $40,000, and, further, the rate on additional income (the marginal percentage rate) is 80 percent! Married people earning $100,000 per year would keep only about $25,000.

Take-home income is of course not the only income a person receives. All citizens are provided with free public services such as medical treatment. Unemployment characterized only 2.5 percent of the labor force in 1981. All individuals have a right to employment in the public sector. Certainly the Swedish welfare state produces welfare for some people, and the economist would count these welfare goods and services as benefits. But what are the costs?

Many economists claim that the high tax rates on Swedish individuals and corporations inhibit work effort, private investment, and formation of capital. If the emergence of the welfare state has in fact had these effects, they should show up in the distribution of employment between the public and the private sector and in the growth record of the Swedish economy. Consider in Table 3–2 how the composition of the public and private labor force changed in percentage terms between 1970 and 1981. Over this decade employment in all private-sector activities either failed to grow or declined. But employment in national and, especially, local government mushroomed. More than 30 percent of all people employed in Sweden now work for the national or local government.

Raw employment percentages in private and public enterprise do not reveal possible growth in the productivity of labor because technological changes may have

TABLE 3–1   Swedish Individual Income Tax Rates, 1977

Pretax Income (U.S. dollars)	Single Persons			Married Couples (one wage-earner)		
	Tax (U.S. dollars)	Percent of income	Marginal rate (%)	Tax (U.S. dollars)	Percent of Income	Marginal rate (%)
5,000	1,129	22.6	32	775	15.5	32
10,000	2,844	28.4	39	2,490	24.9	39
15,000	5,292	35.3	56	4,938	32.9	56
20,000	8,344	41.7	68	7,990	39.9	68
25,000	12,109	48.4	78	11,755	47.0	78
40,000	24,119	60.3	80	23,765	59.4	80
50,000	32,577	65.2	85	32,223	64.4	85
75,000	53,832	71.8	85	53,477	71.3	85
100,000	75,086	75.1	85[a]	74,732	74.7	85

[a]Maximum rate

Source: Svenska Handelsbanken, Sweden's Economy in Figures (1982), courtesy of the Swedish Embassy, Washington, D.C.

TABLE 3–2   Swedish Employment by Sector, as Percentage of Labor Force

	Occupation	1970 (%)	1981 (%)
Private Sector	Agriculture, fishing, forestry	8.2	5.3
	Mining, manufacturing	26.6	22.0
	Electricity, gas, water works	0.7	0.7
	Construction	9.3	7.2
	Wholesaling, retailing, restaurants	14.5	13.3
	Other private and business services[a]	18.9	18.4
	Total	78.2	66.9
Public Sector	National government	6.2	7.3
	Local governments	14.1	23.3
	Total	20.3	30.6
	Unemployed	1.5	2.5
	Total labor force (thousands)	3,910	4,330

[a]Transport, communications, banking, insurance, property management, and so on.
Source: "The Swedish Economy" (Stockholm: The Swedish Institute, 1982).

made labor more productive in both private and public activity. If such were the case in Sweden, we would expect the results to show up in economic growth rates in gross domestic product. Consider in Figure 3–8 the annual percentage average growth in the volume of GDP between 1960 and 1981. By 1982, Sweden had the lowest growth rate of twenty-four Western member nations of the Organization for Economic Cooperation and Development, an international trade and finance organization.

Many economists argue that these results are not surprising. Incentives to work and invest have been progressively reduced with high individual and corporate tax rates and with promises of public employment and guaranteed bureaucratic income. Low and even negative growth rates are the result of reduced private capital investment. This is not to say that the emergence of the welfare state in Sweden has not produced benefits. Economists, however, note that these benefits have come at the cost of work effort, investment, and economic growth. Sweden may find itself with less and less to redistribute as its welfare state grows larger.

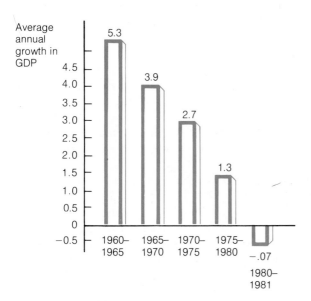

FIGURE 3–8   **Average Annual Percentage Growth of Swedish Gross Domestic Product**

Source: Adapted from "The Swedish Economy" (Stockholm: The Swedish Institute, 1982).

Question

The country of Atlantis provides welfare for all of its unemployed citizens. The law presently states that if welfare recipients find work they will lose an equal amount of welfare benefits. How does this affect the desire to find work? What would happen if the law were changed so that all welfare benefits would not be removed with increases in earned income?

Markets and Prices:
The Laws of Demand
and Supply

S imple specialization began to take place in primitive cultures as individuals recognized unique abilities in themselves and in others. As we have seen, increasing specialization led to organized markets where people bought and sold goods, to the use of money, and to increasingly large groups of buyers and sellers. Economists define these organized markets—such as bazaars, the stock exchange, or Saks Fifth Avenue—as places or circumstances that bring together demanders (buyers) and suppliers (sellers) of any goods or services.

What are the motivations of demanders and suppliers in the exchange of goods and services in these organized markets? How and why do some things get produced and sold at certain prices while other items are not produced or sold at all? Products such as tape cassettes were unknown to our grandparents. How and why do they get produced and sold to us for a certain price at a local music store? What happens to demand and supply when governments or other agencies attempt to intervene in organized markets through policies such as agricultural price supports and rent controls? Can the appearance of surpluses or shortages be predicted? These questions can all be approached through the single theory of supply and demand—the economist's basic tool.

Every day the news media provide dramatic evidence of the workings of the market system. The price of silver rises by 45 percent in one day as massive investments in silver futures by two of the world's richest oil magnates trigger a bandwagon effect. Leak of a technological breakthrough in a certain computer firm creates a frenzy of stock buying, driving up the price of the firm's stock overnight. Crude oil prices fall sharply as a result of the dissolution of the OPEC cartel.

In the familiar economic transactions of everyday life, we too enter markets where buyers and sellers congregate to buy and sell a great variety of products and services. The typical American supermarket sells thousands of products, and as we wander through the store we can view a price system in action. In the produce section, for instance, quantities and prices of fruits and vegetables depend on the quantities consumers want and on the season. Early crops usually bring in the highest prices. It is not uncommon for watermelon to sell for more than two dollars a pound in March but only fifteen cents a pound by the Fourth of July.

What determines who will get the early melons or how they will be rationed among those who want them? Why are prices and quantities constantly rising and falling for millions of goods and services in our economy? How do new products find their way to places where buyers and sellers congregate? The answers are simple. In a market society, the self-interest of consumers and producers, of households and businesses, determines who gets what and how much. To paraphrase Adam Smith, it is not to the benevolence of the butcher and the baker that we owe our dinner but to their self-interest. The primary way that consumers and producers express their self-interest is through the economic laws of supply and demand. Sticking a price tag on a product does not imply price-setting power, as anyone who has run a garage sale knows. In a market system, demand and supply determine prices, and prices are the essential pieces of information on which consumers, households, businesses, and resource suppliers make decisions. High melon prices in March will encourage suppliers and discourage demanders, whereas low prices in July will encourage demanders and discourage suppliers. Before investigating the mechanics of these laws of supply and demand, we consider a simple overview of the price system.

An Overview of the Price System

As market participants, households and businesses play dual roles. Businesses supply final output of products and services—rock concerts, bananas, hair stylings—but also must hire or demand resources to produce the outputs. Households demand rock concerts, bananas, and hair stylings for final consumption but also supply labor and entrepreneurial ability as well as quantities of land and capital to earn income for the purchase of products and services.

Products market: The forces created by buyers and sellers that establish the prices and quantities of goods and services.

Resources market: The forces created by buyers and sellers that establish the prices and quantities of resources such as land, labor, and capital.

As Figure 4–1 shows, businesses and households are interconnected by the **products** (outputs) **market** and by the **resources** (inputs) **market.** Each market depends on the other; they are linked by the prices of outputs and inputs. The particular mix of goods and services exchanged in the products market depends on consumer demands in that market plus the cost and availability of necessary resources. For example, the groups that are featured at a rock concert will depend on what the targeted audience wants to hear plus the ability of that audience to pay the price demanded by the groups and the groups' availability on the chosen date. Similarly, the particular mix of resources available at any one time or through time is determined by what households are demanding—subject also to the availability of the resources. If land suitable for banana growing is available, it is most likely to be sold for banana plantations if households are demanding a lot of bananas, thereby making it possible for banana growers to pay landowners handsomely for their land.

Prices are the impulses of information that make the entire system of input and output markets operate. Take the prices of fad goods: At times certain

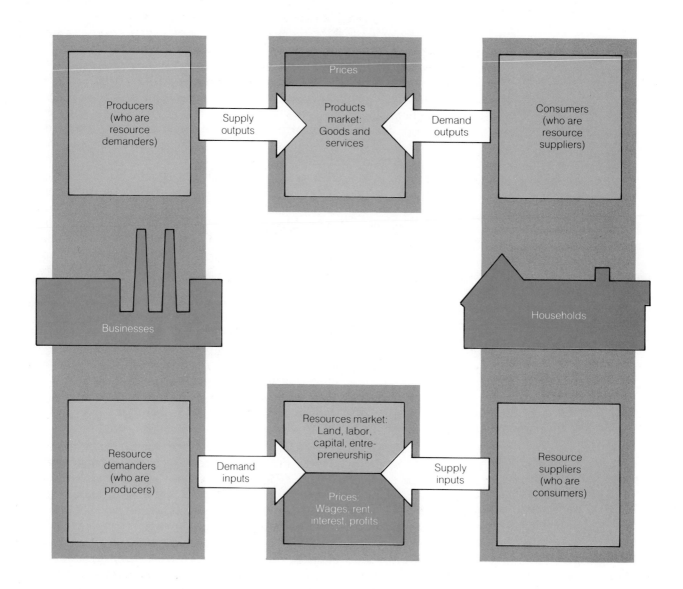

FIGURE 4-1   The Product and Resource Markets: A Circular Flow

Businesses play a dual role in the market economy: They are the suppliers of goods and services as well as the demanders of resources. Households also have a dual role: They are both demanders of goods and services and suppliers of resources.

goods or services—such as Rubik's cubes and hula hoops—have quickly appeared and then disappeared. When a fad begins to catch on, prices tend to be high because there is a high and rapidly growing demand and because resources necessary to produce the good may be scarce and command a high price. It takes time to adapt resources to the production of the fad good.

Machinery must be developed to mold plastic into Rubik's cubes. Labor must be drawn from other uses into the new production. Marketing channels must be established so that sellers and buyers can converge at convenient points of trade. Demand for the fad good is transmitted through businesses to the factors of production through a system of prices. Initially high prices signal the scarcity of goods and resources. As more entrepreneurs and businesses perceive the profit opportunities associated with the fad and as more resources are discovered or developed to produce it, prices of both the inputs and the outputs change accordingly.

The price system reacts similarly with goods that remain in the market for a longer time. Consider the development of computers and computer technology in the 1960s. Initial investments by producers were substantial as businesses rushed to introduce computer systems. In particular, the wages of computer programmers and technicians were high because there was a relative scarcity of workers possessing the skills necessary to use and produce computers. High demand and relative scarcity mean high wages. However, high wages are also an excellent piece of information that may encourage changes that ultimately lower wages. From the 1960s to the present, many schools teaching computer technology have emerged, reducing the scarcity of this resource. Demand for computer technology, however, has grown over time. The wages for people supplying services essential to production depend on consumers' demand for computers and on price (wage)-signaled supply conditions in the market for these services.

The informative signals of a price system work whether goods have short lives (Rubik's cubes) or long lives (computers). The prices formed in both product and resource markets reflect the relative desires of consumer-demanders for particular goods and services as well as the relative scarcity of the resources required to produce them. The very fact that a product or service bears a price means that scarcity exists. Supply and demand in all markets is at the core of scarcity and, therefore, of economics. These critical notions must be understood with the greatest possible clarity. We begin with demand.

The Law of Demand

Law of demand: The price of a product and the amount purchased are inversely related. If price rises, the quantity demanded falls; if price falls, the quantity demanded increases.

Quantity demanded: The amount of any good or service consumers are willing to purchase at some specific price.

Demand curve: A graphic representation of the quantities of a product that people are willing and able to purchase at all of the various prices.

What determines how much of any good—Rubik's cubes, cassette tapes, or hair styling—consumers will purchase over some time period? Economists have answered that question for hundreds of years in the same manner—by formulating a general rule, or law of demand. The **law of demand** states that, other things being equal or constant, the quantity demanded of any good or service increases as the price of the good or service declines. In other words, **quantity demanded** is inversely related to the price of the good or service in question.

The relation between price and quantity demanded is a fact of everyday experience. The reaction of individuals and groups to two-for-the-price-of-one sales, cut-rate airline tickets, and other bargains is common proof that quantities demanded increase with decreases in price. Likewise, gas price hikes will lower the quantities of gasoline demanded. The formalization of this inverse relation between price and quantity demanded is called a law because economists believe it is a general rule for all consumers in all markets. Imagine a graphic representation of the law of demand—called a **demand curve**—for two hypothetical consumers.

The Individual's Demand Schedule and Demand Curve

Suppose that we observe the behavior of Dave and Marcia over one month. These two music lovers own tape players and thus are willing to purchase, or demand, cassette tapes. To determine Dave's and Marcia's demand for tapes, we need only to vary the price of tapes over the month, assuming that all other factors affecting their decisions remain constant, and observe the quantities of tapes they would demand at those prices. This information is summarized in Table 4–1.

Table 4–1 shows a range of tape prices available to Marcia and Dave over the one-month period and the quantities (numbers) of tapes that each would purchase, all other things being equal. Given factors such as their income and the availability of other forms of entertainment, neither person would choose to purchase even a single tape at $10 per tape. Dave, however, would buy one tape at $9 and two at $8. Marcia would not buy her first tape until the price was $7. Each would purchase more tapes as the price falls. Thus, Dave's and Marcia's tape-buying habits conform to the law of demand.

We obtain the individuals' demand curves by simply plotting or transferring the information from Table 4–1 to the graphs in Figures 4–2a and 4–2b. The prices of tapes are given on the vertical axis of each graph, and the quantities of tapes demanded per month are given on the horizontal axis. The various combinations of price and quantity from Table 4–1 are plotted on the graphs. Each demand curve is then drawn as the line connecting those combinations of price and quantity. For both Dave and Marcia, the demand curve slopes downward and to the right (a negative slope), indicating an increase in quantity demanded as the price declines and a decrease in quantity demanded as the price rises.

Factors Affecting the Individual's Demand Curve

Factors affecting demand: Anything other than price that determines the amount of a product that people are willing and able to purchase.

In addition to the price of a good or service, there are dozens, perhaps hundreds, of other factors and circumstances affecting a person's decision to buy or not to buy. These **factors affecting demand** include income, the price of related goods, price expectations, income expectations, tastes, the number

TABLE 4–1   Two Consumers' Demand Schedules

While individuals' demand schedules may differ, they do not violate the law of demand. For both people, the quantity demanded increases as the price falls.

Price of Cassettes (dollars)	Quantity Demanded (one month)	
	Dave	Marcia
10	0	0
9	1	0
8	2	0
7	3	1
6	4	2
5	5	3
4	6	4
3	7	5
2	8	6
1	9	7
0	10	8

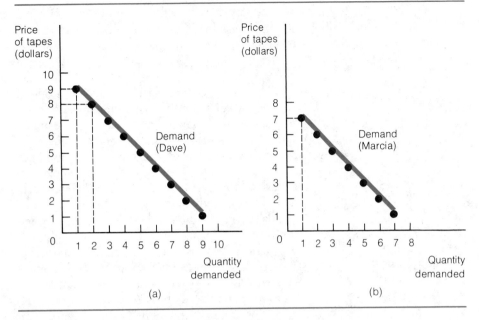

FIGURE 4–2   Demand Curves for Two Consumers

A consumer's demand for a product is the quantity that he or she is willing to purchase at each price. The demand curve is downward sloping for both Dave (a) and Marcia (b): As the price falls the quantity demanded increases, and as the price rises the quantity demanded decreases.

of consumers, and time. Even though this list of other factors is limited, it is too difficult to handle simultaneous variations among all factors in expressing a demand curve or schedule.

Holding Factors Other Than Price Constant. To isolate the effect of price on demand, the nonessential factors must be stripped away. We want to know what quantity of tapes Dave and Marcia would choose to purchase in a month at various possible prices, given that other factors affecting their decision do not change. This condition is called **ceteris paribus** ("other things being equal") by economists. It is essential to the development of any economic theory of model dealing with real-world events since all events cannot be controlled. Economists hold factors such as income and the price of related goods constant when constructing a demand schedule or curve.

This does not mean that these factors do not change, but if they do change, the demand schedule or curve must be adjusted to account for them. Laboratory scientists are in a better position than economists to hold conditions constant. Chemists can perform controlled experiments, but economists, like weather forecasters, deal with a subject matter that can rarely be controlled. But like their scientific counterparts, economists must use scientific methods to organize real-world events into theories of how things work. The economists can use these theories to predict some of the effects on the demand schedule of changes in factors other than price.

Changes in Demand. A simple but crucial distinction exists between a change in Dave's or Marcia's *demand* for tapes and a change in their *quantity demanded* of tapes. Other things being equal, a change in the price of tapes will change the quantity demanded of tapes, as we have seen. A change in

Ceteris paribus: All other things held constant.

Change in demand: A shift of the demand curve or a situation in which different quantities are purchased at all previous prices.

any factor other than the price of tapes will shift the entire demand curve to the right or left. Economists call this a **change in demand.** We consider some possible changes, why they take place, and how they affect the demand curve.

Change in Income. Marcia's or Dave's income may change, and such a change would necessitate a redrawing of the entire demand curve for cassettes. For most goods, a rise in income means a rise in demand. For instance, if Marcia's income increases from $500 to $800 a month, she will demand more tapes at every price because she can afford more. Figure 4–3 shows that, given a new, higher income, Marcia's demand curve shifts to the right for *every* price of tapes. When demand increases, quantity demanded is increased at every price.

Normal good: A product that an individual chooses to purchase in larger amounts as income rises or smaller amounts as income falls.

Inferior good: A product that an individual chooses to purchase in smaller amounts as income rises or larger amounts if income falls.

Although the theory that rising income means greater demand for goods holds true for most goods, it does not apply to all goods. Economists make a distinction between normal goods and inferior goods. **Normal goods** are those products and services for which demand increases (decreases) with increases (decreases) in income; the demand for **inferior goods** actually decreases (increases) with increases (decreases) in consumers' income. Joe's demand for Honda automobiles may decrease as his income increases; a Honda automobile is an inferior good to him. Beth purchases less Häagen-Dazs vanilla ice cream as her income falls, indicating that Häagen-Dazs ice cream is a normal good for her.

The terms *normal* and *inferior* contain no implications about quality or about absolute standards of goodness or badness. Indeed, a good or service that is normal for one consumer in a given income range may be inferior for another consumer in the same income range. It is even possible for a good to be normal for an individual consumer at certain levels of income and inferior at other levels. As one's income rises, for example, hamburger or compact cars may change from normal to inferior. This distinction should not detain us here (we will discuss it in more detail later), but it is important to note that a change in income will produce a shift in the demand curve.

Substitutes: Products that have a relation such that an increase in the price of one will increase the demand for the other or a decrease in the price of one will decrease the demand for the other.

Complements: Products that have a relation such that an increase in the price of one will decrease the demand for the other or a decrease in price of one will increase the demand for the other.

Price of Related Goods. Suppose that the price of a good closely related to tapes—such as records or cassette tape players—changes during the month for which Dave's and Marcia's demand curves are drawn. What happens? Clearly, one of the assumptions about other things being equal has changed, and the demand curve will shift right or left depending on the direction of the price change and on whether the closely related good is a **substitute** for or a **complement** to the product under consideration.

Suppose that other forms of entertainment can substitute for tapes in Dave's or Marcia's budgets. If the ticket price of local movies or record prices decline during the month, the demand curve for tapes for both consumers would shift to the left, that is, the demand for tapes would decline. For every price of tapes, the quantity demanded would be lower. This shift is represented in Figure 4–4. If the price of a substitute good increased during the month, the demand for tapes would increase.

If the price of a good or service complementary to tapes rises or falls, the demand curve for tapes would shift. Such a complementary good or service might be cassette tape players for home or auto, cassette tape carriers, or a new music store opening near Dave or Marcia. If the price of the complement increases, the demand for tapes would decrease (shift left). If the price of the complement decreases, demand for tapes would increase (shift right).

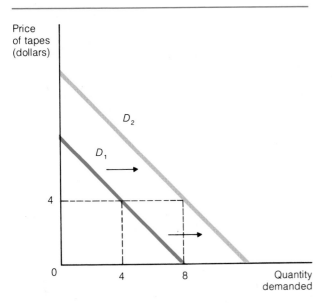

FIGURE 4-3   Increase in Demand

A change in any factors affecting demand causes a shift to the right or left in the demand curve. In this case, Marcia's income increases, causing an increase in her demand for tapes at every price. The demand curve shifts to the right. At a price of $4 per tape, Marcia previously would purchase 4, but given her increased income, she would now purchase 8 tapes. In this representation, D = demand.

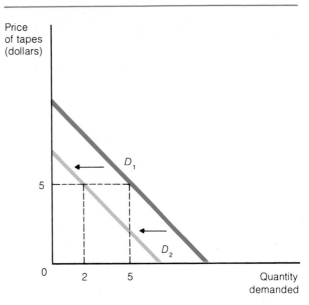

FIGURE 4-4   Decrease in Demand

A leftward shift of the demand curve indicates a decrease in demand. In this case, the price of a substitute good decreased, causing Dave to demand fewer tapes at every price. The demand curve shifts to the left. Before the price change in the substitute good, Dave would purchase 5 tapes at $5 each, but given the change in the substitute good, he would purchase only 2 tapes at $5 each.

Other Factors Shifting the Demand Curve. A number of factors other than income and the price of related goods can cause a shift in the demand curve. Among these are consumers' price and income expectations, consumers' tastes, and the time period. If the price of tapes or the income of consumers is expected to change in the near future, the demand for tapes during the month will be altered. Marcia may discover during the month that she will receive an inheritance sometime in the future. The basis under which her original demand curve was derived changes because she anticipates a change in income. Or back-to-school tape sales might be announced in the middle of July, causing an increase in demand during the time period. Likewise, any alteration in the time period under examination—changing from a month to a day, week, or year—will alter the construction of the curve. A purchase of four pizzas per month at $6.95 would be represented differently than a weekly consumption of one pizza. A change in the time period requires a redrawing of the demand curve.

Reviewing the Law of Demand

The demand curve expresses an inverse relation between the price of a good and the quantity of the good demanded, assuming a number of constant factors affecting demand. As price rises, quantity demanded falls; as price falls, quantity demanded rises. When any one of the non-price factors affecting demand changes, we must reevaluate the demand schedule and curve. In general, we identify only the most important factors affecting demand curves. If we have missed some important factor affecting demand—

Dave's carburetor unexpectedly burns out in the middle of the month, for example—that factor must be accounted for in analyzing demand.

Economists predict that individuals (and collections of individuals), all things being equal, will purchase more of any commodity or service as its price falls. To verify this prediction, the individuals in question do not even have to be fully aware of their behavior; they need only act in the predicted manner. Individuals' response to sales of any kind—they buy more when price declines—is evidence that a general and predictable law of demand exists. That law is a fundamental tool of economists' analyses of real-world events.

From Individual to Market Demand

Market demand: The total demand for a product at each of various prices, obtained by summing all of the quantities demanded at each price for all buyers.

While an individual's demand curve is sometimes of interest, economists most often focus on the **market demand** for some product, service, or input such as automobiles, intercontinental transport, or farm labor. Market demand schedules are simply the summation of all individual demand schedules at alternative prices for any good or service. An increase in the number of consumers increases the market demand curve and a reduction decreases market demand. The key is to add up the quantities demanded by all consumers at alternative prices of the good or service in question.

We can use the tape demand example to understand market demand. We constructed individual demand schedules for Dave and Marcia by varying the price and observing the quantities of tapes that they would buy at those prices, other things being equal. To determine the market demand schedule we simply observe the behavior of all other consumers in the same market situations.

Table 4–2 begins with the data on Dave's and Marcia's demand from Table 4–1. The table also contains a summary of quantity demanded for all other consumers at every price and, finally, the total quantity demanded for

TABLE 4–2   Market Demand Schedule

The total market demand for a product is found by summing the quantities demanded by all consumers at every price.

Price of Tapes (dollars)	Quantity Demanded (one month)			
	Dave	Marcia	All Other Consumers	Total Market Demand
10	0	0	0	0
9	1	0	39	40
8	2	0	78	80
7	3	1	116	120
6	4	2	154	160
5	5	3	192	200
4	6	4	230	240
3	7	5	268	280
2	8	6	306	320
1	9	7	344	360
0	10	8	382	400

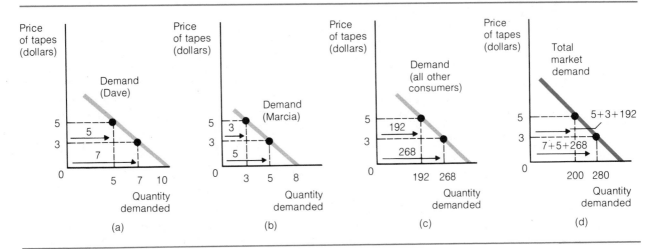

FIGURE 4–5   Market Demand Curve

The total market demand curve for a product is obtained by summing the points on all the individual demand curves horizontally. This is accomplished by selecting prices and summing the quantities demanded by all individuals to obtain the total quantity demanded at each price.

all consumers, or the total market demand. At a price of $10 no one, including Dave and Marcia, wants to buy tapes. At a slightly lower price, $9 per tape, Marcia does not choose to buy, but 40 tapes are sold. The total market demand is, therefore, 40 tapes at a price of $9. Note that actual numbers of tapes sold to all consumers would likely be much higher in any given market. We use low numbers for simplicity. The important point is that the market demand schedule for tapes or any other privately produced product or service is constructed in precisely this manner.

The market demand schedules can be represented graphically as market demand curves (see Figure 4–5). Dave's and Marcia's demand curves are repeated from Figure 4–2. The demand of all other consumers, taken from Table 4–2, is plotted in Figure 4–5c. The total market demand, shown in Figure 4–5d, is simply the horizontal addition of the demand curves of Figures 4–5a, b, and c.

As in the case of individual demand curves, the market demand curve is downward sloping (negatively sloped) and drawn under the assumption that all factors other than the price of tapes remain constant. If the incomes of consumers change, or the price of goods or services closely related to tapes is altered, the market demand would shift right or left, as in the individuals' demand curves. Economists must focus closely on these related factors in any real-world application. Changes in the demand for any product—compact cars, energy, crude oil—will be closely related to factors such as income changes and the price of substitutes and complements.

The important concept of market demand summarizes only half of the factors creating and affecting prices. Like the cutting blades of a pair of scissors, two sets of factors simultaneously determine price. Demand is the first factor. Supply is the second, and not necessarily in that order of importance.

Supply and Opportunity Cost

Indirectly, we have already encountered a supply concept—that of opportunity cost along a production possibilities frontier, which we discussed in Chapter 1. We will briefly review those concepts before shifting attention from the behavior of the consumer to that of the producer.

An opportunity cost—the value of alternative products forgone—is incurred whenever any good or service is produced. Consider again the trade-off between producing oranges and peanut butter in a two-good world, as we did in Chapter 2. See Figure 4–6(a). A given stock of resources is available for the production of these two goods. Society, of course, will choose some combination of the two, and there is an opportunity cost in terms of resources in making a choice of more peanut butter and fewer oranges or more oranges and less peanut butter.

Marginal opportunity costs: The extra costs associated with the production of an additional unit of a product; those costs are the lost amounts of an alternative product.

Consider what the real **marginal** (or additional) **opportunity cost** of producing more oranges is in terms of peanut butter forgone. The marginal cost (short for marginal opportunity cost) of more oranges is simply the amount of peanut butter sacrificed to produce the oranges. This is plotted in Figure 4–6b. From Figure 4–6a we see that society's choice of 4 million tons of oranges is made at a cost of 10 million tons of peanut butter (point G in Figure 4–6b). An additional, or marginal, 4 million tons of oranges can be produced at an additional cost of 15 million tons of peanut butter (point H). Additional oranges will be produced at higher and higher opportunity costs of peanut butter. Figure 4–6b gives the marginal cost curve in terms of real

FIGURE 4–6
Production Possibilities and Marginal Opportunity Cost

The opportunity cost of a product rises as more and more of that product is produced. This is shown in the production possibilities curve as well as in the marginal opportunity cost curve.

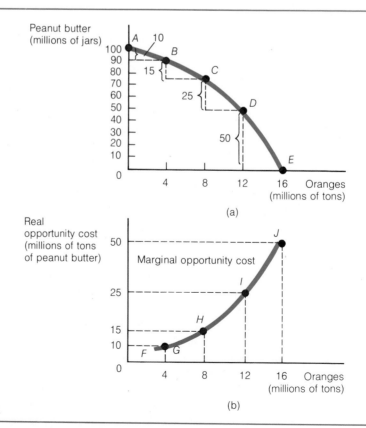

opportunity cost. Note that marginal cost increases as more oranges are produced. Why? Common sense suggests that there are increasing costs of transforming resources from peanut butter production to orange production. At first the resources drawn from peanut butter are easily adapted to producing oranges (that is, there is a low opportunity cost); as more and more oranges are produced, less-adaptable and less-talented resources must be used. The concept of the law of supply, discussed below, is totally analogous to this real opportunity cost except that prices (or opportunity cost in money terms) serve as the proxy for the real output of some good given up (peanut butter) to produce another good (oranges).

Thus, any production involves a real opportunity cost to society. The increasing money costs of producing additional tapes, for example, is merely a reflection of the higher real opportunity costs of drawing resources from other productions. It is crucial to remember that money prices, which represent real opportunity costs, are merely symptoms of the real factors underlying the economy. Scarcity and opportunity cost do not vanish, in other words, in a highly developed economy. Relative resource prices and, consequently, costs of production are but relative signals of scarcity.

The Law of Supply and Firm Supply

Law of supply: The price of a product and the amount that producers are willing and able to offer are directly related. If price rises, then quantity supplied rises; if price falls, then quantity supplied falls.

Quantity supplied: The amount of any good or service that producers are willing to produce at some specific price.

Supply schedule: A schedule or curve that shows the quantities of a product that producers are willing and able to offer at all prices.

The **law of supply** states that, other things being equal, firms and industries will produce and offer to sell greater quantities of a product or service as the price of that product or service rises. There is a direct relation between price and quantity supplied: As price rises, **quantity supplied** increases; as price falls, quantity supplied decreases. The assumption of other things being equal is invoked, as in the case of demand, so that the important relation between price and quantity supplied may be specified exactly. This relation, called a **supply schedule,** shows the quantities of any good that firms would be willing to supply at alternative prices over a specified time period.

The method for constructing an individual firm's (and the market's) supply schedule is identical to the method we used for individual and market demand. All factors affecting supply except the price of the good or service are held constant. The price of the good or service is varied and the quantities that the firm or the industry will supply are specified.

To see how the supply curve is drawn, we turn to the supply side of cassette tapes. Suppose that there are a number of firms supplying tapes in a given geographic area and that the output per month of two typical firms (Grooves Inc. and Joe's Tapes) and all other firms combined is as shown in Table 4–3. The supply schedules of Grooves Inc. and Joe's Tapes are given in the table by the combination of price and quantity supplied. That is, given alternative prices of cassettes and the assumption that all other things are equal, Grooves Inc. and Joe's Tapes specify the quantity of tapes that they would be willing to supply during a one-month period. (Again, numbers are kept arbitrarily low for simplicity.) At a price of $10 per tape, Grooves Inc. would be willing to supply 24 tapes, but if the price falls to $3 per tape, Grooves Inc. will supply only 3 tapes.

As in the case of demand schedules, the information from supply schedules can be graphically expressed as supply curves (see Figure 4–7). The individual supply curves for Grooves Inc. and Joe's Tapes conform to the law of supply—other things being equal, as price rises, the quantity supplied increases and as price falls, quantity supplied decreases. Note that the supply curve for the individual firm is sloped upward and to the right. This is be-

TABLE 4–3   Individual Firm and Market Supply Schedules

Individual firms' supply schedules follow the law of supply: As the price of the product increases, the quantity supplied increases. The total market supply is obtained by summing the quantities supplied by all firms at every price.

Price of Tapes (dollars)	Quantity Supplied (one month)			
	Grooves Inc.	Joe's Tapes	All Other Firms	Total Market Supply
10	24	16	410	450
9	21	14	365	400
8	18	12	320	350
7	15	10	275	300
6	12	8	230	250
5	9	6	185	200
4	6	4	140	150
3	3	2	95	100
2	0	0	50	50
1	0	0	0	0
0		0		

cause the marginal opportunity cost of resources used for increased tape production rises as more tapes are produced. As more tapes are produced, less-adaptable resources are drawn into tape production, just as in the case of peanut butter and orange production. To increase the quantity of tapes supplied, a firm may have to enlarge its quarters by buying and converting buildings formerly used for other purposes, incur the costs of hiring and training workers who have never made tapes before, and perhaps redesign its product to use alternative materials as original materials become more scarce. These

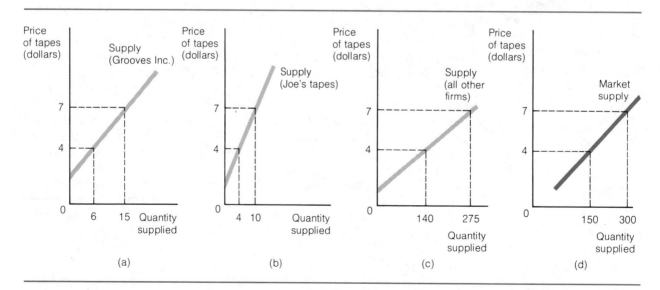

FIGURE 4–7   Market Supply Curves

The market supply, or total quantity supplied, of a product is obtained by summing the quantities that will be supplied by the two individual firms and by all other firms at every price. For example, at a price of $7 the total quantity supplied is 300 (15 + 10 + 275).

increases in marginal costs may make it unprofitable for the firm to supply more tapes unless the price of tapes rises.

Changes in Quantity Supplied and Shifts in the Supply Curve

As in the case of demand, a change in price will alter the quantity that producers are willing to supply, and a change in any other factor will cause a **change in supply,** indicated by a shift in the supply curve either right or left. An increase in price will increase the quantity supplied, but a decrease in price will reduce the quantity supplied. The supply curve is positively sloped—upward and to the right—and, as we saw, the demand curve is negatively sloped—downward and to the right. When **factors affecting supply—** that is, factors other than price—change, the whole curve shifts. Some examples follow.

Changes in Cost of Production. The most important influence on the position of the supply curve is the cost of producing a good or service. The price of resources—labor, land, capital, managerial skills—may change, as may technology or production or marketing techniques peculiar to the product. Any improvement in technology or any reduction in input prices would increase supply, that is, it would shift the supply curve to the right.

Suppose that the price of plastic materials used in cassette tape construction falls or that the wages of salespeople available to tape stores decline. As the production or sales costs to firms producing and selling tapes decline, the quantity supplied increases and the supply curve shifts to the right for every price of tapes. An increase in supply is shown in Figure 4–8a. At price P_0, the firm was willing to supply quantity Q_0 of tapes, and the supply curve was S_0. After the firm (and all other firms) experiences a reduction in costs, the supply curve shifts rightward to S_1, indicating a willingness to supply a quantity Q_1 at price P_0. An improvement in production or sales techniques or a reduction in the price of some resource shifts the supply curve to the right.

Change in supply: A shift in the supply curve or a situation in which different quantities are offered at all of the previous prices.

Factors affecting supply: Anything other than price that determines the amount of a product that producers are willing and able to offer.

FIGURE 4–8

A Shift in the Supply Curve

As factors other than price change, the supply curve shifts. When input costs fall, the quantity supplied increases from Q_0 to Q_1 at the price P_0, shifting the supply curve to the right from S_0 to S_1. An increase in input costs causes a decrease in quantity supplied from Q_0 to Q_2, and the supply curve shifts to the left, from S_0 to S_2.

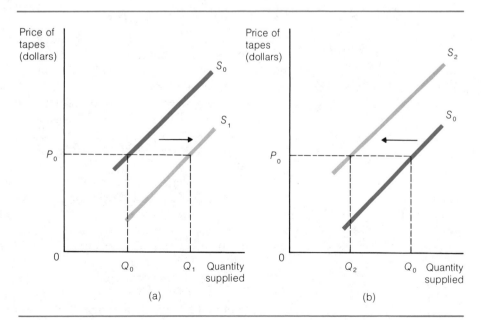

Any decline in production or sales techniques or an increase in any input cost shifts the supply curve to the left, as shown in Figure 4–8b.

Other Determinants of Supply. Factors other than cost of production changes can affect the location of the supply curve. One such factor is changes in producer-seller price expectations. The supply curve, like the demand curve, is drawn for a certain time period. If expectations of future prices change drastically in a market—for example, prices of the good or service are expected to rise suddenly—suppliers would withhold current production from the market in anticipation of higher prices. The current supply would be reduced, causing the supply curve to shift to the left, as in Figure 4–8b.

One other factor deserves mention. The supply curve is drawn for a given time period. A change in the time period (from one month to one week or one year) will alter the dimensions of the supply curve; it must be redrawn if the time period changes.

Market Supply

Market supply is simply the addition of all firms' quantities supplied for every price. If the number of firms in a given market increases or decreases, the supply curve would increase or decrease accordingly because the market supply curve is constructed by adding all the supply curves of individual firms. The market supply of tapes represented by the price-quantity combinations of the first and last columns of Table 4–3 is plotted in Figure 4–7d. The market supply curve is obtained by plotting the total quantities of tapes that would be produced and sold at every price during the time period.

The market supply curve is positively sloped—that is, total quantity supplied increases with increases in price—because the real marginal opportunity cost of tape production rises as more tapes are produced. Resources become more costly as more and more inputs are diverted from other activities into tape production. Remember that money cost is simply a proxy for real marginal opportunity cost.

Market supply: The total supply of a product, obtained by summing the amounts that firms offer at each of the various prices.

Market Equilibrium Price and Output

Market: Any area in which prices of products or services tend toward equality through the continuous negotiations of buyers and sellers.

We now put the concepts of supply and demand together to understand how market forces work to establish a particular price and output. Before proceeding, we must set forth a few terms. A **market** is simply any area in which prices of products or services tend toward equality through the continuous negotiations of buyers and sellers. Competitive (self-interested) forces of both buyers and sellers guarantee this result. All other things being equal, a buyer of dog food will always choose the seller with the lowest price, whereas a seller will choose if possible to sell at higher prices. Buyers will not pay more than price plus transportation costs, and sellers will not take less. Only one price is possible.

Perfect market: A market in which there are enough buyers and sellers so that no single buyer or seller can influence price.

In a **perfect market,** both buyers and sellers are numerous enough that no single buyer or seller can influence price. In addition, buyers and sellers are free to enter or exit the market at any time. In this case of perfect competition, no single seller sells enough of the commodity and no buyer buys enough of the product or service to influence price or quantity. In a perfect market the **law of one price** holds: After the market forces of supply and demand, of buyers and sellers, are at rest or in equilibrium, a single price for a commodity (accounting, of course, for transportation and other costs) will

Law of one price: Exists in a perfect market. After the market forces of supply and demand reach equilibrium, a single price for a commodity prevails.

prevail. If a single price did not prevail, anyone could get rich by buying low and selling at a higher price, thereby driving prices to equality. The self-interested, competitive forces of buyers and sellers acting through supply and demand guarantee this important result.

The Mechanics of Price Determination

Price and output are determined in a market from the simple combination of the concepts of supply and demand already developed in this chapter.

Tabular Analysis of Supply and Demand. Table 4–4 combines the data on the market supply and demand for cassette tapes and contrasts the quantities supplied and quantities demanded at various prices. The numbers used in Table 4–4 come from Table 4–2 (market demand for tapes) and 4–3 (market supply of tapes). The principles discussed here apply to supply and demand functions in any market.

Consider a price of $10 for tapes in Table 4–4. At the relatively high price of $10, the quantity of tapes supplied would be 450 while quantity demanded would be zero. That is, suppliers would be encouraged to supply a large number of tapes at $10, but consumers would be discouraged from buying tapes at that high price. If a price of $10 prevailed in this market, even momentarily, a **surplus** of 450 tapes would exist, that is, would remain unsold on the sellers' shelves.

Surplus: The amount by which quantity supplied exceeds quantity demanded when the price in a market is too high.

These unsold inventories of tapes are the key to what happens if the price of tapes rises above $5 in this market. Such inventories would create a competition among sellers to rid themselves of the unsold tapes. In this competition sellers would progressively lower the price. Consider Table 4–4 and assume that the price is lowered to $7. At $7 the quantity supplied of tapes is 300, while the quantity demanded is 120—a surplus of 180 tapes. Only when price falls to $5 per tape is there no surplus in the tape market.

Shortage: The amount by which quantity demanded exceeds quantity supplied when the price in a market is too low.

Now consider a relatively low price: $3 per tape. As the hypothetical data of Table 4–4 tell us, the quantity demanded of tapes at $3 would far exceed the quantity that sellers would be willing to sell or produce. There would be a **shortage** of 180 tapes; that is, 280 minus 100. Clearly some potential buy-

TABLE 4–4   Market Supply and Demand

The equilibrium price is established when quantity supplied and quantity demanded are equal. Prices above equilibrium result in surpluses; prices below equilibrium result in shortages.

Price (dollars)	Quantity of Tapes Supplied	Quantity of Tapes Demanded	Surplus (+) or Shortage (−)
10	450	0	450 (+)
9	400	40	360 (+)
8	350	80	270 (+)
7	300	120	180 (+)
6	250	160	90 (+)
5	200	200	0
4	150	240	90 (−)
3	100	280	180 (−)
2	50	320	270 (−)
1	0	360	360 (−)
0	0	400	400 (−)

ers would be unable to buy tapes if tape prices remain at $3. In fact, some buyers would be willing to pay more than $3 per tape rather than go without music. These buyers would bid tape prices up—offer to pay higher prices—in an attempt to obtain the product.

As the price bid by buyers rises toward $5, sellers will be encouraged to offer more tapes for sale; simultaneously, some tape buyers will be discouraged (will buy fewer tapes) or drop out of the market. For instance, at a price of $4 per tape, sellers would sell 150 tapes while buyers would demand 240, creating a shortage of 90 tapes. The shortage would not be eliminated until the price reached $5.

Equilibrium price: The price at which quantity demanded is equal to quantity supplied; when this price occurs there will be no tendency for it to change, other things being equal.

Equilibrium price in this market is $5; equilibrium quantity is 200 tapes supplied and demanded. At this price there is no shortage or surplus. A price of $5 and a quantity of 200 are the only price-output combination that can prevail when this market is in equilibrium, that is, where quantity demanded equals quantity supplied. The very existence of shortages or surpluses in markets means that prices have not adjusted to the self-interest of buyers and sellers. *Equilibrium* means "at rest." More precisely, in economic terms, equilibrium is that price-output combination in a market from which there is no tendency on the part of buyers and sellers to change. The free competition of buyers and sellers leads to this result.

Graphic Representation of Supply and Demand. The most common and useful method of analyzing the interaction of supply and demand is with graphs. Figure 4–9 displays the information of Table 4–4 on a graph that combines the market demand curve of Figure 4–5d with the market supply curve of Figure 4–7d.

The interpretation of Figure 4–9 is identical to the interpretation of Table

FIGURE 4–9

Equilibrium Price and Quantity

The equilibrium price is established at the point where quantity demanded and quantity supplied meet. At prices below equilibrium, the quantity demanded exceeds the quantity supplied and the price is bid upward. At prices above equilibrium, the quantity supplied exceeds the quantity demanded and the price is bid downward.

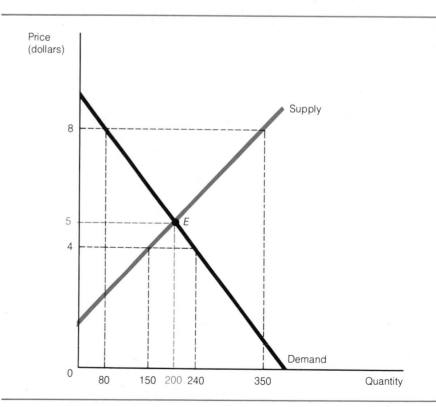

4–4, but the point of equilibrium is pictured graphically. Equilibrium price and quantity for tapes is established at the intersection of the market supply and demand curves. The point where they cross is labeled *E,* for equilibrium. At equilibrium, both demanders and suppliers of tapes are mutually satisfied. Any price higher than $5 causes a surplus of tapes; that is, a higher price eliminates some demanders and includes more suppliers. Any price below $5 eliminates some suppliers and includes more demanders.

The theory of supply and demand is one of the most useful abstractions from the world of events that is available to economists. The following three parts of this text will amplify this statement, but here we will discuss a few of the implications of supply and demand for society and for public policy.

Price Rationing

Prices, which are formed through the interaction of supply and demand, are **rationing** devices. This means that scarce resources are channeled to those who can produce a desired product in the least costly fashion for demanders who most desire the product. Another way of saying this in economic terms is that resources flow to their most highly valued uses. Consider our hypothetical market for tapes again. Suppliers who are able and willing to produce or sell tapes at a cost below $5 (including a profit) are "successful" in that their tapes will be purchased. Demanders who are willing and able to purchase tapes at a price at or above $5 are the successful buyers of tapes. Only the most able sellers and buyers of tapes are successful in this market. High-cost producers (above $5) and buyers with a low preference for tapes (below $5) are eliminated from the market. Tapes and all other goods are rationed by a price system—by the free interplay of supply and demand. In such a system no conscious attempt is made by any organization (such as government) to allocate scarce resources on the basis of factors such as presumed need, eye color, morals, skin color, or ideas of justice. As such, the market system plays no favorites. A price rationing system, in other words, ensures that only the most able suppliers and demanders participate in markets.

> **Rationing:** Prices are rationing devices; the equilibrium price rations out the limited amount of a product produced by the most willing and able suppliers, or sellers, to the most willing and able demanders, or buyers.

Effects on Price and Quantity of Shifts in Supply or Demand

Both individual and market supply and demand functions are constructed by assuming that other things are equal. What happens if other things do not remain equal, if factors other than price change? We can summarize these other factors and indicate their influence on equilibrium price and quantity in any market obeying the law of one price.

Demand Shifts. Any change in a factor other than the price of a good will alter the basis on which a demand curve is drawn—that is, it will shift the curve left or right. (Remember the difference between a change in demand, which is caused by a change in a non-price factor, and a change in quantity demanded, which is caused by a change in price.) A number of shifting factors are summarized in Table 4–5, which indicates the nature of the change, the direction in which demand will shift, and the effects on equilibrium price and quantity. We discussed all of these possible changes earlier in the chapter. Note that when demand changes, price and quantity move in the same direction.

These facts can be seen graphically in Figure 4–10. Factors causing an increase in demand from Table 4–5 will have the effects on price and quantity shown in Figure 4–10a. An increase in demand shifts the whole demand

TABLE 4–5   Factors Shifting the Demand Curve

Changes in factors other than the price of the product will shift the demand curve either to the right or to the left, changing equilibrium price and quantity.

Factors Changing Demand	Effect on Demand	Direction of Shift in Demand Curve	Effect on Equilibrium Price	Effect on Equilibrium Quantity
Increase in income (normal good)	Increase	Rightward	Increase	Increase
Decrease in income (normal good)	Decrease	Leftward	Decrease	Decrease
Increase in income (inferior good)	Decrease	Leftward	Decrease	Decrease
Decrease in income (inferior good)	Increase	Rightward	Increase	Increase
Increase in price of substitute	Increase	Rightward	Increase	Increase
Decrease in price of substitute	Decrease	Leftward	Decrease	Decrease
Increase in price of complement	Decrease	Leftward	Decrease	Decrease
Decrease in price of complement	Increase	Rightward	Increase	Increase
Increase in tastes and preferences for good	Increase	Rightward	Increase	Increase
Decrease in tastes and preferences for good	Decrease	Leftward	Decrease	Decrease
Increase in number of consumers of good	Increase	Rightward	Increase	Increase
Decrease in number of consumers of good	Decrease	Leftward	Decrease	Decrease

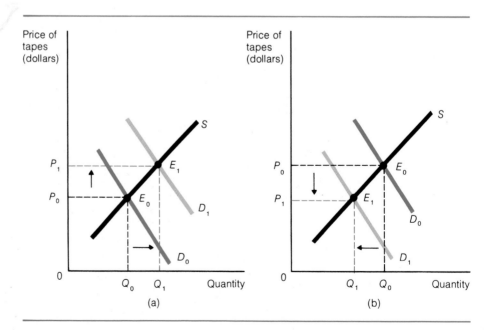

(a)   (b)

FIGURE 4–10   Shifts in the Demand Curve

(a) An increase in demand from D_0 to D_1 will increase the equilibrium price and the equilibrium quantity. (b) A decrease in demand from D_0 to D_1 will decrease both price and quantity.

curve rightward from D_0 to D_1. An excess demand—a shortage—opens up at price P_0. All the self-interested market forces discussed earlier now come into play. Demanders bid prices up to P_1, where additional quantities are supplied by firms and where a new equilibrium, E_1, is established. Thus, an increase in demand has the effect of increasing the equilibrium price *and* the quantity of the product demanded and supplied. A decrease in demand causes the demand curve to shift leftward (shown in Figure 4–10b). This has the effect of reducing prices and quantities demanded.

Supply Shifts. The effects of supply changes are summarized in Table 4–6. Clearly, any increase or decrease in a factor such as resource prices or price expectations will cause increases or decreases in the whole supply schedule. Effects of such changes are shown graphically in Figure 4–11.

Figure 4–11a shows the effects on price and quantity of an increase (rightward shift) in the supply schedule. At price P_0, excess supply—a surplus—opens up, and self-interested competitive firms and buyers bid prices down to P_1. Note, however, that while equilibrium price decreases, equilibrium quantity *increases* from Q_0 to Q_1. A price decline from P_0 to P_1 means that an additional quantity of the product or service will be demanded by consumers. Figure 4–11b shows the effects on price and quantity of a factor that decreases the supply curve (shifts it leftward). A decrease in supply has quantity-decreasing (from Q_0 to Q_1) but price-increasing (P_0 to P_1) effects, as Figure 4–11b shows. In supply shifts, unlike demand shifts, price and quantity change in opposite directions.

Shifts in Both Supply and Demand. Multiple shifts in both supply and demand are also possible, but the results are less predictable. Suppose in both supply and demand in a particular market that the incomes of consumers rise (and the product is a normal good) and that, simultaneously, resource input prices rise. In this case, as a look at Tables 4–5 and 4–6 will verify, demand increases and supply decreases. What will happen to equilibrium price and quantity? An increase in demand and a decrease in supply will clearly have

TABLE 4–6   Factors Shifting the Supply Curve
Factors changing the supply schedule will change equilibrium price and quantity in opposite directions.

Factors Changing Supply	Effect on Supply	Direction of Shift in Supply Curve	Effect on Equilibrium Price	Effect on Equilibrium Quantity
Increase in resource price	Decrease	Leftward	Increase	Decrease
Decrease in resource price	Increase	Rightward	Decrease	Increase
Improvement in technology	Increase	Rightward	Decrease	Increase
Decline in technology	Decrease	Leftward	Increase	Decrease
Expect a price increase	Decrease	Leftward	Increase	Decrease
Expect a price decrease	Increase	Rightward	Decrease	Increase
Increase in number of suppliers	Increase	Rightward	Decrease	Increase
Decrease in number of suppliers	Decrease	Leftward	Increase	Decrease

FIGURE 4–11

Shifts in the Supply Curve

(a) An increase in supply from S_0 to S_1 will lower the equilibrium price and increase the equilibrium quantity. (b) A decrease in supply from S_0 to S_1 will increase the equilibrium price and lower the equilibrium quantity.

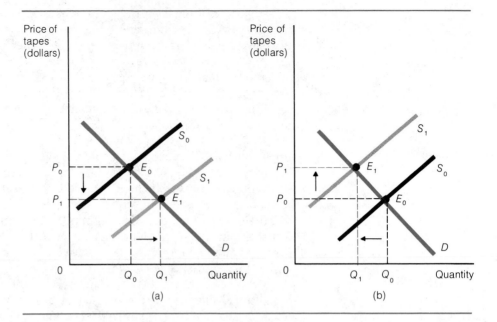

(a)   (b)

price-increasing effects (see Figures 4–10a and 4–11b), but equilibrium quantity is not predictable. Why? Because an increase in demand has a quantity-*increasing* effect, while a decrease in supply has a quantity-*decreasing* effect. Thus, whether equilibrium quantity increases or decreases depends on the magnitude of the relative shifts in the supply and demand curves. More information is needed before the economist can predict an outcome about quantity in such a case.

Simple Supply and Demand: Some Final Considerations

Many real-world events—new entrants into markets, oil price increases, new computer technology—are analyzed using the simple laws of supply and demand. The theory of supply and demand is one of the most powerful tools the economist has to analyze the real world, as you will see throughout the rest of this book. At the outset, however, it is important to understand some limitations and some possible applications and extensions of the simple mechanics described in this chapter.

Price Controls

As we have seen, a major effect of the free interplay of supply and demand is the rationing of scarce goods by a system of prices. One of the best ways to understand price rationing and its usefulness is to examine what happens when government intervenes in freely functioning markets. Such interventions into the natural functioning of supply and demand are called **price controls.** Rent controls, price controls, agricultural price supports, usury laws, and numerous other policies are examples of such "tinkering" in free markets. We deal with many of these matters elsewhere in the book, but simple supply and demand provide the basis for an initial discussion of the rationing of goods and services by nonprice means.

Take the hypothetical market for tape cassettes. The market demand and supply curves are reproduced in Figure 4–12 from the data in Table 4–4.

Price control: Government intervention in the natural functioning of supply and demand.

Price ceiling: A maximum legal price established by government to protect buyers.

Price floor: A minimum legal price established by government.

Suppose the government decides that the price of tapes is too high—"unjust" to consumers. Such is the logic of most state usury laws—"interest rates are too high"—or price controls—"prices are too high." By decree, the government orders the price of tapes to be $4 *or less,* thus establishing a **price ceiling,** a maximum legal price. At a price of $4, quantity demanded (240) exceeds quantity supplied (150), creating a shortage of 90 tapes. Clearly, the desires of buyers and sellers are not synchronized at a price of $4. What will happen? The government will have to police the market, incurring what economists call an enforcement cost. Otherwise, some profit-hungry tape sellers will charge black market prices (above $4) for tapes. Such behavior is perfectly possible since there are buyers willing to pay prices higher than $7 for tapes. Owing to these market forces, such price ceilings seldom, if ever, work. Buyers and sellers find profitable ways to evade them. In most cases consumers would prefer higher prices to non-price rationing which entails waiting lines, the purchase of coupons, or illegal purchases.

Alternatively, suppose the government decides that tape prices are too low and that suppliers must be protected. Many agricultural suppliers have been protected in this manner. In this case a **price floor**—a minimum legal price—is instituted, say at $7 per tape (see Figure 4–12). At a price of $7 the quantity supplied of tapes is 300, while the quantity demanded is only 120, a surplus of 180 tapes. In this case a different (and familiar) problem arises. Assuming that the market is adequately policed—that is, that tape suppliers are disallowed from selling tapes at a price lower than $7—surpluses of tapes build up. The government must stand ready with tax dollars or some other device to buy up and store the surplus tapes to support the price floor. Such policies can become quite costly, as price support programs for butter, milk, cheese, peanuts, and grain have demonstrated. An understanding of

FIGURE 4–12

Price Controls in the Market for Tapes

A price that is fixed either above or below equilibrium will create a surplus or a shortage of a product. Prices set above equilibrium are called *floors* and prices fixed below equilibrium are called *ceilings;* a price floor is a minimum legal price and a price ceiling is a maximum legal price.

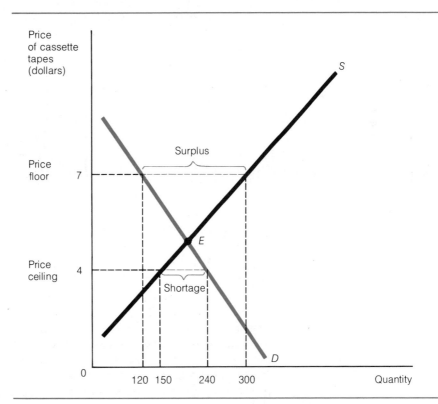

FOCUS   **Usury and Interest Rates:
An Application of Price Controls**

Price controls have been known to all ages and all civilizations. The Roman Emperor Diocletian invoked general price and wage controls prior to the fall of the Roman Empire. During the Middle Ages, church doctrine combined with economic policy to create civil laws related to interest-taking. The laws, called *usury laws*, originated in the belief that an uncontrolled market produced interest rates that were too high. The laws consisted of a legal limit on the amount of interest that lenders could charge or borrowers pay. Such laws survive today in the enactments of certain U.S. state legislatures that wish to protect borrowers and in federal regulations establishing the maximum interest rate allowable on small savings deposits.

The mechanics of the usury laws are simple; we demonstrate them in Figure 4–13. The figure shows the free-market supply and demand for loanable funds. As market interest rates increase, the quantity supplied of loanable funds rises and the quantity demanded declines. As market rates fall, the quantity of loanable funds supplied falls while quantity demanded rises. Market equilibrium occurs at point E, where quantity demanded equals quantity supplied (Q_E) and where the equilibrium price—the market interest rate—is i_E.

What happens when the government (Diocletian or the Ohio state legislature) declares that market rate i_E is too high for the poor and that henceforth the maximum rate will be i_C, a ceiling rate below i_E? At rate i_C the quantity of loanable funds demanded, Q_D (point C in Figure 4–13), exceeds the quantity supplied, Q_S (point B). A shortage of funds, BC, develops at the ceiling interest rate because banks are unwilling to lend the quantity of funds demanded at rate i_C. The available funds, Q_S, must be rationed because of the shortage. In the absence of price rationing, the process of rationing can take other forms. Unscrupulous lenders—"loan sharks"—may charge black market rates for the limited funds. Lenders may begin demanding additional collateral or better credit standing for loans at the ceil-

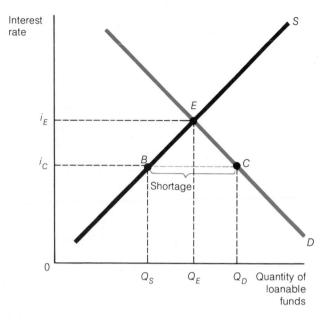

FIGURE 4–13   Interest Rate Ceiling

When a ceiling is placed on interest rates (i_C), a shortage (BC) develops, the difference between the quantity of funds that lending institutions are willing and able to supply (Q_S) and the quantity of funds that borrowers demand (Q_D). Without the ceiling, the forces of supply and demand interact to create equilibrium (E) between interest price and the quantity of loanable funds made available.

ing rate, reducing their risk and making it more difficult for less creditworthy borrowers to obtain funds. Such borrowers are often the poor whom the laws were designed to protect. The poor are often compelled to acquire funds from loan sharks in such circumstances. Usury laws, like all other forms of price controls, have built-in effects that often hurt those who are supposed to be the beneficiaries.

simple supply and demand points out the problems and costs associated with government (or other) interventions in a market system. See Focus, "Usury and Interest Rates," for a further discussion of price controls.

Static Versus Dynamic Analysis

The analysis discussed in this chapter is called *static equilibrium analysis*. This means that (1) time does not enter into the discussion—it is not dynamic; and (2) price-quantity combinations move from one equilibrium to another (E_0 to E_1, for instance, in Figure 4–11) with no indication of the process by which firms and consumers actually move. The *dynamic analysis* of market supply and demand considers the time and process involved in moving from one equilibrium situation to another. How long, for example, does it take for a world oil price increase to influence prices in the various markets using oil as an input? A static analysis cannot tell us. It can indicate only the probable direction of price and quantities. What is the role of the entrepreneur in seeking profit opportunities when demand or supply schedules shift and markets are in disequilibrium, that is, experience shortage or surplus? Again, static analysis provides no method for answering questions about disequilibrium. A number of these issues will be discussed and supply and demand will be analyzed more thoroughly in a real-world context in the following chapters on microeconomics.

Full Versus Money Prices

Money price: The dollar price that sellers charge buyers.

Another distinction must be made about prices resulting from the forces of supply and demand: There is a difference between the **money price** of a product or service and the **full price.** Consider the money price of a haircut for which the customer pays $12.50 in cash. Is this the full price of a haircut? Do consumers react to the money price of products and services or to what economists call the full price?

Full price: The total cost to an individual of obtaining a product, including money price and other costs such as transportation or waiting time.

Money price is often not the only cost to consumers. In the case of a haircut, we must account for the time spent traveling to and from the salon, the time spent waiting, and the time spent with the hairdresser. The economist accepts the truth of the adage "time is money." Time, like diamonds, is a scarce resource, and it bears an opportunity cost. Time costs are often estimated in terms of the wage rate forgone, the leisure time forgone, or, generally, as the opportunity cost of the consumer's next-best alternative. Thus the full price of a haircut or any other good or service includes the price in money terms plus any other resource costs required in the purchase of the commodity. (Focus, "Full Price and a New Orleans Restaurant," gives another example.) Prices as interpreted in the simple model of demand and supply of this chapter are to be regarded as full prices. Consumers react to full prices, not money prices. No one, for example, should take an advertisement for "free puppies" or "free kittens" literally. There is no such thing as a free puppy!

Relative Versus Absolute Prices

Relative price: The price of a product related in terms of other goods that could be purchased rather than in money terms.

In addition to the distinction between money prices and full prices, we make the distinction that consumers react to relative prices, not to absolute prices. **Relative price** is the price ratio in consumption of one product relative to another product or all other products. **Absolute price** is the price of a product measured in terms of money.

Absolute price: The price of a product measured in terms of money.

Many of us come from states that produce some agricultural commodity that varies in quality—citrus fruits from Florida or California, lobsters from Maine, apples from North Carolina or Washington. Residents of such states often complain, "They are shipping the good apples out," or ask, "Why do the high quality apples go to New York?" The theory of supply and demand tells us why.

Suppose that the absolute price of a good apple in Washington State is 10 cents and the absolute price of a bad apple is 5 cents. In this case, the relative price of good versus bad apples in Washington is two to one: A good apple is worth two bad apples. Now assume that the cost of transportation of each apple to New York is 5 cents. In New York, a good apple therefore costs 15 cents and a bad apple costs 10 cents. The relative price of good to

FOCUS   Full Price and a New Orleans Restaurant

New Orleans is considered a restaurant town by gourmets. Restaurants such as Le Ruth's, Antoine's, Brennan's, and Commander's Palace consistently offer some of the best cuisine to be found anywhere. One of our favorite French Quarter restaurants, however, is called Galatoire's. Galatoire's has justifiably earned a reputation for producing some of the highest-quality meals in the city. Menu prices, however, have remained low relative to other famous establishments and, even more important, they have remained low over the past several decades.

The question is, Have both menu prices and full prices remained low at Galatoire's? If not, why not? Finally, what evidence might be offered for a real price increase? How would an economist view the matter?

The economist would focus on elements in the full price of a meal at Galatoire's. It is notable that the restaurant has not enlarged its classic physical plant over the years but that, over time, longer and longer lines form outside the door at many hours during the day and evening. Galatoire's, unlike most of its competition, does not take reservations or accept credit cards—only cash will do. Moreover, Galatoire's enforces a dress code—tie or coat for men (evenings and all day on Sundays) and, until a few years ago, dresses (no pants) for women.

All these factors would tend to increase the full price of dining at Galatoire's. There are opportunity costs to waiting for a table (time is a scarce resource); the time and resources spent dressing for dinner are also applicable to the full price. Unpredictability of the length of the line on arriving at the restaurant can also make the real price higher or lower than anticipated, although diners can form hunches about when the line is apt to be shortest or nonexistent.

These additional costs mean that the full price exceeds the money price printed on the menu. However, the economist must calculate all benefits as well as all costs associated with purchasing products or services. Consider the possibility that some customers get positive benefits from dressing for dinner and from being surrounded by those similarly attired. Others may place a high value on the prestige or satisfaction of dining at the legendary restaurant. Such factors would tend to increase the benefits to these consumers. The full price paid for a meal at Galatoire's varies among consumers, depending on opportunity cost. As we will see in Chapter 6, however, consumers always will marginally balance the perceived costs of buying products and services with the perceived benefits associated with consumption. When price or price formation is discussed in this book, it is therefore the full price, not the nominal or money price, that is being considered.

---

bad apples is therefore different: Good apples are only one and one-half times more expensive than bad apples in New York. Because the relative price of good apples is lower in New York than in Washington, a relatively greater quantity of good apples are demanded in New York. Therefore, the good apples are shipped out of Washington to New York.

The theory of supply and demand discussed in this chapter has not explicitly stressed the distinction between real prices and relative prices. But it is important to interpret even basic supply and demand theory as a theory of relative, not absolute, prices and full, not money, prices. With these distinctions taken into account, the theory of supply and demand forms the foundation for the understanding of how markets function and therefore for the whole science of economics.

Summary

1. The extension of specialization from earlier societies is the modern market society, where individuals and collections of individuals buy and sell—demand and supply—millions of products and services.
2. Demand can be expressed as a schedule or curve showing the quantities of goods or services that individuals want to purchase at various prices over some period of time, all other factors remaining constant.
3. Supply can be expressed as a schedule or curve of the quantities that individuals or businesses are willing to sell at different prices over a period of time, other factors remaining constant.
4. Equilibrium prices and quantities are established for

any good or service when quantity supplied equals quantity demanded.

5. A change in the price of a good or service changes the quantity demanded or quantity supplied. A change in demand or supply occurs when some factor other than price is altered. When these factors change, the demand or the supply curve shifts to the right or left, either raising or lowering equilibrium price and quantity.

6. The market system, through supply and demand, rations scarce resources and limited quantities of goods and services among those most willing and able to pay for them. Products and services, moreover, appear and disappear in response to the market system of supply and demand.

7. Price controls instituted by governments tend to create shortages or surpluses of products. Market forces usually result in some form of rationing other than price rationing under such circumstances.

Key Terms

products market	inferior good
resources market	substitutes
law of demand	complements
quantity demanded	market demand
demand curve	marginal opportunity costs
factors affecting demand	law of supply
ceteris paribus	quantity supplied
change in demand	supply schedule
normal good	change in supply

factors affecting supply	price control
market supply	price ceiling
market	price floor
perfect market	money price
law of one price	full price
surplus	relative price
shortage	absolute price
equilibrium price	
rationing	

Questions for Review and Discussion

1. What happens to the demand for a product if the price of that product falls? What happens to the quantity demanded?

2. What happens to the supply of coal if the wages of coal miners increase?

3. If income falls, what happens to the demand for potatoes? Are potatoes inferior goods?

4. If price is above equilibrium, what forces it down? If it is below equilibrium, what forces it up?

5. What is a shortage? What causes a shortage? How is a shortage eliminated?

6. If the demand for cassette tape players increases, explain the process by which the market increases the production of tape players. What are the costs of tape players?

7. A price ceiling on crude oil has an effect on the amount of crude oil produced. With this in mind, explain what happens to the supply of gasoline if there is a price ceiling on crude oil.

8. What happens to the supply of hamburgers at fast-food restaurants if the minimum wage is increased?

9. What is the full price of seeing a movie? Is it the same for everyone?

10. "Lately the price of gold keeps going up and up, and people keep buying more and more. The demand for gold must be upward sloping." Does this statement contain an analytical error?

⋀⋁ ECONOMICS IN ACTION

When Other Things Do Not Remain Equal: The Pope and the Price of Fish

Perhaps the major lesson about the laws of demand and supply is the importance of holding other things constant. Economists have singled out consumers' income, the price of related goods, and other factors as being of special importance in causing changes in demand (see Table 4–5). Similarly, factors such as the price of inputs or technology create shifts in supply (see Table 4–6). Yet the assumption of other things being equal includes all other factors in consumers' and producers' environment as well. Changes in this environment, in-

cluding institutions and regulations, must sometimes be taken into account when analyzing price and quantity movements. Consider, for example, some effects of lifting the ban on eating meat on Fridays for Roman Catholics in the northeastern United States.

For years, Roman Catholics were required to abstain from eating meat on Fridays. This ban, which can be thought of as a constant over time, helped support the worldwide fishing industry (there are almost 600 million Catholics in the world), including the U.S. com-

TABLE 4–7   Decline in Demand for Fish After Papal Decree

Monthly prices of all seven species of fish declined in the percentages indicated, from a 2 percent reduction for scrod to a 21 percent decline for the price of large haddock. On average, the price of all species of fish declined by 12 1/2 percent.

Species	Percentage Change in Price of Fish After Papal Decree (monthly)
Sea scallops	−17
Yellowtail flounder	−14
Large haddock	−21
Small haddock (scrod)	−2
Cod	−10
Ocean perch	−10
Whiting	−20
All species (average)	−12 1/2

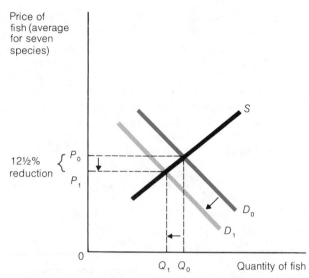

FIGURE 4–14   Papal Decree and the Price of Fish

Following the papal decree lifting the Friday abstinence law, the price of fish declined from P_0 to P_1 and the quantity also declined, from Q_0 to Q_1. The demand curve shifted to the left, from D_0 to D_1.

mercial fishing industry. In February 1966, however, Pope Paul VI issued a decree that allowed local Catholic bishops to end, at their discretion, the rule requiring abstinence from meat on the approximately forty-six Fridays that do not occur during Lent. In December 1966, U.S. Catholics were no longer bound to the rule of abstinence on non-Lenten Fridays.

Economist Frederick W. Bell statistically estimated the impact of the pope's decree on the markets for seven kinds of fish produced and consumed in the northeastern United States.[a] The northeastern United States was chosen because 45.1 percent of the population there is Roman Catholic, more than twice the number in any other area of the United States. To isolate the effects of the decree and estimate a demand function, Bell had to assess the relative importance of all factors affecting fish prices. These factors included many of those discussed in this chapter: the quantity supplied of the particular species of fish; the aggregate personal income for people living in the northeastern United States; cold storage holdings of the particular species of fish; importation of fish into New England; the price of related goods (meat and poultry) and competing fish species; a factor to capture the effects of Lenten and non-Lenten months; and a factor to capture the effects of the papal decree.

In spite of some statistical difficulties, Bell discovered some interesting facts using monthly data. In comparing a period after the decree, December 1966 to August 1967 (excluding Lenten months), with a ten-year

[a]Frederick W. Bell, "The Pope and the Price of Fish," *American Economic Review* 58 (December 1968), pp. 1346–1350.

period from January 1957 to November 1966, Bell found reductions in demand and price after the decree for all seven species of fish, given in Table 4–7.

Figure 4–14 shows graphically the major economic effects of the pope's decree. Demand curve D_0 and supply curve S represent the initial situation. Market forces underlying demand and supply create a price P_0 for all species of fish. After the papal decree, the demand curve shifts to the left at every price, creating an average price decline, from P_0 to P_1, of 12½ percent and a reduction in quantity from Q_0 to Q_1. An exit of labor and capital from the fishing industry could be expected if the price- and revenue-depressing effects of the papal decree persisted. Careful analysis of all factors affecting supply and demand, even those that seem remote, must be made when dealing with changing market prices and quantities.

Question

A coffee bug infests the Latin American coffee producing region. The bug destroys a substantial portion of the coffee crop. What would happen to the supply and demand, the equilibrium price and quantity of coffee? What factors would influence the new levels of output and price? If some consumers chose a substitute such as tea during the crisis, what would you predict about the coffee market when the coffee bug and its effects are totally eliminated?

Adam Smith and Karl Marx:
Markets and Society

Adam Smith

Karl Marx

ADAM SMITH, one of the most important figures in the history of economics, actually began his career as a lecturer in moral philosophy at Scotland's Glasgow College in 1751. Moral philosophy in Smith's time encompassed a wide range of topics, including natural theology, ethics, jurisprudence, and the field we now know as economics. In 1776, coincident with the Declaration of Independence, Smith published his second book (his first was a treatise on moral behavior), *An Inquiry into the Nature and Causes of the Wealth of Nations*, known usually by the shorter title *Wealth of Nations*. The book, which won Smith much attention from scholars of the day, brought together most of what was then known about the workings of the market system. Smith's insights are still being taught today, more than two hundred years later.

Smith was born in Kirkcaldy on the east coast of Scotland in 1723 and lived most of his life in his native country. Although known for his brilliant lectures (and his many eccentricities), Smith did not devote his entire career to teaching. In 1778 he accepted a well-paying job as commissioner of Scottish customs, a post he kept until his death in 1790.

KARL MARX "*looked* like a revolutionary," writes Robert Heilbroner in *The Worldly Philosophers*. "He was stocky and powerfully built and rather glowering in expression with a formidable beard. He was not an orderly man; his home was a dusty mass of papers piled in careless disarray in the midst of which Marx himself, slovenly dressed, padded about in an eye-stinging haze of tobacco smoke."[a]

Coauthor with Friedrich Engels of the *Communist Manifesto*, which predicted the inevitable downfall of capitalism and the triumph of communism, Marx spent

most of his life in difficult circumstances. His activities as a radical in the communist movement caused his exile from his native Germany as well as from Belgium and France. In 1849, a year after the publication of the *Manifesto*, he settled in London, where he and his family survived through the benevolence of Engels and where Marx researched and wrote *Das Kapital*, a theory and history of capitalism and its ills. Marx died in 1883 in London at the age of sixty-five.

As economic theorists, Smith and Marx represent opposite views. Smith was one of the most eloquent defenders of free markets and the promise of capitalism. Marx, who wrote in response to the miseries of the European working class during the Industrial Revolution of the late eighteenth and nineteenth centuries, argued for a new social order and for the overthrow by the working class of capitalists.

The "Invisible Hand"

Smith's views of the free-market system are summarized in a passage from the *Wealth of Nations* in which he writes that individuals pursuing their own self-interest are "led by an invisible hand to promote an end which was no part of [their] intention."[b] Smith believed that by freely exchanging goods and services across markets, individuals contribute to the public good—the aggregate wealth of society—even though they act from purely self-interested motives. In other words, markets cause individuals to benefit others even though they intend only to benefit themselves.

To Smith, voluntary market exchange coordinated the decisions of consumers and producers and generated economic progress. Producers strive in competition

[a]Robert L. Heilbroner, *The Worldly Philosophers* (New York: Simon and Schuster, 1953), p. 131.

[b]Adam Smith, *An Inquiry into the Nature and Causes of the Wealth of Nations*, ed. Edwin Cannan (1776; reprint, New York: Modern Library, 1937), p. 423.

with one another to satisfy consumers with the most appropriate and cheapest goods and services, not out of the goodness of their hearts, not because government planners instruct them to do so, but simply because they maximize the profits of their enterprises by doing so. Markets coordinate supply and demand by way of the price system. Consumers express their relative preferences in their decisions about what to buy; producers attract customers by producing goods at the least cost. In this system of coordination without command, individuals pursuing their own interests are led "as if by an invisible hand" to mesh their interests with those of other individuals trading across markets.

Smith was not opposed to government but argued that its proper role in society was to provide a legal framework—police and courts—within which the market could operate as well as to provide certain other services (including national defense, highways, and education) that he felt the market itself would not supply or would tend to supply in inadequate amounts. Smith also felt that government should provide welfare services for the poor. But he strongly believed that government could best assist the market economy achieve growth by stepping out of the way, that is, by not engaging in most forms of regulation and by ending grants that gave monopoly privileges to favored groups and individuals.

In Smith's view, income was distributed in a market economy by the production of wealth. An individual's income was a strict function of the value of his or her output. Smith did not feel that inequality in income by itself was unfair because the invisible hand ensured that individuals' wealth (or lack of it) was a measure of how much their efforts benefited society as a whole. Anyone can increase his or her income in a free market by serving the consumers in a new, better, or faster way.

The "Anarchy of Production"

Karl Marx rejected Smith's view of the market process. In *Das Kapital* he argued that Smith's writing represented merely the interests of the ruling capitalist class. To Marx, the market process was a system of exploitation by which owners of capital robbed their employees by paying them wages less than the worth of their labor (a situation he termed the "alienation of labor").

The alternative social system that Marx thought would eliminate this exploitation and at the same time greatly increase the efficiency of production was a "general organization of the labor of society . . . [that] would turn all society into one immense factory."[c] He viewed the market economy as one of general disorga-

nization. Its main feature was "the anarchy of production" where producers overproduced and consumers were forced to accept goods they neither wanted nor needed. He claimed that the market process, left to itself, could not coordinate diverse individual plans.

In Marx's view, Smith's "invisible hand" was a euphemism for describing the economic system in which "chance and caprice have full play in distributing the producers and their means of production among the various branches of industry. . . . the division of labour within the society [the theme of much of Smith's *Wealth of Nations*] brings into contact independent commodity-producers, who acknowledge no other authority but that of competition, of the coercion exerted by the pressure of their mutual interests. . . . the same bourgeois mind which praises division of labour in the workshop [as a conscious organization that increases productivity] denounces with equal vigour every conscious attempt to socially control and regulate the process of production."[d] Smith's "invisible hand" was not only invisible but also unbelievable. Only the central planning of economic activity by society (government), which owned all means of production, could coordinate the needs of consumers and producers and eliminate the wastefulness of capitalism. To Marx it was nonsense to describe the market as organized economic activity because there was no "organizer"—coordination of economic activity requires the conscious, centralized control of the economy.

While Adam Smith described the emerging market economy of his day and offered reforms (most of which could be summarized by the phrase "less government"), Marx offered a vision of economic organization that did not exist at the time he wrote but that he maintained was the inevitable wave of the future. In a sense he was proven correct. Followers of the teachings of Marx and his admirer Lenin (who filled in many of the details of what a central planning system would look like in practice) imposed avowedly Marxist-socialist, centrally planned economies on Russia (1917), China (1949), most of the countries of Eastern Europe, and some African and Latin American nations. Today about one-third of the world's population lives in economies organized in accordance with the ideas of Marx, each economy intended to resemble an "immense factory."

In another sense Marx's teachings appear to have failed. The centrally planned economies seem to function poorly—providing low per capita income and poor rates of economic growth—relative to the modern versions of the capitalist economies whose central principle of market organization was so clearly seen by Smith.

---

[c]Karl Marx, *Das Kapital*, ed. Max Eastman (1867; reprint, New York: Modern Library, 1932), p. 83.

[d]Marx, *Das Kapital*, p. 83.

II

The Microeconomic Behavior of Consumers, Firms, and Markets

5

Elasticity

The laws of supply and demand help economists organize their thoughts about real-world problems. But it is not enough to understand that an inverse relation exists between price and quantity demanded or that a decrease in supply causes price to rise in a market. For supply and demand theory to explain and predict economic events, economists must be able to say how much the quantity demanded or supplied of a product will change after a price change. The laws of supply and demand tell us nothing about how responsive quantity demanded or quantity supplied is to a price change. But if, for example, you owned a fast-food restaurant, you would be very interested to know how responsive your customers would be to a hamburger price discount. The law of demand tells you that you would sell more hamburgers, but it does not tell you how many more.

The deteriorating conditions of U.S. highways and bridges led the federal government in 1983 to levy an additional five-cent-per-gallon tax on gasoline. The Reagan administration intended that the tax would generate sufficient revenues to allow the government to make the necessary repairs. But will the tax have the hoped-for effect? The answer depends on the responsiveness of gasoline consumers to the price increase. If they are *very* responsive—if they reduce gasoline consumption a great deal in response to the price rise—tax revenues may be insufficient to finance the repair project.

The Drug Enforcement Administration initiates a crackdown on the import of illegal drugs such as cocaine and marijuana. With stepped-up enforcement, a smaller quantity of drugs is smuggled into the country by organized crime. As a result, the street price of these drugs goes up steeply. Will the revenues of organized crime increase or decrease? There will be some reduction in quantity demanded—that is, fewer drugs will be sold at higher prices. The important question is, how much of a reduction? Clearly, the answer will depend on the responsiveness of drug users to the increase in price.

You are contemplating two business alternatives: buying a gourmet delicatessen or opening a new travel agency. Best estimates tell you that consumers' incomes are expected to rise by 5 percent per year over the next eight years. On these grounds, which business should you enter? How much will increases in consumers' incomes drive up the demand for gourmet foods relative to the demand for travel and leisure?

Elasticity, the topic of this chapter, helps us answer such important, practical questions. The responsiveness of consumers to changes in price (elasticity of demand) is the best-known application of elasticity, but **elasticity** is a far more general concept. It refers to the responsiveness of any "effect" to any change in "cause."

Elasticity is always calculated in percentage terms, as a ratio of the percent change in the effect (quantity demanded or quantity supplied of gasoline, drugs, TV sets, and so on) to the percent change in the cause (income change, price change, or any other change). Any calculation of elasticity is therefore independent of absolute numbers such as quantity, income, or prices.

Consider the relation between egg production and the weather. Assume that egg production falls with increasing henhouse temperatures. The elasticity of egg production with respect to changes in temperature is calculated by dividing the percentage change in production by the percentage change in temperature. If, according to a farmer's records, a 10 percent increase in henhouse temperature resulted in a 30 percent reduction in egg production, we say that the chickens are very responsive to temperature changes; if a 10 percent increase in temperature reduced egg production by only 2 percent, we say that the chickens are not very responsive to temperature changes.

All kinds of elasticities are measured in the same way. The five-cent tax on gasoline will increase the price of gasoline by some percentage and it will reduce the amount sold by some other percentage, large or small, according to gas buyers' response to the price change. The degree of responsiveness is measured by simple division of the two percentages; the amount of tax revenues will depend on this response. More drug enforcement reduces supply and increases prices by some percentage. Drug purchases are reduced by some percentage depending on buyers' responsiveness to the price change, but, given the effects of addiction on drug users, the decrease in purchases will probably be small. Percentage increases in income will, similarly, cause percentage increases in gourmet food and travel consumption. The percentages of these changes can be estimated from similar past experiences, and businesses can use the information to make decisions.

> **Elasticity:** A measure of the responsiveness of one variable caused by a change in another variable; the percent change in a dependent variable divided by the percent change in an independent variable.

Price Elasticity of Demand

Elasticity, as these examples suggest, is a general and wide-ranging concept with many applications. The most common and important applications relate to demand.

Formulation of Price Elasticity of Demand

The formal measurement of demand elasticity, like the generalized concept itself, is simple and straightforward. **Price elasticity of demand** is the percent change in quantity demanded divided by the percent change in price:

> **Price elasticity of demand:** A measurement of buyers' responsiveness to a price change; the percent change in quantity demanded divided by the percent change in price.

$$\epsilon_d = \frac{\text{Price elasticity of demand coefficient}}{} = \frac{\%\text{ change in quantity demanded}}{\%\text{ change in price}}.$$

Price elasticity of demand is expressed as a number. If, for example, a 10 percent reduction in the price of jogging shoes causes a 15 percent increase in the quantity of jogging shoes demanded, then the ratio, called the **demand elasticity coefficient**, ϵ_d, is

$$\frac{15\%}{10\%} = 1.5.$$

This elasticity coefficient or ratio, 1.5, and the percent changes in price and quantity demanded from which it was calculated, are independent of the absolute prices and quantities of jogging shoes. If a 10 percent rise in the price of jogging shoes causes a 2 percent decline in sales, the price elasticity coefficient is calculated in the same way:

$$\frac{2\%}{10\%} = 0.2.$$

The price elasticity of demand coefficient (or, simply, the demand elasticity coefficient) always measures consumers' responsiveness, in terms of purchases, to a percent change in price.[1] To determine the effect of price changes alone, all other changes—such as differences in the quality of goods—must be held constant.

Elastic, Inelastic, and Unit Elastic Demand

The size of the elasticity coefficient is important because it measures the relative consumer responsiveness to price changes. If the number obtained from the elasticity calculation is greater than 1.00, we say that demand is **elastic** over the price and quantity range; the percent change in quantity demanded is greater than the percent change in price. Above a coefficient of 1.00, degrees of elasticity vary. A demand elasticity coefficient of 1.5 for jogging shoes means that consumers are somewhat responsive to a price change. A coefficient of 6.0 for pizza means that buyers of pizza are much more responsive to a change in price (four times more responsive, in fact). The larger the demand elasticity coefficient is above 1.0, the more elastic demand is said to be.

Unit elasticity of demand means that the elasticity coefficient equals 1. In this case, a given percent change in price is exactly matched by the percent change in quantity demanded. If, for example, a 2 percent increase in the price of candy bars causes a 2 percent reduction in purchases, demand would be of unit elasticity. The same would be said if an 8 percent decrease in the price of Volkswagens caused an 8 percent increase in quantity demanded.

An **inelastic demand** coefficient is a number less than 1, meaning that a percent change in quantity demanded is less than the percent change in price that caused the change in quantity. If the price of salt increases by 5 percent and the quantity demanded decreases by 2.5 percent, the demand elasticity coefficient would be 0.5, placing it in the inelastic category. An elasticity number lower than 0.5 for salt would mean that demand is relatively more inelastic.

Demand elasticity coefficient: The numerical representation of price elasticity of demand: $(\Delta Q/Q) \div (\Delta P/P)$.

Elastic demand: A situation in which buyers are very responsive to price changes; the percent change in quantity demanded is greater than the percent change in price; $\epsilon_d > 1$.

Unit elasticity of demand: A situation in which the percent change in quantity demanded is equal to the percent change in price: $\epsilon_d = 1$.

Inelastic demand: A situation in which buyers are not very responsive to changes in price; the percent change in quantity demanded is less than the percent change in price: $\epsilon_d < 1$.

[1]Notice that in actual calculation, the demand elasticity coefficient will always be negative owing to the inverse relation between price and quantity demanded (the law of demand). If price goes up, quantity demanded for a good or service goes down, and vice versa. Unless otherwise noted, this point is irrelevant to the interpretation of elasticity. We will accordingly eliminate use of a negative sign before the demand elasticity coefficient.

The various categories of price elasticity of demand can be shown with demand curves. Figure 5–1 depicts three responses to a 20 percent increase in pizza prices over a given time period in a small college town. In Figure 5–1a, a 20 percent increase in price causes a 5 percent reduction in quantity demanded. This means that pizza consumers are not very responsive to a change in price: ϵ_d is inelastic—it equals 0.25, which is less than 1. In Figure 5–1b, a 20 percent increase in pizza price reduces pizza consumption by exactly 20 percent, meaning that demand is of unit elasticity: $\epsilon_d = 1$. Figure 5–1c shows an elastic demand—a 20 percent price increase causes a 50 percent reduction in the quantity of pizzas demanded: ϵ_d is 2.5 (greater than 1).

The elasticity of two special forms of the demand curve are also of interest in analyzing economic problems. Figure 5–2 shows a completely inelastic and a completely elastic demand curve. Total or complete price inelasticity means that consumers are not responsive at all to price changes. Increases or decreases in price leave quantity demanded unchanged—the demand curve is vertical, as in Figure 5–2a. One might think of the demand for heroin or other addictive drugs—at least for certain price ranges and certain levels of use—as being completely inelastic. The price elasticity of demand coefficient is zero along such a curve: $\epsilon_d = 0$.

A completely elastic demand curve—along which an infinitely large or small quantity is demanded at a given price—is shown in Figure 5–2b. Such a demand curve occurs in a competitive market situation, discussed fully in Chapter 9. In a competitive market, producers are so numerous that they are unable to affect prices. The existence of a huge number of wheat farmers, for example, means that a single small producer is unable to affect the mar-

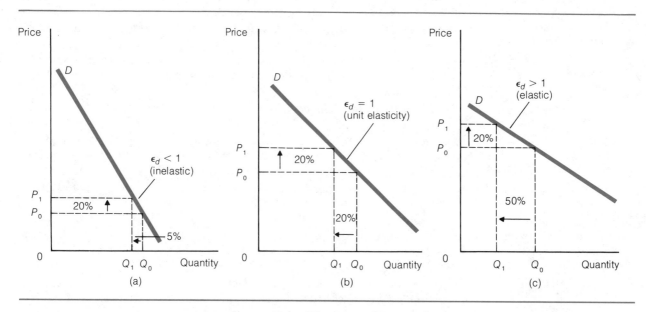

FIGURE 5–1   Price Elasticity of Demand

(a) The percent reduction in quantity demanded (from Q_0 to Q_1, 5%) is less than the percent increase in price (from P_0 to P_1, 20%). Demand is inelastic, $\epsilon_d < 1$. (b) The percent reduction in quantity demanded equals the percent increase in price. Demand is unit elastic, $\epsilon_d = 1$. (c) The percent reduction in quantity demanded is larger than the percent increase in price. Demand is elastic, $\epsilon_d > 1$. The same relations hold for reductions in price or increases in quantity demanded.

FIGURE 5–2

Demand Curves May Be Totally Inelastic or Totally Elastic

(a) Quantity demanded is unresponsive to price changes (totally inelastic). (b) Price is unaffected by smaller or larger quantities demanded (totally elastic). Such a situation occurs in a competitive market.

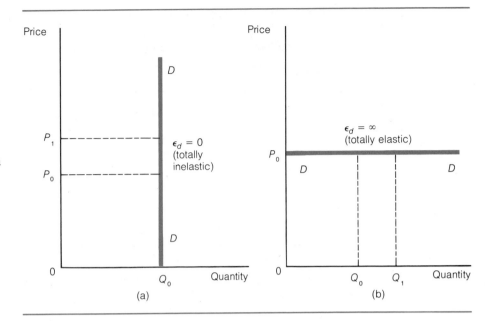

(a)   (b)

ket price by altering production. If the wheat farmer raises his or her price by even a small amount, buyers will not purchase any of the farmer's production. Buyers can get all the wheat they want at the (lower) prevailing market price. In these competitive circumstances, a single wheat farmer faces an infinitely elastic demand curve ($\epsilon_d = \infty$). Quantity demanded is supersensitive to price increases. The farmer can sell as much or as little as he desires at the market price.

To summarize, an elasticity coefficient less than 1.0 means that demand is inelastic, and a coefficient greater than 1.0 means that demand is elastic over a given price range. A coefficient equal to 1.0 indicates unit elasticity of demand. The demand curve can, under certain circumstances, be completely elastic or completely inelastic.

Elasticity Along the Demand Curve

Our discussion of demand elasticity has contained a crucial qualification: that the coefficient (whether elastic, inelastic, or unitary) has meaning only over certain (or relevant) price ranges. We now investigate exactly what this means. To do so requires that we be even more specific about the basic elasticity concept and its algebraic formulation.

Table 5–1 and Figure 5–3 show the market demand function—or price-quantity pairs—that constitute the demand schedule for tapes. (Table 5–1 and Figure 5–3 reproduce some of the information presented in Table 4–2 and Figure 4–5d relating to the market demand curve.)

Is it ever correct or meaningful to ask, "What is the elasticity of this or any other demand curve?" Some simple calculations from the market demand function of Table 5–1 or Figure 5–3 will tell us immediately that the answer is no. Remember that demand curves are downward sloping because quantity demanded drops as price rises. Every negatively sloped demand curve will, in general, contain portions that are elastic, unit elastic, and inelastic. The elasticity coefficient will vary along any straight (linear) or curving (nonlinear) demand curve in all but a few special cases. To verify

TABLE 5–1   Market Demand Schedule for Tapes

As always, the price of tapes is inversely related to
the quantity of tapes demanded.

Price of Tapes	Quantity Demanded
10	0
9	40
8	80
7	120
6	160
5	200
4	240
3	280
2	320
1	360

this fact consider the simple elasticity expression and its algebraic counterpart once more:

$$\epsilon_d = \frac{\% \text{ change in quantity demanded}}{\% \text{ change in price}} = \frac{\Delta Q/Q}{\Delta P/P}.$$

In the algebraic expression, Q simply means quantity, Δ means "a change
in," and P means price. In the numerator, percent change in quantity demanded is determined by dividing the change in quantity demanded by the
initial quantity demanded ($\Delta Q \div Q$). The same is done in the denominator
for initial price and change in price to determine the percent change in

FIGURE 5–3

**Differing Elasticities
Along a Market Demand
Curve**

The elasticity of demand is
different between various
points along a downward-
sloping demand curve. If
prices fall by \$1 between A
and B, demand is elastic. If
prices fall by \$1 between C
and D, demand is inelastic.

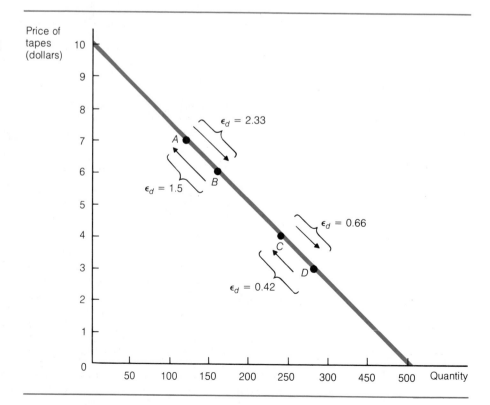

price. Then elasticity can be expressed as the percent change in quantity demanded ($\Delta Q/Q$) divided by the percent change in price ($\Delta P/P$). This simple formulation can be rearranged thus:

$$\frac{\Delta Q/Q}{\Delta P/P} = \frac{\Delta Q}{Q} \times \frac{P}{\Delta P}.$$

We will use this expression to make some calculations.[2]

Suppose from Table 5–1 and Figure 5–3 that the price of tapes declines from $7 to $6. Quantity demanded would increase from 120 to 160 tapes. How would elasticity be calculated? Returning to our simple formula, we can calculate the elasticity across this price range as follows:

$$\epsilon_d = \frac{\Delta Q/Q}{\Delta P/P} = \frac{\Delta Q}{Q} \times \frac{P}{\Delta P} = \frac{40}{120} \times \frac{7.00}{1.00} = 2.33.$$

The number of tapes sold increased by 40 over the original quantity of tapes, those purchased at $7 (120). Thus, $40/120$ is multiplied by the ratio of the original price, $7, to the change in price, $1. The elasticity coefficient is 2.33, which means that demand is elastic over this range of prices. Consumers are responsive to a price reduction from $7 to $6.

Now assume that tapes are selling at $4 and that the price is reduced to $3. Using the same method, we see that

$$\epsilon_d = \frac{\Delta Q/Q}{\Delta P/P} = \frac{\Delta Q}{Q} \times \frac{P}{\Delta P} = \frac{40}{120} \times \frac{4.00}{1.00} = 0.66.$$

The coefficient indicates that for a decline in price over the price range from $4 to $3 demand is inelastic. Consumers are far more responsive to a one-dollar price reduction from $7 to $6 than they are to a one-dollar price reduction from $4 to $3.

This example illustrates that elasticity varies all along a demand curve. It is not correct to say that the demand for anything—salt, cigarettes, tapes—is elastic or inelastic without identifying some specific price range.

The knowledge that elasticity changes along demand curves also points out another difficulty in accurately calculating elasticity. Since elasticity varies at all points along the demand curve, it also varies along the arc, or length, A to B or C to D. This is called **arc elasticity.** In other words, there will be different elasticity coefficients for price changes *between* $4 and $3, such as a price increase from $3.25 to $3.75. Indeed, we will even get a different elasticity coefficient for a price *increase* from $3 to $4, as we verify in the following calculation (also see Figure 5–3):

Arc elasticity: A measure of average elasticity across all intermediate points between two points along a demand curve.

$$\epsilon_d = \frac{\Delta Q/Q}{\Delta P/P} = \frac{\Delta Q}{Q} \times \frac{P}{\Delta P} = \frac{40}{280} \times \frac{3.00}{1.00} = 0.42.$$

Whereas a price decrease from $4 to $3 yielded an elasticity of 0.66, a price increase from $3 to $4 yields an elasticity of 0.42, clearly a more inelastic consumer response. Why the difference? The size of the price change means that widely different initial prices and quantities are being expressed in the simple formula.

[2]Note that this expression for elasticity can be reorganized again to equal ($\Delta Q/\Delta P$) $\times$ (P/Q). This expression tells us immediately that elasticity and slope are different concepts. The slope of the demand curve is ($\Delta P/\Delta Q$); it shows how price changes with unit or other changes in quantity. Elasticity is the inverse of the slope ($\Delta Q/\Delta P$), that is, the slope "turned upside down," multiplied by the ratio of some specific price and quantity (P/Q) along the demand curve.

One compromise solution to the problem of determining the exact elasticity of demand over such a price range is to take the average elasticity within the arc between the two price-quantity combinations.[3] There are other methods of calculation as well. Naturally, the smaller the price change, the more precise is the simple formula $\epsilon_d = (\Delta Q/Q) \div (\Delta P/P)$ at estimating the true elasticity between two points.

While there are several means of calculating elasticity and while these calculations (as with many other representations of actual economic data) are approximations, elasticity is often very useful in assessing real-world problems and policies. We will continue to use the simple formula as our approximation.

Relation of Demand Elasticity to Expenditures and Receipts

As indicated at the beginning of this chapter, it is often necessary to estimate the effect of a proposed increase or decrease in price on total revenue—how much the government will make (or lose) by raising gas taxes or a fast-food chain will make (or lose) by offering a half-price sale on hamburgers. Estimates of elasticity from previous experience can make such projections possible. Or, if the price change and consumer response have already occurred, we can look at what happened to determine the elasticity of demand by examining either what customers have spent for a good or what businesses have received for it.

Common sense tells us that an industry's revenues are the same as consumers' expenditures for the industry's product. The number of items bought by all consumers multiplied by the price paid for each item is the same as the number of items sold by all producers multiplied by the average price charged for the items. Elasticity, therefore, is related both to total expenditures of consumers and to **total revenues** or receipts of businesses.

To understand how elasticity is related to consumer expenditures, we return to the example from Chapter 4 of a single consumer purchasing cassette tapes. Table 5–2 reproduces Dave's demand and total expenditures (quantity times price) for tapes, and Figure 5–4 reproduces his demand curve with the associated elasticities. Dave's total expenditures on tapes begin to rise as the price of tapes falls below $10. As the price falls to $9, $7, and $5, Dave's quantity demanded *and* his total expenditures on tapes rise. Notice that, for decreases in price, if elasticity is greater than 1, total expenditures will increase. Why? When demand is elastic, there is an inverse relation between price and total expenditures because the percent change in quantity demanded dominates (is larger than) the percent change in price. But as price is reduced below $5, Dave's total expenditures begin to decline, even though his consumption of tapes continues to increase. In the price range in which

Total revenue: Price times quantity sold; total expenditures.

[3]To calculate the average elasticity in the $3 to $4 price range for tapes, the prices $3 and $4 and the quantities 240 and 280 are given equal weight. To express the formula algebraically, we can call one quantity Q_1 and the other Q_0. The price at Q_1 is P_1; the price at Q_0 is P_0. The average elasticity of their relation can then be calculated as follows:

$$\epsilon_d = \frac{\dfrac{Q_1 - Q_0}{(Q_1 + Q_0)/2}}{\dfrac{P_1 - P_0}{(P_1 + P_0)/2}} = \frac{Q_1 - Q_0}{Q_1 + Q_0} \times \frac{P_1 + P_0}{P_1 - P_0} = \frac{280 - 240}{280 + 240} \times \frac{3 + 4}{3 - 4} = 0.53.$$

The average elasticity coefficient is thus between those calculated for a price increase and for a price decrease. This is one method of handling the difference in elasticity between two points.

TABLE 5–2   Elasticity and Consumer Expenditures

The average elasticity of demand is greater at higher prices and falls as price falls. At the midpoint (near $5), the elasticity is equal to 1 and total expenditures are at a maximum.

Price of Tapes (dollars)	Quantity Demanded	Total Expenditure	Average Elasticity of Demand
10	0	0	19
9	1	9	5.66
8	2	16	3
7	3	21	1.85
6	4	24	1.22
5	5	25	0.81
4	6	24	0.53
3	7	21	0.333
2	8	16	0.176
1	9	9	0.05
0	10	0	

FIGURE 5–4

An Individual's Elasticity of Demand

The elasticity of demand varies along a downward-sloping demand curve. As price falls, the elasticity falls.

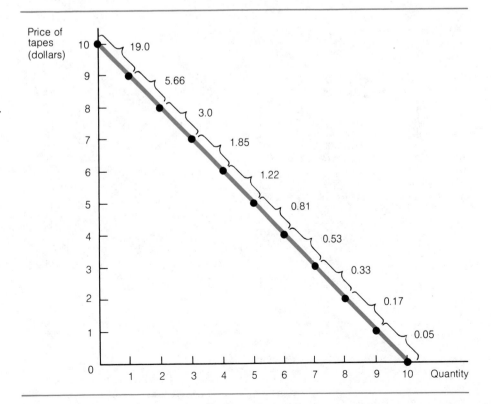

demand is inelastic (a coefficient of less than 1) total expenditures decrease as price falls. Here the percent decline in price dominates the percent increase in quantity, and total expenditures fall. In other words, prices and total expenditures change in the same direction if demand is inelastic.

The same relation between price and total expenditures is found when prices increase. If demand is inelastic, expenditures rise as prices increase from some low level, say \$2, to a new level or over a range. If demand is elastic for that price range, total expenditures actually fall as prices increase. At some price-quantity combination, demand is unit elastic, $\epsilon_d = 1$. For Dave in Figure 5–4, unit elasticity of demand occurs around \$5, when Dave's total expenditures for tapes will remain constant whether price is increasing or decreasing by some minuscule amount. It is at this point that Dave's expenditures are at a maximum.

Being able to estimate the ways that price elasticity of demand affects the total expenditures of consumers is of obvious value to businesses seeking to maximize revenues.

If total expenditures are known, the relations between the direction of the price changes and total expenditures can be used as a "back-door" method for determining elasticity, though not of calculating elasticity coefficients. The relations between the direction of the price change, the elasticity, and total expenditures (or revenues) are summarized in Figure 5–5. As the price falls from P_1, where consumption of the good is zero, to P_2, the midpoint on the demand curve, demand is elastic, and total expenditures (TE) and total receipts (TR) are rising. At P_2, price elasticity of demand is unitary, and total receipts and expenditures remain constant as the price rises or falls around this price. As the price falls below price P_2 (where quantity Q_2 is demanded), the demand elasticity coefficient falls below 1. Below P_2, total

FIGURE 5–5

Total Expenditures and Total Receipts Along a Demand Curve

As the price falls ($P \downarrow$) along the demand curve, both total receipts (TR) and total expenditures (TE) rise, indicating that elasticity is greater than 1. At the point of unit elasticity, TE and TR remain constant. As the price falls from the midpoint, TE and TR decline owing to inelasticity of demand. Rising price ($P \uparrow$) has the opposite effects on expenditures and receipts.

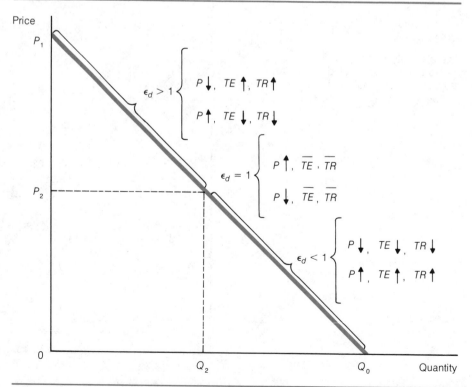

expenditures and receipts fall, and demand is inelastic, becoming more inelastic as the price approaches zero.

The major point to remember is that elasticity varies along any ordinary demand curve. It makes no sense to say, for instance, that salt is an inelastically demanded commodity without reference to some price range. The demand for salt may be elastic over some price range and inelastic over another. In the next section we consider what makes the demand elasticity coefficient greater than, equal to, or less than 1 and what factors determine consumer responses to price changes.

Determinants of Price Elasticity of Demand

There are three major determinants of price elasticity: (1) the number and availability of substitutes, (2) the size and importance of the item in the consumer's budget, and (3) the time period involved. Since all three factors interact, the condition of other things being equal must be invoked to determine the specific effect of any one factor.

Number and Availability of Substitutes

By far the most important predictor of demand elasticity is the ability of consumers to find good substitute products. If the price of one brand of toothpaste rises, many consumers may respond by switching to a different brand. But this substitution will happen only when alternatives are available.

Food is a vital commodity. Everyone must eat; elasticity of demand for nutrients in general is therefore approximately zero over relevant price ranges—completely inelastic. For most of us, there is no substitute for food. But it is possible to substitute between kinds of food. While the demand for food as a whole tends to be inelastic, the consumer may substitute between broad food groups such as meat and seafood and also between foods within the broad groups. Consumers of meat, for instance, have a wide choice of substitutes, such as chicken, beef, pork, duck, possum, or alligator. Consumers are far more sensitive to changes in the price of beef when other substitutes are available. Elasticity of demand for meat itself is lower than that for kinds of meat.

What about the elasticity of demand for beef versus beef products such as hamburger, sirloin steak, beef tails, and so on? The consumer can substitute among beef products. When the price of sirloin rises relative to hamburger, the consumer can substitute hamburger for sirloin. The elasticity of demand for hamburger is therefore higher than the elasticity of demand for meat. By now, the general rule must be apparent: The broader the product or service group, the lower the elasticity because there are fewer possibilities for substitution. We expect the elasticity of demand for Budweiser beer to be more elastic than the demand for beer, just as the demand for a Ford or Mercedes-Benz is more elastic than the demand for automobiles. Ordinarily, the wider the selection and substitutability of similar products and services, the larger the elasticity of demand for some product.

In Chapter 9 we will see that elasticity is a crucial factor in determining how competitive markets are structured. One view of competition is that it exists when products are perfectly substitutable—one seller's wheat is identical to another seller's, for example. A number of closely substitutable products—such as hair stylists in a large city—would also indicate a very competitive market. When goods are demanded and produced with few or no close substitutes, the market changes from competition to monopoly.

The Importance of Being Unimportant

The size of the total expenditure within the consumer's budget is another determinant of elasticity. Ordinarily, the smaller the item in the consumer's budget, the less elastic the consumer's demand for the item will be over some price ranges. A salt user may be insensitive to price increases in salt simply because expenses for salt are a small part of his or her budget. If the price of salt were to rise too high, the user might substitute alternatives, such as artificial salt or lemon juice. Thus, in calculating elasticity it is important, at some point, to analyze both substitutability and the size of the item in the consumer's budget.

On the other hand, the effects of size in one's budget and substitutability may offset each other in determining elasticity of demand. Take, for instance, the demand for electricity by the poor and the aged living on Social Security or other transfer payments. Inflation and rising energy costs have caused electricity rates to soar over the last decade. How would such consumers respond to rising electricity prices, which obviously take up a large portion of their total budget? Using the "importance of being unimportant" criterion, we might be tempted to say that their demand for electricity is highly elastic. But economic theory and common sense tell us that electricity consumers will not be very responsive to price increases because there are few if any viable substitutes for electricity (except perhaps for intolerable house temperatures or highly expensive conversion to oil heat, when electricity is used for heating and cooling). Thus, substitutability—the second determinant of elasticity—outweighs the importance of being unimportant.

The condition of other things being equal clears up our understanding of the elasticity determinants. We can correctly state that, given some constant degree of substitutability, the more unimportant the commodity is to a consumer, the lower demand elasticity will be. This simply indicates that tastes, substitutability, and importance of the commodity in the consumer's budget must all be examined in gauging the demand elasticity of products over given price ranges.

Time and Elasticity of Demand

Time is the final factor affecting the demand curve. We alluded to these effects at the beginning of this chapter, but we can now understand them explicitly.

As a simple example, a local market for tennis rackets is represented in Figure 5–6. (For simplicity we may neglect the supply function.) Assume that the initial price is P_0 and the initial quantity demanded is Q_0. What happens to elasticity of demand if tennis rackets go on sale; that is, if that price falls to P_1 per unit, a 15 percent decline?

The answer depends on how long it takes for consumers and potential consumers to adjust to the new price. If the sale is totally unpublicized, there may be no immediate reaction to the new price. As consumers gain information, however, increases in demand ensue. Thus we may think of the demand curve as rotating around a point (A in Figure 5–6) as news of the sale becomes more widespread. Greater quantities (Q_1, Q_2, and so on) will be sold *through time*. For the given price change, the elasticity of demand will be different at different times—that is, it will depend on whether it is calculated on the first day of the sale ($\epsilon_d = 0.33$) or one week ($\epsilon_d = 0.66$) or one month later ($\epsilon_d = 1.33$). Percent changes in quantity demanded are greater as time passes, meaning that the demand elasticity coefficient is

FIGURE 5–6

Elasticity of Demand Over Time

As time goes by, a given price change (ΔP) may be associated with larger and larger changes in quantity demanded (ΔQ). Since the percent change in quantity demanded grows over time, the elasticity of demand grows as well.

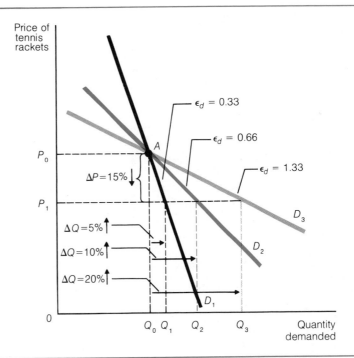

larger and larger (up to some limit, of course). The general rule holds: Elasticity of demand increases the longer any given price change is in effect.

The time period of adjustment is a crucial factor in calculating or estimating all types of actual elasticities. Not only do tastes, substitutability, and size within the consumer's budget affect elasticity, but the ability of consumers to recognize and adjust to changes also plays a part. (Focus, "Time, Elasticity, and the U.S. Demand for OPEC Oil," is an example of the affect of time on elasticity of demand.)

Other Applications of Elasticity of Demand

The concept of elasticity is not restricted to percent changes in quantity demanded to percent changes in price, that is, to price elasticity of demand. In general, we can calculate an elasticity of any dependent variable (effect) to a change in any independent variable (cause). We will consider two other important applications of this versatile economic concept: income elasticity of demand and cross elasticity of demand.

Income Elasticity of Demand

Income elasticity of demand: A measure of buyers' response to a change in income in terms of the change in quantity demanded; the percent change in quantity demanded divided by the percent change in income.

As you will recall from Chapter 4, a consumer's income is an important determinant (independent variable) of the demand for goods and services. It is often very informative to inquire about consumers' **income elasticity of demand.** Producers of all kinds are interested in the magnitude of consumption changes as incomes rise or in consumption habits within various income groups. Budget data compiled by the government and other sources can be used to calculate recent and historical trends in changing consumption patterns as incomes change.

The mechanics of income elasticity of demand are identical to those in-

FOCUS   Time, Elasticity, and the U.S. Demand for OPEC Oil

For a fast-driving, freewheeling, high energy-using country like the United States, the successful embargo of oil supplies to this country by the OPEC oil cartel in 1973–1974 was a sobering experience. Since OPEC began restricting the supply of oil, the effects on prices of oil and oil products, including gasoline, have been dramatic, especially between 1973 and 1978 (see Figure 5–7).

As the figure shows, OPEC's control over supply had the effect of shifting the supply curve to the left. Initially the price of oil rose from P_0 to P_1 along the demand curve (assumed constant) for oil (and implicitly for oil products). Short-run elasticity for oil was quite low, indicating that Americans did not instantaneously or even rapidly adjust to the reduced supply.

The effects of time on elasticity of demand were very different, however, as the ability to find substitute sources of energy grew. Americans, both consumers and producers, began to find ways to economize on high-priced oil. Consumers reduced auto travel, participated in car pooling, and began buying smaller cars. Substitutes for private auto travel, such as urban transit systems, began to gain support in the cities. All in all, these effects produced a 7 to 10 percent reduction in gasoline consumption by 1983. Consumers and producers also economized on home and industrial uses of oil-based energy. Improved home insulation and use of alternative fuel sources, such as solar energy, were common substitutions. Producers substituted alternative resources, including other forms of energy, for high-priced oil.

What were the effects of such substitution over time on the demand for oil and oil products? By the early 1980s elasticity of demand had risen significantly, opening up an excess supply of oil (*AB*) at price P_1. After time adjustments, the demand curve for oil became more elastic (D_2). The prices of oil and oil products fell to lower levels, such as P_2. Along with and, indeed, partially because of the rising elasticity of demand for oil through time, the OPEC cartel lost its

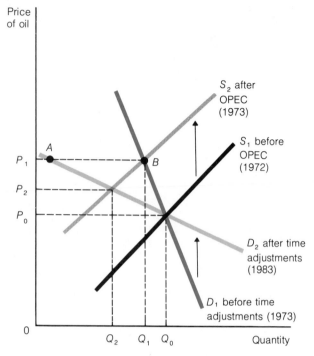

FIGURE 5–7   Effect of the OPEC Cartel on Demand Elasticity over Time 1973–1983

The OPEC cartel's restriction of output effectively shifted the whole supply curve to the left, resulting in higher prices for American consumers. As consumers reduced quantity demanded for OPEC products, however, the demand elasticity for oil rose substantially, creating a reduced quantity demanded (Q_2) and a lower price (P_2).

punch. Disputes among OPEC nations became not so much political as economic in differences over elasticity projections. The bottom line is that elasticity of demand is profoundly affected by time and the ability to substitute.

volved in the calculation of price elasticity; only the independent variable changes. With price held constant, income elasticity is the percent change in quantity demanded resulting from (divided by) a given percent change in income. It is expressed as ϵ_y, with Y representing income:

$$\epsilon_y = \frac{\%\ \text{change in consumption of a good}}{\%\ \text{change in income}} = \frac{\Delta Q}{Q} \div \frac{\Delta Y}{Y}.$$

The income elasticity coefficient, ϵ_y, may be positive or negative depending on whether the good is normal or inferior. (Recall from Chapter 4 that a good is normal if an increase in income results in greater consumption and inferior if an increase in income results in a reduction in consumption of the product.) When applying the income elasticity formula, if ϵ_y is greater than zero, the good is normal; if ϵ_y is less than zero, the good is inferior. For the moment we discuss only normal goods.

If the elasticity coefficient ϵ_y is greater than 1, demand (or consumption) is said to be income elastic; if ϵ_y is less than 1, the product is income inelastic; and if ϵ_y equals 1, income is unit elastic. Suppose that income in the United States rises by 10 percent in 1986 and that the quantity of new automobiles consumed over the year increases by 8 percent. Clearly, new automobiles are a normal good (since their consumption increases along with the increase in income). But what is the income elasticity? The simple computation is the following:

$$\epsilon_y = \frac{8\%}{10\%} = 0.8.$$

The income elasticity of demand for automobiles in this hypothetical calculation is less than 1 but greater than zero. What does this mean for the auto industry? Other things being equal (such as price and tastes), the demand for automobiles will rise but at a slower pace than income. This fact is obviously important to groups in society such as auto manufacturers, investors, boat dealers, and airlines.

Income elasticity for a particular good may be determined for any individual consumer or for consumers as a group. The practical importance of this calculation is undeniable. An individual deciding between opening a gourmet food shop and a travel agency during a period of rapidly rising income would be very interested to know, for example, that income elasticity for gourmet foods is perhaps 0.2, while the same coefficient for Mediterranean vacations is 6.2. However, to achieve accuracy in actual use, all other factors, such as substitutability and the price of the product, must be kept constant. If these factors vary, as they often do in the real world, their impact on consumption must be determined and integrated into the analysis.

Cross Elasticity of Demand

Cross elasticity of demand: A measure of buyers' responsiveness to a change in the price of one good in terms of the change in quantity demanded of another good. The percent change in the quantity demanded of one good divided by the percent change in the price of another good.

Substitutes: Two goods whose cross elasticity of demand is positive.

Cross elasticity of demand simply reveals the responsiveness of the quantity demanded of one good to a change in the price of another good. As such, a cross elasticity coefficient can define either substitute or complementary products or services. In more general terms, cross elasticity is an extremely useful economic tool for identifying groups of products whose demand functions are related.

First there are substitute products. If the price of one good rises and, other things being equal, the quantity demanded of another good increases, those products are **substitutes.** If Dorothy's demand for Bayer aspirin rises 85 percent following a 10 percent rise in the price of Bufferin, her cross elasticity of demand for aspirin is +8.5. The coefficient of cross elasticity of demand is calculated thus:

$$\epsilon_c = \frac{\%\ \text{change in quantity demanded of one good}}{\%\ \text{change in price of another good}} = \frac{85}{10} = +8.5.$$

For substitute commodities, then, the cross elasticity coefficient is positive because an increase in the price of one good causes an increase in the quantity demanded of the other. The larger the elasticity coefficient (in absolute terms), the more substitutable the products or services are.

Next there are items that are complements in consumption, such as bacon and eggs; left shoes and right shoes; gasoline and automobiles; light bulbs, lamps, and electricity. Goods are **complements** when an increase in the price of one good results in a decrease in the quantity demanded of the other. Gin and vermouth are the two essential ingredients (besides olives) in a martini. If there is a 4 percent increase in the price of vermouth and a 16 percent reduction in the quantity consumed of gin, other things being equal, we can call gin and vermouth complementary products. Note that the cross elasticity coefficient is negative:

Complements: Two goods whose cross elasticity of demand is negative.

$$\epsilon_c = \frac{\% \text{ decrease in quantity demanded of one good}}{\% \text{ increase in price of another good}} = \frac{-16}{4} = -4.0.$$

A negative number is obtained for the cross elasticity coefficient for complementary goods since the increase in the price of one is always associated with a decrease in the demand for the other.

In calculating cross elasticity, therefore, a positive sign indicates that goods are substitutes and a negative sign indicates that they are complements. The absolute size of the coefficient, moreover, tells us the degree of substitutability or complementarity. A coefficient of -28.0 indicates a greater degree of complementarity than one of -4.0, for example.

Elasticity of Supply

The versatile concept of elasticity which we have so far related to demand can also be applied to problems related to supply. We will encounter many of these important concepts throughout the book, but the main theme of elasticity of supply and two variations are introduced here. **Elasticity of supply** is the degree of responsiveness of a supplier of goods or services to changes in price or some other variable.

Elasticity of supply: A measure of producers' or workers' responsiveness to price or wage changes; price elasticity of supply is the percent change in quantity supplied divided by the percent change in price.

A price elasticity of supply coefficient can be mechanically calculated in the same manner as all other elasticities. To determine the relation between a change in quantity supplied and a change in price, the simple formula can be applied:

$$\epsilon_s = \frac{\% \text{ change in quantity supplied}}{\% \text{ change in price}} = \frac{\Delta Q_s}{Q_s} \div \frac{\Delta P}{P} = \frac{\Delta Q_s}{Q_s} \times \frac{P}{\Delta P}.$$

Such a simple coefficient or some more elaborate "average elasticity" can be calculated for any supply curve at any instant in time just as for the elasticity of demand described earlier. Elasticity of supply can also be applied to any kind of supply curve, from beets to ball bearings. We turn to an input supply curve—a labor supply curve.

The Elasticity of Labor Supply

A labor supply curve, such as the one shown in Figure 5–8, simply shows the amount of work (measured in number of hours) that a laborer, whom we will call Sam, would be willing to supply at alternative wage rates. Factors such as the worker's wealth, tastes, preferences, and other factors affecting work decisions are held constant. Each person's labor supply function differs because of these factors.

Ordinarily the labor supply curve is positively sloped, like the supply curve for any commodity or service. An increase in the wage rate from W_0 to W_1 in Figure 5–8 causes Sam to increase the number of hours he is willing to work from L_0 to L_1; likewise, a wage increase from W_2 to W_3 increases Sam's

FIGURE 5–8

Elasticity of Labor Supply

Starting from a relatively low wage rate (W_0), a 10 percent increase in Sam's wage leads him to supply 25 percent more work, an elastic response. A 10 percent rise at a higher wage (W_2) increases Sam's hours worked by only 3 percent, an inelastic response.

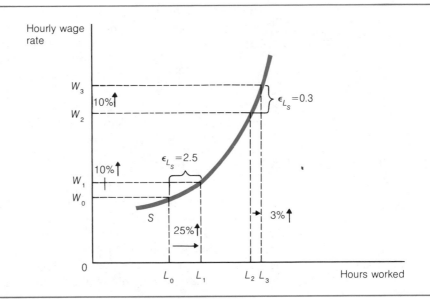

supply of work from L_2 to L_3. The question is, What is the elasticity of the labor supply schedule and what does it tell us?

The elasticity of labor supply shows the responsiveness of labor supply to a percentage change in the wage rate. It is calculated as follows:

$$\epsilon_{L_s} = \frac{\% \text{ change in quantity of labor supplied}}{\% \text{ change in the wage rate}} = \frac{\Delta L}{L} \div \frac{\Delta W}{W} = \frac{\Delta L}{L} \times \frac{W}{\Delta W}.$$

We can make an elasticity calculation for two different segments of Sam's supply curve. Starting from a relatively low wage rate, W_0, an increase in wages of 10 percent will produce a 25 percent increase in the number of hours worked. The coefficient ϵ_{L_s} is 2.5, which indicates an elastic supply of work effort (because it is greater than 1).

But see what happens to an identical wage rate increase (10 percent) starting from a higher wage, W_2. Sam increases the number of hours worked by 3 percent, producing an elasticity coefficient, ϵ_{L_s}, of 0.3—clearly an inelastic response. Why the drastic change in Sam's response to wage increases?

Some personal reflection will give us clues to Sam's work behavior. Work and leisure are two ways of using the scarce resource of time. The more hours we work, the less time we have for leisure. As we work more hours per day in response to wage rate increases, the relative opportunity cost of working (leisure forgone) begins to increase. At some point, laborers start to substitute leisure for work and income. Elasticity of labor supply is one method of calculating this trade-off (although other factors such as income level also shape labor supply). Other things being equal, an inelastic labor supply coefficient for a given wage rate change means that the laborer demonstrates a preference for leisure over work (nonmarket time over market time).

Time and Elasticity of Supply: Maryland Crab Fishing

An extremely important issue, which we have already related to demand, concerns the time dimension over which the economist calculates supply elasticities. A time dimension exists in all facets of life and human activity, and it is important in the economic activities of suppliers and producers.

A supply elasticity coefficient would not ordinarily capture the full response of suppliers over time to a given price change. To illustrate, a crab fisherman from Chesapeake Bay daily brings in a catch and offers it for sale. The supply curve for crabs on any given day would be totally inelastic. On any given day, the quantity supplied of crabs would be completely unresponsive to price changes.

How then does the market establish a price? Price is determined by the interaction of supply and demand. If the demand for crabs on some particular day happens to be D_0 in Figure 5–9, price will settle at P_c and the entire quantity of crabs (a commodity that we assume, unrealistically, is not storable) will be sold.

Now suppose that owing to a change in consumers' taste, Maryland crabs become more desirable, and the demand curve shifts permanently to D_1. The fisherman's good fortune is revealed to him when he brings in his usual catch of Q_0 crabs. All of a sudden he finds that his catch brings a higher price P_0. How will the fisherman respond? If the price P_0 is higher than his average production costs, meaning a higher-than-normal profit, the fisherman will adjust by shifting available resources into crab fishing as soon as possible. If he has idle boats or nets and if there is plenty of labor available, all will be put to use. The act of producing more crabs takes time because resources are not instantly adaptable to crab fishing.

During ensuing days, weeks, or months, more crabs will be offered for sale, resulting in a more elastic supply curve for crabs. Such a supply curve over the initial adjustment period may look like S_1 in Figure 5–9. The price may temporarily fall to P_1 given that the demand for crabs remains stationary at D_1. Comparing the quantity Q_1 sold at price P_1 with the previous quantity of crabs sold, Q_0, shows us that over some adjustment period the supply curve for crabs is not completely inelastic but that quantity supplied is responsive to the initial price change. The response of the fisherman to price changes on any given day will still be nil; but over time, he will adjust resources so as to increase the amount of crabs he offers for sale.

If price P_1 is still abnormally profitable, the fisherman will continue to

FIGURE 5–9

Time and Elasticity of Supply

If the demand for crabs shifts from D_0 to D_1, then price rises quickly to P_0. As time goes by and crab fishermen are able to adjust inputs, the price begins to fall to P_1. After a long period of time and all adjustments are made to the increased demand, price falls to P_2. Over time, elasticity of supply increases.

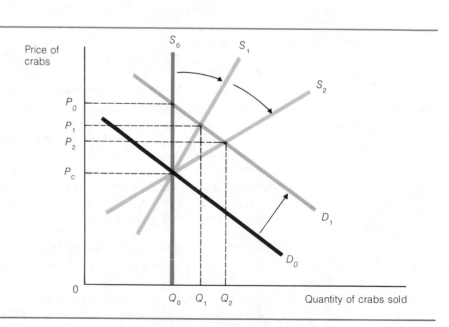

FOCUS   Gasoline, Cigarettes, and Elasticity: The Effects of an Excise Tax

In 1983 the Reagan administration levied a 5-cent-per-gallon excise tax on gasoline in an effort to acquire tax revenues for highway and bridge repairs. The effects of such a tax are of interest not only to consumers and producers of gasoline but also to legislators. By how much will the price of gasoline increase? By how much will the amount of gasoline purchased decrease? Will tax revenues be as great as expected? The elasticities of the demand for and supply of gasoline are the keys to answering these and other questions.

An excise tax is a simple per unit tax on the sale of a particular item. Tax collectors determine the amount of the taxable good sold by a retail or wholesale firm and require that the firm pay the amount of the tax times the quantity sold.

To gauge what may happen under the gasoline excise tax, consider the mechanism of what happens under a similar situation: the levying of an excise tax on cigarettes. The equilibrium effects of an excise tax on cigarettes are shown in Figure 5–10. The hypothetical supply and demand for cigarettes before the excise tax are points S_1 and D. The equilibrium price, E_1, is 70 cents per pack and the quantity is 50 million packs per day. The effect of a 30-cent-per-pack excise tax can be shown by shifting the supply curve vertically upward by 30 cents to S_2. For each quantity along S_2 the price at which producers were willing to offer that quantity has now increased by 30 cents. Producers were previously willing to offer 50 million packs for 70 cents each, but now they must receive $1 per pack to offer 50 million packs.

The new equilibrium quantity, E_2, is now lower, at 40 million. This change occurs because buyers respond to price changes; their elasticity of demand for cigarettes is greater than zero. As the excise tax puts upward pressure on the price, a smaller quantity is demanded. Producers are willing to offer 40 million packs at a price of 55 cents before the tax. The posttax equilibrium price to consumers, P_c, is 85 cents. From this price, producers must pay the 30-cent excise tax and receive the net price, P_p, of 55 cents per pack.

The total revenues received by the government from the tax may not be as great as some politicians expected. Before the tax was instituted, the quantity bought was 50 million. A simple multiplication of 30 cents times 50 million would yield an overestimate of tax revenues. The actual tax revenues are 30 cents times 40 million, the new equilibrium quantity. For any excise tax—including that levied on gasoline—the elasticity of supply and demand must be taken into account before a projection is made for tax revenues.

The elasticities of supply and demand also determine the relative burden of the excise tax. Who actually pays the tax—producers or consumers—is shown by the

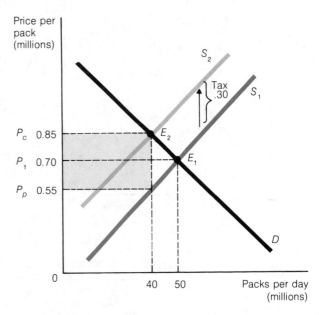

FIGURE 5–10   Effect of an Excise Tax on Cigarettes

When a 30-cent excise tax is added to the price of a pack of cigarettes, the entire supply curve for cigarettes shifts from S_1 to S_2. At the new equilibrium price, 85 cents, the elasticity of demand is greater than zero; specifically, consumers will reduce their consumption of cigarettes by 10 million packs and total tax revenues will equal 40 million times the tax. The shaded area represents government tax revenues, the 30-cent excise tax per pack multiplied by the number of packs of cigarettes sold after the tax is imposed.

change in price to the buyers and sellers. In this example, the price to consumers increased 15 cents and the net price to producers fell 15 cents. Here the burden is shared equally by consumers and producers, but this is not necessarily the case.

If demand had been relatively less elastic, then price would have increased more to consumers than it fell to producers. Figure 5–11a shows that with less elastic demand more of the burden is shifted to consumers. Also, Figure 5–11b shows that if supply had been less elastic more of the burden would have been shifted back to the producers. Figure 5–11c shows that a more elastic demand also shifts more of the burden to producers.

To summarize the effects of an excise tax: The lower the elasticity of demand, the greater the price increase or tax burden to consumers. The lower the elasticity of supply, the greater the net price decrease or burden to the producers. And the greater the elasticity of supply or demand, the lower the total tax revenues.

What then are the effects of a 5-cent excise tax on

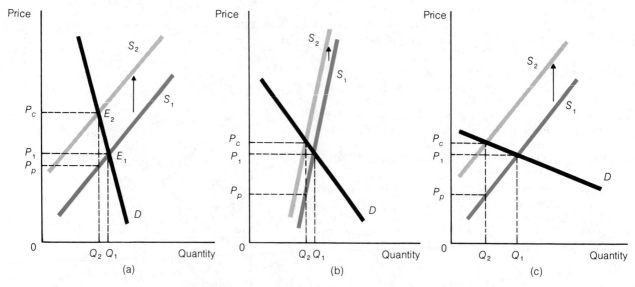

FIGURE 5–11   Elasticity and the Relative Burden of a Tax

(a) With less elastic demand, more of the burden of an excise tax is shifted to consumers: at Q_1 consumers pay the equilibrium price P_1, but at Q_2 following a tax, consumers must pay P_c. (b) With less elastic supply, more of the burden of a tax is shifted to producers. (c) More elastic demand also shifts more of the burden to the producers.

gasoline? The price rises, but not by the full 5 cents. The producers and consumers share the burden of the tax. However, the burden may not be shared equally. The tax revenues are the amount of the tax times the quantity sold *after* the tax is instituted. The change in price, the burden of the tax, and the level of tax revenues are all determined by the elasticity of supply and demand.

shift resources into crab production by purchasing new boats and equipment and by training new workers. Again, such activity takes time, but the effect will be to increase crab supply (perhaps to S_2). The end result in the crab market will depend on the adaptability and availability of resources for crab fishing and on the cost of producing new inputs. But the general rule is that elasticity of supply will tend to increase over time. The longer the time period of adjustment to an initial change in price—whether price is rising or falling—the more elastic supply schedules will be. This principle applies to all supply curves, including market or industry supply curves. In the case of market supply curves, the adjustment to a change in demand and price will depend on how long it takes to draw on unused or idle capacity or on resources from other industries; in the case of a permanent price drop, the adjustment will depend on the time it takes to decrease the use of resources. The full impact of economic policy changes—levying taxes on industries, for example—often hinges on the elasticity of supply over time.

Real-world events are seldom so isolated and data so accurate as to provide the economist with means to calculate elasticities precisely. It is nonetheless important to a great number of economic and practical problems, some discussed in this chapter, that elasticities can be estimated. To estimate elasticities, the economist must apply precise analytical tools to often complicated policy situations. In Focus, "Gasoline, Cigarettes, and Elasticity: The Effects of an Excise Tax," we provide an example of two such situations. In Economics in Action at the end of the chapter we apply the elasticity principle to the U.S. farm industry.

Summary

1. Elasticity is the ratio of the percent change in effect to the percent change in some cause. If changes in quantity demanded (the effect) are caused by changes in price, elasticity of demand is the percent change in quantity divided by the percent change in price. This ratio is called the demand elasticity coefficient (ϵ_d).

2. A demand elasticity coefficient greater than 1 means that consumers are responsive to price changes; demand is elastic. When $\epsilon_d = 1$, demand is unit elastic. When ϵ_d is less than 1, consumers are not very responsive to price changes; demand is said to be inelastic.

3. Elasticity can also be derived (in a general, shorthand manner) by examining total expenditures (or total revenues) as price rises or falls. If total expenditures rise (fall) as price rises (falls), ϵ_d is less than 1. If total expenditures remain constant as price rises or falls, $\epsilon_d = 1$. If total expenditures fall (rise) as price rises (falls), the demand elasticity coefficient is greater than 1, that is, it is elastic.

4. There are three major determinants of demand elas-ticity: (1) the number and availability of substitutes; (2) the size and importance of the item in the consumer's budget; and (3) the time period over which the coefficient is calculated.

5. Other applications of the elasticity concept related to demand are the relation between income changes and changes in quantity demanded (income elasticity of demand) and between price changes for one good and changes in quantity demanded for another complementary or substitute good (cross elasticity of demand between two goods).

6. An elasticity coefficient can be calculated for all kinds of supply curves. A supply elasticity coefficient is simply the percent change in quantity supplied divided by the percent change in price. The concept can be applied to all input and output supply curves, including labor supply.

7. Time is at the center of most important economic calculations, including elasticity. A general rule regarding time and elasticity is that the elasticity of demand and supply increases with the time that a change (in price, for instance) is in effect.

Key Terms

elasticity
price elasticity of demand
demand elasticity coefficient
elastic demand
unit elasticity of demand

inelastic demand
arc elasticity
total revenue
income elasticity of demand

cross elasticity of demand
substitutes
complements
elasticity of supply

Questions for Review and Discussion

1. The formula for elasticity is

$$\frac{\%\Delta Q_d}{\%\Delta P} = \frac{\Delta Q_d/Q}{\Delta P/P} = \frac{\Delta Q_d}{Q} \times \frac{P}{\Delta P}.$$

Using this formula derive the elasticity of demand for a product when the price changes from $1 to $1.50, $2 to $2.50, $3 to $3.75, and the change in quantity demanded is 15 to 10, 25 to 5, 50 to 30, respectively.

2. What are the three determinants of consumers' sensitivity to a change in price? Do these always work in the same direction?

3. Is it feasible to talk about *an* elasticity all along a single demand or supply curve? Why or why not?

4. Given that a price change remains in effect over a period of time, will elasticity increase or decrease? Why?

5. What is meant by cross elasticity? How is it algebraically different from elasticity of supply and elasticity of demand, which we calculated earlier?

6. What do cross elasticities indicate about relations between two goods?

7. What might reports about the link between salt consumption and high blood pressure do to the elasticity of demand for salt?

8. Is the demand elasticity coefficient for large industrial consumers of electricity larger or smaller than that for residential use? If electric companies lower the price to both groups, would total revenues from each group change in the same direction?

9. Suppose a friend has an allowance of $10 per week. She spends all of her weekly income on banana splits. What is her elasticity of demand for banana splits?

10. A college town pizza parlor decides to offer a back-to-school two-for-one special, in effect cutting the price of a pizza in half. More pizzas will be sold according to the law of demand. Will the total receipts of the parlor increase, decrease, or remain the same if pizza consumption increases by 30 percent? By 80 percent?

ECONOMICS IN ACTION   The U.S. Farm Problem

Contemporary American farmers are experiencing severe problems related to elasticity and to long-run trends in technology. The annual income of farmers fluctuates greatly, and the relative price of agricultural products has been steadily declining while farmers' costs have risen. In spite of government programs to alleviate the problem, many farms go out of business every year. Elasticity and the longer-run economic environment under which these businesses operate help explain their current plight.

Both the demand for food and the supply of food are relatively inelastic over short time periods. Price inelastic demand means that small changes in farmers' supply of food result in large changes in price. If weather conditions are particularly favorable, a small increase in production (supply) can result in a large decrease in price. Also with inelastic demand, total revenue falls as price falls. On the other hand, a natural disaster (a late or early ice storm, a hurricane, or an insect plague)

causes a decrease in farm output and raises prices considerably, causing total revenue to rise. These situations are presented graphically in Figure 5–12. The production and demand conditions of individual crops—oranges, wheat, avocados—change continually. Frequent booms and busts caused by changing market conditions create short-run income-maintenance problems for individual farmers.

A problem of greater concern is the long-run trend. Every year more and more farm land is being converted to housing, highways, and shopping malls. Since 1880 the number of farms has decreased from about 4 million to 2½ million. Does this mean that agricultural products are becoming more scarce?

In the early 1800s Thomas Malthus earned economics a reputation as "the dismal science" by suggesting that the human population is doomed to a subsistence wage. This would result from a rate of population expansion greater than the rate of growth in agricultural

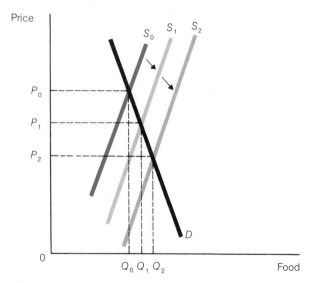

FIGURE 5–12   Short-Run Fluctuations of Food Prices

The demand for agricultural goods is relatively inelastic, and short-run supply fluctuations (shifts to S_0, S_1, or S_2) affect price and revenues dramatically. If farmers' output rises from Q_1 to Q_2, then price falls substantially along with total revenue. If output falls from Q_1 to Q_0, price and total revenue rise substantially.

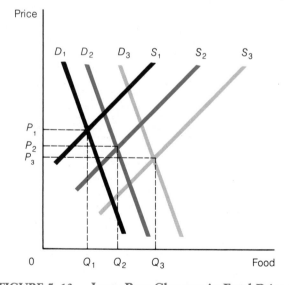

FIGURE 5–13   Long-Run Changes in Food Prices

In recent years the relative price of agricultural products has been falling. Supply has been shifting to the right owing to tremendous strides in food production technology (S_1 to S_2 to S_3). These supply shifts have outstripped increases in the demand for agricultural goods (D_1 to D_2 to D_3) caused by growth in population.

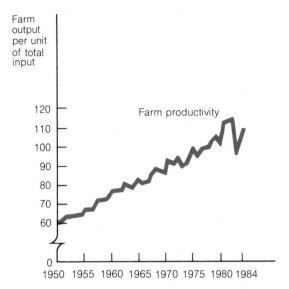

FIGURE 5–14   Farm Productivity, 1950–1984

Productivity—measured as farm output per unit of farm input—has doubled since 1950 on U.S. farms. Rapid technological advance has made this dramatic growth in productivity possible.

Source: U.S. Government Printing Office, *Economic Report of the President* (1985), p. 339.

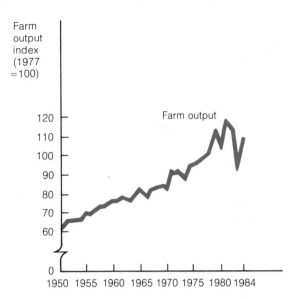

FIGURE 5–15   U.S. Farm Output, 1950–1984

American farm output has grown steadily over the past thirty-five years. With the exception of small and temporary downturns in selected years, farm output has doubled over the period.

Source: U.S. Government Printing Office, *Economic Report of the President* (1985), p. 339.

products. His prediction was that the demand for food would increase faster than the supply. With the increase in food prices, a larger and larger percentage of people's incomes would be spent on food. People's urge to procreate would force population increases, but starvation, disease, and wars would ultimately limit population growth to the growth in the food supply.

Malthus's grim prediction has not panned out for the United States. Even though there are fewer farms and a smaller percentage of U.S. land devoted to crop production, agricultural products have not become more scarce. Instead, the supply of food has been increasing faster than the demand for food. In fact, the price of food relative to other goods and the percentage of income spent on food have fallen steadily through the years. See Figure 5–13.

Tremendous improvements in technology are the reasons for these trends. The U.S. Department of Agriculture and many universities have developed new production techniques, disease and insect control, and high-yield hybrid plants.

Figures 5–14 and 5–15 show farm productivity and the actual growth in total output between 1950 and 1984. A major measure of productivity change is farm output per unit of farm input. Figure 5–14 shows that in spite of a declining number of farms and farmers, output per unit has almost doubled between 1950 and 1984 from an index of 60 in 1950 to an index of 109 in 1984. This statistic indicates that farm technology has been advancing at an extremely rapid pace in the

last thirty-five years. The net result of these factors is faster growth in total output which is seen to be the case in Figure 5–15.

This increase in technology has allowed output prices to fall while input prices have increased, and farmers' income has fallen as a result. As Figure 5–16 shows, current dollar net farm income (in terms of 1967 dollars) has fallen from about 19 billion dollars in 1950 to 8 billion dollars in 1984. This decline is the result of a number of long-run forces discussed above— farm failures due to falling farm prices caused by vastly improved technology and lower growth in the demand for farm products.

Under normal unrestricted competitive conditions, gains in productivity would create only short-run problems for farmers. As fewer farms were needed, competitive forces would drive some farmers out of business. In the long run, the surviving farms would earn a normal profit. Several government programs have been implemented to help support farm incomes and to encourage farmers to remain in the agricultural industry in spite of economic losses. These programs alleviate the risks of farming and ensure an overabundance of agricultural products for the U.S. consumer, but they also contribute to long-run problems for farmers. Rather than encouraging farmers to stay in an unprofitable industry, government programs could be instituted to encourage farmers to leave the industry. This would possibly eliminate both the losses in the industry and the need for government income support programs.

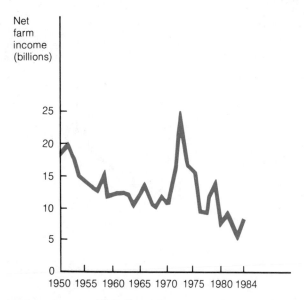

FIGURE 5–16   Net Farm Income, 1950–1984 (in 1967 dollars)

With the exception of a few years, farm income has tended to decline dramatically between 1950 and 1984.

Source: U.S. Government Printing Office, *Economic Report of the President* (1985), p. 338.

In summary, the low income of farmers is the result of improved productivity, low short-run elasticities of supply and demand, and government price-income support programs. These effects have resulted in too many farmers and an unnecessary burden on taxpayers in the form of government support to farmers. All of us who consume agricultural products have benefited from fall-ing relative prices. In addition, the gain in comparative advantage has allowed a substantial increase in exports of agricultural products to foreign countries. The industry's output and efficiency are growing at a much faster rate than in most other industries, but many individual farmers have problems. Low elasticities of demand and supply over short time periods are a large part of the problem.

In 1985, an increasing number of U.S. farmers were experiencing business failures. The strong dollar in international exchange markets encouraged imports of farm products from abroad and discouraged exports of U.S. farm commodities. The problem of the typical farm family has also been exacerbated by falling land prices. In the early 1980s, farmers borrowed heavily to purchase land, only to see the value of that land diminish rapidly. In the first half of 1985 many small farmers threatened with bankruptcy petitioned Congress and President Reagan for additional aid. But the push for a balanced budget has made large increases in support unlikely. The family farm may be a vanishing part of the American experience.

Question

In addition to oil, Iran was a major supplier of pistachio nuts to the United States. In the Khomeni era, U.S. imports of the nuts fell to zero (in 1980). Analyze the possible short-run and long-run effects of demand and supply elasticity on pistachio prices and quantities traded in the United States. What are the predicted effects upon California pistachio nut growers?

6

The Logic of Consumer Choice

We have seen in Chapters 4 and 5 that the law of demand is an accurate and useful generalization about human behavior. Other things being equal, individuals will consume more of a good when its price is lower. Knowing that the law of demand works, however, does not tell us why it works. In this chapter we seek to understand why individuals increase their purchases of a good or service when its price falls. The heart of the analysis involves how consumers deal with the problem of scarcity. In other words, given that a consumer has a fixed budget or income, how does he or she allocate the budget among goods and services to obtain the most satisfaction from consumption?

This chapter's analysis of consumer choice is presented in two alternative forms. In the body of the chapter, individual choice is analyzed in terms of marginal utility; in an appendix at the end of the chapter, the same analysis is cast in terms of indifference curves. Both approaches to the analysis of individual choice-making behavior are useful and introduce new tools and insights with which to study economic behavior in general.

Utility and Marginal Utility

Utility: The ability of a good to satisfy wants; the satisfaction obtained from the consumption of goods.

Why do we demand anything at all? The obvious answer is that we get satisfaction or pleasure from consuming goods and services. A vacation at the beach, a ticket to the big game, dinner at a fine restaurant all give us pleasure. Economists call this pleasure or satisfaction **utility.** Economics, unlike psychology, does not provide any fundamental answers to questions such as why some people prefer red rather than blue shirts or why so many people like chocolate. Economics analyzes the economic results of people's preferences—the observations that people will demand or pay for those things that give them utility.

Marginal utility: The change in total utility that results from the consumption of one more unit of a good; the change in total utility divided by the change in quantity consumed.

Principle of diminishing marginal utility: As more and more of a good is consumed, eventually its marginal utility to the consumer will fall, all things being equal.

Total utility: The total amount of satisfaction obtained from the consumption of a particular quantity of a good; a summation of the marginal utility obtained from consuming each unit of a good.

Although economics cannot offer an answer to questions such as why people like chocolate, economics is interested in the intensity of consumer desires for goods. Why will individuals pay $1.50 for a magazine and $500 for a vacation at the beach? In other words, why is the intensity of demand, as reflected in the prices people are willing to pay, greater for some goods than for others? Economists address questions like these with the aid of the **principle of diminishing marginal utility.** This principle is a simple proposition: As people consume a good in greater and greater quantities, eventually they get less and less extra utility from further increases in consumption.

Although the utility derived by a consumer from some good—shoes, books, raspberries—depends on many factors such as past experience, education, and psychological traits that we may not be able to explain, the principle of diminishing marginal utility allows us to predict certain bounds for consumption behavior. Other things being equal, an individual will not pay more for additional units of a good. And with increasing consumption, sooner or later the consumer will begin to pay less because the good has diminishing marginal utility.

The principle of diminishing marginal utility is based on common sense. At lunch you might be ravenously hungry from skipping breakfast. The satisfaction, or utility, you get from eating the first hamburger in the cafeteria would likely be great. But after one giant burger, or certainly after two or three, the additional satisfaction you experience from one additional hamburger must decline. The amount of money that you would be willing to pay for additional burgers—amounts that reflect their marginal utility for you—would also decline.

The principle of diminishing marginal utility can be numerically and graphically illustrated, as in Table 6–1 and Figure 6–1. Table 6–1 indicates the relation between the quantity of ice cream consumed, the **total utility** for this consumer of all the ice cream eaten, and the marginal utility of each additional scoop for the ice cream eater. For convenience, we construct imaginary units termed "utils" with which to measure quantities of utility going to the ice cream consumer. Note that as the number of scoops consumed increases, total utility—a measure of the total satisfaction gained from the entire amount consumed—increases, but at a decreasing rate. Marginal utility—the satisfaction gained from each additional scoop—declines as additional scoops are consumed within a limited time framework: one visit to the ice cream parlor.

TABLE 6–1   Total Utility and Marginal Utility

In one sitting, an ice cream lover can consume anywhere from one to five scoops of ice cream. The satisfaction the consumer gains from ice cream is measured in imaginary units called utils. Total utility increases as the consumer devours more and more ice cream, but marginal utility—the satisfaction associated with each additional scoop—declines.

Scoops of Ice Cream	Total Utility (utils)	Marginal Utility (extra utils per scoop)
1	6	6
2	11	5
3	15	4
4	18	3
5	20	2

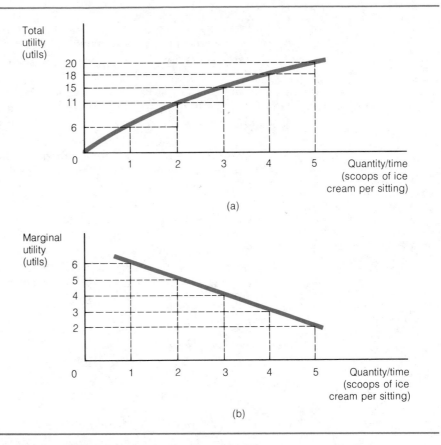

FIGURE 6–1   Total Utility and Marginal Utility

(a) Total utility increases at a decreasing rate as more and more ice cream is consumed in one sitting. (b) Marginal utility declines with each additional scoop of ice cream at one sitting.

In Figure 6–1a, the step line represents the increase in total utility generated by the increased consumption of ice cream. A smooth curve has been drawn through the steps to show the relation between units of consumption and total utility. Notice that total utility increases at a decreasing rate. Figure 6–1b shows the corresponding change in marginal utility. As more ice cream is consumed, the marginal utility of each additional scoop declines.

Consumer Equilibrium: Diminishing Marginal Utility Put to Work

If the principle of diminishing marginal utility applies only to a consumer's behavior in an ice cream store, then it is not very useful. To be useful, the concept of diminishing marginal utility must help us explain behavior in a world in which many different goods are consumed.

Basic Assumptions

To use diminishing marginal utility to understand consumer behavior, four postulates of consumer behavior need to hold.

1. Each consumer desires a multitude of goods, and no one good is so precious that it will be consumed to the exclusion of other goods. Moreover,

goods can be substituted for one another as alternative means of yielding satisfaction. For example, the consumer good of exercise can be satisfied by jogging, playing basketball, hiking, swimming, or a number of other activities. Dinner offers a variety of choices—steak, chicken, pizza, and so on.

2. Consumers must pay prices for the things they want. This seems obvious, but for the purposes of the following analysis, it is important to assume that consumers face fixed prices for the things they consume.

3. Consumers cannot afford everything they want. They have a budget or income constraint that forces them to limit their consumption and to make choices about what they will consume.

4. Consumers seek the most satisfaction they can get from spending their limited funds for consumption. Consumers are not irrational. They make conscious, purposeful choices designed to increase their well-being. This does not mean that consumers do not make mistakes or sometimes make impulsive purchases that they later regret. But gaining experience over time as they deal in goods and services, consumers try to get the most possible satisfaction from their limited budgets given their past experience.

Balancing Choices Among Goods

Armed with these postulates, we can better understand the behavior of consumers in the real world. To see how diminishing marginal utility works in a practical setting, assume that you—a consumer—are deciding how to choose between two goods: sweatpants and socks. For each additional dollar you spend on sweatpants, you experience an increase in utility. Simultaneously, you forgo the utility you could have experienced by spending the dollar on socks. Because you seek to maximize the satisfaction you get from your consumption expenditures (postulate number 4), you will spend your additional dollar on the good that yields you the largest increase in utility, that is, the good with the greatest marginal utility for you.

According to the principle of diminishing marginal utility, whether the utility of one good is greater than the other is a function of the amount of each good consumed. If you already have a relatively large sock collection but relatively few pairs of sweatpants, the marginal utility of socks will tend to be low and that of sweatpants high for you. You are therefore more likely to spend an additional dollar on sweatpants than on socks.

In this simple world of two goods, you will continue to spend additional dollars on sweatpants until the marginal utility you receive from an additional dollar spent on sweatpants is equal to the marginal utility of an additional dollar spent on socks. This process of reaching equality in the two marginal utilities per dollar occurs naturally because the marginal utility of sweatpants declines relative to the marginal utility of socks as you purchase more sweatpants. When the marginal utility per dollar of consuming the two goods, or whatever number of goods are available to you, is equal, you have maximized the total utility of your purchases, subject to the resources you have to spend. In other words, at the point of equality of marginal utility per last dollar spent, you will have no incentive to alter the pattern of your choices between socks and sweatpants. This condition of balance in consumer purchases, from which there is no tendency to change, is called **consumer equilibrium.**

How do economists take the differing goods—as well as their differing marginal utilities—into account? Note that consumer equilibrium occurs

Consumer equilibrium: A situation in which a consumer maximizes total utility within a budget constraint; equilibrium implies that the marginal utility obtained from the last dollar spent on each good is the same.

where the marginal utility of a dollar's worth of socks is equal to the marginal utility of a dollar's worth of sweatpants. To accommodate the differing goods, the following equation is used to describe the condition of consumer equilibrium:

$$\frac{\text{Marginal utility of socks}}{\text{Price of socks}} = \frac{\text{Marginal utility of sweatpants}}{\text{Price of sweatpants}}.$$

Equilibrium is reached when the marginal utility of the last pair of socks purchased divided by the price of a pair of socks is equal to the marginal utility of the last pair of sweatpants purchased divided by the price of a pair of sweatpants.

This equation is just a representation of consumer equilibrium. To see what it means, assume that a pair of socks costs $1 and a pair of sweatpants costs $10, and that you, the consumer, are initially not in equilibrium. This means that the above equation does not hold for you, that is,

$$\frac{\text{Marginal utility of socks}}{\$1} \neq \frac{\text{Marginal utility of sweatpants}}{\$10}.$$

Suppose that you begin by buying a pair of sweatpants because their marginal utility to you is initially higher than that of socks:

$$\frac{\text{Marginal utility of socks}}{\$1} < \frac{\text{Marginal utility of sweatpants}}{\$10}.$$

As you purchase more sweatpants, the marginal utility of an additional pair declines relative to the marginal utility of an additional pair of socks. At each point you are trying to get the most possible satisfaction from your expenditures on the two goods—you are balancing the marginal utility of spending a dollar toward an additional pair of sweatpants against the marginal utility of spending a dollar on another pair of socks.

At the end of this process, as the ratio of the marginal utility of sweatpants to the price of sweatpants declines relative to the ratio of the marginal utility of socks to the price of socks, equality between the two ratios is reached. Consumer equilibrium is restored at the point where an additional pair of sweatpants adds 10 times as much utility as a pair of socks. Put another way, you will adjust your consumption of the two goods until the marginal utility of a dollar's worth of sweatpants is equal to the marginal utility of a dollar's worth of socks. This is what the equation of consumer equilibrium means and how it is reached by a consumer.

The two-good case of gym clothes is a simplification of the consumer's actual choices among thousands of goods. To express the general condition of consumer equilibrium, the equation given above is written with MU_x, MU_y, and so forth standing for the marginal utilities of different goods and P_x, P_y, and so forth representing the corresponding prices of the goods:

$$\frac{MU_x}{P_x} = \frac{MU_y}{P_y} = \frac{MU_z}{P_z} = \ldots .$$

The simplicity of this neat equation is deceptive. The only numbers we have to put in the equation are the prices of the goods. We do not know—or need to know—what the subjective marginal utilities of the consumer are. The important point about the equation is that it is a proposition about individual behavior. It outlines how consumers will achieve balance or equilibrium in allocating their incomes among goods and services. In a sense,

FOCUS   The Diamond-Water Paradox

Adam Smith formulated the diamond-water paradox when he observed:

> Things which have the greatest value in use frequently have little or no value in exchange; and on the contrary, those which have the greatest value in exchange have frequently little or no value in use. Nothing is more useful than water; but it will scarce purchase anything, scarce anything can be had in exchange for it. A diamond, on the contrary, has scarce any value in use; but a very great quantity of other goods may frequently be had in exchange for it.[a]

Water is useful but cheap; diamonds are not useful but expensive. This seems paradoxical. But using the principle of diminishing marginal utility, what errors can we find in Smith's reasoning about the paradox?

First, Smith failed to grasp the importance of the relative scarcity of a commodity in determining its value in use, or marginal utility. He compared a single diamond with the total supply of water. Had he com-

[a]Adam Smith, *An Inquiry into the Nature and Causes of the Wealth of Nations,* ed. Edwin Cannan (1776; reprint, New York: Modern Library, 1937), p. 28.

pared the marginal utility of a single diamond with a single gallon of water, no other water being available, the paradox would have disappeared, for the scarce water would be considered quite valuable. As economic theorists later discovered, water commands little in exchange because its supply is so abundant relative to the intensity of consumer desire for it. Diamonds are scarce relative to consumer desires and therefore command much in exchange. There is no diamond-water paradox when one focuses on the value of the marginal unit of supply to the consumer. Water is plentiful and cheap; diamonds are scarce and expensive.

Second, Smith makes a personal judgment of utility when he suggests that diamonds have no value in use. Many wearers and investors in diamonds would disagree on this point. Modern utility theory does not allow judgments to be made that some preferences are good and others bad.

Of course, Adam Smith did not have modern utility theory available to him when he wrote his statement. Indeed, it was in trying to resolve this simple paradox that modern utility theory was developed by economists in the late nineteenth century in Europe and England.

---

consumers assign their own utility numbers and arrange their pattern of consumption to achieve equilibrium. Consumers solve the equation for themselves as an expression of their rational choices.

Is consumer choice always rational? In describing the famous diamond-water paradox (see Focus, "The Diamond-Water Paradox"), Adam Smith pondered what seemed to be an irrational willingness of consumers to spend vast sums of money on "useless" goods like diamonds. The principle of diminishing marginal utility, developed many years after Smith's writings, helps us understand the rationality behind such choices.

From Diminishing Marginal Utility to the Law of Demand

The marginal utility equation helps us understand the relation between diminishing marginal utility and the law of demand. Assume that you, the consumer, are in equilibrium with socks priced at $1 and sweatpants priced at $10. The equation of consumer equilibrium then looks like this:

$$\frac{MU_{socks}}{\$1} = \frac{MU_{sweatpants}}{\$10}.$$

If the price of sweatpants falls to $5, this equality will be upset. The resulting disequilibrium will be temporary, however, for you will act to restore equality by consuming more sweatpants. As you consume more pairs of sweatpants, the principle of diminishing marginal utility tells us that the marginal utility of sweatpants for you will fall in proportion to their lower price. When the marginal utility of sweatpants has fallen by the same proportion that their price has fallen, consumer equilibrium will be restored.

Note the relation of this behavior to the law of demand. Because the price has been reduced, you will consume more sweatpants until a dollar's worth of sweatpants generates no more utility for you than a dollar's worth of socks—or anything else. The law of demand says that consumers will respond to a fall in the relative price of a good by purchasing more of that good. Diminishing marginal utility thus provides a behavioral basis for the law of demand. When price falls, consumption increases until consumer equilibrium is again established.

The Substitution Effect and Income Effect

Substitution effect: The change in the quantity demanded of a particular good that results from a change in its price relative to other goods.

Consumers' tendency to buy more of a good when its price drops in relation to the price of other goods is called the **substitution effect.** The relatively cheaper good is substituted for other now relatively more expensive goods. Sweatpants are substituted for socks when the price of sweatpants drops. Technically, the substitution effect refers to that portion of the increase or decrease in quantity demanded of a good that is the direct result of its change in price relative to the price of other goods.

Real income: The buying power of money income; the quantities of goods and services that may be purchased with a given amount of dollar income.

Price, however, is not the only relevant factor affecting the quantity demanded of a good. Changes in wealth or income will also tend to shift the demand curve for goods, as we saw in Chapter 4. The larger an individual's income, the more he or she will demand of most goods.[1] Other things being equal (primarily the prices of other goods remaining the same), the fall in price of a particular good will raise the **real income,** or buying power, of the individual consumer. This change in real income means that the consumer can buy more goods under the same budget constraint as before. This **income effect** can be stated more technically. It is that portion of the change in the quantity demanded of a good that is the direct result of the change in the individual's real income that resulted from a price change.

Income effect: The change in quantity demanded of a particular good that results from a change in real income, which has resulted in turn from a change in price.

Marginal Utility and the Law of Demand

The principle of diminishing marginal utility and the law of demand are thus closely related. A price change leads to a change in quantity demanded, which is composed of a substitution effect and an income effect. When the price of a good falls relative to the prices of other goods, the principle of diminishing marginal utility, acting by means of the substitution effect, tends to increase the quantity of the good that the consumer demands. The income effect is a separate and distinct influence that generally reinforces the substitution effect; that is, the income effect tends to increase the quantity demanded of the good whose price has fallen. Though there are cases where the income effect can pull in the opposite direction from the substitution effect, as in the case of an inferior good, economists generally predict that the strength of the substitution effect in such cases will outweigh the strength of the income effect, leading to an increase in the quantity demanded of a good whose relative price has fallen.

[1]Increases or decreases in quantity demanded as a result of an increase in income will depend upon whether the good in question is normal or inferior. Goods for which consumption increases as the consumer's income rises, such as medical care, are called normal goods. When the consumption of a good falls with increases in income, we say that such goods are inferior goods. Used cars may be an example of an inferior good. See Chapter 4 for a discussion of these terms.

Sometimes people claim that they have found exceptions to the law of demand, but the exceptions turn out, on close examination, to be based on simple mistakes in reasoning. For example, it sometimes happens that a fall in price of a good seems to result in a decline in the quantity demanded of that good. A department store may actually sell fewer record albums after it lowers record prices. If we look closer at such cases, we may find that, in the calculations of consumers, the relative price of records has not fallen. Consumers may judge that the store is about to go out of business and, as a result, expect that record prices will soon be even lower. What is crucial to consumers is that, relative to the expected future price, the present "sale" price of records has actually risen. This is the law of demand in action—relative and not absolute prices matter to consumers.

One more of the many possible examples of an apparent, but not real, exception to the law of demand is the case of what is sometimes referred to as a "prestige" good. The demand for caviar, for example, is sometimes said to be much higher as an expensive item than it would be if its price per pound were comparable to, say, tuna fish. The high price of caviar is said to actually increase the demand for caviar, contrary to the law of demand. But this notion is obviously confused. Would the demand for caviar continue to increase as the price rose to $1000, $10,000, or $1,000,000 per pound? Of course it would not. Although prestige goods are usually expensive goods, they do not defy the law of demand. If they did, we would see their prices rise without limit, which does not happen.

Although there may be exceptions of the law of demand, they are extremely hard to find. You might at first think of exceptions that are probably not really exceptions after all. The law of demand seems to describe the way people behave in virtually all cases.

Consumers' surplus: The benefits that consumers receive from purchasing a particular quantity of a good at a particular price, measured by the area under the demand curve from the origin to the quantity purchased, minus price times quantity.

Although the law of demand does seem to explain consumer behavior, what might not be so clear at this point is the potential usefulness of utility analysis. How can subjective magnitudes such as marginal utilities be of any practical relevance? To help answer this question, Focus, "Consumers' Surplus," shows the relation between the important economic concept of **consumers' surplus** and utility analysis. In addition, Economics in Action at the end of the chapter applies marginal utility analysis to the act of voting. These two examples illustrate some ways that utility analysis can be applied to real-world problems.

Some Pitfalls to Avoid

The principles of diminishing marginal utility and marginal utility analysis are useful constructs for analyzing individual behavior. As we have seen, they help explain individual choice behavior and they offer a richer understanding of the law of demand. In using this analysis, however, some fairly common misunderstandings must be avoided:

1. The individual wishing to maximize utility is motivated by personal self-interests, but the principle says nothing about what those interests may be. They may be based on purely selfish, greedy goals or on humanitarian concern for others. Marginal utility analysis does not specify the goals or desires of individuals; it is useful only in understanding and analyzing individual behavior once those goals have been specified.

2. It is important to remember that only individuals make economic decisions. When we refer to a government agency, a corporation, or a snor-

FOCUS   Consumers' Surplus

One stock item in the economist's analytical tool kit is an application of utility theory—the concept of consumers' surplus. Consumers' (or consumer's) surplus is the difference between the amount of money that consumers (or a consumer) would be willing to pay for a quantity of a good and the amount they actually pay for that quantity.

Figure 6–2 shows a market demand schedule for tomatoes. If the price of tomatoes is P_0, a quantity of Q_0 of tomatoes would be purchased, and consumers would spend a total of P_0 times Q_0, or the area P_0BQ_00, on tomatoes.[a] But consumers would be *willing* to spend a lot more money for Q_0 of tomatoes. For some lower quantity, Q_1, tomato consumers would be willing to pay price P_1, higher than P_0. For still lower quantities, some consumers would pay still higher prices. Thus, we may identify the whole shaded area under the demand curve in Figure 6–2 as a surplus of utility that consumers receive when the price of tomatoes is P_0 per pound. They would be willing to pay higher prices, but they only have to pay P_0.

If the surplus utility of individual consumers and of consumers in general can be measured and compared in terms of dollars, the consumers' surplus is a way of measuring the net utility or welfare produced by the consumption of goods. Such an application of utility analysis clearly can be useful. When government decides to produce goods and services that will not flow through a market, such as a nuclear-powered submarine, some means must be found to evaluate the economic welfare created by such production. Cost-benefit

[a]Throughout this book, whenever we identify a geometric area in a graph, we proceed clockwise, beginning from the northwest corner.

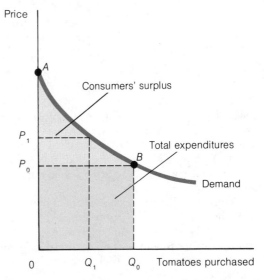

FIGURE 6–2   Marginal Utility and Consumers' Surplus

At a price of P_0, consumers buy Q_0 tomatoes and spend a total of P_0BQ_00 in the process. They would have been willing, however, to spend up to ABQ_00 for the tomatoes. This means that the consumer earns a surplus, in this case equal to $(ABQ_00 - P_0BQ_00) = ABP_0$. The triangle above the market price and below the demand curve is called consumers' surplus. Under certain conditions it measures the net benefit to the economy of the production of a good.

analysis, discussed in detail in Chapter 18, is often used in an attempt to apply the concept of consumers' surplus to evaluating the effects of government programs on the economy.

---

keling club as making a decision, we are only using a figure of speech. Take away the individual members who compose these organizations, and nothing is left to make a decision. This reminder is a principle of positive economics, necessary to avoid confusion.

3. Economics is concerned with the effects of scarcity on the lives of people. It does not address the issue of human needs. A need is not an object that can be measured in a way that everyone can agree on (like the size of a desk or the height of a building). Needs are subjective, just like love, justice, and honor. Economics is neutral with respect to subjective judgments. Marginal utility analysis and the law of demand allow us to discuss resource allocation without having to argue about what "real needs" are.

4. Marginal utility analysis is an explanation, not a description, of individual choice behavior. Economists do not claim that individual consumers actually calculate marginal utility trade-offs before they go shopping. In-

deed, most consumers, if asked, would probably deny that they behave in the way that marginal utility analysis suggests they do. The proof is obviously in the pudding. Individuals behave so as to generate the same outcomes they would generate if they actually did calculate and equate marginal utilities. Marginal utility analysis explains the outcomes we observe rather than describing the mental process involved.

Summary

1. As individuals, we demand goods and services because the consumption of material things gives us utility, or satisfaction.
2. The principle of diminishing marginal utility says that the more we consume of a good, the less utility we get from consuming additional amounts of that good.
3. Consumer equilibrium occurs when individuals obtain the most possible utility from their limited budgets. Their consumption pattern is in balance, and they cannot increase their total utility by altering the way they allocate their given budget among goods.
4. In equation form, consumer equilibrium is

$$\frac{MU_x}{P_x} = \frac{MU_y}{P_y} = \frac{MU_z}{P_z} = \ldots$$

   In other words, the ratio of marginal utility to price is equal among all the goods a consumer consumes.
5. The principle of diminishing marginal utility underlies the law of demand. In the above equation, if the price of x falls, the equality of the marginal utility ratios is upset. To restore equality, the consumer will consume more of the good that has fallen in price, driving its marginal utility down proportion-ate to its fall in price. This is equivalent to the type of behavior predicted by the law of demand. Other things being equal, more of a good is consumed when its price falls.
6. The increase in consumption of a good when its price falls is caused by two effects. More is consumed because the price of the good has fallen relative to the prices of other goods. This is called the substitution effect. In addition, more is consumed because the consumer has more real income owing to the fall in price. This is called the income effect.
7. Generally, the substitution and income effects taken together lead to increased consumption of a good when its price falls.
8. The law of demand is a widespread empirical regularity. We observe its action everywhere. Supposed exceptions to the law are usually just misapplications or misunderstandings of basic economic principles.
9. Marginal utility analysis does not specify or judge goals, address issues of social needs, refer to group behavior, or describe the mental process by which consumers make choices; it merely explains the observed patterns of consumers in making those choices.

Key Terms

utility

marginal utility

principle of diminishing marginal utility

total utility

consumer equilibrium

substitution effect

real income

income effect

consumers' surplus

Questions for Review and Discussion

1. What is meant by *utility?* Is it calculable? Why do economists not speak of *economic needs?*
2. Define *marginal utility.*
3. What are the four postulates of consumer preference?
4. Express consumer equilibrium algebraically in terms of marginal utility. Explain what your expression means in words.
5. "I love seafood so much that I could never get enough of it." Why would an economist disagree with this statement? Could there ever be a time when the person making this statement might ac-tually receive negative utility from consuming more seafood? Give your answer in a graph and in words.
6. Explain the difference between substitution effects and income effects. Which would we expect to dominate during a period of changing demand? During economic downturn and deflation? During economic prosperity and inflation?
7. Oil is essential to the economic life of the nation. Our machines and homes cannot run without it. Yet a quart of oil is cheaper than a ticket to the Super Bowl. How can oil be so precious and yet so cheap? Resolve this oil–Super Bowl paradox.

ECONOMICS IN ACTION   Is It Rational to Vote?

You decide to vote in a presidential election. Typically, more than 60 million people cast their votes for their favorite candidate—Democratic, Republican, Socialist, Libertarian, or other. Your vote therefore has only $\frac{1}{60,000,000}$ worth of influence on the outcome of the election.

We also know that voting is not costless. To make a special trip to the polling place, voters must pay for transportation, take time off from work or leisure, and so forth. Moreover, prior registration often means another trip to the voter registration office. Thus, voting is not free—it clearly places opportunity costs on the voter.

Suppose, in a radical simplification, that all these opportunity costs equal $1; that is, it costs you $1 to vote. Given that you are 1 out of 60 million voters who vote and that it costs you $1 to vote, what must your vote be worth to you to make it rational for you to go to the polls on election day? Since you have only a tiny effect on the election, the outcome would have to have an enormous impact on you to make it worth your while to vote. This concept can be represented algebraically by the formula

$$\frac{1}{60,000,000} \times \text{Benefits} = \$1,$$

which gives the large value of $60 million necessary in benefits to you to justify the cost of voting. In terms of this analysis, it is hardly reasonable to expect the outcome of any given election to be worth $60 million to the average voter. Therefore, no one should vote, and democracy would fall on its face.

Yet herein lies a paradox. A great number of people do vote, and the question is, Why? Economics provides a simple explanation for the conditions under which it is rational to vote. Obviously, voters are smarter than to view themselves as only 1 vote out of 60 million. They have identified—before the election and during the campaign—with parties and individual candidates, and they will make their own estimates of the prospects of the candidates in the election.

Economics says that voters will behave in terms of the marginal utility and marginal cost of voting. Notice that the above example was stated in terms of the *average* influence of a voter on the election. The fact that

the average influence of any one voter is low does not tell us anything about his or her marginal influence on the election. (As we stress repeatedly, the distinction between average and marginal is a crucial one in economics.) While we cannot directly measure the marginal influence of a voter in an election, we can indirectly gauge the marginal benefit from voting. The marginal influence of a vote will be greater the closer the election is predicted to be. In other words, your vote counts for more in a close election than in a lopsided election. Therefore, economic theory leads to the following prediction: Voting turnout will be positively related to expected closeness of the election. In other words, voter turnout will be heavier the closer the election is expected to be.

This is precisely what Barzel and Silverberg found in studying gubernatorial elections in the United States. The closer an election was predicted to be, the larger the voter turnout was.[a] So the simple economic proposition that individuals behave according to marginal cost and marginal benefit provides an explanation for the conditions under which voting is rational. This is not to argue that the only reason people vote is to obtain narrow personal benefits. Benefits can be construed in a wide variety of ways, such as patriotic duty, citizenship, support for an ideological cause, voting for a friend, and so forth. The point is not what the benefits are but that, depending on the closeness of the election, these benefits are likely to be higher or lower relative to costs. Thus, if your friend is running for Congress, you are more likely to vote if he or she is in a tight contest than if your friend is a shoo-in.

Question

Logically, how should each of the following situations affect voter turnout on election day?
a. Rain
b. An international crisis
c. A close race
d. Television coverage

[a]Yoram Barzel and Eugene Silverberg, "Is the Act of Voting Rational?" *Public Choice* (Fall 1973), pp. 51–58.

APPENDIX
Indifference Curve Analysis[2]

Suppose that we wanted to know how a consumer felt about consuming various combinations of two goods, such as vanilla and strawberry ice cream. One approach would be to ask the person how much utility she gets from consuming four scoops of vanilla and nine scoops of strawberry per week. She would perhaps answer, "50 utils." This would not be a very helpful answer because we do not know how much 50 utils represent to this person.

A second approach would be to observe how the individual chooses among various combinations of vanilla and strawberry over a specified time. We could then see, through her **revealed preferences,** how she ranks the various combinations. For example, does she prefer a combination of four vanilla and nine strawberry scoops to a combination of nine vanilla and four strawberry? That is, when offered a choice of the two combinations, does she choose one over the other? Proceeding in this way, we are more likely to derive useful information about preferences. This type of approach yields the concept of an indifference curve.

Revealed preference: A consumer's ordering of combinations of goods demonstrated through observations of the consumer's actions.

Indifference Curves

Indifference set: A group of combinations of two goods that yield the same total utility to a consumer.

Indifference curves are based on the concept of **indifference sets.** Indifference sets can be easily illustrated. Suppose that four possible combinations of two goods confront an individual—say four possible combinations of quantities per week of vanilla and strawberry ice cream. Four combinations are given in Table 6–2. Assume that the consumer, through her revealed behavior, obtains the same total utility from consuming 10 scoops of vanilla and 3 scoops of strawberry as she would from consuming 7 scoops of vanilla and 5

[2]We present indifference curve analysis in an appendix because it is not necessary to know the technique to understand any of the basic points about economics presented in this book. Indifference curve analysis is, however, an integral part of more advanced courses in economics, and many students who plan to do more work in economics will want to learn the technique.

TABLE 6–2   Combinations of Ice Cream Yielding Equal Total Utility

The data represent the consumer's observed behavior. The consumer derives as much satisfaction from consuming combination *A,* 10 scoops of vanilla and 3 scoops of strawberry ice cream per week, as she does from consuming combination *D,* 4 scoops of vanilla and 9 scoops of strawberry. The total utility of each bundle, *A, B, C,* and *D,* is equivalent.

	Scoops Per Week	
Combination	Vanilla	Strawberry
A	10	3
B	7	5
C	5	7
D	4	9

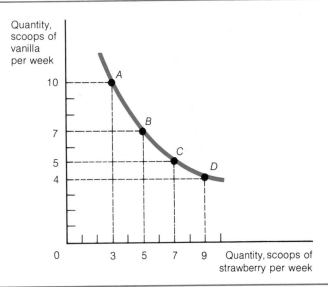

FIGURE 6–3   An Indifference Curve

This indifference curve is based on the data in Table 6–2. The curve shows the various combinations of vanilla and strawberry among which the consumer is indifferent. The marginal rate of substitution of one good for the other is given by the slope of the indifference curve. The indifference curve is convex, which means that its slope decreases from left to right along the curve. This behavior of the marginal rate of substitution is due to the principle of diminishing marginal utility.

scoops of strawberry, and so on through the table. Each combination is equivalent to the others in terms of the total utility yielded to the consumer. For example, if we compare combinations C and D, the extra scoop of vanilla in combination C exactly makes up for the utility lost by consuming 2 fewer scoops of strawberry. The consumer reveals herself to be indifferent among these various combinations of the two goods; each yields the same level of utility to her. An indifference set, then, is all combinations of the two goods among which the consumer is indifferent.

With this information about the individual's evaluation of the combinations, we can draw an indifference curve, as in Figure 6–3. Each point on the graph represents a combination of scoops of vanilla and strawberry ice cream per week. The points representing combinations A through D are shown on the graph. These points and points in between them are joined by a smooth curve because they are members of an indifference set. The curve is the **indifference curve,** a curve made up of points that are all members of a particular indifference set. Any point on the indifference curve is equally preferred to any other by the consumer; all yield the same level of total utility.

Indifference curve: A curve that shows all the possible combinations of two goods that yield the same total utility for a consumer.

Characteristics of Indifference Curves

Indifference curves have certain characteristics designed to show established regularities in the patterns of consumer preferences. Five of these characteristics are of interest to us.

1. Indifference curves slope downward from left to right. This negative slope is the only one possible if the principles of consumer choice are not to

be violated. An upward-sloping curve would imply that the consumer was indifferent over the choice between a combination with less of both goods and another with more of both goods. The assumption that consumers always prefer more of a good to less requires that indifference curves slope downward from left to right.

2. The absolute value of the slope of the indifference curve at any point is equal to the ratio of the marginal utility of the good on the horizontal axis to the marginal utility of the good on the vertical axis. In Figure 6–3, the slope of the indifference curve between A and B is about -1.5, or simply 1.5 as in absolute value. This absolute value tells us that in the area of combinations A and B, the marginal utility of strawberry is approximately one and one half times that of vanilla. In this region, about three scoops of vanilla can be substituted for two scoops of strawberry without lowering the consumer's total utility. For this reason the slope of the indifference curve is called the **marginal rate of substitution,** in this case substitution of strawberry for vanilla ice cream. The marginal rate of substitution tells us the rate at which one good can be substituted for another without gain or loss in utility.

Marginal rate of substitution: The amount of one good that an individual is willing to give up to obtain one more unit of another good.

3. Indifference curves are drawn to be convex: The slope of an indifference curve decreases as one moves downward to the right along the curve. This convexity reflects diminishing marginal utility. As the quantity consumed of one of the goods increases, the marginal utility of that good declines. Hence, the ratio of the marginal utilities—the slope of the indifference curve—cannot be the same all along the curve. The slope must decrease as we move farther down and to the right because the quantity of strawberry ice cream consumed is increasing and the marginal utility of strawberry is therefore decreasing, while the quantity of vanilla ice cream is decreasing and the marginal utility of vanilla is therefore increasing.

In Figure 6–3, the marginal rate of substitution of strawberry for vanilla falls from 1.5 between points A and B to 0.5 between points C and D. As more strawberry is consumed, its marginal utility (measured by how much vanilla the consumer is willing to give up to get strawberry while remaining on the same indifference curve) falls. Less vanilla is consumed, and its marginal utility rises. Between A and B, the consumer is indifferent between three scoops of vanilla and two of strawberry. Between C and D, fortified with more strawberry, the consumer is indifferent between one scoop of vanilla and two scoops of strawberry. The marginal utility of vanilla has risen, and the marginal utility of strawberry has declined.

4. A point representing any assortment of consumption alternatives will always be on some indifference curve. Figure 6–4 presents a graph of combinations of bacon and eggs. We can select a point on the graph—such as A, representing 10 eggs and 1 bacon strip, or D, representing 9 bacon strips and 1 egg—and that point will have a corresponding indifference curve through it showing other bacon and egg combinations that generate the same level of total utility for the consumer. We can see from the graph that this particular consumer is indifferent among 10 eggs and 1 bacon strip (A), 5 eggs and 4 bacon strips (B), and 3 eggs and 8 bacon strips (C) per week.

Moreover, we know that the consumer will prefer any point on the ABC indifference curve, I_1, to any point on the DEF indifference curve, I_2, because points on I_1 contain more of both goods than do points on I_2. The farther from the origin the indifference curve lies, the higher the level of utility the individual will experience.

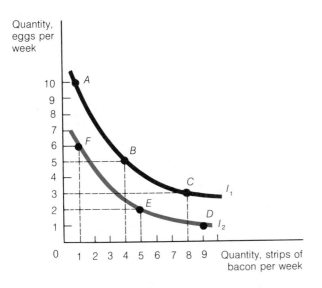

FIGURE 6–4   An Indifference Map

This consumer indifference map illustrates combinations of bacon and eggs. Any point on the graph will be associated with an indifference curve. Since a large number of indifference curves can be placed on the graph, it is called an indifference map. Indifference curves that are farther from the origin, such as I_1, represent higher levels of utility for the consumer because greater quantities of both goods are consumed.

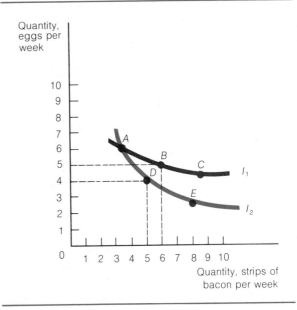

FIGURE 6–5   Indifference Curves Cannot Cross

Because preferences are transitive, point A cannot lie on two different indifference curves. If it did, the consumer would be indifferent among the combinations of bacon and eggs in point A, point B with higher quantities of both, and point D with lower quantities of both.

Indifference map: A graph that shows two or more indifference curves for a consumer.

Transitivity of preferences: A rational characteristic of consumers that suggests that if A is preferred to B and B is preferred to C, then A is preferred to C.

We can draw as many indifference curves on the graph as there are indifference sets confronting the consumer. However many we draw—two or two hundred—the resulting graph is termed an **indifference map.**

5. Indifference curves, which are always drawn for only one individual over a given time period, never cross. The reason for this fact is called **transitivity of preferences,** which simply means that if an individual prefers carrots to squash and squash to artichokes, he or she will also prefer carrots to artichokes. Indifference curves that cross would violate this assumption. Figure 6–5 illustrates this point.

In this graph two indifference curves *do* cross. The point where they cross (point A) lies on both indifference curves, I_1 and I_2. Since, by definition, all points along one indifference curve are equally preferred by the individual, the consumer will be indifferent among choices A, B, and C on I_1 and A, D, and E on I_2. This implies that the consumer is indifferent among all these points. But we can see that this is impossible—if the consumer were indifferent between point B on I_1 and point D on I_2, this would mean that having more of both bacon and eggs (point B) made him or her no better off than having less of both (point D). Other things being equal, consumers always prefer more to less, so indifference curves cannot intersect without implying a type of irrationality that we do not observe in the real world.

The Budget Constraint

Although we know that an individual consumer prefers a point on an indifference curve that is farther from the origin to a point closer to the origin, this knowledge does not enable us to establish which indifference curve represents the best an individual can achieve with a limited budget. We can solve this problem by introducing a **budget constraint,** represented on the graph by a budget line.

In Figure 6–6a, we have assumed that the prices of two goods—apples and oranges—are the same, $1 per pound. If the consumer has a weekly budget of $10, the budget line will be a straight line running from the $10 level of apples (on the vertical axis) to the $10 level of oranges (on the horizontal axis). This line simply reflects the fact that the consumer can allocate his budget between apples and oranges in any way he sees fit—$10 on apples and zero on oranges, $10 on oranges and zero on apples, or any other combination that adds up to $10, such as point A: $5 for each.

How does a change in price affect the budget line? Figure 6–6b shows that the individual's budget remains the same ($10) but that the relative prices of the two goods have changed. While the price of apples remains $1 a pound, the price of oranges has risen to $2 a pound. Although the consumer can still allocate his entire income to the purchase of either apples or oranges, $10 spent on apples would purchase 10 pounds of apples, while the same amount spent on oranges would yield only 5 pounds of oranges.

Budget constraint: A line that shows all the possible combinations of two goods that an individual is able to purchase given a particular money income and price level for the two goods; budget line or consumption opportunity line.

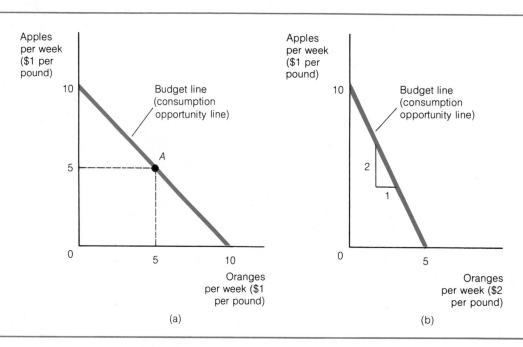

(a) (b)

FIGURE 6–6   Budget Constraint

The budget line, or consumption opportunity line, depends on the relative prices of two goods. The absolute value of the slope of the budget line is the ratio of the price of the good on the horizontal axis to the price of the good on the vertical axis. (a) The budget line represents all possible combinations of apples and oranges at $1 per pound each within a budget constraint of $10 per week. The absolute value of the slope of the line is 1. (b) The budget line, still within the budget constraint of $10 per week, has changed to reflect an increase in the price of oranges to $2 per pound. The absolute value of the slope of the line is 2.

FIGURE 6–7

Determining Consumer Equilibrium with Indifference Curves

The consumer attains maximum satisfaction at point *A*, where the budget, or consumption opportunity, line is tangent to the highest possible indifference curve. At *A* the slope of the indifference curve, or the ratio of the marginal utilities of the two goods, is equal to the slope of the budget line, or the ratio of the prices of the two goods. This relation reflects the same condition for consumer equilibrium as that suggested by marginal utility analysis.

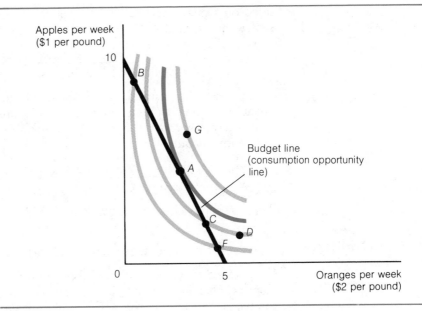

The budget line is thus drawn to reflect any combination of prices for the two goods. The budget line is also called the consumption opportunity line because it shows the various combinations of goods that can be purchased at given prices with a given budget. The absolute value of the slope of the budget line is equal to the ratio of the price of the good on the horizontal axis to the price of the good on the vertical axis. As illustrated in Figure 6–6b, where the price of oranges is $2 a pound and the price of apples $1 a pound, the absolute value of the slope is 2.

Consumer Equilibrium with Indifference Curves

The combination of an individual's indifference curves and budget line allows us to represent consumer equilibrium in a way that is equivalent to the method of marginal utility analysis. This alternative method is illustrated in Figure 6–7, which combines the budget line from Figure 6–6b with a set of indifference curves.

We can see that the consumer represented here will prefer point C to point F because C lies on a higher indifference curve. She will also prefer point A to point C. All three of these points are actual opportunities confronting the consumer; she can afford them, given her budget constraint. Other things being equal, she would prefer point G to point A. However, point G is beyond the limit of her budget line. The best point she can achieve—the point on the highest indifference curve that can be reached within her budget constraint—is point A. This point represents consumer equilibrium—the combination yielding maximum utility from a given budget.

At A, the relevant indifference curve is exactly tangent to the budget line. This means that the slope of the budget line is equal to the slope of the indifference curve. We therefore know that in equilibrium the ratio of the marginal utility of the good on the horizontal axis (oranges) to the marginal utility of the good on the vertical axis (apples), indicated by the slope of the indifference curve, is equal to the ratio of the price of oranges to the price of apples, indicated by the slope of the budget line. Or, expressed somewhat differently:

$$\frac{\text{Marginal utility of oranges}}{\text{Marginal utility of apples}} = \frac{\text{Price of oranges}}{\text{Price of apples}} .$$

With terms rearranged, this formula is exactly the same result we arrived at earlier in the chapter when we discussed consumer equilibrium in marginal utility terms without the aid of indifference curves. That is, the above expression is equivalent to

$$\frac{\text{Marginal utility of oranges}}{\text{Price of oranges}} = \frac{\text{Marginal utility of apples}}{\text{Price of apples}} .$$

The two approaches thus yield similar predictions about consumer behavior.

Indifference Curves and the Law of Demand

Indifference curve analysis can also be used to demonstrate the law of demand. Demand curves are obtained by allowing the price of one of two goods to change and finding the new equilibrium quantities demanded of the two goods. If we allow the price of oranges to fall from $2 a pound to $1 a pound, the budget line shifts outward as in Figure 6–8. With the new budget line, a new equilibrium will occur. The new indifference curve tangent to the new budget line will be farther from the origin, indicating that the lower price of one good increases the level of total utility. As shown in the figure, the lower price of oranges also results in a larger quantity purchased. The consumer equilibrium obtained through indifference curve analysis thus yields downward-sloping demand curves, as predicted by the law of demand.

The increase in the quantity demanded of oranges from O_1 to O_2 is composed of a substitution effect and an income effect. The substitution effect is reflected in the lower relative price of oranges as the budget line shifts

FIGURE 6–8

Indifference Curves and the Law of Demand

As the price of oranges falls from $2 a pound to $1 a pound, the budget line shifts to the right from B_1 to B_2. The consumer reaches a higher level of utility and maximizes satisfaction by obtaining the highest indifference curve, I_2 rather than I_1, possible. As the price of oranges falls, the consumer purchases a larger quantity, from O_1 to O_2. This is precisely what the law of demand predicts—that, other things being equal, quantity demanded varies inversely with price.

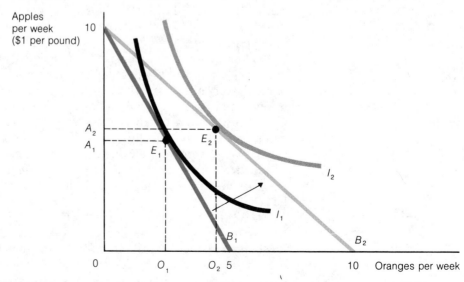

outward. This lower relative price leads to increased orange consumption, just as the law of demand predicts. The income effect measures the degree to which the consumer's real income has increased owing to the fall in the price of oranges. In other words, the shift of the budget line outward means that the consumer is now wealthier. Part of this increased wealth is spent on oranges and part on apples, the consumption of which rises from A_1 to A_2.

Summary

1. The method of indifference curves is an alternative way to study the process of individual choice and consumer equilibrium. It is based on the idea of observing how an individual chooses among consumption alternatives.
2. An indifference set is a group of consumption alternatives that yield the same total utility to the consumer.
3. An indifference curve is a graphical representation of an indifference set. Each point on the curve represents the same total utility to the consumer.
4. Indifference curves have negative slopes, are convex, and do not intersect.
5. The slope of an indifference curve is called the marginal rate of substitution of one good for another.

6. An indifference curve is associated with any combination of goods selected by the consumer. An indifference map consists of a set of possible indifference curves of the consumer.
7. The budget line shows the relative prices of the goods an individual consumes and the amounts of the goods that he or she can consume without spending beyond the budget constraint.
8. In the indifference curve approach, consumer equilibrium occurs at the point where the consumer's budget line is tangent to the highest possible indifference curve. This condition is the same as that derived for consumer equilibrium using marginal utility analysis, namely, $MU_x/P_x = MU_y/P_y$.

Key Terms

revealed preference
indifference set
indifference curve

marginal rate of substitution
indifference map

transitivity of preferences
budget constraint

Questions for Review and Discussion

1. Explain why indifference curves cannot intersect.
2. What is the relation between the principle of diminishing marginal utility and the marginal rate of substitution?
3. Suppose that the consumer feels that each combination of x and y in the following table yields the same total utility. What comment would you make about the individual's choice process?

4. What are the characteristics of indifference curves? What is the rationale for each characteristic? What would be the consequences of violating each of these characteristics?

Combinations	Goods	
	x	y
A	10	7
B	15	8
C	20	9
D	25	10
E	30	11

7

The Firm

We are now at the point in our study of microeconomics where the concept of the business firm comes into play. If you look back briefly at the circular flow of income diagram in Chapter 3, you will see that firms occupy a critical role in a market economy. In fact, there are over 16,000,000 business firms in the United States, ranging in size from the small corner grocery store owned by one individual to the immense multinational corporations owned by millions of stockholders.

What do all of these firms have in common? Why are they organized in different ways? Who controls these firms and to what end? This chapter will explore each of these important questions.

Firm: An economic institution that purchases and organizes resources to produce desired goods and services.

In broad terms, the economic function of a business **firm** is to combine scarce resources (factors of production) to produce goods and services demanded by consumers. Firms that perform this function well make profits and survive; firms that perform this function poorly experience losses and fail. This brief definition of a firm, however, touches only the surface of the matter. To understand the role of firms in a market economy, we need to explore how a firm organizes resources.

Market and Firm Coordination

Production in a market economy is based on the principle of the **division of labor:** Individuals specialize in different productive activities, all of which are coordinated by the price system. The division of labor in a market economy is a vast network of activities based on such factors as the different levels of skill possessed by members of the labor force. Some individuals find their calling in farming, others in medicine. But what coordinates the divi-

Division of labor: An economic principle whereby individuals specialize in the production of a single good or service, increasing overall productivity and economic efficiency.

Market coordination: The process that directs the flow of resources into the production of desired goods and services through the forces of the price mechanism.

Firm coordination: The process that directs the flow of resources into the production of a particular good or service through the forces of management organization within a firm.

Manager: An individual or group of individuals that organize and monitor resources within a firm to produce a good or service.

sion of labor so that we can be assured of a supply of food and a supply of medical services? What is the mechanism by which the production of goods and services by millions of individuals is coordinated so that we can get what we want when we want it?

Basically, there are two methods of economic coordination in a market economy: market coordination and firm coordination.

Distinguishing Market and Firm Coordination

Market coordination is what we have studied so far in this book. It is the use of the price system to provide incentives to suppliers and demanders to produce and consume goods and services in the appropriate amounts. Market coordination refers to the myriad daily economic activities that are guided by the "invisible hand" of the market. Our study of demand and supply is the study of how the market coordinates activities.

Not all activities in a modern economy are coordinated by the market, however. Part of the division of labor is carried out within firms. Automobile assembly-line workers do not sell their output directly in a market. Rather, they supply labor to the auto company, which employs managers to direct and coordinate the uses of their labor within the firm. **Firm coordination** is a productive process that depends on managerial rather than market direction.

In some respects a business firm is like a command economy within a market economy. Decision making in the firm is centralized and performed by **managers.** Managers decide what and how to produce, just as central planners do in a command economy such as that of Russia or Cuba. Within the firm, resources are not bought and sold but are transferred at the command of a manager. The firm, in short, seems to be a contradiction in the marketplace; it appears to conduct its production activities outside the price system.

The analogy of the firm and a command economy is somewhat misleading. A firm is a voluntary institution in which individuals cooperate through free contractual relationships. The manager acts as a specialist in overseeing production and makes decisions that can only figuratively be called commands. Employees agree voluntarily to follow the manager's directions when they join the firm.

The distinction between market and firm coordination is not absolute either. Resource allocation within the firm is not really outside the price system although it might superficially look that way. Managers transfer and allocate resources within the firm in ways that are efficient given the prices for equivalent resources on the market outside the firm. The manager's decision to transfer a quantity of a particular resource within the firm is based on prices, even though resources are not actually bought and sold within the firm.

To understand this point, consider the example of a hypothetical gourmet restaurant, the Greenhouse. It has a luncheon and dinner trade five days a week, it serves brunch on Sundays, and it runs a catering business. After two years of operation, luncheons and Sunday brunch are very crowded, while there are few customers in the evening. In response, the manager of the restaurant will probably transfer resources such as labor, linens, and food from nighttime to daytime use. The transfer is within the firm, but the manager will make the decision by judging the marginal cost of transferring resources—linens, buspersons, waiters, and so on—as if each were being hired or purchased afresh outside the firm. Transfer of resources within the firm is therefore subject to the laws of supply and demand just as if the resources were newly contracted from outside the firm.

If market and firm coordination are so similar, what accounts for the existence of firms? In actuality, much production is undertaken by independent individuals, and it is common for individual demanders to contract with individual suppliers for particular goods and services. For example, Saks Fifth Avenue, Tiffany's, and Neiman-Marcus contract with individual artists and artisans for their wares. Little, Brown and Company—the publisher of this book—hires some work out to freelance editors and art directors rather than having all editorial services performed by in-house staff. Economic coordination between contracting individuals in such cases is achieved directly through the operation of the price system. So an interesting question arises: Why is all production not undertaken by freelance individual operators? Why will an owner of capital goods hire employees for long periods of time instead of contracting out for the performance of specific tasks as they arise?

Least Cost and Most Efficient Size

Ronald H. Coase was the first economist to pose and answer this important question.[1] His answer was simple and profound: Firms exist when they are the least costly form of economic coordination; their size is determined by what is most efficient (least costly) for production. We do not use the market to organize all production because it is not costless to use markets. Using the market necessitates finding out what prices are, negotiating and enforcing contractual agreements with suppliers, going to court when promises are not kept, paying transactions costs, and engaging in various other costly and resource-using activities. For these reasons, not all production in a market economy is coordinated through market exchange. It costs less in many cases to organize production in a firm where resource allocation is directed and coordinated by managers.

Recall the Greenhouse restaurant and some of the reasons that it organized as a firm. The production of food services—luncheons and dinners—requires labor and many other inputs such as refrigeration, stoves, and silverware. It would be too costly to hire cooks or waiters sporadically for specific tasks—catering a wedding or simply serving the regular dinner hours. Using the market afresh to contract for every single input or task required would be costly and inefficient. Instead, a firm is organized to manage and coordinate a variety of inputs. Waiters and cooks are hired on a long-run basis, and some (but possibly not all) capital is leased or purchased through similar long-run arrangements.

In firm production, the inside of the firm will be run by managerial coordination; the firm will deal with the outside world through market coordination. Which activities the firm chooses to organize internally will be related to the cost of organizing the same activities through market exchange. Automobile manufacturers can buy tires in a contractual arrangement with tire companies, or they can make their own tires. The relative costs of these alternatives will determine how the auto company gets its tires. The restaurant may go into the direct market to obtain bookkeeping services or lease, in the short term, specialized capital such as silver serving trays for special catering jobs. It may be too costly to purchase silver trays for one or two occasions a year or to hire a permanent bookkeeper or accountant.

Firms exist in a market economy because they represent a least-cost means

[1]Ronald H. Coase, "The Nature of the Firm," *Economica* (November 1937), pp. 386–405.

of organizing production. However, firms will not grow without limit because at some point the marginal cost of organizing a task within the firm will exceed the marginal cost of organizing the same task through market contracting. This limit on firm size will evolve naturally because the managerial cost of organizing and keeping track of inside production rises along with the number of tasks the firm undertakes. There will be a natural division of production in an economy between firms and markets, determined by the relative costs of producing goods and services in each way.

Suppose the automobile company in its early years produces many of its car parts, such as the seat covers and the spark plugs, because buying these inputs from independent suppliers entails too many risks. This future supply may be uncertain, perhaps because the courts do not strongly enforce supplier contracts for failure to supply these inputs. If, over time, the courts become more concerned about this type of problem and supply contracts come to be more strongly enforced in the event of a failure to supply inputs, the automobile firm can begin buying the auto parts it needs from independent suppliers rather than making them in-house. In the new legal environment, the future supply of seat covers is more certain. In principle, this is how Coase's theory works over time.

Team Production

Team production: An economic activity in which workers must cooperate, as team members, to accomplish a task.

It will further our understanding of why firms exist and how they are organized if we look at the internal operations of a firm in more detail. A fundamental principle involved in the operations of many firms is **team production,** by which several people work together to accomplish a task. None of the individual team members produces a separate product. The classic example is the assembly line invented by Henry Ford.

Production by means of a team is sometimes more efficient than production by individuals working separately. For example, two workers making

"We see ourselves as a team. Wes may discover you have a radiator that won't make it through the summer, while Smitty may decide your transmission needs work, and at the same time Jamie just may come to the conclusion that your whole front end has been twisted out of shape."

Drawing by Booth; © 1984 The New Yorker Magazine, Inc.

jogging shoes—one stamping out the material and the other stitching it together—may be able to produce more pairs of shoes than the two could produce if each were independently assembling whole shoes. Or it may be possible to repair more automobile engines per day if two or three mechanics work on one engine together than if each mechanic concentrates on repairing a separate engine.

Monitor: An individual who coordinates team production and discourages shirking.

Management of the firm can be seen as a device that organizes and **monitors** team production. Management as such functions as a substitute for the discipline of market competition. Management is the firm's solution to a problem that would otherwise limit productive efficiency in group efforts: shirking.

The Problem of Shirking

In individual production, a person's output can easily be measured, but in team production, only the total output of the team as a whole is observable. The contribution of each team member to the total output is harder to detect. Since only the total output of a team is observable, each team member has an obvious incentive to shift the burden of work onto the rest of the team. Such behavior is called **shirking.** This term is not used in a negative sense. We are not saying that workers are naturally lazy; we are simply saying that shirking can be a rational form of behavior when the worker is part of a team production process. Shirking reduces the total output and productivity of the team, but the individual worker who shirks does not bear the full consequences of his or her reduced effort; part of the consequences are shifted to the other team members. Anyone who has ever played basketball, worked in an office or a fast-food restaurant, or participated in any team production process is aware of this problem.

Shirking: A sometimes rational behavior of members of a team production process in which the individual exerts less than the normal productive effort.

Where team production does not exist, individual producers are disciplined by competition from other producers and other products. The individual producer can dawdle or take frequent work breaks if he or she chooses, but the individual bears the cost of reduced production directly in lower income.

In a team setting, it is costly to monitor the performance of each team member. When the increased productive efficiency of team production exceeds this cost of monitoring, team production will replace independent production by individuals. In other words, when teams can produce products at a lower cost than individuals can, firms will exist.

Enter the Manager

The manager is that individual (or group) assigned by the firm to monitor team members and direct them to perform their tasks in a manner that enhances team productivity. The manager is more than just a specialist relieving individual team members of the responsibility for monitoring each other's performance. He or she coordinates the production process in an effort to minimize the costs of production. Managers have the power to hire or fire team members and to renegotiate wage contracts and redirect work assignments. As a result, they are able to discipline shirking and to reward superior productive contributions by team members more effectively than would be possible in a system in which every member monitors every other.

Return momentarily to the Greenhouse restaurant example and suppose that an elaborate banquet is planned for sixty people. The chef, kitchen crew, waiters, and buspeople are alerted. Without management, all members of the team would be responsible for monitoring each other's performance; all would have an incentive to shirk work, placing responsibility on other members of the team. If the food does not appear on time or if it is cold and unpalatable, the chef may blame the kitchen help, the kitchen personnel may blame the waiters, and the waiters may blame the buspeople. It might be more efficient for a manager to step in to organize team activity and to monitor the effort of each member of the food-producing and food-serving team. The manager-monitor would discipline shirking and reward outstanding performance by individual members of the team. The manager, in other words, would attempt to minimize the costs of producing the banquet.

Monitoring the Manager

Of course, within any firm there remains another monitoring problem: Who will monitor the monitor? That is, what is to prevent the manager from shirking? Here competition serves to discipline shirking. More specifically, the threat of losing his or her job to a competing manager will force a manager to avoid shirking. Inefficient managers can and will be fired when more efficient managers are available.

Another incentive to avoid shirking is positive rather than negative: making the managers **residual claimants** who share in the profits of team production. Extra profits that result from more efficient management can be used as a reward to managers. It is common for managers to be given a share of the profits so that they will manage the company effectively, thereby increasing their own wealth. For example, in 1982 the president of Federal Express, a company that provides overnight mail delivery, received a salary of $413,590 and stock gains of $51,129,126 (*Forbes Magazine*, June 6, 1983). This is an astounding amount of profits for a corporate president to receive. Further down the scale was the president of Revco Drugs, who received $30,000 in stock gains.

Voluntary Acceptance of Managers' Commands

What incentive is there for team members to accept the manager's dictates? The manager's power is entirely contractual; team members have voluntarily agreed to accept the manager's commands within the firm. Team members will enter into such contracts because they are better off with a monitor than they would be without one. At first this appears paradoxical. How can employees be better off if they are made to work harder or more efficiently? Why would they voluntarily contract to be disciplined? Even though each team member might be better off if he or she alone could shirk, if every team member shirks, each will bear the full burden of lowered production in reduced income. Hence, every team member will prefer that the entire team be monitored, even if this reduces his or her own opportunities for shirking.

Residual claimant: The individual or group of individuals that share in the excess of revenues over costs, that is, profits.

Size and Types of Business Organizations

As we have seen, firms evolve when team effort can produce goods and services at a lower cost than individual effort. This occurs because of the increased efficiency that may be gained when labor and capital are coordinated through managerial talents. In the following sections, we consider the varying sizes of firms and the advantages and disadvantages of various ways that firms can be organized.

Scale of Production

The most efficient size for firms (the size that performs at the lowest cost of production) varies greatly. The most efficient **scale of production** can be as small as a hot dog stand on a street corner or as large as a General Motors assembly plant.

A firm's scale of production affects the amount of capital equipment it needs. Owners of firms that require large-scale production are forced to make large expenditures on buildings, inventories, machines, and other tools of production. These purchases can amount to millions of dollars. On the other hand, owners of firms with a small scale of production may spend only a few thousand dollars on capital equipment.

When an individual decides to start a business, the amount of money required is determined by the desired scale of production. The larger the scale of production, the larger the amount of funds needed to purchase capital equipment. Often an individual is not able or willing to provide all of the necessary capital. Fortunately, there is more than one way to finance a new firm. An individual may seek other people with complementary talents, ambitions, and funds to start a new business enterprise, sharing the responsibilities, risks, and profits. The way in which a firm's ownership is organized depends on many things—the scale of production, the required capital funds, and the individual talents of owners.

Categories of Ownership

Generally, a firm is organized to fit one of three legal categories: a proprietorship, a partnership, or a corporation. The nature of ownership is different in each category.

Proprietorships. A **proprietorship** is a firm that has a single owner who is liable—or legally responsible—for all the debts of the firm, a condition termed **unlimited liability.** The sole proprietor has unlimited liability in the legal sense that if the firm goes bankrupt, the proprietor's personal as well as business property can be used to settle the firm's outstanding debts.

More often than not, the sole proprietor also works in the firm as a manager and a laborer. Obviously, most single-owner firms are small. Many small retail establishments are organized as proprietorships.

The primary advantage of the proprietorship is that it allows the small businessperson direct control of the firm and its activities. The owner, who is the residual claimant of profits over and above all wage payments and other expenses, monitors his or her own performance. The sole proprietor faces a market price directly. It is up to the sole proprietor to decide how much effort to expend in producing output. In other words, the sole proprietor can be his or her own boss.

The primary disadvantage of the proprietorship is that the welfare of the firm largely rests on one person. The typical sole proprietor is the chief stockholder, chief executive officer, and chief bottle washer for the firm. Since there are only twenty-four hours in a day, the sole proprietor faces problems in attending to the various aspects of the business.

Nevertheless, as Table 7–1 shows, proprietorships are the dominant form of business organization in the American economy. In 1979, they constituted 76 percent of all firms. Though plentiful, they generated only 8 percent of total business revenue. The small business sector is composed of many small firms.

Scale of production: The relative size and rate of output of a physical plant that may be measured by the volume or value of firm capital.

Proprietorship: A firm that has a single owner who has unlimited liability for the firm's debts and who is the sole residual claimant.

Unlimited liability: A legal term that indicates that the owner or owners of a firm are personally responsible for the debts of a firm up to the total value of their wealth.

TABLE 7–1   Types of Firms and Total Receipts, 1979

While the absolute number of sole proprietorships is greater than the number of corporations, their volume of business (total receipts) is much lower.

	Number of Firms	Percent of Total Firms	Total Receipts (millions)	Percent of Total Receipts
Proprietorships	12,330,000	76	$ 487.8	8
Partnerships	1,300,000	8	$ 253	4
Corporations	2,557,000	16	$5,136	88
Total	16,187,000	100	$5,876.8	100

Source: *Statistical Abstract of the United States,* 1982–1983, Table 878, p. 529.

Partnership: A firm that has two or more owners who have unlimited liability for the firm's debts and who are residual claimants.

Partnerships. A **partnership** is an extended form of the proprietorship. Rather than one owner, a partnership has two or more co-owners. These partners—who are team members—share financing of capital investments and, in return, the firm's residual claims to profits. Jointly they perform the managerial function within the firm, organizing team production and monitoring one another's behavior to control shirking. A partnership is a form well suited to lines of team production that involve creative or intellectual skills, areas in which monitoring is difficult. Imagine trying to direct a commerical artist's work in detail or monitoring a lawyer's preparation for a case. If the lawyer is looking out the window, is he or she mentally analyzing a case or daydreaming? By making lawyers residual claimants in their joint efforts, the partnership ensures self-monitoring. (Economics in Action at the end of this chapter analyzes the law firm partnership in more detail.)

The partnership also has certain limitations. Individual partners cannot sell their share of the partnership without the approval of the other partners. The partnership is terminated each time a partner dies or sells out, resulting in costly reorganization. And each partner is considered legally liable for all the debts incurred by the partnership up to the full extent of the individual partner's wealth, a condition called **joint unlimited liability.** Because of these limitations, partnerships are usually small and are found in businesses where monitoring of production by a manager is difficult.

Joint unlimited liability: The unlimited liability condition in a partnership that is shared by all partners.

As Table 7–1 shows, partnerships constitute a small percentage of firms and business revenues in the U.S. economy. One should not underestimate the significance of partnership firms, however, since many important services, such as law, accounting, medicine, and architecture, are organized in the partnership form.

Corporation: A firm that is owned by one or more individuals who hold shares of stock that indicate ownership and rights to residuals but who have limited liability.

Corporations. While **corporations** are not the most numerous form of business organization in the United States, they conduct most of the business. In 1979, corporations accounted for just 16 percent of the total number of firms in the economy but 88 percent of total business revenues. This means that many large firms are corporations and that this form of business organization must possess certain advantages over the proprietorship and the partnership in conducting large-scale production and marketing.

Share: The equal portions into which the ownership of a corporation is divided.

In a corporation, ownership is divided into equal parts called **shares** of stock. If any stockholder dies or sells out to a new owner, the existence of the business organization is not terminated or endangered as it is in a proprietorship or partnership. For this reason the corporation is said to possess the feature of continuity.

Share transferability: The power of an individual shareholder to sell his or her portion of ownership without the approval of other shareholders.

Share Transferability. Another feature that makes the corporation radically different from other forms of business organization is **share transferability—** the right of owners to transfer their shares by sale or gift without having to obtain the permission of other shareholders. For many large organizations, shares of stock are traded on a stock market such as the New York Stock Exchange. Most corporations, however, are smaller, and their shares are traded so seldom that they are not even listed on formal stock exchanges. The shares of these firms are traded by independent stockbrokers on the over-the-counter market.

Share transferability is the most economically important feature of the corporation. It allows owners and managers to specialize, increasing efficiency and profitability in the firm. Owners of stock in a corporation do not need to be concerned with the day-to-day operations of the firm. All that owners need to do is observe the changing price of the firm's shares on the stock market to decide whether the company is being competently managed. If they are dissatisfied with the performance of the company, they can sell their stock. Managers, on the other hand, specialize in reviewing the day-to-day operations of the corporation.

Accumulation of Capital. Specialization in ownership means that relatively large amounts of capital can be accumulated because ready transferability of shares renders investment by individuals feasible. Because stock can be traded at the individual owners' discretion, the risk of owning it is reduced. If the firm's performance sags, dissatisfied investors can sell their shares with a minimum of loss.

Many people who have neither the time nor the inclination to bear the burden of management in a firm nevertheless want to invest in firms operated by skilled specialists and to share in the firms' profits. The resources provided by this investment enable the firms to operate on a larger scale. The stockholders primarily bear the changes in value for those resources—both increases and decreases—and thereby partially free the managers from bearing those risks.

Limited liability: The legal term indicating that owners of corporations are not responsible for the debts of the firm except for the amount they have invested in shares of ownership.

Limited Liability. Another feature of the corporation that distinguishes it from other forms of business organization is **limited liability.** This means that corporate shareholders are responsible for the debts or liabilities of the corporation only to the extent that they have invested in it. For example, if an investor is a millionaire and has invested $10 in one share of XYZ corporation, under no circumstances will he or she risk the loss of more than the invested $10, even if XYZ declares bankruptcy, leaving large unpaid debts. (By contrast, sole proprietorships and partnerships are characterized by unlimited liability; sole owners or partners are legally responsible for the firm's debts up to the amount of their entire personal wealth.) Many investors prefer investments in which their risk of personal loss is strictly limited; the amount of direct investment in corporations is therefore increased as a result of the limited liability involved.

The Separation of Ownership and Control. Not all observers see the corporation as the goose that lays the golden egg. Critics of the modern corporation often claim that it is inefficient because ownership and control are separated: Shareholders are the owners of the firm, but the control of the firm is vested in professional managers. Except for the cases in which managers are also large stockholders in the firm, the problem of separating ownership and control is that management has different objectives for the firm than shareholders. Shareholders are interested in increasing profits and raising the value of their shares. Incumbent managers often have other objectives. They may want to be captains of industry and heads of large firms; they may seek job security or plush corporate headquarters; or they may want to work with friendly rather than competent colleagues. To the extent that managers can pursue such goals at the expense of firm profitability, the separation of ownership and control is a real problem, and shareholder wealth is reduced as a consequence.

This problem may be more apparent than real, however. The corporate shareholder is not held in bondage by the corporation's management—he or she can sell shares at any time. Because the shareholder has this salable right in the capital value of the firm, changes in the behavior of managers that produce losses will provide the inducement to replace inefficient managers with more efficient ones. Such a change may come through a corporate takeover in which another firm or group of investors buys controlling interest in the firm, a stockholder rebellion, or some other means. The point is simply that mechanisms exist to pull errant managers back into line. A subsequent rise in stock price would provide shareholders a gain from the managerial replacement. This potential gain provides a powerful incentive for stockholders to ride herd on management by monitoring the value of their shares (but not by monitoring managerial activities in detail). (See Focus, "Corporate Takeovers and 'Greenmail,'" for a recent and well-publicized method of monitoring corporate managers.)

Other Types of Enterprises

The proprietorship, the partnership, and the corporation are not the only possible types of firms in the economy. Other types of firms are not generally numerous, nor do they account for a large portion of economic activity in the United States, but we mention some briefly.

Labor-Managed Firms. Most firms in Yugoslavia are owned and managed by the employees. The employees share in the profits of the enterprise in addition to earning their normal wage. This form of profit-sharing enterprise is gaining popularity in West Germany and exists to a limited degree in the United States. Profit-sharing programs in U.S. firms allow laborers to share in the performance of the firms. While this is not strictly labor management of the firm, since the workers do not make managerial decisions, it is perhaps a step in the direction of labor management. Other examples include the phenomena of workers buying a plant when management decides to shut it down, as happened recently at a Bethlehem Steel plant in Johnstown, Pennsylvania, and the addition of union leaders to the boards of directors of corporations. The most notable example of this is in the automobile industry, where the president of the United Auto Workers sits on the board of the Chrysler Corporation. Also in this vein are numerous cooperatives around the country in which individuals band together to produce simple clothing or craft goods such as quilts, furniture, and Christmas tree ornaments.

Labor-managed firm: A firm that is owned and thus managed by the employees of the firm, who have the right to claim residuals.

Labor-managed firms are very attractive to employees. Any increase in profits resulting from increased productivity accrue to the workers, which adds an incentive to be efficient and avoid shirking. Increases in wages and strikes only serve to decrease profits and thus occur less often. Also, since

FOCUS   Corporate Takeovers and "Greenmail"

Corporations account for the largest share of total receipts among the various forms of the firm. The principal manner in which large corporations acquire capital to begin business or to make large capital expansions is the issuing of stock shares that are tradable on stock exchanges such as the New York Stock Exchange or the American Stock Exchange. Stocks are supplied and demanded just as any other commodity or service (computers or automotive oil changes). The price of stocks reflects the forces of supply and demand, the demand being determined primarily by the stock purchasers' assessments of the current and future value of the corporation.

Occasionally, the value of a corporation's assets is not accurately represented by the price of stock shares. This sometimes happens when managers are incompetent—that is, when they fail to maximize the wealth of the shareholders. Some investors may come to realize that the stock prices of the firm are too low and do not reflect the value of the corporation's assets. These investor-entrepreneurs may engage in what is called a *takeover bid strategy*: An entrepreneur (usually one already possessing and in control of large-scale assets) finds a corporation whose stock prices are undervalued with respect to the value of the assets of the corporation they represent. He or she then offers to buy up stock in that corporation, thereby driving the value of the shares upward. The entrepreneur is usually followed by other investors in these purchases. Existing shareholders are then faced with a choice: either to sell their stock at the higher prices or to stick with existing management. When sufficient shares are acquired to gain voting control of the corporation, the entrepreneur may attempt to take over the corporation entirely. The takeover entrepreneur in effect gives management of the corporation a choice: Buy back the stock at the new higher price or submit to new corporate control and possible firing. Management may try to prevent stockholders from selling by using all kinds of legal maneuvers, or it may attempt to stave off the "raiders" by selling assets to buy up stock.

Such takeover strategy has become common. In 1984 and 1985, T. Boone Pickens twice attempted to take over the Phillips Petroleum Company of Bartlesville, Oklahoma. The Phillips Company was able to buy out his bid by purchasing Pickens' stock. The al-

T. Boone Pickens

ternative was to submit to new corporate control. Pickens made $89 million on the deal! Such "blackmail" of the corporation is appropriately called "greenmail."

What are the economic implications of takeover bids? While such entrepreneurial activities may appear to carry a "robber baron" flavor, they merely reflect the free market at work. In order for investors to be able to make rational choices between stocks and other assets, shares of stock must reflect the real values of the assets underlying the prices of all assets, including stocks. If a company is being poorly or improperly managed, society is not obtaining maximum output and investors are not receiving maximum returns from their investments. In this sense, the takeover entrepreneur who observes such disparities and acts on them is merely bringing the value of the tradable stock assets to par with the value of the corporation. If the entrepreneur makes a mistake and acquires a corporation whose value is below the price he or she paid for it, the entrepreneur will suffer the consequences.

---

the workers manage the firm, they can determine their own working conditions, safety standards, and fringe benefits.

This type of firm faces one important constraint, as do all other kinds of firms. For a firm to survive in the long run it must be efficient with respect to the costs of production. To compete with other firms, the labor-managed firm must keep its costs as low as possible. Excessive wages, fringe benefits, or costly programs such as office beautification could force firms out of business. Any management decisions by labor must be as efficient as decisions made by traditional managers of other firms. For this reason the working conditions and incomes of employees of successful labor-managed firms do not and cannot differ significantly from those in other firms in a market economy.

In addition, management by labor does not completely eliminate incentives to shirk. When an individual worker's extra effort or lack of it are shared equally with a large number of other employees, then the rewards or costs of these activities to the individual are severely diminished.

In the long run the survival of firms is in the hands of owners and managers whose incomes depend on their special talents and efficiency. Labor-owned and -managed firms must provide this talent and efficiency as well as the labor.

Nonprofit Firms. Nonprofit firms—such as churches, country clubs, colleges, cooperatives, and mutual insurance companies—do not have a group of owners to whom a residual return accrues. No one owns a college in the sense of making a profit from its operation. Generally, funds raised by such organizations must be spent to further the purposes of the organization. Since the reward for good performance is spread among members of the organization, the individual incentive to do a good job is much reduced. Theoretically, this means that workers in nonprofit institutions may have greater incentives to shirk and to pursue their own interests on the job than do workers in profit-making institutions.

Nonprofit firm: A firm in which the costs of production and revenues must be equal and which does not have a residual claimant.

Publicly Owned Firms. Governments sometimes own and operate basic public services, such as railroad, airline, and water companies. One issue in these cases is whether government is a more efficient provider of these services than privately owned enterprises. The answer seems to be no. Note the distinction between publicly owned firms and private firms that are subject to public regulation. (Publicly regulated firms are treated in detail in Chapter 17.) Government water costs more than private water; government garbage collection costs more than private garbage collection; government airlines are less efficient than private airlines. Why? The answer appears to be that operators of government enterprises do not face the test of profit or loss directly as do their private counterparts. Losses in a publicly owned firm can be made up from general tax revenues, for example. Without the discipline of profit and loss, costs can be higher in public than in private enterprises. (We will take up the issues of profits and losses in succeeding chapters.)

Publicly owned firm: A firm owned and operated by government.

The Balance Sheet of a Firm

The financial status of a business enterprise is represented by a **balance sheet,** a statement in which the dollar value of all the firm's **assets** corresponds to an equal total value of **ownership claims.** This correspondence is necessary and exact. An asset is a resource owned by the firm; an ownership claim identifies each party who has a property right or a claim to the firm's assets or wealth. A **liability** is simply a debt of the company—what the company owes to various creditors.

Balance sheet: An accounting representation of the assets and liabilities of a firm.

Asset: Anything of value owned by the firm that adds to the firm's net worth.

TABLE 7–2   Balance Sheet of Joan Robinson, a Sole Proprietor

The net worth or equity of the firm to Joan Robinson is equal to total assets minus total debts. In a sole proprietorship the single owner has complete claim to all of the equity.

Assets		Debts	
Cash holdings	$ 4,000	Accounts payable	$ 8,000
Equipment	20,000	Mortgage payable	20,000
Warehouse inventory	14,000	Total debts	28,000
Land and building	100,000	Joan Robinson's equity	110,000
Total assets	$138,000	Total debts plus equity	$138,000

TABLE 7–3   Balance Sheet of Smith and Ricardo, a Partnership

The balance sheet of a firm that is a partnership is very much like that of a sole proprietorship. In this case, the firm's equity is divided equally between the partners.

Assets		Debts	
Cash holdings	$ 60,000	Accounts payable	$ 11,000
Equipment	10,000	Mortgage payable	40,000
Warehouse inventory	52,000	Total debts	51,000
Land and building	120,000	Adam Smith's equity	95,500
		David Ricardo's equity	95,500
Total assets	$242,000	Total debts plus equity	$242,000

Ownership claims: The legal titles that identify who owns the assets of a firm.

Liability: Anything that is owed as a debt by a firm and therefore takes away from the net worth of the firm.

Net worth: The value of a firm to the owners, determined by subtracting liabilities from assets; also called *equity*. For corporations, net worth is termed *capital stock*.

Underlying the balance sheet is a fundamental identity:

Value of assets − Value of total claims to ownership = Value of liabilities (the amount owed) + Value of owned property in the firm.

Another familiar way of expressing this identity is

$$\text{Assets} = \text{Liabilities} + \text{Net worth.}$$

Net worth is

$$\text{Net worth} = \text{Assets} - \text{Liabilities.}$$

Net worth, also called equity, is the amount of a firm's wealth left over for the owner(s) after all liabilities are met from the firm's assets. If liabilities exceed assets, the firm's net worth is negative.

A simplified hypothetical balance sheet of a sole proprietorship is presented in Table 7–2. In the sole proprietorship, all of the firm's assets are owned by a single individual who is personally liable for all of the firm's debts. The proprietor's own stake in the business is his or her net worth, or equity; this is the difference between the firm's assets and the firm's debts. In this case, Joan Robinson's net worth is $110,000.

Table 7–3 reproduces a simplified balance sheet for a hypothetical partnership. Note that it is identical to the sole proprietorship's balance sheet except that the net worth of the firm is divided equally between partner Smith and partner Ricardo. Equity in partnerships need not always be divided equally between the two (or more) partners. However they divide the equity, each of the partners is individually liable for the firm's entire debt.

TABLE 7–4   Balance Sheet of Robert Malthus, Inc., a Corporation

The capital stock of a corporation is the firm's net worth. This equity is divided among the shareholders, the owners of the firm.

Assets		Debts	
Cash holdings	$ 60,000	Accounts payable	$ 13,000
Equipment	10,000	Mortgage payable	40,000
Warehouse inventory	52,000	Total debts	53,000
Land and buildings	120,000	Capital stock	189,000
		Total debts and	
Total assets	$242,000	capital stock	$242,000

In Table 7–4 we present a simplified balance sheet for a hypothetical corporation. The corporation is legally defined as an individual separate and distinct from its shareholders. Hence, all of its assets are held in the name of the firm. The firm's assets minus its liabilities equals its net worth, termed the *capital stock*. Shareholders hold claims to this capital stock, which represents their equity in the firm. This equity may be paid out to shareholders in the form of dividends or plowed back into investments in the firm in the form of retained earnings. As we have seen, shareholders are legally liable for the firm's debts up to, but not exceeding, the extent to which they have invested in the firm.

The balance sheet for a firm can be likened to a snapshot of the firm's operations taken at an instant in time. The balance sheet may look one way today and radically different next week or next year.

The firm is a fascinating and complex institution. It arises and continues because of its efficiencies relative to simple exchange within markets. Some people see the modern corporation as an unchained beast that seeks to rule the world. The analysis in this chapter leads to a more benign view of the corporate form. It is basically a powerful instrument for economic good, which is not to say that corporations should be completely unfettered from public control. We explore such matters in Chapter 17 on economic regulation. For the present we turn our attention in the next chapter to the conditions of real production and costs in the business firm.

Summary

1. The basic function of a business firm is to combine inputs to produce outputs.
2. Economic coordination in a market economy can take place through prices and markets or within firms. Market coordination relies on price incentives; firm coordination relies on managerial directives.
3. Firm coordination of economic activities exists because there are costs to market exchange and sometimes efficiencies in large-scale production.
4. Team production takes place in firms. In team production, group output can be observed but individual output cannot.
5. Team production makes it natural for individual team members to shirk, that is, to reduce individual effort in achieving the team goal. It is the function of managers in the firm to control shirking.
6. A proprietorship is a single-owner firm whose owner faces unlimited liability for the contractual obligations of the firm. With unlimited liability, the personal wealth of the owner, above and beyond what he or she has invested in the firm, can be drawn on to settle the firm's obligations in the event of bankruptcy.
7. A partnership is owned by two or more individuals who face shared unlimited liability for the contractual obligations of the firm.
8. A corporation is owned by shareholders who have limited liability. Shareholders are liable only for the amount they have invested in the corporation.

9. Corporations are owned by shareholders and typically controlled by professional managers. This potential division of interests has been a source of criticism of the modern corporation.
10. A balance sheet is a financial statement of the economic well-being of a firm. Net worth, which can be positive or negative, is a measure of the wealth of a firm's owner(s). It can be derived from a balance sheet and is defined as assets minus liabilities.

Key Terms

firm	shirking	corporation	balance sheet
division of labor	residual claimant	share	asset
market coordination	scale of production	share transferability	ownership claims
firm coordination	proprietorship	limited liability	liability
manager	unlimited liability	labor-managed firm	net worth
team production	partnership	non-profit firm	
monitor	joint unlimited liability	publicly owned firm	

Questions for Review and Discussion

1. Describe the workings of a balance sheet as presented in this chapter; be sure to include definitions for assets, liabilities, and net worth.
2. Compare and contrast limited liability and unlimited liability. Which of them characterizes (a) a sole proprietorship, (b) a partnership, (c) a corporation?
3. What are the major advantages and disadvantages of the corporation as a form of business organization?
4. What is the concept of team production? Does shirking mean that workers are naturally lazy? Give some examples of team production.
5. What type of firm is your college or university? How would you expect it to operate compared to a private corporation?
6. Is it inefficient for the control and ownership functions of large corporations to be separated?
7. What is the single most efficient way to extract effective performance from managers?
8. Since it would be easy for members of a team to shirk on their work, why does team production exist, from the firm owners' point of view?
9. What types of business activity should be organized as partnerships? Why?
10. A football team is by definition a team production process. Determine three ways that the coaches at your school work to monitor and control individual players' performances on the football team. Do they, for example, film practices and games or do they give differential rewards for superior performance?

ECONOMICS IN ACTION   The Economics of Law Firms

A law firm is an example of team production in action. Most law firms are organized as legal partnerships; partners are mutually liable for commitments made on behalf of the firm by any partner. This liability does not end with the size of the partner's investment in the firm; it extends to the limit of each partner's personal assets (including his or her house). In the event that the partnership goes into bankruptcy, creditors can attach the personal assets of the partners to settle claims against the firm. That is what unlimited liability means legally; what does it mean economically?

As a result of unlimited liability, partners will be

Source: Arleen Leibowitz and Robert Tollison, "Free Riding, Shirking, and Team Production in Legal Partnerships," *Economic Inquiry* 18 (July 1980), pp. 380–394.

very careful about how they conduct their business. In particular, they will screen candidates for partners in the firm very carefully. A young lawyer is first given an associate status in the firm and must learn and compete with other young associates to become a partner after five or six years. Only the brightest and most trustworthy young lawyers rise through the ranks to become partners in the firm. This screening and training process is clearly motivated by the unlimited liability condition—the cost of making a mistake in the selection of a partner can be very high.

Unlimited liability also means that each partner has to keep track of what the other partners are doing. The natural effect of this consideration is that most law firms are small. It is more costly to keep track of the activities of more partners. Costs are also easier to hold down in a smaller partnership. The data on law firms tend to support these arguments. In the mid 1970s, for example, the average law partnership contained only 3.4 partners. Indeed, most of the firms (52 percent) providing legal services consisted of single practitioners.

Although the typical law firm is small, partnerships seem to be growing more popular in the legal services industry than single practitioners. One possible explanation for this trend is that lawyers are now allowed to advertise. Previously, lawyers in small towns were isolated from competition because advertising about price was not allowed. Now that lawyers can advertise, larger firms and legal clinics in urban areas can put competitive pressure on small-town lawyers who provide standard legal services such as wills, divorces, and house sales.

When organized as a partnership rather than a proprietorship, a law firm is a little society of profit sharers. This seems necessary because it is difficult to conceive of paying a lawyer a wage and monitoring his input activities. A lawyer is engaged in what might be loosely called artistic production. How would a monitor know if a lawyer was working on a case when he or she was staring out the window or walking down the street? An external observer cannot tell whether the lawyer is goofing off or considering a new legal angle on a case. Making the lawyer a partner with shared responsibilities in the firm and a share of the firm's profits is a way to make sure that the lawyer devotes his or her time to cases and not to shirking. Thus, one rationale for the partnership is to promote efficient behavior by artistic producers such as lawyers, accountants, architects, and commercial artists.

Question

Although legal advertising has helped many law firms grow, a number of lawyers are vehemently opposed to advertising legal services. What economic motives might these lawyers have? Does this opposition to legal advertising have anything to do with the partnership structure of law firms?

8

Production Principles
and Costs to the Firm

*T*he typical business firm buys inputs at fixed prices, produces a product or service, and sells the product or service at the prevailing market price. The firm's use of inputs leads to costs, and its sale of outputs leads to revenues. The firm has a clear incentive to make revenues exceed costs by as much as possible. When it is successful, the firm earns a profit.

In this chapter we develop concepts related to the firm's costs. Economists do not define costs as the explicit outlays for inputs. Instead, they stress the idea of opportunity costs, or what must be paid to or for a resource to keep it employed in its present use. This amount depends on the value of the next-best alternative use for the resource. As we build on this concept, we will introduce many new terms that economists use in isolating specific costs and analyzing their relations to outputs produced. These ways of pinpointing costs lead to ways of predicting whether it would cost more or less per unit for a firm to increase or decrease its amount of capital or plant size. The tools we develop here are therefore important for understanding how firms maximize profits.

Types of Costs

As we have seen, consumers have virtually unlimited wants, but resources to satisfy their wants are limited. Costs of production arise from the fact that resources have alternative uses. The same resources that are used to produce a good to satisfy consumers' demands can also be used to produce other goods. For resources to be drawn into the production of a particular good, they must be bid away from their present uses. Production of a newspaper might require, among other things, that wood pulp be turned into newsprint rather than notebook paper, that photographers be enticed to work as photojournalists rather than as portrait photographers, and that youngsters be hired to distribute newspapers rather than earn money by baby-sitting or mowing lawns. The expenditures necessary to do all of these things are called **costs of production.** Costs of production are therefore the value of resources in their next-best uses. Focus, "The Opportunity Cost of Military Service," illustrates this point in more detail.

Explicit and Implicit Costs

The opportunity costs of production include both explicit and implicit costs. Take, for example, a small, owner-operated dry cleaning firm organized as a sole proprietorship. The dry cleaner has to purchase labor, cleaning fluids, hat blocks, bagging machines, insurance, accounting and legal services, advertising, and so on, to produce dry cleaning services. The costs of these resources, entered onto the firm's balance sheet, are called **accounting,** or explicit, **costs.** They are so named because they are quite visible; they are the wages and bills the sole proprietor must pay to conduct business.

The firm does not make explicit payments for all the resources it uses to produce dry cleaning services. The owner of the firm may not enter a wage for his own services on the balance sheet. Omitting his salary from the balance sheet, however, does not mean that the owner's services are free. They carry an **implicit cost,** valued by what the owner could have done with his time instead of working in his firm. Implicit costs are the opportunity costs of resources owned by the firm. They are called implicit costs because they do not involve contractual payments that are entered on the firm's balance sheet.

The main implicit cost that is not entered on the firm's balance sheet is the **opportunity cost of capital.** Individuals who invest in firms expect to earn a normal rate of return on their investment. For instance, they could have placed their capital in money market certificates and earned at least a 10 percent rate of return. In this case, the 10 percent rate of return represents the opportunity cost of capital invested in business ventures. Unless investors earn at least the opportunity cost of capital, they will not continue to invest in a business.

The **total cost of production** is the sum of explicit and implicit costs. Total cost is the value of all the alternative opportunities forgone as a result of production of a particular good or service.

Economic Profits Versus Accounting Profits

Businesspeople must compare their total costs with their total revenues (all the money they take in) to determine whether they are making a profit. **Economic profit** is the difference between total revenue and total cost. It exists when the revenue of the firm more than covers all of its costs, explicit and implicit. Economic loss results when revenue does not cover the total cost. A firm's economic profit is said to be zero when its total revenue equals its total cost. Zero economic profit does not mean that the firm is not viable or about to cease operating. It means that the firm is covering all the costs of its operations and that the owners and investors are making a normal rate of return on their investment.

Since accounting costs do not usually include implicit costs—such as the value of the sole proprietor's time or the opportunity cost of capital invested in the firm—accounting costs are an understatement of the opportunity costs of production. This means that **accounting profits** will generally be higher

---

Costs of production: Payments made to the owners of resources to ensure a continued supply of resources for production.

Accounting costs: Payments that a firm actually makes, in the form of bills or invoices; explicit costs.

Implicit costs: The value of resources used in production for which no explicit payments are made; opportunity costs of resources owned by the firm.

Opportunity cost of capital: The value of the payments that could be received from the next-best alternative investment; the normal rate of return.

Total cost of production: The value of all resources used in production; explicit plus implicit costs.

Economic profit: The amount by which total revenues exceed total costs.

Accounting profit: The amount by which total revenues exceed accounting costs.

∧∧∧   FOCUS   The Opportunity Cost of Military Service

Recent attempts to establish an all-volunteer army have been based on the concept of opportunity costs. That is, to induce people to volunteer for military service, the U.S. government offers them benefits comparable to those they would forsake in civilian careers.

Assume that potential recruits have civilian jobs and that economics is the only consideration in their decision to join the army. If government can offer recruits a package including salary, room and board, clothing, education, health care, and travel opportunities that is equal to or greater than their opportunity cost—that is, what they would earn as civilians—then perhaps people will sign up in numbers considered sufficient for national defense.

If the opportunity cost to new recruits is not balanced by the army's offer, people will not volunteer in sufficient numbers. In this case, the government may return to the draft to meet its quota for national defense. When this happens, forced conscription can be seen as a tax—"the draft tax." The draft tax is equal to what an individual could have earned as a civilian minus any compensation he or she gets from serving in the military. If a man was earning $10,000 as a civilian but the military pays him only $5,000 (including the value of benefits such as food and shelter), his draft tax is $5,000.

This example assumes that only money wages matter to the individual. If the individual is not indifferent about the choice between civilian and military work—if he or she strongly prefers to work as a civilian no matter what the pay for military service is—the financial cost of $5,000 understates the real opportunity cost of the draft to the individual.

---

than economic profits because economic profits are calculated after taking both explicit and implicit costs into account.

Private Costs and Social Costs

Private costs: The total opportunity costs of production for which the owner of a firm is liable.

We have so far discussed a world of **private costs,** in which the firm pays a market price for the resources it uses in production. The dry cleaner's costs of production were private costs, equal to the sum of explicit and implicit costs. The key aspect of the concept of private costs is that someone is responsible for them; that is, someone pays for the use of the resources. The owner of the dry cleaning firm incurred both explicit and implicit costs in his business, and he paid these costs to engage in business.

But what if the dry cleaner dumps his used cleaning fluid into the town lake adjacent to his plant? In this case he is using the lake as a resource in his dry cleaning business—as a place to dispose of a by-product of his production process—but he pays no charge for the use of the lake as a dumping site. No one owns the lake for purposes of dumping, and it is therefore unpriced for this use. The private costs of the dry cleaner therefore do not reflect the full costs of his operation. The cost of the use of the lake is not reflected in the dry cleaner's costs of production. Such a cost is called an

External costs: The costs of a firm's operation that it does not pay for.

Social costs: The total value of all resources used in the production of goods, including those used but not paid for by the firm.

external cost because the cost is borne by others.

When external costs arise from a firm's activities—such as polluting the air or water or contributing to erosion or flooding—the firm imposes **social costs.** These are the total of both private and external costs; the total cost to society of the firm's production includes the costs of using environmental assets such as the town lake.

Economics in Action at the end of this chapter discusses the concept of social cost in more detail. In Chapter 18, we further analyze external costs, including ways in which society can correct problems arising from external costs. The important point here is that private costs of production do not always reflect the full opportunity costs of production; the use of unpriced resources, such as the environment, may also carry a hidden price to society.

Economic Time

Time is fundamental to the theory of costs. Economic time is not the same thing as calendar time. To the economist analyzing costs, the firm's time constraints are defined by its ability to adjust its operations in light of a changing marketplace. If the demand for the firm's product decreases, for example, the firm must adjust production to the lower level of demand. Perhaps some of its resources can be immediately and easily varied to adapt to the changed circumstances. In the face of declining or increasing sales, an automobile firm can reduce or increase its orders of steel and fabric. Inputs whose purchase and use can be altered quickly are called **variable inputs.** The auto firm with declining sales cannot immediately sell its excess plant and equipment or even reduce its labor force, however. Such resources are called **fixed inputs;** a reduction or increase in their use takes considerably more time to arrange.

There are two primary categories of economic time: the short run and the long run. The **short run** is a period in which some inputs remain fixed. In the **long run** all inputs can be varied. In the long run, the auto firm can make fewer cars not only by buying less steel but also by reducing the number of plants and the amount of equipment and labor it uses for car production. The time it takes to vary all inputs is different for different industries. In some cases, it may take only a month, in others a year, and in still others ten years. In August 1983, for example, the Chrysler Corporation announced that it planned to be able to respond to an increase in sales by boosting its output by 50 percent within twelve months. The relative brevity of this period needed to increase output was made possible by its purchase of a former Volkswagen assembly plant, an increase in work shifts, the return to production of resources left idle during earlier cutbacks, and a sharp increase in purchases of new capital stock.

Variable input: Factors of production whose quantity may be changed as output changes in the short run.

Fixed input: Factors of production whose quantity cannot be changed as output changes in the short run.

Short run: An amount of time that is not sufficient to allow all inputs to vary as the level of output varies.

Long run: An amount of time that is sufficient to allow all inputs to vary as the level of output varies.

Short-Run Costs

In the short run, some of the firm's inputs are fixed and others are variable. Building on this distinction, the short-run theory of costs emphasizes two categories of costs—fixed and variable.

Fixed Costs and Variable Costs

Fixed costs: Payments made to fixed inputs.

A **fixed cost** is the cost of a fixed input. In the short run, the fixed cost does not change as the firm's output level changes. Whether the firm produces more or less, it will pay the same for such things as fire insurance, local property taxes, and other costs independent of its output level. Fixed costs exist even at a zero rate of output. Perhaps the most important fixed cost is the opportunity cost of the firm's capital equipment and plant; that is, the next-best alternative use of these resources. This cost exists even when the plant is idle. The only way to avoid fixed costs is to shut down and go out of business, an event that can occur only in the long run.

Variable costs: Payments made to variable inputs that necessarily change as output changes.

Total cost: All the costs of a firm's operations, including fixed and variable costs.

Variable costs are costs that change as the output level of a firm changes. When the firm reduces its level of production, it will use fewer raw materials and perhaps will lay off workers. Consequently, its variable costs of production will decline. Variable costs are expenditures on inputs that can be varied in short-run use. A third concept, **total cost,** is the sum of fixed and variable costs.

It is useful to determine the cost of producing each unit of output (each

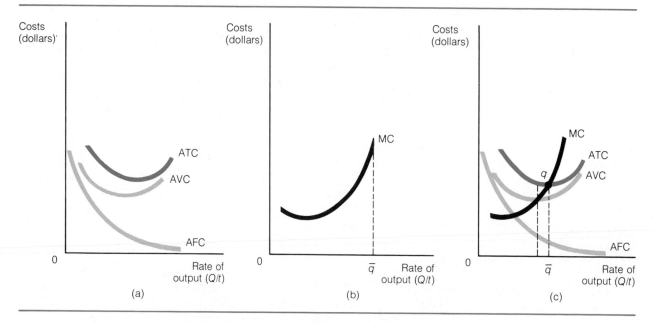

FIGURE 8–1   Short-Run Cost Curves of the Firm

(a) The behavior of average fixed cost (AFC), average variable cost (AVC), and average total cost (ATC) as output changes. ATC is the sum of AVC and AFC. (b) The behavior of the firm's marginal costs (MC) as output increases. MC ultimately rises as the short-run capacity limit of the firm is approached at $\bar{q}$. (c) All the short-run cost curves. Notice in particular the U shape of the ATC curve. ATC is high either where the firm's plant is underutilized at low rates of output or overutilized at high rates of output.

Average fixed cost: Fixed cost divided by the level of output.

car, backpack, or ton of soybeans) in the short run as the level of output is increased or decreased. The fixed cost per unit is called **average fixed cost** (AFC) and is found by dividing the total fixed cost by the firm's total output of the good (referred to as q throughout this chapter). Since fixed costs are the same at all levels of output, AFC will decline continuously as output increases. This relation is shown graphically in Figure 8–1a.

Average variable cost: Variable costs divided by the level of output.

Average variable cost (AVC) is found by dividing total variable cost by the firm's output. It is also drawn in Figure 8–1a; it ordinarily has a U shape.

Average total cost: Total costs divided by the level of output, or average fixed cost plus average variable cost; unit cost.

Average total cost (ATC)—the actual short-run cost to the firm of producing each unit of output—is total cost divided by total output. Average total cost can also be found by adding average fixed costs and average variable costs. Thus, in Figure 8–1a, the ATC curve is shown as the sum of the AFC and AVC curves. Average total cost is sometimes referred to as *unit cost*.

Although the average total cost curve provides useful information, economic decisions are actually made at the margin. In considering a change in the level of production, firms must determine what the marginal cost of the change will be.

Marginal cost: The extra costs of producing one more unit of output; the change in total costs divided by the change in output.

Marginal cost (MC) is the cost of producing each additional unit of output; it is found by dividing the change in total costs by the change in output. The result of such calculations is a marginal cost curve, shown in Figure 8–1b. Marginal cost first declines as output rises, reaches a minimum, but then rises because it becomes increasingly hard to produce additional output with a **fixed plant.** Extra workers needed to produce additional output begin to

Fixed plant: A situation in which the firm has a given size of plant and equipment to which it adds workers.

get in each other's way; inefficiencies proliferate as workers must take turns using machines; storage room for inputs and outputs becomes filled. At some point the firm's maximum output from its fixed plant will be reached (shown at $\bar{q}$).

Figure 8–1c brings all the short-run cost curves together. Notice that the ATC curve is U shaped. At low levels of output a firm does not utilize its fixed plant and equipment very effectively. The firm does not produce enough relative to the fixed costs for its plant, causing the average total cost to be high. At high levels of output the firm approaches its capacity limit ($\bar{q}$). The inefficiencies of overloading the existing operation cause marginal costs to rise, raising the average total cost. These two effects combine to yield a U-shaped average total cost curve. Either low or high utilization of a fixed plant leads to high average total cost. The minimum point on the ATC curve, q, represents the lowest average total cost—the lowest cost per unit for producing the good. This represents the best short-run utilization rate for the firm.

Table 8–1 provides a shorthand reference to all these concepts, and later in the chapter we will discuss them in more detail with a numerical example and more graphical analyses.

Diminishing Returns

Law of diminishing marginal returns: A relation that suggests that as more and more of a variable input is added to a fixed input, the resulting extra output decreases, eventually to zero.

The behavior and shape of the short-run marginal cost and average total cost curves can be understood more completely with the **law of diminishing marginal returns.** The law of diminishing marginal returns states that as additional units of a variable input are combined with a fixed amount of other resources, the amount of additional output produced will start to decline beyond some point. The returns—additional output—that result from adding the variable input will ultimately diminish.

Though the language sounds formidable, the law of diminishing marginal returns is little more than formalized common sense. Suppose that the dry cleaner in our previous example has a plant of a given size that we will view as his fixed input. He begins operations by adding workers one at a time and observes what happens to the resulting output of dry cleaning. The first few additional workers are very useful. One specializes in cleaning, another in pressing, another in bagging, and so on. The output of dry cleaning increases

TABLE 8–1   Short-Run Cost Relations

Terms	Symbols	Definition
Average fixed cost = Fixed cost ÷ Total output	$AFC = FC/q$	A fixed cost does not vary with output, but the average fixed cost per unit declines as output rises.
Average variable cost = Variable cost ÷ Total output	$AVC = VC/q$	A variable cost changes as output changes.
Average total cost = Total cost ÷ Total output = Average fixed cost + Average variable cost	$ATC = TC/q$ $ATC = AFC + AVC$	Total cost is the sum of fixed and variable costs. Average total cost per unit forms a U-shaped curve when graphed from very low to very high output levels.
Marginal cost = Change in total cost ÷ Change in output	$MC = \Delta TC/\Delta q$	Marginal cost rises as the short-run capacity of a firm's fixed plant is approached.

TABLE 8–2   Law of Diminishing Marginal Returns

If units of a variable input (labor) are added one at a time to production with a fixed plant (growing soybeans with no change in technology on a fixed plot of land), the total product, marginal product, and average product all increase at first. But eventually, as more workers are added, they begin to get in each other's way. Total output gained by adding more workers drops, as predicted by the law of diminishing marginal returns.

Variable Input (units of labor)	Total Product (tons)	Marginal Product (tons)	Average Product (tons)
0	0		0
		10	
1	10		10.00
		12	
2	22		11.00
		14	
3	36		12.00
		13	
4	49		12.25
		11	
5	60		12.00
		10	
6	70		11.67
		5	
7	75		10.71
		3	
8	78		9.75
		−2	
9	76		8.44
		−6	
10	70		7.00

as the laborers are added. But this process cannot go on forever. At some point the output provided by one additional laborer will begin to diminish because capacity of the existing physical plant will be reached.

The law of diminishing marginal returns is a fact of nature. Imagine how the world would work if it were not true: All of the world's dry cleaning could be done in a single plant. And if diminishing returns did not exist in agriculture, all of the world's food could be grown on a fixed plot of land, say an acre.

The law of diminishing marginal returns can be illustrated numerically and graphically. A standard example comes from agriculture; we experiment by adding laborers to a fixed plot of land and observe what happens to output. To isolate the effect of adding additional workers (the variable input), we keep all inputs except labor fixed and assume that the technology does not change.

Table 8–2 is a numerical analysis of the results of such an experiment. The first column shows the amount of the variable input—labor—that is used in combination with the fixed input—land. The simplified numbers for **total product** and **marginal product** represent real measures of output such as tons of soybeans grown.

Marginal product is the change in total product caused by the addition of each additional worker. At first, as workers are added, total product expands rapidly. The first three workers show increasing marginal products (10, 12, and 14 extra tons of soybeans with the addition of each worker). Diminish-

Total product: The total amount of output that results from a specific amount of input.

Marginal product: The extra output that results from employing one more unit of a variable input.

ing marginal returns set in, however, with the addition of the fourth worker. That is, the marginal product of the fourth worker is 13 tons, down from the 14-ton marginal product gained by adding the third worker. Marginal product continues to decline as additional workers are added to the plot of land. This squares with our earlier definition of the law of diminishing marginal returns: As additional units of a variable input are added to a fixed amount of other resources, beyond some point the amount of additional output produced will start to decline. In fact, the addition of the ninth and tenth workers leads to reductions in total product. Marginal product in these cases is negative, meaning that the addition of these workers actually reduces output. It becomes more and more difficult to obtain increases in output from the fixed plot of land by adding workers. At some point the workers simply get in each other's way.

Average product: The output per unit of a variable input; total output divided by the amount of variable input.

The last column in the table introduces the concept of **average product,** which is total product divided by the number of units of the variable input. It is the average output per worker. Average product increases as long as marginal product is greater than average product. Thus, average product increases through the addition of the fourth worker. The marginal product of the fifth worker is 11, which is less than the average product for five workers, and beyond this point the average product falls.

Figure 8–2 graphically illustrates the law of diminishing marginal returns

FIGURE 8–2

Total, Average, and Marginal Product Curves

(a) Total product curve, plotting data from Table 8–2. (b) Average product and marginal product curves. Marginal product is the rate of change of the total product curve. The area of diminishing marginal returns begins when marginal product starts to decline. Average product rises when marginal product is greater than average product, and it falls when marginal product is less than average product.

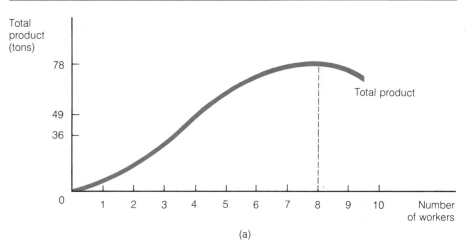

(a)

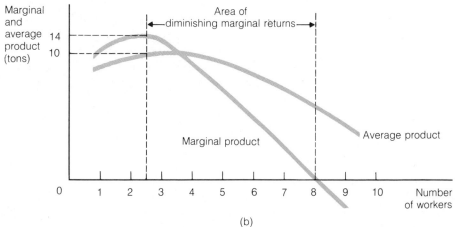

(b)

with the data from Table 8–2. Figure 8–2a illustrates the total product curve. Total product increases rapidly at first as the marginal product of the first three workers increases. Diminishing returns set in with the fourth worker, and thereafter the total product increases less rapidly. Beyond eight workers, the total product curve turns down and starts to decline. The maximum total product is reached with eight workers. The marginal product and average product curves are shown in Figure 8–2b. The marginal product curve is simply the slope of the total product curve. It rises to a maximum at three workers and then declines as diminishing returns set in. Eventually, it becomes negative at eight workers, indicating that total product is decreasing. The average product curve rises when marginal product is greater than average product, and it declines when marginal product is less than average product.

How can the law of diminishing marginal returns be applied to economic decision making? Consider the thinking of the farmer who owns the plot of land in this experiment. How many workers will he or she choose to hire? Before the area of diminishing marginal returns is reached, additional workers have an increasing marginal product. The farmer will add these workers and will go no further in hiring than the point where the marginal product falls to zero. Even if labor were free, the farmer would not go beyond this point because there would be so many workers on the land that the marginal product of an additional worker would be negative and the total product would decline. This implies that it is rational for the farmer to operate somewhere in the area of diminishing marginal returns.

Precisely how many workers will maximize profits? The answer depends on the cost of the workers. Assume that each worker is paid a wage rate equivalent to 4 units of output. The farmer will compare the marginal product of each worker with the wage rate. The fourth worker adds 13 units of output and costs the farmer the equivalent of only 4 units. In fact, the marginal product of the first seven workers exceeds their wage rate. However, the marginal product of the eighth worker is 3 units of output, which is less than the wage rate. The rational farmer will therefore hire seven workers. This is the law of diminishing marginal returns in action. It points the farmer toward the rational utilization of factors of production.

Diminishing Marginal Returns and Short-Run Costs

What does the law of diminishing marginal returns imply about the behavior of the firm's short-run cost curves? Once diminishing marginal returns set in, more and more of the variable input—in our example, labor—is needed to expand output by an additional unit. If the price of the variable input is fixed, the firm's marginal costs will rise as a reflection of diminishing returns. Adding more workers to the plot of land at a fixed wage rate eventually leads to both diminishing returns and rising marginal costs. These are two ways of looking at the same thing.

To see this point more clearly, recall the definition of marginal cost from Table 8–1:

$$MC = \frac{\text{Increase in total cost}}{\text{Increase in output}}.$$

Since the increase in total cost is equal to the increase in the number of units of the variable input times the price of the variable input (which we assume is constant), the expression can be rewritten as

$$MC = \frac{\text{Increase in quantity of variable input}}{\text{Increase in output}} \times \text{Price of variable input.}$$

We know that marginal product is defined as the increase in output associated with an increase in the quantity of the variable input. Thus the first term in the MC expression is the reciprocal of marginal product. MC can further be rewritten as

$$= \frac{1}{\text{Marginal product of variable input}} \times \text{Price of variable input}$$

$$= \frac{\text{Price of variable input}}{\text{Marginal product of variable input}}.$$

This means that marginal cost is inversely related to marginal product. That is, if the price of the variable input is constant, then increases in marginal cost are associated with decreases in marginal product. The law of diminishing marginal returns is equivalent to increasing marginal cost.

Table 8–3 presents a numerical illustration of how the law of diminishing marginal returns affects a firm's short-run costs. Columns (2), (3), and (4) show how total costs behave as output increases in the short run, and columns (5), (6), and (7) show the behavior of the corresponding average cost concepts.

In this example, we keep the numbers low for simplicity. Total fixed cost is constant at $10 per day; average fixed cost per unit is the total fixed cost divided by output. Fixed costs must be paid in the short run regardless of whether the firm operates. The level of average fixed cost falls continuously as output is increased.

Total variable cost is the sum of the firm's expenditures on variable inputs. Average variable cost is the total variable cost divided by output. Notice that total variable cost first increases at a decreasing rate and that after nine units of output it increases at an increasing rate.

Total cost is the sum of total fixed and total variable costs. Average total cost is the sum of average variable and average fixed costs, or simply total cost divided by output.

Marginal cost, shown in column (8), is the change in total cost that results from producing one additional unit of output. The behavior of marginal cost reflects the law of diminishing marginal returns. In this example, marginal cost falls through the production of the ninth unit of output and rises thereafter. The rising portion of the marginal cost schedule reflects diminishing marginal returns for additions of the variable input.

The data in Table 8–3 are plotted in Figure 8–3, which graphically demonstrates the behavior of short-run cost curves. Figure 8–3a illustrates the concepts of total cost, and Figure 8–3b presents the corresponding average and marginal cost concepts. Notice in part b that MC intersects AVC and ATC at their minimum points. This is because marginal cost bears a definite relation to average variable cost and average total cost. In the case of average total cost, marginal cost lies below average total cost when average total cost is falling and above average total cost when average total cost is rising. Marginal cost is equal to average total cost at the point where average total cost is at a minimum. In Table 8–3, this occurs between the thirteenth and fourteenth units of output. At lower levels of output, the MC values in column (8) are less than the ATC values in column (7). ATC therefore declines. At higher levels of output, the MC values are greater than the ATC values,

TABLE 8–3   Short-Run Cost Data for a Firm

The various concepts of short-run costs are expressed in both total and average terms. Columns (1) through (4) present the firm's total cost data; columns (5) through (8) present the cost data in average terms. Notice especially that short-run marginal cost ultimately rises because of the law of diminishing marginal returns.

Total Cost Data (per day)				Average and Marginal Cost Data (per day)			
(1)	(2)	(3)	(4)	(5)	(6)	(7)	(8)
Output (units)	Total Fixed Cost	Total Variable Cost	Total Cost (2) + (3)	Average Fixed Cost (2) ÷ (1)	Average Variable Cost (3) ÷ (1)	Average Total Cost (4) ÷ (1)	Marginal Cost $\Delta(4) \div \Delta(1)$[a]
0	$10.00	$ 0	$10.00	$ 0	$0	$ 0	
1	10.00	1.60	11.60	10.00	1.60	11.60	$1.60
2	10.00	3.00	13.00	5.00	1.50	6.50	1.40
3	10.00	4.35	14.35	3.34	1.45	4.78	1.35
4	10.00	5.70	15.70	2.50	1.43	3.93	1.30
5	10.00	6.90	16.90	2.00	1.38	3.33	1.20
6	10.00	8.00	18.00	1.67	1.33	3.00	1.10
7	10.00	9.00	19.00	1.43	1.29	2.71	1.00
8	10.00	9.90	19.90	1.25	1.24	2.49	.90
9	10.00	10.20	20.20	1.11	1.13	2.24	.30
10	10.00	11.85	21.85	1.00	1.19	2.19	.65
11	10.00	13.15	23.15	0.91	1.20	2.11	1.30
12	10.00	14.65	24.65	0.83	1.22	2.05	1.50
13	10.00	16.40	26.40	0.77	1.26	2.03	1.75
14	10.00	18.45	28.45	0.71	1.32	2.03	2.05
15	10.00	20.90	30.90	0.67	1.39	2.06	2.45
16	10.00	23.80	33.80	0.62	1.49	2.11	2.90
17	10.00	27.20	37.20	0.59	1.60	2.19	3.40
18	10.00	31.15	41.15	0.56	1.73	2.29	3.95
19	10.00	35.70	45.70	0.53	1.88	2.41	4.55
20	10.00	40.95	50.95	0.50	2.05	2.55	5.25

[a]Δ = change in.

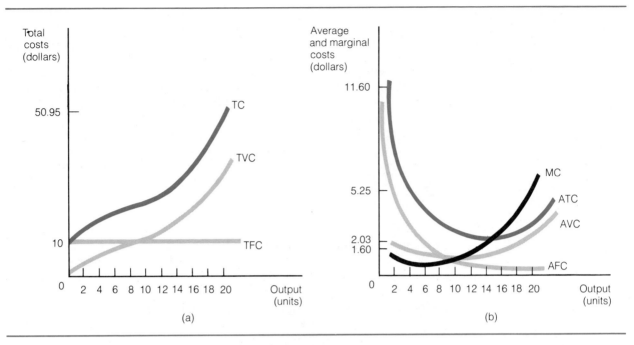

FIGURE 8–3   Short-Run Cost Curves

(a) Total cost data plotted from Table 8–3. (b) Average cost and marginal cost data. Notice the U-shaped average total cost curve. At low levels of output, ATC is high because AFC is high, and the firm does not utilize its fixed plant efficiently. At high levels of output, ATC is high because MC is high as the firm approaches the capacity limit of its plant. These effects give ATC a U shape.

and ATC rises. The same analysis holds for the relationship between MC and AVC.

This type of relation holds for marginal–average series in general. If a baseball player is batting .400 and goes 3 for 4 today (.750), his batting average rises because the marginal figure (.750) lies above the average (.400). If he goes 1 for 4 today (.250), his average falls because the marginal figure (.250) lies below the average (.400). An example can be found closer to home. Suppose that your average grade in economics to date is 85. If you score 90 on your next exam, your average will rise. If you score 80, it will fall. When the marginal figure is above the average, it pulls the average figure up, and vice versa. This same relation holds for average total cost and marginal cost, which both Table 8–3 and Figure 8–3 verify.

In sum, the firm's short-run cost curves show the influence of the law of diminishing marginal returns. First, the firm has certain fixed costs independent of level of output. These costs correspond to what the firm must pay for fixed inputs, and they must be paid whether or not the firm operates. Second, assuming the price of variable inputs is constant, marginal costs reflect the behavior of the marginal product of the variable input. As output rises, marginal costs first decline because marginal product is increasing and it requires less and less of the variable input to produce additional units of output. At some point, however, diminishing marginal returns set in, and it takes more and more of the variable input to produce additional units of output. When this happens, marginal costs start to rise. Third, marginal cost will eventually rise above average variable cost and average total cost, causing these costs to rise as well, which results in a U-shaped average total cost curve.

Long-Run Costs

In the short run, some inputs are fixed and cannot be varied. These fixed inputs generally are characterized as a physical plant that cannot be altered in size in the short run. Short-run costs therefore show the relation of costs to output for a given plant. In the long run, however, all inputs are variable, including plant size. As economic time lengthens and the contracts that define fixed inputs can be renegotiated, a firm owner can adjust all parts of his or her operation. In fact, the scale of a firm's operations can be adjusted to best fit the economic circumstances that prevail in the long run; an owner can choose to arrange the organization in the best possible way to do business. Stated in terms of plant size, the owner will seek the plant size that minimizes long-run costs of producing the profit-maximizing output.

Adjusting Plant Size

The choice of plant size affects production costs. An owner might have the choice of four plants, such as those depicted in Figure 8–4. The ATC curves are short-run average total cost curves that correspond to four different-sized plants. Which plant is best from the owner's point of view? The answer depends on the profit-maximizing level of output. If the firm wants to produce less than q_1, the plant represented by ATC_1 is the lowest-cost plant for those levels of output. For outputs between q_1 and q_2, ATC_2 is the lowest-cost plant. For outputs between q_2 and q_3, ATC_3 is the lowest-cost plant. For even larger outputs—beyond q_3—the plant represented by ATC_4 is the lowest-cost plant.

In effect, the owner's best course of action is depicted by *PLAN* in Figure 8–4. Given that inputs can be varied to build any of these plants, the owner will move along a path such as *PLAN*, gradually expanding output by expanding plant size, seeking the lowest-cost way to produce the profit-maximizing output in the long run.

To determine the optimum long-run plant size, we introduce a new concept: **long-run average total cost** (LRATC). This measure shows the lowest-

---

FIGURE 8–4

Adjusting Plant Size to Minimize Cost

The figure shows short-run average total cost curves for four plant sizes. If these are the only alternative plant sizes available, the long-run average total cost (or planning) curve will be given by *PLAN*. That is, starting with plant size *P*, the rational owner expands plant size to *L, A,* and then *N* in the long run. Each point represents movement to a larger and lower-cost plant size.

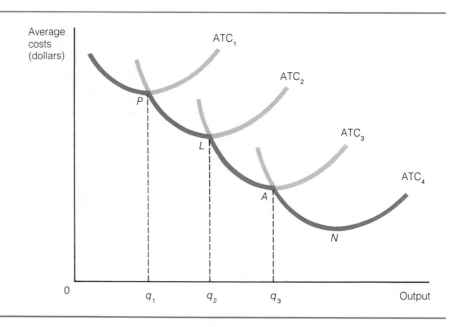

FIGURE 8–5

Long-Run Average Total Cost Curve

LRATC is the planning curve of the firm owner. It shows how plant size can be adjusted in the long run when all inputs are variable. The plant size shown at *q* represents the lowest possible unit cost of production in the long run. Economies of scale prevail before *q* on the curve, and diseconomies of scale prevail past *q*.

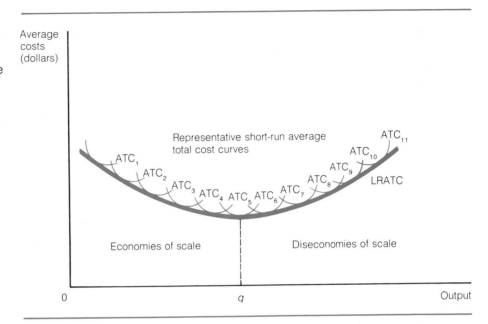

Long-run average total cost: The lowest possible cost per unit of producing any level of output when all inputs can be varied.

cost plant for producing each level of output when the firm can choose among all possible plant sizes. In Figure 8–4, *PLAN* is the long-run average total cost curve when only four plants are possible. In the long run, however, more than four plant sizes are possible. Indeed, the number of possible plant sizes is unlimited. In effect, the owner sees the world of possibilities in terms of a long-run average total cost (or planning) curve such as that drawn in Figure 8–5. The long-run average total cost curve is smoothly continuous and allows us to see the full sweep of a firm owner's imagination. On its downward course, the long-run average total cost curve is tangent to each short-run average total cost curve *before* the point of minimum cost for each given plant size. On its upward course it touches each short-run average total cost curve *past* the point of minimum cost. Only at the bottom of the long-run U shape do the minimum points of the long-run and short-run average total cost curves coincide. In effect, the long-run average total cost curve is an envelope of short-run average total cost curves.

Economies and Diseconomies of Scale

Why does long-run average total cost have a U shape, falling to *q* and rising thereafter? The answer involves two new concepts: economies and diseconomies of scale.

Economies of scale: The relation between long-run average total cost and plant size that suggests that as plant size increases, the average cost of production decreases.

The initial falling portion of the LRATC curve is due to **economies of scale.** To a certain point, long-run unit costs of production fall as output increases and the firm gets larger. There are a number of reasons why a larger firm might have lower unit costs: (1) A larger operation means that more specialized processes are possible in the firm. Individual workers can concentrate and become more proficient at more narrowly defined tasks, and machines can be specially tailored to individual processes. (2) As the firm grows larger and produces more, workers and managers gain valuable experience in production processes, learning by doing. Since workers and managers of a larger firm produce more output, they acquire more experience. Such experience can lead to lower unit costs. (3) Large firms can take advantage of mass production techniques, which require large setup costs. Setup costs are

FIGURE 8-6

Alternative LRATC Curves

Not all LRATC curves are U-shaped. (a) Constant returns to scale. The lowest possible production costs per unit exist within a large range between output levels q_1 and q_2. Both small and large firms can have the same long-run unit costs between q_1 and q_2. (b) Increasing returns to scale. The more output the firm produces, the lower its long-run average total cost.

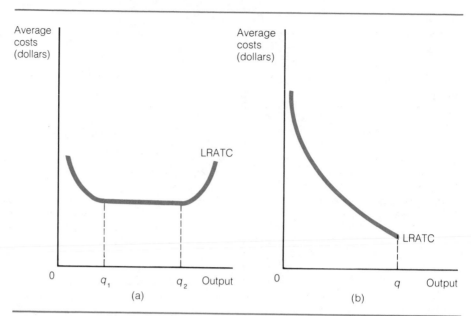

most economical when they are spread over a large amount of output. Production techniques such as the assembly line used by large automobile manufacturers would result in very high unit costs if used by a small producer of specialized cars.

The fact that the LRATC curve rises after q is due to **diseconomies of scale.** In this range of outputs, the firm has become too large for its owner to control effectively. Managers do not have the monitoring technology to hold costs down in a very large firm, and bureaucratic inefficiencies creep in. If such bureaucratic problems did not exist, firms would be much larger. Indeed, a single firm could produce all the world's output of a particular good or service. However, diseconomies may be hard to observe in the real world because it is in the firm's self-interest to correct its operations to keep costs down. Large firms will reorganize, spin off component parts, hire new managers, and in general seek ways to avoid diseconomies of scale.

Diseconomies of scale: The relation between long-run average total cost and plant size that suggests that as plant size increases, the long-run average total cost curve increases.

Other Shapes of the Long-Run Average Total Cost Curve

The LRATC curve in Figure 8–5 shows a unique ideal plant size at q. There is only one minimum point on this U-shaped curve. This means that there is a small range of plant sizes that are efficient in a particular industry, and plant sizes in the industry will tend to cluster at the level of q. The fact that most discount department stores are approximately the same size illustrates this point.

Figure 8–6 shows two other possible shapes the LRATC curve can take. Figure 8–6a illustrates **constant returns to scale.** In certain industries an initial range of economies of scale prevails up to a minimum efficient size of q_1. Beyond q_1, a wide variation in firm size is possible without a discernible difference in unit cost. Small firms and large firms can operate with the same unit costs over this range of outputs. This flat portion of the LRATC curve shows constant returns, or the same unit costs, for a range of output levels from q_1 to q_2. Beyond q_2, diseconomies of scale begin. This is apparently a very common LRATC curve in the real world because we observe both small and large firms prospering and surviving side by side in many industries, such as publishing and textiles.

Constant returns to scale: The relation that suggests that as plant size changes, the long-run average total cost does not change.

Increasing returns to scale: The relation that suggests that the larger a firm becomes, the lower its long-run average total costs are.

Figure 8–6b shows an LRATC curve that exhibits economies of scale over its whole range; this is called **increasing returns to scale.** In such cases, the larger the firm, the lower its costs. This type of LRATC curve is representative of such industries as utilities and telephone service.

Shifts in Cost Curves

Our analysis of cost curves has been based on the familiar *ceteris paribus*, or other things constant, assumption. In other words, we held certain factors constant in the discussion of short-run and long-run costs. What factors did we hold constant, and how do they affect cost curves?

Resource prices have been held constant. If resource prices rise or fall, the firm's cost curves will rise or fall by a corresponding amount. If the price of gasoline falls, the cost curves of a trucking firm fall; this is illustrated by the fall from ATC_1 and MC_1 to ATC_2 and MC_2 in Figure 8–7.

Taxes and government regulation have been held constant. If government increases the excise tax on gasoline, the trucking firm's cost curves will rise. In fact, the average and marginal costs of the trucking firm will rise by the amount of the tax. Similarly, if government imposes more stringent highway weight limits for trucks, the costs of trucking firms will increase.

Technological change has been held constant. Advances in technology make it possible to produce goods and services at lower costs. The invention of the diesel engine, which is more durable and less expensive to operate than the conventional gasoline engine, shifted the cost curves of trucking firms downward. Trucking services can now be produced with fewer resources because of this technological improvement.

The Nature of Economic Costs

Opportunity cost is a forward-looking concept. Opportunity costs are incurred when the decision maker decides about the future use of resources; they are the expected costs of possible alternatives available to an owner.

FIGURE 8–7

The Effect of a Decrease in Resource Prices on Costs

As resource prices decrease, the cost curves of the firm fall from ATC_1 and MC_1 to ATC_2 and MC_2. If resource prices increased, the cost curves of the firm would rise.

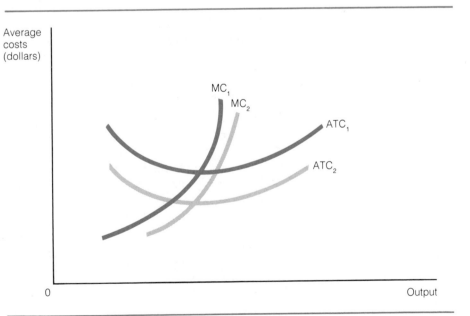

In our discussion of the long-run average total cost curve, we saw that the owner's choice was among potential plants of different sizes. The plants do not yet exist—they are only in the planning or blueprint stage of development. The owner, of course, will base his or her estimate of what the LRATC curve looks like on past experience in the industry or in other industries, so the past is of some help in estimating what will happen in the future. The choice of plant size, however, involves looking into the future and making a decision under uncertainty. To select a plant size means forgoing certain future alternative plant sizes.

Historical costs: Costs of production from the past.

In contrast, accounting costs are **historical costs,** or costs that have been incurred in the past. While historical costs can be useful in estimating economic costs such as short-run marginal cost and long-run average total cost, historical costs look backward; opportunity costs look forward. Historical costs are often poor guides to opportunity costs. Consider the case of sunk costs. **Sunk costs** are incurred as a result of an earlier decision; they are historical costs. For better or worse, earlier decisions cannot be reversed. For this reason economists argue that sunk costs are irrelevant to current decisions. The following examples exemplify this principle and will convince you that it is true.

Sunk costs: Past payment for a presently owned resource.

A jeweler purchases some diamond rings and necklaces. The price of diamonds doubles after her purchase. At what price will she sell the diamonds—the old price of diamonds plus her usual markup or the new price? The amount she paid for the diamonds is a sunk cost. The current price of diamonds determines their worth in the market. She will sell the diamonds at their current price plus her normal markup.

Manhattan, proverbially, was purchased for $24. Suppose you own an acre of Manhattan today. Would your selling price have anything to do with the original purchase price?

You buy a car for $8000. After driving it for a day and imposing only negligible wear and tear, you decide that you do not like it. You put it on the market, and your best offer is $6000. Should you refuse to sell the car for $6000 because you paid $8000 for it?

Each of these examples involves the fallacy of sunk costs. The point is that, viewed rationally, past costs exert no influence on current decisions. This does not mean that we do not learn from experience; it simply means that current decisions are based on current or expected costs and benefits, not on past costs and benefits.

Costs and Supply Decisions

Costs help explain the supply or output decisions of firms. Firms compare their expected costs and expected revenues in deciding how much output to produce. In the short run, the relevant comparison is between marginal cost and expected revenue. If the latter exceeds the former, the firm will supply additional units of output. In the long run the owner of a firm decides whether to enter an industry or to expand output within an industry by comparing long-run average total cost with expected revenue. Again, if expected revenue exceeds marginal cost, the firm enters or increases production, and industry output is expanded. There is a critical link between firm costs and the industry supply curve; we examine this link in detail in the next chapter.

Summary

1. Costs result from the fact that resources have alternative uses. An opportunity cost is the value of resources in their next-best uses.

2. Explicit costs are like accounting costs; they are the bills that the firm must pay for the use of inputs. Implicit costs are the opportunity costs of resources owned by the firm. The total cost of production is the sum of explicit and implicit costs.

3. Private costs are payments for the use of inputs by the firm. External costs arise when the firm uses an input without paying for its services. The sum of private and external costs is social cost.

4. The short run is a period of economic time when some of the firm's inputs are fixed. The long run is a period over which all inputs, including the size of the firm's plant, can be varied.

5. The law of diminishing marginal returns states that when adding units of a variable input to a fixed amount of other resources, beyond some point the resulting additions to output will start to decline.

6. The law of diminishing marginal returns implies that the marginal cost curve of the firm will ultimately rise. As the firm adds more variable inputs to its fixed plant, diminishing marginal returns set in at some point, and marginal cost will start to rise as the firm approaches its short-run capacity.

7. Short-run costs are (a) fixed costs—costs that do not vary with the firm's output; (b) variable costs—the costs of purchasing variable inputs; (c) total costs—the sum of fixed and variable costs; and (d) marginal costs—the change in total cost with respect to a change in output. These costs can be expressed in total or average (unit) terms. The short-run average total cost curve is U shaped. When marginal cost is below average total cost, the latter falls. When it is above average total cost, the latter rises.

8. Long-run average total cost shows the lowest-cost plant for producing output when the firm can choose among all possible plant sizes. It is the planning curve of the firm; it helps the firm pick the right-sized plant for long-run production.

9. A U-shaped long-run average total cost curve results from economies and diseconomies of scale. Economies of scale cause long-run unit costs to fall and diseconomies of scale cause long-run unit costs to rise as output is expanded.

10. Economic or opportunity costs are the expected future costs of forgoing alternative uses of resources. Accounting or historical costs are costs that have been incurred in the past. Sunk costs are historical costs and not relevant to present decisions.

Key Terms

costs of production
accounting costs
implicit costs
opportunity cost of
  capital
total cost of
  production
economic profit
accounting
  profit

private costs
external costs
social costs
variable input
fixed input
short run
long run
fixed costs
variable costs
total cost

average fixed cost
average variable cost
average total cost
marginal cost
fixed plant
law of diminishing
  marginal returns
total product
marginal product
average product

long-run average
  total cost
economies of scale
diseconomies of
  scale
constant returns to
  scale
increasing returns to
  scale
historical costs

sunk costs

Questions for Review and Discussion

1. Explain the difference between explicit and implicit costs.

2. State and explain the law of diminishing marginal returns.

3. What is the definition of marginal cost? How is it linked to the law of diminishing marginal returns?

4. Define economies and diseconomies of scale and give an example of each.

5. How long is the long run? How does it compare to the short run?

6. When is the marginal cost curve below the average total cost curve? When is it above the average total cost curve?

7. An engineering student invests in obtaining an engineering degree. Let us say that his degree cost him $50,000 in opportunity costs. Five years later, he decides he wants to be an artist. Is the $50,000 investment in the engineering degree relevant to his decision to become an artist?

8. Derive the long-run cost model discussed in this chapter using the short-run model. Be sure to include short-run average total cost (ATC), marginal cost (MC), and long-run average total cost (LRATC). Where is the single most preferred operating position in the long-run model?

ECONOMICS IN ACTION   Costs to the Firm and Costs to Society: The Case of Pollution

Costs to a private firm can sometimes differ from full social costs when the firm's production activities create unwanted by-products—external costs. Suppose an iron-smelting firm belches foul-smelling smoke into the air and dumps toxic wastes into streams. The same process that produces goods and services demanded by society also imposes costs on society in health hazards and pollution. What, if anything, is to be done about the costs to society of reducing or eliminating the costly pollution?

It will help to visualize the difference between firm costs and social costs, which we present in Figure 8–8.

Economists have proposed a number of solutions to the externality problem. Cambridge economist A. C. Pigou, in his 1920 work *The Economics of Welfare*, argued that taxes could be levied upon the perpetrators of negative externalities.[a] In a theoretical sense at least, such taxes would have the effect of making the costs faced by the firm include all social costs. Implementation of Pigou's general proposal could take a number of forms: (1) The public could build a treatment plant. (2) Pollution standards could be (and have been) imposed. (3) A tax could be levied on units of waste discharged. All of these proposals to alleviate the effects of externalities have advantages and disadvantages, but all contemplate government intervention into private markets.

In 1960, Ronald Coase challenged Pigou's analysis of externalities.[b] He argued that externality relations can be thought of as bilateral. Stream polluters would not create an externality if there were no downstream population. Cigarette smokers would not impose costs on nonsmokers if the latter did not position themselves in the way of smoke. It takes two to make an externality. Coase also argued that society itself might "internalize" many externalities. Restaurants, for example, can offer smoking and nonsmoking sections. In these cases, the market would adjust to the externality without governmental interference. Coase argued further that if the judicial system assigns liability to the party in the externality relation who would incur the least cost in correcting the situation, market forces may be sufficient to generate efficient solutions to these problems. Perhaps, for example, a scheme that assigns the responsibility for workplace safety to either employers or employees rather than a government regulatory pro-

[a]A. C. Pigou, *The Economics of Welfare* (London: Macmillan, 1920).

[b]Ronald Coase, "The Problem of Social Cost," *Journal of Law and Economics* 3 (October 1960), pp. 1–44.

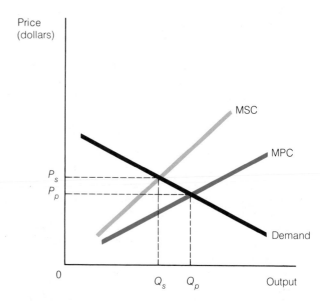

FIGURE 8–8   Costs to the Firm, Costs to Society

In the case of an iron-smelting firm, the marginal social costs (MSC) exceed the marginal private costs (MPC) of production. The marginal social cost curve includes all of the firm's private costs of production plus the costs to society of the firm's pollution. In the case of the polluting firm, MSC is greater than MPC, and "too much," Q_p, of the good is produced at "too low" a price, P_p. If all costs are accounted for, including external or social costs, a lower output, Q_s, would be produced at a higher price, P_s.

gram—such as that administered by the Occupational Safety and Health Administration—is the most efficient way to promote on-the-job safety. These are clearly important matters for debate and analysis, and we return to them in more detail in Chapter 18.

Questions

1. Assume that shrimping firms who fish the Gulf of Mexico "overfish," depleting the stock of shrimp. Such overfishing causes the quantity of shrimp to be lower in the future. This means that the marginal social costs of shrimping are greater than the marginal private costs facing the shrimping firms. How might this problem be reduced or eliminated?
2. Suppose that your neighbor creates a beautiful flower garden. Your neighbor pays all of the costs of the garden. Does he or she receive all of the benefits?

9

The Competitive Firm and Industry

C ompetition exists in virtually every aspect of life. Students compete for grades, animals compete for habitat, college football teams compete for national championships, government agencies compete for budget appropriations, firms compete for customers. In each case, scarcity causes the competition. Funds available to support government agencies are limited; defense and social agencies therefore compete for scarce budget dollars. Likewise, there can be only one national champion in college football. College teams therefore compete for this scarce distinction. If resources were not scarce but freely available, there would be no need for competition. Everyone could have all they wanted of whatever they wanted.

This chapter introduces the economist's model of how firms and industries behave under circumstances of pure competition. It is the first of three chapters devoted to analyzing the impact of industry structure on price and output. In these chapters we look at four models of industry behavior: pure competition, monopoly, monopolistic competition, and oligopoly. In this chapter we discuss the usefulness of the abstract model of pure competition and its relevance for real-world problems. We also use new analytical techniques relating the costs and output decisions of individual firms to the output of an industry as a whole.

The Process of Competition

Most people naturally think of competition as the process of competition or the conduct of competitors. Seen in this way, the important question in determining whether firms are competing with one another is whether they are exhibiting rivalrous, competitive behavior. Do they compete hard for customers? Do they try to outperform one another? Do they use a variety of methods—persuasive advertising, a carefully chosen location, an attractive price—to win and keep customers? Do they seek the best managerial talent available? Are they forward-looking and innovative? Do they seek to eliminate waste and inefficiency in their company? The list easily could go on, but the point is already clear: Competition is normally thought of as a process of rivalry among firms.

This process of rivalry leads to better and cheaper products for consumers. Business firms compete to make profits, but the competitive process actually forces firms to meet consumer demands at the lowest possible level of profit. As Adam Smith observed more than two hundred years ago, firms' self-interest is harnessed by the competitive process to promote the general well-being of society:

> It is not from the benevolence of the butcher, the brewer, or the baker, that we expect our dinner, but from their regard for their own interest. We address ourselves, not to their humanity but to their self-love, and never talk to them of our own necessities but of their advantages.[1]

The process of competition channels the pursuit of individual self-interest to socially beneficial outcomes, and for this reason economists put the study of the competitive process at the heart of their science.

An important point about the process of competition is that it takes place in an uncertain world. Individuals have to make conjectures as they make resource commitments for the future. Some conjectures will turn out to be correct, and individuals who forecast correctly will survive and prosper. Individuals whose conjectures turn out not to be correct will not fare so well in the future. The concept of competition as a process of rivalry under uncertain conditions is especially important for understanding entrepreneurial behavior in the economy, a topic we return to in Chapter 15.

Although competition generally benefits consumers by harnessing the self-interest of producers, producers have incentives to subvert the competitive process to increase their profits. Adam Smith also recognized this tendency:

> People of the same trade seldom meet together, even for merriment and diversion, but the conversation ends in a conspiracy against the publick, or in some contrivance to raise prices.[2]

Independent action among competitors is by no means guaranteed. Indeed, firms may have incentives to avoid competition and independent action in favor of collusion and higher profits. This tendency leads to monopolies of one sort or another, which we discuss in more detail in Chapter 11. Here we simply note that society can influence the rules under which firms compete. Price fixing may or may not be allowed; property rights may or may not be defended; business taxes may be high or low. The process of competition is affected by government regulations, which may be more or less restrictive. Indeed, as we will see, this is a two-way street. Government affects business, and business affects government. Chapter 17 is devoted to an analysis of government regulation of business.

A Purely Competitive Market

In contrast to the concept of the competitive process just described, pure competition is an abstract model of competitive behavior that emphasizes the importance of industry structure. In particular, it stresses the number of

[1] Adam Smith, *An Inquiry into the Nature and Causes of the Wealth of Nations*, ed. Edwin Cannan (1776; reprint, New York: Modern Library, 1937), p. 14.
[2] Smith, *Wealth of Nations*, p. 128.

independent producers in an industry. The difference between the competitive process model and the pure competition model can be compared with the behavior of runners in a race. To determine how close a race will be, the competitive process approach would ask how intensely the runners are competing. Are they striving hard to win? The pure competition approach would ask how many equally qualified runners are in the race. The process view is clearly a richer, more natural interpretation of competition. Nonetheless, the model of pure competition is a useful abstraction that, interpreted carefully, can help us understand the competitive behavior of producers in real-world industries.

Market: A coming together of buyers and sellers for the purpose of making transactions.

The model of pure competition is based on the concept of a **market.** As you recall from Chapter 4, market prices are subject to the laws of demand and supply, and within a market prices of a good or service will tend toward equality. Market forces even transcend geographical boundaries, given free flow of information. Buyers at point x will not pay more for a commodity than the price at point y plus transportation costs. Buyers at point y will not pay more for a commodity than the price at point x plus transportation costs. Price deviations will quickly be spotted by buyers, restoring the market's tendency to one price. Suppose that the price at point y fell below the price at point x plus transportation costs. What would happen? Buyers would shift their purchases from point x to point y, decreasing demand and lowering price at point x and increasing demand and raising price at point y. This shift of buyers would restore the tendency to equality of price.

The behavior of sellers in a market is equally predictable. In fact one way to determine if two commodities are in the same market is to check whether price reductions or increases for the commodity in one area are matched by price reductions or increases in other areas, accounting also for transportation costs. If the price of gasoline goes down in Atlanta, does it also go down in Birmingham? In Charlotte? In Washington, D.C.? In New York? In Chicago? In Seattle? In other words, do sellers in other areas of the country respond with competitive price reductions? To the extent that they do, we can delineate the market for gasoline from the tendency of all sellers to adjust prices in the same direction.

Building on this concept of a market, we can define four conditions that characterize a **purely competitive market.**

Purely competitive market: A coming together of a large number of buyers and sellers in a situation where entry is not restricted.

Homogeneous product: A good or service produced by many firms such that each firm's output is a perfect substitute for the other firms' output, with the result that buyers do not prefer one firm's product to another firm's.

1. Firms in a purely competitive market sell a **homogeneous product.** That is, they sell identical or nearly identical products that are perfectly substitutable for one another. This means that advertising does not exist in a purely competitive market. Why would one firm want to advertise the advantages of its product if it is perfectly substitutable for the product of competitors?

2. A large number of independent buyers and sellers exist in a purely competitive market. This assumption rules out the possibility of collusion among buyers or sellers to affect price and output in the industry. Moreover, the large number of buyers and sellers ensures that the purchases or sales of any one buyer or seller will not affect the market price. Each buyer and seller is small relative to the total market for the commodity and exerts no perceptible influence on the market.

3. There are no barriers to entry or exit in a purely competitive market. Features of economic life such as control of an essential raw material by one or a few firms or government regulation of firms' behavior in the market do not exist under pure competition.

Perfect information: A condition in which information about prices and products is free to market participants; combined with conditions for pure competition, perfect information leads to perfect competition.

4. A perfectly competitive market also offers **perfect information** to buyers and sellers. Everybody in the market has equal, free access to information about the location and price of the product.

These four conditions are the assumptions on which the theory of pure or perfect competition is based. This theory is designed to explain the behavior of many independent buyers and sellers, none of whom has any perceptible influence on the market. The assumptions may seem unrealistic. Products, for example, are rarely homogeneous, and industries with a large number of sellers are rare in a modern economy. But in economics, the realism of the assumptions is not the point of this or any other model. The point is the empirical relevance of the model. Does it explain behavior in real-world markets? As we will see, the purely competitive model helps us analyze important actual markets and industries such as the stock market and agriculture. (See Focus, "The Stock Market," for one application of the concept of pure competition.) Moreover, the purely competitive model gives us an analytical framework for what might be loosely described as the "ideal" working of an economy. In this sense the model provides a benchmark against which other industry models, such as pure monopoly, can be measured.

The Purely Competitive Firm and Industry in the Short Run

How then do firms behave within a purely competitive industry? We focus first on the purely competitive firm's decisions about how much output to supply in the short run; that is, over a period of time when it cannot adjust its plant size.

The Purely Competitive Firm as a Price Taker

In a purely competitive market an individual firm cannot influence the market price for its good or service by increasing or decreasing its output. Because each seller is only a small part of the total market, its actions have no perceptible influence on the market. The competitive firm is called a **price taker:** It must accept the going market price for its product.

Price taker: An individual buyer or seller who faces a single market price and is able to buy or sell as much as desired at that price.

As an example of price-taking behavior, consider the cotton market in 1978–1979. World cotton output was estimated to be 65,337,000 bales in 1978. Of this total, assume that the Commonwealth of Virginia produced a minuscule amount—100 bales in 1978 and 200 in 1979. Further assume that, over a given price range, the price elasticity of demand for cotton worldwide (e_m) is 0.25, which means that the market demand curve for cotton is inelastic.

First consider what happened to the world price of cotton as a result of the 100-bale increase in Virginia's cotton production between 1978 and 1979. Other things being equal, the percent change, in this case a decline, in world price is given by

$$\frac{\text{Percent change in world output}}{e_m} = \frac{100 \text{ bales}/65,337,000 \text{ bales}}{0.25}$$

$$= 0.0000061 = 0.00061\%.$$

(Remember that e_m equals percent change in world output divided by percent change in world price, so percent change in world output divided by e_m equals percent change in world price.)

FOCUS   The Stock Market: An Example of Perfect Competition

The stock market—the market in which shares of ownership of firms are traded—might at first seem like an implausible example of a competitive market. The popular press typically insinuates that the stock market is basically little more than a gambling casino, with ticker-tape machines replacing one-armed bandits. Alternatively, the stock market is portrayed as a kind of free-for-all of combat among an elite few investment tycoons.

But closer observation dispels these popular misconceptions. The stock market is a highly competitive market. Many thousands of firms engaged in hundreds of distinct lines of business offer their shares to many thousands of potential buyers every day. Entry into the market on both sides of potential transactions is practically unrestricted. The shares firms offer for sale are highly substitutable as potential investments to buyers, making them essentially homogeneous goods.

The highly competitive nature of the stock market is obscured by the fact that only a few hundred of the largest corporations are traded in the organized stock exchanges (in the United States, these exchanges are collectively referred to as "Wall Street"). Since the shares of large corporations are traded frequently, it is efficient for trade to occur among exchanges organized for this purpose. The organized stock exchanges radically reduce transaction costs, benefiting both buyers and sellers. But it is misleading to confuse the overall stock market with the organized stock exchanges. Many thousands of other, smaller firms offer stock shares for sale in the so-called over-the-counter market, outside of the larger organized exchanges. All these firms—those participating in the organized exchanges and those trading outside—are competing for the funds of potential investors. Entry into the market is open even to the tiniest firm that wants to offer its shares to the public.

On the buyers' side, although large institutional investors such as well-known mutual funds like the Dreyfus Fund receive much publicity, there are many small investors because the costs of entry into the market—the transaction costs—are very low, both in the organized exchanges and in the market generally. In 1982, for example, General Motors had 1,122,000 common stockholders, General Electric had 502,000 shareholders, Eastman Kodak had 221,000, and Gulf Oil 302,000. In each case, the overwhelming majority of these shareholders were individuals with relatively small investments.

The price of the firm's shares reflects, at any given moment, the sum of all economically relevant information that the market possesses with respect to that firm's past, present, and expected future performance. That price is the result of bidding in the market, which in turn reflects the judgment and relevant knowledge of all interested parties. Hence, stock prices are extremely sensitive to information about any change that affects the firm and its future prospects. An accidental oil field explosion in Sumatra will probably be reflected in the price of oil company stock in seconds, as will be the influence of a flash flood in Alaska on the price of shares in a civil engineering firm. The stock market's first and foremost function is as a communications mechanism: It maintains a flow of such information among buyers, sellers, and firms.

In short, the stock market, rather than representing a kind of floating financial dice game, is probably one of the most efficient and competitive markets imaginable.

Now we can derive the elasticity of demand facing Virginia producers of cotton (e_v). Since we know the 1978–1979 percent change in Virginia output (100 to 200 bales, or 100%) and since we have just computed the percent change in world price caused by the increase in Virginia production, we have all the information we need to calculate the price elasticity of demand facing Virginia producers:

$$e_v = \frac{\text{Percent change in Virginia output}}{\text{Percent change in world price}}$$

$$= \frac{100\%}{0.00061\%}$$

$$= 163,934.$$

The Virginia producers face an extremely elastic demand curve for their output of cotton. If we took only one Virginia producer of cotton as a representative purely competitive firm, the elasticity result would be dramatically larger.

The Demand Curve of the Competitive Firm

From this example it is easy to see why the competitive firm is called a price taker. Expansions or contractions in an individual firm's output within a competitive market have virtually no effect on the market price. To capture this condition analytically, we draw the firm's demand curve as a straight line at the level of market price in Figure 9–1.

Figure 9–1a shows the demand curve (d) facing Virginia cotton producers, and Figure 9–1b shows the world market demand (D) and supply (S) curves for cotton. The demand curve of the Virginia producers is perfectly elastic with respect to the price of cotton; it is drawn as a flat line at the level of market price (P). A perfectly elastic demand curve means that the firm can sell all it wants to sell at the prevailing market price. If it tried to sell its output at a slightly higher price, demand for its product would vanish because buyers can purchase cotton at the lower market price in whatever quantities they choose. If the firm is trying to maximize its profits, it has no reason to sell its product for less than the market price.

The scale for price is the same in both parts of Figure 9–1. They obviously are not the same for quantity because of Virginia's minuscule proportion of world output. Keep this in mind as you interpret the diagrams in this chap-

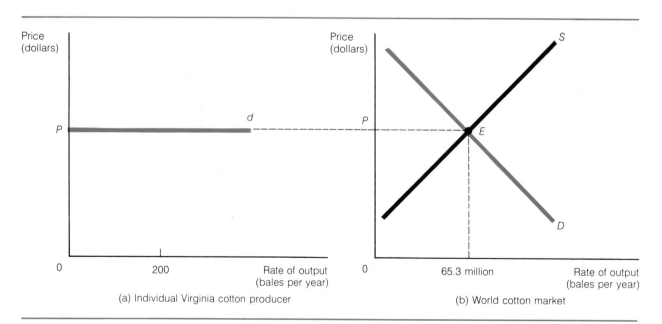

(a) Individual Virginia cotton producer

(b) World cotton market

FIGURE 9–1   Firm and Industry Demand in a Purely Competitive Market

In a purely competitive industry, demand for an individual firm's product is perfectly elastic. The perfectly elastic demand curve facing one individual Virginia cotton producer is shown in (a) at the level of P, which is also the equilibrium price for the world market in (b) at an output of 65.3 million bales per year. Note that the scale for price is the same for both the individual producer and the world market but that the scale for quantity is much different for the two.

ter. Always check to see if the horizontal axis represents *firm* or *industry* quantity.

To summarize: Every firm in a purely competitive industry is a price taker and faces a demand curve such as that of the Virginia cotton producers in Figure 9–1. And every purely competitive firm produces a tiny proportion of industry output.

Short-Run Profit Maximization by the Purely Competitive Firm

The purely competitive firm is a price taker and faces a given price represented by a perfectly elastic demand curve. Under these circumstances how does the firm decide how much of its product to produce? The simple answer is that the firm compares the costs and benefits of producing additional units of output. As long as the added revenues from producing another unit of output exceed the added costs, the firm will expand its output. By following this rule, the firm is led to maximize its **profits** in the short run.

Profits: The amount by which total revenue exceeds total cost.

The added cost of producing an additional unit of output is the marginal cost. We know from our analysis in Chapter 8 that the short-run marginal cost curve of a firm will eventually rise. This is due to the law of diminishing marginal returns, which comes into play as the firm uses its fixed plant more intensively in the short run.

Marginal revenue: The change in total revenue that results from selling one additional unit of output; the change in total revenue divided by the change in amount sold.

The extra revenue from producing an additional unit of output is **marginal revenue** (MR). Simply, marginal revenue is the addition to total revenue from the production of one more unit of output:

$$MR = \frac{\text{Change in total revenue}}{\text{Change in output}}.$$

In this expression, the denominator is simply 1, a one-unit increase in output, and the numerator is always the market price (P) because the demand curve facing the competitive firm is perfectly elastic. Therefore, marginal revenue equals market price, MR = P, for the purely competitive firm. In fall 1984, the market price for used newspaper was $10 per ton, so paper recyclers could assume that their marginal revenue for each additional ton of paper they collected and sold would be $10.

The purely competitive firm will decide how much to produce in the short run by comparing marginal cost with marginal revenue. To maximize its profits, the firm will produce additional units of output until marginal cost and marginal revenue are equal. Figure 9–2 illustrates this process.

The firm's perfectly elastic demand curve (d) in Figure 9–2a is drawn at the level of market price (P) from part b. As we have just seen, d is also the marginal revenue curve for the firm. Revenue will increase by the market price of one unit each time output increases by one unit. In other words, MR = P. We have also drawn the marginal cost (MC) and average total cost (ATC) curves of the firm.

The owners of the firm want to make as much money as possible, so their problem is to find the level of production or output that yields the largest profit. Suppose that the firm is producing hot pepper sauce. It presently produces a certain quantity q_1. Should the owners expand their level of production? At this point, the answer is yes because additional units of pepper sauce add more to revenue than to costs. In other words, marginal revenue exceeds marginal cost at q_1, as you can see by comparing points *mr* and *mc* at quan-

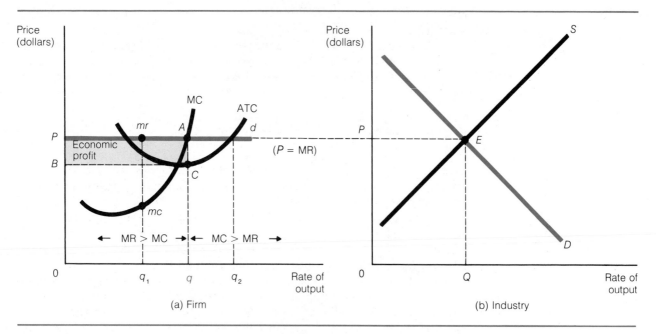

FIGURE 9–2   Short-Run Output Choice and Profit Maximization

(a) The short-run choice of output confronting the individual firm. Before the rate of output q, additional production adds more to revenue than to cost. Beyond q, additional production adds more to cost than to revenue. Therefore, the purely competitive firm will produce at q. At this rate of output the firm earns an economic profit equal to the difference between total revenues and total costs: $PAq0 - BCq0 = PACB$. (b) Prevailing market conditions in the industry.

tity q_1. Since the owners want to maximize their profits, they will therefore expand output beyond q_1 as far as q.

Suppose instead that the owners are operating at an output such as q_2. What will their profit-maximizing reaction be in this case? At this level of production, an additional unit of output adds more to cost than it adds to revenue. That is, marginal cost exceeds marginal revenue. The rational profit-maximizing response of the owners will be to lower production. Through a process of trial and error, they will find that their best level of output is q. At a market price of P, they can do no better for profits than to produce q jars of hot pepper sauce. The short-run equilibrium level of production for the purely competitive firm is therefore defined by the condition $P = MC$. Since we know that price and marginal revenue are the same for the purely competitive firm, we can also write $P = MR = MC$.

Using the average total cost curve in Figure 9–2, we can see exactly how well the hot pepper sauce firm fares by producing at level q. The total revenue of the firm equals its sales, which are the level of output q times the price P for the output. **Total revenue** at q therefore equals Pq, represented by the area $PAq0$ in Figure 9–2. Total cost is the level of average total cost (C) times the level of output q. Total cost at output q thus equals Cq, represented by $BCq0$. Total revenue exceeds total cost in this case: $PAq0 - BCq0 = PACB$. This firm's short-run economic profit is shown by $PACB$, the return in excess of the total cost of production.

In the real world, businesspeople do not spend a lot of time trying to draw marginal cost and marginal revenue curves for their firms. Moreover, they

Total revenue: The total amount of money received by a firm from selling its output in a given time period; price times quantity sold.

operate in an uncertain environment where future costs and prices cannot be known with certainty. Despite these considerations, however, the $P = MR = MC$ rule for profit maximization may have predictive power. It is a rule based on common sense. If additional production promises to add more to revenue than to cost, most businesses will try to increase their production. If additional units will probably add more to cost than to revenue, most businesses will cut back production. Such behavior leads to the $P = MR = MC$ result even when businesspeople know nothing about the economic rule involved.

A Numerical Illustration of Profit Maximization

Another way to understand how the choice of output levels allows a competitive firm to maximize profits is to examine the specific costs and revenues at each level of production. Table 9–1 presents a numerical schedule for the hot pepper sauce firm. The firm's output level is given in column (1). Column (2) shows that the firm confronts an unvarying market price of $2.40 per jar for its output. Price equals marginal revenue ($P = MR$) since this is a purely competitive firm. The marginal cost and total cost data in columns (3) and (5) are taken from Table 8–3 (p. 168), where these cost concepts were introduced and discussed. Total revenue in column (4) is the sales of the firm, or simply price times output, column (1) × column (2). Profit in

TABLE 9–1   Profit Maximization for a Purely Competitive Firm

This numerical schedule of the costs and revenues the hot pepper sauce firm faces at each level of output provides data for determining the maximum profit level in two ways. One way is to find the greatest positive difference between total revenue and total cost, reflected as profit in column (6). The other way is to compare the marginal revenue and marginal cost columns, (2) and (3). The highest profit occurs at the output level where the two columns coincide—between 14 and 15 jars.

(1) Rate of Output (jars per day)	(2) Price (= Marginal Revenue)	(3) Marginal Cost	(4) Total Revenue (1) × (2)	(5) Total Cost	(6) Profit (4) − (5)
1	$2.40	$1.60	$ 2.40	$11.60	− $9.20
2	2.40	1.40	4.80	13.00	− 8.20
3	2.40	1.35	7.20	14.35	− 7.15
4	2.40	1.30	9.60	15.70	− 6.10
5	2.40	1.20	12.00	16.90	− 4.90
6	2.40	1.10	14.40	18.00	− 3.60
7	2.40	1.00	16.80	19.00	− 2.20
8	2.40	0.90	19.20	19.90	− 0.70
9	2.40	0.30	21.60	20.20	1.40
10	2.40	0.65	24.00	21.85	2.15
11	2.40	1.30	26.40	23.15	3.25
12	2.40	1.50	28.80	24.65	4.15
13	2.40	1.75	31.20	26.40	4.80
14	2.40	2.05	33.60	28.45	5.15
15	2.40	2.45	36.00	30.90	5.10
16	2.40	2.90	38.40	33.80	4.60
17	2.40	3.40	40.80	37.20	3.60
18	2.40	3.95	43.20	41.15	2.05
19	2.40	4.55	45.60	45.70	− 0.10
20	2.40	5.25	45.00	50.95	− 2.95

column (6) is the difference between total revenue and total cost, or (4) − (5).

There are two ways to find the profit-maximizing rate of output in Table 9–1. First, we can examine the difference between total revenue and total cost, given in column (6). Notice that at low and high rates of output the firm's economic profits are negative, ranging from a loss of $9.20 per day because of the high cost of producing only 1 jar of sauce to a loss of $2.95 for trying to produce 20 jars a day with an overcrowded fixed plant. The maximum profits the firm can earn in the short run occur at a rate of output of 14 jars of sauce per day and are equal to $5.15 per day, the largest dollar amount in column (6).

Figure 9–3a illustrates the total revenue–total cost approach to maximizing profits. The figure graphs columns (4) and (5) with respect to the output of the firm. The maximum profits possible in the short run occur where the total revenue line (TR) exceeds the total cost curve (TC) by the largest vertical difference. As Table 9–1 indicates, this takes place between 14 and 15 jars of sauce.

The second way to find the profit-maximizing rate of output using the numerical schedule in Table 9–1 is to compare marginal cost and marginal revenue, columns (2) and (3). As long as marginal revenue is greater than marginal cost, it pays to produce additional units of output. Clearly, marginal revenue ($2.40) is greater than marginal cost up to and including an output of 14 jars of sauce. This means that the point of maximum profit occurs between 14 and 15 jars per day. If the firm can turn out jars only one

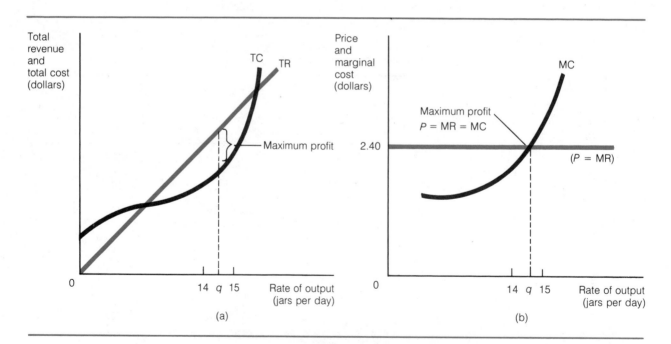

FIGURE 9–3   Profit-Maximizing Output for a Purely Competitive Firm
(a) Total revenue–total cost approach to profit maximization. The profit-maximizing rate of output occurs where the vertical distance between total revenue and total cost is greatest. (b) Marginal revenue–marginal cost approach. Profit maximization occurs where price, or marginal revenue (which are identical for the purely competitive firm), equals marginal cost (P = MR = MC). Both approaches result in the same profit-maximizing rate of output.

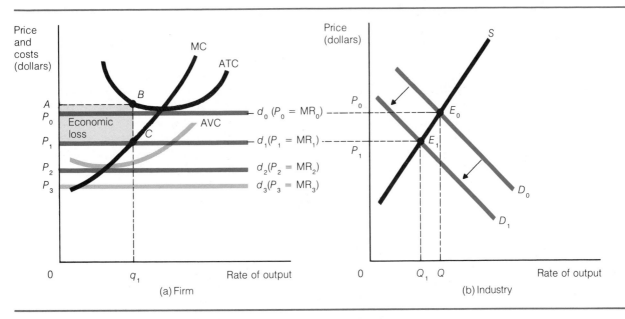

FIGURE 9-4   Short-Run Loss Minimization

(a) Firm's behavior in the face of economic losses caused by (b) decreased industry demand from D_0 to D_1 and a price drop from P_0 to P_1. At P_1, the firm will minimize losses by continuing to operate in the short run because it covers all of its variable costs and some of its fixed costs by operating. At P_2, the firm may either operate, covering its variable costs, or shut down, eliminating its variable costs. At P_3, the firm will shut down in the short run because the price is too low to cover even its variable costs. It minimizes losses by shutting down and paying only its fixed costs, which must be paid whether the firm operates or not.

at a time, the firm will cease producing additional sauce after 14 jars, since the marginal cost of producing the next jar is $2.90 and the marginal revenue is only $2.40. Units of output beyond 14 add more to costs than to revenue; that is, they cause profits to decline. Notice, then, that both the total revenue–total cost and the marginal revenue–marginal cost approaches yield the same answer for the profit-maximizing rate of output.

Figure 9–3b illustrates the marginal revenue–marginal cost approach. Again, note the equivalence of the two approaches by comparing the optimal rate of output, between 14 and 15 jars per day in both graphs.

Economic Losses and Shutdowns

What will the firm do if the short-run situation changes for the worse? What if, for example, demand for hot pepper sauce declines because the government publishes an adverse report on the health consequences of eating too much hot pepper sauce? In this case, the market demand curve for pepper sauce will decrease, as shown in Figure 9–4b, shifting to the left from D_0 to D_1. This results in a price reduction in the hot pepper sauce market, causing the market price confronting the firm to fall from P_0 to P_1. The firm's situation is drawn in Figure 9–4a. Note that the average variable cost curve (AVC) has been added becaue it is now an important consideration.

We see in Figure 9–4a that the firm incurs an economic loss because the new market price, P_1, is below its average total cost curve. The firm's revenues are not sufficient to cover its total costs, and it therefore loses money on its operations. What will the firm do in the face of an economic loss? The firm has two options in the short run.

If it expects the adverse effect on sales to be short-lived, the firm can continue to operate in the face of short-run losses. While the price remains depressed at P_1, the firm can minimize losses by following the same rule it followed to maximize profits: It determines the output level at which price is equal to marginal cost and lowers production to that level, q_1 in the figure. Total cost at an output of q_1 is ABq_10, and total revenue is P_1Cq_10. The firm therefore incurs an economic loss equal to the difference between the two: $ABCP_1$. Why might the firm continue to operate in this case while making a loss? It covers all of its variable costs and some of its fixed costs at a price of P_1. If the firm stopped operating, it would still have to pay its fixed costs. These payments must be made whether or not the firm operates. By continuing to operate at P_1, the firm earns *something* toward the payment of its fixed costs.

Suppose, however, that the market price for hot pepper sauce continues to fall in the industry, say to P_3. At P_3, the firm does not take in enough revenue to cover even its variable costs of production. The loss-minimizing policy for the firm at this point is to cease operations. By shutting down, the firm is ensured of having to pay only its fixed costs. If it tried to operate at a price below its average variable cost, such as P_3, it would not only have to pay its fixed costs but also would incur a deficit in its average variable cost account. The loss-minimizing policy for the firm is to **shut down** when the price falls below its AVC curve.

Shutdown: A loss-minimizing option of a firm in which it halts production in the short run to eliminate its variable costs, although it must still pay its fixed costs.

Suppose that the market price falls to point P_2 where it is just equal to the minimum AVC? What can the firm do in this case? Either operating or shutting down is a reasonable option. If it operates, it can cover its variable costs but none of its fixed costs. If it shuts down, it still must pay its fixed costs. Other things being equal, the firm should be indifferent about whether it operates in the short run under these conditions.

Keep in mind that we are discussing *short-run* policies for the firm. The short run is a period of time in which fixed inputs cannot be varied. Given that fixed costs must be paid, the firm's objective in the short run, when price falls below ATC and losses begin, is to minimize its losses. The firm's long-run adjustment will depend on what it expects to happen to prices and costs in the industry. If it expects prices to rise, the firm may shut down temporarily, keep its plant intact, and plan to reopen at some time in the future. Indeed, the firm may use the shutdown period to reorganize in an effort to have lower costs when production begins again. If the firm expects price to remain so depressed in the industry that losses are likely to continue into the future, it may act to sell off its plant and equipment and go out of business.

Supply Curve of a Purely Competitive Firm

The previous considerations help us understand a competitive firm's willingness to supply products as the market price changes, even when it declines. We saw in Chapter 4 that a supply curve shows the relation between the quantity supplied of a good and its price, other things being equal. What is the supply curve of a purely competitive firm in the short run? We answer the question in Figure 9–5.

We have drawn the firm's marginal cost curve and average variable cost curve and a series of different price levels that the firm faces from a compet-

FIGURE 9–5

Short-Run Supply Curve of the Purely Competitive Firm

The short-run supply curve of the purely competitive firm is its marginal cost curve (MC) above the point of minimum average variable cost. The firm will not operate below P_1 because it cannot cover its variable costs of production. Above P_1, the firm produces where $P = MR = MC$.

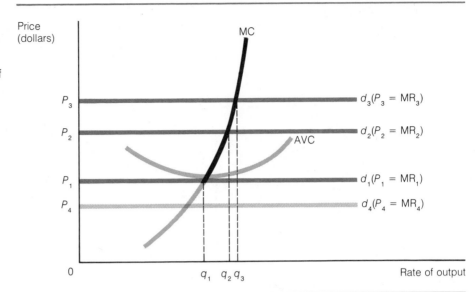

itive market. We know that the general profit-maximizing rule for the competitive firm is to operate at the level of output at which $P = MR = MC$. We also know that the loss-minimizing rule for the firm is to shut down when price falls below the minimum average variable cost. With these two facts, we can derive a supply curve for the competitive firm.

We know that the firm will not produce at a price such as P_4 in Figure 9–5. Since P_4 is below the minimum AVC, the firm will shut down and produce no output at that price. (Notice that no output corresponding to P_4 is given along the horizontal axis.) This decision of the firm applies at any price below P_1, which is the point of minimum AVC. At P_1, the firm will just cover its variable costs and will be economically indifferent about whether or not it operates. We have drawn the firm as producing q_1 units of output at the price P_1. At prices above P_1, the firm will operate in the short run and produce at the output level at which $P = MR = MC$. We have shown two such cases, at P_2 and P_3. In other words, the firm responds to changes in price above P_1 by producing output along its marginal cost curve. The marginal cost curve of the purely competitive firm above the point of minimum average variable cost is the **short-run firm supply** curve; it shows the relation between price and quantity supplied in a period of time when the firm's plant size is fixed.

Short-run firm supply: The portion of the marginal cost curve above the minimum average variable cost.

From Firm to Industry Supply

It is a simple step from the competitive firm's short-run supply curve to the short-run industry supply curve. The **short-run industry supply** curve is the horizontal sum of the individual firms' marginal cost schedules above their points of minimum average variable cost. This process, which is analogous to the way in which we derived market demand curves from individual demand curves in Chapter 4, is illustrated in Figure 9–6.

Short-run industry supply: A summation of all the existing firms' short-run supply curves.

For simplicity, assume that the industry consists of two firms producing hot pepper sauce. At a price of P_0, firm A produces 3 jars of hot pepper sauce per day and firm B produces 2 jars; at these outputs, $P = MR = MC$ for

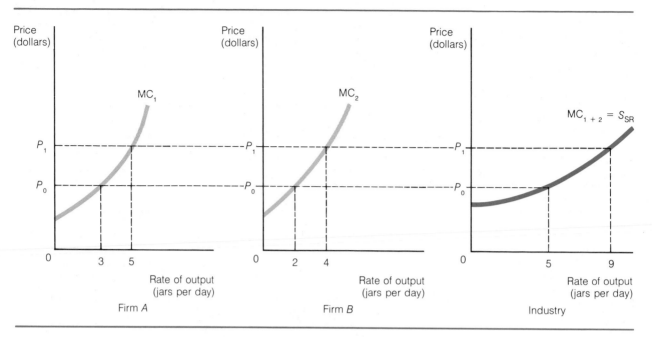

FIGURE 9-6   From Firm to Industry Supply Curves

The short-run supply curve of a competitive industry is the horizontal sum of individual firms' marginal cost curves above their respective points of minimum average variable cost. At price P_0, firm A supplies 3 units of output, and firm B supplies 2 units. Market supply is 5 units at P_0. Other points on the industry supply curve, such as $5 + 4 = 9$ units at P_1, are derived in the same way.

each firm. In each case P_0 is above the firm's minimum AVC. This point on the industry supply curve is 5 jars per day at a price of P_0. That is, 5 jars is the horizontal sum of 3 and 2 jars produced by the two firms. If price rises to P_1, the firms expand production, to 5 and 4 jars, respectively, along their MC curves. The corresponding point for the industry is at 9 jars of pepper sauce. Other points on the short-run industry supply curve (S_{SR}) are obtained in the same manner. The short-run supply curve of a competitive industry is the horizontal sum of all the individual firms' marginal cost curves above their respective points of minimum average variable cost.

Three points should be kept in mind about this discussion. First, a purely competitive industry encompasses many independent producers; the two-firm model used to illustrate the derivation of the industry supply schedule is therefore an abstraction. In a real case of pure competition, thousands of marginal cost curves would have to be summed horizontally. Second, our analysis is conducted in the economic time frame of the short run, in which firms adjust to price changes within the limits imposed by their fixed plant sizes. In the next section we will see what happens to a competitive industry in the long run. Third, the intersection of the industry demand curve and the short-run market supply curve determines the market price for the industry. Both firm and industry are in short-run equilibrium. The firm produces at a level where $P = MR = MC$ for its fixed plant, and industry demand equals industry supply. These two conditions define a short-run equilibrium for the purely competitive model.

The Purely Competitive Firm and Industry in the Long Run

The long run is a period of economic time during which firms can select the lowest-cost plant to produce their output and firms can enter and exit the industry. The result of these adjustments by firms in the face of economic profits or losses is to move the industry toward equilibrium.

Equilibrium in the Long Run

Long-run competitive equilibrium: A situation in an industry in which economic profits are zero and each of the many firms is operating at minimum average total cost.

Zero economic profits: The condition that faces the purely competitive firm in the long run; long-run equilibrium in a competitive industry leads to a condition where $P = MR = MC = LRATC$, which means that firms in the industry earn just a normal rate of return on their investment, or a zero economic profit.

Long-run competitive equilibrium occurs when two conditions are met: (1) quantity demanded equals quantity supplied in the market and (2) firms in the industry are making a normal rate of return on their investments, a situation economists call **zero economic profits.**

Economic profits are returns above and beyond the total (explicit plus implicit) costs to the owner of or investor in a firm. They are returns above the opportunity cost of the owner's capital investment in the firm; that is, they are above the normal return that an owner could expect to make on a capital investment of some other form such as a money market certificate. Economic profits therefore attract the notice of other investors. They are a signal that prods others to try to capture above-normal returns by entering the industry. Economic profits also lead firms already in the industry to seek to expand their scale of operations. Both cases will result in an increase in output in the industry. This causes the short-run industry supply curve to increase and the market price to fall, erasing economic profits and returning the rate of return in the industry to a normal level. This level is referred to as zero economic profits; investors do make a profit, but no more than what their money would have earned through the prevailing rate of return on any other investment.

If firms in the industry are making economic losses, the opposite situation holds. Firms will cut back operations, and some firms will leave the industry. The short-run industry supply curve decreases, causing the market price to rise and restoring a normal rate of return to surviving investors.

A long-run equilibrium state for a purely competitive industry is depicted in Figure 9–7. The two conditions for this equilibrium are illustrated in the diagram: (1) demand equals supply at the industry level of output, and (2) price just equals the minimum average total cost of the firm. In other words, each firm earns a normal rate of return on its investment in the industry.

The Adjustment Process Establishing Equilibrium

Establishment of equilibrium is a continual process in the purely competitive industry. To see how the position of long-run equilibrium is reached, consider a condition of long-run equilibrium in the hot pepper sauce industry, which is depicted in Figure 9–7.

Rightward Shift in Demand. Now imagine that hot pepper sauce not only is given a clean bill of health by government researchers but also is praised as a cure for the common cold by a prominent scientist. The industry demand curve for hot pepper sauce suddenly shifts to the right, as shown in Figure 9–8b, raising the market price to P_1 from the previous equilibrium price P_0.

At first, individual firms in the industry (Figure 9–8a) adjust to the higher price P_1 by expanding output from q_0 to q_1 within their fixed plant sizes, that is, along their MC curves. This increase in firms' output is reflected at the

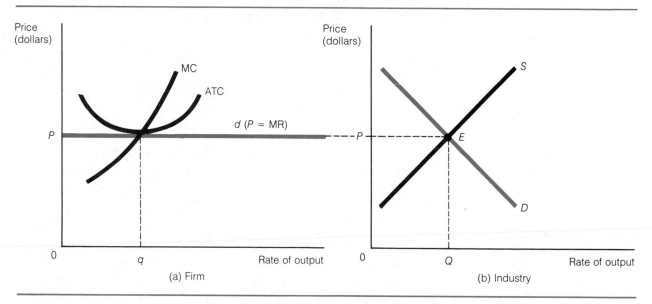

FIGURE 9–7   The Purely Competitive Industry in Long-Run Equilibrium

In a long-run equilibrium state, two conditions are met: (a) the competitive firm makes a zero economic profit, or normal rate of return; (b) industry demand equals industry supply. Price, and therefore revenue, is exactly equal to the minimum total cost, including a normal rate of return for investors.

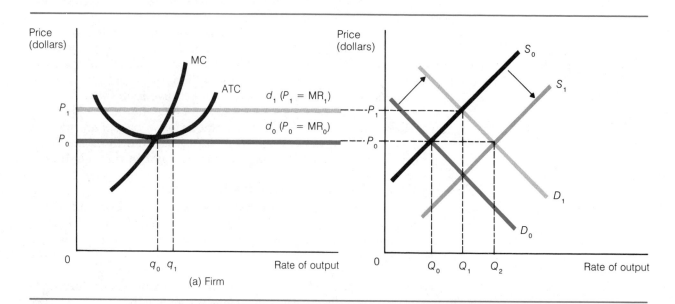

FIGURE 9–8   Industry Adjustment to Long-Run Equilibrium: An Increase in Market Demand

Demand increases from D_0 to D_1 in the industry, and price rises from P_0 to P_1 (b). As demand and price rise, firms now earn profits in excess of their average total costs (a). The opportunity for higher-than-normal profits induces expansion of output by firms in the industry and entry by new firms. As this expansion of output takes place, the industry supply curve shifts to the right until a normal rate of return and the original price P_0 are restored in the industry.

industry level by the movement along S_0 as price rises from P_0 to P_1. Industry output thus initially rises from Q_0 to Q_1. However, P_1 is above the firm's ATC, so firms in the industry are earning economic profits. Other firms will therefore be attracted to the industry, and existing firms will have an incentive to expand their scale of operations to capture more profits. The effect of both adjustments will be to shift the market supply curve to the right, from S_0 to S_1. Eventually long-run equilibrium is restored at the original price P_0, where firms again earn a normal rate of return. In the process, industry output has further expanded, from Q_1 to Q_2. The economy now produces and consumes more hot pepper sauce.

Two points should be noted about this process. First, the long-run adjustment process in this example returned price to the original level of minimum average total cost, that is, to the previously prevailing price P_0. This is a special case of long-run adjustment in an industry in which the prices of resources do not change when industry output is expanded. We discuss this and other long-run adjustment possibilities in more detail later in the chapter.

Second, entry can take place from both without and within the industry. Entry from without is the entry of new firms. Entry from within is the expansion of old firms. Entry from within can take place only if there is a range of long-run firm sizes consistent with minimum-cost production. If the long-run average total cost curve of firms in the industry were U shaped, only one firm size would offer lowest cost in the long run. When this is the case, entry takes place entirely by new firms. The existing firms will not expand their scale of operations because they are already operating with the lowest-cost plant size. An existing firm can expand its scale of operations only if its LRATC curve exhibits a range of constant returns to scale (see Figure 8–6a, p. 172, and its accompanying discussion).

Leftward Shift in Demand. Whereas economic profits lead to the expansion of output in an industry, economic losses lead to a contraction of industry output. Figure 9–9b illustrates a reduction in industry demand from D_0 to D_1. At the resulting lower price of P_1, firms incur economic losses because P_1 is less than firms' average total cost (Figure 9–9a). In the short run, firms' output is reduced, depicted by movement down the MC curve, or firms shut down if P_1 is less than their minimum average variable cost. The short-run drop in production by individual firms causes quantity supplied in the industry to fall to Q_1. In the long run some firms will leave the industry by going out of business and others will reduce their scale of operations. The drop in production (to Q_2) causes the industry supply curve to shift to the left, from S_0 to S_1, putting pressure on the industry price to rise. Ultimately, price will rise enough to restore the original price P_0 and a normal rate of return to firms in the industry.

What are some of the uses of these analyses? First, they depict the desirable effects of competition on the use and allocation of resources. Signals to producers about what consumers want are sent through the market system, and producers respond by adjusting and supplying what consumers want in a way that minimizes the costs of production. Second, the model generates testable propositions, such as the assertion that excess returns should promote entry in a competitive market, from both without and within. An indirect way to test for the presence of a competitive market is to see if excess returns or the presence of economic profits leads to an expansion of output in the industry, or if a decline in output follows economic losses.

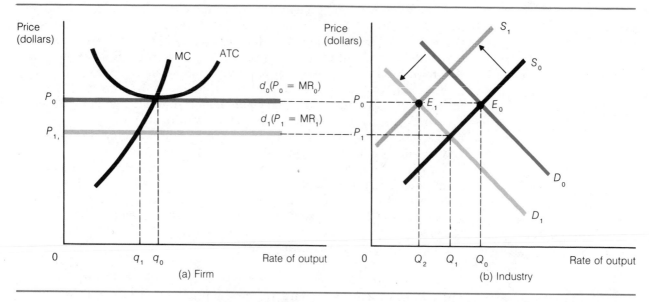

FIGURE 9–9   Industry Adjustment to Long-Run Equilibrium: A Decrease in Market Demand

Industry demand declines from D_0 to D_1, and market price falls from P_0 to P_1 (part b). At P_1 firms incur economic losses and cut back production to q_1. Industry output falls to Q_1. Losses cause some firms to leave the industry and others to reduce their scale of operations. The overall decline in production shifts the industry supply curve to the left, from S_0 to S_1, thus restoring the original price P_0 and a normal rate of return to firms that remain in the industry.

Such an application of this model could be useful to antitrust analysts, for example, whose job is to detect the presence of monopolistic behavior in the economy. Focus, "The Effect of an Excise Tax," describes an application of the competitive model.

The Long-Run Industry Supply Curve

Long-run industry supply: The quantities of a product that all firms are willing and able to offer at all the various prices when the number of firms and scales of operation of each firm are allowed to adjust to the equilibrium level.

The **long-run industry supply** curve represents the minimum price at which firms will supply output to the market over the long run—that is, during a period long enough for entry into and exit out of the industry to occur and for firms to adjust their plant sizes to the lowest-cost level. The long-run supply curve reflects what happens to the prices of inputs as the output of an industry is increased or decreased. In our example of long-run equilibrium, depicted in Figures 9–8 and 9–9, price returned to its original level after first rising or falling with rise or fall in demand. As we noted, this is a special case that does not always apply because the costs of inputs may change as industry output changes. There are three possibilities related to input costs: they may remain constant, increase, or decrease, depending on the nature of the industry.

Constant Cost. In Figures 9–8 and 9–9, long-run adjustments led to the restoration of the original price in the industry (P_0). In the first case the size of the industry expanded, and in the second case the size of the industry contracted. In both cases, price returned to the level of P_0 as entry and exit and adjustments by firms in the industry took place.

FOCUS   The Effect of an Excise Tax on a
Competitive Firm and Industry

Although the model of a purely competitive market is based on abstract assumptions, it is sometimes useful for making predictions about economic events. As we have pointed out, the mark of a good theory is its ability to explain and predict, not the realism of its assumptions. The model of the purely competitive firm and industry can be used to make predictions about the impact of taxation on firms.

Suppose the government decides to place an excise tax of 10 cents per pound on peanut production. Figure 9–10 illustrates the effects of such a tax. The immediate effect of the tax is to raise each peanut farmer's marginal cost curve (MC_0) vertically by the amount of the tax to MC_1. Higher marginal costs lead peanut farmers to cut their output from q_0 to q_1. Output thus falls in the industry to Q_1, reflected in the shift in the industry supply curve from S_0 to S_1. The decrease in industry supply causes the market price of peanuts to rise from P_0 to P_1. As price rises, peanut farmers respond by producing more output, represented along the MC_1 curve. Farmers expand their output to q_2, and industry output correspondingly expands to Q_2. As the market price for peanuts rises, farmers act to restore their output to its original level.

The economic effects of the tax are clear. An excise tax on competitive firms leads to higher costs, lower production, and higher prices.

Notice, however, that the price of peanuts does not rise by the full amount of the tax. The tax was 10 cents per pound, and the price rises by only 5 cents per pound. This means that consumers pay part of the excise tax in the form of higher prices. Who pays the other part? Peanut producers avoid paying part of the tax by reducing their output. This reduction in peanut production leads to a reduction in the demand for factors of production. The other part of the tax is actually paid by the owners of resources, the values of which have fallen because fewer peanuts are now produced. Excise tax payments (pounds of peanuts sold times 10 cents = P_1abP_2) are split between peanut consumers and resource owners in the peanut industry. Excise tax payments in this case are calculated by the amount of peanuts sold at the higher price (Q_2) times the tax (10 cents). In Figure 9–10, this equals $Q_2 \times ab$, represented by P_1abP_2. These payments are split between peanut consumers and resource owners, shown by the shaded areas in the figure.

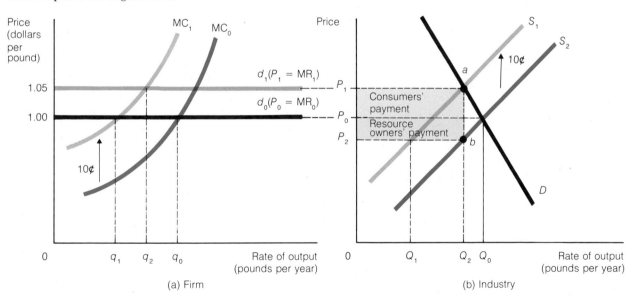

FIGURE 9–10   Effect of an Excise Tax on a Competitive Market

In a competitive market, an excise tax will raise industry price and reduce industry output. The tax is paid partly by consumers in higher prices and partly by resource owners in the industry who experience a decreased demand for their resources. The amount of the peanut tax is P_1abP_2 in part (b) calculated by multiplying the tax rate (10 cents) by the amount of peanuts sold that are subject to the tax (Q_2). Payment of the tax is split between consumers and owners of resources, shown in the shaded areas of part (b).

FIGURE 9–11

Long-Run Price in a Constant-Cost Industry

The long-run industry supply curve (LRS) for a constant-cost industry is a perfectly elastic long-run supply at the level of market price. Expansions in industry output do not cause changes in resource costs to individual firms.

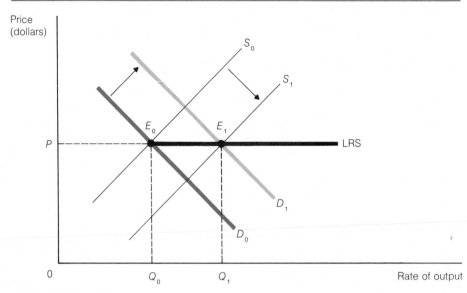

Constant-cost industry: An industry in which the minimum average cost of producing a good or service does not change as the number of firms in the industry changes; an industry for which the supply of resources is perfectly elastic, resulting in a perfectly elastic industry supply.

Figures 9–8 and 9–9 are examples of a **constant-cost industry,** an industry in which expansions or contractions of industry output have no impact on input prices or costs of production. The long-run industry supply (LRS) curve in this case is perfectly elastic—a flat, straight line at the level of the long-run equilibrium price in the industry, as shown in Figure 9–11. Notice that the curve is derived by connecting points E_0 and E_1, the points where long-run industry demand equals supply, that is, in long-run equilibrium.

Constant cost is most likely to occur in industries where the resources used in industry production are a small proportion of the total demand for the resources in the economy. Take, for example, the toothpick industry. A major expansion, say a tripling, of the output of toothpicks would probably not have much impact on the price of the wood used in the industry because far more wood is consumed in other uses, such as building materials, paper, and firewood. The cost of wood to toothpick producers would therefore remain the same as the industry expanded.

Increasing-cost industry: An industry in which the minimum average cost of producing a good or service increases as the number of firms in the industry increases; such an industry has an upward-sloping long-run supply curve.

Increasing Cost. In an **increasing-cost industry,** expansions of industry output lead to higher input prices and therefore higher costs of production for individual firms. This type of industry exhibits the common slope of the long-run industry supply curve because the expansion of most industries puts upward pressure on costs in the industry. An increase in the demand for chicken meat will cause the prices of chicken feed, farmland, and chicken coops to rise. An increase in the demand for newspapers will lead to higher prices for paper, printers, reporters, and so on. These increases in costs lead to a higher long-run price for the goods produced. To obtain more of these goods, consumers must pay higher prices in the long run.

The effects of industry expansion on price and supply in an increasing-cost industry are shown in Figure 9–12. Initially, the firm and industry are in equilibrium at E_0, with a market price of P_0. Demand increases, the market demand curve shifts to the right from D_0 to D_1, and price rises initially to P_1. Entry into the industry and increased production are encouraged by the higher price, and the additional output is represented by a shift in the

FIGURE 9–12
Long-Run Price Changes in an Increasing-Cost Industry

Expansions of industry output in response to higher demand D_1 and resulting higher price P_1 lead to higher resource costs facing individual producers in the industry. Equilibrium is reached at P_2 as costs to producers rise and price falls from P_1 to P_2. The long-run industry supply curve (LRS) for an increasing-cost industry, determined by connecting equilibrium points E_0 and E_1, slopes upward to the right.

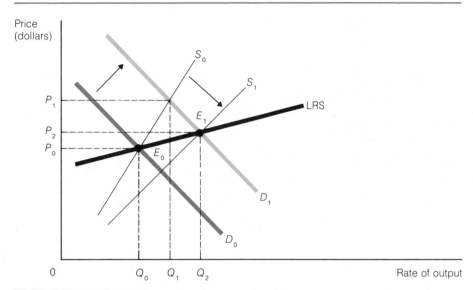

Decreasing-cost industry: An industry in which the minimum average cost of producing a good or service decreases as the number of firms in the industry increases; such an industry has a downward-sloping long-run supply curve.

supply curve from S_0 to S_1. As production expands, however, resource costs to producers in the industry rise, causing the individual firms' cost curves to shift upward. Thus, as entry takes place, price falls from its height at P_1, and costs to producers rise until a new equilibrium is reached at E_1. This new equilibrium price is higher than the initial equilibrium price of P_0. Connecting the two equilibrium points at E_0 and E_1 again yields the long-run industry supply curve. The long-run industry supply curve in an increasing-cost industry slopes upward to the right.

Decreasing Cost. A less typical type of long-run industry supply is demonstrated by a **decreasing-cost industry.** In this industry, expansion of industry output leads to lower input prices and lower costs for individual firms and hence to a lower long-run market price for the product. This case is depicted graphically in Figure 9–13. When industry demand increases and the market demand curve shifts from D_0 to D_1, price first rises from P_0 to P_1. In response, the industry supply curve shifts from S_0 to S_1, and price to consumers falls to P_2. Connecting the two points of industry equilibrium, E_0 and E_1, yields a downward-sloping long-run industry supply curve.

This is an unusual type of long-run industry supply, but a decreasing-cost industry is logically possible. As the clothing industry expands, for example, the costs of certain inputs may fall. Producers of cutting and sewing machines may experience economies of scale, leading to lower prices for their machines. Hence, it is possible for the long-run supply curve of clothing producers to exhibit the decreasing-cost phenomenon. We stress, however, that this is more a logical prospect than a reality.

Pure Competition and Economic Efficiency

The model of a purely competitive market is often used by economists as a benchmark or ideal against which other models of market structure are compared. This point will be more obvious in Chapter 10, where we discuss the market structure of pure monopoly. But we first need to understand in what

FIGURE 9–13

Long-Run Price Changes in a Decreasing-Cost Industry

Though an unusual case, it is logically possible that industry expansion can lead to lower input costs. As demand increases from D_0 to D_1 and the supply curve shifts from S_0 to S_1, prices ultimately settle at P_2, lower than both the initial market price P_0 and the short-run price P_1.

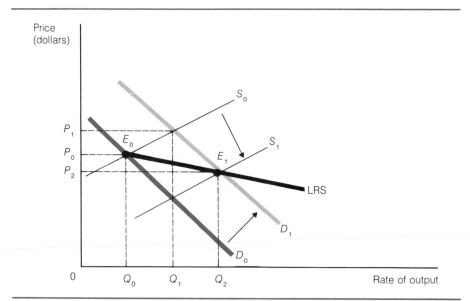

sense the operation of the purely competitive market represents an ideal outcome in the economy and in what sense the model is only an ideal and not a reality.

The Competitive Market and Resource Allocation

Pure competition produces what most economists would agree is an ideal or optimal resource allocation. Purely competitive allocation of resources is optimal or best in two fundamental senses.

First, a competitive industry minimizes the resource costs of producing output. In effect, competition forces firms to produce their output at the minimum long-run average total cost and at a price just sufficient to cover that cost. This **productive efficiency** can be seen in long-run equilibrium in competitive firms, which produce at the point where P = ATC = MC. Consumers benefit because they get the goods they desire at the lowest possible cost. Inefficient, high-cost firms will incur economic losses and will be driven from competitive markets. Surviving will be firms who can provide industry output at the lowest cost. The 1978 lifting of government regulations on the airline industry created a highly competitive situation in which a number of airlines with excess costs suffered severe losses. Some were forced out of business. By contrast, Northwest Airlines thrived because of the leanness of its operations. By avoiding frills such as lavish company headquarters and by promoting values such as employee productivity, Northwest has kept its cost per passenger seat mile 13 percent below the industry average. Such productive efficiency characterizes firms that survive in highly competitive markets.

Second, a competitive market also creates **allocative efficiency,** which simply means that competition causes resources to flow to their most highly valued uses. In a competitive market, consumer demands are met as long as consumers are willing to pay for the production of additional output at a price that is higher than the cost of the additional resources required to produce the output.

This situation is expressed in the now familiar condition P = MC. Price reflects the desire of consumers for additional units of a good. Marginal cost

Productive efficiency: A situation in which the total output of an industry is obtained at the lowest possible cost for resources.

Allocative efficiency: A situation in which the socially optimal amount of a good or service is produced in an industry, given the tastes and preferences of society and the opportunity cost of production.

represents the opportunity cost of the resources necessary to produce an additional unit of the good. If P is greater than MC, consumers will pay more for additional units of production than the cost to produce them. When this is the case, resources in the economy will be reallocated from other uses considered less valuable by consumers. If a chicken farmer, for instance, finds that consumers will now pay more for the relatively scarce rooster hackle feathers prized for fly-fishing than they will pay for relatively abundant chicken meat, the farmer will shift the use of some chicken coops, chicken feed, and farmhands from production of meat birds to production of roosters bred for hackle feathers. Chicken lovers might find fewer chickens for sale.

If P is less than MC, consumers will pay less for additional units of production than the cost to produce them. Resources will be reallocated out of such production. If the price for hackle feathers is less than the cost, then fewer hackle feathers will be available for fly-fishing. If P equals MC, resource allocation is ideal in the sense that the things consumers want are produced by competitive firms in the exact quantities and combination (of feathers and food, for instance) that consumers desire and, as we saw above, at the lowest cost.

The competitive model is really Adam Smith's concept of the "invisible hand" at work. Acting in their self-interest, consumers and producers create a mutually beneficial outcome. Looking out for their own interest, producers seek to maximize profits, and yet in a competitive market the result is that consumers' desires are met in the most efficient way possible. Consumers selfishly demand the goods and services they want, but competitive producers balance those demands against other demands for other goods and services. An incredibly complex process of consumer demand and producer response is put in motion by the behavior of each individual in heeding his or her own interest. No central planner is required, and yet a result emerges that is ideal.

Perfect competition is not perfect in every sense, however. While a purely competitive world leads to an ideal resource allocation, it might not lead to an ideal distribution of income among people. Also, while pure competition accommodates consumers' preferences, these preferences themselves may not conform to anyone's image of the ideal state of human existence. Individuals are free to buy and sell as they see fit. If consumers want gidgets rather than great books, they are free to make this choice.

Reality Versus Pure Competition

In addition to not being ideal in every sense, the model of pure competition is an abstraction rather than a situation that actually occurs in its perfect form. As we turn to the analysis of other models of market structure in the next two chapters, it is useful to point out some of the important ways in which the model of pure competition diverges from real-world markets.

1. In some industries, large firms experience economies of scale and can produce at lower unit costs than small firms. In the long run, therefore, many sellers cannot exist in such industries, and the market structure becomes less than purely competitive.

2. Competition in the real world usually involves more than simply price changes for a homogeneous product. More often than not, competition for consumer purchases involves such factors as location, product design, advertising, and many other aspects of nonprice competition.

3. The model of pure competition emphasizes the establishment of indus-

try equilibrium. But as we all know, the world is generally a place of disequilibrium or change. Technology changes, consumer demands change, owners and investors have new ideas, and so on. The world is rarely at rest. In a world of change, it is the process of adapting to change and moving from one equilibrium state to another that is the most interesting and important way to study competitive behavior.

4. Finally, consumers actually do not desire homogeneous products. Consumer preferences vary widely, encompassing many aspects of products such as their quality, location, color, and design.

To address some of these aspects of more realistic behavior, we turn to the discussion of other models of market behavior and structure. We begin with the model of pure monopoly, which is at the other extreme of the competitive continuum from pure competition. In fact, the model of pure monopoly describes the behavior of industries consisting of a single producer rather than many producers.

Summary

1. Economists study competition in two basic senses: as a rivalrous, natural process among competitors and as an abstract concept described by the model of a purely competitive market.
2. A purely competitive market is characterized by many buyers and many sellers, a homogeneous product, no barriers to entry or exit by firms, and free information. Examples include the stock market and agricultural markets.
3. The purely competitive firm is a price taker: The firm is so small in relation to the total market for its product that its output has no influence on the prevailing market price. The demand curve facing the purely competitive firm is perfectly elastic at the level of the prevailing market price.
4. Marginal revenue is the change in total revenue caused by an increase in output. Since the purely competitive firm faces a perfectly elastic demand curve, each unit is sold at the prevailing market price. Thus, marginal revenue is equal to market price for the pure competitor.
5. In the short run, the pure competitor decides how much output to produce by setting marginal cost equal to market price. If price is above the average total cost, the firm earns an economic profit in the short run. If price is below the average total cost, the firm has an economic loss. If price is below the minimum average variable cost, the firm will shut down in the short run to minimize its losses.
6. The supply curve of a purely competitive firm is its marginal cost curve above the point of minimum average variable cost. The supply curve of a purely competitive industry in the short run is the horizontal sum of individual firms' marginal cost curves above their respective points of minimum average variable cost.
7. Short-run equilibrium for the purely competitive firm and industry occurs when quantity supplied and quantity demanded are equal in the market and each firm is producing at the level where $P = MR = MC$.
8. In the long run, firms in a purely competitive industry can enter and exit the industry and seek the optimal plant size in which to produce their output. The presence of economic profits leads to entry and expansion in the industry, and economic losses lead to the opposite.
9. Long-run equilibrium in a purely competitive industry results when firms in the industry earn zero economic profits and industry demand and supply are equal. In the long run, the purely competitive firm's position is such that price equals marginal cost equals long-run average total cost: $P = MC = LRATC$.
10. The long-run industry supply curve in a purely competitive industry reflects what happens to firm costs as industry output expands or contracts. In a constant-cost industry, resource prices and firm costs are unchanged by industry expansion, and the long-run industry supply curve is perfectly elastic. An increasing-cost industry experiences rising costs as industry output expands; the long-run industry supply curve slopes upward to the right. In a decreasing-cost industry, resource prices fall as industry output expands, and the long-run industry supply curve slopes downward to the right.
11. The working of the purely competitive model represents a benchmark against which the working of the economy can be measured. In this model, resources flow to their most highly valued uses, and output is produced at the lowest cost in terms of resources used. The model of pure competition is useful in some applications but is an abstraction and does not describe most real-world markets.

Key Terms

market

purely competitive
market

homogeneous
product

perfect information

price taker

profits

marginal revenue

total revenue

shutdown

short-run firm supply

short-run industry
supply

long-run competitive
equilibrium

zero economic
profits

long-run industry
supply

constant-cost
industry

increasing-cost
industry

decreasing-cost
industry

productive
efficiency

allocative
efficiency

Questions for Review and Discussion

1. What is a market? What is a purely competitive market?
2. Why are price and marginal revenue the same thing in a purely competitive firm?
3. What is the supply curve of a purely competitive firm? Of a purely competitive industry?
4. Describe the adjustment process between points of long-run equilibrium in an increasing-cost industry.
5. In the peanut excise tax example in the second Focus essay (p. 195), what would have happened if the tax had been applied to one firm in the industry rather than to the whole industry?
6. Suppose that a competitive firm is earning economic profits because its owner figured out a way to lower the costs of production. Since the source of the re-

duced costs is known only to the owner, how can entry take place? Should such information be proprietary (belong to the owner), or should the owner be required to tell potential competitors the reason for the lower costs?

7. Which of the following markets could be analyzed with the competitive model: (a) automobiles; (b) Swiss cheese; (c) blue jeans; (d) cheeseburgers; (e) trash collection; (f) television news; (g) janitorial services?

8. How long will a firm in a competitive industry endure economic losses before it leaves the industry? In your answer assume that price is above minimum AVC but below ATC.

ECONOMICS IN ACTION

Competitive Markets in Disequilibrium

Changes in demand and supply in competitive markets typically lead to a smooth adjustment to a new equilibrium as consumers and producers take appropriate actions in the marketplace. It is possible, nonetheless, that changes in some competitive markets may not follow such a smooth pattern. Consider the case of the *cobweb effect* in agricultural markets.

The cobweb effect arises from the uncertainty of agricultural production and from the fact that farmers do not know when they plant a crop what price the crop will bring after it is harvested. It takes a year or more to grow and harvest most crops, and farmers must decide how much of each crop to plant before they know what next year's price for the crops will be. This process works fine as long as agricultural markets are in equilibrium, but what happens when something knocks a market out of equilibrium?

Figure 9–14 graphically depicts the cobweb effect in terms of the market for Halloween pumpkins. The demand curve, D, represents the relation between this year's price and the quantity demanded in the same year. The supply curve for pumpkins, S, shows the relation between the price of pumpkins this year and the quantity supplied of pumpkins next year. Pumpkin

growers decide how many hills of pumpkins to plant as a function of the level of price this year. The market is presently in equilibrium at a price of P_0. At this price, Q_0 pumpkins are bought and sold.

If the market is in equilibrium, pumpkin growers will bring Q_0 pumpkins to the market each year, and no problem of disequilibrium arises. Suppose, however, that drought destroys a large part of the pumpkin crop one year. In this case, the quantity of pumpkins brought to the market falls to Q_1.

The shortage of pumpkins at Q_1 causes market price to rise to P_1 this year. Farmers respond by planting pumpkins in the higher amount of Q_2 to be brought to the market next year. Next year's *supply* is thus a function of this year's *price*. But next year, when Q_2 pumpkins are brought to the market, they will fetch a price of only P_2 because the higher price of P_1 reflected a shortage that no longer exists. From here, the cycle starts over again. At a price of P_2, farmers will plan to bring Q_3 pumpkins to the market. These pumpkins will sell for P_3, and so on.

The market is in disequilibrium, and it is not returning to equilibrium instantaneously. It is proceeding in cobweb-like fashion, bouncing around the equilibrium

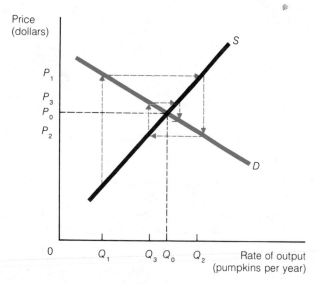

FIGURE 9–14   A Cobweb Process in the Market for Halloween Pumpkins

values of P_0 and Q_0 and approaching these values more closely each year. If there are no further disturbances in the market, equilibrium eventually will be restored. The cobweb process in Figure 9–14 is stabilizing because the old equilibrium values of P_0 and Q_0 will ul-timately be reached if the market is not further disturbed.

The point of the cobweb model is that competitive markets, such as those in agriculture, can exhibit periods of instability in which the adjustment to equilibrium is not smooth. This instability is costly to the economy. In agriculture, farmers may alternately plant too much and then too little of crops.

Of course, we have analyzed the farmer's planting decisions in starkly simple terms. Farmers certainly take more information into account than this year's price for a crop when deciding how much to plant. Regardless of the data they use, however, the fact that they have to plant a crop before they know what price it will bring leads to the potential for cobweb-like behavior in agricultural markets. Indeed, even more complicated and less stable cobweb models—such as the corn-hog cycle in which this year's corn planting is a function of this year's hog prices—can be identified and analyzed.

Question

Is a cobweb process more likely to be at work in the automobile industry or in asparagus production? (Asparagus plants take three years to mature). Is one more likely in turkey production and consumption than in the Christmas tree ornament industry?

10

Monopoly:
The Firm as Industry

*I*n a purely competitive market, each firm's production is such a small proportion of industry output that a single firm has no influence on the market price. As we have seen, the competitive firm is a price taker. The opposite extreme to a purely competitive firm is a pure monopoly: a single seller in an industry. The pure monopoly, in fact, *is* an industry. It alone faces the industry demand curve for its output, and it can affect market price by changing the amount of output that it produces. For this reason, we refer to the pure monopoly firm as a *price searcher:* It must seek the price that maximizes its profits. The ability of the pure monopoly firm to affect market price is not absolute, however. Even a single seller in an industry is subject to competitive pressures from the makers of substitute products.

What Is a Monopoly?

Pure monopoly: An industry in which a single firm produces a product that has no close substitutes and in which entry of new firms cannot take place.

A **pure monopoly** is an industry composed of a single seller of a product with no close substitutes and with high barriers to entry. This definition seems clear, but we must use some care in applying it.

Examples of pure monopoly in the sense of a single seller in an industry are rare. In sixteenth- through eighteenth-century Europe, monarchs granted monopoly rights to individuals for a variety of productive undertakings. These were pure monopolies because each was given the exclusive right to run an entire industry. For example, the king of England created only one English East India Company, which was empowered with a monopoly over the trade with India. This monopoly power was manifested in the high prices the East India Company charged for the goods it brought back from India and sold in England. (See Focus, "Mercantilism and the Sale of Monopoly Rights," for a more detailed discussion of such historical monopolies.)

Mercantilism and the Sale of Monopoly Rights

Mercantilism refers to the form of economic organization that dominated Western Europe from roughly 1500 to 1776. The English and French economies of the time were typical mercantile economies. Both were characterized by monarchies in which a king or queen represented the central government. As practiced by these monarchs, mercantilism involved widespread and detailed regulation of the economy. More often than not, this regulation took the form of creating monopoly rights for favored individuals. The mercantile economies therefore came to be characterized by the existence of pure monopolies in such diverse areas as brewing, mining, trading, playing cards, and so on, endlessly.

The creation and protection of these monopoly rights had a clear purpose—to raise revenue for the "needs" of the sovereign. Such needs included the expenses of the king's court and the resources needed to fight foreign wars. In other words, the monarchies sold monopoly rights to raise revenue.

Why did sovereigns use monopolies rather than taxes for revenue? Fundamentally, tax collection was a relatively inefficient means to raise revenue for the central state because the costs of monitoring and controlling tax evasion were high. Barter and nonmarket production were widespread in the agricultural economy of the times; moreover, commercial record keeping was not highly developed. Tax collection was therefore a difficult and unattractive alternative as a source of revenue.

The granting of monopoly rights as a means to raise revenue did not have the same deficiencies as taxation. Most important, competition among potential monopolists revealed to the state authorities the worth of such privileges. There were no problems of evasion or guessing at taxable values in this case. Those who wanted the monopoly right would come to the king or queen and make an offer for the right. The potential monop-

The coat of arms of the East India Company. The throne of England granted the East India Company monopoly rights over most trade with India and the Far East.

olists were buying the agreement of the ruler to protect and enforce the monopoly privileges that he or she granted.

Thus, pure monopoly had a prominent place in economic history because of its effectiveness at raising revenue for the state. As the state's power to tax has increased in modern times, this revenue-raising role of monopolies has diminished.

Even in these historical cases of monopoly, the monopoly was usually granted to a group of merchants. These monopolies were therefore not technically a single seller; rather, they were a group of sellers acting as a single seller. This is a broader meaning of monopoly, and, in fact, we typically observe monopoly in the real world in groups of sellers acting together. Such associations of sellers are called cartels. The analysis of pure monopoly presented in this chapter is relevant to cartels, but pure monopolies and cartels are not the same. A group of sellers must somehow organize to form a cartel agreement; the single pure monopolist faces no such organizational problem. In this chapter, we discuss pure monopoly in the sense of a single seller facing the industry demand curve. In Chapter 11, we return to the cartel problem and see what extra elements have to be added to monopoly theory to account for cartel behavior.

A second problem in applying the definition of monopoly concerns the criterion that there be no close substitutes. All products exhibit some degree of substitutability. The monopolist's isolation from competition from substitute products is therefore a matter of degree. Consider electricity, usually sold by firms holding local monopolies on its production and distribution. Are there good substitutes for electricity? The answer depends on what the electricity is used for. If it is used for lighting services in residential homes, the few possible alternatives—such as candles and oil lamps—are not very good substitutes for the convenience of electric lighting. If it is used for heating, however, the range of substitute products is greater. The wood stove, oil, and gas industries are competitors with the electric utility company in the market for home heating. The degree of substitutability for the monopolist's product is an important determinant of monopoly behavior. Where there are close substitutes, the monopolist cannot substantially raise prices without losing sales. Where there are no close substitutes, the monopolist has more power to raise prices.

A third consideration in defining monopolies is high barriers to new entry by potential competitors, the basic source of a pure monopoly. There are several types of barriers to new competition: legal barriers, economies of scale, and control of an essential resource.

Legal Barriers to Entry

Sometimes the power of government is used to determine which industry or firm is to produce certain goods and services. Such legal barriers to entry take several forms.

First, it may be necessary to obtain a **public franchise** to operate in an industry. As we noted in the Focus essay, the monarch granted a franchise determining who could sell Asian goods in England in the seventeenth and eighteenth centuries. In modern times, franchises are granted by government for a variety of undertakings. The U.S. Postal Service, for example, has an exclusive franchise to deliver first-class mail. Many universities offer exclusive franchises to firms that provide food service on campus. Similar arrangements for food and gas service are made along toll roads such as the New Jersey Turnpike. The essence of an exclusive public franchise is that a monopoly is created; competitors are legally prohibited from entering franchised markets.

Second, in many industries and occupations a **government license** is required to operate. In most states a license is required to enter occupations such as architecture, dentistry, embalming, law, professional nursing, pharmacy, schoolteaching, medical practice, and veterinary practice. At the federal level, an operating license is required from the Federal Communications Commission to open a radio or television station. If you want to operate a trucking firm that carries goods across state lines, you must obtain a license from the Interstate Commerce Commission. Licensing therefore creates a type of monopoly right by restricting the ability of firms to enter certain industries and occupations.

Third, a **patent** grants an inventor a monopoly over a product or process for seventeen years in the United States. The patent prohibits others from producing the patented product and thereby confers a limited-term monopoly on the inventor. The purpose of a patent is to encourage innovation by allowing inventors to reap the exclusive fruits of their inventions for a period

Legal barriers to entry: A legal franchise, license, or patent granted by government that prohibits other firms or individuals from producing particular products or entering particular occupations or industries.

Public franchise: A right granted to a firm or industry allowing it to provide a good or service and excluding competitors from providing that good or service.

Government license: A right granted by state or federal government to enter certain occupations or industries.

Patent: A monopoly granted by government to an inventor for a product or process, valid for seventeen years (in the United States).

of time. Yet a patent also establishes a legal monopoly right. In effect, the social benefit of innovation is traded off against the possible social costs of monopoly.

Economies of Scale

In some industries, low unit costs may be achieved only through large-scale production. Such economies of scale put potential entrants at a disadvantage. To be able to compete effectively in the industry, a new firm has to enter on a large scale, which can be costly and risky. The effect is to deter entry. Only rarely do new firms attempt to enter the automobile manufacturing industry, for example, because it uses highly automated production techniques on a large scale to keep costs down.

Natural monopoly: A monopoly that occurs because of a particular relation between industry demand and the firm's average total costs that makes it possible for only one firm to survive in the industry.

In a **natural monopoly,** economies of scale are so pronounced that only a single firm can survive in the industry. In such a case competition will not work, and government enacts some sort of regulatory scheme to control the natural monopoly. Public utilities such as natural gas, water, and electricity distribution are examples of natural monopolies that are regulated. The regulation of natural monopoly is discussed in more detail in Chapter 17.

A firm may own all of an essential resource in an industry. In this case new entry is barred because potential entrants cannot gain access to the essential resource. The De Beers Company of South Africa, for example, controls 80 percent of the world's known diamond mines. This makes the company a virtual monopolist in the diamond market.

Monopoly Price and Output in the Short Run

Even though pure monopoly is rare, the theory of monopoly can be applied to a large number of situations in the economy. These range from the behavior of government-created monopolies to natural monopolies, patent rights, and diamond companies. Moreover, as we will see in the next chapter, virtually every firm has some control over the prices it charges. The degree of control varies with the type of market in which the firm sells, but the fact that the firm can choose its price makes the concept of the monopoly firm as price searcher relevant to understanding firm behavior in general. In other words, the analysis of pure monopoly is useful in helping us understand the pricing behavior of all firms, from the corner grocer to IBM.

The monopoly firm, like the competitive firm, faces both long- and short-run production and pricing decisions. In this section we define more carefully what the concept of price searching means, and we examine how the monopoly firm decides how much output to produce and what price to charge in the short run.

The Monopolist's Demand Curve

The demand curve of the pure monopolist is fundamentally different from that of the pure competitor. The purely competitive firm, as we saw in Chapter 9, is a price taker; it accepts market price as given but can sell as much as it wants at the prevailing market price because demand for its product is perfectly elastic. The monopoly firm faces a downward-sloping market demand curve and must find the price that maximizes its profits. This means that to sell more, the monopolist must lower price. The monopoly firm therefore confronts the problem of finding the best price-output combination. For this reason, we say that the monopolist is a **price searcher.** The process of price searching occurs in any firm that faces a downward-sloping

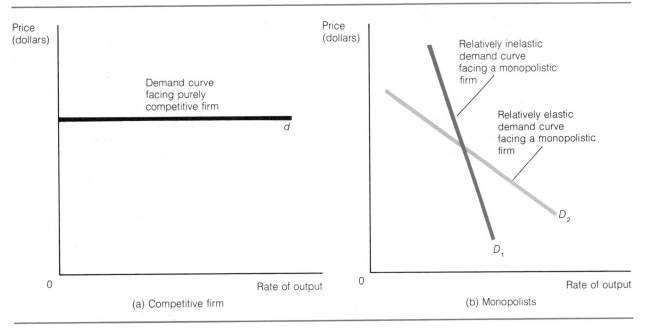

FIGURE 10–1   Demand Curves for Competitive and Monopoly Firms

(a) A perfectly elastic demand curve d facing a purely competitive firm. This firm is a price taker: It can sell all of its product that it wants at the prevailing market price. (b) Two market demand curves facing different monopolists. The firm facing demand curve D_1 has more leeway than the firm facing curve D_2 in choosing price-output combinations because consumers cannot easily find substitutes for the former's product. But both monopolies are price searchers. They face a downward-sloping demand curve: to sell more, they must lower their price.

Price searcher: A firm that must choose a price from a range of prices rather than have a single price imposed on it; such a firm has a downward-sloping demand curve for its product.

demand curve, however slight the downward slope may be. A local dry cleaning firm, for example, has some control over its price and must therefore search for the best price for its service.

Figure 10–1 illustrates the difference in the demand curves facing price takers and price searchers: The former is perfectly elastic, while the latter slopes downward. How elastic or inelastic is the price searcher's demand curve? Economic theory does not give us a single answer to this question. Recall from Chapter 5 that the degree of elasticity along a demand curve depends on a number of variables, including the price and the number of close substitutes for the monopolist's product. We noted earlier that even though monopolists are synonymous with the industries they occupy, they are not immune to competition from producers of substitute products. The cheaper and more numerous these substitutes are, the more elastic the monopolist's demand curve will be. The more elastic a monopolist's demand curve, the less valuable is its monopoly position in setting prices. Figure 10–1b represents the demand curves for two different monopoly industries. The demand curves for these two industries, D_1 and D_2, both slope downward, but D_1 is more inelastic than D_2 because the latter monopoly faces stiffer competition from substitutes.

For this reason a monopoly in tuna fishing, for example, would not be a tremendously valuable monopoly right. While tuna is a unique product, significant price increases for tuna would cause consumers to switch to other protein sources, such as chicken salad rather than tuna salad for sandwiches.

A monopoly over crude oil production, however, where consumers have few viable consumption alternatives in the short run, has proved to be tremendously valuable, as the price-raising success of OPEC testifies.

The basic point to remember about the demand curve facing the monopolist is that it is downward sloping. This means that the monopoly firm must lower price to sell more and must search for the best, or profit-maximizing, level of output. In other words, the monopolist must be able to answer the question, Does the revenue gained from lowering price exceed the revenue lost from lowering price? To answer this question, the monopolist must employ the concept of marginal revenue.

The Monopolist's Revenues

To understand the usefulness of marginal revenue figures in price searching, consider yourself the owner of a monopoly. You have discovered a special kind of rock on your land that, when split, reveals a natural hologram. The mysterious beauty of these rocks is highly appealing to rock collectors and gift purchasers. If your aim is to earn the maximum profit from the production and sale of the rocks, how do you decide what price to charge for them?

By trial and error—for you have no previous experience on which to base your pricing decision—you find that the revenues, costs, and profits you encounter at different prices vary considerably, as illustrated in Table 10–1. Column (1) shows the decreasing prices you must charge to sell increasing numbers of rocks. Together, columns (1) and (2) represent the market de-

TABLE 10–1   Revenues, Costs, and Profits for the Pure Monopolist

These data are for the hologram rock monopolist. Columns (1) and (2) show the components of the demand curve facing the firm. Column (3) gives total revenue, or sales. Column (4) is the important concept of marginal revenue, which is the change in total revenue divided by the change in output. Columns (5) and (6) are the cost data. Maximum profits in column (7) occur where marginal cost equals marginal revenue, at 6 units of output per day.

(1) Rate of Output (units per day)	(2) Price (per unit)	(3) Total Revenue (per day) (1) × (2)	(4) Marginal Revenue $\Delta TR \div \Delta Q$	(5) Total Cost (per day)	(6) Marginal Cost	(7) Profit (per day) (3) − (5)
0	–	–		$ 40.00		− $40.00
			$24.00		$10.00	
1	$24.00	$ 24.00		50.00		− 26.00
			22.00		8.00	
2	23.00	46.00		58.00		− 12.00
			20.00		7.00	
3	22.00	66.00		65.00		1.00
			18.00		10.00	
4	21.00	84.00		75.00		9.00
			16.00		12.00	
5	20.00	100.00		87.00		13.00
			14.00		13.00	
6	19.00	114.00		100.00		14.00
			12.00		15.00	
7	18.00	126.00		115.00		11.00
			10.00		20.00	
8	17.00	136.00		135.00		1.00
			3.50		25.00	
9	15.50	139.50		160.00		− 20.50
			.50		30.00	
10	14.00	140.00		190.00		− 50.00

FIGURE 10–2

The Dual Effects of a Price Reduction on Total Revenues

When the monopoly firm lowers price, it gains revenue from the additional output sold, Q_1 to Q_2, and loses revenue on the output previously sold at price P_1 and now sold at a lower price, P_2. The net effect is a marginal revenue curve that lies below and inside the demand curve.

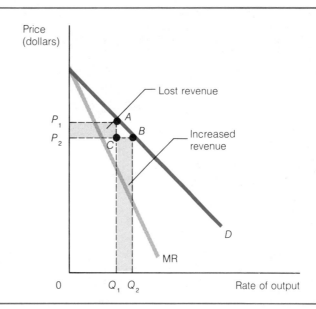

mand curve for hologram rocks, of which you are the sole seller. Your total revenue per day (also known as *sales*) is column (1) × column (2), or price × quantity.

Marginal revenue (MR) is the change in total revenue brought about by a change in output, or $\Delta TR/\Delta Q$. In Table 10–1, column (4) reflects the change in column (3) each time you lower price and produce an additional rock. In Chapter 9, we saw that for the purely competitive firm, MR was equal to price because the firm did not have to lower its price to sell additional units of output (see Table 9–1, p. 185). However, as Table 10–1 shows, MR is not equal to price for you as a monopoly firm, except for the very first unit you sell, because you face a downward-sloping demand curve. When you cut your price, as for example from $22 to $21, two conflicting influences affect your total revenues. You gain additional revenue from the additional units you sell at the lower price. But since the price reduction also applies to output that you were previously able to sell at a higher price, in effect you *lose* revenue on these units. In other words, the three rocks per day you could previously sell at $22 are now priced at $21, a reduction of $3 from your total revenues. Your marginal revenue is therefore $18 in this case.

When demand is downward sloping, MR ($18) is less than price ($21) by the amount of the revenue lost on units that would have sold at the higher price. For price searchers, marginal revenue is less than price (P > MR) because of the need to lower price to sell additional units of output.

Figure 10–2 provides another way of looking at the fact that marginal revenue is always lower than price for the monopoly firm. At a price of P_1, total revenue is $P_1 \times Q_1$. At a price of P_2, total revenue is $P_2 \times Q_2$. Marginal revenue is the difference between these two total revenue rectangles. To see its components more clearly, consider what happens when price is cut from P_1 to P_2. First, additional revenue is generated by the sale of extra units at the lower price. This additional revenue is indicated in the shaded rectangular area CBQ_2Q_1 above increased sales. Second, there is a loss of revenue on units previously sold for P_1 but now sold for P_2. This loss is represented by the shaded area P_1ACP_2. Thus, the marginal revenue de-

FIGURE 10–3

**Changes in Elasticity of
Demand and Total
Revenue as
Price Changes**

The elasticity of the
monopolist's demand curve
is linked to the behavior of
total revenue as price
changes. In the elastic ($\epsilon_d >$
1) portion of the demand
curve, price reductions
cause total revenue to rise.
When demand is unit elastic
($\epsilon_d = 1$), price reductions
leave total revenue
unchanged. When demand
is inelastic ($\epsilon_d < 1$), price
reductions lead to a fall in
total revenue. The point of
unit elasticity in part a,
beyond which marginal
revenue becomes negative—
30 cans per day—is also the
point in part b at which total
revenue stops rising and
starts declining.

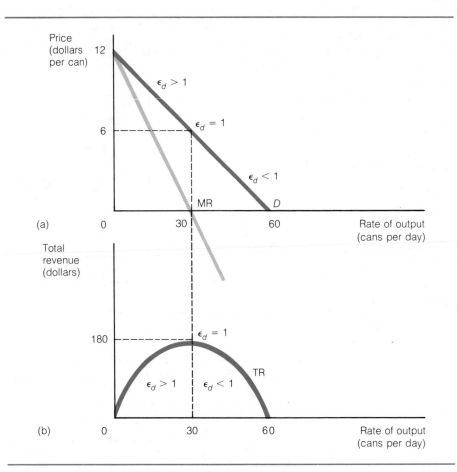

rived from a price reduction by the monopolist is less than the new price
charged. Except for the first unit of output, the MR curve of the monopolist
therefore lies below or inside the demand curve.

Even though the monopoly firm faces no direct competition, it still must
search for its best price. This search for the most profitable price-output
combination explains why the monopoly firm lowers price even though it
has no direct competitors.

Total Revenue, Marginal Revenue,
and Elasticity of Demand

The demand curve facing the monopoly firm shows the amount of output
that can be sold at different prices. It also shows how the firm's revenues vary
as price and output are changed. By observing this, we can learn something
about the elasticity of the monopolist's demand curve over various ranges.
Figure 10–3 illustrates the relation between total revenue, marginal revenue,
and the elasticity of the monopolist's demand curve at various points. Suppose that we are joint owners of a firm with a monopoly on Gas-Saver, a
unique additive that doubles gas mileage if one can of it is added to each
tankful of gas. If we conduct a pricing experiment along our demand curve,
we can observe its effects on our revenues. We start with a high price of $12
per can and lower it gradually to $6. At $12, we will not sell any of the
stuff, for consumers know that a can of it will cost them more than the gas
they save. As we lower the price toward $6, we sell more and more.

Figure 10–3b shows what happens to our total revenue as the price we charge falls. Over the range of sales up to 30 cans per day, our total revenue rises. In Figure 10–3a, this rise is reflected by the positive marginal revenue. As we saw in Chapter 5, a demand curve is elastic when price reductions cause total revenue to rise. Thus, our demand curve is elastic up to the level of 30 units of output and a price of $6. Beyond 30 units of output, however, we observe that as we further lower our price, our total revenue falls, and our marginal revenue is negative. This drop means that we have reached the inelastic portion of our monopoly demand curve. Here, price reductions cause our sales to decline.

The regions of our Gas-Saver demand curve that are elastic and inelastic are labeled as $\epsilon_d > 1$ and $\epsilon_d < 1$, respectively, in Figure 10–3. The point labeled $\epsilon_d = 1$ is where the demand curve is unit elastic. This point occurs where marginal revenue equals zero and where total revenue is at a maximum. It is the dividing line between the elastic and inelastic portions of the monopolist's demand curve. The link between MR and elasticity of demand can be stated thus: MR goes from positive to negative as demand goes from elastic to inelastic.

With this simple model, we can begin to discuss the rational pricing strategy of the monopolist. Suppose that our Gas-Saver monopoly firm has no costs of production. What price should we set to maximize our revenue? Figure 10–3b clearly illustrates that we can do no better than to be at the top of the total revenue curve. At this point, we reach our total maximum revenue: $180 per day. Reading vertically up to part a, you can see that to earn this maximum total revenue, we should produce 30 cans of Gas-Saver per day and sell them for $6 each. The profit-maximizing monopolist with no costs of production will set price along the demand curve where demand is unit elastic, or equal to 1. A monopoly firm will never operate along the inelastic portion of its demand curve because marginal revenue is negative along this portion. Increases in output and reductions in price actually cause total revenue to fall.

This simple analysis helps us understand the pricing behavior of any firm facing a downward-sloping demand curve. Such firms will not set price in the inelastic part of their demand curve because they can earn more revenue by raising price. Soon we will introduce the cost curves of the monopoly firm, which further strengthen this result. For the present we note that this result and the discussion of this section are based on the properties of a straight-line or linear demand curve.[1]

Short-Run Price and Output of the Monopolist

Although for simplicity we did not consider input costs in the previous discussion, we introduce them now, because all firms incur costs of production. The fact that the monopolist possesses a monopoly in the output market says nothing about its position in input markets. In this respect the monopoly firm is like the competitive firm; it is one of many buyers of inputs. The cost curves of the monopolist therefore resemble those of the competitive firm.

By combining the cost and revenue concepts of monopoly, we can analyze the choice of the best, or profit-maximizing, rate of output and price for the monopoly firm in the short run. To maximize profits, the monopolist follows

---

[1]With a linear demand curve, the marginal revenue curve will bisect the horizontal axis and any line parallel to the horizontal axis. In Figure 10–3, for example, MR bisects the horizontal axis at 30 units of output.

the same rule that the competitive firm follows—it sets marginal cost equal to marginal revenue. The logic is identical in both cases. If MR is greater than MC, it pays to produce additional units of output because they add more to revenues than to costs. If MR is less than MC, it pays to reduce production because extra output adds more to costs than to revenues. The monopolist will therefore do best by producing at the output level where MC equals MR.

Look back at the cost and profit data in Table 10–1 for the hologram rock monopoly to check the viability of this rule for profit maximization. Comparing marginal revenue in column (4) with marginal cost in column (6), we see that the first 6 rocks produced per day add more to revenue than to costs; that is, MR > MC over this range of outputs. The sixth unit of output adds $14 to revenues and $13 to costs. It pays to produce that sixth unit. Producing the seventh rock, however, has a marginal cost of $15 and a marginal revenue of $12. The firm would be losing money if it increased operations to 7 rocks a day. The maximum profit occurs at a rate of output of 6 rocks per day and a price of $19 for each. The profitability of this decision is recorded in column (7), which is the difference in total revenue from column (3) and total cost from column (5). The maximum daily profit of $14 occurs when 6 rocks are produced, the point at which MR equals MC.

The same point is illustrated graphically in Figure 10–4, where a monopoly's marginal cost curve is combined with its demand and marginal revenue curves. To determine the profit-maximizing rate of output, the monopolist would find the point along the horizontal axis where MR = MC, in this case Q_m. For outputs less than Q_m, MR > MC, and it pays the monopolist to expand production. In this area, extra units of output add more to revenue than to cost. For outputs greater than Q_m, it pays the firm to reduce output. At Q_m, MR = MC, and the firm can do no better than produce this output. Profits are maximized at Q_m.

The monopolist sets a price for Q_m units of output by reading the market value of this output off the demand curve at a point directly above the point where MR = MC. In other words, the price for this output is read off the

FIGURE 10–4

Profit-Maximizing Price for a Monopoly

The monopoly firm maximizes profits where MR = MC. This means that Q_m is the profit-maximizing output and P_m, read off the demand curve, is the profit-maximizing price for the monopoly firm.

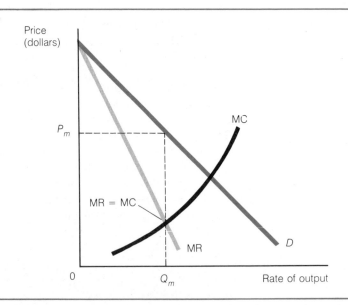

FIGURE 10–5

Monopoly Profits

At point *B*, the monopoly firm would just cover its average total cost. But since it holds a monopoly, it can charge price P_m, creating excess profits equal to $AB \times Q_m$, represented by the area P_mABC. In other words, P_mABC is equal to total revenue, P_mAQ_m0, minus total costs, CBQ_m0.

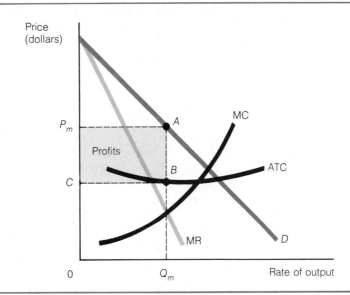

Profit-maximizing price: The price at which the difference between total revenue and total cost is greatest; the price at which marginal cost equals marginal revenue.

demand curve by drawing a straight line from the intersection of MR and MC to the demand curve and over to the price axis, as shown by the dashed lines in the figure. Price is not set where MR = MC on the marginal revenue curve; it is set by reference to the demand curve. In Figure 10–4, P_m is the **profit-maximizing price.**

Figure 10–4 illustrates a principle based on economic behavior, but in real-world firms executives do not calculate MR and MC and set them equal to one another to determine their best price. Business decisions are more complicated, and more often than not businesspeople will be observed using rough and ready approaches to pricing and production decisions. However, through a process of trial and error, price-searching firms will grope toward the MR = MC solution. It may not seem as if they are applying the MR = MC rule, but the practical effect of their decision making ultimately works itself out in these terms.

Note also that although both the purely competitive and the monopolist firm follow the MR = MC rule, the result is fundamentally different in the two cases. For the competitive firm, marginal revenue is the same as price. The competitive firm therefore maximizes profits by producing where P = MC. Marginal revenue is not the same as price for the monopolist; as we have seen, the monopolist produces where P > MC. We will elaborate further on this important difference between competition and monopoly later in the chapter.

Monopoly Profits

Figure 10–4 shows how the monopolist applies the MR = MC rule, but it does not tell us anything about the level of profits the monopoly earns. To discuss this concept we add the monopolist's average total cost to the analysis in Figure 10–5.

Nothing is changed from Figure 10–4. The monopolist continues to set MR = MC, selling Q_m units at a price of P_m. Just as in the case of the purely competitive firm, the monopolist's profits are determined with respect to the average total cost curve. At point C, the monopolist would just cover its average total cost. But its monopoly position and the demand for its product allow it to set price higher, at P_m, on its demand curve, creating profits in excess of its costs. In Figure 10–5, these excess profits are given by the shaded area P_mABC, or the amount by which total revenue, the area P_mAQ_m0, exceeds the total cost of producing Q_m, the area CBQ_m0. Another way to look at monopoly profits in such a graph is to say that the monopolist makes AB profits per unit of output, where $AB \times Q_m = P_mABC$.

This is a very profitable monopoly. These returns are in excess of the total costs of the firm, which means that they are above the opportunity cost of the monopolist's capital investment. Were this a purely competitive firm, these excess returns would stimulate new entry and expansion in the industry. Remember, however, that this is a pure monopoly. Entry cannot occur here, by definition. So, unlike the profits of the competitive firm, which ultimately are restored to a normal rate of return, the profits of the monopolist can persist in the long run.

Not all monopolies make monopoly profits. Some lose money. Such a case is graphed in Figure 10–6. In this case demand is not sufficient to cover the average total cost of the monopolist at the level of output where MR = MC. Total revenue at Q_m is represented by P_mCQ_m0; total cost is represented by the larger area ABQ_m0. Total cost exceeds total revenue by the amount of the shaded area, $ABCP_m$. This monopolist makes an economic loss in the short run. As long as P_m is high enough to cover the monopolist's variable costs, the monopoly will continue to operate in the short run if it expects market conditions for its product to improve. Over time, however, if the losses depicted in Figure 10–6 persist, the monopolist will cease operations and go out of business. The investment of the monopoly firm is not earning a competitive rate of return, and, by closing down, the firm is free to move

FIGURE 10–6

An Unprofitable Monopoly

Not all monopolies make profits. This monopoly makes *BC* losses per unit of output because demand for its product is not sufficient to cover its average total costs. Its total revenue at Q_m is represented by P_mCQ_m0, and its total cost by ABQ_m0. Therefore, its total losses are represented by the shaded area, $ABCP_m$. If the situation does not improve in the long run, this monopoly will go out of business.

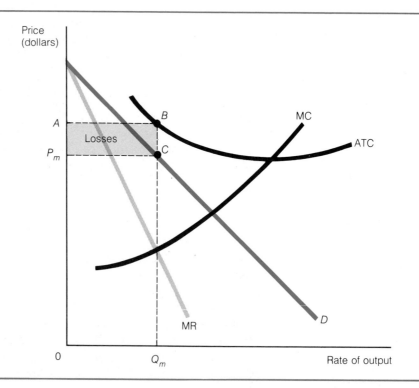

its investment elsewhere in the economy where it can earn higher rates of return.

It may seem curious that monopolists should make losses, but this is a very real phenomenon. Many holders of patents, for example, never market their products because there is no demand for them. The fact that one has a monopoly does not automatically make it a valuable monopoly.

Differences from the Purely Competitive Model

Before we move on to long-run considerations for the monopolist, we note two important differences between the monopoly firm and the purely competitive firm. First, in a pure monopoly no distinction is made between the firm and industry. The pure monopoly firm *is* the industry, so there is no room for a separate theory of industry versus firm behavior, such as the one we presented for pure competition.

Second, recall that the marginal cost curve of a competitive firm (above the point of minimum average variable cost) is its supply curve because the firm sets marginal cost equal to price to maximize profits. Its pricing behavior traces out a relation between quantity supplied and price. No such relation exists for pure monopoly. The monopolist controls the quantity of output produced and sets MR = MC, but, as we have seen, MR does not equal price for the monopolist. Because the monopolist does not set MC equal to price, no unique relation exists between MC and price in the case of pure monopoly. To know what the monopolist will produce at a given price, we need to know more than the firm's MC; we need to know the shape and position of its demand and marginal revenue curves as well. The lesson is simply that the monopolist does not have a supply curve, that is, a unique relation between the amount it produces and the price.

Pure Monopoly in the Long Run

Like all firms, the pure monopolist must face the long run, the period of economic time over which the firm can vary all of its inputs and enter or exit the industry. As we will see, the long run for the pure monopolist is not analogous to the long run for a purely competitive firm and industry. Certain adjustments take place, such as the capitalization of monopoly profits, but entry and other efficiency-enhancing adjustments do not take place. Monopoly and its effects persist in long-run equilibrium.

Entry and Exit

A pure monopoly may not be profitable, as we saw above, and hence may leave the industry. Exit is a distinct possibility. Entry is not. By definition, high and prohibitive barriers to entry exist with monopoly; entry by new firms is barred. Long-run adjustments in the form of competitive entry do not take place in the case of the monopoly.

Adjustments to Scale

The monopoly firm can adjust its scale of operations in the long run. That is, given that it makes a profit, the monopolist can seek to produce the profit-maximizing rate of output in the most efficient plant. Figure 10–7 illustrates the normal plant size result of long-run adjustment for the monopolist.

The monopoly firm produces where MR = MC at Q_m and operates to the left of the point of minimum long-run average total cost. In contrast, the

FIGURE 10–7

Monopoly Plant Size in the Long Run

The monopoly firm can adjust its plant size in the long run along its LRATC curve. Typically, this adjustment results in the selection of an efficient plant size to the left of minimum long-run average total cost, at Q_c, the point at which competitive firms must operate.

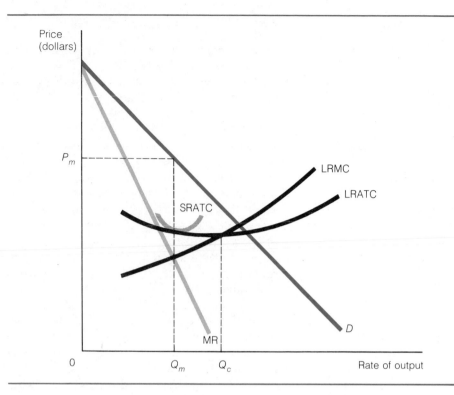

purely competitive firm was forced by competition to produce at the level where long-run average total cost was at a minimum, at Q_c. The monopolist produces Q_m in the most efficient plant that it can, given by short-run average total cost (SRATC) for the lowest-cost plant in which to produce Q_m. But compared with the purely competitive firm, it is operating inefficiently, that is, in a plant size with unit costs greater than minimum long-run average total cost. We discuss this potential efficiency cost of monopoly later in this chapter. This result is by no means unique. It is the standard depiction of the monopolist's selection of a long-run plant size. Other results are possible if the demand and marginal revenue curves are in different positions.

Capitalization of Monopoly Profits

The final long-run adjustment by the monopoly firm concerns what happens to its profits in the long run. We have said that entry is not possible with pure monopoly. But there is no reason that the monopoly cannot be sold to aspiring investors. In this respect, monopoly rights are like any other resource: Where competition is free, resources will flow to their most highly valued uses. Figure 10–8 shows the outcome of this process.

The monopoly is earning profits represented by P_mABC in long-run equilibrium. Suppose that its owners decide to retire and sell the firm. What would they ask for the firm? Rationally, they would include in their asking price the value of the monopoly profits. They would not give these profits away to a buyer. As a result, the average total cost to the buyer would rise to reflect the value of the monopoly profits, from $LRATC_1$ to $LRATC_2$ in Figure 10–8. The price of the monopoly firm is given by the expected present value of P_mAQ_m0, which includes the value of the monopoly profits.

A useful principle has emerged. When monopoly profits are capitalized in this manner, the new owners of the monopoly firm may not earn any mo-

FIGURE 10–8

Capitalization of Monopoly Profits

If the monopoly is sold, the asking price will include the value of the monopoly profits, represented by P_mABC. The selling price is therefore represented by P_mAQ_m0. The average total cost to the buyers is depicted by a shift in the LRATC curve from $LRATC_1$ to $LRATC_2$. The new owners therefore face an LRATC curve of $LRATC_2$; they will not make monopoly profits but will make only a normal rate of return on their investment.

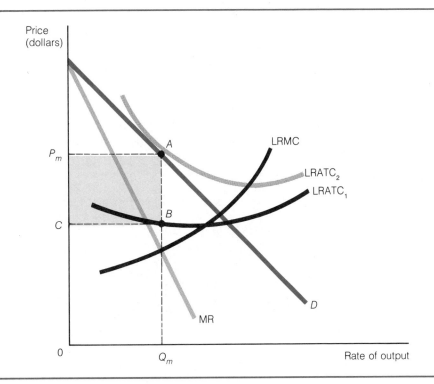

nopoly profits; they will most likely earn a normal rate of return on their investment in the monopoly right. Remember that a normal rate of return is embodied in the long-run average total cost curve. The LRATC curve faced by the new owners of the monopoly is $LRATC_2$, which is just tangent to the demand curve at P_m. At P_m, the new owners just cover their total costs, including a normal rate of return on their investment. The monopoly profits are taken by the original owners, who captured the value of P_mABC when they sold the firm. This does not mean, of course, that the monopoly must cease to exist because it is not profitable. P_m and Q_m continue to prevail in the marketplace, but the current owners make only a normal rate of return on their investment. Why, then, did the new owners buy the firm? Obviously, they think they can run the firm at lower cost than the old owner and make profits through more efficient management of the monopoly.

Price Discrimination

In our discussion of monopoly theory so far, the profit-maximizing monopoly has charged a single price for its product. Each customer who buys the product pays the same price. Under certain conditions, however, a monopoly can make more money by charging different customers different prices for its product. For example, movie theaters and airlines often charge lower prices to children and senior citizens than to others, and utilities charge different rates for electricity to businesses and residences. This method of pricing is called **price discrimination.**

Two points should be kept in mind as we discuss price discrimination. First, price differences that reflect cost differences do *not* constitute price discrimination. For example, large buyers (those purchasing more goods) are often charged less per unit for some goods than are small buyers. The general

Price discrimination: The practice of charging one buyer or group a different price than another group for the same product. The difference in price is not the result of differences in the costs of supplying the two groups.

reason for this difference is that selling costs per unit are lower when dealing with the large buyer. Each sale may require the same paperwork, but the paperwork cost is lower per unit of sales for the large buyer. The lower price charged to the large buyer is not the result of price discrimination; it reflects the seller's different costs of serving the two customers. Second, equality of prices across buyers does not necessarily imply the absence of price discrimination because the costs of supplying the buyers may be different. For these reasons, price discrimination is a tricky concept, and care must be used in asserting its presence in the economy. Some forms of price discrimination are illegal under the Robinson-Patman Act (1936), which is enforced by the Federal Trade Commission.

When Can Price Discrimination Exist?

Price discrimination can occur only under certain conditions. First, the firm must have monopoly power and face a downward-sloping demand curve for its output. In other words, the firm must be a price searcher, not a price taker. A price taker obviously cannot charge different prices to different customers because the price it charges is determined by the prevailing market price, over which it has no influence. Second, the monopolist's buyers must fall into at least two clearly and easily identifiable groups of customers who have different elasticities of demand for the product. Third, the separation of buyers is crucial. Without the ability to separate buyers, the seller will not be able to keep buyers who are charged a low price from reselling the product to buyers who are charged a high price. Such behavior would undermine a price discrimination scheme and lead to a single price for the monopolist's product. The identification and separation of groups of consumers must be possible at low cost to make price discrimination worthwhile for the firm.

A Model of Price Discrimination

Figure 10–9 shows how a monopolist can gain from price discrimination. There are two groups of customers for the monopolist's product. The demand of buyers in market A is relatively more elastic than the demand of buyers in market B.

Two points are crucial in understanding the analysis in Figure 10–9. First, the marginal cost to the monopolist of supplying both markets is the same. The best way to think about this condition is to assume that the output sold to each market is produced in the same plant. MC is thus at the same level in both market A and market B. Second, to maximize profits, the monopolist sets marginal cost equal to marginal revenue.

Following the MC = MR rule in this case, the monopolist sets MC = MR in both markets. In market A this leads to a price of P_1; in market B it leads to a price of P_2. The buyers in market B with the less elastic demand are charged a higher price than the buyers in market A ($P_2 > P_1$). These are the profit-maximizing prices in the two markets. The monopolist benefits from price discrimination to the degree that its profits go up relative to what they would be if it sold its output at a single price to all buyers.

The monopolist must be able to keep the two markets separate to sustain a price discrimination scheme. If it does not, buyers in market A could profitably resell the product to buyers in market B and break down the price discrimination system. In 1984 Apple Computer Inc. offered to sell all students at twenty-four universities its then-new Macintosh computer for less than half of the full price being charged to other customers. Students and profiteers immediately saw the potential for profit in reselling the computers

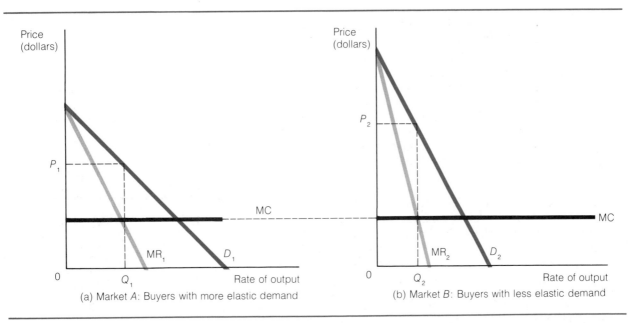

FIGURE 10–9   **Price Discrimination**

The figures demonstrate a case of price discrimination for the same product in two markets. Consumers in market A have a relatively more elastic demand than those in market B. The profit-maximizing monopolist confronts market A consumers with a lower price. That is, $P_1 < P_2$. For a price discrimination scheme to be viable, the monopolist must prevent buyers in market A from reselling to buyers in market B.

to nonstudents at marked-up but still below retail prices. Communication between students with the cut-rate Macintoshes and nonstudents desiring Macintoshes quickly sprang up through newspaper ads and black marketeers.

The seller who wants to practice price discrimination must therefore be able to distinguish and separate customers on the basis of their elasticity of demand. Moreover, the seller must be able to perform this feat without a large expenditure of resources. Thus, most real-world price discrimination schemes are based on general characteristics of customers, such as age, education, income, and sex. For example, children and older people are often charged lower prices for certain products because their age is thought to mean that they have more elastic demand for these products.

Another subtle factor that provides a basis for price discrimination is differences in the opportunity cost of time among customers. The practice of giving discount coupons for grocery purchases, for example, allows sellers to separate buyers into elastic and inelastic demand categories. Buyers in the inelastic demand category will not take the time and trouble to collect coupons and redeem them. These are customers who place a relatively high value on their time. They value other uses of their time and do not want to take the effort to clip coupons. Buyers in the elastic demand group place a lower value on their time and will take the time to clip coupons and convert them into lower prices at the store. Sellers, by using coupons, are therefore able to separate buyers into classes and charge them different prices based on their different elasticities of demand. Focus, "Perfect Price Discrimination," gives another example of price discrimination: the pricing of medical services according to patients' income.

~~~~~ FOCUS   Perfect Price Discrimination

In the example graphed in Figure 10–9, the price-dis-criminating seller separated buyers into two groups and charged each group a different price. Perfect price discrimination occurs when the seller is able to charge each buyer a different price that reflects just what he or she is willing to pay for the product or service.

Perfect price discrimination is difficult to practice because it is obviously hard to separate buyers so that each buyer's willingness to pay can be determined. A possible use is the fee schedule of a plastic surgeon. The surgeon can estimate patients' willingness to pay by such factors as their income or insurance coverage and then charge different prices to each customer for similar services. Resale is out of the question with medical services. Thus, the conditions for perfect price discrimination can be met.

Figure 10–10 illustrates one plastic surgeon's pricing scheme. He charges each patient along the demand curve for plastic surgery just what that patient is willing to pay for the operation. For example, patient 1 is charged P_1, patient 2 is charged P_2, and so on. The cost of performing the surgery, indicated by MC, is assumed to be the same for all patients. All told, the surgeon extracts the maximum possible revenue from patients, represented by the total area $ACQ_0 0$, and a large profit in excess of cost, represented by ACB.

Because it is costly to separate buyers into separate categories and to prevent resale among categories of buyers, perfect price discrimination is rarely found in the real world. But pricing of medical services sometimes fulfills the conditions for perfect price discrimination.

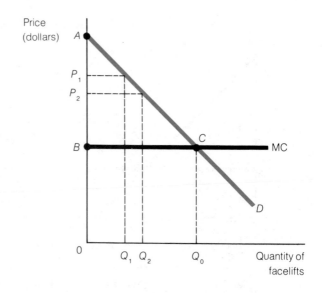

FIGURE 10–10   Perfect Price Discrimination
In this case of perfect price discrimination, the plastic surgeon charges each patient just what he or she is willing to pay for his services. Patient 1 pays P_1, patient 2 pays P_2, and so on. This surgeon's perfect price discrimination can capture the area ACB as a return in excess of costs.

The Case Against Monopoly

Monopoly is generally viewed as "bad" for the economy in comparison to the "good" expected from a competitive organization of markets. The indictment of monopoly by economists is based on arguments that monopolies are lacking in both efficiency and equity or fairness. After considering arguments against monopoly, we will examine the theory that monopolies provide certain social benefits that may offset their social costs.

The Welfare Loss Resulting from Monopoly Power

Contrived scarcity: The action of a monopoly that reduces output and increases price and profits above the competitive level.

The main efficiency argument against monopoly is that this market structure leads to **contrived scarcity,** that is, that a monopoly withholds output from the market to maximize its profits. Remember that the monopolist sets price where MC = MR but where P > MR. At the profit-maximizing output for the monopolist, price therefore exceeds marginal cost. In the case of pure competition, price reflects the marginal benefit that consumers place on additional production, and marginal cost reflects the economic cost of the resources necessary for additional production. When price is greater than marginal cost, this is a signal to producers to increase output. The monopoly firm, of course, will not do this because this approach is not consistent with maximizing its profits. We thus say that the monopolist causes a contrived scarcity, a condition in which the monopoly's product is short in supply and high in price.

This contrived scarcity is a social cost to the economy. Its nature and magnitude can be estimated with an economic model that contrasts the effects of monopoly and competition in a given market. We give such a model in Figure 10–11.

Figure 10–11 shows the industry demand and supply functions, D and S, for some commodity, say cranberry juice. There are no economies of scale in this industry. The average total cost of producing cranberry juice is the same at all levels of output. As defined in Chapter 9, this is a constant-cost industry with a flat long-run supply curve along which MC = ATC.

Suppose that the industry is organized competitively, with many producers of cranberry juice. Long-run industry equilibrium is established at E, where industry demand and supply are equal. The juice output of the various firms is priced according to the marginal cost of producers along the industry supply curve and according to the wishes of the consumers along the demand curve. In other words, price is equal to the long-run marginal cost of production.

Now we ask a special question: At the competitive price-output combination P_c and Q_c, how much surplus do consumers receive? They are willing

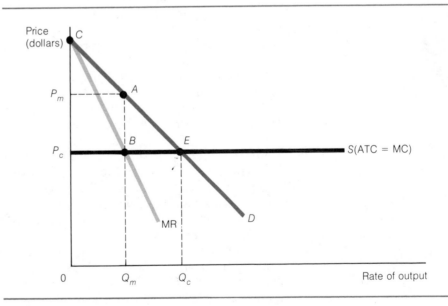

FIGURE 10–11   Welfare Loss Due to Monopoly Power

This graph illustrates the hypothetical case of converting a competitive industry into a monopoly. Consumer surplus at price P_c and output Q_c in a constant-cost industry is represented by CEP_c. When the industry becomes a monopoly, it cuts back output to Q_m, resulting in a decreased consumer surplus of CAP_m and monopoly profits of P_mABP_c. The welfare loss to the economy, that is, the amount of real income that does not reappear for anyone after the monopoly is formed, is represented by AEB. The area of monopoly profits can be considered a transfer from consumers to the monopoly or an additional social cost of the monopoly.

to pay the amounts given by the demand curve, but they pay only P_c. They thus receive a surplus of real income in this case represented by the large area above the industry supply curve and below the industry demand curve, or CEP_c. This triangle represents the concept of consumer surplus discussed in Chapter 6: the amount that consumers would be willing to pay over what they have to pay for a commodity.

Now suppose that this cranberry juice industry is turned over to a single seller, a monopolist. As a result, the monopolist restricts juice output to Q_m (contrived scarcity) and raises price to P_m to maximize its monopoly profits. What has happened to the level of consumer surplus in this market? The consumer surplus before monopoly is given by CEP_c, but consumer surplus after monopoly is only CAP_m. The difference in these two areas is the trapezoid P_mAEP_c.

The area P_mAEP_c is made up of two components. First we consider the rectangle P_mABP_c, depicting monopoly profits, to represent a transfer from consumers to the monopolist. In other words, these dollars do not leave the economy: They are taken out of the pockets of consumers and put into the pocket of the monopolist. (We later discuss both the efficiency and equity aspects of this transfer of wealth.) The triangle AEB is left over. Who gets the real income represented by this triangle? The answer is no one. It simply vanishes when the monopoly is formed. This triangle is therefore called the **welfare loss due to monopoly.** It is a loss because it vanishes when the monopoly is formed. It is a cost to the economy because it does not reappear as income to someone.

Arnold Harberger was the first scholar to attempt to state the total amount of this welfare loss in the manufacturing sector of the U.S. economy.[2] Using some complicated economic formulas, he found that the losses in the U.S. manufacturing sector for 1929 were on the order of 0.1 percent of GNP. In other words, the simple welfare loss to monopoly did not loom large in the economy. Subsequent scholars have used variations of Harberger's technique and have generally found the welfare losses due to monopoly to be low. Thus, even though the welfare loss due to a single monopoly looks large in Figure 10–11, it does not appear to be large as a factual matter in the context of the whole economy.

Welfare loss due to monopoly: The lost consumers' surplus resulting from the restricted output of a monopoly firm.

Rent Seeking: A Second Social Cost of Monopoly Power

The results of Harberger and other scholars suggest that in terms of the welfare loss it imposes, monopoly is not a big problem in our economy. This finding seems paradoxical in an economy such as ours that is dominated by very large and powerful firms. How could the social cost of these giant, powerful concerns be so small?

Gordon Tullock has suggested a solution to this apparent paradox.[3] He considered the area P_mABP_c in Figure 10–11—previously defined as monopoly profits, a simple transfer from consumers to the monopolist—and asked the question, What happens if people compete for these transfers? That is, suppose the monopoly procures its price by hiring lawyers to lobby the government for a monopoly right. The expenditure of resources on lawyers to win the transfer is socially wasteful because these expenditures add nothing

---

[2]Arnold Harberger, "Monopoly and Resource Allocation," *American Economic Review* 44 (May 1954), pp. 77–87.

[3]Gordon Tullock, "The Welfare Costs of Tariffs, Monopolies, and Theft," *Western Economic Journal* 5 (June 1967), pp. 224–32.

to the social product. The opportunity cost of hiring lawyers consists of the productive activities they would otherwise perform. To pay lawyers to take a dollar out of the pockets of consumers and put it into the pocket of the monopolist reduces GNP by the amount of what the lawyers would have produced if engaged in productive pursuits. Not only lawyers but also lobbyists, accountants, executives, and secretaries may be involved in the effort to secure a monopoly transfer. If the competition to capture these transfers is sufficiently strong, resources will be wasted in the process up to the amount of the transfer. This amount, when added to the welfare loss, is the amount represented by P_mAEP_c. When the loss is equal to the trapezoid P_mAEP_c and not the triangle AEB, the social cost of monopolies is no longer small.

This process of competing for transfers is called **rent seeking,** and it is normally associated with the process of seeking monopoly privileges from government. The central proposition of rent-seeking theory is that the expenditure of resources to gain a pure monopoly transfer is a social cost. (We discuss these ideas in more detail in Chapter 15.)

Production Costs of a Monopoly

Recall from our discussion of Figure 10–7 that the typical monopoly firm, after long-run adjustment, does not produce at the point of minimum long-run average total cost. In other words, the monopoly firm does not produce its output in the most efficient plant size in the long run. We learned in Chapter 9 that the competitive firm was compelled by competition to produce at the point of minimum long-run average total cost. By comparison, the monopoly produces its output inefficiently, resulting in another social cost of the monopoly relative to the competitive outcome.

This result assumes that the monopoly and the competitive firm operate along the same long-run average total cost curve. If they do not, a further social cost of monopoly power is possible. This cost was proposed by Harvey Leibenstein, who called it **X-inefficiency.**[4] His reasoning is straightforward. Purely competitive firms must minimize costs of production to survive in the long run. Are monopolists similarly constrained? Leibenstein's answer is no. He argues that the monopolist can afford to live a quiet life; although the monopoly firm gains from reducing costs, there is no competitive pressure to force it to do so. The monopoly firm will not be driven out of business if it fails to minimize costs of production. So the monopolist can take part of its income in terms of a looser, less-efficient organization. The higher production costs caused by this X-inefficiency are also a social cost of monopoly power.

Estimates of the costs of X-inefficiency are hard to derive, so its possible magnitude in the economy is subject to speculation. Moreover, there are forces working to mitigate X-inefficiency costs. As we saw in our discussion of monopoly in the long run, the right to operate a monopoly can be bought and sold like any other asset in the economy. If the monopoly firm is being operated inefficiently, investors can take it over, operate it more efficiently, and make a profit. The competitive pressure of such takeovers may or may not be sufficient to erase X-inefficiency costs, but takeovers surely play a role in keeping such costs to a minimum.

Rent seeking: The activity of individuals who spend resources in the pursuit of monopoly rights granted by government.

X-inefficiency: The increase in costs of a monopoly resulting from the lack of competitive pressure to force costs to the minimum possible level.

[4]Harvey Leibenstein, "Allocative Efficiency vs. X-Inefficiency," *American Economic Review* 56 (June 1966), pp. 392–415.

Monopoly and the Distribution of Income

In addition to the potential efficiency cost arguments against monopoly—welfare loss, rent seeking, and inefficient levels of production—there is an equity argument. In simple terms, the creation of a monopoly transfers wealth from consumers to the monopolist. In Figure 10–11, this transfer is represented by the area of monopoly profits, P_mABP_c. The average consumer gets poorer; the monopolist gets richer. Most people would regard this transfer to the monopolist as unfair, but here we are clearly facing an ethical comparison. What is fair? The truth of the matter is that the economist is no better than anyone else at answering such a question.

We can make some economic points about the transfer of monopoly profits, however. As we saw in the discussion of the capitalization of monopoly profits, it is the original monopolist who receives the transfer. The current owners of a monopoly may be making only a normal rate of return on their investment in the monopoly right. Thus, if the government sets out to break up monopolies in the economy on the grounds that they affect the distribution of income in an undesirable way, government would have to take care to distinguish between current and original owners of monopolies. Also, the concept of rent seeking states that monopoly profits are competed away in wasteful competition to obtain the monopoly right from government. In this sense, the monopoly profits are not transferred to anyone; they are wasted.

The Case for Monopoly

Static inefficiency: A condition, related to the concept of welfare loss due to monopoly power, which is summarized as the production of too little output at too high a price.

We have seen that monopoly leads to economic inefficiency compared with a purely competitive organization of the economy. The monopolist essentially produces too little output at too high a price ($P > MC$). This type of inefficiency, called **static inefficiency,** can be seen as a social cost of the condition that other things are equal.

What if other things are not equal? Specifically, what if the monopoly produces more new ideas and products than the competitive firm? If this is the case, there may be an offsetting **dynamic efficiency** that makes monopoly a desirable form of market organization. One way to look at these counterforces is that the static inefficiency caused by monopoly pushes the economy inside its production possibilities frontier. This loss of production is the cost of monopoly power. But the dynamic efficiency of monopoly pushes the production possibilities curve outward over time, potentially swamping the short-run costs of monopoly power. In other words, monopoly has costs and benefits for the economy, and the benefits (dynamic efficiency) may outweigh the costs (static inefficiency).

Dynamic efficiency: A firm may at first glance impose welfare costs on the economy due to its monopoly power, but may on a closer look be a progressive, innovative firm. In other words, there may be a trade-off in analyzing real firms between static inefficiency and dynamic efficiency, between monopoly power and innovation, and so on.

This view of the potential efficiency of large-scale, monopolistic enterprise was put forth by Joseph Schumpeter.[5] He stressed the advantages of the ability of large firms to finance large research laboratories and hire thousands of scientists. While these firms with well-known, productive research labs—such as Du Pont, AT&T, and IBM—are not typically pure monopolies, they are usually large firms with dominant positions within their industries.

To Schumpeter, the traditional theory that competition spurred innovation therefore seemed wrong. Small competitive firms were at a disadvantage in the innovation process because they could not afford large-scale research. Large firms were the key to innovation and success in the industrial order.

In Schumpeter's theory, the innovative monopolist did not possess a pe-

[5]Joseph A. Schumpeter, *Capitalism, Socialism, and Democracy* (New York: Harper, 1942).

rennial advantage in the marketplace. He believed that innovation would proceed apace over time, and no one large firm would have more than a transitory monopoly in the face of a constant supply of new ideas and innovations by other large firms. The monopoly that any one large firm achieved by being creative was short-lived.

Other famous economists continue to advance Schumpeter's vision of the industrial economy. John Kenneth Galbraith, for example, has argued forcibly in his many works on the economy that large firms are the source of innovation and economic advance.[6] Is the Schumpeter-Galbraith vision of the economy correct? That is, what are the sources of major innovations in the modern economy? Large firms? Small firms? Independent entrepreneurs and inventors?

Numerous studies have been conducted on this issue. Studies of the source of major inventions indicate that a surprisingly large number of innovations have come from the backyard shops of independent inventors. John Jewkes, David Sawers, and Richard Stillerman found, for example, that more than half of sixty-nine important twentieth-century inventions were produced by individual academics or by individuals unaffiliated with any research organization.[7] Among these inventions were air conditioning, the Polaroid Land camera, the helicopter, and xerography. Much invention is individually inspired and created in small-scale, personal labs and workshops.

Other studies have examined the relation between patents as a measure of innovation and firm size and have concluded that in most industries the middle-sized firms have an advantage over both the very large and the very small firms in the innovation process.[8] Although the reasons for this finding are not entirely clear, more patents are issued to medium-sized firms in most industries. Industrial creativity thus seems to stem from something other than size.

Thus, the mass of the evidence suggests that the Schumpeter-Galbraith vision of the economy is not exactly accurate. But, then, neither is the vision that pure competition is best for the economy. It seems to be the case that innovation springs from many sources, from the backyard or basement lab of the inventor to the lab of the large industrial company. Diversity produces innovation, or, better yet, *individuals* produce innovation, whether they are working in their own workshop or a large firm's lab. Perhaps the most typical scenario is that an individual or small firm comes up with a new idea, and a large firm provides the vehicle by which the idea is put into practice.

Finally, where is the balance struck between the costs and benefits of monopoly? It is hard to generalize on this issue. In filling out a report card on firms, the best approach is to be pragmatic. Does the firm possess (monopoly) power? If so, can we judge how large this power is and how much it costs the economy? Does this (monopoly) power carry with it some redeeming merit such as increased innovation? These are not easy questions to answer or to weigh in the balance, but they are the type of issues that must be confronted if the costs and benefits of monopoly are to be understood and its model applied intelligently.

[6]See, for example, John K. Galbraith, *American Capitalism* (Boston: Houghton Mifflin, 1956) and *The New Industrial State* (Boston: Houghton Mifflin, 1967).

[7]John Jewkes, David Sawers, and Richard Stillerman, *The Sources of Invention* (New York: St. Martin's, 1959).

[8]See Fredric M. Scherer, *Industrial Market Structure and Economic Performance* (Chicago: Rand McNally, 1970), Ch. 18, for a survey of such studies.

Summary

1. A pure monopoly is a single seller of a product for which there are no close substitutes in a market characterized by high barriers to new entry.
2. The monopoly firm faces the industry demand curve. When it lowers price, it must accept a lower price for all units previously sold for a higher price. The new revenue at the lower price minus the old revenue at the higher price is equal to the marginal revenue of the monopoly firm. For the monopoly, marginal revenue is less than price.
3. The monopoly selects an output to produce by setting marginal cost equal to marginal revenue. Since marginal revenue is less than price, the monopoly firm sets prices where price is greater than marginal cost.
4. Not all monopolies make profits. In the short run it is possible for the monopoly to incur a loss. If the loss persists in the long run, the monopoly will go out of business.
5. In the long run the monopoly firm can adjust its plant size. Normally, however, the monopoly does not operate at the point of minimum long-run average total cost. Monopolies can be bought and sold in the long run.
6. Price discrimination requires that a seller be a monopoly firm, be able to separate buyers on the basis of their elasticities of demand, and be able to pre-vent buyers in the low-price market from reselling to buyers in the high-price market.
7. Where buyers can be separated into two classes, the monopoly firm will set marginal cost, which is the same in both markets, equal to marginal revenue in each market. This is the definition of price discrimination—price differences that do not reflect cost differences.
8. There are three potential efficiency costs of monopoly: the welfare cost due to monopoly power, which is the lost consumer surplus caused by monopoly; the rent-seeking cost of monopoly, which is the value of the resources used to capture monopoly profits; and the degree to which the monopoly firm has higher costs of production than a competitive firm.
9. Monopoly affects the distribution of income. Where there is no rent seeking, monopoly transfers wealth from consumers to the monopolist.
10. A potential advantage of monopoly is that it allows the large scale necessary to foster innovation. The monopoly or large firm might thus possess a dynamic advantage over a small, competitive firm. The evidence on this issue is mixed, but it appears that the most progressive firms in most industries are middle sized.

Key Terms

pure monopoly	patent	price discrimination	X-inefficiency
legal barriers to entry	natural monopoly	contrived scarcity	static inefficiency
public franchise	price searcher	welfare loss due to monopoly	dynamic efficiency
government license	profit-maximizing price	rent seeking	

Questions for Review and Discussion

1. What is a barrier to entry? Why is it important in the definition of monopoly?
2. Depict the short-run equilibrium of a pure monopolist. Can the monopoly firm make a loss? If so, diagram what the loss situation for the monopoly might look like.
3. What are the conditions that need to be met before a monopoly can price discriminate?
4. What are three efficiency costs of monopoly? What is a potential advantage of a monopoly firm?
5. Insurance companies have started to offer lower rates for life insurance to individuals who do not smoke and who jog. Is this an example of price discrimination? Why or why not?
6. Why is lobbying for an income transfer such as monopoly profits a social cost?
7. Entry is not possible with pure monopoly, but the monopoly can sell its monopoly right. What is the difference?
8. Monopolies and large firms are sometimes seen as the culprits in causing prices in the economy to rise. Does the theory of monopoly presented in this chapter imply that prices under monopoly will rise quickly or will be high? Explain.

State Liquor Monopolies

The wholesale and retail distribution of liquor in the United States takes place in two basic ways. First, some state governments monopolize the sale of alcohol; these states are called control states. In 1980, there were eighteen control states, including Alabama, New Hampshire, North Carolina, and Virginia. In these states, consumers must purchase liquor in the state store. Second, the remaining thirty-two states and the District of Columbia are license states; individual firms can purchase licenses to conduct all phases of the wholesale and retail sale of alcohol. While entry is not completely free and the price of licenses varies across these states, entry is easier than in the monopoly states, where it is impossible. We therefore have a comparative case: states where liquor is sold by state monopolies versus states where liquor is sold by licensed competitors. The theory of monopoly suggests that price should be higher and output lower in the monopoly states. Examine the following 1980 data:

Type of State Control	Nine-Brand Average Price[a]
License	$7.57
Monopoly	7.36

[a]These prices are inclusive of the federal excise tax on distilled spirits, which applies in all states, and the varying excise levies imposed by state and local governments.

The state monopolies, on average, sell liquor at a lower in-store price than the states with competitive suppliers! It would seem that the theory of monopoly has fallen flat on its face.

The key to understanding this case is understanding the nature of "price." We said above that the in-store price was lower in monopoly states. The in-store price

Source: Distilled Spirits Council of the United States, Inc., *Annual Statistical Review 1980: Distilled Spirits* (Washington, D.C., 1981).

is not the real price of a purchase of liquor. The full price of liquor includes such factors as the time it takes to find a store, the hours that the store is open, and the service provided by the store (for example, does it take credit cards and checks or provide help in carrying purchases to the car?).

Now consider the following data:

Type of State Control	Number of Persons per Outlet
License	2,961
Monopoly	29,139

One reason for the great difference in the numbers of potential customers served by each store is that there are simply many more suppliers in the license states, by about 10 to 1. The real price of liquor is therefore lower in these states, for stores are easier to find and are closer on average to the people who use them. In addition, license stores generally are open longer, offer better service, carry a larger selection of brands, accept checks and credit cards, and so on. For such reasons the real price of liquor is lower in competitive than in monopoly states.

Thus, monopoly theory does help us understand real-world monopolies. It predicts that price should be higher and output lower when a market is monopolized, and this is what we find in the case of the state liquor monopolies after taking account of the real price of liquor.

Question

In various cities, states, and nations of the world, such services as airline transportation, water distribution, and garbage collection are provided by government and in private markets. Drawing on the above example of state versus private liquor sales, what might you observe about real price and quality of service with respect to airline transportation, water supply and garbage collection?

11

Monopolistic Competition, Oligopoly, and Cartels

U ntil the 1930s, microeconomic theory analyzed only two basic market structures: pure competition and pure monopoly. These structures are, of course, opposite extremes. In a purely competitive market, prices tend to equal costs in the long run and consumers' economic welfare is theoretically maximized, whereas in a monopoly market the single monopolist may restrict output, raise prices, and capture economic profits. In the 1930s, two economists, E. H. Chamberlin of Harvard University and Joan Robinson of Cambridge University, wrote books developing models of market structures that did not fit the mold of pure competition or pure monopoly. Both Robinson and Chamberlin emphasized that the real world was characterized by market structures that did not easily fit into existing economic theory.

All models of firms and markets other than the extremes of pure competition and pure monopoly are called theories of **imperfect competition.** The word *imperfect* does not imply a value judgment. It simply means that not all the conditions for pure competition are met.

Some imperfectly competitive markets contain large numbers of sellers (as in pure competition) selling slightly different products (unlike in pure competition). These are called monopolistically competitive markets.

Other structures are composed of small numbers of competing sellers who recognize that their actions have an impact on one another's sales, prices, and profits. Under these circumstances, firms may engage or try to engage in various forms of combination or collusion. Such associations are called oligopoly or cartel market structures. They are closer in economic results to pure monopoly than to pure competition. The spectrum of market structures, therefore, runs from pure competition to monopolistic competition to oligopoly and cartels to pure monopoly.

This chapter focuses on monopolistic competition, oligopolies, and cartels. In assessing market structures, we pay particular attention to the contrasts between each of these models and the purely competitive model. Governmental attempts to steer certain industries toward market structures considered more socially beneficial are covered in depth in Chapter 17.

Why is it important to categorize and classify intermediate or imperfect structures, those between competition and monopoly? No real-world market is characterized by the exact characteristics of pure competition or pure monopoly. We may think of the market for ice cream as competitive with many sellers, but in reality every brand or flavor is slightly different. Or, at the other end of the spectrum, we may view our local water company as a monopoly, but there are a few water substitutes available, such as digging a well.

Because real-world markets seldom fit into the neat categories of competition or monopoly, it is very important to develop more realistic models. The economist, moreover, must assess the effects of these real-world market structures on consumers' and producers' welfare and on economic efficiency in general. We begin with monopolistic competition, the economic model closest to pure competition.

Monopolistic Competition

In his search for more realistic models of firms and markets, E. H. Chamberlin developed a model close but not equivalent to pure competition. Since the model contained some monopolistic elements, Chamberlin labeled this theory "monopolistic competition."

Characteristics of Monopolistic Competition

Monopolistic competition: A market model with freedom of entry and number of firms that produce similar but slightly differentiated products.

Monopolistic competition is a market structure in which a large number of sellers sell similar but slightly different products, free from entry and exit barriers. Advertising is the principal tool for differentiating products within monopolistic competition.

Large Number of Sellers. Monopolistic competition, like pure competition, is characterized by a large number of competing sellers. The effect of large numbers is the same in both cases: Collusion to fix prices or other kinds of cooperation is costly—so costly, in fact, that it ordinarily does not occur. The presence of many sellers is a competitive element in monopolistic competition. Unlike pure competitors, however, monopolistic competitors are not always price takers because of another characteristic—product differentiation.

Product Differentiation. Competitive sellers sell a homogeneous or identical product, such as corn. Monopolistic competitors sell products that are highly similar but not identical. Products may be distinguished by brand names, location, services, even differences merely perceived by the consumer. Because the monopolistic competitor is selling a slightly **differentiated product,** the firm will have a degree of control over price, unlike the competitive firm. Product differentiation is an example of **nonprice competition,** a term commonly used to refer to any action other than price cuts taken by a competitor to increase demand for its product.

Differentiated products: A group of products that are close substitutes, but each one has a feature that makes it unique and distinct from the others.

Nonprice competition: Any means that individual firms use to attract customers other than price cuts.

There are many varieties of nonprice competition. One is differences created in the products themselves. Consumers shopping for home computers can now choose among many firms' products, each differing slightly in hardware, software, styling, sturdiness, and user-friendly or state-of-the-art appeal. Sometimes distinctions between brands are more illusory than real. Although aspirin is essentially a homogeneous product, monopolistically competitive firms produce aspirin under many different brand names—Bayer, Excedrin, Cope, and so on—often at very different prices. The reason for the different prices for aspirin is that each brand is differentiated in consumers' minds, often by advertising or some other form of sales effort. Consumers who are convinced that brands of aspirin are actually different will have some allegiance to a particular brand. It is this "brand allegiance" that permits the seller to set price, within limits.

Location may also differentiate a seller's products. There may be ten Exxon gas stations within a city, alike in every respect, including the gasoline they sell. The gas stations each occupy a different location, however, and thus are free to set different prices. The Exxon station nearby the interstate might charge a higher price than the station downtown because travelers in a hurry are willing to pay more. In fact, many businesses have failed because of poor location. You may have noticed a fast-food restaurant go out of business while others selling identical products thrive in locations where parking and accessibility are better.

Service is another form of nonprice competition. Physically identical products may be offered for sale in supermarkets at identical nominal prices. However, one store may be untidy, another may employ rude checkers, and still another suffer slow-moving checkout lines. Markets with well-swept aisles and shelves, automated checkout, and friendly clerks will have differentiated their products by the services they offer.

Subtle product differentiation, whether by packaging, service location, or any other variable, is a common fact of life, but the degree of substitutability among such products is high. While demand is highly elastic under monopolistic competition, it is not infinitely elastic as in perfect competition. A sudden tenfold increase in the price of one brand of aspirin would send buyers scurrying for substitutes. The monopolistically competitive firm, in other words, like the firm in pure competition, is primarily a price taker. It has a small and limited degree of control over price, however, stemming from its ability to differentiate its individual product from those of its many competitors. This limited control over price is the monopoly element in these markets. Chamberlin's ideas are verified by common observation: *There exists practically no market that is not characterized by monopoly elements through some form of differentiation.*

No Barriers to Entry. Just as with competitive markets, entry into and exit from the monopolistically competitive market is free. This fact follows from the high cost of collusion in a market occupied by a large number of sellers. Although monopolistic competition shares the free-entry conditions of perfectly competitive markets, entry itself is not costless in either market structure. There are obvious resource costs and commitments in opening any business: hiring labor, investing capital in fixed items such as buildings, managing or hiring managers, and so forth. Free entry means that there is no government regulation over entry and that essential raw materials are not controlled by one or a few firms.

In addition to the start-up costs associated with entering any new business, entry into monopolistically competitive markets entails the cost of introducing a new product. New product development and market entry through advertising and other sales efforts constitute a cost that the competitive firm does not face. A farmer's entry into the white winter wheat market, for example, will not involve a sales or advertising campaign, whereas entry into the deodorant, paper towel, or restaurant market will. Product differentiation is simply part of the cost of entry under monopolistic competition. Entry is nonetheless assumed to be free, and in this characteristic, monopolistic competition is very much like perfect competition.

The Importance of Advertising. With the general characteristics of the monopolistically competitive structure in mind, we examine a typical firm's demand curve and assess the impact of **advertising** on it. The existence of nonprice competition coupled with a high degree of substitutability means that the firm's demand curve, while highly elastic, is downward sloping. Such a demand curve is drawn in Figure 11–1 for a hypothetical antacid product, Relief. The demand curve for Relief approximates that of a competitive firm, but it is downward sloping because Relief is not the same product as its competitors, Tums, Rolaids, Mylanta, Maalox, and so on.

As with the pure monopoly firm, the monopolistically competitive firm's marginal revenue curve lies below its demand curve, and for the same reason: Additional units can be sold only at a lower price. The main difference between a monopoly firm and a monopolistic competitor is that the monopoly firm sells a unique product—one with no close substitutes—while substitutes abound for the monopolistic competitor's product.

How elastic or inelastic will the demand curve be for Relief? The answer will depend on the size and intensity of consumer preferences for Relief, given the range and price of substitutes. Naturally, the firm will be interested in increasing the size of its market and in intensifying consumers' preferences for its product. The firm accomplishes these goals through product advertising. Advertising is essential to the introduction of products in monopolistic competition. It is also a crucial tool in the manipulation or management of demand.

Advertising is a variable under the monopolistically competitive firm's control, within the limits imposed by its advertising budget. The firm in monopolistic competition will seek, along with all firms who advertise, to equate the addition to revenue from the final dollar spent on advertising to

Advertising: Any communication that firms offer customers in an effort to increase demand for their product.

FIGURE 11–1

Demand Curve Facing Monopolistically Competitive Firm

The demand curve for the monopolistic competitor slopes downward because the firm sells a product differentiated from its competitors. The marginal revenue curve lies below the demand curve because additional units can be sold only at a lower price.

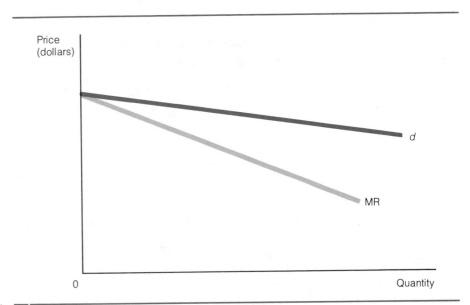

FIGURE 11–2

The Effects of Advertising on the Firm's Demand Curve

The firm advertising its product hopes to shift its demand curve to the right and to reduce its elasticity. In doing so, the firm tries to intensify demand for the product by distinguishing it in the minds of buyers.

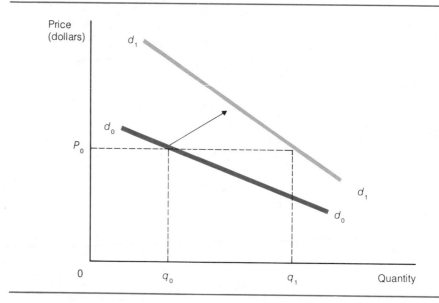

the marginal addition to cost. Naturally, firms seeking to enter markets will often spend large amounts of money to introduce their products. But whether the firm is an established competitor or a new entrant, the purpose of advertising is the same: to increase demand for the firm's product.

In Figure 11–2, demand curve $d_0 d_0$ depicts a monopolistically competitive firm's demand curve. The hoped-for effect of increased advertising is a shift rightward in the position of the demand schedule to $d_1 d_1$ *and* a reduction in the elasticity of the curve. Thus advertising is used not only to increase demand but to intensify demand by differentiating the good or service in consumers' minds so that the firm will be able to raise the price of the product. If the advertising of Relief is able to convince consumers that the product is "easier to swallow," "has no unwanted side effects," and the like, demand becomes more inelastic (less substitutable), and consumers are willing to pay more for it. Focus, "The Pros and Cons of Advertising," discusses advertising in more detail.

Short-Run and Long-Run Equilibrium Under Monopolistic Competition

Now that we have defined the overall conditions for monopolistic competition, we can discuss how firms set prices. The firm in monopolistic competition has some control over price but, as we will see, its control differs in the short run and the long run.

Short-Run Equilibrium. Monopolistically competitive firms, like firms in all other market structures, always equate marginal cost to marginal revenue to maximize profits or minimize losses. In the short run, the firm may enjoy economic profits or endure losses.

Figure 11–3a gives a firm's short-run cost curves, demand curve, and marginal revenue curve. MC intersects, or is equal to, MR at quantity q_0. The firm produces that quantity and sets price P_0 for q_0 units of output. At this price and quantity, the firm will earn economic profits. These profits are

FOCUS   The Pros and Cons of Advertising

Advertising and product differentiation are the hallmark of monopolistic competition. In contrast, competitive advertising is largely unnecessary under either pure competition (since products are homogeneous) or pure monopoly (because there is a single seller with no close substitutes).

No subject in the economics of market structure has been debated as vigorously as advertising. Some economists consider advertising wasteful and "self-canceling." Others judge it to be integral to the competitive process. We briefly consider three of the major arguments.

Is Advertising Informative?

Critics of advertising argue that most ads are tasteless and wasteful assaults on consumers' senses. Since products and services in monopolistically competitive markets are, by definition, close substitutes, product characteristics are difficult to differentiate with *actual* differences. Rather, critics point out that advertising creates only imagined differences. A large number of competitive firms have the incentive to advertise in this manner, creating a confusing array of messages to consumers. In this view, advertising allocates demand among competitors without increasing the total demand for the product. To increase the total demand for a product, such as fast-food hamburgers, cat food, or breakfast cereal, advertising must reduce either consumers' savings or their expenditures on other goods. If advertising is unsuccessful at this task, the critics allege, it is unproductive because it merely allocates demand among competing firms producing goods that are fundamentally alike.

Defenders of advertising argue otherwise. In their view, nonfraudulent advertising offers real information about the existence of products and their characteristics. Such information thus lowers the consumers' cost of searching for goods. Further, actual brand comparisons made possible by a Federal Trade Commission ruling and price advertisement are by nature informative. Defenders of advertising argue that knowledge of any characteristic of a product or service produces information that permits consumers to make a rational choice among competing goods within their budgets. Unsatisfactory products are simply not repurchased. Shifts in consumers' brand allegiance resulting from advertising or any other cause are symptomatic of a working competitive system. According to this viewpoint, critics of advertising's alleged tastelessness are merely trying to substitute their own subjective judgments for consumer sovereignty.

Does Advertising Lower Costs?

Critics argue that advertising expenditures, being largely duplicative and wasteful, increase the price of goods without producing corresponding consumer benefits. Defenders point out that, owing to economies of scale, increased production lowers per unit costs as output increases. Advertising that increases demand might actually lower the average costs of production, resulting in lower prices for consumers.

The argument is clearly empirical. If advertising can reduce unit costs with economies of scale, it could also increase unit costs if the firm is pushed to outputs characterized by increasing average costs. Unfortunately, the economies of scale argument is impossible to assess without some knowledge of the firm's cost curves.

In another variant of the cost argument, defenders of advertising argue that in monopolistically competitive markets advertising is a means of entry. It is part and parcel of the competitive process. Without advertising, existing firms would have greater market power—greater ability to charge higher prices and to exact economic profits. Defenders of advertising point to markets where advertising, especially price advertising, is restricted. For example, physicians, funeral directors, and dentists are all restricted in the amount and type of advertising legally permitted. When such nonprice competition is disallowed, defenders of unrestricted advertising argue, prices to consumers tend to be higher and available outputs tend to be lower than if advertising were free and unregulated.

Does Advertising Create Social Waste?

A third, much-debated advertising issue, originated by economist John Kenneth Galbraith, relates to social waste. Galbraith argues that there is an imbalance between private goods and social goods—that too many automobiles, laundry detergents, and fashions (advertised private goods) are produced relative to dams, highways, and pollution-control facilities (nonadvertised social goods). Many private goods, moreover, are tasteless and ugly, while social goods produce societal welfare and utility.

Defenders of advertising argue that the social balance issue should not be settled by subjective aesthetic opinion but by consumer-voter-taxpayers. If voters are actually in control in a democracy, social wants are filtered through a political process, just as private wants are registered through the dollar votes of consumers. To say that advertising is useless and should be the object of regulation is to deny its potential benefits. Advertising, in the defenders' view, has permitted the rapid in-troduction of technological innovations with great con-sumer benefits. Advertising may make possible the sale of what some consider trash, but it also helps sell Mo-zart, Tolstoy, the ballet, and the New York Philhar-monic. From the perspective of individualism and in-dividual development, then, advertising has fostered a maximum of consumer choice and freedom.

Obviously, these arguments concerning the pros and cons of advertising cannot be settled here. Any stand on a particular issue depends on the facts of the partic-ular case, one's philosophical preconceptions, and so on. However, it is important to understand at least some of the arguments within the context of monopo-listic competition and all other market structures where advertising is observed.

calculated by multiplying price times quantity sold ($P_0 \times q_0$) to get the firm's total receipts, and then subtracting the total cost of q_0 units, $ATC_0 \times q_0$. Profits to the firm are shown in the shaded area of the figure. Notice that the firm's demand curve is highly elastic because of the large number of substitutes but that it is still negatively sloped because, under monopolistic competition, products are differentiated through advertising and other forms of nonprice competition.

Like firms in all other market structures, monopolistically competitive firms do not always make profits. Figure 11–3b depicts a loss-minimizing short-run equilibrium for a monopolistically competitive firm. In the case illustrated, the firm will sell q_0 units at price P_0—the price-quantity combi-nation selected where MC equals MR. This time, however, totol receipts, $P_0 \times q_0$, are exceeded by total costs, $ATC_0 \times q_0$. Losses (the shaded rec-tangle) are incurred. The firm, however, will remain in business in the short run because its total variable costs $AVC_0 \times q_0$ are covered by its total re-

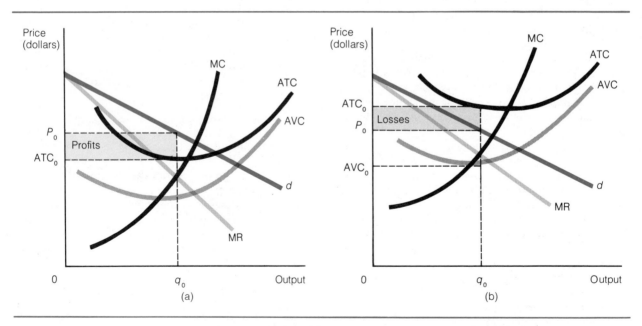

FIGURE 11–3   Short-Run Profits and Losses for the Monopolistically Competitive Firm

(a) The firm earns short-run profits. It sets price P_0 for q_0 units of output. Its profits are the difference between its revenue ($P_0 \times q_0$) minus its costs ($ATC_0 \times q_0$). (b) The firm suffers short-run losses. Costs are greater than revenue. The firm will continue operating, however, because its total variable costs ($AVC_0 \times q_0$) are less than its revenue.

ceipts, so there is some contribution to fixed costs. If these conditions pre-vailed in the long run, the firm (like the competitive or pure monopoly firm) would go out of business. In the long run, expenditures that are fixed in the short run become variable expenses and must be covered.

The monopolistic competitor does have one advantage over the pure com-petitor in the short run: the ability to manipulate demand through advertis-ing and intensified product differentiation. The loss-minimizing firm in Fig-ure 11–3b may increase short-term outlays on advertising or sales effort, thereby shifting demand and marginal revenue curves rightward. Such an increase may permit the firm to break even or to earn profits. Optimum advertising outlay for the firm occurs when one dollar of additional selling costs adds exactly one dollar to the firm's receipts. In this sense, the cost of advertising is treated like the cost of any other input, such as capital or labor.

Long-Run Equilibrium: The Tangency Solution. Long-run equilibrium for the firm in monopolistically competitive markets may take a number of forms. One of them, described by E. H. Chamberlin, is the famous **tangency solution** wherein the firm ends up making zero economic profits; that is, it simply breaks even. The assumptions of this model of the firm's economic behavior can be summarized as follows:

1. The firm operates in an industry composed of many sellers selling substi-tutable products, with the number of sellers constant throughout the pe-riod of adjustment.
2. Each seller assumes that its price actions will not provoke a response from rivals.
3. The degree of product differentiation (and thus advertising budgets) is constant and is determined by all competing firms; firms, therefore, ma-nipulate prices to increase profits.
4. The firm represents all competitors in that their cost curves and demand curves (as well as their reactions) are identical.

A model conforming to these assumptions is shown in Figure 11–4. Two demand curves, d_0d_0 and d_1d_1, are reproduced along with a long-run average

---

Tangency solution: A long-run situation in which the firm's downward-sloping demand curve is just tangent to the average total cost curve, necessarily implying zero economic profits.

FIGURE 11–4

The Tangency Solution

Each firm believes it can increase profits by lowering price. When all firms lower price from P_0 to P_1, however, the demand curve shifts leftward for each firm. In final equilibrium, E, the point of tangency with the LRATC curve, all firms break even.

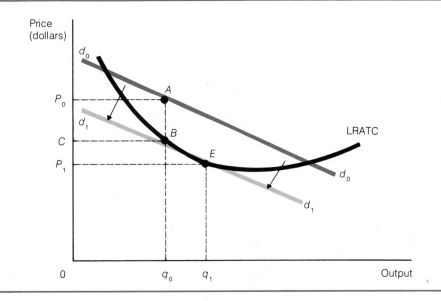

cost curve for the firm. Suppose that the representative firm (out of perhaps 100 firms producing similar products) initially finds itself at point A, charging price P_0 and selling quantity q_0. (The marginal revenue curves associated with d_0d_0 and d_1d_1 are omitted in Figure 11–4 for simplicity.) From the perspective of profitability, the firm, along with all other competitors, is earning economic profits in the amount represented by P_0ABC, B being the long-run average cost of producing quantity q_0. Collusion to maintain profits at this level is impossible because of the large number of producers, so these economic profits cannot be sustained. The initial price and quantity combination selected, P_0 and q_0, will not be permanent in the face of Chamberlin's assumptions about firms' behavior.

Why? Because each firm believes that it can increase profits by lowering its price. In other words, each firm believes that its demand curve is more inelastic than it really is because it believes, erroneously, that rivals will not reduce prices when it does. In fact, rivals usually react to another's price drop by lowering their prices. This widespread lowering of price causes the demand curve for each seller to shift leftward until economic profits are zero, that is, until each firm breaks even. Finally, demand curve d_1d_1 becomes tangent to the long-run average total cost function, LRATC (thus the term *tangency solution* to describe this case).

At point E in Figure 11–4, all firms will produce q_1 units of output and charge price P_1 per unit. Quantities above this amount and prices below P_1 would force firms out of business. Quantities less than q_1 and prices higher than P_1 would provoke the price adjustment responses described above. Under these circumstances, a stable equilibrium characterized by zero economic profits for all firms is generated.

A word of warning: While the tangency solution is perhaps the best-known analytical approach to monpolistic competition in the long run, it is only one of a large number of possibilities. A little reflection on the realism of the four assumptions in Chamberlin's model, listed earlier, will tell us why.

1. The initial existence of economic profits will certainly encourage the entry of new firms into the market, altering the effects of price responses.
2. We cannot assume that firms will continue to think that price reductions will not provoke similar responses from their competitors. Firm managers must have the limited vision of snails to remain unaware of other firms' repeated and identical responses to other firms' price reductions.
3. Monopolistic competitors can use product differentiation and advertising as well as price manipulation to maintain demand and profits. The degree of differentiation is artificially held constant in the model described in Figure 11–4.
4. It is obvious to the most casual observer that all firms are not alike with respect to demand and costs.

Thus the model described in Figure 11–4 reflects but one set of behavioral assumptions among many possibilities. Its value is considerable, however, because it describes a tendency to tangency. Real-world tangency solutions will likely not exist, but as long as the assumptions of Chamberlin's famous model *approximate* reality, an approach to a tangency solution may be observed and predicted in markets displaying the features of monopolistic competition. Additionally, the tangency model of monopolistic competition provides a convenient benchmark for a comparison of the attributes and economic effects of pure competition and those of monopolistic competition.

Does Monopolistic Competition Cause Resource Waste?

A charge commonly leveled at monopolistic competition—unrelated to the alleged wastes of advertising—is that the tangency solution involves, and indeed requires, (1) underallocation of resources in production or (2) generation of excess capacity and prices higher than minimum long-run unit costs of production. This unflattering portrayal of monopolistic competition is possible when it is compared with the long-run competitive model.

Monopolistic Competition versus Pure Competition. Long-run equilibrium in the perfectly competitive model, discussed in Chapter 9, results in the following three conditions:

1. Price = Long-run marginal costs.
2. Price = Minimum long-run average total costs.
3. Total revenue = Total costs.

The first condition means that the consumer's marginal sacrifice (price) just equals the marginal opportunity cost of producing the last unit of a product or service consumed. The second condition means that consumers are getting the product at the lowest possible price at which the product can be produced. The third conclusion means that no competitive firm earns more than a normal profit in the long run. Firms are price takers under competition. Freedom of entry guarantees normal profits in the long run.

Consider, with the aid of Figure 11–5, how monopolistic competition compares to pure competition in these respects. Figure 11–5a shows the de-

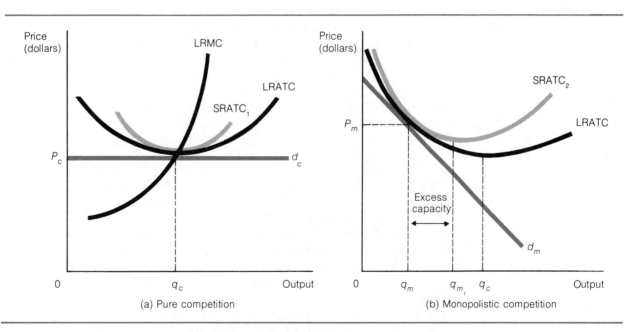

FIGURE 11–5   A Comparison of Pure Competition and Monopolistic Competition

In a technical sense, the monopolistically competitive firm (b) charges a higher price, P_m, than the purely competitive firm (a), P_c. The monopolistically competitive firm sets output at q_m to maximize profits, but the most efficient use of its plant would be an output of q_{m_1}. The firm therefore produces at excess capacity, or underutilization of its scale of plant.

mand curve $P_c d_c$ and the short-run and long-run average cost curves for the typical firm in pure competition. Clearly, both long-run and short-run costs are tangent to the perfectly elastic demand curve for the firms in competition. Quantity q_c is produced at price P_c, which equals long-run (and short-run) marginal cost. Figure 11–5b shows the downward-sloping demand curve d_m of a monopolistically competitive firm in a tangency solution with the firm's long-run average cost curve. Given that the firms' cost curves are identical under both market structures, how do they compare in economic efficiency?

As Figure 11–5 shows, the following features result from monopolistic competition:

1. Price > Long-run marginal costs.
2. Price > Minimum long-run average costs.
3. Total revenue = Total costs.

The first conclusion means that the additional sacrifice (price) that consumers are willing to make for an additional unit of the good is greater than the marginal opportunity cost of producing that additional unit. In other words, resources are underallocated in the production of goods by monopolistically competitive firms, just as they usually are in production by pure monopolies.

The second conclusion is that consumers pay more for goods produced under conditions of monopolistic competition than they do when goods are supplied by perfect competitors. The fact that P_m in Figure 11–5b is greater than P_c in Figure 11–5a means that price is higher and output lower under monopolistic competition than under pure competition. Consumers are not paying the lowest possible unit cost. However, like perfect competitors, monopolistic competitors do not earn economic profits because a tangency solution means that total revenues equal total costs to the firm (or, alternatively, that the firm's average cost equals its average revenue).

Excess capacity: A situation in which industry output is not produced at the lowest possible average total cost, the result of underutilized plant size.

Critics of monopolistic competition raise still another issue: that **excess capacity** is generated under the monopolistically competitive market structure. This criticism ordinarily holds for any less-than-perfectly competitive market. To understand what excess capacity means, consider Figure 11–5 again. The LRATC function is called an envelope curve because it is composed of a series of tangencies of points on the short-run average total cost curves (see Figure 8–5 on page 171). The short-run curves $SRATC_1$ and $SRATC_2$ drawn in Figure 11–5 are two such curves. To utilize fully any scale of plant or to produce at an optimum rate of output in the short run, the existing scale of plant or existing resources invested must be used at the lowest average cost of production. For the monopolistically competitive firm depicted in Figure 11–5b, the output corresponding to that lowest short-run average total cost of production is q_{m_1}, not q_m, the profit-maximizing quantity the firm has chosen. From society's point of view, the firm producing at q_m is wasting, or underutilizing, resources. Excess capacity is the unused capacity of the scale of plant that the firm has built. From the firm's perspective, however, the scale of plant represented by $SRATC_2$ is perfect or optimal in that it produces the quantity q_m more cheaply than any other scale of plant could. Thus critics argue that monopolistic competition creates excess capacity as well as the long-run resource inefficiencies outlined above.

A Defense of Monopolistic Competition. Is monopolistic competition as inefficient as the critics say? In assessing this issue we must be careful to distin-

Economic efficiency: Proper allocation of resources from the firm's perspective.

Economic welfare: The situation in which products and services are offered to consumers at the minimum long-run average total cost of production.

guish between pure **economic efficiency** and **economic welfare.** Economic efficiency means that resources are properly allocated (at minimum cost for a given output) from the firm's perspective. Pure economic welfare means that consumers obtain all products and services at the minimum long-run average total cost of production. Chamberlin, for one, argued that the monopolistically competitive market structure is not wasteful in either sense because product differentiation creates variety and extends the array of consumers' choices.

If price is established close to minimum long-run average total cost, some pure inefficiency might exist with monopolistic competition, but consider what consumers get in return: varied products and a vastly expanded number of products and services from which to choose. If this variety is socially valued, advertising and product differentiation are not necessarily wasteful. Such differentiation, summed across all consumers, means that monopolistic competition may increase social welfare. A society that produces only white shirts may be efficient in a technical sense, but it is apparent that Americans have not chosen such a society. The repeated purchase of other kinds of shirts at higher prices means that consumers are willing to pay for differentiation and that their satisfaction is thereby heightened.

Some excess capacity may also be a benefit and not a cost in terms of consumer welfare. As indicated in Chapter 4, resource costs other than money price are involved in the purchase of goods. Time, for example, is an important cost in consuming goods. Excess capacity in the sale of some goods and services may simply be a means of lowering consumption costs. Taxi cabs typically line up and wait for fares in front of hotels and other busy establishments. This observed "excess capacity" may have a welfare benefit for consumers. A consumer's cost of transportation is certainly reduced by reduced waiting time. And as suggested in Focus, "Why Is There a Funeral Home in Most Small Towns?" even the excess capacity built into funeral homes is socially valued. The conclusion: Excess capacity itself may provide consumer welfare.

How Useful Is the Theory of Monopolistic Competition?

When models of monopolistic competition and pure competition are subjected to real-world observation, is there a dime's worth of difference between them? How useful, in other words, is the model of monopolistic competition?

If the demand curve is very elastic when products are differentiated, the model of monopolistic competition is a close approximation of pure competition. As noted in Chapter 8, the perfectly competitive model is primarily an idealized benchmark with which real-world markets can be compared. Simple observation tells us that most actual markets conform to the set of characteristics we have defined as monopolistically competitive.

Thus, a study of markets with differentiated products and rapid and free entry of substitutes is a study of what is usually called "competition." With the ready emergence of substitutes, the "monopolistic" adjective in monopolistic competition may be downplayed. Firms' ability to set prices is critically restricted by such entry and even by potential entry. Viewed in this light, the model of monopolistic competition, with its emphasis on product differentiation and other forms of nonprice competition, is of great value in helping us understand how dynamic real-world markets function, change, or thrive.

FOCUS   **Why Is There a Funeral Home in Most Small Towns?**

The funeral service business in the United States illustrates the workings of a monopolistically competitive market. Some observers and critics charge that funeral homes operating in small towns have a virtual monopoly on funeral services in their area. These firms are said to be able to price services above marginal costs by restricting output (monopoly power), to thus be able to take advantage of the extremely strained circumstances surrounding death, and to create excess capacity in the process. Is the fact that there is a funeral home in most small towns evidence of a degree of monopoly power and of excess capacity? Or is the market more competitive than commonly believed?

First we consider the information that consumers use in choosing funeral homes. Generally, more than one person goes along in viewing the home. This fact tends to reduce quick, emotional, or irrational choices. Past experience in dealing with firms and the recommendations of family and friends play a part in the selection of a funeral home. If the firm has a reputation of providing poor-quality services, even in small towns, we would expect new firms to enter the market. Indeed, one characteristic of the funeral industry is that of the long-standing firm, implying that quality of service has been at acceptable levels over the years.

But why are these firms in most small towns? One reason is that proximity is a consideration in making the choice. Most funeral directors in small towns also have other means of income because the "supply" of deaths in small towns is not as great as in large cities. Still, the home is there because people want quick service with proximity to where they live. People also like to deal with someone they know, someone who will provide speedy, high-quality, personal services. Waiting would itself be a cost to consumers if funeral service were not quickly available nearby.

Another issue related to monopolistic competition concerns excess capacity. The observed excess capacity in the funeral industry may not be excess at all in an economic sense. If the quality of the service depends on the funeral director's undivided attention, and if each funeral utilizes a major portion of the funeral home's physical plant, such as viewing rooms, then a funeral home may have to build in some excess capacity merely to avoid peak load problems. This is characteristic of many industries, particularly service industries, where the demand for output is unpredictable or variable and where output cannot be stored easily. Thus, these seemingly idle resources have economic value. Because they are idle some of the time, there will be sufficient resources at other times so that funerals can be performed with no delay.

All firms must have ways of dealing with variations

In most small towns, funeral services are usually sold through independent, family-owned businesses. These small firms are able to operate profitably because of the special characteristics of the funeral industry.

in demand, and there are several methods for balancing production and sales flows. Sometimes output can be produced and stored as inventories, which in turn fluctuate with demand and production conditions. Some firms allow queues to form when demand temporarily exceeds capacity. Others hold larger amounts of physical capacity and other resources—excess capacity—and thus have shorter (or no) queues, even during periods of high demand. The cost of these measures has its effect on the price of the product. Consumers, in their choice of firms, balance the price of the service or product against the time they would have to wait for delivery.

In the funeral industry, excess capacity can be explained by the nature of demand. Given the personal nature of funeral services, proximity as well as personal attention of the funeral director may be important qualities of service that the consumer purchases, along with the casket and other items. Even very small towns may therefore have funeral homes, each of which performs a relatively small number of funeral services per year. While this widespread existence of underused facilities appears to be inefficient, it occurs because people are willing to give up some of the benefits of economies of scale for the ability to deal with local, unhurried funeral personnel. In this case a monopolistically competitive market does not meet the strict efficiency criterion of pure competition, but it may produce welfare benefits that more than compensate for the observed waste.

Oligopoly Models

In the remainder of this chapter we turn to the types of market structure between perfect competition and pure monopoly that we have designated as oligopolies and cartels. As we have seen so far, monopolistic competition lies closer to perfect competition than to the pure monopoly model. The demand curve facing the monopolistically competitive seller is downward sloping, but not by very much. The monopolistically competitive firm has some limited control over price, in contrast to the competitive firm. Oligopolies and cartels, however, lie closer to the pure monopoly model. Sellers in these markets face downward-sloping demand curves, but under certain circumstances they can raise price substantially above marginal cost without invoking the entry of new firms.

Characteristics of Oligopoly

Oligopoly: A market model characterized by a few firms that produce either a homogeneous product or differentiated products and entry of new firms is very difficult or is blocked.

Oligopoly refers to a market dominated by a few sellers. Industries with a few sellers are quite characteristic of the modern marketplace: steel, automobiles, large mainframe computers, aircraft, breakfast cereal, and soft drinks are just a few of the products of industries with few sellers. Oligopolies may sell homogeneous products such as steel or differentiated products such as automobiles or soft drinks. In this sense, oligopolies may resemble either pure competitors or monopolistically competitive firms. Later we will consider the special problems of oligopolies selling differentiated products, but first we consider two important characteristics of markets that are dominated by a few sellers: mutual interdependence and barriers to entry.

Mutual interdependence: A relation between firms in which the actions of one firm have significant effects on the actions and profits of other firms.

Mutual Interdependence. The key to the existence of oligopoly is that the small number of sellers leads to interdependence among them. **Mutual interdependence** is a relation among firms such that what one firm does with its price, product, or advertising budget directly affects other firms in the market, and vice versa. If the Chrysler Corporation introduces two new types of automobiles or if the company increases its overall advertising budget by a significant percentage, sales of General Motors, Ford, and foreign import manufacturers will be directly affected. These oligopolistic competitors will know that their sales and profits are affected by Chrysler's actions and will react accordingly. Firms therefore invest resources in keeping track of their competitors' actions.

When mutual interdependence is recognized by competitors, each firm will do the best it can in the market (in price, product innovation, and advertising) depending on what it thinks its rivals will do. Obviously, mutual interdependence can lead to a variety of outcomes, and we will find that the theory of oligopoly consists of a number of plausible models. The variety of possible outcomes arises from the uncertainty facing sellers in this market; each model reflects how competitors expect rivals to react.

Barriers to Entry. Oligopoly firms are normally thought to possess a large degree of **market power.** Subject to the reactions of competitors and to the

Market power: A situation characterized by barriers to entry of rival firms, giving an established firm control over price and, therefore, profit levels.

ability of new competitors to enter the market, oligopoly firms are able to set prices and even make economic profits, much like pure monopolists. But how does a firm achieve market power in the oligopoly market structure?

The source of the market power of oligopolists, as in any market structure, is barriers to entry. The pure monopolist is the sole supplier of a product because of barriers to entry into its industry. In any locality, for instance, there is ordinarily no close substitute for cable TV or residential power service, and there are legal barriers to firms wishing to enter these markets. If entry were free in these markets, new competitors would be attracted by higher-than-competitive prices, and excess profits would be eroded. Entry barriers are likewise essential to the maintenance of market power by oligopolists.

For an oligopoly, an entry barrier is essentially a cost that confronts a potential entrant into the industry but does not affect the incumbent firms. There are both natural or artificial entry barriers in an oligopoly market.

Natural Barriers. Economies of scale are an example of a natural barrier to entry. The economies of large-scale production and distribution mean that long-run average costs decline as output grows. Minimum average cost is not reached until the firm is producing a huge proportion of the total industry output. Economies of scale in the U.S. automobile industry, for example, have promoted the presence of a few firms in the long run—far fewer than the hundreds of automobile manufacturers at the beginning of this century. Other natural entry barriers include high fixed costs, high risk, scarce managerial talent, personal technological knowledge, and high capital costs.

Great care must be used in interpreting natural entry barriers since most such barriers reflect the different efficiencies of firms in producing output. Society benefits when firms achieve economies of scale, produce with the best technology, and allocate scarce managerial talent carefully. The gross national product goes up as a result. If a natural force like economies of scale leads to the possibility of market power, then the firm may be subject to antitrust prosecution. This means that the firm may be subject to laws enforced by the Federal Trade Commission and the Justice Department, which limit or eliminate anticompetitive practices.

Artificial Barriers. Government restrictions—patents, government regulations, licensing, labor union legislation, tariffs, quotas, and other government actions—constitute artificial barriers to entry for ologopolists. Such restrictions are fairly self-evident, but consider some examples. Patent protection is the exclusive right to market an invention for a given number of years. Patent protection is said to promote inventions, but the patent law also creates and protects oligopolies or temporary monopolies. Patents on new drugs and such products as the Polaroid camera create temporary monopolies within oligopoly industries.

Government-sponsored import tariffs on foreign automobiles or steel are an example of government restrictions that encourage and protect oligopolies. Cartels or legal agreements to collude—in such industries as railroads, ocean transportation, communications, TV, and radio—are actually sponsored and enforced by government. Cartel arrangements legalized through regulation are probably the most important artificial barriers to entry established by government. The basic point is that oligopoly theory rests on the concept of barriers to entry, both natural and artificial. Without barriers to entry, oligopoly could not exist.

Models of Oligopoly Pricing

Like other firms, oligopolies seek to maximize their profits. To do so, how-ever, the oligopolist can pursue many different strategies because firms can compete on a number of bases—product development, prices, advertising budgets, and so on—and can form a whole spectrum of opinions about how rivals will react. Many theories of oligopoly based on price or nonprice com-petition have therefore developed. We present some of the models, repre-senting the simplest to the more complex.

Mutual Awareness. A major category of oligopoly model involves the case where a few firms in an industry sell the same product. This is sometimes called *pure oligopoly*, reflecting the assumption of a homogeneous product. In the U.S. economy, the steel and aluminum industries are possible examples of pure oligopoly.

Suppose that there are only two firms in an industry (economists call such a situation a *duopoly*). Both are critically aware of the actions of the other, but they cannot communicate with one another because it is illegal to do so. What market outcome might result?

Obviously, a lot depends on the personalities of the two oligopoly owners. Suppose that a benign lack of competition—"live and let live"—is their characteristic attitude. Figure 11–6 illustrates the simple result.

The firms have the same MC curve and face the same industry demand curve, D. Mutual forbearance in this case will lead to equal shares of the market; d represents one-half of industry demand. The mr curve is drawn marginal to d, and hence d and mr represent the demand and marginal rev-enue conditions facing both firms. Price will be set at P_1, and each firm will produce q_1, where $2q_1 = Q_T$, or total industry output. This is not the same as the profit-maximizing situation for the pure monopolist. The pure monop-olist would set MC = d and would price and produce accordingly, whereas price is lower and quantity higher in a simple mutual awareness model.

This is an example of how two oligopolists might reach an implicit under-standing with one another about market behavior. It is hard to say whether

FIGURE 11–6

A Model of Mutual Awareness

There are two firms in the industry with the same MC curve, and they tacitly agree to share the market equally. The d represents one-half of industry demand, D. Each firm produces q_1, so the industry output $Q_T = 2q_1$. This simple model of oligopoly shows that price is lower and quantity greater than that which would exist under a pure monopoly model.

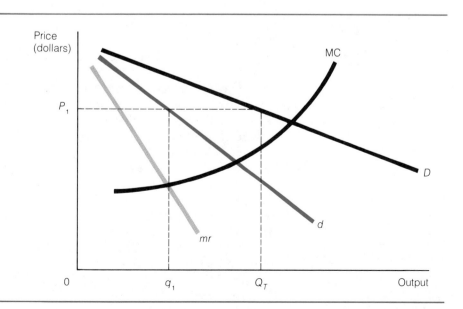

FIGURE 11–7

The Effects of Differing Costs

The duopolists in this case have different costs (MC_A and MC_B), which lead them to prefer different prices (P_A and P_B). Interdependence is harder to achieve, and firm B experiences great pressure to match the price of firm A.

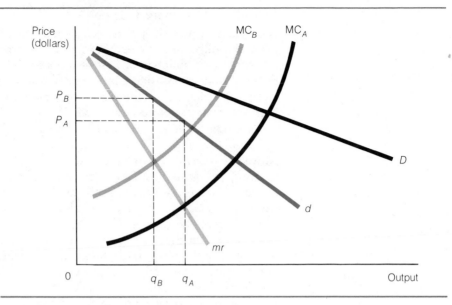

this case represents a likely outcome, but it is surely based on simple assumptions.

Suppose, on the other hand, that the two firms had different costs or faced different demand curves. Figure 11–7 illustrates the case of different costs.

In this case firm A prefers a lower price and a larger output than firm B. One would guess that mutual forbearance would not work very well here. The firms are selling a homogeneous product, but one firm, A, is a more efficient producer than the other. Since the firms cannot explicitly communicate and set a common price-output policy to maximize profits, firm B is in a bad situation. It has little choice except to match firm A's price; it would otherwise lose its share of the market over time. Yet matching A's price may impose losses on B. Such considerations make mutual forbearance difficult when firms have different costs.

Virtually the same analysis holds if the firms have the same MC curve but face different demand schedules. Suppose that the industry is growing faster in the region of firm A than in that of firm B. Again, the pricing strategies of the duopolists will diverge, and it will be more difficult for the two firms to practice interdependent pricing. Indeed, it will be more difficult for the firms to avoid outright price competition. Since we generally expect firms to have different cost functions and to face different demand schedules, we expect interdependent pricing through mutual awareness and forbearance to be hard to implement in real-world markets.

The Kinked Demand Model. The Great Depression of the 1930s provided a setting for the development of a famous model of oligopolist competition. The **kinked demand** theory of oligopolist pricing was developed by Paul M. Sweezy in the 1930s to explain why prices in oligopolist industries with few sellers did not seem to change very much in the face of large-scale reductions in demand during the Depression. Prices were "sticky"—that is, they refused to fall or they fell slowly—over this period.

Figure 11–8 illustrates Sweezy's kinked demand model of oligopolist pricing. The graph looks complicated, but it is not if we understand Sweezy's

Kinked demand: A curve that has a discontinuous slope, the result of two distinct price reactions of competitors to changes in price.

FIGURE 11–8

Sweezy's Kinked Demand Curve Model

Sweezy's basic assumption is that rivals will not follow price increases by a firm but will follow price decreases. This causes MR to be discontinuous over the range *BC*. If MC cuts MR in this segment, there will be no tendency for price or output to change when the level of marginal costs rises or falls between *B* and *C*. Prices will be "sticky."

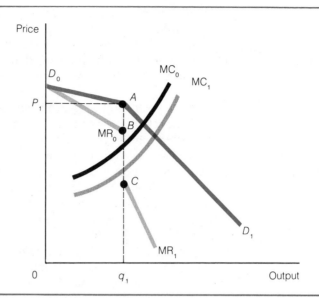

main assumption about oligopoly firm behavior: that rival firms will follow price decreases but will not follow price increases. In Figure 11–8, the price being charged is P_1, and output is q_1. As you can see, the market demand curve facing the firm has a "kink" in it. The slope of the upper portion, D_0A, reflects the assumption that if the firm raises its price, no rival will follow the price increase. D_0A is an elastic portion of the kinked demand curve. The slope of the portion AD_1 is based on the assumption that rivals will match price decreases by the firm. Thus, the firm loses sales if it raises price above P_1 and does not gain sales if it lowers price below P_1. This market situation results in a kink in the demand curve at point A.

The two segments of the kinked demand curve are associated with two marginal revenue curves. MR_0 is the marginal revenue curve corresponding to D_0A, and MR_1 is the marginal revenue curve corresponding to AD_1. Connection of the two differently sloped MR curves results in a discontinuous marginal revenue curve. MR hits the dashed line in Figure 11–8 at B and "jumps" to C, where it begins again as MR_1. The marginal revenue curve has a gap (or discontinuity) over the distance BC. It is useful to think of marginal revenue as consisting of MR_0, BC, and MR_1 in this case.

Now we can see why prices are "sticky" in this model. Applying the MC = MR rule, we see in Figure 11–8 that the MC curve cuts MR in the gap BC. Within this gap MC can rise as high as B or fall as low as C, but applying the MC = MR rule will yield the same price for the firm. Thus, changes in marginal costs do not automatically lead to lower prices or higher quantities. In Figure 11–8, for example, when the firm's MC falls to MC_1, there is no change in its price.

Sweezy's assumption may or may not hold. In fact, the relation of the model to real-world pricing in concentrated markets seems open to question. Careful empirical studies tend to show a reasonable amount of upward and downward price flexibility in concentrated markets. Moreover, the kinked demand model is strictly a short-run theory of oligopoly pricing. It does not tell us how price comes to be set at P_1 or how firms adapt to situations where MC shifts outside the gap of MR.

Price leadership: A pricing behavior in which a single firm determines industry price.

Dominant Firms. Many industries consist of one or a few large firms and many smaller rivals. Examples are the airlines, steel manufacturers, and home computer industry. Pricing in such a market is often characterized by **price leadership,** which means that there is an unwritten agreement that the largest firm sets pricing policy for the industry. When the largest firm announces that it is raising its price by 5 percent, the other firms in the industry follow suit (and usually on the same day!). This unwritten code of industry conduct is one way to solve problems of interdependent pricing.

It is not always the largest firm that is the price leader; sometimes it is the most respected firm in the industry or the firm, called a "barometric" firm, that best reflects average cost, demand, and other conditions. And sometimes the role of price leader shifts among firms. Jones-Laughlin, though not the largest firm in the market, was for years the steel industry's price leader, whereas General Motors, clearly the dominant firm, provided a similar role for the automobile industry.

Figure 11–9 illustrates the dominant-firm model. The diagram looks more complicated than it is. D is the industry demand curve; ΣMC_f is the horizontal sum of the MC curves of the "fringe suppliers" in the industry. These are competitive firms that behave like price takers; they set MC equal to price. D_d is the demand curve facing the dominant firm (or groups of firms), and MR_d is the associated marginal revenue curve. The marginal cost curve of the dominant firm, MC_d, is lower than the marginal cost curves of the fringe suppliers. Its lower costs derive from economies of scale.

How does the model work? The dominant firm takes the existence of the

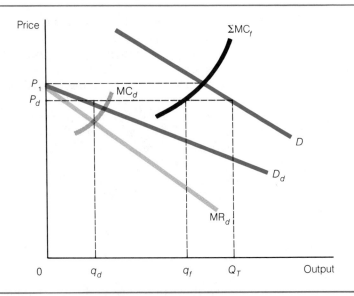

FIGURE 11–9   A Dominant Firm Model

MC_d is the marginal cost curve of the dominant firm, and ΣMC_f is the marginal cost curve of fringe suppliers. At price P_1, the competitive fringe supplies the whole market. At prices below P_1, a residual demand faces the dominant firm ($D_d = D - \Sigma MC_f$). The dominant firm, which has lower MC than the fringe suppliers, sets a monopoly price, P_d, over this portion of the market. The dominant firm produces q_d. The fringe firms accept P_d as given and set MC = P_d, producing q_f; $q_d + q_f = Q_T$. The dominant firm is therefore the price leader in the industry, and the fringe suppliers are price takers.

fringe suppliers as a given, and it prices accordingly. At any given price, the dominant firm derives its demand curve by considering what the fringe firms will do. At price P_1, the fringe suppliers will supply the whole market ($\Sigma MC_f = D$). There is nothing left for the dominant firm to exercise its monopoly power over. At prices below P_1, fringe suppliers cannot supply the whole market; something is left over for the dominant firm to supply. This portion of the market is given by the horizontal difference between ΣMC_f and D at points below where $\Sigma MC_f = D$. In other words, at prices below P_1, the demand curve of the dominant firm is equal to the difference beteen ΣMC_f and D.

Over this residual portion of the market, the dominant firm is a price setter. According to its own MC_d and MR_d, it sets price at P_d and produces q_d units of output. The fringe firms accept P_d as given and produce by equating ΣMC_f to P_d. Fringe suppliers supply q_f. Since $q_d + q_f = Q_T$, market demand equals market supply at a price of P_d.

The dominant firm model mixes elements of monopoly and oligopoly theory. The dominant firm sets price for the industry, and other firms follow this pricing policy, but it sets price by acting like a conventional monopolist over the residual portion of the market not supplied by the fringe firms.

Oligopolists with Differentiated Products. Just as there are many cases of pure oligopoly, there are many cases in which few sellers supply differentiated products. Under such conditions, oligopolists may compete with each other and perhaps try to bar new entrants by advertising, quality variations, or pricing strategies.

Advertising. In oligopoly, as in monopolistic competition, advertising is used by rival firms to increase demand for their products (shift demand curves outward) through the creation of brand loyalty and product identity. Heavy advertising by a few large firms is often held to be an entry barrier to new competition. The argument is that new firms cannot enter because they cannot afford the heavy advertising expenses necessary to establish a presence in the market. But what is at stake in advertising? Basically, what advertising does is establish a firm's brand name, which represents the quality of the firm's products and services. A brand name reflecting reliable performance is quite valuable and reflects what might be called the advertising capital of the firm. However, if firms lie or produce shoddy products, their brand name capital value will fall.

If advertising creates a barrier to entry, we would expect to see firms earning excess profits (greater than a competitive rate of return) on their advertising capital investments in brand names. If they are earning only competitive returns on their advertising capital, advertising is not a barrier to entry. Incumbent firms and potential entrants would face an essentially competitive market in establishing brand names. There is some evidence that the rate of return on advertising investments across large U.S. firms is at the competitive level. This finding would undermine the argument that advertising is a barrier to entry into oligopoly industries.

Quality Variations. Quality variation and product development are other familiar methods of nonprice competition by oligopolists. Just as monopolistic competitors compete in this manner, so oligopolists will alter products to increase demand curve and to make demand more "intense" and inelastic in consumers' minds. Adding new colors to toothpaste, more or less chrome to automobiles, and more service to airline travel are familiar attempts to differentiate products and services.

Not all of this nonprice competition by oligopolists promotes efficiency in the economy. The first Economics in Action, "The Prisoner's Dilemma and Automobile Style Changes," at the end of the chapter illustrates how competition in styling between automobile producers can lead to wasteful results. In addition, this essay illustrates a particularly interesting form of oligopolist interaction called the prisoner's dilemma.

Limit pricing: The price behavior of an existing firm in which the firm charges a price lower than the current profit-maximizing price to discourage the entry of new firms and thus maximize its long-run profits.

Limit Pricing. In oligopoly industries with a differentiated product, incumbent firms will seek to deter potential competition by limit pricing. **Limit pricing** is simply the practice of the incumbent to reduce price as much as necessary to limit entry.

Imagine a situation in which a new firm wants to enter an industry. Since other firms' products have already been differentiated, the new firm will have to incur heavy advertising expenses to become competitive with incumbent firms assumed to be earning economic profits. If a new entrant is successful in its bid for a share of the market, its sales come at the expense of the incumbent firms. To block this challenge, an incumbent firm may lower price and increase output in response to the threat of entry, voluntarily (if temporarily) reducing its profits. In general, the incumbent firm could set a price at which entry would not be worthwhile for any potential rival.

We are talking about oligopoly behavior. The potential entrant must make an estimate of what the incumbent firm will do in the face of entry. If the incumbent firm practices or even threatens limit pricing, entry may not be feasible. Indeed, the most rational thing for the incumbent firm to do is to keep price at profitable levels and at the same time make it clear to potential entrants that it will lower price and increase output if they enter.

Cartel Models

In our discussion of oligopolies, we have assumed that competitor firms will not cooperate to set prices. But what if firms do cooperate? This question brings us to the topic of cartels.

Cartels and oligopolies are close cousins. The distinguishing feature of a **cartel** is an agreement among firms to restrict output in order to raise price and to achieve monopoly power over a market.

Cartel: A formal alliance of firms that reduces output and increases price in an industry in an effort to increase profits.

Characteristics of Cartels

Normally cartels are supported by agreements that limit entry and restrict and segment output among markets. Mutual interdependence—also a characteristic of oligopoly—exists prior to cartel agreements because the purpose of a cartel agreement is to limit or constrain rivals' actions that have effects on the prices and profits of the group of suppliers.

Cartel agreements may be legal or illegal. In countries like the United States with antitrust laws, cartels are illegal, but they sometimes operate covertly. An illegal cartel rests on private, collusive agreements that are out of sight of the antitrust authorities. In the 1960s certain firms made a secret agreement to fix the prices of electrical generators sold to municipalities; this illegal collusion was discovered and prosecuted under antitrust laws, with the guilty firms paying damages. Beyond the threat and reality of antitrust prosecution, such secret arrangements tend to fall apart on their own. Reasons for the fragile nature of cartels will be discussed later in the chapter, but most are related to the incentives for some members to cheat on the agreement.

A legal cartel is one supported by a law limiting entry and restricting competition among members. Such cartels may be made legitimate by the legal and political organization of an industry—communications, electrical distribution, and railroads, for example—under the umbrella of government regulation restricting competition and allowing price fixing. Or cartels may be made legitimate by legislation granting exceptions to antitrust statutes. Such is the case for exporters, labor unions, and farm organizations.

Legal status does not protect such cartels from breaking down in the long run. Technology and competition have all but destroyed the railroad cartel in the United States enforced by the Interstate Commerce Commission since the late nineteenth century. OPEC (the Organization of Petroleum Exporting Countries) was unable to maintain its full cartel status not because it was illegal within the participating countries but because of the lack of adequate enforcement and the unwillingness of certain nations, notably Saudi Arabia, to meet the terms demanded by other members. The Colombian coffee cartel of the 1960s broke down for similar reasons, and the De Beers diamond cartel (see the second Economics in Action at the end of the chapter), though legal, had some shaky times in the early 1980s. Both legal and illegal cartels tend to be fragile, therefore, though agreements enforced by law and the government are less so.

Cartel Formation

Assume for the sake of simplicity that cartels do not violate any law and that we are analyzing a case in which the cartel is to be privately enforced by the cartel members and not by government. The first problem facing any potential cartel or association of sellers is how to get organized. In this respect the cartel is no different from any other group. The usual problem in getting any group to do anything can be roughly summarized as "passing the buck." This is quite a rational attitude for the individual cartel member to take. After all, if somebody else forms the cartel or takes the lead in restricting output, other members can generally enjoy the benefits of the cartel without bearing any of the costs of organization. This behavior is called "free riding."

Of course, if everybody in the cartel were a free rider, the cartel would never be formed. All potential members would hold back and expect to benefit from the organizational efforts of others. Economic theory does not have much to contribute to understanding how groups overcome these free-riding costs and get organized to do things.

Suffice it to say that a cartel has to get organized. Someone has to get the individual suppliers in the industry together and hammer out an agreement to restrict output in order to raise price. The details of this feat are not simple. Which firms will reduce output and by how much? How will the resulting cartel profits be shared? How will deviations from cartel policies be monitored and enforced? And so on.

None of these are easy questions for the cartel manager to answer or to reach agreement on. The problem becomes more pronounced the more firms there are in the industry. As a rule, smaller groups are easier to deal with than larger groups. This is the only general principle that we can enunciate about cartel formation.

FIGURE 11–10

Organizing a Competitive Industry into a Cartel

Price rises from P to P_c as the cartel is formed and output is cut back from Q to Q_c. In the extreme case, the cartel price, P_c, is equivalent to the monopoly price.

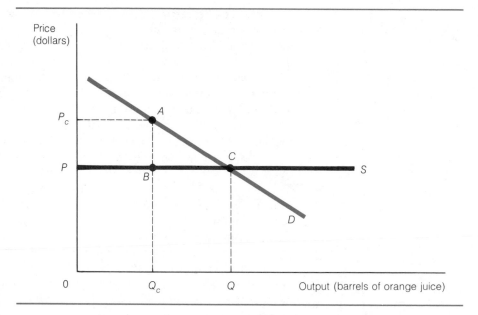

The Cartel in Action

Figure 11–10 illustrates the basic sense of how a cartel operates. Imagine that orange juice producers try to form a cartel. Initially, the orange juice industry is competitively organized, with a number of firms. Orange juice industry demand and supply intersect to yield price P. Industry output is Q.

A cartel manager emerges and strikes an agreement among the competitive producers to cut back production from the industry's Q to Q_c and to raise price from P to P_c. Cartel output Q_c will ordinarily not be equivalent to the output a pure monopolist would select. Why? In an industry such as orange juice production, with a relatively large number of suppliers, the costs of eliminating all competitive behavior—the costs of enforcing the agreement—would be too high. Some competitive behavior will remain. Cartel price will therefore be lower than the profit-maximizing price of a pure monopolist.

The incentive to form cartels is nevertheless clear. Cartel members who were previously earning a competitive rate of return in a competitive industry now earn an agreed-upon portion of P_cABP—an equal share of cartel profits.

Cartel enforcement: An effort by the administrators of a cartel to prevent its members from secretly cutting price below the cartel price.

Cartel Enforcement

Holding the cartel together is not easy. Once the cartel has been organized, has restricted output, and has raised price to P_c, what is the position of the individual orange juice firm in the cartel? If an individual firm in the cartel lowers its price slightly below P_c, it will face a very elastic demand curve for its output. Buyers will prefer the lower price and will switch their purchases to the lower-priced firm. The fact that individual firms in a cartel can significantly increase their sales and profits by secretly lowering their price puts tremendous pressure on the cartel to fall apart. Individual firms have a large incentive to cheat on the cartel agreement.

Of course, to expect such profits to hold up in the long run is unrealistic. Other firms would easily discover what was going on because they would suffer a drastic loss of sales. If they responded in kind with price reductions, the result would be that the secret price cutter would not experience an increase in sales and profits for long. Further spates of secret price cuts and price competition would lead to a return to competitive equilibrium.

Recognizing these possibilities, the individual firm in the cartel must form an estimate of what it can reasonably expect to gain from secret price reductions and compare this estimate to its share of cartel profits. A cartel will not be stable if the former is larger than the latter.

The general case in which a firm might expect to gain by secret price cutting is illustrated in Figure 11–11. Cartel equilibrium is at price P_c and output Q_c, represented by E_c. D/n represents the prorated firm demand curve, where n is the number of firms in the industry, and d_i represents the gains to an individual cartel member from secretly cutting price; d_i is more elastic than D/n. With the more elastic d_i curve, price increases would cause a sharp drop in sales, but price reductions cause a sharp gain in sales and profits at the expense of other cartel members. The gains of a secret price cut to P_s are traced out by d_i below P_c.

The key to cartel enforcement is now apparent. Means must be found to make d_i approximate D/n. Perfect cartel enforcement exists when each cartel member knows that secret price cuts are not possible and will be matched by other members. In this case, the prorated demand curve, D/n, is the relevant demand curve. If a firm cuts price, it will immediately be matched by other cartel members. There is thus no gain to cutting price. Movements will be along the prorated firm demand curve, D/n. Where it is costly to detect secret price cuts, a demand curve like d_i will confront each cartel member. Each firm must decide whether to cheat, depending on the expected profits from cheating relative to the expected profits from staying in the cartel. In general, we can gauge the effectiveness of cartel enforcement by the degree to which the cartel is able to make the cheating demand curve d_i coincide with the prorated firm demand curve D/n. That is, how good is the cartel at detecting cheating?

Cartels will spend resources to control the incentive to cheat. Historically, many devices have been used. Perhaps the most effective device is a common sales agency for the cartel. All production is sold through the

FIGURE 11–11

Cheating in a Cartel

P_c is the agreed-upon cartel price at cartel output Q_c. E_c is therefore the cartel equilibrium. If secret price cuts cannot be detected, the cheating firm will face the more elastic demand curve, d_i, and can gain from cheating on the agreement. If such behavior can be detected, movements will be along D/n, the prorated firm demand curve. The degree to which d_i diverges from D/n reflects the efficiency of the cartel enforcement system.

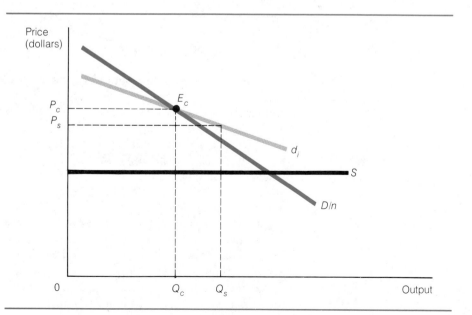

agency, and individual producers do not negotiate sales for themselves. Cheating is not a viable strategy under such a condition. Producers produce their allotted quotas, and the sales agency markets the agreed-upon cartel output (Q_c). Exclusive sales territories and customer lists are other ways to control incentives to cheat and to free ride on the cartel price. For example, exclusive territories are a way to divide markets, and intrusions by other cartel members are easy to spot.

Why Cartels May or May Not Succeed

Collusive cartel agreements are more likely under some circumstances than others. We offer now a menu of factors thought to be conducive to cartel-like behavior. Our list is not meant to be exhaustive, but just illustrative of the types of economic conditions associated with a greater likelihood of cartel activity.

1. A cartel is easier to form among a smaller number of sellers than a larger number. The costs of any group decision-making process rise with the number in the group.
2. It is easier to reach a cartel agreement where members produce a homogeneous product. In such cases, nonprice competition is less of a problem. Sellers do not have to worry that their competitors will secure bigger market shares through advertising.
3. Collusion is easier to effect in growing and technologically progressive industries than in stagnant or decaying industries. In the latter case, excess capacity created by declining demand puts great pressure on individual firms to compete for customers and to cheat on the cartel agreement.
4. Many small buyers reduce the costs of collusion. Large buyers have a clear incentive to seek to undermine collusion among sellers. In effect, large buyers make cheating on the cartel price very worthwhile to individual sellers.
5. Low turnover among buyers makes collusion easier. If sellers cheat on the cartel agreement, frequent turnover of buyers makes it difficult to detect who is cheating.

Other factors could be added to this list of conditions favorable to the formation of private collusive cartels. Indeed, an easy way to find out what makes collusion attractive is to look at the areas of the economy in which the antitrust authorities most typically find such behavior (discussed further in Chapter 17).

At this point, one fact should be clear: Private collusive agreements are hard to sustain. The incentive to cheat on such agreements is very strong, and cartels therefore tend to be unstable. Moreover, where antitrust law prevails, cartels must operate on the sly to avoid law enforcement; this secrecy further reduces their ability to control cheating. Exclusive sales territories, for example, would likely be spotted by antitrust authorities, for customers would complain about high prices and the lack of competitive sales or service alternatives. It is little wonder, then, that cartels rarely last long unless they are supported by law.

Groups that have difficulty sustaining a private cartel agreement can petition government to pass a law making their cartel legal. More than this, the government can enforce the cartel by outlawing entry and price competition among members. So to say that private cartels are not stable and therefore unlikely to side-step the social benefits of competition in the long run is inaccurate if government steps in to reinforce such agreements.

From Competition to Monopoly: The Spectrum Revisited

We are at last in a position to take an overview of the economic characteristics and welfare effects of market structures from pure competition (Chapter 8) to pure monopoly (Chapter 10) to those in between (the present chapter). Table 11–1 organizes the long-run economic tendencies and conclusions of the various structures we have examined.

The table distinguishes the five market structures by number of sellers, barriers to entry, product characteristics, and long-run market tendencies. The number of sellers ranges from many under competition and monopolistic competition to one in the case of pure monopoly. Significant entry barriers characterize oligopolies, cartels, and, of course, monopolies. These three structures are also characterized by the sale of *either* homogeneous or differentiated products, unlike the purely or monopolistically competitive structures.

Most important, Table 11–1 provides a category—long-run market tendencies—that gives us some insight into the consumer and economic welfare created under these alternative structures. As the table reveals, price tends to equal long-run average total cost in only two of the structures—pure and monopolistic competition. Price, moreover, tends to equal long-run marginal cost only in the purely competitive market structure. What does this mean in terms of consumers' economic welfare?

An important concept introduced in Chapter 6 aids us in interpreting the effects of market structure on social welfare. In Chapter 6, we introduced the concept of consumers' surplus, defined as the difference between the amount of money that consumers would be willing to pay for a quantity of a good and the amount that they actually pay for the good.

Figure 11–12 contrasts what happens to consumers' surplus for a given hypothetical market—personal computer software—if the market is organized

TABLE 11–1   Characteristics and Consequences of Market Structures

When static market structures are considered, only pure competition yields an efficient and welfare-maximizing allocation of resources. Resources are underallocated to the production of goods and services under all other structures from consumers' perspective.

Type of Market Structure	Number of Sellers	Barriers to Entry	Type of Product	Market Tendencies
Pure competition	Many	No	Homogeneous	P = MC P = ATC
Monopolistic competition	Many	No	Differentiated	P > MC P = ATC
Oligopoly	Few	Yes	Homogeneous or differentiated	P > MC P > ATC
Cartels	Relatively few, acting as one	Yes	Homogeneous or differentiated	P > MC P > ATC
Pure monopoly	One	Yes	Unique	P > MC P > ATC

FIGURE 11–12

The Effects of Competitive versus Monopolistic Markets

In contrasting the higher price and lower quantity *(P$_m$ and Q$_m$)* produced under a monopolistic market structure with the lower price and higher quantity *(P$_c$ and Q$_c$)* theoretically possible in a purely competitive situation, *ACP$_c$* represents the area of consumer surplus. This is real income available to anyone who can capture it. Under competition, it goes to consumers. Under monopoly, cartels, and oligopolies, part of it *(P$_m$BFP$_c$)* typically goes to producers and part *(BCF)* is lost to society.

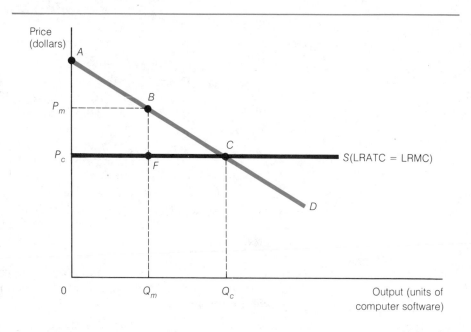

along monopolistic rather than competitive lines. The industry demand curve for this market has the usual negative slope, but the industry has a flat supply curve in this simplified conception. In other words, this is a constant-cost industry, in which additional output may be produced at a constant average cost that is also equal to marginal cost. The supply curve is simply a flat line, as in Figure 11–12.

To verify what happens to consumers' surplus under the different market structures represented in Figure 11–12, look again at the long-run market tendencies of the various models in Table 11–1. If the computer software industry is purely competitive, it will theoretically be producing an output of Q_c at price P_c, which is the lowest possible price because it is equal to both the average cost and the marginal cost of production. In this situation, consumers enjoy a surplus represented by ACP_c. In contrast, price exceeds both average cost and marginal cost in the market models of oligopoly, cartels, and pure monopoly. Economic profits are therefore captured at the expense of consumers in these markets. In Figure 11–12, P_m and Q_m represent the effects of a market structure on the monopoly end of the spectrum; the area P_mBFP_c is redirected from consumers' surplus to producers' profits. Such a redistribution of real income would not worry economists so much were it not that society must also pay a deadweight loss (the area BCF in Figure 11–12) because price exceeds cost. This deadweight or social loss due to monopoly power or the semimonopoly power of cartels and oligopolies causes some economists to condemn such structures as reducers of consumer and social welfare.

As we have seen in this chapter, monopolistic competition is something of a special case. Under monopolistic competition, there is a long-run tendency for price to equal average total cost, although not marginal cost when economies of scale are considered. The benefits of differentiated products—a greater range of choices—also exist in this structure. But the only unambiguously welfare-maximizing structure is that of pure competition. As Table 11–1 and Figure 11–12 indicate, only competition provides the product in this example, computer software, at the lowest possible price (equal to both LRATC and LRMC) and in the greatest quantities possible, Q_c. Only pure competition maximizes society's welfare by maximizing consumers' surplus. Important deviations from this norm are prime subjects for society's economic policy concerns. The following two parts of this book, especially Part IV, discuss a number of important economic issues related to market structure.

Summary

1. Monopolistic competition is a market structure that contains a large number of firms. Each firm sells a product that is similar to but slightly differentiated from the products of other firms in the industry. Freedom of entry and exit exist in this market structure.

2. Advertising is a means by which firms distinguish their products and thereby increase demand for them. Advertising and other forms of nonprice competition characterize monopolistic competition.

3. In the long run under monopolistic competition, a tangency between the firm's demand curve and its average cost curve can exist. If this tangency occurs, then the firm produces a rate of output less than the rate associated with minimum average total cost; that is, excess capacity exists.

4. The excess capacity of monopolistic competition may have some value. The larger number of firms may offer a larger variety of products or a lower real price to consumers than a competitive market with similar production costs.

5. Oligopoly is a market structure characterized by a few firms that sell either homogeneous or differentiated products. Entry into the industry is difficult because of either natural or artificial barriers.

6. In an oligopoly, the small number of sellers leads

to mutual interdependence. The actions of one firm have significant effects on other firms in the industry. For this reason, the behavior of firms in oligopoly is difficult to predict or analyze.

7. The kinked demand curve theory of oligopoly encompasses two distinct reactions of firms to price changes. It is suggested that if one firm increases its price, no other firms will follow the price increase. On the other hand, if one firm lowers its price, all other firms will lower their prices as well. The resulting demand curve can result in a rigid price structure.

8. Cartels are formal agreements among firms within an industry to restrict output or to segment the market in an effort to increase profits. The result is similar to monopoly price and output in the industry.

9. There are potential profits for firms that cheat individually on a cartel. Cartels are therefore difficult to establish and maintain. The costs of enforcing the cartel agreement frequently prevent cartel's survival.

10. Cartels can be legal or illegal. Those created and enforced by the government or those allowed to exist by government sanctions are legal. Otherwise, cartels in the United States are breaches of antitrust legislation.

Key Terms

imperfect competition
monopolistic competition
differentiated products
nonprice competition
advertising
tangency solution

excess capacity
economic efficiency
economic welfare
oligopoly
mutual interdependence
market power

kinked demand
price leadership
limit pricing
cartel
cartel enforcement

Questions for Review and Discussion

1. Select two products or services that you regularly consume. Are these goods produced under monopolistically competitive conditions? What information do you have to have before giving a definite answer to this question?

2. What, exactly, is the role of product differentiation and advertising in monopolistic competition?

3. How many forms of competition can you name and

analyze in addition to price competition? How is location a factor in competition?

4. Given that Coke is the number-one soft drink in the market, is it wasteful for the firm to continue to advertise on so large a scale? Give reasons for your answer.

5. What is the tangency solution to the market model of monopolistic competition? Are half-empty air-

planes on the New York–San Francisco route definite evidence of excess capacity and economic waste? Give reasons for your answer.

6. Compare the results of long-run equilibrium in pure competition and monopolistic competition from the standpoint of both efficiency and welfare.

7. What are the general characteristics of the oligopolist market model? Compare and contrast these to the characteristics of the monopolistic competition model.

8. Saudi Arabia is the dominant firm in the OPEC oil cartel. Describe how world oil prices might have been set in the 1970s.

9. What does the free-rider effect mean? Cite three personal experiences in which you have observed free-riding behavior.

10. Suppose that firms forming a cartel have different marginal costs. Describe, both in words and in a graph, how this situation complicates the various problems that a cartel must solve to be effective in raising prices.

11. Which of the following factors make collusion more or less likely: (a) the purchase of an input in the same market (such as pigs at the stockyard); (b) long-term contracts with buyers; (c) selling to governments; (d) salespeople staying in the same motel on the road; (e) an industry trade association and price list.

ECONOMICS IN ACTION   The Prisoner's Dilemma and Automobile Style Changes

A helpful analytical tool in analyzing the effects of nonprice competition among a small group of oligopolist producers of differentiated products is the concept of the prisoner's dilemma. The prisoner's dilemma originated in a story told by mathematician A. W. Tucker. Two people are caught in a serious crime, but the district attorney has evidence to convict them only for a lesser offense. The D.A. separates the prisoners and attempts to obtain a confession in this manner: Both prisoners are separately informed that (1) if one confesses, the confessor goes free and the other hangs; (2) if neither confesses, both will receive the modest penalty that goes with the lesser crime; (3) if both confess, both will receive a severe penalty short, of course, of hanging. Given the payoffs and the uncertainty, the expected result is a confession from both prisoners.

The prisoner's dilemma can be related to economic behavior. The automobile industry's annual change of style is an example of this type of oligopoly behavior.[a] A style or design change increases automobile production costs (retooling, refabrication, and so on) and at the same time tends to reduce profits. Yet it is consistently chosen by competing firms acting in their own self-interest. Their reasoning is illustrated in Figure 11–13, a hypothetical payoff table, in which two automobile firms are trying to maximize their profits, represented by the varying dollar figures within the boxes.

From the standpoint of the industry, joint profit maximization occurs in box A, where neither firm changes style. Industry profits are $120 here ($65 for Ford and $55 for GM), as opposed to $100 in boxes B and C and $90 in box D. Yet box A is not the profit-maximizing choice for either individual firm.

[a]Harold Bierman, Jr., and Robert D. Tollison, "Styling Changes and the 'Prisoner's Dilemma,'" *Antitrust Law and Economics Review* 4 (Fall 1970), pp. 95–100.

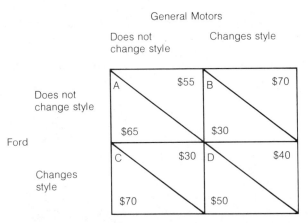

FIGURE 11–13   The Prisoner's Dilemma and Automobile Style Changes

Various possible combinations exist when Ford and General Motors try to maximize profits by deciding whether to change styles. Profits for the firms' decisions are given in millions of dollars. The best choice for the industry is box A, where neither firm changes style, but the best profit-maximizing choice for each firm is to change style.

Take GM's decision. No matter what Ford does, GM's profits are higher if it changes its style. If it changes and Ford does not (box B), GM's profits are $70. If it changes and Ford does likewise, its profit will be only $40 (box D). But if it does not change style and Ford does, GM's profits will drop to the lowest point of all, $30 (box C). Ford management considers its own set of options and reaches a similar conclusion, namely, that Ford will always be better off by changing its style. If they both make decisions that tend to maximize their individual profits, both will introduce style changes and together earn a smaller amount of profit

($90) than they would have earned in the absence of a style change by either ($120).

This is the nature of a prisoner's dilemma. Both parties could be better off if they could communicate and stay in box A. Indeed, perhaps we would all be better off—pay less for cars—if automobile firms did not change styles every year or so. But since firms cannot legally communicate because we fear that they would talk about more than style changes, that is, would collude, we are stuck with the implications of the prisoner's dilemma. The general argument here is also applicable to other forms of nonprice competition among oligopolists, such as advertising.[b]

[b]The classic work applying game theory, of which the prisoner's dilemma is an example, to economic analysis is by mathematician John von Neumann and economist Oskar Morgenstern: *Theory of Games and Economic Behavior* (Princeton: Princeton University Press, 1944).

Questions

1. What would be the pros and cons of allowing auto-makers to meet each year to decide collectively whether or not to change styles?
2. The home computer and computer software industries are both characterized by a rapid rate of technological advance. Assess the potential benefits and costs to consumers when there is (a) open competition in these areas, or (b) a cartel agreement which includes a slower introduction of new technology than would occur under competition.

ECONOMICS IN ACTION

All That Glitters: The De Beers Diamond Cartel

Recession and falling demand have been the ruination of many cartels. During periods of declining demand, the temptation to cheat is great. However, the De Beers diamond cartel—the strongest, longest-running, and most successful cartel of modern times—successfully warded off a threat of dissolution in the recession of the late 1970s and early 1980s.[a] De Beers' maneuvering during this period provides interesting insights into the enforcement powers of a cartel.

Founded by entrepreneur Cecil Rhodes in 1888, De Beers Consolidated Mines is a publicly held corporation that markets more than 80 percent of the uncut diamonds of both gem and industrial quality sold yearly in the world. De Beers owns some South African mines and acts as the marketing agent for virtually all other diamond-producing and -supplying nations, including Zaire and other African countries, Australia, and the Soviet Union. De Beers pays royalties out of its retail sales to these nations for exclusive rights to market their stones. Retailers have no choice but to purchase wholesale boxes of diamonds from De Beers; if they refuse, they are not necessarily invited to purchase diamonds again. Thus De Beers may be thought of as a marketing or middleman type of cartel.

The De Beers cartel system worked well until the

[a]John R. Emshwiller and Neil Behrmann, "Restored Luster: How De Beers Revived World Diamond Cartel After Zaire's Pullout," *Wall Street Journal*, August 7, 1983.

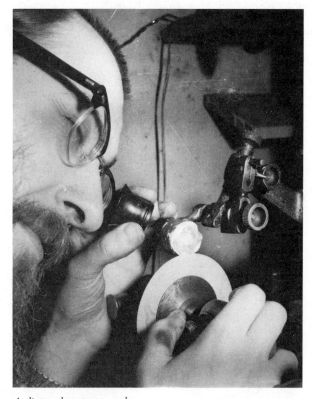

A diamond cutter at work.

late 1970s when several storm clouds appeared to threaten its market power. In 1979, an enormous load of diamonds called the Arggle mine was discovered in Australia. Even worse for the cartel, a worldwide recession created massive reductions in demand for diamonds. De Beers' earnings fell from $673 million in 1979 to $183 million in 1982. Rather than reduce price in the face of declining demand, De Beers' reaction was to remove massive supplies of diamonds from the market by cutting production at its own mines to shore up the cartel price. De Beers also reduced purchases from participating nations in return for lump-sum payments. In all, De Beers added more than one billion dollars' worth of diamonds to its inventory between 1979 and 1982.

Most troublesome of all, Zaire, the world's largest producer of diamonds, pulled out of the cartel in May 1981 and gave its diamond concession to three independent producers for five years. These independents marketed low-priced industrial-quality diamonds (called *boart*) from Zaire's Miba mine. De Beers' reaction was decisive. Though stockpiling diamonds in general, the company poured boart onto the market, causing the price of low-grade industrial diamonds to fall by two-thirds. De Beers was also charged with encouraging illegal diamond smuggling by Zairian citizens. The result: Zaire's revenues and profits fell precipitously. The country was even refused loans from international agencies to upgrade its mines. In March 1983 Zaire returned to the fold, dropping the three independent marketers in midcontract.

The icing on De Beers' cake came when Australia joined the cartel. The two majority stockholders of Australia's Arggle mine, which could produce a full 25 percent of current world production by 1985, had indirect ties to De Beers through a chain of stock ownership. Though there was political opposition and opposition from some of the partners, De Beers' Australian friends helped bring Arggle within the cartel, but with some concessions on the part of De Beers. Australia is now free to market 25 percent of its industrial stones and 5 percent of its gemstones outside the cartel.

Some observers believe that De Beers' concession sets a precedent that could weaken the cartel. Others believe this outcome unlikely. The Soviet Union, for example, could produce and sell enough diamonds to create chaos in world markets but chooses instead to play the cartel game. The Soviet Union apparently thinks that long-run price stability is preferable to possible short-run gains. Zaire, through costly misadventure, learned this lesson all too well.

Question

The De Beers diamond cartel is largely a "private" cartel held together by private enforcement. Compare the enforcement problems that would likely be encountered in cartels such as De Beers to those found in government-supported ("legal") cartels.

POINT-COUNTERPOINT

Alfred Marshall and Joan Robinson: Perfect versus Imperfect Markets

Alfred Marshall

Joan Robinson

ALFRED MARSHALL (1842–1924), an English economist, was the strong-willed son of a harsh father, who early on tried to coerce young Alfred into the ministry. Much to his parent's dismay, Marshall rejected a theological scholarship to Oxford University and entered Cambridge, where he showed early brilliance as a mathematician and received his M.S. in mathematics in 1865. From mathematics, Marshall veered into metaphysics, joined a philosophy discussion group at Cambridge, and lectured for almost ten years in moral science at Cambridge. During this period Marshall became convinced of the overriding importance of economics to individual and social action. In 1877 Marshall married Mary Paley, a former student, and moved to Bristol, where they both lectured on political economy at University College. In 1885, they returned to Cambridge, where Marshall continued his long and illustrious career as lecturer in political economy.

The most important of Marshall's books was the modern bible of microeconomics *Principles of Economics* (1890),[a] which he revised through nine editions. Many of the concepts presented in today's introductory economics courses, including the principles of perfect competition among firms, were first set down systematically by Marshall. His mathematical precision and fondness for lucid example made the text and economics generally accessible and popular. In 1903, while his work on *Principles* continued, Marshall succeeded in establishing economics as a discipline separate from moral science. He is also credited with founding the neoclassical tradition in economics—the modern version of economic principles established by Adam Smith.

JOAN ROBINSON (1903–1983), like Alfred Marshall, spent most of her academic career at the University of Cambridge. She completed her B.A. in economics at Cambridge in 1925, the year after Marshall's death. She began a forty-year teaching career at Cambridge in 1931, at a time when Marshall's influence over the principles and methods of economics study was still great. Despite these close ties, Robinson is perhaps best known for her book *The Economics of Imperfect Competition* (1933),[b] which challenges many of Marshall's conclusions about business organization and his theory of perfect competition among firms.

Robinson's contributions to economics reach well beyond her theory of imperfect competition. In the mid-1930s she and a small group of Cambridge economists helped John Maynard Keynes write his monumental *General Theory of Employment, Interest, and Money*, which ushered in a new era of macroeconomic theory. She also wrote in the areas of economic development, international trade, capital theory, and Marxian economics.

Although Robinson later repudiated her own analysis of imperfect competition, her critique of Marshall's assumptions of how the economy operates presents an interesting chapter in the development of economic thought.

The World of Perfect Competition

To Marshall, the microeconomic behavior of firms and individuals was kept in constant check by the competitive nature of markets. The dynamics of free enterprise, of limitless entry to and exit from markets enforced the norm of perfect competition among buyers and sellers. By his terms, each market is made up of sufficiently large numbers of competing firms buying and selling virtually identical products and services. No

[a]Alfred Marshall, *Principles of Economics*, 8th ed. (London: Macmillan, 1920).

[b]Joan Robinson, *The Economics of Imperfect Competition* (London: Macmillan, 1933).

single demander or supplier of goods can affect the market price because no single firm has a large enough stake in the market. Given these conditions, the market price for goods would always gravitate to a level equal to the firm's costs of production. As a result, economic profits could not persist. Under the rigors of perfect competition, no firm would be able to rise above its equals.

To demonstrate his model of perfect competition, Marshall chose a number of illustrations, including the market for fresh fish. Marshall noted that the ordinary daily price of fish varies according to factors such as the daily intensity of demand along with weather and luck, which affect the size of the catch. He then introduced into this illustration a cattle plague that permanently increased the demand for fresh fish over beef. How would the competitively organized fish market react? (Remember that no freezing or cold storage facilities were readily available in Marshall's day.)

Under competitive conditions, the price per pound of fish would immediately rise. Recognizing that the price rise was not a temporary phenomenon, fishermen would respond to the increased profits (since price would now be greater than costs of production) by using their boats, fishing crews, and nets more intensively. Supply would be increased by these activities as it would by the entry into the market of new fishing firms attracted by profits. Supply increases would ultimately bring price down to the average cost of catching fish.

Marshall's vision of perfect competition comports well with Adam Smith's theory of "the invisible hand." In fact, Marshall's theory of perfect competition describes the invisible hand at work. In his illustration, new fishing firms enter the market and old firms expand not for the benefit of customers but in the hopes of earning more profits. They act only in their own self-interest but in doing so promote the interest of society. The price of fish falls to the costs of production in the long run, and more fish are provided in the market.

Although Marshall did acknowledge the existence of monopolies and oligopolies, he believed that these market structures were a special case. Natural monopolies—those with large economies of scale in production (the phone company, the gas company, and so on)—could exist, as could monopolies due to artificial barriers to entry (government regulation, control of vital resources, and so on), but these were rare exceptions to competitive organization.

Perfect Competition Turned on Its Head

"It is customary," Joan Robinson wrote in The Economics of Imperfect Competition, ". . . to open with the analysis of a perfectly competitive world, and to treat monopoly as a special case. . . . It is more proper to set out the analysis of monopoly, treating perfect competition as a special case." Beginning with these words, Robinson set out to turn Marshall's theory of perfect competition upside down.

Modern industrialized societies, she contended, were dominated by monopolies (a single firm in a market) or oligopolies (a few large firms in a market). To a greater or lesser degree, most markets were influenced by the power of monopoly control. In other words, individual firms in these markets were large enough to affect market prices on their own. Given these assumptions, Robinson pointed out, the prices of goods and services would not gravitate to a level equal to the firm's costs of production. Economic profits could persist, enabling the dominant firms to maintain or even increase their dominance.

Modern examples of imperfect competition include the automobile industry, dominated by industrial giants such as General Motors, or the fast-food industry, dominated by firms such as McDonald's. A single seller or a few sellers of running shoes and athletic equipment in a small college town is an example of monopoly on a smaller scale.

To further illustrate her point that imperfect, rather than perfect, competition was the norm, Robinson pointed to the pervasive evidence of price discrimination in markets. Whereas Marshall contended that only one market price could prevail under perfect competition, Robinson saw many instances in which a single good or service could command several different prices in the same market. Multiple prices for the same product could be possible, Robinson emphasized, only in a world of imperfect competition.

Robinson's theory of imperfect competition was not the only challenge made against Marshall's system of perfect competition. A similar theory was offered by American economist E. H. Chamberlin in his Theory of Monopolistic Competition,[c] also published in 1933.

Other important additions have been made to Marshall's theory of competition. Modern economists stress the manner in which information affects suppliers and demanders, the consequences of potential competition and rivalry on markets, and the impact on Marshall's model of interpreting price as "full price." The new theory of competitive markets has again come to dominate microeconomic analysis as the simple theory of competition did in Marshall's time.

[c]Edward H. Chamberlin, The Theory of Monopolistic Competition: A Re-orientation of the Theory of Value, 8th ed. (Cambridge: Harvard University Press, 1962), originally published in 1933.

III

Microeconomic Principles of Input Demand and Supply

12

Marginal Productivity Theory and Wages

*I*n Chapter 4, we presented an overview of the resource and products markets (see the circular flow model in Figure 4–1, p. 70). We showed business firms as both suppliers of goods and services and demanders of inputs such as land, labor, capital, and entrepreneurship. Households were depicted both as demanders of goods and services and as suppliers of inputs. Up to this point in our study of microeconomics, we have focused on the upper part of the circular flow model and studied how the prices of final products such as psychiatric services, home computers, or diapers are determined. We now turn to the lower part of the model to view ourselves as demanders or suppliers of inputs such as labor services, land, or capital. Specifically, we now inquire how all input prices—wages of day laborers or engineers, lumber, or machine prices—are determined. Once we know the economic process by which input values are determined, we will be able to understand better how and why people earn the incomes they do as owners of their own labor and other resources.

Factor market: The market in which the prices of resources (factors of production, or inputs) are determined by the actions of businesses as the buyers of resources and households as the suppliers of resources; also called resource market.

The markets in which input prices are determined are called **factor,** or **resource, markets.** The prices of factors such as wages, rent, and interest are determined through the interplay of supply and demand. Indeed, the theory of factor prices is just the familiar theory of demand and supply applied to factors of production.

We begin this chapter by presenting the marginal productivity theory of demand for one factor of production, labor. Although our example is labor, the marginal productivity theory is equally applicable to other factors of production, as we will see in Chapters 14 and 15, where we apply the theory to land and capital inputs. In the second part of this chapter, we present the theory of labor supply. Since wages are intimately linked to the level of income earned by labor, we devote the final section of the chapter to discussing the relevance of marginal productivity theory to income distribution.

Factor Markets: An Overview

The price and quantity of inputs such as labor and land are determined by the same laws of supply and demand that we have discussed for outputs. To understand why this is so, consider the general characteristics of factor markets.

Derived Demand

Derived demand: The demand for factors of production that is a direct function of the demand for the product that the factors produce.

The demand for labor or any other factor of production is a **derived demand.** This term means that the demand for labor is directly related to the demand for the goods and services that the labor is used to produce. If the demand for computers increases, the derived demand for computer scientists and engineers will increase. If the demand for physical fitness increases, the derived demand for health foods will increase. If the demand for law and order increases, the derived demand for police officers will increase.

Derived demand also obeys the law of demand. All things being equal, if the price of health foods falls, more will be consumed; if the price of paper rises, fewer or shorter books will be published; if the rental prices on office space fall, more offices will be rented. So even though we call the demand for factors of production a derived demand, the law of demand applies to it in the same way that it applies to demand for final output.

Firms as Resource Price Takers

Just as there are different types of output market structures—such as competition, oligopoly, and monopoly—similar differences can occur in resource markets. Table 12–1 summarizes the characteristics of output and input market structures.

There can be perfectly competitive sellers and buyers of labor, for exam-

TABLE 12–1   Characteristics of Output and Input Market Structures

Type of Market Structure	Output Markets	Input Markets
Pure competition	Many sellers; no barriers to entry; homogeneous product	Many sellers; no barriers to entry; homogeneous resource
Monopolistic competition	Many sellers; no barriers to entry; differentiated product	Many sellers; no barriers to entry; differentiated resource
Oligopoly	Few sellers; barriers to entry; homogeneous or differentiated product	Few sellers; entry is difficult; homogeneous or differentiated resource
Cartels; Union	Relatively few sellers acting as one; barriers to entry; homogeneous or differentiated product	Relatively few sellers acting as one; barriers to entry; homogeneous or differentiated resource
Pure monopoly	One seller; barriers to entry; unique product	One seller; barriers to entry; unique resource

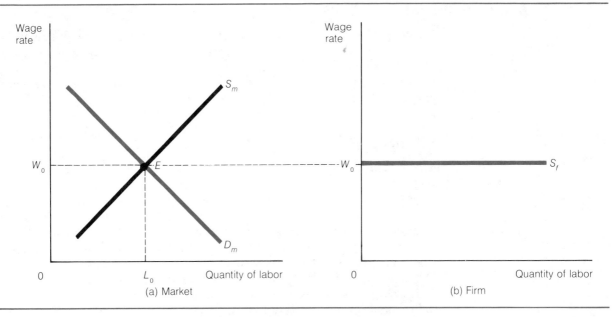

FIGURE 12–1   The Firm as a Wage Price Taker

(a) The equilibrium wage, W_0, is determined in the competitive market by the market supply, S_m, and demand, D_m, for a particular type of labor. (b) The firm must pay wage rate W_0 for workers but is able to hire all it wants at that wage. The supply of labor to the competitive firm, S_f, is infinitely elastic at the prevailing wage.

Competitive labor market: A labor market in which the wage rate of a particular type of labor is determined by the forces of supply by a large number of sellers of labor and demand by a large number of buyers of labor.

Market demand for labor: The sum of, or overall, demand for a particular type of labor by all firms employing that labor; the total level of employment of a particular type of labor at all the various wage rates.

Market supply of labor: The total amounts of labor that all individuals are willing to offer in a particular occupation at all the various wage rates.

ple. In such cases, there are large numbers of both buyers (employing firms) and sellers (workers) of a homogeneous type of labor. There can also be various types of imperfect competition in buying and selling labor, as we will learn in the next chapter. We limit this chapter to the competitive labor market. The following discussion of labor demand examines labor markets in which a large number of firms hire a single type of labor and a large number of people are willing to supply the labor.

In a **competitive labor market,** the wage rate is determined by the familiar forces of supply and demand. The overall or **market demand for labor** is the sum of all firms' demand for the type of labor in question. All of the firms may not be in the same industry—that is, they may not all produce the same product—but they all employ the same type of labor. For example, the demand curve in Figure 12–1a could represent the market demand for electricians. This overall demand would be found by summing the demand for electricians by all buyers of their labor, from AT&T and General Motors to construction companies, cable television companies, and even homeowners.

The overall or **market supply of labor** is obtained by summing the supply of individual workers. As indicated by the supply curve S_m in Figure 12–1a, as the wage rate rises, a larger quantity of labor is supplied. The supply of labor to any particular occupation thus follows the familiar law of supply.

The interaction of supply and demand for labor results in an equilibrium wage W_0. Once this wage is established, the buyers and sellers of labor are price takers. That is, the firms must accept W_0 as the prevailing wage and may hire as many workers as they are willing and able to hire at that wage. Being perfectly competitive buyers of labor, employing firms face an infinitely elastic labor supply curve, as shown in Figure 12–1b. All buyers of any good or service who are price takers face an infinitely elastic supply curve.

Marginal Productivity Theory

We cannot say exactly how many electricians or computer programmers will be demanded within a given time period, but we can present some general rules about how the numbers of people demanded in these occupations might rise or fall. In this analysis, we reencounter many principles from earlier chapters. All apply equally to the demand for resource inputs, including labor.

The Profit-Maximizing Level of Employment

The firm hiring labor in a competitive market faces a market-determined wage. Since the equilibrium wage is established, the point of interest becomes the quantity of labor that an individual firm chooses to hire. If a firm is selling its product in a perfectly competitive product market, then how much labor will it hire at a particular wage rate?

To answer this question, let us say that the firm has many inputs, one of which is labor. Also, let the quantities of all inputs except labor be fixed. Labor, in other words, is the only variable input. (These circumstances, as you may recognize, describe a short-run situation for the firm.) To produce more output, the firm must hire more labor. The firm wishes to maximize profits and must therefore hire the profit-maximizing amount of labor.

Suppose the firm in this instance is a corn farmer who must decide whether to hire a worker to help pick and husk corn. How does the farmer make this decision? Hiring the worker will increase output, and the increase in production will be sold to obtain revenue. At the same time, hiring the worker will increase the cost of production because the farm worker must be paid. The farmer makes a profit-maximizing decision *at the margin*. If hiring the worker increases total revenue more than it increases total cost, then hiring the worker will increase profits. The farmer will always hire a worker if doing so increases profits.

Suppose the farmer is thinking about hiring a second worker. The same decision process is repeated. If the second worker adds more to revenue than to cost, then he or she will be hired. This process continues with every prospective worker. An important principle is in action: the *law of diminishing marginal returns*, which you may recall from Chapter 8. As more and more of a variable input is added to a fixed input, the marginal product of the variable input eventually declines. As more workers are hired, the extra output of each worker eventually falls. Since each additional worker adds less and less to total output, the additional revenues that each worker produces eventually fall below the additional cost of hiring. At some point, the farmer will stop hiring altogether. In short, the profit-maximizing farmer should hire all workers that add more to revenue than to cost but stop hiring at the point where the addition of revenue is just equal to the addition to cost.

The example of a corn farmer can also be expressed in economic terms. The extra output that each additional unit of labor adds to total output is called the **marginal product of labor:**

Marginal product of labor: The change in total output that results from employing one more unit of labor.

$$MP_L = \frac{\Delta TP}{\Delta L},$$

where MP_L is the marginal product of labor and ΔTP is the change in total product brought about by adding one unit of labor, ΔL. As we saw in Chapter 8, adding units of the variable input to other fixed inputs will eventually decrease the marginal product of the variable input in the short run.

TABLE 12–2   A Marginal Revenue Product Schedule for Labor

The marginal revenue product of labor in column (4) falls as more units of labor are employed because the marginal product of labor (column 2) falls. The firm will hire additional units of labor up to the point where the extra revenue these units generate just equals the extra cost of paying for them. At this point, 6 units of labor, MRP = MFC.

(1) Units of Labor (worker-hours)	(2) Marginal Product of Labor MP_L (bushels of corn)	(3) Marginal Revenue MR = P (dollars per bushel)	(4) Marginal Revenue Product $MRP = MR \times MP_L$ $= \Delta TR/\Delta L$	(5) Marginal Factor Cost (dollars per hour) MFC = Wage
1	14	$2	$28	8
2	12	2	24	8
3	10	2	20	8
4	8	2	16	8
5	6	2	12	8
6	4	2	8	8
7	2	2	4	8
8	0	2	0	8

Table 12–2 summarizes the hypothetical choices available to the corn farmer. The first column shows the units of labor, in number of worker-hours. The second column shows the marginal product of the workers—the amount by which total output increases as one more unit, or hour, of labor is added to the production process. MP_L decreases as the amount of labor increases, according to the law of diminishing marginal returns. The first unit of labor adds 14 units of output (bushels of corn), the second adds 12 units of output, and so forth.

When the extra corn produced by each additional unit of labor is sold, the resulting increase in total revenue is called **marginal revenue product** (MRP):

$$MRP_L = \Delta TR/\Delta L,$$

where MRP_L is the marginal revenue product of labor, ΔTR is the change in total revenue, and ΔL is the change in the amount of labor hired.

Another method of expressing MRP_L is

$$MRP_L = MR \times MP_L,$$

where MR is the marginal revenue (the increase in total revenue resulting from selling one more unit of output) and MP_L is the marginal product of labor.

For the firm that sells its product in a perfectly competitive market, marginal revenue is equal to the price of the product. Thus, the marginal revenue product (MRP) is simply the price (P) of the product times the marginal product (MP) of each unit. As shown in Table 12–2, the marginal revenue product of the first unit of labor is found by multiplying its marginal product, 14, by the price of the product, $2, to obtain $28. The MRP_L declines as more workers are hired because MP_L declines.

Regardless of which economic shorthand is used, the profit-maximizing firm will continue adding units of labor as long as the additional revenues the labor produces are greater than the additional costs of the labor. The

Marginal revenue product: The change in total revenue that results from employing one more unit of a variable input.

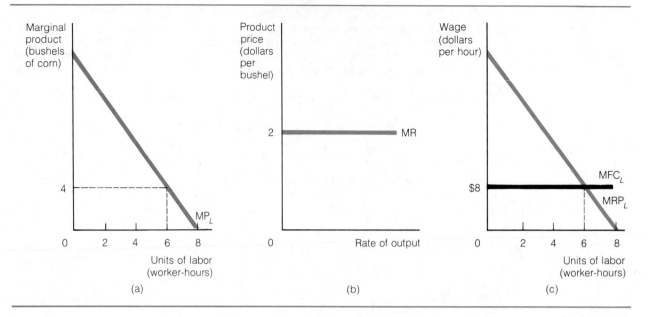

FIGURE 12–2   The Firm's Marginal Revenue Product Curve

(a) The farmer's marginal product of labor (MP_L) curve is plotted from columns (1) and (2) in Table 12–2. (b) The demand curve for the farmer's product is determined by the product price (P = MR in this perfectly competitive market). (c) The marginal revenue product (MRP_L) curve is found by multiplying MP_L by MR. The profit-maximizing firm hires labor up to the point where marginal revenue product of labor equals marginal factor cost of labor (MFC_L). With a wage of $8 and a product price of $2, the firm will hire up to 6 units of labor.

Marginal factor cost: The change in total cost that results from employing one more unit of a variable input.

extra cost of hiring one more unit of labor is called the **marginal factor cost** of labor:

$$MFC_L = \Delta TC/\Delta L,$$

where MFC_L is the marginal factor cost and ΔTC is the change in total cost. Under competitive conditions, the firm may purchase all of the labor it wants at the prevailing wage rate. Each additional unit of labor has a marginal factor cost equal to the wage rate (shown as $8 on the schedule).

The profit-maximizing firm will hire labor to the point where MRP_L = MFC_L. With the numbers in Table 12–2, this equation indicates that it will be most profitable for the farmer to hire 6 units of labor. Each unit from the first to the sixth adds more to revenue than to cost. By hiring 6 units, the farmer adds as much as possible to profit. However, the seventh unit adds more to cost than to revenue and therefore will not be hired.

This process should sound familiar: It is the mirror image of how the competitive firm determines its profit-maximizing output. The firm follows the same process to equate its marginal cost to marginal revenue in hiring inputs as it does in selling its output. In fact, when the firm hires the profit-maximizing quantity of labor, that amount of labor will produce the profit-maximizing level of output.

The numbers used in Table 12–2 are presented in graph form in Figure 12–2. The MP_L curve is drawn in part a by plotting the points from columns (1) and (2). The demand curve for the farmer's corn is shown in part b, where P = $2. The MRP_L is found by multiplying marginal revenue by marginal product; the result is the marginal revenue product curve in part c.

The prevailing wage is \$8, stated as MFC_L. The profit-maximizing quantity of labor is the point where the MFC_L curve crosses the MRP_L curve. At this point, the marginal revenue product equals the marginal factor cost of labor.

Profit-Maximizing Rule for All Inputs

Firms employ many inputs other than labor. The profit-maximizing rule for the level of employment is the same for each resource. The corn farmer may be planning to use fertilizer, for example. If he applies 1 unit of fertilizer—a 100-pound bag—the amount of corn produced will increase. The extra corn is treated as the marginal product of the first bag of fertilizer. The extra revenue that results is the marginal revenue product, and the extra cost of the bag is the marginal factor cost. If the MRP of the fertilizer is greater than the MFC, then the farmer will purchase and use the fertilizer. The profit-maximizing amount of fertilizer is found at the point where the MRP of fertilizer is just equal to the MFC.

This rule applies to all inputs, such as tractors, seed, land, tractor drivers, and water. The profit-maximizing quantity of all inputs is the amount that makes the MRP of each input equal to the MFC of each input:

$$\frac{MRP_1}{MFC_1} = \frac{MRP_2}{MFC_2} = \frac{MRP_3}{MFC_3} = \cdots = \frac{MRP_N}{MFC_N} = 1,$$

where the subscripts 1, 2, and 3, ... indicate the MRP and MRC of the first, second, third, and so on, inputs, and N indicates the total number of inputs used. Each ratio of MRP to MFC is equal to 1 because the MRP of each is equal to the MFC of each. Under this condition, all inputs are employed in the profit-maximizing amounts.

The Demand for Labor

Marginal productivity theory gives insight into the purchasing behavior of firms. In the short run, when labor is the only variable input, the firm is willing to purchase or hire additional units up to the point at which MRP = MFC. The firm's marginal revenue product curve (Figure 12–2c) thus traces the relation between the price of labor (the wage rate) and the amount of labor a firm is willing to purchase. In other words, the marginal revenue product curve is also the firm's short-run demand curve for labor. Figure 12–3 illustrates this. At wage W_0 the firm chooses L_0 units of labor. A higher wage such as W_1 results in less labor hired. For each wage, the firm adjusts the level of employment to maintain the equation $MRP_L = MFC_L$.

In the following sections we examine the demand for labor from several perspectives.

The Short-Run Market Demand for Labor

A single firm's demand for labor in the short run is equivalent to its marginal revenue product of labor. To obtain the overall market demand for labor in the short run, we must make some additional calculations.

Each firm will, of course, demand labor up to the point where $MRP_L = MFC_L$. Thus, at a particular wage, the market demand for labor may be shown as the sum of all the firms' MRP_L curves. Figure 12–4 shows such a curve, MRP_{L0}. This curve represents simply the sum of all the firms' short-run demand curves for a particular type of labor. The curve labeled MRP_{L0},

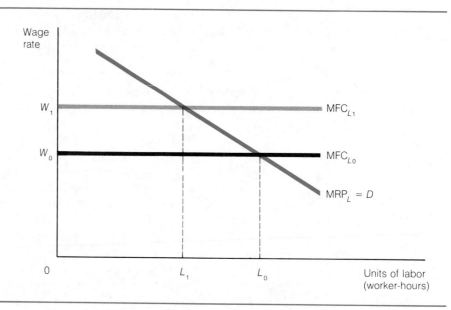

FIGURE 12–3   The Firm's Short-Run Demand for Labor

The firm hires labor up to the point where the wage is equal to the marginal revenue product of labor (MRP_L). The marginal revenue product curve shows the various quantities of labor the firm is willing to hire at all the different wage rates when labor is the only variable input.

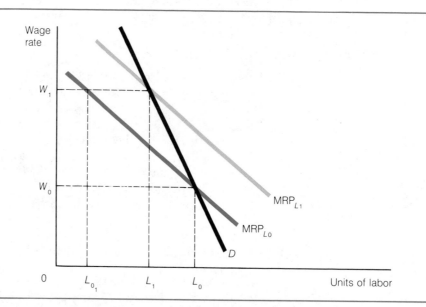

FIGURE 12–4   The Short-Run Market Demand for Labor

MRP_{L0} represents the sum of all the firms' marginal revenue product curves for a particular occupation. When the wage for this occupation is W_0, the firms hire L_0 units of labor. If the wage rises to W_1, the firms initially hire the lower L_{0_1} units of labor. However, the increase in wages causes an increase in the price of the product. The higher product price shifts the sum of the firms' MRP_{L0} to MRP_{L1}. At W_1, when labor is the only variable input, the firms hire L_1 units of labor. D represents the short-run market demand curve for labor because it allows for changes in the wage rate.

however, is not the short-run market demand curve for this type of labor, but it does establish one point on the short-run market demand curve, D.

To establish the whole demand curve, we must examine what happens when the wage rate changes. Imagine the overall demand for corn huskers. Many farmers employ huskers at wage W_0; the total number of corn huskers employed at W_0 is L_0. Recall that the MRP_L curve is obtained by multiplying the price of corn by the marginal product of labor. If the wage rate rises to W_1, then each firm's costs of production will rise. The increased costs of production will decrease the supply of corn and increase the equilibrium price. To obtain the new MRP_L curve (MRP_{L1}), we multiply the marginal product of labor by the new, higher price of corn, and we sum this quantity over all the firms. The new wage, W_1, and level of employment, L_1, establish a second point on the short-run market demand curve. The short-run market demand curve, D, differs from the sum of the MRP_L curves because it allows for increases in the price of the product that result from wage increases.

Firm's long-run demand for labor: The various quantities of labor that a firm is willing to hire when all inputs are variable.

The Firm's Long-Run Demand for Labor

We have so far concentrated on the short-run demand for labor, in which labor is the only variable input. In the long run, many adjustments occur when wages change. Allowed enough time, firms may vary all inputs. Also, at the industry level, not only does the price of products change directly with wages, but the number of firms changes as well.

Suppose that corn huskers demand a raise in wages. What happens in the long run? First, farmers will adjust to the higher price of labor by hiring fewer corn huskers; given enough time, they may turn to substitute inputs for this type of labor, perhaps by buying combines rather than hiring human huskers. Second, the increase in the price of labor increases the farmers' costs of production. With the increase in costs, the supply of the product decreases, and its price rises. Finally, the number of firms will adjust, establishing a new equilibrium. The number of firms, in the end, is likely to decrease. When wages rise, a larger number of workers lose their jobs in the long run than in the short run; or, if wages fall, a larger number of workers are hired. The long-run market demand is more elastic than the short-run market demand.

Changes in the Demand for Labor

In applying the law of demand to labor, we have seen some adjustments at the firm level and at the industry level in response to wage rate changes. An increase in wage rate changed the price of the product, the employment of other inputs, and the number of firms. However, in establishing the demand curve for labor, many other things were held constant, such as the demand for the product, the prices of other inputs, and the level of technology. If these things were not held constant, then we would not have been able to establish the relation of wage rate to the amount of labor demanded.

What causes an increase or decrease in the demand for a particular type of labor? Why would the demand for chemical engineers suddenly increase? Or why would a restaurant suddenly hire more waiters and cooks? Such questions may be answered by examining the origins of the demand for labor and the behavior of profit-maximizing firms.

The demand for labor is derived basically from two things: the demand for the product the labor produces and the marginal product of the labor itself. If either of these increases or decreases, then the demand for labor will increase or decrease in the same direction. Understanding this relation simplifies things; however, further details are of value.

1. *If the demand for the product that labor produces changes, then the demand for labor will change in the same direction.* Remember that for a competitive firm, the marginal revenue product is equal to marginal revenue (the output price) times marginal product. If the demand for the good increases, then its price increases and the demand for labor (MRP) and all other inputs also increases. The demands for electronic technicians and computer components have therefore increased since the demand for personal computers has increased.

The opposite situation occurs when the demand for a product decreases. If the price of a product falls, then the demand for inputs used to produce that product falls. When the demand for American-made cars decreases, the demand for domestic autoworkers decreases (along with the demand for tires and steel).

2. *If the marginal product of labor changes, then the demand for that labor will change in the same direction.* The productivity of a particular type of labor depends to a very large extent on the firm's employment of other inputs in the production process. Additional expenditures on other inputs often increase the productivity of labor. For example, the marginal product of farm workers is influenced by the amounts of other inputs that the farmer buys, such as land, equipment, seed, fertilizer, and water. If such inputs increase the workers' productivity, their use will also increase demand for the workers' labor.

Complementary inputs: Two or more inputs with a relation such that increased employment of one increases the marginal product of the other.

Substitute inputs: Two or more inputs with a relation such that increasing the employment of one decreases the marginal product of the other.

The relation between the quantity of nonlabor inputs and the productivity of labor can be direct or inverse. In other words, increasing the amount of one input may either increase or decrease the productivity of labor. If increasing the quantity of one input increases the marginal product of labor, then this input and labor are **complementary inputs.** For example, if the farmer uses more fertilizer, then corn huskers can produce more output; their marginal product increases. Increases in inputs that complement labor will increase the demand for labor.

If increasing the amount of another input decreases the marginal product of labor, then that input is called a **substitute input** for labor. For corn pickers and huskers, a combine that picks and husks is a substitute input. If the farmer uses more combines, then the marginal product of human huskers and pickers will decrease, and so will the demand for them.

Why would a farmer suddenly choose to purchase more fertilizer or combines? A firm makes decisions on the basis of expected profits. For example, if the price of fertilizer falls, then it is in the farmer's interest to purchase more fertilizer. If the price of combines falls, then the farmer will purchase more combines. Thus, if the price of a complementary input falls, the demand for labor rises. If the price of a substitute input falls, the demand for labor decreases.

In addition to changes in the quantity of inputs, technological advances may shift the demand for labor. New discoveries about production techniques may increase or decrease the demand for labor in a particular occupation. In fact, this is generally what technological change does: It increases the marginal product of some workers and decreases that of others. When robots replace factory workers, the marginal product of the factory workers falls, and the demand for their labor decreases. At the same time, the marginal product of the workers who make robots goes up, and the demand for their labor increases.

The Elasticity of Demand for Labor

We have shown how to find the firm's demand for labor, how to find the market demand for labor, and how the demand curve for labor can shift. We still need to say something about the **elasticity of demand for labor.** Remember that elasticity measures the responsiveness of buyers to changes in price. Specifically, the elasticity of demand for labor is equal to the percent change in the quantity of labor demanded divided by the percent change in the wage rate.

Keep in mind that we are not talking about labor in general. We are talking about the derived demand for a specific type of labor. For example, we want to understand why the level of employment of corn huskers decreases by 30 percent when the wage rate increases by 10 percent; or why the level of employment of nuclear physicists falls by only 10 percent when their wages increase by 30 percent. There are three useful rules to apply when analyzing the elasticity of demand for labor.

1. *The elasticity of demand for labor is directly related to the elasticity of demand for the product that labor produces.* If the demand for a product is highly elastic, then a small percent increase in price will decrease the quantity demanded by a large proportion. If the increase in price is caused by an increase in wages, the resulting large decrease in production will cause a large decrease in the quantity of labor demanded. Suppose that the demand for houses is very elastic, and the wage of carpenters increases. The increase in the price of houses that results from the higher wages will decrease the quantity of houses purchased by a relatively large amount. The decreased production of houses will then decrease the quantity of carpenters demanded by a relatively large amount.

The greater the elasticity of demand for corn, the greater the elasticity of demand for corn huskers. The less elastic the demand for nuclear reactors, the less elastic the demand for nuclear physicists. The elasticity of demand for labor is in part derived from the elasticity of demand for the product that labor produces.

2. *The elasticity of demand for labor is directly related to the proportion of total production costs accounted for by labor.* The total amount of money that a firm pays for labor is a percentage of its total costs. The larger the percentage of total costs accounted for by labor, the greater the elasticity of demand for labor. This relation occurs because of the effect of wage increases on the price of the product. If wages rise by $2 an hour and labor represents a large percentage of total costs, then total costs and thus product price will increase by a large amount. The large price increase will decrease the amount of the product purchased and will decrease greatly the quantity of labor demanded. On the other hand, if labor accounts for a very small percentage of total costs, the same $2-an-hour wage increase will have little effect on overall cost and price. The decrease in the amount of the product demanded will be very small, making the decrease in labor demanded also small.

If the percentage of corn huskers' wages in the farmer's total costs is very large, then an increase in the huskers' wages will result in a relatively large decrease in the number of workers that the farmer hires. Conversely, if the percentage of huskers' wages in the farmer's total costs is very small, then the same wage increase will result in a relatively small decrease in the number of huskers hired.

3. *The elasticity of the demand for labor is directly related to the number and*

Elasticity of demand for labor: A measure of the responsiveness of employment to changes in the wage rate; the percent change in the level of labor employed divided by the percent change in the wage rate.

availability of substitute inputs. The demand for labor is highly elastic if there are good, relatively low-priced substitutes for labor. If the wage rate rises and there is another comparably priced input that can do the job of labor, then the relative amount of labor hired will decrease significantly. If there are no close substitutes for labor, or the substitutes are of a much higher price, then a wage increase will reduce employment by a lesser amount.

Once again, we see that the behavior of firms is similar to the behavior of consumers. The determinants of the elasticity of demand for labor and other inputs parallel the determinants of the elasticity of demand for goods by consumers. A larger percentage of total expenditures for purchases of some item—labor by firms or a good by consumers—creates greater elasticity of demand. A large number of substitutes for an item creates a highly elastic demand.

Monopoly and the Demand for Labor

We have spoken so far of demand for labor only under conditions of competition. What if the firm demanding labor enjoyed a monopoly in its output market but was still a purely competitive buyer of labor? For example, what would the farmer's demand for labor be if he were the only producer of corn? The farmer's decision rule for hiring labor would include a minor twist. Recall from Chapter 9 that the demand curve for the monopolist's product is downward sloping. This suggests that as the firm hires more workers and increases output, the price of the product must fall for the firm to sell the extra output. As the firm hires more labor, not only is the marginal product of labor falling but the product price is falling as well.

Even though the price of the product is falling as more output is produced, marginal revenue product, MRP, still describes the change in total revenue that results from hiring one more unit of labor. As before, MRP is equal to marginal revenue times marginal product. The difference is that the marginal revenue of the competitive seller is equal to the market price, whereas the monopolist's marginal revenue is a downward-sloping curve, derived from its demand curve.

A graphical comparison of the purely competitive firm's demand for labor and the monopolist's demand for labor is shown in Figure 12–5. The competitive firm's demand for labor is MRP_c in Figure 12–5c, obtained, as before, by multiplying MR_c from part a by MP_L, the marginal product of workers in part b. The monopolist's demand for labor, MRP_m, is obtained by multiplying MR_m in part a by MP_L in part b. The marginal product of workers, MP_L, falls at the same rate whether they are working for monopolists or for purely competitive firms. But the monopolist's MRP diminishes more quickly than the pure competitor's because not only is the marginal product of labor falling, but the monopolist's marginal revenue is falling as well.

A firm that enjoys a monopoly in its output market does not necessarily have any advantage in resource markets. For example, your local electric company may have exclusive rights as the only seller of electricity in your area. But this monopoly position does not imply that the electric company is the only employer of electricians, computer programmers, or bookkeepers. For these occupations, the monopoly is a purely competitive buyer of labor and must pay a competitive market wage for its labor. The marginal factor cost of labor is equal to the wage rate because the supply of labor to the firm is infinitely elastic.

In the end, the monopolist's decisions in hiring labor require the same

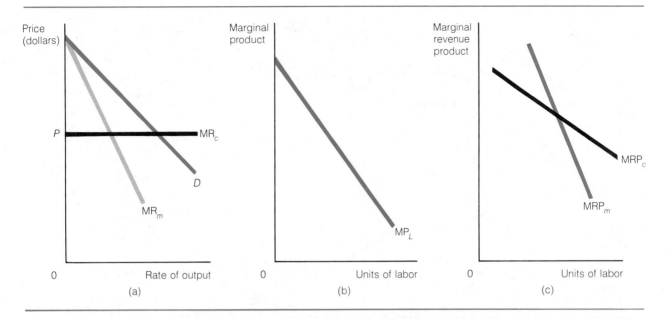

FIGURE 12–5   The Monopolist's MRP versus the Competitive Firm's MRP

(a) The marginal revenue curves for a monopoly (MR$_m$) and for a competitive firm (MR$_c$). (b) The marginal product of labor for both types of firm. (c) The marginal revenue product for each firm, obtained by multiplying marginal revenue by marginal product. The monopolist's marginal revenue curve is downward sloping, and the competitive firm's is equal to a given and constant product price; therefore, the monopolist's demand for labor, MRP$_m$, diminishes more rapidly than the competitive firm's MRP$_c$.

marginal analysis as for any other firm. The firm will hire any unit of labor when its marginal revenue product is greater than its marginal factor cost and will continue to hire labor up to the point where the MRP$_L$ is equal to MRC$_L$. The only difference for a monopoly firm is that its marginal revenue falls along with the labor's marginal product. Focus, "The Marginal Revenue Product of Professional Baseball Players," discusses these ideas as they relate to another market.

The Supply of Labor

Now that we have some insight into the demand for labor, the next logical category is labor supply. We know that labor supply in competitive markets is highly elastic. But this fact does not answer two important questions: What determines the overall, or aggregate, supply of labor and how are wage rates and labor supply interrelated?

Individual Labor Supply

Individuals in free markets are able to choose whether to offer their labor services in the market, and individuals' labor market decisions are based on the principle of utility maximization. The individual must choose between time spent in the market earning the wage that he or she may command and time spent in **nonmarket activities,** such as going to school, keeping house, tending a garden, or watching television.

Nonmarket activities: Anything that an individual does while not earning income from working.

Fortunately, the individual does not have to make an all-or-nothing

FOCUS   The Marginal Revenue Product of Professional Baseball Players

The extra revenue that a firm obtains from hiring a worker is the marginal revenue product; if this is higher than the extra cost of the worker's wages, then the firm adds to its profits by employing the worker. Baseball players' salaries have increased sharply in the last few years, especially for superstars. Are they worth it? Are their MRPs greater than their wages?

To answer this question, Gerald Scully calculated the extra revenues that team owners received from hitters and pitchers. He first found that a player's contribution to his team's win-loss record was best measured for pitchers by the percent each added to the team's strikeout-to-walk ratios and for batters to the contribution of each to the team's batting average. Using this information, Scully then estimated the total team revenue where revenue is simply the number of tickets sold per season times the average ticket price. After adjusting for the size of the city in which the team played its home games, the age of the stadium, the team's league, and other factors affecting revenue, he calculated that every point added to the team batting average by a batter increased the team's revenue by $9,504. Similarly, every .01 point added to the pitching staff's strikeout-to-walk ratio increased revenue $9,297.

Once he estimated player MRPs, it was a straightforward calculation for Scully to determine whether the salary paid to a particular player was commensurate with the amount the player added to team revenue. For example, star hitters added from $250,000 to $383,700

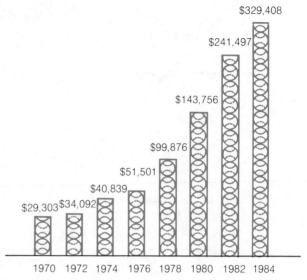

The average salary of major league baseball players has increased from $29,303 in 1970 to $329,408 in 1984.

Source: Major League Baseball Players Association.

Source: Gerald W. Scully, "Pay and Performance in Major League Baseball," *American Economic Review* 64 (December 1974), pp. 915–30.

to team revenues. Star pitchers contributed even more, from $321,700 to $479,000. However, during the 1968 and 1969 seasons, no player in Scully's sample was paid more than $125,000. He concluded, therefore, that players' salaries were well below their MRPs, with the greatest deficiency being the superstars' salaries. Scully's results explain in part free agency and the huge increases in players' salaries in the 1970s and 1980s.

---

choice. No one chooses to work twenty-four hours a day, seven days a week. People can divide their time between market and nonmarket activities. How many hours would one choose to work during an average week? This question cannot be easily answered, but we can speculate about the effect that the wage rate may have on an individual's labor supply choices.

A supply curve relates the quantity supplied of a commodity to its price. Applied to labor supply, it is a schedule that relates the quantity of work offered and different wage rates. The labor supply curve of an individual is given in Figure 12–6. The first thing to observe about this person's labor supply curve, S_i, is that below w_1 she would prefer not to work at all; she chooses all nonmarket activity. This cutoff occurs because wages below w_1 do not meet her opportunity cost; she receives more utility from nonmarket activities than she would from the income received from working. Wage w_1 may not be high enough to encourage her to forgo fishing, painting the house, or attending school, for example. At wages immediately above w_1, the hours of labor she supplies respond positively to increases in pay. As her

FIGURE 12–6

Individual Labor Supply

The individual labor supply curve, the number of hours offered at various wages, may be positively or negatively sloped. If the substitution effect is greater than the income effect, the curve is upward sloping, as shown from w_1 to w_3. The individual substitutes hours of work for hours of nonmarket activities. As the individual's wage rises beyond w_3, his or her demand for nonmarket time increases. The income effect is greater than the substitution effect, and the curve is negatively sloped above wage rate w_3.

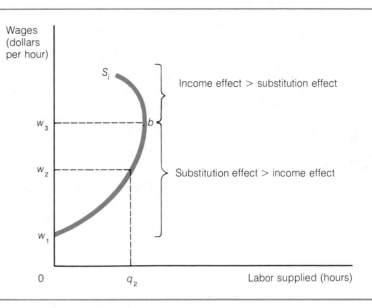

wage rate rises from w_1 to w_2, the hours she is willing to supply increase accordingly, to q_2.

The value of one hour of market time (one hour spent working) is equal to the wage rate. This means that for each hour an individual chooses to enjoy nonmarket activities, he or she is forgoing an hour of pay. In other words, the opportunity cost or the price an individual must pay for an hour of leisure time is the wage rate. As the wage rate rises, the price of nonmarket time also rises. An individual's demand for nonmarket time is downward sloping, just as it is for any other good. If the wage rate rises, the individual is encouraged to purchase less nonmarket time; that is, he or she is encouraged to substitute hours of work for hours of leisure.

The substitution effect, however, may be offset by an income effect. As the wage rate rises, the individual's income rises. If nonmarket time is a normal good, as we might expect, then the demand for nonmarket time will increase as income rises. This type of income effect will encourage the individual to enjoy more nonmarket time and less work.

As the wage rate rises, the two effects pull the worker in opposite directions. The effect that dominates will determine whether higher wages encourage more or fewer hours of work and determine whether the labor supply curve is positively or negatively sloped. If the substitution effect is greater than the income effect, higher wages will bring more hours of labor. In Figure 12–6, the substitution effect is greater than the income effect from w_1 to w_3, so the slope is positive. However, the income effect is greater than the substitution effect at wages above w_3. In this range, higher wages bring fewer hours of labor, so the curve is negatively sloped. At point b, where the income and substitution effects are equal, the supply curve begins to bend backward.

The actual shape of any individual's labor supply curve cannot be determined theoretically. It can be positively sloped, negatively sloped, or vertical through any range of wages. The wage level that initially entices people to enter the labor market also varies.

Since the shapes of individual labor supply curves are difficult to deter-

FIGURE 12–7

The Aggregate Supply of Labor

The aggregate labor supply curve, S_A, representing the numbers of hours of labor supplied by all individuals, is likely to be upward sloping. As wages rise, more people join the labor force, and many people work more hours.

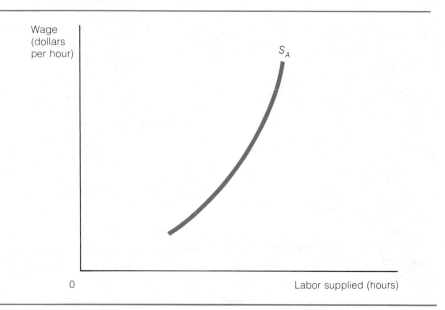

mine, the slope of the total or aggregate labor supply curve is difficult to predict with accuracy. The aggregate labor supply curve is theoretically obtained by summing individuals' labor supply curves horizontally. The aggregate curve shows the total amount of labor supplied at all the different wages. Even though some individuals have backward-bending labor supply curves, we expect the aggregate labor supply curve to be positively sloped, in Figure 12–7, for two reasons. First, as wages rise, many workers will work more hours. Second, as wages rise, more people will enter the labor force.[1]

Although the wage rate has an influence on the number of hours people may choose to work, individuals may also choose their wage rate, within limits. They do so by choosing an occupation. Some occupations must pay higher wages than others to induce people to join.

Human Capital

Some occupations, such as medicine, law, and nuclear physics, require many years of training. The fact that the wages in these skilled professions are higher than the wages of unskilled labor is not coincidental. Before people are willing to endure the years of training to become highly skilled professionals, they must be reasonably sure that their investment of time and other resources will pay off in the long run. People frequently choose to invest in some form of training to make themselves more productive and to enhance their income-earning potential. While a person is going to college, attending trade school, or gaining on-the-job training, he or she is building **human capital.**

Human capital: Any nontransferable quality an individual acquires that enhances productivity, such as education, experience, and skills.

Gaining human capital requires an investment period, such as four years in college. This investment involves a large opportunity cost. A college student loses the next-best alternative when attending college. For many students, the lost opportunity is the income they would have earned if working. Figure 12–8 portrays a simplified version of two alternative lifetime income streams for an individual.

[1]The analysis of aggregate labor supply is in Chapter 26.

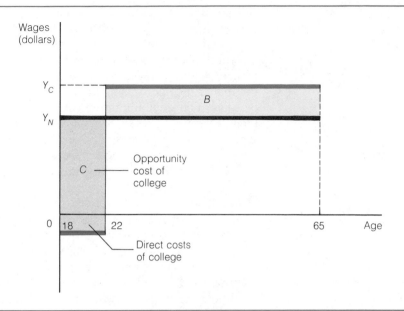

FIGURE 12–8   The Economic Costs and Benefits of College

An individual may choose to earn income Y_N from age eighteen to retirement at age sixty-five. Or an individual may attend college, lose income from age eighteen to age twenty-two, pay the direct costs of college, and then earn income Y_C from age twenty-two to retirement. The total costs of college are represented by the shaded area C, the lost income and direct costs; the benefits of college are shaded area B, the extra income earned with a college degree. In monetary terms it pays to go to college if B is greater than C.

After graduating from high school at age eighteen, individuals have a choice. They may enter the labor market immediately and earn income stream Y_N from that time until retirement at age sixty-five. This income stream could rise through the years as the individuals acquire skills, but for simplicity we let it remain constant.

Some individuals may go to college. From age eighteen to twenty-two, they do not earn income. The opportunity cost of going to college—that is, the income the individuals forgo by choosing not to work—is represented by the shaded area above the x-axis between ages eighteen and twenty-two. In addition, they must pay the direct costs of college—tuition, books, and so on. The shaded area below the x-axis between ages eighteen and twenty-two represents the direct costs of college, a negative income. The total cost of going to college is the sum of both the opportunity cost and the direct costs of college, the entire shaded area C. After graduation, these individuals enter the labor market and begin earning positive income. Figure 12–8 indicates that their starting pay at age twenty-two after attending college is higher than it would have been had they not gone to college. They earn negative income for four years and then Y_C as a lifetime average from age twenty-two until retirement.

Which is the best choice? If the deciding factor is money, then one determines the relative values of the two income streams and chooses the higher. One method of doing so is to compare the area C (the investment period) with area B. Area C represents the total cost of going to college: the direct costs plus the lost income. Area B represents the benefits of going to college; it is the extra income earned with a college degree. If area B is greater than area C, then it pays to go to college.[2]

Some investments in human capital require more time and some require less time than others. But regardless of the length of the investment period, if there is pure competition in the market and if enough individuals choose the higher income streams, then in the long run the income streams will all be equal. (This equilibrium requires that the wage rates for occupations with long investment periods be greater than the wage rates for occupations with short investment periods. The difference in wages is what equalizes the lifetime income streams.) Focus, "Does It Pay to Go to College?" elaborates on the economic decision-making process of whether to go to college.

Other Equalizing Differences in Wages

A long investment period is not all that discourages individuals from entering a particular occupation. Other characteristics make it necessary for some occupations to offer higher wages to induce workers to enter. People will evaluate their alternative work possibilities in both monetary and nonmonetary terms. Pay will be important, to be sure, but so too will be working conditions, location, co-workers, risk, length of contract, personality of the boss, and myriad other factors. Although some of these factors are nonmonetary or psychological, this does not mean that they are not income. They are a part of workers' pay. The **total compensation** of a worker in a given occupation consists of the wage plus any nonmonetary aspects of the job.

The principle of **equalizing differences in wages** works in the following way. In a competitive labor market, laborers choose the occupation with the highest total compensation. This brings the wage down until the total compensation for the occupation is on a par with that of other occupations. If total compensation is too low in some occupation, then people leave that occupation until the wage rises enough to equalize the total compensation. When total compensation is equalized across occupations, the competitive equilibrium is achieved. If this condition is not met, workers will move around and change jobs until it is met.

One student who spends the summer in an air-conditioned office may make a lower wage but more nonwage pay per hour than another student who spends the summer on a construction crew. As students compete for summer jobs, total pay between jobs will be equalized. This is the principle of equalizing differences at work in a competitive labor market.

The basic point is that observed wage differences across occupations may reflect differences in nonmonetary aspects of employment. Of course, labor markets may not always be competitive as we have assumed here. Some jobs may pay more and have more attractive working conditions if there are barriers to entry into the occupation. However, under competition, wage differences can exist. The following are just a few of the many reasons for wage differences other than those caused by human capital differences.

Wages will vary directly with the disagreeableness of a job. The more uncomfortable the job, the higher the pay will be. To induce workers to accept jobs that create discomfort—such as tarring roofs in the heat of summer—a higher monetary reward must be offered. Workers will not offer their labor below a certain wage, given that they have alternatives.

The more seasonal or irregular a job, the higher the pay will be. To induce

Total compensation: The lifetime income that an individual receives from employment in a particular occupation, including all monetary and nonmonetary pay.

Equalizing differences in wages: The differences in wages across all occupations that result in equality in total compensation.

[2]The values of C and B must be discounted for time, as is the case for any investment; this point is discussed in Chapter 14.

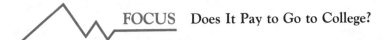

FOCUS   Does It Pay to Go to College?

The decision to go to college is a decision to invest in human capital with alternative payoffs for different levels of investment (see Table 12–3). As with all investments, there are costs and benefits. The monetary costs are the forgone income one could earn by working and the direct costs such as tuition and books. The monetary benefits are the years of extra income that one may earn with a degree. Is college a good investment? Is the rate of return on an investment in education higher than that of other investments?

Richard Freeman calculated the rate of return on a college degree over an individual's life cycle. Freeman's findings were:

Year	Rate of Return
1959	11.0%
1969	11.5%
1972	10.5%
1974	8.5%

These rates of return represent the interest rates that individuals receive on their investments in education. From 1969 to 1974 there was clearly a downward trend.

Freeman investigated whether the trend has continued into the 1980s. He found that through the 1970s the rate of return to college education fell considerably, suggesting that Americans are overeducated. However, the trend seems to have bottomed out. Freeman predicts that during the 1980s the market for college grad-

Sources: Richard Freeman, "Overinvestment in College Training," *Journal of Human Resources* 10 (Summer 1975); and "The Overeducated American in the 1980s: A Report to the National Commission on Student Financial Assistance," *Higher Education Marketing Journal* 11:9 (Summer 1983).

uates will improve and increase the rate of return on investment in a college education.

TABLE 12–3   Estimated Lifetime Earnings for Men and Women by Educational Level (1981 dollars)

Expected earnings from additional education rise for both men and women with more education, but less than is commonly supposed. A dramatic increase for both men and women comes with the high school diploma and with a college degree (versus only some college work). However, the figures probably underestimate actual earnings since they do not include human productivity changes over lifetimes. Also note the obvious disparity in estimated future income for men and for women, a difference due partially to sex discrimination which we fully discuss in Chapter 16 on income distribution.

Category	Men	Women
Less than 12 years education	$ 845,000	$500,000
Completed 12 years education	1,041,000	634,000
Completed 1 to 3 years college	1,155,000	716,000
Completed 4 years college	1,392,000	846,000
Some graduate work	1,503,000	955,000

Source: U.S. Department of Commerce, Bureau of the Census, *Consumer Income* (February 1983).

individuals to supply labor to irregular employments, the wage must compensate them for the likelihood of being laid off frequently. For example, housing construction workers have frequent periods of unemployment between jobs. To have readily available workers, construction employers must pay a higher wage.

Jobs that require trustworthiness carry a higher wage. Take the jobs of bank teller, armored truck driver, and blackjack dealer. One reason that such occupations pay more is that being trustworthy is both an extra burden and a form of human capital; employers must compensate such workers for their extra efforts. Since workers have a choice between stealing and thereby losing future wage income and not stealing but keeping the job, the wage must be high enough to induce continued honesty in supplying labor.

Jobs with greater risk to health will have higher pay. Occupations with higher than average probabilities of death or disability must compensate employees for taking a risk. People are not willing to risk lost future income because of injuries unless they can receive higher income in the present. Steelworkers who construct frames for one-hundred-story skyscrapers must be paid more than those who work on the ground.

Jobs that carry the possibility of tremendous success will have a lower wage. Acting is a good example of such a profession. Many young people aspire to be actors and are arguably attracted into acting by the sumptuous lifestyles of movie stars. Most never make it that big; the average salary of all actors is quite low. Yet some people are willing to take the plunge, feeling that they are good enough to be big winners. Attitudes toward risks are therefore important in determining relative wages. The wages for most actors will be lower because some successful actors make startlingly high pay, a possibility that induces many young people to enroll in acting school.

Regardless of the equilibrium wage, the supply of labor to each occupation is highly elastic. In the short run, wages may deviate from equilibrium because of changes in demand, but in the long run new people entering the labor force and competing for jobs will force the difference in wages that yields equal total compensation across occupations.

Marginal Productivity Theory in Income Distribution

The profit-maximizing tendency of competitive firms to set marginal revenue product (MRP) equal to marginal factor cost (MFC) leads to a wage that equals the value of the marginal product. This result applies to the prices of all factors of production in competitive markets. Each resource—be it labor, land, or capital—is paid the value of its marginal product.

The marginal productivity theory of factor prices is a positive, not a normative, theory. It is not meant to be an ethical theory of income distribution even though it is sometimes attacked on such terms. It is meant as a demand and supply theory that explains the behavior of input prices and the allocation of factors of production to different employments.

However, the marginal productivity theory has been criticized even as a positive theory. Labor markets are not perfect (as discussed in detail in the next chapter), and resources do not seem to flow around the economy in a smooth, frictionless way to equalize rates of return in factor markets. Labor and physical capital are particularly difficult to characterize as easily movable. People and machines and buildings generally like to stay in the same place. In this sense the marginal productivity theory may not always offer a realistic explanation of short-run factor prices. At any time we might observe different wage rates in a market for equally skilled workers, which marginal productivity theory may have difficulty in rationalizing.

Over a longer period of time, marginal productivity theory does offer a good explanation of factor prices. In the long run, resources will relocate to take advantage of higher rates of return. For this reason, marginal productivity theory is alive and well as a scientific explanation of factor prices. Given proper time, it is the best explanation we have of factor prices.

Another criticism of marginal productivity analysis is that business managers could not possibly estimate the marginal revenue products of their inputs. How hard it is to estimate MRP is an interesting question, but the fact is that business managers are forced to make some sort of estimate of what their inputs are capable of doing on an ongoing basis. For example, firms must decide whether to promote from within or hire managers from other firms. Why do firms typically promote from within? The answer is that it is less costly to form an estimate of the MRP of inside candidates. The firm has observed the performance of the inside candidate but must take letters of recommendation and other types of indirect evidence about the outside candidate.

It should now be clear that the theory of marginal productivity implies many things about income distribution. Where there is competitive voluntary contracting for labor services, labor's share of national income will be directly related to its marginal product. "To each according to what he or she produces" perhaps best describes what happens in marginal productivity theory. Of course, factor markets are not perfect, and various other forces affect income distribution in a society. These issues will be carefully addressed over the next several chapters. Our argument in this chapter is not that the marginal productivity theory is necessarily a good normative theory of how income should be distributed (although you might believe it so) but rather that it provides an objective basis for understanding the economic behavior of demanders and suppliers of inputs.

Summary

1. The demand for labor is a derived demand; it is derived from the demand for the product that labor produces.

2. The demand curve for labor is a function of its marginal product and the demand for the product it produces. In the short run, demand is equivalent to marginal revenue product: $MRP_L = MP_L \times MR$.

3. The competitive firm hires the profit-maximizing amount of labor at the point where MRP equals the wage rate. The short-run market demand curve for labor shows the amount of labor that firms hire when labor is the only variable input and the number of firms is constant, but it allows for changes in product price as wages change.

4. The long-run market demand curve for labor shows the amount of labor hired at different wages when all inputs are allowed to vary and the firms are in long-run competitive equilibrium.

5. The demand for labor changes when the demand for the product changes or when the marginal product of labor changes. The elasticity of demand for labor is directly related to the elasticity of demand for the product it produces, the percentage of total cost ac-

counted for by labor, and the number and availability of substitute inputs for labor.

6. The relation between the wage rate and the number of hours of labor an individual supplies depends on the substitution and income effects. If the substitution effect is greater than the income effect, then the individual's labor supply curve is positively sloped; the curve is negatively sloped if the income effect is greater than the substitution effect. The aggregate labor supply curve is the sum of individual labor supply curves and is usually positively sloped.

7. Human capital is anything an individual acquires that increases productivity. In an occupation that requires human capital, the longer and more costly the investment period, the higher the wages.

8. The total compensation in all occupations in competitive markets is equalized by differences in wages. The long-run supply of labor to each occupation ensures the equilibrium difference in wages.

9. In competitive markets, resources are paid a price that is equal to their marginal product. Labor's share of the overall income distribution is related to its marginal product.

Key Terms

factor market
derived demand
competitive labor
   market
market demand
   for labor

market supply of labor
marginal product of
   labor
marginal revenue
   product
marginal factor cost

firm's long-run demand
   for labor
complementary inputs
substitute inputs
elasticity of demand
   for labor

nonmarket activities
human capital
total compensation
equalizing differences
   in wages

Questions for Review and Discussion

1. Suppose that the demand for each of the following goods and services increases. What derived demands will increase as a result? (a) Automobiles, (b) candy, (c) government regulation, (d) physical fitness, (e) education.
2. Explain why an individual's supply of labor curve may bend backward at some sufficiently high wage level.
3. What does the law of diminishing marginal productivity have to do with the demand curve for labor?
4. What is the difference between the demand curve for labor of a monopolist and of a perfectly competitive firm in the output market?
5. Apply the principle of equalizing differences to explain why the relative wage of the following occupations is high or low: (a) politicians, (b) school-teachers, (c) morticians, (d) actors, (e) brain surgeons.
6. In what sense is the marginal productivity theory of wages a positive theory? In what sense is it a normative theory?
7. What activities will frustrate the equalization of total wages across jobs in the labor market? Does this mean that there is such a thing as nonequalizing wage differences?
8. Are tractor drivers substitutes for human cotton pickers? What happens to the demand for tractor drivers if cotton pickers' wages fall?
9. Is the demand for toothpicks elastic or inelastic? What about elasticity of demand for a machine that makes toothpicks?

ECONOMICS IN ACTION   Regional Wage Differences

In the long run, wages may differ across occupations for many reasons. Some occupations require large investments in education or training, some are risky, and others have random periods of unemployment. Economists in the past have also noticed that geographical factors influence wage differences within the same occupations. Why do we observe higher wages in the North than in the South or higher wages in San Francisco than in Oklahoma City? Can we expect these wage differences to persist in the long run?

Wages in any region are determined by supply and demand. However, if wages in market A exceed the wages in market B, then we expect to see labor flow from B to A. The shifting short-run supply brings wages into equilibrium. Wages fall in A and rise in B.

The equality of wages is the result of individuals' actions. Furthermore, an individual's decision to move is based on the costs and benefits of doing so. Relocation requires resource expenditures that must be offset by extra income. Indeed, the decision to move to a higher-paying location within the same occupation represents an investment in human capital.

Suppose a worker could stay in location B and earn wage W_B or could move to location A and earn a higher wage W_A. The direct costs of moving are the expenses incurred in transporting the worker, her family members, and their personal belongings. There are also additional opportunity costs. The act of moving requires time, and so might finding a job in A. This is time that could have been spent earning income in B. The lost income is a cost of relocation, as shown in Figure 12–9.

At time 0 the individual must choose between earn-ing wage W_B from T to retirement and spending time moving so as to earn wage W_A. The time $0T$ is the amount of time spent moving from B to A plus the amount of time spent seeking employment in A. The total cost of moving is the shaded area between 0 and T above and below the x-axis, which represents the income forgone plus direct moving expenses. The benefit of moving—represented by the shaded area between T and retirement—is the extra income that may be earned after the move. It pays to move if the area of extra benefits is greater than the area of costs.[a]

If benefits outweigh costs, then many people will choose to move to A—the move is a good investment in human capital. However, the shifting labor supplies will increase wages in B, where workers are now scarcer, and decrease them in A. The maximum amount by which W_A can exceed W_B in the long run is the amount that forces the area of benefits to just equal the area of costs. Therefore, the lower the costs of moving, the smaller the difference in wages needs to be.

In fact, there are many reasons why we expect little variation in wages across the country within the same occupation in today's markets. First, the duration of lost income for most movers is very small. Indeed, it is quite possible to work until 5:00 P.M. on Friday in New York City and start work at 9:00 A.M. on Monday in Los Angeles. Many people will not move unless they have already found employment in the new location. If

[a] As in the earlier example of the economic costs and benefits of a college education, these values must be discounted for time, as explained in Chapter 14.

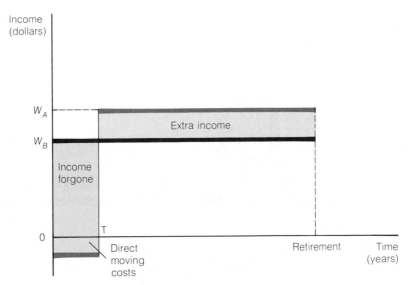

FIGURE 12–9   Wage Differences Across Regions

An individual must choose whether to remain in location B and earn wage W_B from time 0 to retirement or move to location A and earn wage W_A from time T to retirement. The decision to move to A necessitates costs in terms of the time 0T spent moving and income forgone while making the move, represented by the shaded area from 0 to T above the x-axis. After the move, the individual earns extra income represented by the shaded area from T to retirement. If this area of extra income is greater than the area of total costs of moving, the individual will choose to move.

the amount of time it takes to travel across the country today were as great as it was two hundred years ago, then wage differences could be very large.

Another reason that regional differences in wages are small is that young workers have small costs and large benefits in moving. Young workers typically have small families or no families and few material possessions to move. Also, the years of receiving extra benefits are longer for young people. Thus, the difference in wages can be smaller to induce them to move.

Another important aspect of regional wage differences involves the demand side of the labor market. Firms will be encouraged to expand operations in areas where there are low costs of production and decrease operations where costs are high. Firms will invest more capital in low-wage area B and decrease investment in high-wage area A. Capital moves around the country just as labor does, and this movement tends to equalize wages. In fact, all resources that are mobile move to the locations that yield the resource owners the highest income.

In spite of this equalizing tendency, we observe that

wages in the North are about 10 percent higher than in the South in the same occupations. Don Bellante suggests that the difference occurs only in money wages and not in real wages.[b] The cost of living in the South is about 10 percent lower than in the North. This difference in costs equalizes the real wage, just as we expect in the long run.

Question

The City of Houston recently advertised for police men and women on Atlanta TV stations offering 30 percent (on average) higher wages than those paid by the Atlanta police department. Some personnel moved to Houston. What effect would this movement have on regional wage differences for police services between Texas and Georgia? What might we expect the average age of the migratory police workers to be?

[b]D. Bellante, "The North-South Differential and the Migration of Heterogeneous Labor," *American Economic Review* 69 (March 1979), pp. 166–175.

13

Labor Unions

*T*he previous chapter discussed demand and supply conditions in a competitive labor market, where the price of labor tends to equality. Although we see a great deal of wage discrepancies in competitive markets, many of these can be attributed to circumstances such as the amount of risk a particular job involves or the amount of training it requires. We naturally expect brain surgeons to make more than dishwashers in a competitive environment because of the investment a brain surgeon must make in order to perform services. The mobility of a work force also affects its wage. The wages of workers will often vary because many of us are simply unable to move to fill available jobs.

All labor markets, of course, are not strictly competitive. In general, factor markets can assume the same kinds of imperfect structures that we have already studied in Chapter 11. The imperfection in labor markets can occur on the supply side, the demand side, or both. On the supply side, labor unions account for the largest source of imperfect competition.

In this chapter we will look at how imperfectly competitive labor markets work. Specifically, we will look at the effects of labor unions.

The imperfection in labor markets can occur on the supply side, the demand side, or both. On the supply side, labor unions account for the largest source of imperfect competition.

Types of Labor Unions

Labor union: A group of workers who organize to act as a unit in an attempt to affect labor market conditions.

A **labor union** is essentially a group of workers who organize collectively in an effort to increase their market power. By acting as a collective unit, they are able to exert a greater influence over working conditions or wages. In this sense, a labor union is very similar to a cartel (discussed in Chapter 11). The sellers of labor agree not to compete among themselves but to act as a single seller of labor.

The first labor unions in the United States started as workers' guilds in the late 1700s and early 1800s. These organizations of workers within the same trade—carpenters, cordwainers (shoemakers), hatters—began meeting to set standards and prices for their output. These loosely organized trade unions were typically short-lived but were the beginnings of the American labor movement.

The Industrial Revolution brought about new opportunities for labor unions. Large manufacturing plants employed many workers with similar skills and interests who could organize to pursue their common goals. The early stages of the new industrial growth also brought the buyer of labor—the factory owner—considerable market power, the effects of which further encouraged workers to unify to enhance their own power. Today there are three major types of labor unions: craft unions, industrial unions, and public employees' unions.

Craft Unions

Craft union: Workers with a common skill who unify to obtain market power and restrict the supply of labor in their trade; also called trade union.

In 1886, the American Federation of Labor (AFL) was started in an effort to organize craftspeople into local unions. **Craft unions** organize workers according to particular skills—such as electricians, carpenters, or plumbers—regardless of the industry in which they work. The main function of the guilds is to advance their members' economic well-being. When workers act as a unit, their power in the market can limit competition and raise total compensation above the competitive level.

Trade or craft unions increase members' total compensation by decreasing the supply of skilled workers. They do so by excluding potential workers from membership. Frequently, trade unions require high initiation fees, monthly dues, and long apprenticeship programs in an effort to discourage potential entrants. Existing members enjoy the higher wage brought about by the limited supply of workers in their trade. Figure 13–1 shows the impact of this decreased supply on wages: Wages rise from the competitive level, W_c, to the unionized level, W_u.

Trade unions are very much like cartels in the sense that supply is artifi-

FIGURE 13–1

A Craft Union's Effect in the Labor Market

Craft unions decrease the supply of workers in a particular craft by limiting the number of people who may join the union. When a union organizes, the supply of labor decreases to S_u from the competitive supply, S_c. Wages rise from the competitive level W_c to the unionized level W_u. Employment falls from L_c to L_u.

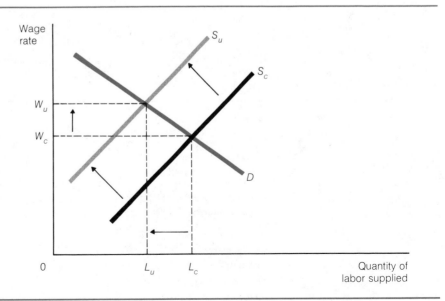

cially restricted so as to command a higher price. As such, craft unions face the same problems as cartels. Just as cartel members have an incentive to cheat on the cartel price, workers outside the union who offer their labor at a wage lower than the union wage can obtain jobs at the expense of union members.

Industrial Unions

Large manufacturing plants first appeared during the late 1800s and early 1900s. A large firm in the textile, automobile, steel, or coal industry would frequently be the only employer in a small, isolated town. As a single buyer of labor, such a firm would be able to keep wages below the competitive level. Under these circumstances, workers often found it in their interest to organize. Rather than organize individual craft unions according to different skills, all workers within the same industry would organize a single industry-wide union. In 1938, John L. Lewis formally organized the Congress of Industrial Organizations (CIO), which unified workers first at the firm level and then at the level of the industry as a whole. Contemporary examples of these **industrial unions** include autoworkers' unions and steelworkers' unions.

For an industrial union to be effective, it must unionize all firms in an industry. Otherwise, the lower-cost, nonunionized firms would prosper while the higher-cost, unionized firms would dwindle. Thus, industrial unions often encourage membership rather than restrict it.

Public Employees' Unions

Public employees' unions are organizations of government workers. Such unions cover a wide variety of jobs—such as firefighting, police work, teaching, and clerical work—and include both blue-collar and white-collar workers. This sector of the union movement has been one of the fastest growing in recent years. The American Federation of State, County, and Municipal Employees (AFSCME) is now among the ten largest unions in the country, as indicated in Table 13–1. Indeed, membership in AFSCME almost tripled over the 1968–1980 period.

Industrial union: Workers within a single industry who organize regardless of skill in an effort to obtain market power.

Public employees' union: Workers who are employed by the federal, state, or local government and who organize in an effort to obtain market power.

TABLE 13–1   The Ten Largest Unions, 1980

The larger unions are industrial unions, which usually encourage membership. The smaller craft unions (electricians, carpenters, and so on) typically restrict new membership.

Type of Union	Membership
Teamsters	1,891,000
Autoworkers (UAW)	1,357,000
Food and commercial workers	1,300,000
Steelworkers	1,238,000
State, county, and municipal employees (AFSCME)	1,098,000
Electrical workers (IBEW)	1,041,000
Carpenters	832,000
Machinists	745,000
Service employees (SEIU)	650,000
Laborers (LIUNA)	608,000

Source: Statistical Abstract of the United States, 1984, p. 440.

Union Activities

Regardless of the means of organization, when workers successfully unite to act as a single seller of labor, the union gains monopoly power in its labor market. The union and its members are therefore no longer wage rate takers. There is a downward-sloping demand curve for the members' labor, and the union may seek any wage along this curve. However, the level of employment for union members is inversely related to the wage rate. Higher wages are gained at the expense of fewer jobs. In the face of this fact, a union's activities will depend on its choices among conflicting objectives, the elasticity of demand for its workers' product, and the union's effect on demand for its workers' labor.

Union Goals

The ultimate goals of the union may depend on many competing objectives. Understandably, the union's elected, policy-making officials suggest that members want improved economic well-being: higher wages, more jobs, greater job security, safer jobs, more retirement pay, more fringe benefits, and so on. But relating union policy to a particular objective is difficult. For example, the union objective may be to maximize any number of options, some of which may be inconsistent, such as the utility of the union officials, the utility of members, the wage rate, the level of employment, the level of membership, the total wage income of members, the total wage income of senior members, and so on. The ramifications of three of these objectives are discussed below.

Employment for All Members. One objective of a union may be to achieve full employment for all members. To do so would require a wage rate that ensures that the quantity of labor demanded by firms is equal to the quantity of labor supplied by existing members. Suppose that the demand for the union's labor can be represented by curve D_L in Figure 13–2. If the amount of labor offered by the union is L_1, then wage W_1 will ensure full employment of union members.

Maximizing the Wage Bill. Although wage W_1 means that everyone in the union is working, full employment may not be the union's goal. Unions frequently have members sitting on the unemployment bench. Any wage higher than W_1 will leave some members unemployed, but there may also be some benefits to a higher wage. For example, it may increase the total wage bill. The **total wage bill** to firms hiring union workers is the wage rate times the total amount (in person-hours) of labor hired. This total wage bill is not only the total cost of labor to the firms; it is also the collective wage income of all the union members. Since workers prefer more income to less, all things being equal, the union may choose to maximize the wage bill. If the full employment wage occurs where the elasticity of demand is less than 1 (inelastic demand), then an increase in the wage will increase total union income ($W \times L$).

To understand this possibility, recall from Chapter 5 the relation between price, total revenue, and elasticity. Total revenue is maximized at the price where the elasticity of demand is equal to 1. If elasticity is less than 1, then an increase in price increases total revenue; if elasticity is greater than 1, it takes a price decrease to increase total revenue. In analyzing labor demand, the total wage bill, the wage rate, and the elasticity of demand have the

Total wage bill: The total cost of labor to firms, equal to wage times total quantity of labor employed; the total income of all workers.

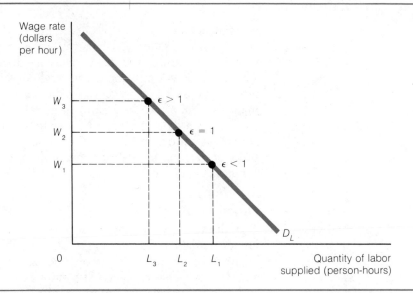

FIGURE 13–2   The Union's Wage Goals

In setting its wage rate, a union must choose among many goals. If there is L_1 amount of laborers in a union, then wage W_1 must be chosen to achieve full employment of all members. If the elasticity of demand at W_1 is less than 1, then an increase in wages up to W_2 will increase the total wage bill while bringing about union unemployment in the amount $L_1 - L_2$. Still higher wages such as W_3 will decrease total union income, although those union members who are working enjoy greater incomes. In trying to maximize total wages, the union will face some degree of unemployment, according to this model.

same relation. If the union's objective is to maximize the total union income, then the wage is set where the elasticity of demand is equal to 1.

In Figure 13–2, wage W_2 maximizes the wage bill. However, with L_1 amount of labor in the union, unemployment in the amount $L_1 - L_2$ results. This may not be a problem because the total union income ($W_2 \times L_2$) is now higher than at W_1. All members can still be better off at a higher wage if the working members provide unemployment compensation to unemployed members. On the other hand, a trade union may choose to limit union membership to L_2.

Maximizing Income for Limited Members. While wage W_2 would maximize income for the union as a whole, a higher wage such as W_3 in Figure 13–2 would increase income only for those who remain fully employed. L_3 workers would be hired, and their incomes would be higher so long as they are working the same number of hours. The workers on the unemployment bench would receive no income. For example, senior members usually retain employment under a seniority system. Thus, wages above W_2 would probably increase senior members' incomes at the expense of younger members. These unemployed members could be phased out of the union so that remaining members could retain higher incomes.

Union Objectives and the Elasticity of Demand

A union's ability to accomplish its goals is limited by the elasticity of demand for its members' labor. Increases in wages decrease the level of employment, but the percent decrease in employment is determined by the elasticity of demand. Unions would like to see large increases in wages with minor effects on employment; in other words, they would like the demand for their members' labor to be as inelastic as possible. The elasticity of demand for union labor is determined by many factors (discussed in Chapter 12). Unions themselves often engage in activities that decrease the elasticity of demand for union members. Such a change in elasticity of demand is shown in Figure 13–3 as a rotation of the demand curve from D_1 to D_2. Wages are increased without causing so much unemployment. Following are some familiar examples of this process.

Elasticity of Demand for the Product. The more inelastic the demand for the good or service that the union produces, the more inelastic the demand for the union's labor. While a union would like the demand for the good or service its workers produce to be inelastic, it cannot do much to decrease this elasticity. Union workers are mostly at the mercy of the market, although in some circumstances they can and do affect elasticity of demand. For example, an item's elasticity of demand is determined in part by the number and availability of substitutes. If a union can limit the amount of substitutes for the product they produce then, the elasticity of demand for the union workers' labor decreases. We see this happening in the automobile industry. Through the political process (see Focus, "The Political Power of Unions") U.S. autoworkers encourage import quotas, which have the effect of limiting the supply of foreign autos and decreasing the elasticity of demand for U.S.-made cars. With import quotas, increases in autoworkers' wages have less effect on employment than the same increases in wages if quotas are not imposed.

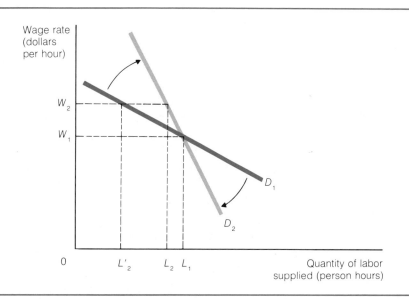

FIGURE 13–3   Decreasing the Elasticity of Demand for Union Labor

Unions typically attempt to decrease the elasticity of demand for their members' labor. For example, unions support immigration laws that retard or block the flow of competitive foreign labor into this country. The effect of such activities is represented by a rotation of the demand curve from D_1 to D_2. This rotation of the demand curve indicates that future wage increases result in smaller employment decreases. If wages rise from W_1 to W_2, employment falls only to L_2 rather than L'_2.

FOCUS   The Political Power of Unions

One way to increase the demand for union labor is to lobby for various types of protective legislation, such as tariffs and minimum-wage laws. Over the years, unions have been quite successful in garnering political support for legislation favorable to their membership. The unions' political effectiveness results primarily from their ability to control a supply of votes.

Unions often attempt to persuade members to work in campaigns and to vote in a bloc for candidates who support union positions. The incentive for a union member to vote can be stronger than that of the average voter because the union member knows that a lot of other members will vote the same way. In other words, individual union members know that their votes will count, unlike average voters, who sometimes feel that their votes are insignificant. Union leaders, in turn, broker union votes among political candidates to win votes for union positions in legislative struggles.

Labor union institutions such as the union hall and the union boss are designed to mitigate the problem of free riding—the tendency of individuals to let others work in campaigns and then reap the benefits of their efforts without bearing any of the costs. Unions overcome this problem through their organizing and monitoring of members' activities.

Availability of Substitute Inputs. The fewer substitute inputs for union labor, the lower the elasticity of demand for union labor. Unions would prefer that there be no substitute inputs for their members' labor. If this were true, then wage increases would have minimal effects on employment. However, there are almost always some inputs that may be substituted for union labor. Typically, nonunion labor is a good substitute. For this reason, labor unions usually form contracts with employing firms in which the firms agree to hire only union labor. Unions also take direct actions to limit the amount of nonunion labor available. For example, laws that restrict immigration to the United States are supported by unions in an effort to decrease the availability of immigrant workers who could substitute for their members in the work force.

Union Labor as a Proportion of Total Costs. The smaller the percentage of firms' total costs accounted for by unions, the more inelastic the demand for union labor. If labor represents a very small percentage of total costs and if wages increase substantially, then the increase in total cost and thus product price will be very small. For this reason, the level of employment will fall by only a very small percent. In such a situation, union goals conflict. If unions attempt to maximize the total wage bill and increase the number of unionized occupations, then their ability to affect wages might decrease.

Increasing the Demand for Union Labor

Regardless of the goals of unions, the members are always in favor of an increase in the demand for their labor. With increased demand, wages, employment, or both are increased. The curve representing demand for labor shifts to the right, as shown in Figure 13–4, when one of two basic things happens: an increase in the demand for the product or an increase in the marginal product of workers.

Unions can and do engage in activities that affect the demand for their members. Some of these demand-increasing activities also decrease elasticity of demand for labor.

FIGURE 13–4

Increases in Demand for Union Labor

Labor unions encourage increases in the demand for their members' labor. Increasing the demand from D_1 to D_2 allows an increase in wages from W_1 to W_2 with a constant level of employment at L_1. Alternatively, employment can increase from L_1 to L_2 with wages constant at W_1. Or there can be an increase in both wages and employment to W_3 and L_3.

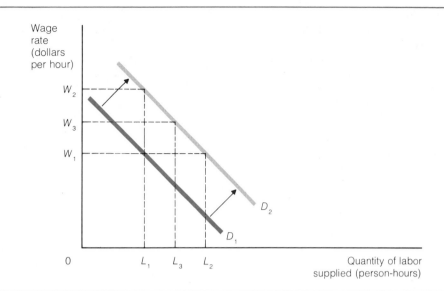

Increasing Product Demand. Unions frequently attempt to increase the demand for the product they produce. Garment workers advertise the union label and encourage consumers, as well as union members, to buy union-made clothes. Autoworkers strongly encourage union members to buy domestically produced cars. As we have seen, unions sometimes try to influence lawmakers to decrease the supply of foreign products that are substitutes for union-made goods. Import tariffs or quotas increase the demand for U.S.-made products.

Increasing Substitute Input Prices. If unions can increase the relative price of inputs that are substitutes for union labor, then the demand for union labor will increase. Unskilled nonunion workers using machines can substitute for skilled union labor, for example. Unions have therefore supported increases in the minimum wage. As the relative price of unskilled nonunion labor rises, the demand for union labor increases.

Increasing Productivity of Members. Unions prefer that their members be very productive. If the marginal product of new members joining the union is relatively high, then the demand for all members increases. Apprenticeship programs offered by unions train new entrants in an effort to increase overall productivity.

Monopsony: A Single Employer of Labor

Having examined unions' attempts to build the economic power of laborers, we now turn to the nature of the firms they face as employers. Before the Industrial Revolution, large manufacturing plants were not common. In most towns or regions there were several potential employers for most workers. As a matter of fact, most skilled craftspeople were self-employed. However, as technology changed, the benefits of mass production increased. Large manufacturing plants such as steel mills, textile mills, and coal mines became

FIGURE 13–5

Monopsony in the Labor Market

A monopsonist faces an upward-sloping supply curve of labor, S_L. It maximizes profit by hiring the amount of labor that equates its marginal factor cost (MFC) to its marginal revenue product for labor (MRP_L). The monopsonist's levels of employment and wage, L_m and W_m, fall below the competitive levels, L_c and W_c.

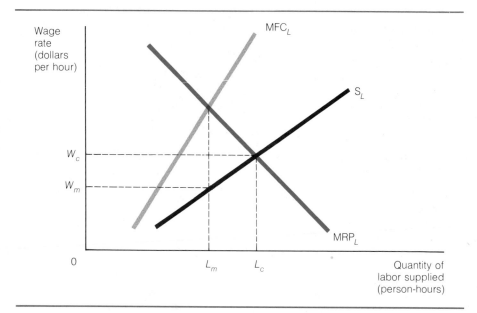

Monopsony: A single buyer of a resource or product in a market.

the principal employers of labor in some regions. In some areas only one firm employed workers.

A single buyer of a resource in a market is known as a **monopsony.** The monopsonist buyer has market power because rather than purchase all that it is willing and able to purchase at a market-determined price, the monopsony faces an upward-sloping supply curve. From along this curve it seeks the price that yields the profit-maximizing quantity of resource employment.

Consider Figure 13–5. The firm sells its product in a competitive market; its marginal revenue product curve for labor is MRP_L. Since it is the only buyer of labor in the market, it faces an upward-sloping supply curve of labor, S_L. To hire more labor, the firm must offer higher wages. Under these circumstances, the marginal factor cost of labor is no longer equal to the wage rate. The increase in total cost that results from hiring one more unit of labor is greater than the price paid for one more unit. And to attract one more unit of labor, the firm must offer a higher wage to all units of labor. For this reason, the MFC_L curve has the same relation to the upward-sloping supply curve as the marginal cost curve has to the average cost curve (discussed in Chapter 8). Indeed, the supply curve of labor represents the average factor cost curve.

The firm will hire all units of labor that add more to revenue than to cost. Thus the monopsonist firm will maximize profit by hiring labor up to the point where MFC_L equals MRP_L. (The monopsonist makes the same marginal analysis as the competitive buyer of labor.) The monopsonist hires L_m amount of labor and must pay wage W_m to attract that amount of labor.

The monopsonist wage and level of employment may be compared to those of pure competition under the same circumstances. If MRP_L in Figure 13–5 had been the summation of many firms' demand curves for labor, then competitive forces would have forced wages to W_c and employment to L_c. Under the same supply and MRP_L conditions, the presence of monopsony results in lower wages and lower employment.

The fact that the monopsonist pays a wage rate less than the MRP_L is known as **exploitation of labor.** If individuals are not being paid the value of their marginal product, then the monopsonist is extracting value from them.

Exploitation of labor: A situation in which the wage rate is less than the marginal revenue product of labor.

In the small regional markets of early days, monopsonists were actually exploiting the immobility of labor. As single employers, firms were able to pay lower-than-competitive wages in isolated areas because the cost of transportation was very high. It was very difficult for individuals to move around the country to work for other employers.

The more mobile an employee, the greater the number of potential employers and the more elastic the labor supply. Today, in spite of the seemingly high price of gasoline, the costs of transportation are relatively low. The opportunity cost of moving from one side of the country to another or commuting to all firms within a sixty-mile radius is lower today than it was a hundred years ago. It is extremely rare to find an individual who has only one potential employer.

Monopsony and the Minimum Wage

The monopsony model represents one of the few cases where a minimum-wage law may have a productive impact on the economy. To see why, ask yourself what would happen in Figure 13–5 if the government set a minimum wage at the level of W_c. In this case, the firm would be a wage rate taker when hiring labor up to L_c units. The supply of labor would be infinitely elastic at w_c from zero to L_c; the supply would be upward sloping from L_c on. With infinitely elastic supply, MFC_L equals the wage. Thus the firm would hire labor up to the point where MRP_L equals the wage. The monopsonist would increase employment to L_c and pay the minimum wage. A minimum wage can increase employment in a monopsonistic market as long as it is not higher than the competitive wage.

Bilateral Monopoly and the Need for Bargaining

If there is a monopoly in a market, then the single seller has the power to determine price. If there is monopsony in a market, then the single buyer has the power to determine price. But in some markets, **bilateral monopoly**—a single seller and a single buyer—could exist. In such a case, what determines the price?

Bilateral monopoly: A market in which there is only one buyer and one seller of a resource or product.

Theoretically, the price in a bilateral monopoly market cannot be determined. No competitive forces determine a single price, and the seller has a preferred price that is higher than the buyer's preferred price. However, if the maximum price the buyer is willing to pay is greater than the minimum price the seller is willing to accept, then a bargain can be reached. Bargaining between two parties is the process by which price is determined.

Consider a labor market with bilateral monopoly. Suppose a monopsony exists in an isolated mining town with only one employer, a mining firm. The firm must rely on the town's supply of labor. As Figure 13–6 illustrates, the mining firm maximizes profit by hiring L_m workers and by paying wage W_m. The workers, though, are unhappy with wage W_m and decide to form a union. They manage to sign up workers of all skill levels throughout the community to the union, maximizing its bargaining strength.

The wage most preferred by the union is not clear. If the union's goal is to maximize employment, then W_c, which coincides with the competitive wage, would be most preferred. However, a higher wage such as W_u may result in the maximum total wage bill.

There is a maximum wage that the union can obtain. Any wage above the maximum would result in no union employment at all because of the high production costs that would be imposed. If wages are too high, then the monopsony may actually go out of business, turn entirely to substitutes

FIGURE 13–6
Bilateral Monopoly

When a monopsonist chooses lower-than-competitive wage W_m and a monopoly union chooses higher-than-competitive wage W_u, the wage theoretically cannot be determined. But it can be determined by bargaining, and it will be between W_u and W_m.

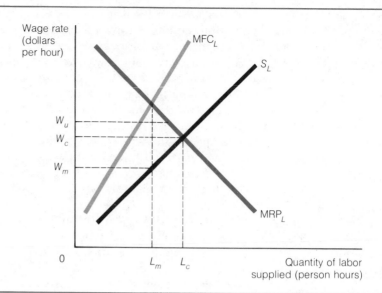

for the union labor, or relocate in a lower-cost area. The union's ability to extract value from the firm is limited by the mobility of the firm, just as the firm's ability to extract value from the workers is limited by the mobility of the workers.

Collective Bargaining

When labor bargains collectively with a monopsonist, the final wage will be somewhere between the wage chosen by the monopsonist (W_m in Figure 13–6), and the wage most preferred by the union (W_u in the figure). The party that comes closest to establishing its goal is the one with the best bargaining strength. In **collective bargaining,** all buyers or all sellers of labor act as a unit to enhance their bargaining power.

Bilateral monopoly brings forth the need to negotiate. The absence of competition on both sides of the market creates a situation in which neither employers nor workers are able to dictate a wage rate. Indeed, there may be no need for bargaining unless a bilateral monopoly exists. Yet each year collective bargaining contracts are negotiated that cover the pay and working conditions for six to nine million American workers. Unions definitely limit wage competition among their members in many markets, but are there that many monopsonies?

In some industrial unions, such as the autoworkers' union, the union representatives bargain with a single firm. Under these circumstances, a firm that employs a very large percentage of a union's members does have some monopsony power. Wages cannot be dictated by the union. On the other hand, craft unions' members are employed by many firms. Hundreds of firms in different sectors of the United States employ carpenters, for example. It might seem that a union would not have to negotiate with an employer that has no monopsony power. However, the potential employers of craft union members may form a coalition, a buyer's cartel, to negotiate with unions as a single buyer of labor. Thus, when labor bargains as a unit to gain market power, previously competitive employers would do the same. Any time that wage contracts are determined by bargaining negotiations there is, in effect, a bilateral monopoly in the market.

Collective bargaining: The determination of a market wage through a process in which sellers or buyers act as bargaining units rather than competing individually.

Strikes

The ability of a union to achieve higher wages depends on its bargaining strength, and its strongest weapon is the strike. When firms fail to meet labor's demands for a higher wage or any other demand, the union can withhold the labor services of its members. Such a **strike,** or refusal to work, can severely limit or even eliminate the firm's ability to continue production. By striking, a union withholds not only its members' labor but also that of other unions who respect the picket line. Under some circumstances, members of nonnegotiating unions also withhold their labor from the negotiating firm. A firm's failure to concede to labor's demands can therefore be very costly. Of course, conceding to higher wages is costly as well.

Strike: A refusal to work at the current wage or under current conditions.

Yet the strike is a double-edged sword. Not only does it cost firms income, but union members lose income as well. A strike is not attractive to either party, but it is the best weapon labor has to force firms to accept the union wage. The union must decide if the lost income is worth its members' increased future income.

Similarly, firms have the power to reject employment at the union's desired wage. By doing so, a firm in effect refuses to hire labor at a particular wage or any higher wage. In a sense, the firm strikes against employing labor at too high a wage rate.

The potential costliness of a strike leads to serious bargaining by both management and labor because both sides have significant incentives to arrive at a new labor contract without a strike. In fact, negotiation is the general pattern of collective bargaining in the United States. Each year, some 120,000 labor contracts are renegotiated in collective bargaining processes. In more than 96 percent of these cases, labor and management arrive at a new agreement without recourse to a strike.

Table 13–2 provides some data about strikes since 1950. Perhaps the most relevant statistic in the table is the percentage of working time lost to strikes.

TABLE 13–2   Strikes: 1950–1983

While there are thousands of strikes each year, the percentage of working time lost is relatively small.

Year	Number of Strikes	Number of Workers Involved (thousands)	Number of Worker-Days Idle (thousands)	Percentage of Working Time Lost
1950	4,843	2,410	38,800	0.33
1955	4,320	2,650	28,200	0.22
1960	3,333	1,320	19,100	0.14
1965	3,963	1,550	23,300	0.15
1969	5,700	2,481	42,869	0.28
1973	5,353	2,251	27,948	0.14
1975	5,031	1,746	31,237	0.16
1977	5,600	2,300	35,822	0.17
1979	4,827	1,727	34,754	0.15
1980	3,885	1,366	33,289	0.14
1981	2,568	1,081	24,730	0.11
1982[a]	96	656	9,061	0.04
1983[a]	81	909	17,461	0.08

[a]After 1981, data on work stoppages involving less than one thousand workers ceased to be collected.
Sources: U.S. Bureau of Labor Statistics, *Handbook of Labor Statistics,* 1983, p. 380; and *Monthly Labor Review* (November 1984), p. 103.

As the data make clear, this figure has fallen consistently over time (1969 is an exception) and generally constitutes less than 0.5 percent of total working time, a figure far less than the time lost to worker absenteeism each year. The fact that not much working time is lost to strikes does not mean that strikes are not a powerful union weapon. The threat of a strike may be sufficient to make collective bargaining work smoothly.

Sometimes, if negotiations are not successful, both parties can agree to **binding arbitration.** Such an agreement brings in a third party acceptable to labor and management to make a decision that both sides must abide by. Binding arbitration is often used for public sector union disputes because public employees usually are forbidden by law to strike. Public sector strikes in basic service activities—such as commuter transportation, garbage collection, teaching, and fire and police protection—have the potential to bring the economy to a halt. If essential services provided by government cease to be performed because of a strike, much private economic activity will cease as well.

In private sector strikes, lost output can be made up to some extent if workers work overtime after the strike is settled. Moreover, substitute products and services are available to consumers during a private sector strike. Purchasers of new cars can buy used cars or drive their present cars until an autoworkers' strike is over, for example, and trips planned in taxis can be made by rental cars or postponed.

Political Influence in Bargaining

The bargaining power of unions and management is affected by legislation, which in turn is affected by the political environment. Legislation can favor either management or labor in the negotiating process. Until the 1930s, the struggle between management and labor in the United States was tilted toward management. The strike was not a strong labor weapon because employers could obtain a court order, called an injunction, against a strike. Employees suspected of union sympathies could be fired or roughed up by company thugs. New employees could be required to sign a "yellow-dog" contract, which required that a potential employee not join the union to get the job.

All this changed in the 1930s with the New Deal. The Norris–La Guardia Act (1932) and Wagner Act (1935) denied management the use of anti-union tactics and granted labor the right to organize and engage in collective bargaining in a legal framework regulated by the National Labor Relations Board (NLRB). The NLRB has five members appointed by the president for five-year terms. Its functions include ruling on unfair labor practices, settling jurisdictional disputes among unions over bargaining rights, and calling for plant elections at the request of workers. These legal changes lent a tremendous impetus to the growth of the union movement and gave unions a favorable environment for pursuing their ends.

The post–World War II era saw much industrial unrest. Unions showed little self-restraint or discipline, so Congress tilted the balance between labor and management back toward management. The Taft-Hartley Act, passed in 1947, allowed states to enact **right-to-work laws,** which forbid unions from coercing workers into their ranks. (Focus, "The Right-to-Work Controversy," discusses these laws in more detail.) It added several other constraints on union behavior, such as outlawing strikes by government workers and secondary boycotts, by which the union would set up picket lines against other suppliers of the company being struck. After the Taft-Hartley Act,

Binding arbitration: An agreement between employers and labor to allow a third party to determine the conditions of a work contract.

Right-to-work law: A law that prevents unions from forcing individuals to join a union as a prerequisite to employment in a particular firm.

charges of union corruption and ties to organized crime led to further legislation to regulate union behavior. The Landrum-Griffin Act, passed in 1959, contained various provisions to ensure that unions were honestly managed. Among other things, this law required the filing of financial reports by union officers and the auditing of union finances.

Even in the face of such restraining legislation, the labor union is important and powerful in American life. Unions are now an accepted fact in our economy. No longer are unions viewed as weak associations of workers fighting for a better standard of living for their members. In some respects, unions are best thought of as strong special-interest groups that are proficient at obtaining special legislative favors such as protective tariffs. On the other hand, union political power is sometimes used to promote general-interest legislation. Civil rights and welfare laws, Social Security, and similar programs have been strongly supported by organized labor.

Union Power Over Wages: What Does the Evidence Show?

As we have seen, unions may have an effect on the relative wages of their members, depending on a number of conditions. But what effect have the unions actually had? Several studies have attempted to measure the influence of unions using data on wages and employment in the economy.

H. Gregg Lewis, who made the pioneering study of this issue, estimated that the average wage of union members was 10 to 15 percent higher than that of nonunion members who had about the same marginal productivity as the union members.[1] Some unions made more than this, on average, while others made less. Strong unions, such as the airline pilots' union, were able to raise members' wages by more than 25 percent compared to the wages of nonunion pilots with the same skill level. Weaker unions included those in the textile and retail sales industries, where unionization has had little perceptible influence on members' wages. Craft unions exhibited more strength than industrial unions. Lewis's results square with the economic theory of union power as we have presented it.

Lewis based his estimates on data for the 1940s and 1950s. Similar empirical research on more recent data suggests that unions' power to raise relative wages may be increasing. Holding the characteristics of employees constant, Frank Stafford estimated that the earnings of unionized craft and semiskilled workers were 25 percent higher than that of their nonunion counterparts. Michael Boskin estimated that the union-nonunion wage difference was 15 to 25 percent on average in 1967.[2] Other estimates have shown even higher differentials between union and nonunion members. It is not completely clear why union power has grown over time, but one factor is the increasing political power of the union movement. As unions have become more important in politics—for instance, as major suppliers of votes to political parties—their power to increase the demand for union labor and to restrict competing labor supply has increased.

[1]H. Gregg Lewis, *Unionism and Relative Wages in the United States* (Chicago: University of Chicago Press, 1963).

[2]Frank P. Stafford, "Concentration and Labor Earnings: Comment," *American Economic Review* 58 (March 1968), pp. 174–181; and Michael J. Boskin, "Unions and Relative Wages," *American Economic Review* 62 (June 1972), pp. 466–472.

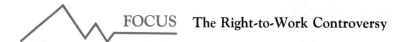

FOCUS   The Right-to-Work Controversy

A union shop provision in a contract between labor and management makes it mandatory for a new employee to join the union after a short probationary period. However, Section 14b of the Taft-Hartley Act allows states to enact right-to-work laws. These laws ban the union shop provision and let individuals decide voluntarily whether to join a union. Such laws are a thorn in the side of organized labor. Union leaders feel it is unjust for workers to refuse to join a union and pay dues to support it while at the same time benefiting from the bargaining gains achieved by the union. Opponents of compulsory unionism perceive a need to check the growth of union power and its negative impact on economic growth and employment. Periodically, labor tries to have Section 14b of the Taft-Hartley Act repealed.

As of 1982, twenty states had right-to-work laws on their books. These states are listed in Table 13–3. Historically, there have been as many as twenty-four states with such laws, but over time some states have repealed their right-to-work laws and others have added them. The thirty states in which labor union membership was compulsory in 1982 are listed in Table 13–4.

TABLE 13–3   Net Change in Manufacturing Employment in Right-to-Work States: 1970–1982

State	Employment Gain
Texas	325,900
Florida	79,300
North Carolina	65,100
Arizona	61,000
Louisiana	37,100
Virginia	31,700
Georgia	29,900
Kansas	29,000
Arkansas	27,200
South Carolina	21,900
Mississippi	21,400
Alabama	9,800
South Dakota	9,200
Georgia	6,400
North Dakota	5,000
Tennessee	4,200
Nebraska	2,000
Wyoming	1,800
Nevada	−2,400
Iowa	−8,300

Source: U.S. Bureau of Labor Statistics, *Handbook of Labor Statistics,* 1983, pp. 175–176.

TABLE 13–4   Net Change in Manufacturing Employment in Non-Right-to-Work States: 1970–1982

State	Employment Gain (+) or Loss (−)
California	370,000
Colorado	62,000
Washington	50,400
Oklahoma	49,500
Minnesota	27,600
New Hampshire	20,500
Oregon	13,300
New Mexico	12,600
Vermont	8,600
Idaho	7,500
Alaska	2,500
Maine	−3,000
Hawaii	−3,100
Delaware	−3,200
Montana	−3,800
Rhode Island	−3,800
Wisconsin	−4,600
Kentucky	−8,400
Massachusetts[a]	−18,500
Connecticut	−25,700
West Virginia	−27,600
Missouri	−42,100
Maryland	−59,200
Indiana	−126,900
New Jersey	−133,000
Michigan	−197,400
Illinois[a]	−223,700
Ohio	−306,300
Pennsylvania	−361,700
New York	−398,800

[a]Massachusetts and Illinois data cover 1970–1981.

Source: U.S. Bureau of Labor Statistics, *Handbook of Labor Statistics,* 1983, pp. 175–176.

Also listed in Tables 13–3 and 13–4 are some data on the change in the level of employment in right-to-work and non-right-to-work states between 1970 and 1982. In general, employment has grown in right-to-work states and decreased in non-right-to-work states. These data would seem to indicate that compulsory unionism retards state and local economic growth and employment. Such a conclusion must be drawn with care, however. The level and rate of change in employment in a state is a function of many factors, including the levels of state and local taxation, the abundance and availability of natural resources, and the quality of the local labor force. Many states that have right-to-work laws are also attractive to industry for other such reasons. It is the sum of these factors that determines employment changes across states. In this context we can say that while the weakness of unions in right-to-work states helps to promote employment and growth in these states, it is not the whole, and maybe not even the most important, part of the story of economic growth in the states.

The Impact of Unions on Labor's Share of Total Income

The studies just mentioned measure the impact of unions on relative wages; that is, the wages of union members versus those of nonunion members. A broader question concerns the impact of unions on the share of national income that goes to all labor, not just union labor.

Primarily, a general increase in wages requires an increase in the productivity of workers. An increase in productivity can occur in a variety of ways. The quality of tools and other physical capital used by workers can increase, workers' skill levels can be raised through education and experience, innovation or better management can increase worker productivity, and so on. Higher real wages for all workers can be realized only if the output of goods and services in the economy is increased. It would seem to follow that unions can increase the wages of all workers only if they increase the general level of productivity in the economy. While such an impact of unions is not out of the question, it is hard to imagine how it might happen.

Who pays for the benefits that unions obtain through bargaining power, legislation, and other means? The naive answer is that, of course, employers pay, but the evidence suggests otherwise. In fact, economic analysis indicates that consumers and nonunion employees pay for the benefits that unions obtain for their members.

Consumers pay because higher union wages mean higher costs for firms that employ union labor. If these firms produce in a competitive market for their output, output prices will rise to reflect the higher cost of union labor. Basically, the same result holds if the firm that hires union labor has monopoly power, although the monopolist may bargain with the union over the level of monopoly profits earned. In either case, higher costs caused by union wages lead to higher prices for consumers. Consumers thus pay part of the tab for higher union wages.

Nonunion employees pay because higher union wages lead firms to produce less output and to substitute nonhuman capital for union labor. Both these effects cause the level of employment in unionized industries to fall. In other words, as we saw in the models of how unions increase the relative wages of their members, these are cases where union bargaining power leads to a lower employment level for labor in the unionized industry. The workers who do not make it into the union must seek employment elsewhere in the economy. This increases the supply of labor to alternative, nonunion employments and drives down nonunion wages. Thus, nonunion employees also pay part of the tab for higher union wages.

It can be argued on the basis of economic theory that employers do not generally pay for all union gains. Consumers and nonunion employees share the burden because the price of union-made goods rises and the wage rate for nonunion labor falls. The latter effect can be substantial. Lewis estimated

TABLE 13–5   Labor's Share of National Income

The percentage of the nation's real output that labor receives has increased slightly since World War II. This increase is attributed to a decrease in small owner-managed firms rather than to union activities.

Year	Total Employee Compensation (Including Employer Contribution for Social Insurance) (percent of national income)	Total Employee Compensation plus Self-Employment Income (percent of national income)
1935	71	84
1940	68	83
1945	68.2	82.4
1950	71.8	82
1955	72.2	82.6
1960	74	81.8
1965	74.3	82.3
1970	74.2	82.8
1975	74.3	83.4
1980	73.9	82.2
1981	74.6	81.4
1982	76.1	82.5
1983	74.9	79.6
1984[a]	73.4	78.6

[a]provisional

Source: U.S. Department of Commerce, *Survey of Current Business,* various issues.

that nonunion wages are 3 to 4 percent lower than they would be without unions.[3] Since this effect on nonunion labor covers about three-fourths of the labor force, it is clear that the resource allocation effects of unions can be substantial in the aggregate.

Table 13–5 provides some data about the behavior over time of the share of national income that goes to labor. Two ways of measuring labor's share are shown. The second column measures total employee compensation (wages and salaries) plus the employer's share of the Social Security tax as a percentage of national income. This series shows a slight upward trend since World War II, a rise that has been attributed to a decline in the number of self-employed workers over this period, primarily in agriculture.

The third column in the table adds the income of self-employed persons—such as business proprietors, lawyers, and accountants—to employee compensation. Self-employment compensation is clearly income that goes to labor. When we look at this more inclusive concept of labor's share of national income, we see that it has been virtually constant for about fifty years. Thus, during an era when labor union strength grew in the United States, we do not find any evidence that labor's share of national income rose. While labor unions have clearly increased the wages of their members relative to the wages of nonunion workers, there is no evidence to suggest that they have made all workers better off in terms of their share of national income.

The fact is that higher real wages for workers can come about only through increases in worker productivity. Workers in the United States earn high wages because their productivity is high. Their wages can rise only if

[3]Lewis, *Unionism and Relative Wages,* pp. 1–308.

the output of goods and services in the American economy is increased. All things being equal, it is the level of productivity that drives real wages.

Conclusion

The goals of unions are easy to understand. The members want more income, greater income security, and more pleasant working conditions than the market provides. Their ability to achieve these goals is limited by many factors: the supplies of competing inputs, the monopsony power of employers, and the political power of the unions. It is clear that unions have provided their members with higher wages and better job security, but they also provide something more.

The union's role extends beyond an economic wage and employment analysis to the plant, where the union provides workers with a set of rules and representation to settle disputes with employers. This presence of the union on the job cannot be discounted as a primary source of nonpecuniary benefits to union members. Such benefits accrue to workers in the form of greater job security, a feeling of power and dignity at work, and less alienation from their work. These are important aspects of work in modern factories and offices, and the union's role in providing workplace representation is a powerful force for creating greater worker satisfaction and thus greater productivity. (See Economics in Action, "Toward a New Theory of Labor Unions," at the end of this chapter for more on the productivity-enhancing aspects of labor unions.)

There is much concern today over the meaningfulness of work. Workers are said to be alienated and bored by assembly-line types of jobs. In this respect, the role of the union as an arbitrator of workplace rules and procedures will probably become increasingly important. Workers may be willing to trade off some wage gains to obtain changes in the way their work is done. And if workers seek more meaningful jobs and factory arrangements, the union will surely be at the forefront in helping promote such arrangements in future labor contracts.

Summary

1. A labor union is a group of workers in a craft, industry, or government job who organize to gain market power.
2. Labor unions attempt to improve the economic welfare of their members by increasing wages, decreasing the elasticity of demand for their services, and increasing the demand for their members' labor through a variety of activities.
3. Monopsony exists when there is a lack of competition in the employment of labor. A single buyer may force wages below the competitive level.
4. In a bilateral monopoly in a labor market, the wage is determined by the relative strength of the bargainers.
5. A union's ability to achieve its goals is in part determined by its bargaining strength. A union's ability to strike is its strongest negotiating tool.
6. It is estimated that unions have increased the wages of their members relative to nonunion workers in similar occupations.
7. Unions have not increased all labor's share of total income. Gains made by union labor have generally been at the expense of nonunion labor and total production in the economy.

Key Terms

labor union
craft union
industrial union
public employees' union
total wage bill
monopsony
exploitation of labor
bilateral monopoly
collective bargaining
strike
binding arbitration
right-to-work law

Questions for Review and Discussion

1. How do craft unions increase wages for their members? What happens to the people who are not allowed into a craft union?
2. If industrial unions allow anyone who wishes to enter the union, how can it increase wages? Could an industrial union in an isolated mining town increase both wages and employment?
3. Suppose the United States exports beef to Japan and imports cars from Japan. If autoworkers persuade Congress to impose a tariff on Japanese cars, what happens to the incomes of cattle ranchers and ranch hands?
4. Why do unions want to decrease the elasticity of demand for their members? How do they do so?
5. What is a monopsony? Does the existence of labor unions promote the existence of monopsonies?
6. It is frequently suggested that firms exploit women in the labor force. Monopsonists' power to exploit is based on the immobility of labor. Are women less mobile than men?
7. In some countries such as the Soviet Union, there is only one potential employer—the government. Do these countries have the power to exploit their workers? How can their workers avoid exploitation?
8. Why are unions and union employers forced to negotiate wages while nonunion wages are determined without negotiation?
9. Are union wages closer to the monopsony wage or to the wage that maximizes the wage bill?
10. Suppose that a union is successful at organizing only part of the laborers in an industry. Do you predict that the union will be strong or weak? Why?
11. Why do you think unions favor minimum-wage laws?

ECONOMICS IN ACTION

Toward a New Theory of Labor Unions

Two economists from Harvard University have developed a new theory of the role of labor unions in the U.S. economy. In *What Do Unions Do?* Richard B. Freeman and James L. Medoff argue that two conflicting views of unions dominate the thinking of economists. On one side is the standard or orthodox view that unions are basically monopolies in the labor market whose main effect is to raise the wages of members, which in turn has the effect of decreasing the efficiency of the economy. On the other side are those who defend unions as increasing labor productivity by improving worker morale, assisting in the development and retention of skills, pressuring management to improve its efficiency, and protecting workers against arbitrary management decisions through collective bargaining.

In their study, Freeman and Medoff argue that the truth is somewhere in between. They present detailed findings in their book, but five basic results stand out:

1. Although unions provide their members with a significantly higher-than-competitive wage, there is no single union/nonunion wage differential among all socioeconomic groups. The effect of a union on wages is greatest for less-educated, younger, and low-seniority workers in heavily organized and regulated industries. Further, Freeman and Medoff

Source: Richard B. Freeman and James L. Medoff, *What Do Unions Do?* (New York: Basic Books, 1984).

state that "the social costs of the monopoly wage gains of unionism appear to be relatively modest, on the order of .3 percent of gross national product, or less" (p. 2).
2. Unions reduce the cost of their members' pensions as well as the cost of their life, accident, and health insurance.
3. Unions on balance increase the equality of income distribution among workers.
4. Unions increase the stability of the work force by providing various services such as grievance and arbitration proceedings and seniority clauses.
5. In many industries, unionized establishments actually have a higher rate of productivity than nonunion establishments.

The main contention of Freeman and Medoff's book—that unions provide services to the work force that are essentially unrelated to monopoly wage differentials and that, taken by themselves, increase economic efficiency—would receive the assent of most economists. Critics of their work have emphasized that Freeman and Medoff underestimate the social costs of union monopoly and that the efficiency-increasing services offered by unions are basically unrelated to their efficiency-decreasing monopoly aspects. In other words, it is possible to have unions provide benefits to workers without simultaneously engaging in restrictive labor market practices such as the closed shop.

Question

If we assume that unions actually produce higher than average wage rates for younger or less educated workers in certain fields, is there reason to believe that future unemployment will increase in these areas? Will businesses faced with such unions attempt to substitute relatively lower-priced machinery for the now relatively higher-priced labor? Comment.

14

Capital and Interest

Wealthy societies such as our own consume, or use, goods and services at ever-increasing rates. Think, for instance, of the rate at which automobiles have become available to nearly every American. Little more than sixty years ago, relatively few Americans owned cars, and most of these proud owners were happy to keep the same car for several years. Today autos are bought and sold in the millions. Many of us own more than one car; few of us are happy to hold onto the same car for more than four or five years. Since consumption is possible only through production, the basis of a wealthy society is not its ability to consume but its ability to produce, to transform raw materials into more useful goods and services through the application of human skills and physical capital such as tools and factories. Societies that are better at this transformation are able to reach higher levels of wealth and to consume more. This chapter is about the role of capital in determining the ability to produce and the level of wealth in a society.

Roundabout Production and Capital Formation

Economic behavior is forward-looking. The availability of the food we eat today is due to the foresight of farmers in the past who undertook the appropriate productive actions. The food we will eat tomorrow depends on actions farmers take now. A simple analogy of a shipwrecked sailor illustrates this point.

Imagine Robinson Crusoe trying to survive alone on a windswept island in the middle of the ocean. Crusoe's economy is primitive. He lives by fishing with a simple wooden spear that he found on the beach. With his spear Crusoe can catch 5 pounds of fish in about 10 hours of fishing. Suppose that one day Crusoe decides to improve his fishing techniques by weaving a fishing net.

Crusoe takes a day off to build the net. His opportunity cost of building the net is the fish that he would have caught during this day by fishing with his spear. Assuming that it takes 10 hours to build the net, the opportunity cost of the net to Crusoe is 5 pounds of fish.

Crusoe's purpose in building the net is obviously to catch more fish in the future. That is, he hopes the net will enable him to catch more than 5 pounds in 10 hours. It is the anticipation of more production and therefore more consumption in the future that drives Crusoe to the more **roundabout** method of catching fish.

Of course, Crusoe must have some way to support himself during the period that he stops fishing to weave the net. He provides for his needs during this time by **saving.** If he consumes 5 pounds of fish every 10 hours and it takes 10 hours to make a net, Crusoe must set aside 5 pounds of fish to support himself while making the net. To do so, he could fish extra hours for several days, or he could eat only 4 pounds each day for 5 days. The hallmark of saving is some form of current sacrifice—extra hours worked mean less hours for current leisure, or fewer fish eaten in the present mean fewer calories consumed—in anticipation of a higher level of future consumption. However Crusoe manages to save fish, it is clearly the act of current abstinence that leads to saving and ultimately to the greater yield of fish.

What is Crusoe's incentive to abstain? Suppose that with the net he can catch 25 pounds of fish in 10 hours and that the net will last indefinitely. By abstaining from spear fishing for 10 hours, at a cost of 5 pounds of fish, Crusoe is able to raise his daily catch by 20 pounds. The increase is even more significant in long-range terms: For an investment of a mere 5 pounds of fish Crusoe raises his weekly catch from 35 pounds to 175 pounds, an increase of 140 pounds per week. This particular act of saving and investment of resources to produce more capital is especially profitable. Indeed, even if the net wears out every so often, Crusoe will be able to afford to take 10 hours to make a new net in anticipation of similar rates of return in the future.

This simple example has relevance to a modern economy. In a modern economy individuals specialize in their activities. They don't spend part of the day fishing and part of the day making rods and reels. But this specialization does not change any of the essential features of what is called capital formation.

The term **capital formation** refers to the process of building capital goods and adding to the capital stock in the economy. Capital goods are things such as machines and implements that are used to produce final goods and services. The capital stock of an economy is the amount of capital goods that exists at a given point in time. To increase capital stock, producers must resort to methods of roundabout production, the use and production of capital goods as a means of achieving greater output in the future. Saving, or the abstinence from present consumption, is required for capital formation to take place. Moreover, the act of saving requires entrepreneurship, a concept involving risk taking, which we discuss in Chapter 15. In the case of capital production, the future is uncertain. Crusoe did not know whether a net would increase his daily catch. He took a chance that it would, and it paid off.

Another important feature of capital is that it does not automatically re-

Roundabout production: The production and use of capital goods to produce greater amounts of consumption goods in the future.

Saving: The act of forgoing present consumption in an effort to increase future consumption.

Capital formation: The use of roundabout production to increase capital stock.

Depreciation: The wearing out of capital goods that occurs over a period of time.

produce itself. Instead, capital goods are automatically **depreciating;** they decrease in productivity or value over time. This wearing out of capital goods cannot be stopped. If Crusoe does not repair or replace his net, he will ultimately return to spear fishing and to a lower standard of living. The capital goods or capital stock of a society must be repaired and replaced over time if the society is to continue to grow and experience high levels of consumption. This is the primary problem of economics: maintaining and increasing the ability to produce.

Capital consumption: The loss of capital that occurs because the rate of depreciation is greater than the rate of capital formation.

Capital consumption is the opposite of capital formation. Present consumption is increased temporarily at the cost of a reduction in the future rate of consumption. Suppose that a lumber company owns trees ranging in age from 1 to 25 years. Assume that it cuts down one thousand 25-year-old trees and plants one thousand seedlings each year. As long as external circumstances do not change, this forest can yield one thousand 25-year-old trees annually. For a while, however, the yield from the forest can be increased by harvesting some of the 24-year-old trees in addition to the regular harvest, then dipping into the stock of 23-year-old trees the next year, and so on. Cut in this manner, the amount of timber harvested will increase for a while. Still, the time will come when the total output of the forest will necessarily decline. Capital in the form of younger trees has been used up to increase current output, so the future output of timber must fall. This process is called capital consumption.

The Rate of Interest

Saving and capital formation depend on the willingness of individuals to pass up current consumption to achieve greater consumption in the future. In other words, individuals must abstain from current consumption to provide the flow of saving that is used to produce capital goods. Economists stress that an individual's willingness to abstain from current consumption is related to his or her **rate of time preference.** Time preference is the degree of patience that an individual has in forgoing present consumption in order to save. A person with a high rate of time preference has a strong preference for current rather than future consumption. Most individuals have positive rates of time preference; that is, they prefer present to future consumption. Since people prefer to consume now rather than later, they must be paid a price for waiting. This price is called **interest.**

Rate of time preference: The percent increase in future consumption that is necessary to induce an individual to forgo some amount of present consumption.

Interest: The price a borrower pays for a loan or a lender receives for saving, measured as a percentage of the amount; the price of not consuming now but waiting to consume in the future.

The concept of interest has two related meanings. In the first sense, interest refers to returns on investments. A person who saves $100 in a passbook account earns 5¾ percent interest on saving. A firm that invests $1000 in a new piece of machinery earns $200 more in revenue because of the resulting improvement in productivity. The firm's investment thus yields 20 percent interest, the firm's reward for roundabout production. In these terms, interest is the amount individuals are willing to *receive* in order to sacrifice current consumption; it is a payment for their abstinence or waiting to consume later.

Interest is also the price that individuals are willing to *pay* to obtain a good or service now rather than later. A consumer who wants a car today might pay 13 percent interest for a loan from the bank. A firm that needs a new plant might pay 10 percent interest to get the necessary financing.

Keep in mind that interest is not just a monetary phenomenon. Whether paid or received, interest is based on the fact that people prefer to consume and invest now rather than later. Moreover, the rate of interest reflects the rate of time preference in the economy. The more impatient people are, the higher the rate of interest must be to induce them to save and to create capital goods.

Demand and Supply of Loanable Funds

The rate of interest is determined in the market for loanable funds. It is, in effect, the price of loanable funds. Like other markets, the market for loanable funds has a demand side and a supply side.

Demand for loanable funds: A curve or schedule that shows the various amounts of money that people are willing and able to borrow at all interest rates.

The **demand for loanable funds** arises from two sources. Consumers demand loanable funds because they want to consume more now than their current incomes will permit. Loans to these individuals are called *consumption loans*. They are made for myriad reasons. Individuals may want to take a vacation now and pay for it on time, borrow to tide themselves over a temporary decline in income, or borrow to buy cars or household appliances.

The second source of demand for loanable funds is desire for *investment loans*. Investors borrow in order to finance the construction and use of capital goods and roundabout methods of production that are expected to be productive. Firms borrow funds to build new plants and to purchase new equipment in the expectation that such investments will increase profits.

The sum of the demand for consumption loans and investment loans equals the total demand for loanable funds. The law of demand applies to the demand for loanable funds just as it does to any other commodity (see Figure 14–1). The price one must pay to borrow funds for consumption or investment loans is the interest rate. According to the law of demand, the amount of funds demanded is inversely related to the interest rate. As the interest rate falls, the cost of borrowing money to finance the earlier availability of consumption and capital goods falls. Other things being equal, we thus expect to see more borrowing when the interest rate falls. At a lower rate of

FIGURE 14–1

Determination of Interest Rates

D and *S* are the demand and supply curves for loanable funds. The market for loanable funds reaches equilibrium at *i* and *Q*, where the plans of borrowers and lenders are compatible. That is, for a given interest rate, the amount that borrowers want to borrow and the amount that lenders want to lend are equal.

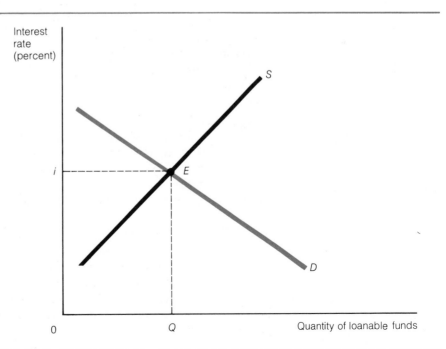

interest, consumers will expand current consumption, and more investment projects will appear to be profitable.

In Figure 14–1, the **supply of loanable funds,** S, is provided by savers. These are individuals or firms who are willing to consume less than their present earnings in order to set aside something for the future. As the interest rate—in this case, the return on investment—rises, more and more individuals and firms will be enticed to save; that is, to forgo current consumption in order to consume even more in the future. Thus, the quantity supplied of loanable funds varies positively with the interest rate offered to savers.

Supply of loanable funds: A curve or schedule that shows the various amounts of money that people are willing and able to lend (save) at all interest rates.

The intersection of the demand and supply curves for loanable funds in Figure 14–1 yields an equilibrium interest rate and the equilibrium level of loanable funds. At i and Q, the market for loanable funds is in equilibrium, and the plans of borrowers are compatible with the plans of lenders. At interest rates above i, there will be an excess supply of saving, putting pressure on the interest rate to fall. Below i, there will be an excess demand for loanable funds, putting pressure on the interest rate to rise.

Variations Among Interest Rates

The previous discussion of the supply and demand for loanable funds might suggest that there is one interest rate in the economy. In reality there is a multiplicity of interest rates: the prime interest rate given to businesses with excellent credit ratings, mortgage rates offered to home buyers, credit card rates, consumer loan rates, and so on. These interest rates tend to be different. For instance, the interest rate on government bonds is generally lower than the interest rate on corporate bonds. What are some of the reasons that interest rates differ?

Risk: The probability of a default or a failure of repayment of a loan.

Risk. The **risk** associated with particular borrowers—their likelihood to default on repaying a loan—is an important source of the differences in interest rates. Creditors go to great lengths to ascertain the degree of this risk, and they adjust the interest rate they charge accordingly. Loans to a government agency will carry a low interest rate because the risk of default is very low. Government can use its power to tax to repay loans. On the other hand, loans to unemployed workers will carry high interest rates because the risk of default is high.

Cost of Making Loans. The cost of making loans differs. Large loans and small loans may require the same amount of accounting and bookkeeping work. The large loan will therefore be less costly to process per dollar loaned; for this reason it will carry a lower interest rate than the small loan. This distinction means, for example, that loans to large companies will carry a lower interest rate than loans to small companies.

Time. The length of time over which a loan is made will affect the rate of interest charged. The longer the term of the loan, the more things that can go wrong for the borrower. Because the risk of default rises with the length of the loan, longer-term borrowers must pay a premium for this rise. Long-term loans will carry higher interest rates than short-term loans, other things being equal. In addition to the risk factor, borrowers are willing to pay a premium for the longer availability of funds.

Nominal rate of interest:
The price of loanable funds measured as a percentage of the dollar or nominal amount of the loan.

Real rate of interest: The nominal interest rate minus the rate of inflation; the price of loanable funds measured as a percentage of the real buying power of the amount loaned.

Nominal and Real Rates of Interest. The **nominal rate of interest** is the interest rate set in the market for loanable funds. With inflation in the economy, the nominal rate of interest can be a misleading measure of how much borrowers pay for consumption and investment loans. Suppose that the inflation rate, which is a rise in the average level of all prices in the economy, is 12 percent per year and that the nominal rate of interest is 17 percent. A borrower who borrows $100 will have to repay $117 in a year. However, during the year the average level of prices increases by 12 percent. The $117 paid to the lender at the end of the year will not buy the same amount of goods as the $100 made available to the borrower one year earlier. Since prices have risen by 12 percent, the $117 will buy only 5 percent more goods for the lender after a year. Thus, when the rate of inflation is factored in, the **real rate of interest** is 5 percent, or 17 percent minus 12 percent.

Lenders and borrowers will not generally be fooled by inflation. The real rate of interest will adjust to account for the expected inflation rate. The nominal rate of interest will include a premium to compensate lenders for the expected depreciation of the purchasing power of their principal and interest. Lenders will have to be compensated for expected inflation, or they will reduce the amount they are willing to lend. Borrowers will also recognize that they will be repaying loans with dollars of less purchasing power and will adjust the amount of interest they are willing to pay to obtain loans. In the above example, if both borrowers and lenders fully anticipate a 12 percent inflation rate, the nominal rate of interest will adjust to 17 percent. As the inflation rate changes, the nominal rate of interest will also change to reflect the level of expected inflation.

The real rate of interest is the nominal rate of interest minus the expected inflation rate. In the latter part of the 1970s, the nominal rate of interest rose to over 20 percent on some types of loans, but since the expected inflation rate was on the order of 15 percent, the real rate of interest was only 5 percent.

Interest as the Return to Capital

So far, we have discussed interest as the price of loanable funds. Interest may also be viewed as the return that goes to capital as a factor of production. In other words, interest is the return to capital as a productive input in the circular flow model of the economy.

If an entrepreneur buys a machine for $100 and makes $25 a year by using it in production, the entrepreneur earns 25 percent interest on the investment in the machine. Using this concept of interest, we can show how the market for capital equipment works in Figure 14–2.

The demand curve, D, is the marginal revenue product curve of capital. It represents the entrepreneur's estimate of how much each unit of capital equipment will add to the firm's revenue. In effect, D is the derived demand curve for capital equipment. We assume that the short-run supply of physical capital—machines, building, tools, and the like—is fixed and cannot be augmented except over the long run. S_1 and S_2 are therefore drawn as vertical straight lines in Figure 14–2. The point where demand and supply intersect determines the rate of interest that entrepreneurs earn on each level of fixed capital investment.

What is the link between the market for loanable funds and the market for physical capital? If the rate of interest being earned on physical capital,

FIGURE 14–2

Return on Investment in Capital Goods

D is the derived demand curve for capital goods. S_1 and S_2 are short-run fixed supply curves for capital goods. Investment in new plant and equipment will take place if the interest earned on such investment (i_1 and i_2) exceeds the interest rate on loanable funds. New investment shifts the supply curve of capital goods from S_1 to S_2.

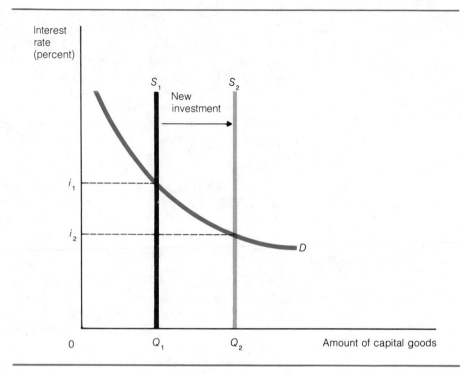

such as i_1, is higher than the rate of interest on loanable funds, it will clearly pay entrepreneurs to borrow funds at the lower rate of interest to purchase additional physical capital. Over time, the given supply of capital will shift to the right as entrepreneurs invest in plant and equipment. The movement from S_1 to S_2 represents such a shift. If the rate of return on physical capital falls below the loanable funds rate, the opposite process will take place, with the level of capital investment falling over time and the supply of physical capital shifting to the left.

Generally, there is a tendency for the rate of return on capital goods to fall over time and for an associated decline to exist in the demand for loanable funds. Offsetting this tendency in the economy are forces such as technological change that increase the productivity of capital goods and shift the demand curve for capital goods to the right.

The Nature of Returns to Owners of Capital

We have seen that individuals invest in capital goods because they expect to make a rate of return in excess of the cost of capital. The elements of this rate of return can be broken down into three categories: pure interest, risk, and profits and losses.

Pure Interest

Pure interest: The interest obtained from a risk-free loan.

Capital investment yields a return called **pure interest,** the interest rate that must be paid to induce saving when the lender bears little or no risk. Think of pure interest this way: You have a sum of money, and you're trying to decide whether to invest in a piece of capital equipment or a savings account at the bank with a virtually guaranteed return of 5 percent. The 5 percent is a pure interest yield on your money; in a year you will have earned a 5 percent return on your investment in the savings account. Unless you are confident that the alternative investment in the equipment will yield at least 5 percent, you have no incentive not to put your money in the bank. Investments in capital equipment must pay at least the pure rate of interest, or investors will not provide funds for investments. Again, the pure rate of interest is the rate that must be paid to induce individuals to abstain from current consumption and to save.

Risk

Suppose that you are confronted with the following offer: You can have $100 for sure or you can have a 50-50 chance of receiving either $200 or nothing. In fact, the expected value of each alternative is the same. In the second alternative, the probability of receiving $200 is 0.50, and the probability of receiving no money is 0.50. Thus, 0.50 ($200) + 0.50 ($0) = $100. Yet the individual decision maker is not indifferent between these alternatives. The latter alternative involves risk, while the former does not. Individuals generally view the bearing of risk as a cost and have to be compensated to bear risk.

Risk has a lot to do with capital investment. There is no guarantee of a handsome rate of return on investments in capital goods. Market conditions can change, making the capital goods obsolete or making the things that the capital goods produce less popular with consumers. In other words, capital investments involve risks, and to induce individual investors to bear risks, they must be paid a premium on their investments. This risk premium is above and beyond the pure rate of interest, and it is paid to induce investors to bear the risk of capital investment.

Profits and Losses

Investors in capital goods can earn economic profits or losses. The opportunity cost of fixed capital is the price that must be paid to keep it committed to its present use. This cost consists of the pure interest and risk premium discussed above. Yet something unexpected can happen in the economy to make the fixed capital more valuable. For example, the demand for the product produced by fixed capital may increase many times over what was expected. In such a case the owners of capital earn economic profits on their investment in addition to pure interest and a risk premium. In a competitive market, these returns will be competed away as other investors become aware of the excess returns and make capital investments of their own in the industry. Keep in mind too that economic profits can be negative; that is, investors in capital goods can incur large and unexpected losses.

These profits and losses are analogous to capital gains and losses to investors in the stock market. A capital gain, for example, derives from an unexpected increase in the value of a firm, its capital stock, and what the capital stock produces.

Summarizing, the rate of return to capital investment embodies three components: a pure interest return, a premium for bearing risk, and a residual component reflecting unexpected changes in the value of capital goods. Each of these components has an important allocative function in the economy. Pure interest induces people to save. The risk premium leads individuals to invest in risky but valuable ventures. Economic profits are the spur to entrepreneurship and innovation in the economy. We have more to say about profits and entrepreneurship in Chapter 15.

The Present Value of Future Income

Present value: Today's value of a payment received in the future; future income discounted by the rate of interest.

Investment in capital goods and saving balances the needs of tomorrow with the needs of today. Economists find it useful to look at this balancing act in a precise mathematical formula called present value. **Present value** refers to the value today of some payment that will be received in the future. This value depends on the rate of interest and the length of time between the present and future.

How, for instance, would the prospect of making $100 in one year or in two years compare with the prospect of making $100 now, if the rate of interest is 10 percent? The $100 made now would by definition have a present value of $100. But what about the $100 made in one year? What would be the equivalent amount of money made now? That is, what amount invested now at 10 percent would be able to generate a fund of $100 in one year? The answer is $90.91 with simple interest, computed by the formula

$$PV = \frac{\text{Receipts in one year}}{1 + \text{Interest rate}} = \frac{\$100}{(1 + 0.10)} = \$90.91.$$

What sum invested now at 10 percent and compounded annually would produce $100 after two years? The answer is $82.64, which is the present value of $100 in two years at 10 percent compounded annually. This amount can be computed from the formula

$$PV = \frac{\$100}{(1 + 0.10)^2} = \$82.64.$$

The logic of this formula is that $82.64 will increase to $90.91 after one year, which in turn will increase to $100 after two years.

By extension of this procedure, the present value of any amount at any time in the future can be computed. The formula for doing this is

$$PV = \frac{A_n}{(1 + r)^n},$$

where A_n is the actual amount anticipated in a particular year in the future, r is the rate of interest, and n refers to the particular year in the future.

Note two aspects of present value computations. First, the lower the rate of interest used in making these computations, the higher will be the present value attributed to any given amount of income to be made in the future. Table 14–1 illustrates this point clearly. The present value of $100 one year from now is $97.10 at a 3 percent interest rate and $89.30 at a 12 percent interest rate. Second, the present value of a given amount to be received in the future declines as the date of receipt advances further into the future. In Table 14–1, for any given interest rate the present value of $100 is less the further it is to be received in the future. To see this relation, simply read down any column in the table. The point is that the present value of future revenues or costs is inversely related to the rate of interest and the distance of the date in the future when payment will be received. Focus, "Present Value and Calculating the Loss from Injury or Death," presents an application of the concept of present value.

TABLE 14–1   The Present Value of $100 at Various Years in the Future

The further in the future the $100 is received, reading vertically down the table at any interest rate, the less it is worth now. Reading horizontally across the table, the higher the interest rate the less is the value of $100 at any year in the future.

Years in the Future	3%	5%	7%	10%	12%
1	$97.10	$95.20	$93.50	$90.90	$89.30
2	94.30	90.70	87.30	82.60	79.70
3	91.50	86.40	81.60	75.10	71.10
4	88.80	82.30	76.30	68.30	63.60
5	86.30	78.40	71.30	62.00	56.70
6	83.70	74.60	66.60	56.40	50.70
7	81.30	71.10	62.30	51.30	45.20
8	78.90	67.70	58.20	46.60	40.40
9	76.60	64.50	54.40	42.40	36.00
10	74.40	61.40	50.80	38.50	32.20
.	.	.	.	.	.
.	.	.	.	.	.
15	64.20	48.10	36.20	23.90	18.30
.	.	.	.	.	.
.	.	.	.	.	.
20	55.40	37.70	25.80	14.80	10.40
.	.	.	.	.	.
.	.	.	.	.	.
30	41.20	23.10	13.10	5.73	3.34
.	.	.	.	.	.
.	.	.	.	.	.
40	30.70	14.20	6.70	2.21	1.07
.	.	.	.	.	.
.	.	.	.	.	.
50	22.80	8.70	2.13	.85	.35

Investment Decisions

The estimation of present value is important to firms because investment decisions involve current expenditures for plant and equipment in the expectation of future revenues from the goods and services produced by the plant and equipment. Two aspects of investment decision making are paramount. First, the future revenues that an investment project will yield are estimated and converted to present value terms. This process provides the firm with an estimate of how much a project is worth now. Second, the firm must know what its cost of loanable funds is; that is, it must know on what terms it can borrow or lend money. This cost is the interest rate that the firm will apply to investment decisions. A simple decision rule follows from these two steps: If the estimated present value of a project exceeds the cost of loanable funds, a profit-maximizing firm will undertake the project. If not, it will turn down the project. For example, if a firm estimates that a project will yield an 8 percent return but money market certificates are currently paying 15 percent, the firm will reject the project and place its funds in the money market.

FOCUS   Present Value and Calculating the
Loss from Injury or Death

In many lawsuits, the formula for present value is used to estimate the loss of earnings that an individual suffers from an incapacitating injury or premature death. Economists are retained by law firms across the country to make such estimates and to present them to judges and juries.

Consider an instance of premature death. The cost of premature death is treated as the loss of income that would otherwise have resulted over the remaining working lifetime of the deceased. For someone who earned $50 a day and worked 240 days per year, the loss of income would be $12,000 per year. If this person died at age sixty but had expected to work five more years before retiring, the total loss of income for the missing five-year period is $60,000. This amount would not be the cost of the premature death. Rather, the cost would be some lesser amount, reflecting the fact that a loss of $100 in the future is not equivalent to a loss of $100 now.

The formula for present value (PV) given in the text is $PV = A_n/(1 + r)^n$. In the case of a person who earned $12,000 per year, A would be $12,000. At a 10 percent rate of interest, what would be the present value of the premature death at age sixty of someone who earned $12,000? Applying the general formula for present value, the answer would be found by solving

$$PV = \frac{\$12,000}{(1.1)^1} + \frac{\$12,000}{(1.1)^2} + \frac{\$12,000}{(1.1)^3}$$
$$+ \frac{\$12,000}{(1.1)^4} + \frac{\$12,000}{(1.1)^5} = \$45,521.$$

While a total of $60,000 of income is lost, the present value of this loss is $45,521. For a person who dies prematurely at age sixty, the cost of this death, labeled as premature because it occurred before age sixty-five, would therefore be estimated at $45,521.

The concept of present value thus plays a role in the courtroom. It helps lawyers, judges, and juries establish reasonable bounds for settling lawsuits based on the loss of future income through injury or death.

Here is an example of this reasoning. A firm is contemplating an investment in a $50,000 machine. Its managers estimate that after expenses for maintenance and repair the machine will add $10,000 per year for the next six years to firm revenues. After six years the machine no longer has any value; it has no scrap value. Assuming that the firm can borrow the funds to purchase the machine at a 12 percent rate of interest, it will discount the estimated future revenues at a 12 percent rate. The present value calculations are shown in Table 14–2.

By taking the appropriate rate of interest, commonly called the **discount rate** when used in investment planning, from Table 14–1 and applying it to the estimated future revenue at the end of each year, we derive in the last column of Table 14–2 how much the flow of future income from the machine for each year is worth today. As you can see, the total present value of the estimated future income produced by the piece of equipment is $41,200 (the sum of the figures in the last column). This total present value is considerably less than the cost of the machine ($50,000), and so the firm will rationally reject this investment proposal.

Whether an investment project will be undertaken is obviously sensitive to the interest rate that the firm faces. As we have seen, lower interest rates lead to higher present values. If the discount rate in this example had been 5 percent, the present value calculations would have indicated that estimated revenues would be greater than $50,000. At a 5 percent interest rate, the firm would have undertaken the project.

Will the rates of return on profitable investments differ greatly in a competitive economy? Ignoring differences in risk of different investments, competition will tend to equalize the rate of return to investment projects everywhere in the economy. Where there are profits from investment projects,

Discount rate: The interest rate that a firm uses to determine the present value of an investment in a capital good; the best interest rate that a firm can obtain on its savings.

TABLE 14–2   The Discounted Present Value of $10,000 for Six Years

Year	Estimated Future Revenue (received at year-end)	Discounted Value (12% rate)	Present Value of Income
1	$10,000	0.89	$8,900
2	10,000	0.80	8,000
3	10,000	0.71	7,100
4	10,000	0.64	6,400
5	10,000	0.57	5,700
6	10,000	0.51	5,100
	Total $60,000		Total $41,200

entry by competing investors in similar projects will occur. For example, in the 1950s Ray Kroc pioneered the concept of a fast-food restaurant by opening McDonald's restaurants across the country. The profits that he earned on this investment attracted many competitors—Burger King, Wendy's, and so on—to make similar investments. Such entry (or exit in the case of losses on investment projects) will drive the rate of return on capital investment to equal its cost. In simple terms, competition will tend to equate the expected future revenues from capital investment to the current cost of the investment. In such a way competition equalizes the rate of return on capital investments in a competitive economy.

The Benefits of Capital Formation

Saving means abstaining from consumption now in the expectation of being able to consume more in the future. Saving thus provides the resources needed to increase the stock of capital goods in the economy. A large stock of capital goods raises the rate of consumption that can be sustained in the future. Saving clearly benefits the saver, but it also benefits those who don't save. Indeed, the benefits from saving are diffused throughout the whole economy. Most of us benefit from advances in computer technology, though few of us contributed to the saving that made such advances possible.

Saving and capital formation also lead to higher wages and incomes for workers. In a competitive economy wages are a function of the productivity of workers, and this productivity is in turn a function of the amount and quality of the equipment and tools that workers use. A worker with a tractor is far more productive than a worker with a mule-drawn plow. Since more productive workers are paid more, some of the benefits from capital formation are spread to workers who did not necessarily contribute to the saving that made the capital formation possible.

Indeed, those who save and who build capital goods are among the chief benefactors of society. As Ludwig von Mises put it:

> Every single performance in this ceaseless pursuit of wealth production is based upon the saving and the preparatory work of earlier generations. We are the lucky heirs of our father and forefathers whose saving has accumulated the capital goods with the aid of which we are working today. We favorite children of the age of electricity still derive advantage from the original saving of the primitive fishermen who, in producing the first nets and canoes, devoted a part of their working time to provision for a remoter future. If the sons of these legendary fishermen had worn out these intermediary products—nets and canoes—without replacing them by new ones, they would have consumed capital

and the process of saving and capital accumulation would have had to start afresh. We are better off than earlier generations because we are equipped with the capital goods they have accumulated for us.[1]

The reverse is also true: Policies that reduce the incentive to save and to produce capital goods, while appearing to harm the suppliers of saving and the owners of capital goods, actually hurt all of us.

Summary

1. Wealth can be defined as a high level of sustainable consumption in an economy. The elements of a wealthy society are production, resources, knowledge, capital, technology, and institutions.
2. Roundabout production is the process of saving and investing in capital goods production in order to produce more in the future.
3. Time preference is a measure of an individual's concern for present versus future consumption.
4. Interest is the price that individuals must be paid for waiting to consume later, or the price that individuals are willing to pay in order to consume now rather than later.
5. The rate of interest is determined in the market for loanable funds by the demand and supply of loanable funds.
6. In reality there are a multiplicity of interest rates—such as prime rates, government rates, and credit card rates—determined by the degree of risk, cost, and time allowed for payment. Taking the rate of inflation into account, the real rate of interest is lower than the nominal rate of interest.
7. The returns to capital are composed of pure interest, a risk premium, and economic profits or losses.
8. Present value is the amount of money in the present that is equivalent to a certain amount of money in the future.
9. The discount rate is the interest rate used to discount future amounts of revenues and costs in order to obtain present values.

Key Terms

roundabout production	capital consumption	supply of loanable funds	pure interest
saving	rate of time preference	risk	present value
capital formation	interest	nominal rate of interest	discount rate
depreciation	demand for loanable funds	real rate of interest	

Questions for Review and Discussion

1. Give some examples of roundabout production. Describe why going to college is a roundabout method of production.
2. Suppose that you are thinking about investing in a state of the art typewriter. The typewriter costs $2000 and will last for four years. You expect to earn $800 a year typing papers part time with the typewriter. The bank will lend you the money to buy the typewriter at 19 percent interest. Should you buy the typewriter? Suppose the bank's interest rate was 12 percent?
3. Why is the concept of time preference important in the discussion of interest rate? Explain.
4. How would the following events affect the rate of interest in the United States?
   (a) The threat of war in Central America
   (b) An increase in the inflation rate
   (c) The discovery of massive domestic oil resources
   (d) Greater impatience in the general population
   (e) More capricious government intervention in the economy
5. If the interest return that entrepreneurs can earn on capital investments is less than the interest rate paid for loanable funds, how will entrepreneurs behave? What happens to the level of investment in the economy?
6. J. M. Keynes in *General Theory* suggests that the marginal product of capital can be driven to zero in one generation by increasing the production of capital. Would this ever happen in a free market? If it did, what would this imply about time preference? What would happen to the nation's output? What would be the interest rate?
7. What is the formula for computing present value?

[1]Ludwig von Mises, *Human Action*, 3rd ed. (Chicago: Henry Regnery, 1966), p. 492.

How is present value affected by changes in the interest rate used for discounting?

8. What is the present value of your entire future income? Would you consider yourself a millionaire? How much of your future income would a bank be willing to lend you now?

ECONOMICS IN ACTION Stocks and Bonds

At the back of most daily newspapers are tables listing the current prices of stocks and bonds. Such tables are good evidence of how the market for loanable funds works, coordinating the supply of funds, primarily held by investors and investor groups, with the demand for funds, primarily the desire by corporations to raise capital for future production.

Corporations can raise capital by two means: They can sell shares of stock or issue bonds. By selling shares of stock, a corporation is effectively selling property rights to itself. The shareholders of a corporation are its owners; shares usually convey a voting right: Each share of stock permits its holder one vote at meetings that determine the firm's future management. In this way corporate policy is directed by the will of the holders of the majority of the firm's shares—whether one or many individuals. Shareholders do not make operating decisions for the firm, but they are legally responsible for hiring the managers—the board of directors—who do. Usually when shareholders are pleased with the existing management of a firm, they assign their votes to the management, who then act as a proxy for the shareholders in voting on corporate policy.

The stock market is an organized exchange where corporate shares are bought and sold. In the stock market the prices of shares issued by different firms reflect the market's estimate of the future profitability of the firms in question. From the shareholders' standpoint, investing in the stock market means bearing the risk resulting from the uncertain prospects of any given firm. A share of the stock's future value is not guaranteed and can rise or fall with the firm's fortunes, or because of general market conditions such as interest rates.

Corporations and governments also issue another type of obligation called a bond. Unlike shares of stock, bonds do not confer ownership rights; they simply represent IOUs. The firm or government body agrees to pay the bondholder a fixed sum (the principal that represents the loan to the firm) either on a specific date or in installments over a specified period. Most bonds pay, in addition, a fixed return per year, usually expressed as a percentage of the face value of the bond. For example, a corporation may issue a $10,000 bond (which, if sold, represents a $10,000 loan to the corporation for some specified period) at an interest rate of 5 percent. This bond would pay the holder $500 per year until maturity, when the entire $10,000 is returned to the investor.

52 Weeks High	Low	Stock	Div.	Yld %	P-E Ratio	Sales 100s	High	low	Close	Net Chg.
			– A–A–A –							
39⅞	30¾	Alcoa	1.20	3.8	16	x2283	32	31⅜	31⅜	– ½
25⅝	15½	Amax	.20	1.1	..	523	17⅝	17⅜	17½	– ⅛
42¼	32½	Amax	pf 3	8.8	..	6	34	34	34	– ½
52½	22¾	AmHes	1.10	3.5	16	3877	32¼	31½	31⅝	+ ½
2¾	1¼	AmAgr		..	..	70	2⅛	2	2⅛	.....
19⅝	15⅛	ABakr		..	8	10	18½	18¼	18½	+ ⅛
70	53	ABrand	3.90	5.9	9	709	67⅞	66⅛	66⅜	– 1½
27¾	24⅝	ABrd	pf2.75	9.9	..	166	27¾	27⅜	27¾	+ ⅜
115	55¾	ABdcst	1.60	1.5	16	899	108⅜	108	108	.....
26¼	19½	ABldM	.86	3.4	13	12	25½	25½	25½	.....
27⅛	20⅛	ABusPr	.64	2.5	15	6	26	25¾	25¾	– ⅜
55¾	40⅛	AmCan	2.90	5.5	11	805	53½	53	53⅛	– ⅜
48	36	ACan	pf 3	6.5	..	54	46⅝	46¼	46¼	– ¼
110	103	ACan	pf13.75	12.	..	16	u110¾	109½	110¾	+ 1¾
19⅞	16¾	ACapBd	2.20	12.	..	85	19⅜	18⅞	19	– ⅛
33	25⅛	ACapCv	2.51e	9.0	..	13	27⅞	27¾	27¾	– ⅛
11	6½	ACentC		..	11	32	8⅝	8½	8½	– ¼

The photo above shows a listing of stocks, their prices, and other information from the New York Stock Exchange in the Wall Street Journal. To the right of the first two columns of numbers, an abbreviation of the stock's name is found. ABdcst, for example, stands for the American Broadcasting Company. In columns one and two, the high and low price of the stock in the previous year (ending at the close of trading the day before) is given. Price information from the latest day is in the last four columns on the right. The latest day's high and low price is reported along with the last (closing) price paid for it. If the high or low price paid for a stock on that day is also a 52-week high or low, a footnote (u for up, d for down) is attached to the day's high or low price and will be reflected in the following trading day's data. The final column gives the net change in a stock's value from the previous closing price. Thus the price of Alcoa (the Aluminum Company of America) stock fluctuated between a high of $32 and a low of $31⅜, also closing at that value on this day. The net change (from the value of the previous day's closing price) was down by 50¢ a share. Other important stock data is also given. The annual cash dividend is given in the column right of the stock's name. It is based on the rate of the last quarterly payout, and extra dividends are indicated by footnotes following the cash amount. The next column gives yield of the stock by dividing the cash dividend by the closing price. This valuable information permits you to compare the yield with other kinds of financial instruments. The P-E, Price-Earnings ratio, is found by dividing latest closing price of the stock by latest available earnings per share, which is (for the most recent four quarters) the profit of the company divided by the number of shares outstanding. The P-E ratio is an indicator of the stock's performance. The "Sales 100s" column lists the number of shares traded. Thus the number 6 means that 600 shares were traded that day.

Although bonds tend to be issued for long terms (thirty-eight years in the case of some U.S. Treasury bonds), bonds are commonly bought and sold prior to the date of maturity. The bond market is like the stock market in that it represents an organized exchange specializing in the buying and selling of bonds.

Interest rates affect the prices of shares of stock indirectly by their impact on the activities and hence the profitability of firms, but interest rates bear a direct relation to the price of bonds. Take, for example, a bond issued with a face value of $10,000 that pays an annual rate of interest of 6 percent. Until maturity, whoever holds the bond is entitled to $600 annually. Suppose, however, that the market interest rate rises to 10 percent. This means that purchasers of newly issued $10,000 bonds can earn $1,000 annually, other things equal. What happens to the price of the 6 percent bonds? Their market value must fall to reflect the new, higher interest rate. No one, in other words, would invest in a bond paying less than 10 percent since they can earn 10 percent on newly issued bonds. The price of the 6 percent bond must fall to $6,000 to reflect the new, higher rate of 10 percent. Individuals will not buy the original bonds unless their price falls so as to produce a 10 percent return on a given investment ($600/$6,000). The purchaser of the original bond (assuming he or she still held it after the interest rate increase) experiences a capital loss of $4,000 ($10,000 − $6,000). The bond will still pay $10,000 at maturity (the firm must pay back what it borrowed), but the sales value of the bond right now is $6,000.

The general point to remember about bond price is that the price of a bond varies inversely with the interest rate. When the interest rate rises, the price of a bond falls; when the interest rate falls, the price of a bond rises.

This U.S. Treasury bond, with a face value of $100,000 and 3½% interest, is dated October 30, 1960, and will mature on November 15, 1998, a term of thirty-eight years. The present sale value of a bond like this is now below $100,000 because interest rates have risen considerably above 3½% in recent years. Current sale prices of bonds are reported in daily newspapers.

Question

In what ways are stocks and bonds similar? In what ways are they different? How does the sale of old, previously issued securities (both stocks and bonds) contrast to the issuance of new securities?

15

Rents, Profits, and Entrepreneurship

T hus far in our study of the factors of production, we have looked at labor and capital and their corresponding factor payments, wages and interest. In this chapter we look at two more factors: land and the special form of labor called entrepreneurship. In discussing these two factors, we will present the economic theories of rents and profits.

The concept of rent has a broader economic meaning than simply the payment for use of land. In theory, any factor of production, including labor, can earn rents. Therefore, after we have examined the relation of land supply to rent, we will look at other types of rents and discuss the concept of rent seeking, which refers to the competition for rents among individuals and firms.

In economic terms, rent and profit are closely allied. Profits are the residual payment to capitalist entrepreneurs after all other factor payments—including rents, salaries, and interest—have been made. Every firm and every individual seeks to maximize profits and minimize losses. Competition for profits, or risk taking, involves decision making under uncertainty. We will explore how this process takes place.

Types of Rent

In everyday language, rent is simply payment for the use of something. A farmer pays rent to a landowner for the use of an acre of land. A student leases an apartment and pays rent to the landlord each month. A sales representative rents a car to call on customers in a distant city. These types of rents are all represented in the circular flow of income as the income to property and landowners, a broad category that includes all types of rent payments made according to lease agreements. In 1984, such payments to-

Rent: A payment to a factor of production in excess of its opportunity cost.

taled $62.4 billion, or about 2.1 percent of national income. In economic analysis the concept of rent has a broader meaning than the payment for the lease of a factor such as land. For the economist, **rent** refers to payment to any factor of production—land, labor, or capital—beyond its opportunity cost, or the forgone value of its next most productive use. Nearly every factor you can name has alternative uses. An acre of farmland could be used as the site of an industrial park. An engineer could be put to use typing correspondence. The economic theory of rents, therefore, has a much broader application than simply the payment for lease.

Pure economic rent: The payment to a factor of production that is perfectly inelastic in supply.

Economists distinguish among various types of economic rent. A **pure economic rent** is the return to a factor of production that has a perfectly inelastic supply curve. In such a case the price of the resource is determined solely by the level of demand for its services because supply is fixed and does not change as price changes. Land in the aggregate is the classic example of a resource whose supply is given and whose return is characterized as a pure economic rent.

Inframarginal rent: A type of rent that accrues to specialized factors of production.

Another type of economic rent is paid to factors of production with rising supply curves. These rents are called **inframarginal rents**, and they represent payments to a factor of production above and beyond its opportunity cost as reflected in its supply price. Virtually all factors of production earn inframarginal rents.

Finally, there are categories of rents that accrue to firms. Quasi-rents are returns to a firm owner in a competitive industry above the opportunity cost of the owner's invested capital. Quasi-rents are thus the short-run economic profits of competitive firms, which we analyzed in Chapter 9. Monopoly rents are the returns that accrue to the owner of a monopoly firm when the firm restricts output and raises price. Monopoly rents are the profits of the monopolist, which we studied in Chapter 10. Competitive and monopoly profits are recast in this chapter as quasi-rents and monopoly rents to illustrate the relation between economic rent and economic profits.

Land Rents

How is land priced? The obvious answer is that the price of land is determined where its demand and supply curves intersect. While this is correct, there are some special aspects of the price-determination process in the case of land. These aspects concern how the supply curve of land is defined. Although the focus in our discussion will be on the supply of land, keep in mind the role of the demand curve for land. As we learned in Chapter 12, this demand curve is based on the marginal productivity of land. In this regard the demand curve for land is the same as the demand curve for any factor of production—it is based on the expected marginal revenue product of land as a factor of production. When land is rented, rent is paid to the landowner. The greater the demand for the land, the larger the rent will be. In equilibrium, where the demand and supply of land are equal, the rent to the landowner is equal to the marginal revenue product of the land to the renter.

Pure Economic Rents

To pursue the relation of the supply of land to its price, assume that land is leased by the acre per year. We are concerned with the rent that must be paid to use the flow of services yielded by land for a year. We are not concerned with the related question of the price of a given stock or amount of land, although the two ways of looking at the pricing of the services of land amount to practically the same thing. Land is a resource that yields a flow of services over time. The price of a stock of land will reflect the estimated present value of this future flow of services. Alternatively, the rental price of land will reflect the estimated value of the flow of services for a given period of time. Since our interest is in land rent, we speak in terms of the price of the flow of services from land for a given period of time.

A distinction must also be made between the fixed supply of land and its potential allocation to varying uses. In Figure 15–1a the supply curve of land is drawn for all land, or the **aggregate supply of land.** The curve is drawn as a vertical line, indicating that land in the aggregate is in fixed supply. In general, the amount of land on earth is given by nature and cannot be changed. Actually, common sense tells us that the assumption of a perfectly inelastic supply of land is not completely true. The supply of usable land can be increased through reclamation of swamps and marshes or reduced through erosion. Nonetheless, it is useful to think of the aggregate supply curve of land as perfectly inelastic, as drawn in Figure 15–1.

What are the economic implications of this vertical supply curve for all land? First, the perfectly inelastic nature of land in the aggregate means that the supply of land is unresponsive to price. Whether land is priced at $1 per acre per year or $1 million per acre per year, the supply of land in the aggregate will not change. In economic terms this fixed quantity, Q in Figure 15–1a, means that the rental price of land will be determined solely by the level of demand. Thus, the level of rent on the vertical axis depends on where the demand curve for land intersects the fixed supply curve.

Aggregate supply of land: The amount of land available to the entire economy at various rental rates.

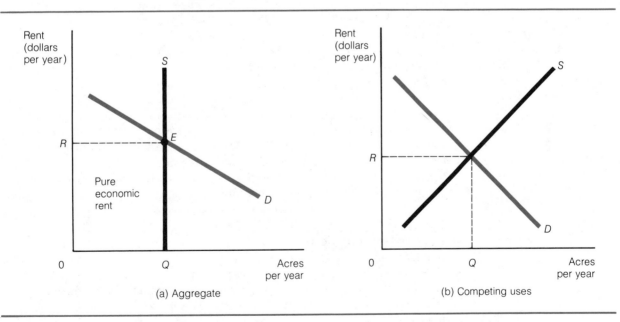

FIGURE 15–1   Pure Economic Rent and the Supply of Land

(a) The supply curve for land in the aggregate is perfectly inelastic with respect to rent, which means that the return to land in the aggregate is a pure economic rent.
(b) The supply curve of land to competing uses appears as it is normally drawn—sloping upward to the right. This slope reflects the fact that land has competing uses and hence that the price of using land or rent plays an important role in determining the uses to which land is put.

The return to land in the aggregate is labeled a pure economic rent. To understand this concept, recall that economic rent is defined as a payment in excess of the opportunity cost of a factor of production. Recall also that a factor's opportunity cost is reflected in its supply schedule because a supply schedule indicates what it takes to bid resources away from their next-best alternative use. A vertical supply curve such as in Figure 15–1a means that land in the aggregate has no opportunity cost; nothing is given up for its supply, for it has no alternative uses. Very simply, this means that the whole return to land in the aggregate is a pure economic rent. In Figure 15–1a, land garners $REQ0$ in pure rent.

The concept of pure economic rent is normally associated with the supply curve of all land. However, it is possible that other resources will exhibit a range of their supply curve that is perfectly inelastic and that they therefore stand to earn pure economic rents. This characteristic may be true of uniquely talented performers in the world of sport and art, a subject to which we will return later in the chapter.

The Supply of Land to Alternative Uses

Since the return to land in the aggregate is a pure economic rent, a price does not have to be paid to call forth a supply of land. The supply of land is fixed and given at Q, and it is not affected by the level of rent, R. Since the supply of all land is independent of its price, we say that price has no allocative function in this case.

If rent does not have to be paid to call forth a supply of land, then why pay it? Or why not tax all rents away as an unearned surplus to landowners? The attractiveness of such a tax is apparent—it raises revenue without affecting the available supply of land. In Figure 15–1a, for example, a tax equal to the area of pure rents, $REQ0$, would leave the supply of land unaffected at Q.

In 1879, in his immensely popular book *Progress and Poverty*, Henry George proposed a single tax of 10 percent on land rent on just these grounds. George argued for a single tax on land because he felt that landowners contributed nothing to the land's productivity. Rising land values were determined by general economic growth and by increases in the demand for land. Thus, a tax on land would not affect the amount of land available to the economy and would prevent landowners from getting rich from windfall profits while nonlandowners remained poor. Moreover, George thought that a single tax on land rents could finance all the government of his day, eliminating the need for other types of taxes.

George's proposal rests on the assumption of a perfectly inelastic supply curve of land. He was essentially assuming that land has no alternative uses. But this assumption concerns land only in the aggregate. While land in the aggregate has a vertical supply curve, the supply of land to alternative uses does not. Consider what happens, for example, as a city expands. Farmland is converted to use as sites for houses and shopping centers as people bid for the right to use the land in other ways. The market process reallocates the land from less valued to more highly valued uses, from farmland to city land.

The supply curve of land for particular uses slopes upward to the right, as shown in Figure 15–1b. To rent land for a particular use, one must bid it away from competing uses. As more desirable land is rented, the rental rate rises. With a normally sloped supply curve, the rental price paid for land serves to allocate land to competing uses. Those who value the use of a parcel of land more highly will offer higher rents than others. Focus, "Land

FOCUS   Land Rent, Location, and the Price of Corn

The rental rate of land varies greatly from one location to another depending on the land's productivity. While the fertility of the soil may be identical in two different locations, the rental rates may not be equal because the physical locations of two plots may account for differences in productivity. To understand this point, consider a straight highway that leaves a large city and extends far into a rural area. Along this highway are farms of equal size that produce the same product, say corn.

Each farm transports its corn to the city and incurs the costs of transporting each bushel. The distance from each farm to the city is of course different, and the cost of transportation increases as the distance increases. Thus, farms nearer the city have lower average and marginal costs than farms farther from the city.

Figure 15–2 shows cost curves of four farms at 50-mile intervals from a city. Farm A is closer to town than Farm B, B is closer than C, and C is closer than D. The costs of production are lower at locations closer to the city; this land can be described as more productive.

Since each farm is a price taker in a competitive market in town, all farms face the same price, P. Profits are shown by the shaded area in each panel. The land

closer to town is more profitable because of its locational advantage and thus is in greater demand. The demand for land decreases as the location is farther from town. Farm D makes no profits; there would be no demand for land by farmers beyond this point. The varying profits bring forth varying demand and thus varying rental rates. The rental rates are bid up by potential farm owners until a normal profit is obtained at each location. The shaded area of profits at each location actually becomes the rental rate. The advantages of a productive location yield income in the form of rents to the suppliers of the land.

What happens to rents if the demand for corn, and thus its price, increases? The higher price for corn will increase profits. Even the land at greater distances than farm D will now be brought into production. With higher profits, the demand for land will increase, and thus rental rates will again rise. After rental rates have risen, normal profits will result at each location, given competitive markets. A decrease in the demand for corn and a reduction in its price will lead to the opposite results. Profits will fall, some distant farms will go out of business, and rental rates will fall, restoring a normal rate of return to farming.

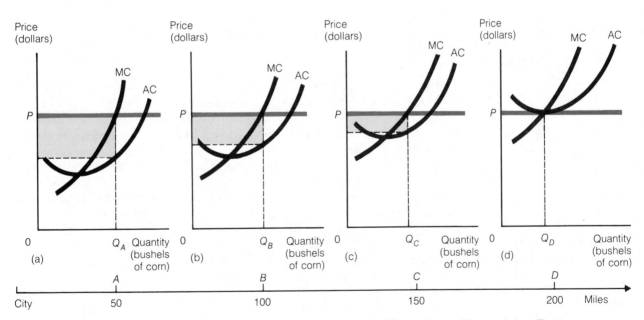

FIGURE 15–2   The Importance of Location in Determining Rents

Graphs a–d show the cost curves for four different farms at four different locations. Farm A is closest to the city, farm D is farthest away. Since costs of transportation increase the farther a farm is from the city, farm A will enjoy lower costs and higher productivity than any of its three competitors.

Rent, Location, and the Price of Corn," discusses this process in more detail. In this sense, rent functions like any other price in signaling the intentions of buyers and sellers in the market for land. Rent for land is not a pure surplus in this case. It serves a useful purpose in the price system, and to tax it away could impose unnecessary shortages and other costs on the economy.

The proper distinction to keep in mind is that while the supply curve of land in the aggregate is approximately vertical, the supply curve of land to alternative uses is positively sloped. Since the latter concept of the supply of land is far more relevant to real-world situations than the former, a proposal such as Henry George's could do much damage to the economy. It was never implemented. Although George became a popular politician and ran for mayor of New York City twice as the candidate of the Labor and Socialist parties, he lost his first race and died during the second, in 1897. Whether land is viewed in the aggregate or as a resource of varying desirability and alternative uses, rent does serve a function: It rations land among available bidders.

Land and Marginal Revenue Product

Like the demand curve for any other factor of production, the demand curve for the services of land is a derived demand curve. It is derived from the value of the output that the land, used in conjunction with other factors of production, can produce. Entrepreneurs will therefore employ land in their production processes as long as the marginal revenue product of land exceeds its rental rate. Where the demand and supply for land are equal, as shown in Figure 15–1b, rental payments equal the marginal revenue product of the land. As we stressed in Chapter 12, the principles of marginal productivity theory apply to any factor of production, including land.

Specialization of Resources and Inframarginal Rents

The concept of economic rent is more general than a rent payment in a lease agreement. An economic rent is defined as a payment to any factor of production—land, labor, or capital—above and beyond its opportunity cost. Rent, in other words, is an excess payment to a resource owner. This concept can easily be applied to resources other than land. For example, a worker provides labor services to a firm. The firm pays the worker the prevailing market wage. The opportunity cost of the worker—her evaluation of her next-best employment prospect—is less than the wage she is paid. The worker thus earns an economic rent in her present job, for she is paid more than it takes to induce her to continue to work in her present employment.

Resource specialization: The devotion of a resource to one particular occupation that is based on comparative advantage.

Note that rents are caused by the specialization of resources. **Resource specialization** simply means that a resource or factor of production is relatively better at working in one occupation or industry than in others. A worker, for example, may be specialized as a welder because he went to technical college to develop his welding skills. As a welder, he will supply his specialized labor to those industries that employ welders.

In general, we can detect the degree of resource specialization by observing the supply curve of the resource. The supply curve of a factor of production measures the willingness of its owner to supply its services at various prices. It therefore gives a measure of the opportunity cost of the resource at different prices.

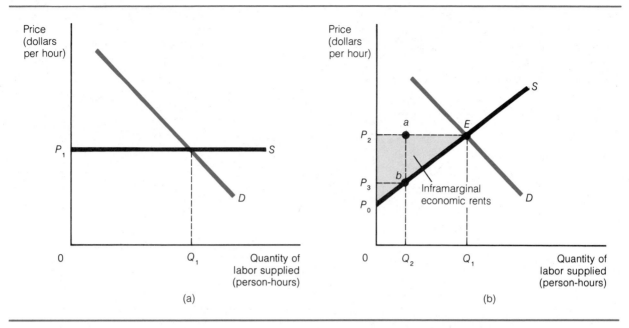

FIGURE 15–3   Unspecialized and Specialized Resources

(a) The supply curve of an unspecialized resource. Such resources earn no economic rents. (b) The normal case of economic specialization and an upward-sloping resource supply curve. Inframarginal rents accrue to the resource owners in the amount represented by P_2EP_0.

Examples are given in Figure 15–3. In part a, the supply curve of the resource is flat or perfectly elastic. This means that the resource is completely unspecialized with respect to its use in this industry. Any amount of the resource can be purchased at the same price, P_1. The resource's price is determined solely by the level of the supply function. This is the opposite extreme from the case of the perfectly inelastic supply curve of land given in Figure 15–1a. Whereas the return to land in the aggregate was a pure rent, the resource depicted in Figure 15–3a earns no rents. Completely unspecialized resources earn no rents.

Many categories of manual labor and jobs that do not require extensive training or experience can be characterized as unspecialized. The supply, for example, of paper carriers, yard workers, and baby-sitters is probably a flat line at the prevailing wage for these services. That is, given the level of demand for such services, the supply of labor to fulfill them is perfectly elastic.

In Figure 15–3b we have drawn a more normal-looking supply curve for a factor of production. In this case, we can say that the upward-sloping supply curve indicates degrees of economic specialization. If the factor is word processors, for instance, some people will be highly specialized at word processing but untrained for anything else. Their next-best alternative would be some low-paying job that requires little skill or training. Such people theoretically would be available as word processors at a relatively low wage—such as P_3 in Figure 15–3 because their opportunity cost is low. At the upper end of the supply curve are people trained as word processors who are also highly skilled in other areas, such as data programming or legal secretarial work. They can be bid into word processing jobs only by relatively high wages such

FIGURE 15–4

Inframarginal Rents for Volunteer Soldiers

S and *D* are the demand and supply curves for volunteer soldiers. Because of the upward-sloping nature of the supply curve, volunteer soldiers earn inframarginal economic rents in the amount represented by *WEA*.

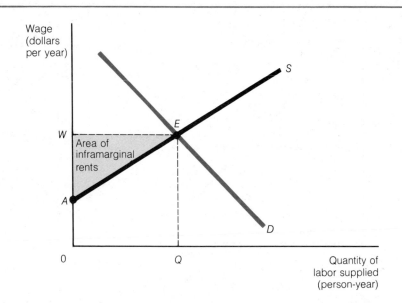

as P_2 or an even higher wage because the cost of their forgone opportunities—jobs as legal secretaries or data programmers—is high.

Although workers with varying degrees of specialization would thus accept different wages, interaction of supply and demand tends to set an equilibrium wage, P_2, paid to all people or other resources doing the same job. At this price, the more specialized resources on the lower end of the supply curve—those who are relatively better at the work for which they are hired than at anything else—will earn economic rents. Look at the word processors or other resources represented by Q_2 in Figure 15–3, for example. Their opportunity cost is P_3, which is the price just sufficient to attract them into this industry. Yet because of this market's equilibrium, all units of the resource receive the same price, P_2. The difference between P_2 and P_3, or *ab*, is an economic rent that accrues to the resource depicted by Q_2. It is a payment above and beyond the opportunity cost of the resource. In fact, the whole area below P_2 and above the supply curve, P_2EP_0, represents economic rents earned by suppliers of this resource. The last unit, Q_1, is paid just its opportunity cost, P_2, so we say that Q_1 is the marginal unit of the resource that earns no rents. All the previous units hired earn economic rents by the amount of the difference between their supply prices and their market rate of compensation. Economists call this kind of economic rent an inframarginal rent.

As another example of inframarginal rents, examine the supply curve of volunteer soldiers given in Figure 15–4. We assume that the army wants to hire *Q* soldiers per year to maintain its force structure. To do so it finds that it must offer a wage of *W*. All soldiers hired are paid *W*, but it is clear from the upward slope of the supply curve that some would have volunteered for lower wages. For example, the very first volunteers would have signed up for a wage slightly above *A*. The volunteer army therefore leads to the receipt of inframarginal economic rents by volunteer soldiers. In Figure 15–4, these rents are given by the area *WEA*.

The concept of inframarginal rents has general applicability to economic life. Factors of production by their very nature are different, and these differences lead to economic specialization, which in turn leads to inframarginal rents. Yet the prevalence of rents does not suggest that there is something unwholesome about the market process. While it is true that these are "excess returns" that do not have to be paid in order to allocate resources to their most valued uses, it would be a virtual administrative impossibility to tax such returns away. Think of the complexity of determining the opportunity cost of each resource. Moreover, it is the ability to capture such returns that leads to the process of resource specialization in the first place. These rents provide incentives for resource owners to behave in certain ways that are good for the economy as a whole.

The Economics of Accountants and Superstars

It is possible that the same resource can attract both inframarginal and pure economic rents. Such a situation may be created temporarily by a sudden surge in demand for a resource. Consider what happens to the rents earned by accountants if the government unexpectedly passes a complicated tax law that can be understood and applied only by those with special accounting skills. Prior to the passage of this law, the market for tax accountants would be in equilibrium at a wage of W_1 and a quantity of Q_1 accountants, as indicated in Figure 15–5. The supply curve of accountants slopes upward to the right, and at the usual wage W_1 accountants earn inframarginal rents represented by W_1BA.

After the passage of the new tax law, the demand for the services of tax accountants increases to D_2. Yet at this point in time, the supply of tax accountants is fixed at Q_1. There are just so many tax accountants and their supply cannot be increased overnight. The supply curve of accountants therefore becomes vertical at B for some specified time period. As in the case of aggregate land supply, a vertical supply curve means that the accountants

FIGURE 15–5

Two Kinds of Economic Rent Earned Simultaneously

When the demand curve for accountants shifts from D_1 to D_2, the supply of accountants remains constant in the short run at Q_1. Over this period accountants earn both pure economic rents of W_2CBW_1 and inframarginal rents of W_1BA.

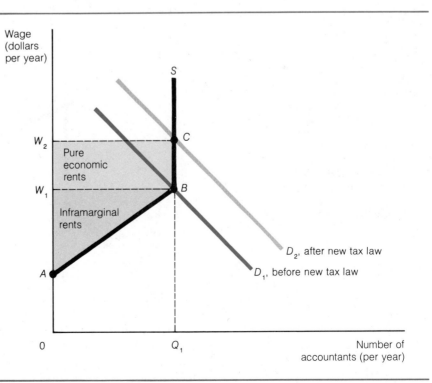

will earn pure economic rents as the demand for their services increases. In Figure 15–5 these rents are given by the area W_2CBW_1. In such a situation, accountants thus earn two types of economic rent: pure economic rents equal to W_2CBW_1 and inframarginal economic rents equal to W_1BA.

The pure economic rents in this example are only temporary. Over time, these returns in excess of opportunity cost will attract entry into the accounting profession. Entry of new accountants will shift the supply curve for accountants to the right, reducing the price of accounting services and the economic rents earned by accountants. During the period in which entry does not occur and the accountants earn pure economic rents, a rising wage for accountants serves to ration their services to the highest bidders.

A supply schedule like that drawn for accountants in Figure 15–5 can also apply to the income of superstars, such as Barbra Streisand or basketball star Julius Erving. Figure 15–6 shows a possible supply curve for a superstar. In this case the next-best alternative to being a superstar is reflected in a price for the star's services of P_1. The job earning P_1 as an alternative to superstardom might be something as ordinary as working in a bank. The supply curve of the superstar becomes vertical at P_1 because he or she possesses unusual abilities that cannot be copied by anyone else. Only Dr. J has certain moves on a basketball court; only Streisand can offer her unique musical sound. Thus the supply curve of the superstar above P_1 indicates that he or she is earning pure economic rent. As in other cases of vertical supply curves, price here serves to ration the services of the superstar among the competing demands of music producers or team owners.

We must be careful, however, in defining economic rents in the case of a superstar. If we define the next-best alternative of a star as being a truck driver, the economic rents that he or she earns will be large. Only if the price for his or her superstar services falls below P_1 in Figure 15–6 would the

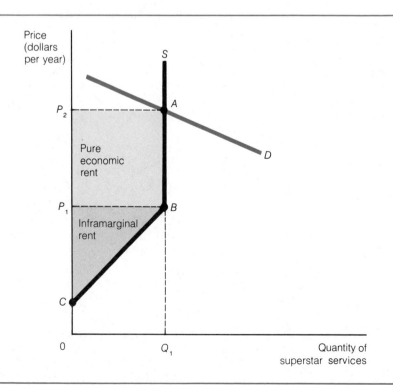

FIGURE 15–6

A Superstar's Supply Curve

The superstar's unique talent cannot be replicated, so it is represented by the vertical portion of the supply curve. The next-best alternative of the superstar is given by P_1. The difference between P_2 and P_1 means that the superstar's income includes a pure economic rent in the amount of P_2ABP_1 as well as an inframarginal rent of P_1BC.

star switch to such an alternative job. Suppose, however, that the next-best alternative of the basketball star who earns $1 million per year is to play for another team who would pay him $900,000. In this case economic rent would be $100,000:

$$\underset{\substack{\text{Resource owner's} \\ \text{income}}}{\$1,000,000} - \underset{\substack{\text{Resource owner's} \\ \text{opportunity cost}}}{\$900,000} = \underset{\substack{\text{Economic} \\ \text{rent}}}{\$100,000.}$$

The measurement of economic rents hinges on how the next best alternative or the opportunity cost of the resource is defined.

Do the rents that superstars earn affect the future supply of talent in their professions? Are their pure economic rents transitory, like those of the accountants, as new stars rise to compete with them and capture some of their rents? No one knows the answers to these questions exactly, but we can surmise that the answers are yes. Surely, it is the lifestyle and general level of ability and wealth of superstars that provide the incentive for thousands of youngsters to practice hard and to develop extraordinary talents. Over a sufficient period of time, new entry may erode the pure economic rents of superstars.

Rents and Firms

Firms may also earn rents, and these rents are analogous to the firms' profits. In fact, the concept of economic rent as a return in excess of the opportunity cost of a factor of production is just another way to think about the profits earned by competitive and monopolistic firms. The purely competitive firm can earn quasi-rents, and the pure monopolist can earn monopoly rents.

Quasi-Rents

Quasi-rent: The short-run payments to owners of capital in a competitive industry that exceed the opportunity cost of capital.

Quasi-rents are the return to the owner of a competitive firm in the short run. They are called quasi-rents because they are a short-run or temporary return. Figure 15–7 illustrates a short-run equilibrium for the competitive firm and industry. The firm earns an economic profit in the short run, which we have labeled as quasi-rent. This return is a rent because it exceeds the firm owner's opportunity cost (which by definition is included in the average total cost curve). Quasi-rents are temporary; entry of new firms into the industry will erode these returns over the long run. Keep in mind that quasi-rents can also be negative instead of positive, for the firm can suffer economic losses in the short run. The rents themselves are a return to the owner of the firm for decision making under conditions of risk and uncertainty.

Monopoly Rents

Monopoly rent: The payments to owners of capital in a monopolized industry that exceed the opportunity cost of capital.

Figure 15–8 shows the case of a monopolist who makes profits of *PABC*. These profits are sometimes called **monopoly rents** because they are a return in excess of the monopolist's opportunity cost, which is reflected in the average total cost curve. Unlike the quasi-rents earned by the competitive firm, the monopoly rents earned by the monopolist will persist over time. They are permanent, short of an event that would threaten the monopolist's control of an industry. But what happens if the monopolist sells the monopoly? Primarily, the monopolist will sell for the present value of future monopoly rents, as discussed in Chapter 10. Monopoly rents are therefore captured by the original owner of the monopoly; buyers of the monopoly right may earn only a competitive rate of return on their purchase of the monopoly.

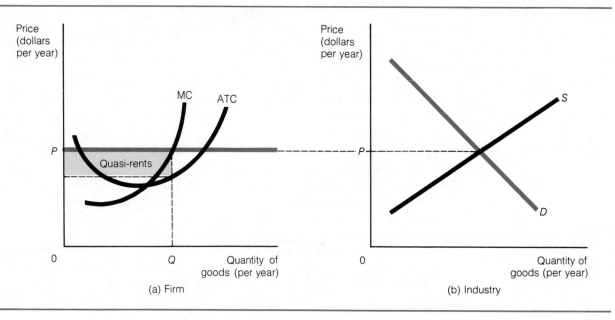

(a) Firm

(b) Industry

FIGURE 15–7   Quasi-Rents

Quasi-rents are illustrated as the economic profits earned by the owner of a
competitive firm in the short run. They are rents because they are a return in excess
of the owner's opportunity cost, included in the ATC curve. They are called quasi-
rents because they are temporary and will be eroded by the entry of new firms in the
long run.

FIGURE 15–8

Monopoly Rents

Monopoly profits can be
called monopoly rents. They
are returns in excess of the
monopolist's opportunity
costs, which are reflected in
the ATC curve. Unlike quasi-
rents, they persist over time
because of the monopolist's
dominant position in the
market.

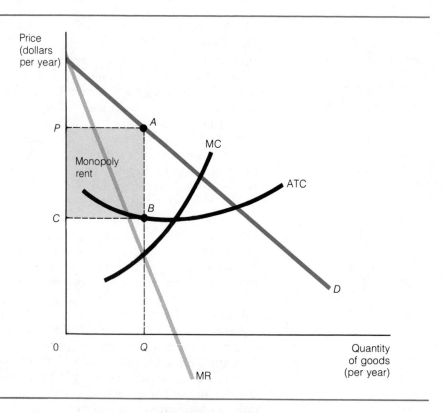

Rent Seeking

When competition is viewed as a dynamic, value-creating, evolutionary process, economic rents play a crucial role in stimulating entrepreneurial decisions and in prompting an efficient allocation of resources. Profit seeking is a normal feature of economic life in a competitive market. The returns of resource owners will be driven to normal levels by competitive profit seeking as some resource owners earn positive rents that promote entry into the industry and others earn negative rents that cause exit. In this basic sense, profit seeking in a competitive setting is socially beneficial.

When the setting in which profits are pursued is changed, sometimes through government intervention, problems can arise. Suppose the government wishes to grant a monopoly right in the production of playing cards. In this case an artificial scarcity is created by the state. As a consequence, monopoly profits or rents are present to be captured by monopolists who seek the government's favor. In a naturally occurring monopoly these rents are thought of as transfers from playing card consumers to the monopolist. Yet this can be the case only if the aspiring monopolists employ no resources to compete for the monopoly rents. To the extent that resources are used to capture monopoly rents by lobbying the government, these expenditures create no value from a social point of view. This activity of wasting resources in competing for artificially contrived transfers is called **rent seeking**. If a potential monopolist hires a lawyer to lobby the government for the monopoly right, the opportunity cost of the lawyer (such as the contracts that he or she does not write while engaged in lobbying) is a social cost of the monopolization process.

Rent seeking: The process of spending resources in an effort to obtain an economic transfer.

The economic wastefulness of rent-seeking activity is difficult to escape once an artificial scarcity has been created. At one level the government can allow individuals to compete for the playing card monopoly and waste resources through such activities as bribery. Such outright corruption is perhaps the simplest and most readily understood level of rent seeking. At a second level the state could sell the monopoly right to the highest bidder and put the proceeds at the disposal of government officials. In such a case the monopoly rents may show up in the wages of state officials; to capture rents at this level individuals will compete to become civil servants. At still another level, the monopoly right may be sold to the highest bidder and the resources dispersed through the state budget in terms of expenditure increases and/or tax reductions. Even in this case, rent-seeking costs will be incurred as individuals seek to become members of the groups favored by the tax-expenditure program. Rent-seeking costs are incurred in any of these options, and the form that such costs take is influenced by how the government transacts its business in creating artificial scarcity.

Profits and the Allocation of Resources

As we saw in the discussion of quasi-rents and monopoly rents, rents and profits are closely related concepts. The word *profits*, however, is the term that we usually apply to the earnings of owners of firms. Profits are the last category of input returns in the circular flow process that we analyze. As we will see, profits are a residual return to those individuals in the economy who provide economic foresight and leadership in an uncertain environment. Though we discuss the role of profits in the resource allocation process last, this does not mean that their role is small. To the contrary, profits are a very important reason that the competitive price system works to allocate resources to their most highly valued uses.

Profit Data

For individual firms, industries, or the economy as a whole, profits are not computed directly but are determined by what is left over from revenues after payments have been made to land, labor, and capital. It is in this sense that the profit earner, or the productive factor to which profits are imputed, is referred to as the *residual claimant*. This view of profits as a residual contrasts sharply with the incorrect notion that employees and other input suppliers are paid only after the entrepreneur has secured profits.

Have levels of profits earned in the economy as a whole grown over the years? Table 15–1 traces after-tax corporate profits and proprietors' income for selected years between 1929 and 1984. Even when taking inflation into account, the absolute level of corporate profits has risen quite steadily over this period, but when measured as a percent of GNP the corporate profit level has remained remarkably stable, ranging between 3 and 5.5 percent

TABLE 15–1   Corporate Profits and Proprietors' Income, 1929–1984

Year	Corporate Profits After Tax (billions of 1972 dollars)	Percent of GNP	Proprietors' Income After Tax (billions of 1972 dollars)	Percent of GNP
1929	22.3	7.1	45.8	14.5
1933	0.4	.2	23.5	10.6
1939	16.5	5.2	41.5	13.0
1940	21.0	6.1	44.7	13.0
1945	20.0	3.6	83.9	15.0
1950	40.3	7.5	72.3	13.5
1955	35.8	5.4	70.5	10.7
1960	29.8	4.0	68.7	9.3
1965	51.1	5.5	76.5	8.2
1966	53.2	5.4	78.8	8.0
1967	48.8	4.8	77.4	7.7
1968	47.9	4.5	77.5	7.3
1969	41.7	3.8	77.2	7.1
1970	32.6	3.0	72.4	6.7
1971	37.1	3.3	72.3	6.4
1972	43.0	3.6	76.9	6.5
1973	53.0	4.2	88.7	7.1
1974	55.1	4.4	77.2	6.2
1975	52.6	4.3	71.7	5.8
1976	62.3	4.8	71.2	5.5
1977	67.3	4.9	74.0	5.4
1978	71.3	5.0	78.0	5.4
1979	76.0	5.1	80.8	5.4
1980	68.0	5.0	73.6	5.0
1981	57.98	3.8	69.4	4.6
1982	35.15	2.6	53.57	3.6
1983	42.17	2.8	56.52	3.7
1984[a]	50.45	3.1	69.25	4.2

[a]preliminary

Source: Economic Report of the President (Washington, D.C.: U.S. Government Printing Office, 1985), pp. 246, 256.

since the mid-1950s. In contrast, proprietors' income exhibits a strong and continuous downward trend as a percent of GNP, while showing little change from year to year in constant dollar terms. Overall, the Table 15–1 data fail to confirm the widely held perception that profits have been growing steadily over time.

Rate of Return on Capital

The profit data given in Table 15–1 are not relevant to the decisions of firms concerning investments in plant and equipment. What matters for such purposes is not the absolute level of profits but the economic **rate of return on invested capital.** That is, at the margin it will pay to buy an additional dollar's worth of equipment only if the firm expects to earn more than a dollar from the sale of the resulting output from the equipment.

The economic rate of return on capital is approximated in Table 15–2 with accounting data by dividing the amount or stock of capital valued at its current replacement cost into profits earned during the year. The data indicate that after-tax rates of return on capital have fallen over time. The average annual rate of return between 1955 and 1969 was 10.1 percent, with

Rate of return on invested capital: Profits that are measured as a percentage of the costs of capital.

TABLE 15–2   Rates of Return on Depreciable Assets, Nonfinancial Corporations, 1955–1982

Year	Rate of Return (percent)	
	Before Tax	After Tax
1955	19.8	9.8
1956	16.8	7.9
1957	15.2	7.4
1958	12.8	6.5
1959	16.4	8.5
1960	15.0	8.0
1961	15.1	8.2
1962	17.4	10.3
1963	18.8	11.2
1964	20.2	12.5
1965	22.1	14.0
1966	21.8	13.7
1967	19.3	12.4
1968	18.9	11.3
1969	16.5	9.7
1970	12.8	7.9
1971	13.5	8.5
1972	14.3	9.1
1973	14.3	8.7
1974	11.0	6.1
1975	11.9	7.7
1976	12.9	7.9
1977	13.7	8.6
1978	13.3	8.2
1979	12.3	7.6
1980	10.7	6.9
1981	11.0	7.7
1982[a]	9.5	7.5

[a]preliminary

Source: Economic Report of the President (Washington, D.C.: U.S. Government Printing Office, 1983).

the highest rates occurring during the 1960s. By contrast, the average annual rate of return on capital between 1970 and 1981 was 7.9 percent. This decline in rates of return has two important implications. The first involves incentives to make capital investments. For investments in capital to be profitable, not only must the rate of return be positive, but the rate of return on capital must exceed the expected return on alternative investments. Since a larger number of competing uses of funds—buying bonds, for example—become attractive as the rate of return on capital falls, the data in Table 15–2 suggest that incentives to invest in plant and equipment during the 1970s were lower than in earlier years. Second, the data provide further evidence that the popular view of large and ever-increasing corporate profitability is mistaken.

Accounting Versus Economic Profits

Accounting profit: Total revenue minus total explicit money expenditures.

Economic profit: Total revenue minus total opportunity costs.

It is important to note that **accounting profits** such as those in Table 15–2 tend to overstate the level of **economic profits** because accountants by necessity take a less theoretical view of costs than do economists. Specifically, accountants count as costs only those business expenditures involving direct money outlays—wages and salaries, raw materials purchased, rental payments on land and buildings, borrowing costs, advertising expenses, and so forth. To economists, costs are opportunities forgone, whether explicit or implicit. For example, what is the economic cost of the company-owned land on which corporate headquarters sits? It is the revenue forgone from putting the land to its next-best use, perhaps by leasing it to a shopping center developer. But the accounting cost of company-owned land will generally be zero because no explicit money payments are presently made for the land by the firm. Thus, because the accounting concept of costs tends to understate economic costs, accounting profits overstate economic profits.

Accounting profits also tend to be misleading because cost is a subjective, forward-looking concept. That is, insofar as they influence economic decisions, costs are based on anticipations about the future that may or may not materialize. The cost of a particular decision will therefore vary depending on when in time cost is measured. Moreover, what the decision maker thinks is being sacrificed by a choice cannot be directly observed by someone other than the decision maker. We are left with the conclusion that accounting profits, based as they are on a restricted concept of costs, provide only a limited guide to the actual level of economic profits in the economy.

Profits and Entrepreneurship

Entrepreneur: An individual who organizes resources into productive ventures and assumes the uncertain status of a residual claimant in the resulting economic outcome.

Assuming that economic profits are capable of being measured, what role do they play in the economic process? Economic profits are the return to a particular factor of production, **entrepreneurship.** As such, profits are the residual income accruing to the entrepreneur as a return on the services he or she brings to the productive process. These services include technical abilities in organizing the other factors of production into combinations appropriate for the efficient manufacture of goods. More important, though, is the entrepreneur's alertness to the existence of potential profit opportunities in the economy. In this sense entrepreneurship consists of linking markets by perceiving the opportunity to buy resources at a lower total cost than the revenue obtainable from the sale of output or by recognizing that sellers in one market are offering to sell output at a price lower than buyers in some other market are willing to pay.

Put another way, entrepreneurship consists of offering the most attractive opportunities to other market participants—perceiving unfulfilled demands, offering higher prices to sellers or lower prices to buyers, improving existing goods or making them more cheaply, finding more effective means of communicating to consumers the availability and attributes of goods. In all of these activities the entrepreneur takes advantage of opportunities that exist because of the initial ignorance of other market participants. But as the result of the entrepreneur's actions, markets move closer to the prices and quantities emerging in equilibrium, the plans of buyers and sellers more closely dovetail, and the knowledge of economic data held by market participants is increased.[1]

In addition to alertness and organizational skills, the entrepreneur brings to the production process a willingness to act in the presence of uncertainty. (We distinguish risk from uncertainty in that risk involves events that occur with known probabilities—the toss of a coin, for example—while uncertainty entails outcomes whose probabilities cannot be specified with precision—for example, war or peace, long-run weather forecasts, and so on.) There is no guarantee that perceived profit opportunities will in fact materialize. Because it takes time between the purchase of inputs and the sale of output—between the expenditures on resources and the receipt of revenue—intervening events may cause the entrepreneur's plans to be either overambitious or underambitious when evaluated with the advantage of hindsight. Entrepreneurship thus entails the bearing of responsibility for incorrect anticipations. Profits can be viewed as a return to uncertainty-bearing.

In summary, the entrepreneur is the prime actor in the economic process, and entrepreneurship is characterized by technical skills in organizing production, alertness to the appearance of profit opportunities, and a willingness to act under uncertain conditions. Profits are just the payment to another factor of production, but they are different in that the entrepreneur is "paid" last, claiming the residual after the other factors have been compensated. Profits are the incentive spurring entrepreneurial activity in the economy. Focus, "Venture Capitalism," discusses a specific type of entrepreneurship.

General Equilibrium

Our study of the essential parts of microeconomic theory is now complete. The basic tools of economic analysis—demand and supply, elasticity, marginal utility, and costs of production—were introduced in Chapters 1 through 8. The pricing and production of goods and services in a variety of market structures ranging from pure competition to pure monopoly were covered in Chapters 9 through 11. In Chapters 12 through 15, the elements of the theory of input markets were presented. We have analyzed each part of the economy in isolation from the other parts. This type of economic analysis is called *partial equilibrium analysis*. Changes in other parts of the economy are held constant while the functioning of a particular market or industry is analyzed. We are now in a position to gather some understanding of how all the parts work together. This study of the whole economy, where no part of the system is held in isolation from the others, is called *general equilibrium analysis*. The appendix to this chapter is devoted to a simple exposition of general equilibrium theory.

[1]The entrepreneur's role in the economic process is detailed in Israel Kirzner, *Competition and Entrepreneurship* (Chicago: University of Chicago Press, 1973).

FOCUS   Venture Capitalism

Entrepreneurs are people with new ideas about how to make money. They see profit opportunities where others do not—in new products, different locations, and an almost endless number of other ways. An entrepreneur, in essence, makes conjectures about the economic future, and these conjectures typically entail very risky undertakings. Since not all entrepreneurs can finance their dreams, who provides the financial backing for their proposals?

The answer is a venture capitalist, a person who supplies capital to fund new investment ideas by entrepreneurs. Some venture capitalists are wealthy individuals, and others are subsidiaries or divisions of large firms or financial institutions. Each year, billions of dollars are invested by venture capitalists in the new ideas of entrepreneurs who cannot find support for their projects through alternative channels.

Venture capitalists must exercise great care and caution in evaluating the proposals that come to them. For

Source: Wall Street Journal, December 19, 1983, p. 27.

example, Citicorp Venture Capital Ltd. has $100 million in capital to invest and each year receives about 3,000 proposals. The proposals range from the interesting (a new concept in toilet seat sanitizing) to the vague (a proposal with no details except the address to which to send $150,000 to open a supermarket). Obviously, most venture projects are quite risky, and the venture capitalist seeks a high rate of return on the investment in such projects as well as a large measure of control in the undertaking.

There are no systematic data on the type of projects that tend to be funded, but it appears that a ripe source of entrepreneurial proposals come from operating executives in established firms who are seeking to establish their own businesses. These individuals can typically provide some of the finance for their proposals, and they have credibility with the venture capitalist because of their industrial experience.

Venture capitalism is an important part of the economy. In a way, it finances the progress that the economy makes over time.

Summary

1. An economic rent is a return in excess of the opportunity cost of an owner of a resource.
2. Land in the aggregate is fixed and has a perfectly inelastic supply curve. Its entire return is a pure economic rent determined by the level of demand for land as a factor of production.
3. The supply of land to competing uses is not fixed. Land must be bid away from competing uses, so the supply curve of land to a specific use is upward sloping.
4. Proposals to tax land rents away would be costly to the economy. Land is like any other factor of production. It has alternative uses, and it is used in the production process by entrepreneurs and firms up to the point where its marginal revenue product equals its marginal cost.
5. Specialized resources earn economic rents. The degree to which the supply curve of a factor of production is upward sloping reflects the degree to which it is specialized. Where all units of a resource earn the same market price, the more specialized units will earn economic rents. These rents are called inframarginal rents because all units of the resource except the very last, or marginal, unit earn rents.
6. Superstars earn economic rents because their talents are rare gifts that cannot easily be duplicated.

The supply curve of superstar services is vertical, and a great deal of the income of a superstar will consist of pure economic rents.

7. Another way to look at the profits of competitive and monopolistic firms is through the concept of economic rent. Competitive firms earn quasi-rents, or returns in excess of the entrepreneur's opportunity cost. Monopolists can earn monopoly rents, which are equivalent to monopoly profits.
8. Rent seeking is the process of competing for artificially contrived rents normally created through government intervention in the economy. From society's point of view, expenditures to capture such transfers are wasted.
9. Economic profits are a residual that the entrepreneur receives after the other factors of production have been paid. Profits can be positive or negative (or zero), and the receivers of profits are called residual claimants. Although the absolute level of profits in the economy has grown over time, the rate of return on capital has not. It has declined in recent years. The rate of return on capital is the crucial determinant of firm profitability.
10. Accounting profits differ from economic profits. Typically, accounting procedures understate economic costs and thus overstate economic profits.
11. Profits are the force in the economy that leads to

entrepreneurship. They are a return to such activities as decision making under conditions of risk and uncertainty. The concept of entrepreneurship has many dimensions, but perhaps the most basic characteristic is that of being alert to economic opportunities.

Key Terms

rent

pure economic rent

inframarginal rent

aggregate supply of land

resource specialization

quasi-rent

monopoly rent

rent seeking

rate of return on invested capital

accounting profit

economic profit

entrepreneur

Questions for Review and Discussion

1. Distinguish between pure and inframarginal economic rents.
2. At what point in the production process do residual claimants receive their return? Can their return be negative?
3. How does rent seeking lead to the waste of resources?
4. What are the essential things that an entrepreneur does?
5. There is a fixed supply of oil in the world. Is the return to oil producers therefore a pure economic rent? Explain carefully.
6. "Land rent should be completely taxed away." Evaluate.
7. Suppose that your whole wage was a rent. What implications would this situation have for your behavior where you work? Would you be cantankerous with your superiors or docile and easy to get along with? Why?
8. Explain why a firm making zero economic profit will probably stay in business.
9. Are you earning economic rents as a college student? For example, suppose your college raised tuition. Would you drop out of school or transfer to another school?
10. Suppose several cable television companies were each attempting to obtain a franchise from your city government. Only one company would obtain the exclusive rights to provide cable services. How much would each company be willing to spend in an effort to obtain the franchise? What is the term that describes their activity, and what is the term that describes the winning firm's profits?
11. Suppose a firm is earning economic profits that are greater than the profits of most other firms in the industry because it has a superior business manager. In the long run what happens to the manager's salary? Also, what happens to the firm's production costs and profits?

ECONOMICS IN ACTION   Computer Entrepreneurs

In the view of many economists, entrepreneurs play a vital role in the growth of an economy. By their willingness to take risks, their inner drive to succeed, and their ability to identify new products and markets, entrepreneurs often spark sudden breakthroughs in technology and reap large fortunes as a result.

The computer revolution of the 1970s and 1980s is a case in point of the role of the entrepreneur. Many of the men and women who saw early on the potential of the personal computer for use in the home, office, or classroom are now managing corporations that provide goods and services undreamed of fifteen years ago. But not everyone who took risks in developing computer products succeeded. As sizeable profits attracted new entrants into computer markets, competition intensified and pricing and marketing strategies became the only means of survival for many. By the early 1980s, news of bankruptcies in America's "silicon valleys" became as common as the new products themselves.

Steven Jobs and Jack Tramiel are two of the most celebrated computer entrepreneurs. Jobs, a 1972 college dropout who spent several years tinkering with computer technology in his parents' garage with his friend and fellow "hacker" Stephen Wozniak, found the means to raise money and market the Apple computer, the first successful personal computer. Jobs's first capital—$1300—came from the sale of his Volkswagen bus. Today Jobs is the Chairman of the Macintosh division of Apple Computer, Inc., whose revenues top $1 billion. He and Wozniak have both become multimillionaires.

Jack Tramiel's career spans many more years than

Steven P. Jobs

Jack Tramiel

Jobs's, and he has overcome many more obstacles to success. Often described as a somewhat ruthless, intensely driven man, Tramiel survived the Nazi holocaust in Poland, emigrated to the United States, opened a typewriter repair shop, and built it into Commodore International, another billion dollar company that today dominates the home computer market. (Apple's more expensive models have been much more successful with small businesses and schools.) Tramiel has seldom been afraid of taking risks. In 1968 he became one of the first marketers of hand-held calculators, but he soon lost most of his business to Texas Instruments, as prices for the calculators dropped almost tenfold in less than 5 years. He retaliated by introducing the PET tabletop computer (later the VIC-20) and became so aggressive in marketing and pricing his products that he drove several competitors, including Texas Instruments, out of these markets. In 1984 Tramiel abruptly resigned his post at Commodore and purchased the moribund Atari company, another computer manufacturer that had suffered huge losses in competition with Commodore. He is now Chairman of the Board at Atari. Many observers believe that Tramiel is bent upon removing his former company, Commodore, from its profitable niche.

Strong egos and intense competitive desire are common traits of entrepreneurs such as Jobs and Tramiel.

(Tramiel is often quoted as saying, "Business is not a sport. It's a war.") Aside from their fascinating psychological profiles, however, entrepreneurs are important in a strictly economic sense. In *The Spirit of Enterprise*, a book about entrepreneurial behavior, author George Gilder writes,

> The capitalist is not merely a dependent of capital, labor, and land; he defines and creates capital, lends value to land, and offers his own labor. . . . He is not chiefly a tool of markets but a maker of markets; not a scout of opportunity, but a developer of opportunity; not an optimizer of resources, but an inventor of them; not a respondent to existing demands but an innovator who evokes demand; not chiefly a user of technology but a producer of it. He does not operate within a limited sphere of market disequilibria, marginal options, and incremental advances.[a]

Question

How, specifically, would you characterize the role of the entrepreneur in society? How is the role of the entrepreneur related to risk taking? Does the entrepreneur provide only economic benefits for herself or himself?

[a]George Gilder, *The Spirit of Enterprise* (New York: Simon and Schuster, 1984), pp. 16–17.

APPENDIX
Putting the Pieces Together: General Equilibrium in Competitive Markets

At this point each of the many parts of microeconomics has been introduced and explored. Studying each individually can add a great deal to our understanding of many events that occur in the economy. However, bringing all the parts together can create an overall picture that may help answer the "what," "how," and "for whom" questions introduced in the first chapter of the book. To illustrate, we can describe a situation of general equilibrium and then create a change in an economic variable, tracing its effects throughout the economy.

Consider an economy that has only two goods, wheat and coal. All citizens spend their income on these two goods in a manner that maximizes utility. With limited incomes and diminishing marginal utility, downward-sloping demand curves result. In Figure 15–9a and 15–9b, which represents supply and demand conditions in the output markets for wheat and coal, D_{w1} and D_{c1} are the current demand curves for wheat and coal, respectively.

Firms that maximize profits and have diminishing returns will have upward-sloping supply curves for the two goods. S_{w1} and S_{c1}, reflecting the current supply curves for wheat and coal.

The supplies of farm workers and coal miners are determined in part by each individual's maximizing of utility in occupational choice. In Figure 15–9c and 15–9d, the equilibrium wage rates W_{f1} and W_{m1} for farm workers and coal miners are determined by competitive forces. There may be a difference in these two wage rates for a variety of reasons. For example, if coal mining is a skilled craft that requires a long training period or if coal mining is especially dangerous, in the long run the wages of coal miners must be greater than the wages of farm workers. The current quantities of labor employed in each industry are L_{f1} and L_{m1}.

P_{w1}, W_1, P_{c1}, and C_1 represent the equilibrium prices and quantities of wheat and coal. In this long-run competitive equilibrium, profits are zero, each firm is forced to operate at minimum average cost, and price is equal to marginal cost. In other words, there is an optimal allocation of resources.

From this initial equilibrium, we can examine a disturbance in the economy. Suppose that consumers permanently decide to spend a larger portion of their income on wheat and less on coal. The demand for wheat shifts to the right to D_{w2} in Figure 15–9a, and the demand for coal shifts to the left to D_{c2} in Figure 15–9b. This action will lead to a series of short-run changes and eventually to a new long-run equilibrium.

In the short run the price of wheat will rise to P_{w2} and the price of coal will fall to P_{c2}. These shifts will generate short-run profits for firms in wheat production and losses for firms in coal production. The demand for farm workers increases to D_{f2}, and the demand for coal miners decreases to D_{m2}. The wages for farm workers rise to W_{f2}, while the wages of coal miners fall to W_{m2}. If wages do not adjust to these new levels, there will be a shortage of farm workers and unemployment for coal miners.

As time goes by, firms and resources adjust to the changed conditions.

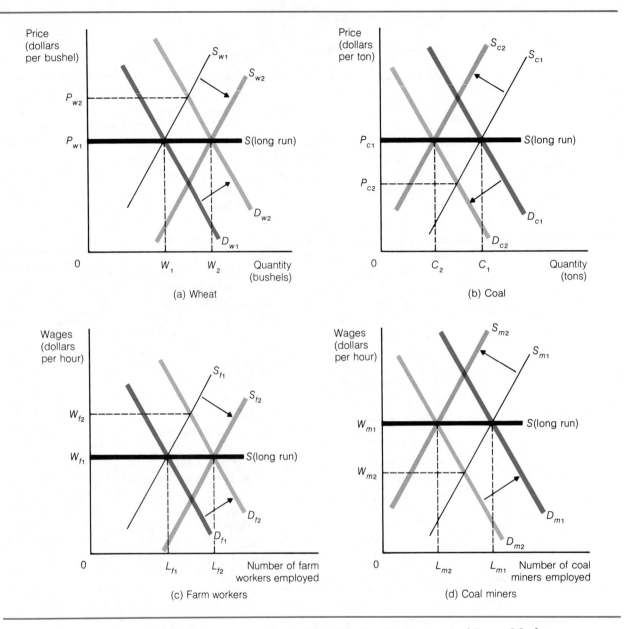

FIGURE 15–9   General Equilibrium in Output and Input Markets

The current demand curves for wheat and coal are given in (a) and (b) by D_{w1} and D_{c1}. If consumers decide to spend more of their income on wheat and less on coal, the demand curve for wheat shifts to the right, from D_{w1} to D_{w2}, and that for coal shifts to the left, from D_{c1} to D_{c2}. The short-run price of wheat will rise from P_{w1} to P_{w2}, and the short-run price of coal will fall from P_{c1} to P_{c2}. In the long run the economy adjusts, and the equilibrium price and quantities become P_{w1}, P_{c1}, W_2, and C_2. The equilibrium wage rates W_{f1} and W_{m1} for farm workers (c) and coal miners (d), respectively, are determined by competitive forces when the demands for wheat and coal shift as in (a) and (b). The demand for farm workers increases, and the demand curve shifts from D_{f1} to D_{f2}. The demand for coal miners decreases, and the demand curve shifts from D_{m1} to D_{m2}. The wages of farm workers rise from W_{f1} to W_{f2} and those of coal miners fall from W_{m1} to W_{m2}. When general equilibrium is reached in the economy, as in (a) and (b), the equilibrium quantities of farm workers and coal miners are L_{f2}, W_{f1}, L_{m2}, and W_{m1}.

The higher economic profits in the wheat industry will attract new firms. The losses in the coal industry will encourage firms to leave the industry. The supply of wheat will begin to increase, while the supply of coal decreases.

In the resource markets, shown in Figure 15–9c and 15–9d, let there be only one input, labor. Workers may choose to enter the wheat industry as farm workers or the coal industry as coal miners. The demands for farm workers and for coal miners are derived from the demands for wheat and coal. The higher wages in the wheat industry will attract workers, and the lower wages in the coal industry will discourage the entry of workers. The supply of farm workers increases while the supply of coal miners decreases.

If resources are able to shift, as they are in this case, the new long-run equilibrium will result in the former prices and wages. Profits will return to zero in both industries. The only permanent changes are in the quantities of wheat, coal, farm workers, and coal miners, W_2, C_2, L_{f2}, and L_{m2}, respectively. The long-run supplies of wheat, coal, and labor are infinitely elastic.

This example demonstrates the way in which resources are demand-directed. As consumers' tastes change from coal to wheat, resources shift from coal to wheat. Thus it is said that resources flow to their most highly valued use.

An understanding of what makes this process work is important. Self-interest is the motivating force of individuals; it is also the impelling force that drives the competitive market to the desired result. Self-interest is what attracts firms to profitable industries and leads them away from unprofitable industries. Self-interest is what attracts workers to occupations with high wages and leads them away from those with low wages. Without self-interest the market system would not produce the products that are most highly valued by society.

In actuality, there are many obstacles to long-run competitive equilibrium. Imperfect competition, externalities, or government intervention may inhibit the free flow of resources to the most desired goods.

In the previous example, what would be the result if there had been a monopoly in the wheat industry or a labor union for farm workers? Or what would have happened if government had taxed away the profits of wheat-producing firms and subsidized the coal industry? The most desired quantities of wheat and coal would not have been produced. While both natural and artificial barriers to optimal resource allocation do exist, this example should provide the basis for an understanding of the overall workings of the microeconomic system.

16

The Distribution of Income

A re you rich or poor? In what sense? Diane, who is passionately interested in art, may be devoting a long and happy life to painting seascapes, earning no more than five or six hundred dollars a month. John, a stockbroker, may regard himself as one of the walking dead in spite of a six-figure income. Is Diane rich? Is John poor? Surely the question of whether one is rich or poor in an economic sense is different from whether one feels rich or poor in emotional well-being.

While most people, including economists, might argue that it is far better to be rich and unhappy than poor and unhappy, happiness cannot be measured in objective terms or easily be related to income. The equation of rich with happy or poor with unhappy, though tempting, is inadmissible in any scientific sense. Economists cannot measure happiness, but they can measure and analyze differences in income.

In studying income distribution, economists look at how and why personal revenue is measured and divided among members of a society—without any implications for the members' relative happiness. Economists seek to explain a given pattern of income distribution as the product of such factors as individual choice, socioeconomic discrimination, and government policies. More important, they go beyond the facts and look to the consequences of policy changes. What are the economic effects of a distribution of income determined by free-market forces? If society, working through the political process, determines that distribution to be unjust, what are the effects on incentives and economic growth of altering the distribution produced through market forces? These and similar important questions are discussed in this chapter. Before turning to these larger issues, consider the simpler question, Why are individuals rich or poor?

The Individual and Income Distribution

Individual income: The sum of labor income, asset income, and government subsidies minus tax payments.

The economic question of why one is rich or poor has many facets. Individuals accumulate income from their labor and entrepreneurial skills, from the ownership of assets such as land, or from the receipt of cash or government subsidies called transfer payments. At the same time, most individuals lose a certain portion of income through taxes. We will analyze each factor of **individual income** briefly.

Labor Income

Labor income: The payments an individual receives from supplying labor, equal to the individual's wage rate times the number of hours of labor supplied.

Labor income—income received from supplying labor—is the wage rate received multiplied by the number of hours worked. The supply of labor is by far the largest source of individual income. The supply of labor ordinarily varies according to wage rates, but individuals also differ in their choice of work hours over leisure hours, and wages themselves vary for a number of reasons.

Choices of Work Versus Leisure. Different people with the same skills will often supply different quantities of labor at identical wage rates. The same wage rate affects people differently because they can make choices in the trade-off between work and leisure. Harpo, who has the same skills as Gummo, chooses to play his harp six hours a day for no pay and to work two hours a day for pay. Gummo chooses work and income over leisure, working twelve or fourteen hours a day. The terms *lazy* and *industrious* are certainly relative, but these personality traits do not totally explain different attitudes toward work among those with identical skills.

What else accounts for differences in attitudes toward work? No human being is a machine who can supply labor without limit. At higher and higher wage rates and longer and longer hours worked, we all become leisure lovers (see Chapter 12 on this point). As we choose more work, the costs of losing more and more leisure rise and the rewards of earning more income fall.

The explanation for income-leisure trade-offs also relates to differences in incentives to work. Incentives to work—willingness to supply labor at a given wage rate—are ordinarily thought to result from different doses of the Protestant ethic of industriousness. The economist must accept this ethic as given and point out that, for some, leisure is freely chosen over income. Unfortunately, the choice of income or leisure is not freely open to all. Disabled persons, the sick, the old, and the unskilled may have few or no choices. Society often compensates by making transfer payments to such individuals, providing them with an income. We will have more to say about such government activities later in the chapter.

Differences in Wage Rates. Wage rates vary widely among occupations. The wage rate you earn is largely a function of the human capital you have built up through education and investments in other skills. Human capital is a major reason why you are ultimately rich or poor. Level of schooling attained (a college degree, a high school diploma) is a ticket to enter some occupations. Given government enforced regulations of the American Medical Association, entry into the physicians' market without a medical degree would almost certainly land you in jail. It is worth noting that if education level and achievement are related to the income class one initially comes from, income differences may be somewhat self-perpetuating over time.

Luck also pays a role in the wage rate. At the beginning of the energy crisis in the early 1970s, those entering the labor market with degrees in petroleum engineering received much higher wages than under previous conditions. (The situation was short-lived, of course, since more and more college students were attracted to petroleum engineering by high wages, and, as usual, supply expanded and wages fell.) Others are natually endowed with rare talents or physical attributes and are rewarded accordingly. P. T. Barnum's midget Tom Thumb was paid a handsome income simply for being abnormally small. For all of us, it is far better to be "lucky than good."

Individuals' or families' incomes are also determined by their stage in the life cycle. Law students at Harvard, Stanford, or the University of Michigan may be poor—students everywhere tend to be poor—but they will not always be poor. Income or earnings tend to rise through the life cycle, ordinarily reaching a peak between the mid-40s and the mid-50s and falling off thereafter. Other factors also influence the wage rate an individual receives. Without question, race and sex discrimination have been ugly practices affecting wage rates. We look at this matter in more detail later in the chapter.

Broadly speaking, then, an individual's income is determined by his or her choices in the trade-off between work and leisure and by the individual's stock of human capital, current life cycle period, special skills, luck, and degree of discrimination. These factors are major determinants of whether one is rich or poor, but there are other factors related to income from savings and government benefits or taxation.

Income from Savings and Other Assets

Asset income: The income received from savings, capital investment, and land, all of which require forgone present consumption.

The other major source of income is the return from savings or other assets one owns or has accumulated. We may save as an ingrained habit. Our parents or culture may instill the ethic of frugality in us to "save for a rainy day" or to accumulate wealth or assets for other reasons. Motives for the sacrifice of current consumption are numerous: security, expected future gaps in income, education, retirement, large consumer expenditures, and so forth. In any case, the sacrifice of current consumption for future consumption is rewarded by an interest return or **asset income.**

Many people save for future generations. Some current income earners set aside assets for their children's later use, which brings us to another major source of income. Inheritance of money or other assets is related to the luck of having had wealthy parents. Historically, free societies have permitted inheritance and protected the rights of income owners to bequeath gifts at death or while still alive. Most societies have nevertheless taxed the lucky recipients of inheritance on the grounds that such wealth was not due to their own productivity.

The returns from both savings and other wealth and asset accumulations also depend on one's skill at investing. Interest returns on all sorts of investments—monetary, real capital, or land—are determined by the costs of investing and the incentives to maximize income from investments. Small savers and investors often receive lower returns than large-scale investors partly because of the high costs of gathering information about markets.

Taxes and Transfer Payments

Ex ante distribution: The distribution of income before the government transfer payments and taxes are taken into account.

A final source of income is government tax collections and benefit distributions. Economists distinguish between the **ex ante** distribution of income, the before-tax and transfer payments distribution, and the **ex post,** or after-tax and transfer payments, **distribution.** We all receive goods, services, and benefits from all levels of government, and we all pay taxes in one form or

Ex post distribution: The distribution of income after the government influences the disposable income of individuals with taxes and transfer payments.

Transfer payment: The transfer by government of income from one individual to another; it may take the form of cash or goods and services such as education, housing, health care, or transportation.

another. National defense, roads, and the local public swimming pool are some of the benefits; property taxes, sales taxes, and income taxes are some of the costs.

Most individuals are either net taxpayers or net benefit receivers. The determination of how rich or poor we are *ex post*, in other words, depends partly on our position with regard to taxes and benefits. The receipt of government services and **transfer payments**—such as direct welfare payments, food stamps, subsidized public housing, Social Security, and unemployment compensation—must be added to privately earned income. Transfer payments are all money or real goods or services transferred by government from one group in society (taxpayers) to other groups (transfer recipients). Tax payments of all kinds must likewise be subtracted to help explain *ex post* why we, as individuals, are rich or poor. We will see that most government statistics on income distribution reveal only *ex ante* money distributions of family or household income, but economists have attempted to make some *ex post* calculations as well.

To summarize, we are rich or poor as individuals because of three major factors: our wage income, our income from savings and other assets, and our position with respect to government taxes and benefits. Part of our position in income distribution depends on luck (health, genetic and financial inheritance, the socioeconomic position of our family), part depends on choice (incentives, investments in human capital and skills), and part depends on political decisions—taxes and public benefits—about the justice of *ex ante* income distribution. With these important distinctions in mind, we turn to the issue of how the government measures income and to the economic tools used to explain income distribution and how it changes over time.

How Is Income Inequality Measured?

Family income: The sum of incomes earned by all members of a household.

The money income of families (often termed *households*), or **family income,** is calculated in *Current Population Reports* published by the U.S. Bureau of the Census. Table 16–1 shows family income in the United States for selected years between 1960 and 1982. Total family income is broken down into the percentage received by each fifth of the total number of families over these years, ranked from the lowest, or poorest, fifth to the highest, or

TABLE 16–1   Money Income of Families, Percent of Aggregate Income: 1960–1982

These data show an amazing uniformity in pretax income distribution over a twenty-year plus period. Special care must be used in interpreting the data, however.

Families by Quintile	Percent of Aggregate Income					
	1960	1965	1970	1975	1979	1982
1. Lowest fifth	4.8	5.2	5.4	5.4	5.3	4.7
2. Second fifth	12.2	12.2	12.2	11.8	11.6	11.2
3. Middle fifth	17.8	17.8	17.6	17.6	17.5	17.1
4. Fourth fifth	24.0	23.9	23.8	24.1	24.1	24.3
5. Highest fifth	41.3	40.9	40.9	41.1	41.6	42.7
Highest 5%	15.9	15.5	15.6	15.5	15.7	16.0

Source: U.S. Bureau of the Census, *Current Population Reports*, in *Statistical Abstract of the United States,* various issues.

richest, fifth. (A fifth of a total distribution is called a *quintile.*) In addition, Table 16–1 gives the percentage of total money income earned by the top 5 percent of income earners between 1960 and 1982.

Interpretation of Money Distribution Data

We are struck immediately by a feature of Table 16–1: *The distribution of reported money income did not change significantly between 1960 and 1982.* Reading horizontally across the table, we see that the shares for each quintile have stayed roughly the same. The lowest quintile still receives little more than 5 percent of aggregate income, while the highest quintile consistently gets the lion's share: over 40 percent. Before accepting these figures at face value, we should note some limitations of the statistics in Table 16–1:

1. Money income statistics reported by the Bureau of the Census do not conform to the economist's definition of *income.* Income, in the economist's sense, is what an individual accumulates from labor, assets, or transfer payments minus what is paid in taxes. By contrast, the *Current Population* statistics on income omit all transfer payments, all assets such as capital gains, and all income and payroll taxes.
2. The government statistics report family, not individual, income. Therefore, an increase in the number of family units, income remaining the same, will alter the reported income distribution. In the period 1960–1982, the number of family units increased significantly through later marriages, high divorce rates, and increasing independence of the elderly. The government's statistics do not take this increase into account.

These two deficiencies in the Census data require great care in drawing conclusions about income distribution from the raw statistics. Refinements must be made before concluding anything about the distribution of economic welfare. Before actually making some of these refinements, we turn to the tools with which economists deal with income distribution.

Economists' Measures of Income Inequality

Economists have developed useful methods for describing and analyzing income distribution. The primary tool for measuring income distribution is called a Lorenz curve, developed in 1905 by M. O. Lorenz. A second device derived from the Lorenz curve is also named for its inventor: the Gini coefficient, developed by Corrado Gini in 1936.

Lorenz curve: A graph that shows the cumulative distribution of family income by comparing the actual distribution to the line of perfect equality.

The Lorenz Curve. A **Lorenz curve** plots the relation between the percentage of families receiving income and the cumulative percentage of aggregate family income. To understand the Lorenz curve, consider a hypothetical income distribution for some country in the year 2020, as shown in Table 16–2. Table 16–2 tells us that the lowest 20 percent of family income earners receive only 5 percent of total income. The next 20 percent of families get 15 percent, and the lowest 40 percent receive a cumulative percentage share of 20 percent. The middle 20 percent get 20 percent of total income, indicating that the lowest 60 percent of income earners receive only 40 percent of total income. Obviously, 100 percent of income recipients receive a cumulative total of 100 percent of income.

The data of Table 16–2 may be translated into graphic form. The vertical axis of Figure 16–1 represents the cumulative percentage of family income, and the horizontal axis shows the percentage of income-earning families. If all families earned equal incomes, the Lorenz curve would be a straight line.

TABLE 16–2   Hypothetical Income Data for the Year 2020

The second column shows the percentage share of total income received by each quintile of families in this hypothetical country; the third column adds these percentage shares cumulatively. For instance, since the lowest fifth receives only 5 percent of total income and the second fifth receives 15 percent, together these lowest two fifths receive a cumulative share of only 20 percent of total income. The lowest four fifths receive a cumulative share of 65 percent. Data from the third column are plotted on a Lorenz curve in Figure 16–1.

	Aggregate Income		
All Families by Quintile	Percent Share (year 2020)	Cumulative Percent Share (year 2020)	
A. Lowest fifth	5	5	20%
B. Second fifth	15	20	40%
C. Middle fifth	20	40	60%
D. Fourth fifth	25	65	80%
E. Highest fifth	35	100	100%

Ten percent of all families would earn 10 percent of income; 80 percent of families would earn 80 percent; and so on. Such a curve would actually be a diagonal cutting the square in half, as represented by the line of perfect equality in Figure 16–1.

A Lorenz curve is constructed from the hypothetical data of Table 16–2. Point A in Figure 16–1 corresponds to line A in Table 16–2—it is the

FIGURE 16–1

A Hypothetical Lorenz Curve for the Year 2020

The Lorenz curve shows the cumulative percentage of family income earned by percentages of families. The Lorenz curve indicates that 40 percent of families earned only 20 percent of total income, while 80 percent earned a cumulative 65 percent. This distribution is unequal and falls short of the line of perfect equality. The shaded area in the diagram indicates the degree of income inequality. The farther the Lorenz curve moves away from the line of perfect equality, the less equal the income distribution.

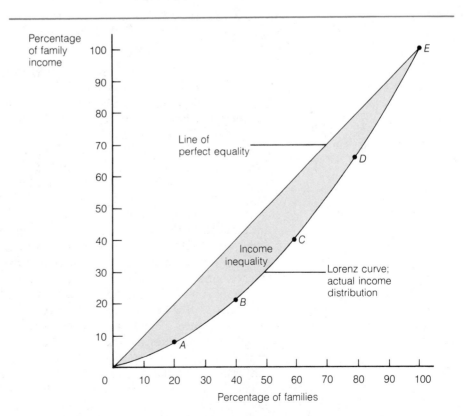

intersection of the lowest 20 percent of families and their cumulative percentage share of total income. Point *D* in Figure 16–1 shows that 80 percent of families (the fourth fifth in Table 16–2) received 65 percent of total income. The combination of points *A*, *B*, *C*, *D*, and *E* forms a Lorenz curve.

The distance of the Lorenz curve from the line of perfect equality is a measure of income inequality. The shaded area of Figure 16–1 is a measure of the degree of income inequality for our hypothetical society. Should the shaded area become larger or more bowed, money income would be more unevenly distributed. A smaller area or a flatter curve would signal a more equal income distribution.

The Gini Coefficient. An economic tool related to the Lorenz curve is the Gini coefficient. A **Gini coefficient** is the ratio of the shaded area of Figure 16–1 to the total area under the line of perfect equality. The Gini coefficient is simply a number between zero and one. The smaller the number—the closer it is to zero—the more equal the income distribution. A larger Gini coefficient, resulting from a more bowed Lorenz curve, means that income is less evenly distributed. A Gini coefficient of 0.30 for a certain year would indicate the presence of greater income inequality than a year with a coefficient of 0.20. Care must be taken in interpreting both Lorenz curves and Gini coefficients.

Gini coefficient: A numerical estimation of the degree of inequality of the distribution of family income.

Problems with Lorenz and Gini Measures

We have already seen some of the limitations associated with government statistics of money income distribution. Such data are not adjusted for taxes, capital gains, or many kinds of income transfers from government to families. Since the Lorenz and Gini measures are based on these statistics, we must be especially careful in using these instruments. In addition, the Lorenz and Gini measures have limitations of their own:

1. The Lorenz and Gini income distribution measures are simply mechanical devices and not precise indicators of the relative level of wealth in society. Economists are particularly interested in the effects of policies and programs on income distribution. Yet a change in the Gini coefficient from, say, 0.30 in 1990 to 0.28 in 1995 may not give an accurate enough indication of how distribution has changed. While the lower number does mean more equality overall, it may be that the poorest 20 percent of families received a lower percentage of income and the richest 20 percent received a greater percentage. If economists are concerned with policies to raise the economic welfare at the lower end of the distribution, the mechanically calculated Gini number is a very misleading indicator of social welfare changes. The actual Gini ratios of 0.376 in 1947 and 0.364 in 1977 tell us little of what the economist wants to know about the component shares in income distribution.

2. Important microeconomic effects may be hidden or neglected by a measure such as the Lorenz curve. Suppose, for example, that policies designed to reduce income inequalities at the top 20 percent are instituted through higher taxes for the rich. The initial *ex ante* effect would be a reduction in incomes of executives and others earning very high salaries and wages. The Lorenz curve would move closer to the line of perfect equality. But *ex post* factors must also be considered. High income earners might decide to work less, and the labor supply in these areas of the economy would diminish. A reduced supply, given demand conditions,

yields higher future returns to workers. The new high returns also signal new entrants into these areas of the economy. These factors may create a new *ex post* bowing out of the Lorenz curve, meaning that the original redistribution policies did not have the ultimate effect of reducing inequalities in income distribution at the upper end.

While changes in both the Lorenz and the Gini measures are suspect for reasons such as these, they remain two of the essential economic tools to measure economic well-being. They are useful so long as we keep their limitations in mind. More important, some of their limitations have been lessened by refinements attached to them by economists interested in problems of economic welfare. We turn to some of these refinements in the following section.

Income Distribution in the United States

We have already looked at family income distribution for selected years between 1960 and 1982. We now construct Lorenz curves with the data for the years 1960 and 1982. In Table 16–3, the cumulative percent share of all families by quintiles is calculated from the simple shares for 1960 and 1982. In 1960, for example, the lowest 60 percent of families received 34.8 percent of total income and in 1982 received 33.0 percent. In 1960, the lowest 20 percent received 4.8 percent of money income, but 4.7 percent in 1982, and so on.

Lorenz curves are calculated for these two years in Figure 16–2. Two important facts may be concluded from these Lorenz curves. First, it appears that money income distribution has changed very little over the twenty-year period, as shown by the minute differences between the Lorenz curves. For the lowest, middle, and fourth highest quintiles of income receivers, the distribution of income remained about the same. The second fifth lost one percent and the richest 20 percent gained one and a half percent, although this movement may be insignificant in a statistical sense. (Focus, "Do the Very, Very Rich Get Richer?" analyzes income distribution further.)

What can we conclude from this money income data? Has overall distribution become more unequal over the twenty-year period 1960–1982? Have

TABLE 16–3   **Share and Cumulative Share in Income Distribution of All Families, 1960 and 1982**

The data show insignificant changes in the lowest, middle, and fourth highest quintiles in income distribution, with a loss of one percent in the second fifth and a gain of about one and a half percent in the highest 20 percent of income receivers.

All Families by Quintile	Percent Share 1960	Percent Cumulative Share 1960	Percent Share 1982	Percent Cumulative Share 1982
Lowest fifth	4.8	4.8	4.7	4.7
Second fifth	12.2	17.0	11.2	15.9
Middle fifth	17.8	34.8	17.1	33.0
Fourth fifth	24.0	58.8	24.3	57.3
Highest fifth	41.3	100.0	42.7	100.0

Source: U.S. Bureau of the Census, *Current Population Reports,* in *Statistical Abstract of the United States,* 1984.

FIGURE 16–2

Lorenz Curves for 1960 and 1982

The Lorenz curves of income distribution for 1960 and 1982 show little change in before-tax income distribution in the United States.

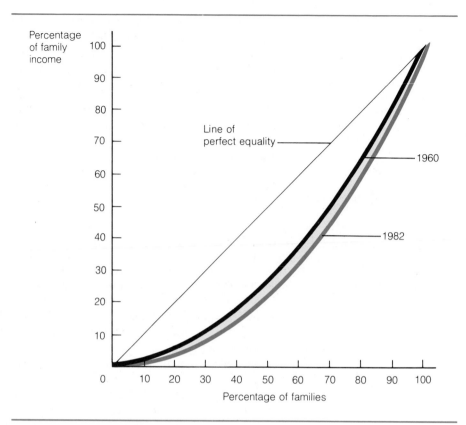

the lowest 60 percent of income recipients on balance become worse off? Have the rich gotten slightly richer?

The answers to these questions lie in the taxes and transfers that are not reported in the *ex ante* money income statistics from which the Lorenz curves of Figure 16–2 were constructed. We have already discussed some of the problems associated with the construction of simple Lorenz curves and Gini coefficients. Consider what happens to income distribution when adjustments are made for such things as income and payroll taxes, capital gains, in-kind transfers, money transfer or welfare payments, and other factors.

Adjustments to Money Income

Two important factors related to money income data must be considered before any conclusions are drawn about recent trends in income distribution in the United States. First, while the reported money income data do reflect the impact of indirect business taxes—excise taxes, property taxes, and corporate income taxes—since these items are paid before money factor payments are distributed, the data in Tables 16–1 and 16–3 do not account for payroll and income taxes. Since the U.S. income tax system is progressive, meaning that the higher-income recipients pay higher tax rates, **adjusted income** distribution is affected in the direction of greater equality.

Adjusted income: The income of an individual after taxes are subtracted, transfer payments are added, and other items are accounted for.

A second and increasingly important set of items that is not included in the raw money income statistics is transfer payments. These include **in-kind transfers** of benefits other than money—such as food stamps, public housing, Medicare, Medicaid, and other subsidies received largely by lower-income groups—in addition to money transfers or welfare payments. Such in-kind subsidies also tend to equalize income distribution across the various classes.

FOCUS   Do the Very, Very Rich Get Richer?

F. Scott Fitzgerald, the Jazz Age American writer, once remarked to Ernest Hemingway, "You know, Ernest, the very very rich are different than you and I." Hemingway's reply: "I know, they have more money." How rich are the very rich?

The rich in America are not as rich as they were sixty or a hundred years ago. Gone are the good old days when Alva Vanderbilt could spend $250,000 1883 dollars on a fancy dress ball or when door prizes consisted of five minutes with a shovel in a sandbox filled with diamonds, emeralds, and rubies. Today, we do not seem to observe huge family accumulations such as those of the Rockefellers, Mellons, Carnegies, or Vanderbilts, although there are some exceptions (such as the Hunts of Dallas). But the rich are still among us, as the following table shows.

Year	Percentage of Personal Wealth Held by Top 1%
1922	31.6
1929	36.3
1933	28.3
1939	30.6
1945	23.3
1949	20.8
1953	24.3
1954	24.0
1956	26.0
1958	23.8
1962	22.0
1965	23.4
1969	20.1
1972	20.7

Source: 1922–1956: Robert J. Lampman, *The Share of Top Wealth-Holders in National Wealth*, National Bureau of Economic Research, 1962; 1958: James D. Smith and Staunten K. Calvert, "Estimating the Wealth of Top Wealth-Holders from Estate Tax Returns," *Proceedings of the American Statistical Association*, Philadelphia, 1965 (copyright); 1962–1972: James D. Smith, unpublished estimates, the Urban Institute, Washington, D.C., and the Pennsylvania State University (copyright).

The percent share of total U.S. wealth (not just income) held by the wealthiest 1 percent of all Americans has changed a good deal since the 1920s. Over the 1920s, wealth holdings concentrated at the top, reaching a high of 36.3 percent in 1929. Since World War II, however, wealth holdings have in general become less concentrated. Tax laws, especially inheritance laws, have probably reduced the concentrations of income in the hands of a few.

Composition of the wealthiest class may have changed, too. Gross statistics on wealth holdings give no indication of who the wealth holders are. With more economic and social mobility in this century than in the last one, membership in the top 1 percent was probably different in 1949, 1972, and 1984. *Forbes* magazine's 1983 poll of the four hundred richest Americans ($125 million and above) is very revealing in this regard.[a] Of a list of the thirty richest Americans compiled in 1981 only seven or eight names from the same families made it in 1983. A full 18 percent of the names on the 1983 list were different from 1982's roster. Rather than being frozen in inheritance or traditional economic power, wealth originated in a large variety of occupations, settings, and backgrounds. Women make up one-third of the 1983 list. It is also the case that high-tech society provides a ripe setting for rags to riches—and riches to rags—experiences. Given the large amount of economic mobility, tax laws, and incentives of inheritors, the likelihood of families going from "shirtsleeves to shirtsleeves in three generations" increases.

Ambition, incentive, and intelligence are still winning attributes in U.S. society, in spite of tax laws tending to reduce wealth concentrations. Americans still believe, in the sentiments of the late Duchess of Windsor, that there are two things a person can never be—"too thin or too rich."

[a] "The Richest People in America: The Forbes Four Hundred, 1983 Edition," *Forbes* (Fall 1983).

In-kind transfer payments: Transfers of benefits other than money from government to citzens, such as food stamps, public housing, and Medicare.

Economist Edgar K. Browning has attempted to develop a measure of income that conforms more closely to economists' definition. Browning estimated the effects of compensation for taxes, welfare transfers, in-kind transfers, and other factors for the year 1972, which we reproduce as Table 16–4. The table shows the distribution of money income in 1972 in billions of dollars with adjustments for in-kind transfers and other factors added in and with income and payroll taxes subtracted from the raw money income data.

Consider the additions to money income shown in Table 16–4. In-kind

TABLE 16–4   Distribution of Adjusted Income in 1972

Adjusted income distribution statistics for any single year require the addition of in-kind transfer benefits and other items and the subtraction of income and payroll taxes. In 1972, a more equal distribution resulted from these adjustments, as is evident in the comparison between the adjusted and unadjusted distribution figures at the bottom of the table.

Income Item	Families by Quintile (shown in billions of dollars)					
	Lowest	Second	Third	Fourth	Highest	Total
Unadjusted money income	37.1	81.7	120.1	164.1	284.2	687.2
Plus						
Benefits in kind, capital gains, potential additional earnings, and other adjustments	58.6	54.8	54.1	50.5	102.9	320.9
Minus						
Income and payroll taxes	1.3	5.8	15.5	26.6	67.5	116.6
Equals						
Adjusted money income total	94.4	130.7	158.7	188.0	319.6	891.4
Adjusted percent distribution	12.5%	15.8%	17.9%	20.4%	33.3%	100.0%
Unadjusted percent distribution	5.4%	11.9%	17.5%	23.9%	41.4%	100.0%

Source: Edgar K. Browning, "The Trend Toward Equality in the Distribution of Net Income," *Southern Economic Journal* 43 (July 1976), p. 914.

benefits are received largely by the poorest classes in society, whereas capital gains on stocks, bonds, and houses—which are not reflected in money income data—are six to seven times greater for the highest quintile of income recipients than for the lowest.

After these items are added to money income, taxes must be subtracted to obtain an estimate of net income. Since the United States employs a progressive income tax system, the highest burden of income taxes—in both percentage and total amount—falls on the highest 20 percent of income recipients. In 1972, as the income and payroll taxes line in Table 16–4 shows, this group paid about $20 billion more in income taxes than all other income earners combined.

What is the net result of these adjustments to money income? Table 16–4 reports the results in terms of total income adjustment and in unadjusted and adjusted percent distribution. Total income rises for each class after additions to and subtractions from money income are made. But the important point is that the percent distribution among the quintiles is affected. The poorest 60 percent of income recipients all receive increased shares, with the lowest 20 percent receiving a full 7 percent increase in adjusted income.

The effects indicated in the data of Table 16–4 are summarized in Figure 16–3 as Lorenz curves. Figure 16–3 shows a set of three Lorenz curves for

FIGURE 16–3

Lorenz Curve Including Income Adjustments, 1972

When taxes are subtracted from before-tax money income and in-kind transfers and other factors are added, the Lorenz curve moves closer to the line of perfect equality.

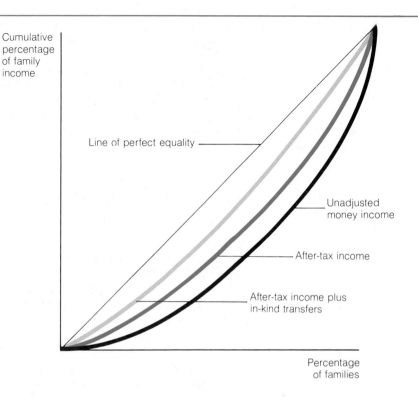

Source: Figure based on conclusions of Browning, "The Trend Toward Equality in the Distribution of Net Income," *Southern Economic Journal* 43 (July 1976), pp. 912–923.

1972. The most bowed curve is the unadjusted distribution of money income. Figure 16–3 also shows a Lorenz curve that is adjusted for payroll and income taxes. The tax adjustment makes the curve less bowed and moves income distribution closer to the line of perfect equality. This shift indicates that a progressive tax system does effect greater equality in income distribution. The inclusion of in-kind transfers and other adjustments pushes income distribution to still greater equality, as the third Lorenz curve shows. These curves indicate the importance of social programs such as Medicare, food stamps, and public housing in creating greater income equality across the various segments of society.

Historical Changes in Adjusted Income

While a study of adjustments to money income over a given year leads to the conclusion that after-adjustment income is distributed more equally, what about long-range trends? Edgar K. Browning has calculated adjusted net income for 1952, 1962, and 1972 to answer this question. His results are reported in Table 16–5.

Table 16–5 indicates that a significant change took place in income distribution over the twenty-year period 1952–1972. Gains were concentrated in the bottom two quintiles, largely due to the growing importance of in-kind transfers. These benefits grew in importance over the period, and the benefits were concentrated in the poorest 40 percent of families. The growth in educational expenditures also had an equalizing effect.

The relative positions of the richest two quintiles also changed. A loss of 5.8 percentage points for these groups balanced the gain of 6.1 in the bottom two quintiles. Middle-income families did not fare significantly better or

TABLE 16–5   Adjusted Net Income for 1952, 1962, and 1972

The data spanning two decades from 1952 to 1972 show gains in the poorest 40 percent of income earners and losses in the highest 40 percent. Middle-income families (the third quintile) received about the same proportion of total income.

| | Families by Quintile (percent share) | | | | |
Year	Lowest	Second	Third	Fourth	Highest
1952	7.8	14.8	18.8	23.3	35.3
1962	9.0	15.1	19.1	22.9	34.0
1972	12.6	16.1	18.4	20.9	31.9

Source: Browning, "The Trend Toward Equality in the Distribution of Net Income," *Southern Economic Journal* 43 (July 1976), p. 919.

worse over the twenty-year period. These data suggest that a redistribution did in fact take place from richer to poorer families, creating a more equal income distribution between 1952 and 1972. The growth of in-kind transfer payments, moreover, appears to have been the principal means of the redistribution from rich to poor.

Age Distribution and Income

Another useful factor in understanding income distribution is the age structure of the income-receiving population. Take your own situation as an example. Perhaps you are a college student with a part-time job. If so, you are now probably part of the poorest quintile of the population in calculated annual income distribution. But you are not always likely to be poor. Education and training alone will give you an edge in earning income later. The relative mean income of heads of households tends to peak between ages forty-five and fifty-five and perhaps even later. Therefore, annual income figures will either overstate or understate multiyear inequality in income distribution, depending on the age structure of the population.

If the population is heavily populated by young people in the college age group, annual data will understate actual lifetime or multiyear inequality in income distribution. There is in fact less inequality over multiyear distributions in the United States than in any given year. Economists' estimates vary on how much more equal distribution is over the long run, but all agree that there is more equality when age distribution is considered.[1]

Rags to Riches Mobility

Economic mobility: The ability or ease with which an individual may move from one income range to another.

Economic mobility—the opportunity for movements up or down in the quintiles of income distribution—also influences the question of how seriously annual data should be taken. How, for example, do we know that a family in the poorest quintile of income receivers in 1960 was not in the richest quintile in 1979 or vice versa? The answer is that we do not know. To the extent that there is upward (and downward) mobility among income classes, greater equality is achieved over time.

The United States, as a nation of rugged individualists, has historically championed the idea that a hard-working and frugal boy or girl, though poor,

[1]See Alan S. Blinder, "Inequality and Mobility in the Distribution of Wealth," *Kyklos* 29 (1976), pp. 607–638. Also see an estimate by Morton Paglin, "The Measurement and Trend of Inequality: A Basic Revision," *American Economic Review* 65 (September 1975), pp. 598–609.

could climb to the top. Such economic mobility can promote equality over the long run. Most studies indicate that there is a high degree of mobility in the United States, especially in the lower quintiles, but there are important statistical problems with identifying mobility over long-run periods.[2] Such problems also extend to measuring age distribution factors and multiyear inequality. While it is safe to say that age distribution factors and economic mobility undoubtedly reduce income distribution inequality over time, economists and policymakers do not yet have the data or the wherewithal to judge the exact amount of the reduction. Annual statistics are nonetheless useful as indicators of certain important trends and characteristics of income distribution. Let us consider some of these important characteristics, especially those related to poverty in the United States.

Some Characteristics of Income Distribution

The data collected in population surveys and census reports, though imperfect, reveal a great deal of information about ourselves as a nation.

Family Income Characteristics

Consider Table 16–6, for example, which shows the total, average or mean, and per capita incomes of all families, black families and white families, by certain characteristics for one year, 1981. The data of Table 16–6 illustrate a number of the points made earlier about income distribution and also show some persistent and disturbing trends.

One important fact that glares out of the data is that the **per capita** (per person) **income** of white families is almost $4,000 greater than that of black families. Such income differences may result from differences in educational attainment, skills, and training, but it is a sad and undeniable fact that some of the differences in black and white income result from economic discrimination. **Mean**—or average—black **incomes** are persistently only 60 percent of white incomes, and, as we will analyze later, discrimination is a large part of the reason.

Another look at Table 16–6 gives us other important information. **Age-distributed income** (statistics broken down by ten-year periods, such as 15 to 24 years, 35 to 44, and so on) rises through all age categories. Maximum per capita income is not achieved until the decade 55–64 years (in 1981 at least) for all groups in society, black or white. Note, however, the persistent per capita income differentials between black and white families at all age levels.

Table 16–6 also tells us something about income distribution by sex. Note the differences in average and per capita incomes of families with a male married householder (head of family) and those with a female married householder. There is little difference between incomes of male householders whether the wife is absent or present, but for female householders, incomes are three times as great when the husband is present than when he is absent. Furthermore, widowed, separated, divorced, and single men receive more than twice the average and per capita income as their female counterparts. In other words, males did better than females on practically all counts. The

Per capita income: The income per individual, found by dividing total income by the total number of people.

Mean income: The income of the average income earner, found by dividing the total income by the number of income earners.

Age-distributed income: The distribution of income over the various age brackets of the population, such as ten-year intervals.

---

[2]See Alan S. Blinder, "The Level and Distribution of Economic Well-Being," in *The American Economy in Transition*, ed. Martin Feldstein (Chicago: University of Chicago Press, 1980), pp. 450–454. This entire essay is an excellent nontechnical and modern introduction to the many problems of income distribution.

TABLE 16–6   The Money Income of Families by Family Characteristics, 1981

The table shows total number of families, their aggregate income, average income, and per capita income by certain characteristics such as race, age, occupation, and family size.

Characteristic	All Races				White				Black			
	All Families (1000)	Aggregate Income (billions of dollars)	Mean Income (dollars)	Per Capita Income (dollars)	All Families (1000)	Aggregate Income (billions of dollars)	Mean Income (dollars)	Per Capita Income (dollars)	All Families (1000)	Aggregate Income (billions of dollars)	Mean Income (dollars)	Per Capita Income (dollars)
Age of householder												
15 to 24 years	3,621	56.2	15,528	5,632	3,030	49.7	16,399	6,075	522	5.5	10,487	3,446
25 to 34 years	14,449	330.2	22,855	6,757	12,205	290.9	23,837	7,114	1,846	30.0	16,256	4,616
35 to 44 years	13,083	380.9	29,114	7,332	11,359	344.9	30,361	7,707	1,399	26.8	19,172	4,617
45 to 54 years	10,710	343.5	32,070	8,861	9,363	313.6	33,495	9,445	1,089	21.6	19,812	4,906
55 to 64 years	9,752	287.6	29,492	10,585	8,802	268.9	30,547	11,336	794	14.4	18,093	5,005
65 years and over	9,403	178.1	18,945	8,171	8,511	166.8	19,599	8,696	763	8.8	11,566	3,913
Marital status of householder												
Male householder, total	49,513	1,388.3	28,039	8,507	44,968	1,283.5	28,542	8,792	3,442	74.1	21,531	5,788
Married:												
Wife present	47,527	1,343.4	28,266	8,509	43,326	1,244.8	28,732	8,784	3,168	69.3	21,872	5,777
Wife absent	274	6.6	24,108	8,691	236	5.9	24,863	9,283	32	.6	(B)	(B)
Separated	227	5.2	23,063	8,234	194	4.6	23,928	8,851	29	.5	(B)	(B)
Widowed	364	8.7	23,819	8,020	295	7.7	26,018	9,002	59	.8	(B)	(B)
Divorced	681	15.9	23,298	8,846	588	14.1	23,980	9,300	78	1.5	19,288	6,379
Single	666	13.7	20,621	8,218	524	11.0	21,010	8,722	104	1.9	18,394	6,759
Female householder, total	11,506	188.3	16,364	5,327	8,301	151.3	18,226	6,320	2,971	33.0	11,096	3,105
Married:												
Husband present	2,103	58.8	27,947	8,699	1,680	49.7	29,591	9,783	366	7.6	20,640	5,191
Husband absent	1,859	18.7	10,078	2,934	1,161	12.2	10,498	3,276	668	6.2	9,345	2,442
Separated	1,651	16.2	9,808	2,847	996	10.0	10,066	3,139	628	5.9	9,382	2,452
Widowed	2,462	43.4	17,621	5,996	1,895	36.4	19,212	7,009	514	5.9	11,051	3,174
Divorced	3,478	50.3	14,475	4,890	2,838	42.7	15,049	5,264	576	6.8	11,875	3,451
Single	1,604	17.1	10,629	3,651	726	10.3	14,127	5,617	846	6.4	7,569	2,321
Occupation of householder												
White-collar workers	21,107	751.6	35,609	10,710	19,345	703.2	36,353	10,994	1,285	31.2	24,310	7,192
Blue-collar workers	16,383	421.4	25,724	7,272	14,489	378.1	26,097	7,493	1,565	35.1	22,414	5,663
Farm workers	1,411	22.4	15,904	4,631	1,318	21.3	16,146	4,784	74	.8	(B)	(B)
Service workers	3,784	74.2	19,603	5,744	2,856	58.6	20,534	6,258	812	12.9	15,936	4,182
Size of family												
Two persons	24,426	552.3	22,612	11,251	22,072	519.4	23,530	11,726	2,024	26.7	13,203	6,460
Three persons	14,079	370.3	26,305	8,682	12,248	337.4	27,545	9,093	1,550	26.2	16,886	5,565
Four persons	12,594	365.0	28,978	7,232	10,998	331.9	30,177	7,530	1,264	23.3	18,407	4,604
Five persons	5,971	173.7	29,092	5,802	5,020	153.7	30,617	6,108	757	14.7	19,450	3,866
Six persons	2,409	71.6	29,723	4,948	1,900	59.8	31,456	5,221	406	8.3	20,535	3,468
Seven persons or more	1,539	43.7	28,359	3,662	1,031	32.7	31,698	4,169	412	7.8	19,048	2,365
Total families	61,019	1,576.6	25,836	7,941	53,269	1,434.8	26,934	8,444	6,413	107.1	16,696	4,571

Source: U.S. Bureau of the Census, *Current Population Survey,* in *Statistical Abstract of the United States,* 1984.

numbers strongly suggest that sex discrimination might be at work, a subject for later discussion.

Not surprisingly, families of two persons enjoy higher per capita income in all categories than larger families. White-collar workers earn significantly more than blue-collar workers on the average, again in all categories. Black farm workers earn the lowest per capita income of all, and white blue-collar workers earn more on average and per capita than black white-collar workers.

In summary, these numbers suggest some features about income distribution in 1981 that are fairly characteristic of all years surveyed: (1) that income is very much distributed along racial and sex lines and (2) that poverty, in a relative sense and perhaps also in an absolute sense, is still very much a feature of generally affluent American life. In the following two sections these interrelated matters are analyzed together with some proposed economic interpretations and solutions.

Poverty

Poverty and economic discrimination are interrelated within our economic system. One cannot be understood without understanding the other. First we examine poverty, its definition, and some of the connected reasons why some families are poor.

Poverty: A term describing family income below a defined level when other things such as size of family, location, and age are considered.

Poverty, in the relative sense that some people have lower incomes than others, is always going to be with us. As long as the perfectly adjusted Lorenz curve is not diagonal, some families or persons will be better off relative to others. Moreover, an equitable distribution of income does not necessarily mean perfect equality. Clearly, however, some degree of absolute poverty persists in the United States: A number of families in the lower quintiles of income distribution are even worse off than the lowest living standards considered acceptable in this society.

Government measures of poverty are based on a poverty index first devised by the Social Security Administration in 1964 and modified by a Federal Interagency Committee in 1969. The index defines a "poverty level" that varies according to such factors as the size of family, inflation, age and sex of the head of family, number of children under age eighteen, and farm or nonfarm residence. Accepting the government's definition of poverty—income below the poverty level—we may observe aspects of poverty over time and in different groups.

Table 16–7 provides an overall poverty profile for persons by race, family status, and sex of householder for selected years between 1959 and 1981. In 1959, almost 40 million people in the United States were poor by government standards, but that number fell to barely 32 million in 1981. More important, the number of poor in percentage terms fell significantly over the twenty-year period. Clearly progress has been made, but disturbing problems remain.

The percentage of black persons below the poverty level, though declining since 1959, remained at 34 percent in 1981. Likewise, the total number of poor persons in families with female householders rose by 5 million between 1959 and 1981, though the percentage dropped from 50.2 to 35.2. The 1981 comparison between poor persons in families with female heads and the percentage of poor in all other families is stark (35.2 percent versus 8.8 percent).

These gross statistics tell us that the low end of income distribution is made up, disproportionately, of blacks and women householders. Other groups in our society, not shown in Table 16–7, fare poorly also. The num-

TABLE 16–7   Poverty by Family Status, Race, and Sex of Householder, 1959–1981

While poverty has declined in absolute numbers and in percentage terms over time, the relative position of blacks and female householders is still a matter of concern.

Family Status, Race, and Sex of Householder	Number Below Poverty Level (millions)					Percentage of Persons Below Poverty Level				
	1959	1966	1969	1975	1981	1959	1966	1969	1975	1981
All Persons	39.5	28.5	24.1	25.9	31.8	22.4	14.7	12.1	12.3	14.0
White	28.5	19.3	16.7	17.8	21.6	18.1	11.3	9.5	9.7	11.1
Black	11.0	8.9	7.1	7.5	9.2	56.2	41.8	32.2	31.3	34.2
Families with female householder, no husband present	10.4	10.3	10.4	12.3	15.7	50.2	41.0	38.4	34.6	35.2
All other families	29.1	18.3	13.7	13.6	16.1	18.7	10.8	8.0	7.8	8.8

Source: U.S. Bureau of the Census, *Current Population Reports,* in *Statistical Abstract of the United States,* various issues.

ber of Hispanic poor tends to fall between that of poor whites and poor blacks. The over-sixty-five age group, as a whole, contained about 3.5 million persons below the poverty level in 1979, for example. Some 35.5 percent of this number were black and 26.1 percent were of Spanish origin, but only 13.2 percent were white. Improvements have been made since 1960 in the poverty status of the elderly, but the improvements have been among the white over-sixty-five age group.

Poverty also has a geographic dimension. The highest percentage of persons below the poverty level is found in the southern states (15.3 percent in 1975) with the lowest in the northeastern section of the United States (8.9 percent in 1975). Alaska and Connecticut tied in 1975 for the lowest percentage of poor (6.7 percent), while 26.1 percent of persons in Mississippi were defined as poor in government statistics.

Poverty During Recession

One important factor in analyzing poverty is that poverty levels can change drastically when the total real output and unemployment in society increase or decrease because of changing total demand. A recent and dramatic example of this feature of poverty is revealed in the recessionary conditions of declining production and employment that took hold of the U.S. economy in 1981–1982.

Table 16–8 focuses on the Census Bureau's estimate of the poverty experience of alternative groups—Hispanics, blacks, the elderly, and so on—between 1979 and 1982. Table 16–8, moreover, gives two definitions of poverty: an official definition counting only the cash income of families in the various groups and an alternative definition that counts cash income plus the market value of noncash benefits such as food stamps, school lunches, public housing, Medicaid, and Medicare. A family of four was classified as poor if it had cash income of less than $7,386 in 1979 or less than $9,862 in 1982. The official poverty level, in other words, is adjusted for inflation, as are the market values of noncash benefits.

According to either measure, poverty was on a serious upswing between 1979 and 1982. Counting only cash income, there were 26.1 million poor people in 1979 but 34.4 million in 1982. But if noncash benefits are included, there were 15.1 million poor in 1979 and 22.9 million in 1982. The

TABLE 16–8   Poverty Rates and Recession, 1979–1982

Reductions in total output and employment, especially between 1981 and 1982, created increases in poverty rates for many groups in society. When cash income only (the official definition) is considered or when the market value of noncash benefits is added to cash income (alternative definition), poverty rose between 1979 and 1982.

Category	1979		1982	
	Official Poverty Definition	Alternative Definition	Official Poverty Definition	Alternative Definition
Total	11.7%	6.8%	15.0%	10.0%
White	9.0	5.6	12.0	8.3
Black	31.0	14.9	35.6	21.5
Hispanic	21.8	12.0	29.9	20.5
Children under six	18.2	11.3	23.8	17.2
Elderly	15.2	4.3	14.6	3.5
Married couples	6.1	3.9	8.9	6.4
Families headed by women	34.9	16.6	40.6	24.8

Source: U.S. Bureau of the Census, *Statistical Abstract of the United States,* 1984.

inclusion of noncash benefits significantly reduces the absolute number of poor people, but note that, in relative terms, there was a higher percent increase in poverty using the noncash benefits calculation than using the official cash income definition.

With the exception of the elderly, recessionary conditions have increased the poverty level of all other groups considered by the Census Bureau, using either definition of poverty. Declining expenditures on all goods and services and the unemployment conditions created by the recession are clearly the major reasons for the observed increase, but there are several other possible explanations. The rate of increase in some noncash benefits was reduced by Congress in 1981. More important, there was a 33 percent increase in the consumer price index between 1979 and 1982 (primarily felt in 1979 and 1980), an increase that reduced the average market value of the noncash benefits received by the poor by more than 10 percent. In sum, poverty levels undoubtedly increased over the recession, but it is important to re-member that decreases in poverty levels, both absolute and in percentage terms, are reduced during periods of economic expansion and recovery.

The overall statistics on poverty since 1960 do show improvement, but the undeniable fact is that there are poor among us and that poverty is often related to race, sex, and family status. What are the reasons for the facts of poverty? Factors are many and often interwoven. As noted earlier, some people choose to be poor because they choose more leisure. Others are un-lucky. Still others are not able-bodied; they are simply unable to work. But these factors cannot explain the existence of all poverty.

Lower incomes also exist for clearly identifiable economic reasons. The poor tend to have lower skills and lower stocks of human capital. These lower stocks of human capital are the result of less education, inferior education, and lower job-related training. Lower productivity means unemployment and lower wages, lower wages mean lower incomes, and lower incomes mean poverty. Environmental factors may make poverty a vicious cycle: Children of the poor often receive less education and training. Moreover, many wage and income differences are the result of outright race and sex discrimination. Indeed, differences in training and education may be related to past and present economic discrimination.

Race and Sex Discrimination in Wages

One might object that poverty, race, and sex discrimination are not necessarily related. After all, a black woman may earn $75,000 per year in legal practice but still endure discrimination with respect to her white male counterparts. It is more than coincidence, however, to find such groups as blacks, Hispanics, and women ranking consistently high in poverty statistics.

We must be careful to distinguish between social discrimination and economic discrimination that leads to significantly lower wages and income, although the two are often related. Groups such as Jews, gays, and the handicapped may meet with social discrimination in seeking housing and club memberships but be little affected by economic discrimination.

Economic wage discrimination exists when individuals of equal ability and productivity in the same occupation earn different wage rates. The bald economic facts, already observed in the statistics discussed earlier in this chapter, are that the median income of blacks is only 60 percent of that of whites (in all income classes) and that females earn only about 60 percent of median male income for full-time work. Empirical studies show that about half of this difference is related to educational, productivity, and job-training differences.[3] The remaining half is the apparent result of economic discrimination or other unidentified factors.

Economic wage discrimination: A situation in which an employer pays individuals in the same occupation different wages, the wage difference based on race, sex, religion, or national origin rather than productivity differences.

The Practice of Discrimination. Economists do not pretend to know the causes of discrimination. Some observers have argued that sex discrimination may originate from historical roles of women that are matters of cultural tradition. Women's biologically unique potential for childbearing may also be a factor in differences in wage rates. Employers may not want to invest in as much job training or pay high wages for women whose work is to be interrupted or terminated by childbirth. Both women and blacks receive less formal education and on-the-job training than men and whites, a factor that may be related to past discrimination.

Economic discrimination involves costs to those who discriminate as well as to those against whom discrimination is practiced. An employer who refuses to hire women or blacks of equal productivity to men or whites has what economists call a "taste" for discrimination. If the wages of the preferred groups are driven up by these actions, employers must pay a premium for men or whites. In a competitive system, discriminating firms will be at a disadvantage in the marketplace. Employers who do not discriminate in hiring can acquire equally productive labor at lower wage rates and, in the end, will tend to drive discriminating competitors out of business.

Even if employers do not discriminate, consumers may. In a competitive

---

[3]See Dwight R. Lee and Robert F. McNown, *Economics in Our Time: Concepts and Issues* (Chicago: Science Research Associates, 1983), chapter 10, for an excellent analysis of economic issues related to discrimination. Some of the present discussion is drawn from their work.

Consumer-initiated discrimination: A circumstance in which people prefer to purchase a good or service produced or sold by individuals of a particular sex, race, religion, or national origin; such consumers are willing to pay a premium to indulge their taste for discrimination.

system, **consumer-initiated discrimination** is costly to the discriminator as well as to those discriminated against. Consumers who will deal only with whites or men (avoiding, say, black or female lawyers, doctors, or interior decorators) must pay a premium for their prejudice. Prices of goods and services of favored sellers will be higher than prices for the same goods or services available to those who have no taste for discrimination. In sum, the competitive system makes it costly for employers or consumers to discriminate. This economic deterrent has failed to eradicate discrimination in our economy, however, as have government regulations explicitly forbidding discrimination.

Economic Restrictions and Discrimination. The law is clear on one point—employers are to give equal pay for equal work without regard to race, creed, or sex. But employers may simply avoid the hiring of women or blacks. Laws establishing quotas (minimum numbers) of females or racial minorities in employment or educational situations may not have the fully intended results. Reverse discrimination cases have resulted in some instances of quota imposition. In other cases, employers may subtly hide discrimination by reducing the amount of on-the-job training while paying the same money wages. The full wage—both money and training—may not be paid to those against whom discrimination is practiced.

Although there are encouraging statistics on the entry of women and blacks into male- and white-dominated occupations in the past twenty years, a number of factors remain that make the rapid and total elimination of discrimination unlikely. As suggested above, a fully competitive system discourages discrimination, placing extra costs on discriminating employers and consumers. Where, then, do we observe the most discrimination? In markets where regulations or restrictions prevent the competitive process from working. According to the research of economist Thomas Sowell, high degrees of economic discrimination have traditionally been found within labor unions, in regulated industries, and in the professions where legal and government-sanctioned entry restrictions are permitted.[4] In such industries and occupations, nonmarket characteristics—such as color and sex—may be used in deciding whom to employ. Such discrimination may be detected in medicine, law, and dentistry, for example, as well as in those historically regulated areas of our economy such as railroads and the postal service. Government itself has practiced a good deal of discrimination in this century as, for example, in the U.S. military up to World War II and the Korean conflict.

In sum, discrimination appears to be most pronounced and most long-lived where government-granted regulations and restrictions substitute for freely competitive market forces. Many economists feel that a movement toward a deregulated system of competition would go far in removing much sex and race discrimination. Where economic incentives are placed before employers and consumers to avoid discrimination, less discrimination may be confidently predicted.

Programs to Alleviate Poverty

Poverty is the result of a number of factors—lower productivity, inferior educational opportunities, and economic discrimination. Until full equal economic opportunity for all is a reality, some type of redistribution of wealth is likely to take place.

[4]See Thomas Sowell, *Markets and Minorities* (New York: Basic Books, 1981), pp. 34–51.

Contemporary welfare programs (see Table 16–9) center on the relative needs of the poor. In-kind benefits such as food stamps, Medicare, housing subsidies, and outright money transfers (such as Aid to Families with Dependent Children) are allotted to the poor on the basis of a government agency's determination of need. Benefits, moreover, vary from state to state and among local governments. Without questioning need or the existence of poverty, economists might analyze the costs of fairly and accurately administering such programs. But more important, economists are concerned with the effects of such programs on the incentives of the poor.

While there will always be poor unfortunates, who, for health or other reasons, cannot work, it is clear that the contemporary welfare system discriminates against most of the poor in one respect: It discourages any effort to better themselves by earning income additional to their subsidy and perhaps acquiring job training in the process. Take a highly stylized example: Rita, a widow with three small children, could earn $4,000 a year to supplement her modest welfare subsidy of $7,000 per year. Such work would surely not be in her interest, however, since under the welfare system in some circumstances her subsidy would be taxed at a rate of 100 percent. By "taxed" we mean that Rita would lose her welfare benefits if she works. In other words Rita would have less incentive to work if her subsidy were reduced by an amount equal to what she earned and still less incentive if she were to lose the entire subsidy by working.

Negative income tax: A progressive income tax that allows for a negative tax rate (income subsidy) for income below a particular level. As income rises, the subsidy gradually diminishes to zero.

Consider another possibility, one supported by a large number of economists interested in public policy relating to poverty: the **negative income tax.** The negative income tax originates from the observation that welfare recipients such as Rita will react to economic incentives if such incentives are provided. Under a negative income tax system Rita would be taxed for working but not at a rate of 100 percent.

Suppose that Rita and her children are guaranteed a minimum income subsidy of $7,000; an income of $14,000 is determined by some agency calculation to be adequate for her situation—a widow with three children to support (the figures given here are wholly hypothetical). Further suppose that

TABLE 16–9   **Summary of Principal Welfare Programs**

U.S. welfare programs include both cash subsidy and in-kind benefits. Welfare programs, moreover, are aimed at various segments of society and are administered at all levels of government.

Program	Level of Administration	Purpose	Beneficiaries
Aid to Families with Dependent Children (AFDC)	Federally mandated; administered by state and local governments	Income maintenance, often used in conjunction with in-kind subsidies	Poor families with unemployed heads of household
Unemployment Compensation	State governments	To aid the unemployed for limited periods	Directly benefits the unemployed and their families
In-Kind Benefits, Housing Subsidies, Food Stamps	Federal, state, and local governments	Supplements to nutrition and to the quality of housing of the poor	Poor families and the children of poor families
Social Security, including Medicare and Medicaid	Federal government	To provide retirement insurance	Insurance for survivors, welfare for the aged poor, medical care for the elderly (Medicare) and the poor (Medicaid)

TABLE 16–10   Negative Income Tax for a Hypothetical Welfare Recipient

If Rita could earn $4,000 per year from working, she would be taxed only at a rate of 50 percent, giving her a total income of $9,000. At a marginal or additional negative income tax rate of 50 percent, she would have an incentive to work and earn additional income. After she earns an income greater than $14,000, Rita becomes a net taxpayer.

Rita's Income from Working or Other Sources	Rita's Subsidy (Negative Income Tax)	Rita's Total Income (Private Sources plus Subsidy)
$ 0	$7,000	$ 7,000
4,000	5,000	9,000
8,000	3,000	11,000
14,000	0	14,000
16,000	−1,000	15,000
20,000	−2,000	18,000

rather than being taxed at a rate of 100 percent, as in the current welfare system, Rita is taxed at a lower rate, say 50 percent of all additional private earnings. What happens when Rita works and earns income?

As Table 16–10 shows, Rita now has an incentive to enter the marketplace. If she earns $4,000, she does not lose $4,000 in subsidy but instead gains $2,000. In other words, if Rita earns $4,000, she pays only 50 percent of it in taxes, which, in effect, reduces her subsidy by $2,000. Her total income would climb to $9,000. Should Rita be able to earn $8,000 by job advancement, she would pay in taxes only $2,000 of the additonal $4,000 earned. Her subsidy would be reduced, in effect, to $3,000, giving her a total income of $11,000. Should she be able to earn $14,000, her minimum acceptable income, Rita would receive no subsidy (paying, in effect, an additional tax of $3,000). If Rita earns an income over $14,000 per year she would become a net taxpayer.

The point is that Rita would have positive incentives in this system to work and be self-supporting. While the numbers in this example—such as a 50 percent tax rate—are completely hypothetical (progressivity and regressivity may be built into the plan), the principle is clear. If the poor could improve their lot by working, economists would predict that such incentives would encourage work effort. While a negative income tax plan has not received much political support, many economists have strongly urged its implementation in preference to much of the contemporary welfare system. In an economic view at least, it would help make the welfare system a tonic to the poor rather than a sedative, as the present system can be regarded.

The Justice of Income Distribution

Economists, who have been writing about income distribution for more than two hundred years, have attempted—not always successfully—to avoid advocating any given distribution of income or any redistribution of income. In other words, economists have tried to avoid normative statements about what should be and have taken a more positive stance—what income distribution is and how it might be changed to produce greater incentives to work and produce while maintaining some politically derived concept of justice. The proposal that a negative income tax plan be instituted in place of the

Pareto Movements and Income Distribution

Are there any rules governing the ability of an economist to make statements concerning income distribution? Practically speaking, can an economist say anything about whether a housing subsidy or a new dam would make people better off, all things considered?

A famous turn-of-the-century economist, Vilfredo Pareto (1848–1923), had something to say on these matters. His view, sometimes called Pareto optimality, was that the economist could pronounce one distribution of income or wealth as preferable to another if and only if the change made one group better off without leaving another group worse off. "Better off" is expressed as satisfaction or utility, so judgment depends on the ability of the economist to measure welfare.

It is almost impossible to think of distribution changes that do not help one group without simultaneously making another group worse off. The provision of food stamps obviously aids the recipients, but it re-

duces the welfare of those who pay taxes to support such expenditures. While it is tempting to argue that a dollar's worth of something provides more satisfaction to a poor person than it takes away from a rich person, the positive economist cannot make this leap because utility cannot be measured. Support for a progressive income tax on this basis is invalid for the same reason. It would involve what economists call an interpersonal utility comparison—a judgment about the relative utility of, say, a dollar to a rich person and to a poor person. As noted in this chapter, the economist can measure rich or poor but cannot measure happy or unhappy. Thus the economist depends on the political process to decide on a particular distribution to be made through taxes and expenditures. The economist is then concerned with the effects of specific taxes and specific expenditures on economic efficiency.

contemporary system of welfare transfers is an example of a positive economic solution related to income distribution. Focus, "Pareto Movements and Income Distribution," considers further the role of economists.

Justice itself is not part of the positive economist's vocabulary, but the effects of some concept of justice are within the economist's purview. For example, some concepts of justice could create extreme income redistributions that would make income distribution more equal at the expense of working and producing members of society. This notion of extreme redistribution is condemned in some popular literature as putting too many people in the wagon with too few pulling the wagon. In economic terms, such redistributions and the high tax rates involved would probably create disincentives to work. If so, total real output of goods and services available to all citizens would decline. As a result, a shrinking segment of a smaller pie would be available to all classes of economic society, including the poor.

More than a hundred years ago, the classical economist John Stuart Mill observed that two kinds of equality are involved in questions of distribution: *ex ante* equality and *ex post* equality. *Ex ante* equality is a situation where "all start fair"—where, through educational opportunities, the absence of discrimination, social and economic mobility, and so on, every individual is given an opportunity to maximize his or her potential. *Ex post* equality is a guarantee that through redistributive policies all end up at the same place. While Mill championed *ex ante* equality, he believed that differing incentives, innate talents, luck, and so on meant there would and must be differences in income distribution in the end. To guarantee all citizens *ex post* equality would stultify incentives, which are the very mode of economic progress. Enforced *ex post* equality would surely shrink the total economic pie.

Modern economists concerned with income distribution face the same problem. The trick is to evaluate the contemporary system and to propose means that provide that all start fair (akin to golfing or bowling handicaps) but that do not force redistributions that significantly reduce incentives to produce. The modern economist is as much on a razor's edge as were past writers dealing with the questions of income distribution. The issues of fairness, justice, and equality will always be matters of dispute. The economist, as a maker of positive statements and proposals, will always be a contributor to this important debate.

Summary

1. Economists deal with positive and not normative aspects of income distribution, for the economist's view of "who should receive what" is only as good as anyone else's. The economist therefore does not deal with questions of justice or fairness but focuses instead on the facts of income distribution and on the possible effects of alternative welfare programs.

2. Money income distributions differ for a number of reasons: Some persons choose leisure over work and income while others earn lower incomes because of bad luck, poor health, lower educational opportunities and productivity, and race or sex discrimination.

3. A Lorenz curve and the related Gini coefficient are measures of income inequality. Ordinarily, the more bowed the Lorenz curve or the larger the Gini coefficient, the more inequality there is in a society's income distribution.

4. Lorenz curves showing money income distribution for a given year overstate actual inequality when in-kind benefits, income and payroll taxes, and other factors are considered. Age distribution of the population and income mobility also create greater long-term equality in income distribution.

5. Poverty and economic discrimination are interrelated features of our economic system. Economic discrimination exists when individuals of equal productivity are paid different wages on the basis of sex or racial differences.

6. Economic discrimination exists against blacks and women in our society in the sense that only half of the difference between black-white and male-female earnings is explainable on the basis of productivity differences.

7. Economic discrimination cannot be practiced by employers or consumers without costs in a competitive environment. Government-sanctioned regulations and restrictions on competitors appear to account for a large amount of observed economic discrimination in the United States.

8. Contemporary welfare programs based on need do not provide positive incentives for the poor. Many economists support a negative income tax structure in place of the current system to provide work incentives to the poor.

Key Terms

individual income	transfer payment	in-kind transfer payments	poverty
labor income	family income	economic mobility	economic wage discrimination
asset income	Lorenz curve	per capita income	consumer-initiated discrimination
ex ante distribution	Gini coefficient	mean income	negative income tax
ex post distribution	adjusted income	age-distributed income	

Questions for Review and Discussion

1. What are the components of an individual's income? Which of these are matters of choice by the individual?

2. How does an individual obtain asset income?

3. What does a Lorenz curve show? What does it mean if the Lorenz curve moves closer through time to the line of perfect equality?

4. What is the difference between income as measured by government and adjusted income as reported by some economists?

5. What is economic wage discrimination? Give an example of wage discrimination that fits the description.

6. What are the problems associated with the use of Gini coefficients to report income distribution?

7. Why do young people have lower incomes than older people? If average lifetime incomes rather than annual incomes could be used to construct Lorenz curves, would the Gini coefficient be lower?

8. If every married couple in the United States obtained a divorce, would the Lorenz curve move away from the line of perfect equality? Would this alteration change the actual distribution of income?

9. Suppose all people were born with the same

amount of natural abilities and the same amount of inherited money wealth. Under these circumstances, what would the sources of income differences be?

10. Do income redistribution programs affect the supply of labor? How does a negative income tax compare to other transfer programs in its effect on work effort?

ECONOMICS IN ACTION

Who Are the Economic Minorities in the United States?

Who are the economic minorities in the United States, and on what does their relative position in income distribution depend? Economist Thomas Sowell has suggested answers to these questions in an extensive investigation of the characteristics of a number of minorities in America.[a] Sowell's thesis is that discrimination is not a good single explanation for why some minority incomes are above or below the national average. Other factors—some of them brought out in this chapter—must figure prominently in any explanation of income differences.

A part of Sowell's intriguing explanation is revealed in the following table. The table ranks ethnic origin (or ethnicity) with the ethnic group's average income compared to the average per capita income of the U.S. population as a whole. WASPs' incomes—the average income of white, Anglo-Saxon, Protestant Americans—is not reported in the table but amounts to 104 percent of national average income.

Ethnicity	Relative Income (percent of national average)	Median Age	Children per Woman
Jewish	172	46	2.4
Japanese	132	32	2.2
Polish	115	40	2.5
Chinese	112	27	2.9
Italian	112	36	2.4
Irish	102	37	3.1
Mexican	76	18	4.4
Puerto Rican	63	18	3.5
Black	62	22	3.7
American Indian	60	20	4.4

Source: Adapted from Thomas Sowell, *Markets and Minorities* (New York: Basic Books, 1981), Tables 1.1, 1.2, and 1.3, pp. 8, 11, and 16.

What factors explain the differences? According to Sowell, genetic color differences are not much of an answer. The income of the average black West Indian American (also not reported in the table) is 94 percent of the mean U.S. income, and the average Japanese American earns 132 percent as much as the average

[a]Thomas Sowell, *Markets and Minorities* (New York: Basic Books, 1981).

American. Contrast these relatively high earnings for some dark-skinned minorities with the earnings of light-skinned Puerto Ricans: only 63 percent of the mean income. Color alone cannot be a big factor explaining ethnic income differences.

According to Sowell, at least three factors explain differences in average incomes: (1) median age in the ethnic group, (2) locational concentrations, and (3) the fecundity of women (children per woman) within the ethnic group. The table above shows, in general, that the older the median age and the fewer children per woman in the ethnic group, the higher the median income of that group. We have seen in this chapter that in all groups, income earned varies with age differences. Income earnings generally peak between the ages of forty-five and fifty-five and perhaps at higher ages. We would thus expect those ethnic groups with the lowest median ages to earn less income than those with higher median ages. This age differential in part explains, for example, why Mexican Americans—with a median age of eighteen—earn 63 percent of average national income and why Jewish Americans—with a median age of forty-six—earn 172 percent.

In these age distribution statistics nevertheless lie hidden signs of hope for the condition of ethnic minorities. Attitudes toward minorities are clearly changing, but the effects of attitude changes on equal education and job opportunities are being felt largely by the younger members of ethnic minorities. As Sowell reports, blacks over forty-five years of age earn less than 60 percent of the sum earned by their age-peers in the population, but blacks aged eighteen to twenty-four earn 83 percent of their age-peers' income.

The number of children per woman is also a factor explaining the relative economic position of minorities. In general, those with lower relative incomes have a higher fecundity rate. The number of children can also have an impact on differences between family and per capita incomes among minorities. In family income terms, for example, Mexican Americans earn significantly more than black families. But in per capita terms, when the number of children per woman becomes a factor, the more fecund Mexicans earn lower incomes than blacks. Age and fecundity factors lead to other surprising statistics. For example, half of all Mexicans and Puerto Ricans in the United States are below the age of eighteen.

Geographic differences also account for relative income differences. As we saw in this chapter, greater poverty exists in the southern part of the United States. The same fact pertains to comparisons of ethnic incomes and poverty. As Sowell notes and as indicated in the table, Mexican Americans and Puerto Ricans earn more than blacks nationally, but blacks outside the South earn more than both those groups. The place one lives also makes a difference. American Indians living in Chicago or New York earn more than twice as much as those living on reservations. Mexican Americans living in Detroit earn more than twice the amount of those living in cities in the Rio Grande Valley of Texas such as Brownsville.

With arguments such as these, Sowell questions the extent to which discrimination is a function of others' sins. A number of factors explain the relative position of minorities, including age, number of children, and geographic differences. Discrimination cannot explain all differences. How, for example, could we explain Jewish and Japanese relative incomes in spite of rampant discrimination against these groups in the past and to a certain extent in the present as well?

Question

Explain why most generalizations about income distribution and minority incomes must be carefully analyzed and dissected.

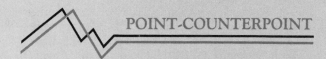

POINT-COUNTERPOINT   **Thomas Malthus and Gary Becker:**
The Economics of Population Growth

Thomas Malthus

Gary Becker

THOMAS MALTHUS (1766–1834) was born in Surrey, England, during the Industrial Revolution. It was widely believed at this time that the effects of industrialization, such as increased trade and specialization of labor, would eventually improve the quality of human life. When anarchist and pamphleteer William Godwin published his utopian outlook for society in his book, *Political Justice*, in 1793, an argument developed between Malthus and his father. Malthus's father was inclined to agree with Godwin's views, but Malthus believed that society was caught in a trap in which population would increase more rapidly than the food supply, leaving the standard of living at a subsistence level at best. In 1798, the same year he became a minister of the Church of England, Malthus published his views of population and the economy in his treatise, *An Essay on the Principle of Population*. It was this pessimistic treatise on the future of humanity that led essayist Thomas Carlyle to label economics "the dismal science."

Malthus went on to become the first professor of political economy in England at the East India College in 1805. His work on population influenced his friend and critic David Ricardo as well as Charles Darwin.

GARY BECKER (b. 1930), a current University Professor of Economics at the University of Chicago, gained international recognition for his work in applying microeconomic theory to areas such as marriage, crime, and prejudice. His book, *The Economics of Discrimination* (1971), introduced prejudice and discrimination as forces that can be analyzed and measured in their effect on the economy. In *The Economics of Human Behavior* (1976), Becker argues that all human behavior, even selection of a marriage partner, is based on economics.

In 1965, Becker authored his famous treatise on the concept of human capital. In *Human Capital*, Becker treats the individual as a "firm" that makes investment decisions (such as education or on-the-job training) on the basis of rate of return. Soon after the publication of *Human Capital*, Becker was awarded the John Bates Clark medal by the American Economic Association for excellence in research by an economist under the age of forty.

Viewed simply in economic terms, population growth represents sustained demand for children on the part of parents. Both Malthus and Becker have analyzed the decision to have or refrain from having children on the basis of rational self-interest. Their widely different conclusions offer insight into this timely issue.

The Check of Misery

Malthus's theory of population growth is strongly pessimistic. In *An Essay on the Principle of Population*, Malthus wrote that the world's population will increase at a rate that will ultimately test the limits of our available food supply and other subsistence goods. As the supply of food increases through additional labor and agriculture, population will increase as well, with the result that per capita income—the fruits of labor—will never exceed bare subsistence standards for the population as a whole. The threat of widespread famine will persist indefinitely. Only the lucky will survive.

An alternative way of stating Malthus's proposition is that children are what economists call normal goods, demand for which rises and falls in response to changes in income. Any increase in income to parents will increase their demand for children, other things being equal. At the point when an additional child will actually reduce living standards below subsistence—a point Malthus called the "check of misery"—parents will cease reproducing.

Given this gloomy scenario, it is not surprising that Malthus, a parson, would urge "moral restraints" on parents. Such restraints included abstinence and postponing marriage. It is interesting to speculate about whether Malthus would urge modern forms of birth control. The morality of contraceptive use, forced sterilization, and abortion is, of course, one of the burning issues of our times.

Malthus's theory of population growth seems today most applicable to the exploding populations of nations such as China, Mexico, Bangladesh, and other poor, developing countries. In the industrialized nations, including the United States, the last century has seen actual declines in average family size, while average family incomes have risen well above the level of bare subsistence. Many economists, including Gary Becker, have sought reasons why the theory has not, fortunately, become generally true.

The Price of Children

Becker's theory of human capital helps explain both the power and the limits of the Malthusian theory. For Becker, the choice of whether to reproduce has a second rationale, one beyond a simple change in income. Parents' demand for children will be influenced not only by changes in income but also by the relative "price" of children. This price is partly the direct costs of raising the child, but it also includes the opportunity costs that additional children represent. As the parents' income increases, the opportunity cost of children will also tend to increase. In other words, the sacrifices borne by increasing the size of a family will increase with income. To take one example, if an executive forsakes a high-paying job to stay home with the children so that his or her spouse can pursue a career, the cost of that decision is the wages and income the executive receives now and the higher income he or she could expect to receive in the future.

As the costs of children increase, the quantity demanded will decrease, other things being equal. As a result, parents may substitute quality for quantity in their decisions to raise children. Instead of feeding an additional child, parents might decide to spend additional income on housing, education, or any of a host of goods that improve living standards for their children.

Becker's approach to the costs of children helps explain why parents in more developed countries, with relatively high incomes, have fewer, better-educated children than parents in undeveloped countries with relatively low incomes. In undeveloped, largely subsistence agricultural countries, children represent direct labor inputs—even small children can do useful work on a subsistence farm—thus lowering the price of children. Where the price of children is lower, we expect to see more children produced, other things being equal. Thus we see the law of demand at work in a novel way.

According to Becker, then, Malthus went wrong by failing to take into account relative prices in his theory of population. While rats or horses may naturally tend to breed up to the limit imposed by the available food supply, the behavior of human beings will also reflect the influence of opportunity costs and the rational choices such costs inspire.

IV

Microeconomics and Public Policy

17

Market Structure and Public Policy

O ur market economy has grown in ways that Adam Smith could never have imagined. Smith, you may recall, believed that market forces, operating under a laissez-faire form of government, would eventually bring us to a point of perfect competition among firms, what Smith termed "a nation of shopkeepers." Our economy, however, has followed a different path, and its market structures range from nearly pure competition to nearly complete monopoly. In the process, a few of our "shopkeepers" have managed to capture huge shares of their markets. In 1882, for example, John D. Rockefeller's Standard Oil Trust controlled 95 percent of the oil refineries in the United States, virtually eliminating all competition. Such amassing of market power may temporarily allow lower prices through economies of scale and coordination of production—characteristics often necessary to undermine competitors. But, once such a monopoly is created, some believe that it can dominate the industry at the consumers' expense. Responding to public alarm over this potential for abuse, the U.S. government began late in the nineteenth century to pass laws designed to thwart the growth of monopolies. These laws have slowed, but not eliminated, the trend toward monoply power. When U.S. authorities intervened and broke up Rockefeller's Standard Oil Trust, it created several firms in its place. Today one of those firms, Exxon, is still the largest industrial corporation in the world, with income in 1984 of over $5½ billion.

Is big business bad for the economy? Should the government intervene in private markets to regulate the size and monopoly power of firms? How does the govenment determine whether a firm has monopoly power? Which government policies make sense from an economic point of view? All of these questions form the focus of this chapter. The overall question of monopoly regulation is certainly one of the major microeconomic issues of our time.

Measuring Industrial Concentration

Before it can intervene to break up, regulate, or even personally operate a monopoly in the public interest, the government must be able to determine whether a monopoly exists. The number of firms in an industry is a major indicator of the degree of competition. We examine two rough estimates of the degree of concentration of market power within an industry—aggregate concentration and industry concentration ratios. We also question why some industries exhibit high concentration ratios while others don't, noting that changing demands or costs will change the number of firms in an industry and thus the potential for monopoly concentration.

Aggregate Concentration

One method of gauging the extent to which production assets are dispersed among different firms is to examine the **aggregate concentration** of firms in the economy as a whole. These statistics measure the percentage of all business assets that are owned by the largest one hundred or two hundred firms in the country. This approach ignores the percentage of individual industries' assets that firms own and instead focuses on the percentage of the nation's total assets held by the largest enterprises. Instead of determining the degree to which General Motors controls the assets of the automobile industry, aggregate concentration is a measure of the degree to which General Motors, Allied Chemical, and other manufacturing giants control assets among all manufacturing firms.

Aggregate concentration: A measure of the percentage of total productive assets held by the largest one hundred or two hundred firms in the economy.

Table 17–1 shows trends in aggregate concentration for firms in the manufacturing sector between 1925 and 1977. The share of total manufacturing assets held by the top one hundred and top two hundred companies increased by about ten percentage points over the past fifty years, but the aggregate concentration has remained remarkably stable since 1958. Table 17–1 suggests that there is little cause for concern that the productive assets of this country are becoming increasingly concentrated in the hands of fewer and fewer firms.

Industry Concentration Ratios

Industry concentration ratios: An estimate of the degree to which assets, sales, or some other factor is controlled by the largest firms in an industry.

An alternative method of looking at business structure is to calculate **concentration ratios** (CR) for various industries. Concentration ratios are determined by ranking firms within an industry from largest to smallest and then calculating the share of some aggregate factor such as sales, employment, or assets held by the largest four or eight firms. Sales is the most common factor used in these calculations.

Figured by this method, the concentration ratio will lie between zero and one hundred. If an industry is totally monopolized, one firm will have 100 percent of industry sales, CR = 100. If an industry has several firms but the four largest have 50 percent of the market sales, then CR = 50. If a very large number of companies each have a very small market share, then CR will be close to zero, suggesting the existence of pure competition.

Table 17–2 displays four-firm and eight-firm concentration ratios for a small sample of manufacturing industries. A brief look at the concentration ratio data indicates that quite a bit of variation in business structure was hidden within the aggregate concentration percentages shown in Table

TABLE 17–1   Trends in Aggregate Concentration in the Manufacturing Sector, 1925–1977

To determine aggregate concentration, we measure the percentage of assets owned by the largest firms in the country. Since 1958, the percentage of assets held by the top one hundred and top two hundred firms in the U.S. manufacturing sector has remained remarkably stable.

Year	Percentage of Assets Held by the Largest Firms	
	100 Largest	200 Largest
1925	34.5%	—
1929	38.2	45.8%
1935	40.8	47.7
1939	41.9	48.7
1948	38.6	46.3
1950	38.4	46.1
1958	46.0	55.2
1963	45.7	55.5
1967	47.6	58.7
1972	45.3	56.5
1974	46.1	57.6
1975	46.0	57.3
1976	46.1	57.3
1977	45.6	56.6

Source: Richard Duke, "Trends in Aggregate Concentration," *FTC Working Paper No. 61,* (June 1982), p. 18.

17–1. Some industries are apparently very highly concentrated. In 1977, for example, the top eight producers of flat glass accounted for 99 percent of industry shipments, indicating that the remaining flat-glass manufacturers were very small businesses indeed. In contrast, the eight largest firms supplying ready-mix concrete had only an 8 percent share of total sales in 1977, suggesting that this industry is made up of a large number of relatively small companies.

Table 17–2 also reveals trends in concentration. Concentration ratios that have risen over the years indicate that the leading firms have grown relative to other industry members, either by internal expansion, merger, or decreases in the number of other firms. Declining concentration ratios imply that more and more competitors have made inroads into the industry and thereby reduced the dominance by leading firms. While examples of each pattern appear in the table, concentration ratios remain fairly stable in general.

One problem with using concentration ratios to determine the degree of competition in an industry is that the data can obscure quite vigorous competition. Since the ratios do not identify individual firms, the concentration ratio can remain stable over time even though the leading firms rapidly turn over. For example, if the first and fourth companies in meat packing change places, the four-firm concentration ratio would remain unchanged.

A further difficulty with the concentration ratios in Table 17–2 is that they are based on the Commerce Department's Standard Industrial Classification (SIC) groups. The SIC codes group firms that produce similar products, but these codes do not account for substitutability by buyers. For example, plastic panes may frequently be substituted for flat glass, but the SIC codes compare glass manufacturers only with other glass manufacturers. Thus, the concentration ratios do not show the competition between groups

TABLE 17–2   Selected Concentration Ratios in Manufacturing

Concentration ratios measure the share of sales held by the largest four or eight firms in an industry. The data shown here represent a selected group of manufacturing industries for the years 1947, 1963, and 1977. In some industries—such as men's and boys' suits—the concentration ratio increased through the years, suggesting that the largest firms obtained a larger share of total industry output. In some cases—such as meat packing—the ratio fell, suggesting a more competitive situation.

| | Concentration Ratios | | | | | |
| | 1947 | | 1963 | | 1977 | |
	Four-Firm	Eight-Firm	Four-Firm	Eight-Firm	Four-Firm	Eight-Firm
Meat packing	41	54	31	42	19	37
Fluid milk	–	–	23	30	18	28
Cereal breakfast foods	79	91	86	96	89	98
Distilled liquor	75	86	58	74	52	71
Roasted coffee	–	–	52	68	61	73
Cigarettes	90	99	80	100	–	–
Men's and boys' suits and coats	9	15	14	23	21	32
Women's and misses' dresses	–	–	6	9	8	12
Logging camps and contractors	–	–	11	19	29	36
Mobile homes	–	–	–	–	24	37
Pulp mills	–	–	48	72	48	76
Book publishing	18	29	20	33	17	30
Pharmaceutical preparations	28	44	22	38	24	43
Petroleum refining	37	59	34	56	30	53
Flat glass	–	–	94	99+	90	99
Ready-mix concrete	–	–	4	7	5	8
Blast furnaces and steel mills	50	66	48	67	45	65
Metal cans	78	86	74	85	59	74
Electric lamps	92	96	92	96	90	95
Radio and TV receiving sets	–	–	41	62	51	65
Motor vehicles, car bodies	–	–	–	–	93	99
Jewelry, precious metal	13	20	26	33	18	26
Pens and mechanical pencils	–	–	48	60	50	64

Source: U.S. Department of Commerce, Bureau of the Census, Census of Manufactures, 1977.

that results from actual market structures. Moreover, the SIC categories include sales data for the entire nation and therefore do not give information about concentration levels in different geographic regions. The four- and eight-firm concentration ratios were replaced in 1982 by the Herfindahl index, described in Focus, "New Merger Guidelines," but the Herfindahl index is subject to the same problems.

Differences in the Number of Firms

Why are there so few firms, sometimes only one, in some industries, while other industries have several hundred? The answer lies in the relation between the industry demand and the firm's average cost of production.

For simplicity's sake, consider a hypothetical industry in which all firms have identical products and costs of production. Let the curve AC_1 in Figure 17–1 represent one firm's average costs. The output that minimizes average costs, q_1, is the equilibrium output for each firm. In other words, as we recall from Chapter 8, the optimal size for the individual firm is the output level that allows the lowest possible per unit cost of production. On an industry-wide level, D represents the entire demand in the industry. The amount

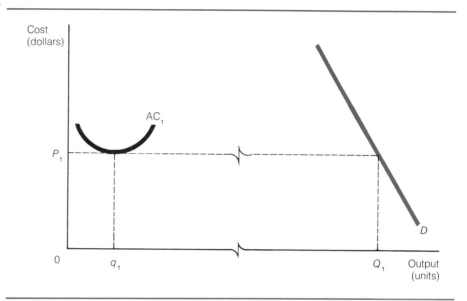

Cost (dollars)

AC_1

P_1

D

0   q_1 Q_1 Output (units)

FIGURE 17–1   Determination of the Number of Firms in an Industry

The equilibrium number of firms is determined by the magnitude of industry demand relative to the firm's average cost of production. The average cost curve for one representative firm is given as AC_1. The minimum average cost determines the optimal rate of output and plant size for each firm. The total number of firms is equal to the output per firm, q_1, divided into the amount demanded in the industry, Q_1, at the price that is equal to minimum average cost. If $q_1 = 100$ and $Q_1 = 10,000$, then the equilibrium number of firms in this industry would be 100.

Natural monopoly: Market conditions that allow for the survival of only one firm in an industry.

demanded in the industry at price P_1 is Q_1. The equilibrium number of firms in the industry is therefore Q_1 divided by q_1. If the optimal output for each firm is 100 units and the amount demanded in the market is 10,000 units, then 100 firms can exist in this industry.

The greater the industry demand relative to the firm's output at minimum average cost, the greater the number of firms will be. Highly concentrated industries result from industry demand that is insufficient to support a large number of firms of optimal size. In fact, a **natural monopoly** exists when industry demand is just great enough relative to a firm's average cost to support only one firm.

Figure 17–2 shows two types of natural monopolies. In Figure 17–2a, a natural monopoly develops because demand is insufficient to support more than one firm. The monopolies held by small-town newspapers provide a familiar example: Demand for local news in small towns often is not great enough to support more than one newspaper. In such a case, D_A represents the industry demand—the demand for local news in the small town; Q_a represents the quantity of newspapers demanded at price P_a—a quantity that can be produced profitably by one firm. AC_a and MC_a represent this newspaper firm's average and marginal costs. In this industry, only one firm can survive because industry demand is not great enough to support two firms. If another firm enters the market, industry sales will be divided by two. For example, at P_a the total industry output demanded is still Q_a, and thus each firm's sales are $Q_a/2$. Indeed, the marginal revenue curve represents each firm's demand curve when the industry demand is shared by the two firms. Since the individual firm's demand curve lies below AC_a at all levels of

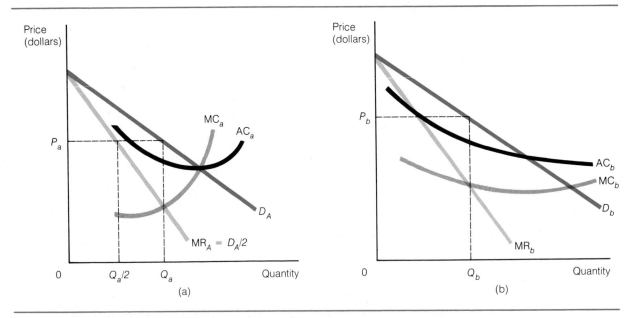

FIGURE 17–2   **Demand-Cost Relations of Natural Monopolies**

When industry demand is not great enough to support more than one firm of optimal size, then a natural monopoly results. (a) Only one firm can survive because half of industry demand, MR_A, the demand that each firm faces when there are two firms in the industry, does not rise above the firms' average cost. (b) A single firm survives because one large firm, through economies of scale, can achieve lower average cost than two smaller firms.

output, two local newspapers cannot profit and survive; only one firm can exist under these market conditions.

Figure 17–2b demonstrates a natural monopoly created by economies of scale. A single television company might dominate the market in a given area because it can supply the entire existing demand among customers at lower cost than two or more smaller firms. One large TV company serving many customers can therefore offer lower commercial prices than several firms serving fewer customers. The key feature of the model shown in Figure 17–2b is that the industry demand intersects the average cost curve at a point where average costs are still declining owing to economies of scale. Even though two firms could survive in this industry because their demand curves lie above AC_b at some points, we expect that price competition would eliminate one firm. The reason is that a single firm with a larger output can produce the entire industry output at a lower cost than two smaller firms. The existence of economies of scale yields a cost advantage for a single large firm.

Comparing the industry demand to the firm's average cost is a useful approximation of the number of firms that can exist in an industry. This method does have some limitations, however, for its assumptions may be unrealistic. For one thing, not all firms have identical costs. Firms in different parts of the country are confronted with different input prices and a different economic environment, for example. Consequently, the optimal size of the firm varies from one location to another, and the number of firms becomes difficult to predict. The assumption of identical products may also be false. If firms in an industry produce products that are not perfect substi-

 FOCUS   New Merger Guidelines

The revised merger guidelines issued by the Department of Justice in 1982 list conditions under which the department's Antitrust Division will "more often than not" challenge a proposed merger or acquisition. The conditions, which focus on the pre- and post-merger concentration levels, are expressed in terms of the Herfindahl index.

In brief, the Herfindahl index seeks to summarize in one number the extent to which a market is "concentrated" or dominated by a few firms. The index is calculated by squaring and then summing the individual market shares of all firms in the market. That is,

$$H = s_1^2 + s_2^2 + \ldots + s_n^2,$$

where, for example, s_1 is firm 1's percentage market share (usually its percentage of industry sales) and there are n firms in the market.

H will always be between zero and 10,000. If the industry is monopolized, one firm will have 100 percent of the market, so $H = 100^2 = 10,000$. In contrast, if each of a very large number of firms has a very small market share, H will be close to zero.

The advantage of the Herfindahl index over the more traditional four- and eight-firm concentration ratios is that it gives information about the dispersion of firm size within a market. The Herfindahl index does this by giving greater weight to large market shares than to small rather than treating all market shares sizes equally. A disadvantage of the Herfindahl index is that its calculation requires the gathering of information on the market shares of every firm in the industry.

The 1982 Justice Department merger guidelines label a market as concentrated if H is equal to 1,800 or more. (This situation would occur, for instance, if four firms in an industry accounted for approximately 50 percent of sales.) In such cases a merger that causes H to increase by 100 will generally be challenged by the Antitrust Division. In unconcentrated markets (where H is 1,000 or less), mergers leading to a change in H of 200 or more will usually trigger antitrust action.

Suppose that an industry is made up of three equally sized firms. The Herfindahl index would work out to

$$H = 33^2 + 33^2 + 33^2 = 3267.$$

A merger between any two of the firms would raise H by 1178:

$$H = 66^2 + 33^2 = 5445.$$

Since the industry was already concentrated by Justice Department standards, the merger would undoubtedly draw legal action.

In contrast, suppose that one firm accounts for 25 percent of sales in an industry and that one hundred other sellers each have a 0.75 percent market share:

$$\begin{aligned} H &= 25^2 + 0.75^2 + 0.75^2 + \ldots + 0.75^2 \\ &= 683.25. \end{aligned}$$

If the large firm merges with one of the smaller firms, H rises accordingly:

$$\begin{aligned} H &= 25.75^2 + 0.75^2 + 0.75^2 + \ldots + 0.75^2 \\ &= 719.56. \end{aligned}$$

This merger would generally go unchallenged.

Note that both the Herfindahl index and the merger guidelines are based on the assumption that firm size is an indication of market power. That is, concentration is considered bad because markets dominated by one or a few firms are thought to exhibit prices and profits exceeding the competitive level. Of course, firms can be large because they are efficient and serve consumers well. Thus, one should look beyond simple concentration levels before concluding that antitrust action is warranted.

---

tutes for one another, then industry demand and product price will be difficult to determine.

Changes in the Number of Firms

Since the number of firms in an industry is determined by the magnitude of industry demand relative to each firm's average cost, any change in demand or cost can lead to a change in the equilibrium number of firms. If the demand in a particular industry increases, then we expect to see more firms enter the industry. As demand for computers has soared, IBM's previous dominance of the computer market has decreased. On the other hand, if the price of inputs changes such that the average cost curve shifts to the right, then the optimal plant size increases. Such a shift will lead to a decrease in the number of firms. Changes in technology can also affect the optimal plant size and thus the number of firms. Such changes have led to a decrease in the number of automobile manufacturers in the United States.

Since the economy is dynamic, we expect fairly frequent changes in production cost and industry demand. The number of firms in various industries will therefore change through time. New firms enter and old ones go out of business. The concentration ratio of an industry may thus change because of events external to the firms.

Figure 17–3 gives an indication of the turnover rate in U.S. markets between 1929 and 1980. The turnover rate is the rate at which new businesses open and older ones that are failing close their doors. The first point to note in Figure 17–3 is that many new enterprises are formed and quite a few fail in any given year. In 1980, for example, more than 500,000 new firms were incorporated, and nearly 12,000 businesses went bankrupt. The business failure rate during the Great Depression was very high. Since then the rate of failure has slowed considerably, but failure rates are still significant. Starting a new business continues to be a high-risk proposition. There appears to be no shortage of entrepreneurs willing to take those risks, however, in hopes of succeeding.

Monopoly Profit Seeking

Entrepreneurs organize and operate businesses in an effort to obtain profits. The profit motivation is not socially undesirable, for businesses earn profits by producing products that are most highly valued by society while simultaneously using the fewest and least expensive resources. However, the level of profits is limited in purely competitive industries. When new firms are able to enter profitable industries and act independently in seeking profits, economic profits are forced to zero. On the other hand, firms that are natural monopolies may enjoy positive economic profits—a greater than normal return on their investments.

Even though profits in competitive industries are no greater than normal, there is a potential for obtaining greater profits by decreasing the amount of competition. If all the firms in a competitive industry could agree not to compete and simultaneously could prevent new firms from entering the market, monopoly-level profits could be obtained.

As explained in Chapter 10, converting a competitive industry into a monopoly results in loss of part of the consumer surplus. Consumer surplus is the amount that consumers would be willing to pay for a good above the price actually charged for it. This difference translates into real income to consumers. In Figure 17–4, a competitive market reaches equilibrium price and quantity for a certain good—perhaps lime juice—at price P_c and quantity Q_c, with D and S indicating industrywide demand and supply conditions. At the competitive price and quantity, consumers enjoy a surplus equal to FCP_c, the area beneath the demand curve but above the supply curve.

If competition is eliminated in the lime juice industry, firms acting with monopoly power can create contrived scarcity by restricting output to Q_m and raising the price of lime juice to P_m. At this level, consumer surplus shrinks to the area FAP_m, shifting the profits represented by P_mABP_c to the pockets of the lime juice monopolists. The area ACB is simply lost to everyone. It is neither transferred to monopolists in the form of profits nor re-

FIGURE 17–3

Business Formation and Business Failures, 1929–1980

During the Great Depression, the rate of business failures was quite high. The chances for survival have improved considerably since that time, yet business formation remains risky. The number of new incorporations has increased steadily since 1950, indicating that many individuals are still willing to take the risk.

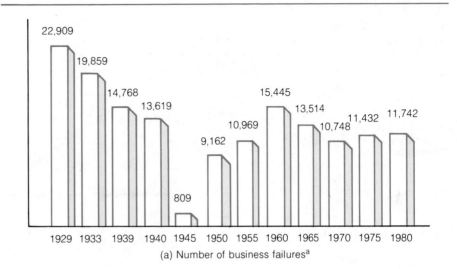

(a) Number of business failures[a]

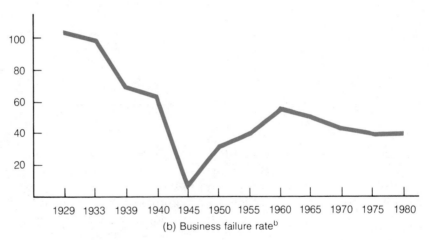

(b) Business failure rate[b]

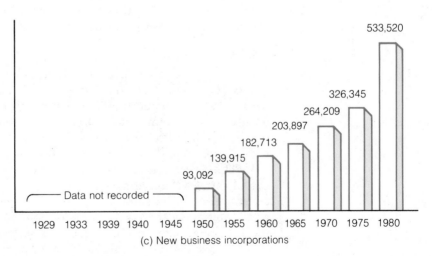

(c) New business incorporations

[a]Commercial and industrial failures only. Excludes failures of banks and railroads and, beginning in 1933, of real estate, insurance, holding and financial companies, steamship lines, travel agencies, and the like.
[b]Failure rate per 10,000 listed enterprises.
Source: Economic Report of the President (Washington, D.C.: U.S. Government Printing Office, 1982), p. 338.

FIGURE 17–4

Welfare Loss Resulting from Monopoly

With pure competition, the demand D and supply S intersect at C and the price and quantity at P_c and Q_c. The consumer surplus is FCP_c. If this industry is monopolized, then the profit-maximizing price and quantity are P_m and Q_m. Profits are the area P_mABP_c, and consumer surplus is decreased to area FAP_m. The welfare loss is the lost consumer surplus (ACB) that is not transferred to the monopoly in the form of profits.

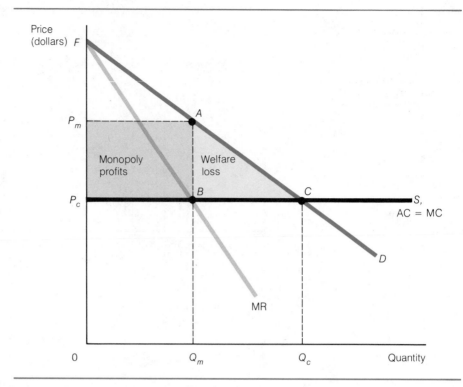

Welfare loss: The consumer surplus that is lost to consumers but not transferred to the monopoly in profits when a competitive industry is monopolized.

tained by consumers as consumer surplus. This area is therefore treated by economists as a **welfare loss** to society—to both consumers and producers. It is the cost to society of having a monopolist produce a product rather than having the good supplied by a competitive industry.

To capture monopoly profits, firms may engage in a variety of activities designed to decrease competition. As described below, anticompetitive behaviors include the formation of cartels, mergers, and trusts, and collusive agreements and government franchises.

Cartels

Cartels, as described in Chapter 11, represent a formal alliance among producers of a good. The firms agree not to engage in price competition and agree to restrict output in order to extract monopoly profits. To be successful, a cartel must prevent member firms from charging a price below the monopoly price. The cartel must also be able to prevent entry of new firms—otherwise, profits are dissipated by the entry of competitive firms.

OPEC (Organization of Petroleum Exporting Countries) was a fairly successful cartel. OPEC firms are prevented from cheating on the cartel by the governments of the countries in the organization. New firms are prevented from entering by the endowments of nature—new firms cannot produce oil unless they acquire an oil deposit. These features explain the longevity of OPEC. Most cartels that are not able to police cheating and limit entry are short-lived as OPEC's experience in the early 1980s proved.

The endowments of nature are unknown until exploration becomes profitable. Indeed, OPEC's initial success made exploration and development profitable to firms that formerly would not have entered. Consider the new production from the North Sea, Mexico, China, and the North Slope of

Alaska. These new "firms" have entered, and OPEC is now much weaker in part as a result of their entry.

Trusts

Trust: An institution that organizes firms in the same industry in an effort to increase profits by decreasing competition.

In the late 1800s, during the Industrial Revolution, the optimal size of firms—the size allowing the lowest possible per unit cost of production—began to increase. This trend lowered the number of firms in some industries and increased concentration ratios. When there are only a few firms in an industry, organizing them to act as a single firm becomes less costly. Firms in some industries joined holding companies called **trusts** that monitored the activities of members. For example, John D. Rockefeller capped his concentration of power in the oil-refining industry by forming the Standard Oil Trust, an organization to which stockholders of almost all oil-refining companies turned over controlling amounts of their stock in return for dividend-paying trust certificates. Such unifying organizations allowed firms to act as a cartel. Antitrust legislation, to be discussed later, has since been instituted to prevent most such organizations.

Collusion

Collusion: An explicit or implicit agreement between firms in the same industry not to engage in competitive behavior.

When there are very few firms in an industry but antitrust laws prevent formal organization, businesses sometimes engage in informal agreements to restrict output or raise prices, a form of **collusion.** Price competition can be diminished through explicit or implicit agreements. Firms may simply agree to charge a monopoly price and restrict output. The success of simple collusion is particularly rare since price cheaters and new entrants cannot be controlled informally.

Mergers

Merger: The joining of two or more firms' assets that results in a single firm.

One way that concentration ratios may rise is through **mergers.** If there are only a few firms in an industry, the combination of two or more firms into one may decrease the degree of competition. However, the types of mergers vary and thus their effect in the market varies. There are three basic types of mergers: horizontal, vertical, and conglomerate. Each type is distinguished by the relation between the products produced by the merging firms.

A horizontal merger occurs when the merger partners produce the same product. Examples include mergers between two beer producers, two manufacturers of glass containers, two retail grocery stores, and so forth. Horizontal mergers usually decrease competition because they have the direct effect of raising industry concentration ratios and moving the affected market closer to monopoly.

Mergers are characterized as vertical if the product of one firm serves as an input for the other firm, that is, the merging companies are at different stages of the production process. A vertical merger would occur if a company producing crude oil acquired a petroleum refinery, if a steel producer purchased iron ore deposits, or if an automobile manufacturer acquired a string of retail car dealerships. Vertical mergers do not raise industry concentration levels and are usually motivated by the prospect of lower costs. Accordingly, vertical mergers do not normally decrease competition.

Conglomerate mergers occur when the products of the firms are unrelated: for example, when an insurance company buys a firm that produces bread or if a cigarette producer merges with a soft drink company. The usual explanation for such mergers is that they are motivated by a desire to spread the risks associated with economic ups and downs across products. However, spreading risks can be accomplished much more simply by, for example, purchasing shares of a variety of businesses. Rather, conglomerate mergers can better be explained by the desire to diffuse managerial expertise across firms. That is, if an efficient corporate management team believes that their executive talents can be profitably applied to some other firm, they will have an incentive to acquire another company regardless of the product produced by the other firm. Since conglomerate mergers do not raise industry concentration levels, they do not generally decrease competition.

Government-Managed Cartels

Given the problems associated with organizing and policing a cartel in an industry with a large number of firms, it is not surprising that firms turn to the power of government when seeking profits. Sometimes firms are able to entice the government to enforce a higher-than-competitive price or a lower-than-competitive output or to prevent new firms from entering their otherwise competitive industry. Long before OPEC, for example, the Texas Railroad Commission limited the rate at which oil could be pumped from wells in Texas. This action bolstered price above the competitive level.

Public Policy

Purely competitive industries represent the ideal in market performance. In industries characterized by imperfect competition, the ideal is not obtained—profits may be greater than normal, production costs are not necessarily minimized, and welfare losses can occur. If these losses to society are significant, then there is a call for social action in the form of government regulation of some businesses' activities.

Public policies regulating business performance take many forms. Variety in policies is desirable because the imperfections in markets are caused by many diverse conditions. For example, the market conditions that create a natural monopoly differ significantly from those that promote a trust company or cartel. Even though the conditions vary, the final consequences of these market structures are indeed the same—monopoly price, restricted output, and welfare loss.

Public solutions to the monopoly problem take four basic forms:

1. Antitrust laws basically designed to prevent the creation of monopolies.
2. Price regulations designed to allow the existence of monopolies but to control prices and profits.
3. Government ownership that removes the firm from the private sector and thus places its performance directly in the hands of the public.
4. Laissez-faire policy that allows the market to determine the fate and performance of firms with no government interference.

Antitust Law

Antitrust policy: The laws and agencies created by legislation in an effort to preserve competition.

Antitrust policies are aimed at preventing firms from engaging in anticompetitive activities. During the 1870s and 1880s, an antitrust movement sprang up primarily among American farmers. Facing falling prices for their products and apparently stable prices for the goods they bought, they became fearful that monopolies or trusts were wielding unwarranted economic power. Through organizations such as the National Grange and the National Anti-Monopoly Cheap Freight League, the farmers were successful in influencing the two major political parties to add antimonopoly planks to their 1888 election platforms. Antitrust legislation was enacted soon after, with subsequent revisions to strengthen the government's ability to fight monopolistic practices in court.

Sherman Antitrust Act. The Sherman Antitrust Act of 1890, the first antimonopoly law passed by Congress, provided that "every contract, combination in the form of trust or otherwise, or conspiracy, in restraint of trade or commerce among the several states, or with foreign nations, is hereby declared to be illegal. . . ." The Sherman Act also declared "every person who shall monopolize, or attempt to monopolize . . . guilty of a misdemeanor, and subject to fine or imprisonment." The statute did not spell out what would constitute an illegal restraint of trade. Its main purpose was to place the antitrust issue under federal law and to provide a means of penalizing monopoly whenever it was discovered.

Clayton Antitrust Act. Believing that the Sherman Act was ineffective in preventing monopoly, Congress passed the Clayton Act in 1914. This act sought to limit a firm's acquisitions of the stock of another firm "where in any line of commerce in any section of the country, the effect of such acquisition may be substantially to lessen competition, or to tend to create a monopoly." Unfortunately, the Clayton Act was silent on the legality of mergers through the acquisition of physical assets, leaving a large loophole for avoiding antitrust prosecution.

The Clayton Act also enumerated some specific restraints of trade that could be challenged if their effects tended to lessen competition substantially or create a monopoly. These included (1) price discrimination—selling the same product to different customers at different prices not related to cost differences; (2) tying arrangements—making the sale of one good contingent on the purchase of some other goods; (3) interlocking directorates—the same person's serving on the boards of competing companies; and (4) exclusive dealing—selling to a retailer only on the condition that the firm not carry rival products.

Federal Trade Commission Act. Section 5 of the Federal Trade Commission (FTC) Act, also passed by Congress in 1914, contains the broadest statutory language, declaring illegal "unfair methods of competition in commerce." The act established a five-member commission independent of the executive branch, with the idea that the FTC would be a repository of economic and antitrust expertise not available to the federal courts.

Other Antitrust Statutes. The Clayton Act has been amended twice. In 1936, the Robinson-Patman Act revised the provisions relating to price discrimination and added language prohibiting predatory pricing, the act of selling goods below cost as a method of destroying competitors (see Focus, "Cutthroat Competition"). In 1950, the Celler-Kefauver Act closed the Clayton Act loophole relating to mergers through the acquisition of physical assets.

The FTC Act has been revised even more often. First, in 1938 the Wheeler-Lea Act added "unfair or deceptive acts or practices in commerce" to the behavior declared illegal. In the late 1970s, the Hart-Scott-Rodino Antitrust Improvement Act established a premerger notification system by which firms contemplating a merger would notify the FTC and the Department of Justice prior to consummating the acquisition.

FOCUS   Cutthroat Competition

Predatory pricing was declared illegal in the Robinson-Patman amendment to the Clayton Act. Predatory pricing—sometimes known as "cutthroat competition"—is said to exist when a firm lowers its price below its competitors' cost of production with the intention of driving competitors out of business. If the firm is able to do this, it may eventually become the only firm in the industry. It can then raise prices, and its subsequent monopoly profits can compensate for the losses it incurred during its predatory pricing period.

This activity was made illegal to prevent monopolies from arising in industries where competition could presumably exist. However, economic theory suggests that predatory pricing will not result in a monopoly unless the market conditions for a natural monopoly already exist. The only way that a single large firm can survive with no entry of new firms is to have lower average cost of production than the potential entrants. Otherwise, it must continually incur losses to prevent new firms from entering.

When a natural monopoly evolves in a market, one firm becomes larger, its price falls, and smaller firms go out of business. The term "predatory pricing" seems to describe these circumstances fairly accurately, but making cutthroat competition illegal does not change the market conditions that create a natural monopoly. Antitrust laws do not change the cost of production or the demand for the product.

Laws prohibiting predatory pricing are at best temporary hindrances to the emergence of a natural monopoly. Such prohibitions may indeed postpone the emergence of monopoly long enough for market conditions to change. Otherwise, outlawing price competition can have undesirable effects. For example, if a firm is prevented by law from charging a price that is below its competitors' average cost, then the law in effect may be protecting inefficient firms. Price competition is what forces the survival of only low-cost firms and thus low prices for consumers. Preserving competition by protecting firms from low prices ensures high prices for consumers.

Antitrust Agencies. The United States has a dual system of antitrust enforcement whereby two government agencies—the Federal Trade Commission and the Justice Department's Antitrust Division—share responsibility for policing the antitrust laws. Both agencies can enforce the Clayton Act, and, since 1948, the FTC has had the power to bring charges against behavior that violates the Sherman Act.

In addition to the two government agencies, complaints charging Sherman and Clayton antitrust violations can be brought by private parties. Private plaintiffs can sue for treble damages; that is, guilty defendants can be required to pay penalties of up to three times the value of the actual injury caused by their illegal conduct. Because private suits offer such potentially large rewards for successful plaintiffs, in any given year the number of private antitrust actions is many times higher than the number of cases brought by government.

The Value of Antitrust Policies

The antitrust laws and agencies have provided a means to control anticompetitive behavior in industries. Initially they were promulgated to prevent firms from organizing into trusts or cartels designed to increase members' profits by decreasing price competition. Antitrust policies have since been extended to prevent other activities that decrease consumer surplus, such as price discrimination and tying arrangements. Under most circumstances, the free market prevents these forms of competition. However, when concentration of market power is high, the market can fail to prevent such behaviors. Antitrust laws and agencies have therefore extended government control over industry with the goal of preventing increases in concentration.

Many factors external to firms can increase concentration, however. As we have seen, a fall in industry demand or a change in either input prices or technology that increases the optimal plant size can cause increases in concentration. These changes are beyond the control of the firms in the industry, and a decrease in the number of firms is indeed socially optimal under these circumstances. Thus, antitrust policies designed only to promote low concentration can be detrimental to economic efficiency.

Enforcement of Antitrust Policies

Antitrust policies are carried out in the courtroom. Antitrust officials, policymakers, and court officials face difficulties in distinguishing between behavior that is a conscious intent of firms to decrease competition and behavior that is the result of firms' responding to external changes. For this reason, the FTC treats every market situation as unique. Before the FTC takes action in altering market activities, it examines the individual firms and their activities. Firms judged to be in violation of antitrust laws may be ordered to stop their anticompetitive practices or to divest themselves of some of their assets by splitting the monopoly into several smaller competing firms. Sound economic theory is not always well represented in these court decisions. We will consider some specific cases that have been tried and the consequences of the decisions.

Cartel Policies. In 1927, the Supreme Court in the Trenton Potteries Case[1] held that price-fixing conspiracies were per se illegal under the Sherman Act even though the act did not specifically prohibit price fixing. That is, attempts by competitors to fix prices would be found illegal regardless of whether the fixed price was actually above the competitive level.

The case involved the activities of the Sanitary Potters Association (SPA), a Trenton, New Jersey trade association whose twenty-three member firms produced 82 percent of the U.S. output of bathroom sinks, tubs, and commodes. The SPA published price lists that included suggested discounts and surcharges for six geographic regions; members following the SPA lists charged identical prices.

The question addressed by the Supreme Court was whether a lower court judge was correct in directing the jury not to consider the reasonableness of the particular prices charged. The Court held that "uniform price-fixing by those controlling in any substantial manner a trade or business in interstate commerce is prohibited by the Sherman Law, despite the reasonableness of the particular prices agreed upon."[2] It seems that economic theory would support the court's decision. If a large percentage of the firms in an industry agree on price, then it is likely that price competition has been abolished.

Merger Policy. As we noted earlier, the Clayton Act, as amended by the Celler-Kefauver Act, seeks to limit corporate acquisitions that "tend to create a monopoly." Under the Hart-Scott-Rodino premerger notification system, most large firms contemplating a merger are required to apprise both the FTC and the Justice Department's Antitrust Division of their intentions. The agency chosen to handle the merger must decide whether to challenge it by seeking a preliminary injunction in federal court or to allow the merger to proceed without opposition. In the case of relatively small mergers, the

---

[1]*United States v. Trenton Potteries Co.*, 273 U.S. 392 (1927).
[2]See Phillip Areeda, *Antitrust Analysis: Problems, Text, Cases*, 3rd ed. (Boston: Little, Brown, 1981), p. 165.

antitrust authorities may become involved after the fact. In particular, a firm that is later found to have acquired another company in violation of the Clayton Act may be required to divest itself of the acquired assets in order to restore the premerger status quo.

An important aspect of merger law enforcement is the use of guidelines to establish which particular acquisitions will be challenged. For example, vertical and conglomerate mergers are not often challenged, but horizontal mergers that significantly decrease competition are. The merger guidelines issued by the Justice Department contain numerical standards that identify markets thought to be "unconcentrated" or "concentrated" prior to a merger. The guidelines set limits on the amount that concentration will be allowed to rise in each case before antitrust action is initiated. In 1968, concentration ratios were established for this purpose; in 1982, these ratios were supplanted by the Herfindahl index (described in the Focus essay earlier in the chapter).

In 1966, before these indexes were in use, the Supreme Court considered the legality of a merger between two Los Angeles retail grocery chains, Von's Grocery Company and Shopping Bag Food Stores.[3] At the time of the acquisition, 1960, Von's was the third largest grocery chain in Los Angeles; Shopping Bag was the sixth largest. Together, the two grocery retailers accounted for only 7.5 percent of sales in the Los Angeles area, however. Despite the relatively small market shares involved in the merger, the Court held that Von's had violated the Clayton Act. The court decision placed great weight on the fact that the number of owners operating single grocery stores in Los Angeles had declined from 5,365 in 1950 to 3,818 in 1961. This decline, the Court alleged, indicated a trend toward concentration.

In his dissenting opinion, Justice Potter Stewart called the Court's decision "a requiem for the so-called 'Mom and Pop' grocery stores . . . that are now economically and technologically obsolete in many parts of the country." He went on to say that the Court,

> through a simple exercise in sums, . . . finds that the number of individual competitors in the market has decreased over the years, and, apparently on the theory the degree of competition is invariably proportional to the number of competitors, it holds that this historic reduction in the number of competing units is enough under Section 7 of the Clayton Act to invalidate a merger . . . with no need to examine the economic concentration . . . , the level of competition . . . , or the potential adverse effect of the merger.[4]

If the number of firms in an industry is declining because of technological changes, economic theory suggests that the optimal size of firms is increasing. This change will naturally lead to an increase in concentration. According to economic theory, therefore, the merger probably should not have been prevented, especially since the two firms constituted such a small percentage of the market.

Price Discrimination. Charges of price discrimination may also require subtle distinctions. In a private suit, the Utah Pie Company charged Continental Baking Company,[5] Carnation Company, and Pet Milk Company with price discrimination in the sale of frozen pies. Utah Pie, located in Salt Lake City, alleged that its three California-based competitors sold pies shipped to

---

[3]*United States* v. *Von's Grocery Co.*, 384 U.S. 270 (1966).
[4]Areeda, *Antitrust Analysis*, pp. 961–964.
[5]*Utah Pie Co.* v. *Continental Baking Co.*, 386 U.S. 685, 699 (1967).

Salt Lake City at prices below those charged for pies sold nearer their own plants.

The evidence before the Supreme Court pointed to a highly competitive pie market in Salt Lake City. The price discrimination evidence appeared superficially correct in the sense that, given the cost of transportation, one would not expect the prices of the California firms to be lower in Salt Lake City. Accordingly, the Supreme Court found in favor of Utah Pie.

The Court's decision does not square with some of the facts, however. For one thing, the prices charged by Utah Pie were consistently lower than the prices of its rivals throughout the price discrimination episode. Second, Pet Milk Company suffered substantial losses on its Salt Lake City sales. Third, Utah Pie consistently increased its sales volume and continued to make a profit while facing the alleged anticompetitive practices of its rivals. Finally, prior to the entry of Continental, Carnation, and Pet, Utah Pie enjoyed a 67 percent market share in Salt Lake City, and while the new competition cost Utah pie a portion of its market, at the end of the price discrimination period the Salt Lake City firm still accounted for 45 percent of the pies sold in the area.[6] It appears that Utah Pie was using the antitrust policy to eliminate price competition and to gain a larger share of the market.

Natural Monopoly and Antitrust Cases. Antitrust policies may also be inappropriately applied to natural monopolies. The Aluminum Company of America (Alcoa) was involved in antitrust litigation as early as 1912 on allegations that it had violated the Sherman Act by, among other charges, monopolizing deposits of bauxite (the crucial ore in aluminum production), conspiring with foreign aluminum firms to fix world prices, and entering into exclusive contracts with power companies guaranteeing that the power companies would not supply electricity to any other aluminum producers. Finally in 1945, the U.S. Circuit Court of Appeals rendered its decision on charges brought by the government that Alcoa had monopolized the production and sale of virgin aluminum ingot.

The appeals court decision, written by Judge Learned Hand, is one of the most celebrated in American antitrust history.[7] Judge Hand, in overturning a lower court decision favoring Alcoa, wrote that even though Alcoa's monopoly was "thrust upon it" and the firm "stimulated demand and opened new uses for the metal," the firm should not increase production in anticipation of increases in demand. He contended that Alcoa prevented new firms from entering by increasing its capacity as demand increased and that because of Alcoa's lower cost, no new firms could effectively compete.

Although the court found Alcoa guilty, it refused to dissolve "an aggregation which has for so long demonstrated its efficiency." The "problem" of lack of competition in the aluminum industry was solved at the end of World War II, when the government sold to Reynolds Metals and Kaiser Aluminum the aluminum production plants it had set up for the war effort.

It is clear from this evidence that Alcoa had a natural monopoly and that **divestiture**—breaking the company into several smaller firms—was undesirable. Indeed, antitrust policies are generally ineffective when a natural monopoly exists. Antitrust legislation is best suited to stopping cartels, trusts, and conspiracies from eliminating competition when competition is the natural order.

Divestiture: A legal action that breaks a single firm into two or more smaller independent firms.

[6] Areeda, pp. 1070–1075.
[7] *United States* v. *Aluminum Co. of America,* 148 F.2d (2nd Cir. 1945).

An Economic View of Antitrust Cases

How well enforcement of the antitrust laws agrees with economic principles is a complex question with no simple answers. On the basis of our brief sketch of laws, agencies, and court interpretations, we can draw the rough conclusion that economic theory has not made a large impression on legal thinking. In the words of George Stigler, "Economists have their glories, but I do not believe that the body of American antitrust law is one of them."[8]

In particular, the courts seem to equate the degree of competition with the number of firms in an industry and to seek to protect competition by protecting competitors. This focus on the number of firms in an industry ignores the other dimensions of competition (number of buyers, entry and exit conditions, degree of product homogeneity, and so forth). More important, the numbers game neglects the possibility that firms can grow relative to their rivals not because they use unfair methods of competition or expand by buying out their competitors but because they can serve consumers better at lower cost.

Regulation

Antitrust policies are basically designed to maintain competition when competition can exist. Therefore, the antitrust policies are not very effective when a natural monopoly exists. How does the public avoid the potential problems of monopoly when only one firm can survive in an industry? Price regulation is offered as a solution to some of the problems that can arise under these circumstances. Governments may attempt to regulate prices to keep them low for consumers' sake or may even establish agencies controlling the activities of specific industries. But like antitrust litigation, these regulatory efforts may have unwanted economic effects.

Price Regulation. Public utilities, such as firms that provide local electricity, telephone, gas, and water services, are often considered natural monopolies because their provision requires economies of scale. In this case, Figure 17–5 illustrates the relation between demand for and cost of these services.

With no regulation, a firm holding a natural monopoly chooses the profit-maximizing price and output P_m and Q_m, which result in profits. If the local government chooses to regulate the firm to eliminate profits, then it may force the firm to charge a price that just covers its average cost of production.

This price regulation is achieved through an agreement between the firm and the local government—the government allows the firm to exist as the sole provider of services and, in return, the firm agrees not to allow profits to rise above a normal rate of return and also agrees to meet the total demand for the product. The resulting price and output, P_{AC} and Q_{AC}, occur where the industry demand crosses the average cost curve. Economic profits are zero since price equals average cost.

Average cost pricing: A form of price regulation that forces price equal to average cost and thus economic profits equal to zero.

This **average cost pricing** is indeed a very popular form of price regulation. It seems to eliminate one of the problems of monopoly. However, it does not result in the optimal level of output. Many economists believe that price should instead be set equal to marginal cost, in what is called **marginal cost pricing.** The price and output that result from the intersection of the demand curve and the marginal cost curve, P_{MC} and Q_{MC}, are theoretically optimal. That is, when price equals marginal cost, the benefit to society of one more

[8]George Stigler, "The Economists and the Problem of Monopoly," *American Economic Review* 72 (May 1982), pp. 1–11.

FIGURE 17–5

Two Forms of Price Regulation

An unregulated natural monopoly might charge the profit-maximizing price P_m, at which it would sell the quantity Q_m. With average cost price regulation, the price would be set equal to the firm's average cost, P_{AC}, increasing the quantity sold to Q_{AC}. Marginal cost pricing would result in even lower price and higher quantity: Price P_{MC} and output Q_{MC} result where the demand curve intersects the marginal cost curve. But such a policy would force the firm to operate at a loss, for the marginal cost is lower than the firm's average cost of production.

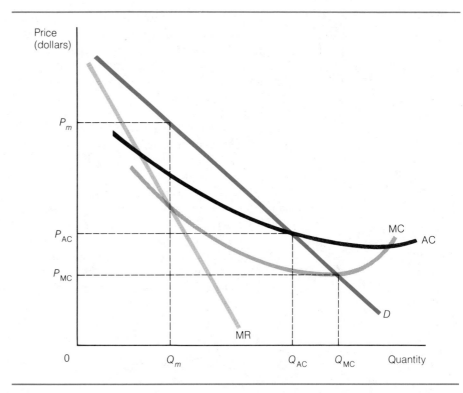

Marginal cost pricing: A form of price regulation that forces price equal to marginal cost and results in optimal allocation of resources.

unit of output is just equal to the extra cost of that unit. However, as in the case shown in Figure 17–5, marginal cost pricing may result in losses for the firm. One of the basic problems of price regulation is that monopolies cannot be forced into the ideal competitive solution where $P = AC = MC$. Since marginal cost pricing may result in losses, average cost pricing may be the preferred form of price regulation.

Another problem with price regulation is that when firms are forced to earn only a normal rate of return, their incentive to achieve efficiency is diminished. Maintaining low costs of production is not rewarded with higher profits. For this reason we may see public utilities allowing costs to rise.

In spite of the potential problems of price regulation, direct control of industry prices became popular in the late 1800s. It was believed that government regulation beyond antitrust legislation could promote more efficient behavior than could market forces.

Government as a Cartel Manager. Many government agencies have evolved through the years in an effort to regulate particular industries. In 1887 the Interstate Commerce Commission was established to regulate rates and routes of the U.S. railway industry, which was considered a monopoly. Since that time, the ICC has extended its control to the interstate trucking industry and the inland waterway transportation industry.

There are many more federal regulatory agencies intended to protect the public interest in specific industries. The Securities and Exchange Commission (SEC) regulates stock issues, brokers, and the major trading exchanges; the Federal Communications Commission (FCC) licenses radio and television broadcast rights; the Food and Drug Administration (FDA) controls the introduction of new drugs and regulates the purity of food and cosmetics; and the Federal Power Commission (FPC) regulates natural gas prices. Regulation is not confined to the federal level, however. State and local governments set standards and issue licenses for barbers, medical doctors, architects, teachers, and other professionals. Various state and federal agencies regulate public utilities (telephone, water, and electricity suppliers). The list could go on and on.

These agencies frequently were formed in an effort to control monopoly profits, but many have extended into competitive markets. In fact, we observe these agencies actually eliminating competition by forcing firms to charge a minimum (cartel) price. Their regulatory activities also prevent the entry of new firms into the industry. In a sense, the existing firms use the regulatory agency to police a cartel that is beyond the jurisdiction of antitrust legislation. Before the deregulation of the airline industry, for example, the Civil Aeronautics Board (CAB) prevented airlines from charging air fares lower (or higher) than those set by the agency.

Government regulation is often justified as a consumer-protection measure when, in fact, it usually ends up being a protection device for the producer. The prevalence of regulatory intervention in everyday life requires one to keep in mind that whether one believes regulation is good or bad from a social point of view, its economic effects are often to restrict output and raise prices.

Government Ownership

Socialization: Government ownership of a firm or industry.

An alternative to antitrust policy or regulation is government ownership, or **socialization.** When monopolies exist and their performance is less than socially optimal, government may choose to own and operate the firm in the interests of the public. We see this happening occasionally, more often at the state or local level than at the national level. For example, many public utilities such as electric services, water, and natural gas are owned and operated by local governments.

Socialization of firms is a solution to the monopoly problem, but is it an efficient solution? Ideally, the firm should approximate the competitive equilibrium. It should minimize its costs of production and charge a price equal to marginal cost. However, the performance of the firm depends on its managers. The managers, of course, must operate within the bounds allowed by the politicians who employ them, the voters, and the buyers of their services.

The efficiency of publicly owned firms is questioned by many economists. Many suggest that there is a lack of incentive for efficiency because there are no residual claimants. That is, no one may receive any profits acquired by the socialized firm, so managers have no reason to ensure low costs of operation or product quality.

To judge the efficiency of publicly owned firms, it is possible under some circumstances to compare their price, costs, and quality of service to those of privately owned firms in the same industry. For example, we can compare a locally owned electric firm's price and service to a privately owned electric firm's price and service. Many industries have both privately owned and publicly owned firms, such as education, medical care, telecommunications, television, garbage services, and employment agencies. However, in most situations we must compare government-owned firms with government-regulated firms, as in the case of electricity.

The U.S. Postal Service is a socialized firm that we cannot compare directly to a private firm, for the government prohibits the delivery of first-class mail by anyone except the U.S. Postal Service. However, there are many firms wishing to enter the mail delivery market, and occasionally they do so illegally. The suggestion that private firms could make a profit at the same price whereas the U.S. Postal System does not is an indication that the Postal Service is run inefficiently. Indeed, many government-owned firms must be subsidized by tax dollars.

Despite its lack of incentives for efficiency, socialization is still considered by some supporters to be an appealing alternative to trust busting or regulation in monopoly-prone industries. The price of medical services has been rising, and socialized medicine has been offered ás a solution. AT&T was recently divested through an antitrust case, but an alternative could have been to socialize telephone services. The possible consequences of such actions can be predicted by observing the prices and quality of services of similar industries that have been socialized in other countries.

Laissez-Faire

Antitrust policies, regulations, and government ownership are not the only alternatives to the monopoly problem. Under many circumstances, the forces of the market are capable—given enough time—of eliminating the undesirable performances of monopolies, cartels, and trusts. Leaving an industry alone so that it can clear itself of monopolist inefficiencies and high prices is called **laissez-faire** (allow to do) policy.

Laissez-faire: A government policy of not interfering with market activities.

Natural Shifts in Market Power. To understand how markets can be self-correcting, let us examine the circumstances facing a natural monopoly.

Natural monopolies exist because of the special relation between industry demand and the firm's average cost. In an industry in which the demand curve intersects the firm's average cost curve at a point where average cost is still declining due to economies of scale, as in Figure 17–6, only one firm will survive, and profits will exist. These monopoly profits motivate competition from other firms.

New firms will want to reap the potential profits, and their profit-seeking activities can limit the welfare loss due to monopoly. For example, many entrepreneurs may attempt to discover and produce substitutes for the monopolist's product, or they may attempt to create a new technique of production that allows them to produce the same product on a smaller scale of production with equally low cost.

Without the threat of entry, the monopolist may charge price P_M, the typical profit-maximizing price. However, if new firms are on the verge of entry, then the monopolist may engage in **limit-entry pricing.** That is, the monopolist may drop to a price lower than P_M, such as P_E in Figure 17–6. The entry-limiting price P_E must be low enough so that new firms are not able to survive. Though designed to squeeze out potential rivals, this activity in effect allows the monopoly to maximize its long-run profits since demand is infinitely elastic at price P_E in the long run. If the firm raises prices above P_E, it will lose customers through time to new entrants.

Limit-entry pricing: A pricing policy by a firm that discourages entry of new firms by selling at a price below the short-run profit-maximizing price.

The monopoly power of a firm may diminish in the long run anyway. The monopoly exists because of the particular demand-cost relation. Through time we expect demand and costs to change. New firms can enter and survive if industry demand increases or if prices of inputs and technology change in such a way that the optimal plant size decreases. Because the economy is truly dynamic, we expect to see the optimal number of firms and concentration ratios rise and fall in many industries, including monopolistic industries.

FIGURE 17–6

Limit-Entry Pricing

The monopoly with no threat of entry charges price P_M and sells quantity Q_M. However, the threat of entry of new firms may be diminished if the monopoly charges a price, such as P_E, below the potential entrant's average cost. The entry-limiting price is still above the monopoly firm's average cost, which is kept low through economies of scale. Limit pricing in effect maximizes the firm's long-run profits and diminishes the adverse social effects of a monopoly.

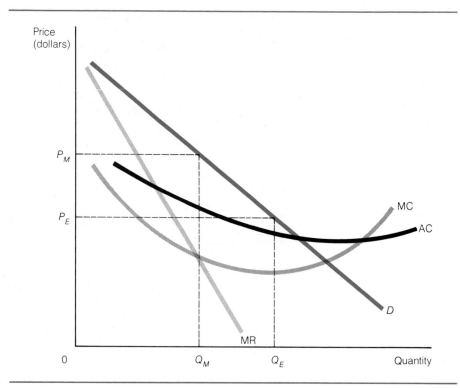

Just because a monopoly exists now does not imply that it will be there forever. According to laissez-faire thinking, there may be no need to intervene in certain monopolized industries because the monopolies may eventually disappear anyway. Focus, "What Happened to the IBM and GM Monopolies?" gives two recent examples of declining monopolies: General Motors and IBM.

The situation illustrated in Figure 17–6, with the demand curve intersecting the average cost curve where there are economies of scale (AC is falling), represents the usual natural monopoly. The figure also represents the situation believed to exist for public utilities such as electric companies. If there are economies of scale, then firms with low rates of output, such as electric companies in small towns, should have higher average cost and thus higher prices than electric companies in larger towns with high rates of output. If we do not observe this relation, then natural monopolies no longer exist, and the firms are monopolies only because the local authorities grant them monopoly rights. Indeed, it is difficult to find a free-market monopoly; that is, one that exists without monopoly rights granted by government.

Deregulation. In addition to simply leaving monopolized industries alone, a laissez-faire policy might dictate removing previous regulations—**deregulation** of the industries. As one would expect, firms in regulated industries often attempt to use the regulatory agencies to promote the firms' profits. Government regulation of industry often serves to protect the very companies that antitrust laws and regulations are intended to suppress. In particular, regulatory limits on the entry of new firms into an industry and on the prices that are charged by existing companies can give these firms a degree of market power they could not otherwise achieve.

In view of the fact that antitrust laws and regulations have been misused,

Deregulation: A situation in which government ceases to regulate a previously regulated industry in an effort to improve the performance of that industry.

FOCUS   What Happened to the IBM and GM Monopolies?

In the 1960s IBM was essentially the only U.S. supplier of computers. For many years IBM enjoyed a lack of competition, which of course resulted in higher-than-normal profits and a call for antitrust action by IBM's customers and its few small competitors.

Similarly, in the 1960s, General Motors gained a disproportionately large share of the U.S. auto market. The firm was large and getting larger because it was making the most popular cars at the most popular prices. Through the 1950s and 1960s several smaller auto companies such as DeSoto and Packard went out of business. It seemed that GM was gaining a great deal of monopoly power. Antitrust authorities were contemplating drastic actions.

Customers, competitors, and antitrust lawyers suggested that IBM and GM should be divested of some of their market power by breaking each firm into several smaller firms. For example, an antitrust action could have made Chevrolet, Buick, Pontiac, Oldsmobile, Cadillac, and GMC Trucks all independently owned companies rather than divisions of General Motors. Such an action would have increased the number of firms and lowered concentration in the automobile industry. A similar action could have divided IBM into several smaller companies.

However, as we so frequently observe, market conditions changed. Under a laissez-faire government policy, both IBM and GM have lost their market power. Each now has many competitors. New firms were able to enter the market because demand or cost conditions changed. In the computer industry, the demand for computers increased along with tremendous changes in technology. IBM is no longer the only name in computers. The market thus solved the computer monopoly problem.

Competitors from abroad have of course decreased the market strength of General Motors. The expansion of production and technology in both Asia and Europe cured the domestic monopoly problem. Under some circumstances, free international trade is the ultimate weapon in destroying monopoly power. In fact, public policy has flip-flopped from an effort to diminish GM's profits to an effort to bolster them with import quotas.

The experience with IBM and GM leads us to believe that the market itself can decrease the monopoly power of firms through time. Industry concentration fluctuates through time and a "leave it be" policy can benefit the public in the long run.

it is not surprising that we have observed a trend toward deregulation of industry. Two recent examples of limited deregulation have occurred in the interstate trucking and airline industries. Until 1980 the Interstate Commerce Commission severely restricted entry of new firms into the trucking industry as well as price competition between firms. Since the Motor Carrier Act of 1980 decreased the ICC's ability to restrict new entry, the number of firms has increased by several thousand and price has fallen by 20 percent and more. In the airline industry, the Civil Aeronautics Board set passenger fares and restricted entry until the Airline Deregulation Act in 1978. From that time until the CAB's demise in late 1984, the CAB's ability to restrict price competition and entry was severely limited. The results of deregulation in the airline industry have been similar to those in trucking. The number of firms has increased and prices have fallen. Since deregulation, some trucking and airline firms have also gone bankrupt. One would expect that competition would eliminate the inefficient firms. Nevertheless, concentration has decreased in both industries.

Conclusion

A purely competitive free market is impersonal and unforgiving. It forces efficient behavior by eliminating firms that are not able to achieve the minimum cost of production. No special-interest groups can reap profits because a competitive market acts as an impartial arbitrator in the task of resource allocation. However, as we have seen, certain market conditions can lead to the survival of only a few firms. In such cases, pure competition cannot exist, and firms are not forced into the socially optimal solution.

We have examined several forms of public policy that are aimed at preventing poor market performance. Each of these—antitrust, regulation, socialization, and laissez-faire—has its merits, but there is no public policy that can achieve the efficiency generated by a competitive market. When imperfect competition exists, our best solutions are themselves imperfect.

Not only are our best solutions theoretically imperfect, but actual government policies appear to be schizophrenic. One action attempts to increase competition and reduce profits while other actions promote monopolies and their profits. The reason for these contradictions is that public policies are formulated in a political arena and not by impartial economic theorists or forces. An understanding of public choice theory—to be explored in Chapter 20—can tell us more about public policies in monopolized industries.

Summary

1. Aggregate concentration ratios have not increased significantly in the past several decades. This suggests that big businesses have not been gaining greater control over the nation's productive assets. Industry concentration ratios change through time because the optimal number of firms changes as demand and cost conditions change.

2. The number of firms in an industry is determined by the number of optimal-sized firms that industry demand is able to support. A natural monopoly exists when industry demand is not great enough to support two or more firms of optimal size.

3. Firms seek to increase profits by increasing their share of industry demand, by decreasing production costs, and by decreasing the number of competitors. They may achieve these goals in a variety of ways—cartels, trusts, mergers, collusion, and government grants of monopoly power.

4. The social welfare loss due to monopoly calls for public policy to end anticompetitive practices. Active public policies include antitrust legislation, regulation, and socialization.

5. Antitrust laws are basically designed to promote and maintain competition when competition is possible. Thus their main goal is to decrease concentration, a goal that is not always desirable.

6. Price regulation allows the existence of monopoly but attempts to decrease the social loss by forcing a particular price. However, price regulation may decrease the firm's incentive to maintain efficiency. Regulation of competitive industries has resulted in government-managed cartels.

7. Government ownership as an alternative to monopoly can prevent higher-than-normal profits. However, a government-owned business is not forced by the market to maintain efficiency.

8. A laissez-faire system has been offered as an alternative to active public policy. Under some circumstances, this policy to leave the market alone is clearly preferable. Under other circumstances, concentration could increase under laissez-faire policy.

9. All public policies have their shortcomings. When competition does not and cannot exist, then no public policy can force the competitive solution.

Key Terms

aggregate concentration	trust	divestiture	laissez-faire
industry concentration ratios	collusion	average cost pricing	limit-entry pricing
natural monopoly	merger	marginal cost pricing	deregulation
welfare loss	antitrust policy	socialization	

Questions for Review and Discussion

1. Suppose that the output of an entire industry were produced by one firm but there were no barriers to entry. Should the antitrust authorities be concerned about this industry?

2. What does market power mean in antitrust analysis? How can this concept be used to assist antitrust enforcers in deciding where to look for monopoly power in the economy?

3. What is the Herfindahl index of concentration? How is it used by government to screen mergers?

4. Do a few large firms control all the wealth in the economy? Does the identity of these firms remain constant over time?

5. In the United States there is dual antitrust enforcement by the Federal Trade Commission and the Department of Justice. Discuss the pros and cons of having two antitrust enforcement agencies.

6. What is a conglomerate merger? What is the rationale for conglomerate mergers?

7. Suppose that a merger reduces the costs and increases the market power of two firms. Can you show this shift graphically? How should the antitrust officials decide such a case?

8. Why does government regulation sometimes act to enforce a cartel? Why do voters and consumers put up with such programs and policies? What effect would deregulation have on such an industry?

9. Blue laws outlaw Sunday sales in a given locality. From the point of view of economic regulation, explain who gains and who loses from such legislation.

10. How does rate-of-return or profit regulation affect firm behavior? Does this regulation promote production efficiency? Do we achieve the optimal output with average cost pricing?

11. Are local television companies natural monopolies? How could the answer be determined? If they are, would they want to be regulated?

12. Can public policy change an industry characterized as a natural monopoly into a competitive industry that results in competitive equilibrium?

ECONOMICS IN ACTION   The Breakup of AT&T

For most of the twentieth century, the American Telephone and Telegraph Company (AT&T) was the regulated quasi-monopoly supplier of both local and long-distance telephone service in the United States. In 1974, for example, AT&T supplied almost 80 percent of U.S. telephone service.

In January 1982, AT&T struck a deal with the U.S. Justice Department ending an antitrust suit that the department filed in November 1974. By the terms of the agreement, the suit was dropped, and AT&T agreed to divest itself of its local telephone service subsidiaries. The twenty-two local Bell Telephone operating companies will continue much as before, organized into seven large regional holding companies. The "new" AT&T after January 1984 (when the divestiture took effect) has assets on the order of $35 billion and continues to supply long-distance telephone service, but in competition with other suppliers. Competing non–Bell long-distance services have been gradually entering the market over the last several years. The AT&T divestiture has eliminated some barriers to entry facing these firms, mostly by mandating "equal access" to local telephone facilities for all suppliers of long-distance service.

Meanwhile, the local telephone companies "spun off" from AT&T will still be subject to state and federal regulations, but their holding companies will be able to invest in areas previously prohibited. Some of these companies, for example, have entered the computer business.

AT&T voluntarily agreed to the divestiture, but one of its main goals in doing so may have been to end its expensive legal battle with the Justice Department. As

An advertisement for one of AT&T's competitors in the long-distance phone market. Although AT&T still dominates the industry, many new companies have entered the market in recent years.

things stand, it would be inaccurate to claim that divestiture has in and of itself significantly increased the degree of competition within the telephone industry. By the end of the 1960s certain forms of long-distance communication in addition to those supplied by AT&T had begun to be approved by the Federal Communications Commission. These developments accelerated during the 1970s, and it is not clear that divestiture itself significantly lowered entry barriers in the long-distance telephone service industry. Furthermore, local telephone service continues to be supplied by local companies that are effectively regulated monopoly suppliers, even though they are now independent concerns. Pacific Telephone in California and New England Telephone are now each a part of independent companies, but each has an effective monopoly in the supply of local telephone service in its respective operating area. In other words, divestiture has not introduced competition in the local telephone service market, which remains basically controlled by a set of regionally regulated monopoly firms.

It will be some time before a final assessment of the economic impact of forced divestiture in the case of AT&T can be made. One important fact is that the cost of a long-distance call has gone down. Beyond this, the aspects of the breakup likely to have the most significant positive effects are those relating to the relaxation in operating restrictions on both the "new" AT&T (now permitted to enter the fast-growing information processing industry) and the regional holding companies. It remains to be seen whether these consequences will outweigh any losses in efficiency the "old" AT&T may have enjoyed due to economies of scale.

Question

Why do you think AT&T went along with forced divestiture rather than continue to fight it in the courts? What will happen to local phone rates as a result of the divestiture? Why?

18

Market Failure and Public Policy

Market failure: A situation in which a private market does not provide the optimal level of production of a particular good.

Chapter 17 addressed some of the ways that private markets can fail to achieve ideal economic results. Monopoly was the primary culprit in this discussion, with its attendant restriction of output and high prices causing a reduction in the economic welfare of society. Monopoly is a form of market failure, that is, a failure of private markets to provide goods and services in the right quantities at the right prices. It is not, however, the only type of market failure that is possible in an economy.

Market failure typically arises from problems of incomplete or nonexistent ownership rights to basic resources. Owners of private property bear directly the economic results of the use of property and are therefore motivated to use resources efficiently. In contrast, when property is held or used by everybody, users do not bear the full results of its use. Common ownership creates a conflict between the pursuit of self-interest and the common good.

Consider some examples. We are all familiar with the problem of litter in public parks or along city streets. Littering occurs in such places because each individual is not directly assigned property rights for the use of common property. We do not ordinarily allow litter to pile up indefinitely in our living rooms or automobiles because we are forced to bear the costs of our actions. We are not completely forced to bear the costs of public littering, however. As we know, fines and threat of arrest have failed, in many cases, to significantly reduce unsightly litter from our environment. Stream or water pollution by manufacturers or refiners, noise pollution by aircraft around airports and elsewhere, and overfishing in coastal waters are also examples of market failure. In some cases government regulation and intervention can be applied to correct market failures; other market failures may simply require a redefinition of who owns the rights to basic resources, after which the private market will work fine. Still other complex intermediate solutions emphasize public-private interaction to control market failures.

In this chapter we review all these approaches to market failures. We do not live in a perfect world, and to improve on situations of market failure we must carefully weigh the costs and benefits of alternative policies. Our discussion begins with an analysis of the problem of commonly owned resources and how this problem is approached in two concrete cases: the common stock of fish in international waters and the oil pool.

The Economics of Common Property

Common ownership: Lack of clearly specified ownership rights of a resource for whose use more than one person competes.

The fundamental economic disadvantage of **common ownership**—the right to the use of a resource by anyone or by competing users—is the lack of incentive to invest in the productivity of the common property. Rather, each person with access to the resource has an incentive to exploit the resource and neglect the effects of his or her actions on productivity. This principle applies to both renewable resources such as fish and nonrenewable resources such as oil.

Fishing in International Waters

In international waters anyone may acquire exclusive ownership of a fish by catching it, and no one has rights to the fish until it is caught. This system causes each fisher to ignore the effect of the well-known biological law that the current stock of a species of fish determines its reproduction rate. Thus, fish harvested today reduce today's stock and thereby affect the size of tomorrow's stock and tomorrow's harvesting costs and revenues. Overfishing results because no single fisher has an incentive to act on this bioeconomic relation. If all fishers would cooperate to restrain themselves now, tomorrow's stock would be larger and future harvesting cheaper and more profitable. However, under competitive conditions, each individual knows that if he or she abstains now, rivals will not, and much of the effect of abstention is lost. Further, the reduction in future costs would accrue to everyone, not just to the abstainer. Hence each person has little reason to abstain, since the major effect is to lower others' future costs at some immediate present cost to the abstainer.

Sole ownership removes this dilemma. If there is only one person fishing, he or she need not worry that rivals will not abstain. Alternatively, all the people involved could negotiate an agreement to abstain, and all would benefit. The fact that such agreements are not usually successfully negotiated is due to the costs of dealing with all current and potential people who will fish and the difficulty of ensuring that all abstain as agreed. A third method of dealing with this dilemma is government regulation.

Such regulation usually sets an annual catch quota. To allocate the quota, regulators rely on restrictions on technology or simply close the season when the quota is taken. Both means create difficult enforcement problems and potential economic waste. If the season is closed when the quota is taken, for example, excess profits are dissipated in competition among fishermen to buy bigger, faster boats and thus get a larger share of the quota. The season progressively shrinks, and resources stand idle or are devoted to inferior employments.

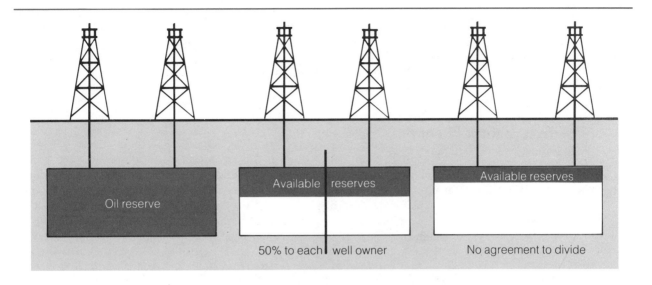

FIGURE 18–1   The Problem of Common Ownership in Oil Drilling

If a fixed reserve of oil is owned in common, and the co-owners cannot agree to share the reserves, then each owner will likely try to pump as much oil as possible from the reserve. Doing so reduces the available reserves at a rate disadvantageous to both owners.

Drilling for Oil

Oil presents another version of the same common ownership dilemma (see Figure 18–1). Technically, oil, unlike fish, is nonrenewable. However, oil does move about, so pumping one well lowers the entire reserve and affects the pressure below other wells. These characteristics create a problem when more than one person has rights to the same oil reserve. For example, if the land above a reserve of oil is owned by two people, each has the right to drill for oil. But the only way to own the oil is to pump it to the surface. Oil in the ground belongs to the landowner only when that person pumps it. Herein lies the crucial lack of private ownership rights to oil reserves.

It may be more economically efficient to exploit an oil reserve with a single well than with two wells not only because it costs more to drill two wells than one but also because the existence of one well interferes with the other's technical ability to pump. If one person owned all the property rights, he or she would probably use only a single well. Likewise, if two landowners with rights to a common oil pool negotiate, both can be made better off by agreeing to split the proceeds of a single well rather than both drilling.

If they cannot negotiate, perhaps because of legal restrictions, or if a larger number of owners causes difficulties in communication, each may go ahead and drill. Once two wells are in production, there is an economically efficient rate of pumping for each well that a single owner would adopt to maximize his or her wealth. Nevertheless, if the two owners cannot agree on efficient exploitation, each will be tempted to pump oil at too high a rate, a tendency reinforced by the fear that the other is doing the same thing. In fact, even if negotiations are successful, it may be costly to monitor each other's pumping, and the agreement might dissolve in mutual cheating. The costs of negotiating and enforcement probably rise in proportion to the numbers of owners, as do the incentives for each to overpump because others are likely to do so.

The lessons to be learned about common property resources are straightforward: What belongs to everyone belongs to no one, and the deterioration and rapid exploitation of common property resources are clear implications of economic theory.

Externality Theory

Externality: Spillover costs or benefits.

When resources are not owned privately, a market failure called externality can occur. An **externality** is an economic situation in which an individual's pursuit of his or her self-interest results in costs or benefits to others. The costs and benefits are external to the individual who caused them. Obviously, any economic entity can create an externality, including firms, government agencies, and individuals.

Negative and Positive Externalities

A common example of externality is a factory's emission of smoke as a by-product of production. If the factory is able to avoid responsibility for the consequences of its smoke—such as the expenses for painting soot-covered buildings nearby or laundering sooty clothes—the output of the factory will be excessive in that the value of additional output is less than the cost of the damage. If the factory were held responsible for the damage done by its smoke, its costs would rise and its product would become more expensive. This example illustrates a **negative externality,** a situation in which production entails costs for others, such as those living near a factory. Other potential examples of negative externalities are the exhaust emissions of automobiles, a neighbor who plays a stereo too loud, low-flying jet airplanes landing at an airport, cigarette smoke in a crowded elevator, drunk drivers, the contamination of underground water by toxic chemical dumps, and so on.

Negative externality: A situation in which the social costs of producing or consuming a good are greater than the private costs.

Positive externality: A situation in which the social benefits of producing or consuming a good are greater than the private benefits.

Positive externalities are social benefits accruing without costs to the beneficiaries. Suppose a beekeeper and the owner of an apple orchard have adjoining land. The beekeeper's bees fly into the orchard and pollinate the apple blossoms, making the orchard more productive. If the beekeeper were to reap the consequences of the increased value of the orchard, the yield on his investment in bees would rise, and he would raise more bees. But as it is, his bees' social service is provided free. For both illustrations the point is identical: Externalities arise when one person's activities affect the well-being of others, either positively or negatively. Some other examples of positive externalities are the impact of public education on literacy rates, personal hygiene and health care, inoculation against contagious disease, charity, invention, and so forth.

Determining the Need for Government Intervention

To promote the general economic well-being, an ideal government would intervene to promote positive externalities and to discourage or prevent negative externalities. In the preceding examples, government might intervene to reduce the amount of smoke emitted by the factory or to increase the number of bees kept by the beekeeper. Intervention is not always called for, however. Individuals might settle such matters privately, because the costs of intervening might be prohibitive.

To enhance the bees' contribution to the orchard's productivity, the orchard owner could contract with the beekeeper to supply more bees.[1] Alternatively, the orchard owner could raise his own bees. As long as the beekeeper's marginal costs of raising additional bees are less than the value the bees contribute to the orchard owner, it will be possible for both parties to reach an agreement that will leave each better off. Such private contracting in effect eliminates the apple-bee externality and makes public intervention unnecessary.

In addition to the possibility of private settlements, a second consideration when determining the need for government intervention in externalities is whether the externality is irrelevant or relevant. Externalities are trivial when the situation is not worth anybody's effort to do anything about it. The world is full of trivial externalities. You may not like the way a friend dresses but not enough to say anything about it. Some people's unfailingly gracious and friendly behavior creates external benefits, and yet they are rarely complimented or rewarded for such behavior. The same goes for an attractive lawn. Neighbors reap external benefits from a well-kept lawn but not so much that they would offer a subsidy to support the yard work.

Externalities are termed irrelevant when the externality generates a demand, but not a sufficient demand, by the affected party to change the situation. Your neighbor's messy yard may disturb your sense of propriety. You would be willing to pay her $50 a year to clean up the mess, but her price to clean it up is $600 a year. In this event her yard will not be cleaned up, and the externality will persist. The messy yard is a nuisance but not enough of a nuisance to cause sufficient demand to get the yard cleaned up. In principle, the best thing to do with an irrelevant externality is to leave it alone because it costs more than it is worth to correct the problem.

A relevant externality creates sufficient demand on the part of those affected by it to change the situation. The neighbor's yard is messy, and the neighbor will supply a cleanup for $10. You are willing to pay as much as $50. In this case you can make an effective offer to correct the externality and reach a deal with your neighbor. With relevant externalities, it is worthwhile to correct the situation because the benefits of correction exceed the costs.

Correcting Relevant Externalities

Economic theory suggests that correction of relevant externalities can be approached in a number of different ways. In general, these can be categorized as defining property rights, taxing negative externalities or subsidizing positive ones, selling rights to create an externality, and establishing regulatory controls.

Establishing Ownership Rights. Externality problems generally persist because of the presence of some element of common ownership. The beekeeper's and the apple grower's ownership rights are clearly established and easily transferable. In such a situation it will be in the interest of both parties to conclude an agreement that places resources in their most highly valued uses. But in the case of the polluting factory, does the factory have the right to use the air as it pleases, or do the people in the surrounding community have the right to clean air? Air-use rights are indefinite and nontransferable: It is generally impossible for people to buy and sell rights to the use of air. In this

[1]See Steven N. S. Cheung, "The Fable of the Bees: An Economic Investigation," *Journal of Law and Economics* 16 (April 1973), pp. 35–52.

case there is no reason to believe that the pursuit of individual interest will promote the use of air in its most highly valued manner.

The establishment of ownership, then, is one way of eliminating problems associated with externalities. Since nonownership is a source of market failure, the creation of ownership is a means of correcting market failure.

An important analysis of the economics of establishing ownership rights was first introduced by R. H. Coase.[2] Coase posed the question, Does the legal assignment of property rights to one party or the other in an externality relationship make any difference to the observed market outcome? To keep the analysis simple, Coase assumed that the costs of bargaining and transacting among the parties was zero.

We will apply the Coase analysis to an example of air pollution—a factory belching smoke over a nearby community.[3] Figure 18–2 illustrates the analysis. Firm output per unit of time is given along the horizontal axis. The marginal benefits (MB) to the firm of producing this output are given in dollar terms on the vertical axis. For the sake of this analysis, marginal benefits can be thought of as the net profits of producing additional units of output. The MB schedule therefore declines with increases in output because the rate of return on additional production generally declines (see Chapter 8). The marginal cost (MC) curve represents the externality caused by the firm's production and is also given in dollar terms along the vertical axis. The MC function measures the additional cost created at each output level by additional smoke. (Assuming that the amount of smoke the firm produces is directly related to its rate of output, more production will cause more smoke.) The MC curve rises as output increases.

Now we apply the Coase analysis. As an experiment, we give the ownership right to clean air to the homeowners in the surrounding neighborhood. Assuming that bargaining costs nothing, we can predict what will happen with the help of Figure 18–2. For rates of output up to Q, we observe that MB > MC. That is, the firm's profits on additional units are greater than the additional pollution costs borne by the neighborhood. This means that the firm would be willing to buy and the neighborhood willing to sell the right of using the air to produce these units. Since the externality caused by this output is represented by the area under the MC curve, 0EQ, a bargain can be struck between the firm and the neighborhood to produce Q by agreement on how to divide the surplus marginal benefit, AE0. Beyond Q, MC > MB. A bargain to produce these units cannot be reached between the firm and the neighborhood because the firm would not be willing to bid enough to obtain the right to produce these units. So Q, where MB = MC, is the equilibrium outcome when the neighborhood is given the transferable right to use of the air.

Suppose that a judge decided instead to award the firm the right to use the air as it chooses. What happens in this case? We go through the same analysis, but now we see that the neighborhood must pay the firm to produce less output. Up to Q units of output, MB > MC, which means that the neighborhood cannot offer a large enough sum of money to induce the firm

[2]R. H. Coase, "The Problem of Social Cost," *Journal of Law and Economics* 3 (October 1960), pp. 1–44.

[3]More subtle, widespread damages such as the loss of fish and trees to acid rain and the loss of work time to health impairment are not dealt with here. For the sake of simplicity, we consider only obvious costs to the local community. But if claimants organize and the source of pollution can be clearly pinpointed, the same analysis can be applied to the more subtle costs of air pollution.

FIGURE 18–2

The Coase Theorem

In this graph depicting the Coase theorem, MC represents the externality imposed by the firm's pollution, and MB represents the profits to the firm from producing various levels of output (Q). The Coase theorem states that regardless of who is assigned ownership rights to the air, costless bargaining between the parties will result in the same market outcome at Q.

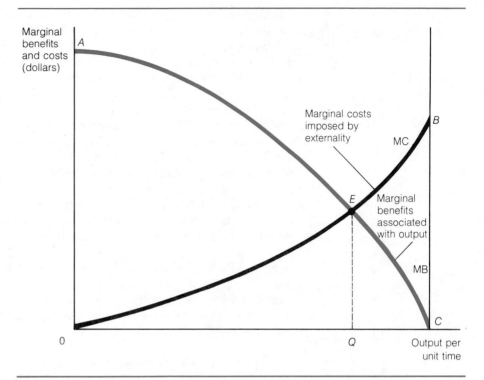

Coase theorem:
Externalities will adjust to the same level when ownership rights are assigned and when the costs of negotiation are nonexistent or trivial, regardless of which party receives the rights.

to reduce its output and its smoke. Beyond Q, the situation is changed: The firm would be willing to accept an offer of money to reduce its output. Beyond Q, the smoke causes EBCQ in damage to the neighborhood, and the benefits to the firm of this output are only ECQ. The neighborhood is therefore willing to make it worthwhile to the firm to reduce pollution to the point where MB = MC at Q.

Provided that we have been logical, we have shown a simple but powerful result: No matter who has the legal right to the use of the air, the amount of pollution is the same. When the firm must pay to pollute, it produces Q. When the firm has the unfettered right to pollute, it produces Q. This is the central insight offered by the **Coase theorem:** In a world in which bargaining costs nothing, the assignment of legal liability *does not matter.* A certain equilibrium level of output and its resulting level of pollution will exist regardless of whether firms or consumers "own" the air.[4] Under the stated conditions of the Coase theorem, government intervention—in the form of pollution guidelines, tax penalties, and the like—cannot improve upon a settlement negotiated by those parties who are directly involved with the externality problem.

The value of the Coase theorem is that it provides a benchmark for analyzing externality problems. It shows what would happen in a world of zero bargaining and transaction costs. Many real-world externality problems can be analyzed with the Coase theorem as long as the bargaining and transaction costs are low.

[4]The Coase theorem also marked the beginning of a new field in economics called law and economics, which studies the impact of legal rules and institutions on the economy; the Coase theorem, for example, analyzes the role of legal liability assignment in an externality problem. Many law schools have specialized fields in law and economics, and much of the literature in this area is published by the *Journal of Law and Economics.*

Nevertheless, using the establishment of ownership to correct externalities cannot always be relied upon to eliminate the difficulties of market failure because it may be exceedingly difficult if not impossible to establish ownership in some instances. Many externality problems do not fit the assumption of zero or low transaction and bargaining costs. In the factory smoke example, the neighbors in the area surrounding the plant are likely to be numerous, difficult to organize, and diffuse in their interests with respect to the air pollution problem. The widely dispersed sufferers of the smoke damage would face insuperable costs if they had to organize to purchase the agreement of the factory to reduce its emissions. If the firm had the right to pollute, it is not likely that the neighborhood would be able to overcome these problems, get organized, and offer the firm money to cut pollution back to Q in Figure 18–2. Looked at the other way around, if the neighborhood owned the air space, the firm would find it costly to track down all homeowners and arrange for the purchase of their consent to produce up to output Q.

No matter who held the right to use of the air, bargaining costs would be high, and the likelihood is that nothing would be done to correct the situation. The costs of defining and enforcing a system of ownership can be so high that some system of taxation or administrative control would be more effective than reliance on the definition of property rights and contractual arrangements.

Taxing or Subsidizing Externalities. A seemingly simple way of dealing with externalities is to tax activities such as pollution that create negative externalities and to subsidize activities such as beekeeping that create positive externalities. Many complexities surround such a simple-sounding prescription, however. How could a negative externalities tax be levied? Against the product the factory produces? Against the amount of smoke it emits? Either solution might create burdens on the industry and on consumers that are not necessary in cleaning up the environment. Similarly, how would a subsidy for positive externalities be worked out? Based on the honey produced by the beekeeper? Based on the number of bees kept? Despite such complexities, the essential idea of taxing the negative and subsidizing the positive has a rich tradition in economic theory.[5]

To understand the effects of taxation, consider the pollution problem in terms of output decisions facing individual firms. We have learned that the competitive firm, in deciding what output to produce, compares marginal cost and marginal revenue. Marginal cost in this context means **marginal private costs** (MPC), the extra costs that the firm must pay to increase its output. These costs do not include externalities. By definition, an externality imposes costs on others that are not reflected in the private costs of the firm. The full costs to society of an increase in the firm's operations are summarized as **marginal social costs** (MSC). MSC are equal to MPC plus the externality costs (E): MSC = MPC + E. When there is no externality at the margin (E = 0), then MSC = MPC.

This analysis can be applied to the competitive industry as well. Consider the situation depicted in Figure 18–3. These are the same types of graphs that we used in Chapter 9. Figure 18–3b shows that an industry is initially in equilibrium at E_1 where industry demand (D) and supply (S_1) are equal. Under these conditions the firm (Figure 18–3a) reacts as a price taker to P_1

Marginal private costs: The increase in a firm's total costs resulting from producing one more unit.

Marginal social costs: The increase in total costs to society (the firm plus everyone else) resulting from producing one more unit.

---

[5]The tax-subsidy approach to externalities is often called *Pigovian*, after A. C. Pigou, author of *The Economics of Welfare* (London: Macmillan, 1932).

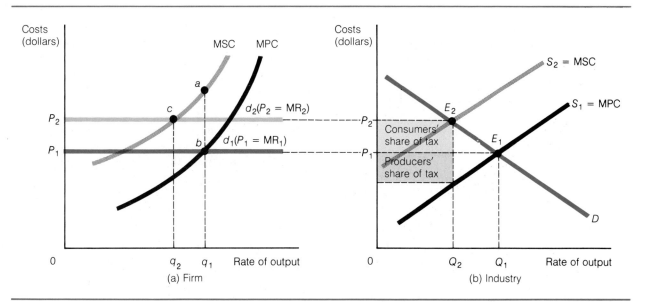

FIGURE 18–3   The Effects of a Tax on an Externality

(a) A pollution excise tax of *ab* per unit is placed on the output of the firm to correct the externality. The marginal private costs facing the firm are given by MPC, and marginal social costs are given by MSC. The tax raises the MPC curve to become identical with the MSC curve. In such a way the firm is made to account for the externality in its decision about how much output to produce. (b) At the industry level, the pollution tax reduces industry supply to S_2, and, according to our earlier analysis of the incidence of an excise tax in Chapter 9, the pollution tax is paid by consumers and producers in the proportions indicated in the diagram.

and sets its MPC curve equal to P_1 ($= d_1 = MR_1$). The competitive firm, acting on the basis of marginal private costs, produces q_1 units of output.

Notice, however, that the firm's MSC curve lies above the MPC curve at all levels of output; for example, it exceeds MPC by *ab* at q_1 units of output. The MSC curve traces the externality that the firm's operations impose, such as the amount of damage that the firm's air pollution causes in the surrounding neighborhood.

Our question is, What can government do to improve the situation? Basically, government must get the firm to behave as if its marginal cost curve is MSC, not MPC. In this way the competitive firm would equate MSC to price, and the externality would be "internalized" in the firm's output decision; that is, the firm will produce the output where MSC = P, which is the correct output from society's point of view. (As we saw with the Coase theorem, the optimal level of pollution is not zero. At some point, clean air costs more than it is worth.)

This solution can be obtained by imposing a tax on the industry's output that reflects the degree to which MSC and MPC diverge. Such a tax is illustrated in Figure 18–3 as an excise tax imposed on the firm of *ab* per unit. The impact of the pollution tax is to shift the marginal cost curve of the firm from MPC to MSC. This change is reflected at the industry level by a shift in the industry supply curve from S_1 to S_2. The tax forces the competitive firm to account for the externality in choosing its output level. It thus chooses the socially efficient output level where MSC = P_2, or where marginal social costs are equal to marginal private benefits.

Who pays the pollution tax? As we learned in Chapter 9, part of the tax will be paid by producers and part by consumers. Consumers will face a higher price for the industry's output, but the price will not rise by the amount of the tax. The producer thus absorbs some of the pollution tax.

Selling Rights. A simple alternative to the tax example is to establish a market for externality rights such as the right to pollute. In this approach government sets an allowable level of pollution and sells the right to pollute. In theory, this market works like any other market. Those who value the pollution rights most highly will buy and hold them. Firms will make decisions about whether it is less costly to install pollution-control equipment or to buy pollution rights. Government does not tell industry how to clean up the air. It sets the allowable level of pollution and lets firms decide how to control their pollution. This marketlike approach to pollution control has the virtue of ensuring that the acceptable level of pollution is reached at least cost.

Difficulties arise in such a scheme, however. Indeed, whether it chooses to tax pollution or sell pollution rights, government must somehow estimate what the "optimal" level of pollution is. The relevant knowledge in this case is economic rather than technical. The question is not, What is the technical damage of degrees of air pollution? but rather, How is this technical damage valued by individuals? Such knowledge is not easy to come by. Moreover, both a tax scheme and a pollution rights scheme must be monitored and enforced by government. Without such actions tax evasion and pollution without permit will be serious problems.

Setting Regulations. For such reasons government has most often not adopted an economic approach to control of externalities such as pollution. More typically, government has sought to control pollution and other externality problems through direct regulation of industry. In the case of pollution, regulation generally consists of detailed rules about the technology that firms must adopt to control pollution. Firms in certain pollution-prone industries are required to install special pollution-control equipment as a condition of being able to stay in the industry. This is a practical way to control pollution. Each firm must clean up its emissions in the prescribed manner. In contrast to the pollution rights approach, however, direct regulation does not allow firms to choose the most efficient means of staying below the allowable level of pollution. Firms cannot choose to pollute or not pollute or select among different types of pollution-control technologies. They must follow the rules laid down by government. Moreover, these rules can affect different firms in an industry in different ways. Large firms, for example, may be better able to adapt to the technology required by government than small firms. Some small firms may be driven out of the industry as a result, and the large firms will benefit through higher prices and profits.

Each approach to the problem of externality control that we have discussed—the establishment of property rights, taxation or subsidy, selling of rights, and direct regulation—has costs and benefits in different applications. In some cases, the establishment of property rights may resolve the externality. In other cases, like automobile pollution, some form of government taxation or regulation may be used to control the externality of pollution. No approach will yield a perfect solution to these problems. There will still be disputes among property owners about where one's property ends and another's begins. And government will not always act to resolve an externality in the most cost-effective way. There may, for example, be very little air pollution in Wyoming, which has fewer cars or industries, but new cars purchased there will nonetheless be required to have the same pollution control equipment as new cars purchased in smog-ridden Los Angeles or New York. Actual policies to control externalities are not perfect, as Focus, "Saving Endangered Predators," illustrates.

Public Goods Theory

As we discussed in Chapter 3, government affects the economic well-being of a society in two essential ways. Government provides and maintains a system of laws to protect and enforce property rights and to permit the free flow of goods through markets. And government provides various goods and services that are not ordinarily provided through the free-market system.

Public good: A product that is noncompetitive in consumption and nonexclusive; a good that, once produced for one individual, is available to everyone.

These government-provided goods and services fall into the broader category of **public goods**—goods such as radio broadcasts and highways that, once provided to one person, are available to all on a noncompeting or nonrivalrous basis. It is difficult or impossible to exclude anyone from the use of a public good, and its use by one person does not prevent its use by another. A nuclear submarine, for example, provides national defense as a public good for all members of a society. One member's consumption of national defense does not preclude another's consumption. The use of a lighthouse beam by one ship does not stop other ships from using the same beam. The consumption of a public good by one consumer does not impose an opportunity cost on other consumers; all can consume the public good at the same time.

A private good is consumed exclusively by the person who buys it. Its consumption by one individual precludes consumption by others. When Benny eats a cheeseburger, Barbara cannot eat the same cheeseburger. The concern here is not with the fact that Benny bought the cheeseburger and owns it in the sense that he has legal title to it. The concern is with a technical characteristic of consumption—that when one person is consuming a private good, it cannot be simultaneously consumed by other people. The consumption of a private good carries an opportunity cost: Consumption of the same good is denied to other consumers.

A mixed good is a good that embodies attributes of both private and public goods. An outdoor circus, for example, would seem to be a pure public good. Two people can watch the circus at the same time without interfering with each other's consumption. If the crowd gets large enough, however, the circus can become more of a private good because congestion detracts from each person's view. In other words, rather than being equally available to all, consumption becomes rivalrous; some people's consumption of the circus detracts and may even prevent other people's consumption. We say then that the outdoor circus is a mixed good. It has characteristics of both private and public goods.

It is important to keep in mind the difference between public goods and goods that are publicly provided. The concept of a public good refers to the attributes of the good and not to whether it is produced by government. Indeed, the public sector produces many private goods such as first-class mail delivery, and the private sector produces many public goods such as beautiful architecture. In other words, the supply of public goods comes from both the public and the private sector. As we will see, it is largely the costs of exclud-

FOCUS   Saving Endangered Predators

The preservation of endangered predators, such as the bald eagle, the mountain lion, and the timber wolf, is a pressing environmental issue. Because these predators cause an enormous amount of damage to ranchers (in 1979, approximately 1.3 million sheep were lost to predators), the wild animals and birds are hunted, trapped, and poisoned without regard to their endangered status. Environmentalists worry about the possible extinction of the predators themselves.

In the case of bald eagles, the present solution is a federal law that stipulates a $10,000 fine, a year in prison, or both for killing or possession of bald or golden eagles. This penalty appears to be ineffective. There is evidence that ranchers are willing to pay bounties of $25 for dead eagles, even in the face of a $10,000 fine and a year in jail. Why? People apparently do not take the fine and prison penalty seriously because they do not expect to get caught. The expected penalty (the probability of being caught times the fine and jail sentence) is very low; people rationally ignore the penalty and continue to kill eagles.

An obvious solution to the problem would be to raise the penalty. If the expected penalty were raised to equal the expected gain from killing an eagle, eagle killing would recede. Such a solution has been avoided out of concern that this fine would probably have to be so high that the unlucky rancher who got caught would have to go to jail for the rest of his life or pay a fine that would bankrupt him.

Source: R. C. Amacher, R. D. Tollison, and T. D. Willett, "The Economics of Fatal Mistakes: Fiscal Mechanisms for Preserving Endangered Predators," *Public Policy* (Summer 1972), pp. 411–441.

The economic theory of externalities suggests another solution: Establish property rights in eagles. Rather than considering eagles unowned as they are now, the "property rights" to eagles and other endangered predators could be transferred to conservation groups and their supporters. Then the damage done by the predators would become the liability of those who care about them. In assigning liability to conservation groups, it would be possible for a market for eagles to evolve. Farmers could take their damage claims to conservation groups (perhaps state by state), and conservation groups could finance whatever level of these claims they wanted to support out of their membership fees. As bargaining took place between farmers and conservationists, some idea of the "optimal" eagle population would emerge.

Needless to say, such a proposal would have many kinks to be ironed out. Farmers would seek to "overclaim" damages; conservationists would "underestimate" compensation. To get around such problems, independent arbitrators could perhaps be employed to settle damage claims.

Another possibility would be for the government to protect predators by compensating farmers for any livestock killed. Such a move would remove the burden of compensation from the few who care strongly and distribute it across the general population through taxation. In a modified version of this approach, the state of New Mexico is supporting a program that breeds Mediterranean sheepdogs and then leases the dogs to ranchers for $120 a year, the approximate price of one ewe. Subsidizing the breeding of sheepdogs is an alternative to outright compensation for lost sheep.

---

ing nonpayers that determine whether a public good is privately or publicly provided.

Public Provision of Public Goods

Abraham Lincoln's dictum on government's proper place in the economic order still has merit: "The legitimate objective of government is to do for a community of people whatever they need to have done, but cannot do at all, or cannot do so well themselves, in their separate and individual capacities. In all that the people can individually do as well for themselves, government ought not to interfere."[6] The theory of public goods follows Lincoln's dictum. Just as Lincoln pointed out, there are some things that a free-market system fails to provide members of a society, usually because doing so would be difficult if not impossible.

[6]As quoted in Peter G. Sassone and William A. Schaffer, *Cost-Benefit Analysis: A Handbook* (New York: Academic Press, 1978), epigraph.

How, for example, can a free-market system provide national defense? Once a defense system is provided, it is equally available to all consumers or citizens. Since individuals cannot be excluded from the benefits of military defense, who would voluntarily contribute to its provision? Most individuals would hold back in the expectation that someone else would provide the necessary funds, and then they would reap the benefits of military defense for free.

Lighthouses provide another illustration of market failure to provide public goods. A lighthouse's beam is available to all ships passing by, regardless of whether a particular ship pays for the light. Since it is difficult or impossible to exclude nonpayers from using the beam, individual shipowners are likely to refrain from contributing to the lighthouse even though it provides an obviously valuable service for all shippers. To overcome this market failure, government assumes responsibility for the lighthouse. Through its power to tax, government can raise enough revenue to build and maintain the installation of the lighthouse.[7]

Public health services also illustrate the need for government to overcome market failure to provide a public good. At certain times when the public health seems in danger, such as the swine flu outbreak in the late 1970s, the government provides inoculations against disease. Individuals who are inoculated protect themselves against the disease. Their inoculation also protects those who are not inoculated. As a person becomes inoculated, the sources of contagion for the disease are reduced, so person A's inoculation reduces the odds that B and C, who have not been inoculated, will catch the disease. B and C theoretically owe A something for the protection, but would they pay for it? Obviously not, since they cannot be excluded from A's protection. Government's providing such a public health service is a way of overcoming the market failure in this case.

Not all goods provided to a number of people at once require government provision. In some cases the costs of excluding noncontributors are low enough that private firms will produce goods enjoyed on a nonrival basis by many people. For example, a movie, a football game, and a concert are goods that are jointly provided to many consumers at once. In this respect these goods are like the lighthouse and national defense. The difference is that it is feasible for the producers of these goods to exclude those who do not pay for their provision, and so they can be supplied by private producers within the free market.

Free Riders

Free rider: An individual who is able to receive the benefits of a good or service without paying for it.

Sellers of private goods can exclude nonpayers from using the goods by charging a price for their use. Once a public good is produced, however, it is equally available to all consumers, regardless of whether they contribute to its production or not. Individuals who do not contribute to production and yet enjoy the benefits of goods are called **free riders.**

The problem facing producers of public goods is that free-riding behavior makes it difficult to discover the true preferences of consumers of a public good. Individuals will not reveal their true preferences for public goods because it is not in their self-interest to do so. For example, if a neighborhood tried to organize a crime-watch group, many neighbors might not contribute

[7]While the lighthouse has long served as an archetype of a public good, the private provision of lighthouses is a historical fact. See R. H. Coase, "The Lighthouse in Economics," *Journal of Law and Economics* 17 (October 1974), pp. 357–376.

but would still appreciate any protection provided, thus concealing their true preferences for the service. On the other hand, some who do not contribute voluntarily might be totally disinterested in the service. The possibility of free riding makes it difficult to determine the true level of demand for the public good.

If everyone free rides and no one contributes to the production of public goods, then everyone will be worse off as a result. No public goods will be produced, even though each individual would like some provision of public goods. Herein lies a rationale for government provision of public goods. Government has the power of coercion and taxation, and it can force consumers to contribute to the production of public goods. Citizens cannot refuse to pay taxes and thus contribute to the costs of public sector output. When the cost of excluding nonpayers is high, government can produce public goods through tax finance.

Pricing Public Goods

Should a price be charged for use of public goods? In a sense, the correct price of a public good is zero. Take the example of a bridge. Once the bridge is built, the marginal cost of one more car's going over it is approximately zero. If we follow the rule that we learned in Chapter 9 for optimal pricing, we should set price equal to marginal cost. Since the marginal cost of an additional bridge crossing is effectively zero, price should be zero according to the $P = MC$ rule. According to this logic, public goods should be free.

This rationale is fine as far as it goes, but it does not go far enough. In rationing the use of the good among demanders, zero is the correct price. Except for cases of traffic congestion, a public good presents no rationing problem; by definition, additional individuals can consume a public good at zero marginal cost. But rationing is not the only function of price. A second important function of price is to achieve the right quantity of the public good by relating quantity to demand. Without information about what people are willing to pay to cross the bridge, authorities are left in the dark about the desired amount of bridge construction to undertake. After all, public goods involve the expenditure of scarce resources, and prices for public goods help public officials develop information about the demand for these goods.

How then should a price be set for a public good? Suppose a defense alliance is formed by two countries, A and B.[8] The alliance provides for a common defense policy through a treaty agreement. National defense is a public good for the alliance, and the problem facing the two countries is how to reach an agreement on the amount of national defense to be produced by the alliance. Figure 18–4 illustrates the problem.

We assume that each country faces the same marginal cost curve (MC) for providing defense. MC rises because additional scarce resources devoted to defense come at an increasing cost to the two countries.

Each country has a demand curve for national defense D_A and D_B. Recall that the consumption of private goods is mutually exclusive. When A gets a unit of a private good, there is one less unit available for B to consume. In Chapter 4, this condition meant that to derive a market demand curve for a private good, the demand curves of individual consumers had to be summed

[8]See Mancur Olson and Richard Zeckhauser, "An Economic Theory of Alliances," *Review of Economics and Statistics* 47 (August 1966), pp. 266–279.

FIGURE 18–4

Sharing the Costs of Defense

MC is the marginal cost curve for production of military protection under an alliance. D_A and D_B are the individual country demand curves for national defense. The appropriate "prices" for national defense for each country are P_B and P_A. $D_A + D_B$ is the vertical sum of D_A and D_B. Optimal output for the alliance is where $D_A + D_B = MC$, or at Q_T.

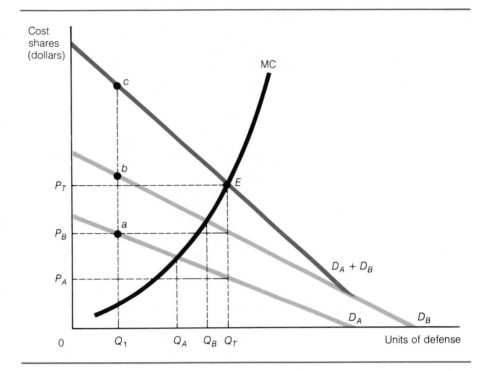

horizontally. National defense, however, is not a private good. It is a public good, and this makes the nature of the "demand curves" in Figure 18–4 quite different.

Consumption of a public good is not mutually exclusive among consumers. All the units along the horizontal axis in Figure 18–4 are equally available to both demanders. Both A and B receive benefits from, for example, Q_1 units of military protection. Thus, the total protection benefits offered by Q_1 to A and B are equal to the vertical sum of their demand prices for this quantity or output. At Q_1 the benefit that each country places on alliance protection is read vertically off its demand curve. A derives aQ_1, B derives bQ_1, and both countries together place a value of cQ_1 on Q_1 units of alliance output.

This point applies to all public goods. The market demand curve for a public good is obtained by vertically summing all individuals' demand curves. The demand curves are summed vertically rather than horizontally because all demanders can consume the same units of output along the horizontal axis. In Figure 18–4, then, market demand is given by $D_A + D_B$, which is the vertical sum of D_A and D_B.

At points along the horizontal axis less than Q_T, marginal benefit exceeds marginal cost ($D_A + D_B > MC$), and production will be expanded. At points beyond Q_T, marginal benefit is less than marginal cost ($D_A + D_B < MC$), and production will be cut back. Equilibrium output is at E, where $D_A + D_B = MC$. The equilibrium defense output in the alliance is Q_T, and the alliance defense budget is $P_T \times Q_T$.

How should this budget be divided between the two countries? One way to apportion the costs is in proportion to the benefits received. P_B represents the benefit that accrues to country B, and P_A is the benefit that goes to country A. Country B pays more because it benefits more from alliance output. *B*, for example, may be a larger country, with more income, capital, and population to defend.

This is, of course, an idealized presentation of alliance behavior. For example, what happened to the concept of free riding? Basically, we ignored it in order to depict a perfect outcome for the alliance. But free riding will be a thorn in the alliance's side. Look at the individual equilibrium positions of the two countries in Figure 18–4. The small country prefers Q_A as its defense output; the large country prefers Q_B. When the treaty to form the alliance is signed, *A* gains access to the benefits of *B*'s production. It is apparent that *B*'s production more than satisfies *A*'s demand for national defense. In group alliances, small countries like *A* will therefore have strong incentives to reduce their defense spending and to free ride on the defense expenditures of larger countries such as *B*. This tendency will lead to much bargaining and many debates about contributions to alliance output with the large countries complaining about the efforts and contributions of the small countries. The problem of free riding will be more pronounced the more countries there are in the alliance.

Cost-Benefit Analysis

Cost-benefit analysis: A process used to estimate the net benefits of a good or project, particularly goods and services provided by government.

Externalities and public goods require government intervention in the economy. As we saw, unless private ownership can be established, externalities call for government regulation or taxation of private activities that create externalities. The need for public goods calls for government provision of certain goods and services such as national defense. Up to this point we have assumed that government officials are able to obtain the necessary information about how to correct externalities and to provide public goods in more or less the correct quantities. This, of course, is a questionable assumption. Government decision makers face tremendous practical difficulties in obtaining information to guide their decisions. For example, since consumers will rationally not reveal their true preferences for a public good, how does government decide how much of the good to produce? This is not an easy question to answer in practice, but the technique of cost-benefit analysis has been applied to help government officials answer such questions.

Cost-benefit analysis is just what the term implies: It is a procedure by which the costs and benefits of government programs are weighed so that decisions can be made about which programs to undertake. Clearly, a wise use of public resources is to rank programs from highest to lowest in their benefit-cost ratios and undertake those programs with the highest ratios.

Cost-benefit analysis emerged as a result of efforts to find more systematic information on which to base government decisions about investment in physical resources. While cost-benefit analysis can be traced to the 1920s, its development truly got under way after World War II. In this postwar period various water resource investments under the aegis of the Army Corps of Engineers and the Bureau of Reclamation provided the subject matter for cost-benefit analysis. Military applications became important in the early 1960s in the Department of Defense, although these applications were typically referred to as systems analysis. Eventually, cost-benefit analysis came to be viewed as appropriate for application throughout the government budget process. Investments in human resources and proposed changes in legislation were added to investment in physical resources as reasonable subjects for cost-benefit analysis.

Government cost-benefit analysis can be divided into two main categories, according to its purpose. In the case of public investment, the goal of cost-benefit analysis is efficiency—to help choose more efficient over less efficient dams, for example. The other main category is the use of cost-benefit analysis to determine the merit of government intervention in private sector activity. The application of cost-benefit analysis in such areas as penalizing alcohol abuse, determining health effects of smoking, breaking up monopolies, regulating drugs, discouraging disposable containers, analyzing food additives, and promoting automobile safety are illustrations. The task of cost-benefit analysis in such instances is to provide evidence about the desirability of particular government interventions or public policies.

Although cost-benefit analysis is equally applicable to firms and governments, its development has been inspired mainly by an interest in promoting efficiency in government. Careful comparisons between the costs and benefits of a proposed dam, for example, help discipline government decisions, since market forces are not present to reward good or discourage bad government decision making. Estimating anticipated costs and benefits resulting from a particular government program helps decision makers select the most efficient programs from among those that could be adopted. Cost-benefit analysis is forward looking; it seeks to assess the consequences of various proposed courses of action.

Prior to conducting any cost-benefit analysis, planners must agree on a desired objective and array the various means of attaining the objective. Cost-benefit analysis does not propose a course of action. It is simply a technique for helping to evaluate proposed courses of action. The proposals themselves come from outside the framework of the analysis.

Once the goals of a government program are stated and the possible ways to reach these goals described, the primary task of cost-benefit analysis is to measure the anticipated benefits and costs of the project. The relevant benefits fall into two categories: direct benefits and indirect benefits. Direct benefits accrue to users of the project after its completion, such as the benefits that a new subway system provides to its users. Indirect benefits might occur if the subway system reduces congestion on the highways and highway users save travel time. Needless to say, estimating the benefits, even the simple direct benefits, of a program is a difficult task. It requires much judgment because hard evidence of real benefits is often impossible to develop.

In addition to measuring or estimating benefits, a cost-benefit analysis must measure or estimate the anticipated costs of a proposed project. The cost of a project is the value of the benefits from other potential uses of resources that must be sacrificed to undertake the project. If the costs of a dam are determined to be $10 million, this $10 million represents the value of the other services that could have been produced in place of the dam. This is why economists typically use the term *opportunity cost* as synonymous with *cost* since cost represents a sacrificed opportunity.

As with benefits, there will be direct and indirect costs associated with public programs. Moreover, the same inherent problems that plague benefit estimation plague cost estimation. As in the case of benefits, prices of programs often are not available to estimate the value of alternative uses of public resources. And what is forgone when specific alternatives are adopted is not easily observable. Cost and benefit estimations are therefore likely to be only crude approximations of real values even under the most favorable circumstances.

A Role for Government

The presence of externalities and public goods means that government has an important and productive role to play in the economy. Externalities lead to a regulatory role for government. Public goods imply a role for government as a direct producer of some goods and services. In this case, in order to pay for such public goods and services, government levies taxes on citizens to raise revenue. It is this important area of government activity—taxation—that we turn to in the next chapter.

Summary

1. Market failure occurs when the private market system fails to allocate resources in an optimal or ideal fashion. Externalities and public goods are two categories of market failure.

2. Owners of private property bear the economic results of its use. When property is held in common, users do not bear the full results of its use. Common or undefined ownership creates a conflict between the pursuit of self-interest and the common good. The implication is that common resources will be overused and exploited.

3. An externality occurs when an economic unit does not bear all the value consequences of its actions. Externalities can be positive or negative, trivial, irrelevant, or relevant. A relevant externality creates an effective demand to correct the externality.

4. The Coase theorem assumes that in a world of zero transactions and bargaining costs, the amount of an externality such as pollution is independent of who has legal liability for pollution damage. Under the stated conditions, government cannot improve upon what individuals negotiate.

5. Where transaction and bargaining costs are high, other routes to externality control become feasible. A per unit pollution tax would force a firm to act as if its marginal cost curve were identical to its marginal cost curve plus the cost of the externality, or the marginal social cost curve. Other approaches include establishment of a market for pollution rights and direct regulation of firm technology.

6. Public goods are goods that are nonrival in consumption, with high costs of excluding nonpayers. Whereas a private good is consumed exclusively by one individual, a public good can be simultaneously consumed by all consumers. A mixed good has both private and public characteristics.

7. To free ride means to reap the benefits of a public good without bearing its costs. If everyone free rides, no public good is produced by the private sector even though each individual values its production. Such behavior means that public goods typically have to be produced by government through its power to tax.

8. Public goods are not free; they cost real resources. In terms of rationing a public good among consumers, a zero price is the correct price because additional consumers can consume a public good at near-zero marginal cost. However, a zero price for a public good offers no guidance for public officials about how much of a public good to produce. There is thus a trade-off regarding whether public goods should be priced.

9. Individual demand curves for public goods are summed vertically. Optimal output of a public good occurs where the vertically summed individual demand curves intersect the marginal cost curve. Optimal prices for this output are read off the individual demand curves.

10. Cost-benefit analysis is an analytical tool for evaluating and promoting efficiency in government programs and policies. Cost-benefit analysis does not set goals; it evaluates the costs and benefits of attaining goals in alternative ways.

Key Terms

market failure
common ownership
externality
negative externality

positive externality
Coase theorem
marginal private costs
marginal social cost

public good
free rider
cost-benefit analysis

Questions for Review and Discussion

1. Are whales an example of a common property resource? If so, what sort of behavior would you predict among whale hunters?
2. What is an externality? Name the various types of externality. Why is an externality considered a market failure?
3. Suppose that pollution is controlled with a pollution tax on polluting firms. Who will pay the tax? Firms? Consumers?
4. Define a public good. What is the difference between a private and a public good? How will individuals behave when asked how much they will contribute toward payment for a public good? What does this mean for the nature of the demand curves for public goods?
5. "Museums should be free." Evaluate.
6. Describe how you would apply cost-benefit analysis to the evaluation of an addition to the football stadium at your college or university.

ECONOMICS IN ACTION   Is Urban Mass Transit Worth the Cost?

Several U.S. cities have new public transit systems in construction or in planning. Between 1984 and 1989 these cities (and the federal government) will spend roughly $14 billion for construction and repair in an effort to overcome traffic congestion, air pollution, and the economic decline of downtown metropolitan areas. Los Angeles, for example, will soon begin an 18.6-mile, $3.5-billion subway, and Washington, D.C., now operates a partially completed 47-mile subway system that will cost $6 billion by the time it is finished.

For the most part, urban mass transit systems represent a large subsidy to their ridership, who typically bear only a fraction of a system's true cost in the fares they pay. It was estimated that in 1982 fares in the New York City subway system failed to cover $1.3 billion in total operating expenses. But mass transit advocates argue that investment in these systems is nevertheless justified because of the positive externalities these systems offer, which more than make up for typical operating losses. The question faced by voters and politicians is quite straightforward: Is urban mass transit worth its cost?

Backers of mass transit point to three main categories of positive externalities generated by large-scale bus and subway systems. First, population density is increased in urban areas as a result of the availability of transit service because new residents are attracted by the reduced costs of transportation. This increase in population brings more tax income to hard-pressed local governments. The second positive externality is economic development. Businesses will concentrate in or near the denser population centers to reduce the costs of business transactions and to take advantage of

Source: David M. Stewart, "Rolling Nowhere," *Inquiry* (July 1984), pp. 18–23.

The Jefferson Park station on the O'Hare line in Chicago.

a large, accessible labor supply. The Bay Area Transit in San Francisco has been credited with stimulating $1.4 billion of construction since it opened. Third, mass transit may reduce traffic congestion and pollution levels as commuters switch from cars to buses or the subway. For these three broad reasons, mass transit proponents argue that the positive externalities generated by such systems are high enough to equal or exceed the costs of construction and operation.

Critics who oppose the large public investment in mass transit argue that the positive externalities are exaggerated and that the costs of construction and operation of these systems exceed their benefits. For example, some maintain that reduced business costs would tend to be canceled by increased property values and tax liabilities. Others point to the fact that in those areas with large mass transit systems, traffic congestion and pollution have remained approximately constant or have even increased. In Washington, D.C., for instance, rush-hour trips downtown by car actually increased after the transit system began operating. In short, the mass transit controversy hinges on weighing positive externalities against the costs of construction, maintenance, and operation.

Question

Houston, Texas is a large and rapidly growing urban center. Voters in that city, however, have turned down several proposals for mass transit systems over the past decade. Is this proof that the costs of such systems outweigh the benefits to the city of Houston? Upon what economic considerations should such decisions be made?

19

Taxation

W hen government finances the production of public goods or services—a local sewer system, a research program looking into the cause of acid rain, a nuclear warhead, a presidential limousine—it uses taxes to help pay the bill. In the United States about 82,000 governmental bodies at federal, state, county, city, school district, municipal, and township levels have the power to establish and collect taxes.

Unpopular as they may be, taxes are a necessary part of life in a mixed economy. Every tax has a purpose; directly or indirectly, tax revenues finance government expenditures. As we saw in the previous chapter, the market fails to provide the optimal amounts of public goods such as national defense, police and fire protection, and education or to deal with externalities through programs such as environmental protection. Traditionally, federal, state, and local governments have taken an active role in providing or influencing the production of such goods and services. **Public finance** is the study of government expenditures and revenue-collecting activities.

Public finance: The study of how governments at federal, state, and local levels tax and spend.

Government Expenditures

In Chapter 3 we reviewed some of the major categories of government expenditures and gave some statistics on the rate of government growth. There are two basic types of government expenditures: direct purchases of goods and services and transfer payments, the redistribution of income from one group of people to another. Direct purchases involve spending on items such as defense, fire and police protection, and wages and salaries of government employees. Transfer payments include such expenditures as Social Security payments, Aid to Families with Dependent Children, and unemployment insurance.

TABLE 19–1   Federal, State, and Local Government Expenditures

All government expenditures have increased in recent decades, with state and local expenditures increasing at about the same rate as federal expenditures. This increase was not simply a function of population growth; notice that the expenditures per person increased threefold between 1950 and 1984.

Fiscal Year	Expenditures (billions of 1983 dollars)			Total Expenditures	
	All Governments	Federal	State and Local	Per Household	Per Capita
1950	264.2	168.4	95.8	6,068	1,752
1960	511.6	329.0	182.6	9,692	2,864
1965	642.4	406.3	236.1	11,183	3,332
1970	837.9	523.9	314.0	13,215	4,134
1975	999.4	607.6	391.8	14,052	4,660
1976	1057.3	659.5	397.8	14,511	4,881
1977	1083.9	686.1	397.8	14,620	4,956
1978	1099.9	705.0	394.9	14,466	4,977
1979	1123.3	722.9	400.4	14,526	5,028
1980	1172.5	752.7	419.8	14,516	5,190
1981	1229.8	795.0	434.8	14,931	5,384
1982	1246.7	818.5	428.2	14,926	5,408
1983	1300.3	864.8	435.5	15,345	5,587
1984	1359.4	883.2	476.2	16,035	5,788

Source: Tax Foundation's *Monthly Tax Features*, (June/July 1984). Conversion to constant dollars provided by authors.

The size of all government expenditures is shown in Table 19–1 for selected years from 1950 to 1984. Total government expenditures have increased dramatically since 1950; they are expected to rise even higher in the future.

In Table 19–1, total expenditures are broken down into federal government expenditures and state and local government expenditures. Notice that federal expenditures (including grants-in-aid to state and local government) exceed state and local expenditures by a fairly constant proportion. This has not always been the case. Around the turn of the century, for example, local government expenditures exceeded the sum of federal and state expenditures. Also, the per household and per capita government expenditures have increased in the last few decades.

Not only has the absolute level of expenditures increased; the level of government expenditures as a percentage of GNP has increased. Government expenditures at all levels increased from 21.3 percent of GNP in 1950 to 26.9 percent in 1960, 31.3 percent in 1970, and 35.5 percent in 1982.

Federal government expenditures have increased most in the area of transfer payments. This trend began in the 1960s under the Great Society policies of the Lyndon Johnson administration (see Chapter 3 for a review of these trends). Direct purchases of goods and services by the federal government have also increased in absolute terms but have decreased as a percentage of total expenditures.

Direct purchases at state and local levels exceed those of the federal government. This may not be surprising when you consider the programs offered. Most streets, roads, and highways, police and fire protection, hospitals, education, and sewage and garbage disposal are provided by state and local governments. Indeed, state and local governments have a greater influence on our daily lives than does the federal government.

Tax Revenues

The two basic methods of financing public expenditures are taxing or borrowing. State and local governments usually borrow by selling bonds to finance special projects such as highways, schools, or hospitals. The federal government has come to use borrowing as a routine means to finance its burgeoning deficit. Nevertheless, the largest portion of government expenditures are still financed by taxes. Table 19–2 shows the estimated total tax receipts of the federal government from 1982 to 1986 from each source. The largest single source of federal revenues is the personal income tax, followed closely by Social Security taxes.

Property taxes and sales taxes are the two largest contributors to state and local government revenues, which are broken down in Table 19–3. The statistics reveal a recent trend in state and local financing: Property tax revenues have not been rising as rapidly as sales and income tax revenues. This indicates that property owners are being taxed relatively less over time than other groups. Focus, "Tax Freedom Day," discusses a little-known effect of taxation on individuals.

Theories of Equitable Taxation

Given the government's need for tax revenues, how should taxes be levied among the people? Our society attempts to find an equitable way of distributing the tax burden, but there are no easy formulas for determining what is truly equitable. In theory, however, two basic ways are used to determine who should be taxed and how much: the benefit principle and the ability-to-pay principle.

TABLE 19–2   Actual and Estimated Tax Revenues of the Federal Government

The federal government obtains most of its tax dollars from personal income and Social Security taxes. Notice that total revenues are expected to rise by more than $200 billion between 1982 and 1986.

	Tax Revenues (billions of dollars)				
	Actual	Estimate			
Source	1982	1983	1984	1985	1986
Individual income taxes	197.7	285.2	295.6	317.9	358.6
Corporate income taxes	49.2	35.3	51.8	60.5	74.0
Social Security taxes and contributions	201.5	210.3	242.9	275.5	304.9
Excise taxes	36.3	37.3	40.4	40.8	74.8
Estate and gift taxes	8.0	6.1	5.9	5.6	5.0
Customs duties	8.9	8.8	9.1	9.4	9.7
Miscellaneous receipts	16.2	14.5	14.0	14.5	14.8
Total budget revenues	617.8	597.5	659.7	724.3	841.9

Source: U.S. Office of Management and Budget.

TABLE 19–3   State and Local Tax Revenues

Unlike the federal government's revenues, state and local governments' largest share of tax revenues is collected in the form of property taxes. Property taxes are falling in relative importance, while sales and income taxes are rising.

	Tax Revenues (billions of dollars)			
	1977		1982	
	Absolute	Percent	Absolute	Percent
Level of Government				
State	106,111	58	165,116	60.3
Local	76,661	42	108,814	39.7
Total	182,772	100	273,930	100
Type of Tax				
Individual income	30,852	16.8	52,742	19.2
Corporation net income	9,709	5.3	13,494	4.9
Property	64,164	35.1	86,811	31.7
General sales and gross receipts	38,740	21.2	61,910	22.6
Motor fuel sales	9,365	5.1	10,597	3.8
Tobacco product sales	3,684	2.0	4,210	1.5
Alcoholic beverage sales	2,309	1.2	2,909	1.0
Motor vehicle and operators' licenses	4,961	2.7	6,712	2.4
All other	18,988	10.3	34,545	12.6

Source: Tax Foundation's Monthly Tax Features.

Benefit Principle

Benefit principle: The notion that people who receive the benefits of publicly provided goods should pay for their production.

According to the **benefit principle,** individuals who receive the most benefits from government-produced goods should pay the most for their production. Many taxes arise from this principle.

When nonbuyers can be excluded, government can directly tax the users of its goods and services. People who ride on public transportation facilities such as buses or subways are charged a tax in the form of a fare. Electrical power companies owned by local governments charge individuals according to the amount of electricity they use and thus the amount of the benefits they receive. Highway users may be taxed directly for their use of a road by a toll. Certain other taxes attempt to force users to pay indirectly for categories of benefits they enjoy. For example, the revenues collected from excise taxes on gasoline, tires, automobiles, and auto batteries are used to build highways, roads, and bridges.

The benefit principle is not easily applied to all taxes, however. For one thing, benefits that people receive are not always obvious. With some goods, such as highways and education, direct users receive the greatest benefits, but others receive indirect benefits. A Milwaukee resident who has no car may still buy fruits trucked across highways from California, Texas, and Florida. Families with no children to educate benefit from an educational system that increases worker productivity and raises the standard of living.

FOCUS **Tax Freedom Day**

The average American worker spends a great deal of time earning income just to pay taxes. Not only does the federal government tax the worker directly in the form of income tax, but state and local governments collect taxes in many other forms. How much time does the average worker spend earning money to pay all these taxes?

In 1948, Dallas L. Hostetler of Florida decided to estimate how long the average worker spent each year working to pay taxes, beginning on January 1. Bringing his calculations up to date, the Tax Foundation, Inc. yearly estimates "tax freedom day," the day of the year that the average worker begins to earn take-home pay. Figure 19–1 represents the foundation's estimates.

In 1930, the average worker devoted only 43 days per year to pay all taxes. In 1985, 121 days—one-third of a year's work—were required to pay taxes.

The Tax Foundation also reports the number of hours per day that the average worker supports the public sector. Figure 19–2 breaks down the hours and minutes the average individual worked to support federal and state and local government.

Notice that the federal government now requires more of the worker's day than state and local governments combined. The total hours required for taxes represent one-third of a worker's day.

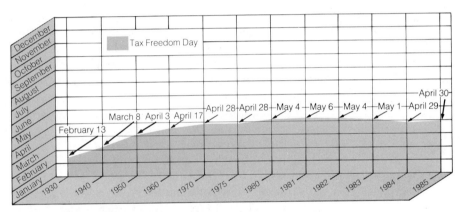

FIGURE 19–1   Tax Freedom Day

"Tax freedom day" is the day of the year on which the average worker is able to begin earning take-home pay.

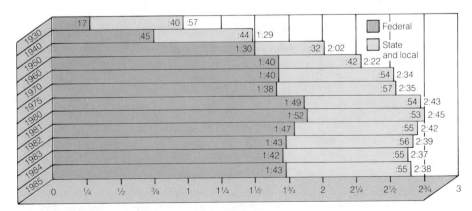

FIGURE 19–2   Supporting the Public Sector

The average individual worker spends a part of each work day supporting federal and state and local governments.

A second problem with the benefit principle is that individuals who receive government benefits cannot always afford to pay an equitable share of taxes. For example, welfare recipients are not likely to have the ability to pay for all the benefits they receive. For these reasons, many taxes are based on the ability-to-pay principle.

Ability-to-Pay Principle

Ability-to-pay principle: The notion that tax bills should vary directly with income levels or the capacity of individuals to pay taxes.

According to the **ability-to-pay principle,** individuals who are more able to pay should pay more taxes than those less able. Under this principle, levels of an individual's income, wealth, or expenditures are measures frequently used for determining the level of tax obligations. The higher the level of income, wealth, or expenditures, the greater the tax liability, regardless of the benefits an individual receives. For example, income and property taxes result in higher tax payments for people with a greater ability to pay.

Horizontal equity: A tax structure under which people with equal incomes pay equal amounts of taxes.

Vertical equity: A tax structure under which people with unequal incomes pay unequal taxes; people with higher incomes pay more taxes than people with lower incomes.

When taxes are levied according to the ability-to-pay principle, the equity of taxation is theoretically measured in two dimensions. The ideal of **horizontal equity** suggests that people with equal abilities to pay should pay equal amounts of taxes. For example, all people who earn $10,000 income would pay the same amount in taxes—perhaps $2000. The ideal of **vertical equity** suggests that people with greater ability to pay must pay a higher absolute amount of taxes than people with less ability to pay. For example, if one individual earns $10,000 and pays $2000 in taxes, anyone who earns more than $10,000 must pay more than $2000 in taxes. Ideally, an equitable ability-to-pay tax has both horizontal and vertical equity.

This ideal is not easily met, unfortunately. One of the problems with establishing an equitable tax is defining ability to pay. Income is the most common measure used to determine ability to pay, but two families with the same income may have dramatically different abilities to pay. One family might have four children, huge medical bills, and no stored wealth. Another family with the same income might have no children, few medical bills, and a great store of wealth. Wealth itself is sometimes proposed as a better measure of ability to pay, but wealth is very hard to measure: Is it the value of a house, of a work of art, of a portfolio of stocks and bonds? A family with enormous wealth may find itself with very little income.

There is general agreement that the wealthy should pay more taxes than the poor, but the question of how much more is not easily resolved. Three ways have been used to relate taxes to ability to pay. The following paragraphs discuss these three ways with reference to an income tax, but they can also be applied to other taxes. Figure 19–3 graphically illustrates the effects of the three types of taxes.

Proportional income tax: A tax based on a fixed percentage of income for all levels of income.

Proportional Tax. A tax that requires individuals to pay a constant percentage of their income in taxes is a **proportional income tax.** If the proportion is 10 percent, a taxpayer earning $50,000 a year would pay $5000 in taxes. If a taxpayer earned $5000, his or her tax bill would be $500. A proportional income tax can result in both horizontal and vertical tax equity. Some states have a proportional income tax.

Progressive income tax: A tax based on a percentage of income that varies directly with the level of income.

Progressive Tax. A **progressive income tax** requires that a larger percentage of income be paid in taxes as income rises. If the tax structure is based on this principle, an individual with $15,000 income may pay only 5 percent of income in taxes while an individual with $30,000 income pays 15 percent. A progressive income tax is also capable of vertical and horizontal equity.

FIGURE 19-3

Progressive, Proportional, and Regressive Taxes on Individual Income

With a progressive tax, individuals pay a larger percentage of their income in taxes as their income rises. A proportional tax takes a constant percentage of income for taxes. A regressive tax takes a smaller percentage of income as income rises.

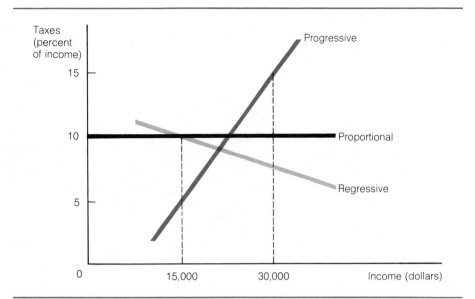

Regressive income tax: A tax based on a percentage of income that varies inversely with the level of income.

Regressive Tax. With a **regressive income tax** a lower percentage of income is paid in taxes as income rises. An individual with $15,000 income may pay 10 percent in taxes while an individual with $30,000 income may pay 8 percent in taxes. While a regressive tax results in a lower percentage of income paid in taxes as income rises, people with higher incomes may pay higher taxes in absolute amounts. According to our example, the individual with $15,000 income pays $1500 in taxes, while the individual with $30,000 income pays $2400 in taxes. For this reason, regressive taxes can be vertically equitable. Further, these taxes can be horizontally equitable so long as people with equal incomes pay equal taxes.

Types of Taxes

Dozens of different taxes are imposed by national, state, and local governments on individuals and businesses. These taxes are usually based on income, wealth, or particular activities.

Personal Income Tax

The federal government receives more tax revenues from personal income taxes than from any other single source. An individual's taxable income is found by subtracting tax exemptions and deductions from total income. The remainder is subject to a progressive tax rate that is used to determine total tax liability. Figure 19-4 illustrates some of the details for determining personal **taxable income.**

Taxable income: The amount of income that is subject to income taxes; total income minus deductions and exemptions.

Tax liability is usually determined by applying tax tables with a progressive tax rate to taxable income. Table 19-4 shows a sample tax table for an unmarried taxpayer. Taxable income in column (1) is divided into the tax obligation in column (2) to yield the **average tax rate** in column (3)—the percentage of income paid in taxes.

Average tax rate: The percentage of income that is paid in taxes; total tax liability divided by total income.

The **marginal tax rate** in column (6) is the percentage of a small increase in income that must be paid in taxes. It is the change in taxes (column 5) divided by the change in taxable income (column 4). For example, if a

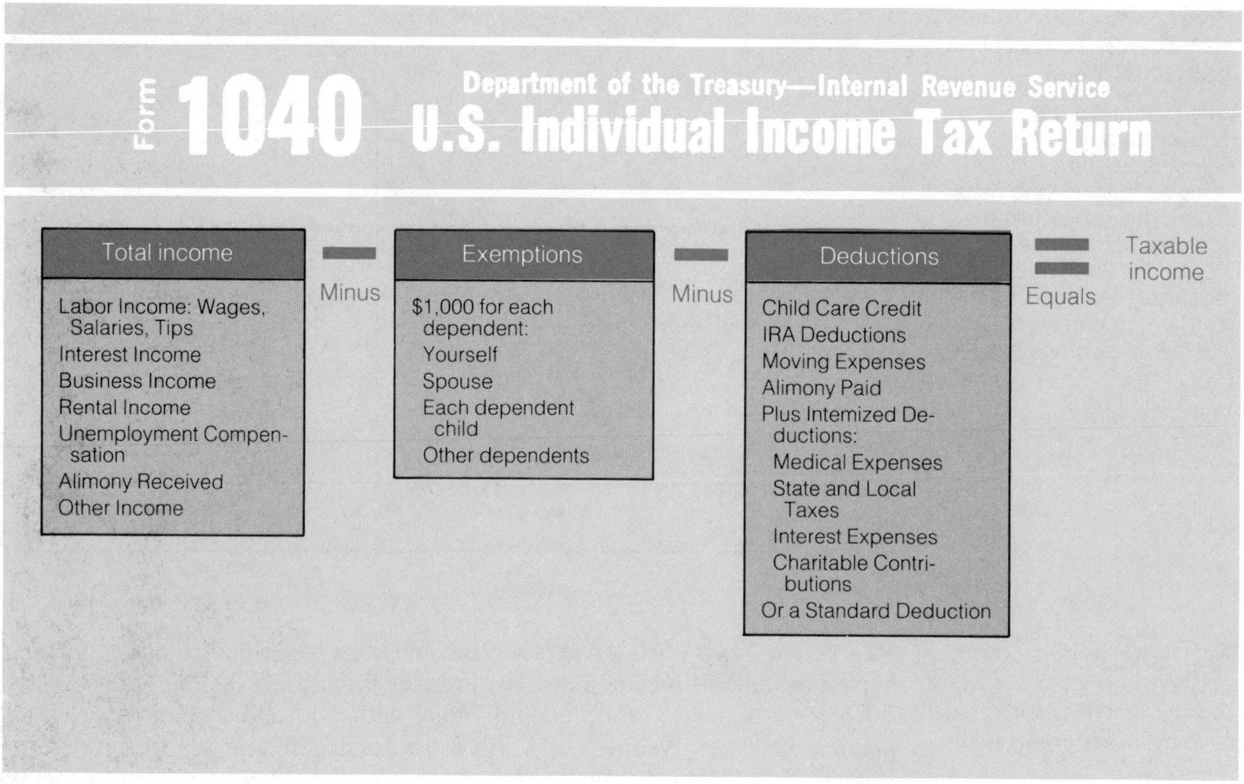

Form 1040 — Department of the Treasury—Internal Revenue Service
U.S. Individual Income Tax Return

Total income		Exemptions		Deductions		Taxable income
Labor Income: Wages, Salaries, Tips Interest Income Business Income Rental Income Unemployment Compensation Alimony Received Other Income	Minus	$1,000 for each dependent: Yourself Spouse Each dependent child Other dependents	Minus	Child Care Credit IRA Deductions Moving Expenses Alimony Paid Plus Itemized Deductions: Medical Expenses State and Local Taxes Interest Expenses Charitable Contributions Or a Standard Deduction	Equals	

FIGURE 19–4   Estimating Taxable Personal Income

This chart demonstrates the basic method of determining personal income subject to federal income taxes. Notice that total income is not taxable. Individuals are allowed to exempt some income for each dependent. Also, certain other expenses are deducted from total income regardless of whether the taxpayer itemizes expenses.

Source: Abbreviated from IRS Form 1040, 1983.

Marginal tax rate: The percentage of an increase in income that must be paid in taxes; the change in tax liability divided by the change in taxable income.

worker normally earns $23,500 a year but has the opportunity to earn an extra $5300 in overtime pay, then 32 percent of that extra income would be paid in taxes; that is, $1696 of the $5300 would be used to pay taxes and only the remaining $3604 would be take-home pay. Notice that the marginal tax rate is greater than the average tax rate and that it rises as income rises. The maximum marginal tax rate is 50 percent. For every dollar earned over $55,300, one-half is taken in taxes.

Some economists believe that the federal income tax system does not result in vertical or horizontal equity. Some people with higher incomes pay lower taxes than others with lower incomes. Also, people with equal incomes do not always pay equal taxes.

Income tax inequities occur because many people engage in tax avoidance. This is a legal activity in which individuals are able to reduce their tax liability by using what are commonly called tax loopholes. One means of avoiding taxes is obtaining income from nontaxable sources. For example, the interest income received from municipal bonds is nontaxable, whereas the interest income received from bonds issued by businesses is taxable.

Individuals may also avoid taxes by spending their money in a particular way. For example, interest paid on borrowed money is deducted from total

TABLE 19–4   Average and Marginal Income Tax Rates

As an individual's taxable income rises, the average tax rate rises. The average tax rate is equal to taxes paid divided by income. The marginal tax rate is the change in income from one tax bracket to the next divided by the change in taxes. With a progressive income tax, the marginal tax rate is greater than the average tax rate. The 50 percent bracket is the highest marginal tax rate.

(1) Taxable Income (dollars)	(2) Taxes (dollars)	(3) Average Tax Rate (%) (2) ÷ (1)	(4) Change in Taxable Income (dollars) (1)	(5) Change in Taxes (dollars) (2)	(6) Marginal Tax Rate (%) (5) ÷ (4)
2,300	0	0	—	—	—
3,400	121	3.5	1,100	121	11
4,400	251	5.7	1,000	130	13
8,500	866	10.2	4,100	615	15
10,800	1,257	11.6	2,300	391	17
12,900	1,656	12.8	2,100	399	19
15,000	2,097	14.0	2,100	441	21
18,200	2,865	15.7	3,200	768	24
23,500	4,349	18.5	5,300	1,484	28
28,800	6,045	21.0	5,300	1,696	32
34,100	7,953	23.3	5,300	1,908	36
41,500	10,913	26.3	7,400	2,960	40
55,300	17,123	31.0	13,800	6,210	45
69,100	24,023	34.8	13,800	6,900	50

Source: IRS Form 1040, Schedule X, 1983.

income. Therefore, mortgaging a house rather than renting can lower tax obligations. Deductions and exemptions such as these may result in tax inequities.

Corporate Income Tax

The federal government charges incorporated businesses an income tax on their accounting profits. The corporate income tax is progressive, with a maximum marginal tax rate of 46 percent for profit income over $100,000. The after-tax profits can be distributed to the shareholders (owners) of a corporation in the form of dividend payments. The shareholders' income is also subject to personal income taxes. Thus corporate profits are taxed twice, once from corporations and once from shareholders.

Social Security Taxes

Social Security taxes, the second largest source of federal tax revenues and the fastest-growing revenue source, represent taxpayers' contribution to federal programs that support the elderly, ill, and disabled. The Social Security tax is a payroll tax in which the employer and the employee each pay 6.7 percent of the employee's income up to $35,700 (as of 1984). The marginal tax rate is zero for income above $35,700. Nonlabor income is not subject to Social Security taxes. These features lead to a vertically and horizontally inequitable tax. For example, an individual with wage income of $35,700 pays $2,191.90 in Social Security taxes, and an individual who earns $50,000 in wage income pays the same amount because income over $35,700 is not taxed. Since unequals are not treated unequally, vertical inequity results. As currently structured, the Social Security tax is also horizontally inequitable: Equals are not treated equally because interest income is not taxed. An individual who earns $35,700 in interest from bonds pays no Social Security tax.

Property Taxes

The largest source of state and local government tax revenues is the property tax. Owners of land and permanent structures are charged a tax based on the assessed value of their property. The assessed value is usually a percentage of the market value; that is, the selling price of the buildings and land.

A property tax is actually a tax on wealth rather than on income. Individuals who hold their wealth in the form of property pay more taxes than people who hold their wealth in other forms such as gold, art, or bonds. For example, people who hold bonds are taxed on the income from bonds—just as property owners are taxed on rental income—but they are not taxed on the value of the bonds, as property owners are taxed on the value of their property. From this perspective, the property tax may be regarded as inequitable.

Sales and Excise Taxes

Sales and excise taxes also account for a large portion of state and local tax revenues. They are levied on the purchase of consumer goods. A sales tax is a specific percentage of the price of a good; an excise tax is a per unit tax. For example, a sales tax may be 5 percent of the price of a good regardless of the price. Under a sales tax, as the price of a good rises, the amount of the tax per unit rises. On the other hand, an excise tax may be 5 cents per unit of the good. The amount of the tax per unit does not change as the price rises or falls.

Sales taxes levied by state and local governments vary from state to state. In many locations, items such as food and medicine are exempt from sales taxes. Excise taxes are levied both by national and by state and local governments but only on selected items. For example, liquor, tobacco, gasoline, and certain sporting goods have excise taxes. Frequently these tax revenues are earmarked for specific government projects.

The purchase of nonconsumption goods such as stocks and bonds is not subject to a sales tax. This feature of sales taxes may create a horizontally inequitable tax distribution when people with the same income spend their money in different ways. An individual can pay less in sales taxes by spending more on investment goods than on consumption goods. Furthermore, if higher-income families invest more of their income than lower-income families, the sales tax is a regressive tax.

The Effects of Taxes

The nature of the U.S. tax system effectively eliminates total tax avoidance—almost all goods and many resources are taxed. Indeed, this situation is desirable. Taxes are intended to eliminate free riders and thus should be unavoidable. Yet the burdens of taxes and the attempt to avoid them do influence individuals' activities. Altering economic behavior is sometimes intentional and desirable for society as a whole; other times it leads to undesirable economic inefficiency.

Neutral tax: A tax that has no effect on the production or consumption of goods.

Taxes that have no effect on behavior are called **neutral taxes.** Taxes are rarely neutral. More typically, by affecting price, they affect the quantity of goods supplied and demanded and the allocation of resources. Furthermore, the burden of taxes is frequently shifted away from the group that was intended to pay the taxes. Taxes are necessary to obtain desired public goods, but taxation can have a detrimental effect on the overall productive capacity of the economy.

Basic Effects on Price and Quantity

To understand how taxation affects price and quantity, consider a simple excise tax. Let D and S in Figure 19–5 represent, respectively, the demand and supply of gasoline in the United States with no taxes. P_0 and Q_0 represent the original price and quantity. Then assume that the federal government imposes an excise tax of t cents per gallon for every gallon sold. This tax effectively increases the cost to the suppliers of producing and selling gasoline.

What happens? The supply of gasoline decreases. The supply curve shifts upward by the amount of the tax. Before the tax, producers were willing to offer quantity Q_0 at price P_0; after the tax, the producers are willing to offer Q_0 but only at a higher price, P_0 plus the tax t. The tax results in a new equilibrium E_t. The price increases, as we would expect, but probably not by the full amount of the tax. In this example, if the original price was $1.30 and the tax is 20 cents, then the price to consumers, P_c, rises by only 10 cents to $1.40. In this case, the price that producers receive after paying the tax is $1.40 minus 20 cents, or $1.20. The price the producers receive, P_p, may be found on the original supply curve at the new reduced quantity, Q_t.

The tax thus has two primary effects in the market for the taxed item. First, it changes the relative prices that sellers and buyers face. In fact, it lowers the price that sellers receive and raises the price that buyers pay. Second, as we would expect, the quantity exchanged falls—the lower price

---

FIGURE 19–5

Effects of an Excise Tax

An excise tax shifts the supply curve upward by the amount of the tax. This shift increases the price paid by consumers from P_0 to P_c and lowers the net price received by the sellers from P_0 to P_p. The tax also lowers the amount produced and consumed from Q_0 to Q_t. The shaded areas represent tax revenues for the government and the welfare loss to society, discussed later in this chapter.

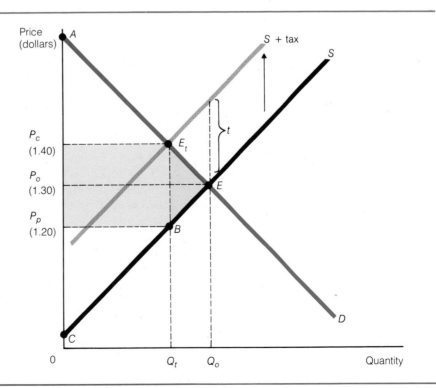

to sellers encourages them to offer fewer goods or services, and the higher price to buyers encourages them to buy fewer goods or services. The decrease in quantity depends on the size of the tax and on the elasticities of supply and demand.

If supply or demand for a good or service is perfectly inelastic, a tax will not affect quantity and is therefore neutral. This is rarely the case. As we have seen, elasticity varies at different price levels. The higher the tax-increased price, the more likely that the tax will affect quantity. With the Social Security tax, employers and employees turn over a fixed percentage of the employees' paychecks to the government. This tax increases the cost of labor to the firms and decreases the after-tax wage that workers take home. If the demand and supply of labor are less than perfectly inelastic, as is the case with most goods and services, the tax will encourage firms to hire fewer workers and discourage people from entering the labor force or encourage them to work fewer hours. If the supply of labor is relatively inelastic or if the percentage of wages paid in Social Security tax is lower, then the tax has relatively little effect on the level of employment. As the tax rises, more people seek ways to support themselves without earning heavily taxed wages, from fixing their own cars and growing their own food to investing in tax-exempt enterprises.

Although the burden of the Social Security tax is theoretically shared equally by employers and employees, who really pays taxes levied on items such as consumer goods? The distribution of the tax burden between buyers and sellers is known as tax incidence.

Tax Incidence

Tax incidence: The burden of a tax or fiscal resting place.

At a basic level, we know that the buyers of goods in competitive markets cover the entire cost to producers—production costs plus any and all taxes. However, we can determine the actual burden of the tax—the **tax incidence**—by observing the relative change in price to producers and consumers. Notice that in the excise tax example in Figure 19–5, the price to producers fell by 10 cents ($P_0 - P_p$) and the price to consumers increased by 10 cents ($P_c - P_0$) after imposition of the tax. These changes indicate that the tax incidence was shared equally by the buyers and sellers—each faced a price change of 10 cents.

The incidence of a tax is not always shared equally by buyers and sellers—indeed, we suspect that an equal incidence is a rare event—but it *is* most often shared to some degree. It is most unusual for the entire burden of a tax to be shifted forward to the buyers or backward to the sellers. In other words, the price to consumers does not usually increase by the full amount of the tax or decrease to producers by the full amount.

Tax shifting depends on the elasticity of demand and supply. In general, the more elastic the demand for a product, the greater the incidence of tax shifted backward to the sellers. Conversely, the more elastic the supply, the greater the incidence shifted forward to the buyers. In some circumstances the tax incidence may be shifted entirely to one group or the other. In particular, if demand is perfectly inelastic or if supply is infinitely elastic, the entire burden is shifted to the buyers. The entire burden is shifted to the sellers if demand is infinitely elastic or if the supply is perfectly inelastic. Such extremes of elasticity are not common.

Regardless of whether a tax is shared or is borne entirely by the buyer or the seller, it drives a wedge between the price that buyers pay and the price that sellers receive. This chunk claimed by the government occurs whether the tax is an excise tax, a sales tax, a payroll tax, a property tax, or an income tax. For example, the income tax drives a wedge between the price that employers pay for labor and the price employees receive for their services.

The Allocative Effect of Taxes

Unless taxes are neutral, they affect the allocation of resources. Taxes change relative prices, so the relative quantities of goods produced tend to change. In Figure 19–5, the imposition of the tax caused the equilibrium quantity to fall from Q_0 to Q_t. The quantity falls in such a case for two reasons: The price to consumers increases and the price to producers falls. The decrease in production and consumption represents a loss to society that can be measured in terms of consumer and producer surplus.

The consumer surplus at the original price is represented by the triangle AEP_0, and the producer surplus by P_0EC. When the tax causes an increase in the price to consumers, consumer surplus falls to AE_tP_c; when the tax lowers the price to producers, the producer's surplus falls to P_pBC. The loss to producers and consumers is represented by the area $P_cE_tEBP_p$. The area $P_cE_tBP_p$ represents tax revenues for the government (t times Q_t), so revenue is not lost to society as a whole, but the area E_tEB is lost. It is a deadweight loss to society that is similar to the welfare loss due to monopoly.

In some cases, a decrease in quantity is intentional because it is thought to be socially desirable. For example, if government wants people to consume less liquor or tobacco, an excise tax may discourage drinking and smoking. Or government may encourage the consumption and production of some goods by not taxing these goods and taxing all others. If government wants to increase consumption of agricultural products relative to other goods, for example, then agricultural products may be exempted from a general sales tax.

Tax policies may have unintended allocative effects. For example, property taxes may cause geographical shifts in population and industry. The aggregate supply of land, many people believe, is perfectly inelastic. However, the current property tax system is not neutral in its effects because demand for land is elastic. State and local governments tax land according to market value, which varies according to demand. Owners of land in the center of cities and other strategic locations must therefore pay higher taxes than owners of less valuable land. These high property tax rates may drive firms and homeowners out of such areas in search of lower taxes, other things being equal.

To summarize, the allocative effect of a tax depends on the size of the tax and the elasticity of supply and demand for the product. In general, the larger the tax, the greater the reduction in the production and consumption of the good. Also, the more elastic the supply or demand for the good, the greater the reduction in production due to the tax. Conversely, when demand or supply is perfectly inelastic, the tax has no effect on the quantity of the taxed good; the tax is said to be allocatively neutral.

The Aggregate Effect of Taxes

By changing quantities demanded and supplied and thus reallocating resources, taxes have a strong impact on the economy. Some economists are concerned that the impact of taxes may stunt the growth of the economy. As we have indicated, for example, payroll taxes lower real wages to workers and thus may decrease the participation of workers in the labor force (a phenomenon discussed in Economics in Action at the end of this chapter). If this occurs, then the nation's total output will fall because of the tax.

Taxes on savings can also affect the overall performance of the economy. Interest obtained from bonds or savings accounts is subject to the personal income tax. The tax lowers the interest rate received from savings, so people tend to save less. With a decrease in savings, fewer funds are available for business purchases of capital. The tax on interest income can therefore diminish the rate of capital formation.

To encourage growth, the government sometimes cuts taxes on certain activities. The government may, for example, encourage productive behavior by lowering the tax rate for businesses or individuals engaged in activities that increase the supply of resources. For example, the investment tax credit lowers the tax liabilities of firms that invest in new production equipment. Moving expenses are also deductible when people relocate for new jobs. Such tax credits and deductions can improve the productive capacity of the nation and thus increase real output.

The overall effect of taxes on production in the United States is not certain. However, the burden of taxes is growing, and the notion that their effect is detrimental to economic performance is gaining popularity. Many people are calling for decreased taxes and for tax reform. Some economists even believe that when tax rates get too high, tax revenues to the government actually fall (see Focus, "The Laffer Curve"). The important issue is whether the net effect of taxes encourages or discourages productive efficiency, or, more specifically, whether we have the optimal amounts of public goods and whether our methods of taxation result in the most efficient use of resources.

Tax Reform

The U.S. tax system has become so complex that the federal tax code, the textbook of federal tax regulation, is more than 40,000 pages long. Individual tax liabilities are difficult to determine without the help of specialists. Each taxpayer is also subject to a maze of sales, excise, and property taxes. And an unknown portion of the incidence of business taxes is passed on to taxpayers in the form of higher prices or lower wages.

Taxpayers are beginning to voice their displeasure with the current level and nature of taxes. People have always wanted a lower tax incidence. The current tax system itself may be inequitable and inefficient. If so, tax reform could improve the equity of the tax incidence and the overall productivity of the economy.

The Need for Tax Reform

Much of the concern about our present tax system centers on the income tax. The nature of the current personal and corporate income tax tends to divert resources away from the production of desirable goods and services. For example, the personal and corporate income taxes induce people to devote resources to tax avoidance rather than to productive activity.

The income tax system allows a maze of deductions and exemptions that change through time. Households and businesses employ resources, both labor and capital, in an effort to minimize their tax liability. Keeping records and receipts throughout the year and staying abreast of new tax loopholes is not a trivial task. Several hundred million dollars are spent on professional

---

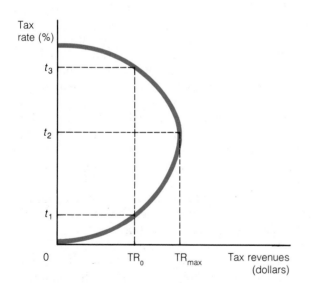

FOCUS   The Laffer Curve

In the early 1980s, the Laffer curve, named for economist Arthur Laffer, became a favorite illustration of those pressing for cuts in the personal income tax. Essentially, the Laffer curve represents the relation between the tax rate and tax revenues. It predicts that as the tax rate rises, total tax revenues will rise at first and then begin to fall. The importance of the Laffer curve for tax cut advocates is its suggestion that the government may actually lose revenues by increasing tax rates too much (see Figure 19–6).

The Laffer relation can be demonstrated with a simple excise tax. In Figure 19–7, S represents an infinitely elastic supply curve for a good produced by a constant-cost industry. When a simple, per unit excise tax is imposed on the good, the supply curve shifts upward by the amount of tax. Demand for the product is represented by D, and total tax revenues are shown in the shaded areas (the tax rate, t_1, t_2, and so on, times the amount sold, Q). For example, with excise tax t_1, the total tax revenues are t_1 times Q_1, or area $(P + t_1)AEP$. A higher tax rate will increase tax rev-

enues. With excise tax t_2, tax revenues are represented by the shaded area $(P + t_2)BFP$. Notice that the excise tax t_2 maximizes tax revenues in this model. A higher tax, t_3, actually results in lower tax revenues, area $(P + t_3)CGP$. $(P + t_2)$ is identical to the price that would maximize profits if this industry could become a monopoly or a cartel. Indeed, with an excise tax the government can restrict output and raise price in an industry. The profits, however, are received by government in the form of tax revenues.

In this example, the elasticity of demand determines which tax will maximize revenues. Debate over the Laffer curve and the advantages of cutting income taxes has focused on this issue of elasticity. In theory, if the supply of or the demand for labor is perfectly inelastic, then higher tax rates would always yield higher tax revenues. If labor demand or supply is elastic, however, then raising taxes will not necessarily result in higher revenues, and cutting taxes may actually increase revenues.

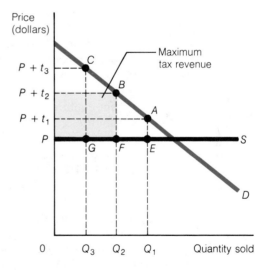

FIGURE 19–6   The Laffer Curve

When the tax rate is zero, total tax revenues are zero; but when the tax rate rises, revenues rise at first and then fall. This suggests that there is a tax rate that maximizes tax revenues.

FIGURE 19–7   Tax Revenues at Different Tax Rates

Total tax revenues are a function of the tax rate and the elasticity of demand and supply. Excise tax t_1 results in tax revenues represented by the area $(P + t_1)AEP$. Maximum revenues occur at rate t_2. A higher rate such as t_3 results in lower tax revenues because of the high elasticity of demand.

FIGURE 19–8

The Effect of Tax Reform

An inefficient tax system can depress the supply of resources. Point *A* shows a low level of production with an inefficient tax system. If tax reform results in a more efficient tax, then there would be greater production of both public and private goods—point *B* on the shifted production possibilities curve.

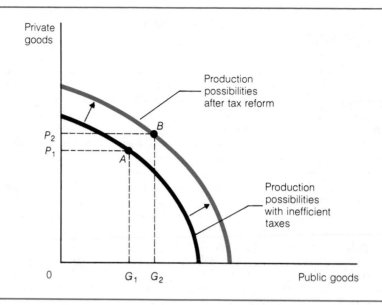

Inefficient tax: A tax that decreases the overall productive capacity of the country; a tax that decreases the supply of resources or decreases the efficiency of resource use.

Flat-rate tax: A proportional income tax with no exemptions or deductions.

tax services each year. Not only is labor employed in bookkeeping for tax purposes, but many lobbyists are employed by industry in an effort to obtain more loopholes from Congress. While these resources are employed in the production of a service that is valuable to the recipients of the tax reductions, they are not necessarily employed in the production of goods and services that benefit all of society.

Figure 19–8 shows the effects of an **inefficient tax** system on the productive capacity of the nation. Point A shows the level of output of publicly produced goods, G_1, and privately produced goods, P_1, with an inefficient tax system; that is, one that suppresses the supply of resources to the production of goods and services. Tax reform that encourages a more efficient tax system could increase the supply of resources and thus shift the production possibilities curve outward. More of all goods—public and private—could be produced with a more efficient tax system.

Suggested Reforms

A tax system that is allocatively neutral, equitable, and efficient is certainly desirable. Although no known tax is both neutral and equitable, some economists believe there are more efficient and equitable tax systems than the current U.S. system. One radical reform would be a flat-rate tax on personal income. A milder version of the flat-tax concept, embodied in President Reagan's 1985 tax reform bill, is currently making its way through Congress.

Flat-Rate Tax. Some economists support the idea of replacing our present progressive tax with a **flat-rate tax,** a simple proportional tax with no exemptions or deductions. For example, a 20 percent tax on income for every level of income with no loopholes would generate an equitable tax. Everyone with the same income would pay the same amount in taxes, and people with higher incomes would pay more in taxes than people with lower incomes. The flat-rate tax would probably eliminate some of the costs of bookkeeping and the uncertainty of tax liability. Also, if Congress were not allowed to institute deductions or exemptions for special interests, lobbying efforts would probably decrease. Furthermore, the flat rate would eliminate the high marginal tax rates that can discourage work efforts. Problems emerge with implementing this proposal, however, including political opposition from those whose taxes would rise and the possible burden it would place on those with little ability to pay.

Reagan's Tax Reform Plan of 1985. The flat-tax concept has been modified by the Reagan administration into a sweeping legislative proposal to simplify the Federal tax code. Instead of a single tax rate for individuals, President Reagan proposes three tax brackets: a top rate of 35 percent (for all single incomes over $42,000), a middle rate of 25 percent (for single incomes between $18,000 and $42,000), and a bottom rate of 15 percent (for single incomes above $2,900). Anyone earning less than $2,900 would pay no income tax at all. Joint incomes would also be taxed in three brackets, but the income ranges would proportionately higher in each bracket.

The Reagan proposal eliminates many but by no means all of the deductions available to taxpayers. Deductions for interest payments on home mortgages, charitable contributions, and medical expenses would still be allowed. Wealthy taxpayers, however, would find fewer "loopholes" with which to reduce their tax bill. For their part, businesses would bear a higher percentage of the total tax bill. Corporate taxes would increase, in some estimates, by 23 percent.

It is too early to say whether the reform will achieve its primary purpose of reducing the enormous complexity and cost associated with the present system. If tax reform, as promised, results in a more efficient tax structure, then economic growth will be stimulated. At the same time, tax reform will have many effects on the incentives for individuals to invest. If, for example, homeowners can no longer claim credit on their tax bill for spending on energy-saving devices, the demand for solar heating panels and other such innovations will decline. If investors cannot deduct the interest for "second home" mortgages, the housing market will feel the effects.

As with any legislative proposal, the effort to simplify the tax code will depend as much upon political realities as upon economic costs and benefits. Who wins and who loses with tax reform? The myriad decisions involved rest with elected officials, whose behavior — the subject of our next chapter — is a matter of economic concern as well.

Summary

1. Taxes are a means of financing public goods. Although some taxes are levied directly on the users of public goods, in many cases beneficiaries cannot be explicitly defined, so general taxes must therefore be used to diminish free ridership.
2. Taxes are levied according to two basic principles: the benefit principle and the ability-to-pay principle. The benefit principle suggests that the people who receive the greatest benefit from public programs should pay the most taxes. The ability-to-pay principle suggests that people with the greatest ability to pay taxes should pay the most.
3. The ideal of equity in taxation has both a horizontal and vertical dimension. An equitable tax takes greater revenues from higher-income groups than from lower-income groups and also taxes people with equal incomes equally.
4. In general, the relation of taxation to level of income falls into one of the following categories: A proportional tax takes a constant percentage of income regardless of the level of income. A progressive tax takes a larger percentage of income as income rises. A regressive tax takes a smaller percentage of income as income rises.
5. The marginal tax rate is the increase in tax liabilities that results from an increase in income. With a progressive income tax, the marginal tax rate is greater than the average tax rate.
6. The U.S. government imposes a progressive income tax. Because of loopholes—exemptions and deduc-

tions—the federal income tax has both horizontal and vertical inequities.

7. Most existing taxes effectively drive a wedge between the price buyers pay and the price sellers receive for a product or resource. The burden of a tax is usually shared by the buyers and sellers of the taxed item.

8. Taxes usually reallocate resources and may decrease the supply of resources. Taxes that decrease the productivity of the economy are inefficient. A tax that is neutral has no effect on the allocation of resources.

9. Tax reform such as the proposed flat-rate tax or a tax simplification scheme could lead to a more equitable tax incidence and to an increase in overall productivity and economic growth.

Key Terms

public finance	proportional income tax	marginal tax rate
benefit principle	progressive income tax	neutral tax
ability-to-pay principle	regressive income tax	tax incidence
horizontal equity	taxable income	inefficient tax
vertical equity	average tax rate	flat-rate tax

Questions for Review and Discussion

1. Why do we have taxes? What principle of taxation does a user tax follow? What principle does a property tax follow?

2. Which principle of taxation does the U.S. income tax follow? Is the U.S. income tax horizontally and vertically equitable? If a tax is not horizontally equitable, can it be vertically equitable?

3. When is a regressive income tax vertically equitable? With a regressive income tax, is the marginal tax rate greater than the average tax rate?

4. Draw the demand curve for a product with a supply curve that is perfectly inelastic. Show the effect of an excise tax. Who bears the burden of the tax? Is the tax neutral?

5. Would a property tax system that makes taxes identical at every location in the country be more efficient than the present property tax method?

6. Does the government use valuable resources to prevent tax avoidance? When state and federal agents attempt to detect and prevent bootlegging, are they attempting to protect a tax base? If there were no tax on liquor, would the government prevent bootlegging?

7. Does an income tax decrease the amount of labor supplied? If the supply of labor were perfectly inelastic, would an income tax decrease the level of employment? Who would bear the entire tax burden?

8. If consumption expenditures rather than income were taxed, would the after-tax wage still fall? Would a consumption tax decrease the amount of labor supplied? Would a consumption tax decrease the amount of capital supplied?

9. What would happen to the amount of capital in the United States if the interest income from savings were not taxed? Describe a tax that could increase the supply of capital.

10. Some states have a competitive market for alcoholic beverages in bottles, but they charge an excise tax. Other states have state-owned and -operated liquor stores that are profit-making monopolies. Under which of these two systems would a state maximize revenues? Could a state make more money by selling the monopoly right to the highest-bidding firm?

11. What are the benefits of eliminating tax loopholes? How can tax loopholes be used to improve the productivity of the economy?

12. How the Reagan tax plan affect tax equity? How would it affect economic growth?

⋀⋁⋯ ECONOMICS IN ACTION   **If You Are Taxed More, Will You Work Less?**

Just about everyone has heard someone complain about high tax rates, especially in regard to earning extra income such as overtime pay. The percentage of extra income taken in taxes is determined by the marginal tax rate. With a progressive income tax, the marginal tax rate rises, so extra income is taxed at a higher rate than the rate at which the first dollars earned are taxed. As we have suggested, this system may make people who have the opportunity to earn extra income resent the higher tax rate, but does it actually affect their work decision? Specifically, does a progressive tax rate have a more detrimental effect on the supply of labor than a proportional tax rate, such as the proposed flat-rate tax?

Any tax based on labor income effectively decreases a worker's wage. A proportional tax decreases the wage by a constant amount regardless of the hours worked. For example, if a worker earns $10 per hour, a 20 percent income tax would lower his or her after-tax wage to $8 per hour regardless of the number of hours worked. A progressive income tax results in a lower hourly wage as the number of hours worked increases. For example, if the worker chooses to work 15 hours in a week, he or she may be taxed at a 10 percent rate, whereas 25 hours of work may be taxed at a 20 percent rate and 40 hours at a 30 percent rate. Figure 19–9 shows the relative effects of progressive and proportional tax rates. Both lower the after-tax income and thus lower the wage rate, but by different amounts at different numbers of hours worked.

What effects do these two patterns of taxation have on the amount of labor supplied? As we suggested in Chapter 12, a decrease in wages has two effects on labor supply decisions: an income effect and a substitution effect, which work in opposite directions. The income effect suggests that a wage decrease lowers real income and thus encourages individuals to work more. The substitution effect suggests that a wage decrease lowers the price of leisure and encourages individuals to buy more leisure; that is, to work less. Because these two effects work in opposite directions, the combined effect of a wage decrease on the amount of labor supplied cannot be determined. However, we may suggest the following: If per hour wages decrease because of the income tax, workers with an upward sloping supply curve of labor will work fewer hours, and workers with a negatively sloped supply curve will work more hours. Furthermore, if the aggregate supply of labor is upward sloping, an income tax will decrease the overall amount of labor supplied.

The relative effects of proportional and progressive tax rates on the supply of labor can also be determined. In Figure 19–9, the two tax rates result in equal tax revenues at 40 hours of work per week. At this level, the income effect of the two tax rates is therefore identical, for they both decrease income by the same amount. However, the substitution effects are not the same under both tax systems. The progressive income

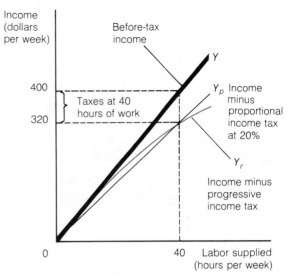

FIGURE 19–9 Before- and After-Tax Income with Proportional and Progressive Tax Rates

The Y curve represents the level of income earned before tax withholdings. The Y_p curve shows after-tax income with a proportional tax rate, and Y_r shows after-tax income with a progressive rate. As workers increase the number of hours they work, their after-tax hourly wage falls. The decline is steady with a proportional tax, but not with a progressive tax. In this example, both tax schedules result in identical tax revenues at 40 hours of work per week: $80 for the government, with $320 left as take-home pay.

tax has a stronger substitution effect than the proportional tax because it lowers the wage at the margin by a greater amount. This difference suggests that the progressive income tax will have a more detrimental effect on the amount of labor supplied than the proportional tax when both taxes bring in the same amount of tax revenues.

Question

Consider that after graduation you can earn, over a period of years, from $25,000 to $80,000 per year. What factors would your income and substitution effects depend on? Given these factors, would your own work effort be reduced at any given annual income?

20

The Theory of Public Choice

Public choice: The economic analysis of politics; the branch of economics concerned with the application of economic principles to political decision making.

*T*he discussions of public goods and externalities in Chapter 18 and taxation in Chapter 19 were based on the concept of market failure. When or where markets do not produce ideal outcomes, government can and often does play a role in correcting the situation. In Chapter 18 we more or less assumed that government is capable of making these corrections. The overall capability of government to compensate for market failure, however, is itself an issue. Democratic governments act through a unique set of institutional arrangements, such as majority voting, representative democracy, political parties, bureaucracies, and special-interest groups. The theory of **public choice,** introduced in this chapter, is an attempt to apply economics to understand and to evaluate how democratic governments operate. A central precept of public choice theory is that government is not perfect; like markets, governments can fail to achieve ideal outcomes. Appropriate solutions to economic problems in the economy are frequently a choice between markets that are not perfect and government programs that are not perfect. Appropriate public policy cannot be decided simply by saying "Let the market work" or "Let government do it." Rather, the costs and benefits of alternative institutional arrangements must be carefully examined case by case.

The fundamental premise of public choice is that political decision makers (voters, politicians, bureaucrats) and private decision makers (consumers, brokers, producers) behave in a similar way: They all follow the dictates of rational self-interest. In fact, political and economic decision makers are often one and the same person—consumer and voter. The individual who buys the family groceries is the same individual who votes in an election. If the premise of public choice is correct, we can learn a lot about issues such as

Drawing by Mankoff; © 1984 The New Yorker Magazine, Inc.

why people take the time and trouble to vote by applying the same logic to the voting decision that we have applied to the grocery-buying decision. As we do so, we should keep in mind that the institutions and constraints facing political decision makers are different from those facing private decision makers. For example, the benefits of buying groceries are fairly obvious, but what are the benefits of voting? We will be careful to specify such differences between private and public choices throughout our analysis.

Public choice, like the rest of economics, has both a normative and positive side. **Normative public choice** involves value judgments about the desirability of certain political situations. In issues such as the design of voting procedures, for example, how does majority rule compare with other voting rules in reflecting the true preferences of voters? Is there a better voting rule or process through which to make political decisions? Normative public choice looks at the way political institutions work from the standpoint of how we might make them work better.

Positive public choice seeks to explain actual political behavior. Why do high-income individuals vote more often than low-income individuals? How do committees and legislatures work? What is the impact of special-interest groups on government? In this chapter we will examine both normative and positive aspects of public choice after first exploring how self-interest affects decision making.

Normative public choice: The study of shortcomings and possible improvements of political arrangements, such as voting rules.

Positive public choice: The analysis and explanation of political behavior.

The Self-Interest Axiom

The public choice approach to politics is based on the idea that political actors are no different from anyone else: They behave in predictable ways. We do not seek to serve the public interest when we vote or the private interest when we buy a car. We seek our self-interest in both cases.

Nevertheless, market behavior and political behavior differ in the constraints facing decision makers. The market is a **proprietary** setting—one in which individuals bear the economic consequences of their decisions which either enhance or decrease their wealth. In the market economy, a firm that produces a new product stands to make profits or losses depending on the quality of its efforts. The political arena is a **nonproprietary** setting—one where individuals do not always bear the full economic consequences of their decisions. For example, the political entrepreneur who comes up with a new

Proprietary: Relating to private ownership and profit seeking by private owners.

Nonproprietary: Relating to public ownership.

political program does not bear the full costs of the program if it is a failure and does not reap all the benefits if it is a success. Behavior will therefore differ in market and in political settings, not so much because the goals of behavior are different but because the constraints on behavior are different.

We can distinguish the economic constraints at work in a market or proprietary setting and in a political or nonproprietary setting by analyzing the roles of agents and principals in each case. In both settings the agent, whether a firm or a politician, agrees to perform a service for the principal, whether a consumer or a voter. Because the agent and the principal are both self-interested, it is likely that the agent will not always act in the interest of the principal, particularly if the behavior of the agent is costly to monitor.

The agent-principal problem has been analyzed primarily in a private setting. For example, the corporate manager is an agent, and stockholders are principals. How do the stockholders get the manager to act in their interest? They do so primarily through economic incentives. Managers of private firms have incentives to control costs in their firms because increased costs cause a decrease in the firm's profitability. Stockholders, with such mechanisms as stock options and takeover bids, can discipline managerial behavior toward maximizing wealth.

Managers of political "firms," or bureaucracies, do not face a similar incentive to control costs. They cannot personally recoup any cost savings that they achieve for their agencies, and the means available to voters to curtail poor performance by political managers are minimal and costly to implement. This does not mean that public officials can do anything that they want to do. Like any other economic actors, they are constrained by costs and rewards.

The main point about the agent-principal problem is that political agents face different constraints on their behavior than do private agents because the principals in the two cases face different incentives to control the behavior of the agents. This difference is the focus of public choice analysis.

Normative Public Choice Analysis

Normative public choice theory analyzes the performance of political institutions in serving voters. In this section we illustrate the usefulness of this approach by examining the process of voting by majority rule and by introducing the problem of constitutional choice.

Majority Voting[1]

In a democracy, public choices are typically based on a majority vote: If candidate A wins one more vote than candidate B, then candidate A is elected to office. In some cases, particularly when more than two choices are available, majority rule does not always accurately reflect voter preferences.

Imagine a three-person nominating committee that must choose among three alternatives—Smith, Jones, and Tobin—as the nominee for club president. Each committee member ranks the three candidates in order of preference (first, second, third), and the winner is to be selected by majority vote (two out of three committee members in favor). Table 20–1 shows the rankings of the three candidates by the committee members (A, B, and C). Member A, for example, prefers Smith to Jones and Jones to Tobin. Note

---

[1]The inventor of this analysis was Kenneth Arrow, a Nobel Prize winner in economics. See his *Social Choice and Individual Values* (New York: Wiley, 1951).

TABLE 20–1   **A Voting Problem**

The committee must choose among Smith, Jones, and Tobin. The candidate receiving a simple majority (2 votes) wins. When each candidate is paired against another (Smith versus Jones, Jones versus Tobin, Tobin versus Smith), no clear winner emerges. Smith beats Jones, Tobin beats Smith, but Jones beats Tobin. This is a problem of simple majority voting on more than two candidates two at a time. Individual voters can clearly rank candidates, but majorities cannot. Majority voting in this case leads to a repetitive cycle of winners.

Committee Member	Smith	Jones	Tobin
A	1	2	3
B	2	3	1
C	3	1	2

that each member has consistent preferences; that is, each member is able to rank the three candidates in order of preference.

Now let the voting begin. To find the preferred nominee, the committee pairs each candidate against one of the other two until a winner is found. Suppose they start with Smith against Jones. Who wins? *A* votes for Smith, *B* votes for Smith (*B* prefers Tobin overall but between Smith and Jones he prefers Smith), and *C* votes for Jones. Smith wins by a majority vote of 2 to 1. Can Smith now beat Tobin? *A* votes for Smith, *B* votes for Tobin, and *C* votes for Tobin. Tobin wins that round. Now, how does Tobin fare against Jones? Since Smith previously beat Jones, we would also expect Tobin to dominate Jones because Tobin dominated Smith. But look what happens: *A* votes for Jones, *B* votes for Tobin, and *C* votes for Jones. This time Jones wins! Thus in three rounds of voting, a different candidate wins each round.

The committee's stalemate points up a problem with majority voting. Under majority voting each individual voter can have consistent preferences among multiple candidates or issues, and yet voting on candidates two at a time can lead to inconsistent collective choices. In other words, individuals can make clear choices among Smith, Jones, and Tobin, but majorities cannot. Majorities will choose Smith over Jones, Tobin over Smith, and Jones over Tobin, and on and on in a repetitive voting cycle until a way is found to stop the voting. Either Smith, Jones, or Tobin will win when the voting is stopped, but only because the voting stopped at a particular point, not because one candidate is a clear winner over the other two.

This simple proof strikes at the rationality of majority voting procedures. It says that although individual choices will be clear and rational, majority choices will be inconsistent and cyclical. Where does such a paradoxical result lead us? First, it leads us to look for a better voting procedure than the simple majority rule when voting on two issues or candidates at a time. We explore this possibility in Focus, "A National Town Meeting," which discusses a proposal for national computer voting, and in Economics in Action at the end of the chapter, which examines how voting may be designed to protect minority interests. Second, we note that the problem of voting cycles is mitigated to some extent by the institutions of democracy. Our legislatures and committees do not cycle endlessly among alternative proposals without reaching a decision. Votes are taken and decisions are made. Attempts to reach agreement do not go on forever. Powerful committee chairpersons, for example, can call for votes on legislation and see to it that the legislature

FOCUS   **A National Town Meeting**

Direct or participatory democracy, in which each citizen has the opportunity to cast his or her vote on all public issues, is often held up as the ideal method of reaching a political consensus. The use of such a voting process has been limited historically—the Athenian agora and the New England town meeting are the prime examples—chiefly because of the costs of organizing large groups. But the advent of small computers makes conceivable in the near future the direct voting system proposed by James C. Miller III.

In conjunction with cable television or other data transmission systems, Miller envisions personal computers being used to register and tally the votes of millions of individuals on legislative issues now decided by elected representatives. One can imagine that voters would spend a portion of each evening sitting in front of their computers examining the legislation that had been brought up for consideration, casting their votes, and soon after learning the results of the night's balloting.

Although the computer would vastly reduce the costs of registering and counting votes, it would not appreciably lower the cost to the public of becoming informed about issues. Miller's proposal consequently allows voters to assign proxies to experts or specialists. The proxy could be a general one, giving the proxy holder permission to cast a vote on all proposed legislation, or it could be confined to particular topics. For

example, a voter could assign a proxy to an economist to vote on economic matters, to an expert in military affairs for votes on defense issues, and so forth. Voters would assign proxies to individuals who would be expected to cast ballots in the same way the voter would if the voter were informed about the proposed legislation. The assignment of a general proxy would be similar to electing a congressional representative or senator except that proxy holders would represent the views of each voter and not just the views of the majority. Moreover, the speed of computers would allow proxy assignments to be made for as long as each voter wished—perhaps only an hour—instead of for fixed terms of two to six years.

Under Miller's system a role would exist for a legislative assembly to propose bills or to debate important issues. He suggests that the "house of representatives" be composed of the 400 individuals wishing to serve there who have the largest number of proxies and that the "senate" be constituted from the 100 largest proxy holders. The possibility of quick recall would assure that each representative would exercise the proxy exactly as voters wish.

The three main elements of Miller's proposal—direct registering of citizen votes, proxy assignments, and a bicameral legislative body—are not far removed from the current system established by the U.S. Constitution. The main difference is that technological development makes it possible for the democratic ideal of a small New England town meeting to operate on a national scale.

Source: James C. Miller III, "A Program for Direct and Proxy Voting in the Legislative Process," *Public Choice* 6 (Fall 1969), pp. 107–113.

---

produces new laws. While such legislative procedures do not solve the problem of inconsistent choices in majority voting, they do keep the tendency toward voting cycles in check.

Constitutional Choice

In a constitutional democracy there are two levels of political decision making: day-to-day decisions about running the government, which are made under given rules and procedures, and decisions about rules and procedures themselves. The first level of decision making takes place under given political constraints. The second involves the choice of the constraints themselves, which amounts to a constitution of the set of rules under which day-to-day government operates. For example, the choice of whether to employ simple majority voting or some other system would be a constitutional choice, while voting on current issues with the chosen voting system would be part of the day-to-day process of running the government.

Public choice theory can be applied to an analysis of constitutional deci-

FIGURE 20–1

Constitutional Choice

Curves D and E represent the expected costs facing the constitutional decision maker. D represents the costs of collective decision making; as the number of voters increases, D increases at an increasing rate. E is the cost of a decision rule requiring less than unanimity. Summing the two curves ($D + E$) and taking the minimum point gives the optimal voting rule chosen by the constitutional decision maker. N_1/N represents the optimal number of individuals in the society of N persons who must agree before a collective decision is made. N_1/N may be any number. Here it is a simple majority, $(N/2) + 1$.

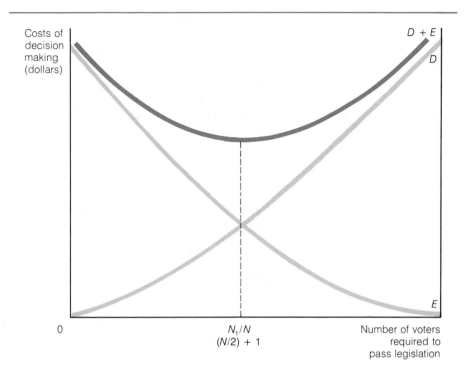

sion making.[2] As an illustration, we will analyze the choice of a voting rule for the legislature. We assume that the constitutional decision makers are unbiased and impartial (a large assumption). Figure 20–1 illustrates the analysis.

The horizontal axis gives the quantity of N, all the members of a hypothetical society; N can also be thought of as all the voters in the society. The vertical axis is a measure in dollars of the expected costs of the various voting rules that could be selected for the society. Examples of these costs will be presented shortly.

Constitutional decision makers face the following problem: How do they choose the voting rule that minimizes the expected cost of collective decision making in this society? The answer to the question can be found by examining the D curve and the E curve. The D curve measures the costs of collective decision making. As the voting rule is made more inclusive—requiring, for example, a two-thirds vote rather than a simple majority for approval or defeat of an issue—the costs of reaching collective decisions increases. In fact, as more people are required for agreement, the D curve increases at an increasing rate. This is a logical result in any group decision process. More people require more discussion, larger meetings, more bargaining, and so on. Agreement is more costly to obtain in a larger group.

The E curve reflects the potential costs that a collective decision can impose on the individual affected by the decision. If very few people are required to agree on a measure, agreement will occur more frequently and E-type costs will be high. In other words, a small number of people could get together and pass measures that benefit them at the expense of other voters.

[2]This analysis was first developed by J. M. Buchanan and G. Tullock, *The Calculus of Consent* (Ann Arbor: University of Michigan Press, 1962).

If a unanimity rule prevails, however, requiring that all members of the legislature must agree before a tax can be passed, then E-type costs to any individual will be zero. Under these conditions, no collective decision can be made unless every voter consents. Unanimity means that one person can block a collective decision.

We have, then, the two major categories of costs related to the choice of a voting rule by constitutional decision makers. The minimum of the two costs can be found by vertically summing the D and E functions and picking the minimum point on that curve. This occurs at N_1/N in Figure 20–1. In this society, the two types of voting costs are minimized where N_1 persons are required to agree before a collective decision is made. N_1/N may be any number. It may, for example, be a simple majority voting rule—$(N/2) + 1$ in Figure 20–1—or it may be a stricter voting rule in which more than a simple majority agreement is required before a collective decision is made. The analysis in Figure 20–1 leads to a more inclusive voting rule than simple majority. Though stricter voting rules involve higher D-type costs, the prospect of lower E-type costs could easily make stricter voting rules an important route to the improvement of collective decision procedures. What would happen, for example, if every proposal for a public expenditure had to carry with it a proposal for the taxes to finance it and, moreover, to pass the legislature with a majority of seven-eighths?[3]

Positive Public Choice Analysis

Positive public choice analysis is like positive economics; it consists of the development of models of political behavior that can be subjected to empirical testing. A positive theory can be wrong; it can fail to be supported by the evidence. A positive theory can never be proved; it can just fail to be refuted by the evidence. In this case a positive theory will hold its ground against alternative theories until a theory comes along that offers a better explanation of the process. This is the spirit in which we outline positive public choice theory in this section. We present some positive propositions about government and political behavior that seem to be supported by evidence from the world around us.

Political Competition

Have you ever wondered why by election time the positions of the two major parties' candidates sound like Tweedledum and Tweedledee, and you cannot see any difference in the views espoused by either? Perhaps more important, why do we have only two major political parties in the United States? If we had only two firms in every market, antitrust authorities would become alarmed and might intervene to increase competition. The following model helps us answer these and other questions about political competition in a democratic setting.

Assume that voter preferences on issues can be distributed along an imaginary spectrum running from radical left (perhaps the Socialist Workers) to radical right (perhaps the John Birch Society), as in Figure 20–2. Further assume that voters vote for the candidate closest to their ideological position. The normal distribution of voters holding different ideological positions is split down the middle by M. We call the voter at M the median voter. If

---

[3]A famous Swedish economist named Knut Wicksell proposed such a scheme more than ninety years ago.

FIGURE 20–2

Ideological Distribution of Voters

The distribution of voters in this example is single-peaked, ranging from radical left to radical right. *M* represents voters at the middle or median of the distribution. Two-party political competition leads to middle-of-the-road positions, similar to those held by voters at *M*.

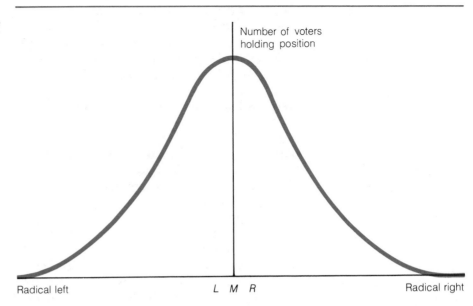

Number of voters holding position

Radical left   L M R   Radical right

either candidate in a two-party race adopts a middle-of-the-road position such as M's position, he or she is guaranteed at least a tie in the election; he or she will receive at least 50 percent of the vote. If the other party's candidate adopts any other position, such as *R*, he or she will get less than 50 percent of the vote. (Remember: Voters follow the rule of voting for the candidate closest to their ideological position; fewer than half the voters in Figure 20–2 are closest to *R*.) The best strategy for each candidate is to adopt position M and hope that random error (pulling the wrong lever) by voters will make him or her a winner. M, the median position, is the vote-maximizing position for both candidates; by election time, therefore, both candidates are virtual carbon copies of each other.

What happens if a third candidate enters the race? If he or she adopts a position such as *R* while the other two are at M, he or she would get all votes to the right of *R* and half of those between M and *R*, thereby beating both of the look-alike candidates at position M. Such an entry would probably induce one of the other two candidates to move away from M, perhaps to *L*. The remaining candidate at M would now be isolated with a small fraction of the vote. This candidate at M would have an incentive to move outside of the *LR* segment, thereby trapping one of the other candidates, and so on. Starting from a position in which all candidates are at the center, at least one candidate in a three-party or larger system will have an incentive to move away from the center. But this spreading-out process has its limits. As long as the peak of the distribution of voters remains at M, a candidate can increase his or her percentage of the vote by moving toward M. With three candidates at *L*, M, and *R*, the two outside candidates can increase their votes by moving toward M.

This model of political competition is oversimplified, but it does offer some insights into political behavior. Where the distribution of voter preferences is single-peaked, as at M in Figure 20–2, there are strong tendencies for median outcomes to be produced. This holds true for virtually any type of collective decision process, from a committee consensus to a public election. Middle-of-the-road policies are vote-maximizing; therefore, we expect middle-of-the-road candidates and policies to be winners. In addition, the analysis underlying Figure 20–2 gives some clues as to why we have two parties in the United States. If parties can enter on either the left or right flank of parties at M, such entry jeopardizes the control of government by the center parties. We can expect that the M-type parties will do their best to place barriers to entry in the way of noncenter parties, such as making it very costly and difficult for a new party to get on the ballot for elections.

Logrolling

Logrolling: The exchange of political favors, especially votes, to gain support for legislation.

Positive public choice also takes into account the behavior of officials once they are in office. A typical political behavior is **logrolling,** a term used to describe the process of vote trading. Representatives in a legislature are constantly making deals with one another. One wants a dam in his district; the other wants a new courthouse in hers. They agree to trade votes on issues: You vote for my dam; I'll vote for your courthouse.

There are many beneficial aspects of logrolling. Its main benefit is that it enables representatives to register their intensity of preferences across issues. For example, a minority representative who feels strongly about animal rights can trade his votes on issues about which he does not feel so intensely for support on the animal rights issue. Logrolling also helps mitigate the problem of indifferent majorities winning over intense minorities. In essence, logrolling or vote trading is a form of exchange, and economists usually view exchange as a productive and efficiency-enhancing process.

We must be careful, though, in assessing logrolling. Its general efficiency depends on the political setting. Most legislatures, for example, are formed on the basis of geographic representation, under which a representative's constituents have one thing in common: They all reside in the same area. Any bill the representative can get through the legislature that benefits the members of her district should win her votes at election time. In effect, a geographically based system provides the legislator with incentives to represent local interests in the national legislature at the expense of broader national issues.

Most individuals in a community share an interest in the vitality of the community's economy. In many instances a bloc of voters from a single voting district may receive income from a single firm or industry, as in a company town or in geographically concentrated industries like steel, automobiles, lumber, defense, and regional agricultural industries. When economic activity is concentrated, it is possible for representatives to win political support by serving the economic interests of their home districts. Tariffs, industry- and company-oriented tax concessions and subsidies, local public works projects, and defense contracts are all examples of issues that often are decided, in part, on the basis of their economic impact on certain regions.

Representatives therefore trade their support on national issues, such as air pollution control, for support on amendments or separate bills that serve their local interests. National legislation often becomes a vehicle for local support, and the result is the "pork barrel" type of legislation that carries rich rewards for specific locales. Logrolling, which attempts to redistribute income toward certain regions and industries, generally does not lead to a more productive economy. Rather it leads to legislation such as individual industry tariffs, unnecessary and costly public works legislation, and special-interest tax "reform" bills. This form of logrolling at the national level is unproductive not because it acts as a means for revealing relative intensities of preferences on national issues but because in its most blatant forms it is used to reveal relative intensities of preference on essentially local issues. When restricted to national issues, however, logrolling can be a beneficial means for revealing voter preferences. For example, one senator may prefer more foreign aid for Latin America and another a reform in the Social Security program. Through exchange of their votes on these two national issues, both may be passed as a result.

Voting

In Chapter 6 we discussed whether voting was rational. Simply put, one voter out of many has very little impact on any election. Therefore, even if the marginal costs of voting—the time it takes to register and go to the polls—are very low, they may make it too costly for the average voter to vote. Of course, the voter is rational; he or she will vote when the marginal benefits exceed the marginal costs of voting. If an election is predicted to be close, the marginal benefit of voting will rise, and voter turnout will rise.

In this section we examine the act of voting in more detail by discussing some of the relevant differences between the market and voting as means to allocate resources. We compare consumer behavior within market institutions and voter behavior within political institutions.

In markets, consumers make marginal choices; in politics, voters must evaluate package deals. When you buy carrots or hot dogs, you can buy one more or one less; that is, you engage in marginal decision making. When you vote for a politician or a party, you vote not for a single issue but for a package or a platform—all-or-none decision making. Politicians do not offer voters a little more or a little less of this or that public program. They offer package deals. This feature of voter choice clearly makes voting complex and costly. Moreover, when people are forced to make package choices, they are likely to end up choosing a lot of items that they really do not want in order to get a few things that they do want.

Elections generally involve the choice of numerous candidates and issues. In addition to the package nature of political choice, the voter has to make a variety of choices at the same time; many candidates and issues are on the ballot. To some extent, of course, this is efficient. It is better to vote on many things at once than to vote on a lot of individual matters in separate elections. Clearly economies of scale accrue in voting. But having numerous candidates and issues on the ballot also leads to greater complexity and cost in voter choice.

Consumption is frequent and repetitive; voting is infrequent and irregular. The consumer goes grocery shopping weekly or daily; the voter votes every year or every two or four years. This infrequency makes it more difficult for the voter to find reliable policies and candidates. Imagine how the grocery-buying decision would be altered if you had to buy groceries once a year. Your ability to discard bad products and try new ones would be reduced dramatically.

The primary difference between voting and the market, deriving from the above three factors, is that voters have little incentive to be informed. Voting is more complicated than market choice. This means that voters have little incentive to gather information about their public choices. In market choice, as we have seen, consumers have direct and strong incentives to gather information and to search for useful, reliable products. In public choice, voters have difficulty evaluating candidates and issues. The costs of being informed are high. Suppose the Defense Department argues for an expensive new missile system on the grounds that it is needed to deter a Soviet threat. How can the average voter hope to obtain the information needed to make a rational choice? The same is true for proposals for dams, foreign aid, welfare, relations with China, money supply, jobs, and so on. Voters are quite rationally uninformed about such matters; they estimate that the costs exceed the benefits of being fully informed. Indeed, they may free ride by not gathering information, not voting, and letting those who do vote make the choices for them. If the choices of other voters happen to be beneficial to nonvoters, the nonvoters benefit without bearing any of the costs. This behavior leads to a fifth and final implication.

Since voters are not well informed, the information transmitted to politicians by elections is not very useful in helping politicians determine what voters want. Politicians predictably have great difficulty assessing the implications of an election with respect to what voters want them to do. Of course, politicians who are good at reading the public's pulse will survive and be reelected. Nevertheless, the transmission of information is not direct. In markets, if consumers want more carrots, they can effectively transmit this information to producers and get more carrots. In voting, if consumers want more butter and fewer guns, it is costly and hard to make this preference known to politicians through voting.

In sum, private and public choices differ in important ways. Political markets are more imperfect than private markets. This is a positive proposition derived from the manner in which our political institutions are designed.

Interest Groups

Interest groups have received much attention in positive public choice analysis. The behavior of interest groups helps explain a significant amount of government activity. To see how interest groups affect government policy, consider the process by which a statewide group of barbers organizes to advance their interests.

The most fundamental problem that the barbers face is how to get organized. Specifically, each barber has incentives to stay outside the interest group, for if the interest group is successful in, say, obtaining higher prices for haircuts, a free-riding barber can benefit without having incurred any costs. Free-riding behavior makes the formation of interest groups difficult but by no means impossible. For such reasons, interest groups seek ways to limit the benefits of their activities to members only.

Assume that the barbers organize. The group's representatives then go to the state legislature and seek a legally sanctioned barbers' cartel. The purpose of this cartel is to protect barbers from new competition by erecting barriers to entry into barbering, thereby allowing barbers to raise the price of a haircut above the costs of providing a haircut. In other words, the barbers petition the state government to raise their wealth at the expense of the consumers of haircuts. Put another way, they are demanders of a wealth transfer from the state.

Who are the potential suppliers of wealth transfers? Clearly, the suppliers are those—in this case consumers of the barbers' product—who do not find it worthwhile to organize and to resist having their wealth taken away by the barbers. This is an unusual concept of supply, but it is a supply function nonetheless. The attitude that will prevail among suppliers is that it would cost them more to protest the barbers' proposal than any benefits they would derive from protesting. For example, the average consumer might have to spend $100 to defeat the barbers' proposal, the net result of which would be a saving of $10 in haircut costs. Why spend $100 to save $10?

This example describes a "demand" and a "supply" of wealth transfers in the case of barbers. How does the market for wealth transfers work? Politicians are brokers in this market and in the market for legislation generally. This means that politicians get paid for pairing demanders and suppliers of wealth transfers. If they transfer too much or too little wealth, they will be replaced at the next election by more efficient brokers.

The outcome for the barbers is shown in Figure 20–3. The barbers start out as a purely competitive industry, producing Q_c haircuts per year at a price of P_c. When their interest group forms, they persuade the legislature to grant them cartel status. This persuasion is not very difficult because the consumers who are harmed by the cartel do not find it worthwhile to protest to the legislators. The barbers are the only ones to be heard, and perhaps by putting their arguments for a cartel in public-interest terms ("higher prices mean higher quality," "barbers must be licensed to ensure an orderly marketplace," and so on), they carry the day with the legislature.

The legislators assess the costs and benefits of creating a barber cartel. In Figure 20–3, the barbers seek P_gABP_c in cartel profits at the expense of consumers' welfare; haircut consumers incur an additional loss equal to ACB. (Area ACB represents the welfare loss due to monopoly discussed in Chapter 10.) The per capita gains to barbers, who are a relatively small number of producers, exceed the per capita costs to the more-numerous haircut consumers. The result: The legislature passes a law giving the barbers a cartel.

Note, however, that this discussion is highly stylized. Interest groups do not always work as producers versus consumers. Technically, any group can organize and act as an interest group, and we find in fact that interest groups represent many overlapping segments of society. There are labor unions, women's rights groups, religious groups, trade associations, organizations for

FIGURE 20–3

How an Interest Group's Demands Are Met

P_c is the competitive price that the industry, in this case barbers, charges, and P_g is the price established by the group after it is granted cartel status. After the cartel is formed, it stands to gain P_gABP_c in profit from consumers, and consumers stand to lose ACB in welfare. Generally, the former wealth transfer dominates the latter welfare loss, and a law is passed sanctioning the cartel.

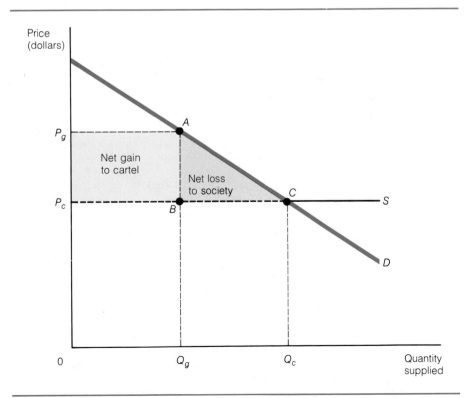

doctors, lawyers, hunters, environmentalists, farmers, and so on. Moreover, not all of these groups are small. Many large groups, such as farmers, are effectively organized as an interest group. The basic point is that you begin explaining why a certain law is passed if you try to answer two questions: Who wins? and Who loses?[4] Focus, "Interest Groups and the British Factory Acts," presents a historical example of how the demands of interest groups result in legislation.

Bureaucracy

While legislatures enact laws and authorize levels of public spending, the actual implementation of those laws is delegated to a variety of agencies, commonly referred to as bureaus. Bureaus are created by the legislature and, understandably, are generally supportive of the legislature's interests. Employees of bureaus generally fare better as the bureaus' budgets expand because, among other things, opportunities for promotion expand and, with them, salaries. With this expansion bureaus become more effective as vehicles for imposing their version of the public interest on the citizenry. At the same time legislators generally have particularly strong interests in the enactment of legislation that will be implemented by bureaus. Consequently, bureaus and the legislative committees with which they deal generally have similar interests.

As organizations engaged in the supply of services, public bureaus differ from private firms in two important respects. These differences in turn lead to important differences between the conduct of bureaus and the conduct of firms.[5]

1. Bureaus derive their revenue from legislative appropriations, which in turn come from tax collections. By contrast, firms derive their revenue from the voluntary buying decisions of customers.
2. There are no transferable ownership rights to public bureaus. In bureaus, there is no status comparable to that of stockholders, the residual recipients of differences between revenues and expenses. This difference does not mean that there are no profits in the operation of bureaus, but only that these profits do not show up as residual income to residual claimants or owners. Rather than being converted to personal use, any differences between revenue and expense disappear through added expenditure. For example, at the end of each fiscal year in Washington, bureaus with funds left over try to find ways to spend the funds. Many large research grants are made, for instance, at the end of the fiscal year.

The difference in incentives between private and public decision making yields many familiar results. It is often more difficult to register a car, get a passport, or buy alcoholic beverages in a state that has state-franchised stores than it is to buy automobiles, plane tickets, or wine from private suppliers. Under government operation, a bureau's hours of operation are usually shorter than those of private businesses and, hence, less convenient to customers. There is also a reluctance on the part of bureaus to accept checks and credit cards.

[4]For more analysis of interest groups see R. E. McCormick and R. D. Tollison, *Politicians, Legislation, and the Economy* (Boston: Martinus Nijhoff, 1981).

[5]See, for example, Ludwig von Mises, *Bureaucracy* (New Haven: Yale University Press, 1944); Gordon Tullock, *The Politics of Bureaucracy* (Washington, D.C.: Public Affairs Press, 1965); Anthony Downs, *Inside Bureaucracy* (Boston: Little, Brown, 1967); and William A. Niskanen, *Bureaucracy and Representative Government* (Chicago: Aldine-Atherton, 1971).

FOCUS   Interest Groups and the British Factory Acts

The interest-group theory of government evaluates legislation in terms of who benefits and who loses. Not only is this framework a useful way to analyze present-day laws and regulations, but it is also helpful in the study of economic history. As an illustration, we can apply the theory to legislation passed in England more than a hundred years ago.

The British Factory Acts of 1833–1850 placed restrictions on the employment of children and women in the English textile industry. Most historians have analyzed this legislation by accepting at face value the rhetoric of those who agitated to get it enacted. This rhetoric was couched in humanitarian terms. Those who sought these laws said they were doing a great favor for women and children and for society in general; they said they were acting in the "public interest."

Suppose that we go behind what the reformers said and apply the interest-group theory to this legislation. Who won and who lost as a result of the Factory Acts?

Two primary groups stood to gain from the restrictions on children and women in the labor force. Male workers, especially skilled male workers, stood to gain because children and women were competitors for their jobs. If competing labor could be outlawed, male wages

Sources: G. M. Anderson and R. D. Tollison, "A Rent-Seeking Explanation of the British Factory Acts," in David Colander, ed., *Neoclassical Political Economy* (Cambridge, Mass.: Ballinger, 1984), pp. 187–201; Howard P. Marvel, "Factory Regulation: A Reinterpretation of Early English Experience," *Journal of Law and Economics* 20 (October 1977), pp. 379–402.

would rise. There is, in fact, evidence that male workers were the primary agitators for the factory legislation, as the interest-group theory suggests they should have been. Of course, they presented their lobbying efforts in terms of doing a favor for women and children and serving the public interest.

The other group that may have gained from the restrictions on women and children was the owners of steam textile mills. Mills in the 1830s were water- or steam-powered. Steam mills could run all the time, but water mills were subject to droughts. The water mills typically made up for time lost during droughts by operating for long hours when water power was available. The restrictions on women's and children's hours of work made this catching-up process more difficult. Textile prices rose as a consequence, and the wealth of the owners of the steam mills increased because their operations were less hampered by the law.

Who lost? Obviously, the women and children who lost jobs and income suffered to a degree from the legislation. One would have to count against this loss the extent to which the law led to better working conditions, especially for children. Women were particularly vocal about their loss, but since women could not vote, politicians could ignore their wishes at low cost. The other losers were the owners of water-powered mills.

So interest-group analysis goes beyond what people say they want and evaluates the costs and benefits that actually result from government action. Seen in this light, the Factory Acts appear to have been primarily an example of wealth transfer and sexism.

Carding, drawing, and roving in a Lancashire, England, cotton mill, 1834.

In addition to these common observations, empirical studies have shown that public bureaus have higher costs of doing business than private firms. To cite a few examples, Roger Ahlbrandt has estimated that the cost of providing the same type of fire protection is about 88 percent higher when it is done by public bureaus than when done by private firms. David Davies has estimated that a private airline in Australia was 104 percent more productive in carrying freight and mail and 22 percent more productive in carrying passengers than a public airline. Robert Spann estimated that refuse collection is 43 percent more expensive when done by public bureaus than when done by private firms.[6]

Without the profit-maximizing incentives of private business, there are numerous ways that the indirect appropriation of excess funds through higher expenditures can take place within bureaus. Different public bureaus face different opportunities for appropriating profits. A public hospital, for instance, may overinvest in expensive equipment that is underutilized. A highway department may award contracts without competitive bidding and end up paying higher prices than necessary. Since excess funds cannot be appropriated directly, they are appropriated indirectly as the profits are dissipated through expenditure.

As we have noted, one main difference between a private firm and a public bureau lies in the identification of the customer. For private firms, the people who use the good or service are the customers whose continued favor is essential for the success of the firm. The public bureau's customers, however, are not the people who queue up for space at a public campground or who try to get efficient service at the motor vehicle department. Although the bureau provides services to those people, the public bureau's customer actually is the legislature because the legislature is responsible for the bureau's existence through legislation and funding. To remain in existence and to be successful, the bureau must please the legislature.

To some extent, of course, the legislature reflects the interests of citizens. But we must distinguish between the special interests of particular citizens and the interests of citizens in general. In cases where a bureau's performance pursues special interests, those special interests may represent a quite different set of people from those who use the bureau's service. For legislation to reward some special interests, it must penalize other, general interests. The often-poor performance of public bureaus is another feature of a special-interest approach to government, and this performance may not be so poor when it is viewed from the perspective of the bureau's true customers—the special interests that demanded the legislation.

The Growth of Government

A particularly important problem for public choice analysis is to explain why government changes size over time. As Table 20–2 shows, the size of government in the United States has grown substantially over the period from 1929 to 1984. As the data clearly show, this growth has occurred in both absolute

---

[6]Roger Ahlbrandt, "Efficiency in the Provision of Fire Services," *Public Choice* 16 (Fall 1973), pp. 1–16; David G. Davies, "The Efficiency of Public Versus Private Firms: The Case of Australia's Two Airlines," *Journal of Law and Economics* 14 (April 1971), pp. 149–165; and Robert M. Spann, "Public Versus Private Provision of Governmental Services," in *Budgets and Bureaucrats*, ed. Thomas E. Borcherding (Durham, N.C.: Duke University Press, 1977), pp. 71–89.

TABLE 20–2   The Growth of Government Expenditures in the United States Since 1929[a]

Year	GNP (billions of dollars)	Total Government Expenditures (billions of dollars)	Percentage of GNP
1929	103.4	10.3	10.0
1934	65.3	12.9	19.7
1939	90.9	17.6	19.3
1944	210.6	103.0	48.9
1949	258.3	59.3	23.0
1954	366.8	97.0	26.4
1959	487.9	131.0	26.9
1964	637.7	176.3	27.6
1969	944.0	286.8	30.4
1974	1434.2	460.0	32.1
1976	1718.0	574.9	33.5
1978	2156.1	681.8	31.6
1980	2626.5	868.5	33.1
1982	3069.3	1090.1	35.5
1984	3661.3	1258.1	34.3

[a]Dollar amounts have not been adjusted for inflation.

Source: Adapted from Richard E. Wagner, Public Finance (Boston: Little, Brown, 1983), p. 13. Original data from Facts and Figures on Government Finance, 21st ed. (Washington, D.C.: Tax Foundation, Inc., 1981), p. 36.

and relative terms. The relative size of government at all levels rose from 10 percent of GNP in 1929 to 34.3 percent in 1984. Moreover, this growth continued under the Reagan administration, which came into power on a platform of reversing the trend.

Explanations for the growth of government are easy to find in some periods. Expenditures for World War II clearly explain the growth of government over the period 1939–1944. However, the general upward march of government expenditures in other periods has many complex causes. For public choice theorists, the growth of government poses both positive and normative issues.

On the positive side is the fundamental question, Why does government grow? Part of the answer rests with the various forces that we have discussed in this chapter. Those who pay for and elect government officials do not pay critical attention to what government is doing; interest groups and government expand in size and scope as a consequence. Politicians, of course, are happy to accommodate the interest groups and bureaus so long as voters remain uninterested. Such things can come to an end, however. Government and independent bureaus in government can become so large or controversial that they start to attract the voters' attention. In this case, the force of voter or citizen reaction can lead to a reduction in government or bureau size. It is also important to recognize that government growth is not inevitable. Our basic argument has been that government is a more imperfect mechanism for allocating resources than the market is. This does not mean that government is out of control; it just means that it takes longer to effect changes in the size of government.

On the normative side, scholars have addressed themselves to the issue of how to contain government expansion. The basic thrust of developments has been to suggest fiscal constraints on government. That is, in recognition of the incentive of politicians and bureaus to expand government, the way to check such expansion seems to be to impose a system of fiscal discipline on political decision makers. In this approach the Constitution would be amended to require that the annual budget be balanced or that government expenditure not exceed a given percentage of gross national product. This approach recognizes the incentives of politicians and bureaucrats, and it seeks to design constitutional rules that will enhance the degree to which the political process reflects the underlying desires of voters.

Summary

1. Public choice analysis is the study of government with the tools of economics. It treats government decision makers like private decision makers—as rationally self-interested actors.
2. Government officials maximize their individual interests, not the "public interest." In the process they face different constraints than private decision makers, and so they behave differently.
3. Normative public choice analysis evaluates the effectiveness with which political institutions represent the preferences of voters, focusing on the evaluation of voting rules.
4. Voting on issues two at a time by majority rule can lead to a voting cycle where there is no clear winner even though each individual voter knows clearly who he or she prefers.
5. Constitutional decision-making analysis considers the choice of rules in the political arena. In other words, it asks how disinterested decision makers design such things as a voting rule for legislatures.
6. Positive public choice analysis seeks to present testable theories of government behavior. It is analogous to positive economics applied to politics.
7. In political competition the median voter is the controlling force. Parties and candidates compete by taking middle-of-the-road positions.
8. Logrolling, or vote trading, is useful in revealing relative intensities of preference across issues. When combined with geographic representation, however, logrolling can lead to an overexpansion of local public projects.
9. Voting and the market are two different means of allocating resources. For a variety of reasons, voters have little incentive to be informed in making voting decisions. As a result, voting is an imperfect mechanism for allocating resources.
10. Interest groups seek wealth transfers from government. Successful interest groups win transfers at the expense of general efficiency in the economy.
11. Bureaucracy is the production and management side of government. Unlike private production, government production takes place in a not-for-profit environment. As a consequence, bureaucrats behave differently from private decision makers.
12. Government grows because in general the costs of stopping growth appear to voters to exceed the benefits to them. There has been some interest in placing a constitutional constraint on the size and fiscal operations of government.

Key Terms

public choice	positive public choice	nonproprietary
normative public choice	proprietary	logrolling

Questions for Review and Discussion

1. What is a voting cycle? What conditions cause a voting cycle?
2. What is logrolling? What are the benefits and costs of logrolling?
3. Name three differences between voting and the market as mechanisms for allocating resources.
4. What is the difference between a public bureau and a private firm as a productive unit?
5. Discuss three experiences you have had in dealing with government that can be explained by the not-for-profit nature of government production.
6. Suppose that the size of government is declining relative to the size of the economy. Apply public choice theory to explain how this might happen.
7. Using the median voter model of voter preferences discussed in the chapter, explain why Democratic presidential candidates are more liberal in the primaries than in the general election.
8. Why do we not observe voting cycles in the U.S. House of Representatives?

ECONOMICS IN ACTION

In this chapter we explored the problems of *majority* voting. What about the interests of *minorities?* How do they fare under majority rule?

Imagine the simplest of all democratic processes: a local referendum to increase the property tax by a relatively small amount and to use the funds to build a new school. This situation can easily lend itself to a problem of neglected minority interests. For example, a majority of voters may not have children and may be slightly opposed to the measure because of the small tax increase that accompanies it. The parents of school-age children may be intensely in favor of the tax-school package because of the poor condition of the existing school. Despite their strong concern, the parents lose to a relatively indifferent majority in the defeat of the referendum. Conversely, one could envisage a situation in which the parents are in the majority, the proposed tax increase is substantial, the present school is in good condition, and the nonparents feel tyrannized by the passage of the tax-school referendum because it promises more costs than benefits for them. In one person/one vote majority rule, unless all voters have an equal expected gain or loss from the outcome of an issue, voting may not accurately reflect the underlying intensities of voter preferences.

To clarify the example further, look at the distribution of potential gains for two voters over a set of issues, shown graphically in Figure 20–4. The height of each curve is a measure of the potential gain to a voter from a favorable outcome on a particular issue (a favorable outcome may represent either the passage of a desired bill or defeat of an undesired bill). To avoid the problem of minority interests under majority rule, each voter must have exactly the same utility curve. For example, all voters must experience a Y gain from the outcome on issue X. If 51 percent of the voters have a utility curve like A and 49 percent have a curve like B, the A's will win on every issue even if the B's feel more strongly in each case. Majority rule is disadvantageous to minorities in this case and does not lead to voting outcomes that reflect the preferences of all the voters.

The problem is that no mechanism exists by which minorities and majorities can vote according to their differing intensities of feelings on issues. When voters have only one vote per issue, this vote represents a different underlying amount of gain or loss to majority and minority voters.

Source: D. C. Mueller, R. D. Tollison, and T. D. Willett, "Solving the Intensity Problem in Representative Democracy," in R. C. Amacher, R. D. Tollison, and T. D. Willett, *The Economic Approach to Public Policy* (Ithaca, N.Y.: Cornell University Press, 1976), pp. 444–473.

Voter Preferences and Majority Voting

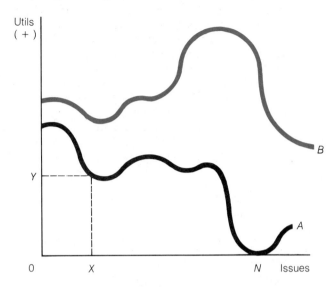

FIGURE 20–4   The Minority Voting Problem

Utility curves *A* and *B* measure the intensity of preference of two voters or two groups of voters across issues. The *B* group feels more strongly about each issue than the *A* group. A one person/one vote majority rule does not register this different intensity of feeling in voting outcomes. Point voting, when each voter is given a certain number of votes to allocate among all issues, may be a solution to this problem.

Suppose, however, that instead of having one vote on each issue, each voter is given 1000 votes and is asked to allocate them in proportion to his or her preferences over a wide number of issues. If issue *J* provides 7 times as much benefit as issue *K*, the voter allocates 7 times as many votes to it. Voting would consist of each voter's filling out a form indicating how the 1000 votes are to be assigned to each issue and whether the voter casts them for or against each issue. The issues would be decided by totaling the votes over all voters and passing all the issues that receive more yes votes than no votes.

This *point voting system* gives accurate information about the relative intensities of each voter's preferences over the various issues and makes it unnecessary for all voters to have the same utility curve. Stated in economic terms, point voting reveals individuals' marginal rates of substitution among issues, analogous to marginal rates of substitution among private goods. In other words, point voting for public issues is like the process of consumer equilibrium discussed in Chapter 6. The consumer allocates his or her fixed number of dollar votes in such a way as to establish equal marginal rates of substitution across goods. Point voting provides the voter with the opportunity to do the same with public issues.

Point voting is a possible solution to the problem that minority interests are not registered under one person/one vote majority rule. No voting system is perfect, however. The voting scheme that accurately reflects the preferences of all voters in final voting outcomes has yet to be invented. One of the key jobs of the public choice analyst is therefore to analyze voting systems and seek improvements in them.

Question

It is well known that different cities provide different levels of public goods such as education, street repair, police and fire services, and so on. Individual citizens, however, do not value different "packages" of public goods in the same manner. One response to the situation is to move to other cities when those other cities offer a more satisfactory package. Describe conditions of local public goods supply under which you might prefer one city over another.

POINT-COUNTERPOINT   A. C. Pigou and Ronald Coase:
Solving the Problem of Social Costs

A. C. Pigou

Ronald Coase

ARTHUR CECIL PIGOU (1877–1959), an English economist, succeeded Alfred Marshall (see page 259) in the
chair of political economy at Cambridge University in
1908. Like Marshall, Pigou had been drawn to economics because of its social importance. Both men
were deeply interested in improving the living standards of the poor. Both had great faith in the power of
economics to improve the welfare of all members of
society. Pigou and Marshall disagreed, however, on the
proper role of government. Like his neoclassical peers,
Marshall was an advocate of laissez-faire policies,
whereas Pigou felt that government's power to tax and
redistribute wealth among all members of society was a
crucial ingredient of general economic prosperity.

Pigou held his prestigious position at Cambridge for
more than thirty-five years. During this period he wrote
several books, including *Wealth and Welfare* (1912),
The Economics of Welfare (1920), and *The Theory of
Unemployment*. His students at Cambridge included
Joan Robinson (see page 259) and John Maynard
Keynes (see page 590). Pigou continued to be productive after his retirement from Cambridge in 1943. For
many, his death in 1959 marked the end of the English
neoclassical tradition in economics, a tradition begun
in the 1870s.

RONALD COASE (b. 1910) has earned international
recognition for his papers, "The Nature of the Firm"
(*Economica*, November 1937) and "The Problem of Social Cost" (*Journal of Law and Economics*, October
1960). Modern theories of the firm and of externalities
owe much to the work of this English-born economist.
Coase's novel economic analysis of institutions, particularly the law of business, has helped spawn a new area
of economic inquiry, much of which has challenged
traditional views of how government and government
regulations affect the behavior of firms. Coase himself

has written extensively on utilities, broadcasting, and
other government-regulated monopolies. He is generally critical of the effects of regulatory law, believing
that solutions to market failure can and do exist outside
of government intervention. Coase is currently retired
Professor Emeritus of Economics at the University of
Chicago. He has also held posts at the University of
Buffalo and the University of Virginia.

For almost half a century, the theory of social cost was
based on A. C. Pigou's original insights into the problem of market failure. Coase's challenge to Pigou's theory and his own approach to market failure offer a good
illustration of how economics is evolving in our time.

Taxing the Smoke

Pigou pioneered the modern microeconomic theories of
externalities and market failure. In *The Economics of
Welfare* he distinguished between private costs of production—the operating and maintenance costs of producers—and social costs of production—the expense or
damage to society that results from the producer's activity. There are many possible illustrations of social
costs. A nuclear power plant's social costs include the
potential danger of radiation to the surrounding community. A railroad's social costs might include the danger of fire caused by sparks from its wheels. Pigou's own
example of private and social costs is a smoke-belching
factory. The factory owner's private costs derive mainly
from the people and machines that make up the factory. The social costs of the factory derive from the
smoke the factory produces as a by-product of its operation. The smoke from the factory soils clothing and
property nearby, poses a health risk to the community,
and creates a noxious odor.

Pigou pointed out that there is nothing in the economic forces of market supply and demand to prevent the factory owner from continuing and even increasing the pollution. Since the market, in essence, permits the owner to shift part of the costs (the smoke) onto others, the owner will likely protect this advantage.

Pigou also sought to prove that market failure resulted in inefficiency of resource allocation. To do so, he looked at the factory's costs and benefits in marginal terms. The owner's marginal private costs are the expenses to produce one more unit. The marginal social costs are the damages suffered as a by-product of producing one more unit. The marginal private benefit to the owner is simply the price of the product, or unit, and the marginal social benefit is whatever society gains from the product.

According to economic theory, the factory owner will try to manage production so that the marginal private costs of the factory equal the price of the unit. Pigou pointed out that by doing so, however, the owner is not responding to the true costs of production. Since the owner's private costs are less than the full (private plus social) costs of production, the owner will produce "too much" output. If, however, the owner managed production so that private plus social costs equaled price, the excess output would be eliminated, and all resources would be allocated according to their true opportunity cost.

To remedy the inequities and inefficiencies of market failure, Pigou proposed a tax on polluters and other sources of social costs. In the case of a smoke-belching factory, the tax would depend on the amount of smoke produced and would force the owner to restrict pollution to the point at which the benefits of production were equal to its true costs. Such a tax, of course, would also likely restrict output. In this way, both producers and consumers would share the costs of the tax.

Assigning Liability

In his 1960 article, "The Problem of Social Costs," Ronald Coase challenged Pigou's major assumption that social costs always move in one direction—from producer to society. In fact, Coase argued, the issue of social cost poses potential harm to both parties. On the one hand, society may be harmed by the unwanted by-products of production, and on the other hand, producers may be harmed by society's attempt to correct the market failure through taxation. By the terms of Pigou's example, should the factory owner be allowed to pollute the local community, or should the local community be allowed to drive up the factory's costs of production by forcing it to cease polluting? The issue is especially pertinent today in the debate over acid rain and the problem of toxic waste.

Coase argued that an efficient solution to the problem of social costs would involve negotiation between the polluter and those affected by the pollution. If it were possible for both sides to negotiate on even terms, then a market might be created between them. The two sides could "trade" for the right to the pollution: Either the factory would pay the local community for the right to pollute or the community would pay the factory not to pollute. Coase showed that the actual outcome of such negotiations would be the same regardless of which side had the "right" to use the atmosphere as it saw fit.

In the real world, such negotiations between parties help solve a variety of potential social cost problems. For example, restaurant customers have influenced restaurants to segregate smokers from nonsmokers, eliminating the social cost of tobacco smoke. Home buyers shopping for a house near an airport can negotiate for a lower price to offset the social cost of airplane noise.

Often, however, one or both sides in the dispute face enormous transaction costs to enter such negotiation. A community, for instance, might find it very costly to organize into a bargaining unit with a nearby factory. The millions of people affected by acid rain and the hundreds of producers who might be responsible for the pollution could hardly transact freely. In such instances Coase suggested that the parties look to the courts, not the legislature, for a solution. Courts could determine the relative costs involved and could assign liability, or responsibility, to the party whose costs of adjusting to the social cost are lower.

For example, who should be responsible for injuries and death resulting from the use of defective products? For a long time, the rule of *caveat emptor,* or "let the buyer beware," prevailed among courts, and the consumer was fully responsible for the safe use of products. Such a rule worked reasonably well in times when products were relatively simple in design and use. Now, however, as products are becoming more complex, consumers may not be able to judge their safety and reliability. As a result, the courts have gradually increased the range of cases in which producers are held liable for shoddy and dangerous merchandise. This shift in the legal posture of the courts is an illustration of the Coase idea at work: The legal system alters liability assignment in response to the relative costs of adjusting to social costs.

V

Private-Sector and Public-Sector Macroeconomics

21

Macroeconomics: Contemporary Problems and Issues

*T*oday's local newspaper, this week's *Time* or *Newsweek,* or the evening television newscast is likely to contain much economic news. There might be stories on conditions in microcomputer markets, on the migration of agricultural labor from Minnesota to South Dakota, or, if filler is needed to take up space or time, on a temporary price increase in the sesame seed market. But by far the most consistently reported economic news relates to the overall health of the economy.

The media pay careful attention to monthly and quarterly statistics on such measures as the inflation rate, the unemployment rate, and the growth rate in gross national product (GNP). During periods when one or all of these economic barometers portend trouble, we can expect to see in-depth stories devoted to unemployment and the human suffering it causes, to skyrocketing prices for such basic commodities as food, or to the GNP's sluggish performance and the prospects for economic recovery. Government economic policies to remedy macroeconomic problems are also heavily reported.

Topics such as the real rate of growth in the economy may seem abstract and distant to some of us, but they certainly have an immediate and a lasting impact on our individual economic well-being. Economists have therefore spent more than two hundred years seeking to understand how the macroeconomy—the large-scale economy—shapes our lives.

Think for a moment about how the major macroeconomic issues—inflation, unemployment, and growth—are likely to influence your future. Will the price of textbooks and, indeed, of a college education, soar beyond your reach? Will you enter the job market during a time of high unemployment? Will your hopes for a high-paying job, a comfortable home, a promising future for your children be realized? The study of macroeconomics certainly cannot answer these questions for you, but it can better equip you to plan for the future, whatever the future may hold.

The Goals of Macroeconomics

In Chapter 1 we defined *macroeconomics* as the analysis of the economy as a whole. When working with macroeconomic theory, the focus is on whole quantities, or what economists call aggregate quantities. Macroeconomics concerns not just one market but all markets; not just one price change but all price changes; not just one firm's employees but all employment.

As with any discipline, the field of macroeconomics is made up of many specialized areas, competing theories, and ongoing debates. Consistent throughout macroeconomics, however, is a shared set of goals. Both macro-economists and the politicians who heed (or disregard) their advice are interested in achieving three separate, sometimes conflicting objectives: full employment, price stability, and economic growth. We discuss each of these goals separately, but you will soon see they are closely intertwined.

The Full Employment Goal

Full employment is a primary goal of any economic society for obvious reasons. The more fully resources are employed, the greater the levels of output of goods and services, and the higher the prosperity.

Social Concern over Unemployment. Not only does high unemployment threaten to bring poverty to millions of citizens, but it also can lead to political upheaval. During the Great Depression of the 1930s, for example, signs of social and political unrest appeared in the United States. Farmers marched on Washington, D.C., makeshift camps of the poor and out-of-work dotted the nation's cities, and mass migrations of the jobless from the Dust Bowl of the Midwest to the Promised Land of California took place. When World War II mobilized the economy, full employment was restored before serious political and social upheaval could erupt.

Other nations of the world have not been spared similar upheavals. Witness the frequent changes of government and political systems in less-developed nations, where poverty—much of it caused by widespread unemployment—is rampant. Increased crime and even violent revolution can result from failure to attain the macroeconomic goal of full employment.

In recognition of the social and political implications of unemployment, the U.S. government passed the Employment Act of 1946 on the heels of the Depression and World War II. The act recognized and enshrined maximum employment as a macroeconomic goal of the federal government. The meaning of the term *maximum employment* has been modified in the intervening years, but policymakers still respect the intent, if not the letter, of the law. Unemployment rates of 10 percent and more, for example, which characterized the U.S. economy in 1982 and 1983, brought great concern to the U.S. Congress and to the Reagan administration. Unemployment rates also ran high in Europe during the early 1980s, reaching over 13 percent in England and close to 10 percent in Italy by 1984. These rates, while relatively high, were not even close to those commonly existing in less-developed countries. In all nations, the goal of full employment is critical. Social programs such as unemployment compensation and food stamps ease the burden of unemployment, but they cannot substitute for the economic benefits of full employment.

Defining the Level of Employment. Some unemployment in specific markets is always expected. In a dynamic economy, demand grows for some goods and services and declines for others. In recent years, for example, the demand for high-technology outputs such as microcomputers, lasers, and fiber optics has outpaced demand for the products of heavy industry, and temporary unemployment in heavy industries such as steel and machine tools has been very high. Some workers may develop new skills, however, and temporarily unemployed resources may flow into new areas of consumption.

Temporary unemployment caused by the natural shifts of jobs between producers in a dynamic economy is called **frictional unemployment.** Full employment does not mean that everybody has a job all the time. But if rigidities—such as a lack of training or aptitude for the new technology or inability or unwillingness to move where the jobs are—prevent the rapid adjustment of employment to changed conditions, **structural unemployment** is created in the economy. As the structure of the job market changes, some displaced workers—such as automobile assembly-line workers replaced by robots—may be unable to find suitable alternative employment for a long period because they have no skills currently in demand by employers.

What, then, is full employment? A meaning relevant for policymaking is difficult to determine, but the actual measuring of employment levels is easier to understand. Data on employment and unemployment are collected by the Bureau of the Census through a sampling procedure. A random sample of 56,000 households is surveyed each month to determine whether civilians over sixteen years of age who are not in an institution such as a hospital or a prison or serving in the military are employed, unemployed, or not in the labor force. Each person interviewed is placed in one of those three groups. The civilian labor force is defined as those who are employed plus those who are officially defined by the Bureau of Labor Statistics as unemployed. Those classified as not in the labor force are individuals such as nonworking students and persons working in the home without pay.

Basically, individuals are counted as employed if they did any work for pay during the survey week. If they were on vacation or absent from work because of illness, they are counted as employed. Unemployed workers are those who were available for work during the survey week and who had actively looked for work in the past four weeks but did not have a job.

Official unemployment statistics, in the view of some observers, seriously underestimate the true U.S. unemployment picture. The official unemployment statistics do not count workers who, for one reason or another, drop out of the ranks because they believe jobs to be unavailable in their line of work. These discouraged workers do not include individuals such as students or the permanently disabled who choose not to work. Although discouraged workers are not counted in the official unemployment statistics, the Bureau of Labor Statistics estimates their number. At the end of 1983, for example, their number was estimated at more than one million people.

Using the data compiled by the Bureau of the Census, the Bureau of Labor Statistics each month reports the total number of employed and unemployed workers in the United States for the previous month. Figure 21–1 shows the growth of the civilian labor force and the fluctuations in employment and unemployment during the 1950–1984 period.

The **unemployment rate** is the percentage of the labor force accounted for by unemployed persons. It is a ratio: the number of unemployed persons divided by the labor force. Table 21–1 shows the unemployment rate, unemployment, and the civilian labor force for the 1975–1984 period.

Frictional unemployment: Unemployment that results from typical changes in the demand for goods and services.

Structural unemployment: Unemployment that occurs when there are major and permanent changes in the demand for goods or technology that affect the demand for labor.

Unemployment rate: The number of people that are unemployed divided by the number of people in the labor force; the percentage of the labor force that is unable to find jobs.

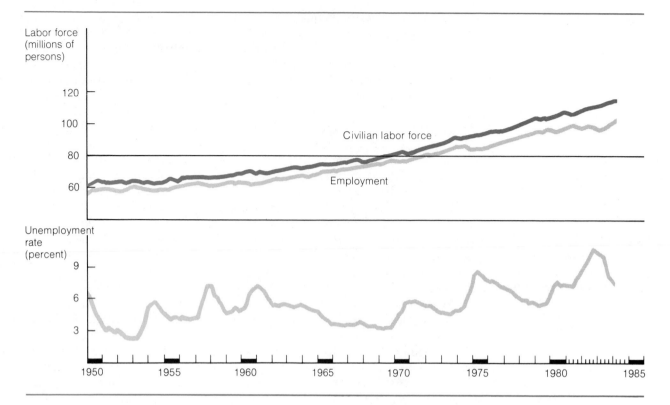

FIGURE 21–1   The Labor Force, Employment, and Unemployment

The civilian labor force is the sum of the employed and the unemployed. The labor force has increased steadily because of population growth, but the level of unemployment has fluctuated with changes in the overall economy.

Source: U.S. Board of Governors of the Federal Reserve System, *Historical Chart Book* (Washington D.C., Board of Governors of the Federal Reserve System, 1984).

The Bureau of Labor Statistics also breaks down the unemployment rate into rates for various demographic groups to show the impact of unemployment on particular groups in the economy. Figure 21–2 shows the total unemployment rate in the U.S. economy for the 1950–1984 period and the corresponding unemployment rates by race, sex, and age. Some facts about the composition of the unemployment rate are apparent: (1) Minority groups have experienced a higher unemployment rate than whites, (2) teenagers in the 16–19 age group have experienced a higher unemployment rate than any other age group, and (3) women over twenty years of age have generally experienced higher unemployment rates than men of the same age though this was not the case in 1984.

Full employment: A situation in which the optimal level of employment is reached with no underemployment and no excessive unemployment.

Despite the presence of unemployment, **full employment** remains a political goal. As we have seen, the Employment Act of 1946 charged the federal government with the responsibility of promoting maximum employment. In 1978, Congress passed the Full Employment and Balanced Growth Act, which committed the government to full employment, defined as an unemployment rate of 4 percent. The goals of the act have been modified in the intervening years, with the goal for 1986 being a 5.9 percent unemployment rate.

How can full employment be defined as allowing for any unemployment? Some economists feel that there is a **natural rate of unemployment** (or em-

TABLE 21–1   Unemployment, Unemployment Rate, and the Civilian Labor Force, 1975–1984

Year	Unemployment Rate (percent)	Unemployment (thousands of persons)[a]	Civilian Labor Force (thousands of persons)[a]
1975	8.5	7,929	93,775
1976	7.7	7,406	96,158
1977	7.1	6,991	99,009
1978	6.1	6,202	102,251
1979	5.8	6,137	104,962
1980	7.1	7,637	106,940
1981	7.6	8,273	108,670
1982	9.7	10,678	110,204
1983	9.6	10,717	111,550
1984	7.5	8,539	113,544

[a]Age sixteen and over.

Source: Council of Economic Advisers, *Economic Report of the President* (Washington, D.C.: U.S. Government Printing Office, 1985), pp. 226 and 271.

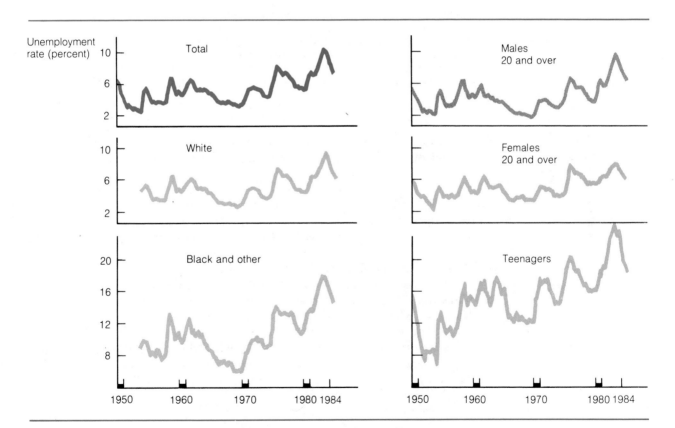

FIGURE 21–2   Unemployment Rates

The unemployment rate (the number of persons unemployed divided by the labor force) rises and falls with other economic activity. However, blacks and teenagers consistently have higher rates of unemployment than other groups.

Source: U.S. Board of Governors of the Federal Reserve System, *Historical Chart Book* (Washington, D.C., Board of Governors of the Federal Reserve System, 1984).

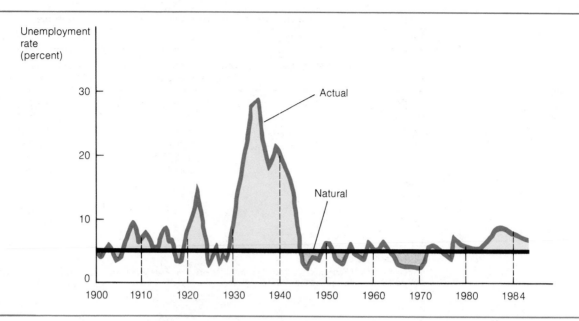

FIGURE 21–3   Rate of Unemployment, 1900–1984

The actual rate of unemployment has fluctuated widely in the twentieth century. The unemployment rate rose to dangerous levels during the 1930s but has stayed within narrower limits since World War II. The actual unemployment rate has fluctuated around the natural rate of unemployment, a theoretical concept that depends on the situations and institutions surrounding the demand and supply of labor.

Source: Robert J. Gordon, *Macroeconomics,* 3rd ed. (Boston: Little, Brown, 1984), p. 23.

Natural rate of unemployment: The rate of unemployment that occurs along with economic stability or equilibrium in labor markets.

ployment), the rate that would exist under long-run equilibrium conditions because of the time needed for changes in the labor market, the lag in matching vacancies and workers, the costs of hiring and firing and of changing jobs, regulations affecting structural changes in labor markets, and so forth. Natural forces, not individuals, determine the natural rate of unemployment at any given time. The actual rate of unemployment can be compared to the theoretical concept of the natural rate as in Figure 21–3. Note that the natural rate has risen slightly since 1950, while the actual unemployment rate has fluctuated around the natural rate, with cycles of business activity producing unemployment below the natural rate in some periods and above the rate in other periods. In the most direct sense, macroeconomics deals with why the actual rate of unemployment differs from the natural rate and, of course, with what causes the natural rate to change over time.

Economists debate, sometimes hotly, what the appropriate level of full employment is. In fact, the unemployment rate has not fallen below 5 percent since 1975, and many economists now think that the full unemployment rate should be in the 5–7 percent range, a range close to the theoretical concept of the natural rate of unemployment.

The Goal of Price Stability

The second major macroeconomic goal is **price stability**—the absence of inflation or deflation in overall prices. **Inflation** is not simply a rise in prices; rather, it is a percentage increase in price experienced over some time period (day, month, or year) for some representative basket of goods or for all

Price stability: A situation in which there is no change in the overall price level—the average price of all goods, services, and resources.

Inflation: Sustained increases in the overall price level.

Rate of inflation: The rate at which the overall price level of goods, services, and resources has increased over a specified time period; the percent increase in the average price of all items.

Deflation: A sustained decrease in the overall price level.

Real income: The purchasing power of income; the quantity of goods and services that an individual, a household, or a nation can consume with the money received from all sources of income; income adjusted for inflation or deflation.

Nominal income: The amount of money or its equivalent received by income earners; money income.

goods. A 2 percent increase in food prices in March means that, on the average, food will cost the consumer 2 cents more per dollar spent on food than in February. Never mind that bacon prices actually fell in March and that potato chip prices rose by more than 2 percent. Inflation percentages are expressed as an average of some representative bundle of particular goods—food prices, fuel or energy prices, and so on—or of an average of a representative bundle of goods consumed at some level—such as the retail, consumer, or wholesale level. If all prices increased by an average of 2 percent on all items consumed each month of the year, the annual inflation rate would be 2 percent multiplied by 12 months, or 24 percent per year. Some base period (prices in the previous month or the previous year) is always used in calculating inflation. Such calculations are used to determine the consumer price index (CPI), to be discussed further in Chapter 22. The CPI is the government's measure of the monthly **rate of inflation** in prices for typical consumer goods.

Deflation is simply a decrease in the general price level. Like inflation, it is always calculated from some base period. Since World War II, price instability has been due to inflation rather than deflation. During the 1970s, for example, the United States experienced double-digit inflation rates reaching almost 14 percent per year in 1980.

Certainly we have plenty of evidence that inflation can get out of hand. Like severe unemployment, runaway inflation can wreak havoc in a society, causing social and political disintegration. A classic example occurred in post–World War I Germany when the inflation rate climbed thousands of percentage points *per day*. The German mark became worth more as paper— it was actually used as wallpaper—than as money. As a result, goods disappeared from markets and people starved. Looking further back in time, economic historians claim to have evidence that runaway inflation was partly responsible for the breakdown of the Roman Empire's economic and political institutions.

The inflation problem has been particularly acute in the United States during postwar periods. The post-Vietnam years of the 1970s were beset by high inflation rates, but during the early 1980s inflation subsided, as indicated in Figure 21–4. Other nations have not been so fortunate. In 1982, France and Italy experienced inflation rates of 10 and 16 percent, respectively. Developing countries such as Mexico, Brazil, and Argentina have tried, with varying degrees of success, to hold the inflation rates between 100 and 200 percent per year in the 1980s. Actual (official) inflation in Bolivia reached 2,700 percent in 1984!

Even when the rate of inflation stands at relatively modest levels, it can pose problems for the economy. First of all, inflation has peculiar redistributive effects. That is, it enhances some people's real incomes at the expense of others'. **Real income** is the quantity of goods and services that can be bought with an individual's **nominal,** or money, **income.** In other words, real income is the real purchasing power of one's nominal income. For example, if a person was earning $10,000 per year in 1972 and continues to do so in 1985, a modest inflation rate of 5 percent per year from 1972 to 1985 would mean that the person's real income, or purchasing power, would be more than halved despite the fact that the nominal income remained the same. This loss of real income is all too familiar to people on fixed money incomes—those on fixed pension plans, for example. When prices of the goods and services consumed by these groups rise, but their nominal incomes remain the same or do not rise as fast as prices, the real income of these

FIGURE 21–4
The Consumer Price
Index (CPI),
1951–1984

The CPI is calculated for a representative market basket of goods and services using 1967 as the base year. Inflation rates have varied greatly over the period, although the trend has been toward overall higher rates. Note the downward trend in the inflation rate since 1980.

Source: U.S. Board of Governors of the Federal Reserve System, *Historical Chart Book* (Washington, D.C., Board of Governors of the Federal Reserve System, 1984), p. 37.

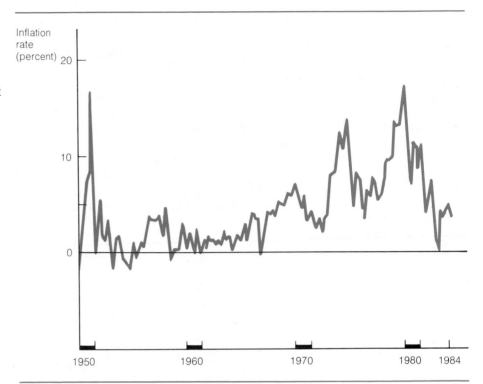

consumers falls. Naturally, consumers whose income is rising at a faster rate than prices are better off during inflationary periods. Under such circumstances, arbitrary redistributions take place, with those on fixed incomes bearing the costs.

Uncertainty about future prices also results from price instability. Debtors and creditors, for instance, must guess at future prices and charge or pay interest rates that may or may not cover the real change in the value of money. Inflation also disrupts decisions of producers and consumers. Consumers may rush to "buy now" because prices may be higher tomorrow, and producers may withhold output of goods and services for the same reason.

Inflation may also mean an increasing tax bill for taxpayers and a redistribution of real spending power and income to government. Since the U.S. tax structure is based on a progressive rate of taxation, individuals pay higher taxes for each marginal increase in income. When money incomes rise, total tax payments rise. But when increases in money income do not keep up with rises in the price level, real purchasing power falls, partly because consumers are pushed into higher income tax brackets. In the past, tax rates have been levied on money income received, not real income earned. Thus another cost of inflation is an arbitrary, unvoted increase in taxes. A program called indexation, which partly alleviates this tax burden by indexing the Federal tax tables to changes in the inflation rate, went into effect in 1985.

The Objective of Economic Growth

Economic growth: An increase in overall production of an economy over time.

The third major macroeconomic issue is growth in the economy. **Economic growth** refers to any increase in the productive capacities of the economy, whether as a result of an increase in the labor supply, an increase in the productivity of labor (the output per worker), or a net increase in the quality or quantity of the nation's capital stock, the wherewithal of production.

The labor supply grows through increases in population or in the number of people willing to work. Increased productivity of labor in output per worker is achieved through improvements in education and human capital or through a higher quantity and quality of capital stock supplied to labor. Writers and secretaries, for instance, may increase their productivity by switching from typewriters to word processing systems. Additions to the nation's capital stock are made through new investment in capital goods—word processors in offices, robots in factories. This investment arises from another macroeconomic variable, the rate of growth in private saving. To save, individuals must forgo present consumption for future consumption. Under favorable economic conditions, when the two goals of full employment and price stability are achieved, the rate of private spending is apt to be higher, generating in turn a high rate of new investment, new capital formation, and growth in actual and potential output.

Gross national product (GNP): The dollar value of all final goods and services produced in a year.

Real economic growth in terms of **gross national product,** or GNP, a measure of the final output of goods and services produced in the economy over a year, has averaged about 3 percent per year in the United States over the past hundred years. During the decade of the 1970s, however, the U.S. economic growth rate fell below the average. Potential growth exceeded actual growth during these years chiefly because of the presence of unemployed resources and a high degree of uncertainty about the inflation rate.

Nominal GNP: Total output or total spending expressed in current dollars.

Figure 21–5 shows two measures of aggregate output growth between 1950 and 1984. The black line gives GNP in current dollars; it is a measure of **nominal GNP.** The colored line gives GNP adjusted for price changes; it is a measure of **real GNP,** calculated according to 1972 prices.

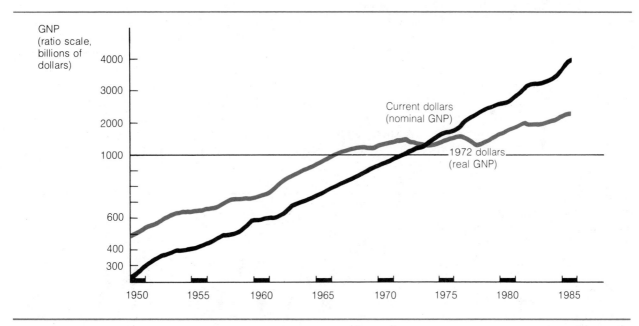

FIGURE 21–5   Real and Nominal GNP

The black line shows the rising level of GNP. Most of this increase is caused by rising prices. The colored line shows the trend of real GNP by removing the effects of inflation on prices. GNP, in other words, rises less rapidly in terms of constant 1972 dollars.

Source: U.S. Board of Governors of the Federal Reserve System, *Historical Chart Book* (Washington, D.C., Board of Governors of the Federal Reserve System, 1984), p. 12.

Real GNP: Total output or total spending after adjustments for inflation or deflation.

Why is the distinction between nominal and real GNP important? To get a picture of economic growth, we measure over time the changes in the aggregate output of goods and services. Changes in prices cloud the picture. For example, prices can rise with no change or even with a fall in output. Expressing GNP in constant-dollar terms by way of a price index (a procedure discussed in detail in Chapter 22) adjusts for changes in prices over time to give us a clearer picture of the course of the aggregate output of goods and services. For example, look at the period between 1973 and 1975. Real GNP declined over this period even though nominal GNP showed a steady increase. This divergence between the two lines gives us a clue to its cause. If real aggregate output was falling and nominal GNP was rising, then inflation must account for the divergence. In fact, 1974 and 1975 were years of high inflation.

It is important to remember that the macroeconomic goals of full employment, price stability, and economic growth are very much intertwined. Attainment of any or all of these goals will have both current and future effects on the economy.

Economic Stability and Business Cycles

Economic stabilization: When aggregate variables such as the price level, the unemployment rate, and the economic growth rate are at acceptable levels, varying only slightly and temporarily from desired and achievable goals.

The overall goal of macroeconomic policy is the achievement of **economic stabilization.** In a condition of economic stabilization, price changes (deflation or inflation) and unemployment do not cause much variation in the plans of buyers and sellers and do not prevent maximum output in the economy. Obviously, many prices in individual product and resource markets change daily in response to changes in the supply and demand for goods and services. Such price changes and some level of unemployment are necessary and important in any free-market system. However, an overall rate of change in the price level—deflation or inflation—or excessive unemployment may have disastrous effects on the economy by disrupting overall consumption and production and by limiting growth.

Economic stability thus means the achievement of full employment under inflationless or near inflationless conditions to attain maximum economic growth in the present and future. Stabilization is a tall order in a modern economy where economic activity is subject to fluctuations. These fluctuations, called **business cycles,** are the result of severe variations in the plans

FIGURE 21–6

Cycles of Real Annual Growth Rates in GNP in the U.S., 1950–1984

Periods of economic expansion and contraction have varied a great deal in this century. Note the roller-coaster path since 1960, punctuated by sharp recessions in 1974–1975, 1980, and in 1981–1982.

Source: U.S. Board of Governors of the Federal Reserve System, *Historical Chart Book* (Washington, D.C., Board of Governors of the Federal Reserve System, 1984), p. 12.

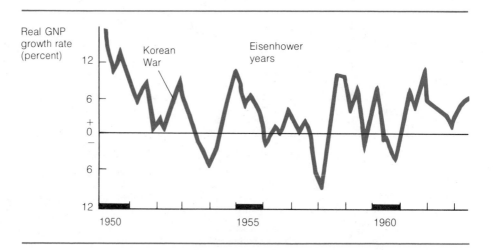

FIGURE 21–7

A Typical Business Cycle

Most business cycles vary in length and intensity. A typical cycle includes an expansionary phase of rising business activity and output, a peak of activity, a contractionary phase of falling activity, and a low point, usually referred to as a recession or depression.

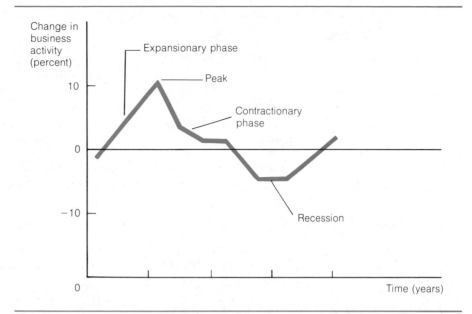

Business cycles: Recurrent fluctuations in the level of business activity, as measured by real GNP.

of buyers and sellers beyond those variations necessary for changes and improvements in production and consumption.

The extremes of business cycles are called peaks and recessions or depressions. At the peak of a business cycle, the economy is expanding rapidly. In a recession or depression, resources, especially human resources, are grossly underused. The expansionary phase of a cycle, a period often characterized by increased inflation as well as increased business activity and employment, is often contrasted to the contractionary phase characterized by stable or falling prices and excess production capacities and unemployment. Figure 21–6 shows cycles of business activity as measured by real GNP growth rate changes since 1950; Figure 21–7 illustrates a typical business cycle.

The goals of macroeconomic policy are to even out or counterbalance the opposing forces of the business cycle. Such policy is therefore called countercyclical policy. Its role is to counter the business cycle to produce inflationless economic growth with full employment. The role of macroeconomic and monetary theory is to understand the causes of changes in business activity; that is, the causes of the business cycle.

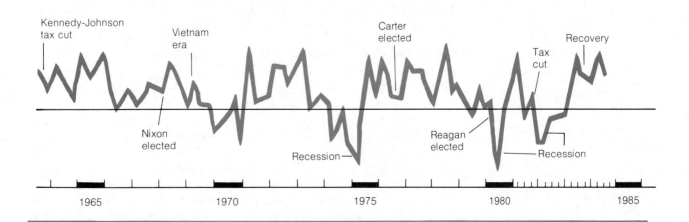

Aggregate Demand, Aggregate Supply, and Macroeconomic Policy

Identifying goals is only the first step. The real questions are, How can full employment be reached? How can inflation be tamed? and How can economic growth be sustained? Questions such as these require both theoretical and political answers. In this section we briefly introduce two of the most important macroeconomic theoretical tools—aggregate demand and aggregate supply—and the two most important types of macroeconomic policy.

Aggregate Demand and Aggregate Supply

As we saw in Chapter 4, the laws of demand and supply are fundamental to economic analysis. In macroeconomic terms, demand and supply are always spoken of as aggregate demand and aggregate supply.

It is impossible to conceive of all the millions of decisions made on a given day or in a given year to spend or consume resources. In the abstract, **aggregate demand** represents all such willingness to spend: consumption, investment, government spending, and foreign expenditures. Aggregate demand is always expressed over a specific period of time, such as a year. Like demand expressed by individual consumers, aggregate demand varies inversely with the price level of goods and services.

In the macroeconomy, **aggregate supply** represents the millions of individual goods and services available over a specified period of time, such as a year. The aggregate supply of goods and services varies directly with the price level of goods and services.

Although the analogy between demand and supply in an individual market (such as the market for cassette tapes) and the aggregate demand and supply of all goods and services is not perfect, the concepts are nonetheless similar. Consider Figure 21–8, which shows a common conception of aggregate demand and aggregate supply. The aggregate demand curve is negatively sloped just like the demand curve for tapes or candy bars: As the price level declines, the quantity of goods and services demanded (aggregate output) rises. The curve for the aggregate supply of goods and services is somewhat different from the individual supply curve. Simply stated, the aggregate supply curve is positively sloped up to the full employment of resources, Y_F in Figure 21–8, and is vertical thereafter, indicating that higher price levels will not force more resources and production out of the economy.

Equilibrium in the economy occurs where aggregate demand equals aggregate supply and where, simultaneously, maximum output and employment as well as price stability are achieved. This occurs in Figure 21–8 where aggregate output or productivity is Y_F and the price level is P_0.

Fiscal and Monetary Policy

The concepts of aggregate demand and aggregate supply recur throughout our discussion of macroeconomics. While all macroeconomists agree that, to one degree or another, the economy's performance depends on these two concepts, they rather sharply disagree over exactly how this is so. We will have much to say about these varying interpretations as we move to the basic macroeconomic theories. For now we briefly address the two types of policy meant to influence aggregate demand and aggregate supply.

Basically, aggregate demand is affected in two ways: through fiscal policy and through monetary policy. **Fiscal policy** entails alterations in tax policies and in spending by government to influence the aggregate demand for goods

Aggregate demand: Spending, including consumption, investment, and government and foreign spending, over a specified period at specific price levels.

Aggregate supply: All the goods and services available over a specified period at specific price levels.

FIGURE 21–8
Aggregate Demand and Aggregate Supply Curves

Like the demand and supply curves for a single good, the aggregate demand and aggregate supply curves for all goods and services determine a price level and a level of output for all goods produced and consumed. Equilibrium is achieved at the point where full employment of resources and maximum output, Y_F, as well as price stability, P_0, are achieved.

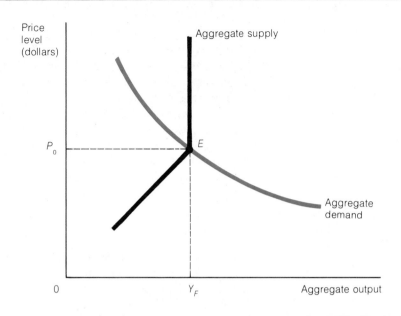

Fiscal policy The use of taxation and government spending to effect changes in aggregate economic variables.

Monetary policy: The act of changing the money supply or the rate of change in the money supply to create changes in aggregate economic variables.

and services. **Monetary policy** is set by a central bank (in the United States by the Federal Reserve System). The goal of monetary policy is to alter the aggregate demand for goods and services by changing the supply of money or by adjusting interest rates.

Suppose that the economy is in a recession or depression with widespread unemployment of labor and other resources. Such a situation is depicted with aggregate demand and supply curves AD_0 and AS, respectively, in Figure 21–9. At the intersection of these curves, a level of output Y_0 is produced at a price level P_0. Note that Y_0 is less than the full-employment output level Y_F. The goal of both fiscal and monetary policies is to cause aggregate demand

FIGURE 21–9
The Goal of Fiscal and Monetary Policies

Fiscal and monetary policies are used to shift the aggregate demand curve to attain full employment Y_F and price stability P_1 at a full-employment level of production.

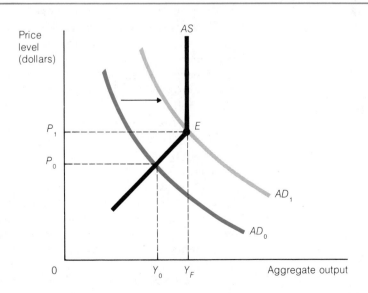

to shift rightward to a full-employment equilibrium intersection with aggregate supply. As aggregate demand shifts rightward, resources become scarcer, causing the price level to rise also. If fiscal and monetary policies are successful, equilibrium is finally achieved at output-production level Y_F and at a higher price level P_1. A high rate of inflation due to excessive aggregate demand would, of course, require fiscal and monetary policies to restrain or shift demand leftward (a situation not shown in Figure 21–9).

Use of both fiscal and monetary policies can also affect aggregate supply over the long term, as we will see in Chapter 26. Tax policies affect work incentives, for example, changing the willingness of laborers to supply productive resources. In other words, fiscal and monetary policies may approach goals from the supply side as well as from the demand side.

Macroeconomic Issues: Achieving Goals Is Not Easy

Unfortunately, many problems are associated with implementing policies to achieve full employment, price stability, and economic growth. Macroeconomic issues are controversies about how to use and design policies or institutions to achieve goals. Issues are always more visible than either theory or goals, and there seems to be no end to conflicts concerning the achievement of macroeconomic goals.

Trade-Offs

Trade-off: A situation in which the attainment of one desirable goal necessarily implies the loss of another desirable goal.

One problem is that economic goals may seem to conflict with each other, especially in the short run. Unfortunately, the solution of one economic goal, such as full employment, may not be compatible with the fulfillment of other goals, such as price stability. In fact, there may be a **trade-off** or a contradiction between full employment and price stability, at least over the short run. This trade-off occurs because an increase in the aggregate demand for goods and services will increase the employment of all resources, including human resources, and will lead to full employment. As resource supplies become scarcer, prices of inputs begin to rise. Further, as full employment is approached, less adaptable resources are used. For these reasons, attainment of the goal of full employment may create higher inflation rates.

On the other side, inflation control may lead to increased unemployment. Aggregate demand may be restrained through either monetary or fiscal policy to hold back upward pressures on prices. Such monetary and fiscal restraint, often accompanied by high interest rates, may create a recession in output and in resource employment. The high inflation rates of the late 1970s were in fact followed by high unemployment rates in the early 1980s as monetary policies were applied to curb inflation.

Other desirable goals may also conflict. Attaining high growth rates in real output may be thwarted by rules and regulations to clean up the environment. Looser environmental controls, on the other hand, may mean a faster growth rate but more pollution and environmental decay. Quality and quantity may not be achievable simultaneously.

Another trade-off occurs because expenditures on government-provided goods such as national defense and social programs tend to increase tax requirements. Increased taxes, however, may tend to discourage work efforts, labor productivity, and the amount of private spending and savings. Lower private savings, as noted earlier in this chapter, may reduce net investment in new capital equipment, decreasing the country's growth prospects. And these are only a few of a number of possible trade-offs between economic goals facing policymakers.

Can Discretionary Macroeconomic Policy Work?

Discretionary policy: An attempt to use temporary fiscal or monetary policy to reestablish economic stability when short-run disturbances have changed aggregate economic variables.

A more fundamental macroeconomic issue relates to the advisability of using **discretionary policies** (monetary or fiscal adjustments in the economy). Some economists question whether the economy can be managed at all by fallible public officials who possess imperfect information. Fiscal policy—taxing and spending policy—is controlled by the Congress and president of the United States. Monetary policy—control over the growth in the money stock—is controlled by the politically appointed Federal Reserve Board. The issue is whether these groups could ever know enough about the economy to direct and control it in discretionary fashion. If mistakes are possible, could not discretionary policies create so much uncertainty and confusion that they might sometimes have the opposite effects to those intended?

Some contemporary macroeconomists believe that, for reasons such as conflicting goals and imperfect information, discretionary fiscal and monetary policies have built-in side effects that make them worse than useless. These economists call for the use of policy rules such as a predetermined growth in the money supply and balanced federal budgets instead of discretionary control of the macroeconomy. Rules and balanced budgets would, in the view of some, take policy out of error-prone human hands, leading to more certain expectations among consumers, producers, and investors in the economy.

Underlying a faith in establishing rules rather than discretion in enacting policy is the belief, sponsored by early classical writers and upheld by their followers today, that the economy, if left undisturbed by government, will automatically achieve the goals of full employment, price stability, and maximum economic growth. Much of the policy controversy discussed in ensuing chapters centers on issues such as these. Can discretionary policy work? How automatic are free-market forces in achieving macroeconomic goals? Are rules preferable to continual alterations in fiscal and monetary policies?

Demand-Side Versus Supply-Side Policies

Most contemporary macroeconomic theory and policy, originating in the 1930s in the writings of John Maynard Keynes, is directed toward manipulating aggregate demand, either in discretionary fashion or with rules designed to produce full employment, price stability, and growth. This approach is now referred to as **demand-side policy.**

Demand-side policy: Fiscal or monetary policy that directs or changes the overall level of spending to create changes in macroeconomic variables.

Supply-side policy: Fiscal or monetary policy that directs or changes the overall productive capacity of the economy by affecting work effort, savings, or investment.

Some economists, notably those associated with the Reagan administration, have focused instead on some of the factors related to output response or to aggregate supply changes. As noted earlier, output response is largely determined by factors related to resource supply and to investment and capital accumulation. A **supply-side** view of the macroeconomy has developed that emphasizes the possible effects of taxes on work effort, labor supply, and investment. High tax burdens on consumer-workers and on investors mean that the economy's output response is constricted. A lower output response creates rigidities in the economy, limiting the attainment of macroeconomic goals. Institutional changes surrounding the work decision—such as growth in the number of women in the work force—plus government social programs and minimum-wage laws may also play a role in influencing aggregate supply.

A focus on aggregate supply is simply a matter of emphasis. In no way does the emphasis connote that aggregate demand theory or policy is unimportant. As we will see, both aggregate demand and aggregate supply play critical roles in explaining success or failure in obtaining macroeconomic goals.

Modern Macroeconomics: A Mixed Approach

A brief look at the confusing and often conflicting issues surrounding the attainment of macroeconomic goals should not discourage us. Modern economists, as we will see, can lay claim to a sound and practical understanding of how the aggregate economy functions.

Economists, moreover, often take different policy positions based on alternative outcomes—and different magnitudes of outcomes—predicted by different theories. We will discover the rather unsettling fact that there is not one single theory that tells us everything we want to know. Rather, there is some truth in a number of theoretical approaches that may be viewed as alternative or complementary to each other.

Summary

1. Macroeconomics studies the effects of the interplay between aggregate demand and aggregate supply on inflation, unemployment, and economic growth, whereas microeconomics is concerned with the functioning of supply and demand in specific markets.

2. The overall goal of macroeconomic theory and policy is economic stabilization. Specifically, macroeconomic policy aims at an inflationless and fully employed economy with maximum economic growth. Some unemployment is expected in any fluid, dynamic economy as individuals change jobs and other resources adapt to changing market conditions.

3. Inflation and unemployment are the enemies of economic growth because they create an environment within which private saving and new investment are reduced. A reduction in new investment means lower capital formation and reduced growth prospects.

4. Runaway inflation can have disastrous social and economic effects in an economy. Redistributions of real income from those on fixed incomes, often the

poor and elderly, extreme variations in buyers' and sellers' plans, and arbitrary increases in taxes required by government are among the problems created by inflation.

5. Fiscal policy and monetary policy are the two major means for manipulating aggregate demand and aggregate supply. Fiscal policy is the alteration in tax or spending activities by government. Monetary policy is control of the money supply and interest rates.

6. Trade-offs may exist between the achievement of two or more economic goals such as inflation and unemployment, especially over the short run.

7. Economists have differing opinions about whether discretionary macroeconomic policy is stabilizing or destabilizing in promoting economic goals. In the view of some economists, rules should be substituted for authority in the quest for economic stabilization.

8. Supply-side economists have shifted attention to the factors affecting output response in the economy, such as the effect of taxation on work effort, new investment, and labor productivity.

Key Terms

frictional unemployment	rate of inflation	real GNP	monetary policy
structural unemployment	deflation	economic stabilization	trade-off
unemployment rate	real income	business cycles	discretionary policy
full employment	nominal income	recession or depression	demand-side policy
natural rate of	economic growth	countercyclical fiscal policy	supply-side policy
unemployment	gross national product	aggregate demand	
price stability	(GNP)	aggregate supply	
inflation	nominal GNP	fiscal policy	

Questions for Review and Discussion

1. What is price stability? Is a constant rate of inflation at 10 percent less harmful to people than an inflation rate that randomly fluctuates between 2 percent and 8 percent?

2. What are the three major goals of macroeconomic policy? Why have politicians in the past been less concerned with policies aimed at achieving sustained long-run growth than achieving the other two goals?

3. Who controls fiscal policy and monetary policy? Does fiscal policy have a greater microeconomic impact in markets than monetary policy, that is, does it change the quantities of individual products and resources more than monetary policy does?

4. Why do economists disagree on macroeconomic policies? Does this imply that economists do not understand macroeconomics?

5. What is discretionary-demand management policy? Can discretionary policy eliminate swings in the business cycle?

6. The aggregate supply curve shown in this chapter indicates that aggregate output can rise when there are unemployed resources but that once full employment is reached, output cannot be expanded. Does this have any relation to what you have learned about production possibilities curves?

7. What is necessary to achieve economic growth? Do inflation and unemployment hinder this process?

8. Can discretionary-demand management cause aggregate supply problems? What other trade-offs exist for policymakers?

9. What is stagflation? What causes it?

10. Which do you suppose is better for the economy, an increase in aggregate demand or an increase in aggregate supply? What is the difference?

22

Measuring the Macroeconomy

*I*n Chapter 21, we presented recent data on the American economy, including statistics on gross national product, unemployment, and inflation. In this chapter we look at these measures in more detail, examining the components of gross national product, the relation between GNP and national income, and the methods through which economists derive these aggregate data. We also look at the methods for measuring inflation in the economy and at how a price index is compiled. Statistics such as GNP and the rate of inflation are the bread and butter of macroeconomic study and of policy debate.

National income accounting—and macroeconomics itself—was born in the seventeenth-century endeavor to develop a "political arithmetic." Chief among the writers in this tradition were Sir William Petty (1623–1687) and his follower Charles Davenant (1656–1714). "By Political Arithmetick," wrote Davenant, "we mean the art of reasoning by figures upon things relating to government. . . . the art is undoubtedly very ancient. . . . But Petty first gave it that name and brought it into rules and methods."[1] Petty tried to estimate the national income of England as early as 1665, and economists have attempted to measure the economy's economic performance ever since.

Although economists have long been interested in aggregate economic statistics, it was not until the 1930s that Congress instructed the Department of Commerce to collect and report such data. Simon Kuznets, who won a Nobel Prize in economics in 1971 for his work in this field, was the father of modern **national income accounting,** statistical measuring of the nation's

National income accounting: A statistical measuring of the nation's aggregate economic performance.

[1]Charles Davenant, quoted in Joseph A. Schumpeter, *History of Economic Analysis* (New York: Oxford University Press, 1954), pp. 210–211.

economic performance.[2] Kuznets's work in conjunction with the Department of Commerce led to the publication in 1934 of estimates of national income for the U.S. economy. Such estimates continue to be published by the Department of Commerce and other government agencies.

Gross National Product

Before we examine the process of national income accounting, let us examine the concept of gross national product (GNP) in more detail. GNP is the most comprehensive national income statistic and the one that usually gets the most public attention.

Gross national product (GNP): The market value of all final goods and services produced in an economy over a given period.

Gross national product is the market value of all the final goods and services produced in an economy over a given period. As we shall see in more detail later, the word *gross* means that the total market value of all goods sold, including capital goods in the economy, is included in GNP. GNP therefore does not account for the wear and tear on physical assets—called depreciation—that takes place during a given period. The word *national* means that GNP covers the whole economy. The word *product* refers to the market value of final goods and services. We cannot sum physical amounts of goods because no data are kept on the total number of many goods—such as hamburgers and haircuts—sold in a given year. We can, however, sum the money or market values of different goods and services because businesses keep records of their receipts for tax purposes. GNP is thus the sum of the final market values of all goods and services sold.

Two other elements of the definition are important. First, only sales of all *final* goods and services are counted in GNP. To do otherwise would be to double-count. The value of McDonald's or Burger King's purchases of raw meat would be added to consumers' purchases of hamburgers and thus be double-counted if all purchases of goods and services sold in the economy were used to calculate GNP. The flow of inputs and outputs in the economic system that creates double-counting problems also permits an alternative means of calculating GNP, as we shall see presently.

A second point involves the time period. GNP ordinarily refers to final sales over the period of a year. However, monthly and quarterly growth rates in GNP and other national income accounts are also calculated and discussed. When GNP or any other national income account is being expressed as a rate or percentage (3 percent, 6 percent) per period (month, quarter, year), economists are utilizing a flow concept. A stock concept is being expressed when we say that GNP (or any other account) was 750 billion dollars or 1 trillion dollars in some year or period. Flows are processes taking place through time measured in some unit per time, whereas stock refers to a measurement at some specific time. To say that GNP grew at a rate of 3 percent per quarter in 1985 uses a flow concept; to say that GNP was 3.8 trillion in 1985 is to state GNP as a stock.

What GNP Does Not Count

Like any measure or approximation, GNP is not perfect. Some goods, services, or costly activities are not counted by GNP accounting procedures. Some of the important exclusions are the following.

[2]Economists use the term *national income* both as a generic term encompassing all of the national income accounts (gross national product, net national product, national income, and so on) and as a specific term referring to the total compensation of all the factors of production (labor, capital, and so on) in the economy over some period.

GNP does not count the value of goods that are produced but not sold in any market. Home production, for example, is not counted in GNP. This category includes such activities as backyard vegetable gardening, do-it-yourself activities such as home repair, and the myriad activities of homemakers. The point is not that such activities are not productive; they most surely are. The point is that such activities are not directly bought and sold in a market so that national income accountants can place a value on them to include them in GNP. Rather than guess at the magnitude of these activities, national income accountants leave them out of their estimates of GNP.

GNP does not count the value of illegal goods and services or "paperless" transactions. Illegal drugs and prostitution, for example, are not counted in GNP because data on illegal sales are obviously hard to obtain. Transactions made in cash and not reported to tax authorities cannot be counted in GNP estimates either. Where there is no paper trail, there is no source of data for the national income accountants. This means, of course, that there is more real production in the economy than is measured by GNP. Economists have estimated "underground transactions" to be as much as 20 percent of GNP.

GNP does not include the sale of used goods. GNP is a measure of current production. Goods produced in previous periods are not counted in this period's production. The sale of a used car or an older home, for example, does not get counted in the calculation of current GNP, for the production of used cars and older homes was counted when they were first produced.

GNP does not count the value of financial transactions and government transfer payments. Exchanges of stocks and bonds are not counted in GNP because such transactions represent an exchange of assets rather than the current production of assets. Government transfer payments, such as veterans' benefits and Social Security payments, are not a payment to the recipient for current production and hence are not counted in GNP. Interest paid on the government debt is also treated like a transfer payment; that is, it is not viewed as a payment that leads to the current production of government goods and services.

GNP does not include the value of leisure. A significant shortcoming of GNP estimates is that they fail to account for the value of leisure. In the United States, for example, the length of the work week has fallen significantly in this century. The additional leisure that is consumed by workers is a valuable good, but it is not counted in GNP.

GNP does not account for social costs. GNP does not account for many of the costs of economic growth. Water and air pollution, for example, have gone hand in hand with a rising GNP in the United States and other industrialized countries, and yet since such costs are not reflected in market transactions, they do not get counted against or deducted from GNP. Because GNP does not account for such side effects of economic growth, it tends to overstate the real output of economic goods and services of an economy.

GNP does not measure happiness or economic welfare. A word is in order about the relation between GNP and the overall health of an economy. While GNP is a standard index of output that is used, for example, to compare the performance of national economies, it is not and was never intended to be a measure of the happiness or well-being of a people. GNP is a measure of the production of goods and services without any indication of who gets the goods and services or what the goods and services are. GNP is indifferent about whether handguns or operas are being produced. However, some economists have attempted to compute a **measure of economic welfare** or well-being (see Economics in Action at the end of this chapter).

Measure of economic welfare: A measurement of social and economic well-being that accounts for the production of all goods, not just marketed goods. These include leisure, pollution, and certain government services.

Final-Goods Versus Value-Added Approaches

GNP may reflect either market values of all final goods and services or the value added at each stage of production because both sums are identical. **Final goods** and services are end products in the economy; **intermediate goods** go into the production of end products. As suggested above, Mc-Donald's meat purchases to produce hamburgers are purchases of intermediate goods. In the final-goods approach, production of intermediate goods is excluded from GNP to avoid the problem of double counting. For example, a number of stages are involved in making a cigar. If we counted the value of the tobacco sold by the farmer to the cigar maker and the value of the cigar sold to the cigar store by the cigar maker, as well as the price consumers pay for the finished cigars, we would have counted the value of the tobacco three times. This is an example of double counting or, we should say, triple counting. The value of the tobacco should be counted only once. One way to avoid double counting is to count only the price of the final good; the other is to count only the value added at each stage of production. **Value added** is the difference between the selling price of goods at a stage of production and the costs of the inputs used to make the good.

To better understand value added, consider the making of a cigar in more detail. Suppose that the farmer sells a quantity of tobacco to the cigar maker for $3. The value added by the farmer is thus $3, which covers the farmer's costs of producing the tobacco (wages, rent, interest, and profit). In the second stage, the cigar maker sells a certain number of cigars, produced from the quantity of tobacco, to the cigar store for $8. Since the cigar maker paid the farmer $3 for the tobacco, the value added at the second stage by the cigar maker is $5 ($8 − $3). In the final stage, the cigar store sells the number of cigars to smokers for $15. Since the cigar store paid $8 for the cigars, the value added in the third stage is $7 ($15 − $8). Table 22–1 illustrates the value-added approach to GNP with the example of cigars.

Summing the value added at each stage of production, we find that this sum, $15, equals the market price of the finished good. If we had instead added up the sales price at each stage of production ($3 + $8 + $15), we would have double-counted. The correct measure of GNP is $15, not $26. The value-added approach is logically equivalent to using the market price of final goods and services to estimate GNP. The final-goods approach is easier to calculate, but the value-added approach serves as a useful check on other methods of estimating GNP.

Final goods: Goods at the final stage of production that are not used to produce other goods.

Intermediate goods: Goods that are inputs in the production of final goods.

Value added: The increase in the value or worth of a good at each stage of production.

TABLE 22–1   Value-Added Approach to GNP

When GNP is estimated by the value-added approach, the extra cost at each stage of production is summed. At the first stage, the farmer adds $3 in value; at the second stage, the cigar maker adds $5 in value; and, finally, the cigar store adds $7 in value. The sum of the three values added is $15, which is identical to the final sale price.

Stage of Production	Seller	Buyer	Sale Price of Good	Value Added
First	Farmer	Cigar maker	$3	$3
Second	Cigar maker	Cigar store	$8	$5
Third	Cigar store	Smoker	$15	$7
				$15

National Income Accounting

Although we have thus far spoken only of the market value of goods, aggregate value in the economy can actually be derived either by summing all expenditures for final goods or by summing all income received. The following sections explain why these constructs should yield identical results and what factors must be taken into account to arrive at identical sums.

The Circular Flow of Economic Activity

National income (GNP) can be measured in two ways: the flow of expenditures approach and the flow of earnings, or the income approach. The **flow of expenditures** approach looks at national income as the total amount spent on final goods and services. The **flow of earnings,** or income, approach looks at national income as the amount earned by factors of production (land, labor, capital, and the entrepreneur).

Logically, the two approaches have to yield the same figure for national income. The circular flow diagram of economic activity indicates why. Figure 22–1 is a circular flow diagram for an economy with only two sectors—a household, or consumer, sector and a firm, or business, sector. In the inner loop of the upper part of the diagram, firms (suppliers) produce goods for sale to consumers (demanders). The flow of goods to households is matched by a return flow of expenditures, indicated by the outer flow line. Businesses use the revenue from final sales to cover their costs of operation.

The inner part of the lower loop shows businesses or firms (demanders) employing factors of production from the household sector (suppliers). The factors of production are paid incomes that reflect their productivity in producing final output. Firm revenues are exhausted by factor payments because any left over, or residual, portion accrues to the entrepreneur or owners of the firm in the form of profit, which is a type of income.

The upper and lower loops of the circular flow diagram illustrate the identity of the two approaches to measuring national income. The revenues from producing final goods and services are transformed into payments to cover the costs of producing the output. Total revenues from producing goods are decomposed into profits, wage payments, interest, and land rent. By necessity, the dollar value of the expenditure flow on goods and services equals the dollar value of the income, or earning, flow to the factors of production. The important point is that national income can be measured with either approach, and the results will be identical. The flow of expenditures approach is formally identical to the flow of earnings or income approach.

Flow of expenditures: The total amount of money that consumers, businesses, and government spend on final goods and services in a particular period.

Flow of earnings: The total amount of money earned by resource suppliers in a given period; as an approach to measuring GNP, also called income approach.

Leakages and Injections: Some Complications

Before using these equivalent approaches to national income, consider some complications. The simple case described by Figure 22–1 excludes such features as saving and other important sectors of the economy such as government or the foreign sector. Without them the flow of income to factors of production matches the flow of expenditures on final goods and services to firms. But since the excluded features and sectors are part of the economy, Figure 22–1 does not account for certain leakages and injections out of and into the simple flow of expenditures and receipts. A leakage occurs when spending is diverted from the income stream. Household or business saving is a leakage from the expenditure stream. When processed through banks and other financial institutions or intermediaries, such saving becomes investment, an injection or addition to the expenditures flow.

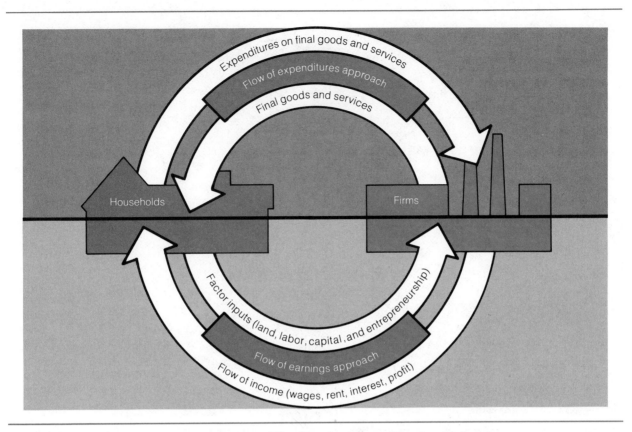

FIGURE 22–1   The Circular Flow of Economic Activity

The circular flow diagram is portrayed with only a business and a household sector. The upper loop is the flow of final output and expenditures on final output. The lower loop is the flow of input factors of production and the return flow of factor incomes. By definition, the value of the upper and lower loops must be equal. Therefore, GNP as measured by the flow of expenditures approach and GNP as measured by the flow of earnings approach must be identical.

Likewise, government expenditures are injections into the simple flow of expenditures, since the government, like households and businesses, makes expenditures on goods and services. Government transfers income to recipients through welfare and other programs as well. All such expenditures are injections into the circular flow. Financing such expenditures, however, requires the taxation of households and businesses. Taxation represents a leakage from the expenditure stream of the private economy.

The U.S. economy is also heavily involved in international trade and exchange, but the simple flows shown in Figure 22–1 do not include exports and imports. Exports—sale of goods and services produced in the U.S.—are an injection into the circular flow of expenditures and income, while imports—expenditures on goods and services produced abroad—are leakages from the simple income-expenditure flow.

In short, the simple income-expenditure model of Figure 22–1 does not account for certain leakages and injections. Government expenditures and exports must be counted as injections into the expenditure stream of private household and business spending. Likewise, household and business savings, along with government taxes and imports, are leakages from the simple circular flow of economic activity (see Figure 22–2).

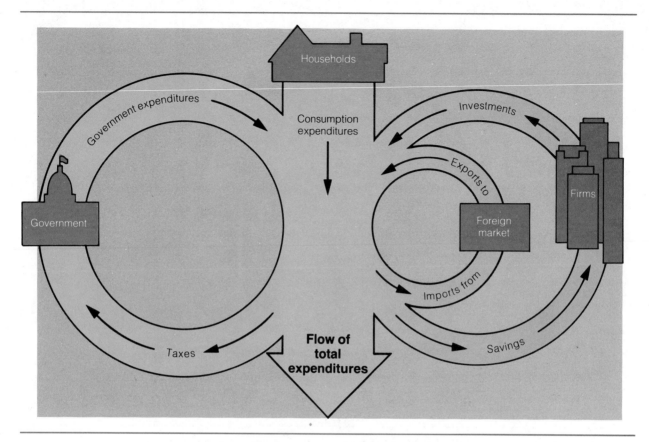

FIGURE 22–2   Leakages and Injections into the Circular Flow
The simple two-sector economy portrayed in Figure 22–1 does not include the
leakages represented by savings, taxation, and imports and injections represented
by investment, government expenditures, and exports. When these additional
concepts are included, economists speak of a four-sector economy, including firms,
households, government, and foreign markets.

The Flow of Expenditures Approach

National income accounting puts into practice the principles of the simple,
two-sector circular flow diagram as amended by injections and leakages. The
national income accountants apply the flow of expenditures approach to
measuring GNP by breaking the economy down into four sectors: household,
business, government, and foreign. Individuals in the economy are assigned
among the four sectors, and GNP is estimated by adding up the amount
spent on final output by individuals in each sector. Table 22–2 gives the
1984 data on spending in each sector. Figure 22–3 compares the flow of
expenditures to the flow of earnings, which is discussed beginning on page
484.

Personal consumption
expenditures (C): The total
amount of money that
households spend on final
goods and services.

Personal Consumption Expenditures. Expenditures by individuals and nonprofit institutions are called **personal consumption expenditures (C).** These
expenditures are broken down into spending on durable goods, on nondurable goods, and on services. Durable goods are items such as refrigerators that
are expected to last for more than a year. Nondurable goods are such things

TABLE 22–2   Gross National Product 1984: The Flow of Expenditures Approach

Total expenditures in the economy are broken down into four basic groups: private consumption expenditures; private domestic investment expenditures by businesses; government expenditures; and net exports (foreign purchases of U.S. goods minus U.S. purchases of foreign goods).

	Expenditures (billions of dollars)	
Personal consumption expenditures (C)		
Durable goods	318.4	
Nondurable goods	858.3	
Services	1,165.7	
Total		2,342.3
Gross private domestic investment (I)		
Fixed investment	580.4	
Change in business inventories	56.8	
Total		637.3
Government expenditures (G)		
Federal	295.5	
State and local	452.4	
Total		748.0
Exports and imports of goods and services		
Exports (X)	363.7	
Imports (M)	429.9	
Net total (X − M)		−66.4
Total expenditures in GNP		3,661.3

Note: GNP is in current, nominal dollars. Subcategories may not total because of rounding.

Source: Council of Economic Advisers, *Economic Report of the President* (Washington, D.C.: U.S. Government Printing Office, 1985), pp. 232–233.

as food and clothes that are not expected to last for more than a year. Services are intangible items such as travel, car repair, entertainment, and medical care. Personal consumption expenditures represent the largest expenditure component of GNP ($2,342.3 billion in 1984).

Gross private domestic investment (I): The total amount of money that private businesses spend on final goods, including capital goods.

Gross Private Domestic Investment. Expenditures on final output by private business firms (including resource-owning households) are called **gross private domestic investment (I).** Investment expenditures include spending on capital goods such as machinery and warehouses, new residential construction, improvements to existing houses, farm investments, and inventories. These expenditures were $637.3 billion in 1984. The qualifier *gross* is used because these expenditures include spending on new plant and equipment as well as on the replacement of worn-out plant and equipment. Business inventories are an investment in the holding of finished goods, semifinished goods, or raw materials by business firms. Investment in business inventories can vary greatly. That is, inventories can be built up or drawn down. For example, firms often draw down the level of their inventories in a period when business is bad, such as a recession. They hold fewer goods in inventory in the anticipation of decreased demand for the goods by consumers. Conversely, businesses will add to inventories in anticipation of sales increases and prosperous times.

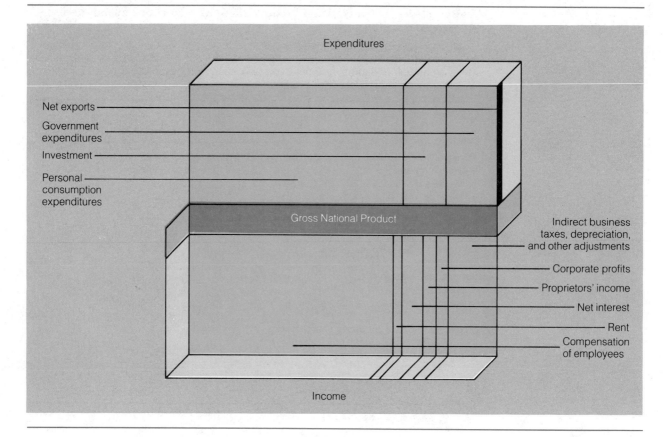

Net exports

Government expenditures

Investment

Personal consumption expenditures

Expenditures

Gross National Product

Indirect business taxes, depreciation, and other adjustments

Corporate profits

Proprietors' income

Net interest

Rent

Compensation of employees

Income

FIGURE 22–3   Total Expenditures Equals Total Income Plus Adjustments

In any given year total expenditures, composed of consumption, investment, government, and net foreign spending must equal the income to the factors of production plus corrections. This income is composed of wages, rents, interest, proprietors' income, and corporate profits. The flow of total expenditures must equal the flow of total earnings.

Government expenditures (G): The total amount of money that federal, state, and local governments spend on final goods and services.

Government Expenditures. The third component in the expenditure approach to GNP is **government expenditures (G)** on final goods and services at the federal, state, and local levels. Government spending was $748.0 billion in 1984. Government transfer payments, such as Social Security and veterans' benefits, are not included in this figure because such payments are not made to individuals for current productive activities.

Net exports (X − M): The value of domestic goods and services purchased by foreigners minus the value of foreign goods purchased by domestic citizens.

Exports (X): Domestic goods purchased by foreigners.

Imports (M): Foreign goods purchased by domestic citizens.

Net Exports. The final category of spending in the expenditure approach to GNP is **net exports (X − M)**. **Exports (X)** are domestic goods purchased by foreigners. **Imports (M)** are foreign goods purchased by U.S. citizens. The purpose of GNP accounting is to measure current production in the economy. Thus, we must add the value of domestic goods purchased by foreigners and subtract the value of foreign goods purchased by Americans in calculating GNP by the expenditures approach. In other words,

$$\text{Net exports} = \text{Total exports } (X) - \text{Total imports } (M).$$

Net exports can be negative or positive. Obviously, if we spend more on imports than we sell in exports, net exports are going to be negative. In

1984, net exports were negative. We imported more than we exported by $66.3 billion.

In sum, the flow of expenditures approach to measuring GNP is quite simple. It says that GNP can be estimated with the following equation:

$$GNP = \text{Personal consumption expenditures } (C)$$
$$+ \text{Gross private domestic investment } (I)$$
$$+ \text{Government expenditures } (G)$$
$$+ \text{Net exports } (X - M), \text{ or}$$
$$GNP = C + I + G + (X - M).$$

By this method, GNP was $3,661.3 billion (over 3 trillion dollars!) in 1984, as shown in Table 22–2. Keep in mind that the expenditures approach is the national income accounting measure of the upper loop in the circular flow model of economic activity.

The Flow of Earnings Approach

The flow of earnings approach should give us the same value for GNP as does the expenditures approach. Table 22–3 gives the categories used in the flow of earnings approach. (Refer also to Figure 22–3.)

National income (NI): The income earned by suppliers of resources.

National Income. The least inclusive figure in the flow of earnings approach to GNP is specifically called **national income (NI).** NI is determined by adding up the income earned by factors of production (land, labor, capital, and entrepreneurship) used to produce final goods and services during a given period. Table 22–3 shows national income of $2,959.4 billion for 1984, the sum of income items 1 through 5: compensation of employees, proprietors' income, rental income of persons, corporate profits, and net interest.

Compensation of Employees. Compensation of employees is the largest component of national income. It is the sum of wages and salaries paid to

---

TABLE 22–3   Gross National Product, 1984: The Flow of Earnings

GNP can be calculated by summing the incomes received by suppliers of resources. Labor income, interest income, rental income, and profits are all summed—with adjustments for indirect business taxes and depreciation—to obtain the same level of GNP as found by the expenditures approach.

	Earnings (billions of dollars)
Compensation of employees	2,172.7
Proprietors' income	154.7
Rental income of persons	62.5
Corporate profits	284.5
Net interest	285.0
National income (NI) (total of 1–5)	2,959.4
Indirect business taxes and other adjustments	299.0
Net national product (NNP)	3,258.4
Capital consumption allowances (depreciation)	402.9
Gross national product (GNP)	3,661.3

Note: GNP is in current, nominal dollars. Subcategories may not total because of rounding.
Source: Council of Economic Advisers, *Economic Report of the President* (Washington, D.C.: U.S. Government Printing Office, 1985), pp. 254, 256, and 257.

employees plus employer contributions to Social Security and employee benefit plans. In 1984, this category amounted to $2,172.7 billion.

Proprietors' Income. Proprietors' income is the net income earned by sole proprietorships and partnerships. The latter are mostly individuals such as self-employed professionals—doctors, lawyers, accountants, and so forth—and the former are small businesses across the country that are not incorporated. The 1984 figure for proprietors' income was $154.7 billion. (For more discussion of the role of proprietors' income and corporate profits in the economy, see Focus, "How Much of National Income Goes to Profits?")

Rental Income of Persons. Rental income of persons is the income of individuals from renting property, like a house or a car, as well as returns to individuals who hold patents, copyrights, and rights to natural resources, such as an oil or timber lease. National income accountants also impute a rental value to owner-occupied houses. In other words, home ownership is treated like a business that produces a service that is sold to the owner. All these forms of rental income yielded earnings of $62.5 billion in 1984.

Corporate Profits. Corporate profits are the net income of private corporations, including profits on foreign operations. The total of corporate profits was $284.5 billion in 1984.

Net Interest. Net interest consists of the interest received by U.S. households and governments minus the interest paid by these households and governments. Some interest payments are not counted as a part of national income because they are not considered payment for current production. Net interest totaled to $285.0 billion in 1984.

From National Income to Gross National Product. The figure for 1984 national income in Table 22–3—the sum of income items 1–5—does not equal the 1984 figure for gross national product in Table 22–2 figured by the flow of expenditures approach. Yet we know by the principle of the circular flow model that both approaches to measuring GNP must yield the same result. The problem is that there are two items—items included in total expenditures that are not income to anyone—that must be added to national income to obtain GNP. Let us examine these adjustments in detail.

Indirect Business Taxes and Other Adjustments. Indirect business taxes and other adjustments are the largest single item that must be added to national income to determine GNP because such taxes are part of the total expenditures on goods and services but are not received by anyone as income. These include sales and excise taxes paid by purchasers of goods and services; property taxes; business transfer payments such as corporate donations to charitable institutions; government subsidies; and other minor statistical adjustments.

Capital Consumption Allowances. Capital consumption allowances are an adjustment made for the wearing out of capital goods such as plant and equipment during the current production period. In Table 22–3, depreciation is equal to $402.9 billion. With adjustments for indirect business taxes and capital consumption allowances, GNP figured by the flow of earnings approach is $3,661.3 billion, the same estimate derived from the flow of ex-

FOCUS   How Much of National Income
Goes to Profits?

The typical small business hires inputs at given prices in order to produce goods or services. It sells its output at the prevailing market price. The difference between its costs and revenues accrues to the owner(s) of the firm as a residual return—what is left over after all expenses have been made. The owner of the firm is called a residual claimant. He or she earns profits when the residual return is positive and suffers losses when it is negative.

It is important to carefully measure what is being called a profit here. In an owner-operated business, the opportunity cost of the time and effort of the owner must be counted as a cost against firm revenues when calculating profits. In other words, the owner's efforts are not free; the owner could be working at the next-best alternative and being compensated accordingly. This same opportunity wage must be imputed to the owner's efforts in his or her own firm. This means that most of the "profits" observed in small businesses are really just a return to the time and effort of small businesspeople.

In a corporation, the picture is a little clearer. Stockholders are the residual claimants who earn returns such as dividends. Stockholders' returns are paid only after all the other factors of production have been paid. Managers are paid a salary for their services, though they are sometimes also paid in stock options to provide them with an incentive to increase the value of the firm.

If we examine the role of profits in national income, it does not appear to be large. In Table 22–3, taking proprietors' income and corporate profits as representing returns to residual claimants, we see that the ratio of these returns to national income was about 14.8 percent ($439.2 billion divided by $2,959.4 billion) in 1984. As we just discussed, a large part of proprietors' income is simply wages for the time and effort of small businesspeople; the share of residual claimants in 1984 national income is therefore lower than 14.8 percent. In general, a good guess is that the share of economic profits in national income is ordinarily less than 10 percent.

Less than one-tenth does not seem to be a very large component of national income, given the importance that economists normally attach to profit-seeking behavior. After all, profit seeking is the driving force behind the free-market system. It is the profit seekers who arrange for productive activities to take place. If profits are such a small part of national income, why are they so important?

The residual claims reported in the national income accounts are the actual result of profit-seeking activities. As such, they only approximate what entrepreneurs and firms had in mind when they first engaged in productive activities or announced their goals for the year. At the earlier stage, entrepreneurs and firms attempted to make the difference between their costs and revenues as large as possible. It is the *attempt* to make profits that is crucial to explaining economic behavior. The results, as reported in the national income accounts, are secondary, at best, to the continuous search for sources of profits in the economy. Profit-seeking refers to before-the-fact decision making by firms and individuals; the national income accounts record after-the-fact results of this important activity.

penditures approach given in Table 22–2. Using the flow of earnings method of accounting, GNP is, in short, the sum of returns to the factors of production (wages, proprietors' income, rents, interest, and corporate profits) plus indirect business taxes and depreciation.

Net National Product

Net national product (NNP): GNP minus depreciation; the value of the nation's output minus the capital used to obtain it.

Another useful national income concept can be derived from the statistics of Table 22–3. Net national product is equal to GNP minus depreciation, the value of the capital consumption allowances. In Table 22–3, the net national product for 1984 can also be determined by adding to the national income of $2,959.4 billion the indirect taxes and other adjustments of $299.0 billion. Net national product in 1984 was $3,258.4 billion. **Net national product (NNP)** is the net market value of goods and services produced in any economy in a given period. It tells us the value of all cars and houses and such produced during a certain period minus the value of the capital, plant, and equipment used in this production. Since such capital is

TABLE 22–4   National Income, Personal Income, and Disposable Personal Income, 1984

National income minus income not received (items 1, 2, 3) plus income received (items 4, 5, 6, 7) yields personal income. Subtracting personal taxes from personal income yields disposable personal income.

	Value (billions of dollars)	
National income		2,959.4
1. Minus: Undistributed corporate profits[a]	−284.5	
2. Minus: Net interest	−285.0	
3. Minus: Contributions for social insurance	−305.9	
4. Plus: Government transfer payments	+399.5	
5. Plus: Personal interest income	+434.8	
6. Plus: Personal dividend income	+ 77.7	
7. Plus: Business transfer payments	+ 17.3	
Personal income		3,013.2
8. Minus: Personal taxes	−435.1	
Disposable personal income		2,578.1

[a]Includes inventory valuation and capital consumption adjustments.

Note: Disposable personal incomes in current nominal dollars.

Source: Council of Economic Advisers, *Economic Report of the President* (Washington, D.C.: U.S. Government Printing Office, 1985), pp. 255 and 260.

constantly being used up or transformed into current production of goods and services, it must be subtracted to get the net value of current production.

Other National Income Concepts

Macroeconomics is often concerned with the amount of income that goes to the household sector in an economy. By adjusting the figure derived for national income—income earned by factors of production—we can arrive at the concepts of personal income and disposable personal income.

Personal income: The total amount of income actually received by individuals in a given period.

Personal Income. Personal income is the amount of income that households receive during a given period. There is a difference between income earned and income received. A writer, for example, may not receive royalties on the great American novel until years after writing the book because of the time lags in the production and marketing processes. Current production does not necessarily translate instantly into current income. The major categories of situations in which household earnings do not coincide in time with household income are given in items 1, 2, and 3 in Table 22–4.

To determine personal income, items 1, 2, and 3 must first be subtracted from national income because they represent income earned but not received by households in the current period. Undistributed corporate profits are retained and reinvested by corporations and do not flow to households as dividends in the current period. Net interest may also be earned by households but may not be received in the current period. Contributions for social insurance (− $305.9 billion)—such as payroll deductions for Social Security— are also excluded from personal income because they are earned but not received by workers in the current period.

Items 4 through 7 are received by households in the current period and are therefore counted as part of personal income. These include government transfer payments, such as Social Security payments, personal interest income, dividends, and business transfer payments. These statistics, not included in national income, are added to it to determine personal income.

Depending on the magnitudes of the minuses and pluses, national income can be larger or smaller than personal income. In a period of declining business activity, for example, national income can fall below personal income because corporate profits go down while government transfer payments go up. Personal income in 1984 was $3,013.2 billion, a greater amount than national income.

Disposable personal income: Individuals' income less taxes, that is, spendable income.

Disposable Personal Income. Disposable personal income—what income receivers have left to spend or save after taxes—is found in Table 22–4 by subtracting personal taxes from personal income. Personal taxes include income, property, and sales taxes to federal, state, and local governments—individuals' personal shares in the expense of government. In 1984, disposable personal income was $2,578.1 billion.

Personal Saving

Personal saving: The portion of disposable personal income that consumers choose not to spend on final goods and services.

Personal saving, another important economic statistic, represents the funds provided for investment by the saving of individuals and unincorporated businesses in the economy. Their choice to refrain from spending all they receive is a crucial ingredient in economic growth. Table 22–5 shows the calculation of personal saving as $156.9 billion in 1984.

The Measurement of Price Changes

Was GNP in 1985 greater than in 1980? To answer this question, we might simply compare GNP for 1980 and 1985. Yet changes in prices make such comparisons difficult. Think of a simple world in which only chocolate fudge is produced. The fact that total expenditures on fudge rose from $5 million in 1980 to $10 million in 1985 tells us very little. Total expenditures will increase if (a) more fudge is produced, (b) the price of fudge rises, or (c) both (a) and (b) happen at the same time. The same facts hold for changes in GNP. GNP can change over time because total output increases, prices increase, or both happen at the same time. Thus, since we are interested in comparing differences in aggregate output over time when using national income statistics, a means has to be found to adjust GNP data for changes in prices. (The more difficult issue of changes in the type and quality of goods over time is treated in Focus, "GNP and Economic Change.")

TABLE 22–5   Personal Saving, 1984

Disposable personal income may be saved or spent by households. In 1984, households chose to save $156.9 billion and spend $2,421.2 billion of their disposable personal income.

	Value (billions of dollars)
Disposable personal income	2,578.1
Minus: Personal outlays	−2,421.2
Personal saving	156.9

Note: Personal savings in current, nominal dollars.
Source: Council of Economic Advisers, *Economic Report of the President* (Washington, D.C.: U. S. Government Printing Office, 1985), p. 260.

FOCUS   **GNP and Economic Change**

Comparisons of GNP over time are even more complicated than controlling for price changes with a price index. Such comparisons are fundamentally difficult because entrepreneurs introduce new and improved goods and services to the economy. For example, how useful is it to compare GNP in 1884 with GNP in 1984? The United States at these two time periods comprised two different economies. In 1984, there were cars, fast-food restaurants, televisions, computers, and liver transplants. These goods and services did not exist in 1884. On the other hand, there was a great deal of open land, a relatively low crime rate, and little or no acid rain in 1884. These conditions largely no longer existed in 1984.

The goods available to consumers—that is, the final outputs of the economy—change over time. Even someone with a lot of money in 1884 could not have purchased what the average middle-income family purchases today. The point is simple but profound. An economy changes across time, and comparisons of GNP therefore lose much of their relevance.

So while we know that the U.S. economy has grown remarkably over its history, the fact that the bundle of goods available to consumers has changed over time makes the measurement of this growth difficult. For example, real per capita GNP in 1950 was $3535. In 1980, it was $6645. Does this mean that workers in the United States produced roughly twice the amount in 1980 as they produced in 1950? The problem is that GNP cannot give a clear answer to this type of question. The goods produced in each period have changed. We do not now produce twice the amount of the *same* goods produced in 1950. Our capacity to produce goods and services has increased, no doubt, but the exact amount of this growth is not measured by the growth of real GNP.

The further apart in time that comparisons of GNP are made, the more severe is the problem caused by the differences in economics in the two periods. However, even over a short period, GNP comparisons are difficult because of changes in the quality of existing goods. In a span of a little more than ten years, consumers in the United States have gained access to improved goods such as cable television and more fuel-efficient cars. Do these improvements in quality mean growth in the value of the economy? If so, how could it possibly be measured?

Finally, exactly the same point can be made about the difficulty of comparisons of GNP across countries. Countries obviously produce different bundles of goods, and for the reasons just given, cross-country comparisons of GNP are difficult and subject to careful interpretation. None of this argues that comparisons of GNP are meaningless. The point is that such comparisons should be made carefully and conclusions drawn about economic growth only after a careful study of all the relevant information about the economy or economies concerned.

Price level: The weighted average of the prices of all goods and services.

Price index: A device used to estimate the average percent change in the prices of a particular bundle of goods.

The **price level** is the weighted average of the prices of all goods and services. It is not to be confused with relative prices, which show the price of one good in terms of another.

Economists and national income statisticians approach the problem of unstable prices by constructing **price indexes.** The most popular index uses a base year as the year against which subsequent changes in output are measured. In this method of measuring price change, called the Laspeyres index for its nineteenth-century inventor,[3] base year quantities are used as weights for the price

[3]An understanding of the Laspeyres index calculation can be reached with a simple example. Assume that the consumer price index is composed of only two goods, beer and pizzas, and that the relevant data on prices, outputs, and expenditures are as follows.

Hypothetical Data for the Construction of a Price Index

Commodities	Price (per unit)	Quantity sold	Total Expenditures
Beer (six-packs)			
1984	$2.00	2 million	$ 4 million
1985	$3.00	3 million	$ 9 million
Pizza (each)			
1984	$4.00	3 million	$12 million
1985	$5.00	5 million	$15 million

indexes of ensuing years. Other theoretical approaches can measure price changes, but the Laspeyres method underlies many of the price indexes used presently by government. The important point is that the function of a price index is to control for price changes so that we can obtain a more accurate picture of the behavior of aggregate output over time.

Government statistics calculate several price indexes for the American economy. The underlying principle of these indexes is the same: They are generally calculated by the Laspeyres method. The details of how each is constructed differ, however.

The Consumer Price Index

Consumer price index (CPI): The device used to estimate the percent change in the prices of a particular bundle of consumption goods.

The **consumer price index (CPI)** measures price changes in a typical market basket of goods purchased by urban wage earners and clerical workers. This "market basket" was developed from a survey of about 20,000 families, who provided information on their buying habits. The market basket includes food, housing (rental costs), apparel, transportation, health and recreation, and miscellaneous services. To measure price changes in the market basket, the Bureau of Labor Statistics at the Department of Labor collects data every month from a wide variety of retail stores and service establishments. The CPI is issued monthly, and its base year is 1967.

Figure 22–4 shows the movement of the CPI for the period 1971–1984. Notice that the CPI for all items has been broken down into four categories: food, rent, transportation, and apparel. Each of these categories fluctuates over time, with some rising and others falling during a given period.

The Producer Price Index

Producer price index (PPI): The device used to estimate the percent change in the prices of a particular bundle of inputs used by manufacturing businesses.

The Bureau of Labor Statistics also publishes statistics known as the **producer price index (PPI)**, previously known as the wholesale price index. This index excludes consumer prices and instead covers about 2800 industrial commodities, from raw materials to finished goods. For the PPI the Bureau of Labor Statistics collects about 10,000 price quotations each month from producing companies.

The Implicit Price Deflator

A final major index is the **implicit price deflator,** often called the GNP price deflator. This index goes to the heart of the problem of measuring output changes over time. As we have seen, it is impossible to add up the different

---

Assume that the base year is 1984 and that we are interested in how prices have changed in 1985—that is, has there been inflation or deflation? A Laspeyres base year quantity weight index for beer and pizza can be expressed as

$$\text{Price index} = \frac{q_b^{84} p_b^{85} + q_p^{84} p_p^{85}}{q_b^{84} p_b^{84} + q_p^{84} p_p^{84}}.$$

The superscripts denote the relevant year (1984 or 1985), and the subscripts denote beer (b) or pizza (p). Price is p and quantity is q. Using this price index formula with 1984 as the base year, the calculation of price change is equal to 1.31. That is,

$$\text{Price index} = \frac{\$(6 + 15) \text{ million}}{\$(4 + 12) \text{ million}} = 1.31.$$

Since the value of the index is always equal to 1.0 in the base year, the rate of inflation is relative to the prices or quantities of the base year. The indexed change in prices in this example is found by dividing 21 by 16. Calculated with this price index, prices rose by 31 percent in the beer-and-pizza economy.

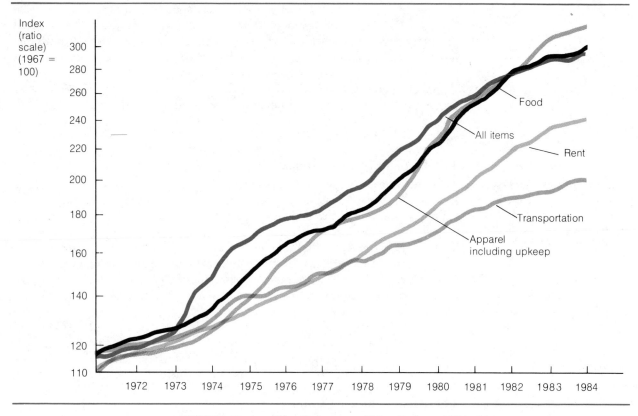

FIGURE 22–4   The Consumer Price Index, 1971–1984

The consumer price index measures the average change in price of a bundle of consumer goods. The prices of goods in various categories, such as food or rent, change at different rates. However, the overall CPI has increased fairly consistently over the past decade. The base year for constructing the consumer price index on the vertical axis is 1967.

Source: Board of Governors of the Federal Reserve System, *Historical Chart Book* (Washington, Board of Governors of the Federal Reserve System, 1984), pp. 40–41.

Implicit price deflator: A price index that attempts to use the entire national output as the bundle of goods and thus to generate the most comprehensive price index and the best measure of inflation; also called GNP price deflator.

outputs of goods over time. Rather, we add up the dollar amounts spent on commodities. But since both prices and quantities of commodities can change over time, the comparison of GNP at different points in time is difficult. Remember that the economist is interested in measuring the economy's real output. Therefore, the GNP statistician must be careful to distinguish current or nominal GNP, which is evaluated in current dollars, from real or constant GNP, which is evaluated in constant dollars. The implicit price deflator is a means to convert nominal GNP values into real GNP values.

Nominal GNP is converted into real GNP by evaluating or deflating current GNP in terms of 1972 prices, which are used as a standard and set equal to 100 in the index. The implicit price deflator applied to current GNP is found by dividing the current, nominal value of GNP by its value in terms of 1972 prices. The nominal value of GNP in 1980 was $2631.7 billion; at 1972 prices the value of this GNP was $1475.0 billion. The implicit price deflator was thus 2631.7/1475.0 = 178.42 percent. In other words, over the 1972–1984 period, prices of final output rose by 78.4 percent.

To understand the relation between GNP in current dollars and the im-

TABLE 22–6   Current GNP, GNP Deflator, and Real GNP, 1972–1984

The GNP deflator is used to convert current GNP into constant GNP. By dividing any year's GNP by that year's GNP deflator, one can estimate real GNP in terms of 1972 dollars.

Year	Current GNP (billions of dollars)	GNP Deflator (1972 = 100)	Constant GNP (billions of dollars)
1972	1185.9	100.0	1185.9
1973	1326.4	105.75	1254.3
1974	1434.2	115.08	1246.3
1975	1549.2	125.79	1231.6
1976	1718.0	132.34	1298.2
1977	1918.3	140.05	1369.7
1978	2163.9	150.42	1438.6
1979	2417.8	163.42	1479.4
1980	2631.7	178.42	1475.0
1981	2957.8	195.60	1512.2
1982	3069.3	207.38	1480.0
1983	3304.8	215.34	1534.7
1984	3661.3	223.38	1639.0

Source: Council of Economic Advisers, *Economic Report of the President* (Washington, D.C.: U.S. Government Printing Office, 1985), pp. 234, 236.

plicit price deflator, divide current GNP by its implicit price deflator. In the 1980 example, this means that

$$\text{Constant GNP} = \frac{2633.7}{178.42} \times 100 = \$1475.0 \text{ billion.}$$

Constant GNP shows what today's output would have been worth had it been sold in 1972.

Table 22–6 shows GNP in both constant and current terms, using the varying implicit price deflator for GNP for the years 1972 to 1984. One interesting aspect of these data is that between 1973 and 1975 and again between 1981 and 1982, GNP in current terms was rising but constant GNP was falling. In other words, growth in the GNP was due to inflation in prices rather than to increases in output. The proper measure of the real performance of the economy is the movement of real GNP over time.

Summary

1. National income accounting is concerned with the measurement of the aggregate performance of the economy.
2. Gross national product (GNP) is the market value of all the final goods and services produced in an economy during a given period. It includes only currently produced goods and services, and it excludes such items as illegal transactions and household production.
3. GNP is not a measure of economic welfare. It does not count social costs such as pollution, which lowers economic welfare, or such goods as leisure, which raises economic welfare.
4. Gross national product can be measured by the value-added approach, the final-goods approach to expenditures, or the income paid to factors of production to produce final outputs. The latter two methods are the most useful.
5. Gross national product can be computed in two ways: by adding up expenditures on final goods and services produced during a given period or by summing the earnings or incomes of the factors of production used to produce final goods and services during the period. The two approaches, subject to a statistical discrepancy, yield the same result for GNP.

The Expenditures Approach

Personal   Gross private   Government
consumption   domestic   purchases of
expenditures   investment   final goods
    (C)   +   (I)   +   and services
                                     (G)

Net exports
+ (X − M) = GNP

The Flow of Earnings Approach

            Proprietors'   Net
Wages +   income   + Rents + interest

            Indirect
  Corporate business
+ profits + taxes   + Depreciation = GNP

The flow of expenditures approach focuses on buyers' evaluations of goods produced during a year. The flow of earnings approach concentrates on the cost of production of goods and services. The circular flow of economic activity, therefore, ensures that the two approaches are identical. That is,

$$\frac{\text{Dollar spending}}{\text{on final outputs}} = \text{GNP} = \frac{\text{Dollar costs of}}{\text{producing final outputs.}}$$

6. Other important national income accounting concepts are personal income, disposable personal income, and personal saving. Each provides important information for the macroeconomist in analyzing economic stability and growth.
7. To compare GNP in different time periods, some method must be used to control for price changes. The government uses various price indexes, including the consumer price index, the producer price index, and the implicit price deflator for GNP.

Key Terms

national income   flow of expenditures   exports (X)   personal saving
  accounting   flow of earnings   imports (M)   price level
gross national product   personal consumption   national income (NI)   price index
measure of economic   expenditures (C)   net national product   consumer price index
  welfare   gross private domestic   (NNP)   (CPI)
final goods   investment (I)   personal income   producer price index
intermediate goods   government expenditures (G) disposable personal   (PPI)
value added   net exports (X − M)   income   implicit price deflator

Questions for Review and Discussion

1. Why does GNP count only the production of final goods and services? Why aren't intermediate goods and services counted?
2. What are the relations among gross national product, net national product, national income, personal income, and disposable personal income?
3. List three reasons why GNP should not be considered as an indicator of society's well-being.
4. What are the three approaches used to estimate GNP?
5. American society has become more urbanized and industrialized since 1930. Other things being equal, do you think this means that our measurement of GNP is more or less precise? Why?

6. What is GNP in constant dollars and how is it calculated?
7. Why might GNP be a misleading statistic with which to compare the economy of the United States with that of the Soviet Union?
8. Which of the following are counted in the calculation of this year's GNP? (a) the services provided by a homemaker; (b) the wage paid to a maid; (c) Sam's purchase of an antique desk; (d) Joan's purchase of ten shares of stock; (e) Social Security checks received by the elderly; (f) Social Security taxes paid by workers.

ECONOMICS IN ACTION   GNP, Leisure, and Social Welfare

As indicated in this chapter, GNP is not a perfect measure of economic well-being. GNP statistics exclude, for example, various aspects of economic existence such as home production and leisure. Some scholars have tried to overcome these problems in measuring GNP and to propose measurements of an economy's performance that would more closely reflect the underlying welfare or happiness of its people. William Nordhaus and James Tobin have proposed a tool called *measure of economic welfare* (MEW). Their measure would modify the traditional measurement of GNP in the following ways:

1. Measures of the value of leisure time and household production and consumption would be included in GNP.
2. Social costs such as pollution, litter, and congestion would be excluded from GNP.
3. Expenditures on intermediate goods produced by government such as national defense and police pro-

tection, which are judged to be activities that are really inputs to other end-product activities yielding utility, would be excluded from GNP.

The net result of their calculations indicates that since 1965 MEW has grown less rapidly than GNP. Thus, the GNP of recent years tends to overstate our economic progress relative to MEW. This conclusion is understandable since many of the costs of our economic growth such as pollution became paramount over this period, undermining the quality of our lives at the same time that we have seemed to be advancing in material wealth.

MEW is new and controversial among national income scholars. It represents, however, a step toward better and more accurate measurement of the macroeconomy, a step that will attract the energies of economists and statisticians toward further refinement.

Source: William Nordhaus and James Tobin, "Is Growth Obsolete?" in *Economic Growth,* Fiftieth Anniversary Colloquium (New York: National Bureau of Economic Research, 1972).

Question

In what sense (or senses) is the MEW measure of national well being subjective in nature? How, for example, could one measure the value of leisure?

23

From Classical to Keynesian Theory

M easuring the rise and fall of macroeconomic variables such as the rate of inflation, GNP, and unemployment is only a first step toward understanding why these economic indicators fluctuate and what measures, if any, are needed to maintain stability and growth.

The next four chapters survey some of the major theoretical tools of macroeconomics to discover the conditions under which the economy can achieve stability and growth. We concentrate on both aggregate demand—the spending side of the economy—and aggregate supply—the producing half of the economy. This chapter and Chapter 24 develop economic principles pertaining to aggregate demand. Chapters 25 and 26 deal with the interaction of aggregate demand and aggregate supply.

Some simplifications are necessary to understand how aggregate demand and supply interact to sustain full employment, economic growth, and low levels of inflation. Chapter 23 concentrates on two views of consumer and business spending—the private or nongovernment part of aggregate spending. The first view is the classical notion of a self-regulating aggregate economy; the second is the perspective of J. M. Keynes, who focused attention on how spending changes cause prosperity and depression. To simplify our development of aggregate demand and supply theories we at first ignore important but complicating variables such as the money supply. Major attention to money is postponed until Part VI.

Classical Theory

Can the economy remain as close as possible to its production possibilities frontier? Can it manage to keep all resources, human and nonhuman, fully employed? More fundamentally, can the economy automatically produce full employment, a maximum GNP, and price stability? If not, what actions are necessary to maintain these important goals?

Historically, economists have given many different answers to these difficult questions. **Classical macroeconomic theory** stems from Adam Smith's pioneering work, *An Inquiry into the Nature and Causes of the Wealth of Nations*, published in 1776. Smith and all who followed in the classical tradition believed that, given laissez-faire government policies and enough time, the economy would achieve the goals of price stability, full employment, and economic growth through its own ability to correct short-run unemployment and inflation. In other words, the classical theorists believed that unemployment and inflation were temporary phenomena. In the long run (a period of time that is quite hard to specify) the economy would remain close to or on its production possibilities frontier and enjoy a stable level of prices. This classical tradition lasted for well over 150 years (the last great neoclassical economist, A. C. Pigou, died in 1959). The theory that grew from Smith's faith in a self-adjusting market mechanism was the work of many economists who wrote at different times in response to different conditions. The following simplified discussion of classical macroeconomics is a composite of various individual contributions.

Classical macroeconomic theory: The school of economic theory dominant during the late eighteenth, nineteenth, and early twentieth centuries. Classical theory viewed the macroeconomy as a self-adjusting mechanism capable of generating and sustaining maximum employment without government intervention.

Cornerstones of Classical Theory

The classical economist's belief that the economy self-adjusts in response to short-run disturbances and produces full employment and economic growth without government interference rests on four cornerstones:

1. Say's law
2. Interest rate flexibility
3. Wage-price flexibility
4. Quantity theory of money

The first three cornerstones deal with the flexibility of the macroeconomy in maintaining full employment. The fourth concerns the determination of the price level and the control of inflation. We encounter the first three in this chapter. We return to the fourth, the quantity theory of money, in Chapter 30.

Say's law: An assumption of classical economists that the supply of goods and services in the economy creates a corresponding demand for those goods and services.

Say's Law. **Say's law** is an economic principle first attributed to Jean Baptiste Say (1767–1832). According to Say's law, diagrammed in Figure 23–1, the act of supplying goods, or total real output, is the equal but opposite side of demanding goods, or total real expenditures. Say's law implies that full employment is a permanent, built-in feature of the macroeconomy. Resource unemployment is impossible because the act of producing goods is the same act as demanding goods. In other words, supply creates its own demand.

Say's law is most easily understood in a barter economy, where goods are exchanged for goods. Suppose, for example, that a chicken farmer supplies eggs and demands auto parts. If a deal can be worked out with an auto parts supplier, the farmer's act of supplying eggs is simultaneously the act of demanding auto parts. Likewise, the auto parts supplier's demand for eggs is reflected in his supply of auto parts. What is true for the farmer and the auto parts supplier is also true for all traders in the economy.

Say's law also applies to an economy using money. When money is used as a common denominator or as a medium of exchange for all goods and services, the egg supplier does not deal directly with the auto parts seller, but he or she does barter goods indirectly through the use of money. Money

FIGURE 23–1

Say's Law

In the circular flow of income and output, the act of supplying goods and services is necessarily equal to the act of demanding goods and services.

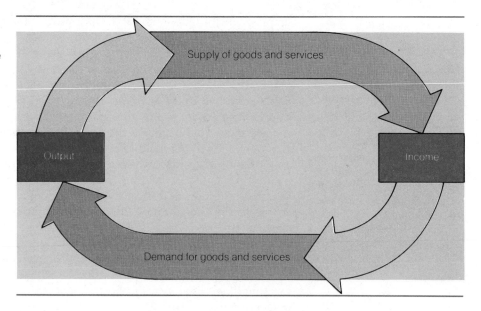

in this system is merely a veil that hides the real workings of the economic system.

Say believed that in a money economy the aggregate demand for goods and services is financed through the earning of income, as in the circular flow diagram in Figure 22–1. How do productive factors earn wages, rents, interests, and profits? They do so by producing goods and services. In this system, as in a barter system, the act of producing specific goods and services results in the demand for other goods and services.

Say's law seems perfectly sensible, except for one important issue: What happens when consumers choose to save part of their income? The act of

Jean Baptiste Say

saving—the sacrifice of present income for future consumption—would disrupt the perfect balance between income and output.

Saving is a leakage, or withdrawal, from the circular flow of income and spending. If this withdrawal from spending is not matched by an injection—another form of spending to compensate for the saving withdrawal—underconsumption and unemployment would result. Investment spending, as we saw in Chapter 22, is one form of injection into the circular flow to make up for a savings withdrawal. But how do private saving—the leakage—and investment—the injection—become linked? The answer to this riddle, the classical economists thought, was in the mechanism of interest rates.

Interest Rate Flexibility. People save, or postpone current consumption, because of a reward for doing so. Interest paid on savings is their reward, enabling savers to consume more goods later. This important relation may be expressed as

$$S = s(r),$$

or saving (S) is a function of the real rate of interest (r). The rate of interest is both the nominal percentage that savers earn annually on their savings and the percentage that borrowers must pay to use funds deposited in savings institutions. The **real rate of interest** is the nominal rate of interest minus the annual rate of inflation. The real rate reflects the fundamental forces of saving and investment in the economy.

Real rate of interest: The nominal rate of interest minus the rate of inflation; a rate that reflects the true incentive of savers and investors to save and invest.

Classical economists believed that saving was positively related to the real interest rate: A rise in the interest rate increased the amount saved. For example, when the real interest rate increases from r_0 to r_1, depicted in Figure 23–2, consumer-savers in the economy are encouraged to save more for future consumption and to consume less now, assuming that attitudes toward thrift are constant. Any increase in attitudes encouraging saving would cause the S curve in Figure 23–2 to shift to the right, indicating more saving at all rates of interest. If attitudes toward thrift become discouraged, of course, the whole curve would shift to the left.

Keep in mind that increases in saving are accompanied by simultaneous decreases in current consumption. In other words, according to the classical model, aggregate demand falls by the amount that saving increases. The important question is, therefore, Will an increase in saving mean that aggregate supply will not call forth enough aggregate demand? Will goods remain on shelves, inventories pile up, and thousands of workers be laid off? The classical economists said no, if the economy is allowed to reestablish its equilibrium. Not only does the interest rate paid to savers determine their choices, but the interest rate charged on loans determines the behavior of investors. Investment is a flow of expenditures to repair or replace capital goods or to make additions to the capital stock of the nation. Investment spending, just like consumption spending, generates income and employment.

In classical theory, a lower rate of interest makes more investment projects—such as expenditures for warehouses and machinery—profitable. Investors, in other words, spend more and more on capital goods as the interest rate falls. Technically, this relation can be expressed as

$$I = i(r),$$

or investment (I) is a function of the real rate of interest (r). However, unlike the relation between saving and interest, investment and the real rate

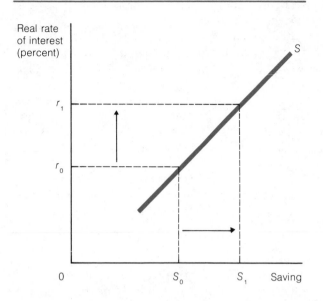

FIGURE 23–2   The Classical Concept of Saving

Saving and interest rates are positively related. As the real rate of interest rises, from r_0 to r_1, the level of saving in the economy rises from S_0 to S_1. An increase in the amount saved means that present consumption, or demand, is diminished by an equal amount.

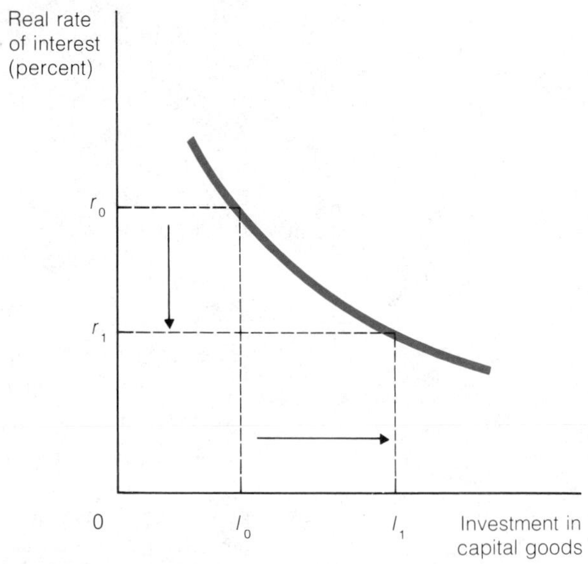

FIGURE 23–3   The Classical Concept of Investment

The amount invested (I) and the real rate of interest (r) are negatively related, given a constant rate of return on investment in capital goods. As the rate of interest falls from r_0 to r_1, businesses invest more in capital goods. The increase in investment expenditures from I_0 to I_1 represents a net increase in demand.

of interest are negatively related. Rises in r cause declines in the amount of I, and reductions in r provoke increases in the amount I.

This inverse relation is depicted in Figure 23–3. As the real rate of interest declines, the investment spending represented by curve I increases. Conversely, when the rate of interest rises, fewer and fewer investment projects become profitable, and investment spending declines. Increases in the productivity of capital—through inventions or improvements in technology—would increase the level of investment at every rate of interest, shifting the whole curve to the right.

According to the classical system, saving and investment are balanced by means of the interest rate. Figure 23–4 reproduces the savings and investment curves already discussed. An initial economy-wide equilibrium is established where saving equals investment at interest rate r_0. At interest rate r_0, in other words, the amount that individuals in society wish to direct from consumption to savings is S_0. Real interest return r_0 is the savers' reward for their thrift. At interest rate r_0, moreover, investors find I_0 worth of investment projects to be profitable—no more, no less. Given curves S and I, no other interest rate is compatible with equilibrium. Why? Because an interest rate lower than r_0 would create a shortage of investment funds, and one higher than r_0 would mean a surplus of funds. In the former case, according to the law of supply and demand, investors' demand would force the interest rate up to r_0; in the case of a surplus of funds, savers' demand would force the interest rate down to r_0. To put the matter formally, macroeconomic

FIGURE 23–4

The Relation Between Rate of Interest, Saving, and Investment in the Classical System

Initial equilibrium E_0 is established at interest rate r_0 with saving and investment curves S and I. An increase in thrift will cause the saving function to shift rightward from S to S_1, lowering the rate of interest from r_0 to r_1. As the interest rate declines, the amount of investment spending increases. Reduction in the amount of consumption spending brought about by increased saving is therefore accompanied by a counterbalancing increase in the amount of investment spending.

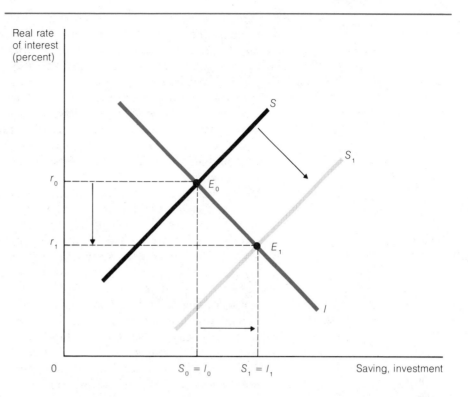

equilibrium requires that the amount of saving equals the amount of investment, or, in terms of the above expressions, that

$$S = I.$$

What happens if the whole saving curve or investment curve shifts, reflecting a sudden change in society's underlying attitudes toward thrift (increased saving for retirement or expectations of hard times) or a technological breakthrough that increases the productivity of investments? Suppose, for example, that individuals wish to save more at every rate of interest, shifting the savings function rightward from S to S_1, as shown in Figure 23–4. A surplus of saving develops at interest rate r_0, forcing the rate of interest down. As the rate falls, investors take advantage of new, profitable investments. Finally, at the new equilibrium interest rate, r_1, a new, higher quantity of savings, S_1, equals a new, higher level of investment expenditures, I_1.

If Say's law is true, all leakages from the income stream must be replaced by injections. According to classical economics, increases in saving (leakage) are replaced by increases in investment (injection) through the mechanism of the real rate of interest. Aggregate demand for goods and services would fall short of aggregate supply were it not for the mechanism of interest rates. In other words, the classical economists believed that Say's law is valid if the interest rate is able to adjust freely upward or downward without regulations or restrictions. Consumption spending plus investment expenditures will be sufficient to purchase all of the output produced by the economy.

Price-Wage Flexibility. The validity of Say's law requires another cornerstone of classical thought: price-wage flexibility. To understand this second

Hoarding: Storing idle cash; accumulating money that is neither spent nor saved.

concept, suppose that future economic conditions appear very uncertain. There might be expectations of war, of a coming depression, or even of an overall decrease in prices. In response to this uncertainty, individuals seek greater immediate economic security. Rather than increase their savings, they begin **hoarding** their income, cutting their demand for goods and services in the process. In the event of a sudden surge of hoarding, aggregate demand would be insufficient to carry off the aggregate supply of goods and services. Unemployment would increase disastrously, and the economy might collapse. According to classical theory, however, any such reduction in demand would also cause prices and wages to decline in proportion to the size of the reduction.

As the prices of shoes, pizzas, hair stylings, and all other goods and services decline, additional quantities of goods will be demanded, and all excess production will quickly be bought up by consumers. Wages of laborers and the prices of all other inputs will also decline, but laborers' real wages and incomes will be the same as before the price decline. The flexibility of prices and wages in a competitive market guarantees full employment even in the event of a sharp and protracted reduction in aggregate demand.

It must be emphasized that the classical cornerstone of price-wage flexibility applies when aggregate demand is insufficient to maintain full employment for *any* reason. Natural disasters such as wars and famines were no threat to full employment, so long as prices and wages were allowed to adjust. The purely competitive market system—an assumption of classical economists—is the fail-safe mechanism through which unemployment over long periods of time was thought to be impossible. For the system to operate properly, of course, there could be no restrictions on prices (such as price controls) or wages (such as a minimum wage) in either product or resource markets.

Classical Theory and Policy: A Self-Adjusting Economy

The cornerstones of classical macroeconomic theory imply that self-adjusting forces in the economy will, over some unspecified period, guarantee full employment. Market forces will establish full employment given the validity of Say's law, the theory that supply will create its own demand, given interest rate flexibility and price-wage flexibility. Long-run unemployment is simply not possible if the economy is allowed to function as the theory predicts.

These conclusions sound almost too good to be true, but they provide the bases for practically all of the following classical policy recommendations.

Balance the Budget. Classical economists emphasized the need for a balanced government budget for a number of reasons. If government is allowed to spend more than it receives in taxes and so on, the government would be forced to compete with investors to borrow available funds. Such competition would tend to increase real interest rates, choking off private investment, capital formation, and economic growth. Financing deficits by printing money—an attractive option when governments have access to the presses—would have equally bad effects on the private economy. As we will discuss in Chapter 30, increases in the money supply create inflation, which is tantamount to a tax on the private sector.

Keep Government Small. High levels of taxes necessary to finance big government reduce incentives to private saving from which new investments are made. Big government, in the classical view, short-circuited the mainspring to progress.

Moreover, large-scale government included domestic or international regulations that tended to reduce trade, productivity, and consumers' well-being. In the classical laissez-faire view, the government should be restricted to providing national defense, a legal system, and few other functions. Over the past two centuries, a number of policymakers have pursued these classical economic conclusions, as described in Focus, "Classical Macroeconomics Under William Gladstone and Ronald Reagan."

Laissez-faire. The classical cornerstones of macroeconomic theory were part of a long-run view of economic activity. Yet the classical writers never stated the actual time period needed for the economy to self-adjust. They merely emphasized that any short-run government tinkering with the private economy in either a macroeconomic or a microeconomic sense would have negative long-run effects. In their view, the best long-run economic hope for all members of society—laborers, households, consumers, businesspeople, investors, savers—was to let unfettered market forces work in the private economy, unassisted by government.

Recessions, depressions, or periods of high unemployment caused by massive shifts in consumption spending, natural disasters, or wars could all wreak short-run havoc and temporary unemployment in the economy, as all classical economists knew. Society's members, however, would all be better off to suffer the temporary consequences rather than demand that government intervene. In the classical view, nearly all short-run actions of government in the aggregate economy only made problems worse, prolonged recessions or depressions, and created built-in instabilities in the macroeconomic system. Capitalism was not perfect, but society's inability to endure the short-run pains of recession and unemployment (reflected in demands for government to do something) meant that there would be greater economic pain in the long run.

Theory Versus Events: Keynes, Macroeconomics, and the Great Depression

Classical economic theory dominated the economic debate until the early twentieth century, when a series of catastrophic events severely weakened some economists' faith in a self-adjusting economy and the validity of Say's law. After an extraordinarily prosperous Roaring Twenties, the United States and the rest of the world fell into the deepest and most prolonged depression in modern history.

In September 1929, the market for stocks and other securities began to fail. On Black Thursday, October 24, 1929, stock prices plummeted and thousands of investors lost millions of dollars' worth of securities. The stock market crash ushered in (though it did not necessarily cause) steep declines in industrial production, real income, and civilian employment. For more than a decade, between 1929 and 1940, industrial production in the United States failed to exceed its 1929 level. Real income of Americans (expressed in 1929 dollars) was about $90 billion in 1929, a level unmatched until 1939. Most important, civilian unemployment rose to more than 30 percent by 1933.

The exact causes of the Great Depression are still a matter of debate

FOCUS   Classical Macroeconomics Under
William Gladstone and
Ronald Reagan

In 1841, William Gladstone was appointed vice-president of the Board of Trade in England under the Conservative government of Prime Minister Robert Peel. Gladstone exercised a major influence on British economic policy until 1865, and his reign marked the end of mercantilism—an era characterized by abundant government control, regulation, and intervention in England's economy.

Gladstone's policies were designed to eliminate as much government intervention in the private sector as possible. These policies included a reduction in import tariffs and the abolition of the income tax. Before Gladstone, approximately four hundred products were subject to import tariffs, a form of excise tax on imported goods. Gladstone removed the tariffs on all but about fifteen of these items.

Major tariff reductions resulted in a temporary government deficit. Therefore, Gladstone saw it necessary to impose a 10 percent income tax to cover the deficit and balance the budget. However, the income tax was reduced to a lower rate over several years, eventually being eliminated altogether. Hence, the income tax was a temporary measure in the budget and disappeared after its aims were achieved.

Do these arguments sound familiar? They are the essence of President Ronald Reagan's original economic programs, which emphasized less government and lower taxes to increase private saving and investment. In addition, Reagan championed reductions in domestic regulation in such industries as transportation and communications. Reagan's policies, moreover, were oriented to the long run. His midterm slogan in 1982, "Stay the Course," and the so-called trickle-down theory (the idea that prosperity follows a downward course from increased industry to the poor) were aspects of a long-run view that has its origins in classical macroeconomics.

The Reagan administration sought—at least as an ideal—a lower level of total government involvement in the economy. Despite this goal total federal government outlays grew between 1980 and 1984 partly as a result of a large defense buildup. At the same time huge deficits began to worry financial markets and became a campaign issue in 1984.

A key to classical laissez-faire policy is the denial of permanent political access to tax revenues. Though not an exact parallel, the Reagan administration's policies are an attempt to lower the financial base of the federal government and to transfer control over resources from the public to the private sector and from federal to state and local governments. Whether Reagan's policies can achieve, have achieved, or will achieve these objectives remains a matter of great debate.

William Gladstone

Ronald Reagan

An unlucky speculator, one Walter Thornton of New York,
offering to sell his roadster on October 30, 1929.

among macroeconomists, but the prolonged economic chaos is a matter of fact. The Depression and the unemployment that accompanied it brought ruin for millions of Americans and similar hardships around the world. The classical self-adjusting macroeconomic system did not appear to work.

Keynes's Objections to Classical Theory

Economists were among the first to recognize that events appeared to overwhelm theory. In particular, some economists at Cambridge and Oxford universities in England began to question and analyze the cornerstones of classical economics. Economists such as Richard Kahn, Joan Robinson, R. G. Hawtrey, Roy F. Harrod, and John Maynard Keynes met regularly during the early 1930s to discuss reasons for the apparent failings of the classical system.

J. M. Keynes (1883–1946) was particularly vocal. In open letters to President Franklin D. Roosevelt, published in the *New York Times,* Keynes advocated the use of government spending and taxation policies to supplement private spending as a cure for the ailing economy. (Roosevelt, clinging to more traditional thinking, heeded Keynes's advice guardedly and hesitantly.) Keynes was firmly convinced that, contrary to classical theory, private market forces would not be sufficient to regain full employment equilibrium in the depressed economy. He dismissed the long-run self-adjustment theory of the classical economists with scorn. As he said in another context, "In the long run we are all dead." The economy could be stuck at some equilibrium characterized by high levels of unemployment for an extended period of time, as it seemed to be in the early 1930s.

Since the classical theory, in Keynes's view, did not offer rescue from the Depression, Keynes sought a better model from which to interpret events. In 1936, Keynes published his *General Theory of Employment, Interest and Money,* which established a new theory of how the economy functions—a new macroeconomics. In proposing his theory, Keynes had to counter the foundations of the classical macroeconomic system.

Keynes's Evaluation of Classical Theory

Recall that Say's law states that "supply creates its own demand" if savings (a leakage) could be transformed into investment expenditures (a compensating injection) through a flexible interest rate and if prices and wages were flexible when aggregate demand is exceeded by aggregate production. Keynes took exception to Say's law and to the two other cornerstones that support it—the belief that price-wage flexibility and interest rate flexibility will cure any temporary disruptions in the economy.

First, Keynes argued that saving and investment were determined by a host of forces in the economy in addition to the rate of interest. In Keynes's view, savers and investors are different groups with different sets of motivations and interests. Savers, he thought, are more responsive to their amount of personal disposable income than to the rate of interest in deciding how much to save. Higher interest rates, thought Keynes, might even mean less saving if people were saving toward a particular goal—such as some fixed retirement income—since a high interest rate means that less saving is required per period.

Similarly, investment was not very responsive to the short-run interest rate, in Keynes's view. Much investment, especially in large projects, takes place over long periods. Once investment decisions are made, the investment is autonomous, unrelated to the interest rate or other variables. The point is that the different motivations of savers and investors mean that saving and investment plans could become unlinked. A flexible interest rate, responsive to the desires of savers and investors, would not guarantee the saving leakage would automatically be turned into the investment injection. There was no built-in assurance that savings would equal investment at a full-employment, growth-maximizing level of economic activity.

Second, Keynes argued that in reality the internal structure of the economy was not competitive enough to permit prices and wages to fall in response to insufficient aggregate demand. To Keynes, the existence of monopoly and union pressures in the economy had to be taken into account. If demand fell off, monopolies would let output fall rather than accept price reductions. Workers, moreover, would refuse to take cuts in their money wages, thereby creating unemployment and layoffs by businesses. Even if workers did take money wage cuts, the reduction in their income would further reduce the demand for goods and services, probably reducing output and employment even more. The conclusion: *Classical self-adjusting mechanisms in the private economy would not lift us out of depression.*

The Importance of Income

Keynes believed that a prolonged depression proves that the economy can establish equilibrium at less than full employment. The classical economists, as we have seen, placed their confidence in a flexible price system—flexible prices, wages, and interest rates—as the self-adjusting mechanism that would assure full employment when aggregate demand got out of kilter with aggregate supply.

Keynes argued instead that aggregate demand and economic activity in general were determined by income and by changes in income. He thereby developed an **income-expenditures model** of macroeconomics. The income-expenditures model shows that private expenditures (the same as aggregate demand) are primarily determined by the level of national income and that such expenditures, in turn, establish equilibrium (or nonequilibrium) levels of output and employment in the economy. Other factors, as we will see, may help determine expenditures, but they are unimportant as short-run determinants of what people spend.

The important point is that private spending—all nongovernment spending—may create some level of output and employment that does not fully

Income-expenditures model: The theory of John Maynard Keynes showing that private expenditures are primarily determined by the level of national income and that such expenditures in turn determine the level of output and employment in the macroeconomy.

utilize society's resources. An equilibrium could exist in the economy, in other words, at some high level of unemployment. The chronic disease of unemployment, Keynes argued, could be effectively treated only by increasing or decreasing the aggregate demand for goods and services. Only government is powerful enough, in his view, to affect aggregate demand by changes in spending or taxing policies to restore the economy to full employment. Before we consider government's role in the Keynesian model, we will examine the Keynesian theory of spending—specifically, the interrelation of income, consumption, saving, and investment—in more detail.

The Income-Expenditures Model

Keynes focused on aggregate demand or spending because he viewed rapid aggregate demand changes as the villain in recessions and depressions. Keynes believed that aggregate supply factors—large-scale changes in productivity, such as new technology and inventions and changing incentives to work and produce—changed slowly and could therefore be neglected when considering short-run macroeconomic problems. (Aggregate supply factors will be considered in Chapters 25 and 26.) In the short run, the time period most relevant to Keynesian macroeconomics, the economy's production or supply of goods and services seemed to react passively to changes in total expenditures.

Assumptions of the Income-Expenditures Model

The Keynesian view of aggregate demand is somewhat complex, and we must use some initial simplifications to understand it. As we proceed to develop Keynes's macroeconomic theory of aggregate demand, our assumptions will be eliminated one by one. The assumptions are as follows:

1. Government spending and taxation as well as foreign trade (exports and imports) are excluded from the basic model of private spending. Taxes, transfer payments (such as Social Security payments), and any other government activities are not considered. Depreciation is also ignored. Income or output is produced and consumed only in the private sector of the economy. This simplification means also that national income and disposable after-tax income are the same.
2. The purely monetary side of the economy is excluded from the basic model. The interest rate and any possible effects of interest rate changes are therefore excluded from consideration. This assumption makes sense if we accept Keynes's basic point that real income, not the interest rate, is the principal determinant of both consumption and saving. (Keynes did believe that money and interest rates play some role in explaining total spending in the economy, but we will defer discussion of these factors for the time being.)
3. We assume that, over the short run, prices and wages remain constant. This assumption means that total income or output will increase or decrease solely in response to changes in aggregate spending and not to price or wage changes. All national income figures are therefore expressed in nominal terms unless otherwise noted.

Private Consumption Expenditures

The classical writers, as we saw earlier, placed primary emphasis on the real rate of interest in determining the relation between saving (income reserved for future consumption) and expenditures (income used in current consumption). One of Keynes's basic criticisms of the classical theory was that the rate of interest could not always enable us to predict how much of current income an individual would choose to save or to spend. Keynes believed that current income was the most reliable and predictable determinant of consumption expenditures. Private saving, the residual of income after consumption, was also a function of, or was explained by, income. The simple Keynesian model of consumption and saving, in other words, does not account for any factors other than current income. Consider how these relations between consumption and income and between saving and income might be expressed and analyzed.

Consumption and Saving Functions and Schedules

Keynes constructed consumption and saving functions and schedules based on income. The consumption function and the saving function can be expressed simply as

$$C = C(Y) \quad \text{and} \quad S = S(Y).$$

In other words, consumption (C) is a function of income (Y), and saving (S) is a function of income, which simply means that consumption and saving are related to income in some way. Both consumption and saving, as we will see, are positive functions of income, meaning that an income increase will increase both consumption and saving, and an income decrease will decrease both consumption and saving.

Consumption and saving relations can also be regarded as schedules. A consumption-income schedule shows the amount that households would desire or plan to consume at every level of income. Likewise, a saving-income schedule shows the desired or planned level of saving that households would undertake at various levels of income.

In the basic Keynesian model, income can be disposed of by households in only two ways: They may consume it or save it. This observation means that,

$$Y = C + S,$$

and it also means that consumption and saving are related in a unique way.

These relations can be uncovered by the construction of budget studies; that is, studies of exactly how households consume and save over some period, usually a year. Consider such a study for the year 1983, reproduced in Table 23–1. Table 23–1 shows the average income of households estimated for 1983, along with the actual average consumption and average saving by income receivers within each income group. Both the consumption and the saving data tell us that consumption and saving rise with increases in income, but we can gather still more information about exactly how they are related.

Marginal propensity to consume (MPC): The percentage of an additional dollar of income that is spent on consumption; the change in consumption divided by the change in income; 1 − MPS.

The Marginal and Average Propensity to Consume and Save. Macroeconomists are often interested in knowing how much consumption or saving will change with a change in income. The two concepts that give us this answer are the marginal propensity to consume and the marginal propensity to save. The **marginal propensity to consume (MPC)** is defined as a ratio of the change in consumption (ΔC) to the change in income (ΔY) that caused the change in consumption, or

TABLE 23–1   Household Budget Data on Consumption and Saving, 1983

Estimated budget data for 1983 reveal rises in both consumption and saving as income increases. Average consumption (C), however, falls with increases in income, and average saving (S) rises with increases in income. Further, the data reveal a general tendency to consume less out of each increase in income, as indicated in the MPC column, and to save more out of the same income increases, as indicated in the MPS column.

Income Bracket	Average Income (Y)	Average Household Consumption (C)	Average Household Saving (S)	Average Propensity to Consume (APC = C ÷ Y)	Marginal Propensity to Consume (MPC = ΔC ÷ ΔY)	Average Propensity to Save (APS = S ÷ Y)	Marginal Propensity to Save (MPS = ΔS ÷ ΔY)
$ 3,000–$5,999	$ 4,560	$ 5,298	− $738	1.16		−0.16	
					0.92		0.08
$ 6,000–$8,999	7,536	8,037	−501	1.07		−0.07	
					0.90		0.10
$ 9,000–$11,999	10,515	10,716	−201	1.02		−0.02	
					0.78		0.22
$12,000–$14,999	13,482	13,038	+444	0.97		0.03	
					0.84		0.16
$15,000–$17,999	16,455	15,537	+918	0.94		0.06	
					0.71		0.29
$18,000–$22,499	20,151	18,159	1,992	0.90		0.10	
					0.72		0.28
$22,500–$29,999	25,665	22,146	3,519	0.86		0.14	
					0.66		0.34
$30,000–$44,999	35,283	28,521	6,762	0.81		0.19	
					0.46		0.54
$45,000+	66,432	42,819	23,613	0.64		0.36	

Source: U.S. Department of Labor, Bureau of Labor Statistics, *Consumer Expenditures and Income, Total United States, 1960–61,* BLS Report No. 237–93 (Washington, D.C.: U.S. Government Printing Office, 1965), p. 16. Updated and adjusted for estimated price level changes by the authors.

$$\text{MPC} = \frac{\Delta C}{\Delta Y} = \frac{\text{Change in consumption}}{\text{Change in income}}.$$

Marginal propensity to save (MPS): The percentage of an additional dollar of income that is saved; the change in savings divided by the change in income; 1 − MPC.

The **marginal propensity to save (MPS)** is the ratio of a change in saving (ΔS) to a change in income, or

$$\text{MPS} = \frac{\Delta S}{\Delta Y} = \frac{\text{Change in saving}}{\text{Change in income}}.$$

Since households in the simple Keynesian model dispose of income only by consuming or saving it, the MPC and the MPS must add up to 1:

$$\text{MPC} + \text{MPS} = 1.$$

Average propensity to consume (APC): The percentage of total income that is spent on consumption; total consumption expenditures divided by total income; C/Y; 1 − APS.

Economists are also interested in how much, on average, households consume and save at various income levels. The **average propensity to consume (APC)** is the proportion of income consumed at any income level, or C/Y. Likewise, the **average propensity to save (APS)** is the ratio of saving to income at any level of income, or S/Y. Since consumption and saving are the only two ways that households can dispose of income, a unique relation exists between the APC and the APS:

Average propensity to save (APS): The percentage of total income that is saved; total saving divided by total income; S/Y; 1 − APC.

$$1 = \frac{\text{Consumption}}{\text{Income}} + \frac{\text{Saving}}{\text{Income}}$$

or

$$1 = \text{APC} + \text{APS}.$$

Dissaving: The act of spending more than saving or borrowing against future income. Dissaving results from present consumption expenditures that are greater than present income.

What Do the Budget Data Reveal? The data from Table 23–1 establish some important facts related to consumption and saving behavior.

A break-even level of income occurs where consumption expenditures equal income. Below this level (within the $12,000–$14,999 income bracket) households **dissave;** that is, they consume more than their income. Households dissave by drawing down on past saving or by borrowing from banks or other financial institutions. Naturally, households with incomes greater than the break-even point, where C = Y, engage in positive saving.

As income rises, the average propensity to consume (APC) declines and the average propensity to save (APS) rises. This relation simply means that we tend to consume smaller portions and to save larger portions of higher incomes, a fact that is ordinarily verified by individual behavior.

As we observe changes from one income level to another, increases in consumption are smaller and smaller. As Table 23–1 verifies, the marginal propensity to consume (MPC) tends to decline with increases in income. The tendency to consume less and less out of increases in income is matched by the tendency to save more and more out of such increases, so the marginal propensity to save (MPS) rises as income increases.

Keynes's Theoretical Model of Consumption and Saving

The features of actual household consumption and saving behavior can be translated into a theoretical model of consumption and spending for the entire economy: how the country as a whole would choose to spend or save at different levels of national income. (Using hypothetical data for the entire economy, we abstract essential information from the actual behavior of households as reported in Table 23–1.) To make this model easier to comprehend and with no important loss in accuracy, we will assume that the consumption-income and saving-income relations are in the form of straight lines. This assumption means that, in contrast to the data from budget studies, the MPC and the MPS are constant. This simplifying assumption does no damage to our conclusions, and it greatly facilitates understanding of these crucial ideas.

Hypothetical information about economy-wide consumption and saving is given in Table 23–2. The hypothetical data of Table 23–2 show, in billions of dollars, what private consumption and private saving would be at alternative levels of income. Consider, for instance, a year in which national income is $100 billion. How would households plan to divide this income between purchases of consumption goods and saving? Consumption expenditures at a $100 billion level of national income would exceed income by $20 billion. This −$20 billion would be dissavings. To consume $120 billion at an income level of $100 billion, in other words, households would have to draw down $20 billion in savings. A special interpretation is given to dissaving at a zero income level. Note that all households would consume at the level of $40 billion even if no income were earned; this $40 billion is independent of income. Such **autonomous consumption** expenditures are independent of income. At levels of income higher than zero, consumption expenditures increase until they are equal to income. If national income were $200 billion, all income would be devoted to consumption. At income levels higher than $200 billion, consumption expenditures fall short of current income, and the remainder is devoted to saving.

Autonomous consumption: Consumption expenditures considered independent of the level of income.

Consumption and the 45-degree Line. An important method for understanding how consumption is related to income is to compare historical con-

TABLE 23–2   Hypothetical Consumption and Saving Data for the Economy

The data in the table, for the entire economy, replicate the important facts described about household budgets in Table 23–1. The economy consumes and saves more as income increases. Further, the average propensity to consume declines and the average propensity to save rises with increases in income. For the sake of simplicity, the MPC and the MPS are assumed to be constant.

National Income (billions of dollars) (Y)	Planned Consumption Expenditures (billions of dollars) (C)	Planned Savings (billions of dollars) (S)	Average Propensity to Consume (APC = C ÷ Y)	Marginal Propensity to Consume (MPC = ΔC ÷ ΔY)	Average Propensity to Save (APS = S ÷ Y)	Marginal Propensity to Save (MPS = ΔS ÷ ΔY)
0	40	−40	−	0.8	−	0.2
100	120	−20	1.2	0.8	−0.2	0.2
200	200	0	1.0	0.8	0	0.2
300	280	20	0.93	0.8	0.07	0.2
400	360	40	0.90	0.8	0.10	0.2
500	440	60	0.88	0.8	0.12	0.2
600	520	80	0.87	0.8	0.13	0.2
700	600	100	0.86	0.8	0.14	0.2

sumption levels with a hypothetical relation in which consumption spending is always equal to income. In graphical terms, such a relation would always produce a 45-degree line, reflecting Keynes's basic idea that the more income we have, the more we will consume.

Examine Figure 23–5, on which national income levels are displayed on the horizontal axis and consumption expenditures on the vertical axis. At every point on the 45-degree line in Figure 23–5, total spending—in this case, consumption spending—*equals* total income. The extent to which actual consumption levels for particular periods differ vertically from this 45-degree line indicates the degree to which consumption does not equal income. Points along the **consumption function** above the 45-degree line indicate dissavings; points below it indicate savings. In Figure 23–5, actual consumption is equal to income only at the $200 billion income level. At a national income below this break-even point, society would go into debt or dip into savings; above it, society would save some of its income.

Consumption function: The curve that shows that the level of consumption is positively related to the level of income, holding all other important factors constant.

The Propensity to Consume. Household consumption is the major factor in total spending, spending that creates jobs and production in the economy. Macroeconomists are therefore very interested in how much household consumption will change with changes in national income. The value determined for the marginal propensity to consume tells us how much will be consumed out of an additional dollar—or 100 billion dollars—in national income.

The marginal propensity to consume is of course different for different individuals and for different groups in society. Some of us are more likely than others to spend every dollar we have. In this example, we simplify the variations by choosing an average MPC value of 0.8 for society as a whole and holding this value constant across varying national income levels. In Figure 23–6, an increase in national income from $300 billion to $400 billion will increase the level of consumption spending from $280 billion to $360 billion, an increase of $80 billion. Likewise, an increase from $400 billion to $500 billion would cause consumption spending to increase by

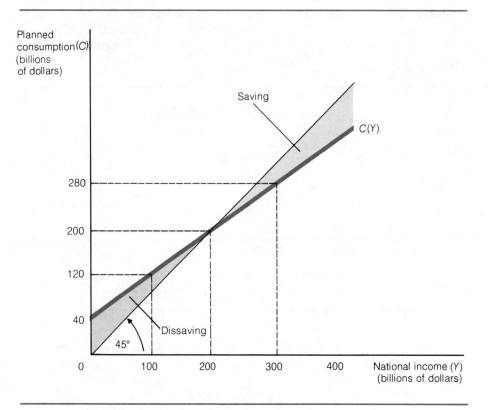

FIGURE 23–5 **The Level of Consumption as a Function of the Level of Income**

The consumption function, $C(Y)$, shows that the level of consumption expenditures rises as income rises. When income is $100 billion, consumption spending is $120 billion; when income is $200 billion, consumption is $200 billion; and when income is $300 billion, consumption is $280 billion. Along the hypothetical 45-degree line, consumption expenditures are exactly equal to income. The degree to which consumption differs from income indicates dissaving or saving by society as a whole. In this example, consumption spending is equal to income only at the $200 billion income level. Along the consumption function and above the 45-degree line, society dissaves; along the consumption function and below the 45-degree line, positive saving takes place.

another $80 billion. The MPC is, therefore, the ratio 80/100 or 80 percent or 0.8, as indicated in Table 23–2.

Though the actual relation of the MPC to national income is a matter of debate, we know that the average propensity to consume changes as income levels change. In general, as income rises, the APC, or the ratio of consumption to income, falls. At a $300 billion income level, for instance, consumption expenditures are $280 billion and the APC equals 280/300 or 0.93. On the average, the population would be spending 93 percent of its income. But if national income rose to $600 billion, households would spend only 87 percent of their income (520/600).

The Saving Schedule. Like consumption, saving is a positive function of income. That is, the more income we have, the more we save; the lower our income, the less we are able to save. In the hypothetical **saving function** in Figure 23–7, at a national income level of $400 billion, desired saving is

FIGURE 23–6

Marginal Propensity to Consume

Along a straight-line consumption function, C(Y), the MPC is constant. In this graph, MPC = 0.8. Each time income rises by $100 billion, the level of consumption rises by $80 billion. Or, for every $1 increase in income, consumers increase expenditures by $0.80.

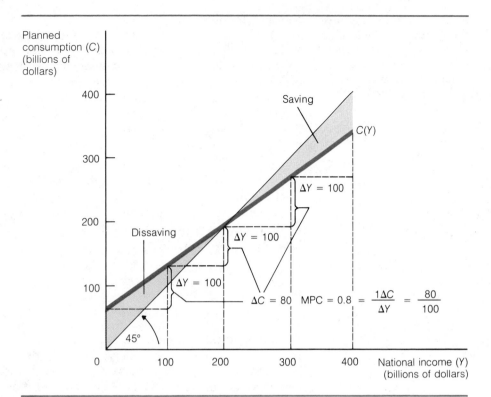

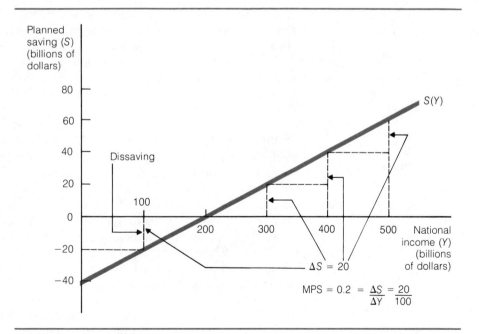

FIGURE 23–7   A Saving Function

The saving function, S(Y), shows that the desired or planned saving is positively related to the level of national income. At a low level of income (below $200 billion) there is dissaving. As the level of income rises from $200 billion to $300 billion, the level of saving rises from $0 billion to $20 billion. At Y = 400, S = 40, and at Y = 500, S = 60. The figure also illustrates the marginal propensity to save. Along a straight-line saving function, the MPS is constant. Here the MPS is equal to 0.2. For each increase in income of $100 billion, the level of saving increases by $20 billion.

$40 billion dollars. In fact, saving becomes positive rather than negative at all levels of income greater than $200 billion in this example.

The marginal and average propensity to save are also illustrated in Figure 23–7. As we saw earlier in the household budget data, the MPS is the ratio of a change in saving to a change in income. As shown in Figure 23–7, incremental changes of $100 billion change saving by $20 billion,

$$\text{MPS} = \frac{20}{100} = 0.2.$$

The average propensity to save is the ratio of savings to income at any level of income. The APS rises as income rises, as you can see in Table 23–2, from which the data in Figure 23–7 are taken.

Assuming straight-line consumption and saving functions and the disposal of all income as either saving or spending, the MPC, MPS, APC, and APS bear the following relations:

$$1 = \text{MPC} + \text{MPS} \quad \text{and} \quad 1 = \text{APC} + \text{APS},$$

or

$$\text{MPS} = 1 - \text{MPC} \quad \text{and} \quad \text{APS} = 1 - \text{APC}.$$

According to these relations, a rise in the MPC will mean a fall in the MPS by the same amount. And since the APC decreases with increases in income, the APS must increase with increases in income.

The values of MPC, MPS, APC, and APS are very important for economic policy. (Keep in mind that the values of MPC and MPS given in Figures 23–6 and 23–7 are based on hypothetical data.) A high MPC means that increases in income will generate a large amount of additional private spending. A low value for the MPC (or, exactly the same thing, a high MPS) means that increases in income will generate only small increases in consumption. These features of the consumption function will have extremely important impacts on the model of total expenditures developed in the following chapters.

Nonincome Factors Influencing Consumption and Saving

Although Keynes's basic argument involved the effects of income levels, income is not the only factor affecting consumption expenditures. A host of nonincome factors also determines the amount of planned consumption spending by all households. If any nonincome determinant of consumption spending changes, the entire consumption function shifts upward or downward. The amount of planned consumption changes at every income level when a nonincome element affects consumption changes. In contrast, a change in income simply changes the amount of income consumed, expressed graphically as a movement along a given consumption function.

In Figure 23–8, assuming an initial consumption function $C_0(Y)$, a change in any nonincome factor affecting consumption will shift the consumption schedule either upward to $C_2(Y)$ or downward to $C_1(Y)$. (The break-even level of income where consumption spending equals income—on the graph, the point where the consumption function intersects the 45-degree line—also changes to Y_2 with consumption C_2 or to Y_1 with consumption C_1.) By contrast, if income increases from Y_0 to Y_2, consumption increases along consumption function $C_0(Y)$ from point A to point F.

In this section, we address seven of the most important nonincome factors that may affect consumption: wealth; the price level; price and income ex-

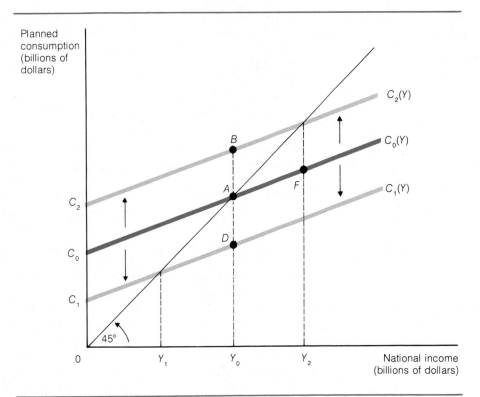

FIGURE 23–8   Shifts in the Consumption Function

Any change in nonincome factors affecting consumption spending will shift the consumption function, $C_0(Y)$, either upward or downward. An increase in consumption means that consumption spending rises for every level of income (from A to B); a decrease in spending creates a fall from A to D in the figure. By contrast, a change in income, all other factors being equal, moves consumers along a given consumption function. For example, increasing national income from Y_0 to Y_2 would move consumption from point A to point F in the figure.

pectations; credit and the interest rate; taxation; age, geographic location, and population; and the distribution of income.

Wealth or the Stock of Assets. People consume not only on the basis of their flow of current income but also on the basis of their previously accumulated stock of wealth. Purchase of real goods—such as precious stones or a painting by Andrew Wyeth—takes place at a given time. As the money value of such assets increases over time, people feel richer and consume more even though their incomes do not increase. Conversely, reductions in the money value of assets reduces wealth and tends to reduce consumption at every level of income.

The Price Level. Certain assets such as money or assets denominated in money terms, such as bonds, change value with changes in the price level. If you hold a given stock of money, a halving of the price level would double the real value of the money assets you hold. Since you would then be richer, your desired consumption would increase. For the opposite reason, rising prices reduce consumption expenditures.

Price and Income Expectations. We all have anticipations about our economic lot in the future. If we expect prices to be higher in some future period, we will tend to increase consumption expenditures in the present to beat price hikes. Expectations of lower prices will reduce present consumption. Likewise, if we expect higher future incomes, our present consumption will rise. Expectations of harder economic times will reduce present consumption and increase present saving.

Credit and the Interest Rate. Two closely interrelated factors affecting consumption are the availability of credit and the interest rate. The number of institutions offering credit in the economy has undergone phenomenal growth since World War II. The overall effect of increased credit availability, along with improved terms of credit (longer payback period, lower down payment), has been to increase consumption over time.

Along with credit availability and improved credit terms, an important factor affecting consumption is the cost of credit, determined by the interest rate. As the interest rate rises or falls, other things such as credit terms being equal, consumers will reduce or increase their present consumption expenditures. A higher rate of interest, moreover, will mean a higher level of saving in the present because reduced consumption will lead to increased saving. As indicated earlier, Keynes did not believe that the interest rate had predictable effects on consumption and saving. His view was in direct contrast to the classical writers, who thought that the interest rate was central to explaining saving and consumption behavior.

Taxation. As we will see more fully in Chapter 24, taxation affects spending. The existence of an income tax, for example, reduces the amount of income, thereby reducing consumption spending. Tax reductions increase consumption. A change in taxation affecting households will always affect the position of the consumption function because it affects the amount of households' spendable income.

Population, Age Distribution, and Geographic Location. An increase in population will obviously increase consumption over time. The age distribution of the population—the proportion of young, middle-age, and older citizens to the total population—is a slowly changing but important factor affecting consumption. The young and the old have a tendency to spend a larger portion of their income than do people in their middle years. An age distribution shift to the middle years will therefore tend to reduce consumption and increase saving.

Factors such as geographic location may also affect consumption spending. A large population movement from rural to urban areas may increase present consumption, for example, if city dwellers have a higher tendency than country dwellers to consume their income.

Distribution of Income. A final factor affecting planned household consumption is how income is distributed among groups of different income levels. Low-income, middle-income, and high-income households are thought to have different average propensities to consume and to save. Specifically, low-income families may have generally higher APCs than high-income families. In the Keynesian analysis of consumption, a permanent redistribution of income from high-income families to low-income families may cause an increase in total consumption; that is, it may shift the consumption curve upward.

Although Keynes envisioned differences in the APC between income groups, empirical evidence on actual consumption functions over the long run suggests that a long-run proportionality exists between consumption and income; society seems to consume the same proportion of income at low as well as at higher levels of income, indicating a more or less constant APC. A number of economists have attempted to reconcile data showing short-run nonproportionality with long-run proportionality although explanations vary. At bottom, it is not clear how a redistribution of income from rich to poor families would affect aggregate consumption spending at any one time or over a long period of time.

In sum, consumption spending is affected by a large number of nonincome determinants. A change in any one of them will shift the consumption function either upward or downward. A single consumption function, however, is constructed by holding all the factors constant and varying income.

Investment Expenditures

Investment spending: The level of expenditures by businesses on capital goods and additions to inventories, by households on home construction, and by government on structures such as schools, roads, and housing.

We have so far looked at household consumption and its relation to income. Private spending in the economy also includes investment spending. In a macroeconomic context, **investment spending** means spending by the private sector—mainly businesses—on capital goods. Public investment in goods such as schools, water reclamation projects, or dams is not included in private investment expenditures. Public investment expenditures are categorized as government spending and will be covered in Chapter 24. Rather, private investment refers to the national income category gross private domestic investment, defined in Chapter 22 as including fixed investment in plant and equipment, all private-sector residential and nonresidential construction, and changes in business inventories (increases or decreases in stocks of finished goods, semifinished goods, or raw materials).

Clearly, there are two effects on the capital stock of the nation at any given time: Some of it is being used up or depreciated, and new net investment is creating additions to it. Businesses usually replace their worn-out capital stock and add to their stock of capital with new investments.

Autonomous Short-Run Investment Spending

Keynes argued that in the short run, investment expenditures could be viewed as autonomous. They are independent of the level of income, profit expectations, the interest rate, and all of the other factors possibly affecting investment. The idea that investment could be autonomous in the short run stems from the fact that most investment expenditures in any current period have been determined by past investment decisions.

Businesses, in fact, make investment decisions from a long-run perspective. The building of new plants and the installation of sophisticated equipment typically take a number of years to complete. During the investment period, planned investment expenditures are often carried out regardless of business conditions. Planned investment expenditures are those that businesses desire to undertake over the period of a year. (Planned investment expenditures may not be the same as actual expenditures, as we will see later in this chapter.) These investment expenditures are then independent of the level of income.

Autonomous investment: Investment expenditures considered independent of the level of income.

The concept of **autonomous investment** spending can be analyzed within the framework of the simple Keynesian model of private spending. If I stands for investment spending by businesses, autonomous investment is simply described as some constant level, $\bar{I}$, or

$$I = \bar{I}_0,$$

where the bar on top of I means that these expenditures are autonomous and the subscript 0 means that they are constant at some specified level—$20 billion, $50 billion, $10 billion—during the period under consideration.

If we assume that autonomous investment expenditures are constant at $20 billion, we can express investment simply as:

$$I = \bar{I}_0 = \$20 \text{ billion}.$$

Figure 23–9 shows the relation between income levels and autonomous investment. Autonomous investment is constant at every level of income. An increase or a decrease in autonomous investment shifts the flat **investment function** upward or downward, respectively, by the amount of the increase or decrease.

Investment function: A curve that shows the relation between the level of investment and the level of some other variable such as income.

In short, investment expenditures, while volatile and changing from period to period, may be assumed constant over any given short-run period. They provide, in addition to consumption spending, the means of arriving at the Keynesian concept of total expenditures in a purely private economy.

Long-Run Factors Affecting Investment. While investment is considered autonomous in the Keynesian context over some short-run period relevant for policymaking, its volatility over longer periods is subject to a number of rapidly changing factors, including the rate of interest, the cost of capital, and current and expected sales.

Rate of Interest and the Cost of Capital. Ultimately, it is profitability—or expectation of future profits—that determines the amount of investment that businesses will undertake. Profitability is, in simple terms, the difference between a firm's revenues and its costs.

FIGURE 23–9

Autonomous Investment Expenditures

The horizontal investment function, *I*, shows that the level of planned investment expenditures is independent of the level of income. As income changes, the level of investment remains constant. An increase in investment caused by factors such as a change in interest rates is shown as a horizontal shift upward.

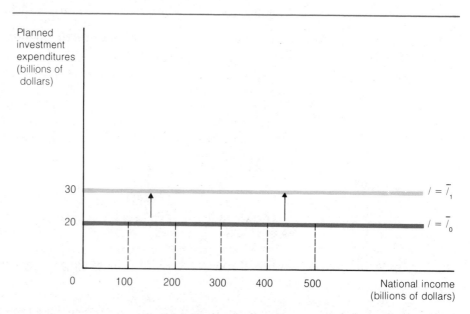

The cost of capital investment is heavily dependent on the rate of interest. Whether a business uses its own internal funds or borrows to make an investment, it incurs an opportunity cost in the form of interest lost. An increase in the interest rate, other things being equal, will reduce the amount of investment; a reduction in the interest rate will increase investment. The cost of the capital equipment itself is also important in determining the amount of desired investment. An increase in costs, other things being equal, reduces the amount of investment.

Current Sales and Expected Sales. The other side of the profitability of business investment is sales. If a large capacity to produce already exists, an increase in current sales will not encourage new investment expenditures very much. If businesses are operating close to the limits of their ability to produce, an increase in current sales will cause current investment spending to rise.

Expectations are also a central feature of Keynes's theory of investment. Sales expectations require estimates of future business conditions and tend to vary greatly and quickly over time. Optimism concerning future sales and business conditions will mean increases in capital investment in the present. Pessimism will have the opposite effect.

As we will see in later chapters, the volatility of profit expectations is a principal cause in the creation of unstable income levels, unemployment, and cycles of business activity. Before turning to these issues, however, we must first understand how income equilibrium is defined in a purely private economy.

Private Income-Expenditures Equilibrium

The classical economists argued that full employment would be achieved automatically simply through the activity of supplying goods in an essentially private economy, an economy theoretically free of government intervention. Should difficulties in adjustment to this full-employment equilibrium take place, prices, wages, and interest rates would change to bring the aggregate output of goods into line with aggregate expenditures.

For reasons already outlined, Keynes did not agree that wages, prices, or interest rates would change in any predictable way to bring the demand for goods into line with the supply of goods. How could a depression economy reach equilibrium with high unemployment? Keynes argued that levels of private spending (consumption and investment) determine the output of goods produced in the economy. In other words, businesses react to any level of total expenditures by producing the quantities demanded. If households want more kitchen appliances, they are produced; if businesses demand less computer software, less is produced. Since it obviously takes more or less labor and other resources to produce more or less total output, the level of employment is also affected.

Keynes's theory of output and income determination can be expressed in symbolic terms using the tools of private spending analysis developed above. Since total private expenditures are a function of income, and since these total private expenditures consist of consumption and investment spending, we can express the Keynesian equilibrium as follows:

$$Y = C + I = \text{Total expenditures.}$$

Private-sector equilibrium, or equilibrium output, occurs when $C + I$ equals

TABLE 23–3   Planned Total Expenditures Determine Total Output and Equilibrium Income

For equilibrium in the private economy, planned total expenditures, $C + I$, must equal total income or output, Y. In this example, equilibrium is reached at a national income (or output) level of $300 billion. At national incomes less than $300 billion, total expenditures exceed total income. At national incomes higher than $300 billion, total expenditures are less than total income.

	Planned Consumption Expenditures (C)	Planned Investment Expenditures $(I = \bar{I}_0)$	Desired or Planned Total Expenditures $(C + \bar{I}_0)$	National Income or Output (Y)	Differences Between Total Expenditures and Total Output $(C + I_0 - Y)$
	120	20	140	100	+40
	200	20	220	200	+20
Equilibrium	280	20	300	300	0
	360	20	380	400	−20
	440	20	460	500	−40

Note: All values are in billions of dollars.

Private-sector equilibrium: The state of equality between private expenditures and income or output; a condition of macroeconomic stability from which there is no tendency for change.

income. In other words, equilibrium output production in the economy takes place when total expenditures equal what is produced.

An alternative way to define the establishment of equilibrium output involves the relation of income to expenditures. We know that income received by all factors of production may be disposed of in two ways, or

$$Y = C + S.$$

That is, income may be used for consumption spending or for savings, the residual of consumption spending. Equilibrium will occur when income received is equal to total expenditures:

$$C + S = C + I, \quad \text{or} \quad S = I.$$

Total Expenditures Equal Total Output

Keynes's theory—that equilibrium income (or, its mirror image, equilibrium GNP or output) is a function of total expenditures—can be expressed numerically and graphically using tools already at our disposal. All of our previous consumption, investment, and income figures are reproduced in Table 23–3. Total expenditures, shown in the third column, are equal to the sum of consumption and investment expenditures. Recall that planned consumption depends directly on the level of national income, reported in the fourth column. According to the circular flow model, the fourth column represents both the total output of goods and services in the economy as well as national income. Total expenditures may be greater or less than total production or output. The amount by which income or output exceeds expenditures is shown in the last column.

At an income level of $100 billion, for example, total expenditures are greater than total output by $40 billion. This situation could be possible only if businesses draw down on inventories of goods. Producers react to such a situation by producing greater quantities of goods and services, thus generating additional income to workers and to the other factors of production. The resulting increases in income will change consumption plans. At incomes higher than $100 billion, households will spend greater amounts on goods and services, indicated in the first column of Table 23–3 and in Figure 23–10, which merely reproduces the information of Table 23–3 graphically.

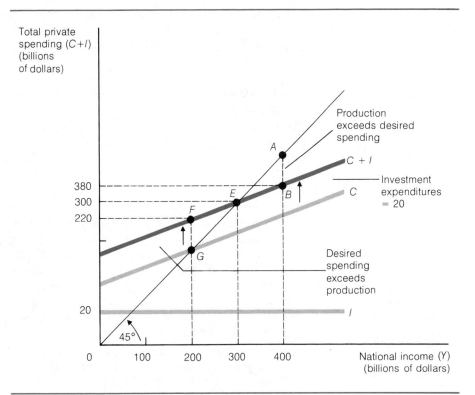

FIGURE 23–10   Private-Sector Equilibrium

The equilibrium level of income occurs when total private expenditures are just great enough to purchase the total output of the economy. At equilibrium, expenditures and output will meet at a point along the 45-degree line because they are exactly equal. At income level $Y = 300$, the economy is at equilibrium, *E*. At an income or output level of 400, desired total private expenditures at point *B* are insufficient to purchase the total output generated at point *A*. Inventories of goods will pile up, causing reductions in the rate of output and a lowering of income. These reductions will take place until income and output again equal the equilibrium level of 300. At this level total expenditures equal total output produced in the economy.

(Note in Figure 23–10 that investment is added vertically to the consumption function at an autonomous level of $20 billion.)

A new, higher level of output produced and income received at $200 billion creates higher total spending, but there is still an excess of total expenditures over output (amount *FG* in Figure 23–10). Inventories continue to be drawn down and production stepped up until an **equilibrium level of national income** of $300 billion is reached. Only at this level of income will production plans be in line with the plans of households and businesses to consume and invest.

What would happen at an income level greater than $300 billion, say $400 billion? If an income level of $400 billion were temporarily established, total output would temporarily exceed total expenditures by $20 billion, or *AB* in Figure 23–10. Unwanted inventories of goods would pile up unsold, and producers would cut back the rate of production. Since a cutback in production would reduce income, consumption plans of households would be revised downward. Equilibrium would again be achieved when the level of income dropped to $300 billion.

It is always worth remembering that the simple Keynesian model of total

Equilibrium level of national income: The income level that exhibits no tendency to change. In the income-expenditures model, the level at which total expenditures equal total output.

private spending is a short-run model. Prices, wages, and interest rates do not adjust as they do in long-run classical theory. The only factors that are adjusting are the real quantities of goods and services produced in the economy.

Desired Saving Equals Desired Investment: An Equivalent Method

An equivalent method for determining private-sector equilibrium is to equate private saving and private investment. This method is based on the fact that private saving is what is left over after planned private consumption is determined for various income levels. Saving, like investment, is the other side of consumption.

Desired saving and desired investment concern the plans of households and businesses. Over any period and at alternative income levels, businesses plan to invest some amount—in the above model, $20 billion worth of investment at every level of income. Households, likewise, plan to save some amount out of income for every level of income; but unlike business investment, household saving varies with income. Desired saving at an income level of $400 billion in our numerical example is a positive $40 billion. But it is a negative $20 billion at an income level of $100 billion—that is, $10 billion will be dissaved.

In other words, at an income level of $100 billion (see Table 23–4), businesses plan or desire to invest $20 billion but because total expenditures draw down inventories by $40 billion, they actually end up with −$40 billion worth of unplanned investment. What do they actually invest? Actual investment is the sum of desired (planned) investment and unplanned investment. In the numerical example of Table 23–4, actual investment is −$20 billion. In this state of affairs, businesses will increase output so that planned investment will equal actual investment.

At alternative income levels, therefore, actual investment may diverge from desired or planned investment, but whenever this occurs, producers will change their rate of production. The important point is that actual investment will always equal the actual amount of savings. But when the plans or desires of savers and investors differ, income and output will increase or decrease. The only possible equilibrium will take place when the plans of

TABLE 23–4   Equilibrium Income

The actual levels of investment and saving are accountings of the difference in total expenditures and total income. If total expenditures are greater than total income, then actual investment is below desired investment. In such cases, businesses increase output until their actual investment equals their desired investment. If total expenditures are less than total income, then actual investment is greater than total income, and businesses decrease output. Equilibrium occurs where desired and actual saving equals desired and actual investments.

	National Income (Y)	Desired Saving	Desired Investment	Unplanned Investment	Actual Investment	Actual Saving	Income Adjustment
	100	−20	20	−40	−20	−20	increase
	200	0	20	−20	0	0	increase
Equilibrium	300	+20	20	0	+20	+20	—
	400	+40	20	+20	+40	+40	decrease
	500	+60	20	+40	+60	+60	decrease

Note: All values are in billions of dollars.

FIGURE 23–11
Saving-Investment Equilibrium

When $C + I = Y$, saving is equal to investment. The equilibrium level of income occurs where the saving function intersects the autonomous investment function ($\bar{I}_0$), in this case at 300.

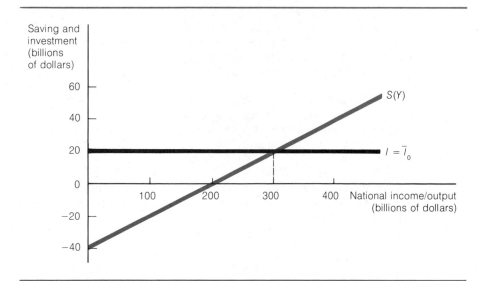

both household savers and business investors are identical. As Table 23–4 and Figure 23–11 show, this equilibrium income occurs when $300 billion of output is consumed by consumers and investors.

Only where desired saving comes into line with desired investment does the process of adding to (or subtracting from) inventories stop. Only at an income level of $300 billion does desired saving of $20 billion equal desired investment of $20 billion. At this level, and only at this level, producers' motivations to increase or decrease production will cease. That is why it is called the equilibrium level of income. Only at an income of $300 billion is the behavior of savers or, more properly, consumer-savers, consistent with that of investors. Not only does $S = I$ in the actual accounting sense, but the desired economic behavior of savers and investors is mutually consistent. Thus, for equilibrium,

$$\text{Desired } S = \text{Desired } I = \text{Actual } S = \text{Actual } I.$$

Private Equilibrium in the Keynesian System

Unlike the classical economists, who believed that the economy would self-adjust until full employment was reached, Keynes thought that the economy could adjust to an equilibrium level with widespread unemployment. This unfortunate state of affairs, in Keynes's view, either was permanent or would last a long time. In either case, some action was demanded in the economy. That action, as we will see, was a role for the government.

Summary

1. Classical macroeconomics consisted of four cornerstones: Say's law, price-wage flexibility, interest rate flexibility, and the quantity theory of money. From these foundations, classical economists concluded that the economy would self-adjust to reach a full-employment level of production and income.

2. Say's law indicated that the act of supplying goods created an automatic demand for the goods. The act of saving or hoarding did not mean that overproduc-

tion would occur. If prices and wages were flexible downward, price and wage declines would ensure that all units of production would be sold. Further, the linkage between the rate of interest and saving and investment guaranteed that what was removed from the income stream (saving) would be returned in a different form of spending (investment).

3. Theory clashed with events when, in the 1930s, a decade-long depression in output and employment

took hold of the United States and other Western countries. The response of economists was to provide a new version of how the macroeconomy functions. The principal developer of the new macroeconomics was the English economist John Maynard Keynes.

4. Keynes criticized classical theory as being an inadequate description of events in the real world. For example, Keynes did not believe that prices, wages, and interest rates were flexible enough to create a permanent and sufficiently rapid tendency to private-sector full-employment equilibrium.

5. Keynesian economics is predicated on a model of total expenditures. In the simple private-sector model, consumption and investment spending determine equilibrium output and income.

6. Keynes argued that consumption was primarily a function of current income and that the marginal and average propensities to consume were related to income in a stable and predictable manner. Investment, the other kind of private spending, was related to expectations and to a number of factors but could, in the short run, be considered autonomous of income.

7. Income and output equilibrium occurs in the simple Keynesian model of private spending when total expenditures ($C + I$) equal the total output of goods and services. Alternatively and equivalently, equilibrium occurs when desired and actual saving equal desired and actual investment.

Key Terms

classical macroeconomic theory
Say's law
real rate of interest
hoarding
income-expenditures model
marginal propensity to consume (MPC)

marginal propensity to save (MPS)
average propensity to consume (APC)
average propensity to save (APS)
dissaving
autonomous consumption

consumption function
saving function
investment spending
autonomous investment
investment function
private-sector equilibrium
equilibrium level of national income

Questions for Review and Discussion

1. According to Say's law, under what circumstances would the quantity of goods demanded be less than the quantity of goods supplied?

2. Would people be willing to save money when the real rate of interest is equal to zero? What is the real rate of interest on a passbook savings account if the inflation rate is 7 percent and the bank pays 5½ percent interest?

3. According to the classical view, why does an increase in the level of saving not decrease the total level of spending? Would this be true if interest rates were not flexible?

4. Why did classical economists recommend small balanced budgets for government? Why did they suggest that large budget deficits would lead to lagging economic growth?

5. How does Keynes's view of saving differ from the classical view of saving? How does Keynes's view of wage-price flexibility differ from the classical view?

6. Define marginal propensity to consume and average propensity to consume. What is the essential difference between the two?

7. If the MPC is 0.75 and national income increases by $200 billion, then by how much will consumption expenditures increase?

8. Can you think of any nonincome factors not listed in the text that might affect consumption decisions?

9. What is autonomous investment spending? Does it change as the level of income changes? Do you think that this is an accurate interpretation of how businesses make investment decisions?

10. When does equilibrium occur in the Keynesian income-expenditures model? What adjusts to obtain equilibrium?

11. Why does the level of actual investment always equal the level of actual saving but the level of desired investment may not equal the level of desired saving?

12. Consider a basic income-expenditures model (with no government sector and where $Y = C + \bar{I}$) in which Y is real national income, C is real consumption expenditures, and $\bar{I}$ is real autonomous investment expenditures. Let $C = \$40$ (billion) $+ (MPC)Y$ and $I = \bar{I}_0$. If $\bar{I}_0 = \$50$ billion, and the MPC $= .5$, compute the equilibrium levels of national income Y; consumption, C; and saving, S.

ECONOMICS IN ACTION   Is Thrift Beneficial to Society?

The notion of the wisdom of thrift has been part of our society for hundreds of years. Indeed, the Protestant ethic, which in part extols saving and other conservative financial behavior, became a large part of the philosophical underpinning of laissez-faire economic policy. To the classical economists, saving—that is, abstaining from present consumption—was the surest means to economic growth because when saving was transformed into new productive investment via the interest rate, capital accumulation and growth took place.

Keynesian economics takes issue with the classical view of saving. While it might be true for the individual that "it's not what you consume that makes you rich but what you save that counts," Keynes believed that saving could be disastrous from society's perspective. Thus, the so-called paradox of thrift: Individuals may be made better off by saving, but society is made worse off. And, further, the attempt by members of society to save more creates a situation where saving more becomes impossible.

The Keynesian analysis of saving and investment discussed in this chapter can be used to analyze this paradox. Figure 23–12 shows an initial saving-investment equilibrium for the economy. Initially, planned saving and investment are equal at an income level of Y_0. Suppose, however, that individuals in society suddenly become more thrifty and the saving schedule shifts upward at every level of income. Any nonincome determinant of desired or planned consumption and saving—such as an increase in the interest rate on saving—could have this effect.

The effect of the increase in desired saving, however, will be to reduce desired consumption by an equal amount. Since desired saving exceeds desired investment at income level Y_0 by the amount BA in Figure 23–12, total desired spending $(C + I)$ will be less than actual output. Inventories will build up, and businesses will cut back the rate of output, which will generate lower incomes. An equilibrium between spending and production, or between saving and investment, will not be reattained until the new level of desired saving again equals desired investment. As indicated in Figure 23–12, the new saving-investment equilibrium does not occur until income has declined from Y_0 to Y_1. But at income level Y_1, total saving is the same amount as it was at income level Y_0. Society's attempt to save more

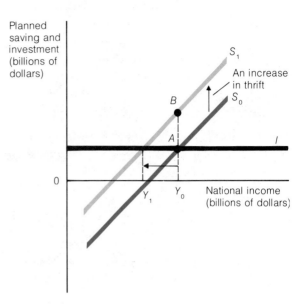

FIGURE 23–12   The Paradox of Thrift

An increase in thrift could cause an increase in desired saving of society in the amount BA for every level of income. The increase in thrift, however, reduces the consumption component of total spending and actually results in a reduction in equilibrium income from Y_0 to Y_1. Saving at this new lower income level, Y_1, is no greater than it was at the higher level of income.

creates a reduction in total spending, which forces society to save the same amount but at a lower income level. The implications of the paradox, if it holds true in the economy, are of great importance. Contrary to classical thinking, saving may be injurious to the economy because it lowers employment of labor and all resources.

Question

Can you think of any government policies that tend to discourage saving? One example may be the taxation of unearned income. People must pay income taxes on the interest they earn from savings accounts. Are there other examples? What sorts of policies toward saving would classical economists advocate?

24

Output Fluctuations and the Public Sector

*T*he last chapter focused on how equilibrium in the private economy is reached, both in the classical model and in the Keynesian model. In the classical view, equilibrium was conceived from a long-run perspective, a period of time that could encompass short-run swings in income and employment. From a short-run perspective, however, equilibrium seems an ephemeral concept, for we commonly observe fluctuations in real GNP, employment, and inflation whenever the government releases its latest statistics on the economy. An essential part of the Keynesian model and Keynes's critique of classical economics was an explanation of how and why the economy in the short run increases or decreases its output in response to changes in spending. Keynes sought, in other words, the reasons why the economy suffers short-run instability and what can be done to correct it.

In this chapter we see how changes in private expenditures actually affect changes in output (and income) by multiples of the initial expenditure change. This process is important in understanding how spending changes are translated into fluctuations in the employment and production levels in the economy—what are known as business cycles of economic activity. Increases in private expenditures that put pressure on the economy's available resources are apt to create inflation in the economy, while cutbacks in private spending may result in economic contractions and widespread unemployment.

Keynes argued that spending in the private economy would normally be insufficient to produce an economy characterized by full employment and little or no inflation. He felt that the public sector—the government—must pursue spending and taxation policies to ensure full employment and economic growth.

Fluctuations in the Private Sector

Before we analyze the mechanics of the Keynesian model, a brief review is helpful. The private-sector equilibrium discussed at the end of Chapter 23 was based on Keynes's central assumption that total expenditures determine output and income. (Keep in mind that the terms *output* and *income* are used interchangeably in macroeconomics. In the circular flow of goods and services, one person's expenditures on output always represent another person's income. The final or total value of output must therefore equal the level of income in the economy.)

In the Keynesian model, output and income can increase or decrease in response to any change in the components of total expenditures. For example, if autonomous consumption or investment increases, income will increase. If the economy-wide marginal propensity to consume increases, income will increase. Table 24–1 summarizes possible changes in private expenditures and their effects on income and output.

To see how changes in these factors actually affect income and output, consider an increase in autonomous investment. Recall that autonomous investment spending is independent of income and depends instead on a number of factors such as the cost of capital, expected sales, and business profits. What happens to income when investment spending increases? In the example illustrated in Figure 24–1, the initial level of income of $300 billion is determined by private spending level $C + I_0$. Assume that spending then increases by $20 billion because of an increase in expected future profits. The total spending curve shifts upward to $C + I_1$ by the amount $\Delta I = \$20$ billion. Desired spending then exceeds the output of goods and services by $20 billion (distance AB in Figure 24–1). To meet the demand, inventories are drawn down, producers step up the rate of output, and income rises. The question is, What effect does a $20 billion increase in autonomous investment have on equilibrium income? Does income increase by $20 billion exactly or by a greater or lesser amount? As you may have noticed in Figure 24–1, income increases by $100 billion with an investment spending increase of $20 billion. Why does such a relatively small increase in investment result in such a relatively large increase in income?

The Investment Multiplier

The answer to this question is given by the value of the multiplier, a number found by dividing the change in income by the change in autonomous spending that caused the change in income. In Figure 24–1, we are looking at a

TABLE 24–1   The Sources of Income and Output Changes in the Private Economy

Increases or decreases in autonomous consumption or investment or in the marginal propensity to consume (and save) will cause increases or decreases in the level of income and output in the economy.

Component Change	Change in Income and Output
↑ Autonomous consumption	Increase
↓ Autonomous consumption	Decrease
↑ Marginal propensity to consume	Increase
↓ Marginal propensity to consume	Decrease
↑ Autonomous investment	Increase
↓ Autonomous investment	Decrease

FIGURE 24–1
The Effects of an Increase in Investment Expenditures

When the level of autonomous investment increases by $20 billion, from 60 to 80, *AB* in the figure, and the marginal propensity to consume is 0.8, equilibrium income rises by $100 billion, in this case from 300 to 400. The multiplier effect causes the level of income to rise by more than the amount of the increase in investment.

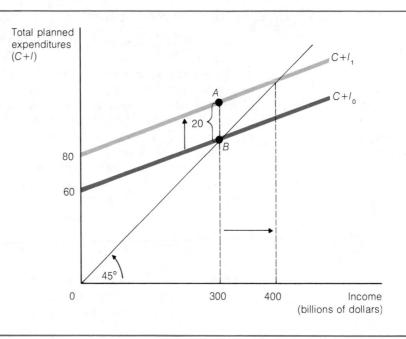

Investment multiplier: A number that is multiplied by the change in autonomous investment to reveal the change in income; 1/MPS.

change in autonomous investment. Thus the **investment multiplier** is found by dividing $100 billion by $20 billion or, if k_I represents the investment multiplier,

$$k_I = \frac{\Delta Y}{\Delta I} = \frac{100}{20} = 5.$$

If autonomous investment increases by ΔI, or $20 billion, the initial expenditure first generates an addition to income of $20 billion ($\Delta Y$) because expenditures on investment goods create returns (or income) to resources. However, the process does not end here, for all resource owners, laborers, holders of capital or land, and so on, are consumer-savers. New income will cause consumer-savers on average to spend a certain part of their income, determined by the MPC, and to save the other part of their income, determined by the MPS. Their expenditures on goods and services then create additional income out of which another set of income recipients spend a portion and save a portion. In this process, the initial increase of $20 billion is multiplied at each stage.

This multiplier process is illustrated in Table 24–2. The initial injection of $20 billion in autonomous investment creates an addition to income (ΔY) of $20 billion. Assuming that MPC is 0.8 and MPS is 0.2, income recipients will spend 80 percent and save 20 percent of the new income. The new consumption thus creates an addition to total income of $16 billion on the first round of consumption spending. The recipients of the $16 billion consumption spending in turn spend part of it — 80 percent of $16 billion, or $12.8 billion — and save part of it — 20 percent of $16 billion, or $3.2 billion. The multiplier process goes on until the total change in income equals $100 billion, of which $80 billion is composed of new consumption and $20 billion of new saving. It is no accident that the new saving is exactly equal to the initial amount of new autonomous investment. Why? Because only when the increase in desired investment is matched by an equal amount of additional desired saving will the disequilibrium caused by the new investment be corrected and equilibrium be restored.

The Multiplier Formalized

The value of the multiplier is determined by the MPC or MPS. A quick method for determining the multiplier is therefore simply to divide 1 by the MPS as follows:

$$k_I = \frac{1}{(1 - \text{MPC})} = \frac{1}{\text{MPS}},$$

or, in the example,

$$k_I = \frac{1}{(1 - 0.8)} = \frac{1}{.2} = 5.$$

Any change in income caused by a change in an autonomous expenditure may then be determined by multiplying the autonomous expenditure change by the multiplier or, in this case,

$$\Delta Y = k\Delta I = \frac{1}{\text{MPS}}(\Delta I) = 5 \times 20 = 100 \text{ billion.}$$

A change in the MPS, which also means a change in the MPC, will obviously change the value of k, the multiplier. The multiplier is the reciprocal of the MPS. If the MPS is 0.1, the multiplier is 10. Under these circumstances, a $20 billion injection of autonomous expenditures would increase income by $200 billion. Or, if Americans saved one half of every dollar received in income, the multiplier would take on a value of 2. (On a piece of scratch paper, demonstrate to yourself why this is so.) In this case a $20 billion increase in investment spending would raise equilibrium income by only $40 billion.

TABLE 24–2   **The Workings of the Investment Multiplier**

Any change in autonomous spending changes equilibrium income by some multiple, depending on the value of the MPS. In this example, MPC = 0.8 and MPS = 0.2. The simple multiplier is the reciprocal of the MPS, in this case 1/0.2 = 5. In a simple Keynesian framework, the change in equilibrium income is equal to the multiplier times the original change in spending, or 5(20) = 100.

Round of Spending	Increase in Autonomous Spending (ΔI)	Increase in Income (ΔY)	Increase in Consumption (ΔC)	Increase in Saving (ΔS)
1	20	20	0.8(20) = 16	0.2(20) = 4
2		16	0.8(16) = 12.8	0.2(16) = 3.2
3		12.8	0.8(12.8) = 10.24	0.2(12.8) = 2.56
4		10.24	0.8(10.24) = 8.19	0.2(10.24) = 2.04
.		.	.	.
.		.	.	.
.		.	.	.
All others		40.96	0.8(40.96) = 32.96	0.2(40.96) = 8.19
		100	0.8(100) = 80	0.2(100) = 20

Note: All values are in billions of dollars.

Consumption multiplier: A number that is multiplied by the change in autonomous consumption to reveal the change in income; 1/MPS.

The Consumption Multiplier

The **consumption multiplier** applies to the autonomous portion of consumption, that portion of consumption spending independent of income and determined instead by variables such as thrift attitudes and the rate of interest on savings.

If autonomous consumption increases, income also increases by some multiple. If, for example, the autonomous component of consumption increases from $40 billion to $50 billion in a society with a marginal propensity to save of 0.2, equilibrium income would rise to $450 billion from its previous level of $400 billion. The change in income can be expressed as follows:

$$\frac{\Delta \text{Autonomous consumption}}{\text{MPS}} = \frac{10}{0.2} = 50 \text{ billion.}$$

Thus the consumption multiplier, like the investment multiplier, is the reciprocal of the MPS (in this example, $1/0.2 = 5$). The higher the MPC (or the lower the MPS), the higher the consumption multiplier and the higher the income change from a given change in autonomous expenditures.

To summarize, income will rise or fall by an amount greater than the expenditure change because of a multiplier process. The multiplier, which is simply the reciprocal of the MPS, multiplies the total amount of autonomous expenditures to create equilibrium income. Stated another way, changes in equilibrium income are calculated by applying the multiplier to changes in autonomous expenditures.

These concepts are of great importance for practical policymaking. The most significant effect of the multiplier is that very small changes in spending may be greatly magnified in resulting income and employment changes. For example, a small change in private spending could precipitate a relatively large economic contraction or reduction in employment and business activity. As we will see later in this chapter, Keynes recommended that government's fiscal actions be used to counteract the instabilities generated by the multiplier effects of private spending on GNP, employment, economic growth, and inflation.

The Multiplier and Private Income Equilibrium

Before turning to more elaborate models of the macroeconomy and the impact of prices in the Keynesian framework, it is useful to reflect on some crucial assumptions about the multiplier process and about the basic Keynesian private-sector model in general.

The multiplier process just described is a simplification. No time horizon is placed on the rounds of spending, and the process requires that income receivers who spend and save do so in the exact amounts given by the MPC and the MPS. In the simple model there are no other deductions from income except saving. Since there is still no role for government in the model, income receivers do not have to pay taxes on income received or on income spent, such as excise or consumption taxes. (As we will see later in this chapter, taxes are a leakage from income that has the effect of reducing the numerical value of the multiplier.) Further, there is no hoarding, or holding idle income (not to be confused with saving), in the simple multiplier model.

In addition, the simple private model does not consider spending injections into the macroeconomy other than autonomous consumption and investment. As we will see in the next section, investment can be accelerated by changes in income, and there are other injections such as government spending and inputs to account for. Such additional spending tends to increase equilibrium income. In general, then, leakages—deductions from income such as saving and taxes—reduce equilibrium income, while injections—additions to spending—increase equilibrium income. Most of these factors are fully integrated into the basic macroeconomic model later in this chapter.

The Multiplier, the Accelerator, and the Business Cycle

The income-expenditures approach to macroeconomic theory establishes an explanation for equilibrium levels of GNP and employment. As we have seen, the multiplier process explains how changes in autonomous consumption or investment spending change the equilibrium level of income. These models are extremely useful in organizing our understanding of how GNP and employment of resources are determined. However, data on unemployment, GNP fluctuations, and inflation suggest that our economy undergoes cycles of business activity. As indicated in Chapter 21, the business cycle consists of observed upswings and downswings in the production of goods and services and in the employment of resources to produce goods and services. Keynes left the business cycle unexplained, although Nobel laureate Sir John R. Hicks sought an explanation in an extension of Keynes's theory of the simple investment multiplier.[1] He called it the accelerator principle.

The Accelerator Principle

Investment activity is closely linked to overall business activity. As we have seen, changes in autonomous investment spending produce greater changes in income via the multiplier. There is also reason to believe that the reverse is true—changes in sales or output also produce changes in investment. The **accelerator principle** relates investment to rates of change in consumption or output. To better understand this important relation, let's look at the investment process more closely.

Why do businesses invest? As we noted in Chapter 23, the costs of investment projects, interest rates, expected sales, and actual sales all help determine the amount of investment spending in the economy. But think about the possible responses of businesses to increases or reductions in the demand for their products. In the short run, businesses with a relatively fixed capacity must either add labor and other variable resources when sales increase or lay off labor and other resources when sales decline. In the short run, businesses have relatively few options. Decreases in sales will create unemployment of labor and other resources as well as unused capacity; increases in output demand may lead to pressure on labor and other resources, accompanied possibly by a high rate of inflation. The result: In the short run, businesses may alter the rate of output of their plants, but actual increases in output are severely constrained because of the impossibility of adding capacity by building warehouses and plants or purchasing sophisticated equipment over a short-run period.

Accelerator principle: Changes in the level of investment expenditures (and thus GNP) take place because capital requirements change with changes in output rates.

---

[1] John R. Hicks, *A Contribution to the Theory of the Trade Cycle* (London: Oxford University Press, 1950). The idea of an accelerator was expressed very early, though in primitive form, by American economist J. M. Clark in "Business Acceleration and the Law of Demand," *Journal of Political Economy* (March 1917), pp. 217–235.

Capacity to produce can be changed in the long run, however. New plants may be built or existing plants may be expanded in response to increased demand. Plants may also be scrapped when demand for businesses' output is persistently down. These changes in capacity are changes in capital investment and are made in response to long-run changes in sales.

The real question is, By how much will an increase in output or sales increase physical capital requirements assuming that we are initially using our existing capital stock to the fullest extent? A certain amount of capital is needed to produce a certain amount of consumer goods, from hamburgers to cars and diamond necklaces. A ratio exists of capital to output for the entire economy. This **capital/output ratio** is a measure of the capital investment generated by a change in sales. A capital/output ratio of 0.8 would mean that 80 cents in new capital investment is facilitated by every dollar's increase in sales or output.

Capital/output ratio: The ratio of the amount of capital to the amount of output, or the ratio of investment expenditures to total expenditures.

Investment is therefore generated by two factors: (1) the ongoing autonomous investment that takes place no matter what sales or output happen to be in the short run and (2) an investment component that is directly related to the rates of growth or decline of sales of goods and services in the economy in the long run. We may express these two components for a hypothetical year 1988 as follows:

$$\begin{matrix} \text{Investment} \\ \text{expenditures} \\ (1988) \end{matrix} = \begin{matrix} \text{Autonomous} \\ \text{investment} \end{matrix} + \frac{\text{Capital}}{\text{Output}} \left[\begin{matrix} \text{Rate of} \\ \text{change} \\ \text{in output} \\ (1988) \end{matrix} - \begin{matrix} \text{Rate of} \\ \text{change} \\ \text{in output} \\ (1987) \end{matrix} \right].$$

When output or sales rates are changing, say from 3 percent in 1987 to 5 percent in 1988, investment expenditures will be accelerated.

Note that is it the rate of change in output that causes the accelerator effect. The special significance of the accelerator is that it can explain a decrease in investment when output is increasing at a decreasing rate. For example, if output rises at a rate of 3 percent in one year and only 2 percent the next, investment expenditures will actually fall. Such an occurrence would give rise to a business cycle, the kind and size of which would depend on the numerical value of the capital/output ratio.

The Business Cycle

Business investment in any particular period is determined by the level of autonomous expenditures and by the growth or decline of sales, as discussed above. If sales are growing at a steady and constant rate (for example, 3½ percent per year), only autonomous factors will affect investment. (To clarify, suppose the hypothetical output rates in 1987 and 1988 are both 3.5 percent. Were this the case, the second term in the right-hand side of the above expression would equal zero.) What happens if there is a shift in autonomous investment expenditures caused by a change in future sales expectations or some other nonincome factor affecting investment?

Figure 24–2 shows the growth rate of real GNP through time with the straight line depicting a constant and steady growth rate in output. When sales are assumed to be growing at a steady rate, as between 1983 and 1988 in Figure 24–2, income also grows at a steady rate. But what happens in 1988 when autonomous investment rises? A cycle process begins, with a shift in the autonomous portion of investment. As income is increased through the multiplier process, investment expenditures are accelerated in the system because of the increased growth rate of output. Additional capital invest-

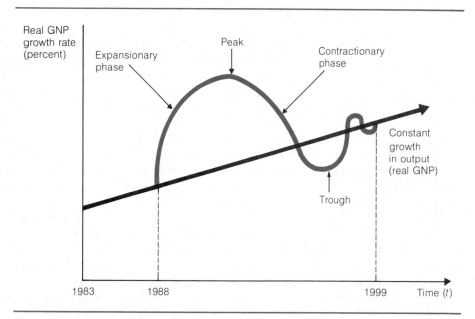

FIGURE 24–2   The Accelerator Effect and the Business Cycle

A jump in autonomous investment in 1988 creates a business cycle as output rises owing to the multiplier effect. As output rises, investment expenditures are accelerated because of the capital requirements of producing additional output. A peak is reached as the rate of output increase slows, creating a contractionary phase of the cycle. A trough is finally reached when the decline of output slows, creating another upturn or expansionary phase of the cycle. In the real world a return to the equilibrium growth path is unlikely since expectations about future sales and business conditions tend to change frequently, setting up a continuous business cycle.

ment takes place as the capital/output ratio creates an expansionary phase of the business cycle. The expansionary phase is characterized by the struggle to hire more labor and other inputs in order to expand business activity and production. Higher prices may accompany this phase (which is why the vertical axis shows *real* GNP growth).

With no other influence, the rate of growth in real GNP eventually slows owing to resource scarcities, causing a slowdown in new investment expenditures. A peak of the cycle is reached, and business activity, measured in real GNP, begins to decline. Workers are laid off and plant production is reduced in this contractionary phase of the cycle. The rate of decline in real GNP finally slows down, and a bottom or trough of the cycle is experienced. The process then starts over again.

Figure 24–2 shows the cycle of output swings getting smaller, ultimately reaching stable equilibrium in 1999. However, is this outcome likely in the real world? The answer is no when uncertainty, sales expectations, and other factors affecting investment and income are brought into play. The likely scenario is an economy in a perpetual state of flux with recurring cycles of expansion (and possibly inflation) and contraction.

Contractionary and Expansionary Gaps

Failure of the private economy to provide stable growth (such as the upward-sloping growth curve of Figure 24–2) and predictable, stable levels of full employment was Keynes's fundamental criticism of classical economics.

Though Keynes himself did not discuss a business cycle, in his view the private economy was erratic and subject to long-run periods of large-scale unemployment and low production levels. In terms of Figure 24–2, the contractionary phase of the cycle would engender more and greater pessimism about future business conditions, keeping the economy in a contractionary phase of the cycle for a very long period of time. Only accident, not dependable market forces as the classical economists had maintained, would produce a stable equilibrium at full employment with no inflation.

Contractionary gap: The amount by which total expenditures fall short of full-employment income; recessionary gap.

Expansionary gap: The amount by which total expenditures exceed the amount necessary to achieve full employment; inflationary gap.

For our more familiar income-expenditures approach to the macroeconomy, Keynes envisioned the economy as reaching, and remaining in, either a contractionary gap or an expansionary gap. A **contractionary gap** (sometimes called a recessionary gap) is a situation where equilibrium settles at a less than full employment level owing to insufficient total expenditures at the full employment income level. An **expansionary gap** (sometimes called an inflationary gap), the opposite situation, means that total private expenditures exceed those expenditures necessary to produce a full-employment level of income without inflation.[2] The excess of expenditures, in other words, puts pressure on the employment of resources. Since the economy cannot push output beyond its production possibilities frontier, inflation is created.

These gaps can be illustrated in graphic terms. Figure 24–3a and Figure 24–3b show the two possible equilibrium positions for the private economy. In Figure 24–3a, private spending results in an equilibrium at point E. Private consumption and investment spending produce an equilibrium income (and level of employment) Y_E. Income level Y_E is not, however, a full-employment level of income. Workers are laid off and other resources are unemployed, as indicated by the spending or contractionary gap represented at income level Y_F, the level needed for full employment without inflation. The contractionary gap in the private economy (AB in the figure) presents the amount of additional consumption and investment spending that would be necessary to bring the economy to a full-employment equilibrium.

Figure 24–3b depicts the opposite situation. Consumption and investment spending produce equilibrium and a level of employment at Y_E. Full employment without inflation occurs at the lower income level Y_F, however. Spending in excess of what is needed for the full-employment income level Y_F creates pressures on the economy to expand. Because resources are limited, these pressures usually create inflation, a general rise in prices. An expansionary gap exists in the amount represented by FG at full-employment income level Y_F. This gap represents the amount of excess consumption and investment that causes inflation at full employment.

Government Spending, Taxation, and Foreign Trade

Keynes believed that the invisible hand—those forces that would return the economy to full-employment equilibrium without inflation—was seriously arthritic. Even if such forces on private spending did exist, in Keynes's opinion they would probably be unendurably slow in restoring equilibrium. Keynes

[2]The expansionary gap is also called an inflationary gap, but the amount of inflation produced by excess spending depends on how close the economy is to maximum capacity. When the economy is on its production possibilities frontier, no additional output is possible and the excess spending produces pure inflation.

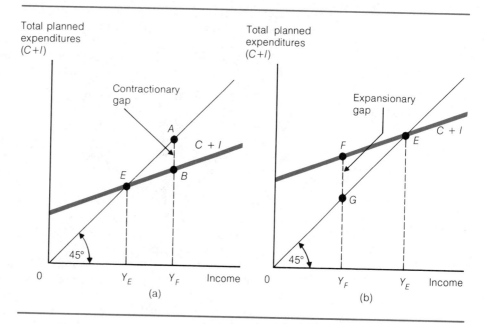

FIGURE 24–3   Contractionary and Expansionary Gaps

(a) A contractionary, or unemployment, gap exists when private spending is insufficient to bring the economy to full employment without inflation. Y_E is the equilibrium income reached by private spending and investment. Unemployment is high and production is low, however, with a contractionary gap AB represented at Y_F, the level needed for full employment without inflation. (b) An expansionary gap is the result of private spending exceeding the resource capabilities of the economy. Such excess spending creates inflation. Equilibrium at Y_E is in excess of the level Y_F needed for full employment without inflation. The excess spending creates pressure on the economy to expand, usually causing inflation.

argued that society should not have to endure contractionary or expansionary gaps of long duration. Since automatic forces were unreliable, he saw discretionary government interventions as necessary. Keynes felt that macroeconomic management should be entrusted to the central government, which theoretically has the ability to adjust aggregate spending in line with national economic goals by the use of its two major budget weapons: government spending and taxation. Changes in one or both of these areas constitute fiscal policy, the object of which is the manipulation of total spending in the economy. In this section we will examine the macroeconomic effects of government spending, taxation, and foreign trade; in the final section of the chapter, we will analyze how leakages and injections to total expenditures can be manipulated to control the whole economy.

Government Expenditures

In the context of the basic Keynesian income-expenditures model, government spending (G) is an increase in total spending, represented graphically by raising the C + I curve to account for the new source of spending. (In adding government spending, we are relaxing one of the simplifying assumptions of our basic model discussed in Chapter 23.)

Government spending can take many forms and have many (sometimes conflicting) objectives. For example, in one sense the government behaves like a business, purchasing inputs used in its daily operations, such as labor, machinery, and buildings. In another sense, the government acts like a consumer, purchasing products for its own use, such as missile systems, tanks, ammunition, and airplanes. Another kind of government spending that is more extensive and more controversial does not result in the outright purchase of goods and services; rather, it is made up of transfer payments and redistributes income among various groups in society—from rich to poor, from able to disabled, from healthy to sick, from employed to unemployed. As a first approach, we will not include transfer payments; instead, we will confine the meaning of G to government purchases of goods and services.

One important point to keep in mind is that a government's decision about how much to spend is somewhat different from private decisions to consume or invest. A major difference is that the government is a not-for-profit sector of the economy. Its decisions are often not motivated by private profit or economic efficiency. Furthermore, its spending is not constrained by its income in tax receipts. Congress can, and often does, spend far more than it takes in in the form of taxes, issuing government bonds or creating money to cover the difference. Because government spending is not constrained by government income, enormous budget deficits have built up in recent years, causing much concern among the electorate and in Congress. One reflection of this concern is the broad support for a balanced budget amendment to the Constitution (see the Economics in Action at the end of this chapter). It is dangerous to ignore consequences of budget deficits, but in this first approach we will do so to keep the model simple.

Autonomous government expenditures: A level of government spending that does not change as income changes.

To further simplify the analysis, we will assume that government expenditures to purchase goods and services actually are **autonomous expenditures;** in other words, they do not depend in any predetermined way on national income. Thus we may write, as we did previously for investment, $G = \overline{G}_0$ to indicate that government expenditures are autonomous of income although, for simplicity, we will use G_0 instead of $\overline{G}_0$ below and in the figures. Aggregate demand can now be redefined as follows:

$$\text{Aggregate demand} = C + I + G = \text{Total expenditures.}$$

Graphically, an increase in autonomous government spending, G, results in a parallel upward shift of the aggregate demand function. In Figure 24–4a, the $C_0 + I_0 + G_0$ curve is higher than the $C_0 + I_0$ curve at each level of national income by whatever amount of G_0 Congress decides to spend. Notice that the addition of government spending to the aggregate demand curve has the effect of moving the aggregate demand curve to an intersection point higher up the 45-degree line. In other words, government expenditure, taken by itself, is expansionary; it leads to a higher level of equilibrium income.

Government expenditures are also an injection into the expenditures stream. In Figure 24–4b, government expenditures in the amount of G_0 are added vertically to private investment expenditures of I_0. Since there are no leakages in this simple model other than private saving, equilibrium occurs where saving—the leakage from the income-spending stream—is equal to the sum of investment and government spending injections into the income stream. The equilibrium level of income and output is exactly equal to that produced in the total expenditures approach to income determination shown in Figure 24–4a.

The Effects of Taxation

To introduce the concept and effect of government spending, we merely acknowledged the impact of an injection of new spending on the macroeconomy. In fact, government spending decisions are rarely made in a budgetary

FIGURE 24–4

Effects of Government Expenditures on Aggregate Demand

The introduction of government spending increases total spending and the equilibrium level of income. (a) With government spending of G_0 added to autonomous consumption and investment, $C_0 + I_0$, total income rises from Y_0 to Y_1. (b) An equivalent level of income appears in a leakage-injection analysis. The saving leakage, S, is equated to the investment and government spending injections, $I_0 + G_0$, to determine an equilibrium level of income.

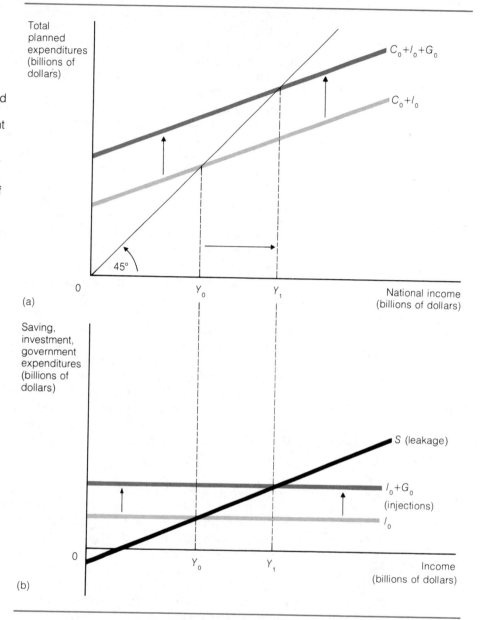

vacuum. Decisions about taxes—both their level and their structure—almost always accompany decisions about government spending. In reality, tax decisions are always more complicated than they are presented here because in the real world of politics questions about taxation and income redistribution are inextricably entwined. And since income redistribution involves gains by one group at the expense of another, power relations and vested interests invariably come into play.

Income redistribution is not the major focus of this chapter. We will therefore disregard changes in the structure of taxes and concentrate on the effects on output and employment of changes in the level of taxes. Taxes represent a leakage from the income-expenditures flow. In this respect they are like saving and for analytical purposes can be treated in the same way as saving. (You might think of taxes as "public saving" in conjunction with the private decision to withhold from spending a part of national income.) The fact that taxes represent involuntary saving does not concern us for the moment because the macroeconomic effects do not depend on whether saving is voluntary or involuntary.

To determine the effect of taxes on the macroeconomy we must ask, Who pays taxes? The answer, of course, is that consumers (and, to a lesser extent, businesses) do. The effect of taxes, therefore, is primarily to lower planned consumption insofar as taxes reduce the disposable income received by households, the income left after taxes. We must refine another assumption in light of government spending. As we saw earlier, so long as government is not a factor in the economy, several of the aggregate income measures are identical to each other. But once we subtract taxes from national income, national income and disposable income are no longer the same. Disposable income is equal to national income minus the amount of taxes paid. Aggregate demand is, therefore, a function of disposable aggregate income once taxes are admitted into the analysis.

Exactly how do taxes affect spending and saving levels? Let us assume for convenience that taxes are paid out of household income only and that businesses are not taxed. In Chapter 23 we saw that any increase in household income will be distributed between future consumption and future saving and that once we know the marginal propensity to save (MPS), we thereby know the percentage of the change in income allocated to each use. What works for injections to household income also works in reverse for leakages from household income. That is to say, facing a higher tax bill (a lower disposable household income), households will finance the higher level of taxes partly at the expense of future consumption and partly at the expense of future saving. This is true of any kind of tax increase. But the effect of the tax on the shape of the consumption and saving curves depends on the type of tax levied.

Lump-sum tax: A fixed level of taxes, or taxes that do not change as income changes; an autonomous tax.

Lump-Sum Taxes. If the tax is a **lump-sum tax,** the effect is simple and straightforward: Every income earner is taxed the same amount, say $100. Each dollar increase in lump-sum tax revenue collected by government results in a dollar lost from household disposable income. An increase in the lump-sum tax therefore shifts the consumption curve downward. But the curve does not shift downward by the full amount of the tax because the saving schedule is also shifted downward. If MPC equals 0.8 and the MPS equals 0.2, then every dollar increase in the lump-sum tax will reduce consumption by 80 cents and reduce saving by 20 cents. The downward shifts in consumption and saving will be parallel shifts because a lump-sum tax means that the tax amount is the same regardless of the level of income. Figure 24–5 shows the shifts in consumption and saving resulting from the imposition of a lump-sum tax.

In Figure 24–5a, the imposition of a lump-sum tax on each household causes household consumption to decline by 80 percent of the reduction of income caused by the tax (MPC $\times \triangle Y$) because the MPC is 0.8. If the lump-sum tax is $100, the shift from $C_0 + I_0$ to $C_1 + I_0$ represents an $80 cut in household expenditures at every level of aggregate income. In Figure 24–5b, the shift of the saving curve from S_0 to S_1 represents a $20 reduction of household saving (MPS $\times \triangle Y$) at each level of aggregate income.

The effect of a tax increase, taken by itself, is contractionary. It causes a reduction in aggregate income and output. This point is readily seen in Figure 24–5a, where the effect of the tax is to reduce national income equilib-

FIGURE 24–5
Effects of a Lump-Sum Tax on Aggregate Household Spending and Saving

A lump-sum tax decreases both consumption and saving. (a) The decrease in consumption from C_0 to C_1 is determined by the MPC. If MPC = 0.8, a lump-sum tax reduces consumption by 80 percent of each tax dollar paid. A lump-sum tax reduces equilibrium income, in this case represented by the shift from income level Y_0 to Y_1. (b) Imposition of the lump-sum tax reduces saving by an amount equal to the MPS times the tax paid, T. The saving function shifts downward from S_0 to S_1. The total tax leakage, T, is then added vertically to saving function S_1, making the total leakage $S_1 + T$. When equated to the investment injection, I_0, equilibrium income shifts to Y_1.

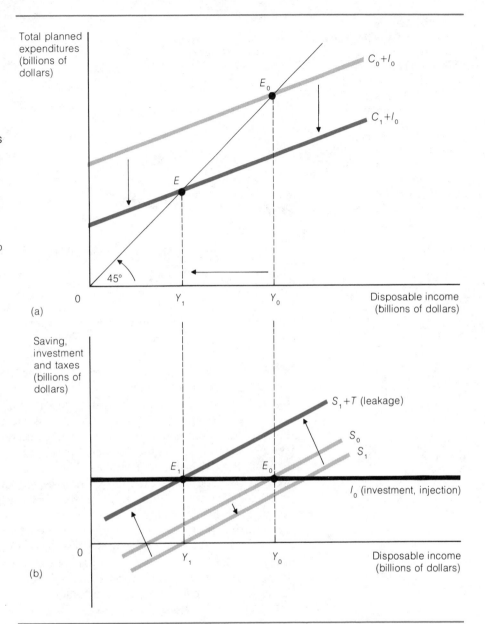

rium from Y_0 to Y_1. Likewise, in the leakages-injection approach illustrated in Figure 24–5b, a new equilibrium is attained where the sum of the leakages, $S_1 + T$, equals the sum of the injections, I (investments being the only injection considered in this example). This equilibrium occurs at income level Y_1.

Income Taxes. An **income tax** differs from a lump-sum tax in that the amount of the tax collected varies systematically with the level of income; the **tax rate** specifies the percentage of income that must be paid to the government as tax. Income taxes can be progressive, regressive, or proportional. Progressive tax rates increase in percentage terms when income increases; regressive tax rates decline as income increases; proportional tax rates remain the same in percentage terms (though not in total amount) as

Income tax: A tax based on the level of income; an income-induced tax.

Tax rate: The percentage of income that is devoted to taxes.

A tax that is based
proportionally on income has
the effect of rotating the
consumption function
downward from $C_0 + I_0$ to
$C_1 + I_0$. Disposable income
falls from Y_0 to Y_1 because of
the tax.

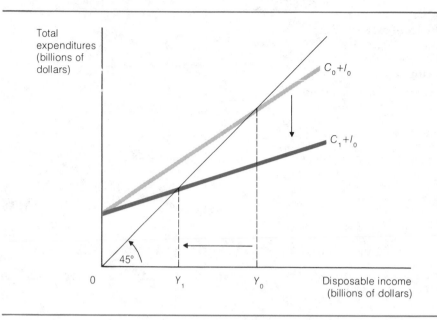

income increases. In most Western economies where it is used, the income
tax is designed to be progressive, although when all taxes are considered
together, the effect is closer to being proportional.

To trace the effects of a proportional income tax as opposed to a lump-
sum tax, let us suppose that 20 percent of household income is taxed regardless of the level of income of each household (such a system, called a flat
tax, has, in fact, been much discussed as a substitute for the U.S. progressive
tax system). The effect on the expenditures curve is shown in Figure 24–6.
As in the case of the lump-sum tax, the curve shifts downward, but instead
of a parallel shift, the curve rotates downward. The reason the aftertax curve
grows farther from the pretax expenditure curve as the level of income increases is that the same percentage of higher and higher income levels extracts higher and higher total dollar amounts of tax revenue. (If the tax
structure were progressive, the divergence between before-tax and after-tax
curves would be even more pronounced.)

A 20 percent income tax reduces the amount of income that gets into the
spending stream by 20 percent. If the MPC is 0.8, the 80 cents of each
dollar that would be spent were it not for taxes is reduced by 20 percent, or
$0.8 - 0.8 (0.2) = 0.64$. With an MPC of 0.8 reduced by a tax rate of 0.2,
only 64 cents of each dollar of national income reaches the expenditure flow.

A quick comparison of the income tax and the lump-sum tax shows that
the former may have a stronger contractionary effect on the economy, other
things being equal. The lump-sum tax claims a smaller percentage of larger
incomes than of smaller incomes.

Foreign Trade

Exports: The goods, or the
dollars received from goods,
that are sold to people in
foreign countries.

The impact of **exports** (X) and **imports** (M) on the macroeconomy is analogous to the impact of government spending and taxation. When the United
States exports goods, the expenditure flow rises by the amount of the purchases made by other countries. Therefore, this amount is a spending injection in the United States and should be added to the aggregate demand

Imports: The goods, or the dollars spent on goods, that are sold by foreigners to domestic citizens.

Net foreign trade: Exports minus imports $(X - M)$; the net amount of dollars spent on imports and exports.

schedule $C + I + G$, making it $C + I + G + X$. In the leakages-injection approach, exports are an injection to the $I + G$ schedule, or $I + G + X$.

When U.S. citizens spend dollars to buy foreign-made imported goods, the value of these imports constitutes a leakage from the domestic income flow. Expenditures on imports should therefore be deducted from aggregate demand or added to the saving-taxation schedule. Total expenditures or aggregate demand now equals $C + I + G + (X - M)$. This expression is a convenient way to account for foreign trade in the Keynesian model we have been studying. The **net foreign trade** balance $(X - M)$ is determined by subtracting imports from exports and adding that number (either positive or negative) to the $C + I + G$ schedule of aggregate demand. Although the net trade balance figure has taken on increased importance in the U.S. economy, the figure is still a relatively small share of total national income, ordinarily accounting for less than 5 percent of GNP. We postpone extensive discussion of the subject of foreign trade until Chapters 35 and 36, where both real and financial aspects of international economic activity are discussed.

Total Expenditures: A Recap

In summary, equilibrium income (and, by implication, the amount of employment hired to produce it) may be determined either by examining total expenditures, $C + I + G + (X - M)$, or by equating leakages to injections, $S + T + M = I + G + X$. Total income, the result of total planned or desired expenditures, can be viewed schematically in Figure 24-7 as sub-

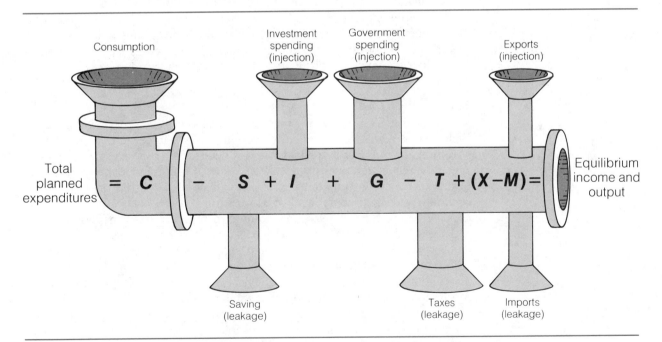

FIGURE 24-7   Total Planned Expenditures Determine Total Income and Employment

The Keynesian model of aggregate demand can be viewed as a series of injections into and leakages from the flow of income. Saving, taxes, and imports are subtracted from private consumption expenditures; investment, government, and export spending injections are added, yielding total planned or desired expenditures. These expenditures determine an equilibrium income and output.

tractions from or additions to, leakages from or injections to, private consumption expenditures. The net effect of these leakages and injections from private consumption is to determine a specific level of total spending, which in turn determines a specific level of income. This level of income is produced by some quantity of resources, including labor usage. Total employment or jobs filled is therefore the product of consumer spending modified by leakages and injections.

Countercyclical Fiscal Policies

Countercyclical fiscal policy: Changes in government expenditures or taxes that attempt to decrease swings in total expenditures which create unemployment or inflation.

Having reached an understanding of how the presence of a government sector influences the economy, we now move the discussion toward a fuller understanding of the Keynesian prescriptions for **countercyclical fiscal policy.** Fiscal policy, as we have already noted, means budget actions: government expenditure changes, tax changes, or a combination of the two. To say that a policy is countercyclical means that the policy seeks to correct whatever phase of the business cycle—contractionary or expansionary—the economy happens to be in. In a recession, for example, aggregate spending, output, and employment are below the full-employment equilibrium level. In such circumstances, countercyclical fiscal policy would seek to raise the level of aggregate demand to the full-employment level. During inflation, a countercyclical fiscal policy must try to induce a lower level of total spending. But knowing the appropriate direction of budgetary changes does not guarantee that the exact equilibrium target will be achieved. To do that, policy must be guided by knowledge of the relevant expenditure multipliers. Expenditure multipliers play a major role in the following analysis. Any autonomous expenditure is subject to the multiplier process; thus it is possible to identify consumption multipliers, investment multipliers, government spending multipliers, export multipliers, import multipliers, and so on.

Fiscal Policy to Deal with a Recession

To see how fiscal policies are used to change the direction of the business cycle, let us first suppose that the economy is in a recession. Imagine that government economic policymakers have determined that the actual level of aggregate income/output is $50 billion below the full-employment level. In Figure 24–8, the government's problem can be seen as that of moving the economy from an income level of $2.50 trillion to one of $2.55 trillion (an increase of $50 billion). What is required in this case is an increase in G or a decrease in T, either of which would lead to an increase in aggregate demand $(C + I + G)$. If government simply increases G by $50 billion, it will overshoot the desired target because the new additional expenditure of $50 billion will be multiplied through various rounds of spending into a much larger increase. What, then, is the appropriate fiscal stimulus?

The answer depends on the size of the contractionary gap, whether households or businesses are the ultimate recipients of government expenditures, and the value of the relevant multiplier. In the following analysis, assume (1) that only households will receive the government expenditures and (2) that MPC equals 0.8. We can determine the size of the contractionary gap, AB, and design a fiscal policy to eliminate it.

Since we know the value of the MPC, we can compute the consumption multiplier (the only relevant multiplier in this case):

FIGURE 24-8

Government Spending to Eliminate a Contractionary Gap in Income

If aggregate spending, $C_0 + I_0 + G_0$, is not great enough to achieve full employment, then an increase in government expenditures may restore full employment. In this example, initial expenditures, $C_0 + I_0 + G_0$, are insufficient to create a full employment level of income. Government expenditures are increased to $C_0 + I_0 + G_1$ and, through a multiplier process, the contractionary gap, AB, is eliminated.

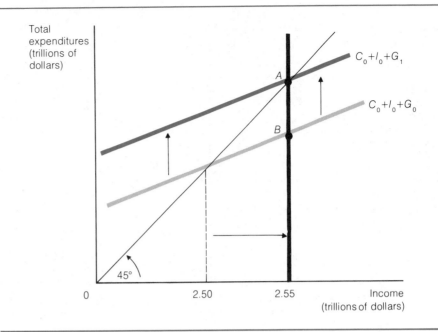

$$k = \frac{1}{1 - MPC} = \frac{1}{0.2} = 5.$$

We know the size of the income gap ($50 billion), and we know that whatever increase in government spending occurs will be multiplied by 5. Therefore, we can easily determine the size of the contractionary gap, AB, by the formula

$$\text{Contractionary gap} = \frac{\text{Income gap}}{k} = \frac{\$50 \text{ billion}}{5} = \$10 \text{ billion.}$$

In this case, an increase in aggregate expenditures of $10 billion will, via the multiplier, produce an increase in aggregate income of $50 billion, thereby moving the economy to its full-employment equilibrium level.

There are two ways that fiscal policy could produce the additional $10 billion expenditure. The most direct way is to increase government expenditures, G, by $10 billion. We assume that this direct injection to the spending stream is not subject to any leakages. Therefore, the full $10 billion enters the expenditure flow at once.

An indirect way to accomplish the same thing is to cut taxes to stimulate consumption spending, as shown in Figure 24-9. Remember, however, that tax changes do not produce expenditure changes of the same dollar amount because taxes are paid partly at the expense of consumption and partly at the expense of saving. Against a contractionary gap of $10 billion, a lump-sum tax cut of $10 billion will only induce additional consumption of $8 billion (MPC $\times$ ΔT). To raise consumption spending by $10 billion requires in this instance a tax cut of more than $10 billion to offset the leakage to saving that occurs when household disposable income increases. A lump-sum tax cut of $12.5 billion, given that MPC = 0.8, would be required to raise aggregate consumption by $10 billion and thereby eliminate the $10 billion contractionary gap.

FIGURE 24–9

Taxation to Eliminate a Contractionary Gap

If total spending, $C_0 + I_0 + G_0$, is not great enough to achieve full employment, then aggregate expenditures may be increased by a tax cut. In this example, if taxes fall, then consumption increases from C_0 to C_1 to achieve full employment. A tax cut increases the disposable income of consumers, causing them to spend more.

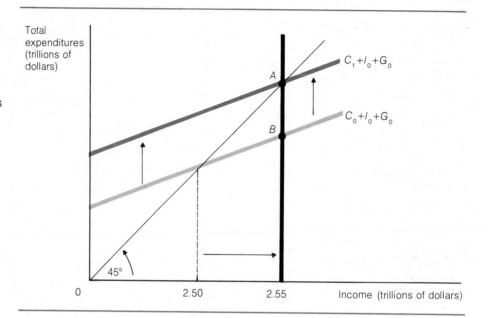

Based on what we have learned in this section, it should be obvious that government expenditure changes are more direct and more powerful, dollar for dollar, than tax changes of the same amount. This recognition must be kept in mind when expenditures and taxes are changed simultaneously (see Focus, "The Balanced Budget Multiplier").

Fiscal Policy to Deal with Inflation

Once full employment occurs, further increases in aggregate demand will tend to drive the price level upward, creating an expansionary gap. The reverse of a given policy that works in one phase of the business cycle works for the opposite phase of the cycle. Thus, if the economy is characterized by inflation and full employment, the appropriate fiscal response would be a decrease in government expenditures or an increase in taxes.

An expansionary gap, as discussed earlier, is the difference between the existing level of total spending and the desired level. The trick is to bring the actual level of aggregate demand, $C + I + G$, in line with the desired full-employment level. As in the case of contractionary gaps, the correction can be achieved in one of two ways: either a change of government expenditures or a change of taxes.

Assume that the economy is operating at a level of nominal output-income too high for the full-employment level. (Since we are talking about the effects of inflation, the concepts of real and nominal come into play.) In what sense could the nominal level of income be too high? "More is better than less" when it comes to economic output, but there are two factors to consider. First, by definition, aggregate real output in the macroeconomy cannot increase after full employment has been reached because the economy is already at a point on the production possibilities frontier. Second, output in the Keynesian model is always measured in dollar terms, so that income measures beyond the full-employment level mean that the same amount of goods and services are valued at higher unit prices, the effect of inflation. When the newspapers say that the economy is "overheating" they mean that a lower level of national income is preferred to the present higher

FOCUS   The Balanced Budget Multiplier

When the government's expenditures equal its revenues, its budget is balanced. Maintaining a balanced budget has long-run economic benefits and is gaining popularity with voters (see Economics in Action at the end of this chapter), but does a balanced budget prevent the use of fiscal policy for managing aggregate demand? Will an increase in government spending in an effort to reduce a contractionary gap be completely offset by an equal increase in taxes?

An increase in government expenditures is an injection into the total spending stream. Meanwhile, an increase in taxes is a leakage from the spending stream. However, an increase of $10 billion in autonomous government spending, along with a $10 billion lump-sum increase in taxes, *will* cause an increase in total spending. This spending increase occurs because not all of the tax increase is at the expense of consumption. Households pay taxes by decreasing both consumption expenditures and saving. Thus, not all of the $10 billion tax increase is removed from spending.

Increasing the size of the total balanced budget, say from $1 trillion to $1 trillion 10 billion, will increase total spending, but by how much? The government expenditures multiplier, $1/(1 - MPC)$, when multiplied by the increase in government expenditures, ΔG, tells us the increase in income is $\Delta G/(1 - MPC)$. Under a balanced budget taxes increase in amount equal to the increase in government expenditures. The reduction in consumption spending and saving from the tax will be ΔT multiplied by the MPC. The multiplier that applies to the tax increase is thus $-MPC/(1 - MPC)$. The reduction in income resulting from the tax increase alone is $-\Delta T(MPC)/(1 - MPC)$. If we add the government expenditures and tax multipliers together we see that $(1 - MPC)/(1 - MPC)$ is equal to 1.

Thus, an increase of $10 billion in the size of the budget is multiplied by 1 to obtain the increase in total spending. The balanced budget multiplier is 1. Under a balanced budget, the multiplier does not come into play. Expenditures are not really multiplied at all.

With a contractionary gap of $50 billion, balanced budget fiscal policy calls for an increase of G and T by a full $50 billion. On the other hand, an inflationary gap of $50 billion would require a $50 billion decrease in the size of government. Countercyclical balanced budget fiscal policy thus requires very large changes in the size of government to obtain the desired goals. While a balanced budget theoretically could be used to fine-tune the economy, large increases and decreases in the level of taxes and the size of government programs might be very disruptive.

---

one. Too high a level of aggregate demand is redundant in that it produces inflation.

The problem posed by inflation, then, is how to reduce the level of nominal income by some amount, say by $50 billion. With a multiplier of 5, a reduction in spending of one-fifth the observed income gap will produce the desired effect. This result can be achieved by cutting government expenditures by $10 billion, leaving taxes unchanged. Likewise, it can be achieved by leaving government expenditures unchanged and raising lump-sum taxes by $12.5 billion (to reduce the consumption component of aggregate demand by $10 billion). The mechanics of these changes are exactly the reverse of those used in the contractionary-recessionary situation. Changes in taxes or government expenditures therefore composed Keynes's arsenal of methods for counteracting severely contractionary or inflationary swings of the business cycle.

Keynesian Economics in Historical Context

Was Keynes right about the ability of government to use countercyclical fiscal policies to counteract extreme phases of the business cycle? Although the policy and theoretical issues are complex, it is probably the case that certain theoretical and political effects unforeseen by Keynes have inhibited the ability of government to accurately control the business cycle and thereby guarantee continuous inflation-free levels of full employment and maximum economic growth.

Keynes was primarily concerned with the economic and human costs of depression (or, if you prefer, an extreme contractionary phase of the business cycle). His advice, though not fully implemented in Western democracies during the Great Depression of the 1930s and early 1940s, was at least partially responsible for the increased economic role of government at the time and, perhaps, for the failure of prolonged and serious depressions to materialize in the post–World War II period. But problems other than depression became paramount in the U.S. economy of the late 1960s, 1970s, and 1980s—particularly high rates of inflation and slowed economic growth. Keynesian economics was not explicitly designed to deal with these problems. Further, U.S. fiscal history since the beginning of the Vietnam War may have contributed greatly to contemporary difficulties.

How could actual fiscal policy have caused these problems? The use of Keynesian countercyclical fiscal policy means that balancing government expenditures and receipts—a balanced budget—takes on secondary importance. The cumulative effect of unbalanced budgets is ignored by this approach. As we have seen, antirecessionary fiscal policy requires raising government expenditures or reducing taxes, either of which tends to produce a budget deficit. For reasons that will be explained more fully in Chapter 32, the history of fiscal policy in the United States has been one in which, since World War II, annual deficits have outnumbered annual surpluses by a margin of almost twenty to one, a policy that Keynes probably would not have supported. A deficit implies that government expenditures exceed its revenues, further implying the need for creation of money and credit. Credit creation, in turn, has implications in the money markets of the macroeconomy, the most important of which is inflation.

A major deficiency of the income-expenditures model we have employed to this point is that it excludes monetary variables, particularly prices. Therefore, it does not tell all of the macroeconomic story. Prices affect both consumption and investment spending, which in turn affect macroeconomic activity. In other words, simple Keynesian economics leaves us ill equipped to adequately analyze recent experience with unemployment and inflation. A more complete Keynesian theory of aggregate demand, one including prices and other monetary factors, will help us better understand today's macroeconomic problems. This expanded theory will be presented in Chapters 25 and 26.

Summary

1. A multiplier is calculated by dividing a change in income by the change in autonomous expenditures that caused the change in income. The multiplier is the reciprocal of the marginal propensity to save, or $1/MPS$.

2. A multiplier may be applied to any change in autonomous expenditures. For example, if the autonomous portion of consumption expenditures increases, income changes by the value of the expenditures multiplied by the reciprocal of the MPS.

3. The accelerator principle means that investment depends on the role of change in output over time.

When the accelerator principle is combined with a multiplier, business cycles—fluctuations in business activity over time—are created.

4. Fiscal policy consists of budgetary action taken by the central government. More specifically, fiscal policy means congressional action to adjust either the level of taxes or the level of government expenditures to achieve a stable level of growth.

5. An increase in government expenditures, G, is expansionary. A decrease in government expenditures is contractionary.

6. An increase in the level of taxes, T, is contractionary. A decrease in the level of taxes is expansion-

ary. Insofar as tax changes affect private spending only indirectly, a tax change is less effective than the same dollar amount of change in government expenditures.

7. Output effects from changes in G or T are multiplied. Although the value of the consumption multiplier (1/MPS) is the same in both cases, the output effects are different because a $1 change in taxes affects consumption less than a $1 change in government expenditures does.

8. Keynes's macroeconomic theory and his remedies for macroeconomic ills, while probably appropriate for periods of contraction and widespread unemployment, omitted crucial monetary variables, especially prices, from primary consideration. A modern and useful theory with a more complete ability to explain inflation and other contemporary macroeconomic problems requires the consideration of such factors.

Key Terms

investment multiplier
consumption multiplier
accelerator principle
capital/output ratio

contractionary gap
expansionary gap
autonomous government
  expenditures

lump-sum tax
income tax
tax rate
exports

imports
net foreign trade
countercyclical fiscal
  policy

Questions for Review and Discussion

1. What effect will a $30 billion decrease in autonomous investment have on total expenditures if the MPC is 0.9?
2. What is the essential implication of the investment multiplier? What is multiplied when autonomous investment changes?
3. How does the consumption multiplier differ from the investment multiplier? Will a $50 billion change in autonomous consumption expenditures have the same effect as a $50 billion change in autonomous investment expenditures when the MPS is 0.2?
4. Suppose autonomous consumption expenditures are $50 billion, autonomous investment expenditures are $20 billion, and the MPC is 0.8. What is the equilibrium level of income? What happens to the equilibrium level of income if autonomous consumption falls to $20 billion?
5. Does the full effect of changes in investment occur instantly? How long should it take?
6. What determines the level of investment expenditures? Does the level of investment expenditures determine the level of income or does the level of income determine the level of investment?
7. What was Keynes's basic dispute with the classical economists? What were his recommendations for smoothing out the business cycle?
8. What fiscal policy would Keynes recommend for large-scale unemployment? What would he recommend for inflation?
9. What can government do to offset a leakage from the income-expenditures flow? What can it do to offset an expansionary injection into the flow?
10. What effect does a lump-sum tax of $20 billion have on total expenditures if the MPC is 0.9? Would it have a greater effect if MPC were 0.8?
11. Does an increase in government expenditures of $20 billion have the same effect on total expenditures as a $20 billion tax cut? Does an increase in both government expenditures and taxes by $20 billion have any effect on total expenditures? Why?

⌐\/⌐_____ ECONOMICS IN ACTION

Should Congress Be Forced to Balance the Budget?

There is much grass-roots support for a constitutional amendment to balance the budget. Since 1977, thirty-one states (four short of the required number) have called for a constitutional convention to consider such an amendment. In August 1982, the Senate approved the amendment by a vote of 69–31 with all senators present, but it was rejected in the House, 236–187, only 46 votes short of the two-thirds majority needed to amend the Constitution. Debate over the issue has not subsided, however, and opinion polls of the American electorate continue to give overwhelming support to a balanced budget amendment. What would be the

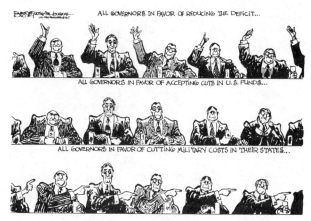

Reprinted by permission of United Feature Syndicate, Inc.

economic effects of such an amendment? Consider the diverse views of three economists, all past members of the Council of Economic Advisers.[a]

William A. Niskanen, member of President Ronald Reagan's council, defends a constitutional amendment as appropriate and workable. It is appropriate, in Niskanen's view, as simply another economic rule like those already enshrined in the Constitution regarding property rights and tax limitations. A balanced budget rule is defensible, moreover, to help prevent the economic excesses of deficits and deficit finance. Niskanen argues that a constitutional amendment would not necessarily lead to a total abandonment of countercyclical fiscal policy by Congress; a supermajority in Congress could override the balanced budget in certain economic conditions, but violation of the balanced budget rule would require broad support in Congress. As Niskanen has argued: "Fiscal rules . . . are like a dam or like a fence—they can be valuable even if they have some openings."

James V. Tobin, Yale University professor and 1981 Nobel laureate in economics who served on President John F. Kennedy's Council of Economic Advisers, argues against the proposed amendment. How, asks Tobin, could the mid-1980 yearly budget deficits of approximately $200 billion be corrected without creating another Great Depression? Aside from the immediate problem of eliminating big contemporary deficits, Tobin believes that the amendment would actually increase economic instability. Escape valves to a strictly

[a]These views are reported in "At Issue: The Balanced Budget Amendment," *Fiscal Policy Forum* (Washington, D.C.: Tax Foundation, January 1983).

balanced budget would be open, but delays could be caused by minorities in Congress, which could be very costly for the economy. Tax receipts fall during business cycle contractions as incomes fall, and they rise during expansions as incomes rise, creating built-in stabilization in the economy. These stabilizers would tend to become ineffective over the business cycle as deficits or surpluses would force Congress to enact procyclical spending and taxing policies. Deficits and surpluses, in other words, are themselves affected by business cycles. Tobin, in sum, thinks that abandonment of discretion in the use of fiscal policy would eliminate the insurance that has guarded the economy against depressions since the end of World War II.

Hendrik S. Houthakker, Harvard professor and economic adviser to President Richard M. Nixon, agrees with Tobin that a balanced budget amendment would be useless in view of the size of present deficits, but he argues that deficits must be harnessed. To Houthakker, broad-based public support for an amendment is merely an indication that the public is disenchanted with fiscal policy as it has been conducted over the past twenty-five years. Constant stimulus of the economy—witnessed by only one budget surplus (in 1969) in the previous twenty-five years—has been self-defeating. Like the excessive prescribing of antibiotics, deficits will not work now when we really need them. As Houthakker says: "We have tried to prime the pump even when it was already working overtime."

The problem, in Houthakker's view, is that the federal budget does not follow sound budgetary principles. For one thing, it makes no distinctions between capital and current accounts. For example, on the assumption that bridges and highways will last forever, no provisions have been made for maintenance of the nation's capital stock. Budget reform is needed with respect to both revenues and expenditures. Both Social Security and Great Society income transfers and entitlements have grown dramatically without adequate provisions for financing. Fiscal reform, in other words, must take place before any sort of constitutional amendment could be successfully implemented.

Question

Would the passage of a balanced budget amendment likely lead to a smaller total size of the federal budget? Would politicians, in other words, be apt to find less support if they voted for policies and programs that must be immediately financed through taxation?

25

Aggregate Demand, Aggregate Supply, and Demand Management

M any of the policies that became the legacy of Keynes and his powerful theories are still in use, although circumstances have changed greatly from Keynes's day to our own. Keynes wrote for a time when the specter of depression and the threat of permanent, widespread unemployment dominated the news. Since his time, and partly because of his contributions, our economy has undergone tremendous growth, despite many fluctuations in GNP and employment. The effects of such growth have not all been positive. Beginning in the late 1960s, price inflation has become an important practical and theoretical problem.

Most of us have experienced some effects of inflation on our consumption spending. In 1980, for example, the average cost for attending public colleges and universities was approximately $2,200 per year. Today, those same costs are approximately $3,000. If inflation remains at close to 5 percent per year until 1990, the cost of attending college then will be $4,020, nearly twice the cost in 1980.

The basic Keynesian income-expenditures model held prices constant. From here on, we will discard this assumption and consider how changes in the price level affect private consumption and investment spending. We will reintroduce the concepts of aggregate demand and aggregate supply, and we will reexamine the effects of government policy in relation to the price level.

Price Changes and Aggregate Demand

We have defined the total expenditure function, $C + I + G + (X - M)$, as aggregate demand. But an important distinction must be made between total expenditures and aggregate demand once the price level is explicitly recognized. The term *aggregate demand* will henceforth refer to the relation

553

between the price level of all output and the total quantity of all goods and services demanded. In other words, from here on, as we discuss contractionary and expansionary phases in the economy and proper macroeconomic policies to remedy such phases, we will be tying total expenditures to a specific price level, a composite index of all prices in the economy. We will then vary the price level to see how such changes affect equilibrium output or income.

The simple Keynesian model we have used up to this point assumes, correctly, that total expenditures are a function of national income. But it is also true that the price of something affects the amount that people choose to buy, so expenditures are determined by prices in addition to income.

Price Changes and Changes in Purchasing Power

Chapter 21 defined the difference between nominal, or money income, the income we receive in dollar terms, and real income, the purchasing power of those dollars. In times of inflation, nominal incomes may increase, yet real incomes may remain the same or even fall. Since inflation erodes the purchasing power of each dollar received, real income, or income adjusted for changes in the price level, is a more accurate measure of how the economy is performing.

Inflation affects more than income, unfortunately. It also affects wealth. The term *income* is usually taken to mean the sum of weekly, monthly, or annual earnings. **Wealth**, on the other hand, is the sum of all assets. Income is a flow variable; wealth is a stock variable. Flow variables are measured per unit of time. Speed, for example, is a flow variable—it is measured in feet per second or miles per hour. By contrast, weight is an example of a stock variable; it is defined independently of time. It is meaningful to say that someone weighs one hundred pounds without making any reference to time.

Wealth may be pecuniary (dollar-denominated, such as a savings account) or nonpecuniary (non-dollar-denominated, such as a house, auto, or Persian rug). The distinction between pecuniary and nonpecuniary wealth is important because the value of each type of wealth is affected differently by changes in the price level. Compare a $10,000 Persian rug with a $10,000 deposit in a savings account. If the price level doubles under 100 percent inflation, what happens to the value of these two kinds of wealth?[1] The higher price level implies that the nominal value of all non-dollar-denominated assets, including the Persian rug, has increased, assuming that prices of collectibles keep pace with the general price level. But a rug is still a rug: In real terms it is worth as much as it was before. It would now fetch $20,000 in the marketplace instead of $10,000, but the higher amount would merely purchase the same amount of goods and services as $10,000 would have previously. In other words, the real value of many nonpecuniary assets (real estate, paintings, jewels) is unaffected by changes in the price level.[2]

Wealth: The stock of monetary and nonmonetary assets.

[1]Note that inflation, strictly speaking, is a flow variable, whereas a change in the price level (or simply the price level) is a stock concept expressed at some point in time rather than as a rate over time. We simplify our discussion by sometimes using the term "inflation" when considering price level changes. The phenomenon of inflation is discussed in flow terms in Chapters 30 and 32.

[2]The real value of nonmonetary assets can change, of course. The belief, for example, that the purchase of Persian rugs would protect investors from inflation might mean brisk real demand increases. If the prices of Persian rugs were rising higher than the inflation rate, their real value would be increasing. Their increased real value would mean that they could be sold to purchase more goods and services than they were worth before the inflation.

Now contemplate the unfortunate effects of the same degree of inflation on the $10,000 savings deposit. Except in the unlikely case that the interest rate has kept pace with the 100 percent rate of inflation, the real value of the deposit will have declined. Assuming *no* interest return, the same $10,000 that would have purchased $10,000 worth of goods and services before inflation will purchase only $5,000 worth of goods and services now. In other words, price level changes bring about changes in the real value, or purchasing power, of pecuniary assets.

The Real Balance Effect

As inflation erodes the purchasing power of the money in our bank account, we have to economize—we buy fewer books and try to keep our old car running instead of buying an expensive new one. This change in our purchasing power that results from a change in the price level is sometimes called the **real balance effect.**

Real balance effect: The change in purchasing power that results from a change in the price level and its effect on wealth.

If we let MS stand for society's entire stock of pecuniary, or dollar-denominated, assets and P for the aggregate price level, then the value of real balances can be expressed as MS/P. The level of household consumption will be determined by the real balance effect as well as by the level of disposable income.

The relation between consumption and real money balances is positive, just as the relation between consumption and income is positive. If prices fall, assuming that MS remains the same, the value of real balances rises and consumers experience an enhanced wealth effect; that is, their money assets will purchase more goods and services than before. Thus, consumption will rise in response to greater real wealth. Graphically, the consumption function will shift upward, establishing a higher equilibrium level of output and employment. Conversely, if prices rise, the value of real balances thereby declines, the consumption curve subsequently falls, and lower output and employment result. Changes in real balances are capable of inducing the same kinds of changes in total expenditures as fiscal policy manipulations (ΔG and ΔT).

We have already encountered this important concept once before. Recall that neoclassical economists before Keynes generally believed that wage-price flexibility would automatically bring about total expenditure changes that would correct the cyclical swings of the macroeconomy.

The logic behind this self-correcting view was as follows: If the economy slides into a recession and unemployment occurs, labor markets will be faced with an excess supply of workers. In competitive labor markets, this excess of workers will create downward pressure on wages. As wages fall, so will the unit costs of production; prices will therefore decline as well. The general decline in prices will sooner or later raise the real value of money balances, thus inducing consumers to spend more. The increased spending will thereby stimulate output and employment, lifting the economy out of the recession. In a similar fashion, the presence of inflation will eventually lead to a reduction in real balances and a consequent decline in total spending, thereby eliminating the inflationary tendencies. For the classical economists, the real balance effect on consumption helped explain the self-adjusting nature of the economy.

This important process can be fixed more precisely in graphical terms. In Figure 25–1 three total spending functions are shown for three alternative price levels. Given some price level P_0 and some constant level of pecuniary assets in the economy, real balances will produce some level of consumer

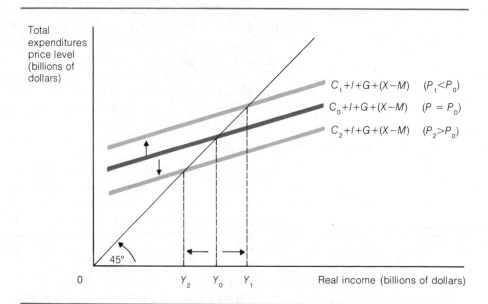

FIGURE 25–1   How Price Changes Alter Consumption Spending and Real Income

Any change in prices from their initial level, P_0, will create a real balance effect on consumption, shifting the total spending curve upward or downward. A lower price level, P_1, is associated with a higher level of equilibrium real income, Y_1, and the total spending curve shifts upward to $C_1 + I + G + (X - M)$. A price level, P_2, higher than the initial one would, for similar reasons, be associated with a lower level of equilibrium income, Y_2, and a spending curve that shifts downward to $C_2 + I + G + (X - M)$.

spending, C_0. Consumer spending level C_0 and given levels of investment, government, and net foreign expenditures, I, G, $(X - M)$, produce an equilibrium level of real income Y_0.

What happens in the event of a price decline? A price reduction will increase the value of pecuniary real balances, causing an increase in the consumption component of total spending. In Figure 25–1, the original total spending function shifts upward to $C_1 + I + G + (X - M)$, at the new, lower price level. This new, higher total expenditures function is associated with a higher level of equilibrium income, Y_1. By the same reasoning, a rise in the price level above the initial level brings a reduction in equilibrium income to Y_2.

The Interest Rate Effect

We have seen that changes in the price level affect the purchasing power of our pecuniary assets and equilibrium income. Price level changes also affect income through the mechanism of interest rates.

Interest rates are always in the news. Consumers are highly interested in whether interest rates are high or low, rising or falling. Interest rates to some extent determine purchasing decisions—whether to buy a new car or to mortgage the purchase of a new home. Businesses are also very watchful of interest rate levels when deciding whether to make capital investments— build a new warehouse, acquire new machinery, expand their inventories.

In general, interest rates depend on the price level. When the price level rises, interest rates tend to rise. When the price level falls, interest rates

FIGURE 25–2

The Interest Rate Effect

A rise in the price level may be related to a decrease in real income and employment via the process of investment spending. Conversely, a fall in the price level can lead to an increase in income. The lower half of the diagram shows that declines in the price level tend to lower the interest rate and expand investment expenditures. Increased production and jobs result.

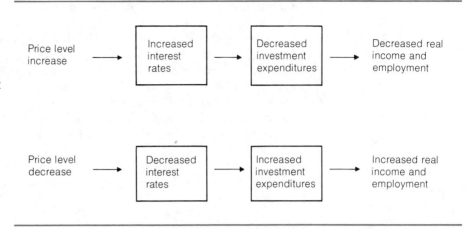

tend to fall. Of course, other factors, including the supply of and demand for loanable funds, also determine interest rate levels, but for the moment we will hold these other factors constant.[3]

As shown in Figure 25–2, changes in the price level cause changes in the interest rates; changes in interest rates cause changes in investment (or consumer) expenditures, which in turn cause changes in income and employment. As prices rise, output and real income decline; as prices fall, output and real income increase.

The **interest rate effect** can be viewed in the conventional income-expenditures model depicted in Figure 25–3. With the price at some initial level, P_0, total expenditures equal $C + I_0 + G + (X - M)$. A decrease in the aggregate price level to P_1 will lower interest rates. Other things being equal, lower interest rates will lower the cost of all new investment projects. The consequent effect in the total expenditures model is to raise the investment

Interest rate effect: The change in investment spending that results from a change in the interest rate.

[3]The details about the relation between price levels and interest rates are explained in Chapter 30. We are simplifying our assumptions here for the sake of clarity.

FIGURE 25–3

Prices and Investment Spending

Decreases in the price level from P_0 to P_1 lower interest rates and increase investment from I_0 to I_1. The higher level of investment increases real income from Y_0 to Y_1. Increases in the price level to P_2 increase interest rates and decrease investment to I_2 and income to Y_2.

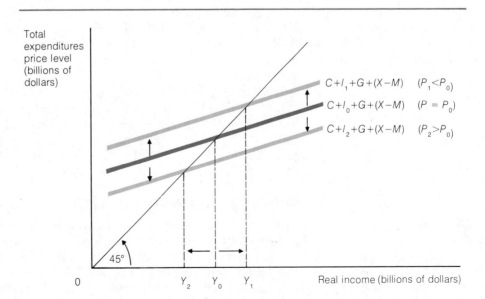

function, thus stimulating the economy to a higher level of output and employment, Y_1.

Conversely, higher aggregate price levels tend to drive up interest rates. Now all new investment projects will be more costly than before, so a decline in the investment schedule can be predicted. In Figure 25–3, the total expenditure schedule will shift downward to $C + I_2 + G + (X - M)$ and establish a lower level of national income at Y_2.

To summarize, changes in the aggregate price level are likely to affect aggregate consumption (the real balance effect) and aggregate investment (the interest rate effect). The macroeconomy, viewed realistically, is a complex structure of relations in which the controlling variable, total expenditures, is determined by the aggregate levels of income, wealth, and prices.

Aggregate Demand Is Negatively Sloped

Aggregate demand curve: Graphical relation showing the level of demand at various price levels.

In Chapter 4, we saw that all demand functions relating prices and quantity demanded are downward sloping. This also applies to the **aggregate demand curve**, which captures the relation between the aggregate price level and real national income. The aggregate demand curve basically tells us that as the price level falls in the macroeconomy, other things being equal, households and businesses tend to buy more. Conversely, as the price level rises, other things being equal, households and businesses tend to decrease their spending. This relation can be shown by using the Keynesian income-expenditures model.

Deriving the Aggregate Demand Curve

Examine Figure 25–4, assuming that the macroeconomy is in equilibrium at Y_0. The $C_0 + I_0 + G + (X - M)$ curve in part a is constructed for a given price level, P_0 in part b. Starting from this point, we can hypothetically vary the price level up and down in part b, observing the corresponding income level produced. Matching pairs of P and Y will trace out the aggregate demand curve. For example, assume the price level falls from P_0 to P_1. The effect of this price decline will be to raise household consumption (the real balance effect) and to raise business investment (the interest rate effect). Thus, total expenditures will shift upward, establishing a higher income level at Y_1. A lower price level, P_1, therefore, matches a higher level of real income, Y_1. This matched pair of variables is plotted in Figure 25–4b as point F.

We can continue the conceptual experiment by raising the price level from P_0 to P_2. Inflation causes the consumption function to shift downward (the interest rate effect), moving the equilibrium income to Y_2. The combination P_2, Y_2 is shown in Figure 25–4b as point B. The aggregate demand curve (AD) is the locus of the three points B, A, F. Note that the curve is downward sloping, as stated at the beginning of this section.

Shifts in the Aggregate Demand Curve

In tracing out the relation between the aggregate price level and the aggregate level of income, certain variables other than price are assumed constant. These constants are numerous. They involve all of those things that shift the demand function either rightward or leftward. The following list of factors capable of shifting aggregate demand is representative but not exhaustive:

FIGURE 25–4

Derivation of Aggregate Demand from Total Expenditures

A decrease in the price level from P_0 to P_1 creates an increase in consumption and investment spending. (a) A decrease from P_0 to P_1 increases real income from Y_0 to Y_1. Increases in the price level decrease C and I, so real income falls from Y_0 to Y_2. (b) The price and income combinations trace out the aggregate demand curve: Quantities of goods and services demanded are greater at lower prices.

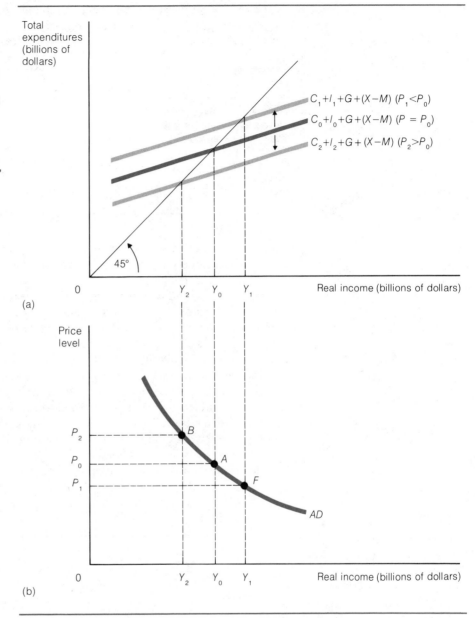

household consumption (C): autonomous component
business investment (I): autonomous component
government expenditures (G)
saving (S): autonomous component
net exports $(X - M)$
the money stock (MS)
taxes (T)

A change in any of these factors will result in a different level of national income at the same level of prices. Graphically, a change in any of these factors will create a rightward or leftward shift in the aggregate demand function. Table 25–1 summarizes the effects on aggregate demand for each change listed, and Figure 25–5 shows the shifts graphically.

To test our understanding of the principles involved, assume some change

TABLE 25–1   Shifts in the Aggregate Demand Function
A change in any of the listed variables will shift the entire aggregate demand function up or down, depending on whether the variable increases (↑) or decreases (↓).

Rightward Shift	Leftward Shift
Autonomous consumption ↑	Autonomous consumption ↓
Autonomous investment ↑	Autonomous investment ↓
Government expenditures ↑	Government expenditures ↓
Net exports $(X - M)$ ↑	Net exports $(X - M)$ ↓
The money supply ↑	The money supply ↓
Saving ↓	Saving ↑
Taxes ↓	Taxes ↑

in total expenditures, such as an increase in the net export balance. Spending increases at every price level, creating higher real income for every possible price level. In Figure 25–5, the aggregate demand curve shifts to the right from AD_0 to AD_1, illustrating that the real income at each price level increases. The same effect would occur with an increase in consumption (a reduction in saving), investment expenditures, or government expenditures.

Leftward shifts in the aggregate demand curve are likewise caused by both private-sector and government-controlled factors. Reductions in consumption (increases in saving), investment, or net exports would cause a leftward shift in the aggregate demand curve, from AD_0 to AD_2 in Figure 25–5, for example. An increase in taxes or a reduction in government spending would also shift the function leftward. These shifts occur due to a reduction in spending at every possible price level, meaning that the real income associated with any given price level falls.

As we have seen, many economic variables are capable of altering aggregate demand in the macroeconomy, thereby also altering the equilibrium levels of output, income, and employment. Some of these variables—such as

FIGURE 25–5

Shifts in Aggregate Demand

Any change in a nonprice factor affecting total expenditures will shift the aggregate demand curve rightward or leftward. An increase in saving, for example, will reduce total expenditures and aggregate demand at each price level, causing a leftward shift from AD_0 to AD_2. A decrease in taxes or an increase in government expenditures will shift the curve rightward from AD_0 to AD_1, reflecting greater spending on goods and services at every price level.

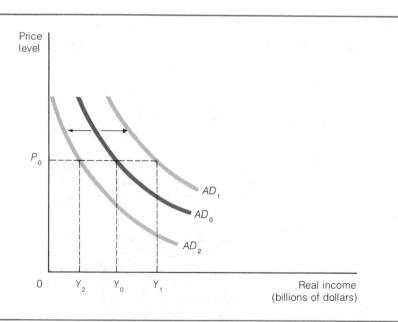

household consumption and business investment—are determined by millions of decentralized, individual decisions. Others are determined by highly centralized government directives—decisions such as the appropriate level of government expenditures and taxes, the money supply, and the interest rate. Among the latter set are various policy actions to combat the ups and downs of the business cycle.

The point of discretionary budget policies—manipulating taxes or government spending—is to stimulate the economy during periods of economic contraction and unemployment and to cool down the economy during periods of rapid expansion and inflation. Such efforts by the government to steer the economy are known as **demand management.**

Demand management: Government action that attempts to determine or change the level of total spending and thus aggregate demand.

Before we can discuss the possible effects of fiscal policy on aggregate demand, however, we must look at aggregate supply, the other major force determining the quantity of goods and services produced in the economy. We cannot determine the actual quantity of output produced through demand factors alone; we must also know what all businesses in the economy are willing to produce at alternative price levels.

Aggregate Supply

Recall from the chart of the circular flow of income in Chapter 3 (page 51) that output, or supply in the economy, is a function of inputs such as labor, land, capital goods, and capital. In Chapter 4 we saw how prices function in the economy to signal what and how much is produced. Other things being equal, higher prices generate more output.

Prices or, more generally, the price level is also the key to aggregate supply. Earlier we asserted that aggregate supply or total output was always equal to expenditures along the 45-degree line in the Keynesian income-expenditures model. But the supply of goods and services is clearly related to the price level. **Aggregate supply** is defined as the aggregate output of goods and services that will be produced at alternative price levels. The **aggregate supply curve** traces the relation between output levels and price levels at any given moment. Figure 25–6 shows three ways in which the price level might affect total output of goods and services. We refer to these three conditions as the Keynesian range, the intermediate range, and the classical range.

Aggregate supply: The level of output produced at all the various price levels in the macroeconomy.

Aggregate supply curve: Curve that traces the relation between level of output and price levels.

The Keynesian Range

In a very real sense, the shape of aggregate supply depends on the state of the economy. In the **Keynesian range** in Figure 25–6, so called because Keynes described an economy in the throes of large-scale unemployment and depression, real output can expand without any increase in prices. Why? Because all inputs, including labor, may be increased without causing price-increasing scarcities in supply. Plenty of resources are available for increases in output, in other words, when the economy is well within its production possibilities frontier.

Keynesian range: An infinitely elastic aggregate supply curve that suggests that output can adjust with no change in the price level because of large-scale unemployment.

The Intermediate Range

In the ordinary state of affairs in the economy, aggregate supply is upward sloping, or in the **intermediate range.** This means that the economy is producing close to its production possibilities frontier. As output is pushed forward, the economy begins to run out of specialized resources, and the price of these increasingly scarce inputs, such as computer technicians or petroleum engineers, rises. The result is an overall rise in the price level, as shown

Intermediate range: An upward-sloping curve that suggests that the price level and output are directly and positively related.

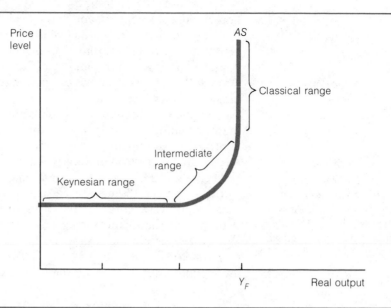

FIGURE 25–6   The Aggregate Supply Curve

The aggregate supply curve, *AS*, shows the relation between real output, measured on the horizontal axis, and the price level, measured on the vertical axis. The slope of the aggregate supply curve depends on economic conditions. The horizontal portion of the curve, called the Keynesian range, represents supply during times of severe unemployment and underutilization of resources. In this range, any increase in output will not be accompanied by an increase in the price level. The upward-sloping portion of the curve, termed the intermediate range, represents supply under normal or near normal conditions. The economy has not reached its production possibilities frontier (represented by Y_F, full employment), but neither is it depressed. In this range, any real increase in output will be accompanied by higher price levels. The vertical portion of the curve, labeled the classical range, represents the economy at full employment. In this range, output cannot increase and supply is perfectly inelastic regardless of price.

in Figure 25–6. Over this range the aggregate supply of all goods and services is positively related to the price level: Additional output can be obtained only with increases in prices.

The Classical Range

Classical range: A perfectly inelastic aggregate supply curve that suggests that the price level can adjust with no change in output because of fully employed resources.

In the **classical range**—so called because the classical economists described a full-employment economy—resources are fully employed and the economy is producing at a point on its production possibilities frontier. All available resources are fully employed, so any attempt to produce more in the economy is totally frustrated. Absolute resource scarcity at any given time means that the prices of these resources will increase in proportion to the increase in demand for them. Aggregate supply is vertical at the point of full employment. No increase in real output is possible; only inflation is possible.

Demand Management

We will return to the aggregate supply curve in Chapter 26. With the simple concepts outlined, however, we can now describe the possible effects of de-

FOCUS   Politics and Demand Management

In any discussion of government economic policy, some allowance must be made for the possibility that policy-makers will not make the "right" response; namely, the policy change consistent with accepted countercyclical economic theory. Experience tells us that in the world of interest-group politics, economic considerations are not always uppermost in the minds of policymakers. There is every reason to believe that elected politicians, like you and I, are self-interested individuals. What is it that the self-interested politician seeks to maximize? Is it the public interest or is it job security (the probability of future reelection by voters)? Most economists now believe it is the latter, so they have come to recognize that knowledge of economic theory on the part of government policymakers does not guarantee decisions justifiable in terms of that theory. Political motives, in other words, frequently dominate demand management decisions.

Consider the accepted "cure" for inflation, according to economic theory. The appropriate fiscal response is a cut in government expenditures, an increase in taxes, or some combination of the two. But how many

elected officials have the courage to vote for a tax increase in an election year? By the same token, reductions in government expenditures are politically unpopular with the interest groups that are the recipients of these expenditures. What are politicians to do? If they wish to curry favor with the electorate—and what politician doesn't?—they will predictably vote against unpopular fiscal measures even though such measures may be economically sound countercyclical actions. Proof of this principle is the fact that the United States has incurred budget deficits in every year but two since World War II. If economic theory had dictated the deficits, the existence of the deficits would imply that the U.S. economy has been in a quasi-perpetual state of recession, with Congress continually trying to spend us out of the slump. Historically, of course, we know that this has not been the case. For many of the postwar years, Congress has actually pursued procyclical rather than countercyclical policy. The reason is that in many of those years, politicians were unwilling to pay the political price for an unpopular though analytically sound economic policy.

---

mand management on the price level. For more detail, also see Focus, "Politics and Demand Management."

Demand Management and Aggregate Supply

Ordinarily, the economy's aggregate supply curve is positively sloped, meaning that, like the supply curves of individual products, higher price levels for all outputs are associated with higher levels of production. In Figure 25–7, such a supply curve is reproduced along with the negatively sloped aggregate demand curve developed earlier in the chapter.

Equilibrium output occurs initially at the intersection of the aggregate demand curve AD_0 and the aggregate supply curve AS along the ordinary, upward-sloping intermediate range of the supply curve. A level of real output or income is produced within the economy in the amount of Y_0 at a price level P_0. However, aggregate demand level AD_0 is insufficient to provide a level of full-employment output, designated Y_F. What kind of macroeconomic policy would be called for? For fiscal policy, the government would attempt to stimulate spending by reducing taxes and thereby increasing consumption, by increasing government spending, or by stimulating consumption and investment spending by increasing the money supply and reducing interest rates. Demand management consists of manipulation of these politically controlled variables to shift AD_0 to AD_1, or to move demand in the opposite direction to combat rapid expansion and inflation. By fine-tuning the economy by shifting aggregate demand either rightward (to stimulate economic activity) or leftward (to cool down economic activity), the government thus manages aggregate demand to achieve economic goals of full employment and maximum economic growth.

FIGURE 25–7

Fiscal Policy Shifts the Aggregate Demand Curve to Combat Recession

Assuming a positively sloped aggregate supply curve, *AS*, an increase in government spending or a decrease in taxes will shift the aggregate demand curve rightward from AD_0 to AD_1. The increase causes prices to rise as real income and employment rise to the full-employment level.

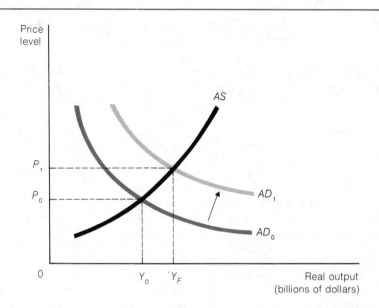

Effects of Demand Management and the Shape of Supply

Notice in Figure 25–7 that an increase in aggregate demand and an increase in real income and employment accompanied an increase in the price level. Whether or not prices will rise with increases in aggregate demand depends on the shape of the aggregate supply curve.

Figure 25–8 depicts aggregate demand shifts under alternative economic conditions summarized in the shape of the aggregate supply curve. With widespread unemployment of resources, as in the Depression of the 1930s, the economy is performing well within its production possibilities frontier. In Figure 25–8 an increase in aggregate demand from AD_0 to AD_1 through government fiscal or monetary policies will have the effect of increasing real output and employment *without* creating inflation by raising prices. Why? Because widespread resource availability permits increased real output without putting pressure on resources and output prices. This situation is close to the one Keynes described for the Depression world of the 1930s and 1940s. Under these extreme conditions, demand management could achieve increases in output and employment without inflation.

The opposite extreme is also possible. When resources are fully and most efficiently employed, the economy is operating at or very near the production possibilities frontier. Given the state of technology, resources, and institutions, there is some maximum possible amount of real output and employment. If the economy is at this maximum point, the aggregate supply curve is vertical, as represented by the vertical section of the supply curve, the "classical case" in Figure 25–8. With a vertical supply curve, an increase in aggregate demand created by monetary or fiscal policies will be inflationary. In an increase of aggregate demand from AD_4 to AD_5, for example, real output would remain at a level Y_4 no matter what sort of discretionary demand management policies are followed. Demand management is unnecessary in the classical case because, as we saw in Chapter 23, the economy is self-adjusting and will produce full employment and maximum output automatically. Economists adhering to classical principles or to modern versions

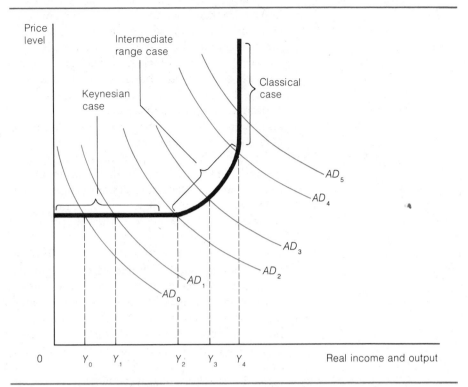

FIGURE 25–8   The Effects of Demand Management and Alternative Shapes of Aggregate Supply

The effects of equal changes in aggregate demand on prices and output will be determined by the shape of the aggregate supply curve. Given the Keynesian case of depression and widespread unemployment, an increase in demand will increase real income with no increase in prices. In a classical world of fully employed resources, an identical increase in aggregate demand will simply cause inflation with no increase in income. In the classical case, output cannot increase in real terms; it can, however, increase in nominal terms; in other words, the economy suffers inflation with no increase in output. In the intermediate case, both prices and real income respond somewhat to aggregate demand increases.

of classical macroeconomics therefore argue that discretionary demand man-agement by government is unnecessary and possibly harmful to the economy.

Between the two extremes lies an intermediate case. Increases in aggregate demand will put some pressure on prices along with some increase in real national income and employment. This situation was depicted in Figure 25–7 and is shown as the "intermediate case" in Figure 25–8. An increase in demand from AD_2 to AD_3 will create some increase in prices, but real output and employment will also rise, from Y_2 to Y_3. The economy's produc-tion possibilities frontier is not reached with increases in demand. However, shortages will develop in some resource markets, creating bottlenecks in pro-duction and price increases. As the economy approaches full employment and the production possibilities frontier, prices will ordinarily rise at a faster rate and output increases will be smaller and smaller as resource shortages become more acute.

In sum, the effectiveness of managing demand by monetary or fiscal means depends in part on the shape of the aggregate supply curve. In the Keynesian case, reflecting depression conditions, demand changes will have their full impact on real income and employment with no effect on prices. In the classical case, demand management is virtually useless in affecting employment or output: Its sole effects are on the price level. The intermediate case contains elements of both Keynesian and classical conclusions. Aggregate demand changes will change real output and employment as well as the price level. The government's ability to adjust spending and aggregate demand to contractionary or expansionary gaps—to control unemployment and inflation and to promote economic growth—is therefore partially limited by the shape of the aggregate supply curve.

The effectiveness of demand management may also be limited by a condition called *stagflation*. As discussed in Economics in Action at the end of this chapter, stagflation (stagnation plus inflation) occurs when aggregate demand first rises and then declines, causing price increases that stick at an inflationary level, coupled with widespread unemployment.

Possible Limits to Discretionary Demand Management

In addition to exteme shapes of the supply curve or the presence of stagflation, demand management by the government may be limited in its effectiveness by the realities of private decisionmaking. About two-thirds of annual total expenditures in the economy come from the private sector—households and businesses. While the private sector may be said to have an ability to spend roughly equivalent to its income and wealth, its desire to spend is not subject to direct government control. Thus it is possible, in general terms, for any government policy designed to accelerate or decelerate total expenditures to be thwarted by cumulative individual decisions to reduce or increase spending. In the context of the simple Keynesian model, for example, an increase in government spending could be partially or totally offset by a decrease in some autonomous component of private expenditures. By and large, the limits to discretionary demand management are related to expectations in the private economy and to the supply behavior of workers and other resource suppliers in the economy.

Rational Expectations. One of the powerful influences on private expenditure decisions that we have ignored up to this point is expectations. Individuals generally base their current spending and saving decisions not only on circumstances they know to exist at the time but also on what they expect to happen in the near term. If people expect inflation in the future, they may purchase more goods now rather than wait until prices rise. Or they may alter their spending pattern to buy more goods capable of retaining their value during inflation. Similarly, if policymakers announce a concerted effort to curb inflation and back it up by a credible fiscal policy, people may believe they will succeed and may lower their inflation expectations accordingly.

Rational expectations hypothesis: A view of economic behavior that suggests that people completely and rationally anticipate changes in economic variables, including those created by government.

The idea that people base their financial decisions on their expectations of what will happen in the future is called the **rational expectations hypothesis.** It assumes that people behave rationally in processing all information available to them. This information includes past inflationary experiences, understanding of the impact of expected policy actions related to deficits, government spending, and fiscal policy, and knowledge of macroeconomic relations within the economy.

The theory of rational expectations has interesting implications for the success of fiscal policy. In the extreme, it leads to the conclusion that publicly announced or accurately predicted demand management policies can have no effect whatsoever on real output or employment. Suppose the economy confronts a contractionary gap and announces huge government expenditures to stimulate the economy. Under the rational expectations hypothesis, households and businesses will anticipate that the deficit created by massive government spending will put upward pressure on interest rates and downward pressure on real balances and the profitability of investments. People will also anticipate future taxation to pay for the deficit. Their response—weighing all possible effects of the deficit—will be to cut back on household consumption and business investment, thereby sharply reducing the effects of the planned countercyclical policy.

At present there is widespread disagreement among economists on the practical significance of the rational expectations hypothesis. Although few economists accept the idea that demand management has no effect on real economic variables unless conducted in secrecy, the idea that you can't fool all of the people all of the time is winning acceptance. It is an aspect of economic policy that should not be ignored, and we return to it in Chapter 31.

Aggregate Supply Impact. One important trend in macroeconomic theory and policy over the 1970s and 1980s was the growing recognition that excessive and prolonged use of demand management might adversely affect aggregate supply. That is, discretionary aggregate demand changes may produce offsetting shifts in aggregate supply. Although we have not yet fully developed the notion of aggregate supply, it is important to note that certain fiscal measures, particularly tax measures, have the potential to alter incentives so that workers may respond to tax changes by producing more or less goods and services.

Supply-side economics: The attempt to stimulate real production by providing incentives to workers and businesses.

The Reagan administration made **supply-side economics**—the attempt to stimulate real production by providing incentives to workers and businesses—a hallmark of its economic policy. The main issue it focused on was the income tax structure. Extremely high marginal tax rates (say 40 percent) mean that income earners are allowed to keep only 60 cents out of each extra dollar earned. Under such taxation, many individuals may choose to work less, especially if the personal value to the worker of time off is more than the extra after-tax income earned. By contrast, lower marginal tax rates may stimulate workers to work more. In the former instance, aggregate supply would decline; in the latter, it would increase.

As supply issues such as the effects of the tax structure indicate, it is dangerous to treat aggregate demand in isolation. For this reason, a more complete discussion of fiscal and monetary policy must be postponed until the supply side of the macroeconomy is developed in detail in the next chapter.

Summary

1. The simple income-expenditures model is limited because it ignores monetary variables and price changes. These factors have become increasingly important since the late 1960s owing to inflation.
2. Price level changes affect household consumption by changing the real value of money balances held by consumers. This real balance effect means that price increases will decrease consumption spending and that price decreases will increase consumption

spending. These spending changes affect the level of real income and employment, meaning that price changes may be related to the aggregate demand for output.
3. Price changes also alter business investment expenditures via the mechanism of interest rates. Investment spending will increase with declines in interest rates and decrease with increases in interest rates. The interest rate effect means that real income and

employment are negatively related to prices through investment spending.

4. The aggregate demand function relates national income and output to changes in the price level; it is a downward-sloping function, like the demand curve discussed in Chapter 4.

5. Changes in both private and public nonprice variables (autonomous consumption, imports, government spending, taxation) cause the aggregate demand curve to shift its position relative to the price and income axes.

6. The effects of demand management by government depend on the shape of the aggregate supply curve. In a Keynesian range, demand increases cause increases in real income but not in prices; in the classical case, demand increases are simply inflationary with no increase in income and employment; in the intermediate case, both prices and real income increase with increases in demand.

7. One of the deficiencies of the simple Keynesian total expenditure model is that it cannot explain price effects because prices are not included in the model. The aggregate demand–aggregate supply model helps cover this gap, permitting explanations for both inflation and stagflation, the combination of inflation and a high level of unemployment.

8. The prospects of successful demand management by government fiscal or monetary policies may be limited if economic participants correctly or rationally anticipate the effects of such policy. Demand policies, moreover, may have supply-side effects by affecting incentives to work and invest. These actual and potential effects became part of economic policy and debate in the late 1970s and 1980s in the United States.

Key Terms

wealth	aggregate demand curve	Keynesian range	supply-side economics
real balance effect	demand management	intermediate range	
interest rate effect	aggregate supply	classical range	
real money supply	aggregate supply curve	rational expectations hypothesis	

Questions for Review and Discussion

1. What components of aggregate demand are omitted in the Keynesian total expenditures model?

2. Explain how a decrease in the price level would restore full employment. Is this explanation consistent with the classical or Keynesian view of the economy?

3. What effect does an increase in the price level have on the interest rate, the level of investment, consumption, and total output? Are these changes induced by the interest rate effect or the real balance effect?

4. In deriving the aggregate demand curve, what macroeconomic variables are being held constant? What would happen to the aggregate demand curve if these variables changed?

5. Explain why you agree or disagree with the following statement: "Keynesian fiscal policies are effective only when there is widespread unemployment, and classical fiscal policies are effective only when there is full employment."

6. When does the Keynesian aggregate supply curve best represent economic conditions? When does the classical aggregate supply curve best represent economic conditions?

7. Can fiscal policy effectively eliminate both unemployment and inflation?

8. If the government attempts to increase aggregate demand, what changes in the economy could render the government's fiscal policies ineffective?

9. Do changes in government spending and taxes affect only aggregate demand? What actions by government could simultaneously increase aggregate demand and decrease aggregate supply? When this occurs, what happens to real output and prices?

10. Does fine-tuning the economy require both demand and supply management? Does the rational expectations hypothesis suggest that the government can achieve this goal?

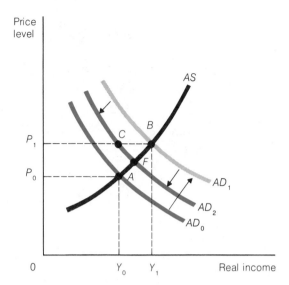

ECONOMICS IN ACTION

Uses of Demand and Supply Analysis: The Problem of Stagflation

Aggregate demand and supply theory has an obvious advantage over the Keynesian total expenditures approach in explaining price inflation. Aggregate demand and supply theory helps predict what effects fiscal policy will have on prices. Assuming that the aggregate supply curve is in the intermediate range (see Figure 25–8), aggregate demand changes will either increase prices along with increasing output and employment or they will decrease prices with accompanying decreases in jobs and production. In other words, there may be a built-in trade-off between inflation and employment. We will discuss this phenomenon in detail in Chapter 32, but consider still another possibility—stagflation, inflation accompanied by unemployment or stagnation. This condition appears to have existed in the United States between 1969 and 1970, 1971 and 1972, 1973 and 1974, 1975 and 1976, and, most recently, 1979 and 1981.

Stagflation can be analyzed in terms of demand and supply curves such as those developed in this chapter. In Figure 25–9, point A represents a recessionary level of output and employment in the economy. In response to high unemployment, government increases spending or decreases taxes in an effort to shift the aggregate demand curve rightward from AD_0 to AD_1. Prices rise, creating inflation, but output and employment also increase. Now assume that some other component of private expenditures—consumption, investment, or exports—decreases, causing a shift leftward in the aggregate demand curve from AD_1 to AD_2.

The result is that prices fall along with output to point F. But what if union or monopoly pressures mean that prices are too "sticky" to fall at all in response to the decreased demand? In this case, prices remain at high levels, and the full impact of the demand decrease falls on employment and output. Production falls as the economy stagnates, but inflationary pressures do not diminish. In terms of Figure 25–9, the economy would move to point C rather than to point F. Thus inflation and joblessness might go hand in hand.

While there are still other interpretations of stagflation (see Chapter 32), the problem illustrates the usefulness of a macroeconomic theory—aggregate demand and supply analysis—that includes prices. In the post-Keynesian period, attention has focused on price infla-

FIGURE 25–9   Aggregate Demand Shifts and Stagflation

The phenomenon of stagflation is produced by an increase in aggregate demand from AD_0 to AD_1 caused by some fiscal action (government spending increase or a tax reduction) and a reduction in some component of private spending (autonomous investment, consumption, or exports). Prices, however, are sticky and fail to fall with the demand reduction (from AD_1 to AD_2), creating stagflation—inflation and unemployment at the same time.

tion as well as on employment in considering aggregate demand management as a means for fine-tuning the economy. The fact that stagflation poses a possible trade-off between inflation and unemployment illustrates some of the problems associated with the use of fiscal policy to control demand and to stabilize the economy.

Question

Does the apparent trade-off between inflation and employment observed in the U.S. in recent decades mean that fiscal policy is inadequate in controlling and stabilizing economic activity?

26

Aggregate Supply

C urrent macroeconomic theory of aggregate supply owes much to earlier classical theories. Classical economists such as Adam Smith believed that the supply of productive resources such as labor and capital determined a nation's prosperity and economic growth. Early macroeconomic theory was based on the premise that "supply creates its own demand" and that economic policy that stimulates productivity and investment would at the same time allow the economy to self-adjust to a level of economic growth with full employment and without inflation.

Today, in the debate over economic policy, more and more attention is being paid to the concerns first expressed by Smith and other classical economists. Supply-side economics, a popular term during the early days of the Reagan administration, is based on the idea that overreliance on aggregate demand management, in the form of taxation and spending policies, has had ill effects on the growth of aggregate supply and has fostered much economic instability.

Our study of aggregate supply is broken into three parts. First, we will look at the various factors in the economy that can cause the aggregate supply curve to shift and thereby raise or lower the economy's productive capacity. The most important such factor is the price of labor, or the wage rate, which is determined through the macroeconomic market for labor.

Next we will look at three different models of how aggregate supply responds to changes in the price level: the long-run model, the short-run model, and the natural rate of unemployment hypothesis, a modern variant of the long-run supply curve. Finally, we will try to bring our analysis of aggregate supply and aggregate demand together in evaluating some of the important policy questions facing politicians and voters.

Factors Determining Aggregate Supply

Output, or supply, in the economy is a function of several inputs, including labor, land, capital, capital goods, and entrepreneurship. When these inputs are relatively scarce, the aggregate supply of goods and services in the economy will be lower than when these inputs are relatively plentiful. In Chapter 4 we saw how prices function in the economy to signal what and how much gets produced. Other things being equal, higher prices generate more output.

Prices or, more generally, the price level is the key variable in aggregate supply. **Aggregate supply** is the total output of goods and services produced at various price levels. The aggregate supply curve traces the relation between output levels and price levels; Figure 25–6 (page 562) shows the aggregate supply curve under three separate economic conditions. Our previous discussion of the aggregate supply curve, however, sidestepped some of the reasons for the alternative shapes of the supply curve and some of the factors that can cause the curve to shift right or left, resulting in more or less output at all price levels. We turn now to some of these factors.

Aggregate supply: The value of the total output of goods and services produced at various price levels.

Labor's Wages

By far the most important factor explaining the position of the aggregate supply curve is the labor market, since labor is the largest cost of production. The supply and demand for labor at any point in time are determined by conditions such as income and corporate tax rates, labor productivity (output per person-hour), aggregate preferences for leisure or work, labor laws such as the minimum-wage law, and the availability of other complementary resources. This supply and demand for labor produce an equilibrium wage rate, the labor cost to all producers. An increase in equilibrium labor costs causes the whole aggregate supply curve to shift to the left (see Figure 26–1), while a decrease in equilibrium labor costs shifts the aggregate supply curve to the right. Since labor costs result from a host of factors determining labor supply and labor demand, shifts in aggregate supply are also determined by these factors. We will explore this mechanism in more detail later in the chapter.

FIGURE 26–1

Labor Costs and the Aggregate Supply Curve

Supply is greatly influenced by labor costs, which depend directly on the real wage paid to workers. An increase in these costs will cause the whole aggregate supply curve to shift to the left from AS_0 to AS_2, whereas a reduction in labor costs increases the aggregate supply curve (shifts it to the right from AS_0 to AS_1).

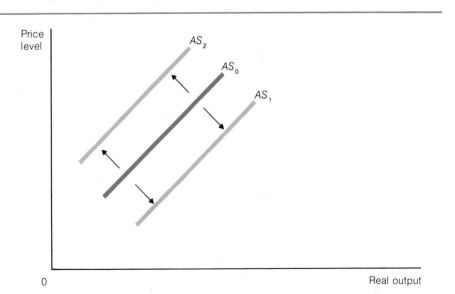

Capital and Other Inputs

The quantity and availability of all factors of production other than labor also determine business costs and therefore the shape and position of the aggregate supply curve. A reduction or expansion in the availability of raw materials—such as energy—will shift the supply curve for all goods and services. Discovery of new or larger quantities of energy resources will have cost-lowering effects on aggregate supply. Likewise, larger stocks of fixed capital assets such as machinery, equipment, buildings, and inventories will increase aggregate supply.

Productivity

Productivity: The average output produced per worker in a given amount of time.

Productivity describes the average output produced per worker per hour of work—the number of shoes, shirts, books, and so forth produced in a given time period divided by the number of workers. Increases or decreases in the productivity of the labor force as a whole shift the aggregate supply curve rightward or leftward. For instance, higher education or more on-the-job training increases labor's productivity. A larger stock of capital goods or natural resources with which labor can work also enhances productivity.

Technology

Technological change, brought on by innovations in productive methods or by the invention of new inputs, also shifts the supply curve. Improvements in technology, such as the use of contour plowing, the discovery of aluminum, or the development of the digital computer, can shift the aggregate supply curve to the right.

Output Restrictions and Institutional Change

Changes in the institutional structure affecting production is also a factor that can change aggregate supply. Increases in monopoly or union power cause output or input restrictions, reducing production possibilities and aggregate supply. Regulations on businesses that increase the costs of production also have that effect. Note, however, that regulations by such agencies as the Environmental Protection Agency (EPA) or the Occupational Safety and Health Administration (OSHA) may increase the quality of output while decreasing its quantity.

The macroeconomic effects of a number of these factors will be analyzed later in this chapter and in the discussion of economic policy in Chapters 31–34. But remember that a change in any of these factors will shift the entire aggregate supply curve either rightward or leftward at all possible price levels. (For a recent example of a major supply shift, see Focus, "Aggregate Supply Shifts: OPEC in the 1970s and 1980s.")

The Market for Labor and Other Resources

Each of the factors listed in Table 26–1 can increase or decrease, causing the aggregate supply curve to shift rightward or leftward. The question before us is what causes an increase or decrease in labor's wages, in productivity, in monopoly power, or in any of the other factors. By analyzing these causes, it is possible to understand the economic roots of aggregate supply.

Delving into the *microeconomic* causes of aggregate supply, however, is an enormous task. We cannot repeat here all of the microeconomic analysis devoted to topics such as labor's costs, monopoly power, and the market for capital. We can, however, briefly explain the supply and demand function

Aggregate Supply Shifts: OPEC in the 1970s and 1980s

When the supply of any resource or the level of technology changes, the production possibilities of the economy expand and the aggregate supply curve shifts. This means, in effect, that for any given level of labor employment, more or less output can be produced. Supplies of resources can change for a variety of reasons. For example, the oil embargo imposed by the Organization of Petroleum Exporting Countries (OPEC) in the 1970s decreased the amount of oil available for the production of goods and services in the United States. Labor and other productive factors suddenly had less energy to work with, indicating that the productive effects of labor and other resources were lower, which in turn reduced the demand for labor. OPEC's action thus caused a "supply shock"—an unexpected, once-and-for-all leftward shift in aggregate supply, as illustrated in Figure 26–2.

A shift in aggregate supply from AS_0 to AS_1 in Figure 26–2, given a stable aggregate demand curve AD_0, meant that the price level rose permanently and output and employment declined. The emergence of the OPEC cartel contributed to the inflationary and recessionary pressures observed in the United States and in the world economy over the 1970s.

The first half of the 1980s witnessed a reversal, though not a complete reversal, of the OPEC supply shocks of the 1970s. Owing partially to price cutting by OPEC members and to OPEC's inability to maintain and police output restrictions, the cartel has lost much of its power. As a result, there is an increased availability of oil at lower prices to United States producers and consumers as represented by the rightward shift in aggregate supply to AS_2 in Figure 26–2. To the extent that the formation of OPEC contributed to the inflationary and recessionary trends of the 1970s, its partial dissolution has helped foster lower inflation rates and higher output levels in the 1980s. The economy may

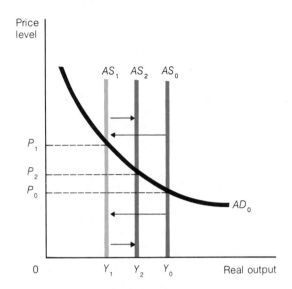

FIGURE 26–2   Reduced Availability of Oil Shifts Aggregate Supply Leftward

The imposition of an oil embargo in the early 1970s reduced the productive capacities of the United States in a supply shock, a sudden leftward shift in the aggregate supply curve. The effects were an increase in the price level from P_0 to P_1 and a reduction in real income and output from Y_0 to Y_1.

of course respond to supply shocks by developing alternative resources, a reaction observed many times throughout history. Part of the rightward shift to AS_2 has undoubtedly been due to the adaptation of production to the initial supply shock. New resource development, such as solar energy, and alternative production techniques would, for example, help reduce the impact of the initial OPEC-induced supply shock.

for the input of labor as an example of how macroeconomics and microeconomics are intimately related. By looking at the price mechanism for the labor market, we will have a clearer idea of how the overall aggregate supply curve is derived.

As indicated above, the labor market plays a primary role in determining aggregate supply. There are two key questions to be answered in analyzing the labor market: Why do firms demand labor? and Why do workers supply labor?[1] In other words, what motives underlie the decisions to hire labor and to perform labor? The conclusions we reach in this section generally apply

[1] These questions were analyzed in Chapter 12, where the microeconomic behavior of labor input demand and supply was discussed in detail.

TABLE 26–1   Factors Affecting Aggregate Supply
When any of these factors increases or decreases, the aggregate supply curve shifts to the left or to the right.

Factor	Effects
Wage rate	An increase in labor costs will shift supply curve to left; a decrease in labor costs will shift supply curve to right.
Capital goods and raw materials	An increase in capital assets will shift curve to right; discovery of new resources such as oil or natural gas will shift curve to right.
Productivity	Tax incentives, better education, and more job training will increase productivity and shift supply curve to the right.
Technology	New inventions, such as the digital computer, will shift curve to right.
Output restrictions; institutional change	Increases in monopoly power will shift aggregate supply curve to left. Regulations that increase costs of production will shift curve to left. Regulations that enhance productivity will shift curve to right.

to all inputs—to nonhuman resources as well as to human resources. We are using labor as an example of the mechanisms that determine aggregate supply.

Why Do Firms Demand Labor and Other Resources?

In a competitive system, firms hire labor and all other resources to make profits; more fundamentally, firms hire labor to produce output, the sale of which persumably will earn a profit. A restaurant, for example, purchases raw materials and hires cooks, waiters, and so on as inputs so that an output—meals—can be produced and sold. Each firm hires a certain number of workers on the basis of labor's productivity—that is, on the basis of how much physical output per hour, week, or month laborers can produce. The real cost of hiring these workers is determined not by the nominal wage rate W but by the **real wage rate:** the nominal wage rate divided by P, the price level. At lower real wages, firms can afford to hire additional workers with lower productivity. At higher real wage rates for more productive workers, fewer laborers will be hired.

This principle makes sense in the labor market and in all other factor input markets, as in the markets for capital or natural resources. The more receipts that a unit of labor or any other input adds to a firm, the more the firm will be willing to pay for the labor or other resource unit. The less income added by a unit of factor input, the less a business will be willing to pay for it.

It is important to emphasize that firms hire quantities of labor on the basis of the real wage and not the nominal wage. If the nominal (or money) wage, for example, is $4.75 per hour and the price level rises, the real wage (the quantity of goods and services that could be bought with $4.75) will fall. Since the real wage has declined, making the hiring of labor cheaper to firms, firms will hire more labor. The same would be true for a constant price level and a declining nominal per hour wage rate: The real wage falls and firms hire more labor. Naturally, if real wage rates rise, due to nominal wage

Real wage rate: The nominal wage rate, W, divided by the price level, P.

increases or price decreases, firms will hire less labor. Now let us consider the other side of the coin—the supply of labor and other inputs.

Why Do We Supply Labor and Other Resources?

We work to earn income, to buy the things that give us satisfaction, and, in the limit, to survive. In other words, it is in our self-interest to work, just as it is in firms' self-interest to hire workers. But our decisions about how much labor to supply are a bit more complex. There are trade-offs between work effort and leisure, for example, or between supplying work and raising a family. Taxes, moreover, may affect the amount of work effort we supply. We may choose higher education to enhance our productivity, but to do so we cannot work much in the present.

In general, however, we are likely to work more if we are offered higher wages. While it is true that at a very high real wage rate, many individuals would be expected to forgo additional income for additional leisure time, ordinarily we do not assume that a whole society of workers has reached this fortunate point. Economists assume that, in general, increases in real wages will provoke more work effort rather than less.

What is true of labor is also true of all other inputs. Higher real returns to owners of all other resources will encourage them to supply larger quantities of the resources. For example, higher real prices of pine lumber will encourage greater production in the forests of Washington or Georgia; higher steel prices will cause more steel to be produced and sold in the United States or elsewhere.

The Importance of Resource Market Equilibrium for Aggregate Supply

To see how these forces determine real returns to labor and to all other factors, we need merely to put the concepts of demand and supply together, concentrating on the labor market. In the labor market, equilibrium occurs when the quantity of labor that workers would supply at some real wage exactly equals the quantity businesses will demand at that real wage. This equilibrium determines the quantity of labor (and all other resources) that will go into production, and thus it determines the aggregate supply in the economy at any one time and under given circumstances.

As we saw earlier in this chapter, a change in conditions affecting the labor market or any other resource market will change demand or supply in these markets. These changes will also affect the real returns to resources, altering costs to producers and shifting aggregate supply (see Figure 26–1).

The important point is that the real wage rate and the equilibrium supply of labor are freely responsive to changes in the desires of workers to supply work effort and to factors such as technology, invention, or education affecting labor's productivity and, therefore, firms' desire to hire labor. These basic factors determine how much labor will be used in the economy at any one time.

Before turning to the impact of wages and other factor costs on the slope of the aggregate supply curve, an important qualification is in order.

Money Illusion: Concepts of Real and Nominal

Both the demand for labor and the supply of labor have been related to the real, not the nominal, wage. Practically, what does this mean? From the demand side, firms are not fooled by nominal wages. They will react to price-level changes by changing the amount of labor they demand. For instance, a decrease in the price level raises the real wage paid, so firms will reduce the quantity of labor they demand. (But if wages are flexible and fall in the same proportion as the price level, the new equilibrium real wage will equal the old equilibrium real wage. As a result, firms will have no reaction to the change in W and P and will continue to demand the same amount of labor as before.)

Likewise, suppliers of labor are ordinarily not fooled by changes in nominal or money values. A rise in the price level, with money wages constant, erodes workers' purchasing power and reduces their real wage. Other things being equal, workers will reduce the quantity of work effort they supply if the real wage falls. (A proportionate increase in money prices and wages that keeps real wages constant has no effect on the supply of labor by workers.)

Money illusion: The assumption that nominal values are actually real values and that changes in nominal wages mean equal changes in real wages.

Later in this chapter we will present some important exceptions to this assumption, exceptions we refer to as money illusion. **Money illusion** among workers means that they will supply labor on the basis of the nominal wage rather than the real wage. Thus, if the nominal wage rises with a simultaneous higher increase in the price level (meaning an actual reduction in the real wage to workers), laborers would supply more, not less, work. They (temporarily at least) perceive themselves as receiving higher wages, but they are actually receiving lower returns. For now, we assume that firms or workers cannot be fooled by nominal changes in either prices or wages. We assume that supply decisions are based on certainty about real wages and that both firms and laborers expect any real wage changes to be permanent. We now turn to an explanation of how adjustments in the market prices for labor and other inputs affect the aggregate supply curve.

The Slope of the Aggregate Supply Curve

The concept of aggregate supply relates the aggregate output of goods and services in the economy to alternative price levels of that output. But that relation will vary over time and economic conditions. Consider economic conditions. We have already seen in Chapter 25 (Figure 25–8, page 565) that during periods of depression the aggregate supply curve is horizontal: Increases in output occur at no increase in wage rates or wage cost because unemployed labor and other resources are plentiful. During periods of prosperity after full employment has been achieved in all markets, aggregate supply is a vertical line at some level of production corresponding to full employment. In an in-between or intermediate case, aggregate supply is an upward-sloping curve: The price level positively relates to the total quantity of goods and services produced and supplied.

The time period under scrutiny and the rapidity with which resource markets achieve equilibrium are also a prime concern in explaining how the aggregate supply curve is shaped at any time or under given economic conditions. We look at how two aggregate supply curves may be developed under alternative time adjustments and economic conditions.

The Classical or Long-Run Aggregate Supply Curve

The classical aggregate supply curve shows the relation between price and the output of goods and services *after all short-run market adjustments of prices and wages have taken place.* For this reason, most macroeconomists regard the classical conception as a **long-run aggregate supply curve.**

What does the long-run aggregate supply curve look like if we assume

FIGURE 26–3

The Classical Aggregate Supply Curve

Classical economists believed that the nation's output is independent of price level. If prices rise or fall, the long-run level of output, Y_F, does not change after price and wage adjustments take place, so the aggregate supply curve is vertical in the long run.

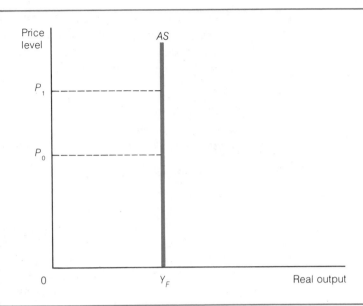

Long-run aggregate supply curve: The relation between price and the output of goods and services after all short-run market adjustments of prices and wages have taken place.

instantaneous or very rapid adjustments to changes in all markets? As Figure 26–3 shows, the long-run notion of aggregate supply is a vertical line. That is, real output and the level of employment remain exactly the same at every price level. Classical macroeconomists believed that changes in the price level will not affect supply if economic forces are allowed to work.

How, exactly, is this unique and unchanging level of supply, designated as Y_F, guaranteed? Suppose that prices rise above the equilibrium level P_0 to P_1. As a result of the higher price level, the labor market is in temporary disequilibrium. The real wage (W/P) *falls* to W_0/P_1, which is lower than the equilibrium real wage. The nominal wage rate W has not adjusted to the rise in prices. In disequilibrium, the quantity demanded of labor at the lower real wage rate will be greater than the quantity of labor supplied at that wage. And, more important, a reduced input of labor means that output or GNP will fall.

In a classical view of the macroeconomic world this situation is quickly corrected; in theory, it is corrected instantly. Since firms demand a larger quantity of labor than is supplied, they will compete for the use of the lower quantity. Such a competition will cause nominal wage rates to rise, creating upward pressure on the real wage. As the real wage rises, more laborers enter the work force and the demand for labor slacks off. Equilibrium is restored when the equilibrium real wage is restored. Note, however, that restoration of the same real wage requires that nominal wages rise so that the new real wage is exactly equal to the old real wage.

Note also that a vertical aggregate supply curve is premised on rapid upward and downward adjustment of money wages and prices. Any disequilibrium pressures bring an adjustment of wages and prices. If a lengthy period of adjustment is required to restore equilibrium, the aggregate supply curve is not vertical during the adjustment period, as we will see in the next section. For this reason, most macroeconomists regard the supply curve of Figure 26–3 as a long-run aggregate supply curve; that is, one that would result after all price and wage adjustments have taken place.

The Short-Run Aggregate Supply Curve

Keynes focused primarily on short-run adjustments on the aggregate demand side of the economy. He assumed that in the short run, prices and wages were more or less inflexible. Keynes's view of the supply side of the economy also included this assumption. For instance, in the face of falling demand, laborers would not accept temporary cuts in their nominal wage rates. (Naturally, workers would be willing to take increases.) Under such circumstances, the aggregate supply curve looks very different from the classical vertical function, for there is a "stickiness," or inflexibility, of wages or prices in the markets for inputs.

If the real wage rises above equilibrium—perhaps because union pressures increase nominal wages or because of price-lowering deflation—laborers would be unwilling to take money wage cuts, according to Keynes. Their unwillingness to accept less money would prevent or postpone the restoration of equilibrium over some fairly long time.

The portion of the aggregate supply curve that results when wages are too sticky to move downward is given in Figure 26–4 as AS_{SR}, along with the long-run classical version of aggregate supply. The upward-sloping short-run part of the curve is explained as follows: A price level of P_0 is associated with a full-employment level of output Y_F. A lower price level P_1 is associated with a lower level of output Y_0. The inflexibility of money wages, in other words, produces a positively sloped short-run aggregate supply curve. Since price level P_1 is associated with income level Y_0, the short-run curve

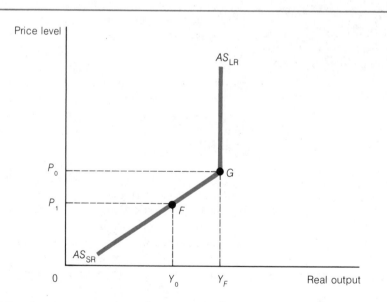

FIGURE 26–4   Short-Run Aggregate Supply Curve

An upward-sloping supply curve occurs in the short run because wages and prices are not instantly flexible but take time to rise or fall. Price level P_0 is associated with a full-employment level of output Y_F. At a lower price level, P_1, however, workers are unwilling to take cuts in their nominal wages, thereby creating a higher real wage rate. The higher real wage creates a reduced demand for labor, lowering employment and output and income. The aggregate supply curve is therefore upward-sloping in the short run, AS_{SR}, and when the economy reaches its employment potential at Y_F, the supply curve becomes vertical in the long run, AS_{LR}.

traced out is FG. The inflexibility of money or nominal wages explains the positively sloped portion of aggregate supply.

We may also think of the AS_{SR} and AS_{LR} curves in terms of the production possibilities frontier. The economy is right on the frontier when the aggregate supply curve is in the vertical (classical) position. Along the long-run curve, according to classical theory, full employment of labor and all other resources is achieved. But when there are rigidities in the economy that temporarily prohibit the full employment of all resources, production and output levels fall below the production possibilities frontier of the economy. Thus, at all points along the upward-sloping curve, AS_{SR}, the economy is not reaching its maximum potential. Over time, these rigidities tend to dissolve, meaning that the aggregate supply curve becomes completely vertical. This is why the upward-sloping portion of the aggregate supply curve in Figure 26–4 is labeled AS_{SR} (aggregate supply, short run) and the vertical portion is labeled AS_{LR} (aggregate supply, long run).

Expectations and Aggregate Supply

Another very useful view of the aggregate supply curve highlights the relation between employment of resources and the price expectations of both resource owners and firms over time. Labor again serves as our example of how this relation works.

The Natural Rate of Unemployment

Full employment: A situation in which all workers who desire employment at the current real wage are employed.

As we have stressed, the quantity of labor supplied depends on workers' perceptions of wages offered. To the classical economists, **full employment** meant that everyone who desired employment at the real wage had a job with no excess demand for labor. In this view, the real wage established by the interaction of the supply and demand for labor is the full-employment wage. Some workers would work for wages higher than the equilibrium real wage, but they are not regarded as "unemployed" because they voluntarily take themselves out of the labor force. All who would accept lower than equilibrium real wages are employed.

Today economists prefer to talk of a natural rate of employment or unemployment. We know that a certain amount of unemployment is natural in an economy at any given time. Because of changing demand conditions, resources are being constantly shifted among producers. As explained in Chapter 21, a certain rate of unemployment arising from the "friction" in the economic system is to be expected, though the precise rate is a matter for much debate among economists, and it has changed over time.

Natural rate of unemployment: The frictional rate plus all unemployment resulting from institutional changes that affect the decisions of workers to work and employers to demand labor.

The **natural rate of unemployment** is easier to define than to measure: It is the frictional rate plus all unemployment resulting from institutional changes that affect the decisions of workers to work and employers to demand labor. For example, changes in minimum-wage laws, tax laws, the number of women working, and the availability of retirement or welfare benefits are conditions that affect workers' incentives to work and the natural rate of unemployment. We will equate full employment with the natural rate of unemployment, estimated by some economists to be as much as 7½ or 8 percent of the civilian labor force.

Temporary Misperceptions of Real Wages

Some economists have argued that employment always fluctuates around the natural rate of unemployment. When combined with a theory of worker and employer expectations, the natural rate of unemployment is used to explain business cycles of inflation and recession. To see how this simple idea can be so powerful, we must integrate the concept of expectations with the workings of the labor market. The single difference between this aggregate supply theory, called the **natural rate hypothesis**,[2] and all of our earlier models of the labor market is that in the earlier models both employers and workers were assumed to instantly perceive changes in real wages and to adjust their demands for labor or supplies of work effort accordingly and instantly.

According to the natural rate hypothesis, however, expectations of workers do not adapt instantaneously to changed circumstances. Instead, their **price expectations**—their expectations of what prices or inflation will be in six months, one year, or two years, for example—may temporarily reflect misperceptions, what we earlier called the *money illusion*. Suppose, for instance, that there is an increase in prices and nominal wage rates. Workers may adapt slowly to the change in prices because they expect prices (or the inflation rate) to maintain their former level or to increase more slowly than they actually do. At the same time, workers may correctly perceive that their nominal wage, W, is rising. If workers expect prices to remain constant or to rise more slowly than their nominal wage and at the same time perceive that their nominal wage is rising, in effect they believe that their real wage is increasing. As argued earlier, an increase in the perceived real wage, even though it is a misconception, will increase work effort, shifting the labor supply curve to the right.

This money illusion may not affect firms, though. Employers who encounter rising prices for their products believe correctly that the real wage they must pay for labor is falling. A drop in the real wage would occur if prices of final output rise faster than the nominal wage rate. For reasons outlined earlier, a fall in the real wage rate means that employers demand a greater quantity of labor.

Consider the effect of this one-sided illusion. Labor suppliers recognize the rise in their money wages but do not yet perceive that prices are rising at a faster rate. In spite of the reality of a lower real wage, workers think that their real wage is higher, and they temporarily supply more, not less, labor. On their part, firms—labor demanders—will hire additional workers because the real wage has actually declined. As a result, the supply curve of labor temporarily shifts rightward, and a higher quantity of laborers is employed. This quantity is greater than the natural level of full employment, a situation that is sometimes termed **overfull employment**.

Overfull employment is not a permanent state of affairs. Workers slowly begin to catch on about rising prices. When the prices of the goods and services purchased by workers rise perceptibly, the workers demand higher money wages. The real wage thus rises once more to its initial (higher) equilibrium rate. As wage expectations of workers are revised upward, the supply curve of labor shifts back to its initial position.

This mechanism also works in reverse. Should prices and therefore nominal wages fall, workers would temporarily believe that their real wage had fallen. If prices were falling faster than nominal wages, however, the real wage would actually be rising, a situation that firms correctly perceive. Labor supply would again shift temporarily, but this time in a leftward direction.

Natural rate hypothesis: Workers do not adapt instantly to changes in the demand for labor or in the wage rate but weigh their responses to changes in light of their expectations of future price and wage conditions.

Price expectations: Anticipations of future prices or inflation.

Overfull employment: A temporary situation in which the quantity of laborers employed is greater than the level of full employment.

---

[2]The natural rate hypothesis was initially developed by Milton Friedman in his presidential address to the American Economic Association: "The Role of Monetary Policy," *American Economic Review* (April 1968), pp. 1–17. Also see Friedman's *Price Theory* (Chicago: Aldine, 1976), pp. 227–229.

A level of employment below the natural rate would be temporarily generated, causing a short-run fall in the level of output and income. A lower price level would therefore be associated with a lower level of employment and output in the short run. When price expectations finally caught up to the actual price decline, workers would be willing to respond to the (actually) higher real wage. The supply curve of labor would shift rightward, reestablishing a natural rate of employment. In Chapter 32, we will see that this behavior in the labor market helps explain business cycles of inflation and recession and the so-called trade-off between inflation and unemployment.

Long-Run Labor Agreements and Contracts

Some economists argue that misconceptions or "fooling" of laborers is not necessary to explain the slow adjustment of nominal wages to changed aggregate demand conditions over a short-run period. In this view, laborers can be "fooled" about price level changes only for short periods of time—a few weeks or months—since price level changes, such as inflation, are felt immediately at the grocery store or at the gas pump. Certainly such news is prominent on nightly TV news. Workers, in short, catch on quickly.[3]

Although workers may be temporarily fooled into supplying more or less labor due to misconceptions of their real wage, other reasons may be stressed for the slowness of nominal wages to adjust to changed demand conditions. To a certain extent, there are forces at work on the part of both firms and workers to avoid frequent or unnecessary job changes. Employees typically spend considerable time and money for on-the-job training specific to their needs. These trained workers are paid a premium (over untrained workers), a premium that workers might lose by moving to another job or occupation. For this reason, during periods of rising demand and lagging nominal wages, workers will hold out and actually work harder in anticipation of future nominal wages adjustments to an increased price level.

Additionally, a large number of industrial workers are covered by union contracts—often lasting up to three years—that also contribute to sticky wages. These union contracts are designed to avoid strikes and costly and frequent negotiations with workers. They remain in force whether aggregate demand rises or falls. Though these contracts often contain escalator or cost-of-living clauses which provide for nominal wage increases over the contract period (often through some tie with an index of prices), they seldom provide for a full increase. Individual firms will not ordinarily agree to fully tie the wages paid their laborers to a price index since the increase in aggregate demand which precipitated the higher price level may not materialize for their own product. A computer firm which made such an agreement, for example, might find itself with higher wage costs and no increase in the nominal demand for computers.

Wage contracts are not tied to aggregate demand conditions, and nominal wages are only partially adjusted to changing demand and price conditions. Aspects of the employer-worker relationship and long-run wage contracting, in addition to short-run misconceptions of workers, help explain the stickiness of nominal wages over short-run periods of time. Just as expectations catch up with actual price behavior, workers will also realize nominal wage increases to match price level increases when union and other bargaining

---

[3]A fuller discussion of these issues may be found in Robert J. Gordon, *Macroeconomics*, 3rd ed. (Boston: Little, Brown and Company, 1984), pp. 209–214.

FIGURE 26–5

Aggregate Supply Adjusted for Expectations

When price level rises from P_0 to P_1, output temporarily rises above the full-employment level to Y_0. When real wages adjust, output returns to the full employment level Y_F, which is also assumed to be the natural rate of unemployment. The short-run aggregate supply curve consequently shifts to the left from AS_0 to AS_1.

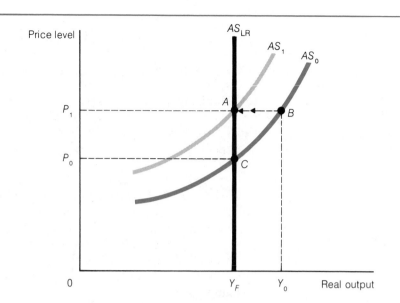

contracts come up for renegotiation, especially during expansionary or boom periods of high output and employment. Workers will also be convinced to negotiate nominal wage cuts during periods of recession and low employment. Ultimately, real wages readjust to equilibrium levels of employment and output.

Expectation-Adjusted Aggregate Supply

Expectations or aspects of long-run labor contracts and agreements will alter an aggregate supply curve because wages will tend to even out in the long run. Suppose, as we did earlier, that prices rise at a faster rate than nominal wages. Further assume that workers do not perceive that prices are rising faster than wages and that they maintain their former expectations. A perceived rise in real wages causes the input of labor to increase, creating increases in output. To obtain this result, employers must perceive before workers that the price increases are causing an actual decline in the real wage. A higher price level, P_1, in Figure 26–5 is therefore associated with a higher overfull level of real output and income, Y_0. This new combination of prices and output produced is shown at point B in Figure 26–5. The economy, in effect, travels up its short-run supply curve, AS_0. In this situation, output exceeds the natural rate of employment by $Y_F Y_0$.

Overfull employment does not last in the system, however. Workers are finally forced to adjust their price expectations upward and to respond by supplying less labor owing to the actual rise in the price level. As the labor supply is reduced, the short-run aggregate supply curve moves from point B to point A, a shift from AS_0 to AS_1. The economy has returned to a level of income and output that corresponds to the natural rate of unemployment, or full employment, Y_F.

Several points are worth remembering about the aggregate supply curve in Figure 26–5. First, the long-run aggregate supply curve is vertical when adjusted for expectations. It is simply a series of intersections of the full-employment level with the short-run supply functions. Each short-run function, moreover, is positively sloped because actual prices in the economy differ from the price perceptions of workers. *Either unemployment or overfull employ-*

ment will result when price expectations differ from actual prices. When perceptions catch up to actual price levels, the economy returns to the vertical long-run supply function. The speed with which workers' expectations adjust will determine the length of time that the economy must endure unemployment (below the natural rate) or overfull employment (above the natural rate). The expectation-adjusted aggregate supply curve will be of great use in understanding the views of modern monetarists in Chapter 30 and the so-called macroeconomic trade-off between inflation and unemployment in Chapter 32.

The Effects of Policy on Aggregate Supply

While we reserve an extended discussion of macroeconomic policy for Part VII of this book, we now turn briefly to the combination of aggregate demand and aggregate supply, highlighting the effect this combination has on economic policies. As we have seen, we may choose among various shapes of aggregate supply functions depending on our assumptions about the flexibility or inflexibility of the labor market in response to price changes.

The Classical or Laissez-Faire Perspective

The vertical classical supply curve was constructed under the assumption that prices and wages were flexible when either an excess demand or an excess supply of labor built up in the labor market.

This classical supply curve can be matched with an aggregate demand curve, as in Figure 26–6. The elegant simplicity of the classical argument is revealed when aggregate demand is either increased or decreased from an initial equilibrium. Suppose, as in Figure 26–6, that the economy experiences a reduction in aggregate demand due to a fall-off in some component of private spending—the autonomous component of consumption, investment, or expenditures on exports. This reduction in aggregate demand is shown as a shift in the demand function from AD_0 to AD_1.

Equilibrium income was established initially at E_0, where the economy was achieving full employment (corresponding to a natural rate of unemploy-

FIGURE 26–6

The Classical Long-Run Self-Adjustment Mechanism

In the classical view, when aggregate demand decreases, as from AD_0 to AD_1, the level of output temporarily falls from E_1 to A, causing the level of employment to fall as well. The price level then falls from P_0 to P_1, and full employment, Y_F, returns. With flexible wages and prices, the economy self-adjusts to full employment.

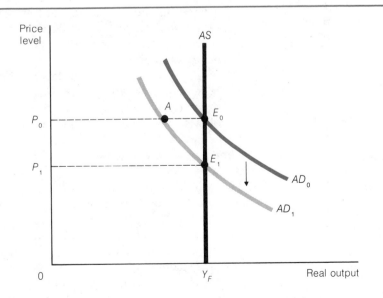

ment) at price level P_0. With the decrease in aggregate demand, we can think of the economy as moving to point A, where supply ceases to create its own demand. Demand is insufficient to carry Y_F worth of output at price level P_0. Contraction of production and employment would set in were it not that prices (and money wages) will be flexible downward. As prices fall from level P_0 to P_1, the quantity demanded for goods and services increases. As noted in Chapter 25, falling prices will increase consumption spending *and* investment spending through increases in the real balances and decreases in the rate of interest.

When spending declines, unemployment is initially created. But in the classical version of events, wages also decline, so that all laborers seeking work find work, but at a lower nominal wage. In terms of Figure 26–6, prices and wages fall, so the economy rapidly ends up at a new full-employment equilibrium, E_1, at Y_F, but at a lower price level, P_1. In fact, the key to understanding the basically noninterventionist policies of the classical writers in macroeconomic affairs is the point that price, wage, and interest rate adjustment take place very rapidly.

Keynesian and Modern Classical Views of Macroeconomic Adjustment

As we will see, Keynesians and the modern classical theorists differ on most policy issues. But they agree on one thing: Macroeconomic adjustment takes time. Consider Figure 26–7. Once more, assume that autonomous consumption, investment, or export expenditures decline, causing the aggregate demand curve to shift from AD_0 to AD_1. Demand is again insufficient to purchase Y_F goods and services and the labor supply necessary to produce them at price level P_0. The economy moves from E_0 to point A or B, depending on the relative, or short-run, inflexibility of wages and prices. Either drop, however, creates a certain amount of **demand deficiency unemployment**, unemployment that results whenever aggregate demand is insufficient to purchase the quantity of goods and services produced under full employment.

Demand deficiency unemployment:
Unemployment that results whenever aggregate demand is insufficient to purchase the quantity of goods and services produced under full-employment conditions.

Under rigid Keynesian assumptions, prices and money wages are totally inflexible, meaning that the economy would move to point A with a relatively large amount of demand deficiency unemployment. If prices and wages were somewhat flexible, however, a movement from E_0 to some point B on the short-run supply curve AS_0 might be predicted, leaving "only" demand deficiency unemployment of Y_0Y_F.

Short-run unemployment of labor and reductions in real output are a fact of life for both modern Keynesians and for modern classical theorists. But, as we will detail in Chapter 30 and Part VII, they differ widely in the kind of policies they would recommend. As we saw at the close of Chapter 25, Keynesians favor the use of discretionary fiscal policy—the use of government expenditures and taxation or, to a lesser degree, discretionary monetary policy—to force the aggregate demand rightward from B to E_0 in Figure 26–7. According to Keynesian analysis, the economy is quasi-permanently stuck at point B, where unemployment exists. Government is required to do something about it.

By contrast, contemporary classical economists believe that attempts to manipulate aggregate demand are doomed to fail. They feel that the discretionary meddlings of government—through fiscal and monetary policy—simply set up unsettling and disruptive expectations about inflation and the economy in general, making the restoration of equilibrium even more diffi-

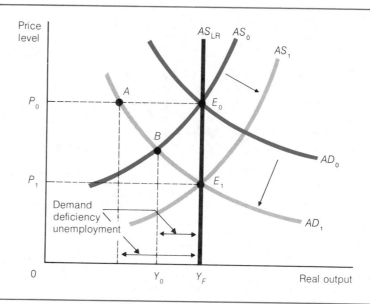

FIGURE 26–7   Demand Deficiency Unemployment: Short-Run Keynesian and Modern Classical Views

If demand falls from AD_0 to AD_1, unemployment will result. The economy moves from E_0 to A or B, in either case causing demand deficiency unemployment. In the short-run Keynesian view, full employment can be achieved only by restoring aggregate demand to the original level, Y_F. In the contemporary classical view, full employment will be achieved at lower aggregate demand, AD_1, by an adjustment in the short-run supply. If the economy moves from E_0 to B, a demand deficiency unemployment of Y_0Y_F still exists in the short run, but the aggregate supply increases from AS_0 to AS_1, and the economy moves to full employment at Y_F, but at a lower price level, P_1.

cult to achieve. They would prefer that government refrain from counter-cyclical fiscal policies except under certain conditions.

From the perspective of the expectation-adjusted long-run supply curve, moreover, any reduction of aggregate demand from AD_0 to AD_1 in Figure 26–7 would be purely temporary. Worker expectations about prices would soon match actual price changes, causing the short-run aggregate supply curves to shift rightward, restoring the equilibrium levels of output and employment. Indeed, an important school of modern classical theorists—the rational expectationists—believes that, on average and after repeated experience, market participants perfectly anticipate the discretionary actions of policymaker-politicians and the results of their actions. The public's ability to anticipate the results of fiscal policy on unemployment, for example, would effectively sterilize any attempts by policymakers to stimulate demand.

Supply-Stimulating Policies

In addition to modern classical and Keynesian policies to affect aggregate demand, recent fiscal policies and proposed policies have been oriented toward the stimulation of aggregate supply. Federal income tax cuts of the early 1980s were designed to provide new work incentives among laborers and incentives to businesses to accumulate capital goods and thus increase productivity. The Reagan administration's proposed institutional changes in labor markets were also supply-oriented. For instance, modifying the minimum-wage laws to permit the hiring of teenagers in need of job experience

FOCUS   Supply-Side Economics in the 1980s

Supply-side economics, some of the principles of which are discussed in this chapter, is more than a theoretical toy of macroeconomists. During the first half of the 1980s, under the sponsorship of President Reagan, a massive attempt was made to put supply-side principles in place in the United States. These include

1. Policies to increase incentives and motivations for businesses and to encourage investment in new plant and equipment;
2. Policies designed to encourage individuals and households to work and save; and
3. Control of inflation to stabilize price expectations and thereby expand incentives to work, save, and invest.

New investment tax credits, along with accelerated depreciation, were offered to businesses to help promote new investment. Restrictions on the interest rate paid to small savers were loosened or removed, thereby encouraging private saving to finance investment projects. The largest income tax cut in recent history was passed in 1980 to increase both saving and work incentives. And, potentially most important of all, the rate of inflation was brought down from 18 percent in 1979 to little more than 6 percent in 1983 and even less in

1984 due to restrictive money supply policies of the Federal Reserve System. These inflation-reducing policies, as implied in our discussions of the price expectation-adjusted labor market and aggregate supply curves, allow more rapid adjustment to the natural rate of employment.

The results of these policies have so far been mixed. A recession in the early 1980s, possibly precipitated by radical inflation-control measures, dampened enthusiasm for President Reagan's policies, although growth rates in real output and employment began to be impressive in 1983 and even more so in 1984. The economic growth rate in terms of real GNP in 1984 was at a two-decade high, although unemployment rate reductions were not as strong. A major problem leading to uncertainty in private markets was the growing federal deficits over the period, signaling to potential savers and investors that inflation and government competition for private funds might soon be on the way.

Regardless of the ultimate outcome of the Reagan program, which contains many political as well as economic trade-offs, policies to remove impediments to work, save, and invest have gained fairly broad approval among economists.

at lower-than-minimum wages is an effort to enhance aggregate supply and to improve long-run labor productivity. Several supply-side policies are discussed in Focus, "Supply-Side Economics in the 1980s."

Many macroeconomists have come to recognize the ill effects that decades of countercyclical demand policies have had on aggregate supply. In this new view, the tendency to deficit spending by governments at all levels and the increased taxation or inflationary tendencies created by deficit finance have had a constraining effect on work effort and on private investment by businesses. Aggregate demand policies by government may work faster than supply policies to alter inflation or unemployment, but they have negative long-run effects on the economy's aggregate output of goods and services. In other words, quick-fix policies do not get at the long-run problems of economic growth and productivity. Aggregate supply policies—especially those relating to labor productivity, long-run business investment, and institutional change—take more time to become effective. But in the view of many economists, choices should fall on the side of longer-run economic strength. They feel that demand policies should not be neglected, but supply issues should be raised to equal importance.

The Importance of Money

Fiscal policy, as we have seen, affects both aggregate demand and supply. But fiscal policy tells only half the story of macroeconomic policy. Money and its regulation in the economy is the other important tool of economic stabilization. There is now a school of macroeconomic thinking that places money at the center of the control of cycles of recession and inflation. The economists of this school—the monetarists—follow classical lines of thought, emphasizing the self-adjusting character of the economy in establishing full employment, maximum economic growth, and stable prices. Furthermore, to these economists, money is the principal determinant of economic health.

Before we can fully understand how money can be characterized as the primary explanation for business cycles, we must understand what money is, how and why money originated, how money is produced, and how it is regulated by the federal government. After understanding money's role in the world of macroeconomic theory we will return, in Part VII, to alternative views of macroeconomic policy. The important matter, of course, is to discover whether or not it is within the power of government to provide full employment without inflation.

Summary

1. An aggregate supply curve relates the total amounts of output of goods and services that would be forthcoming at alternative price levels over some period. Aggregate supply ultimately depends on the incentives of firms and workers throughout the economy to produce goods.

2. All aggregate supply functions stem from an aggregate production function that relates output to input. The output of goods and services is a function of all inputs. However, with all other inputs in the economy held constant, total output may be viewed as a sole function of the input of labor.

3. The shape of all aggregate supply functions is related to the flexibility of prices and money wages in the labor market.

4. In the classical perspective, aggregate supply is displayed as a vertical curve rising at the full-employment level of output. A vertical aggregate supply curve follows from the instantaneous adjustment of prices and wages to disequilibrium conditions in the labor market.

5. Keynes argued that prices and wages tend to be inflexible in downward directions. These inflexibilities impart an upward slope or a short-run portion to the aggregate supply function.

6. When the possibility of diverging price expectations of workers and employers is introduced into aggregate supply analysis, long-run aggregate supply becomes a vertical expectations-adjusted supply curve. When workers' price expectations differ from actual prices, however, a positively sloped short-run aggregate supply curve for the economy is produced.

7. Aggregate supply curves, when combined with demand curves, produce a unique price level and a level of output in the economy. However, classical, Keynesian and modern classical theorists disagree about the ability of private forces in the economy to produce full employment without inflation.

8. Aggregate supply policies include tax policies and institutional changes in labor markets to stimulate productivity, work incentives, and incentives to invest.

Key Terms

aggregate supply	money illusion	natural rate of unemployment	overfull employment
productivity	long-run aggregate supply curve	natural rate hypothesis	demand deficiency
real wage rate	full employment	price expectations	unemployment

Questions for Review and Discussion

1. What does the aggregate supply curve show? What makes it shift to the right?

2. What effect will each of the following have upon aggregate supply: an improvement in the quality of education; an improved and more efficient method of pumping oil from existing wells; an increase in the minimum wage; a regulation by the Occupational Safety and Health Agency (OSHA) that

prevents construction workers from working after dark?

3. If the real wage rises, what happens to the quantities of labor demanded and supplied? What is a real wage as opposed to a nominal wage?

4. What is money illusion? Suppose that workers receive a five percent increase in nominal wages and the price level rises by ten percent. If workers are

subject to money illusion, would they increase the amount of labor supplied? By how much would they think their income had changed? By how much would it actually change?

5. What does the classical aggregate supply curve look like? In order to achieve this shape, what must be assumed about nominal wages? Does this assumption hold in the short run and long run?

6. If the economy is operating at a point on the aggregate supply curve where output is below the full employment level, then where is the economy operating relative to its production possibilities frontier?

7. Suppose that union members are expecting an inflation rate of ten percent over the next two years and they negotiate a two-year contract that calls for an increase in nominal wages of ten percent per year. What happens to their real wages if the ac-

tual rate of inflation falls to only five percent? What happens to their levels of employment and unemployment? What would happen if the actual rate of inflation increased to fifteen percent?

8. Explain what happens in the short run to the price level, the wage rate, and the level of employment when aggregate demand decreases. What happens in the long run?

9. Suppose the government increases taxes and government expenditures in an effort to stimulate aggregate demand. What effect will this increase in taxes have upon after-tax wages, the supply of labor, and the aggregate supply of goods and services? Will the nation's output increase?

10. Should the government attempt to manage the level of aggregate demand or should it manage the level of aggregate supply in an effort to maintain full employment and economic growth?

ECONOMICS IN ACTION

The Opening of the Iron Age: A Supply Shock in Ancient Greece

The long-run reactions of Americans to the energy crisis precipitated by the activities of the OPEC cartel in the 1970s have been numerous. Initially, the supply shock caused a mass movement away from enormous, gas-guzzling American cars to small, fuel-efficient cars. After OPEC lost some of its power to control prices, gas prices stabilized, and larger automobiles were again in favor. Things were not the same after the supply shock, however, and will never be the same again. The new, larger cars are made to be fuel-efficient, solar and other alternative forms of energy have found new uses, and production processes have been adjusted permanently to oil shortages and to the possibility of future shortages. In sum, the doomsdayers who claim that society will have to learn to live with less do not appear to have been very accurate. In fact, OPEC's supply shock may have had extremely beneficial effects on the American economy.

Two economists have recently taken the doomsdayers to task, examining a large number of supply shocks throughout 10,000 years of recorded history.[a] Supply shocks, they argue, never permanently reduce aggregate supply; instead, adjustments to supply shocks produce beneficial effects that reverse leftward shifts in aggregate supply.

Consider, for example, a supply shock in ancient Greece, a shock that marked the transition from the Bronze Age to the Iron Age. Why did our Greek ances-

tors switch from bronze to iron tools and weapons beginning about 1000 B.C.? The reasons often given—that the Greeks didn't invent iron until 1000 B.C. and that iron was a much better metal than bronze—are questionable. People in ancient Greece had a knowledge of iron making dating to 3000 B.C. In addition, hammered bronze tools and weapons are almost as hard as iron ones and will hold an edge almost as well as iron tools. Since neither of these explanations is satisfactory, what other factor may have ushered in the Iron Age?

Consider some conditions existing during the Bronze Age. From trading records, we know that iron was extremely expensive during the Bronze Age: Records from the nineteenth century B.C. indicate that the exchange ratio of iron to silver was 1:40. Bronze is an alloy of about 90 percent copper and 10 percent tin. At the time, copper was abundant, especially relative to iron—the trading ratio of copper to silver was 200:1. Tin was not so abundant and was probably imported into Greece from Iran, for there is practically no tin in the eastern Mediterranean area. During the Bronze Age, tin exchanged for silver at between four and ten units of tin to one unit of silver. Everything considered, the components for a bronze tool cost about 0.05 percent of what an iron tool would have cost. Small wonder the Greeks used bronze.

What happened to alter this situation? We know that around 1000 B.C. tin became extremely scarce in the Aegean. The Greeks experienced a supply shock of enormous severity owing to the wartime disruption of trade caused by the invasion of the Sea Peoples (the

[a]See S. Charles Maurice and Charles W. Smithson, *Doomsday: 10,000 Years of Economic Crises* (Stanford: Hoover Institution Press, 1985), pp. 95–105.

Greek shoemaker's shop, 520–510 B.C. *Black figured amphora. By the early 6th century* B.C., *the use of iron in toolmaking and weaponry had become commonplace.*

Philistines) into the eastern Mediterranean. These invasions, lasting from about 1025 to 950 B.C., led to the collapse of the major Bronze Age civilizations—Mycenean Greece, New Kingdom Egypt, and the Hittite Empire.

With tin no longer available, the price of bronze rose precipitously. Old bronze was melted down for its tin content and, because iron now became cheaper relative to bronze, ancient smiths began to forge iron tools and weapons. The response of the Greeks to the "tin crisis" was to usher in the Age of Iron. Even when trade was reestablished and bronze prices fell, iron continued to be used. The crisis had encouraged the smiths to learn ways to produce iron more cheaply. Iron continued to be used even though the Greeks reverted in part to the use of bronze. In short, when resource crises or supply shocks occur, alternatives become cheaper and more attractive. Leftward shifts in aggregate supply create new circumstances that may mean vastly increased aggregate supply in the future. Within crises are the seeds of undreamed-of future economic possibilities.

Question

Assume that fires and disease destroy three-quarters of America's timber stock. What might be the effects of the catastrophe upon alternative resource development in the short run? Over a longer period of time?

POINT-COUNTERPOINT   J. M. Keynes and Joseph Schumpeter: Intervention or Innovation?

John Maynard
Keynes

Joseph A.
Schumpeter

JOHN MAYNARD KEYNES (1883–1946) was the son of John Neville Keynes, a famous logician and writer on economic method. Educated at Eton and later at King's College, Cambridge, the younger Keynes developed interests in literature, mathematics, and later, in economics. One of his teachers at Cambridge, the great neoclassical economist Alfred Marshall, was much impressed with Keynes's precociousness and strongly urged him to become a full-time economist.

During his years as an undergraduate, Keynes became an integral part of a small coterie of British intellectuals that came to be known as the Bloomsbury Group. Its members included novelist Virginia Woolf and biographer and literary critic Lytton Strachey. The group provided Keynes with an arena for intellectual debate, but he still had to decide on a career. Self-confidently, he wrote to Strachey, "I want to manage a railway or organize a Trust." Neither of these options materialized, and Keynes entered London's India Office of the British civil service in 1907.

Soon bored with civil service duties, Keynes returned to Cambridge and became editor of the prestigious Economic Journal, a post he held for thirty-three years. He joined the British treasury in 1915 as a monetary expert and became a key figure representing Britain at the Versailles Peace Conference at the end of World War I. In 1919, he wrote The Economic Consequences of the Peace, a condemnation of the Versailles Treaty, which brought him international recognition. He went on to write a Treatise on Probability (1921) and to amass a personal fortune in the risky game of speculating in foreign exchange markets.

Through the Bloomsbury Group, he was acquainted with the cultural set in British society, knew Pablo Picasso and Bernard Shaw, and met Russian ballerina Lydia Lopokova, whom he married in 1925. He collected modern art, a practice begun during his work for the treasury when he had acquired valuable paintings

from France for settling that country's debts to Great Britain.

In the late 1920s, Keynes's interest turned increasingly to the theory and practice of macroeconomics. His productivity was enormous: Major works of the period include the Treatise on Money (1930), Essays in Persuasion (1931), and Essays in Biography (1933). In 1936, Keynes published the work for which he is most famous, The General Theory of Employment, Interest, and Money. In this book, whose influence has been compared with Marx's Das Kapital and Smith's Wealth of Nations, Keynes rejected the idea of automatic adjustment in the economy and maintained that public policy and government expenditure are required for the prevention of economic stagnation and excessive unemployment. During World War II, Keynes negotiated lend-lease programs and was a leading figure in plans to restore the international monetary system. He died of a heart attack soon after the war ended.

JOSEPH A. SCHUMPETER (1883–1950) and Keynes were born only a few months apart. Schumpeter was raised in a provincial town in Austria (then Austria-Hungary) and studied law at the University of Vienna, where he also attended seminars on economics led by Carl Menger and Frederick von Weiser, two of the founders of the neoclassical Austrian school of economics. In 1906 Schumpeter earned his law degree and practiced law for a short time before deciding to devote himself to economics. At age twenty-eight he produced a brilliant doctoral dissertation, The Theory of Economic Development (1911), which brought him recognition as a first-rank theorist. After World War I and the breakup of the Austro-Hungarian monarchy, Schumpeter served as Austria's minister of finance. Throughout the 1920s he lectured widely throughout Europe. In 1932, as Fascism began its rise in Central Europe, he emigrated to America and became the senior economics faculty member at Harvard, where he remained until his death.

Schumpeter stands out as an extremely innovative thinker. He rejected many contemporary approaches to macroeconomic theory partly because they were based on pure mathematical insight. Schumpeter preferred to base his theory of economic change on the creative force of the individual, whose social, historical, and psychological dimensions are largely ignored by strict mathematical formulas. Schumpeter is also known for his broadly historical views of the discipline itself, which are presented in the posthumously published *History of Economic Analysis* (1954), edited by his wife Elizabeth Boody Schumpeter. Another work, *Capitalism, Socialism, and Democracy* (1942), is famous for its prediction that capitalism will eventually destroy itself, not because of its failures (as a Marxist would contend) but because of its successes.

Keynes and Schumpeter differ in most respects in their analysis of how the macroeconomy works and what can or cannot be done to make it work. In many ways, their disagreements over the true forces of change within an economy and the proper role of government in economic stability help frame the "great debate" in twentieth-century macroeconomics.

Fine-Tuning the Engine of Demand

Keynes's central work, *The General Theory of Employment, Interest, and Money*, shared one important characteristic with the work of the great neoclassical economist Alfred Marshall: a love for abstraction. Robert Heilbroner calls the *General Theory* "an endless desert of economics, algebra, and abstraction, with trackless wastes of the differential calculus, and only an oasis here and there of delightfully refreshing prose."[a] Beneath the calculus, however, were ideas capable of influencing an entire generation of economists and of affecting the economic fortunes of millions of people.

Keynes's great insight rested upon his central abstraction, aggregate demand, in shorthand $C + I + G + (X - M)$. Unlike his classical and neoclassical forebears, Keynes believed that insufficient demand, or spending, on the part of consumers would leave the economy in disequilibrium, stagnating permanently below full employment. Accordingly, Keynes focused on means to increase demand through government policies and interventions in the economy.

In Keynesian terms, the economy is inherently unstable yet manageable. Guided by economic variables such as national income and business investment, government policymakers can rely on fiscal measures to increase or decrease aggregate demand in amounts sufficient to restore equilibrium. Individual, or microeconomic, decisions to spend, invest, or save could predictably follow whatever course the fiscal and monetary planners design.

Keynes, despite the increased role he recommended for government, was deeply mistrustful of overreliance on central planning. He was not attempting to redraw capitalism but to rescue it. (In the year when the *General Theory* appeared, unemployment in the United States was close to 25 percent.) Responding to criticisms by economist Friedrich Hayek that an overplanned economy represents tyranny, Keynes wrote:

> Moderate planning will be safe enough if those carrying it out are rightly oriented in their own minds and hearts to the moral issue (of tyranny).[b]

Creative Destruction

Whereas Keynes could be said to honor the economist's role in rescuing a stagnant economy, Schumpeter honored the entrepreneur's role. In many respects Keynes viewed the economy from above, from the heights of abstraction. Schumpeter looked from below, from the vantage point of individuals whose risk-taking and profit-seeking behavior spurred innovations and new opportunities for growth. Accordingly, Schumpeter looked for ways to ensure free enterprise, not to manage it.

To Schumpeter, the tendency of an economy to fall below full employment levels resulted from shrinking opportunities for profits. As new breakthroughs in technology or production occur, inspiring new investment and greater opportunities for profit, the economy generates growth. Schumpeter called the process *creative destruction* of profit opportunities, the continual rebirth of production frontiers.

Schumpeter naturally argued against government intervention and central control of the economy. He was much more mistrustful of the results of fiscal management than Keynes and felt deeply that governmental tyranny would be its inevitable result. The first victim of such tyranny, in Schumpeter's mind, would be the great engine of capitalism, the entrepreneurial spirit. As Schumpeter summarized the matter: "The problem that is usually being visualized is how capitalism administers existing structures, whereas the relevant problem is how it creates and destroys them."[c]

[a]Robert Heilbroner, *The Worldly Philosophers*, rev. ed. (New York: Simon & Schuster, 1961), p. 235.

[b]Heilbroner, p. 244.

[c]Joseph A. Schumpeter, *Capitalism, Socialism and Democracy*, 3rd ed. (New York: Harper and Row, 1950), p. 81. Also see the excellent article by Peter Drucker, "Schumpeter and Keynes," *Forbes* (May 23, 1983), pp. 124–128, on which some of the themes of this section are based.

VI

Money:
Its Creation
and Management

27

An Introduction to Money and the Banking System

A s suggested in our discussion of macroeconomic theory, money is a vital link in our economy. This chapter and the three that follow it explore the nature of money and its economic role. We look at how economists define money, how money functions within the economy, and the reasons people hold and use money. Further, we see how money is related to the banking system and how the banking system itself is regulated by the government.

Money and monetary control are intimately related to the major goals of economic stabilization. Conversely, lack of monetary control can contribute to a society's economic downfall. Before we delve into these issues, our goal in the present chapter is to lay the groundwork for an understanding of money.

Familiar quotations point up the fascination—and the suspicion—that surrounds money. Shakespeare wrote that the person who wants money, means, and content is without three good friends. George Bernard Shaw noted, contrary to a certain religious teaching, that the *lack* of money is the root of all evil. And, in a cynical reversal of the notion that money cannot buy happiness, it has been said that happiness cannot buy money either.

Economics, of course, is not interested in the moral questions that money provokes, but it is interested in the nature of money and its role in our individual and collective lives.

Functions of Money

To understand the essential functions of money, try to imagine what a society would be like without it. In a **barter economy** all goods or services would be traded for other goods or services. A visit to a psychiatrist, for example, would require a trade of some good or service for an hour on the couch.

Barter economy: An economy in which all goods and services are traded for other goods and services without the use of a medium of exchange such as money.

Double coincidence of wants: A situation, necessary for barter to be successful, in which the goods and services that an individual has to offer are the goods and services another individual desires.

Medium of exchange: Anything that is used as payment for goods and services, most commonly money. A medium of exchange eliminates the inefficiency and costs of barter.

Commodity money: An item used as money that is also a salable commodity in its own right.

Suppose that you are the psychiatrist and a group of three potential customers are waiting for your services, each with his or her own item of specialization to trade. Sam offers four hours of typing services, Bill three Persian kittens, and Judy, a farmer, one fully dressed hog. Will the demanders of psychiatric services be able to trade with you?

The result would depend on whether you, the psychiatrist, demand those goods or services in exchange for your own skills and on whether a price—a certain quantity of pork or kittens per hour of psychiatry—could be agreed on. In other words, a **double coincidence of wants**—a cat breeder wanting psychiatry and a psychiatrist wanting cats—would be required for a mutually satisfactory exchange to take place. It may be that, not being a cat lover yourself, you know someone who would be willing to exchange stereo tapes for Persian cats. In that event, a more complicated set of exchanges might be arranged. But these sorts of trades would require either luck or a great deal of information gathering. Transaction costs—the costs of getting buyers and sellers together for mutually advantageous exchanges—are astronomical in a barter economy.

It is not surprising that where barter is practiced, specialization and trade tend to be at relatively low levels. The high transaction and information costs associated with barter inhibit specialization and trade. A modern free-trade economy could not be founded on a system of barter because of the system's limiting effects. Money was therefore invented to overcome these limitations. In economic terms, money serves four separate functions: as a medium of exchange, as a unit of account, as a standard of deferred payments, and as a store of value.

Money as a Medium of Exchange

The evolution of money as a **medium of exchange,** or means of payment, came about from the desire to avoid the transaction costs associated with barter and to achieve greater economic efficiency. Clearly the need for a coincidence of wants between traders is avoided by the introduction of money. Once money is generally accepted as a means of payment, the hog or kitten owners sell their output in hog and kitten markets. They are willing to accept money in exchange for their products or services because they know that psychiatrists and sellers of all other items they consume are willing to do likewise. No longer is the psychiatrist required to search out demanders for psychiatric services who are simultaneously suppliers of goods or services that the psychiatrist demands. Use of money as a means of exchange fosters specialization and economic efficiency. Within an economic system, the item that serves as a medium of exchange does so exclusively; anything else is not generally acceptable as money.

Commodity Money. Virtually anything can serve as money if it is generally accepted as a means of exchange or payment. Historically, an incredible assortment of items have served as **commodity money** within different societies, including horses, cowrie shells, elephants, stone wheels, cigarettes, colored beads, slaves, gold and other precious metals, cows, paper, and feathers. A commodity serving as money should be relatively scarce. The feathers of the extremely rare and beautiful quetzal bird would not have made a good medium of exchange in pre-Columbian Central America if the birds and their feathers had been in abundant supply. A sudden growth in the quetzal population would have produced inflation in the price of feathers—too many feathers available to exchange for too few goods. (Indeed, this is what infla-

FOCUS The Gold Standard

Features of precious metals such as scarcity and durability help explain why most early economically advanced societies, including the United States up to the early 1930s, chose precious metals, particularly silver and gold, as their medium of exchange.

Monetary systems in which gold and silver are the medium of exchange are called *gold* (or silver or specie) *standards*. A pure gold standard means that gold in the form of coins or ingots or paper representations of gold, fully redeemable for an equivalent value of gold, circulates as the medium of exchange both domestically and internationally. Under a pure gold standard the government may certify the purity and content of new gold brought into circulation.

Although gold standards were popular among many nations throughout history, they were never without difficulties. The value of gold, and thus the price level, is determined by costs of gold production under long-run competitive conditions. That is, supply and demand determine the price of gold just as they determine the price of any good or service. The price of gold is influenced by the costs of mining and by the amount of gold discovered. If new discoveries of gold are made, the price of gold decreases and the price of other goods relative to gold increases. An increased quantity of gold is available to exchange for the same quantity (or a more slowly growing quantity) of goods and services. Inflation of the prices of the goods and services results.

Or consider the opposite scenario—a persistent scarcity of the precious metal in relation to a growing population and a growing real output of goods and services. More gold could be pumped into the economy only by digging deeper mines, using more capital and labor, and so on, all at increased costs of production. These increased costs and the resulting increase in the price of gold would mean that the price level of all goods and services would decrease—that is, the economy would undergo price deflation.

Under a pure gold standard, then, society is subject to instabilities in prices and economic activity depending on the relative scarcity or plenitude of gold. This problem is inherent in the use of any commodity medium of exchange, from cowrie shells to whale teeth.

tion means—that money becomes overabundant in relation to the goods and services it can purchase.)

Yet another and possibly more important problem associated with all commodity monies, including quetzal feathers, is that there is an opportunity cost to using any commodity as money. A moment's reflection tells us what that cost is. Use of a commodity as a monetary medium of exchange means that it cannot simultaneously serve other uses. If quetzal feathers are used as money, they cannot simultaneously be used in headdresses.

Gold is perhaps the most familiar commodity money. The opportunity cost of using gold as money is particularly important (see Focus, "The Gold Standard"). Most societies have valued gold highly as decoration, as an object of possession, and, more recently, in industrial or medical uses such as dentistry. Gold that is used for money cannot simultaneously be used to produce jewelry or to serve any other purpose. The value of gold in nonmoney uses is precisely the value given up when gold is used as money. When this opportunity cost rises above the value of gold in use as a medium of exchange, gold will be converted—coins melted down—to the uses of higher value.

History is peppered with such conversions resulting from changes in the opportunity cost of using gold or other precious metals as money. And social visionaries, seeking to abolish the "tyranny" of gold as money, have naively sought to demonetize gold by decree. Lenin once said that the first goal of his revolution would be to use gold in the manufacture of toilets. It does not appear that there are any gold-plated toilets in Russia today, and opportunity cost provides the explanation.

Fiat money: Money in the form of paper or some other inexpensive item that is not backed by a commodity such as gold but is legally certified by the government to function as the medium of exchange.

Fiat Money. Two problems—the opportunity cost of using commodities as money and instabilities in the exchange value of the commodity used as money—contributed to the emergence of another alternative, called **fiat money.** Fiat money is paper (or some other inexpensive, low-cost item such as lead or nickel) that is certified by government decree, or fiat, to be money. Dollars, pesos, rubles, and yen are all fiat money. Fiat money is not backed by a commodity such as gold; that is, it is not freely and perfectly convertible into that commodity. A system of paper backed by a commodity functions exactly the same as commodity money: The commodity must be stored for instant convertibility, thereby incurring an opportunity cost.

Even though fiat money is not backed by a commodity, governments have various ways of giving the money a generally accepted value as a medium of exchange. Fiat money sponsored by government is issued under closely enforced monopoly restrictions by the state. Since the value of the fiat commodity, usually paper, is lower than its monetary value, counterfeiting cannot be permitted. Valuing of the paper as a medium of exchange is further encouraged by the government's acceptance of it in payment of taxes and by the declaration of debts as "forgiven" when creditors refuse to accept the fiat currency as payment. Government thus contributes to the general acceptability of fiat money by making it legal tender.

Use of fiat, nonconvertible paper money has one great benefit: It releases valuable resources—gold, silver, or other items—to other uses. But like all commodities, paper is not perfect money. Although it is easily transferable, divisible, and portable, it is less durable than some other forms of money such as gold. Paper money must be periodically replaced by the government, so replacement is a cost of using it. Another important cost is that the government monopoly might overissue paper money relative to the demand for it, thereby creating inflation. However, the widespread use of fiat money indicates that the benefits from using it outweigh the costs.

The reader may object: Is not paper a commodity and is there not an opportunity cost to using paper as fiat money? Paper is a commodity, but the price and thus the opportunity cost of using paper as money is extremely low relative to other possible forms of money. The important point is that the paper money must be generally acceptable as a medium of exchange. If a base metal or anything else meets this criterion, it also can serve as money.

Money as a Unit of Account

Money is preferred to barter partly because money reduces the cost of economic transactions. To illustrate, a leathersmith in a small-scale barter economy produces tanned leather in exchange for blankets, gunpowder, food, alcohol, and metal traps. To rationally allocate exchanges, the leathersmith must consult four separate markets to find out the leather price of blankets, of food, and so on. Each trader must do the same. As the number of traders and the number of goods rises, the number of calculations rises exponentially.

Unit of account: A common measure, such as dollars and cents, that expresses the relative values of goods and services.

The use of money as a **unit of account** helps solve such problems. When money is introduced, all goods are valued in a common measure. Thus, if Jane knows that leather coats cost $4 each and that alcohol costs $2 a bottle, she will also automatically know that her can of gunpowder, worth $8, will trade for two leather coats or four bottles of alcohol or any other combination of goods equaling $8. The function of money as a unit of account means that the number of mental calculations required to determine the relative prices of all traded goods is vastly reduced. This function of money was called the *numeraire* by the French economist Leon Walras (1834–1910). The numeraire is the unit in which the prices of all other goods are stated. Prices are ordinarily stated in the society's monetary unit, such as dollars, sous, or pounds, but there is no reason why any other commodity could not serve as the numeraire. Prices may be reckoned through a leather unit of accounting or through an alcohol standard. The important point is that some thing serves to reduce the number of mental transactions required for rational trade. Money has often served in this capacity.

Money as a Standard of Deferred Payments

Standard of deferred payment: Money's acceptability in contractual arrangements, such as loans, involving future payments.

Money not only helps us keep track of the relative prices of goods, it also serves as a standard for payments that are deferred from the present to the future. When we borrow money for a Caribbean vacation, a home, or a new car, we are using money as a **standard of deferred payment.** Money is used as a standard of deferred payment when a debt or obligation is expressed in terms of dollars, pounds, or any other medium of exchange.

Contracts to buy or sell in the future use money as a standard of deferred payment, and there are potential costs in doing so. One of them is inflation—a reduction in the value of money. When money is used as a standard of deferred payment, there are potential gainers and losers. A contract to pay for a new car two years hence means an inflation risk for the lender. If the inflation rate depreciates the value of money by 10 percent in two years, the lender would receive dollars that would buy 10 percent fewer goods and services. Lenders will therefore try to protect themselves by charging interest to cover the risk of inflation.

Money as a Store of Value

Store of value: A characteristic of money or any other commodity that allows it to maintain or increase its worth over time.

Individuals who want to save must store wealth, and one way to store it is in the form of money. Money's role as a **store of value** is its least exclusive function. Virtually all commodities, including furniture, houses, stamps, Pekinese puppies, and money, serve as temporary storehouses of wealth or value. But there are better and worse keepers of value. Some commodities such as perishable foods are used up relatively quickly. Other commodities, such as land, precious metals, works of art, or buildings, tend to change value relatively slowly.

Liquidity: The ease or speed with which any asset or commodity can be converted into money.

While any commodity is a store of value, not all commodities are instantly salable. The salability of commodities is called **liquidity,** the ease or rapidity with which any commodity can be converted into a medium of exchange. Money is by definition the most liquid asset—the one most readily accepted in exchange for other goods and services. All other commodities take varying amounts of time to be converted into money. When cash is needed, land or warehouses cannot easily be converted into cash, but stocks and bonds can. Land is therefore less liquid than stocks and bonds.

Money has its own advantages and disadvantages as a store of value. It may retain its value over long periods, but during periods of inflation it may be a poor store of value. Under extreme conditions, as when a nation at war is forced to make huge expenditures to defend itself, money may lose its value or purchasing power very rapidly through inflation. Such was the case during the American colonial period, when paper money was known as Continentals. The phrase "not worth a Continental" refers to a rapidly depreciating store of value. In fact, inflation, or currency depreciation, has occurred during or after every American war, including the Vietnam War.

In general, money has served fairly well as a store of value in the United

States, although its performance over the last several decades has been uneven. The holding of money, therefore, is both safe and risky. Money is instantly convertible to other goods and services, but it may lose its value more or less rapidly than other commodities. When a person holds any other asset (an interest-bearing savings account, IBM stock, a Picasso painting), liquidity is sacrificed.

The Official Definition of Money

Now that we have gained some understanding of money by examining its functions, we must define more carefully what we mean by the word *money*. The definition of money is crucial in that it determines how the nation's money supply is measured. As we have said in previous chapters, the nation's money supply is critical in explaining and predicting output, employment, and prices in the economy.

The most common definition of money is "anything generally acceptable as a medium of exchange." We are all familiar with money in this context. It includes coins, currency, and checkbook money, all of which are generally acceptable for goods and services. However, many other definitions of money address the varying degrees of liquidity of assets that might serve as money (see also Focus, "Are Savings Accounts and Credit Cards Money?").

Transactions accounts: Any account, such as a demand or checkable deposit account, against which checks may be written or funds transferred.

Possible confusion caused by the terminology surrounding checkbook money can be avoided by remembering that any accounts from which payments or funds transfers may be made are called **transactions accounts.** These include all demand deposits and checkable deposits managed by banks and other financial institutions. **Demand deposits** are those transactions accounts against which an unlimited number of checks may ordinarily be written. Banks are the primary issuers of demand deposits. **Checkable deposits** generally refer to other types of transactions accounts, such as those available at credit unions, savings and loan associations, and so on. These deposits often carry restrictions on transferrability—a maximum number of checks may be written per month without penalty, and so on. Both demand deposits and checkable deposits are "checkable" in the sense that either may be utilized as a medium of exchange.

Demand deposit: Any account against which funds may be transferred on demand and up to the value of the account.

Checkable deposits: Deposits at financial institutions such as credit unions or saving and loans against which checks may be written, often with restrictions, and funds withdrawn sometimes with interest or other penalties.

Though the "general acceptability" definition is reasonably clear-cut, a number of assets, such as noncheckable savings accounts, may have important effects on the behavior of investors and consumers and, therefore, on the economy. These less-liquid assets, sometimes called "near-monies," are therefore included in some statistical compilations. Thus the nation's money supply is officially measured in a number of ways.

Four major measures of money developed by the Federal Reserve System, the nation's central bank, are described in Table 27–1. Notice that the M-1 measure corresponds to the definition of general acceptability. M-1 is composed of coins and currency in the public's hands and all checkable deposits.

The measures are arranged from M–1 to L in order of decreasing liquidity. Small savings deposits, included in M–2, are ordinarily more liquid than large-denomination time deposits, included in M–3, since convertibility of the latter into cash often requires advance warning to the financial institution. U.S. government securities, included in measure L but not in M–3, are less liquid still, and so on. Conceptually there is no limit to these distinctions (Is the Brooklyn Bridge a more or less liquid asset than the Hoover Dam?), but these three measures plus M–1 are sufficient for most important economic calculations.

**FOCUS   Are Savings Accounts
and Credit Cards Money?**

Numerous recent innovations in the financial system blur the function of money as a medium of exchange. Are savings accounts to be considered money? Savings accounts ordinarily are not checkable; that is, they cannot be transferred from individual A to individual B by check. If they are checkable, then they are money. Likewise, credit cards are an extremely convenient means of paying for goods, but they are not money. The reason is simple: You must meet your obligation to Visa, MasterCard or American Express by a check on your demand deposit account, which *is* money. Credit cards are just what their name implies—temporary credit. Nevertheless, the distinction between "checkable" and "noncheckable" accounts is becoming compli-

cated. Credit unions, commercial banks, and a variety of financial institutions now offer "double-dealing" deposits such as credit union share draft accounts, negotiable orders of withdrawal (NOW) accounts, automatic or electronic transfer services, and so on. As far as the concept of money is concerned, all these innovations simply mean that more types of assets are checkable and generally acceptable as means of payment. The fact that some checkable deposits now pay interest—as savings deposits always have—is totally irrelevant to the concept or definition of money. General acceptability as a means of exchange is the only requirement.

Definitions of *money* and *near-monies* evolve with new developments, like the recent emergence of checkable accounts at nonbank financial institutions such as credit unions. Various items in the Federal Reserve System's definitions are lumped together because they are more alike than other possible groupings. Each definition has its own strengths and weaknesses as a tool for measuring the nation's money stock. M–2, for example, may be used in analyzing economic problems where a broader view of liquidity is appropriate.

Lest the reader get the impression that money is the economist's elusive cat in the hat, calculation of the various measures produces real numbers. Figure 27–1 graphically presents money supply data for November 1984. These figures, calculated in billions of dollars by the Federal Reserve System, correspond to the definitions given in Table 27–1. The figure shows that M–2 is almost four times the size of M–1, and L, the broadest definition of money, is almost six times the size of M–1. In general, however, economists consider "money" or the "money supply" to be M–1, which consists of assets that are generally accepted as a medium of exchange.

Money and the Banking System

The largest components of M–1 are bank deposits. As indicated in Figure 27–1, commercial bank demand deposits were worth $245.5 billion at the end of August 1984. Checkable deposits at other financial institutions accounted for another $139.9 billion. Together these bank deposits are more than two and a half times the size of the holdings of currency (including coins) by the nonbank public. We are all accustomed to accepting currency and coin as money. But how did bank money—that is, checkable deposits—evolve?

The Evolution of Banking

The emergence of banks and checkable deposits was intimately linked with the origins of fiat money. Before the Renaissance and the emergence of a market society, commodity money, usually gold or silver, circulated as a medium of exchange.

TABLE 27–1   Four Measures of the Money Stock

Federal Reserve System classifications of money stock range from M-1 to L in order of decreasing liquidity. The most liquid measure is M-1, which includes currency in the public hands, all checkable and demand deposits at financial institutions, and travelers' checks. Although L is many times greater than M-1, as shown in Figure 27–1, economists often identify money as only assets that are generally accepted as a medium of exchange, that is, M-1.

Measure	Components	Definitions
M-1	Currency, including coins and paper money held outside banks	
	Checkable and demand deposits at commercial banks, savings institutions, and credit unions	
	NOW and super-NOW accounts	NOW (negotiable order of withdrawal) account: Interest-earning account on which owner may write checks; super-NOW account: NOW account with higher interest and additional restrictions on withdrawals.
	ATS accounts	ATS (automatic transfer savings) account: Interest-earning account, the contents of which are automatically transferred to an individual's checking account when the checking account falls to a minimum level.
	Travelers' checks	
	Checkable money market accounts	
M-2	M-1 plus	
	Savings deposits and small-denomination time deposits in both commercial banks and savings institutions	Time deposit: large savings deposit requiring notification (usually 30 days) or interest penalties for early withdrawal.
	Certain money market mutual funds	Money market mutual funds: Investment funds managed by banks or investment companies whereby investors' money is pooled and used to buy or sell bonds or other interest-earning investments; investors generally can write checks against their fund balances under some restrictions.
M-3	M-2 plus	
	Large-denomination time deposits ($100,000 or more) at commercial and savings institutions	
L	M-3 plus	
	U.S. savings bonds	
	Short-term U.S. government securities	
	Bankers' acceptances	Bankers' acceptances: Notes that are issued by businesses to obtain short-term credit and that are accepted by a bank or other financial institution. Acceptance by the bank indicates that the bank stands ready to pay the principal of the note to the holder at maturity. At maturity the bank pays off the note, charges the business issuer's account, and charges a fee for its services.
	Commercial paper	Commercial paper: Short-term (up to six months) financial obligations (notes) of large corporations traded among finance companies and companies dealing in installment loan credit. Commercial paper may be placed in the market by corporations directly or through specialized dealers.
	Eurodollar deposits held by U.S. individuals	Eurodollars: Deposits denominated in U.S. dollars held by individuals and nonbank businesses in banks located abroad (mostly in Europe).

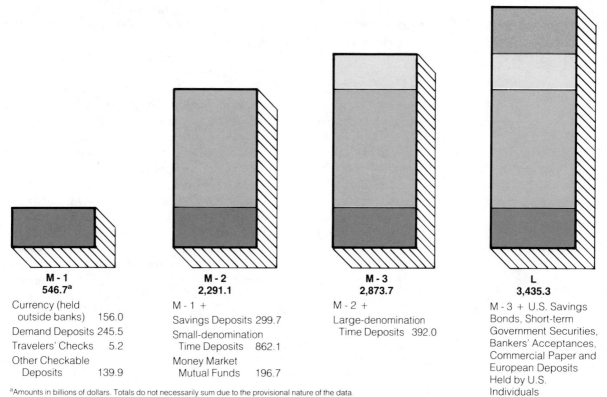

M - 1	**M - 2**	**M - 3**	**L**
546.7[a]	**2,291.1**	**2,873.7**	**3,435.3**
Currency (held outside banks) 156.0	M - 1 +	M - 2 +	M - 3 + U.S. Savings Bonds, Short-term Government Securities, Bankers' Acceptances, Commercial Paper and European Deposits Held by U.S. Individuals
Demand Deposits 245.5	Savings Deposits 299.7	Large-denomination Time Deposits 392.0	
Travelers' Checks 5.2	Small-denomination Time Deposits 862.1		
Other Checkable Deposits 139.9	Money Market Mutual Funds 196.7		

[a]Amounts in billions of dollars. Totals do not necessarily sum due to the provisional nature of the data.

FIGURE 27–1   Numerical Measures of the Money Stock, August 1984

The Federal Reserve System has calculated numerical measures of the money supply for August 1984. As the numbers indicate, L includes the largest numbers and amounts of assets, and the range from M–1 to L is in order of decreasing liquidity of assets. M–1, consisting of currency, all demand and checkable deposits, and travelers' checks, is the most useful measure of money. The *Federal Reserve Bulletin* does not report numerical values of all components of a measure; those reported for M–2, M–3, and L are thus selected portions of the measures.

Use of gold, for all its advantages as a medium of exchange, had some drawbacks in actual trade. As Eastern and Western trade routes expanded and as exchange became less localized, merchants were forced to transport gold overland and overseas. A sale of goods by a merchant in Constantinople to buyers in Venice required a corresponding payment in gold. But traders soon recognized the safety and convenience of not actually transporting gold at all.

Gold "warehouses" developed, often run by people whose job it was to certify the gold or silver content of coins and to stamp the coins accordingly. These warehouses were a primitive though essentially complete form of the modern bank. Goldsmiths stored the precious metal for traders, issuing a receipt representing the quantity of gold on deposit. These warehouse receipts soon became acceptable as the medium of exchange in both local and international commerce.

This acceptability was a giant leap toward the development of nonconvertible, nonredeemable fiat money, although a commodity standard was still in existence at the time since these receipts were instantly convertible into gold or silver. Traders acquired faith in the goodness or ready convertibility of their warehouse receipts into gold, and with general acceptance, the receipts themselves became money. Currency had thus arrived; and modern checkbook money was only a small step away.

Accepting deposits is one of the two essential functions of banks, and the early goldsmith-warehousers originated the practice. A second major role of commercial banks is to make loans, a profitable but sometimes dangerous practice that was also devised by early bankers. In the beginning the warehouser undoubtedly earned some income solely by guarding the deposited gold and charging a fee for risk and insurance. The warehouser's receipts were backed by 100 percent of the deposited gold. Soon, however, the primitive banker recognized that not all depositors presented receipts for payment at the same time. On the average, perhaps half the receipt holders demanded their gold during any given day or week. The ever-present profit motive took over. Gold (or gold receipts) could be lent to borrowers at some rate of interest so long as some "safe" percentage of gold holdings was kept in reserve for conversion on demand. In this way, a **fractional reserve banking system,** in which banks maintain only a fraction of cash reserves against deposits, was born. The early goldsmiths issued more receipts than the actual value of gold in their vault. They kept a fraction of gold reserves in their vaults against the likelihood that depositors would demand their specie. The rest was lent at interest, creating another source of profit.

Fractional reserve banking system: A monetary system in which a bank or other financial institution holds only a portion, or fraction, of its total assets in liquid reserves against its liabilities; also called partial reserve system.

In performing the banking functions of accepting deposits and making loans, early goldsmith-bankers also encountered dangers similar to those inherent in modern fractional reserve banking. Greed in the form of lending too much money for profit—that is, underestimating the fraction of total gold reserves demanded by depositors at any given time—brought disaster. Bank runs occurred when customers became fearful that the goldsmith's deposits were not safe and immediately sought to withdraw their funds.

Two crucial points related to early banking are central to understanding modern banking systems: (1) The practice of lending at interest in a fractional reserve system meant that goldsmiths could actually alter the money supply, expanding it by lending, reducing it by calling in loans; and (2) goldsmiths could place all deposits and banks in jeopardy by misjudging the fraction of depositors' liabilities that would be presented for payment at one time. The first matter—multiple expansion and contraction of checkable deposits by commercial banks in a modern fractional reserve system—is the subject of Chapter 28. The second issue—the safety of the banking system and its regulation by the Federal Reserve System—is treated in Chapter 29. Before investigating these topics in detail, we take a brief overview of the contemporary banking system.

The U.S. Banking and Financial System

Commercial banks: Chartered financial institutions that make commercial and consumer loans and accept various types of deposits, including demand deposits.

There are about 15,000 commercial banks in the United States. A **commercial bank** is a privately owned but publicly regulated financial institution whose primary role is accepting checkable deposits and making business and consumer loans. There are many other nonbank financial institutions (see Table 27–2), some of which accept checkable deposits (money) but which specialize in particular kinds of securities or notes bought and sold. For example, savings and loan institutions have historically specialized in providing long-run mortgages and in accepting long-run time or savings deposits. These institutions are not, strictly speaking, commercial banks, but they are beginning to share the major function of commercial banks—to accept checkable

TABLE 27–2 **Major Types of Financial Institutions in the United States**

Type of Institution	Major Activities	Approximate Total Number	Total Assets (billions of dollars)
Commercial banks	Accepting checkable deposits; making business loans	15,000	1386
Savings and loan associations	Accepting time deposits and NOW accounts; making residential mortgage loans	4500	630
Life insurance companies	Issuing insurance policies; buying corporate bonds and making commercial mortgage loans	2000	470
Pension funds	Issuing pension plans; buying corporate stocks and bonds	Not known	455
Money market funds	Issuing shares in fund; buying short-term liquid securities	670	241.4
Mutual savings banks	Accepting time deposits and NOW accounts; making residential mortgage loans	442	172
Credit unions	Accepting savings deposits and NOW accounts; making consumer loans	20,000	69
Mutual funds	Issuing shares in fund; buying corporate stock	670	64

Source: From George S. Kaufman, *The U.S. Financial System: Money, Markets and Institutions,* 2nd ed., © 1983, p. 121. Adapted by permission of Prentice-Hall, Inc., Englewood Cliffs, N.J. Data from *Statistical Abstract of the United States, 1981* (Washington, D.C.: U.S. Government Printing Office) and Board of Governors of the Federal Reserve System, *Flow of Funds Accounts, 1957–1980,* (September 1981).

Federal Reserve System: The central bank of the United States, established by Congress in 1913, which regulates the nation's banks and other financial institutions.

deposits. In short, our financial system is evolving, and with the development of checkable deposits at nonbank financial institutions the distinction between banks and nonbanks is beginning to lose meaning. Moreover, modern regulations have brought the nonbank institutions under the control of the Federal Reserve System.[1] But it is still most informative to focus on commercial banks in discussing the role of financial institutions in money and money creation.

Both the individual states and the federal government charter, or license, privately owned commercial banks. National bank charters are issued by the Comptroller of the Currency (an official of the U.S. Treasury Department), while state banking commissions or similar bodies issue state bank charters. By law, all national banks must belong to the **Federal Reserve System,** whereas state banks may elect membership in "the Fed" subject to the approval of the Federal Reserve System. The basic role of the Federal Reserve

[1]Do not confuse the Federal Reserve System with the U.S. Treasury Department. The U.S. Treasury handles the budget, issues U.S. government bonds and securities to make up the difference when the budget is in deficit, and issues small amounts of money, mainly coins. The Federal Reserve System does not issue securities and does not act as the fiscal agent for government. It is primarily responsible for control and stability of prices and employment in the economy through control of the money stock.

System, a system of banks within geographic regions of the United States, is to regulate member banks and other financial institutions. About two-thirds of all banks do not belong to the Federal Reserve System, but as we will see, this limited membership does not prevent the Federal Reserve System from overseeing and regulating the activities of all banks.

Statistics on commercial bank assets and liabilities provide an idea of the overall importance of the commercial banking system in our economy. Table 27–3 shows the balance sheet (total assets equal total liabilities plus net worth) of all domestically chartered commercial banking institutions in the United States as of the end of December 1984.

A glance at Table 27–3 reveals that the major functions of the U.S. commercial banking system—accepting deposits and making loans—are identical to those of the early gold warehousers. Modern U.S. banking follows the same tradition, but it is far more complex in that many financial instruments—variants of the early bankers' warehouse receipts—have evolved. Regulations to prevent problems in the banking system have also evolved; in

TABLE 27–3   The Balance Sheet of All Commercial Banks, December 1984

The aggregate balance sheet of all commercial banks highlights the major assets and liabilities of commercial banks. Major assets are cash and income-earning loans and securities. Major liabilities include demand, savings, and time deposits. Double-entry bookkeeping guarantees that total assets equal liabilities plus capital or net worth. Aggregate data include all Federal Reserve System member and non-member commercial banks, mutual savings banks, and non-deposit trust companies in the United States, except branches of foreign banks.

Total Assets (billions of dollars)			Total Liabilities and Capital (billions of dollars)		
Total cash assets		190	Total deposits		1482
Currency and coin	23		Demand	371	
Reserves with the Federal Reserve banks	18		Savings	460	
			Time	650	
Balances with depository institutions	75		Borrowing		216
Cash items in process of collection	73		Other liabilities		117
Loans and securities		1525			
Loans, excluding interbank loans	1095				
U.S. Treasury securities	181				
Other securities	248				
Other assets		253	Residual assets (assets less liabilities)		152
Total assets		1969	Total liabilities and capital		1969

Source: Federal Reserve Bulletin (February 1985).

the United States the Federal Reserve System, which was created by the Federal Reserve Act of 1913, makes and enforces such regulations.

Assets. Many of the assets and liabilities of the commercial banking system are self-explanatory, but some require comment. In Table 27–3, total commercial bank cash assets in December 1984 were over $190 billion. The first but smallest category of cash assets is vault and till cash (currency and coins). The second and third cash assets are currency deposits that commercial banks have in other banks including the Federal Reserve banks. Commercial banks often have financial arrangements with other commercial banks (and other depository institutions such as savings and loans). These are called correspondent relations with correspondent banks. A special type of correspondent relation exists between commercial banks and the Federal Reserve banks: Commercial banks are required to keep a percentage of their deposit liabilities—a fractional reserve—available at all times. Some of this reserve is deposited in the Federal Reserve banks. The last cash asset, cash items in process of collection, refers to temporarily uncleared checks held by banks that, when cleared, will give the banks control over cash assets.

Loans and purchases of government securities and other interest-bearing instruments are the two most important means by which commercial banks earn returns and, presumably, profits. Loans include commercial, industrial, and private loans. Other assets ($253 billion in Table 27–3) include accrued interest on notes and securities, Federal Reserve bank stock (a required purchase by member banks), and real estate—bank premises, furniture and fixtures, and long-run real estate mortgages.

Liabilities. The major liabilities of commercial banks are deposits of several kinds. Demand deposits are checkbook money—generally acceptable orders by one business or individual depositor to pay another. Recall that such checkable deposits are included in M–1 and are therefore money as we have defined it. Savings deposits and time deposits include passbook savings deposit accounts, time deposits of various long-run durations, and interbank deposits. Most time deposits require notification of intended withdrawal or invoke interest penalties for early withdrawal. If savings accounts are checkable or instantly transferable to checkable deposits, they are money in the M–1 sense.

The borrowing category includes borrowings from other banks, financial institutions, or the Federal Reserve System. Commercial banks often borrow from each other and from the Federal Reserve to invest in interest-earning assets such as loans or securities.

The Federal Reserve System

The Federal Reserve System is the basic regulatory agency in the business of commercial banking and other financial intermediaries. Its relation to the banking system is similar to that of the Federal Communications Commission and the radio and television industry. One might assume that the main purpose of the Federal Reserve System is to harness the activities of overzealous bankers, who might act in the manner of overzealous gold warehousers—that is, lending out too much money at interest and thereby being unable to meet commitments to depositors. Bank and depositor safety, however, is not a primary or even essential function of the Federal Reserve.

The Federal Reserve System, or any other modern central bank such as the Bank of England or the Bank of Sweden, has two essential functions.

One is the older function of serving as a lender of last resort to commercial banks and other lending institutions—to respond quickly and adequately to bank runs, panics, or liquidity crises by providing currency or specie to meet withdrawals. The other, more modern function is to actively control the money supply to affect the business cycle and economic activity—to produce significant short-run effects on the rates of employment, inflation, and real income growth. This latter function developed slowly and has come into prominence only since the Great Depression of the 1930s, but it is now the major purpose of the Federal Reserve System. The Federal Reserve also conducts other service functions such as issuing currency and holding deposits of the federal government and its agencies.

The following chapters discuss monetary economics in detail—how the Federal Reserve controls the money supply and how changes in the money supply affect employment, inflation, and real income. As an introduction, we consider now the structure and membership of the Federal Reserve System.

Structure. Failure of the old National Banking System, established in 1863, to act as the lender of last resort led to the creation of the modern Federal Reserve Banking System in 1913. The system began operation with the following purposes: (1) to provide an elastic currency to help eliminate panics and bank runs; (2) to supervise the banking system at the federal level; (3) to provide facilities for the buying and selling of commercial paper, a means of providing funds to banks at an interest rate called the discount rate; and (4) to enlarge facilities for check clearance. A formal structure within which these functions were to be performed was also established. The original structure survives almost unchanged today, although concentrations of power over monetary policy have developed within the system over the years of its existence. These power concentrations will become apparent in our discussion of the formal structure.

The structure of the Federal Reserve System is shown in Figure 27–2. The Federal Reserve System is composed of two basic units: the decentralized twelve Federal Reserve banks (see Figure 27–3 for their locations) and the central node of power, the Board of Governors in Washington, D.C. The Board of Governors consists of seven members appointed by the president, with six governors serving staggered fourteen-year terms and the Chairman of the board serving a four-year term. The board in general and the chairperson of the board in particular are almost solely responsible for the establishment of national monetary policy.

The **Federal Open Market Committee (FOMC)** consists of the seven members of the board plus five Federal Reserve bank presidents; the president of the New York bank is always on the committee because of the amount of major financial activity that takes place in New York City. Virtually all monetary policy is carried out by the board and the Open Market Committee. The board must approve any changes in regulations that directly affect the supply of money.

In contrast to the centralization of power and action in the hands of the board and the FOMC, the actual structure of the Federal Reserve System is elaborate and decentralized. As Figure 27–3 shows, the system is organized into twelve districts with one Federal Reserve bank and a varying number of branch banks for each district. The Federal Reserve bank for the sixth district is in Atlanta, for example, and branch banks for that district are in Miami, Jacksonville, New Orleans, Birmingham, and Nashville. The Federal

Federal Open Market Committee (FOMC): A committee of the Federal Reserve System, made up of the seven members of the Board of Governors of the Federal Reserve System and five presidents of Federal Reserve district banks, that directs the open market operations (buying and selling of securities) for the system.

FIGURE 27–2

The Federal Reserve and Commercial Banking Systems

The commercial banking system and other institutions issuing checkable deposits are regulated and controlled by the Federal Reserve System, which is composed of twelve Federal Reserve banks and twenty-five branch banks. Monetary control is exercised through the Board of Governors and the Federal Open Market Committee in Washington, D.C. The seven members of the board, including its chairperson, are appointed by the president with Senate approval.

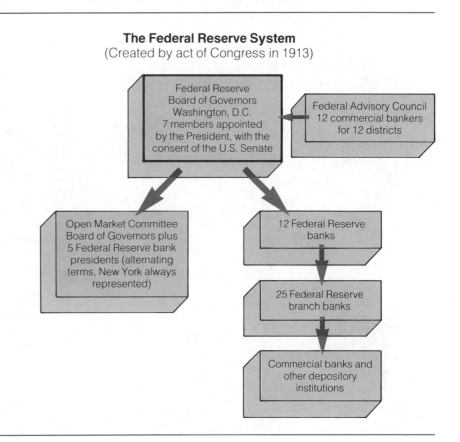

The Federal Reserve System
(Created by act of Congress in 1913)

Federal Reserve Board of Governors Washington, D.C. 7 members appointed by the President, with the consent of the U.S. Senate

Federal Advisory Council 12 commercial bankers for 12 districts

Open Market Committee Board of Governors plus 5 Federal Reserve bank presidents (alternating terms, New York always represented)

12 Federal Reserve banks

25 Federal Reserve branch banks

Commercial banks and other depository institutions

Reserve banks and their branches are geographically dispersed to service and inspect member banks and other depository institutions in their areas.

Membership. Membership in the Federal Reserve System has fluctuated over the years. Assets and deposits of member banks declined between the mid-1960s and 1980, especially during the late 1970s. The major reason for this decline was the increasing regulatory cost of belonging to the federal system in contrast to looser regulations over banks at the state level. Shifts therefore occurred from national bank charters to state charters. In response, Congress gave the Federal Reserve System sweeping new powers over all banks and over nonbank depository institutions with the passage of the Depository Institutions Deregulation and Monetary Control Act of 1980 (discussed in Chapter 29). Despite declining membership in the Federal Reserve System, the system directly controls all checkable deposits in all U.S. financial institutions, the largest component of the money supply. This control, as we will see in the following chapters, might enable the Federal Reserve System to control the overall money supply and greatly influence the course of economic growth.

The Importance of Money: A Preview

Individually we might take the use of money for granted, for some form of money has always been and always will be part of our lives. But most previous and contemporary economists have regarded the control of money as central to sound economic policy. The basis for this belief is the quantity

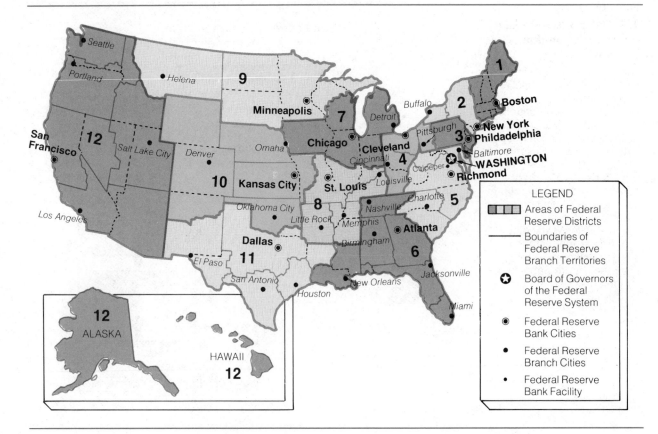

FIGURE 27–3   The Twelve Districts and Branches in the Federal Reserve System

The Federal Reserve System is geographically decentralized to service banks and other financial institutions throughout the United States. Services include the provision of an elastic currency supply to prevent bank runs, the discounting of paper, and check clearing facilities. The major purpose of the Federal Reserve System is to help provide full employment, economic growth, and price stability by control of the money supply.

Source: Federal Reserve Bulletin (February 1985).

theory of money (discussed briefly in Chapter 23), an idea that is well over two hundred years old and is now the basis for modern monetarist thought. Before turning to the intriguing conflicts over monetary policies, we must more closely investigate how bank money is created, how the Federal Reserve System affects money creation, and how monetarists view the world of aggregate economic activity.

Summary

1. Money was invented to avoid the transaction and information costs of barter, which required a coincidence of wants among traders and imposed limits on the division of labor and specialization.

2. Money serves as a medium of exchange, a unit of account, a standard of deferred payments, and a store of value. Money's most important function is

its general acceptability among traders as a means of exchange or payment. Virtually anything may serve as a medium of exchange, and there are costs and benefits attached to any item chosen.

3. Fiat money is paper or other low-cost money that is nonconvertible to gold or other commodities. Fiat money is certified by governments, and its general

acceptability is fostered by its acceptance by the government in the payment of taxes.

4. Money has a number of alternative official definitions, but the one most commonly used by economists is the classification M–1, which consists of all currency and coins in circulation and all checkable deposits at banks and other financial institutions. Currency and checkable deposits are the only two items that are generally acceptable as media of exchange.

5. The major functions of commercial banks are accepting checkable deposits (also called demand deposits) and making loans.

6. The Federal Reserve System was formed in 1913 to regulate and control the banking system. Its functions are to prevent bank collapse and to control the money supply to promote full employment and economic growth and to prevent inflation.

7. The structure of the Federal Reserve System includes twelve geographically dispersed Federal Reserve banks and twenty-five branch banks to serve the commercial banking and financial system. Decision-making power rests with the Board of Governors and the Federal Open Market Committee, both located in Washington, D.C.

Key Terms

barter economy
double coincidence of wants
medium of exchange
commodity money
gold standard

fiat money
unit of account
standard of deferred
    payment
store of value

liquidity
checkable deposits
    (demand deposits)
fractional reserve
    banking system

commercial banks
Federal Reserve System
Federal Open Market
    Committee (FOMC)

Questions for Review and Discussion

1. Explain the functions of money. Does inflation make money lose some of its functions?

2. Explain why a money system of exchange is better than a barter system.

3. For money to lower the costs of transactions and to perform its other functions well, it must have some particular physical characteristics. List and explain the desirable physical characteristics and also explain why paper makes better money than M&M's or stamps.

4. Would using gold as a medium of exchange solve the problems of constant changes in the level of prices? Explain your answer.

5. Define money and then discuss why each of the following is not considered money: IRAs, savings accounts, certificates of deposit, and gold.

6. What is the major feature that distinguishes M–1 from M–2, M–3, and L?

7. Essentially, what is a bank and what is the value to society of banks?

8. What is the major purpose of the Federal Reserve System? How does it achieve this purpose?

9. Why do you think the Federal Reserve is a system of decentralized banks rather than one central bank?

 **ECONOMICS IN ACTION**

The Spontaneous Emergence of Money: The Prisoner of War Camp

Stalag 17, a highly acclaimed play and movie, depicted everyday life in a German prisoner of war camp. Seth, the antihero entrepreneur, ran rat "horse races," sold schnapps fermented in his own distillery, and seemed to be able to obtain almost anything for a price. Money in the form of coins and currency did not exist, but trades were conducted by the prisoners with another form of money—cigarettes. This part of the fictionalized account of prison life has basis in fact as reported by an economic observer, R. A. Radford, who was a POW in Germany and Italy.[a] Radford's report tells us

much about the relation between economic activity and the spontaneous emergence of money.

POWs in Stalag 17 did not do paid work, but they were allotted weekly rations of goods from the Red Cross and other private sources. These rations consisted of such items as tinned milk, jam, butter, biscuits, chocolate, sugar, and cigarettes. Tastes differ among individuals, and with everyone receiving roughly equal amounts of the same commodities, prisoners would barter among themselves for the items they preferred. As time passed, trade expanded and the relative price of one item in terms of others became well known. Trade became increasingly complex, and a generally accepted medium of exchange developed in

[a]R. A. Radford, "The Economic Organization of a P.O.W. Camp," *Economica,* n.s. (November 1945), pp. 189–201.

Prisoners of war playing poker for cigarettes at Stalag Luft III, Germany, 1944.

the form of cigarettes, a commodity in common use. With the camps and their residents becoming semipermanent, prices of items in terms of cigarettes quickly became widely known. An exchange and mart board listing desired trades was set up: a pound of cheese for seven cigarettes, two cigarettes for a drink of schnapps. The use of cigarettes as a standardized commodity greatly reduced the transaction and information costs of trade and thus increased the volume of trade.

Additional economic issues were made easier by the new medium of exchange. Price fluctuations from Mondays when the provisions were handed out until a new shipment of rations came in on the following Sunday were smoothed out through a futures market. Speculators would buy on Mondays when the price of toothpaste and food was low. As supplies diminished through the week and prices rose, speculators would sell their hoarded goods for profit. Relative price differences for goods between Mondays and Sundays were thus narrowed.

The POW camp did not avoid the economic problems of inflation and deflation. Periodic injections of higher quantities of cigarettes produced inflation. Scarcity of cigarettes caused the price levels to fall.

As all economists know, prices are also affected by psychological factors. Air raids, good or bad war news, and the weather all affected the prices because these events clearly changed the nonmonetary demand for cigarettes. The prisoners would smoke cigarettes rather than save them. All commodity money standards are influenced by such factors. Altered demand for quetzal feathers as decoration or for gold in dentistry or industry would clearly affect the price level in economies using feathers or gold as money.

The POW camp experience illustrates the naturalness or spontaneity with which money as a medium of exchange emerges. The liberation of prisoners by the U.S. infantry created chaos for the cigarette standard. When commodities are no longer relatively scarce, there is no need for economic organization or activity. The POW economy, as Radford reports, simply collapsed. The lesson to be learned is that scarcity, coupled with the recognition by traders of the costs of barter, will inevitably lead to the use of money.

Question

Money, in order to function efficiently, must be portable, durable, divisible, easily recognized, stable in supply, and relatively scarce. How well do cigarettes conform to these characteristics? Can you think of any other goods likely to be found in a POW camp that would be preferable to cigarettes as money?

28

The Creation of Money

C hapter 27 defines *money* as anything that is generally acceptable as a medium of exchange—consisting of all currency, coins, and checkable deposits in the hands of the public. The most important part of our generally accepted money is checkbook money, called demand deposits or checkable deposits. In fact, in January 1985, the value of checkable deposits was more than two and a half times the value of dollars and coins in circulation—$398.4 billion in checkable deposits compared to $159.4 billion in currency.

Checkable deposits are money because of our faith that others will accept our checks—our orders to transfer funds—as currency and the faith of others that we will do likewise. This means that when the amounts kept in checkable deposits expand or contract, the economy's money supply also expands or contracts. In this chapter we explore how the process of money creation and destruction works through the commercial banking system. In the next chapter we will look at how the Federal Reserve System regulates the deposits to maintain economic stabilization.

Historically, commercial banks were exclusive sellers of checkbook money, and other nonbank financial institutions, also called thrift institutions—savings and loan institutions, mutual savings banks, credit unions, and the like—limited their services to transferring funds from savers to investors such as home buyers. But nonbank financial institutions began dealing in the checkable deposit business in the 1970s. Commercial banks are still the biggest sellers of checkbook money, but the nonbank institutions are beginning to compete strongly for such accounts. Checkable deposits at nonbank financial institutions grew from $8.4 billion in 1978 to $149 billion in January 1985. The general blurring of the roles of banks and nonbank institutions reflects the dynamic character of the financial side of our economy. For ease of understanding, we discuss money creation from the perspective of commercial banks.

Money Creation by a Single Bank

The major functions of any commercial bank are to accept deposits and to make loans and investments—the same functions as the early banker-warehousers. We present a simplified balance sheet for a hypothetical individual bank in Table 28–1 to show how these roles interact. A balance sheet uses the accounting identity, the convention that the sum of all bank assets must equal the sum of all liabilities plus capital or net worth. The assets of the hypothetical bank are listed on the left side and liabilities and net worth on the right side. Double-entry bookkeeping—in which every asset creates an equal liability and vice versa—guarantees that the balance sheet always balances.

Major bank liabilities are sums that the bank owes to others. These obligations include checkable, or demand, deposits, which depositors can draw on by writing checks, as well as savings, time, and interbank deposits. Liabilities also include the bank's borrowings from other commercial banks and from the Federal Reserve bank. Capital accounts, funds raised from the sale of stocks, are placed on the right side of the balance sheet along with the liabilities.

The assets that balance these figures are all the things the bank owns—cash, notes representing loans to consumers and businesses, securities, bank buildings, equipment, and so on.[1]

Balance sheets give a snapshot of banking activities. To better understand the importance of banks and their role in the expansion and contraction of the money supply, we need to step back for a moment and look at the process of how a bank gets started.

How a Bank Gets Started

Suppose that a group of investors organizes and decides that the town of Show Low, Arizona, needs a new bank and that such a bank would be profitable. They apply for and obtain a state charter and pledge $2 million in

[1]Throughout this chapter we will be using the word *cash* to refer to currency and coin held in the bank's vaults or tills plus cash balances held with the Federal Reserve.

TABLE 28–1   Simplified Bank Balance Sheet

Capital investments of $2 million permit the formation of the Second Bank of Show Low. The left side of the balance sheet gives the bank's assets—cash, notes representing loans to its customers, securities, and buildings and equipment. This balance sheet shows that the bank's entire initial investment was used to purchase the building and furnishings. The right side of the balance sheet gives the bank's liabilities—all deposits (checkable, time, and savings), its borrowings from other banks, and capital invested in it. Double-entry bookkeeping is used, so every asset creates an equal liability and vice versa.

Second Bank of Show Low Balance Sheet			
Assets (thousands of dollars)		Liabilities and Capital Accounts (thousands of dollars)	
Cash	0	Deposits	0
Loans	0	Borrowings	0
Securities	0	Capital	2000
Plant and fixtures	2000		
Total assets	2000	Total liabilities and capital	2000

TABLE 28–2   **Recording a Deposit on the Balance Sheet**

A deposit of $1 million creates liabilities of $1 million for the bank and cash assets of an equal amount. Total assets remain equal to total liabilities plus capital.

Second Bank of Show Low Balance Sheet			
Assets (thousands of dollars)		Liabilities and Capital Accounts (thousands of dollars)	
Cash	1000	Deposits	1000
Loans	0	Borrowings	0
Securities	0	Capital	2000
Plant and fixtures	2000		
Total assets	3000	Total liabilities and capital	3000

capital investment. With these funds they purchase a building, they elect directors, they christen their bank the Second Bank of Show Low. The balance sheet of the bank after the purchase of the plant and equipment is the one given in Table 28–1.

Note that the accounting identity is satisfied in the balance sheet. Total assets are $2 million worth of plant and fixtures (all of the original asset, cash, was converted into the bank building and furnishings). These assets are equal in value to the sum of total liabilities and capital stock because at this point of bank organization, the initial investors have claim against the value of the bank's plant and fixtures. There are no other assets besides the building and no liabilities, but this situation cannot last if investors hope to earn a return on their capital, which they certainly do. The bank must begin to function by accepting deposits.

Assume that an entrepreneur, Josephine Eccentric, is a citizen of Show Low. Josephine has made a fortune as a fast-food restaurant organizer. She deposits $1 million in cash in the Second Bank of Show Low. The bank's new balance sheet appears as Table 28–2.

The Second Bank has now acquired $1 million in cash and a counterbalancing demand deposit liability. As a result of Josephine's deposit, the total money supply has not changed in amount, but it has changed in composition. Josephine gave up currency for a checking account deposit. The checkable deposit is counted as money, however, so Josephine's deposit simply altered the composition, not the amount, of the money supply. The act of putting money in the bank as a deposit to be drawn at a later time allows the bank to create more money by means of the fractional reserve banking system.

The Fractional Reserve System

The *fractional reserve banking system*—in which banks keep only a percentage of funds deposited with them available for withdrawals—is a vital aspect of banking behavior regulated by the Federal Reserve System. We will briefly examine how Josephine's cash in the Second Bank becomes part of the fractional reserve system.

Cash assets of commercial banks held in their own vaults or at the district Federal Reserve bank are called **cash reserves.** The Federal Reserve specifies that banks and other financial institutions must keep a certain percentage of their cash reserves on hand or with a Federal Reserve bank at all times.

Cash reserves: Cash assets of commercial banks held in their vaults or at the Federal Reserve district bank.

Required reserves: A portion of a bank's cash assets that must be kept available at all times, according to the legal requirements of the Federal Reserve System.

Reserve ratio: A percentage of the checkable or demand deposit reserves held by a bank.

Excess reserves: Cash reserves above required reserves.

These **required reserves,** sometimes called legal reserves, can be expressed as a percentage of checkable deposits or as a **reserve ratio:**

$$RR = r(D),$$

where r is the reserve ratio, D is the amount of demand deposit liabilities, and RR is the amount of required reserves for demand deposits. (Different reserve ratios apply to the various kinds of savings and time deposits.) If, for example, the deposit liabilities at a commercial bank are $1,000,000 and r is 10 percent, or 0.10, required reserves must equal $100,000. If r is 5 percent, 0.05, required reserves are $50,000.

Depending on business conditions or expected loan demand, the bank may well decide to hold more than the required percentage of deposits as cash reserves. The total quantity of cash reserves held by a bank or the banking system consists of required reserves and excess reserves. **Excess reserves** are simply cash reserves over and above those reserves required by the Federal Reserve. If the required reserve ratio is 10 percent and deposit liabilities are $1 million, excess reserves would total $900,000 ($1 million − $100,000 = $900,000). Some of these excess reserves could be desirable for the bank—an extra cushion against the possibility of default, of its being unable to meet its depositors' demands for cash. Some excess, however, may be undesirable, for the bank is holding funds that it could use to earn profits through loans or purchases of securities. Generally banks try to minimize the amount of excess reserves they keep on hand.

Assume that the bank in Show Low puts all of its cash on deposit with the district Federal Reserve bank in San Francisco. Our assumption that the bank keeps none of its deposits in its own vaults is unrealistic, of course, but we are simplifying matters for clarity. The balance sheet of the Second Bank of Show Low now appears as Table 28–3. The balance sheet of the San Francisco Federal Reserve bank appears as Table 28–4. The composition of the Second Bank's assets has changed—its cash is now listed as cash reserves—but the structure as well as the total amount of liabilities and net worth remains the same as in Table 28–2. The Second Bank's cash reserves can be instantly converted into cash disbursements as long as the legal reserve requirement is still met. Assuming that the reserve ratio (r) is 10 percent (0.10), legal required reserves against deposits of $1 million are 0.10 ×

TABLE 28–3   Depositing a Cash Reserve in a Federal Reserve Bank

The Second Bank of Show Low deposits cash in the amount of $1 million in the San Francisco Federal Reserve bank. These cash deposits are called cash reserves or simply reserves and are recorded as cash reserve assets rather than cash on the balance sheet. Bank cash reserves are sometimes physically held in the bank's own vaults.

Second Bank of Show Low Balance Sheet			
Assets (thousands of dollars)		Liabilities and Capital Accounts (thousands of dollars)	
Cash	0	Deposits (demand)	1000
Cash reserves	1000	Borrowings	0
Loans	0	Capital	2000
Securities	0		
Plant and fixtures	2000		
Total assets	3000	Total liabilities and capital	3000

TABLE 28–4   The Federal Reserve Bank's Balance Sheet

By accepting the Second Bank's deposit, the San Francisco Federal Reserve bank acquires cash in the amount of $1 million and incurs an equal amount of liabilities to the Second Bank in the form of reserves. Reserves are assets to the Second Bank but are liabilities to the Federal Reserve bank.

San Francisco Federal Reserve Bank Balance Sheet			
Assets (thousands of dollars)		Liabilities and Capital Accounts (thousands of dollars)	
Cash	1000	Claims against reserves (Second Bank of Show Low)	1000

$1,000,000, or $100,000. Excess reserves are $1,000,000 − $100,000, or $900,000.

Table 28–4 shows the change in the San Francisco Federal Reserve bank's balance sheet after accepting the Show Low bank's cash deposit. On the asset side, the cash deposit is an asset to the Federal Reserve bank. The new liabilities are the claims against the reserves that the Federal Reserve bank holds for the commercial bank.

How Checks Clear

What happens within the banking system when a depositor in one bank sends a check to a depositor in another bank? Currency is not trucked from one bank to another; the transaction simply involves bookkeeping shifts known as check clearance, orchestrated by the Federal Reserve district banks and by other clearing facilities provided by commercial banks.

Suppose Josephine, the initial depositor in the Second Bank of Show Low, plans to expand her restaurant business and requires additional inputs and equipment. She has found bargains at a restaurant supply shop in Needles, California, called Roberto's Restaurant City. A purchase of $400,000 worth of restaurant equipment is arranged, and Josephine writes a check in that amount to the order of Roberto's and drawn on the Show Low bank.

After Roberto's receives Josephine's check, it deposits the check in the Needles National Bank—a member of the Federal Reserve System with deposits in the Federal Reserve bank in San Francisco. Figure 28–1 traces the movement of the check. Needles National Bank incurs a new demand deposit liability to Roberto's, but it also acquires a matching asset—cash reserves—after the check is cleared through the San Francisco Federal Reserve bank. When Needles National Bank forwards Josephine's check to the San Francisco bank, the Federal Reserve bank balances its books by increasing its reserve liabilities to Needles National Bank and simultaneously decreasing its reserve liabilities to the Show Low bank. The Federal Reserve bank then "clears" the check by sending it on to the Second Bank of Show Low, which adjusts its own accounts by reducing cash assets by $400,000 and deposit liabilities to Josephine by the same amount. These changes are all summarized in Figure 28–1.

Check clearance is not a mysterious activity. The simple example of Josephine and Roberto's describes what happens every day when millions of checks are written and cleared. The example actually is one of intradistrict check clearance, where the two banks are in the same Federal Reserve district but in different and distant cities. Check clearinghouse facilities often

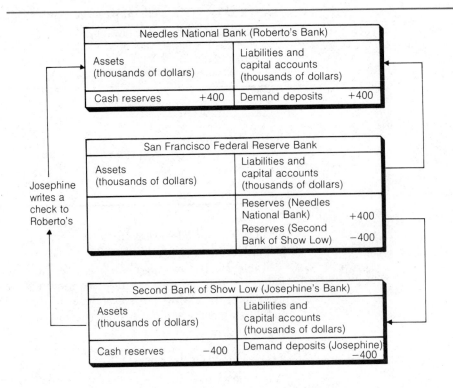

FIGURE 28–1   Clearing a Check

Josephine writes a check to Roberto's in the amount of $400,000, and Roberto's deposits the check in the Needles National Bank. Needles National Bank in turn deposits the check in the San Francisco Federal Reserve bank, obtaining reserve assets. The Show Low Bank loses reserve assets in the amount of $400,000, as does Josephine's account when the check is cleared and all books are balanced.

exist within cities, in which case the services are called intracity clearing. When a check is written to someone who holds an account with the same bank, the check is cleared *intrabank,* or within the bank: Deposit liabilities are simply shifted with a change in the accounting entries on the bank's balance sheet. The principles of clearance are the same whether the writer and the receiver of the check live in the same city or in separate Federal Reserve districts (interdistrict clearing). In the latter case, all Federal Reserve banks have deposits in other Federal Reserve banks, which permits interdistrict clearing. Bookkeeping entries are adjusted without the necessity for cash to change hands. One bank gains cash reserves, another loses them. Note that the two banks involved both gain and lose reserves and deposits in equal amounts. This point is crucial in describing the process of money creation by a commercial bank or by the banking system as a whole.

Loans and Money Creation

Banking, like any business, seeks to maximize profits. Banks earn income by investing in various private and government securities, but the most important means of earning income is making loans and earning interest on them. In a fractional reserve banking system, banks are able to make loans equal to the amount of their excess reserves. This important principle can be expressed symbolically as

Drawing by Levin; © 1984 The New Yorker Magazine, Inc.

$$L = R - rD,$$

where L is the credit or loan creation potential of a single commercial bank, R is the total reserves of the single bank, r is the legal reserve requirement on demand deposits, and D is demand deposit liabilities.

Loans, L, cannot exceed the value of excess reserves because the individual bank faces the real prospect of losing both its reserves and its deposits. Consider the Second Bank of Show Low, whose cash reserves and deposit liabilities have dropped to $600,000 after Josephine's check to Roberto's for $400,000 clears (Table 28–5). If the Federal Reserve bank's required cash reserve ratio (r) is 10 percent, $60,000 of the cash reserves must be kept on hand, leaving excess reserves of $540,000. Suppose that the Show Low bank lends this $540,000 to one borrower, Madge's Greenhouse, in demand deposits (for simplicity, we assume that no currency is withdrawn). Madge is creditworthy and will pay interest on the loan, providing the bank with

TABLE 28–5   Losing Deposits and Reserves on the Balance Sheet

After Josephine's check for $400,000 has cleared, the Second Bank of Show Low loses deposit liabilities in that amount and cash reserves in the same amount.

Second Bank of Show Low Balance Sheet			
Assets (thousands of dollars)		Liabilities and Capital Accounts (thousands of dollars)	
Cash reserves	600	Demand deposits	600
Loans	0	Borrowings	0
Securities	0	Capital	2000
Plant and fixtures	2000		
Total assets	2600	Total liabilities and capital	2600

Second Bank of Show Low Balance Sheet

1. Initial balance sheet

Assets (thousands of dollars)	Liabilities and Capital Accounts (thousands of dollars)
Cash reserves 600	Demand deposits (Josephine) 600

2. Madge's Greenhouse is lent $540,000.

Assets (thousands of dollars)	Liabilities and Capital Accounts (thousands of dollars)
Cash reserves 600	Demand deposits (Josephine) 600
Loans (Madge's Greenhouse) 540	Demand deposits (Madge's Greenhouse) 540

3. Madge draws check in the amount of $540,000.

Assets (thousands of dollars)	Liabilities and Capital Accounts (thousands of dollars)
Cash reserves 600 − 540	Demand deposits (Josephine) 600
Loans (Madge's Greenhouse) 540	Demand deposits (Madge's Greenhouse) −540

4. Balance sheet after Madge's deposit is checked away.

Assets (thousands of dollars)	Liabilities and Capital Accounts (thousands of dollars)
Cash reserves 60	Demand deposits (Josephine) 600
Loans (Madge's Greenhouse) 540	540

FIGURE 28–2   A Single Bank Making a Loan and Creating Money
The Second Bank of Show Low creates money when it accepts something that is not money—a note pledging future payment from Madge's Greenhouse—and provides something that is—a checkable demand deposit. The bank is fully committed with loans when Madge's deposit, created by the loan, is checked away because $60,000 in reserves is required to support demand deposits of $600,000 when the required reserve ratio is 10 percent.

returns and, presumably, profits. No one borrows money to let it sit in the bank. Madge promptly writes checks on her demand deposit account for $540,000. These transactions are summarized in Figure 28–2, beginning with the initial situation of the Second Bank of Show Low.

By lending Madge its excess reserves of $540,000, the bank changes its balance sheet to number 2 in Figure 28–2, with new loan assets and deposit liabilities. Madge then purchases $540,000 worth of goods and services, so in balance sheet number 3 of Figure 28–2, the bank's demand deposits and reserves are "checked away" to another bank or banks. In the final accounting, the balance sheet of the Second Bank of Show Low appears as number 4 in Figure 28–2.

The Second Bank of Show Low has created money in the process of making loans to Madge. It traded a demand deposit, which is money, for a promissory note, Madge's debt, which is not money. The money supply expands by exactly $540,000 (until, of course, the loan is paid off by Madge).

The single bank can lend out only the amount of its excess reserves because it faces the legal reserve requirement as well as the loss of both reserves and deposits. In Figure 28–2, for example, the Show Low bank is fully committed with loans after Madge's check has cleared (the bottom balance sheet). One more dollar loss in deposits and reserves would force the bank below the legal reserve requirement.

Money Creation in the Commercial Banking System

The single commercial bank can lend only up to the amount of its excess reserves, but what of an entire banking system? How much money can be created by all of the depository institutions under control of a central bank such as the Federal Reserve System? To simplify the answer to this important question, we initially assume that there is only one bank to service the entire economy—a monopoly bank, in effect. This assumption is farfetched, but it will help us understand the principles of money creation in a multibank context.

Deposit Expansion by a Monopoly Bank

We call our hypothetical monopoly bank the Universal Monopoly Bank. We initially assume that it is under the control of a central bank and that its balance sheet appears as in Table 28–6. If the reserve requirement imposed by the central bank is 10 percent, then Table 28–6 indicates that the Universal Monopoly Bank is fully committed with loans. Universal Monopoly Bank holds $150 million in cash reserves against deposit liabilities of $1.5 billion; in other words, cash reserves are exactly 10 percent of total deposits. Universal Monopoly Bank has no excess reserves and therefore no lending capacity.

Suppose that because of the interest-earning potential of U.S. government treasury securities, the central bank wishes to buy $50 million of the securities from Universal. (These securities are promises to pay back the principal, or face value, at maturity plus interest over specified time periods.) How does the central bank induce the Universal Monopoly Bank to sell its securities? It goes into the open securities market and increases its demand for securi-

TABLE 28–6   Universal Monopoly Bank's Initial Balance Sheet

With deposit liabilities of $1.5 billion and a cash reserve requirement of 10 percent, the Universal Monopoly Bank is fully committed with loans because it holds exactly $150 million in cash reserve assets. The Universal Monopoly Bank does not face the possibility of losing deposits and reserves to other banks, however, because there are no others.

Universal Monopoly Bank Balance Sheet			
Assets (millions of dollars)		Liabilities and Capital Accounts (millions of dollars)	
Cash reserves	150	Deposits	1500
Loans	250	Borrowings	0
Securities	1000	Capital	500
Plant and fixtures	600		
Total assets	2000	Total liabilities and capital	2000

ties. The increase in demand raises the price of such securities, provoking a sales response from the monopoly commercial bank, which seeks to earn a capital gain. If we assume that Universal Monopoly Bank sells $50 million worth of its previous total of $1 billion in securities, its new balance sheet appears as in Table 28–7.

In Table 28–7, securities have been converted into cash reserves, meaning that the monopoly bank now has excess reserves of $50 million. The question now, is What quantity of demand deposit liabilities will this $50 million support?

If the Universal Monopoly Bank lent out all its excess reserves, loans would expand by $50 million, as would deposit liabilities. But what of cash reserves? Since the bank is a monopoly, it cannot lose reserves to other banks in the system via check clearance, for there are no other banks in the system. Cash reserves therefore will remain at the same level—$200 million—after an initial loan and deposit creation of $50 million. Since reserves cannot be checked away, the Universal Monopoly Bank will still hold excess reserves after the loan. Under a 10 percent reserve requirement, deposit liabilities of $1.550 billion require only $155 million in legal cash reserves. The bank still holds excess reserves in the amount of $45 million ($200 − $155 = $45). Suppose it lends that amount. Lending out $45 million in deposits will create still another income-earning asset (loans), and new deposit liabilities will now be $1.595 billion and required reserves will be $159.5 million (10 percent). Excess reserves still remain in the amount of $40.5 million. If the bank continues to lend excess reserves to new borrowers who hold demand deposits (and no currency), the creation of money through **deposit expansion** will be many times the initial amount of excess reserves. Deposit expansion is the total amount of money created above some initial amount of excess reserves.

The total amount of deposits that can be supported by cash reserves can be conveniently expressed as

Deposit expansion: The total value of bank deposits that an initial amount of reserves can support given some reserve requirement.

$$D = R\left(\frac{1}{r}\right),$$

where D is the total amount of deposits and demand deposit money created, R is the total amount of cash reserves, and r is the required reserve ratio.[2]

The reciprocal of the reserve ratio ($1/r$) is sometimes referred to as the **simple money multiplier**—the number multiplied by reserves to obtain total deposit expansion. In the example of the Universal Monopoly Bank, the money multiplier is 10 ($1/0.10 = 10$). The lower the required reserve ratio, the larger the money multiplier; the higher the ratio, the smaller the multiplier. A value of r of 20 percent would reduce the value of the multiplier to 5 ($1/0.20 = 5$), and so on.

Simple money multiplier: The reciprocal of the reserve ratio.

To see the effects of the money multiplier, consider the final position of the monopoly bank after all excess reserves have been eliminated from the system. The monopoly bank is again fully committed in loans, as shown in

---

[2]There is a clear parallel between the investment (or autonomous expenditures) multiplier discussed in Chapter 23 and the money multiplier. Recall that the autonomous expenditures multiplier is equal to the reciprocal of the marginal propensity to save (MPS). It is the multiple by which income changes when autonomous (investment or consumption) spending changes. The money multiplier is calculated in similar fashion, but it is the reciprocal of the required reserve ratio. It is the percentage by which money expands or contracts with increases or decreases in excess cash reserves in the banking system.

TABLE 28–7   Selling Securities and Creating Excess Reserves

A sale of $50 million worth of securities (such as U.S. Treasury bills) for cash increases Universal Monopoly Bank's reserves by $50 million. These $50 million in cash reserves are excess reserves because they are greater than the amount ($150 million) required against $1.5 billion in deposit liabilities.

Universal Monopoly Bank Balance Sheet			
Assets (millions of dollars)		Liabilities and Capital Accounts (millions of dollars)	
Cash reserves	200	Deposits	1500
Loans	250	Borrowings	0
Securities	950	Capital	500
Plant and fixtures	600		
Total assets	2000	Total liabilities and capital	2000

TABLE 28–8   Increasing Deposit Liabilities and Loans

Initial excess reserves of $50 million can support an additional $500 million in deposit liabilities in the monopoly bank when the money multiplier is 10. A monopoly bank, unlike a single commercial bank, cannot lose deposits or reserves to other banks.

Universal Monopoly Bank Balance Sheet			
Assets (millions of dollars)		Liabilities and Capital Accounts (millions of dollars)	
Cash reserves	200	Deposits	2000
Loans	750	Borrowings	0
Securities	950	Capital	500
Plant and fixtures	600		
Total assets	2500	Total liabilities and capital	2500

Table 28–8. Reserves of $200 million support $2 billion in deposits (money). This value is calculated from the deposit expansion equation, $D = R(1/r)$, or $2000 = $200(10)$. The change in deposits and the money supply can be calculated using a slight variation of the earlier expression:

$$\Delta D = \Delta \text{Excess reserves} \left(\frac{1}{r} \right),$$

or, in the case of the Universal Monopoly Bank,

$$\$500 = \$50 \left(\frac{1}{0.10} \right).$$

By selling $50 million in securities, the monopoly bank actually increased the money supply by $500 million, the amount of deposits that the cash it receives will legally support.

A crucial principle emerges from this example of the monopoly bank: Single commercial banks can lend and create checkbook money only in an amount equal to their excess reserves, whereas a monopoly bank can lend and create money by a multiple $(1/r)$ of its excess reserves. Now we will see how this principle translates into the real world of the U.S. multibank competitive banking system.

Money Expansion in the Multibank System

Just as the monopoly bank cannot lose reserves when deposits are checked away, the banking system as a whole can lose neither reserves nor deposits. Money expansion through the loan process is the same for the fractional reserve multibank system used in the United States as for the simple monopoly bank we just described. Once again, we will analyze the process of money creation in the multibank system with the use of balance sheets.

As with the Universal Monopoly Bank case, we must begin with some simplifying assumptions. First, we assume that the entire banking system is initially fully committed with loans. There are initially no excess reserves in the system. Second, all borrowers at all commercial banks are assumed to want their loans in demand deposits rather than in currency. Third, as in the previous balance sheets, savings deposits are assumed not to exist. Fourth, we assume that banks are solely profit maximizers and do not wish to hold any excess cash reserves. In other words, banks do not hold any reserves in excess of legally required reserves. Finally, to simplify matters even further, we assume that in each transaction, the bank granting a loan lends its entire excess reserves to a single borrower who places the loan in another bank, which then does the same thing. We assume that excess reserves work their way through the banking system in this simplified manner.

Initially we assume that Cut and Shoot National Bank and all other commercial banks are fully committed with loans. Now suppose that Monica, a wealthy miser, deposits $1 million in cash in Cut and Shoot National Bank. The balance sheet change for Cut and Shoot is shown in Table 28–9a. The bank acquires cash reserve assets and demand deposit liabilities (to Monica) of $1 million. Cut and Shoot is now in a position to lend dollar for dollar with its excess reserves. Given the assumptions described and a legal required reserve ratio of 10 percent, excess reserves exist in the amount of $900,000 because total reserves in the bank are $1 million, and required reserves are $100,000 $(R - rD) = \$1,000,000 - [0.10 \times \$1,000,000] = \$900,000$.

Cut and Shoot finds a borrower, Jonathan, for the entire amount of its excess reserves. Table 28–9b reflects the position of Cut and Shoot after it makes the loan. The asset category Loans is enhanced by $900,000, as are deposits of a like amount. Jonathan, however, does not let his demand deposit lie idle. He has been in debt to Yvonne of Altoona State Bank on a business deal and wishes to pay back $900,000 of the debt. Jonathan writes a check to Yvonne, who deposits it in her bank. When Jonathan's check is cleared through the Federal Reserve System's interdistrict clearinghouse, Cut and Shoot loses cash reserves of $900,000 and Jonathan's deposit liabilities of the same amount. Simultaneously, Altoona State Bank gains cash reserves and deposit liabilities of $900,000. Cut and Shoot is left with an interest-earning asset (the loan to Jonathan), but its cash reserves have been checked away to another bank. At a 10 percent legal reserve ratio, Cut and Shoot is fully committed with loans.

Not so at Altoona State Bank. After Yvonne's deposit, Altoona State Bank has excess reserves in the amount of $810,000 ($900,000 − [0.10 × $900,000]). These reserves may be lent to provide income for the bank. A borrower, Chris, wants the entire proceeds of the loan in the form of a demand deposit so that he can pay off a debt to Susan, a creditor who lives

TABLE 28–9   Money Creation in the Commercial Banking System

Money is created as loans and demand deposits are created in the banking system. In this example, as before, the reserve requirement is 10 percent. (a) Cut and Shoot National has excess reserves of $900,000, which it lends to Jonathan (b). When Jonathan checks his new deposit away to Yvonne, a customer of Altoona State Bank, Cut and Shoot loses deposit liabilities and cash reserve assets of $900,000 to Altoona State Bank (c, d). Altoona State Bank, with excess reserves of $810,000, lends the entire amount to Chris, creating new money in the amount of $810,000 in deposits (e). Chris checks his deposits away to a third bank (not shown) leaving Altoona State Bank fully committed with loans. As excess reserves and deposits move through the financial system, money is created.

(a) Cut and Shoot National Balance Sheet

Cash reserves	+1000	Deposits (Monica)	+1000

(b) Cut and Shoot National Balance Sheet

Cash Reserves	+1000	Deposits (Monica)	+1000
Loans (Jonathan)	+900	Deposits (Jonathan)	+900

(c) Cut and Shoot National Balance Sheet

Cash reserves	+100	Deposits (Monica)	+1000
Loans (Jonathan)	+900		

(d) Altoona State Bank Balance Sheet

Cash reserves	+900	Deposits (Yvonne)	+900

(e) Altoona State Bank Balance Sheet

Cash Reserves	+900	Deposits (Yvonne)	+900
Loans (Chris)	+810	Deposits (Chris)	+810

(f) Altoona State Bank Balance Sheet

Cash reserves	+90	Deposits (Yvonne)	+900
Loans (Chris)	+810		

Note: Balance sheet figures in thousands of dollars.

in Buffalo. Susan accepts Chris's check and deposits it in her bank (Buffalo Bank and Trust, not shown in Table 28–9). After Chris's deposit is checked away to Buffalo, Altoona State Bank loses deposit liabilities of $810,000 and cash reserves of the same amount. Altoona State Bank is fully committed with loans (Table 28–9f) when it holds $90,000 in legally required reserves against Yvonne's deposit of $900,000.

Excess reserves make their way through the banking system in the manner described in Table 28–9. Each individual bank can lend only an amount that is within its excess reserves, but part of these reserves becomes excess to some other bank in the system.

Naturally, the process described in the balance sheets of Table 28–9 does not end with Altoona State Bank. Reserves are checked away to Buffalo and may stay in Buffalo, be checked away to some other bank, or return to Cut and Shoot. Given our assumptions about the continuous movement of excess reserves through the system, the money creation process stops only when there are no more excess reserves within the commercial banking system. Table 28–10 summarizes this process. (Also see Focus, "Money and Computers," for a description of an alternative step in the money creation process.)

In the limit, $9 million worth of new loans and deposits are created from the initial $1 million deposit of cash. As the individual commercial banks make loans, they are accepting something from borrowers that is not

TABLE 28–10   Creating Money as Excess Reserves Work Their Way Through the Banking System

The money multiplier, the reciprocal of the required reserve ratio, determines the amount by which the money supply can increase. With a required reserve ratio of 10 percent and initial excess reserves of $900,000, demand deposits and the money supply can be increased by $9 million.

Bank	Reserves and Deposits Acquired	Required Reserves	Excess Reserves	Bank Loans	Increase in Demand Deposits	Increase in Money Supply
Cut and Shoot	1000	100	900	900	900	900
Altoona State	900	90	810	810	810	810
Buffalo Bank and Trust	810	81	729	729	729	729
.	.	.	.	.	.	.
.	.	.	.	.	.	.
.	.	.	.	.	.	.
All other banks	7290	729	6561	6561	6561	6561
Totals	10,000	1000	9000	9000	9000	9000

Note: All numbers are in thousands of dollars.

money—a private note payable to the bank—and are providing the borrower with a demand deposit that is money. Individual banks can lend dollar for dollar with excess reserves only because they face the possibility of losing both cash reserves and deposits, but reserves and deposits cannot be lost to the multibank system as a whole. We may think of this system as closed, just as the monopoly bank could not lose deposits and reserves. In fact, the expression for deposit expansion that applies to the monopoly bank applies to the commercial banking system as well. In both cases, deposit expansion is expressed as

$$\Delta D = \Delta \text{Excess reserves} \left(\frac{1}{r} \right),$$

or, in this case,

$$\$9,000,000 = \$900,000 \left(\frac{1}{0.10} \right).$$

Deposits—and therefore the money supply—expand by $9 million in the multibank system when excess reserves of $900,000 are introduced into the system. Do not forget that the initial cash deposit of $1 million (Monica's deposit) merely altered the form of the money supply from currency outstanding (in Monica's possession before her deposit) to demand deposits. The money supply was not increased by Monica's cash deposit, although Cut and Shoot Bank had to hold a legal cash reserve of $100,000 against it. The money supply increased by $9 million through the process of deposit expansion *after* the initial deposit of $1 million. Thus, the change in the money supply, ΔM, is also expressed as

$$\Delta M = \Delta D = \Delta \text{Excess reserves} \left(\frac{1}{r} \right), \qquad \text{or}$$

$$\$9,000,000 = \$9,000,000 = \$900,000 \left(\frac{1}{0.10} \right).$$

FOCUS   **Money and Computers:**
Banking at Home

Usually a trip to the bank is necessary to withdraw cash or take out a loan, but transferring checkable deposits in payment of bills (or, in the future, transfers like those from Jonathan to Yvonne) and other banking business may be conducted in the comfort and privacy of the home. Big banks such as Chemical Bank, Citibank, and Bank of America, offering services called Pronto, Home Base, and Home Banking, permit individuals to plug in a phone modem to a home computer and conduct a growing variety of banking activities.[a] For between $8 and $12 per month, these big banks offer computerized balance information, fund transfers between accounts (such as from savings to checking), electronic statements, and automatic bill paying. Many banks across the United States are experimenting with such systems, although the sale of such banking services may be tied with other services such as the Dow Jones News and Information Service and other broader computer packages. Technology may alter the physical means and speed through which deposits are shifted from one account to the other, but such shifts are still part of the process of creation and destruction of deposit money and reserves in the banking system.

[a]See Robert A. Bennett, "Banking Goes into the Home," *New York Times* (December 7, 1983).

Advertisement for a home banking service

Money expansion: The increase of the money supply made possible by an initial amount of excess reserves in the banking system.

Money is expanded on the basis of excess reserves, and money expansion takes place at the same rate as the expansion of loans and deposits. In fact, **money expansion** is equivalent to demand deposit expansion. The process of money expansion can also work in reverse. Contraction of the money supply is possible within the fractional reserve commercial banking system.

Money Supply Leakages

Two important sources of leakage from the money expansion process are cash withdrawals and idle reserves. In the previous section, we eliminated both situations for simplicity. Here we examine how either cash withdrawal or idle reserves can slow or even reverse the process of money expansion.

Currency Drain

In the example described in the previous section, if a depositor had withdrawn cash from his or her account in the Cut and Shoot Bank, a chain of deposit destruction could have been brought about. While the act of currency withdrawal by itself does not initially alter the money supply, the bank's cash reserves—the raw material that it draws on to make loans—would be depleted. Especially if Cut and Shoot Bank fell below its legal reserve requirement, it would be forced to call in or not renew loans. This action would cause reductions in cash reserves in other parts of the commercial banking system. In the limit, deposits and, along with them, the money stock could be reduced by

$$\Delta D = \Delta \text{Reserves} \left(\frac{1}{r}\right).$$

A normal demand for cash is to be expected, even in a modern system of electronic transfers and at-home computerized bill payment. Currency and coin is still necessary to accommodate day-to-day transactions—coins for a candy bar or soda from a machine or currency to pay for a subway ride or lunch. Economists express this demand for cash through the **currency-deposit ratio,** which is simply the desired ratio of currency holdings to demand deposit holdings, on average and for all money holders. In the United States, the ratio has fluctuated widely through history, from high levels during bank panics and depressions to lower levels with the advent of federally insured and closely regulated commercial banking as well as with modern computerized credit card payment schemes. In recent years, the ratio has been around $1 in currency desired for every $3 or so in demand deposits.

Currency-deposit ratio: The desired ratio of currency holdings to demand deposit holdings, on average and for all money holders.

Let us assume that the currency-deposit ratio approximates 30 percent. What is the significance of a currency-deposit ratio of 30 percent for money expansion in a commercial bank? Any currency drain from the system is a leakage that reduces the ability of the commercial banking system to create deposits and money. If a bank acquires cash reserves of $1 million and lends to the limit with a 10 percent reserve requirement, the bank will lend $900,000. But if the average borrower's desired currency-deposit ratio is 30 percent, the borrower will want 30 percent of the loan, or $270,000, in cash, taking the remainder, $630,000, in demand deposits. If this currency leakage continues to circulate and the cash does not find its way back into banks—an unlikely event—the deposit- and money-creating potential of the banking system is reduced.

A slightly more complicated expression for deposit and money expansion or contraction gives us an idea of the effect of this currency leakage:

$$\Delta D = \Delta M = \text{Excess reserves} \left(\frac{1 + c}{c + r}\right),$$

where c is the currency-deposit ratio, r is the reserve-deposit ratio (the legal required reserve ratio), and D and M are demand deposits and money created, respectively. If we give a value of 10 percent to r, 30 percent to c, and excess reserves increase to $900,000 in an otherwise fully committed banking system, how might the money supply expand? The result will be

$$\Delta D = \Delta M = \$900,000 \left(\frac{1.30}{0.40}\right),$$

or

$$\$2,925,000 = \$2,925,000 = \$900,000 \ (3.25).$$

Deposits and the money stock expand by $2,925,000, much lower than the $9 million expansion when no currency drain was considered. The money multiplier in this case is $(1 + c)/(c + r)$, or 3.25, whereas it was $1/r$, or 10, when no currency drain was considered. The existence of a currency drain, therefore, lowers money expanison possibilities in the U.S. fractional reserve banking system.

As we will see in Chapter 29, the Federal Reserve System could stem this contraction by supplying supplementary cash reserves to the banking system. Indeed, one of the reasons the Federal Reserve System was created was to stem semihysterical currency drains or bank runs by acting as a lender of last resort. You should be aware nevertheless that the money supply can be shrunk as well as expanded through the banking system.

Idle Reserves

In the earlier examples we assumed that banks did not wish to hold idle reserves above those that were required by the central bank. This assumption is clearly not the case, especially during periods of high economic uncertainty. Banks are privately owned, profit-maximizing institutions, but they also tend to be prudent. As we indicated earlier, a bank totally committed with loans could fall below its legal required reserve if so much as $1 in cash reserves was withdrawn. Banks will not ordinarily let this happen because of the resulting embarrassment and possible repercussions from the Federal Reserve System or from other lenders or clients. Commercial banks therefore usually hold an additional amount in cash reserves. It should be clear that any holding of idle reserves by commercial banks further reduces the possibilities of money and deposit expansion. In other words, the holding of idle reserves, like the currency drain, acts as a brake to increases in the money supply as well as a cushion against decreases.

While other, more complex money multipliers involving savings and government deposits could be considered, it is sufficient here simply to understand the major factors affecting money expansion and contraction. By now you may suspect a deeper theme to money creation and destruction than the simple mechanical exercises of this chapter. Another actor in addition to the commercial banks and the public lurks backstage. The Federal Reserve System pulls the strings that control deposit and money expansion within the commercial banking system. We have only indirectly hinted at the tools of the Federal Reserve in manipulating the money stock. The details of this process will be discussed in Chapter 29.

The Bottom Line: To Make Money by Creating Money

Commercial banks and some other financial institutions are basically no different from firms that sell shoes or hamburgers. They are in business to make a profit. Yet they also perform an essential macroeconomic role in that they create money. The proper management of these institutions is crucial to achieve three interrelated goals: (1) pleasing stockholders with adequate—that is, competitive—returns on their investments; (2) keeping the bank solvent and liquid; and (3) reacting to the implementation of ever-changing government (Federal Reserve) regulation over banking activity.

Returns on Investment

Bank management is tricky business. Think about how banks make profits. Banks and other depository institutions sell services such as savings, checking, and money market savings accounts. As any bank customer knows, these services come in a variety of packages. Some checking accounts are interest-bearing, offering limited checking privileges without cost, while others do not bear interest but offer unlimited checking with minimum balances. NOW accounts and super-NOW accounts available at various financial institutions are examples of interest-bearing checking services.

In addition to the costs of supervising many different kinds of accounts, depository institutions must bear the interest costs associated with savings accounts. Interest must be paid on all passbook and other types of savings and time deposits. These institutions have been allowed to pay unlimited interest on money market savings accounts since 1983, with competition for depositors' money leading to higher interest costs to banks and other financial institutions.

Demand deposits and all other deposits form the raw material from which financial institutions can earn income by lending. These institutions make loans for many purposes, but specialization in certain types of loans has occured. Thrift institutions such as savings and loans and mutual savings banks have traditionally given long-term home and property mortgage loans, while commercial banks have concentrated on short-term consumer and business loans. Many of these institutions also earn interest returns from investments in securities.

The difference between a bank's costs and returns determines the profit to its stockholders. The spread between interest paid and interest received is an indication of the profitability of a bank, but banks incur costs other than interest. Labor costs in the servicing of deposit liabilities is a significant factor, so a spread of 14 percent interest received and 7½ percent interest paid is not necessarily an indication of great profitability. In more specific terms, the profitability of a financial institution depends on management of the bank's portfolios of assets and liabilities. The profit-maximizing structure of the portfolio changes constantly with changing market conditions and Federal Reserve regulations or deregulations.

Solvency and Liquidity

Liquidity: The ability to meet current depositor liabilities in cash.

Solvency: The ability of all assets to cover all liabilities.

As with any other business, the bank or financial institution must maintain both **liquidity,** the ability to meet current depositor liabilities in cash, and **solvency,** the ability of all assets to cover all liabilities. In an attempt to make high returns, an incautious bank manager may develop an unwise or unsafe loan policy, perhaps by lending to speculative or high-risk borrowers. Although longer-term and higher-risk loans typically earn higher interest and income, both liquidity and solvency may be threatened by such loans. To further complicate the problem, variability in deposits and shifts within a bank between types of deposits are not exactly predictable. The possibility of coming up short when depositors demand repayment in cash is always very real for a bank or depository institution. As a result, these institutions have developed sophisticated methods of portfolio management.

Government Regulations

Banks are for-profit, privately owned institutions, but they are also regulated by government. Their major function of creating or destroying money is of the highest importance to economic society. The regulatory umbrella over the entire banking system includes a multiplicity of state and, particularly, federal regulations. The maintenance of legally required reserves is but one of the Federal Reserve System's regulations. Indeed, the U.S. constitutional authority to print and control money is carried out through the Federal Reserve System. The manner in which the Federal Reserve conducts its activities through a privately owned banking and financial system is the subject of Chapter 29.

Summary

1. A bank gets started by accepting capital through selling stock to investors. After purchasing a bank building and equipment, the bank begins to perform the major functions of accepting deposits and making loans.
2. Through double-entry bookkeeping, the deposit of cash by an individual or business means that the bank's cash assets are counterbalanced by a corresponding amount of demand deposit liabilities.
3. The Federal Reserve System requires that all banks and financial institutions accepting checkable deposits must hold a certain percentage, called the required reserve ratio, of cash reserves against deposit liabilities.
4. All reserves above required reserves are called excess reserves. Banks can lend and create demand deposit money on the basis of their excess reserves.
5. An individual commercial bank can lend dollar for dollar with its excess reserves, but a single monopoly bank can lend and create money by a multiple of its excess reserves because its reserves and deposits cannot be checked away to other banks. The commercial banking system, like the monopoly bank, can lend and create money by a multiple of its excess reserves.
6. The multiple of excess reserves by which the banking system can create money is called the money multiplier. The simple money multiplier is the reciprocal of the required reserve ratio ($1/r$). Deposits and the money supply will expand or contract by an amount equal to the multiplier times the change in excess reserves.
7. A currency drain resulting from an increased demand for cash and coin or the maintenance of idle reserves by banks will reduce the value of the money multiplier.
8. Banks create or destroy money in an attempt to make profits. Commercial banks and other financial institutions, though privately owned and motivated to make profits, are nonetheless controlled in the public interest by the Federal Reserve System.

Key Terms

cash reserves
required reserves
reserve ratio
excess reserves

deposit expansion
simple money multiplier
money expansion

currency-deposit ratio
liquidity
solvency

Questions for Review and Discussion

1. Are commercial banks the only creators of money in the market?
2. What are the options for a bank if it finds that its actual reserves have fallen below its required reserves?
3. Explain how checks are cleared within a bank, between two banks in the same Federal Reserve district, and between two banks in different Federal Reserve districts. Look on the backs of some cleared checks. Are there clues as to whether these cleared between banks in the same district or in different districts?
4. By how much can a single bank increase the money supply from a deposit of $1 million of new cash? By how much can the entire banking system increase the money supply from such an injection?
5. If a bank purchases U.S. government securities, does this purchase increase the money supply? Does the money supply change if an individual purchases a government bond?
6. If $2 million were stolen from a bank's vault, would this theft increase or decrease the money supply?
7. What is a currency drain? If the currency drain increases, does it increase the money supply? Why?
8. "Banks do not like to hold excess reserves." Is this statement true? Explain.
9. What is the money multiplier? What does it show?
10. Why would anybody want to own a bank? What is the output of a bank? What is the essential input that banks must purchase to produce this output?

ECONOMICS IN ACTION   How Does a Pub Become a Bank? The Modern Case of Ireland

The proportion of hand-to-hand currency relative to checkbook money is low in most advanced societies. Historically, Americans' use of currency is only one-third the amount of checkable deposits, though the ratio changes over time. Other countries have even lower ratios of currency to demand deposits. Ireland's ratio, for example, was 18 percent in 1966, 15 percent in 1970, and 14 percent in 1976. But Ireland is unique in that its banks close frequently and for longer periods of time than banks of any other relatively advanced nation in the world. In fact, Irish banks closed for varying periods during 1966, 1970, and 1976 because of industrial disputes. For almost seven months (May 1–November 17) during 1970, citizens of the Republic of Ireland were deprived of the services of the Associated Banks, their branches, and their clearing facilities, which controlled virtually all of Ireland's demand deposits. Without 85 percent of its money supply, could Irish society function? Did the Irish resort to barter and to the reduced economic activity associated with barter? In 1979, Antoin E. Murphy of Trinity College, Dublin, presented some intriguing evidence on all of the bank closures, especially the long 1970 closure.[a]

Money did not totally disappear after the bank closures. Irish currency and coin continued to circulate, and some major companies were provided with account

[a]Antoin E. Murphy, "Money in an Economy Without Banks: The Case of Ireland," *Manchester School of Economics and Social Studies* 46–47 (1978–1979), pp. 41–50.

facilities and clearing services by North American and non-Associated banks. The Central Bank of Ireland transferred currency to government departments at the beginning of the closure to pay wages and salaries of government employees and to continue welfare payments, but at the end of the closure there was only a net addition of £4 million to the currency supply.

The increased demand for currency was partially offset by the summer tourist trade. Currency freely circulates in Ireland, and from April to November currency in circulation grew from £5 million to about £40 million. The North American and non-Associated banks provided some demand deposit transactions with means of alternative payment, but the aid was very limited because these banks had no branch facilities and were physically incapable of handling the volume of new business. By the end of May most of these banks refused to handle new accounts. In November 1970, according to Murphy's estimate, there was a total of £52 million in new demand deposits to facilitate consumers' money demands. This was less than one-twelfth of the closed Associated Banks' demand deposit accounts! How, then, did the Irish manage to transact?

People simply continued writing checks against preclosure deposits and against checks received from other parties. During the bank closure, checks were written not against known accounts, but against the value of other uncleared checks along with the check receiver's assessment of the writer's creditworthiness. In such circumstances, default risk increased. Further, there was

A pub in Maghery, County Donegal, Ireland

uncertainty about when the banks would reopen. Credit was therefore undated.

In this situation a personalized transaction system substituted for an institutionalized one. The nature of the Irish economy helped. A high degree of personal contact exists in the Irish population of about 3 million. Where personal information was lacking, credit information often existed at the 12,000 retail shops and at the more than 11,000 pubs in the Republic. One pub exists in Ireland for every 190 citizens over eighteen years of age. A pub keeper does not serve ale to a customer for years, as Murphy put it, "without discovering something of his liquid resources." Thus, pubs and shops provided goods, services, and currency for their customers against undated checks. They in fact formed the nexus of a substitute banking system.

Economic activity actually grew over the period at a somewhat reduced rate. There were no significant differences in retail sales and no significant deflationary trends. The important point is that information was the key to the emergence of a substitute medium of exchange, which reduced the cost of information. The Irish economy did not collapse or even cease growing when deprived of over 80 percent of its money. It simply and naturally fell back on or invented new and alternative forms of transacting.

Question

Suppose all banks closed in the United States. Who would experience more difficulty in finding bank substitutes—big city dwellers or small town citizens? Who would most likely assume the role of banker—your local power company or your local hairstylist?

29

The Federal Reserve System

T he economic power of the Federal Reserve Board is awesome. Changes in Federal Reserve policy or even general comments on the health of the economy by the chairman of the Federal Reserve Board frequently send the stock market plunging or soaring. Fed-watching—anticipating the Federal Reserve's next change in policy—is so important to investors that some major investment banking houses pay experts more than $200,000 a year, three times as much as the salary of the Federal Reserve Board chairman, to monitor Federal Reserve activity.

Where does the Federal Reserve Board get its economic authority? Why is even the president often helpless in directing the course of monetary policy? The Federal Reserve Board uses its position as the nation's central bank and bank regulator to influence the level of fractional reserves in all banks in the country. In Chapter 27 we examined the institutional structure that gives the Federal Reserve this power over banks. Recall that the Federal Reserve System is made up of twelve Federal Reserve banks and twenty-five Federal Reserve branch banks operating in all regions of the country. Overseeing the activities of these banks is the Federal Reserve Board of Governors, appointed by the president and based in Washington, D.C., whose role is to establish and oversee the nation's monetary policy. In this chapter we will see how Federal Reserve policy is carried through and how the nation's central bank controls the fractional bank reserves, the engine of the money supply.

The Federal Reserve System's Balance Sheet

The Federal Reserve System's basic activities in the banking system are concisely illustrated in its balance sheet, shown in Table 29–1. Its assets and liabilities are reported in the monthly *Federal Reserve Bulletin*, which com-

Federal Reserve Bank of Minneapolis, Minnesota

piles statistics on the activities of the central bank and on the financial system in general. The following is a brief description of the major assets and liabilities of the Federal Reserve System, each of which totaled $200 billion in January 1985.

The Federal Reserve's Assets

The first asset listed in Table 29–1 is gold certificates. The Federal Reserve System no longer holds gold, and since 1968 all ties between gold and Federal Reserve notes (dollars or currency) and other deposit liabilities have been abandoned. Before 1968, the U.S. Treasury issued and sold gold certificates to the Federal Reserve, which was required by law to hold a percentage of these certificates against currency and other liabilities. Some of these gold certificates remain on the Federal Reserve books as assets.

TABLE 29–1   Basic Setup of the Federal Reserve System Balance Sheet

Generally, increases in the Federal Reserve's assets will increase member commercial bank reserves, while increases in its liabilities—except for the reserves deposited with the Federal Reserve by banks themselves—will decrease member bank reserves. Changes in bank reserves have important effects on the overall money supply.

Assets	Liabilities and Capital Accounts
Gold certificates	Federal Reserve notes (outstanding)
Loans and securities (earning assets)	Total deposits
	All other liabilities
Bank premises	
All other assets	Capital accounts

The next and largest group of items, loans and securities (earning assets), is the most important for understanding the Federal Reserve System's role in regulating bank reserves and the money supply in general. These assets consist of loans to banks and other depository institutions and securities, including federal agency obligations (various kinds of bills and notes) and U.S. government bonds. The Federal Reserve, as we will see, may purchase these securities from banks or from the nonbank public. Lumped together, the assets loans and securities are sometimes called **reserve bank credit** because increases and decreases in these assets affect member institutions' reserves—the raw material of money creation. We will explain this process later in the chapter.

Reserve bank credit: Loans and securities assets of the Federal Reserve System. Increases and decreases in these assets affect member institutions' reserves.

The Federal Reserve's Liabilities

Table 29–1 shows two major liabilities of the Federal Reserve System. One is Federal Reserve notes (currency or dollars) held outside Federal Reserve banks. The dollar bills in circulation are part of this liability because they represent claims against the assets of the Federal Reserve. Cash issued by the Federal Reserve and kept on hand is neither an asset nor a liability: It is paper.

The liabilities also include deposits, such as member institutions' reserves—the required reserves of banks and other depository institutions discussed in Chapter 28 as well as other commercial bank funds deposited at the Federal Reserve. As liabilities to the Federal Reserve, they are assets to the commercial banks, and any excess reserves may be instantly converted by the banks into cash. Recall that money can be created within the fractional reserve system on the basis of such excess reserves.

Other deposits listed as liabilities in Table 29–1 include checking account privileges to the U.S. Treasury and to foreign countries and their residents. The capital accounts include several liabilities, notably the stock deposits of financial institutions that are required to become members of the Federal Reserve System.

The Monetary Base: Raw Material of Money Creation

Our enumeration of the Federal Reserve's major assets and liabilities is more than an accounting exercise. The Federal Reserve balance sheet illustrates how it attempts to control the money supply.

Basically, the Federal Reserve achieves control over the money supply by manipulating member institutions' reserves. Recall from Chapter 28 that banks can lend by a multiple of their excess reserves, thereby expanding or contracting the money supply. By increasing assets, either its loans to member institutions or its own portfolio of securities, the Federal Reserve makes additional reserve funds available to member banks.

Reserve bank credit outstanding: The difference between the value of the factors increasing bank reserves and those decreasing reserves.

While increases in Federal Reserve assets increase bank reserves, increases in its liabilities other than commercial bank reserves reduce the banks' reserves. Increases in the Federal Reserve's stock of securities would increase bank reserves, for example, while decreases in loans to banks would decrease reserves. The difference between the value of the factors increasing bank reserves and those decreasing reserves is called **reserve bank credit outstanding.** This statistic is deemed so important that it is reported on a weekly basis in the *Federal Reserve Bulletin*. Changes are often viewed as an indicator of whether the Federal Reserve is increasing bank reserves (and therefore increasing the money supply) or limiting bank reserves (and therefore decreasing the money supply).

The Federal Reserve attempts to control the money supply by controlling the **monetary base,** the sum of banks' reserves (including the reserve in their own vaults as well as the reserves deposited with the Federal Reserve) and currency in the hands of the public. Controlling the monetary base is the object of Federal Reserve control because of the direct relation between the monetary base and M-1, the money supply. Naturally, the public decides how much currency it wishes to hold relative to demand deposits. To control the money supply, then, the Federal Reserve must control bank reserves—the stuff of demand deposit money creation—and adjust for the changes in currency holdings of the public, which also affect the ability of banks to create money.

> **Monetary base:** The sum of bank reserves plus currency held by the public.

We know from Chapter 28 that currency withdrawals limit the amount of money that commercial banks can create. Sharp increases or decreases in the public's currency demands could have far-reaching effects on the total quantity of bank reserves in the system and thus on the ability of such reserves to support a certain total money supply. Factors such as currency withdrawals are said to be outside the system's control, which does not mean, however, that the Federal Reserve cannot estimate and predict them. It does mean that overall Federal Reserve control of reserves and the money supply is a complicated business.

Methods of Federal Reserve Control

The Federal Reserve can control the monetary base in a number of ways. Some methods, such as open market operations, affect the monetary base directly. Others, such as changes in the reserve requirement, affect the ability of depository institutions to lend and affect the monetary base indirectly. The major tools of the Federal Reserve, sometimes called **credit controls,** are the following:

> **Credit controls:** The major methods used by the Federal Reserve, such as open market operations or changes in the reserve requirement, to control the monetary base.

1. Open market purchases and sales of securities
2. Changes in the discount rate
3. Alterations in the reserve requirement
4. Changes in the margin requirement and, on occasion, imposition of consumer credit controls

The Federal Reserve can also exercise warnings to banks, an option called moral suasion.

Open Market Operations

The most important and flexible tool available to the Federal Reserve in its attempt to control the monetary base is **open market operations**—the buying and selling of securities on the open market. The Federal Reserve uses this tool daily to directly affect member banks' reserves. Again, when the Federal Reserve increases its securities holdings by purchasing government-issued bonds, the reserves of banks are increased, meaning that banks have greater capacity to lend money, thus increasing the money supply. When the Federal Reserve sells its securities, the reserves of banks are decreased and the

Open market operations:
Purchases and sales of
securities by the Federal
Reserve that determine the
size of the money supply by
affecting the amounts of
reserves in the banking
system; the Federal Reserve's
major method of directing
the money supply and
economic activity.

money supply contracts. Open market operations are used to correct short-run, predictable fluctuations in the money supply or to expand or contract the money supply over longer periods. To make sure that such actions have the desired effect, the Federal Reserve does not announce its intentions; Fed-watchers can only take note of what the Federal Reserve System does and try to guess whether it is making normal short-run adjustments or embarking on a new long-range course.

Some alterations in the monetary base take place owing to seasonal currency drains, such as the withdrawal of cash in December for Christmas purchases. These drains predictably increase the demands of the public for checkable deposit money. The Federal Reserve intervenes to compensate for these predictable drains. Other currency drains are less predictable and would contract the money supply were it not for the Federal Reserve's attempt to compensate by purchasing securities. When currency needs have passed, the Federal Reserve again compensates by selling securities, depending on business conditions. It may also purchase or sell securities in a concerted effort to expand or contract the money supply, to alleviate recession and unemployment, or to control inflation.

The Role of the Federal Open Market Committee (FOMC). The Federal Open Market Committee, introduced in Chapter 27, is the principal operating arm of the Federal Reserve System. It is responsible for Federal Reserve Board decisions to buy or sell securities. This committee ordinarily meets every three or four weeks to set trading policies, which are kept secret to avoid upsetting the plans of buyers and sellers. In these sensitive, closed-door sessions, the FOMC considers such factors as the inflation rate, the economy's growth or real income (GNP), the unemployment picture, the size of excess reserves and borrowings from the Federal Reserve, probable currency drains, and the international balance of payments. It then decides on a monetary base, or reserve target, and a federal funds rate target to shoot for.

Federal funds rate: The
interest rate commercial
banks can charge on
overnight and short-run
loans to other banks.

The **federal funds rate** is the interest rate commercial banks charge on overnight and short-run loans to other banks. When the Federal Reserve targets changes in the federal funds rate, it is, in effect, changing all interest rates, such as the prime interest rate and mortgage rates.

After the FOMC decides how reserves, the money supply, and the federal funds rate are to be altered, this decision is transmitted to the open-market account manager, an officer of the Federal Reserve Bank of New York. This individual, who controls the open market desk at the New York Federal Reserve and deals directly with commercial securities and investment houses on Wall Street, then implements the intentions of the FOMC. According to 1979 rule changes, the account manager has only narrow latitude in affecting monetary aggregates (such as bank reserves, the monetary base, and the money supply), but the account manager is still given a specific target for the federal funds rate.

As of November 1984, the Federal Reserve System held almost $158 billion worth of securities on its balance sheet. In October 1984 it acquired almost $86 billion, while it sold $90 billion. Let us see how these open market transactions affect the condition of the banking system, expanding or contracting the money supply. There are two open market channels through which the Federal Reserve can affect bank reserves and the money supply: It can deal directly with banks or with the nonbank public.

TABLE 29–2   Purchase of Securities from Banks

When the Federal Reserve purchases securities from commercial banks or other depository institutions, it provides them with reserves on which the banks may make new loans, increasing the money supply. The Federal Reserve's purchase of $5 billion in securities from commercial banks changes those assets from a securities entry to a reserves entry in the commercial banks' aggregate balance sheet.

Federal Reserve System		Commercial Banks	
Assets (billions of dollars)	Liabilities and Capital Accounts (billions of dollars)	Assets (billions of dollars)	Liabilities and Capital Accounts (billions of dollars)
+ Securities 5	+ Reserves 5	− Securities 5 + Reserves 5	

Purchase of Securities from Banks. First consider a Federal Reserve purchase of securities (bonds, notes, or bills) from commercial banks. The effects of a $5 billion purchase are summarized in Table 29–2, which shows only changes in accounts. The Federal Reserve acquires $5 billion in assets (+ securities) but creates a new reserve liability to the banks to pay for the securities in exactly the same amount (+ reserves). On the other hand, the commercial banks lose a securities asset on the asset side of their balance sheet but gain an asset—reserves—of equivalent amount. The crucial point is that the monetary base rises as excess reserves—those above required reserves—open up in the banking system. Banks cannot lend on the basis of securities, but they can lend on the basis of excess reserves. After the Federal Reserve's purchase of securities, these excess reserves are ready to be lent. Given a specific reserve requirement, the banking system's desired reserve holdings, and a desired ratio of currency to deposits on the part of the public, money creation may proceed apace. In theory, as the Federal Reserve buys new securities and as the banks begin supplying new loan funds, interest rates tend to fall initially, encouraging consumers and investors to borrow. (A securities sale would have the opposite effect.) In reality, the monetary expansion also depends on general business conditions and expectations for the future.

Purchase of Securities from the Public. Another method by which the Federal Reserve conducts open market operations pumps demand deposits in or out of the monetary system directly. Suppose that instead of purchasing $5 billion worth of securities from commercial banks, the Federal Reserve buys them directly from the nonbank public. The results of such a purchase are summarized in the balance sheets of Table 29–3.

When the Federal Reserve buys securities from the public and pays for them with, in effect, checks written on itself, the public gains demand deposit assets and loses securities assets in equal amounts. Note that the commercial banks gain new reserve assets and new deposit liabilities of $5 billion if the public deposits all of its proceeds from the sale into banks. In such a case, the money supply is increased directly by $5 billion by the very act of the public sale. The extent of the monetary expansion will depend on banks' desired idle reserves, the legal reserve ratio, the public's desired currency holdings, and, more broadly, general economic conditions.

These open market operations take place daily and are the most flexible and efficient tool in the Federal Reserve's arsenal. But how can the Federal

TABLE 29–3   Purchase of Securities from the Public

When the Federal Reserve buys securities from the public and the public deposits the proceeds in banks, the public gains demand deposit money and commercial banks acquire new reserves. Banks may then lend additional money on the basis of their excess reserves.

Federal Reserve System Balance Sheet

Assets (billions of dollars)		Liabilities and Capital Accounts (billions of dollars)	
+ Securities	5	+ Reserves	5

Public Balance Sheet

Assets (billions of dollars)		Liabilities and Capital Accounts (billions of dollars)	
− Securities	5		
+ Demand deposits	5		

Commercial Banks' Balance Sheet

Assets (billions of dollars)		Liabilities and Capital Accounts (billions of dollars)	
+ Reserves	5	+ Demand Deposits	5

Reserve be sure that banks or the public will be willing to sell or buy securities on demand? Simple supply and demand analysis provides the answer. When the Federal Reserve sells securities, it increases the supply of securities on the open market in quantities large enough to affect the interest rate. When the Federal Reserve buys securities, demand for them is increased. An increase in supply lowers the price and increases the interest return on securities, making security holdings an attractive investment for banks or the public. Likewise, when the Federal Reserve places an order to buy on the open market, the demand increase causes security prices to rise and the interest return from holding them to fall. Selling securities to the Federal Reserve then becomes attractive to banks and the public because of the possible capital gains from selling and because of the reduced yield from holding the securities. Supply and demand conditions thus assure the Federal Reserve that there will be a response to its actions in the open market.

Who gets the interest return on the large quantity of securities held by the Federal Reserve? The Federal Reserve itself does. Congress permits the Federal Reserve to use this income to finance its operations. Historically, the Federal Reserve has been able to pay its own way out of its earning assets. Excess income is turned over to the U.S. Treasury. Remember, however, that the Federal Reserve is not supposed to be in business to make profits but to control the money supply for purposes of economic stabilization. The same is true of the other earning assets on the Federal Reserve balance

FOCUS   Bureaucracy and the Federal Reserve System

When economists speak of the Federal Reserve System, they are ordinarily concerned with its performance in managing the money supply and with how it affects the macroeconomic health of the economy. Critiques of the Federal Reserve System's performance are essential to the public's understanding of whether it is doing a good job. Recently, there has been some interest in the bureaucratic behavior of the Federal Reserve. In other words, how do the bureaucratic incentives faced by Federal Reserve officials alter their conduct of monetary policy?

First, how does the Federal Reserve earn income? It does so in the operation of the money supply process. The Federal Reserve produces and sells money through commercial banks. In carrying out such open market transactions, it builds up a portfolio of assets—Treasury bills, notes, certificates, and U.S. bonds—on which it earns interest income. For example, when the Federal Reserve purchases securities in an open market operation, it essentially exchanges money, which it produces at low cost, for securities that generate a stream of interest payments. These interest payments are substantial. During 1981, for example, the Federal Reserve earned $15.51 billion in such income.

Second, what are the expenses of the Federal Reserve System? Its costs are similar to those of a typical commercial bank, including maintenance and depreciation of premises and equipment, the costs of operating an extensive check-clearing network, and costs for supervising member institutions. However, the Federal Reserve's expenses are only a small proportion of its income. As noted above, the Federal Reserve earned $15.51 billion in 1981, out of which it deducted $1.49 billion to fund the operation of the system and the Board of Governors in Washington. The excess of $14.02 billion was returned to the Treasury.

These are the facts of how the Federal Reserve operates. What are the economics? The economics are like that of virtually any other nonprofit bureaucracy. The Federal Reserve and its managers cannot keep the profits that they make in any given year for themselves. That is, these profits cannot be taken home by Federal Reserve managers in increased paychecks. The only way the profits can be used is to pay for the operation of the system. In simple economic terms, this fact means that it is relatively less costly for managers of

the Federal Reserve to "buy" amenities on the job—things that make the job more pleasant—than it would be for managers of a private firm where profits can be taken home as increased wealth. Amenities are cheaper to the Federal Reserve because if it does not spend its profits on expenses, they must be turned over to the Treasury; they cannot be taken home.

For managers, amenities mean such things as content employees, plush physical facilities, good retirement programs, and a growing and expanding number of employees in the organization. The latter is particularly important because the number of employees supervised is often used as a way to measure the importance and prestige of a manager, such as a governor on the Board of Governors.

Does the Federal Reserve behave this way, enhancing its own amenities? Not very surprisingly, the answer is yes. There have been several studies of the relation between the growth in the money supply over time and Federal Reserve employment. These studies uniformly report a strong relation between the two variables, other things being equal. The Federal Reserve makes money when it expands the money supply, and it uses these profits to finance expansions in its own employment and other amenities. Of course, much of this employment expansion is related to the increased duties of the Federal Reserve, such as increases in the area of bank supervision. Nonetheless, there is nothing inherent in the money supply process itself that suggests that Federal Reserve employment would grow as strongly as it has over time. Thus, the conclusion suggests that the Federal Reserve does engage in a significant amount of discretionary spending on amenities.

What is the importance of this analysis? Many discussions of Federal Reserve policies are based on the idea that the behavior of the managers of the system can be changed by pointing out to them what they are doing right and wrong. This approach may be partially effective, but if the managers are guided by the bureaucratic incentives discussed above, such critiques will have little impact on their behavior. Indeed, since the Federal Reserve makes money on its portfolio only when it expands the money supply, bureaucratic incentives could be a factor in explaining an inflationary bias in the money supply process.

---

sheet—loans to depository institutions and acceptances. Like securities, these assets are used to affect monetary policy and are not for profit. (Focus, "Bureaucracy and the Federal Reserve System," discusses the relation between the Federal Reserve's ability to create money and its monetary policy.)

Discount rate: An interest rate charged by the Federal Reserve to depository institutions for loans backed up by some form of collateral.

Loans to Banks: The Discount Rate

The process of lending to banks at some interest rate, called a **discount rate,** is the oldest function of a central bank. The discount rate is an interest rate (expressed as a percentage) charged by the Federal Reserve to depository institutions for loans backed up by some form of collateral—securities, notes, or commercial paper. This loan process was originally related to the status of a central bank as a lender of last resort—an institution that provided liquidity and currency to the banking system in times of sudden currency demands. Many of these bank crises or panics took place in the United States prior to the establishment of the Federal Reserve System in 1913. Yet modern banks are not immune to crises. In 1984 the massive Continental Illinois National Bank and Trust Company of Chicago, the nation's largest business and industrial lender, narrowly avoided closing its doors. The Federal Deposit Insurance Corporation (FDIC) and the Federal Reserve came to the rescue, the latter through loans and through the discounting of notes for the bank.

Initially, the discount rate was perceived to be the Federal Reserve System's major tool to avert crises. In modern times loans to member banks and other financial institutions have performed another function as well—to control expansions and reductions in money and credit. Several discount rates are charged by the Federal Reserve depending on the kind of collateral put up by the banks and the purpose and duration of the loans. The rate for short-run or seasonal credit is lower than that for extended credit to institutions whose loans extend beyond 150 days. These differential rates reflect the Federal Reserve's policy toward banks' motivations for borrowing. A commercial bank may borrow because of seasonal cash drains, to cover very short-run deficiencies in legal reserves, or because it sees the Federal Reserve discount loans as a source of profit. If there is a wide spread between the discount rate paid for borrowing excess reserves at the Federal Reserve and the interest rate banks receive on lending out these excess reserves, banks may seek to borrow from the Federal Reserve to enhance their profits. The Federal Reserve discourages borrowing for this reason by charging higher rates for longer-run credit. In fact, the Federal Reserve may stop this sort of borrowing entirely.

Whatever the reason for borrowing, lending by the Federal Reserve swells the quantity of reserves at the disposal of the banking system. Table 29–4, for example, shows the Federal Reserve balance sheet and that of a single commercial bank, the First Bank of Nome. The process of discounting a note

TABLE 29–4   Lending to the First Bank of Nome

When the Federal Reserve lends at interest to a commercial bank, the commercial bank acquires reserves as a new asset and a new liability to the Federal Reserve. The bank may now create new demand deposit liabilities along with new loan assets, thereby expanding the money supply.

Federal Reserve System		First Bank of Nome	
Assets (billions of dollars)	Liabilities and Capital Accounts (billions of dollars)	Assets (billions of dollars)	Liabilities and Capital Accounts (billions of dollars)
+ Loans (First Bank of Nome) 2	+ Reserves 2	+ Reserves 2	+ Due Federal Reserve 2

for $2 million from the Bank of Nome has the clear effect of increasing the bank's stock of reserves. If any or all of these new reserves are in excess of the required reserves, money creation may proceed in multiple fashion, again depending on banks' required and desired excess reserves, currency drains, and general economic conditions.

The Federal Reserve has several ways of encouraging or discouraging borrowing: (1) it alters the rate or rates charged; (2) it changes the form of collateral required; and (3) it always retains the option to lend or not to lend. Loans to depository institutions and acceptances (another form of short-run lending with different collateral required) make up a relatively small percentage of the Federal Reserve's financial transactions compared to open market operations. But the ultimate importance of the discount rate hinges not so much on the quantity of loans as on the information the rate conveys concerning the intentions of the Federal Reserve. Commercial and consumer interest rates, such as the prime rate (the rate that banks charge to their lowest-risk commercial borrowers), tend to follow the discount rate. Along with the Federal Reserve's changing monetary base targets, the discount rate is seen by the banking and business community as one indicator of whether the Federal Reserve is following an expansionary or contractionary monetary policy. A higher rate indicates a tightening of credit and a contractionary policy. A lower rate indicates the reverse. The stock market, for one, is apt to react sharply to changes in the Federal Reserve discount rate. Thus, while the discount rate is not the major mechanism of Federal Reserve control over the monetary base, it is often considered the principal messenger of the Federal Reserve's policy intentions.

Changing the Reserve Requirement

Changing the legal reserve requirement is potentially the most powerful tool at the disposal of the Federal Reserve, but it is seldom used. To understand this seeming contradiction, remember that an alteration in the legal reserve affects the ability of any given amount of excess reserves to support new money, and it increases or decreases the amount of excess reserves held by depository institutions. A simple example illustrates the importance of the legal reserve requirement for the money supply. If we assume that excess reserves in the banking system are $5 billion and that the legal reserve requirement is 15 percent, the potential expansion in the money supply, neglecting currency drains and other factors, is

$$\Delta M = \Delta D = \text{excess reserves} \left(\frac{1}{r}\right),$$

or

$$\Delta M = \Delta D = \$5 \left(\frac{1}{0.15}\right) = \$33.3.$$

The increase in the money supply M equals the increase in deposit expansion, $33.3 billion. The money multiplier $1/r$ is equal to $1/0.15$, or 6.66. As indicated in Chapter 28, the money multiplier is simply the number by which excess reserves are multiplied to determine potential change in the money supply. It is the reciprocal of the reserve requirement.

A reduction by 5 percentage points in the reserve requirement, from 15 to 10 percent, would have dramatic effects. If excess reserves are still assumed to be $5 billion, the increase in money and deposit expansion would equal $50 billion:

$$\Delta M = \Delta D = \left(\frac{1}{0.10}\right) \$5 = \$50 \text{ billion.}$$

In other words, the quantity of excess reserves increases or decreases with increases or decreases in the required reserve ratio. If total reserves are $100 billion and the reserve requirement is 15 percent, excess reserves are $85 billion. A reduction in the legal requirement to 10 percent raises the quantity of excess reserves by $5 billion. Small wonder that a relatively small change in the legal reserve can have very large effects on the banking system's ability to expand or contract money.

Actually, banks must observe several reserve requirements depending on the size of their deposit liabilities. Table 29–5 provides a summary of reserve requirements in force in January 1985. After implementation of the Monetary Control Act of 1980 (see Focus, "The Monetary Control Act of 1980"), reserve requirements apply uniformly to all depository institutions according to the size of checkable, savings, and time deposit liabilities. As Table 29–5 shows, transaction accounts—all checkable accounts—less than $28.9 million at depository institutions carry a reserve requirement of 3 percent, while 12 percent must be held on deposits over $28.9 million. Some large ($100,000 or more) nonpersonal time deposits, such as certificates of deposit, and savings deposits (less than one and one-half years maturity) now carry a requirement of 3 percent. All in all, these reserve requirements are more uniform and lower than in the past.

Reserve requirements limit the profitability of banking and depository institutions and so are regarded as a necessary evil by the financial system. Changes in the requirements (especially increases) can cause massive disruptions and large-scale adjustments in the portfolios of member banks and other institutions that are required to hold legal reserves. Even small changes might create financial crises for some institutions, depending on their profitability. When banks are fully committed with loans—that is, have low amounts of excess reserves—an increase of only one-quarter or one-half percent in the legal reserve could reduce lending and money creation by billions of dollars, creating tight money and possibly bank failures. The Federal Reserve is therefore reluctant to make sudden large changes in the reserve requirement. Open market operations are far more flexible in operation and predictable in effect. The reserve requirement will always be a part of monetary control, but it will likely be used sparingly.

TABLE 29–5   **Contemporary Reserve Requirements of Depository Institutions**

Reserve requirements have been made uniform for all institutions issuing demand, savings, or time deposits. The total amount of reserves required for an individual bank or savings and loan, for example, depends on the type and amount of deposits issued by the institution.

Net Transaction Accounts	Nonpersonal Time Deposits
3% up to 28.9 million 12% over 28.9 million	3% on less than 1½ years maturity 0% on more than 1½ years maturity

Source: *Federal Reserve Bulletin* (February 1985).

FOCUS   **The Monetary Control Act of 1980**

In the 1970s, a growing number of problems affected the financial system of the United States. Two problems seemed to be paramount: a declining membership of commercial banks in the Federal Reserve System—a "flight from the Fed"; and a growing financial unsoundness of the "thrift institutions," such as savings and loan associations and mutual savings banks. Declining membership in the Federal Reserve System clearly had its roots in the costs and benefits associated with membership. Chief among the costs of belonging to the system was the legal reserve requirement. State banks, which are chartered and regulated at the state level, were often required to hold lower reserves than Federal Reserve members. More money to lend out meant higher profitability. The result: lower Federal Reserve membership and declining amounts of deposits under Federal Reserve control.

The financial problems of the thrift institutions began to intensify as market rates of interest and inflation rates began to rise over the 1970s because thrift institutions, as the adjective implies, depended primarily on savings and time deposits as sources of investment funds. A Federal Reserve regulation, Regulation Q, stated a maximum interest rate that banks could pay on savings and other time deposits. When the market rates exceeded the rates allowed by the Federal Reserve, a process called disintermediation began to take place. Savers took their funds directly to private investments such as mutual funds and money market funds, thus bypassing the financial intermediaries. Many thrift institutions went out of business, but others responded to the crisis by offering accounts called NOW (negotiable order of withdrawal) accounts, ATS (automatic transfer services) accounts, or credit union share draft (CUSD) accounts. These accounts paid interest but could also be used to write checks.

The combination of these two factors—declining Federal Reserve membership and the emergence of new kinds of checkable deposits—led to a lessened Federal Reserve control over the money base and the money stock. In February 1980, Paul Volcker, chairman of the Federal Reserve Board, bemoaned this lack of control before the U.S. Senate Banking Committee. Little more than one month later, on March 31, Congress responded by passing the Depository Institutions Deregulation and Monetary Control Act of 1980. This act contains provisions that promise to ultimately change the structure of banking and financial institutions in the United States forever. The two major provisions of the act are as follows:

1. The application of reserve requirements by the Federal Reserve to all financial institutions after a phase-in period of three to eight years. These requirements will apply uniformly to commercial banks and thrift institutions issuing any form of checkable or time deposits.
2. The gradual elimination of Regulation Q—the interest ceiling on time and savings deposits—for all institutions and the granting of permission to pay interest on checkable deposits (NOW, ATC, and CUSD accounts).

With the stroke of its pen, Congress changed the ground rules under which the financial system had functioned since the Great Depression of the 1930s. How? The requirement that all institutions issuing checkable deposits must hold reserves at the Federal Reserve or in their vaults means that, for all intents and purposes, the distinction between member and nonmember banks disappears. Nonmembers are now even allowed to borrow and discount at the Federal Reserve and to avail themselves of other privileges such as check clearing. Further, the distinction between a bank and a thrift institution has become moot since both accept checkable and savings deposits.

The elimination of Regulation Q has also had far-reaching effects. The emergence of interest payments on checkable deposits created large shifts from traditional checking accounts to the interest-bearing NOW accounts. In 1983, super-NOW accounts were permitted that, given a deposit of $2500 or more, yielded unlimited market rates of interest with checking privileges. Competition between banks and thrift institutions for deposits has become intense, but with some squeeze on profits since interest costs have risen. The future of the financial system is not exactly predictable, but one thing appears to be certain: Banks are becoming more like thrift institutions, and thrift institutions are becoming more like banks. The commercial bank, as society knew it prior to the 1980s, no longer holds its unique niche in the financial structure.

The Margin Requirement and Other Selective Credit Controls

Stock margin requirement: The percentage of the total price of a stock that must be paid in cash.

Other tools sometimes available to the Federal Reserve through congressional action are selective credit controls, which include credit restrictions on stock market purchases and consumer borrowing for durable goods and home purchases. At present, the Federal Reserve controls only the **stock margin requirement,** which is simply the percentage of the total price of a stock that must be paid in cash. A margin requirement of 40 percent means that only 60 percent of the purchase price of stock can be borrowed; 40 percent must be paid in cash. The Federal Reserve acquired the power to demand a margin requirement through the Securities and Exchange Act of 1934 allegedly to stem the rampant stock speculation that was thought to have contributed to the stock market crash of 1929 and the ensuing Depression. The purpose of the margin requirement is thus to provide a brake in what is regarded as unwise stock speculation. Alterations in the required margin selectively affect credit for stock market purchases and, therefore, also affect the volume of such purchases. Since 1968 the rate has varied between 80 percent and 50 percent, the latter rate being in force since 1974.

Finally, during periods of potential and real economic crisis, Congress has in the past given the Federal Reserve powers over the terms of home mortgages and consumer credit for the purpose of buying durables. The rationale for restrictions on the amount of down payment and the period of repayment is simple: During periods of all-out war, resources must flow into military spending. Simultaneously, however, the overfull employment that accompanies a war effort means high nominal incomes, which create huge demands for consumer durables and new housing. Consumer credit controls tend to suppress these demands, having the dual effect of releasing real resources to the war effort and keeping a temporary lid on domestic price inflation. Between 1941 and 1947 and again during the Korean War, Congress authorized the Federal Reserve to impose restrictions on consumer credit. The Federal Reserve does not especially like the discriminatory impact of these controls because they affect only particular markets, such as home building and automobile manufacturing. At present and for the foreseeable future, the only selective control maintained by the Federal Reserve is the stock margin requirement. Under the Credit Control Act passed by Congress in 1969, the president of the United States was given the power to institute consumer credit controls, administered through the Federal Reserve System. With one minor exception, these controls have been unused, probably for some of the reasons given above.

Moral Suasion

Moral suasion: The Federal Reserve's ability to persuade banks that a particular course of action is desirable. Also called jawboning, moral suasion is one of the monetary tools of the Federal Reserve.

Another tool of the Federal Reserve alluded to earlier is **moral suasion,** the ability to persuade banks that some course of action is desirable. For example, during a period of rapid credit and money expansion—deemed undesirable by the Federal Reserve—banks may be warned subtly or not so subtly that continued expansionary behavior will result in implementation of policies that will force banks to cut back on money expansion. As we have seen, these threatened actions could be a raising of the discount rate, the buying of securities, or possibly increases in reserve requirements. To the extent that such suasion achieves the desired results, it could be a tool of the Federal Reserve. Sometimes, however, admonitions do not work, and the threatened action becomes necessary. Thus, the effects of moral suasion, while real, are

TABLE 29–6   Major Federal Reserve Controls and Their Probable Effects

Alterations in any one of the three major credit controls of the Federal Reserve will affect the monetary base of the financial system, bank reserves, and the money supply.

Federal Reserve System Tools	Effects					
	Bank Reserves	Money Supply	Monetary Base	Excess Reserves	Money Multiplier	Currency
Open Market Operations						
Federal Reserve buys bonds	Increase	Increases	Increases	Increase	Remains same	Increases
Federal Reserve sells bonds	Decrease	Decreases	Decreases	Decrease	Remains same	Decreases
Discount Rate						
Raise rate	Decrease	Decreases	Decreases	Decrease	Remains same	Decreases
Lower rate	Increase	Increases	Increases	Increase	Remains same	Increases
Reserve Requirement						
Raise requirement	Decrease	Decreases	Remains same	Decrease	Decreases	Decreases
Lower requirement	Increase	Increases	Remains same	Increase	Increases	Increases

not exactly predictable. The Federal Reserve is always on sounder ground if it can affect the nature of banks' portfolios directly.

Table 29–6 summarizes the Federal Reserve's major tools for money control and their likely effects.

Monetary Control: Reserves, the Monetary Base, and the Money Supply

M-1 money multiplier: The number by which the monetary base is multiplied to obtain the M-1 measure of the money supply.

We have seen that the Federal Reserve has numerous means of altering the stock of reserves and the monetary base. It remains to be shown how the monetary base—the sum of bank-held reserves and currency in circulation—is related to the money supply. In January 1985, for example, the monetary base was $220 billion, while the M-1 measure of the money supply was $563 billion. These figures indicate a money supply on the order of 2.56 times the value of the monetary base. The **M-1 money multiplier** is the number by which the monetary base is multiplied to obtain the M-1 measure of the money supply. This number can change over both short-run and long-run periods. We must understand why and how it changes because the very ability of the Federal Reserve to control the money supply through the monetary base depends on the predictability of the money multiplier.

The Exact Relation Between the Money Stock and the Monetary Base

To understand the forces determining the money stock, M-1, we must come to a more exact appreciation of the money multiplier. In simplified terms the money stock may be expressed as

$$\text{M-1} = \text{Monetary base} \times \text{M-1 money multiplier}.$$

We have already seen that the Federal Reserve controls the monetary base by pumping reserves into or withdrawing them out of the banking system through its credit control powers. But recall that bank reserves are only one part of the monetary base. Currency in public circulation is the other part of the monetary base, and currency holdings by the public are determined by the desires of the public. An individual may, at his or her discretion, convert demand deposits into cash by writing a check at the bank or some other place that cashes checks. The individual's decision to hold cash has an impact on the banking system's ability to expand the money supply. It would reduce the money-expanding capabilities of the system if an individual made a permanent decision to hold more cash than checkbook money.

The money multiplier discussed earlier in this chapter is actually a simplification. It was simply the reciprocal of the reserve requirement. While we will not account for every factor contributing to the multiplier, a more realistic way of determining the money multiplier must take into account the desires of the public to hold currency in relation to demand deposits. The components of a more realistic money multiplier can be defined as follows:

c = Desired currency holdings of the public relative to demand deposits,
r = Holdings of reserves required by the Federal Reserve and desired by depository institutions relative to demand deposits.

If, on average, people wish to hold 35 cents in cash or coin for each dollar in demand deposits, the value of c is 0.35. If the sum of required and desired (or idle) reserves in the banking system is 15 cents for every dollar of demand deposits issued, the value of r is 0.15. For obvious reasons, c is sometimes called the currency-deposit ratio, and r is called the reserve-deposit ratio.

A new and slightly more complicated expression for the money multiplier can be given as

$$\text{Money multiplier} = \left(\frac{1 + c}{c + r}\right).$$

This expression differs from our earlier one by the inclusion of the currency-deposit ratio in both the numerator and the denominator. In practical terms, the addition of c reduces the value of the simple multiplier. Using only the reserve-deposit ratio of 0.15 produces a money multiplier of $1/r = 1/0.15$, or 6.66. Use of a multiplier that includes the impact of the public's desire to hold currency gives a money multiplier of 1.35/0.50, or 2.7.

We are immediately able to understand the actual relation mentioned above for January 1985 between an M-1 of $563 billion and a monetary base of $220 billion:

$$\$563 \text{ billion} = \$220 \text{ billion} (2.56).$$

The monetary base of $220 billion is only one factor in explaining the money supply. The composition of the money multiplier, which equals $(1 + c)/(c + r)$, or 2.56, in our real-world example, is the other factor influencing Federal Reserve monetary control. Why is a more realistic view of the money multiplier important? The truth is that although the Federal Reserve is able to control the monetary base and banks' required reserve holdings, there are outside factors that the Federal Reserve does not control directly in the money supply equation. These factors can hinder the Federal Reserve's control over the money supply.

Pitfalls in the Money Multiplier

The Federal Reserve's ability to control the money stock may be impaired for any of the following reasons.

A major problem related to the money multiplier has already been mentioned. If the money multiplier equals $(1 + c)/(c + r)$, any increase in c—the public's desired holdings of currency relative to demand deposits—will reduce the value of the multiplier. Short-run changes in c are often unpredictable, causing changes in the ability of the monetary base to support a predetermined money stock.

Still another look at the expression for the money multiplier tells us that changes in r—the required reserve ratio or the banks' desired holdings of idle reserves—will affect the value of the multiplier. Should banks become more cautious for any reason and choose to hold more vault cash, the money multiplier would be reduced.

Other factors also affect the monetary base and thus the Federal Reserve's ability to control the money supply. The most important of these factors concerns the gradual removal since 1980 of interest rate ceilings on time and savings accounts and the relatively recent payment of interest on demand and checkable deposits. As Table 29–5 indicates, the reserve requirement on time and savings deposits is significantly lower than that on demand deposits or transaction accounts. A desire on the part of the public to hold greater quantities of time deposits relative to demand deposits would ordinarily cause the money multiplier to fall. Given our definition of money as M-1—currency plus demand deposits—the total quantity of money would decline. A good deal of shifting into and out of transaction accounts has been fostered by the relaxation of interest payment restrictions on time and savings accounts. These changes may have an impact on Federal Reserve control over changes in the money stock.

Real-World Multipliers, the Monetary Base, and the Money Stock

The actual M-1 money multiplier, calculated since 1959, has tended to decline except for a minor turn up since the late 1970s. As shown in Figure 29–1, the real-world multiplier has steadily declined from a value of about 3.1 in 1960 to about 2.8 in 1984. A brief analysis of this general decline gives us some valuable information on trends in our banking system.

In general, both the banks' desired holdings of idle reserves and the required reserve ratio set by the Federal Reserve have declined over the period. The Federal Reserve has very gradually lowered the reserve requirements and has, at the same time, applied them more uniformly to all financial institutions accepting deposits.

Other things being equal, the reduction in the r ratio would have increased the money multiplier rather than reducing it. But other factors have not remained equal. Over the past twenty-five years the ratio of the public's desired holdings of currency to its demand deposits has steadily risen, causing a reduction in the money multiplier. The increase in the public's desired holdings of currency relative to checkable deposits has been accompanied by a shift to money substitutes, notably to the numerous forms of savings and time deposits, many of them new. The Federal Reserve, of course, has been aware of this overall decline and has attempted to adjust the monetary base to fit its money supply policy intentions.

FIGURE 29–1

The Real-World Money Multiplier, 1960–1984

The actual M-1 money multiplier has, in general, declined over the past twenty-five years in the United States. Reductions in the required reserve ratios *(r)* over the period would have increased the multiplier, but increases in the currency-deposit ratio have been the primary cause of the overall decline shown in the figure.

Source: Federal Reserve Bulletin, various issues.

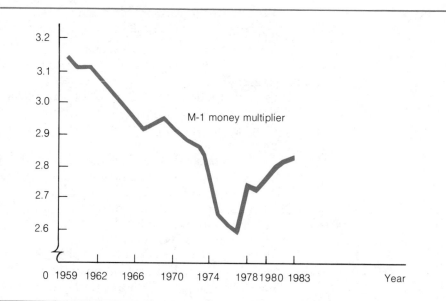

While there is a discernible long-run downward trend in the money multiplier, there are month-to-month and week-to-week changes that render Federal Reserve control of the money stock through changes in the monetary base more difficult. Constantly changing market rates of interest on time, savings, and checkable deposits, as noted earlier, are a devilish factor in the Federal Reserve's calculations. But since 1979, the Federal Reserve has made a concerted effort to control the money supply by controlling a monetary aggregate—the monetary base and bank reserves. Before 1979, an interest rate target—the federal funds rate—was the principal means used by the Federal Reserve to direct money and credit conditions. Despite short-run alterations in the money multiplier, the Federal Reserve is perfectly able to control the money supply within fairly narrow bounds by following monetary aggregate targets. The question that concerns monetarists is whether any discretionary control over the money stock would be adequate medicine for the macroeconomic ills of inflation, unemployment, and lagging growth in GNP. A whole spectrum of issues raised by the monetarists is the subject of Chapter 30.

Summary

1. Major activities of the Federal Reserve System are summarized in its balance sheet. On the asset side, the Federal Reserve holds a portfolio of earning assets such as loans and securities. Increases in these earning assets indicate an increase in the commercial banking system's reserves, the monetary base, and the nation's money supply.

2. Liabilities of the Federal Reserve System include Federal Reserve notes (currency) in the hands of commercial banks and the public and the deposits of commercial banks and domestic and foreign governments. An increase in any liability item, except bank reserves themselves, will reduce the monetary base and the nation's money supply.

3. Reserve bank credit outstanding is an important statistic in that it shows the difference between factors increasing reserves and factors decreasing reserves of the commercial banking system.

4. The monetary base is the sum of banking system reserves and currency in the hands of the public. The Federal Reserve attempts to control the monetary base, which is the basis of expansions and contractions in the M-1 money supply.

5. The monetary base is controlled by the Federal Reserve's use of major credit controls: open market operations, loans or discounts to commercial banks, and changes in the required reserve ratio.

6. Purchases of securities from commercial banks or

the public, lowering of the discount rate, and lowering of the required reserve ratio all have the effect of increasing bank reserves. Sales of securities, raising of the interest or discount rate charged to banks for loans, and increases in the required reserve ratio decrease bank reserves and the monetary base.

7. Less-important controls over bank reserves and lending are sometimes used by the Federal Reserve. Selective credit controls include special powers over consumer credit, mortgage loans, and stock margin requirements. Moral suasion may also be used by the Federal Reserve.

8. Federal Reserve control over the money supply is made trickier by changes in the money multiplier. A change in the money multiplier may be caused, in both the short run and the long run, by changing desires on the part of the public to hold currency relative to demand deposits.

Key Terms

reserve bank credit
reserve bank credit outstanding
monetary base
credit controls

open market operation
federal funds rate
discount rate
stock margin requirement

moral suasion
M-1 money multiplier

Questions for Review and Discussion

1. How does the Federal Reserve System control the money supply through the banking industry? What tools does the Federal Reserve use? How does it control the money supply without using the banking system?
2. What is the monetary base? Is this the same thing as the money supply?
3. "The Federal Reserve can control the minimum amount of reserves that the banking system holds but not the maximum amount." Is this statement true or false? Explain.
4. If the statement in question 3 is true, does it mean that the Federal Reserve can limit increases in the money supply but cannot stop the money supply from falling?
5. What are the things that determine the "powerfulness" of the monetary base?
6. "If the Federal Reserve purchases securities from the nonbank public rather than the banking system, it has a different effect on the money supply." True or false? Explain.
7. Why would a bank want to borrow money from the Federal Reserve and pay the discount rate rather than just accept demand deposits and pay no interest?
8. Is moral suasion an effective monetary tool of the Federal Reserve?
9. Does the reserve-deposit ratio or the currency-deposit ratio have anything to do with the monetary base?
10. "The Federal Reserve is just a big cartel that is operated by the government to make greater profits for the banking industry." Give some arguments that may support this view and some that may refute it.

ECONOMICS IN ACTION

The Quasi-Independence of the Federal Reserve System

The Federal Reserve System is unique among government regulatory agencies in one important respect: The Federal Reserve enjoys relative freedom from the oversight of Congress or the executive branch. This unique status is called quasi-independence and has been a source of much conflict among politicians, economists, and the investment community.

The Federal Reserve system was created by Congress in 1913 to administer and manage the money supply as directed in the Constitution. As an agency of Congress, the Federal Reserve Board represents or is an extension of Congress, but Congress set up the Federal Reserve to be quasi-independent; that is, at some distance from the control of the president and from direct congressional involvement. For example, Federal Reserve Board members other than the chairman are appointed for fourteen-year terms by the president, which means that they can remain on the job regardless of which party controls the White House. Independence is also reflected in the fact that the Federal Reserve is not dependent on congressional and administrative budget appropriations. The Federal Reserve finances all of its own operations, mainly through interest earned on government bonds. This independence from congressional appropriations and from the typical auditing procedures applied to all other U.S. regulatory

Paul Volcker, Federal Reserve Board Chairman, smokes a cigar during his appearance before the Senate Budget Committee on February 8, 1985.

agencies has gone far in placing the Federal Reserve out of the grasp of Congress or the administration.

Is the Federal Reserve too independent? Should it be brought under closer scrutiny of Congress or the ad-ministration? Critics argue that the loose connection between the president and the Federal Reserve System creates discoordination between fiscal and monetary policy. Monetary policy, in this view, should be under the direct control of the executive branch of govern-ment so that it is tied closely to the democratic wishes of voters. Such criticism is especially vocal during pe-riods of high interest rates and tight money.

Defenders of the quasi-independence of Federal Re-serve control argue otherwise. They claim that the Fed-eral Reserve *is* a democratic creature of Congress. Con-gress could, at any time, with the requisite votes modify the Federal Reserve Act to "reform" the actions of the Federal Reserve. The Federal Reserve Board chairman is required to present periodic written and personal reports to Congress relating to the manage-ment of M-1 and interest rates. To go beyond this, de-fenders argue, would give politicians too much author-ity to inflate the money supply as a possible quick solution to budget deficits, unwise government spend-ing, or political ambition.

Question

Suppose that the Federal Reserve's motto becomes "Beat Bolivia" which has an inflation rate in the tens of thousands in percentage terms. Would Congress step in and impose more direct controls over the actions of the Federal Reserve's Board of Governors?

30

Money Demand, Inflation, and Monetarism

M oney clearly matters. The Federal Reserve Board announces weekly changes in M-1, the nation's basic money supply, and that announcement—which can suggest the possibility of loose or tight credit in the future—creates optimism or pessimism along Wall Street. Investors react by buying or selling stock. Buyers and sellers of home mortgages and all other financial instruments also anxiously observe the weekly changes in M-1 in an attempt to predict changes in interest rates and the future profitability of their financial investments.

The main reason for all of the attention paid to M-1 is the relationship—understood by most Americans—between M-1 and price inflation. Inflation has been most aptly described as "too much money chasing too few goods," and although economists have constructed elaborate theories to explain inflation, it is still best understood in these simple terms. When the production of real output (bananas or home computers or hair stylings) does not grow as fast as the M-1 medium of exchange, inflation sooner or later is the result. And most people readily understand that when their nominal money income does not grow at the same rate that prices increase, they are worse off.

In the three preceding chapters we investigated the nature of money, how it is created in the commercial banking system, and how the supply of money is regulated by the Federal Reserve System. In this chapter we take a brief look at monetary theory, specifically at the relation between the money supply, M-1, and the price level. There are several theories of how M-1 is related to prices and to inflation. The key to any process relating M-1 and the price level is to understand how and why money is demanded. After discussing money supply and demand, we examine two theories of how the supply of and demand for money affect the price level: the quantity theory of money and the liquidity preference theory of money. Next we discuss the role of price expectations and a modern variant of the quantity theory of money, called monetarism. Finally, we recap our discussion of monetary theory by once again looking at why inflation occurs and how anti-inflation policies might work.

The Demand for Money

How do economists explain the relation between M-1 and inflation? Economists argue that the value of money, on which the price level of all output depends, rests like the value of other goods on both its demand and supply. We have seen that money is supplied and controlled by the Federal Reserve, but the Federal Reserve does not control the demand for money. People as money holders determine what demand is and what quantities of money they wish to hold. Economists have singled out three reasons why people demand money.

Transactions Demand

People hold or demand money, the medium of exchange, as a simple means of carrying out transactions—to buy chewing gum, dinner at a fancy restaurant, or a home computer. Businesses demand money to pay for labor, materials, and other inputs. The extent of money holdings to accommodate transactions of businesses and consumers is referred to as the **transactions demand** for money.

Transactions demand: The amount of money desired by businesses and consumers to facilitate the purchase of goods and services.

The transactions demand for money is related to the income of consumers and businesses. As incomes rise, the demand for more and more money to carry out transactions also rises. For example, a family with an income of $30,000 will have a greater transactions demand for money than a family with an income of only $15,000. This general relation is true of the economy as a whole as well. As money income rises throughout the economy (either because of a rise in the price level or because of an increase in the real value of GNP), the transactions demand for money will also increase.

Precautionary Demand

A second reason for holding money is the "rainy day" motive. People usually want to keep some money readily available to meet unforeseen emergencies such as illness or car repair. Likewise, firms generally exercise caution in how much money they hold.

Precautionary demand: The amount of money that businesses and consumers want to hold to cover unforeseen expenses.

The **precautionary demand** for money is simply the amount of money that households and firms want to hold to meet unforeseen events. Ordinarily, the amount of money that people wish to hold against unforeseen contingencies rises with income. Precautionary demands may therefore be included with transactions demands as being directly related to income.

Speculative Demand

A final explanation for money demand is the **speculative motive** or **demand.** Like precautionary demand, the speculative motive for holding money is based on the idea of uncertainty: Some individuals will want to hold money to speculate in the markets for bonds or other liquid assets whose price and interest return vary. These speculators have to choose between holding money and holding interest-bearing bonds. To make the choice, they consider the present interest rate and the degree of uncertainty about future conditions.

Speculative motive or demand: The amount of money desired by individuals and businesses to invest at various interest rates.

To money holders, the nominal interest rate is the opportunity cost of holding money. It is the return we give up by holding money and not investing it in some other asset that would yield interest. Economists theorize that high interest rates make people want to hold bonds. The price of a bond and its interest yield are inversely related, so high interest rates mean that bond prices are relatively low. (See Chapter 14 for a discussion of bond prices.) Under these conditions, bonds are a good deal for two reasons. One obvious reason is the high interest return; in addition, if the price of bonds rises in the future, bond holders would experience a capital gain. A capital gain occurs when bond buyers purchase bonds at a low price and then bond prices increase. In the event of rising bond prices—which occurs with falling interest rates—selling bonds becomes more attractive. With high bond prices and low interest rates, people generally prefer to hold more money and fewer bonds. Conversely, low bond prices and high interest rates mean that people will hold less money and more bonds.

The speculative motive for holding money means that money demand is inversely related to the interest rate. As the interest rate rises, bonds are a more attractive asset, and the quantity of money demanded declines. As the interest rate falls, bonds are less attractive and money more attractive.

Figure 30–1 graphically expresses the preceding ideas. The curve labeled L_S is the speculative demand for money. (L is demand for money or "liquidity"). At any time, money holders have formed some expectations about what interest rates and therefore bond prices will be in the future. Given expectations about future bond prices and interest rates, a present interest rate of r_0 means that the quantity of money demanded for speculative purposes is L_{S_0}. A lower interest rate, such as r_1, means that actual bond prices rise and, as discussed above, the amount of money that individuals will want to hold for speculative reasons will rise. In Figure 30–1, the quantity of speculative balances demanded rises to L_{S_1}.

FIGURE 30–1

The Speculative Demand for Money

Individuals hold money for speculative purposes; that is, to take advantage of good deals in the bond market. A low interest rate, r_1, means relatively high bond prices and increased money holdings, L_{S_1}. A higher interest rate, r_0, indicates higher bond holdings and lower holdings of money for speculative purposes, L_{S_0}.

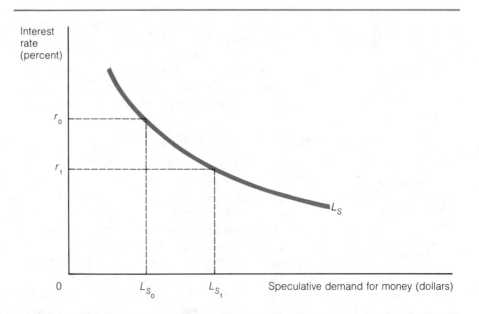

TABLE 30–1   The Three Motives for Demanding Money

Both the level of money income and the rate of interest determine the total demand for money. The quantity of money demanded by all consumers and businesses varies directly with income and inversely with the actual rate of interest, given some level of expectations about future bond prices.

Type of Money Demand	Factor Affecting Demand	Effects on the Quantity of Money Demanded
Transactions To buy and sell goods and services	Rising money income	Higher money demand
	Lowering money income	Lower money demand
Precautionary To save money for unforeseen circumstances	Rising money income	Higher money demand
	Lowering money income	Lower money demand
Speculative To invest in money instruments	Rising interest rates	Lower money demand
	Falling interest rates	Higher money demand

Money Demand: Summary

We now summarize the three influences on the demand for money. Table 30–1 lists the three types of money demand with the chief factors affecting them and the effects of a change in the factors on the quantity of money demanded. As Table 30–1 indicates, the total demand for money is related to income and the rate of interest. As income rises or falls, the transactions and precautionary demand for money rises or falls. If income is held constant, a rise in the interest rate reduces money holdings for speculative purposes. Conversely, a fall in the interest rate produces a rise in the holdings of speculative money.

But this information is only half the story. Just as the price of potatoes is the product of both supply and demand, the price level is a product of both money supply and money demand. On this fact all economists agree. But there are two different theories about how money demand and supply interact to produce a given price level, inflation or deflation, and employment and real income. One theory, the quantity theory of money, places primary emphasis on transactions demand; the other, the liquidity preference theory, highlights the influence of the speculative demand for money.

The Quantity Theory of Money

Quantity theory of money: A hypothesis that there is a direct and predictable relation between the money supply and prices; as one increases so will the other.

One of the essential ingredients of classical economic theory and policy, developed during the century beginning around 1776, was the **quantity theory of money.** This theory maintains that there is a direct and predictable relation between the money supply, M-1, and prices, given a constant transactions demand for money. If the number of dollars demanded to carry out everyday transactions (referred to by the classical economists as the income velocity) remains constant, then an increase in the money supply will increase prices.

The quantity theory is sometimes stated as an identity (an equation that must be true) called the **equation of exchange.** In economic shorthand, the equation of exchange is written

Equation of exchange: An identity stating that the money supply multiplied by the income velocity of money is equal to the final output of all goods multiplied by the price level of all goods ($MV \equiv PQ$).

$$MV \equiv PQ,$$

where

M = the money supply, M-1, consisting of currency in the hands of the public and checkable deposits;

V = the income velocity or the average number of times dollars (the medium of exchange expressed in M-1) are used per year (or week or month) to finance the final purchase of goods and services;

P = the price level of all goods and services; and

Q = the final real output of all goods and services produced and sold in the economy over some period (a year, a month, a week).

Income velocity: The average number of times that a dollar is used to purchase goods and services annually.

The concept of **income velocity** is crucial to our understanding of the quantity theory of money. M-1 in 1984 was approximately $558 billion. In that same year, GNP, which is expressed as the price level P times final output Q was $3661 billion. For $558 billion to pay for $3661 billion worth of output, the average dollar must finance at least $6.56 worth of goods. ($3661 ÷ $558 = $6.56). We say that the average dollar turned over at least 6.56 times in the course of the year, or that the income velocity of one dollar in 1984 was 6.56. If we know the value of GNP and the quantity of money (M-1) at some point in time, velocity is easily calculated from the equation of exchange as

$$V \equiv \frac{PQ}{(\text{M-1})} = \frac{\text{GNP}}{(\text{M-1})}.$$

While the equation of exchange is useful in some respects, it does not help the economist predict how M-1 is related to P and therefore to inflation. Economists need a theory to do that. To make the equation of exchange a theory, we must specify something about V, Q, P, or M in advance. That is, we must make something in the expression variable and then predict cause and effect on the basis of some hypothetical value of the variable or variables. This is exactly what the classical and neoclassical economists did in converting the equation of exchange into a quantity theory of money.

The Simple Quantity Theory

The simple quantity theory is exactly the same as the equation of exchange except that the equivalence sign is not given as $\equiv$, which means "must be equal," but as $=$, which means "is predicted to be equal." The equation of exchange

$$MV = PQ$$

is stated so that it can be falsified—shown to be untrue—and then is therefore a theory.

One of the oldest commonsense macroeconomic propositions is that inflation is "too much money chasing too few goods." The quantity theory explains why that is true. First note that, in general, V and Q were assumed to be constant and predictable. Income velocity—the turnover of money—was thought to be relatively constant over the short run and predictable over the long run. Although the velocity of money or its average turnover time for money holders varied from individual to individual, its average remained constant and stable for all money holders taken together.

Why didn't early quantity theorists worry about the level of Q, the final output of goods and services? They simply assumed that real output was at a maximum. In fairness to these early theorists, we should point out that theirs was a long-run view. Economic disasters created by financial panics, wars, famines, supply shocks, or other disruptions were certainly as much as part of their world as ours, as were positive economic events such as inventions and improvements in technology and labor productivity. These events obviously caused reductions or increases in real output. Early monetary theorists such as Adam Smith, David Ricardo, Alfred Marshall, and Irving Fisher neither were unattuned to real-world events nor were they fools. They knew that there would always be short-run fluctuations in output—what economists call business cycles. Adjustments were always taking place. So the classical economists assumed that output and therefore income was at a maximum because they believed there is a persistent tendency to have economy-wide full employment of resources, implying maximum Q.

As we saw in Chapter 23, the classical and neoclassical economists were believers in Say's law, a principle given expression by Jean Baptiste Say, an early French follower of Adam Smith. Briefly stated, Say's law was that supply creates its own demand. In other words, the act of producing and supplying goods creates the wherewithal to purchase these goods from the market. Gluts of goods or shortages of goods would not ordinarily take place, but if they did, prices of final goods and services and of labor and other inputs would quickly adjust. This classical assumption of price and wage flexibility meant that adjustment time to full employment would be brief. Should prices and wages fail to be flexible, interest rates would adjust to save the day, according to the classical economists.

The significance of assuming V and Q to be constant or predictable is that the quantity theory of money becomes a theory of cause and effect. Money supply is the cause; prices are the effect. In a modern context, if income velocity and real income are constant, increases in M-1 will increase prices; decreases in M-1 will decrease prices. But by how much do prices increase with increases in M-1? And, more pointedly, how exactly do "too many dollars" end up "chasing too few goods"? A feature of the relation between money and prices assumed by many early economists answers both questions.

Money and Relative Prices

Neutrality of money: A proposition stating that the relative prices of all goods and services are independent of overall price level.

Relative prices: The market value of goods or services compared to the market value of other goods or services.

Early economists advanced a proposition called the **neutrality of money,** which means that the **relative prices** of all goods and services (how computer games trade for rock concert tickets, for example) are independent of the overall price level.

The neutrality of money can be easily understood with a simple example. Suppose that, starting from some stable equilibrium money supply that produces an equilibrium price level—that is, where money demand equals money supply—the money supply is suddenly doubled. That is, we wake one morning to find that the banks will redeem every $1 bill with $2 and will double the value of all of our checkable deposits. The quantity theory predicts what will happen:

$$MV = PQ,$$
$$2MV = 2PQ.$$

If consumers' tastes remain the same, the price level will exactly double in the long run when the excess balances created by a doubling of the money supply are spent, but relative prices of goods and services will not change.

What does this mean in practical terms? At first, individuals perceive the

TABLE 30–2   Effect of Doubling the Money Stock in a Three-Good World

If money is neutral—that is, if it does not affect relative prices—a doubling of the money stock doubles the price level from P_0 to P_1, though relative prices and output do not change. The price of beer per case in terms of pretzels is the same before and after the price increase.

Good	Final Output Produced Q	Initial Price P_0	$P_0 \times Q$	P_1	$P_1 \times Q$
Beer	2 cases	$10 per case	$20	$20	$ 40
Pretzels	10 bags	$1 per bag	$10	$ 2	$ 20
Pizzas	5 large	$10 each	$50	$20	$100
Nominal GNP			$80		$160

increase in their money holdings to be a bonanza. Their new money supplies exceed their demand for money. What do these individuals do to restore a balance between money demand and supply? They spend their excess money holdings in the attempt to acquire more goods. If their tastes do not change, their relative expenditures on shoes, candy, and gasoline do not change. But—and here is a major point—there are no more goods to be had. The economy is already at a full employment level of resources and real output. The only effect of these increased money expenditures is to drive all prices up. The price level doubles, but relative prices remain unchanged if money is neutral.

We will investigate this process with a numerical example. Consider a simple, three-good world where quantities of beer, pretzels, and pizzas are produced at certain base prices. Table 30–2 gives data showing the final output of the three goods and an initial set of prices, indicated as P_0. GNP in this simple world is measured by price multiplied by the quantities of goods produced, an initial sum of $80. Now suppose that the money stock doubles and that the new nominal GNP equals $160. If money is neutral, the price of each good produced before the money stock increase must exactly double. This means that relative prices of all the goods traded both before and after doubling of the money stock must remain the same. And, in the example of Table 30–2, they do. One case of beer still trades for ten bags of pretzels and for one large pizza.

Thus, if money is neutral, real relative prices are independent of the price level and of the nominal prices of all goods and services traded in the economy. Neutrality means that money is the oil of trade, not the wheel of trade.

The Simple Quantity Theory: Summary

We summarize the cause-and-effect relation between money and prices in Figure 30–2. Increases in M-1 create an excess supply of money in the hands of consumers and businesses. Their demand for money is unchanged as a proportion of income, but they now hold larger quantities of money. Their spending attempts are met with frustration since real output is already at a maximum. The effects of this attempt to acquire more goods and services simply drive the money price of goods and services upward. The process ends only when nominal GNP has risen to the point where money demanders are again holding balances equal to their real demands for money.

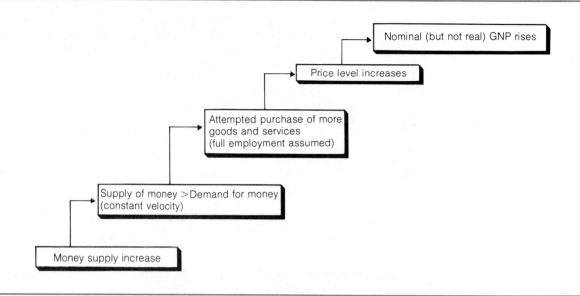

FIGURE 30–2   The Simple Quantity Theory of Money
Increases in the money stock create excess supplies of money balances, causing consumers and businesses to attempt additional purchases of goods and services. Assuming a constant output and full employment of resources, price level increases result.

Money and Prices:
The Liquidity Preference View

We turn now to the second theory about the relation between money supply and prices. Recall from the beginning of this chapter that there are three motives for holding money. The classical economists emphasized transactions demand and, implicitly, the precautionary demand for money. In this view, money demand is related soley to the level of income. The higher the level of income, the greater the amount of money demanded for transactions and precautionary purposes. Money demand was a constant proportion of money income, making velocity constant and predictable.

What about speculative money demands? In the 1930s and later, a theory of money demand was developed on the basis of speculative money demand. The **liquidity preference theory** states that the rate of interest is determined by overall money demand and supply and that economic activity, including price changes, flows from changes in the rate of interest.

Liquidity preference theory: A proposition that the rate of interest is determined by overall money demand and supply and that economic activity flows from changes in the rate of interest.

The Relation of Interest Rates to Money Supply and Demand

Consider the relation between interest rates and economic activity more formally. Total money supply, M-1, may be thought of as being composed of M_T, a portion that goes to satisfy money holdings for transactions and precautionary purposes, and M_S, a portion that is held for speculative purposes. Symbolically,

$$M = M_T + M_S.$$

If we assume a given level of income, the quantity of money going to satisfy transactions and precautionary demand, M_T, is fixed. The amount of money left over for speculative holdings therefore depends on M, the total supply of money, or

$$M_S = (M - M_T).$$

An equilibrium interest rate can then be calculated by setting the speculative demand for money (L_S) equal to the supply of speculative holdings (M_S), as in Figure 30–3. Given that the level of income remains constant, a speculative demand curve for money, L_S, can be drawn by varying the interest rate. In addition, once income is given, we may determine the total amount of money that is available for speculative purposes. In Figure 30–3 this amount is designated by M_{S_0}, which is shown as a vertical line.

Interest rate r_E is the equilibrium rate in Figure 30–3: At r_E, speculative demand L_{S_0} is equal to the quantity of money available for speculative purposes. Any interest rate higher than r_E, such as r_1, would not last because the quantity of money supplied would exceed the quantity of money demanded by the amount AB. At a lower rate, such as r_2, the quantity of money demanded for speculation would exceed the quantity of such funds supplied, by the amount CF. Under these conditions, the interest rate would rise to adjust the discrepancy between the supply and demand of money for speculative balances.

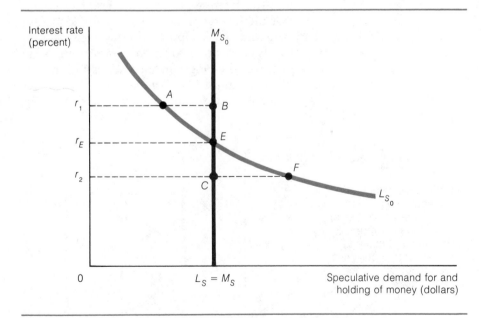

FIGURE 30–3   Determination of the Interest Rate in the Money Market

The rate of interest is partly a product of the interplay of speculative balances demanded, L_S, and speculative balances held, M_S. Demand curve L_{S_0} and money supply M_{S_0} determine an equilibrium interest rate r_E. Given these particular curves, any rate lower than r_E, such as r_2, means that an excess demand for funds, represented by CF, exists, forcing the rate to rise to r_E. A rate higher than r_E creates an excess supply of money, such as AB at rate r_1. Interest rate r_E is the only rate at which the demand for and supply of money are equal.

The Importance of Speculative Demand in the Economic System

We can immediately appreciate the importance of the speculative demand for money by examining the effects of increases in the money supply on the rate of interest. As Figure 30–4 shows, increases in the money supply increase the supply of money available for speculative purposes and drive interest rates down. The higher bond prices that result from lower interest rates cause asset holders to opt for fewer bonds and more speculative money balances. In other words, interest rates have created a preference for the liquid asset, money, over the less-liquid asset, bonds.

The key to understanding the importance of speculative demand and liquidity preference theory is to observe that investment and consumption might be affected by changes in the interest rate, as the classical economists suggested. As discussed in Chapter 23, a rise in the interest rate reduces desired investment and increases desired saving, whereas a fall in the interest rate has the effect of increasing both investment and current consumption.

To the classical economists, lower interest rates ordinarily meant higher spending, higher output, and higher employment. Monetary policy—policies designed to alter the quantity of money—could be used to lower interest rates, to increase consumption and investment spending, and to help bring the economy back to full employment equilibrium if it ever veered from that level.

J. M. Keynes, as developer of the theory of speculative demand for money, did not agree at all with this classical assessment. While he thought that the interest rate was the result of complex forces, he did not believe that interest rate reductions would necessarily have much impact on investment and consumption. In his view, consumption and investment were largely insensitive to interest rate changes, being autonomous or more responsive to changes in income, as indicated in Chapter 23. Monetary policy—alterations in the money stock—was therefore not as effective as fiscal policy to nudge the economy toward full employment. (Focus, "Depression

---

FIGURE 30–4

Increase in the Money Stock

As the money stock available for speculative holding, M_S, increases from M_{S_0} to M_{S_2}, the interest rate falls, given demand curve L_{S_0}. Three different equilibrium rates are shown along demand curve L_{S_0}.

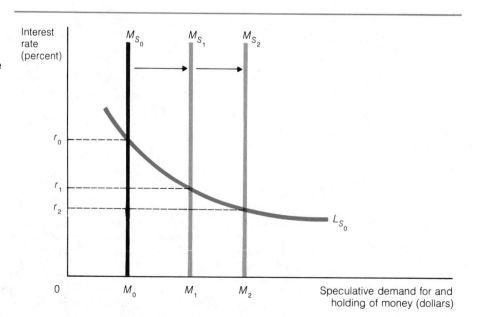

FOCUS   **Depression Economics:**
The Liquidity Trap

The classical economists believed that individuals held money in cash for transactions and precautionary demand. Keynes argued that people hold money to speculate in the bond market, and, on this theory, he constructed the liquidity preference function.

The liquidity preference function, shown in Figure 30–5, unlike the one shown in Figure 30–4, has a characteristic that is specifically Keynesian—the liquidity trap. Keynes argued that the interest rate might fall so low (bond prices rise so high) as to make people believe that bonds were a bad investment. All asset holders would prefer to hold more money and fewer bonds. In Figure 30–5, r_1 is the interest rate where individuals would regard holding bonds as risky and would hold cash instead. The horizontal section of the speculative demand curve L_S is referred to as the liquidity trap.

Keynes argued that the existence of the speculative demand for money meant that the mechanism by which money influenced income and employment in the economic system was not as simple and predictable as the classical economists believed. Specifically, one of the major impacts of money on spending, income, and employment was its effect on interest rates. Lower interest rates, other things being equal, lead to higher levels of investment and consumption.

The figure shows an increase in the nominal money stock, M, with the price level remaining constant. This increase in the money supply from M_{S_0} to M_{S_1} reduces the interest rate from r_0 to r_1, causing income and employment to rise as the lower interest rate spurs economic activity. However, in the liquidity trap portion of the function, the interest rate does not fall because investors as a whole prefer to hold cash rather than interest-bearing bonds, so investment and consumption

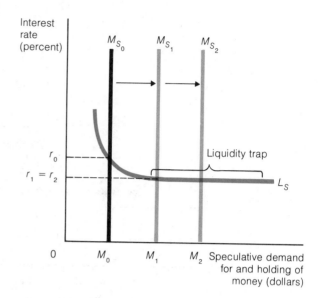

FIGURE 30–5   The Liquidity Trap

When all money holders think bonds are unsafe, the speculative demand curve for money, L_S, contains a horizontal section. After interest rate r_1 is reached, further increases in the money supply no longer result in reductions in the interest rate and will not cause any additional investment or consumption activity, which would normally result from lower interest rates.

are not affected. According to Keynes, given the liquidity trap, monetary policy—which works through the interest rate—is useless in the face of depression and unemployment. Real-world evidence does not verify the existence of a liquidity trap, although the idea was popular among some of Keynes's followers.

Economics: The Liquidity Trap," provides more detail on Keynes's dispute with the classical economists.)

What does Keynes's view mean for the short-run relation between the money supply and prices? The addition of speculative demand to the transactions and precautionary demand for money short-circuits the cause-and-effect relation between the money supply and prices. Increases in the money supply may have little or no effect on the price level if the spending generated by an interest rate decrease is small or nonexistent.

Figure 30–6 illustrates Keynes's scenario. An increase in M-1 creates an excess supply of money. The resulting reductions in interest rates (and concurrent increases in bond prices) cause increased holdings of idle money. Consumption and investment spending increase little, meaning that aggregate demand increases only slightly. Even if resources are fully employed, prices cannot be much affected. If there is widespread unemployment, prices

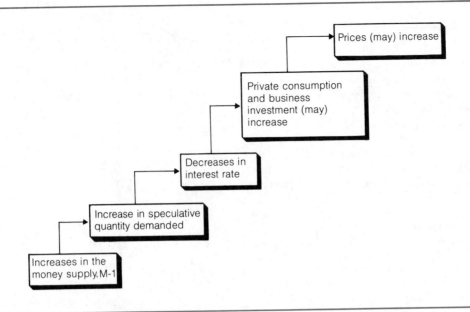

FIGURE 30–6   Money and Prices: Liquidity Preference View

According to Keynes, an increase in M-1 has an uncertain effect on the price level. A declining interest rate, for instance, may not encourage additional total demand. Even if demand increases, prices may not increase if the economy is experiencing widespread unemployment.

are not affected. The liquidity preference theory therefore implies an uncertain and unpredictable relation between the money supply and prices.

Modern Monetarism

The quantity theory of money predicts a long-run relation between the money supply and the price level, whereas the liquidity preference approach explains a possible link between money and prices in the short run through the effects of money supply increases on interest rates. The leading contemporary explanation of inflation and deflation—*monetarism*—combines elements of both views along with a new perspective on the inflation process.

Monetarists are concerned with both short-run and long-run adjustments of the economy to different rates of change in the money supply. The monetarist view of inflation emphasizes the uncertainty people feel about what inflation rates will be in the future and the effects this uncertainty will have on present behavior. As we will see in Chapter 32, these expectations will have a short-run impact on employment and real income as well as on prices. For the present, however, we focus on a simple monetarist explanation of the inflation process, which contains three basic elements: a modern explanation of the demand for money, a distinction between the market or nominal interest rate and the real interest rate, and the impact of people's expectations about future prices and the inflation process.

Money Demand

One of the foundations of modern monetarism is the formulation of a short-run demand for money. Chicago economist Milton Friedman developed a short-run money demand model in the mid-1950s as a direct extension of both the simple quantity theory and the liquidity preference approach. As we saw, earlier economists argued that people hold money for transactions, precautionary, and speculative purposes. Friedman makes no such distinctions, simply accepting the fact that money is held in response to several independent variables. The most important of these variables are income and the nominal interest rate.

Like the quantity theorists and Keynes, Friedman argues that some money holdings for transactions depend partly on income.[1] Also in the manner of Keynes, the monetarists find that the nominal interest rate affects money holdings. While other variables affect money demand in the monetarist model, we neglect them in the following discussion of the monetarist view of inflation and concentrate only on y, current income, and i, the nominal interest rate, as the factors influencing money demand. A shorthand version of money demand can thus be expressed as

$$L_d = L_d(y, i).$$

In this expression the demand for money is determined only by current income and the nominal interest rate.

The Nominal Interest Rate and Price Expectations

Real interest rate: The current rate of interest minus the rate of inflation.

Market or nominal interest rate: The current rate of interest paid at banks and lending institutions.

The classical economists had written extensively about the determinants of the real rate of interest. As we saw in Chapter 23, the **real interest rate**—the current rate paid minus the rate of inflation—is determined by the real and fundamental forces of thrift (saving) and productivity (investment) in the economy. But this fundamental rate is generally not the rate of interest that one would pay to borrow money at a bank. The **market** or **nominal interest rate**—the rate paid at banks and lending institutions—has two components: the real rate of interest and a dynamic element based on price expectations. Consider why the market rate includes both elements.

If you decide on January 1 to borrow $100 from a bank for one year, how will the interest rate be determined? Put another way, on what basis will banks decide to make loans? If the real rate of interest, reflecting basic forces of savings and investment, is 4 percent, the bank will certainly charge at least 4 percent. But will the bank charge more? If the rate of inflation expected over the year is 8 percent, the bank would lose real income on a 4 percent loan. It would be lending at a return of 4 percent, but the value of the money returned at the end of the year will have also lost 8 percent of its purchasing power. Thus, after the fact, the bank would have made a bad bargain at any loan rate below 12 percent.

The nominal interest rate is thus influenced by the expected inflation rate as well as the real interest rate. When inflationary expectations rise, the nominal interest rate also rises. When the inflation rate is expected to fall, nominal rates tend to fall. When actual, realized rates of inflation are different from those expected, borrowers and lenders adjust in the following period. The effect of inflationary expectations on market rates of interest is a stock-in-trade of modern monetarist thought. Through price expectations, a dynamic period-to-period element is brought into modern interpretations of the quantity theory of money.

---

[1]Friedman's assumption rested on permanent income—the income expected over a lifetime—rather than current income. In our simple discussion of inflation, however, we retain current income as a determinant of money demand, neglecting Friedman's more elegant and complex measure of permanent income.

Adaptive expectations theory: The proposition that present price expectations about the future are formed on the basis of actual experiences with prices in the recent past.

How do we form expectations about future prices? A well-known and currently used theory about this matter is the **adaptive expectations theory,** which states that present price expectations about the future are adapted to the public's most recent experience with prices. The basic premise of adaptive expectations is that it takes time for individuals to adjust to changed circumstances. Expectations about occurrences in the future hinge primarily on present knowledge and present experiences. Consistent experiences in the present ("the sun rose in the east every day this week") tend to create the same expectations about future events ("the sun will rise in the east tomorrow"). The adaptive expectations theory simply asserts that our expectations about future price conditions are formed on the basis of actual past price experience, with the most recent experience having the greatest influence. There are other theories about the formation of expectations (some of which will be explored in the following chapters), but the simple adaptive expectations idea goes far in explaining the modern monetarist connection between monetary expansion and the inflation rate.

Modern Monetarism and Inflation

The simple tools of the old quantity theorists combined with a modern version of the demand for money and the adaptive expectations theory provide a monetarist explanation for the process of inflation. Before turning to this concept, it is important to recall exactly what inflation is. *Inflation* is a rate of increase in prices, not a simple rise in the price level. In a static framework, it is sometimes convenient to say, as we did using the simple quantity theory, that a doubling of M produces a doubling of P and to identify this once-and-for-all change as inflation. In a dynamic real-world setting, however, the money supply is constantly being increased or decreased by the Federal Reserve Board at some rate per week, month, or year. Likewise, prices are rising or falling at some rate. Real income is also changing, by values typically expressed in terms of growth rates such as 3½ percent per year, although in the present simple discussion of the inflation process we assume a constant income growth rate. To arrive at the modern conception of inflation, therefore, we must adapt the static quantity theory to a dynamic setting.

The Process of Inflation

The simple process of inflation is best understood by starting from a dynamic equilibrium position in the economy. In this initial position, the economy is characterized by the following conditions:

1. A constant rate of monetary expansion (whether 3 percent or 50 percent growth in the money stock, M-1, per year—the actual percentage makes no difference, so long as it is constant).
2. Expected and actual inflation rates are equal (these rates will be equal to the rate of monetary expansion, and, further, market participants will expect the inflation rate to be exactly what it has been in the present and recent past).
3. The nominal rate of interest equals the real rate of interest plus the constant inflation rate. Since the inflation rate has been constant in the most recent past experience, according to the adaptive expectations theory the inflation rate (and therefore the nominal rate of interest) is expected to be the same in the future.

4. People are actually holding real cash balances equivalent to their desired holdings of money balances. The demand for money is therefore equivalent to the supply of money in real terms, that is, M "weighted" by the level of prices.

5. Real income or real GNP is growing at a constant rate.

If we numerically express the percentage rates of change of the items in the quantity theory, we can calculate a rate of change in prices. If, for example, the monetary expansion rate, which we now designate with a dot on top of M, or $\dot{M}$, is 12 percent, and the constant growth rate in real income, $\dot{Q}$, is 4 percent, and velocity V is constant ($\dot{V}$ is thus zero), the inflation rate $\dot{P}$ is easily calculated. The dynamic quantity theory becomes

$$\dot{M} + \dot{V} = \dot{P} + \dot{Q},$$

or, given the hypothetical numbers,

$$12\% + 0\% = \dot{P} + 4\%,$$
$$\dot{P} = 12\% - 4\% = 8\% \text{ inflation rate.}$$

The inflation rate is thus calculated as 8 percent. Furthermore, the inflation rate equals the rate of monetary expansion, which is the percent increase in the money supply minus the growth rate in income, or $\dot{P} = \dot{M} - \dot{Q}$ (8% = 12% − 4%).

The exercise of casting the quantity theory into rates of change might seem mechanical so far, but it is invaluable in providing a simple understanding of inflation as a process. To initiate the inflation process we need simply to suppose that the Federal Reserve, through its powers over the monetary base, suddenly increases the rate of monetary expansion from 12 percent to 15 percent. Further, the Federal Reserve maintains the rate of monetary expansion at 15 percent from that point forward and forever.

Figure 30–7 illustrates what happens. The new, higher growth rate in the money supply initially creates disequilibrium among money holders. After the expansion, money demanders find themselves holding more real money balances than they want to hold, given the previous state of equilibrium. Consumers and businesses will attempt to rid themselves of these excess balances by spending more on all goods and services.

What, however, is the impact of the increased spending on real income and employment? The answer is "nothing" in real terms. Since output and employment were already growing at a constant full-employment rate, the attempt to purchase more goods will only drive up prices and money income. Thus the actual inflation rate begins to rise. Since it takes time for higher actual inflation rates to be translated into expectations of higher inflation, that is, for the adaptive expectations theory to come into play, higher expected rates will follow the actual rate.

It is easy to predict what will happen next. Since the nominal interest rate equals the real interest rate (which we assume to remain constant) plus the expected inflation rate, nominal rates begin to climb. This rise in the nominal rate affects the desired holdings of money. Since the real demand for money is partially a function of the nominal interest rate (the opportunity cost of holding money), the higher interest rate generated by higher inflation expectations leads to lower desired money holdings. This effect reinforces the attempt of money holders to rid themselves of cash balances. That is, the higher nominal rate further fans the rise in the inflation rate.

When does the process stop? When is a dynamic equilibrium reattained?

FIGURE 30–7

How the Fire of Inflation Gets Fanned

The process of inflation begins with an increased rate of M-1 expansion by the Federal Reserve. Excess money holdings are spent, creating higher actual inflation and higher expected inflation. Higher expected inflation causes higher market interest rates, further increasing spending and inflation. The process continues until the Federal Reserve stabilizes the rate of growth in M-1.

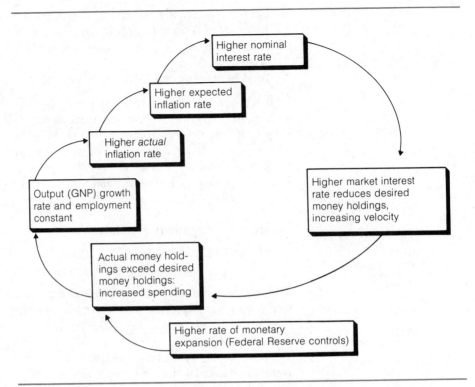

The process will stop only if the Federal Reserve sticks to a constant rate of monetary expansion (15 percent in our example). The process of increasing inflation will continue only if the Federal Reserve continues to increase the rate of money creation. If the Federal Reserve holds the rate to 15 percent, for example, equilibrium will be reattained at a new constant inflation rate, a new nominal interest rate, and a new rate of price expectations equal to the actual inflation rate. In the numerical terms of our example, equilibrium will take place when

$$\dot{M} + \dot{V} = \dot{P} + \dot{Q},$$
$$15\% + 0\% = 11\% + 4\%.$$

Prices must rise, in other words, so that *actual* real money holdings are equivalent to *desired* real money holdings. Spending and the inflation rate will rise until real money demand equals money supply.

Parts of this process are presented pictorially in Figure 30–8, which depicts the growth rate of inflation, money supply, and income on the vertical axis and time on the horizontal axis. As Figure 30–8 shows, the Federal Reserve maintains a monetary expansion rate of 12 percent up to time period t_0, after which it begins increasing the money supply at a rate of 15 percent. As money holders begin to spend these excess balances on the same quantity of goods and services, the actual inflation rate rises, with the expected inflation rate rising shortly thereafter. As expectations begin to adapt and as increases in the nominal interest rate reinforce the actual inflation rate through increased spending, the actual and expected rates oscillate around the new permanent inflation rate of 11 percent. A dynamic equilibrium is reattained only when money holders again reach equilibrium, when expected and actual inflation rates are equal, and when the Federal Reserve holds increases in the money supply to a constant rate.

Some Preliminary Lessons About Inflation

The monetarist view of inflation contains certain explicit and implicit lessons. First and foremost, monetarists argue that inflation is always a monetary phenomenon. Milton Friedman has supported this proposition with massive hundred-year empirical studies of the quantity theory in the United States and England. As we note in Figures 30–7 and 30–8, the villain in the monetarists' inflation scenario is expansionist policies of the monetary authority that controls the money stock.

In the monetarist version of events, for example, high interest rates do not *cause* inflation; rather, they are the *result* of inflation. What we did not mention in our simplified treatment is that one short-lived initial effect of the Federal Reserve's monetary expansion is a temporary decline in market interest rates. Of course, the market rates begin to rise as expectations of higher inflation take hold of borrowers and lenders. Observers and politicians often make the mistake of arguing that the money stock growth rate should be expanded to lower the interest rates. As we have seen, however, the monetarists show that such a short-sighted policy will have the opposite effect: Increasing the expansion of the money supply will soon raise nominal interest rates. Higher interest rates follow higher inflation rates, but higher monetary expansion is necessary for both events to occur.

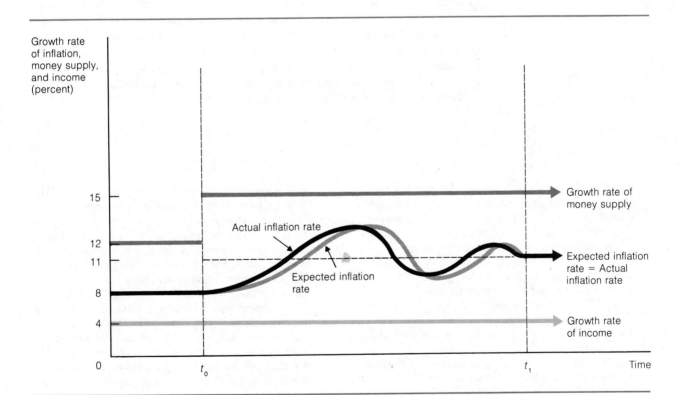

FIGURE 30–8   A Simplified Monetarist Explanation of Inflation

When the Federal Reserve steps up its rate of M-1 expansion from 12 to 15 percent, the actual inflation rate begins to rise. Shortly thereafter, the expected rate rises as well, resulting in new inflationary pressures in the economy and higher nominal interest rates. If the Federal Reserve stabilizes its growth rate of M-1 at 15 percent, the actual inflation rate will ultimately equal the expected rate (at 11 percent in the figure).

Many people also think that inflation is self-generating. Higher nominal interest rates do affect desired money balances and spending. But in the monetarist view, this process must come to an end if the central bank holds the money expansion to a constant rate. It may, of course, take time for inflationary expectations to adjust between consumers and producers, borrowers and lenders, and employers and employees, but when the market adjusts, a dynamic equilibrium will be attained.

Hyperinflations—as sometimes experienced in war-ravaged countries such as post-World War I Germany—can proceed only through increasing rates of monetary expansion. In these cases, money holders will often complain of a lack of liquidity as inflation rates rise, depleting the purchasing power of money balances. In these cases, the central bank may be urged to speed up its supply of money to replenish these balances. But in the monetarist scenario, the effects are all too predictable. Spending merely rises at faster rates, driving the inflation rate still higher. The bottom line of such behavior is economic collapse and reversion to more primitive systems of barter.

The early quantity theorists presented essentially a theory of inflation but without the sophistication of a well-formed theory of money demand and price expectations. Modern monetarists have filled this gap and presented an enriched view of the process of inflation. Implicit in their theory is a criticism of any Federal Reserve monetary policy that they regard as erratic and destabilizing. Although this chapter has focused on the relation between the money supply and inflation, monetarists recognize that Federal Reserve policies also have an effect on short-run income and employment. They agree that in the short run at least money management is important in influencing real factors in our economic system. Modern monetarist thoughts on the critical policy issues of short-run unemployment and economic growth will enter our discussion of macroeconomic and monetary policy in Chapters 31–34.

Summary

1. There are three basic demands for money: transactions demands, precautionary demands, and speculative demands in the bond or securities market.

2. Transactions and precautionary demands are directly related to income. An increase in income, for example, would cause an individual to hold a proportionately greater amount of money for transactions and precautionary purposes.

3. The public also holds money to speculate on the bond market. Expectations of higher bond prices (lower interest returns) would mean that individuals would choose to hold money rather than bonds, whereas low bond prices (higher interest returns) would mean that individuals would prefer to hold bonds rather than money. Speculative money demands are therefore inversely related to the interest rate.

4. The simple quantity theory of money was a macroeconomic theory of how the price level is formed. When transactions demands are considered proportional to income, velocity is also assumed constant. Since real output and income are also assumed to be at a maximum, increases in money supply, M-1, cause proportionate increases in price level.

5. The liquidity preference theory focuses on the speculative demand for money. In this model, the interest rate is determined by the supply and demand for money. An increase in the money supply lowers the rate of interest, possibly encouraging consumption and investment spending. If aggregate demand increases, prices may increase, depending on whether resources are fully employed.

6. The modern monetarist theory of inflation integrates the transactions, precautionary, and speculative demands for money into a dynamic view of the inflation process. In the monetarist conception, money demands depend in part on the nominal interest rate, which in turn depends on both the real rate of interest and inflationary expectations.

7. The monetarists view inflation as a dynamic process originating with higher rates of M-1 expansion by the Federal Reserve System. Increased spending leads to higher actual and expected inflation rates (real GNP being constant) and therefore to higher

nominal interest rates. Higher nominal interest rates lead to lower money demand, more spending, and more inflation.

8. In the monetarist explanation of inflation, the inflationary process does not end until the Federal Reserve stabilizes growth of the money supply.

Key Terms

transactions demand
precautionary demand
speculative motive or demand
quantity theory of money

equation of exchange
income velocity
neutrality of money
relative prices
liquidity preference theory

real interest rate
market or nominal interest rate
adaptive expectations theory

Questions for Review and Discussion

1. Why do people hold money? What effect does the level of income have on the amount of money people hold?
2. What is the cost of holding money? If bond prices fall, does the cost of holding money rise or fall?
3. What is the difference between the quantity theory of money and the equation of exchange? When does the quantity theory of money hold true and when does the equation of exchange hold true?
4. What is the income velocity of money, V? How does V relate to the transactions demand for money?
5. What is liquidity preference? How does this determine interest rates?
6. What is the difference between the nominal interest rate and the real interest rate? If the expected rate of inflation increases, what happens to the real rate of interest and the nominal rate of interest?

7. If the rate of growth in the money supply is equal to 10 percent, the velocity of money is constant, and real output is falling by 2 percent, what will be the rate of inflation according to monetarist theory?
8. What effect does an increase in the rate of monetary expansion have on the total level of spending; the actual rate of inflation; the expected rate of inflation; and the nominal rate of interest?
9. "Higher interest rates on loans at banks are going to cause an increase in the rate of inflation." Explain why this statement is true or false.
10. According to monetarists, what causes inflation? How is inflation stopped?
11. How can interest rate targets by the Federal Reserve lead to inflation? Explain Federal Reserve interest rate targets briefly and tie the concept to monetary expansion.

ECONOMICS IN ACTION

Money Supply versus Interest Rate Targets: The Monetarist Perspective

Market, or nominal, rates of interest, as we have seen, are the product of two factors: the real rate of interest and the expected inflation rate. That is, the interest rate is a product both of real factors of saving and investing in the economy and of the monetary factors of money supply and demand. Before 1979, the target of Federal Reserve policy was primarily to expand or diminish the monetary base to control or to stabilize interest rates. This goal was expressed as "keeping orderly conditions in the money markets." This policy did not mean that the Federal Reserve was unconcerned with the money supply but that it acted to hold the interest rate (specifically, the federal funds rate, the interest

rate at which banks borrow from each other) to narrow variations while controlling the money supply (M-1 and other monetary aggregates) within broader limits.

Monetarists believe that this previous Federal Reserve policy tended to destabilize economic activity because the Federal Reserve did not and could not know why the interest rate increases or decreases. For example, the interest rate may rise in response to an increase in real investment demand. Attempts by the Federal Reserve to compensate for the rise in the interest rate by injecting more money into the system would reinforce short-run increases in total spending and tend to be inflationary. If the increase in the interest rate was due to a short-run increase in the demand for money, the Federal Reserve's attempts to compensate by preventing the interest rate from rising would simply be neutral in effect. The monetarists therefore believe that interest rate control by the Federal Reserve could at best be neutral but at worst have destabilizing effects on economic activity. These arguments, coupled with the contention that inflation or deflation rates and other economic activity in the short run may be closely traced to changes in M-1, led the monetarists to advocate direct control by the Federal Reserve over M-1.

In 1975, the Federal Reserve Board was required by Congress to set money supply growth targets for certain periods. (At present, weekly money supply targets are announced.) To achieve these goals, the Federal Reserve watched closely over measures such as the federal funds rate. In October 1979 the Federal Reserve announced that it would target monetary aggregate goals through tighter daily control over bank reserves, the monetary base, and other aggregates. This announcement meant that the Federal Reserve has promised to hold M-1 and other aggregates within tight bounds while permitting interest rates to fluctuate more widely.

These policies are not to the liking of modern-day Keynesians, who insist that short-run economic activity is influenced, if at all, by money through interest rates. They further argue that money demand is unstable in the short run and that the Federal Reserve should compensate for shifts to prevent monetary factors from troubling the economy. Monetarists, in contrast, believe that the interest rate is a stable and predictable function of a number of variables. Monetarists argue that interest rate concerns should take a back seat to the control of M-1 and the monetary base because M-1 is the best short-run predictor of inflation.

Since 1979, the Federal Reserve has paid more attention to the control of monetary aggregates such as M-1 and less to interest rates. Problems have emerged,

however. It is argued within and outside of Federal Reserve circles that the growth in new forms of checkable deposits—such as NOW accounts, super-NOW accounts, and money market certificates—makes it difficult to know exactly what M-1 is at any time. Since these accounts both bear interest (and may thus be considered investments or savings balances) and are at the same time checkable, it is impossible to know how much is intended for transactions, and how much for other purposes.[a] The critics of monetary aggregate controls thus argue that the Federal Reserve should "go backward when forward fails" and return to interest rate controls.

Monetarists counter by noting the strong relation between changes in the monetary base and M-1 and between M-1 and GNP since monetary aggregates were introduced as a Federal Reserve target. Monetarists within the Federal Reserve System, such as Lawrence K. Roos, president of the St. Louis Federal Reserve Bank, have adamantly defended a predictable relation of M-1 to economic activity. The relation between M-1 and inflation, after an 18 to 24 month lag, is particularly persuasive (see Figure 32–3). Monetarists, therefore, continue to argue that monetary and economic stabilization can only be achieved through targeting monetary aggregates, especially M-1.

Question

"Interest rates are extremely high today. We must increase M-1 in order to reduce them to tolerable levels." Evaluate this statement.

---

[a]Bryon Higgins and Jan Faust, "NOW's and Super-NOW's: Implications for Defining and Measuring Money," *Economic Review* (Federal Reserve Bank of Kansas City) (January 1983), pp. 3–18.

POINT-COUNTERPOINT   Milton Friedman and James Tobin:
How Important Is the Money Supply?

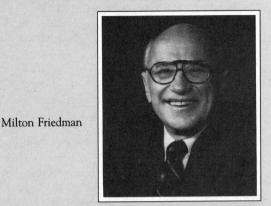

Milton Friedman

James Tobin

MILTON FRIEDMAN (b. 1912) is best known for his strong and eloquent defense of capitalism and the free market and for his arguments against government intervention in the economy.

Born in Brooklyn, New York, Friedman was the son of immigrant parents. His mother was a seamstress, his father the owner of a small retail dry goods store. When his father died, fifteen-year-old Friedman and his mother were left with very little money. The teenager showed great aptitude in mathematics in high school and won a scholarship to Rutgers University in 1929. While a student at Rutgers, Friedman waited tables to support himself, passed exams to become an accountant, and then became increasingly interested in economics. His professor, Arthur Burns, who later became chairman of the Federal Reserve Board, taught him the importance of empirical research in economics, and another professor, Homer Jones, encouraged Friedman to apply for a scholarship to graduate school at the University of Chicago. Friedman went on to receive his M.A. from Chicago and, in 1946, a Ph.D. from Columbia University. During the 1930s he worked as a statistician for the National Bureau of Economic Research, and in 1938 he married economist Rose Director, a fellow graduate student at the University of Chicago.

Friedman taught economics at Chicago until his retirement in 1977, when he moved his scholarly base to the Hoover Institute at Stanford University. He won the Nobel Memorial Prize in Economics in 1976, authored a regular *Newsweek* column, and wrote many highly acclaimed books. In *Capitalism and Freedom* (1962), Friedman argues that market forces are sufficient in both individual and aggregate markets to direct resources and to establish steady growth in the economy. Government fiscal policies, Friedman believes, often bring about results contrary to those that policymakers intend and disrupt planned expenditures in the private sector. In his classic work, *A Monetary History of the United States* (1963), coauthored by Anna Schwartz, Friedman claims that money velocity—related to the demand for money—is stable enough to make the money supply an excellent barometer of inflation, output, and economic growth. His views provided the basis for modern monetarist theory. Friedman is a leading contemporary conservative economist.

JAMES TOBIN (b. 1918), winner of the 1981 Nobel Memorial Prize in Economics, grew up during the Depression with an awareness that many of the world's problems were economic in origin. He made the decision to become an economist while taking an introductory economics course during his sophomore year at Harvard University in 1936. One of his major influences at that time was his teacher, Spencer Pollard, who suggested he read Keynes's *General Theory,* which had just been published. According to Tobin, *The General Theory* "was a difficult book, but when you are 19 you don't know what's difficult and what's not. You just plow into it."

Describing himself as a "very, very shy, noncompetitive individual" during his undergraduate years, Tobin went on to obtain his Ph.D. from Harvard in 1947 and was appointed Sterling Professor of Economics at Yale University in 1957. He faithfully carried Keynesian principles into the public policy arena when he was appointed to President John F. Kennedy's Council of Economic Advisers in 1961. He was one of the architects of the Kennedy-Johnson tax cut of 1964 and has consistently advocated a loose reign on the money supply. He has also played a key role in the recent effort to join Keynesian macroeconomics and neoclassical microeconomics.

Tobin received the Nobel Prize primarily for his analysis of portfolio selection and financial markets. He was president of the American Economic Association in 1971 and continues to support modern activist economic policies.

Friedman and Tobin certainly represent different views on monetary and fiscal policies. For Friedman, control of the money supply is the main determinant of economic stability. For his part, Tobin emphasizes the effects of government spending, taxation, and stable interest rates on economic performance. Juxtaposed, the views of each man illustrate the difference between monetarist and Keynesian approaches to the economy.

Keeping Watch over the Money Supply

Are money supply fluctuations a cause or effect of changes in real income? This question may be impossible to answer definitively, but Milton Friedman has consistently argued that money supply growth is the best barometer we have of economic expansion and contraction. Friedman cites a variety of statistics to back his claim. For example, since the turn of the century, fluctuations in rates of growth in real GNP have strongly coincided with annual growth rates in the nation's money supply. Prior to recession and depression, the money supply growth rate has fallen dramatically; prior to expansions and a booming inflationary economy, the money supply growth rate has shown marked increases.

To explain this statistical marriage of money supply and economic activity, Friedman argues that money velocity is a stable variable (see Chapter 30). Therefore, interest rates, which may vary up or down in response to government spending, do not contribute to economic fluctuations as many economists believe. They are merely a sideshow to the growth patterns in the money supply.

If Friedman's premise is correct, then the government's attempts to fine-tune the economy through spending and taxation policies are largely misdirected. For the sake of stability, Friedman argues, government should concern itself with a stable growth rate in the money supply. By doing so, he believes, the rollercoaster ride of business cycles, of temporary boom followed by temporary bust, can be leveled out and the economy can achieve consistent, measured growth for the future.

There are many complicated issues surrounding Friedman's proposal. For instance, can government officials afford to sit by idly in times of high unemployment, believing that over the long run a stable monetary policy will rescue the economy? Experience in the 1960s and 1970s suggests that Friedman's proposal is perhaps politically unfeasible. A second issue involves the definition of money supply. For Friedman, the only significant money supply definition is M-1—currency plus demand deposits. Recently, however, changes in the banking system have created uncertainties over

what exactly constitutes money in the traditional sense of M-1. If the Federal Reserve policymakers cannot rely on a firm definition of the money supply, efforts to control its growth may be frustrating or even futile. Friedman strongly believes that the concept of M-1, for all its apparent weaknesses, is the Federal Reserve's best measure of money supply. Larger aggregates, such as M-2 or M-3, are far beyond the control of the Federal Reserve System.

Fiscal Activism and Stable Interest Rates

For James Tobin, the issue of what determines ups and downs in the economy is not nearly so straightforward. In essence, Tobin believes that consumption and investment activity is the true determinant of economic performance. Controlling or directing such activity, however, is much more complicated than simply monitoring the money supply. On the one hand, it involves spending and taxation policies to counterbalance the rise and fall of aggregate demand. On the other hand, it involves monetary policies to keep interest rates at acceptable levels.

Unlike Friedman, Tobin believes that the velocity of money is unstable. If this assumption is correct, then interest rates are much more important in economic policy than M-1 growth rates because interest rates determine the shifting demand between different kinds of money assets, leading to greater or lesser consumption and investment activity. Tobin argues that a "strict monetarist regime" he believes to have been in effect at the Federal Reserve between 1979 and 1982 was a failure. He blames the recession and high unemployment between 1981 and 1983 on the Fed's attempt to place limits on M-1 growth.[a] Rather than target M-1 or the money base, Tobin believes the Fed should target broader economic aggregates such as the growth rate in nominal GNP and adjust its monetary policy according to its best estimate of where the economy is headed. (Friedman disputes the idea that monetarism has ever been tried by the Federal Reserve System. See Focus, "1981-1984: Recession and Recovery," in Chapter 33.)

Tobin's position not only admits the need for politicians to do something in times of high unemployment, but it argues for such action. His activist stance, however, can be a two-edged sword. Spending policies can adversely affect interest rates, and therefore fiscal and monetary policies can conflict with one another. The current debate over high government deficits reflects this conflict, for many economists argue that growing deficits drive up interest rates.

[a]For details on Tobin's position see his "Monetarism: An Ebbing Tide?" *The Economist* (April 27, 1985), pp. 23–25.

VII

Monetary and Macroeconomic Problems and Policy

31

From Macroeconomic Theory To Policy: An Overview

*T*he next four chapters deal with policies designed to improve the health of the economy. But as we have seen in previous chapters, economists find it hard to agree on where the economy is heading and what should be done about it. How many times have you heard different predictions of future macroeconomic events by government economic advisers, academics, and business economists? Economists often are equally at odds over the proper policy to correct or sustain economic forces. These differences over policy often arise from differences in theory, particularly whether the economy is seen as inherently stable or unstable. Earlier we discussed how specific theories of the economy, such as early classical theory and Keynesian theory, have led to quite different policy prescriptions. In this chapter we compare and contrast three separate theoretical models of the economy—the Keynesian, the monetarist, and the rational expectationist—and carefully examine the different policy alternatives each theory suggests.

The divisions and categories in this chapter by no means encompass every economic view. There are many self-styled Keynesians who share one or more of the monetarists' views, for instance. As a road map to policy debates, however, the distinctions drawn here are useful.

Policy Alternatives

The purpose of macroeconomic policy is to ensure economic stability—full employment, price stability, and steady growth. There are, of course, various interpretations of exactly what full employment, price stability, and steady growth mean in percentage terms, but most economists and policymakers would agree, for instance, that persistent unemployment rates of 12 or 14 percent or inflation rates of 15 or 20 percent are unacceptable. Likewise, most would agree that unemployment rates of 5½ to 7½ percent, inflation rates of 3 to 5 percent, and GNP growth rates of 3 percent are "in the ballpark" as acceptable macroeconomic policy goals.

In previous chapters on macroeconomic and monetary theory, we saw that there are a number of interacting monetary and nonmonetary factors that affect prices, real income, and employment. Figure 31–1 summarizes these various forces. As the figure shows, the nonmonetary factors are consumption, investment, and net foreign trade expenditures on the private side and government spending and taxation on the public side. The monetary factors are money supply and money demand.

A subset of these variables can be manipulated through discretionary government policy. Fiscal policy—manipulation of government spending or taxation by Congress or the president—is one means to ensure overall stability. Monetary policy—actions of the Federal Reserve Board to alter the money supply or interest rates—is another means to achieve macroeconomic goals. Table 31–1 summarizes these two types of policy.

Fiscal and monetary policies differ according to the amount of time each takes to implement, the magnitude of their effects, and the direction of their effects. For example, by using one of its specific tools the Federal Reserve can act quickly to alter M-1, the money supply, to correct a perceived unemployment downturn or inflation upsurge, but Congress usually requires a bit more time to enact fiscal correctives. Fiscal policies, moreover may have more immediate effects on the economy, but the effects of monetary policies may be longer lasting. In the following sections, we consider different views of these possible effects.

FIGURE 31–1

Determinants of Price Stability, Full Employment, and Economic Growth

Both monetary and nonmonetary variables determine whether the goals of economic policy are achieved. Of these determinants only government spending, taxation, and the money supply may be directly altered in discretionary fashion (also see Figure 31–2).

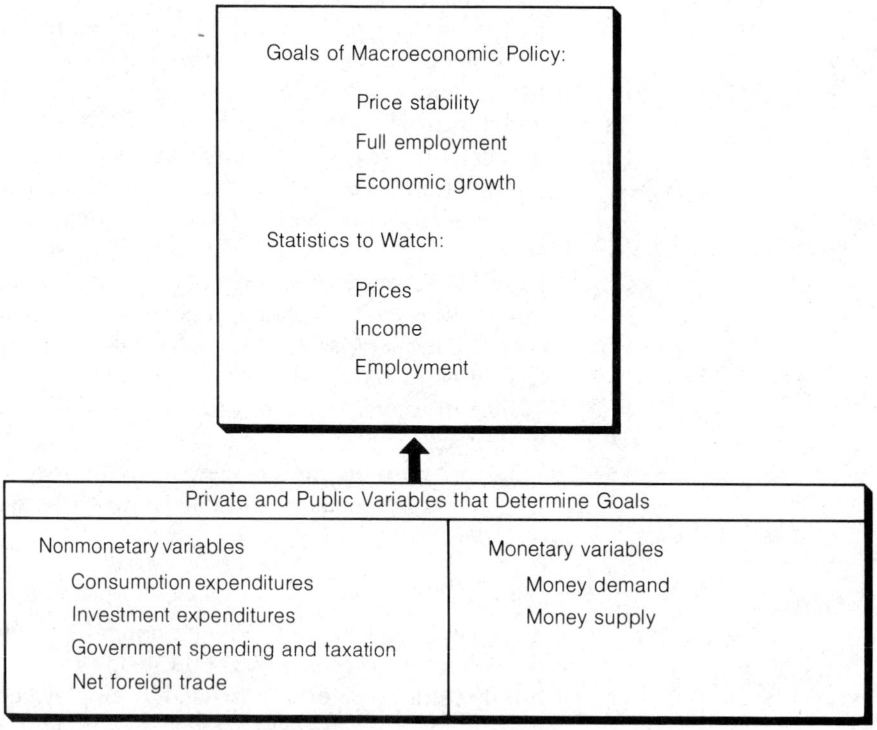

Goals of Macroeconomic Policy:

Price stability

Full employment

Economic growth

Statistics to Watch:

Prices

Income

Employment

Private and Public Variables that Determine Goals

Nonmonetary variables	Monetary variables
Consumption expenditures	Money demand
Investment expenditures	Money supply
Government spending and taxation	
Net foreign trade	

TABLE 31–1   Types of Macroeconomic Policies

Fiscal policy is controlled by the president and Congress. Monetary policy is directed by the Federal Reserve Board.

Type of Policy	Scope	Tools
Fiscal policy	Changes in government taxation, spending	Alterations in income tax rate or other tax rates
		Changes in government spending on goods and services or for transfer payments
Monetary policy	Changes in money supply, interest rates	Open market operations
		Changes in discount rate
		Changes in reserve requirements

Keynesian Theory and Policy

Early Keynesian economics, developed in the 1930s and 1940s, was depression-born and primarily oriented toward the problems of depression and unemployment. Modern Keynesians believe that the economy, left to itself, is unable to adjust to falling aggregate demand and restore the equilibrium level of income and full employment. In this view, fundamental instabilities pervade the economic system. In the broadest sense, the role of policy is to overcome the forces of instability in the economy.

A Nonstabilizing Economy

Sharp fluctuations in GNP, unemployment, and inflation can occur for a variety of reasons. Investment spending, spurred by the accelerator, can initiate cycles of rising and falling business activity. Supply shocks, caused by natural disasters or international upheaval, can disrupt domestic markets. Autonomous consumer spending, magnified by the multiplier, can drive GNP and unemployment up or down.

Our analysis of aggregate demand shifts in Chapter 25 helps illustrate graphically the effects of such disturbances. Figure 31–2 shows aggregate demand and aggregate supply of a hypothetical economy. Along curve AD_0, the economy reaches full employment, Y_F. Assume, however, that investment spending, in response to a sudden change in profit expectations, drops. As a result, unemployment increases and real output falls. Figure 31–2 shows the effect: a shift in aggregate demand from AD_0 to AD_1 and a fall in income from Y_F to Y_1. The gap in employment caused by the shift in aggregate demand is usually referred to as demand-deficiency unemployment. With resources idle and a gap in demand, policymakers are left with the questions, Will the economy be able to self-correct the deficiency? and How long will it take to do so? In the long run, as we have noted, theory predicts that prices and wages will fall in response to a shift in aggregate demand. But what about the short run? Can the economy or the unemployed afford to wait for the natural market forces to restore full employment?

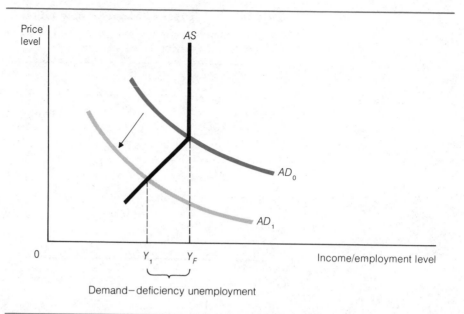

FIGURE 31–2   The Post-Keynesian Stand on Demand-Deficiency Unemployment

A decrease in real expenditures causes the aggregate demand curve AD_0 to shift downward to AD_1, also causing a shift from full employment, Y_F, to Y_1. This gap in employment caused by a shift in aggregate demand is called demand-deficiency unemployment. Keynesians suggest that this unemployment can be eliminated through the use of fiscal policy, returning aggregate demand and employment to their original levels.

Countercyclical Fiscal Policy

In the Keynesian view, time is at the heart of the problem. While the economy might be self-correcting in the long run, the adjustment would be painful and time-consuming. In the short run, prices and wages would not fall sufficiently to restore full employment. Unnecessary economic disruption—instabilities in product and labor markets—would occur if the economy were allowed to establish its own natural rate of income and employment growth. Keynesians' recommended cure for macroeconomic problems is to apply countercyclical fiscal policy. Proper macroeconomic policy in the event of recession and unemployment is to run budget deficits—that is, to keep government expenditures greater than tax revenues. And, though the advice has seldom been taken (see Focus, page 687), inflation calls for budget surpluses—that is, for tax receipts to exceed government expenditures.

Table 31–2 simplifies and summarizes the Keynesian position. During recession or the more severe economic downturn called depression, Congress, the administration, and state and local governments should increase spending and/or reduce taxes. Inflation calls for the opposite policies.

Keynesian Monetary Policy

Note in Table 31–2 that monetary policy is also included in the Keynesian approach to economic instability. Most Keynesians advocate that the Federal Reserve Board respond to inflationary or recessionary pressures in the economy by keeping order in financial markets. That is, the Federal Reserve should use its discretionary power over the M-1 money supply to stabilize

TABLE 31–2   Keynesian Prescriptions to Control the Business Cycle

In the fiscal arena, budget deficits should be employed to cure unemployment and lagging economic growth, while inflation should be controlled by budget surpluses. Throughout all phases of the cycle, stable interest rates should be maintained by the Federal Reserve in money markets.

Economic Problem	Keynesian Corrective	
	Fiscal Policy	Monetary Policy
Recession (unemployment and slow income growth)	Budget deficits ($G > T$) (Government spending greater than taxation)	Increase M-1 and keep interest rates stable
Inflation	Budget surpluses ($G < T$) (Taxation greater than government spending)	Reduce M-1 but keep interest rates stable

interest rates, which, left unchecked, can rise to levels that discourage or even choke off private spending.

An illustration of the Keynesian approach to monetary policy came in the early 1980s. At that time, the Federal Reserve, in response to severe inflation, initiated dramatic cuts in the M-1 supply. Such cuts, many Keynesians felt, were unwise, since they brought about a sharp rise in interest rates. The economy lurched from high inflation to a costly recession. The Keynesians blamed the recession on the Federal Reserve's failure to target stable interest rates while at the same time adjusting M-1 to combat inflation.

Why do the Keynesians prefer to target interest rates? Because such rates help determine consumption and investment spending, which in turn determine the economy's performance. Low interest rates, for example, encourage spending and help reduce gaps in aggregate demand. Discretionary monetary policy therefore should aim for whatever level of M-1 is necessary for stable interest rates.

Direct versus Indirect Policy. Given recurring short-run instabilities in the economy, Keynesians consider fiscal policy a more direct, more immediately effective tool to manage aggregate demand. Monetary policy in general is a less-effective tool. In the Keynesian model, there is no solid link between changes in M-1 and changes in aggregate demand. If M-1 increases, the larger holdings of money by individuals and businesses might not all be spent on more goods and services; some of these funds will be channeled into speculative holdings.

Problems with M-1. In addition, the money supply is less suited to discretionary policy. The Federal Reserve is sometimes unable to keep interest rates within specified targets by manipulating M-1. One reason for this problem is the recent emergence of close money substitutes such as certificates of deposit and money market certificates, which have tended to blur the definition of M-1 money. Without a reliable measure of the money supply, the Federal Reserve is potentially unsure of the effects of its M-1 policy.

For these reasons and others, Keynesians believe that tax and spending policies can give policymakers greater discretionary control over the business cycle than can monetary policies. The effects of monetary adjustments, especially on interest rates, are an "intermediate" mechanism to ensure stability.

More important, perhaps, than the issue of whether fiscal or monetary policy is the preferred tool of policymakers is the issue of whether such policy does what it is supposed to do—correct the instabilities caused by shifts in aggregate demand. Keynesians obviously believe that ignoring such instabilities in the short run would inflict greater potential harm on the economy. As we will see, however, not all economists believe that countercyclical discretionary policy is the best way to ensure stability.

Monetarist Theory and Policy

The monetarist policy position, aspects of which were treated in Chapter 30, differs significantly from the Keynesian view. The foundation of monetarism, as of the classical macroeconomics that preceded it, is that the economy is inherently stable.

A Self-Stabilizing Economy

We can see the self-stabilizing nature of the economy by analyzing what happens when there is a reduction in aggregate demand that leads to reductions in output and to unemployment. Consider Figure 31–3, which analyzes demand-deficiency unemployment from the monetarist perspective. Assume there is a reduction in autonomous consumption or investment, shifting the

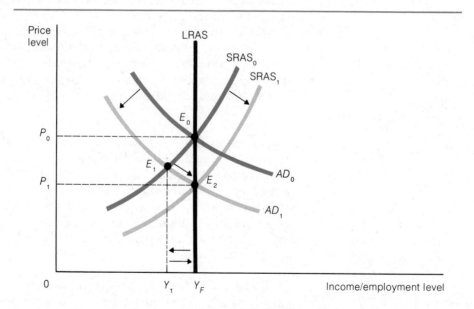

FIGURE 31–3   Monetarist Policy on Demand-Deficiency Unemployment

The economy will self-adjust to the natural rate of employment, Y_F, in the event of a real decline in aggregate demand. Price and nominal wage reductions occur when aggregate demand declines from AD_0 to AD_1, creating increased real balances, which in turn increase consumption and investment spending. Such price reductions also create, through adaptive expectations, rightward shifts in short-run aggregate supply from $SRAS_0$ to $SRAS_1$. The economy finally readjusts to the natural rate of employment, Y_F, at equilibrium E_2 with lower price level P_1.

original aggregate demand curve from AD_0 to AD_1. The reduction in autonomous spending is accompanied by an increase in consumers' and businesses' holdings of real money balances. Initially, the economy undergoes an adjustment from equilibrium income—the natural rate of employment at Y_F—and an equilibrium price level P_0 at E_0 to a new short-run equilibrium, E_1, at income level Y_1. Then the economy self-adjusts to the natural rate of employment, Y_F, but at a lower price level, P_1, and equilibrium, E_2.

How might we explain this important process of movement from E_0 to E_1 and then from E_1 to E_2? With the reduction in demand, the level of prices and, to a lesser degree, the level of income, or wages, begin to fall. Employers correctly perceive the decline in prices as an increase in the real wages of laborers and will respond by reducing the amount of labor employed. Although prices and nominal wages are falling, workers' price expectations are unchanged from price level P_0, so they perceive the reduction in nominal wage rates as a reduction in their real wage rates. They will consequently reduce their real input of labor.[1] The economy will move along the short-run aggregate supply curve $SRAS_0$. The temporary result is that the economy experiences temporary unemployment of resources in the amount $Y_F - Y_1$.

Demand-deficiency unemployment of $Y_F - Y_1$ cannot exist for long at point E_1. In the monetarist view of the macroeconomy, dynamic forces are at work assuring a return to the full employment of resources at some new equilibrium, E_2. If we assume that no fiscal or monetary actions are undertaken by Congress or the Federal Reserve to deal with unemployment, two sets of automatic changes will dominate. First, falling prices will affect the aggregate demand for goods and services through the real balance effect. As we learned in Chapter 25, a decline in the price level will increase real holdings of money, leading to increased consumption and, through lower interest rates, to increased investment.

A second effect will act on the short-run aggregate supply curve. When workers' expectations adapt to the lower price levels, their perceptions of the real wage they receive will change. Specifically, they will adapt their expectations to falling actual prices, which signal an increase rather than a decrease in real wages. Time is required, in short, for workers to perceive the actual decline in prices. They will begin to supply more, not less, labor at every wage level. This adaptation in expectations has the effect of shifting the short-run aggregate supply curve rightward from $SRAS_0$ to $SRAS_1$.

Given pressures on prices, where will equilibrium finally reemerge? After effects from the drop in prices have taken place and after expectations have finally adjusted to the lower prices, equilibrium will reemerge at E_2. The economy moves along aggregate demand curve AD_1 and the relevant short-run supply function will be $SRAS_1$. Due to the initial decrease in consumption and investment spending, prices will fall to P_1, a level to which both employers and workers will again adjust.

In the monetarist view, the interplay between price changes and aggregate demand and between price/wage changes and short-run aggregate supply is sufficient to return the economy to full employment at E_2. However, the monetarist result hinges crucially on the *absence* of interventions by Congress or the Federal Reserve. In the monetarist view, such interventions merely prolong and complicate the natural adjustment process in the economy.

---

[1]You may at this point wish to review sections of Chapter 26 dealing with the concept of the expectations-adjusted aggregate supply curve.

Destabilizing Effects of Fiscal and Monetary Policies

Monetarists adhere to the belief that the market economy will stabilize at full employment. But we observe recession and inflation today. How, then, do monetarists explain the facts of instability?

Like Keynesians, monetarists admit that shifts in aggregate demand can cause waves, or cycles, of unemployment and inflation. Monetarists, however, believe that the primary cause of aggregate demand shifts is not the movement of investment or consumption spending but the actions of government fiscal and monetary policymakers. According to monetarists, the destabilizing element in the macroeconomy is the shifting expectations of market participants—laborers, businesspeople, consumers, and producers—created by erratic alterations in government expenditures, taxation, and money supply targets.

The Effects of Financing Government Expenditures. In the monetarist view, increases in government expenditures—one of the primary tools of fiscal policy—come at the expense of private expenditures. If government spending is financed through taxation, private consumption and investment spending are reduced by the amount of the tax. If government spending is financed through borrowing in private markets, its demand for loanable funds will likely raise real interest rates, and higher interest rates will tend to choke off investment, depending on the responsiveness of investors to the increase. This competition for private funds is termed the **crowding-out effect.** If government spending is financed through the selling of Treasury bonds to the Federal Reserve, the result is potentially inflationary (Chapters 29 and 30 explain why this is so). In sum, the effects of financing government expenditures may offset any gain those expenditures offer a demand-deficient economy. In addition, such financing can create destabilizing effects.

Crowding-out effect: A fall in private investment as a result of an increase in government borrowing to finance expenditures.

Lags in Fiscal Policy. Another problem with fiscal policy often cited by monetarists concerns legislative frailties. Given the nature of public policy and the competing demands of political interests, fiscal policymakers are often slow to recognize and act on economic distress signals. **Recognition lags** are the time it takes the president and Congress to identify impending or existing economic conditions. Predicting economic events is both a science and an art. There are often compelling interpretations offered by both sides to a debate over policy. (Economics in Action at the end of this chapter illustrates the problem.)

Recognition lag: The period between a potentially destabilizing event in the economy and recognition of it by policymakers.

In addition to the difficulties of recognizing a problem, there are administrative lags with fiscal policy. **Administrative lags** are the length of time between politicians' recognition of a business cycle problem and their enactment of legislation to correct the problem. We cite two examples of such lags. The Kennedy-Johnson tax cut of 1964 to stimulate economic growth was proposed as early as 1961 and 1962; its not being enacted until 1964 indicates an administrative lag. Likewise, a jobs bill passed in 1982 to deal with unemployment was debated for months and, when finally passed, was scheduled to go into effect more than a year later. Monetarists emphasize that economic conditions may change before the effects of these legislative acts take place. This would mean that fiscal policy could be procyclical rather than countercyclical in effect, buttressing inflation or recession as the case may be.

Administrative lag: The period between recognition or passage of a proper policy and its implementation.

Once fiscal policy is enacted, it also takes time for its full effect to occur. Estimates vary, but it is thought that tax or spending changes may take from one to two years to have their full impact on income and employment. Recognition and administrative lags make monetarists dubious about the effectiveness of fiscal policy. The unpredictability of how long it may take even well-intentioned fiscal policy to have an effect means that market participants' expectations about prices and future incomes are thrown into uncertainty in the meantime.

Effects of Monetary Policy. Discretionary monetary policy has its own problems in the monetarist view. Like fiscal policy, the attempt to control the money supply in countercyclical fashion is also subject to recognition and administrative lags. In the case of monetary policy, it is the quasi-independent Federal Reserve Board and its Federal Open Market Committee, not politicians, who must recognize macroeconomic problems and then enact and administer a change in policies affecting interest rates or the money supply. While problems of recognizing signs of recession or anticipating unemployment or assessing inflationary pressures may be as great as with fiscal policy, they are somewhat easier to handle because of the small number of Federal Reserve Board or Open Market Committee members. Concerted discretionary recognition and action is probably quicker than with the larger numbers of participants in the fiscal setting.

Despite the quicker response, the time between enactment of monetary policy and when it takes effect is as unpredictable as the time related to fiscal manipulations. Some economists believe that there is as much as a two-year lag between growth in the money supply and the rate of inflation; the effects of monetary growth on income and employment are somewhat more immediate. Other studies appear to show that most of the effects of monetary growth on prices, income, and employment take place rapidly, perhaps in less than six months. A third body of evidence seems to indicate that changes in aggregate demand induced by monetary growth will first have an impact on employment and income and will affect prices later when expectations catch up. Such competing evidence illustrates the present uncertainty over the effects of discretionary monetary policy.

The monetarist conclusion is that, on balance, discretionary fiscal and monetary policy will not be able to contain swings of the business cycle. Indeed, in some regard, business cycle alterations are *caused* by destabilizing attempts to control cycles of inflation and unemployment. Stop-and-go policies that accelerate, decelerate, or otherwise interfere with private saving and spending are seen as the primary causes of aggregate demand instability, leading to prolonged recession and unemployment. The solution to the problems created by discretionary political or quasi-political attempts to control the business cycle is to establish rules through which fiscal and monetary policy is carried out.

Monetary Rules, Government Restraints, and Market Deregulation

In general, monetarists advocate that both fiscal and monetary policy be taken out of the hands of discretionary authorities. On the fiscal side, government's budget should be balanced at all levels—local, state, and federal—through fiscal restraint or through a balanced budget amendment. Along with balanced budgets, the monetarists and other economists urge that government regulations supporting monopoly should be eliminated or modified. Deregulation in some industries and some occupations would create greater price and wage flexibility. Greater price and wage flexibility would in turn encourage a faster adjustment to both demand-deficiency unemployment and to inflation when the economy is disturbed.

Monetary rule: Monetary policy based on a prescribed rate of growth in the M-1 money supply.

Monetarists, led by Milton Friedman, also advocate the use of a **monetary rule** over the money supply growth rate.[2] In a study of the effects of monetary growth, Friedman and Anna Schwartz argue that depressions in the United States and other nations have always been preceded and accompanied by severe reductions in the money stock and that all inflations have been precipitated by sharp increases in the money stock. With regard to the Great Depression of the 1930s, which was accompanied by a one-third drop in the money stock, Friedman and Schwartz conclude:

> The Great Depression in the United States, far from being a sign of the inherent instability of the private enterprise system, is a testament to how much harm can be done by mistakes on the part of a few men when they wield vast power over the monetary system of a country.[3]

Stop-and-go policies of the Federal Reserve Board—raising the monetary growth rate 16 percent one month and lowering it to a negative 6 percent the next—create uncertainty among market participants. The monetarists' solution: Take discretionary policy out of the hands of the Federal Reserve and institute an announced and constant growth rate in the money stock of 3 to 5 percent per year. Why 3 to 5 percent? Because the average rate of growth in the American economy through increases in labor productivity and technology has been on the order of 3 percent per year for more than a hundred years. In the monetarists' view, a growth rate in the money supply of 3 to 5 percent is consistent with a stable price level and a healthy rate of economic development. Moreover, a monetary rule would create much-needed stability and correspondence between expectations and actual price experience.

Critics of the monetarist notion that there should be a rule for monetary policy (see Focus, "Monetarism's Lackluster Support") point out that such a conclusion rests heavily on the short-run and long-run stability of velocity—the demand for money, a process described in Chapter 30. Dissenters also criticize both Friedman's test methods and his results. The outcome of this debate is far from clear. The monetarists' tracking of inflation to changes in M-1 (with a two-year lag) has nevertheless persuaded a number of macroeconomists that the dog (money supply growth or declines) wags the tail (faster or slower inflation rates). We discuss this evidence in Chapter 32.

Rational Expectations: Theory and Policy

The contemporary school of macroeconomics called rational expectations theory goes one step further than monetarism on the matter of the ability of the economy to adjust to the policy actions of Congress or the Federal Reserve.

In Chapter 30 we presented the adaptive expectations theory, which holds that market participants form expectations about the future course of the inflation rate by tracking past rates of inflation. To use a simple example, suppose that people expect inflation this year to be what it was last year. If it was 10 percent last year, people expect it to be 10 percent this year, but

[2]Milton Friedman, *A Program for Monetary Stability* (New York: Fordham University Press, 1960).

[3]Milton Friedman and Anna Schwartz, *A Monetary History of the United States, 1867–1960* (Princeton, N.J.: Princeton University Press, 1963).

FOCUS   Monetarism's Lackluster Support

During the 1970s and 1980s political support for the policies supported by monetarism—strict discipline over the growth rate in M-1 and fiscal restraint leading to a balanced budget—has been lackluster. Indeed, many economists and politicians blame the recession of 1982, the worst economic downturn since World War II, on monetarist theory. Prior to that recession, Federal Reserve Board Chairman Paul Volcker appeared to shift the Federal Reserve's policies of targeting interest rates in favor of narrow and limited growth rates in M-1. As an anti-inflationary strategy, Volcker's vigorous monetary policy certainly had dramatic effects, but the costs in unemployment were severe and politically unacceptable.

Regardless of whether the Federal Reserve had decided to "try" monetarism (monetarists insist that the Volcker action was not consistent with their views), the Federal Reserve soon loosed the reigns over the money supply, and interest rates slowly returned to lower levels. In 1983, President Reagan's chief economists argued that "the monetary authorities should be guided by the principle of keeping money supply growth within a prespecified target range while adjusting these targets when a careful consideration of the evidence indicates that sustained shifts in asset demands have occurred."[a]

In other words, the president's advisers recommended a combination of rules and discretionary authority for the Federal Reserve. For their part, monetarists admit that overreliance on M-1 data is risky because the definitions of savings and checking accounts are constantly evolving. As banks continue to offer variations of interest-bearing checking accounts, depositors have begun shifting their money (in the M-1 sense) into near money (M-2) accounts and back again on a regular basis. Monetarists continue to argue, however, that M-1 growth is the most reliable indicator of inflation. Adopting broad monetary targets and attempting to fine-tune the money supply will only lead to more erratic cycles of inflation and unemployment, in their view. To monetarists, the only problem with monetarism, as G. B. Shaw said of Christianity, is that it has never been tried.

[a]Council of Economic Advisers, *Economic Report of the President* (Washington, D.C.: U.S. Government Printing Office, 1983), p. 25.

---

if the inflation rate turns out to be 5 percent this year, people adapt and expect it to be 5 percent next year. At each period the public adjusts its expectations according to the differences between the actual inflation rate and the predicted inflation rate. It takes time and many readjustments of expectations for people to become convinced that any inflation rate (be it 6 percent or 106 percent) is permanent.

The Rational Expectations Hypothesis

Rational expectations theory: A theory of macroeconomic behavior that assumes people will anticipate the effects of macroeconomic policy and will thereby be in a position to neutralize its intended effects.

A group of economists have developed a new theory of expectations to challenge and amend the adaptive expectations hypothesis.[4] Although much of this work—called **rational expectations theory**—is extremely technical, the major thrust of the idea is simple, straightforward, and intuitively appealing. According to the rational expectations view, all market participants are rational in that they do not throw useful information away. After a time, individuals begin to understand the workings of the economy. For example, they will learn through experience that increases in monetary expansion by the Federal Reserve are followed by inflation, which is followed by higher

[4]Pioneers include macroeconomists Thomas J. Sargent, Neil Wallace, and Robert E. Lucas; see Thomas J. Sargent, "Rational Expectations, the Real Rate of Interest, and the Natural Rate of Unemployment," *Brookings Papers in Economic Activity* 2 (1973), pp. 429–472; Thomas J. Sargent and Neil Wallace, "Rational Expectations and the Theory of Economic Policy," *Journal of Monetary Economics* (April 1976), pp. 169–184; and R. E. Lucas, "An Equilibrium Model of the Business Cycle," *Journal of Political Economy* (December 1975), pp. 1113–1144.

nominal interest rates. Knowing the basic structure of the real economy, individuals will be able to anticipate the most likely outcomes.

Does this mean that market participants will always be correct in their anticipations? In a world filled with uncertainty, outcomes will depend on an enormous number of random occurrences. Individuals will not always be right in their expectations about future economic events, but they will be correct on average.

If people can anticipate discretionary policy and its effects, they are in a position to neutralize the purpose of the policy. For example, if the Federal Reserve Board increases M-1 to increase employment but workers and firms perfectly anticipate the resulting increase in prices, then workers will instantly demand a proportionately higher nominal wage. Firms, anticipating higher nominal revenues, would be willing to pay the higher nominal wage, thus leaving the real wage and the level of unemployment unaffected. In this contest, policymakers are pitted against market participants, and policymakers will not win continuously and certainly will not win over time. Policymakers may attempt to surprise the public, but they cannot do so forever. People catch on to, say, the effects of federal deficits on inflation, anticipate the change, and neutralize or counteract the effects of the change. In the long run, when a model of macroeconomic behavior is learned and actions are correctly anticipated, fiscal or monetary policy is totally ineffective in producing intended alterations on the demand side of the economy. Paradoxically, to have any purposeful effect fiscal and monetary policy would have to be random. If it is not—if policymakers act in any systematic manner—learned behavior over time will neutralize the policy.

Rational Expectations and Monetary Policy

As with Keynesians and monetarists, time is an important factor in rational expectations. Short-run and long-run effects of rational expectations will depend on whether monetary policy—or any event affecting the macroeconomy—is anticipated.

Anticipated Monetary Policy. We can imagine the (non)effects of rational expectations as a result of the increases or decreases in monetary expansion with the aid of Figure 31–4. Note that with *adaptive* expectations, an increase in monetary expansion shifting the aggregate demand curve rightward to AD_1 from AD_0 would increase income and employment above the natural rate, Y_E, because it would take time for the price expectations and perceptions of workers and employers to adjust to the full effects of monetary expansion. In other words, there would be an upward-sloping short-run aggregate supply curve through E_0 in Figure 31–4 as there is in Figure 31–3.

But this is not the case when perfect economic foresight is assumed. Figure 31–4 shows no short-run expectations adjusted aggregate supply curve because all effects are perfectly anticipated by all market transactors. An increase in aggregate demand from AD_0 to AD_1 caused by an increased rate of monetary expansion (or a decrease from AD_0 to AD_2 caused by a reduction in monetary expansion) simply creates a new equilibrium at E_1 (or at E_2) with all adjustments perfectly worked out instantaneously. In this case the inflation or deflation is perfectly accounted for by all market transactors, and the natural level of employment Y_F is left undisturbed by policy changes. Only prices change. This principle also applies when, for example, employment falls below the natural rate, due perhaps to some reduction in private auton-

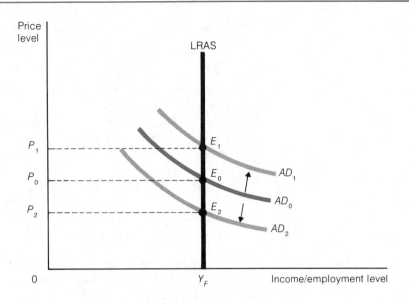

FIGURE 31–4   Rational Expectations and Monetary Policy

According to the rational expectations approach, wages, interest rates, and prices adjust in anticipation of changes in aggregate demand caused by upward or downward adjustment of the money supply. For this reason there is no short-run aggregate supply curve. The economy simply moves along the long-run aggregate supply curve LRAS, with prices rising as the money supply is increased, shifting the demand curve from AD_0 to AD_1, or falling as the money supply is decreased, shifting aggregate demand to AD_2.

omous expenditures that reduces aggregate demand. Any attempt on the part of monetary authorities to remedy the situation will be met with counteracting reactions on the part of market participants, leaving the unemployment rate unchanged.

Unanticipated Monetary Policy and Random Shocks. The preceding conclusions apply as long as policy changes aimed at correcting aggregate demand are fully anticipated and expected. So long as the expectations of market participants (buyers, sellers, lenders, or borrowers) about demand management are fully realized, no change in real income or employment can occur. But the rational expectationist argument also allows for surprises or random events in the short run. If, for example, the Federal Reserve Board changes the M-1 growth rate erratically or if it alters its target, short-run changes in income and employment will occur until all market participants catch on and acknowledge the new policy. Unless Federal Reserve policy is totally random, expectations will adjust in the long run.

Surprises such as crop failures caused by climatic changes or timber losses caused by a volcano eruption will also disrupt production and market plans in the short run. The rational expectations theory predicts, however, that all market participants will adapt and adjust behavior to the new conditions in the long run. In sum, the rational expectations idea does not preclude short-run fluctuations in output and employment due to policy or natural surprises. It does predict that policy changes will be neutralized in the long run unless policy is implemented in a totally random manner.

Rational Expectations and Fiscal Policy

When fiscal policy encounters rational expectations, it may not leave the supply side of the macroeconomy unchanged, especially in the long run. Consider Figure 31–5, for example, and assume that the economy is initially in equilibrium at full employment level Y_F. Further assume that Congress enacts increases in government expenditures financed by increases in taxes or by borrowing in private markets.

In the classical, monetarist, or rational expectationist view of the economy, the increase in government expenditures is closely matched by reductions in private consumption, investment, or imports. Such a reduction would leave the aggregate demand curve of Figure 31–5 stationary. With no further effects, it would appear that equilibrium would remain at E_0 and the natural rate of income and employment would remain constant at Y_F. However, rational expectations theorists assert that fiscal policy would likely have effects on aggregate supply as well as on the public versus private composition of aggregate demand.

Changes in tax rates will have ultimate effects on work incentives. As we saw in Chapter 26, any reduction in work incentives may reduce the total equilibrium labor input, shifting the aggregate supply curve leftward. Moreover, increased taxes on business investment could adversely affect the demand for labor. Both of these possible effects would reduce the natural rate of employment, increasing the natural rate of unemployment. This means that the rate of employment consistent with equilibrium, given all factors (such as taxes) that affect hiring and labor-supply conditions, is lower than before. In Figure 31–5, the aggregate supply curve shifts (or drifts) leftward, reducing equilibrium income with a tendency toward higher prices. Fiscal policy therefore may have effects on aggregate supply as well as on aggregate demand.

At base, the theory of rational expectations reinforces the classical-monetarist position: The private economy, while not perfect in providing full employment and maximum income at all times, is superior to fiscal and monetary authorities in providing a relatively stable economy. But the ra-

FIGURE 31–5

Rational Expectations and Fiscal Policy

If rational expectations exist in the long run, fiscal policy has no effect on aggregate demand. Furthermore, fiscal policy can cause decreases in productivity, causing long-run aggregate supply to shift to the left.

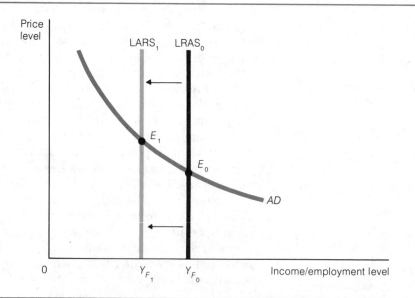

TABLE 31–3   Alternative Positions on Macroeconomic and Monetary Policy

A simplified menu of economic policy choices includes various recommended roles for fiscal and monetary policy. The modern Keynesian position places primary emphasis on fiscal policy, while the monetarist defends a rule for monetary policy and balance in the budget. The rational expectations theory predicts that, over the long run, neither fiscal nor monetary policy will have a positive effect on the aggregate economy.

Policy Position	Role of Monetary Policy	Probable Effects of Monetary Policy	Role of Fiscal Policy	Probable Effects of Fiscal Policy
Keynesian	Maintain orderly and low interest rates through discretionary adjustments in money supply	Destabilizing effects on interest rates if Federal Reserve only attempts to control money supply	Primary tool of macroeconomic stabilization; discretionary changes in government spending and taxation	Discretionary fiscal policy capable of ensuring full employment
Monetarist	Maintain a 3% to 5% growth rate in M-1.	Discretionary policy will destabilize decision making in private markets; monetary rule will stabilize expectations	Provide stable environment for private economy through balanced budgets	Discretionary policy will bring erratic effects on income, employment, and prices; such policy creates an unstable economy
Rational Expectations Theory	Same as monetarist	In the long run, rational expectations will neutralize discretionary policy	Same as monetarist	Possible adverse long-run aggregate supply effects

tional expectations theory goes further and proposes that over time the effects of fiscal and monetary manipulations can be fully anticipated and neutralized. Policies related to aggregate demand will either have no effect or in the longer run will actually produce reductions in aggregate supply. In short, there is little or no room for discretionary macroeconomic policy. Robert Lucas, a leading proponent of the rational expectations hypothesis, has therefore offered this advice: "The Administration and Congress should stop thrashing around pretending to know better than the economy how much can be produced. And the Federal Reserve should move as quickly as possible to a 4 percent monetary growth."[5]

The Major Policy Positions: A Summary

We have now evaluated three major policy positions based on three theories of the functioning of the aggregate economy. Some theoretical or policy views are contrasting and some are complementary, as Table 31–3 reveals.

Modern Keynesians hold that the private economy is inherently unstable and unable to correct itself during periods of inflation and unemployment. In their view, government should take a primary role in managing fiscal and monetary policy for economic stabilization. Of the two major arms of policy, Keynesians advocate discretionary fiscal policy as the primary weapon in controlling unemployment and inflation. Fiscal policies, Keynesians believe, have direct effects on spending in the private economy, whereas monetary policy (control of the money stock) has only indirect results. Monetary pol-

[5]Robert E. Lucas, quoted in "The New Economists," *Newsweek* (June 26, 1978), p. 60. For a nontechnical explanation of the rational expectations idea see "The Rational Expectations Model," *Wall Street Journal* (April 2, 1979).

". . . First I was a Keynesian . . . Next I was a monetarist . . . Then a supply-sider . . . Now I'm a bum . . ."

©Bill Schorr, 1982 *Los Angeles Herald*

icy should maintain the order or stability of interest rates, keeping them low or moderate.

Monetarists and rational expectationists view the economy as essentially stable and self-correcting. Both views envision the private decisions of laborers, employers, money demanders, and consumer-savers and investors as producing a natural equilibrium constituting some natural rate of output and employment (or unemployment). In the monetarist view of this process, discretionary policy of any sort produces instabilities in private markets that distort, exaggerate, or accentuate any natural changes. In the monetarist view, discretionary fiscal or monetary policy contributes to instability in the private economy. That is, discretionary policy either creates inflation and unemployment of itself or it has erratic effects on the private economy's ability to establish economic stability. The solution: Conduct monetary and fiscal policy so as to minimize disturbances of private market participants' expectations and decisions. In the monetarist's policy views, such stability can be achieved by invoking a monetary rule with balanced government budgets.

In the rational expectations variant of macroeconomic theory, the recommended role of fiscal and monetary policy is identical to that espoused by monetarists. Differences between these two positions lie rather in the rational expectationists' view of the probable effects of monetary and fiscal policy. As market participants gain sufficient information about the effects of economic policy through time, policy can have no effects on the private economy. Participants learn the effects of anticipated policy and act more and more quickly to counteract them. The rational expectations approach might be regarded as an extreme view within the monetarist position.

The important point is that macroeconomists have not reached a consensus on exactly how the macroeconomy works. Under such circumstances we expect to see a variety of policy views, as Table 31–3 shows. One of the problems with establishing a consensus among macroeconomists is the difficulty of proving one view with a statistical test that all would accept. The economist, like the meteorologist, is faced with a kaleidoscope of ever-changing conditions; hard-and-fast answers are difficult to come by.

As we will see in the following chapters, macroeconomists have made serious attempts to explain actual happenings, thereby providing some means of choosing between macroeconomic theories. But, as we will also discover, particularly in Chapters 32 and 33, there are yet other views of the causes and effects of macroeconomic events. The effects of aggregate supply changes—discussed in detail in Chapter 26 and briefly in this chapter—will be viewed in Chapter 34 as a major factor in promoting or retarding economic growth. Additionally, the world of politics and its possible effects on macroeconomic well-being must also be considered, in Chapter 33. Before we turn to these matters, however, the macroeconomic explanations for the thorny problems of inflation and unemployment must be more fully discussed.

Summary

1. Macroeconomic policies—those prescribed to solve macroeconomic problems such as inflation and unemployment—rest on alternative versions of macroeconomic theory.
2. The tools of policymakers include changes in taxes, government spending, and the money supply to effect changes in total spending.
3. The modern Keynesians argue that fiscal policies—government spending and taxation policies—have direct and predictable effects on the private economy. Discretionary fiscal policies should be used to cure economic problems such as demand-deficiency unemployment and inflation. Money supply changes may affect spending indirectly through their impact on interest rates.
4. Monetarists advocate a stable, nondiscretionary fiscal and monetary approach to economic stabilization. Monetarists assert that lags in the recognition of economic problems on the part of both fiscal and monetary authorities, as well as lags between policy

implementation and the effects of policies, are potentially disruptive in controlling the macroeconomy. Monetarists advocate a balanced budget on the fiscal side and a money supply growth rule for the Federal Reserve's conduct of monetary policy.
5. Rational expectations theorists argue that, rather than throw useful information away, market participants use all learned information to predict the outcome of policymakers' discretionary decisions. Decision makers with rational expectations are not always right, but they learn through time and are able to predict outcomes over the long run. The net result is that when the discretionary decisions of policymakers are perfectly anticipated, monetary and fiscal policy will have no effect on aggregate demand. Unanticipated policy or events will have short-run effects, which are eventually overcome through natural adjustments. Fiscal demand-side policies, moreover, may result in long-run adverse shifts in aggregate supply.

Key Terms

crowding-out effect   administrative lag   rational expectations theory
recognition lag   monetary rule

Questions for Review and Discussion

1. Explain how fiscal and monetary policies differ with respect to the amount of time necessary to implement them and the amount of time required for their full force to be realized.
2. Under what conditions is Keynesian fiscal policy most effective? What fiscal policies are called for during times of recession? During inflation?
3. What role do the monetary authorities play in Keynesian policies? In the monetarists' view, what is the proper role of the monetary authorities?
4. According to monetarists, what variables adjust to maintain full employment? What effect do discretionary demand management policies have on the economy?

5. According to monetarists, what effect does increased government spending have on aggregate demand? What effect do recognition and administrative lags have on the cycles of inflation and unemployment?
6. Why do monetarists suggest that monetary rules be followed rather than discretionary policies?
7. What is the difference between rational expectations and adaptive expectations? If individuals have rational expectations rather than adaptive expectations, would discretionary monetary policy be more effective in diminishing swings in the economy?
8. According to the rational expectations view, what

are the probable effects of discretionary monetary and fiscal policies?

9. What theoretical conditions would have to exist before discretionary fiscal or monetary policies could completely eliminate the cyclical swings in inflation and unemployment?

10. Some policymakers in the United States adhere to supply-side policies (see Chapter 26 for an overview of these policies). Would they be most closely aligned to the Keynesian, monetarist, or rational expectationist view?

 ECONOMICS IN ACTION

The Council of Economic Advisers Predicts

Provisions of the Full Employment and Balanced Growth Act of 1978 require that the Council of Economic Advisers submit a yearly report to Congress on the state of the economy.[a] Part of this report is a congressionally required prediction of future behavior in key macroeconomic indicators such as employment, inflation, and productivity. In February 1983, the council (then composed of economists Martin Feldstein, chairman; William A. Niskanen, and William Poole) submitted a report on the state of the economy along with a five-year prediction. It will be instructive to compare the council's predictions with the actual values of these crucial economic variables as they became available. How good at prediction are those who should be in the best position to predict?

Table 31–4 shows the council's predictions through 1988 of raw employment statistics (in millions), the

unemployment rate, real GNP, budget outlays, and other key indicators. The council predicted a dramatic decline in the unemployment rate, a dramatic rise in the real GNP growth rate between 1983 and 1984 with sustained high growth afterward, a low and steady rate of inflation, rising real wages (real compensation per hour), and a high and reasonably steady growth rate in productivity (output per worker hour).

Predictions, however, always depend on assumptions concerning certain crucial determinants of economic activity. The council (and citizens of the United States and the world) of course had to deal with a severe recession in 1981–1982, preceded by extremely high rates of inflation between 1976 and 1981. Budget deficits were running well over $150 billion per year in 1982, 1983, and 1984. The supply-side orientation of President Reagan and the Council of Economic Advisers led to the following analysis of the cause of the situation: "A major cause of our present economic ills was the inclination in the past to pursue one economic goal single-mindedly, without adequate attention to the longer-run consequences for other economic objectives.

[a]Council Of Economic Advisers, *Economic Report of the President* (Washington, D.C.: U.S. Government Printing Office, 1983).

TABLE 31–4   **Projected Macroeconomic Indicators**

	1983	1984	1985	1986	1987	1988
Item	Level					
Employment (millions)	101.5	104.2	107.0	109.6	112.3	114.9
Unemployment rate (percent)	10.7	9.9	8.9	8.1	7.3	6.5
Federal budget outlays as percent of GNP (fiscal year basis)	25.2	24.3	24.1	23.9	23.5	23.0
	Percent Change					
Consumer prices	4.9	4.6	4.6	4.6	4.5	4.4
Real GNP	1.4	3.9	4.0	4.0	4.0	4.0
Real compensation per hour	1.2	.6	1.1	1.4	1.6	1.6
Output per hour	2.1	1.9	1.5	1.7	1.6	1.7

Source: Council of Economic Advisers, *Economic Report of the President* (Washington, D.C.: U.S. Government Printing Office, 1983), p. 143.

This administration remains determined to avoid the errors of past policies."[b]

The following are the council's recommendations to achieve the macroeconomic goals predicted in 1983:

1. A continued moderate growth in the money supply, increased reliance on the private sector, and increased domestic and international competition.
2. Encouragement of saving and investment leading to capital formation through reductions in corporate and personal income taxes.
3. Reduction in unnecessary domestic regulation in the form of reductions in entry and price controls and modifications of regulations such as minimum-wage laws affecting the structure of the labor market.
4. Policies to reduce deficits that could choke off new investment and capital formation by keeping interest rates high.

[b]Council of Economic Advisers, *Economic Report of the President*, 1983, p. 144.

Question

Were these optimistic predictions realized? 1984 data (see Chapters 21 and 22), for example, reveal wide differences between actual values and predictions. Can these economic policies be blamed if long-run growth and prosperity do not materialize over the 1980s?

32

Inflation and Unemployment

P erhaps no two problems bedevil the American economy more than inflation and unemployment. During the 1970s, especially between 1973 and 1980, U.S. wage earners experienced nominal wage increases but little or no increase in real income, income adjusted for inflation. Real income actually fell during 1974, 1975, 1980, and in 1982, although recovery produced dramatic increases in 1983 and 1984. For real income to have fallen in the earlier years, price increases must have exceeded increases in nominal income.

Unemployment has been no less a problem. A sharp recession in the United States and the world economy in 1981 and 1982 brought the unemployment problem into sharp focus. Unemployment rates over 10 percent in the United States created an outcry among citizens that something be done, although rates have moderated since then. The problems associated with high unemployment affect some groups more than others. In 1982, the most recent recession, for instance, minority unemployment rates climbed to 30, 40, and even 50 percent in some age groups. Some sectors of the economy— automobiles and textiles were recent examples—are harder hit than others by recession and unemployment, causing massive layoffs and disrupting the lives of hundreds of thousands of people.

Macroeconomists are acutely aware of these economic problems and have devoted special attention to them. In this chapter we will review what we know about inflation and unemployment and present some alternative policy measures designed to prevent their destructive force. One very important issue is whether there is a trade-off between inflation and unemployment, as some economists believe. Does a reduction in unemployment "cost" Americans by driving up the inflation rate? Does a reduction in the inflation rate mean that unemployment must rise? Or are unemployment and inflation separate problems to be dealt with in separate ways? Before addressing these questions, we will look at inflation and unemployment in greater detail.

Inflation in the Economy

As we have shown at many points in this book, it is important to distinguish between the price level at some time—how high or low prices are—and the rate of change in prices (inflation or deflation). Prices can be either high or low and at the same time be in the process of rising (inflating), falling (deflating), or remaining stationary. To say, for example, that prices are 30 percent higher on January 1, 1986, than they were in 1966 is to make a statement about price levels. In Chapter 22, we explained the use of price indexes to measure price change. Such an index uses a base year (often 1967, 1972, or 1977) to calculate the movement of prices.

Defining Inflation

While the level of prices is often of interest to economists, inflation, strictly speaking, means something else. Inflation is always expressed as a rate of price change per year or per month. The statement that prices are higher on January 1, 1986, than in 1977 tells us nothing about the current (1986) rate of inflation. The rate of inflation on January 1, 1986, is expressed as a percentage—5 percent, 9 percent, or −3 percent (deflation). For convenience in analyzing macroeconomic questions, we have described inflation and deflation as once-and-for-all changes in the price level in previous chapters, and we will continue to do so. Inflation, however, is most accurately understood as a flow or process of price change.

A monthly inflation rate is calculated by the Bureau of Labor Statistics. If, in February of 1986, the rate for January 1986 is announced to be ½ percent, an approximate yearly rate can be calculated by multiplying the January rate by 12. In this case, the projected yearly rate is approximately 6 percent if the January rate holds for the remaining eleven months. If the yearly rate is in fact 6 percent, we say that the price level is 6 percent higher on January 1, 1987, than it was on January 1, 1986. If monthly inflation rates differ over the year, each monthly rate is added up to determine the annual rate for the year.

Several measures of inflation are used in the real world to guide policymaking. As we have seen in Chapter 22, a consumer price index and a producer price index are calculated by the Bureau of Labor Statistics. A third major index is the implicit price deflator, a means of converting nominal GNP values into real GNP values. No matter what measure is used, the presence of higher indexes over time indicates a rising price level. Actual U.S. inflation rates are shown in Table 32–1.

The Effects of Inflation

The statistical presence of inflation in the economy tells us little about the costs and other distortions it inflicts there. Runaway inflations of 80 or 2000 percent per day or per month, such as those sometimes experienced in wartorn or underdeveloped countries, can reduce an economy to barter or cause it to collapse altogether. Milder inflation—3 to 15 percent a year—also creates economic costs in any advanced economy. The full effects of increases to inflation rates may take some time to be fully realized, so inflation has both short- and long-run effects.

TABLE 32–1   Two Measures of Inflation, 1960–1984

Two popular measures of inflation used in the United States are the implicit price deflator, a means for converting nominal GNP into real values, and the consumer price index, which bases the inflation rate on price changes in some hypothetical market basket of consumer goods. The numbers given for both indexes are annual rates.

Year	Inflation Rate, as Measured by		Year	Inflation Rate, as Measured by	
------	Implicit Price Deflator	Consumer Price Index	------	Implicit Price Deflator	Consumer Price Index
	Implicit Price Deflator	Consumer Price Index		Implicit Price Deflator	Consumer Price Index
1960	1.6	1.6	1973	5.8	6.2
1961	0.9	1.0	1974	8.8	11.0
1962	1.8	1.1	1975	9.3	9.1
1963	1.5	1.2	1976	5.2	5.9
1964	1.5	1.3	1977	5.8	6.5
1965	2.2	1.7	1978	7.4	7.7
1966	3.2	2.9	1979	8.6	11.3
1967	3.0	2.9	1980	9.2	13.5
1968	4.4	4.2	1981	9.6	10.4
1969	5.1	5.4	1982	6.0	6.1
1970	5.4	5.9	1983	3.8	3.2
1971	5.0	4.3	1984	3.7	4.3
1972	4.2	3.3			

Source: Council of Economic Advisers, *Economic Report of the President* (Washington, D.C.: U.S. Government Printing Office, 1985), pp. 237, 296.

Inflation, Fixed Incomes, and Assets. The most publicized effect of inflation is the redistributive impact it has between specific groups in society. Inflation affects those who can least afford to pay for it—the sick, the poor, and especially the elderly and others on fixed incomes—through erosions of purchasing power. Congress, in recent years of inflation, indexed or "tied" Social Security and other welfare benefits to a price level index, thereby alleviating some of the problems inflation causes for the retired or the aged. Nevertheless, inflation is a constant threat to those who can least afford it.

Inflation also adversely affects holders of assets denominated in nominal, or money, terms as opposed to real terms. Suppose you purchase a house—a real asset—and hold it for five years, over which time the price level doubles because of inflation. What happens to the nominal value of your investment? Excluding the real factors of natural appreciation and depreciation that might take place over the five years (resulting from the location, deterioration of property, and so on), the money value of your house should closely follow the inflation rate over the period, doubling at the end of the five years.

While the value of real assets tends to follow inflation, the value of money-denominated assets may not. Suppose that you lend a friend $10,000, to be paid at the end of a five-year period and that your friend issues you a promissory note attesting to the loan. Again suppose that the price level doubles over the five-year span because of inflation. Aside from any interest paid by your friend, what has happened to the real value of your asset? Although the nominal value of the note remains at $10,000, the real value is only half of what it was at the beginning of the five-year period. You will now be able to purchase only half the goods and services that you could have bought before you made the loan. Capital gains or losses—changes in real value—are always associated with holding any kind of asset in the face of unanticipated inflation. Real assets generally gain in value, whereas money assets—those expressed in fixed money terms—generally lose value.

Asset holders also experience transaction and information costs as they move into and out of inflation-prone assets. It is costly, in other words, to find out which real assets will best follow the price level—a Picasso or a Jackson Pollock, a house or IBM stock.

Disruptive Expectations. A second major effect of inflation is that it distorts expectations concerning prices, wage rates, and interest rates. When the inflation rate is not perfectly anticipated, workers and employers, buyers and sellers of goods and services, and borrowers and lenders will be uncertain of future real values. Since these transactors will generally make mistakes in forecasting the effects of inflation in specific markets, there will be continuous and unanticipated transfers of wealth among workers and employers, buyers and sellers, borrowers and lenders.

Escalator clause: Wage contracts that tie wage increases to a price index so that nominal wages will increase with increases in the price level.

Market participants try to protect themselves, of course. Workers belonging to labor unions often bargain for **escalator clauses** in their contracts, provisions that tie their future wages to some index of the price level as a hedge against inflation. Your banker will set interest rates on the basis of some real rate of interest, reflecting fundamental forces of borrowing and lending, plus his or her expectations concerning inflation over the period of the loan. If the real rate of interest is 4 percent and the banker expects inflation of 8 percent, you will be charged 12 percent to borrow money. If the actual inflation rate turns out to be 10 percent, the banker loses—you will be paying back dollars that are worth less in real terms than the ones you borrowed. If the actual rate of inflation is 6 percent, you lose because you are paying back dollars that are worth more than the ones you borrowed.

Inflation as a Tax. Inflation also amounts to a "tax" on wealth because it taxes holdings of cash balances and forces income receivers into higher tax brackets.

As inflation proceeds, all cash balances—whether in people's pockets or in non-interest-bearing checking accounts—decline in real value. If the erosion of the dollar's purchasing power takes place because of the government's overissuance of money (perhaps to finance government expenditures), inflation represents a tax like any other, except for one feature: No one votes for it. Inflation can also be a tax on one other count—bracket creep—as explained in Focus, "Inflation, Bracket Creep, and Indexation."

Theories of Inflation

Cost-push inflation: Increases in the price level caused by monopoly and/or union pressure or by decreases in the supply of products or resources.

Theories about the cause of inflation can be broadly divided into two categories: cost-push and demand-pull. In brief, **cost-push inflation** results from monopoly or union power pushing prices up. Either unions may respond to rising monopoly output prices by demanding higher wages or output monopolists may respond to union demands for higher wages by raising prices. A variant of the cost-push explanation is the supply shock theory of inflation, which explains inflation as the result of a sudden increase in production costs caused by some resource scarcity, such as that created by OPEC or by crop failure.

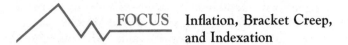

FOCUS   **Inflation, Bracket Creep, and Indexation**

The United States has a progressive income tax. This simply means that as an individual's income rises, the percentage of the person's income paid in taxes increases. A progressive income tax is based on the ability-to-pay principle: People with higher incomes pay a higher percentage of their income in taxes (progressivity) because they can afford to do so. Table 32–2 gives the amount of taxes and the percentage of income paid in taxes for various levels of income in 1983.

During periods of inflation, when wages and prices rise, nominal incomes rise and people are pushed into higher income tax brackets, a process called *bracket creep*. For example, take an individual with a taxable income of $35,000. According to Table 32–2, the individual would pay 26.34 percent of his or her income in taxes. If 15 percent inflation occurred, the person's nominal income would rise to $40,250, and taxes would rise to 28.54 percent of income. Inflation can thus result in an increase in one's tax rate without an increase in real income.

This effect is not necessarily bad for the economy. Inflation and its companion, tax increase, will decrease total spending and reduce inflationary pressures if the government does not spend the extra tax revenues. The automatic stabilizing effects of bracket creep work in the opposite direction as well. If a recession and deflation occur, nominal incomes fall and people drop into lower tax brackets. With constant government expenditures, the cut in taxes will increase private spending and relieve the recessionary pressure.

However, this is not the way it has been working in the United States economy. During the inflation of the late 1970s, the government was collecting more tax revenues as a result of bracket creep. But rather than holding government expenditures constant, Congress has been spending all of the extra tax revenues and then some. Increasing budget deficits may have increased inflation and pushed individuals into still higher tax brackets. In this situation, inflation, coupled with a progressive income tax, is a means of increasing taxes without legislation. In other words, Congress can fund additional government purchases without voting for increased taxes. Even when Congress votes for a tax cut, as it did in 1980, the effects on real income are much reduced because of inflation and bracket creep.

TABLE 32–2   Federal Income Tax Rates

A progressive income tax means that individuals or families with higher incomes pay an increasingly higher percentage of their income in taxes.

Taxable Income (dollars)	Tax (dollars)	Percentage of Income
10,000	1,238	12.38%
15,000	2,337	15.58%
20,000	3,760	18.80%
25,000	5,371	21.48%
30,000	7,182	23.94%
35,000	9,219	26.34%
40,000	11,419	28.54%
45,000	13,831	30.73%
50,000	16,306	32.61%

Source: IRS Form 1040, 1983.

To prevent tax increases caused by inflation, tax indexation was conceived. This plan, scheduled to go into effect in 1985, is a simple means of preventing tax increases without increases in real income. Through indexation, income is adjusted so that the percentage of real income paid in taxes remains constant regardless of the inflation rate. In effect, the individual's nominal income is reduced by the amount of inflation through the use of a price index (GNP price deflator or consumer price index); the person's tax is then calculated based on real income. For example, the person whose income increased to $40,250 during a period of 15 percent inflation would have his or her income adjusted by dividing nominal income by 1 plus the inflation rate (1 + 0.15) to obtain $35,000. The adjusted income is then taxed according to the tax rate table. This method prevents bracket creep. However, individuals whose real income rises—whose nominal income increased at a rate faster than the inflation rate—would move into a higher tax bracket. The progressive tax structure would be preserved, but it would be based on real income. If indexation lives up to its promise, Congress will no longer be able to increase taxes without voting to do so.

Demand-pull inflation:
Increases in the price level caused by increases in the demand for products and resources.

Demand-pull inflation is a rise in prices owing to increases in expenditures on goods and services. When actual holdings of cash balances exceed consumers' and investors' desired holdings, they begin to spend. Such spending drives prices upward until they rise enough to reduce actual real money holdings to desired real holdings. In other words, increases in demand pull prices up.

Monetarists adhere to the demand-pull explanation of inflation. In their view, long-run inflation is the result of the Federal Reserve's pumping the money supply above the level necessary to keep its growth in line with the growth in the public's desired real holdings of money. Most Keynesians view inflation as a combined result of demand-side and supply-side forces in the economy.

Cost-Push Theories of Inflation

Cost-push explanations place the blame for inflation on pressures from monopolies or unions or from sudden shortages of natural resources. The corresponding theories of how these factors can cause inflation are called the monopoly power/wage-price spiral argument and the supply, or resource, shock explanation.

The Monopoly Power/Wage-Price Spiral Theory. Some economists argue that one cause for inflation is that the prices of goods and services are pushed up by monopoly power. There are two variants of this argument, depending on the source of monopoly power—output monopolies in product markets or input monopolies, such as labor unions, in resource markets.

Concentration of monopoly power in input or output markets causes rigid constraints on the ability of prices and wages to move downward in response to falling aggregate demand. Output monopolies such as the dominant oil firms are able to resist cutting prices by controlling output—by supplying lower quantities in an effort to maintain their revenues. Rather than accept lower wages, input monopolies such as giant labor unions curtail labor's supply in an effort to counteract falling demand for labor. The rigidities imposed by output and input monopolies mean that prices and wages are not freely adaptive to changes in demand. When aggregate demand falls, the result is lower output and employment rather than falling prices and wages.

The effects of monopoly or union rigidities occur over time. As industries become more and more concentrated and unions increase their membership, the rigidities and their effects on prices become greater. This process is shown in Figure 32–1, which shows an initial equilibrium between aggregate demand and aggregate supply at E_0. At the initial equilibrium the economy is operating at full employment (Y_F) and the price level is P_0. The aggregate supply curve, AS_0, producing the optimal rate of income and employment, Y_F, is drawn on the assumption that, at a particular point in time, there is a given degree of monopoly and union power in the marketplace. Any particular degree of monopoly and union pressures in output and input markets will constrain labor supply and labor demand.

Further increases in the degree of monopolization or of union membership will create new price and wage rigidities in the economy and, with them, reductions in output and employment. These increases will have the effect of shifting the aggregate supply curve leftward over time. The rate of output and employment will decline and prices will rise.

Another version of this argument is that, at any single point in time, the price demands of monopolies or the wage demands of unions will create a

FIGURE 32–1

Effect of Monopoly Power on the Aggregate Supply Curve

An increase in monopoly or union power shifts the aggregate supply curve leftward from AS_0 to AS_1, causing a reduction in output and employment from Y_F to Y_1 and a rise in prices from P_0 to P_1.

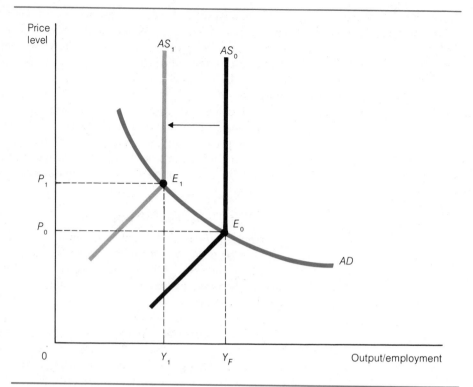

Wage-price spiral: A process in which increased product prices cause increased wages, increased wages cause increased prices, and so on. For the wage-price spiral to continue, the money supply must increase or real output must fall.

wage-price spiral of inflation. If monopolies can set prices and make higher prices stick, the wage demands of unions will escalate; erosions of real purchasing power will cause union members to demand higher nominal wages. Either increased monopoly price demands or union wage demands will reduce the short-run aggregate supply of output because both market activities have the short-run effect of temporarily reducing the hiring of labor.

The process may also be viewed in reverse. Union demands may create cost increases for businesses. These cost increases, where output markets are monopolized, are passed on to the buying public by monopolies in the form of price increases. A wage-price spiral may develop, and the result is the same whether the original cause is unions or monopolies.

In Figure 32–2 assume that the economy is in initial equilibrium at E_0, where the aggregate demand curve, AD_0, intersects short-run aggregate supply curve, AS_0, and long-run aggregate supply curve, LRAS. Suppose that unions demand and receive higher nominal wages in the form of three- or four-year contracts. These nominal wage demands would reposition the short-run aggregate supply curve leftward to AS_1, causing output to fall and prices to rise. If businesses passed these wage increases on to the public, pressures on the leftward shift in the short-run supply curve would intensify. The economy would move to a new equilibrium, E_1, and would experience inflation and some unemployment.

Would such unemployment persist? If wages and prices are assumed to be rigid in a downward direction, output and employment would be reduced below the optimal rate. Such a situation would be correctable only through slow reductions in nominal wages and in prices, if monetary and fiscal restraint were exercised by Congress and the Federal Reserve Board.

Fiscal and monetary authorities might decide, however, that such unemployment is unacceptable and might step in and enact policies to increase aggregate demand. For instance, the Federal Reserve might decide to increase the rate of monetary expansion, thus increasing aggregate demand to AD_1. The economy would then move from E_1 to E_2. We would therefore say that the Federal Reserve had **accommodated** or validated **cost-push inflation**—that is, the increased monopoly or union pressures in the economy, creating new inflationary pressures. Note that, in Figure 32–2, as aggregate demand increases from AD_0 to AD_1, the price level rises.

Under union and monopoly pressures, discretionary monetary and fiscal authorities have a hard choice. They may stand fast, forcing unions and monopolies to endure unemployment until prices and wages fall, bringing the economy back to E_1. Or they may accommodate such pressures by increasing the money stock, thereby validating inflation. Any further pressures by unions or monopolies from E_2 would entail additional hard choices. If the Federal Reserve continues to validate the demands of monopolies and unions, the result is a wage-price spiral, cycles of short-run unemployment and increasing inflation. In Figure 32–2, this spiral moves the economy from E_0 to E_1 to E_2 to E_3, and so on. In this scenario, the quest for higher wages and profits is the source of inflation.

Accommodation of cost-push inflation: An increase in the money supply by the Federal Reserve designed to offset the decrease in output that results from an increase in union and/or monopoly pressures.

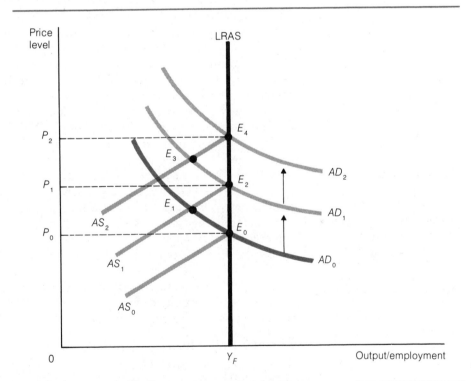

FIGURE 32–2   The Wage-Price Spiral

Though initiated by wage and price demands of unions and monopolies, inflation is accommodated by the Federal Reserve Board's increase of the money supply. The economy is in initial equilibrium at E_0. An increase in union wages causes a leftward shift of the short-run aggregate supply curve from AS_0 to AS_1 as output falls and prices rise. At the new equilibrium, E_1, the economy experiences inflation and unemployment. Monetary expansion by the Federal Reserve then causes an increase in aggregate demand, shifting the aggregate demand curve from AD_0 to AD_1. Prices rise to P_1 and employment reaches the full level again. But further pressures from unions or monopolies for increasing wages would cause the same shifts and, the Federal Reserve Board accommodating, create a wage-price spiral.

Supply shock inflation: An increase in the price level that results from a decrease in aggregate supply, usually caused by a sudden decrease in the availability of an essential resource.

The Supply Shock Theory of Inflation. Another cost-push explanation of inflation is related to **supply shocks**—natural or artificial reductions in the supply of vital natural resources. If, for instance, weather conditions in the Midwest bread basket cause massive shortages in crops one year, the economy undergoes a supply shock. In response to shortages, the costs of producing livestock and other final goods increase. These increases are then passed on to the consumer in higher food prices.

In the 1970s, the OPEC nations progressively monopolized the oil export industry and raised the price of oil and oil products. Since oil-based energy was an important input in U.S. manufacturing and other uses, some observers argued that increases in costs were passed on to consumers in the form of higher prices. OPEC's price demands were therefore seen as ushering in a higher rate of inflation. (Between 1974 and 1980, the average consumer price index inflation rate was approximately 9 percent.)

While it is certainly correct that oil and energy-related prices rose during the 1970s and, further, that the period witnessed high rates of inflation, it is by no means clear that OPEC *caused* inflation, viewed as a continuous rate of price increase. OPEC lost much of its cartel power in the 1980s, moreover, an event which coincides with moderate inflation. Does this mean that OPEC's partial dissolution *caused* the reduction in inflation?

For the supply shock to inflict a continuous increase in prices, other forces must be at work in the economy. For instance, in the 1970s, the Federal Reserve Board steadily increased the growth in the money supply, partly in response to the rising prices of oil-related goods brought about by OPEC. By increasing the money supply, the Federal Reserve not only accommodated OPEC's higher prices but fueled inflation for all other goods as well. As a trading partner with the United States, OPEC nations soon faced higher prices for their imports and responded by increasing the price for their oil. This sequence—higher costs leading to higher prices leading to higher costs, and so on—is a **cost-price spiral,** a variant of a wage-price spiral. Naturally, other forces contributed to the pressure on prices; monopoly and union rigidities complement the effect of a supply shock.

Cost-price spiral: A process in which increased production costs lead to increased product prices, leading to increased production costs, and so on. For the spiral to continue, the money supply must increase or real output must fall.

In determining who was to blame for the inflation of the 1970s, then, some economists might begin with the oil sheiks. Others believe that the supply shock scenario just described would have been impossible without the cooperation of the Federal Reserve's expansionary monetary policy. Not every OPEC-dependent country suffered the same inflationary consequences. Switzerland, for example, whose Central Bank followed a much more conservative policy, seemed to suffer least from the OPEC supply shock.

Demand-Pull Theories of Inflation

The theories we have discussed so far—monopoly power, wage-price, and supply shock—focus on the supply side of inflation. These theories predict inflation on the basis of shifts in the aggregate supply curve.

Inflation, of course, also has its roots in the demand side of the economy. Any rightward shift in the aggregate demand curve, given an upward-sloping supply curve, will create or increase inflationary pressures. Figure 32–1 shows that monetary expansion on the part of the Federal Reserve will shift the aggregate demand curve and drive the price level up. (We know, from previous chapters, that the effects of demand shifts depend on the shape of the aggregate supply curve.) Other factors that can shift the aggregate demand curve include increased government spending, investment spending, and consumer demand.

Money-Supply Growth and Inflation. Monetarists and other economists believe there is a direct relation between expansionary monetary policy on the part of the Federal Reserve and inflation. In other words, inflation—a continuously rising price level—cannot take place without a rising growth rate in the money supply.

This demand-pull theory of inflation is based on a modernized version of the quantity theory of money (see Chapter 30). Briefly stated, the monetarist demand-pull model assumes that monetary expansion on the part of the Federal Reserve will merely increase the cash holdings of individuals without affecting output. If individuals rid themselves of excess money holdings by spending, but output is constant, then prices will go up. Price increases will fuel expectations of higher prices as well as higher nominal interest rates. Higher interest rates will increase pressure on the Federal Reserve to increase the money supply, and so on.

Monetarists believe that this cycle of inflation can be halted only when the Federal Reserve maintains a steady growth rate in the money supply. Such a policy will stabilize money demand, price expectations, and interest rates and permit steady growth in output.

The Relation Between M-1 and Inflation. Does the monetarist demand-pull theory correspond with actual inflation rates? Milton Friedman, the leading proponent of the monetarist position, argues that M-1 growth rates and inflation should be tracked for a number of years (see Figure 32-3). The relation is fairly consistent: Inflation generally follows an increase in M-1 growth with about a two-year lag between the change in M-1 and the change in prices.

The lag between the two variables is explained by the presence of recog-

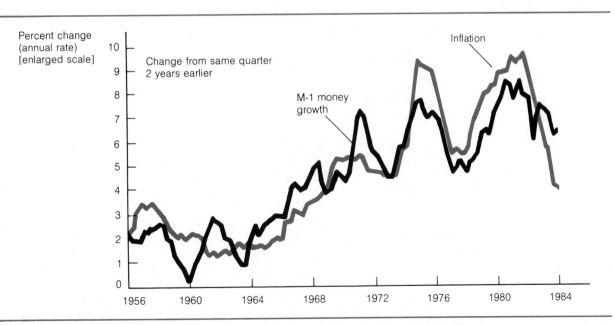

FIGURE 32–3   The Monetarists' Explanation for Inflation

In the monetarist view, inflation rates follow monetary growth rates but with a two-year lag. Erratic inflation rates are caused by the stop-and-go policies of the Federal Reserve System in altering growth rates in the money supply.

Source: Economic Report of the President (Washington, D.C.: U.S. Government Printing Office, 1985), p. 50.

nition and administrative lags on the part of the Federal Reserve and by the time it takes the public to catch up with actual prices and adjust their price expectations.

Inflation and the Deficit. Politicians are fond of blaming high inflation on growing government deficits. Economists, however, are quite unsure of the effects of deficits on the inflation rate. For one thing, deficits are not always caused by increasing government expenditures. They may also result from decreasing revenues. Therefore, the deficit may be increasing at a time when the economy is in a recession and prices are not under inflationary pressure. Deficits can affect inflation, however, if the government must finance them through public and private borrowing. When the government borrows money from the Federal Reserve System, it injects new growth into the money supply, setting off the inflationary spiral. When the government borrows from private sources, it contributes to higher interest rates, another source of inflationary pressure. The relation between federal deficits and inflation is further discussed in Economics in Action at the end of this chapter.

Cures for Inflation

Solutions to inflation—policies that might bring about price stability—are always being hotly debated. Just as there is no single agreed-upon cause for inflation, there is no single accepted cure. When policymakers attempt to resolve the inflation problem, moreover, they may bring about other, perhaps worse, consequences. Many blame the Federal Reserve's anti-inflationary policy of 1979–1980 for the severe recession that followed almost immediately.

Two broad categories of strategies to combat inflation are nondiscretionary policies and discretionary policies.

Nondiscretionary Cures

Those who believe that inflation is directly attributable to monetary expansion urge restraint on the Federal Reserve. This restraint may take several forms. Monetarists urge that the Federal Reserve avoid policies that disrupt the price expectations of the public. Sudden expansions or contractions of the money supply usher in cycles of rising and falling prices. Such changes affect future prices and interest rates and require future discretionary actions. The roller-coaster path of high inflation followed by debilitating recessions is not caused by a destabilized economy but by a destabilizing stop-and-go policy on the part of the Federal Reserve. As we saw in Chapter 31, monetarists urge that the Federal Reserve adopt a money supply growth rule. Such a rule they believe, can allow the economy to reach a proper, long-run equilibrium rate of GNP growth with stable prices.

Discretionary Cures

Discretionary policies to combat inflation are varied. The first step in proper policymaking is accurate diagnosis: Is the inflation a supply-side or demand-side phenomenon? If the inflation is caused by excessive demand, the proper fiscal cures involve budget cuts or higher taxes to dampen the growth in expenditures. Such policies are obviously not very popular with politicians or voters, and inflation-fighting measures are often left to the Federal Reserve. The Federal Reserve's proper monetary policies include reductions in

M-1 growth rates and higher interest rates. Both actions tend to choke off consumption spending and investment, reducing the pressure on prices (and meanwhile risking a recession).

Policies to combat inflation caused by supply factors such as monopoly or union pressures and the wage-price spiral are also varied. We briefly look at two such policies: price and wage controls and tax-induced policies.

Price and Wage Controls: Gain Without Pain? Under price and wage controls, price and wage increases are forbidden by Congress, with penalties for noncompliance. Direct price and wage controls are a familiar, staple policy of governments. From the Roman emperor Diocletian (and probably earlier) through the Nixon years, governments have tried to put a ceiling on price levels by legal means or by the use of price and wage "guidelines," with means of enforcement ranging from fines to death by firing squad. Two basic beliefs underlie the modern advocacy of price and wage controls: (1) Structural rigidities—monopolies and unions—are the cause of inflation, and (2) fiscal and monetary policies are at times insufficient for dealing with the inflation problem.

Price and wage controls usually fail because they cannot be enforced or because they are too costly to enforce. Black markets develop in both output and input markets where there is excess demand for goods or input services (see Chapter 4). Rationing output through coupons and other devices usually fails to overcome these problems. After the OPEC oil crisis, the Carter administration initiated plans for an emergency rationing program for gasoline. The program sparked much controversy and was eventually dropped.

Relaxation of controls opens the floodgates of inflation. After the price and wage controls accompanying World War II were lifted, for example, huge levels of pent-up demand for consumer goods and services—especially durable goods such as autos—were unleashed on insufficient quantities of these goods. The same effects were felt in 1973 after the Nixon price and wage controls were lifted.

Presidential guidelines are another "voluntary" form of price and wage controls. Congressional arm-twisting of businesses and unions to back off from price or wage demands has the same purpose as legally imposed price and wage controls, but with less strict and even unannounced penalties.

Most economists are opposed to price and wage controls because they distort production and economic relations between transactors in the market. If the government is successful as an enforcer of controls, which is itself a costly and difficult job, short-run price expectations may be somewhat stabilized, helping to contain inflationary pressures. Most often, however, a simultaneous and politically popular increase in monetary expansion takes place along with the price and wage controls, turning expectations in the other direction and creating more difficulties. Many economists believe that announced and concrete action on the part of government to control money and fiscal growth has proved to be of greater benefit in containing and stabilizing expectations and, therefore, inflation. Most economists believe that wage and price controls attack the symptoms of inflation but not the cause.

Tax-Induced Policies. Tax-induced policies for fighting inflation are a recent variant of direct price controls. They were first proposed in the United States in the late 1970s. Institution of tax-induced policies would mean the establishment of price guidelines for businesses and labor. These guidelines would not be legally enforced but would carry penalties for violation in the form of higher taxes. Lower taxes would be applied to price or wage changes that did not exceed the guidelines.

The major advantage in using tax-induced policies is that nonprice rationing, such as long waiting lines that discourage consumers, is eliminated. Those willing to pay higher prices or higher wages get the goods or the resources. The disadvantage of tax-induced policies is also a potential disadvantage of price and wage controls. If inflation is not caused by monopoly-union pressures, but rather by increases in the rate of money expansion, imposition of tax-induced policies will not stop inflation.

Unemployment: Concept and Cure

Unemployment is another major effect of instabilities in the economic system, as we have noted many times. Economists are divided on what constitutes the best cure for unemployment. Before examining policies suggested for decreasing unemployment, we review economists' attempts to define full employment.

The Definition of Full Employment

The concept of full employment is not easy to define. What rate of unemployment can it be identified with? As indicated in Table 32–3, yearly statistics show that the rate of unemployment in the United States has varied considerably in recent decades, between 3.5 percent in 1969 and 9.7 percent in 1982.

Full employment clearly does not mean that everyone in the labor force is employed. As we saw in Chapter 21, our market economy is dynamic. Labor is geographically mobile. A number of laborers will be changing jobs

TABLE 32–3   The U.S. Unemployment Rate for all Civilian Workers, 1960–1984

Between 1960 and 1984, the U.S. unemployment rate has varied from a low of 3.5 percent in 1969 to a high of 9.7 percent in 1982. High unemployment rates are associated with recessions—downturns in the economy's use of its productive capacities—although rates of 5 to 7 percent are considered normal and are due to frictional factors such as worker mobility.

Year	Unemployment Rate (percentage)	Year	Unemployment Rate (percentage)
1960	5.5	1973	4.9
1961	6.7	1974	5.6
1962	5.5	1975	8.5
1963	5.7	1976	7.7
1964	5.2	1977	7.1
1965	4.5	1968	6.1
1966	3.8	1979	5.8
1967	3.8	1980	7.1
1968	3.6	1981	7.6
1969	3.5	1982	9.7
1970	4.9	1983	9.6
1971	5.9	1984	7.5
1972	5.6		

Source: *Economic Report of the President* (Washington, D.C.: U.S. Government Printing Office, 1985), p. 271.

at any point in time. But these factors do not help much in determining a meaning of full employment that is useful for policymaking.

Throughout our discussion of macroeconomic theory we have identified full employment with a *natural rate of unemployment*. The natural rate depends on slowly changing factors affecting the labor market: the amount of unionization, the speed with which vacancies and workers are matched, employers' costs of hiring and firing, regulations affecting work and hiring incentives, and so forth. Before 1977 in U.S. government statistics, an unemployment rate of 4 percent was full employment. In 1977, the Council of Economic Advisers suggested that full employment was associated with an unemployment rate of 5 percent. Economists supporting the natural rate hypothesis argue that the natural rate is at least 5 percent and perhaps even 6 percent or more.

Do the data of Table 32–3 provide any evidence on this matter? Is there any indication that the natural rate may be increasing? We attempt to develop a full answer in the discussion of the so-called Phillips trade-off between inflation and unemployment later in this chapter. Consider, however, the fact that the average unemployment rate between 1960 and 1984 is 6.0 percent—not far from the natural rate suggested by economists.

Job Creation Policies Versus Labor Market Flexibility

Economists and politicians often disagree over what to do when the level of unemployment is high. In the midst of recession, when unemployment reaches politically unacceptable levels, legislators tend to recommend specific jobs bills designed to create new jobs for the idle worker force. Jobs bills are usually directed at specific sectors of the economy and can take different forms. The Chrysler Corporation bail-out, for instance, was a direct subsidy to a troubled firm and its workers; protectionist trade legislation has the same effect of increasing employment in specific industries. Another approach is a **jobs creation bill.** Government programs such as CETA (Comprehensive Employee Training Act) provide temporary employment and training to those unable to find work in the private sector. During the seventeen-month recession through December 1982, a number of jobs creation proposals were made in Congress. In late 1982, a $1 billion public employment bill was passed in Congress and signed by President Reagan. Similar bills, such as the Humphrey-Hawkins Act in the 1970s, have been proposed to handle specific unemployment problems or to serve as a permanent stopgap of all unemployment situations. According to the 1978 amendments to the Full Employment Act of 1946, the federal government has the legal responsibility ". . . to use all practical means . . . to promote maximum employment, production and purchasing power." The act is interpreted by many to mean that the role of government is to institute fiscal and monetary policies to help create full employment, not necessarily to provide government jobs.

Economists are divided on the effects of jobs creation programs. Certainly there are short-run positive effects on the unemployed, but since work creation is not permanent, economists question the long-run consequences of tax-financed federal jobs creation. The development of public works to reduce unemployment may create temporary prosperity at the expense of long-run economic goals in society. If long-run full employment is in fact an economic goal, policies that create greater flexibility in labor markets with permanent employment prospects may be more desirable.

The timing of such programs is also an important economic matter. Insti-

Jobs creation bill:
Legislation to increase the demand for labor either in the overall economy or in a particular sector or industry.

tution of a jobs bill that will not go into effect until some months later may backfire in terms of economic stabilization. If a jobs program takes effect at the beginning of an economic recovery, the effects may be inflationary. Miscalculation of fiscal implementation may therefore be quite costly.

Inflation and Unemployment: Is There a Trade-off?

Most vexing for economists is the possibility that there may be a trade-off between inflation and unemployment. If so, achievement of price stability would require unemployment, or full employment would require inflation.

British economist A. W. Phillips published an important paper related to inflation and unemployment in 1958,[1] in which he used empirical data to show that there may be a long-run trade-off between inflation and unemployment.[2] Phillips' data are based on money wage rates; since they follow the inflation rate, money wage changes may serve as a proxy for inflation. Phillips' data showed an almost hundred-year (long-run) inverse relation between inflation and unemployment for the United Kingdom. The data also indicated that the cost of less unemployment is higher inflation rates or, conversely, that the cost of lowering the inflation rate is a higher unemployment rate.

Phillips' study caused a great stir among macroeconomists because an inverse relation between unemployment rates and inflation rates means that the long-run aggregate supply curve is positively sloped, not a vertical line. In other words, higher levels of output and employment are invariably associated with higher price levels. (Alternatively, a long-run curve relating unemployment rates with inflation rates would be negatively sloped.) This means that long-run income and employment growth must come at the expense of inflation. If the Phillips curve is a reality, the promise of achieving one economic goal (price stability or full employment) means that the other economic problem cannot be solved simultaneously. The policymaker must choose between problems in a no-win trade-off.

Phillips Curves: Hypothetical and Actual

Phillips curve: A curve that shows the relation between the rate of inflation and rate of unemployment.

Before analyzing American experience with inflation rates and unemployment, examine a hypothetical **Phillips curve** and what it means. Figure 32–4 depicts the long-run relation between the unemployment rate and the inflation rate that Phillips' long-run data suggested. Note that in place of the level of income or employment, the unemployment rate appears on the horizontal axis. When these data are displayed against the inflation rate, a curve of negative slope results.

Phillips' basic argument is straightforward: A reduction in inflation comes at the expense of additional unemployment. The degree of the inverse relation, however, depends on the initial rate of inflation. At higher rates of inflation, the trade-off is less severe. For instance, as Figure 32–4 shows, a

[1]See A. W. Phillips, "Relation Between Unemployment and the Rate of Change of Money Wage Rates in the United Kingdom, 1861–1957," *Economica* 25 (November 1958), pp. 283–299.

[2]Actually, neoclassical economist Irving Fisher was the first to notice the inverse relation between inflation rates and unemployment rates. See Irving Fisher, "A Statistical Relation Between Unemployment and Price Changes," *International Labor Review* (June 1926), pp. 785–792. Reprinted as "I Discovered the Phillips Curve," *Journal of Political Economy* (March–April 1973), pp. 496–502.

FIGURE 32–4

A Hypothetical Phillips Curve Relating Inflation and Unemployment

According to A. W. Phillips, there is an inverse relation between the inflation rate and the rate of unemployment. At an inflation rate of 16 percent, the unemployment rate is 4 percent. If the inflation rate falls to 14 percent, unemployment rises to 5 percent. At zero inflation, unemployment is very high.

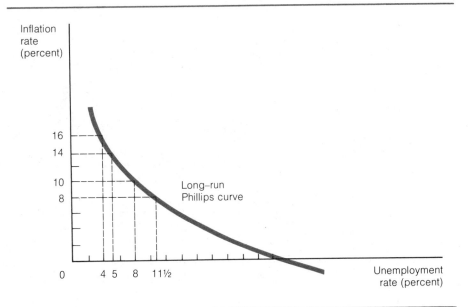

reduction of 2 percent in inflation from an initial rate of 16 percent causes unemployment to rise by only 1 percent (from 4 to 5 percent), whereas a similar reduction at a lower inflation level (from 10 to 8 percent) may force unemployment to rise by a full 3½ percent (from 8 percent to 11½ percent).

In the analysis of the Phillips curve, inflation is a necessary evil. Resource scarcities are such that higher and higher inflation rates are required to coax the resources into use. Likewise, lower and lower rates of inflation are insufficient to put large quantities of labor resources to work. The economy requires inflation to achieve high rates of employment and real income growth. Alternatively, the cost of price stability at low rates of inflation is high unemployment, low income, and lower growth rates.

What is the evidence for a Phillips curve for the United States? Tables 32–1 and 32–3 give us some information about inflation and unemployment that is required to gain an insight into this important question. Figure 32–5 relates the inflation rate to the unemployment rate for the years 1960–1984, using the GNP price deflator reported in Table 32–1 as the measure of inflation. From looking at the raw data of Figure 32–5, it appears that trade-offs are indeed a fact! Note, for example, that a reduction in the unemployment rate between 1967 and 1969 was accompanied by a rise in the inflation rate.

As we get closer to the present, the trade-off still appears to hold for some time periods. Look at the movements between 1972 and 1973, between 1976 and 1979, and between 1981 and 1982. In the latter case, a reduction in the inflation rate from 9.4 to 6.0 percent was accompanied by an increase in the unemployment rate from 7.6 to 9.7 percent, a dramatic increase with all sorts of economic and political consequences.

Consider, however, the relation between the unemployment rate and the inflation rate for other recent time periods. Between 1969 and 1970, 1971 and 1972, 1973 and 1974, 1975 and 1976, 1979 and 1981, and 1982 and 1984 it appears that a positive rather than inverse relation existed between the unemployment rate and the inflation rate. Does this positive relation mean that Phillips' theory is wrong, or does it mean that the sometimes positive relation indicates a movement of the trade-off to higher and higher

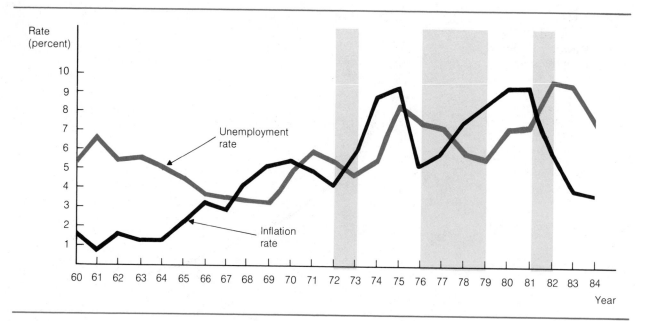

FIGURE 32–5   Combined Unemployment and Inflation Rates, 1960–1984

The Phillips relation—the theory that increases in inflation rates are associated with decreases in unemployment rates and vice versa—appears to hold over selected years in the United States. Additionally, the relation is established at higher and higher rates of unemployment and inflation from 1960 through 1984. Shaded areas represent possible Phillips tradeoffs between inflation and unemployment.

Source: Inflation data, GNP price deflator. All data from *Economic Report of the President* (Washington, D.C.: U.S. Government Printing Office, 1985).

rates of unemployment and inflation? The answer given by some economists—mostly monetarist and rational expectationist—is that the Phillips relation must be viewed from both short-run and long-run perspectives and that expectations play a major role in explaining cycles of inflation and unemployment.

A Phillips Curve Adjusted for Expectations

When the Phillips relation is analyzed from both short- and long-run views, the expectations of consumers and producers regarding the inflation rate and the rate of unemployment become important.

Figure 32–6 reproduces a long-run vertical supply curve, three short-run aggregate supply curves based on certain levels of price expectations by workers and employers, and three related aggregate demand curves. Recall from earlier discussions of the adaptive expectations model in Chapter 26, that along any short-run aggregate supply curve, price expectations of both employers and workers are at given levels but that employers' expectations adjust instantly to new actual price levels while workers are slower to adjust.

In Figure 32–6 the economy is initially in equilibrium at E_1—the price level and the natural rate of employment are in equilibrium where aggregate demand AD_0 and long-run aggregate supply converge. Assume that the Federal Reserve steps up monetary expansion. The result will be a rightward shift in the aggregate demand curve from AD_0 to AD_1. The economy will reach a new equilibrium up the short-run aggregate supply curve at E_2. Why? Because employers correctly perceive a reduction in real wages and hire more

Short-run Phillips curve: A downward-sloping curve that suggests that the rate of unemployment is inversely related to the rate of inflation when the expected rate of inflation is constant.

labor. Laborers are willing to supply more labor input because they believe the rise in their nominal wages to be a rise in the real wage rate. The economy temporarily moves to overfull employment, a point higher than the natural rate. Thus the **short-run Phillips curve** describes an inverse relation between unemployment and inflation, just as Phillips predicted.

But the economy cannot remain at points such as E_2 for very long. Over time, expectations of workers adjust to their actual price experience, and either their wage demands increase or they supply less labor. When that adjustment takes place, the short-run aggregate supply curve shifts leftward from E_2 to E_3. Equilibrium again occurs on the long-run aggregate supply curve, at the natural rate of income and unemployment.

Similarly, consider a reduction in the rate of monetary expansion beginning at E_4 in Figure 32–6. At E_4, aggregate demand curve AD_2 intersects with aggregate supply curve AS_2. When the Federal Reserve reduces monetary expansion, employers correctly perceive the resulting reduced rate of inflation. Workers, however, are slower to realize that actual prices are falling and incorrectly perceive that their real wage is declining. This set of expectations along AS_2 persists for a time, reducing employment below the natural rate to E_5 and lowering prices. Again the Phillips relation holds:

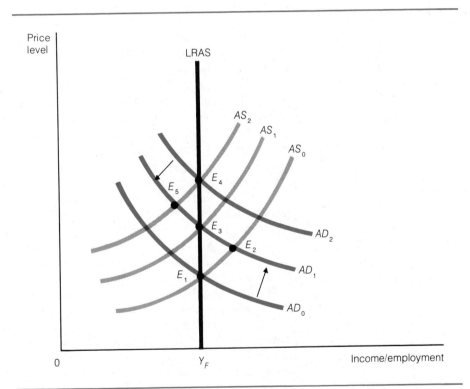

FIGURE 32–6   The Expectations-Adjusted Phillips Relation

With full-employment equilibrium at E_1, an increase in the money supply growth rate will shift aggregate demand from AD_0 to AD_1. Employment and prices rise to equilibrium at E_2. After workers' expectations adjust to higher prices, aggregate supply shifts from AS_0 to AS_1 and the economy returns to the full-employment level. However, in the short run, the Phillips relation holds. The difference between short- and long-run effects also applies to decreases in the money supply growth rate. Starting at equilibrium E_4, a decrease in money growth moves the economy to E_5, but in the long run it moves back to full employment at equilibrium E_3.

Long-run Phillips curve: A vertical line that shows that the rate of unemployment is independent of the rate of inflation when the expected rate of inflation adjusts to equal the actual rate of inflation.

Higher unemployment rates are associated with lower inflation rates. But the situation is only temporary. When workers' expectations adjust to the lower rate of inflation, short-run aggregate supply again shifts to the right, returning the economy to the natural rate of unemployment and income growth at E_3. The conclusion, therefore, is that when adjustment is made for expectations, the **long-run Phillips curve** is vertical—it is the classical, monetarist, and rational expectationist supply curve—although the short-run Phillips curve depicts an inverse relation between unemployment and inflation rates.

Actual data show some evidence of this cycle of expectations, unemployment, monetary acceleration and deceleration, and inflation. In Figure 32-7 we have graphed the inflation and unemployment data from Figure 32-5. This time, however, we have drawn a vertical line representing a natural rate of unemployment at 6 percent, the approximate actual average rate of unemployment between 1960 and 1984. Strikingly, it appears that cycles of unemployment and inflation do fluctuate around this natural rate, just as predicted in the theoretical model shown in Figure 32-6.

Some economists conclude from the expectations-adjusted natural rate supply curve that the short-run Phillips relation would not be observed—certainly not in the extremes of the 1970s and 1980s—if a monetary rule were followed by the Federal Reserve and the budget were balanced by Congress.

Rational Expectations and the Phillips Curve

We can also view the Phillips curve question from a rational expectations perspective. Recall from Chapter 31 that the rational expectationists argue that all individuals (including employees) are rational in that they do not throw information away. While recent price information is important in this theory, as in the adaptive expectations approach, individuals are presumed to act on a much broader base of information, which includes the effects of money supply changes by the Federal Reserve, tax rate changes by Congress, and the like. People are not always right about the future, but they learn and are increasingly right on average.

In the rational expectations view, the path of inflation and unemployment rates is different from that predicted by either Phillips or the adaptive expectationists. If people—all market participants—get smarter as time goes on, expectations will adjust more and more rapidly to economic policy decisions and to prices. The practical effect of this assumption is a decline in the size of the swings or cycles of unemployment and inflation.

In this view, the surprise actions of Federal Reserve Board and congressional policymakers become less and less effective through time. Market participants become better and better at predicting the effects of macroeconomic and monetary policy and adjust to new situations more and more rapidly. In the long run—and with perfect perception of the effects of policy on the part of all market participants—the economy follows the vertical path of the long-run aggregate supply curve and maintains a level of inputs corresponding to the natural rate of income and employment.

While there are numerous theories of the causes of unemployment and inflation in the economy, the debate over cures settles down to a question of whether the necessary adjustments can take place, with or without government intervention. Economists in the Keynesian tradition argue that the economy is incapable of reaching acceptable levels of employment and inflation without discretionary policy on the part of fiscal and monetary authorities. Monetarists and rational expectationists argue that the economy can

FIGURE 32–7

Average Rate of Unemployment and Inflation, 1960–1984

Cyles of unemployment and inflation rates appear to have clustered around the average unemployment rate, 6 percent, between 1960 and 1984. These factors also lend weight to the claim that the natural rate of unemployment is close to 6 percent or possibly higher.

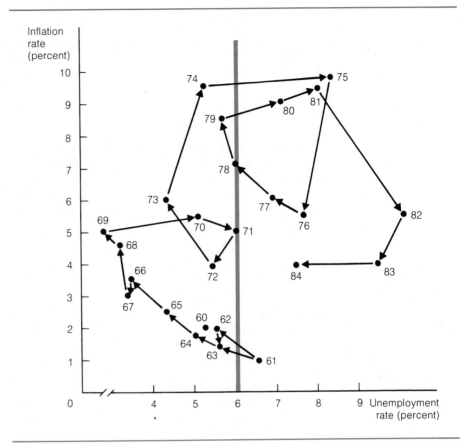

reach stability through internal adjustments. These two schools of thought disagree, however, on the speed of such adjustments. In the past twenty years much has been learned about inflation and employment, but economists are far from settling their debates. In the meantime, economic stabilization remains an elusive goal. In the next chapter we discuss inflation and employment in a broader context and see how economists approach the problem of business cycles.

Summary

1. Inflation and unemployment constitute the most important macroeconomic policy issues of the day. Both have costs and both reduce economic prosperity and economic growth.

2. Inflation is a process of price increase. Inflation arbitrarily redistributes income in society away from those on fixed incomes and those holding dollar-denominated assets. Inflation also disrupts expectations and is a tax that is imposed without having been voted on.

3. Nonmonetary theories of inflation feature structural problems in the economy, such as the existence of or increase in monopoly and union power. In this view, unreasonable demands can create a wage-price spiral that can be stopped by lessening such power

through regulation or through price and wage controls.

4. Monetarists argue that the only ultimate cause of inflation is monetary expansion in excess of the economy's ability to produce goods and services. They feel that only rules pertaining to the conduct of both fiscal and monetary policy can create stability in the macroeconomy.

5. Unemployment is of policy interest when it exceeds the level of unemployment created by job and labor mobility. The natural level of unemployment is thought to be between 5 and 6 percent, although the official measure of full employment currently used by the government means that 5 percent of the nation's work force is without work.

6. The Phillips curve inversely relates the inflation rate and the unemployment rate. In Phillips' view, unemployment rates can be lowered only at the cost of higher inflation rates, and vice versa.

7. One view of the Phillips relation divides the problem into short- and long-run components. Monetar- ists agree that a short-run Phillips-type trade-off exists but argue that there is no trade-off in the long run. Rational expectations theory argues that the short-run trade-off may be corrected by rapid adjustments of expectations.

Key Terms

escalator clause	accommodation of cost-push inflation	Phillips curve
cost-push inflation	supply shock inflation	short-run Phillips curve
demand-pull inflation	cost-price spiral	
wage-price spiral	jobs creation bill	long-run Phillips curve

Questions for Review and Discussion

1. How does inflation decrease individuals' wealth? How can people protect themselves from inflation?

2. In what sense is inflation a tax? Do government tax revenues increase when inflation occurs?

3. How do monopolies or unions cause inflation? Can monopolies cause inflation without the assistance of monetary expansion?

4. Why do wage and price controls fail to stop inflation when monetary expansion continues?

5. In the monetarists' view, what is the ultimate cause of inflation? What do they recommend as a cure for inflation?

6. What causes unemployment to rise above the natural rate? Can it fall below the natural rate?

7. What does the short-run Phillips curve show? Why does this relation between inflation and unemployment occur?

8. What is the relation between the short-run Phillips curve and the short-run aggregate supply curve?

9. What does the long-run Phillips curve show? What would cause it to shift to the right?

10. What causes a shift in the short-run Phillips curve? According to the rational expectations theory, does the short-run Phillips curve shift rapidly?

11. What would a negatively sloped long-run Phillips curve suggest?

12. If labor suppliers expect an inflation rate of 10 percent and the actual rate of inflation is only 5 percent, what would you expect to happen to the rate of unemployment? What would this do to the short-run Phillips curve after workers adjusted to the lower rate of inflation?

ECONOMICS IN ACTION

Budget Deficits and Inflation

The economic effects of federal budget deficits on inflation are complex and controversial.

Before considering economic effects, take a look at budget deficits and surpluses throughout U.S. history[a] (see Figure 32–8). Between the establishment of the U.S. Treasury in 1789 and the year 1985, there have been 195 budgets, of which 92 have been in deficit and 103 in surplus. Between 1789 and 1930, the budget was in deficit 32 percent of the time, but between 1931 and 1985 the government ran deficits more than 84 percent of the time. Onrushing deficits in the post-1931 era

[a]An excellent survey on the effects of deficits is "The Deficit Puzzle: Fitting the Pieces Together," *Economic Review*, Federal Reserve Bank of Atlanta Special Issue (August 1982).

reached an all-time peak in 1983, when the total deficit reached the $178.6 billion level (depending on how it is calculated). The deficit declined in 1984 by about $2 billion, but it is projected to reach $200 billion in 1985 and 1986.

How might these huge deficits matter? Since the deficit is simply the difference between what the government spends and what it takes in as taxes, the difference must be made up somewhere. Roughly, the government has two alternatives to finance the deficit: (1) having the Treasury sell bonds to the Federal Reserve System, which in effect means "to print more money," and (2) selling government bonds on the open market, thus competing with businesses for private savers' funds.

Though it is often difficult to determine the extent

Surplus/deficit
(percentage of GNP)

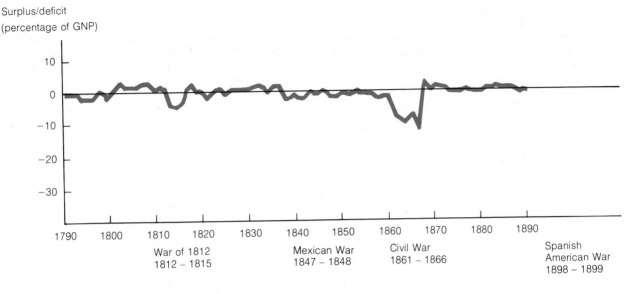

War of 1812
1812 – 1815

Mexican War
1847 – 1848

Civil War
1861 – 1866

Spanish
American War
1898 – 1899

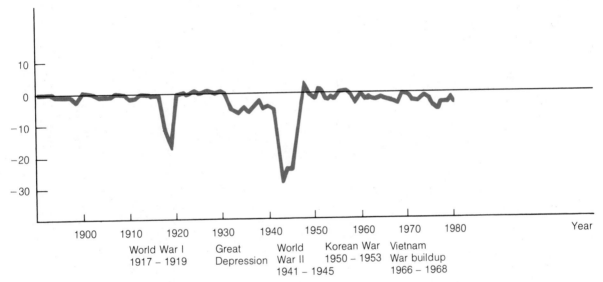

World War I
1917 – 1919

Great
Depression

World
War II
1941 – 1945

Korean War
1950 – 1953

Vietnam
War buildup
1966 – 1968

Year

FIGURE 32–8   Federal Budget Surplus or Deficit as a Percentage of GNP, 1789–1981

Source: James R. Barth and Stephen O. Morrell, "A Primer on Budget Deficits," *Economic Review,* Federal Reserve Bank of Atlanta Special Issue (August 1982), pp. 10–11.

to which the money supply M-1 increases as a result of deficit financing, most economists (and most citizen-taxpayers) believe that such financing is inflationary.

If the government finances deficits by selling newly printed bonds in the private markets, interest rates will rise, causing private investment to decline. The rise in interest rates will also reduce consumption spending. In this view, deficits will crowd out private investment and consumption expenditures, creating a reduction in the production and consumption of private goods and services and permitting an increase in the production of public goods and services. Either method of financ-

ing deficits forces resources from the private sector into production in the public sector.

Critics of this position on the economic effects of deficits point out that there is no clear correlation between deficits and the money supply. The Federal Reserve, after all, is not required to purchase bonds from the U.S. Treasury. Indeed, the Federal Reserve may attempt to neutralize the impact of deficit financing on M-1 through compensatory manipulation of open market operations and other credit controls. Still other economists believe that deficits have no long-run effects if they are financed through the sale of bonds in private markets. Why? Because government bonds are "wealth" to those who buy them, but they are an equivalent and corresponding liability to future taxpayers. In the main, however, most economists and citizens believe that deficits are inflationary and that they crowd out private expenditures.

Question

In what ways do budget deficits inhibit long-run economic expansion? Does the existence of a huge deficit affect your expectations of how high interest rates or inflation rates will be?

33

Business Cycles

*T*he real GNP of an economy fluctuates up and down over time. Chapter 21 defined such fluctuations as business cycles, and in our examination of classical, Keynesian, and monetarist theories of the economy, we sought reasons why an economy can undergo periodic expansions and contractions in output. In this chapter we look at business cycles in greater detail. We examine how business cycles are measured, discuss the tools forecasters use to predict ups and downs in business activity, and survey some theoretical approaches to output fluctuations: investment theories, monetary theories, and theories of a political business cycle. As in Chapter 32, where we looked at inflation and unemployment, some of the theoretical tools used in this chapter will be familiar.

The study of business cycles has great importance for policymakers and business decision makers. Can downturns in the economy be predicted? What factors cause business cycles to differ in their length and severity? Are business cycles caused by instability in private spending? What role does government play in the process? These are some of the questions this chapter addresses.

Defining and Predicting Business Cycles

Over a long period of time, such as a hundred years, economic capacity to produce goods and services tends to increase because of factors such as technological change, growth in labor supply (population), and discovery and development of natural resources. Within the overall trend of long-run growth, economies fluctuate in aggregate output and employment (see Figure 33–1).

A **business cycle** is a downward or upward fluctuation in the real GNP of

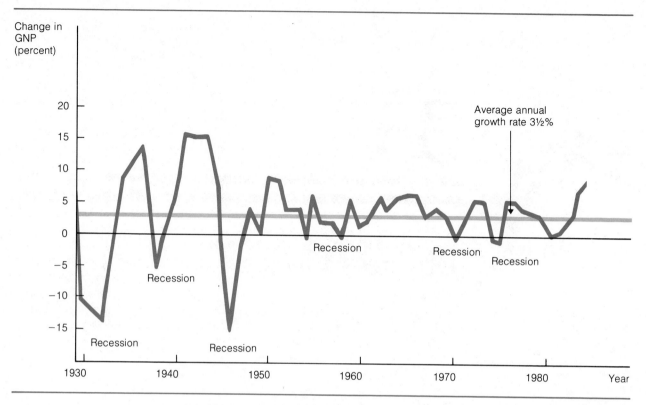

FIGURE 33–1   Annual U.S. GNP Growth Rates, 1929–1984

The percentage change in a country's GNP tends to fall and rise over time; economists refer to these fluctuations as business cycles. Cycles of expansion and recession have characterized the U.S. economy throughout most of its history. Despite these fluctuations, between 1929 and 1984 the average growth rate has been 3.5 percent per year.

Sources: Economic Report of the President (Washington, D.C.: U.S. Government Printing Office, 1966, 1982); *Federal Reserve Bulletin* (September 1984).

Business cycles: Recurrent fluctuations in the level of business activity, as measured by real GNP.

an economy that repeats itself over time. Three points should be kept in mind about this definition. First, business cycles recur over time but not necessarily in a predictable pattern. Recurrent and predictable seasonal variations in business activity—such as the purchase of more firewood in winter than in summer—are not business cycles. For example, everyone is aware of the upsurge in the purchase of goods and services every December. This seasonal change in business activity is a short, predictable fluctuation that is fully accounted for and therefore is not a business cycle.

Second, the term *business cycle* normally refers to overall business activity as measured by real GNP. There can also be cycles in the component parts of GNP. Spending on consumer durables, such as refrigerators and cars, exhibits cyclical behavior over time, whereas consumer spending on nondurable items, such as food, does not. Cyclical patterns are also observed in retail sales, business profits, business incorporations, interest rates, residential construction, and many other reported statistics. For our purposes, however, we take business cycles to refer only to the movement of real GNP.

Third, there can be different types of business cycles, depending on the length of time over which a given business cycle runs. Short cycles last less than two years; intermediate cycles extend five years or more; and long-run cycles run over ten years. Indeed, some students of the business cycle have identified cycles that extend over several decades or even centuries.

The Language of Business Cycles

A business cycle includes expansions and contractions of business activity. An expansion is a cumulative increase in real GNP, while a contraction is a cumulative decrease. Contractions are often called recessions or depressions. A **recession** is a decline in real GNP that lasts for six months or more. **Depression** is a more ambiguous term, but it generally refers to a serious and persistent decline in economic activity lasting several years, accompanied by an unemployment rate of 15 percent or more.

The period, or length, of a business cycle can be measured in two ways: from peak to peak or trough to trough. Figure 33–2 illustrates the alternative measurements. Peaks are upper turning points in a cycle where an expansion finally peters out. Troughs are the opposite of peaks; they are the lower turning points of a cycle where contractions bottom out. The length of a business cycle differs depending on whether it is measured from peak to peak or trough to trough.

The trend line in Figure 33-2 measures a hypothetical long-run growth path of the economy. The extent of deviations from the trend line, at either a peak or a trough, indicates the intensity or amplitude of a business cycle.

Recession: A contraction in business activity, usually defined as a decline in real GNP over a period of six months or more.

Depression: A severe, persistent contraction in business activity that may last for several years and result in 15 percent or more unemployment.

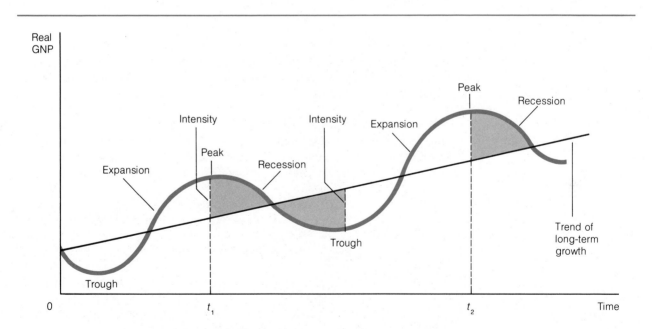

FIGURE 33–2   Hypothetical Business Cycle

The curve represents the four phases of a hypothetical business cycle. One phase is contraction or recession, when aggregate output and employment fall. The trough occurs when this part of the cycle bottoms out. The expansion or recovery phase occurs when the economy exhibits expanding output and employment. The peak occurs where general business activity and employment stop rising. When the next recession begins, the economy starts another business cycle. The period of the cycle is measured from peak to peak or trough to trough. This cycle has a period of t_1t_2, measured peak to peak. The intensity of the cycle is given by the deviation of the peak or trough from the trend line of long-run growth.

TABLE 33–1   U.S. Business Cycles, 1900–1981

Trough	Peak	Length of Cycle (peak to peak in months)
December 1900	September 1902	39
August 1904	May 1907	56
June 1908	January 1910	32
January 1912	January 1913	36
December 1914	August 1918	67
March 1919	January 1920	17
July 1921	May 1923	40
July 1924	October 1926	41
November 1927	August 1929	34
March 1933	May 1937	93
June 1938	February 1945	93
October 1945	November 1948	45
October 1949	July 1953	56
May 1954	August 1957	49
April 1958	April 1960	32
February 1961	December 1969	116
November 1970	November 1973	47
March 1975	January 1980	62
March 1980	January 1981	12

Sources: U.S. Bureau of Economic Analysis, *Business Conditions Digest* (June 1980); *Economic Report of the President* (Washington, D.C.: U.S. Government Printing Office, January 1982).

Business cycles have no set patterns. They recur over time, but each one is slightly different. Each business cycle varies in length and intensity, and within a given business cycle the intensity of an expansion need not equal the intensity of a contraction.

Business Cycles in the United States

Using intensive statistical analysis, researchers have identified numerous business cycles in U.S. economic history. Table 33–1 presents some data on business cycle activity in the United States since the turn of the century. The statistics show that business cycles are recurrent but have different lengths. Nineteen cycles were identified from 1900 to 1981. Measured peak to peak, the average duration of these cycles was approximately 51 months or a little over four years. The shortest business cycle (January 1980–January 1981) lasted 12 months. The longest (April 1960–December 1969) lasted 116 months.

Moreover, the business cycles exhibited varying degrees of intensity. In the two-year expansion from April 1958 to April 1960, real GNP rose by 8 percent. By contrast, in the expansion from March 1980 to January 1981, real GNP rose by only about 2 percent. Each business cycle and its phases possess a unique character.   Since 1981, the business cycle pattern in both recession and recovery has continued. Focus, "1981–1984: Recession and Recovery," discusses how the business cycle continues to be a part of the American economic experience.

Forecasting Business Cycles

Some economists specialize in forecasting business conditions. These individuals are often called **econometricians** because they are highly trained in the use of statistics to predict events in the economy. Some well-known eco-

FOCUS   **1981–1984: Recession and Recovery**

The period between 1981 and 1984 provides an interesting case study of the business cycle in both recession and recovery. The United States experienced a brief recession in 1980 and a longer recession in 1981–1982 (see Figure 33–3). A number of possible reasons might be given for these recessions, but a primary one relates to the high inflation rates of the late 1970s and early 1980s. The annual Consumer Price Index (CPI) rose to 13.3 percent in 1979 and fell only modestly to 12.4 percent in 1980. Wage and price guidelines were instituted in the late 1970s along with price controls on oil and gas but with little effect. The stage for recession was set.

Unstable expectations about future prices, together with rising nominal interest rates, fostered a great deal of nervousness and uncertainty among business investors and consumers and generally reduced most levels of market activity. The Federal Reserve attempted to slow money growth in November of 1978, but because

of a probable one-to-two year lag in effects, inflation continued unabated. Figure 33–4 shows the effects of M-1 money supply growth rates on recent cycles of recession and expansion. (Figure 33–4 incorporates the customary lag in M-1 effects by subtracting one year from M-1 growth rates. For example, what the figure shows as M-1 growth for 1976 is actually 1975's number.) The effects of M-1 contraction on the economy in 1980 were stark. The real income growth rate was negative (see Figure 33–3) and the unemployment rate climbed from 5.6 percent in 1979 to 7.1 percent in 1980 with inflation still running in double-digit figures.

As Figure 33–4 indicates, monetary growth was curtailed considerably in 1981 and in the first half of 1982, a policy that created a severe credit crunch and a short-run increase in interest rates. The result was another recession as shown in Figure 33–3 and in the shaded area (1981–1982) in Figure 33–4. This time the unemployment rate jumped from 7.6 percent in 1981 to

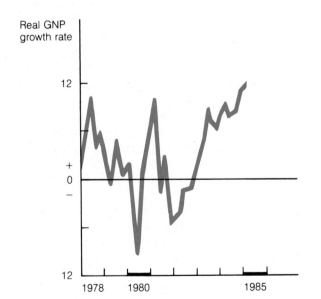

FIGURE 33–3   Annual Growth Rate in Real GNP, 1978–1984

The growth rate in real GNP changed erratically between 1978 and 1984. Two recessionary phases are discernible, in 1980 and a longer one in 1981–1982 with a brief recovery in between. Since 1982 the cycle has been in the recovery phase.

Source: Board of Governors of the Federal Reserve System, *Historical Chart Book* (1984), p. 12.

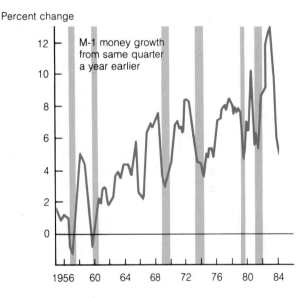

FIGURE 33–4   Money Supply and the Business Cycle, 1956–1984

The M-1 growth rate is one determinant of the business cycle. When M-1 growth rates *one year earlier* are plotted against standard recessions, money growth rates have usually declined before the beginning of a recession and often extended into the recession. Money growth has usually increased before a recovery has begun.

Note: Shaded areas indicate recessions (peak to trough) as defined by the National Bureau of Economic Research.

Source: Council of Economic Advisers, *Economic Report of the President* (Washington, D.C.: U.S. Government Printing Office, 1985), p. 55.

9.7 percent in 1982 with a negative growth in real income once again and for a longer period of time. One benefit came out of the recessionary experience. The CPI declined from 12.4 percent in 1980 to 3.9 percent in 1982. The benefit was obtained, however, at high costs in terms of bankruptcies and the human costs of unemployment.

Between 1982 and 1984 the economy underwent a recovery in terms of rising real income growth and rising employment in all categories. The overall unemployment rate fell from 9.6 percent in 1983 to 7.5 percent in 1984—unsatisfactory by pre-1980 averages, but moving in the right direction nonetheless. Real income growth has been even more dramatic, increasing in 1984 at the highest real rate (6.8 percent) since 1951, using 1972 dollars as a base. There are still important problems to be attacked in the economy, especially the huge federal deficits predicted to be on the order of $175 to $200 billion for the next several years. The financing of these deficits creates competition for private capital formation and contributes to high interest rates which could choke off the recovery. Whatever the outcome, it is predictable that business cycles of some duration will be a part of American economic experience.

Econometrician: One who utilizes economic theory, mathematics, and statistics to analyze and predict economic performance.

nomic forecasters are Lawrence Klein of the University of Pennsylvania, who won the Nobel Prize in economics in 1980; Michael Evans, who has his own forecasting firm; and Alan Greenspan, who was chairman of the Council of Economic Advisers under President Ford and who also had his own forecasting business.

Forecasting is both a science and an art. The science of forecasting involves the construction of statistical models of the operation of the economy and of economic indicators that predict the future level of such variables as GNP. The art of forecasting is the use of past experience and commonsense judgment about what types of variables are important in predicting how the economy will behave, since no two business cycles are precisely alike.

Who are some of the users of economic forecasts? Business decision makers and government policymakers need reliable data on the future course of GNP to plan their activities. When business executives make decisions about investments in plant and equipment that will lead to increased output in the future, they need to have some idea of what the state of the economy will be when this output comes on the market. Overall tax revenues will usually depend on the level of national income and consumer spending in the future. To plan government spending, policymakers need estimates of future revenue.

To produce a forecast, the forecaster seeks information about what changes in the economy today are a good indication of what the state of the economy will be tomorrow. For this purpose economic forecasters have developed **economic indicators,** key economic statistics that provide clues about the state of the future economy.

Economic indicators: Key statistics, such as the net change in inventories, stock price index, and the money supply, that are used to forecast the pattern of business cycles. There are three varieties of indicators: leading, concurrent, and lagging.

Economic indicators come in three basic varieties. Leading economic indicators anticipate a business cycle. They tend to turn down before a general economic downturn and up before an expansion. A coincident indicator runs in step with a business cycle, and a lagging indicator follows changes in the business cycle. Of these three sets of indicators, the most attention is paid to leading indicators because they give decision makers clues about the future course of the business cycle.

There are twelve commonly used leading indicators, listed in Table 33–2. Data on these statistics are published quarterly by the Department of Commerce, and the trend of leading indicators is usually reported in the press. The leading indicators tend to change together over time, and a particularly useful summary statistic is the composite or combined index of leading indicators. The behavior of this index since 1948 is shown in Figure 33–5.

TABLE 33–2   Leading Economic Indicators

Indicator
Common stock prices
Length of the average work week
Net business formation
New building permits
Unemployment
Vendor performance
Orders for new plant and equipment
Consumer good orders
Producer price changes
Real money balances
Changes in the holding of financial assets
Changes in business inventories

Source: Department of Commerce, *Survey of Current Business,* various editions.

How reliable is the composite index of leading economic indicators in predicting business downturns? Compare the path of this index to the occurrence of recessions, represented by the shaded areas in Figure 33–2. Prior to each recession, the index of leading indicators dropped. Sometimes, however, it dropped more than a year before a recession started (as in the 1957–

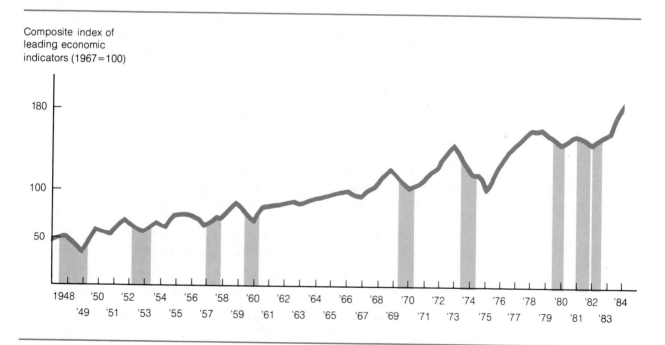

FIGURE 33–5   A Comparison of Business Cycles with the Composite Index of Leading Economic Indicators, 1948–1984

The shaded areas represent business downturns or recessions. The composite index of leading economic indicators dropped before each recession. However, the lag between the drop and the subsequent recession varied. In addition, the index sometimes forecast a recession that did not materialize.

Source: Department of Commerce, *Business Conditions Digest* (December 1984), p. 10.

1958 recession), and at other times it dropped just months prior to a recession (as in the 1953–1954 recession). Moreover, the index dropped several times during the 1960s and no recession ensued.

No predictive model of the future is perfect, but the fact that the leading indicators are widely used by economists, government officials, and businesspeople is a testimony to their usefulness as inputs to economic forecasts.

In addition to forecasting business cycles, economists have tried to explain why they occur. Business cycle theories fall into two broad groups. The first set of theories can be described as investment theories because they suggest that the causes of business cycles can be traced to variations in the level of private investment spending. Recall that Keynesians emphasize destabilizing tendencies in the economy, especially the uncertainty of profit expectations on the part of business investors.

The second group of theories examines a wide variety of causal factors, all of them external to the private economy. For instance, monetary theories emphasize the unpredictability of the money-supply process and the impact of this unpredictability on the economy. Theories of a political business cycle emphasize the effects of political institutions, such as the periodic race for the presidency, on the general economy. The remainder of this chapter will examine the various theories of the causes of business cycles.

Investment Theories

The central proposition of the various investment theories of the business cycle is that any change in the level of private investment spending will have a significant impact on the level of GNP. Investment theories include many of the ideas of Keynes and his followers, particularly the concepts of the accelerator and the investment multiplier, introduced in Chapter 24. Such theories also include several ideas of what initiates business cycles. After we review the concepts of the accelerator and the investment multiplier, we will look at a few possible causes of instability: business expectations, creative destruction of profit opportunities, and supply shocks.

The Investment Multiplier and the Accelerator Effect

Keynes showed that changes in autonomous private investment spending could create exaggerated changes in income and output. The basic reason for the phenomenon is the investment multiplier effect.

The Investment Multiplier. The investment multiplier (described in Chapter 24) is expressed as the reciprocal of the marginal propensity to save:

$$k_i = \frac{1}{1 - \text{MPC}} = \frac{1}{\text{MPS}},$$

Where k_I is the investment multiplier, MPC is the marginal propensity to consume, and MPS is the marginal propensity to save. The investment multiplier enables statisticians and economic forecasters to determine how much a given change in autonomous investment spending, I, will affect an overall change in GNP, Y:

$$\Delta Y = k_I \Delta I = \frac{1}{\text{MPS}} \Delta I.$$

What does this mean in practical terms? Suppose that the economy is initially in equilibrium at an output level of $500 billion and that business investors' expectations about future profits suddenly change for the worse and businesses reduce investment spending by $10 billion. By how much will GNP fall from its initial equilibrium level of $500 billion? By $10 billion or by more than $10 billion?

In Chapter 24 we saw that a reduction in investment spending causes an initial decline in income, which in turn reduces both consumption and saving. The reduction in consumption spending reduces income, which further reduces consumption and saving, and so on until a new, lower level of GNP is reached. The amount of the reduction in GNP depends on the values of the MPC and the MPS. As seen above, the multiplier is equal to the reciprocal of the MPS. If the MPS is one-fourth, or 0.25, the reduction in GNP will be the multiplier (1 ÷ 0.25 = 4) multiplied by the initial reduction in autonomous spending, $10 billion. The initial GNP of $500 billion, will fall to $460 billion, a new equilibrium level.

Keynes, through the investment multiplier, described how a small change in investment spending could change equilibrium output by a larger amount. It was left to two of Keynes's followers, Nobel Laureates John R. Hicks and Paul Samuelson, to show how the multiplier effect would be combined with an accelerator effect to produce business cycles.

The Accelerator Principle. The investment multiplier explains why changes in investment produce greater changes in income. Business cycle forecasters also perceive that the opposite is true: Changes in output can produce dramatic changes in investment. As the economy expands or contracts, these reciprocal forces can generate continuous fluctuations in GNP.

The accelerator principle (described in Chapter 24) relates changes in output or sales to changes in investment. The accelerator is defined as the capital-output ratio in the economy. It identifies, in any given period, the amount of investment capital required on the part of businesses to produce one unit of output. An accelerator of 3, for example, means that businesses require $3 of capital to produce $1 worth of output.

To relate the accelerator to changes in output, we will use a hypothetical example and assume that the textile industry's capital-output requirements match the requirements of the whole economy.

In the textile industry (see Table 33–3) we suppose that it takes $4 of investment capital to produce $1 worth of cloth. In other words, the accelerator for the industry is 4. The industry currently has $10 million worth of capital, which enables it to produce $2.5 million of output.

In any given year, the textile industry will devote part of its investment capital to replacing worn-out machinery. Such expenditure is called depreciation costs. Any investment above and beyond depreciation is termed net investment. Adding net investment and depreciation gives gross investment. The three right-hand columns in Table 33–3 give the breakdown between depreciation, net investment, and gross investment for the textile industry.

Industry investment over an eight-year period shows how sensitive the industry is to changes in demand for its product. In year 2, sales increase from $2.5 million to $3 million, an increase of $0.5 million. Owing to the accelerator effect, however, this increase in sales results in a steady increase in desired capital and gross investment. Capital needs grow $2 million in year 2 (0.5 × the accelerator 4). In year 3, sales increase another $0.25 million, and desired capital increases another $1 million. Note, however, that the smaller increase in sales between years 2 and 3 results in a smaller

TABLE 33–3   The Accelerator Principle in the Textile Industry

Increases in sales, brought about by a rise in autonomous investment spending and the multiplier, create additional investment spending because capital is needed to produce the additional output. As the rate of increase in sales slows up, net additions to the capital stock fall off, which means that investment spending declines, creating a recession. The recession or depression ends when the superfluous capital stock finally wears out and requires replacement.

Year	Sales (millions of dollars)	Desired Capital Stock (millions of dollars)	Replacement (depreciation) (millions of dollars)	Net Investment (new machines) (millions of dollars)	Gross Investment (millions of dollars)
1	$2.5	$10	$1	$0	$1
2	$3.0	$12	$1	$2	$3
3	$3.25	$13	$1	$1	$2
4	$3.25	$13	$1	$0	$1
5	$2.5	$10	$0	$0	$0
6	$2.5	$10	$0	$0	$0
7	$2.5	$10	$0	$0	$0
8	$2.5	$10	$1	$0	$1

gross investment. Net investment in new machines falls because of the falloff in growth of demand. By year 5, the industry has returned to its first year sales, and gross investment falls to $0. After a time, the earlier additions to capital stock are worn out and replacement expenditures are made, stimulating other industries to produce the capital and causing a repeat of the economy-wide cycle.

This single industry has undergone a representative boom and recession, spurred on by relatively small changes in the rate of growth of sales for its product. If this industry's sales experience were proportionately equal to changes in the economy's GNP, then one can understand the meaning of the accelerator principle: If output in the economy grows at a decreasing rate, investment spending will actually decrease. When this decrease, via the multiplier, works its way through the economy, output can fall even further, continuing the accelerator effect. A recessionary trend persists until investment spending and output are rejuvenated through the wearing out of capital and the desire to replace it.

When autonomous investment increases, output begins to increase. As income is generated, consumption spending and sales rise on account of the multiplier. Additional income spurs additional capital investment, and so on. Private spending changes created by the combined actions of the multiplier and the accelerator thus help create business cycles.

The mechanical relation between investment spending and general business activity described by the multiplier effect and accelerator is illustrated in Figure 33–6, which compares year-to-year changes in GNP with year-to-year changes in investment. Note the wide fluctuations in investment accompanying each upturn and downturn in GNP.

Limitations of the Accelerator. The accelerator principle itself is not a theory of the business cycle. Rather, it describes a mechanism by which changes in output demand are translated into changes in investment spending. There are several assumptions implicit in the accelerator principle that limit its usefulness as a model for explaining the business cycle. In the description of the accelerator principle we assumed that the capital-output ratio was a fixed

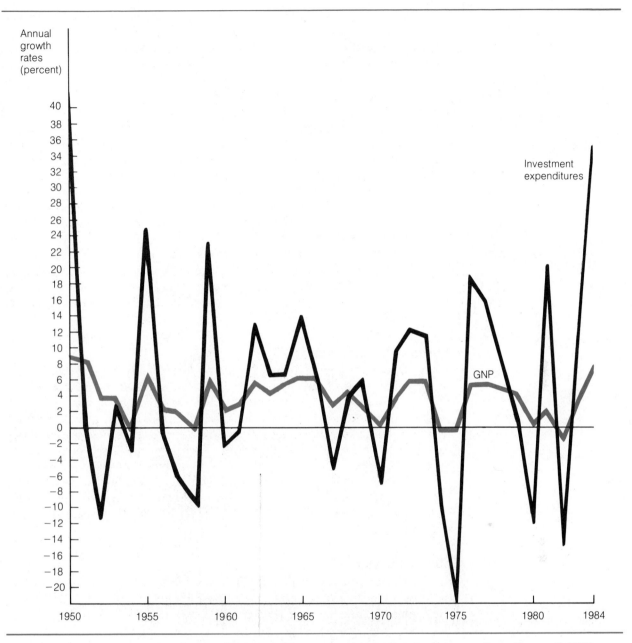

FIGURE 33–6   Annual Growth Rates in Investment Spending and GNP, 1950–1984

The change in private investment spending over time fluctuates more widely than the change in GNP. This relation between changes in investment spending and GNP is predicted by the accelerator principle.

Source: Council of Economic Advisers, *Economic Report of the President* (Washington, D.C.: U.S. Government Printing Office, 1985), pp. 232, 234.

number, such as 4, but this simplification ignores other important factors that affect investment decisions. Among these factors are interest rates, expectations about the future, and technological change. For example, new technology can alter the capital-output ratio by making labor more productive. If such innovation lowers the relative price of the produced goods, the firm will be able to sell more output with the same capital stock. Moreover, the accelerator model assumes that the firm sells all of the goods it produces. Thus, the model fails to account for inventories. In recovering from a downturn, firms can increase sales without increasing capital stock by drawing down previously accumulated inventories of finished goods. A third limitation of the accelerator model is that it assumes that firms operate at full capacity. With room and machinery to spare, firms can increase their output without raising their capital stock; such increases in output can also be accomplished by having employees work overtime or by adding extra shifts. These sorts of adjustments would obviously alter the capital-output ratio.

In addition to the above caveats about the capital-output ratio, the accelerator effect merely provides the basis for a mechanical relation between output and the level of investment in an economy. What it does not do is tell us why investment or other types of autonomous spending are so volatile. The accelerator effect may be set off by any changes in autonomous spending by consumers or the government or by exports or imports. These changes may alter the need for capital in the economy and cause fluctuations in investment spending. The following sections discuss some of the reasons why businesses decide to invest or not to invest, setting the accelerator effect into motion.

Business Expectations

Many economists have concluded that the level of investment spending in the economy depends largely on expectations of future economic conditions. Will the economy grow or stagnate? Will interest rates go up or down? Will inflation increase or decrease? All of these questions weigh heavily on business decision makers. In Chapter 31, we mentioned the concepts of adaptive expectations and rational expectations. Both concepts have been used to explain the investment behavior that leads to business cycles. Pessimistic forecasts naturally lead to less and less investment spending, while optimism about the future leads to more investment. Keynes even spoke of "animal spirits"—unexplained changes in business investors' decisions leading to large overall changes in spending.

Supply Shocks

Sudden changes in the supply of certain resources owing to natural disasters or international conflicts may also initiate business cycles. In addition to their effect on the costs of production, these so-called supply shocks may drastically alter business expectations. The OPEC supply shock in the 1970s is a case in point. The sudden cutback in the availability of oil drove up prices for gasoline and all other oil-related products and affected output in nations around the world. The OPEC crisis also shook investor confidence, contributing to the recessionary conditions that plagued many economies for several years afterward.

Creative Destruction

Another theory focuses on the role of the entrepreneur, one whose new ideas and new products offer expanding opportunities for profit. According to Joseph Schumpeter, the contributions of entrepreneurship to the economic process—the new commodity, the new technology, the new source of supply, the new type of organization—spur economic growth by "destroying" old

Creative destruction theory: A business cycle theory that emphasizes the role of the entrepreneur in supplying new ideas and products and thereby creating profit opportunities that are emulated by competitors; such creativity and its emulation are hypothesized to occur in waves, leading to a business cycle.

ways of doing things.[1] Innovators who are successful in cutting costs or in fulfilling previously unmet demands earn economic profits, and these profits attract entry by new firms that rush to emulate the discovery. In the scramble to obtain a share of the profits, however, overinvestment by new entrants occurs, and the market becomes saturated. Some of the firms go bankrupt, and a period of general retrenchment ensues, precipitating a downturn in economic activity. In microcosm, the recent upheaval in the computer industry in which a temporary boom has been followed by lagging profits and a rash of bankruptcies illustrates the sequence. The cycle finally turns up again as new ideas and new profit opportunities arise. To Schumpeter, the business cycle—a cycle of creation and destruction of profit opportunities—is a necessary accompaniment to capitalist development. This point of view has been called the **creative destruction theory.**

Monetary Theories

Monetary theories of the business cycle are based on the idea that business cycles arise from attempts by government to fine-tune the macroeconomy. In the case of monetary policy, decisions by the monetary authority (the Federal Reserve Board) made in the presence of implementation lags, in error, or in a conscious attempt to influence political outcomes destabilize the economy in one direction. A period follows during which the Federal Reserve takes necessary corrective action. The business cycle is thus thought to be generated by a stop-and-go monetary policy cycle.

To see how monetary policy might generate a business cycle, we use the concept of a Phillips curve, introduced in Chapter 32. The short- and long-run Phillips relations are depicted in Figure 33–7. In the short run there is a trade-off between inflation and unemployment, and in both parts a and b this trade-off is represented by a family of negatively sloped Phillips curves, each corresponding to a particular expected rate of inflation.

The long-run Phillips curve is a vertical line at the natural rate of unemployment, U_F. When the expected rate of inflation is equal to the actual rate, the economy will be in long-run equilibrium with unemployment equal to U_F. In other words, there is no trade-off between inflation and unemployment in the long run, and many possible inflation rates are consistent with long-run equilibrium.

Consider now the effects of an unexpected increase in the rate of monetary expansion. In Figure 33–7 the economy was initially in equilibrium at E_1 on the short-run Phillips curve $SRPC_1$ with unemployment at the natural rate and an inflation rate of P_1. If individuals form their expectations about inflation adaptivity, they will continue to expect an inflation rate of P_1 for a time. Because unexpected monetary expansion pushes the actual inflation rate higher than P_1, however, some workers will believe that their real wage has risen, while employers will begin hiring additional workers in the mistaken belief that the demand for their products has increased. As a result, individuals will accept jobs for which the real wage will eventually turn out to be too low. Unemployment nevertheless falls, and the economy moves leftward along $SRPC_1$.

As workers and employers begin to recognize the rising price level, they

[1]Joseph A. Schumpeter, *Capitalism, Socialism, and Democracy*, 3rd ed. (New York: Harper and Row, 1950), esp. pp. 81–86.

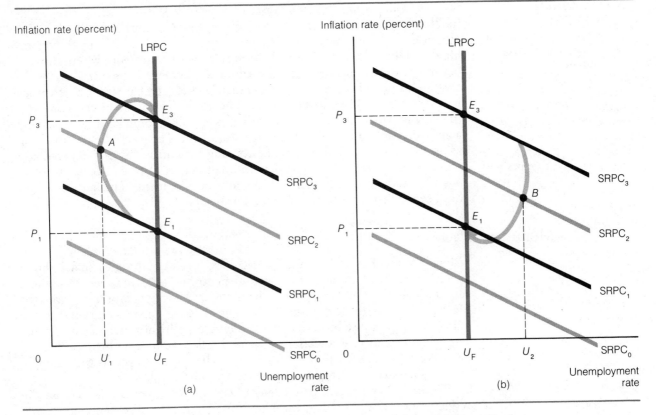

FIGURE 33–7   Monetary Policy and the Business Cycle

The economy's long-run Phillips curve is LRPC, and several short-run Phillips curves (SRPC) are also shown. (a) An unexpected increase in money supply causes the unemployment rate to fall for a while from U_F to U_1, but ultimately its impact is to increase the rate of inflation that is consistent with the natural rate of employment. (b) An unexpected decrease in the money supply causes the opposite process of adjustment. Through these processes, uneven performance by money supply authorities can lead to cyclical behavior in the economy.

revise their expectations about inflation, and as a result the short-run Phillips curve begins shifting to the right to $SRPC_2$.[2] At some point, say A on $SRPC_2$, unemployment reaches its lowest level at Y_1 but then begins to rise again as individuals' expectations about inflation begin to catch up. A new equilibrium is eventually reached at E_3 when the expected rate of inflation is equal to the higher actual rate, P_3. The unemployment rate returns to the natural rate. In the long run, the expansionary monetary policy did nothing more than raise the inflation rate from P_1 to P_3. In the short run, however, what looks like the expansionary phase of a business cycle was generated.

Suppose that the inflation rate P_3 is viewed as too high, inducing the Federal Reserve to take corrective measures. In particular, suppose that the Federal Reserve adopts a policy of reducing the rate of growth in the money supply. The adjustment process now operates in reverse. Beginning from E_3 in Figure 33–7b, the inflation rate starts to fall, but workers and employers continue to expect inflation rate P_3. Some workers will be fooled into thinking their real wage has fallen as employers begin laying off workers in response to a

---

[2]See Chapters 30 and 32 for a more detailed discussion of the adaptive expectations theory and how it relates to the labor market.

mistaken belief that demand has fallen. Individuals will refuse job offers for which the real wage will eventually turn out to have been acceptable. Unemployment rises, and the economy moves rightward along $SRPC_3$.

Downward revisions in the expected inflation rate shift the short-run Phillips curve to the left. Unemployment reaches its highest level at B on $SRPC_2$ and then falls again until a new equilibrium is attained at E_1 when inflationary expectations catch up. When that occurs, the unemployment rate is again at the natural rate, U_F, and the expected and actual inflation rate have fallen from P_3 to P_1. The short-run adjustment process looks like the contractionary phase of a business cycle. Using historical data, monetarists have argued that mistaken money supply adjustments by the Federal Reserve Board have been responsible for business cycle swings including the Great Depression, as described in Focus, "The Monetary Explanation of the Great Depression."

It is important to keep in mind that some degree of surprise is required for monetary policy changes to generate a business cycle. As discussed in Chapter 31, if individuals form their expectations about the effects of monetary policy rationally rather than adaptively, then monetary policy will have a milder impact on the levels of the real variables in the economy. Consumers, workers, savers, and investors will adjust if expectations are rational, but cycles of unemployment and income may be generated until expectations are fully adapted to new conditions.

Political Business Cycles

Political business cycle: The idea that incumbent politicians manipulate the economy, increasing employment before the election and fighting inflation after the election, to win reelection.

Some economists have suggested that there is a **political business cycle.**[3] Incumbent politicians pursue economic policies that promote their reelection, and as a consequence the economy goes through a political business cycle every four years. In an attempt to enhance their reelection prospects, incumbent politicians promote expansionary policies prior to election day—tax cuts, increased government spending, and greater money supply growth. These policies have politically popular consequences in the short run: lower unemployment and interest rates along with increased real income (output). Immediately after the election, the politicians reverse course. To limit the higher inflation rates that the preelection strategy fosters, they raise taxes and cut spending, and money supply growth is reduced. The result of these political moves is a business cycle whose length is roughly equal to the interval between elections.

Do Political Business Cycles Exist?

Three main arguments have been raised against the concept of political business cycles. First, the model is said to be naive because it assumes that voters in a presidential election care only about inflation and unemployment. The critics note, correctly, that election results turn on a variety of issues including foreign relations and the personalities of the candidates, among others. A second criticism raises the question of timing. Monetarists in particular argue that monetary policy lags are long and unpredictable. Since the repercussions of a change in the rate of monetary expansion are not felt immediately but occur over a period of perhaps two years, it may be difficult for

[3]The political business cycle idea is usually attributed to William Nordhaus, "The Political Business Cycle," *Review of Economics and Statistics* 42 (April 1975), pp. 169–190.

FOCUS   The Monetary Explanation of the Great Depression

The Great Depression, occurring over the decade of the 1930s in the United States and in other major countries of the world, was a period of record low levels of employment, industrial production, and real income.

Monetarists Milton Friedman and Anna Schwartz lay the blame for the Great Depression on a series of blunders by the Federal Reserve Board.[a] In particular, between August 1929 and March 1933, the Federal Reserve did little to prevent a 35 percent decline in the stock of money. On the contrary, the policy decisions made by the Federal Reserve Board of Governors during this period seemed to have been designed to prolong the economic crisis. For example, the board allowed the New York Federal Reserve Bank to lower the discount rate—a proper countercyclical policy—only if the district bank ceased its purchases of securities on the open market, a decision that prevented the New York Federal Reserve Bank from using its primary tool for increasing the money supply. Moreover, the Board of Governors voted in early 1933 to reduce the monetary base by $125 million, and twice in 1937 the board raised the reserve requirements it imposed on member banks. The latter action was sufficient to choke off an economic recovery that had begun in late 1934. Friedman and Schwartz cite additional evidence to support their main point that the actions of the monetary authority during the 1930s helped prolong and deepen the worst economic contraction in U.S. history.

Other, less technical, reasons might be offered for the passivity of the Federal Reserve System during the banking panics of the early 1930s. One possible reason is that the Federal Reserve Board did not understand the critical impact that bank failures might have on the money supply and economic activity. Bank failures feed on themselves by lowering confidence in all banks. The rush to acquire "safe" currency on the part of money holders will increase the currency/deposit ratio (see Chapter 29) and further reduce the ability of the entire banking system to expand loans and the money supply. The effect is to strongly discourage business activity and consumer spending. The Federal Reserve System chalked off small bank failures to "bad management," not realizing that such failures put big city banks in

[a]Milton Friedman and Anna Schwartz, *A Monetary History of the United States, 1867–1960* (Princeton: Princeton University Press, 1963).

Depositors besiege the Merchants Bank of Passaic, New Jersey, which was closed by the state of New Jersey in 1929.

jeopardy. Another reason offered for the apparent failure of the Federal Reserve System to soften the jolt of the Great Depression was internal jealousy between the Board of Governors in Washington, D.C. and the Federal Reserve Bank of New York. Until 1928 the New York Bank was the dominant political force within the System. The New York Bank strongly argued for large open-market purchases of securities to stave off bank panics (a policy which might have been effective), but to no avail. The Board of Governors, in a fight to obtain control of the monetary system, resisted the policy.

Citing such poor historical performance by the Federal Reserve, Friedman has argued that a rule governing the rate of growth of the money supply over time would be superior to the present system of discretionary monetary management. Friedman's policy proposal for a monetary rule is discussed in Chapters 31 and 32.

incumbents to chart an economic course that maximizes their political support within the confines of a four-year election cycle.

The most important question, of course, is whether voters can be systematically fooled. If individuals are in fact short-sighted, then the political business cycle hypothesis would seem to be plausible. Under the rational expectations thesis, however, no such policy cycles can be manufactured because voters will foresee the consequences of preelection macroeconomic manipulation. Individuals would quickly revise upward their expectations about inflation when confronted with a higher rate of monetary expansion, thus deflating the political gain the incumbent had hoped for.

A number of attempts have been made to test various aspects of the political business cycle theory. Most of the direct tests consist of searches for patterns in unemployment rates or other macroeconomic variables during the periods surrounding congressional or presidential elections. Ryan Amacher and William Boyes grouped U.S. presidential elections between 1952 and 1975 according to whether the incumbent party won or lost[4] and then examined the behavior of the unemployment rate for the eight quarters prior to the elections and for similar postelection periods. Amacher and Boyes found that during the periods preceding elections that the incumbent party won unemployment rates fell significantly. In contrast, there was no clear preelection unemployment rate pattern when the incumbent party lost.

Other studies have failed to confirm a link between politics and the macroeconomy. In a recent paper Bennett McCallum found no instance between 1949 and 1974 where elections explained a significant proportion of the variation in unemployment rates.[5] McCallum's technique was to first estimate the relation between the unemployment rate in one quarter of a year and the rates in each of the three preceding quarters. He then added an election variable representing one of a variety of possible cyclical patterns (for instance, unemployment falls continuously for the eight quarters prior to a presidential election, then rises for the next eight). McCallum tried six different patterns; none added to the explanatory value of his original equation. He concluded that the unemployment rate in any quarter is better explained by previous unemployment rates than by the nearness of an election.

The studies concerning the existence of a political business cycle have so far been inconclusive. There is evidence on both sides of the issue, and the possibility of a link between politics and the macroeconomy continues to be of interest to economists. In this regard, Economics in Action at the end of this chapter offers a unique perspective on the Federal Reserve's possible involvement in the political business cycle.

The Austrian Theory of the Political Business Cycle

Economists who work in the tradition of the Austrian economic theorists of the late 1800s (Menger, Böhm-Bawerk, Von Weiser) are referred to as the neo-Austrian school of economics. These modern Austrians include Friedrich A. Hayek, who won the Nobel Prize in economics in 1972.[6]

Following the lead of Hayek, the modern Austrians have presented a spe-

---

[4]Ryan C. Amacher and William J. Boyes, "Unemployment Rates and Political Outcomes: An Incentive for Manufacturing a Political Business Cycle, *Public Choice* 38 (1982), pp. 197–203.

[5]Bennet T. McCallum, "The Political Business Cycle: An Empirical Test," *Southern Economic Journal* 44 (January 1978), pp. 504–515.

[6]Friedrich A. Hayek, *Monetary Theory and the Trade Cycle* (New York: Harcourt Brace, 1932).

Austrian theory of the political business cycle: The argument that politically inspired monetary manipulation leads to costly distortions in investment decisions in the economy and ultimately to a business cycle.

cial theory of the political business cycle. They stress two basic points. First, the inspiration for a political business cycle is microeconomic, not macroeconomic. Politicians do not manipulate aggregate economic variables such as the unemployment rate; rather, they seek to influence specific markets and industries, a process that in turn has aggregate economic consequences. To take just one small example, a subsidy to the domestic steel industry benefits steelworkers directly by providing jobs. It also represents a large expenditure on the part of government. If such legislation is carried out for many industries or for many separate economic groups, the government will incur huge costs leading to a large deficit and subsequent macroeconomic problems.

The second neo-Austrian point is that real economic distortions and reductions in income result from politically inspired monetary expansions and contractions. In essence, the Austrians argue that politically inspired expansions and contractions in the money supply upset production plans by changing the nominal rate of interest. The process works in the following way: An unexpected and politically motivated expansion in the money supply creates an artificially low rate of interest. In the division of labor between production for present and future consumption, the interest rate, which reflects the forces of real saving and investment, is the price that coordinates production plans efficiently. A low interest rate leads to an overinvestment in capital goods in the economy. In effect, the lower interest rate misleads investors; it provides them with a signal that is not compatible with the real plans of consumers and savers in the economy. The overinvestment in capital goods leads to huge losses on the part of investors and then to an economy-wide reduction in income and employment, or to what we normally call a recession. The recession will not end until the plans of producers of capital goods and producers of consumer goods are coordinated again, that is, until they plan according to the same relative price and interest rate data. The cycle bottoms out when the excess capital equipment caused by the overinvestment is depreciated and the accompanying economic losses cease.

The essential problem that the Austrian theory of the business cycle addresses is the distortion in relative prices and economic efficiency that results from a political manipulation of the money supply. The aggregate economic effects of this manipulation, such as inflation, are merely incidental to its impact on relative prices.

Moreover, the Austrian theory of the political business cycle focuses on groups of voters, not on voters in the aggregate. The question is not whether voters in general will respond positively or negatively to the inflation rate; it is how certain groups in the electorate will react to specific changes in public policies.

It is safe to say that there is no single theory of the business cycle. None of the hypotheses we have reviewed in this chapter provides a completely satisfactory explanation for why there are recurrent fluctuations in the general level of economic activity. At most, economists have developed a rich statistical and factual knowledge of business cycles. It remains to find a generally acceptable theory to explain these facts.

Summary

1. A business cycle is a recurrent but not predictable fluctuation in GNP. No two business cycles are exactly alike.
2. The four phases of a business cycle are expansion,

peak, recession, and trough. The length of a cycle can be measured from peak to peak or from trough to trough. The intensity of an upturn or downturn in business activity is measured by the degree to which a peak or trough deviates from the long-run growth path of the economy.

3. Economic forecasters try to predict economic upturns and downturns. Economic indicators are an integral part of the forecaster's tools. Leading economic indicators, economic variables that tend to turn up or down before a general business upturn or downturn, are key statistics for forecasters.

4. Investment theories of the business cycle stress the instability of private investment spending as a primary cause of economic fluctuations. Among those theories, the accelerator principle plays an important role. This principle stresses that small changes in the output of the economy cause large changes in the desired capital stock and hence in net investment spending.

5. Other investment theories of the business cycle are the creative destruction theory of Joseph Schumpeter and those relating to business expectations. The expectations of businesspeople are also emphasized as an important determinant of investment spending in Keynesian theories of the business cycle.

6. The monetary theory of the business cycle sees the uneven and unpredictable management of the money supply by the Federal Reserve Board as the basic cause of economic fluctuations. Concern over Federal Reserve behavior has led some monetarists, such as Milton Friedman, to propose that discretionary authority over the money supply of the Federal Reserve be replaced with a rule to govern the growth of the money supply.

7. Some economists have identified the possibility that business cycles are politically inspired. Given short-sighted voters, incumbent politicians win reelection by inflating the economy prior to an election, driving down unemployment, and then allowing unemployment to rise and inflation to recede after the election. Each four years, the cycle is repeated, and so we say that there is a political business cycle. The evidence for this type of manipulation of the economy by politicans is inconclusive.

8. The Austrian theory of the political business cycle is that politicians intervene in specific markets and industries to affect specific groups of voters. Relative prices and interest rates in the economy are thus distorted, leading to an overinvestment in capital goods industries. This overinvestment must subsequently be wrenched out of the economy. These cyclical effects are caused by political manipulation of the money supply.

9. Economics has developed a rich factual understanding of business cycles, but no single plausible theory of the cycle has yet emerged.

Key Terms

business cycles	econometrician	political business cycle
recession	economic indicators	Austrian theory of the political business
depression	creative destruction theory	cycle

Questions for Review and Discussion

1. Why do we say that business cycles are recurrent but not predictable?
2. To which theory of the business cycle would you attribute cycles caused by the following events:
   a. The Federal Reserve Board increases the money supply dramatically.
   b. Sales in the garmet industry fall.
   c. The French suddenly change governments and close their doors to American goods.
   d. Major new oil reserves are discovered in Alaska.
3. Why does the accelerator principle imply that investment spending will be unstable? How do excess capacity and inventories affect the operation of the accelerator principle?
4. How can the Federal Reserve Board cause a business cycle?
5. In a modern economy, information about the motives and activities of politicians is relatively easy to obtain, much easier than it was 100 years ago. Does the availability of information mean that a political business cycle is presently more or less likely?
6. Do changes in leading economic indicators actually cause the changes in GNP?
7. What would macroeconomic policymakers need to know in order to eliminate the business cycle?

ECONOMICS IN ACTION

Can Economics Help You Predict the Next Presidential Election?

Political pollsters, TV commentators, special-interest groups, and especially political candidates would like to know what it takes to win a presidential election. Many believe that presidential elections are won through political organization, TV advertising, good luck, and charisma. Others are more likely to view economic variables such as personal income and inflation as critical variables. Some have tried to determine which economic variables are the most reliable indicators of success or failure for an incumbent candidate.

Ray Fair found that the only issue of real concern to voters in a presidential election is the real growth rate in per capita GNP.[a] Incumbents who can boast of strong growth are usually kept in office. Those who cannot are voted out. In Fair's view, no other single

[a]Ray C. Fair, "On Controlling the Economy to Win Elections," Cowles Foundation Discussion Paper No. 397, 1975.

issue correlates strongly with the outcome of an election.

Another predictor of election outcomes may be the Federal Reserve's handling of the money supply, M-1, before the election. Although the Federal Reserve was designed as a semi-autonomous branch of government, some economists and political observers believe that the Federal Reserve behaves politically to protect its own interests. David Meiselman has presented evidence that monetary expansions and contractions are possibly politically inspired. Figure 33–8 shows the rate of growth in M-1 prior to the last seven presidential elections. The vertical axis in each graph shows the six-month rate of change in M-1. The rates shown are annual percentages.

In each case, except in 1960, when Vice-President Richard Nixon lost to John Kennedy, and in 1976, when incumbent Gerald Ford lost to Jimmy Carter, the

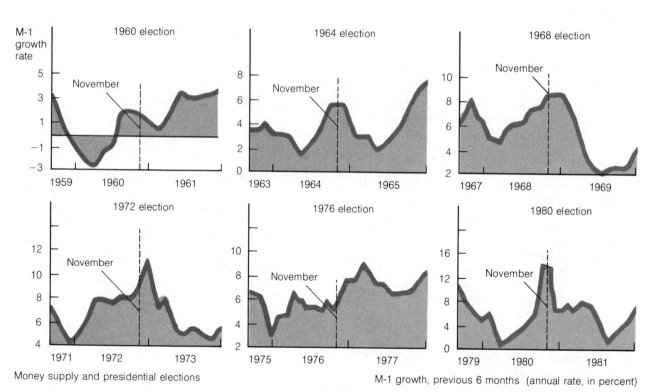

Money supply and presidential elections

M-1 growth, previous 6 months (annual rate, in percent)

FIGURE 33–8   Money Supply and Presidential Elections

These graphs show the relation of the growth rate of the money supply, M-1, to presidential elections between 1960 and 1980. Notice that in most cases, the money supply increased sharply before the election. This finding is evidence that there may be a link between monetary policy and elections.

Source: David I. Meiselman, "The Political Monetary Cycle," *Wall Street Journal* (January 10, 1984) p. 32.

money supply was expanded prior to the election and contracted soon after.[b] Notice too that the scale on the vertical axis increases over time. Monetary growth was 2 percent when John Kennedy was elected and 15 percent when Jimmy Carter was defeated in 1980.

Tracking economic variables such as the growth rate in per capita GNP and the rate of M-1 growth may not always predict the outcome in an election, but it provides interesting evidence of how closely our economic health is related to the four-year cycle of presidential elections.

Question

How might the unemployment rate be a prediction of election outcomes? Would it likely be more important in congressional contests or in presidential elections?

[b]David I. Meiselman, "The Political Monetary Cycle," *Wall Street Journal* (January 10, 1984).

34

Economic Growth and Productivity

A full-employment economy without inflation is the major short-run goal of macroeconomic policy. But achievement of this goal does not necessarily mean that an economy is growing over the long run. Full employment and price stability affect long-run economic growth prospects, but other factors such as population, natural resources, capital growth, and improvements in technology ultimately determine the economy's future.

Long-run growth has been the key to the United States's prosperity. The **standard of living,** usually defined as the quantity of real goods and services that the average citizen is able to consume per year, has risen by an average of 3 to 4 percent per year for more than 150 years in America. Other nations have not experienced such good fortunes. Many Third World nations in Asia, Africa, and Latin America, for example, are stymied in their growth prospects because of a lack of productive labor or capital equipment or because their population is growing as fast as or faster than their output of real goods and services.

Unfortunately, there is some evidence of a slowdown in long-run growth rates in the United States. Is this slowdown temporary or is it a permanent state of affairs owing to the fixed or dwindling supplies of some of our natural resources? Will future generations of Americans suffer the effects of a shrinking economy, or are there economic policies that might ensure steady prosperity? Before we can begin to address these important questions, we need to define what we mean by economic growth and examine its primary causes.

Standard of living: A measure of the value of real goods and services consumed by the average person.

The Meaning of Economic Growth

Economic growth: Increases in real GNP or real GNP per person through time.

Economic growth actually has two definitions. It may simply refer to increases in real GNP, the economy's annual output of goods or services adjusted for inflation. While this measure certainly tells us whether the overall standard of living has improved, the real GNP growth rate tells us nothing about whether the average citizen is better or worse off than at some previous time.

Real GNP per capita: Output per person; real GNP divided by the population.

Increase in **real GNP per capita** is a more accurate measure of economic growth. If real GNP increases by 4 percent over some period and population by 2 percent, real economic growth per capita has taken place. By contrast, a population increase of 6 percent with 4 percent growth in real output means that the average citizen is worse off, that economic decline has occurred in per capita terms.

Both measures are useful for understanding growth processes. For example, a statement that Mexico's real GNP grew at a rate of 3 percent last year tells us that some of the prerequisites for growth exist in Mexico. Likewise, to say that per capita GNP grew at only 1½ percent last year indicates that Mexico may be having population problems since per capita income growth depends on population growth. Both measures, however, are merely statistical concepts; economists generally recognize that neither measure, however useful it may be for comparison purposes, can tell us anything about the quality of life or the distribution of wealth in a society.

The Causes of Growth

The causes of economic growth are complex. We have already pinpointed some of the important short-run requirements of economic growth: Adequate employment of productive resources, price stability, efficient allocation of resources, and other such conditions generally ensure a rising standard of living.

To sustain long-run growth in real GNP, however, an economy must enjoy steady growth in population, natural resources, productivity of labor and resources, capital formation, human capital, and technological improvement, encouraged by a free-market environment.

Population. Population may be detrimental to growth in real per capita GNP in many modern nations, but under certain conditions, an expanding population may be essential to growth. During the industrialization of the United Sates in the nineteenth century, rapid population growth, much of it achieved by immigration, was essential to the phenomenal economic growth rate the United States experienced over that time. On the supply side, growth in population and labor supply led to increased efficiency in production through higher productivity in the use of machinery and natural resources. Mass production such as assembly-line techniques greatly improved the productivity of labor through increased specialization. On the demand side, higher population fanned increased demand for all goods and services, making mass production and greater consumption possible for the United States. Population growth is therefore an ingredient in growth, but increases in population by themselves are insufficient for sustained economic growth. As we will see later in this chapter, rampant or unrestrained population increases may retard, halt, or even reverse any progress in per capita GNP.

Natural Resources. Every society is endowed with some quantity of land and natural resources such as freshwater, forests, and minerals. The United States, the Soviet Union, and many less-developed countries have large quantities of both. Possession of land and natural resources may make growth easier, but it will not guarantee growth; nor will lack of natural resources prevent growth. Japan has a small quantity of land and natural resources, yet Japan's post-World War II growth rate has been greater than that of most other countries.

By themselves, natural resources are valueless. They must be developed by labor and capital. Improvements in technology, moreover, may make resources more productive—as in fertilizing land—or even contribute to the development and invention of new resources. Aluminum, for example, is an invented alloy. It is a common misconception that the quantity of resources such as oil is absolutely fixed or, in economic terms, absolutely scarce in supply. Technological change and growth in the capital stock continuously act on the productivity and even on the existence of resources, altering their relative scarcities.

Productivity of labor:
Output per worker.

Productivity of Labor. Economic growth is very closely tied to the **productivity of labor**—how much output per hour, week, and so on, results from labor input. In fact, output per worker per time is a direct measure of the productivity of labor. Labor's productivity depends on a number of factors, including the quantity (population size) and quality (degree of education and skill) of the labor supply, the stock of capital and other resources each laborer has to work with, and the technology available for production. Educational development and on-the-job training of laborers—growth in human capital—is a rather obvious requirement for productivity increases, but productivity increases are also closely linked to capital formation and technology.

Capital Formation. A fourth ingredient of growth is growth in the size and quality of the capital stock—the quantity of fixed assets consisting of buildings, machinery, inventories, and equipment used in production. Labor productivity—output per worker—is enhanced when the stock of capital available to labor grows faster than the labor supply. When capital is growing faster than the labor supply, capital is said to be deepening. **Capital deepening** has occurred throughout this century in the United States as tools, machinery, and high-technology equipment have grown at consistently faster rates than the labor supply.

Capital deepening: An increase in the total stock of capital relative to the amounts of other resources supplied.

Increases in capital stock are costly, however. Resources, including time, must be sacrificed to produce capital goods that are not directly consumable, so capital growth requires saving or abstention from current consumption. For new investments in machinery, buildings, and equipment to grow, society's ability to save must grow. The reward for saving is greater future growth in either consumable goods or in more capital stock. In the latter sense, economic growth is cumulative: It feeds on itself by using capital to produce more capital.

Improvements in Technology. Technological growth and invention—improvements in the methods by which goods and services are produced and sold—is another key to economic growth. The Industrial Revolution of the seventeenth, eighteenth, and nineteenth centuries in Western countries ushered in the age of mechanization, increasing the productivity of labor and natural resources. The result has been the highest standards of living for the greatest numbers of people in the recorded history of humanity. Such living standards would be unthinkable without the invention of items such as the steam engine, the spinning jenny, assembly-line production, and indoor plumbing. Who could have imagined twenty or thirty years ago, for example, that a high-speed digital computer no larger than a breadbox would be within the reach of the average American to calculate income taxes, store recipes, or play games on? Technology and invention have been responsible for nothing less than the modern world.

The Market Environment. Most Western economists believe that the market environment surrounding the growth in natural resources, labor productivity, capital formation, and so on, is a large contributing factor to economic growth. A competitive market system, many believe, encourages invention and rapid innovation. All private and public restraints on competition, such as excessive or unnecessary regulation or tariffs or quotas on goods exchanged in international trade, will tend to reduce the rate of growth in real per capita GNP.

The Production Possibilities Curve Revisited

All the fundamental elements of long-run economic growth may be summarized in terms of a production possibilities frontier. A production possibilities frontier, introduced in Chapter 2, shows the choices available to society in producing two goods given a fixed quantity of resources—labor, capital, and natural resources—and assuming a given state of technology. If society is positioned on the production frontier, it is fully employing and efficiently allocating all resources and utilizing technology to its best advantage. However, most societies, even highly efficient ones, are at times positioned inside the frontier because of unemployment or underemployment of resources. Societies cannot advance beyond the frontier because of absolute limits at any given time in the quantity and quality of resources.

Figures 34–1a and 34–1b show production possibilities frontiers for two economic societies, country X and country Y, respectively, both of which must choose production levels for consumption goods and capital goods. Each country is assumed to possess a similar resource base and the same technology. Country X's initial choice between capital goods and consumer goods is marked A in part a; country Y's choice in part b is marked A_1. Country X has devoted a higher proportion of its resources to capital goods production in year 1 than has country Y. The net result—and the key to understanding the principles underlying economic growth—is that country X's choice results in a higher production possibilities frontier in some future year, say year 2. In year 2, societies X and Y will again choose some capital-consumer goods combination on a new frontier, but if both countries remain fully employed, country X's economy will have grown to a higher level of output and presumably to a higher level of real GNP per capita.

Increases in economic growth rates, in other words, are determined by society's choice of future goods over present goods. This fact does not mean that the capital stock is the only element in growth. As we noted above, capital combines with labor and natural resources to increase society's growth possibilities, even with a constant state of technology.

Improvements in technology will also shift the production possibilities frontier out from a given resource base or from a resource base growing at a constant rate. Figures 34–2a and 34–2b show the effects of several forms of technological change in computer hardware and food production. If society, for simplification, consumes only these two goods, a neutral improvement in technology—one that affects both goods—would increase potential production of the two goods equally, as in Figure 34–2a. A nonneutral change—one affecting either computer technology or food production—would in-

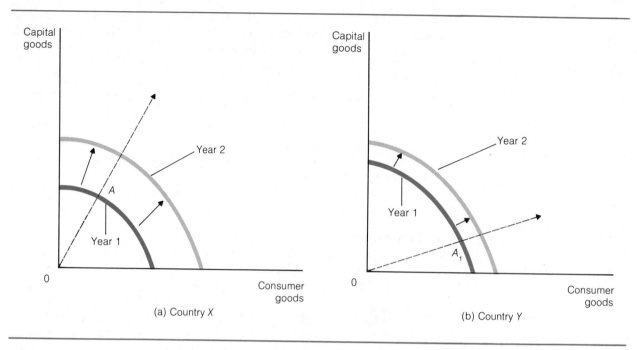

FIGURE 34–1   Economic Growth and the Production Possibilities Frontier
Growth is assumed to result from increases in the stock of capital goods. (a) If an economy forgoes a relatively large amount of consumption goods to obtain more capital goods, then the production possibilities frontier will shift outward by a larger amount than the economy that does not make the sacrifice. (b) Country Y's preference for consumer goods in year 1 results in significantly less growth by year 2.

crease either the potential to produce computers or the potential to produce food. Whether society takes advantage of the increased potential, of course, depends on whether resources are fully employed and efficiently allocated.

Different forms of resource growth would have similar effects on the production possibilities frontier. An equal growth rate in all resources, or in all nonspecialized resources, would shift the curve out as in Figure 34–2a, whereas faster growth rates in resources specialized in computer hardware or in food production would enhance the production possibilities of either computers or food, as in Figure 34–2b.

The Principle of Diminishing Marginal Returns

As always, the law of diminishing marginal returns (or of increasing marginal cost) applies. The law states that as more and more additional units of any variable resource are combined with some fixed amount of another resource or resources, successive applications of the variable input will eventually increase total output at a smaller rate. The law may be understood from your own experience: Your second or third hour of study for an economics quiz is ordinarily more productive than your eleventh or twelfth hour of studying. Why? You study the toughest or most important material first, that most likely to get you a higher grade. Your brain may also be more receptive and able to better absorb new material earlier on. As detailed in Chapter 8, the law of diminishing marginal returns explains the shape of short-run production and cost curves to any individual firm producing any sort of output or service. The law also explains some macroeconomic aspects of growth.

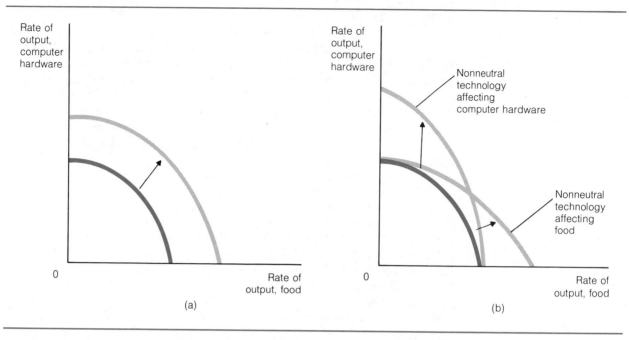

FIGURES 34–2   Technological Change and the Production Possibilities Frontier

The hypothetical economy is based on just two products—computer hardware and food. An improvement in technology shifts the production possibilities frontier to the right. (a) A neutral change in technology—one that affects both products equally—causes an increase in output of both products. (b) A nonneutral change—one that affects either one product or the other but not both equally—increases the output of one but not the other.

The law of diminishing marginal returns clearly explains the typical shape of any given production possibilities curve. As we know from experience, resources, talents, and training are not all alike and are not perfectly adaptable to alternative uses. Given any state of technology and level of knowledge, one good cannot be transformed with equal facility into another good. As we learned in Chapter 2, orange groves cannot be converted into peanut farms at constant opportunity costs. The same applies to any two goods or classes of goods: The most adaptable resources are transformed first. As less adaptable resources are transformed, total additions to output get smaller and smaller. This fact explains the concave shape of the typical production possibilities curve.

The law of diminishing returns also applies to growth in the following way: In the event that any factor affecting growth is fixed or grows at slower rates than the others, output possibilities will increase, but at slower and slower rates. Successive incremental applications of expanding resources to fixed production factors will result in smaller and smaller additions to output. If labor supply is fixed, for example, progressive increases in all other resources, excluding technology, will increase output potential, but at slower and slower rates. Technological change may enhance possible increases in output by "economizing" on the necessity to use labor. The gain would only be temporary, however. Given any state of technology, some point of diminishing marginal returns would be reached.

Comparative Growth Rates

Countries' economies are growing at different rates, for various reasons. Diminishing marginal returns when one or more resources is fixed or growing slowly is a major feature of theories featuring limits to growth. For instance, bottlenecks may be predicted when all arable land and freshwater resources in a country are already being used intensively for food production. Regardless of whether there are absolute or relative limits to long-run economic growth, it is clear that growth is an extremely complex matter. Further, it is clearly the case that even small differences in growth rates matter a great deal over time.

Table 34-1 presents average growth rates in real GNP between 1975 and 1980 for eight industrialized nations. (Real GNP is converted to U.S. dollars for January 1983.) The real growth rate varies from 5.1 percent per year for Japan to −1.2 percent for Sweden. These growth rates are applied to a hypothetical $500 billion in GNP in 1980 and projected for the years 1985, 1990, 1995, and 2000. If these average growth rates in real GNP held through the year 2000, the relative differences in the economic well-being of countries would be dramatic. In 1990 Japan would enjoy twice the real income of Sweden, and by the year 2000 Japan would produce twice the real output of the United Kingdom and four times the real output of Sweden. Relatively small differences in the growth rate will produce huge differences in real income. A growth rate of only about 3½ percent will double real income in only twenty years.

A warning about the determinants of growth in different nations: Many cultural and institutional factors heavily influence actual growth rates. Significantly different growth rates even in highly developed nations may ultimately be explained only in terms of a particular society's cultural, government, and institutional structures, areas where the economist may not be able to analyze key differences. Why, for example, has the Japanese economy grown recently at a faster rate than that of most other industrialized nations

TABLE 34-1   Cumulative Effects of GNP Growth

Different growth rates applied to the same hypothetical GNP in 1980 will produce dramatic changes in real GNP in the future. GNP will double in twenty years, with an approximate growth rate of 3.5 percent per year.

Country	Actual Growth Rate of Real GNP, 1975–1980 (annual percent change)	Hypothetical Real GNP, 1980 (billions of dollars)	Projected Real GNP, 1985 (billions of dollars)	Projected Real GNP, 1990 (billions of dollars)	Projected Real GNP, 1995 (billions of dollars)	Projected Real GNP, 2000 (billions of dollars)
Japan	5.1	500	641	822	1054	1352
Italy	3.9	500	605	733	888	1075
United States	3.7	500	600	719	862	1034
West Germany	3.6	500	597	712	850	1014
France	3.2	500	585	685	802	939
Canada	2.9	500	577	665	768	886
United Kingdom	1.6	500	541	586	634	687
Sweden[a]	−1.2	500	471	443	417	393

[a]Swedish growth rate reported for 1978–1983.

Sources: U.S. Department of State, *Economic Growth of OCED Countries 1970–1980*; U.S. Bureau of the Census, *Statistical Abstract of the United States*, 1981, p. 879; Federal Reserve Bank of St. Louis, *International Economic Conditions* (April 1984).

of the world? The economist can focus only on measurable quantities such as output per worker per time, the rate of savings, and population growth. Other causes, such as the Japanese work ethic, a complex worker-employee social relationship, property rights, and government factors, all lie behind the measurable quantities.

To understand the economic development of any nation, we must look to cultural and institutional factors as well as the quantifiable features of growth.

The Classical Dynamics of Growth

More than any other economists before or since, the classical writers, in the period 1776–1848, were interested in why economies grow, stagnate, or decline. Of the classical writers, David Ricardo (1772–1823) and Thomas Robert Malthus (1766–1834) were notable for their predictions about growth. Both Ricardo and Malthus predicted that society would reach some stationary state because of diminishing marginal returns in agriculture and severe population pressures.

Diminishing Marginal Returns and the Malthusian Principle

Malthus and Ricardo wrote about a hypothetical purely agrarian economy with fixed quantities of land and natural resources and only one output, food. They also assumed that technology was fixed. They reasoned that as more and more capital and labor were applied to land of any given quality, marginal output of food would decline, for two reasons: (1) diminishing marginal returns—as more capital and labor were applied to any given quantity of land, marginal output would decline; and (2) scarcity of fertile land—as capital and labor were applied to agriculture, lower and lower grades of land would be used, yielding smaller quantities of food per acre.

Malthusian population principle: Population has the capability of increasing geometrically while food production can increase only arithmetically; the population can grow faster than the economy's ability to feed it.

At the same time that the marginal productivity of land was falling, society's demand for more and more food to feed a growing population would be rising. The **Malthusian population principle,** developed by Malthus in 1798 in his book *On Population*, was that population increased geometrically over successive generations (1, 2, 4, 8, 16, 32, and so on), while the food supply was capable of increasing only at an arithmetic rate each generation (2, 4, 6, 8, and so on).

Related to the population principle was another Malthusian idea—that real wages are always tending toward the subsistence level. A **subsistence wage** was defined by the classical economists as a wage sufficient only to reproduce oneself (to bring two children to adulthood). A higher-than-subsistence wage would cause larger family sizes over a generation, reducing the higher real wage to the subsistence level again.

Subsistence wage: A wage that is just enough to sustain the current population.

The population-subsistence wage mechanism is presented in simple terms in Figure 34–3, which shows labor supply and demand curves in initial equilibrium at E_0. We assume that the population, as well as the input of labor, is at level N_0 and receives a subsistence wage W_S. If the demand for labor increases from D_L to D_{L_1}, a new equilibrium is established along the labor supply curve S_L at E_1, and a new wage rate W_0 is established that is higher than subsistence. This shift occurs slowly—it may take a generation or more to reach the higher wage rate. If, as Malthus assumed, individuals react to a higher-than-subsistence wage rate by increasing family size and population to N_1, the supply curve of labor will shift rightward, bringing

FIGURE 34–3

The Classical Population–Subsistence Wage Mechanism

Malthus's analysis starts from equilibrium E_0, where the quantity of labor N_0 calls for the subsistence wage W_S. An increase in the demand for labor increases the wage to W_0, a level above the subsistence wage, and equilibrium is established at E_1. This surplus for workers, however, is eventually squandered through increased population. As the population increases from N_0 to N_1, the labor supply increases, and in the long run the wage returns to the subsistence level, at E_2.

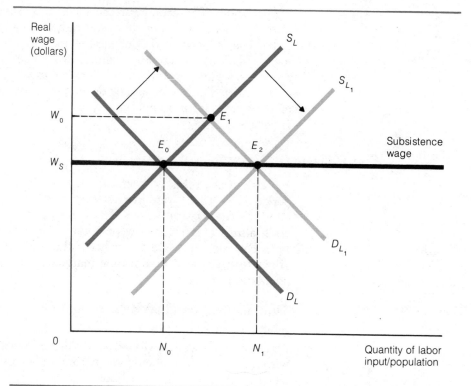

society back to the subsistence wage. If the increased economic well-being of workers $(W_0 - W_S)$ is squandered in increased population, there is little hope for the progress of society. According to Malthus's theory, the subsistence wage will always tend to prevail over the long run. If unrestrained, population will grow from N_0 to N_1, returning the per capita income of workers to subsistence.

Malthus also considered a worse possibility. If population and labor supply overreacted to the high wage rate W_1 and increased beyond S_L, wages would temporarily fall below the subsistence wage. Starvation would result for some, and Malthus predicted all sorts of dire consequences, including famine, plague, wars, and urban crowding.

Deficiencies in Classical Growth Theory

While this simple classical theory is a marvelous simplification of the growth process, it suffers from several oversimplifications. The first is the Malthusian population theory. Does the population mechanism operate as Malthus described it? Do individuals squander increases in real income by increasing population growth rates? Evidence since the days of Malthus does not give a great deal of support to his idea, especially in the industrialized countries of the United States and Western Europe. Massive growth in real GNP per capita has been made possible largely because increases in real income did not create an urge to populate. As Table 34–2 suggests, growth rates in the U.S. population have remained at relatively low levels since 1940 and have actually declined since the early 1960s. This does not mean that some developing nations have not had population problems. Indeed, Malthus's

TABLE 34–2   Population Growth in the United States, 1961–1983

The U.S. growth rate has shown a dramatic decline since the baby boom of the 1950s, when population growth averaged between 1.7 and 1.8 percent per year.

Year	Population (millions)	Percent Change	Year	Population (millions)	Percent Change
1961	183.7	1.7	1973	211.9	1.0
1962	186.6	1.5	1974	213.9	0.9
1963	189.2	1.4	1975	216.0	1.0
1964	191.9	1.4	1976	218.0	1.0
1965	194.3	1.3	1977	220.2	1.0
1966	196.6	1.2	1978	222.6	1.1
1967	198.7	1.1	1979	225.1	1.1
1968	200.7	1.0	1980	227.7	1.2
1969	202.7	1.0	1981	229.8	0.9
1970	205.1	1.2	1982	232.1	1.0
1971	207.7	1.3	1983	234.2	0.9
1972	209.9	1.1			

Source: U.S. Bureau of the Census, *Statistical Abstract of the United States*, 1984, p. 6.

model may be more applicable to countries with rapidly expanding populations, such as India or Ethiopia.

A second major deficiency in the classical theory of economic growth is the classical theorists' mistaken belief that technology, especially related to food production, was a relatively constant factor in growth. In the early nineteenth century, one farm worker could feed only three or four nonfarm workers. Because of technological change, a contemporary U.S. farmer can feed seventy, eighty, or more nonfarm people. This dramatic increase in productivity has been made possible by the growth in capital—in farming techniques—that assists farm labor. The tremendous increase in technology and invention in the nineteenth and twentieth centuries has been the mainspring of real per capita GNP growth, and it has accrued on an unprecedented scale. The steam engine in transportation, highly sophisticated mass production processes, modern electronics, and a host of other technological breakthroughs have all helped forestall what Malthus perceived as an inevitable decline in growth. (The arguments of the "no-growthers," an opposing school of thought, are presented in Economics in Action at the end of this chapter.)

Growth and Modern Macroeconomic Theory

Modern economic theory has much to say on the issue of growth. Foremost in current analysis is the importance of capital formation and improvements in technology. According to modern analysis, growth potential also depends on the ability of an economy to absorb growth.

The Demand Side of Growth

To **absorb growth** means to keep up with the expanding productive capacity of the economy through increasing total expenditures. Recall that aggregate demand must always be sufficient to create full employment at any point in

Growth in Total Expenditures Required for Economic Growth

In this simple model, increases in total expenditures must keep up with expanding resources and population to bring about maximum economic growth. If the economy grows from an output capacity of Y_0 in 1970 to Y_1 in 1980, the aggregate demand curve must shift from TE_0 to TE_1 by 1980 to keep resources fully employed. Total output equals total expenditures along the 45-degree line.

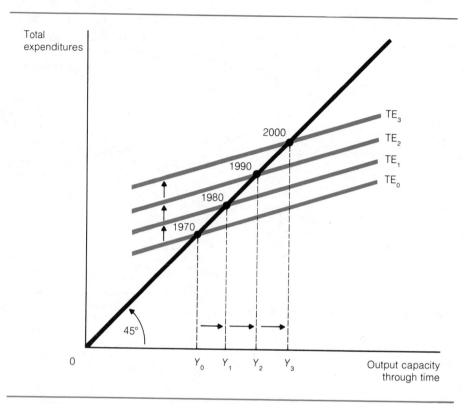

Absorption of growth: An increase in total spending that is great enough to purchase the extra output realized by an increase in productive capacity.

time. The same must hold true to create full employment and maximum economic growth over time.

Figure 34–4 shows total expenditures on the vertical axis and the output capacity of the economy over time on the horizontal axis. Along the 45-degree line, total expenditures equal the total output capacity of the economy. A central requirement for sustainable economic growth is that total expenditures, private and public, must keep the economy on or near this 45-degree line through time; otherwise, unemployment and slower growth will result. If output capacity is Y_0 in 1970, total expenditures must be at the level of TE_0 to achieve maximum income and income growth. A given level of net investment and capital formation in 1970 creates a larger output capacity in 1980, Y_1, which in turn requires a rise in total expenditures in 1980 to realize the growth potential of the 1980 economy.

Ensuring adequate aggregate demand for maximum growth is one of the goals of short-run macroeconomic theory. As we saw in Chapter 31, modern Keynesians argue that the economy should be managed in discretionary countercyclical fashion to keep aggregate demand at full-employment levels. Monetarists and rational expectationists argue that adequate growth in aggregate demand can be achieved only through established rules for Congress and for the Federal Reserve System. But however one looks at the matter, policies to ensure adequate aggregate demand are essential in maintaining a sustained high rate of economic growth.

The Supply Side of Growth

The other facet of economic growth involves the factors that enhance or enlarge the productive capacities of the economy. All the factors—capital growth, technological improvements, growth in natural resources—discussed

FIGURE 34-5

Economic Growth and Rightward Shifts in Long-Run Aggregate Supply

To meet an increase in aggregate demand, the level of technology or the supply of resources may increase, shifting the long-run aggregate supply curve to the right and increasing full-employment income.

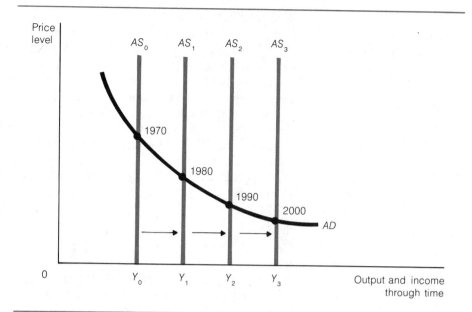

earlier in this chapter create growth in supply. We call these supply-side factors because they have the effect of shifting the aggregate supply curve of real output to the right.

In Figure 34–5 we assume that the economy's demand for goods and services will be sufficient to absorb the increased output of goods and services made possible by growth. The question we ask instead is, What factor or factors underlie rightward shifts in the aggregate supply curve? As we have seen in a number of previous chapters (Chapter 26 in particular), the long-run supply of goods and services is influenced by variables—capital accumulation, the productivity of labor, taxation, and so on. For purposes of discussion, however, there are two major economic factors to watch in predicting the growth rate in the productive capacities of any economy:

1. Growth in private saving and investment, which leads to capital formation; and
2. Growth in the size of the labor supply, in laborers' participation in the work force, and in labor's productivity.

Any variable, policy, or institutional factor that affects these two rates will affect economic growth through its effect on the long-run aggregate supply curve. Consider some recent trends in the United States of these components of economic growth.

Recent U.S. Economic Growth and Productivity

The various growth rates in real GNP and per capita GNP in industrialized nations reported in Table 34–1 have numerous causes. Short-run economic stabilization—the ability of an economy to provide full employment of resources—will naturally affect long-run growth prospects. But the supply-side features of economic growth are of greatest importance in explaining U.S. growth trends.

Capital Formation

Capital formation: An increase in the stock of capital.

Consider recent investment and **capital formation**—increases in the stock of capital—in the United States. Figure 34–6 reports two useful measurements of capital formation between 1950 and 1982: investment as a percentage of nominal GNP and the percent change in capital expenditures per worker. Both of these measures show a large decline since the late 1960s. Net investment expenditures on capital above and beyond replacements—the source of capital formation—was 4 percent of GNP in 1967 but fell to about 3 percent by 1982. The effect of this decline can be seen in the second statistic: the rate of growth in capital per worker, which has also declined. Labor's productivity is closely related to the amount of capital that it has to work with.

Savings

Domestic savings are the principal source of new investment and capital formation in the United States. Table 34–3 shows net domestic savings as a percentage of GNP for periods between 1951 and 1981 from federal, state, local, and private sources. Negative figures represent dissaving created by budget deficits. On average between 1951 and 1981, private citizens, corporations, and government saved 6.7 percent per year more than was necessary to replace the loss of capital stock caused by depreciation. In real terms—after adjustments are made for inflation—private savings have declined since the 1966–1970 period to 4.4 percent of GNP in the 1976–1980 period and

FIGURE 34–6

Two Measures of Capital Formation for the United States, 1950–1982

Sharp declines in investment as a percentage of GNP and in the amount of capital per worker since 1967 explain some of the reduced real growth rate in GNP over the 1970s and early 1980s.

Source: Council of Economic Advisers, *Economic Report of the President* (Washington, D.C.: U.S. Government Printing Office, 1983), p. 80.

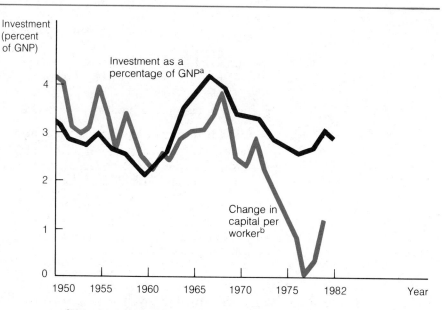

aNet private nonresidential fixed investment as a percentage of GNP, five-year centered moving averages.

bPercent change in real net private nonresidential fixed capital stock per worker in the business sector, five-year centered moving averages.

TABLE 34–3   Net Domestic Savings as a Percentage of GNP, 1951–1981

Domestic savings as a percentage of GNP declined between 1971 and 1981.

Period	Savings as a Percentage of GNP			
	Total	Federal	State and Local	Private (Individual and Corporate)
1951–1955	6.7	0.9	−0.1	5.9
1956–1960	6.9	1.1	−0.1	5.9
1961–1965	7.4	0.2	0.2	7.0
1966–1970	7.5	0.6	0.4	6.5
1971–1975	6.4	−0.3	1.1	5.6
1976–1980	5.8	−0.2	1.6	4.4
1951–1981	6.7	0.4	0.6	5.8

Source: Economic Report of the President (Washington, D.C.: U.S. Government Printing Office, 1983).

to only 3.6 percent in 1981, although real savings rates rose significantly in 1983 and 1984.

To finance their expenditures, governments must compete with private investment for private savings in the markets for funds. When government expenditures require massive borrowing, this drain effectively reduces the funds available for capital formation. Focus, "Changing the Way Americans Save," provides more details about this aspect of saving and capital formation.

Productivity

The declines in net investment and in savings have combined to reduce U.S. productivity in recent years (see Table 34–4). The percentage change in output in the nonfarm business sector of the economy has risen only by an average 1.2 percent between 1978 and 1982. Output actually declined between 1980 and 1982. Growth in output per hour over the five-year period averaged *zero!*

Evidence of a lagging U.S. growth rate relative to other industrialized nations is found by comparing the U.S. figures to investment and productiv-

TABLE 34–4   Output and Productivity Growth Rates in the Nonfarm Business Sector, 1978–1984

Lagging U.S. growth is emphasized by sharply declining growth rates in output and in output per worker over the period 1978–1982. Growth rates in both measures turned up sharply in 1983 and 1984 causing the averages between 1978 and 1984 to more than double the averages between 1978 and 1982.

	Growth Rates (percent)							1978–82 (average)	1978–84 (average)
	1978	1979	1980	1981	1982	1983	1984		
Output per hour	0.6	−1.5	−0.7	1.5	0.2	3.5	3.1	0.02	0.95
Output	5.7	2.2	−1.4	2.1	−2.6	5.0	8.5	1.20	2.78

Source: Council of Economic Advisers, *Economic Report of the President* (Washington, D.C.: U.S. Government Printing Office, 1985), p. 279.

FOCUS   Changing the Way Americans Save

The recent rate of growth in U.S. real per capita income can be attributed to many things, but clearly the single most important cause is the lack of capital stock formation. For the stock of capital to grow, the rate of gross investment in new capital equipment must be greater than the rate at which old capital wears out or becomes obsolete. That is, net investment (gross investment minus depreciation) must be positive before we realize an increase in the stock of capital.

In classical theory, the level of gross investment is determined in the market for loanable funds. The demand for loanable funds represents the actions of borrowers, people who obtain funds for the purchase of capital equipment, houses, or consumption goods. The supply of loanable funds represents money that people save through bank deposits, bond and stock purchases, retained earnings, or a variety of other means. Thus, the level of savings in part determines the amount of money available for gross investment.

In recent years, the level of savings and investment has not been great enough to yield the growth in real output enjoyed by previous generations. At the same time, many of the government's tax policies have discouraged savings and investment. For example, the income earned from interest on savings is subject to income tax. But interest paid on loans to purchase consumption goods, such as cars and houses, is tax-deductible. Households are thus encouraged to borrow rather than to save. In addition, the Social Security program has taken away some of the basic need for saving for retirement. Also, small savers have been discouraged by interest rate ceilings on passbook savings accounts.

While these fiscal policies and regulations have discouraged investment, they are fortunately reversible. Indeed, saving has recently been encouraged through Individual Retirement Accounts (IRAs); taxes are not paid on the income or interest until retirement. Also, the interest ceiling on savings accounts is being repealed.

It should be remembered that the importance of saving is that it supports investment and capital formation. The economic recovery of 1982 through 1984 is a case in point. Private investment is financed by private saving, government saving (which is the negative

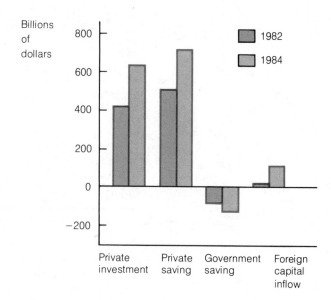

FIGURE 34–7   Trends in Saving and Investment, 1982–1984

The economic recovery experienced between 1982 and 1984 was partially supported by increases in private saving and investment, with the latter facilitated by large capital inflows from abroad.

Source: Council of Economic Advisers, *Economic Report of the President* (Washington, D.C.: U.S. Government Printing Office, 1985), p. 102.

of government borrowing), and by capital inflows from abroad. Consider investment and the changes in these other amounts between 1982 and 1984 (Figure 34–7). Private saving increased by approximately $150 billion between 1982 and 1984 in order to help finance a $220 billion increase in private investment. An increase in negative government saving or positive government borrowing partially offset the increase in investment spending, but the large increase in capital inflows, as shown in Figure 34–7, enabled private investors to finance almost $90 billion in additional capital goods and equipment. The main point is that saving, which enables productive investment, is a major springboard of economic growth.

ity growth in other nations. As Figure 34–8 shows, the United States has been bested by the United Kingdom, Italy, Germany, France, and especially Japan in both these factors over the 1970s.

Small differences in productivity growth rates and in the rates of investment and capital formation that cause them may seem inconsequential. Over

FIGURE 34-8

International Comparison of Capital Formation and Productivity Growth, 1971–1980

Of six major industrialized nations, the U.S. had the smallest growth rate in capital formation, measured here by net fixed investment as a percentage of GNP. The U.S. also showed the smallest growth rate in productivity, measured by manufacturing output per hour. Japan led all other developed nations in these two categories of growth.

Source: Council of Economic Advisers, *Economic Report of the President,* (Washington, D.C.: U.S. Government Printing Office, 1983), p. 82.

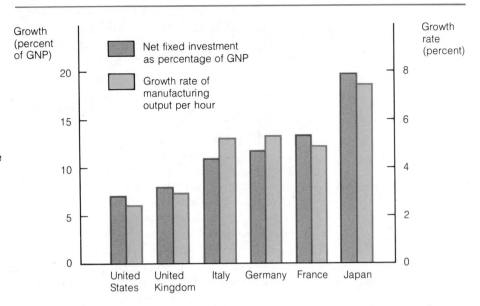

the long run, however, they are critical to growth. The Council of Economic Advisers points out:

> The consequences of reduced productivity growth for our standard of living over the long run are greater than those of any other current economic problem. In 1981 the American economy produced approximately $12,780 worth of output per capita. Had productivity growth continued at the 1948–67 rate during the 14 years subsequent to 1967, output per capita would have reached $16,128 in 1981, 26 percent higher than the actual value.[1]

Small changes in growth rates in productivity, investment, and capital formation can have a major impact on economic growth. Increasing the annual productivity growth rate by only 2 percent, for example, would more than double the standard of living in terms of goods and services by the year 2020.

Denison's Studies

This disappointing picture of recent growth in the U.S. economy has many interrelated causes. Unfortunately, economists cannot pinpoint the exact causes of the slowdown. More important, once the causes of the slowdown in productivity are identified, are policies available to reverse the trend?

The slowdown in labor productivity has been attributed to numerous causes: lower research and development expenditures; a transition to a high-technology economy; higher energy prices; swings in the business cycle; increased regulation affecting saving, investment, and the labor markets; changing composition of the labor force; and changing attitudes of workers. A good explanation places major responsibility for labor productivity declines on the reduction in capital formation.[2] Discrimination against private capital formation has taken a number of forms, including tax policies, destabilizing

[1]Council of Economic Advisers, *Economic Report of the President* (Washington, D.C.: U.S. Government Printing Office, 1983), p. 83.

[2]See, for example, John A. Tatom, "The 'Problem' of Procyclical Real Wages and Productivity," *Journal of Political Economy* 88 (February 1980), pp. 385–394.

inflation created by expanionist monetary policies, and the competition for investment funds at the federal level (such as through government deficits). Public policies of the Reagan administration, for example, have been aimed at countering some of these trends in hopes of increasing capital formation and thus promoting growth. Large increases in both saving and investment in 1984 (see Figure 34–7) offer some promise for the future, although the large federal deficits could choke the recovery.

A more comprehensive accounting for productivity declines and for slower overall growth must be found in the identification of the sources of economic growth itself. Since the early 1960s, economist Edward F. Denison has sought to account for the relative importance of all of the major ingredients of economic growth by using index numbers. Denison has converted trends in total output (national income) and in various factor inputs (labor, capital, legal and human environment, and so on) for the period 1929–1976 into index numbers (see Figure 34–9).

According to Denison's estimates, U.S. national income rose at an average annual rate of 2.98 percent per year over the period 1929–1976. Of the factors contributing to this rise, increases in factor inputs—labor and capital combined—were most significant. Increase in the labor supply consisted partly of increased education and partly of an increased rate of employment. Recall that new net capital growth is the result of growth in savings and in net investment. The remaining major factors contributing to growth, in order of importance, are advances in technology (knowledge), economies of scale, and improvements in resource allocation.

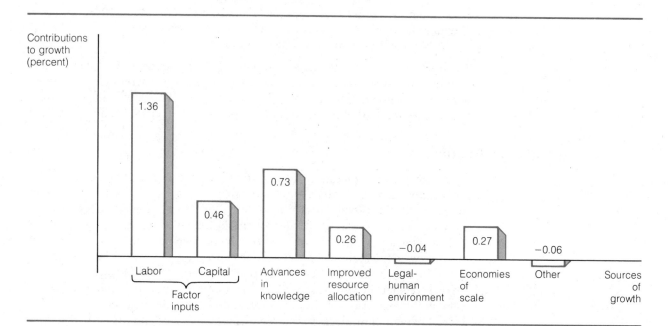

FIGURE 34–9   Denison's Estimates of U.S. Growth Factors, 1929–1976

In addition to the growth in factor inputs themselves, Denison estimated that advances in technology or in knowledge accounted for the largest contribution to economic growth. Environmental restrictions reduced the output rate by 0.04 percent, but the quality of output may have increased.

Source: Edward F. Denison, *Accounting for Slower Economic Growth* (Washington D.C.: Brookings Institution, 1979), p. 104.

Note that changes in the legal and human environment actually reduced economic growth. The latter statistic makes sense because investments in environmental protection, for example, do not increase the quantity of the output of goods and services but do increase the quality. To the extent that quality is important in assessing economic growth and well-being, output per worker is understated because capital investments in environmental and pollution production do not ordinarily result in more real salable output but simply in higher-quality output.

Looking Ahead

Denison's careful studies of the sources of growth point up the complexities of trying to identify causes of growth and then trying to stimulate them. Factor growth and, with it, technological change—the importance of which was missed in classical economics—are the two essential causes of growth. The long-run benefits and the possible short-run costs of rapid technological change have received a great deal of attention from politicians, economists, and the media in recent years, creating a rather muddy picture.

It is clear that the United States is experiencing a gigantic transformation from traditional smokestack industries (coal, steel, heavy machinery) toward high-technology industries (computers, communications). This transformation was clearly evident in the 1982 recession, when there were 25,346 business bankruptcies but 566,942 new companies opened. There were massive worker layoffs in the automobile, steel, and shipbuilding industries and in most heavy industries, while engineers and computer scientists were able to choose between high-paying jobs.

The political blame for much of this economic change has been placed on foreign competition. The United States is no longer a self-sufficient economy; it imports more than twice as many goods now as in 1970. For example, the United States now imports 28 percent of its autos, 18 percent of its steel, 55 percent of its consumer electronic products, and 27 percent of its machine tools.

The two major reasons for the foreign threat are the lower wage rates and the swift spread of technology in foreign countries. Labor costs in South Korea and Taiwan are about one-fifth of those in the United States. Japan, relying on automation, uses approximately six times as many industrial robots as the United States. Some economists and business leaders fear that more automation in the United States will lead to higher unemployment, while other economists feel there will just be a shift into other jobs.[3]

Service industry: An industry whose major product is customer service such as medical services, legal services, food services, entertainment, and the like.

According to projections, new jobs are being created in the service industries. The Labor Department has predicted employment in 1995 of 28.5 million people in **service industries**—those offering teaching, health care, and food service, for example, rather than manufactured goods—compared to 19 million in 1983. There has been a 55.3 percent increase in the number of service industry workers since the early 1970s. By contrast, the number of employees in manufacturing has dropped 1.6 percent since 1973, and predictions are that only 22 million workers will be employed in manufacturing industries in 1995, considerably fewer than the 28.5 million service workers. An example of the increase in service industries compared to heavy industries is the fact that McDonald's now employs more workers than U.S. Steel.

[3]See, for example, "The New Economy," *Time* (May 30, 1983), pp. 62–70, from which the discussion in this section is developed.

The United States is not alone in its transition from heavy industry to high-technology and service industries. European industrial nations are also having difficulty with changing economies. Even Japan is experiencing some decline in heavy industry, but Japan, unlike Western countries, encourages changes in the economy. Japanese companies diversify greatly in areas of production. One company produces products ranging from foodstuffs to nuclear power plants. This diversity allows workers whose jobs become unnecessary to move into other branches of production within the same company. The setup requires large-scale retraining programs within the companies.

Rather than following the Japanese model, politicians, businesspeople, and workers in the United States have demanded protection for the old smokestack industries against foreign competition. During the late 1970s and early 1980s, the United States has experienced the worst outbreak of protectionism since the 1930s. The United States has tried to reduce imports and place quotas on autos, steel, and many other products, as we will see in Chapter 35. However, trade restrictions could well do more harm than good to the United States. The United States now exports more than twice the quantity of products it exported ten years ago, and retaliation by other nations is always a possibility. Such reactions inevitably reduce economic growth and well-being among all traders.

It is not likely that the United States will meet the challenge of economic growth by increasing restrictions and regulations on competition or on capital formation but rather by adapting to new technology through competition and human capital development. The management of change in a complex economy is a difficult matter. It often amounts to a balancing of the present interests of workers and other resource suppliers against the overall interests of consumers and economic growth. American prosperity is clearly the result of earlier technological challenges that were met head-on within the context of vigorous and open competition with due regard for the temporary distortions which new trade and technology might create.

Summary

1. Economic growth is defined as growth in real GNP or in real GNP per capita.
2. The sources of economic growth are complex, but they certainly include growth in population, labor productivity, capital accumulation, and technology.
3. The classical writers envisioned an economy progressing to a stationary (no-growth) state. A Malthusian population mechanism with a long-run tendency to a subsistence wage was a cause of the stationary state, as was a diminishing return to food production.
4. The classical economists also viewed technology as being constant, but modern economists have pointed to growing technology as the mainspring of the economic growth of the Western world.
5. Economic growth requires increases not only in the productive capacity in the economy but also in aggregate demand, which must keep pace with increases in potential output.
6. An essential feature of slower U.S. economic growth in the 1970s has been the decline in the productivity of labor—in output per worker. Though numerous factors caused the decline, the major factor appears to be a reduction in capital formation.
7. A lower savings rate, which has led to a lower growth rate in investment expenditures, is a major factor in explaining the lower rate of capital formation.
8. The United States is faced with a high-technology challenge to economic growth—a movement away from heavy industries such as steel, automobiles, and shipbuilding to industries based on electronics and other advanced technologies. America's response to this challenge will determine growth prospects for many decades to come.

Key Terms

standard of living
economic growth
real GNP per capita
productivity of labor

capital deepening
Malthusian population principle
subsistence wage
absorption of growth

capital formation
service industry

Questions for Review and Discussion

1. What is the difference between growth in real GNP and growth in real GNP per capita? Would it be accurate to suggest that the percent change in real GNP per capita is equal to the percent change in real GNP minus the percent change in the population?

2. Do increases in the population cause increases in real GNP per capita? Would an increase in the labor force participation rate (the percentage of the population that is in the labor force) increase real GNP per capita?

3. Would it be accurate to suggest that a less-developed country with an abundance of natural resources has the potential for a high growth rate? What must occur in such countries before a high rate of growth can occur?

4. What is the cost of capital? What happens to real GNP per capita when the number of machines increases?

5. What changes occurred during the Industrial Revolution that resulted in large increases in real GNP per capita? Are similar changes necessary to obtain growth in less-developed countries?

6. Explain what economic growth means in terms of production possibilities curves. What causes this growth? Is this identical to increases in aggregate supply?

7. What is the difference between a neutral and non-neutral change in technology? Can society enjoy increased consumption of all goods with a nonneutral improvement in technology?

8. Suppose that 1000 tractors were added to a nation's stock of capital every year. Would total farm output rise by the same amount every year? Why or why not?

9. What did Malthus and Ricardo predict about the real GNP per capita? Explain why this prediction has not occurred in the United States. Do their theories appear accurate for some countries?

10. If growth causes an increase in aggregate supply, what must happen to the price level to achieve the higher level of output? Does this suggest that increases in aggregate demand must accompany increases in aggregate supply to maintain price level stability and full employment?

11. What has happened to the rate of growth in real GNP per capita in the United States since the late 1960s? What has caused or accompanied this trend?

12. If you had to choose between price and employment stability and sustained growth in real GNP per capita, which would you choose?

ECONOMICS IN ACTION   A No-Growth Economy?

A number of contemporary economists, alarmed at population growth, rapid depletions of nonrenewable natural resources, and growing pollution, have proposed curtailment by government of traditional kinds of production and consumption in the economy.[a] Although these proposals take various forms, they all end up proposing zero economic growth (ZEG) or zero population growth (ZPG). Predictions of some ultimate doomsday when resource levels fall and pollution rises

precipitously, forcing population declines and sharp reductions in real standards of living, are part and parcel of the idea of ZPG and ZEG. To those envisioning this scenario of the future, the ecological and environmental damage caused by rapid economic growth strongly outweighs the larger quantities of goods and services provided by strict economic efficiency and rapid growth.

Is unlimited economic growth desirable or even possible? The classical writers, as we saw in the present chapter, were themselves doomsdayers. Following Malthus and Ricardo, many classical economists emphasized that all economies grew and matured into a sta-

[a] See E. J. Mishan, *The Costs of Economic Growth* (New York: Praeger, 1967); and J. K. Galbraith, *The Affluent Society* (Boston: Houghton Mifflin, 1958).

tionary state, at which population was at a maximum sustainable level, real income and output were at their highest possible amount, and net investment in new capital was zero. Analytically, this seemingly unhappy state of affairs was caused by an absolutely limited supply of fertile lands (and thus diminishing marginal returns to capital and labor) and by the unrestrained tendency to population increase. To one classical economist at least, John Stuart Mill (1806–1873), the prospect of a stationary state of ZEG was not unhappy at all. Mill said:

> I cannot . . . regard the stationary state of capital and wealth with the unaffected aversion so generally manifested towards it by political economists of the old school. I am inclined to believe that it would be, on the whole, a very considerable improvement on our present condition. I confess I am not charmed with the ideal of life held out by those who think that the normal state of human beings is that of struggling to get on.[b]

Why would the stationary state or economic stagnation be a desirable goal? For Mill, the stationary state was a kind of stable utopia in which wealth redistribution, equality of women, the rights of labor, consumerism, and education could all be advanced. Though Mill staunchly defended, on grounds of productivity and efficiency, every person's right to the fruits of his or her labor, he denounced large concentrations of wealth and supported high and progressive inheritance taxes to reduce inequalities in wealth. He felt that the best place to reduce these inequalities and to encourage the moral and intellectual progress of society was in a state of ZEG.

[b]John Stuart Mill, *Principles of Political Economy*, ed. W. J. Ashley (1848; reprint, New York: A. M. Kelley, 1965).

How are these arguments, old and modern, to be evaluated? Is economic growth desirable? For most economists, economic growth in both advanced and developing nations is the foundation of all progress—economic, intellectual, and moral. In this view, contemporary doomsday economists have made mistakes similar to the classical writers in grossly underestimating technological advance. They have also underestimated the potential of the free-market system. Resource shortages do develop, but the price system responds by creating higher prices for the scarce resources. Producers substitute other inputs for the scarce resources and/or invest in developing new technologies to minimize the effects of the scarcities.

While pollution is a by-product of economic growth, undesirable levels of pollution can be reduced by redefining legal responsibility for pollution or by taxation and other schemes to eliminate pollution problems. It is clear, moreover, that pollution would not be eliminated with ZEG—pollution of the environment would merely slow to a constant rate.

Finally, and perhaps most important, the intellectual and moral development of society, including the desire for equality of which Mill spoke, has taken place primarily within the context of rapid growth in real incomes, at least in the United States and Western Europe. ZEG would increase poverty, unemployment, and inequality, not diminish it. Those advocating ZEG, in short, must come to grips with the social problems that would be created within such an economic environment.

Question

Could a policy of ZEG be carried out under a capitalist system? A democratic political system?

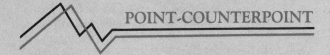

POINT-COUNTERPOINT   **Paul Samuelson and Robert Lucas: Do We Need Discretion or Discipline in Policy?**

Paul Samuelson

Robert E. Lucas, Jr.

PAUL SAMUELSON ". . . has done more than any other contemporary economist to raise the level of scientific analysis in economic theory." So announced the Swedish Academy of Sciences in 1970, when Samuelson (b. 1915) became the first American recipient of the Nobel Memorial Prize in economics. Although he has been frequently called on by presidents and Congress for his advice on economic matters, Samuelson has never held an official policy role in any administration. He has worked principally in the service of his profession: as a professor of economics at M.I.T., as president of the American Economic Association, and as a frequent contributor to journals and magazines. Born in Gary, Indiana, Samuelson received his Ph.D. in economics from Harvard in 1941. His doctoral dissertation, titled *Foundations of Economic Analysis*, was published in 1947 and became one of the definitive technical treatises on neoclassical economics. Today Samuelson is one of the most widely respected Keynesian economists. He is known as a policy activist, one who recommends the use of fiscal and monetary policy to counter recessionary and inflationary cycles of economic activity.

ROBERT E. LUCAS, JR. (b. 1937), credited with introducing rational expectations theory into macroeconomics, has helped change the way economists view the role of macroeconomic policy. Born in Yakima, Washington, Lucas received his Ph.D. in economics in 1964 from the University of Chicago, where he gained a firm background in Keynesian economics. One of the texts he found particularly influential at that time was Paul Samuelson's *Foundations of Economic Analysis*, which he believes is "a great book for first year graduate students."

After graduate training, Lucas became increasingly disillusioned with Keynesian interpretations of the macroeconomy. His article "Econometric Testing of the Natural Rate Hypothesis," published in 1972, is one of many that discuss the implications of rational expectations theory. This theory assumes that people will make rational and intelligent economic decisions based on their knowledge of economic policy, their past experiences, and their expectations of future events. In theory, such rational behavior can distort or even neutralize the effects of well-publicized fiscal and monetary policies. According to Lucas, he and his early followers were initially regarded as "very far out" by colleagues for their nontraditional views. However, appreciation of rational expectations theory has grown so that Nobel Memorial Prize winner George Stigler has suggested that Lucas himself may soon share the award. Lucas has held academic posts at Carnegie-Mellon University, the Ford Foundation, and the University of Chicago, where he currently teaches.

One of the most important debates in economics today centers on the proper role of government policy and its effects on economic growth over the short run and the long run. Activist policy, the heir to Keynes's theories of macroeconomic equilibrium, calls for concerted effort on the part of government to counteract fluctuations in the business cycle. Nonactivist policy, the prescription of monetarists and rational expectations theorists, calls for strict guidelines such as a balanced budget amendment and a money supply growth rule in place of discretionary policy. Samuelson and Lucas speak for the two sides of this debate.

The Case for Activism

In 1964 the unemployment rate stood at 5.2 percent and GNP was $638 billion. The economy was near the end of a bitter recessionary period. Government policy advisers, following Keynesian theory, called for direct intervention in the form of a massive tax cut. Congress responded by passing the 1964 Revenue Act, which slashed personal income taxes by almost 20 percent and corporate taxes by almost 8 percent.

The tax cut, aided by a strong surge in the money supply, ushered in several boom years. Over the next four years, GNP grew over 11 percent and unemployment dropped to 3.8 percent. Policy activists such as Paul Samuelson cite the 1964 tax cut as good evidence of the usefulness of well-planned fiscal and monetary actions. Although the evidence since the 1960s has been somewhat less clear-cut, the case for activism rests on the assumption that the economy cannot easily readjust to full employment without some form of countercyclical policy, whether in the form of monetary growth, monetary restraint, fiscal cutbacks, or fiscal spending. The fiscal actions are generally preferred by activists because they can exert a direct impact on the flow of expenditures. Monetary policy works somewhat more slowly and less directly on expenditures, through the mechanism of interest rates.

Defending the need for activism, Samuelson and other economists focus on the terrible social costs of economic downturns and sudden shocks to the system. Left to correct itself, a shrinking economy could inflict insufferable damage on millions of people before prices and aggregate demand readjust to full-employment levels.

The record of policy activism gives some support for this view. For the past thirty years, the United States has suffered several recessions, but depressions seem to be a thing of the past. "Another depression on the order of the 1930s just doesn't seem possible," says Samuelson. The government nowadays "will do what it has to do" to avert economic disorder. "There's no longer the sense that we must somehow sweat these things out."[a]

Although activist policy has eased the severity of business cycles, it has not found a means to avoid them altogether. Indeed, there are many who argue that the economy's boom-and-recession pattern since World War II has been more the effect of discretionary policy than the cause. Also, there are many economists who point to the accelerating inflation suffered over these years and the failure of government to apply the appropriate countercyclical fiscal medicine: budget surpluses.

The Case Against Activism

For Robert Lucas, the attempts on the part of policymakers to unbend the business cycle through discretionary, countercyclical policy are based on some unexamined assumptions. For such policy to work, individuals and firms must respond cooperatively; when the government tries to close a recessionary gap, they must respond as though such actions will have no effect on prices.

Lucas believes that expectations about prices will always direct behavior, whether one is speaking in microeconomic or macroeconomic terms. If individuals and firms believe that government policy will drive up prices, they will act to protect themselves: Wage earners will demand higher wages, investors will seek higher interest rates, and so on. When such expectations are considered, Lucas deems most countercyclical policy as futile and counterproductive. For instance, if policymakers try to reduce unemployment through deficit spending, wage earners who have experienced the effects of such spending in the past will anticipate higher inflation. They will naturally try to keep the wages they have and increase them. In the face of such demands, employers will find it more and more difficult to hire additional labor, and unemployment will remain despite the increase in aggregate demand. Rather than a bend in the recessionary cycle, the policymakers have managed to foster inflationary pressures, perhaps ushering in high unemployment and high inflation, or what is known as stagflation.

The checkered performance of macroeconomic policy in the early 1980s has given Lucas plenty of ammunition to blame discretionary policy for a variety of economic ills. To Lucas, appropriate macroeconomic policy would try neither to fool nor to ignore the expectations of individuals and firms. Instead, Lucas calls for some fixed limits for policy that recognize the long-run patterns of growth in the economy, which have been in the range of 3 to 4 percent for GNP over the past one hundred years. He also advocates a monetary rule proposed by Milton Friedman that would keep the growth rate of the money supply to approximately 3 percent.

Economists who accept the rational expectations hypothesis have begun the large task of accumulating evidence in its behalf. In the meantime, many find the rational expectations model extreme in its assumptions. How is it, they ask, that citizens can adopt an accurate set of expectations about the effects of policy when even competent observers of the economy are often misjudging the rate or even direction of economic change? Moreover, what happens when fiscal or monetary policies are overwhelmed by a supply shock? Will government then be required to overcome the instability? Questions such as these have brought lively debate over the future course of policy. The issue of activism versus nonactivism in policy is certain to occupy economists for years to come.

[a]Quoted in "Economists Don't See Threats to Economy Portending Depression," Wall Street Journal (October 12, 1984), p. 1.

VIII

International Trade and Economic Development

35

International Trade

We do not often stop to ponder all the ways in which imported goods enrich our lives and improve the material well-being of all nations of the world. On a typical day, for example, an American student's consumption patterns are clearly global. She awakes on sheets made in England and prepares breakfast on a hot plate made in Taiwan. Donning a cotton dress made in India and shoes made in Mexico, she catches the morning news on a TV made in Japan and drives to school in a German-made Volkswagen. Her art history text was printed in the Netherlands, and the movie she sees that evening was imported from France.

In all but a few previous chapters and sections of this book, we have simplified our analyses by treating the national economy as closed—as isolated and with no international interchange. In the next three chapters, we focus exclusively on international trade, finance, and development, with attention to U.S. involvement.

Today's revolutions in technology, especially in telecommunications, are enabling nations to become more and more economically interdependent despite the often explosive political tensions that have divided nations for centuries. In a sense, the whole world is now an economic system. International economics is the study of this world economy with the tools of economic theory. The concepts of supply and demand and production possibilities introduced earlier in the book are quite useful in discussing worldwide trade, the international finance system, and the problems facing the less-developed economies of the world. In these chapters we also examine the organization and functioning of national economies such as that of the Soviet Union that are operated by means of central planning by government rather than by private property and markets.

International economics is composed of both microeconomic and macroeconomic elements. This chapter focuses essentially on the microeconomics of trade, including the reasons for and the advantages of specialization and trade. The effects of artificial interferences with trade, such as tariffs and quotas, are emphasized in an economic defense of free trade. Historical and contemporary trade policies in the United States are highlighted to provide a perspective on this issue of primary importance to our economic well-being.

The Importance of International Trade

All nations have particular talents and resources; like individuals, whole nations are able to specialize in one or many activities. For example, the islands of the Caribbean have abundant sunshine and good weather year round, and so these islands specialize in tourism. Specialization enables nations to emphasize the activities at which they are most efficient and at the same time gain certain advantages through trade.

Examples of international specialization and trade abound. France, a country with a favorable climate and specialized land for wine growing, exports wine to Colombia and the United States and imports Colombian coffee and U.S. machinery. Likewise, both Colombia and the United States specialize and trade products that best utilize their qualities and quantities of resources.

Trade and specialization take place within a given economy as well as among economies. The former type of trade is called intranational or interregional trade; the latter is called international trade. In principle, the two types of trade are the same. As we saw in Chapter 2, Texas may specialize in beef production and Idaho may specialize in potato production. Each meets its need for the commodity it does not produce through intranational or interregional trade. This process is not fundamentally different from the trade of U.S. wheat for Japanese television sets.

In practice, though, some differences exist between interregional and international trade. Resources such as climate, fertile land, and work forces with specialized skills cannot easily be moved between countries. A nation typically trades with the products and services of its resources, not with the resources themselves. Beyond this basic point, countries have different currencies and political systems and values. For such reasons international trade is different from interregional trade and therefore merits special study by economists.

National Involvement in International Trade

The overall magnitude of international trade and countries' shares in this trade can be measured in a variety of ways. The aggregate value of goods in international trade is given in Table 35–1 by major areas of the world. In 1984, world trade totaled more than $2 trillion, measured in exports or imports.

Table 35–2 lists the foreign trade of thirteen nations as a percentage of their gross national products for 1982. A wide range of difference occurs in the percentages. Exports account for almost 50 percent of Saudi Arabia's total output but less than 10 percent of Brazil's or the United States' output. Imports are about one-fourth to one-third of the real consumption of such countries as South Africa, West Germany, Sweden, and Switzerland.

Percentage figures do not tell the whole story, of course. In absolute terms the United States was the world's largest trader in both exports and imports in 1982. The value of exports totaled $261 billion and the value of imports

TABLE 35–1   World Trade: Exports and Imports, 1984

Area	Exports (billions of dollars)	Imports (billions of dollars)
Developed countries[a]	1288.9	1405.0
Developing countries[b]	470.6	432.5
Communist countries[c]	244.6	220.2
Total	2004.1	2057.7

[a]Includes United States and other major developed nations such as Japan, West Germany, and so forth.
[b]Includes OPEC countries and other developing nations.
[c]Includes Soviet Union, China, and Eastern European countries.

Source: *Economic Report of the President* (Washington, D.C.: U.S. Government Printing Office, 1985), p. 352.

TABLE 35–2   Exports and Imports as a Percentage of GNP

Country	Exports as a Percentage of GNP	Imports as a Percentage of GNP
United States	8.5	9.4
United Kingdom	27.0	24.8
Switzerland	33.4	33.5
Saudi Arabia	49.3	26.4
Sweden	33.3	34.1
Mexico	12.5	14.3
Japan	16.9	16.0
Italy	24.4	27.3
West Germany	33.5	31.2
Canada	27.6	23.4
Brazil	9.0	9.3
South Africa	28.8	28.9
Venezuela	25.5	27.8

Source: *International Financial Statistics*, International Monetary Fund (August 1983).

was over $290 billion. The United States trades with virtually every nation of the world, with the primary sources of imports being Canada (16 percent in 1981), Japan (8 percent), and Saudi Arabia and Mexico (5 percent each). The main destination of U.S. exports in 1981 was Canada (15 percent), Japan (8 percent), Mexico (6 percent), and the United Kingdom (5 percent). In terms of products, the three major U.S. imports are crude oil, motor vehicles, and food, while our three major exports are machinery, motor vehicles, and grain.

A more detailed account of U.S. exports and imports for 1982 is presented in Table 35–3. More than 25 percent of U.S. imports consisted of petroleum and other fuels. Machinery and manufactured goods were also large import items. While food and raw materials were only about 6 percent of total imports, food and raw materials constituted about 12 percent of U.S. exports in 1982. The export of machinery and transport equipment, the largest single export, was almost 40 percent of the total.

Why Trade Is Important: Comparative Advantage

The reason for all this trading is that all nations benefit from specialization and trade. Suppose two nations produce computer components and food with an equal expenditure of time and resources. As we saw in Chapter 2, if

TABLE 35–3   U.S. Imports and Exports, 1982, by Major Classification

Imports	Value (billions of dollars)	Percentage of Total	Exports	Value (billions of dollars)	Percentage of Total
Food and live animals	14.5	5.9	Food and live animals	24.0	11.6
Beverages and tobacco	3.4	1.4	Beverages and tobacco	3.0	1.5
Mineral fuels	65.4	26.8	Mineral fuels	12.7	6.1
Chemicals	9.5	3.9	Chemicals	19.9	9.6
Machinery and transportation equipment	73.3	30.1	Machinery and transportation equipment	87.1	42.1
Crude materials	8.6	3.5	Crude materials	19.2	9.3
Manufactured goods	61.2	25.1	Manufactured goods	32.7	15.8
Total imports	244.0		Total exports	207.2	

Source: U.S. Department of Commerce, *Statistical Abstract of the United States*, 1984 pp. 838–841.

Absolute advantage: The ability of a nation or a trading partner to produce a product with fewer resources than some other trading partner.

Specialization: Concentration in a single task in an effort to increase productivity.

Comparative advantage: The ability of a nation or a trading partner to produce a product at a lower opportunity cost than some other trading partner.

nation *A* can absolutely outproduce nation *B* in food and *B* can outproduce nation *A* in computer components, we say that *A* has an **absolute advantage**—the ability to produce a good at lower input cost—in producing food, and *B* has an absolute advantage in producing computer components. Trade between the two countries will likely emerge because each can specialize at what it does best—emphasizing the production at which it is most efficient—and trade with the other country for its requirements of the other good. As we will see, both countries will be better off because **specialization** and trade lead to increases in production and therefore to increases in the attainable consumption levels of both goods in both countries.

According to the principle of **comparative advantage,** countries will specialize in producing those goods and services in which they have lower opportunity costs than their trading partners. For example, a hilly, rocky country will not be able to raise as many sheep per acre as a country with fertile grasslands, but the rocky land cannot support any production other than sheep raising, whereas the grassland will support more lucrative cattle production. Even though the grassland is absolutely more efficient at producing both sheep and cattle, the rocky land has a comparative advantage in sheep growing because the opportunities forgone are nearly worthless. The rocky country will therefore tend to specialize in sheep, the grassy country in cattle.

Consider the simplicity and power of the idea that each country does what it can do best. Countries that have relatively lower opportunity costs of producing certain goods and services have a strong incentive to produce those goods and services. Production across the world will thus come to reflect the principle of comparative advantage at work. Indeed, the rough statistics in Table 35–3 provide a glimpse of America's comparative advantage. The large amounts of fertile U.S. farmland combined with advanced farm technology have made the American farmer the most productive in the world. America has a clear comparative advantage in farm products and raw materials. America's skilled labor force and high rate of technological advance through huge investments in private and public research and development have helped make the United States relatively efficient in producing highly sophisticated machinery and equipment. Trading partners of the United States, especially

Japan, have specialized in routinized productions such as automobiles and steel. Energy regulations and increased demands have forced the United States to import immense quantities of petroleum. The pattern of United States imports and exports therefore shows the process of economic specialization and comparative advantage at work.

Production Possibilities Before Specialization. To demonstrate more precisely how all trading partners benefit by exercising their comparative advantages, we must look at what happens to production possibilities before and after specialization and trade. Suppose that both the United States and Japan, in isolation, produce just two goods: computer components and food. Further assume that the pretrade production possibilities schedules facing the United States and Japan are depicted by the numbers shown in Table 35–4, which gives the alternative combinations of food and computer components that could be produced in the United States and Japan if resources are fully employed. The United States, for example, may choose to produce 60 units of food and no computer components, 40 units of food and 10 of computer components, 20 units of each, or 30 units of computer components and no food. With fully employed resources, Japan may choose between 45 units of computer components and no food at one end of the production spectrum and no computer components with 30 units of food at the other, or combinations between these extremes, as shown in Table 35–4.

These production alternatives can also be expressed in terms of the trade-offs among production possibilities. For the United States, the choice is between producing 60 units of food and 30 units of computer components. This equation of the possibilities can be expressed as 60F = 30CC, where F represents food and CC represents computer components. Within the United States, 1 unit of food therefore exchanges for ½ unit of computer components. Alternatively, 1 unit of computer components exchanges for 2 units of food. The equivalent expressions are as follows:

$$60F = 30CC,$$
$$1F = \tfrac{1}{2}CC,$$
$$2F = 1CC.$$

TABLE 35–4   U.S. and Japanese Production Possibilities Schedules for Food and Computer Components

Prior to trade between the United States and Japan, 2 units of food exchange for 1 unit of computer components within the United States. ⅔ unit of food trades for 1 unit of computer components within Japan. The relative opportunity costs of food production is lower in the United States, and the relative opportunity cost of computer component production is lower in Japan.

	Production Possibilities (at full employment)			
	1	2	3	4
United States				
Food	0	20	40	60
Computer components	30	20	10	0
Japan				
Food	0	10	20	30
Computer components	45	30	15	0

These ratios measure how many units of food must be given up to obtain a unit of computer components, or how many units of computer components must be given up to get an additional unit of food. In either case, what is being measured is the relative opportunity cost of producing food or computer components in the United States; that is, what must be given up to increase the production of food or computer components.

Exactly the same analysis can be applied to Japan. For Japan, the alternatives range from no food and 45 units of computer components to 30 units of food and no computer components. Japan's trade ratio is therefore $30F = 45CC$, meaning that in Japan 1 unit of food exchanges for $1\frac{1}{2}$ units of computer components, or that 1 unit of computer components exchanges for $\frac{2}{3}$ unit of food. These alternative but equivalent expressions can be written as follows:

$$30F = 45CC,$$
$$1F = 1\frac{1}{2}CC,$$
$$\frac{2}{3}F = 1CC.$$

Gains from Specialization and Trade. The concepts of production possibilities and opportunity costs provide a perspective for understanding the benefits of trade. Continuing our example of trade between the United States and Japan, first look at the U.S. and Japanese production possibilities curves in Figures 35–1a and 35–1b. Notice that these production possibilities curves are straight lines. In practical terms, this means that resources can be transformed from food production to computer component production at constant opportunity cost, that resources are perfectly adaptable to one production or

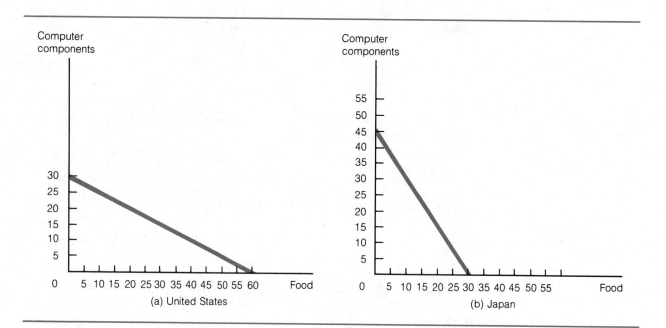

FIGURE 35–1   Prespecialization Production Possibilities Curves

The straight-line production possibilities curves for the United States and Japan indicate that resources can be transferred from computer component production into food production at constant opportunity cost. The differences between the two curves provide the basis for both countries to benefit from specialization and trade.

the other. This assumption is simplistic, but it does not change the argument for specialization and trade.

The important part of the argument for gains from specialization turns on the pretrade opportunity costs of producing the two goods that face Japan and the United States. Before trade, consumers in the United States must sacrifice 2 units of food to get 1 unit of computer components, while Japanese consumers must sacrifice only 2/3 unit of food to get the same amount of computer components. On the other side of the bargain, American consumers must sacrifice only 1/2 unit of computer components to obtain 1 unit of food, while their Japanese counterparts must give up 1 1/2 units of computer components to get 1 unit of food.

Both countries can gain from trade. Since the United States produces food at an opportunity cost of 1/2CC and Japan produces computer components at an opportunity cost of 2/3F, we know that the Americans would gain if trade gave them more than 1/2CC for 1 unit of food, and the Japanese would be happy with anything more than 2/3F for a single unit of computer components. These differences in relative opportunity costs of producing food and computer components yield a basis for specialization and trade between the two countries.

Terms of trade: The ratio of exchange between two countries, based on the relative opportunity costs of production in each country.

At what price will trade take place? Clearly, the **terms of trade**—the ratio at which two goods can be traded for each other—will settle somewhere between the relative opportunity cost ratios for each country. For example, terms acceptable to both Japan and the United States will fall between 2/3F = 1CC (Japan) and 2F = 1CC (United States). We might say loosely that the two countries bargain to set the real exchange rate at which trade takes place, and the bargaining range is determined by each country's internal opportunity cost trade-off. For simplicity, let us assume that the rate settles at 1 unit of food for 1 unit of computer components, or 1F = 1CC.

Figures 35–2a and 35–2b illustrate the potential outcomes of this agreement in terms of the posttrade production possibilities curves. If the countries choose total specialization, the United States can completely specialize in food, producing 60 units, and Japan can completely specialize in computer components, producing 45 units. Given that each country can trade for the product it no longer produces at a price of 1F = 1CC, the production possibilities frontier for each country shifts outward to the right. In contrast to its domestic opportunity cost of 60F for 30CC, the United States now enjoys a situation in which production of 60F can be traded for production of 60CC. Japan enjoys a similar expansion of production possibilities through specialization and trade. By specializing and trading, both countries are made better off. Specialization does not even have to be complete for the two countries to gain from trade. We have used an example of complete specialization here just for illustration.

The production possibilities curves actually illustrate all possible combinations of both goods—food and computer components. If we assume that U.S. consumers originally purchased the bundle of food and computer components labeled US in Figure 35–2a, we can note the improvement in their well-being by considering the new possible consumption bundle US_1 on the posttrade production possibilities frontier. More of both commodities is available after specialization and trade. A similar consumption possibility exists for Japan if we compare hypothetical pre- and post-specialization and trade points J and J_1. Specialization and trade enlarge the consumption possibilities of the trading nations so that both countries can consume more of both commodities after trade.

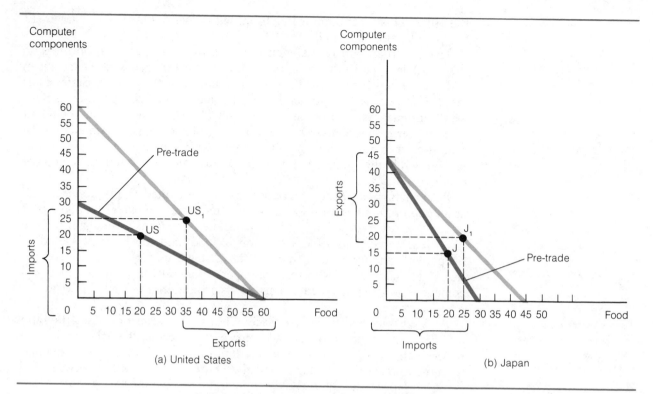

FIGURE 35–2   Posttrade Production Possibilities Curves

The pretrade production possibilities of the United States (a) and Japan (b) are given by the black straight-line curves. Specialization and trade permit the United States to specialize in food, producing 60 units, and Japan to specialize in computer components, producing 45 units. The production possibilities curve for each country shifts to the right (colored straight-line curve) because each country can trade 1F for 1CC: Japan can trade its 45 units of computer components for 45 units of food, whereas before specialization it could produce only 30 units of food for 45 units of computer components. Likewise, the United States can trade 60 units of food for 60 units of computer components instead of its pretrade production of 60 units of food for 30 units of computer components.

It is crucial to remember that the beneficiaries of trade are the consumers in all trading nations. American consumers benefit from the import of such goods as oil, automobiles, and shoes because they can be produced at lower real opportunity cost and therefore lower prices elsewhere. Likewise, consumers of other nations benefit from the United States comparative advantage—a lower real opportunity cost—in producing food, high-technology machinery, and products like Coca-Cola.

Our discussion has been based on a simple example of how the principle of comparative advantage works. In discussing two-good trade between Japan and the United States we assumed that each country was absolutely more efficient than the other at producing food or computer components. What if one of the countries has an absolute advantage at producing both goods? That is, for an equal expenditure of time and resources, one of the countries can outproduce the other in both industries. Is trade still possible? Will each nation stand to gain from specialization and trade? The answer to these critical questions is yes. The principle of comparative advantage still works in such a case. Although one country may be absolutely more efficient than the other at producing both goods, trade is possible so long as the relative opportunity costs of producing the two goods differ in the two countries.

Barriers to Trade

Despite the great advantages of specialization and trade, there are certain barriers that make trade less than free. Some are natural costs of exchange; others are artificial barriers imposed by governments.

There are natural costs to all forms of exchange, from purchases at the local supermarket to international trade. These exchange costs decrease the possible gains from trade because they represent an increase in the real opportunity costs of trade. Contracting costs, negotiating costs, and transportation costs are examples of such natural trade barriers. We might think of them as reducing the size of the outward shift in the posttrade production possibilities curve. Means such as cheaper transportation and more efficient methods of trade negotiations are constantly evolving to reduce the magnitude of such barriers. Natural trade barriers, however, will always exist in one form or another.

Tariff: A tax on imported goods designed to maintain or encourage domestic production.

Quota: A limit on the quantity of an imported good.

Other barriers to the gain from trade are artificial in the sense that they are contrivances of governments designed to raise revenues or protect domestic industries from foreign competition. These artificial trade barriers fall into two major classes: tariffs and quotas. A **tariff,** or import duty, is simply a tax levied on particular imported goods. For example, the United States imposes a tariff on automobiles and steel imported into this country. A **quota** is a partial or absolute limitation on the quantity of a particular good that can be imported. For example, until 1972 the United States imposed a quota on the importation of foreign oil and permitted oil refiners to import only the allowed amount of oil. Both tariffs and quotas reduce specialization and the gains that consumers might obtain from trade. Quotas are generally more protective than tariffs because with tariffs, goods are at least admitted into the country and consumers can decide whether to pay the added amount imposed by the tariff. A more specific type of trade barrier within the United States is discussed in Focus, "State Protectionism."

The Effects of Artificial Trade Barriers

Though tariffs and quotas are adopted by governments presumably out of national self-interest, they may not benefit most citizens in the long run. Tariff duties on imports were an essential source of U.S. government revenues from the Revolution until the late nineteenth century, and the use of tariffs to protect domestic industries is still being promoted. The effects of tariffs and quotas have long been debated, however. Adam Smith made the definitive economic statement on the matter in his *Wealth of Nations* in 1776:

> To give the monopoly of the home-market to the produce of domestic industry, in any particular art of manufacture, is in some measure to direct private people in what manner they ought to employ their capitals, and must, in almost all cases, be either a useless or a hurtful regulation. If the produce of domestic industry can be bought there as cheap as that of foreign industry, the regulation is evidently useless. If it cannot, it must generally be hurtful.[1]

[1]Adam Smith, *An Inquiry into the Nature and Causes of the Wealth of Nations,* ed. Edwin Cannan (New York: Modern Library, 1937), pp. 423–424.

FOCUS   State Protectionism

The United States, like the European Common Market, can be regarded as a huge free-trade zone with no tariffs or other trade restrictions permitted in interstate commerce. ·Or can it?

Part of a state's police power is the authority to prevent agricultural pests from crossing the state's boundaries. But restrictions or absolute prohibitions imposed for safety or on "scientific grounds" also amount to partial prohibition quotas. For years, trucks and passenger cars entering California have had to submit to a fruit check conducted by the state Department of Agriculture, which restricts or confiscates incoming fruit. Regardless of whether these regulations prevent epidemics (inspections failed to prevent entry of the Mediterranean fruit fly in the late 1970s), producers benefit from reduced imports, and consumers lose.

The Kansas laws respecting the licensing and regulation-inspection of plant nursery operators provide another example of protectionist state quotas. The Kansas secretary of agriculture is empowered to license nursery operators within the state and to supervise inspections and certifications of nursery stock entering the state.[a] The instigators of such regulations would like the public to believe that they are sound and well intentioned—that, for example, such regulations improve the quality of the nursery industry in Kansas. But economists remain skeptical. A trade barrier is always a curious way to protect consumers.

[a]"Nursery Laws Governing Kansas," *Southern Florist and Nurseryman* (June 10, 1983), pp. 17–21.

Smith knew that individual traders, such as tailors and shoemakers, could gain from specialization and trade. But he went further and argued that the principle applied no less to nations:

> What is prudence in the conduct of every private family, can scarce be folly in that of a great kingdom. If a foreign country can supply us with a commodity cheaper than we ourselves can make it, better buy it of them with some part of the produce of our own industry, employed in a way in which we have some advantage. The general industry of the country, being always in proportion to the capital which employs it, will not thereby be diminished.[2]

Free trade: The free exchange of goods between countries without artificial barriers such as tariffs or quotas.

With very few exceptions (to be discussed later in the chapter), **free trade**—trade without artificial barriers—leads to maximum welfare among nations. Free trade and specialization expand production possibilities and deliver goods to consumers at the lowest possible costs. Yet given all the benefits of free trade, the world does not often seem to allow it to work, at least not completely. The reason is that the welfare of consumers is not always the single most important goal of a country. Other pressures are brought to bear on governments, such as the pressure by producer groups for protection from foreign competition. Sometimes these groups win political favors, such as tariffs or quotas, to reduce the competitive threat. In this section we use economic theory to show how these interferences with free trade reduce the economic welfare of a country.

Trade and Tariffs: Who Gains? Who Loses?

To understand who really gains and loses from a tariff, consider Figure 35–3. It shows two possible prices facing domestic suppliers of automobiles: the free-market world price of autos, P_w, and the free-market world price plus some per unit tariff on autos imported to the United States, P_{w+t}.

Imposition of a tariff on imported autos clearly benefits domestic automobile manufacturers. Since imported cars are now more expensive, domestic

[2]Smith, *Wealth of Nations*, p. 424.

FIGURE 35–3

Domestic Producers' Supply Schedule for Automobiles

Imposition of an import tarriff on imported automobiles allows U. S. producers to raise the prices of their automobiles from the free-market world price, P_w, to $P_{w + t}$. The higher prices encourage them to increase the quantity of automobiles they produce from Q_0 to Q_1. Their profits are therefore increased by the shaded area $P_{w + t} E_1 E_0 P_w$.

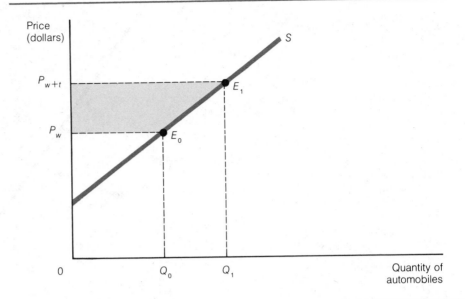

suppliers can raise their prices, without any increase in costs. The higher price $P_{w + t}$ encourages domestic auto producers to increase output from quantity Q_0 to quantity Q_1, increasing domestic suppliers' profits by the shaded area $P_{w + t} E_1 E_0 P_w$. Naturally, the tariff also increases domestic employment by an amount necessary to produce the additional cars.

What if the auto tariff were reduced? A reduction in the tariff would reduce producers' profits as well as domestic employment in the auto industry. But before we can present an economic assessment of tariffs, we must investigate their effects on consumers.

The effects of a tariff on consumers are shown in Figure 35–4. The demand curve for automobiles shows an initial price, P_w, and an initial quan-

FIGURE 35–4

Effects of Tariff on Consumer Demand

A tariff on imported automobiles causes the price to increase from P_w to $P_{w + t}$, which in turn causes a reduction in the quantity of automobiles demanded by consumers from Q_0 to Q_1. Benefits to consumers are reduced by the amount represented by the shaded area $P_{w + t} E_1 E_0 P_w$.

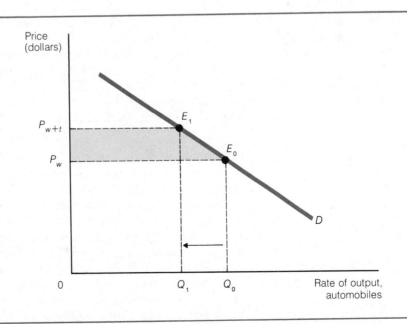

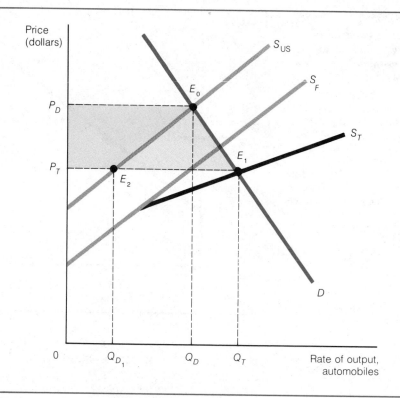

FIGURE 35–5   The Benefits of Free Trade

The total supply curve S_T is the horizontal summation of domestic (S_{US}) and foreign (S_F) supplies of automobiles. In the pretrade situation, when consumers purchase only domestic automobiles, equilibrium E_0 is established at price P_D and quantity produced Q_D. Under free trade, imports lower prices from P_D to P_T and increase quantity consumed in the domestic market from Q_D to Q_T, with equilibrium established at E_1. Consumers benefit in the amount represented by $P_D E_0 E_1 P_T$, the amount they would be willing to pay over the amount they do pay, P_T.

tity of autos, Q_0, purchased by consumers. Imposition of a tariff on imported cars increases the price to P_{w+t} and reduces consumption of autos from Q_0 to Q_1. A reduction in benefits to consumers accompanies the increase in price. (In earlier chapters, we have equated these benefits with consumers' surplus.) The shaded area $P_{w+t} E_1 E_0 P_w$ represents the reduction in benefits to consumers caused by the tariff on automobiles. This concept will be elaborated on in the following sections.

The Benefits of Free Trade

Now we can see the benefits of free trade in the automobile industry. Consider Figure 35–5, which shows the U.S. domestic supply curve for automobiles, S_{US}, the foreign supply curve for automobiles that faces U.S. customers, S_F, and the American demand curve for automobiles. The domestic supply curve and the foreign supply curve for automobiles are added together horizontally to produce a total supply curve, S_T.

Equilibrium is established at the intersection of the demand curve for all automobiles, D, and the total supply curve, where the price to consumers is P_T and the total quantity purchased is Q_T. If no trade took place, the price of automobiles in the United States would be higher, at P_D, and the quantity purchased would be lower, at Q_D. This is because the higher-priced domestic cars force some buyers out of the market. When free trade is introduced, a new price P_T prevails, domestic production falls from Q_D to Q_{D_1} because part of the domestic supply of cars is replaced by lower-cost foreign production. U.S. imports of autos, as shown in Figure 35–5, now occur in the amount of $Q_{D_1}Q_T$. Total consumption rises from the pretrade quantity Q_D to the posttrade quantity Q_T.

Who gains from free trade and who loses? In this simple example, domestic producers lose profits after trade in the amount represented by $P_DE_0E_2P_T$. U.S. consumers of automobiles gain benefits in the amount represented by $P_DE_0E_1P_T$. This is the consumers' surplus, the amount that consumers would be willing to pay for the additional automobiles they buy (Q_DQ_T) over what they did pay for them, P_T. The net gain in U.S. benefits is equivalent to the difference between domestic producers' losses and domestic consumers' gains, or to the area $E_0E_1E_2$.

In assessing the effects of trade, it is also worth noting that imports of foreign automobiles release the resources formerly needed to produce $Q_D - Q_{D_1}$ cars in the domestic market. In other words, opening or expansion of imports creates temporary displacement or unemployment of labor and other inputs, a fact that helps explain some domestic opposition to free-trade policies.

Welfare Loss from Tariffs

We have now seen how trade creates net benefits to domestic consumers. Let us now look at the effects of a protective tariff on the society as a whole, using a hypothetical market for television sets.

Figure 35–6 represents the domestic market for TV sets, including the domestic supply of the product and the U.S. demand. Prior to the imposition of a tariff, American consumers buy Q_4 units at the world price P_w. Of this quantity purchased, Q_1 are sold by domestic manufacturers and Q_1Q_4 are imported from abroad.

Assume that a tariff is imposed in the amount t, causing the price of TVs to American consumers to rise to $P_{w + t}$. American TV producers gain additional profits in the amount represented by $P_{w + t}E_2BP_w$ because of the increased domestic output permitted by the higher price. Imports are reduced to Q_2Q_3, and the total number of sets sold decreases to Q_3. Consumers lose total benefits in the amount of $P_{w + t}E_3E_4P_w$. At the same time, the tariff on imported TV sets creates government revenues. These revenues are composed of the per unit amount of the tariff multiplied by the units of TV sets imported after the tariff is imposed (area E_2E_3AB in Figure 35–6).

Economists typically assume that the government revenue is used in a manner that produces benefits equivalent to those lost by consumers of TV sets. However, the tariff causes a net loss in benefits to society, of which TV consumers are a part, represented by the two areas E_2BE_1 and E_3E_4A. The sum of these triangles represents a loss that is not counterbalanced by the sum of producers' gains and government revenue resulting from the tariff. Economists call this loss a welfare loss to society.

Why Are Tariffs Imposed?

If tariffs create a welfare loss to society, why are they imposed? Certainly tariff revenue to the federal government is minuscule when compared to other sources of government revenue. More than two hundred years ago, Adam Smith identified the real cause of protective tariffs:

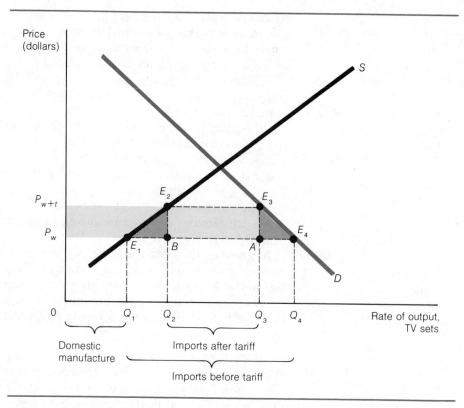

FIGURE 35–6   The Effects of a Tariff

Gains in producer profits and government revenue do not equal consumer losses from tariff imposition. Net consumer losses are shown in the shaded triangles in the figure. Before a tariff on TV sets, consumers purchase Q_4 units at price P_w. The tariff causes the price to rise to P_{w+t}, imports to decrease from Q_1Q_4 to Q_2Q_3, and number of TVs sold to decrease to Q_3. Consumers' total loss is represented by $P_{w+t}E_3E_4P_w$, but producers gain $P_{w+t}E_2BP_w$ and government revenues increase by E_2E_3AB. The net loss to society is therefore the sum of the areas E_2BE_1 and E_3E_4A.

Merchants and manufacturers are the people who derive the greatest advantage from this monopoly of the home-market . . . . Manufactures, those of the finer kind especially, are more easily transported from one country to another than corn or cattle. It is in the fetching and carrying manufacturers, accordingly, that foreign trade is chiefly employed. In manufactures, a very small advantage will enable foreigners to undersell our own workmen, even in the home market . . . . They [merchants and manufacturers] accordingly seem to have been the original inventors of those restraints upon the importation of foreign goods, which secure to them the monopoly of the home-market.[3]

The source of tariff protection is to be found in the urgings of "merchants and manufacturers" today as it was in Smith's day. Consumers of TV sets and of any goods that actually bear a tariff could conceivably convince both producers and the government not to impose tariffs because, as we have shown, consumers lose more than producers and government gain from tariffs. Practically, however, the world does not work this way. Consumers, the gainers from free trade, are widely dispersed and costly to organize in any fight for free trade. Interest in fighting a tariff or quota is also apt to be low

---

[3]Smith, *Wealth of Nations*, pp. 426, 429.

among consumers because their pro rata share of losses from artificial barriers to trade is generally small. To an individual consumer, for example, the tariff is a small proportion of the total price of an automobile. The incentive to organize and fight protective tariffs is therefore small.

Not so for producers. The pro rata share of the effects of tariff protection is much higher among manufacturers, for they are far fewer in number than consumers. Manufacturers thus have stronger incentives than consumers to form an interest group, and their costs of organizing are lower since they are a smaller group. For such reasons, the consumer interest in free trade is often thwarted.

The Case for Protection

Throughout history numerous arguments have emerged in defense of protection from free trade. Some arguments are well constructed; others are but thin veils for producers' interests. All the arguments—both the well-constructed and the questionable ones—deserve careful scrutiny.

National Interest Arguments for Protection

The two oldest and best-formed arguments in defense of protection and against free trade are the national defense argument and the "infant industries" argument. Both arguments contain a grain of truth, but each should be closely scrutinized before serving as a base for protectionist policies.

National Defense. The oldest argument for protection—and the one with superficially the best justification—is that such restrictions are necessary for national defense. As military technology has changed, steel, gunpowder, manganese, uranium, and a host of other inputs have all been commodities essential to making war at various times, and they have been the subject of tariffs and quotas to keep domestic production of these commodities strong.

The logic of protection for defense is clear: Since the ultimate function of government is national defense, any possible threat to national defense, such as the unavailability of some resource during a crisis, must be avoided. A loss in consumers' benefits from protection is thus justified for a greater benefit—the availability of essential materials to be prepared for war. Artificial barriers to trade in these materials are used to protect domestic industries considered essential for national defense.

The national defense argument is plausible for some industries and some products, but consider that cheese, fruit, and watch manufacturers and other, non-defense-related industries have all resorted to the argument. History seems to prove that patriotism is the last and best refuge of a producer seeking protection from foreign competition. The national defense argument has seen double duty as an argument for maintaining unprofitable routes on railroads and protecting truck, air, and railroad companies by setting legalized cartel rates. The telephone industry was long protected on similar grounds.

Other ways can be devised to handle the national defense issue besides imposing tariffs and quotas. A tariff is nothing but a hidden **subsidy** or cash grant to domestic producers. Thus, if an industry is to be protected for national defense reasons, an explicit subsidy may be more straightforward than tariff protection. That way the national defense issue is made clear, and voters know how much defense actually costs them. If, for example, it is deemed vital to national security that the United States have a certain capacity to produce iron and steel, an alternative public policy would be to

Subsidy: A government cash grant to a favored industry.

subsidize domestic iron and steel producers rather than to protect them against foreign competition with a tariff or quota.

Protection of Infant Industries. Possibly the most frequently head argument for protection, especially in less-developed nations, is the so-called **infant industries** argument. Discussed by Adam Smith and supported by Americans such as Alexander Hamilton, the infant industries argument reached its highest expression in the mid-nineteenth-century writing of German nationalist Friedrich List (1789–1846). List sought to unify and develop the German states against the incursions of British imports. He argued that protective tariffs were necessary in the transition from an agricultural-manufacturing to an agricultural-manufacturing-commercial stage of development. His reasoning was clear: Specific tariffs were necessary to maintain fledgling industries until they could compete with foreign imports on their own. After the "infancy" period of industrial development was over, protective tariffs would be lifted, forcing the businesses to meet the rigors of competition.

> **Infant industry:** A new or developing domestic industry whose average costs of production are typically higher than those of established industries in other nations.

A number of less-developed nations use this argument today in their quest for unilateral tariff protection. One-crop economies whose foreign earnings are heavily dependent on a single export seek to protect and diversify their economy with an umbrella of protectionist policies.

As much as one might be concerned for the economic plight of poor nations, great care must be exercised in applying the infant industry argument to them. At best, the argument is one for buying time, for allowing domestic industry to gain a foothold in international competition. Protection for any reason, as we have seen, means lost benefits for consumers. Obviously, any gains from ultimate independence must be set against the costs of lost consumer benefits over the period of protection.

A final question that should be applied to the infant industries argument concerns the vagueness of the goal of removing protection "when the industry grows up." It is hardly legitimate to apply the infant industries argument to the present-day steel industry in the United States and Western Europe, but the argument is still being used. Entrenched protectionist interests will always attempt to prolong the "infancy" of any industry. The dangers of giving protectionists a general legal foothold negate any merit the argument might have in limited and specific cases.

Industry Arguments for Protection

A number of other arguments for protection crop up from time to time, usually put forward by domestic firms and industries seeking protection for protection's sake. We characterize these arguments as wrong because they are assaults on the very principle that gains may be realized from trade. They are all variations on the theme that special-interest groups deserve protection from international competition.

The "Cheap Foreign Labor" Argument. A common argument for tariff or quota protection is that labor or some other resource is cheaper abroad, enabling foreign manufacturers to sell goods at lower prices than U.S. manufacturers. The defenders of this argument are often producers or workers displaced by imports and foreign competition. During the early 1980s, for instance, U.S. producers and workers bitterly complained through their

FOCUS   The Effects of
Textile Quotas

Import quotas ordinarily are more damaging to consumer welfare than are tariffs. Tariffs permit consumers to import as much as they want so long as they pay the duty. Quotas place absolute limits on the foreign supply of a commodity. Quotas are imposed in two ways: (1) A limit on imports may be established, with profits going to the lucky sellers who are able to buy within the quotas on a first-come-first-served basis, or (2) the government may sell import licenses to sellers, with the profits going to government coffers. Either way, consumers lose because government has barred their access to foreign supplies.

In mid-1983, pressure from U.S. garment workers' unions, mills, and manufacturers led the U.S. Department of Commerce to provide "relief" to the industry through quota restrictions on eighty-nine categories of imported apparel and fabrics. Wool items, with American products of vastly inferior quality, were especially hard hit. Only about 20 percent of ordered imported wool skirts and men's and boys' wool coats got through

to U.S. wholesalers and retailers. Prices rose on apparel and fabrics by 10 to 70 percent.

Domestic textile workers and manufacturers obviously gained, but who lost? Arnold Schmedock, executive director of the Ladies Apparel Contractors Association, said, "The only one that will lose out are the retail stores and importers because they won't have the markups and profits they've had."[a] Schmedock forgot to mention the other big losers: American consumers of wool and textile products. There have been other losers in the textile trade war as well. In a Chinese retaliation to the textile quota, U.S. exports to China in May 1983 fell to $75.2 million from $241.7 million in May 1982. The big losers were soybean and wheat producers: Retaliation punished American farmers as much as protection punished American buyers of textile products.

[a]Quoted in "We'll Pay More for Clothes," *USA Today* (July 21, 1983).

union representatives that imports were creating unemployment. (See Focus, "The Effects of Textile Quotas," for a discussion of quotas in one industry.)

While the argument may be correct—foreign producers may be more efficient combiners of resources—this truth in no way denies the benefits to free trade. Indeed, the request for protection from cheap labor turns the gains from trade position on its head: The reason *for* trade becomes a reason to limit or restrict trade. Open trade is beneficial precisely because resources may be cheaper or combined more efficiently elsewhere. The opening or extension of trade creates a temporary disruption of markets, including the unemployment of resources. To impose tariffs or quotas on the grounds that some groups of U.S. laborers are temporarily thrown out of work or that some U.S. stockholders are losing wealth is to subsidize special interests at a greater cost to all U.S. consumers of the subsidized product or service. Moreover, the domestic economy may actually gain from the movement of domestic resources into new fields of comparative advantage. For example, resources released from domestic steel production may be reallocated to computer production.

Side issues arise in the cheap foreign labor argument. One common complaint is that foreign governments subsidize the production of exported goods to shore up their domestic industries and prevent unemployment. As illustrated by Economics in Action at the end of this chapter, this complaint is used as a plea for protection from cheaper imports. But to argue for tariffs on these grounds is to look a gift horse in the mouth. The benefits from trade are independent of the reasons that imports are cheaper. If governments choose to subsidize their exports, they in effect tax their own citizens to benefit foreign consumers.

An activity called dumping is also related to the cheap foreign labor ar-

Dumping: The selling of a product in a foreign nation at a price lower than the domestic market price.

gument for protection. **Dumping** is simply the selling of goods abroad at a lower price than in the home market. The practice, an example of price discrimination (see Chapter 10), means that foreign buyers gain greater benefits from consuming the commodity or service than domestic consumers. Special interests in the favored country often complain that dumping constitutes "unfair competition," but the argument for free trade remains intact nonetheless. Consumers in the favored nation are able to purchase goods at lower prices and overall welfare in the consuming country is enhanced.

It should not matter why foreign prices are low as long as the foreign supplier who is dumping goods in U.S. markets has no monopoly power. If the producers of Japanese television sets undersell U.S. producers with an eye to putting them out of business and subsequently raising their price to a monopoly level, dumping does pose an issue for public concern. This is not usually the case, however. More often, dumping by foreign producers is simply a reflection of their lower costs of production and puts competitive pressure on U.S. producers.

The "Buy American" Argument. The "buy American" argument, a call to patriotism, means to keep money at home. More specifically, buy American means that imports should be restricted so that high costs or inefficient producers and their employees may be protected from foreign competition.

This well-known argument is fallacious on several counts. First, it asks consumers to pay higher prices for goods and services than are available to them through trade, thereby negating the potential expansion of trade benefits. Second, when money is kept at home, foreign consumers are unable to purchase domestic exports, reducing the welfare of domestic export producers and their workers and other input suppliers.

"Terms of Trade" Advantage. A final argument applies to countries that have monopoly or monopsony power in international markets. It suggests that a country employ its power to increase its share of the gains from international trade. Thus, if the United States is the world's largest supplier of computers, putting restrictions on computer exports will drive up the price of computers in the international market. Although the United States will then sell fewer computers, it will get a higher price per computer and possibly higher computer sales revenues. In effect, by restricting computer exports, the United States would shift the terms of trade, or the exchange rate, in international transactions in its favor.

Terms of trade argument: The use of export restrictions on goods in an effort to increase a country's monopoly or monopsony power in international markets.

The **terms of trade argument** therefore suggests that countries exercise monopoly and monopsony power where possible. Doing so is a means for a country to increase its revenues from international trade. The argument founders, however, on a simple point: the possibility of retaliation by other countries. If all countries seek terms of trade advantages, all countries will be worse off. It is thus hard for a country to win terms of trade advantages on a unilateral basis and have its trading partners sit idly by.

Perhaps the greatest flaw in all arguments for protection lies in the implicit assumption that other nations will lose export markets and will not retaliate. U.S. history is peppered with examples of tariff wars. One of the greatest tariff wars in history took place in the midst of the Great Depression of the 1930s, discussed in the next section. Tariffs or other forms of trade barrier retaliation can only create a reduction in worldwide economic welfare and massive and inefficient allocations of resources. Whatever the initial reasons for protective tariffs, consumers ultimately suffer the costs.

U.S. Tariff Policy

U.S. tariff history has been punctuated by cycles of free trade and of protectionism, with a move toward free trade in the past fifty years that may be more apparent than real. The average tariff rates for the years between 1821 (the year of the first good statistics) and 1982 is shown in Figure 35–7 along with some of the highlights of our tariff history. Import duties as a percentage of the value of all imports subject to duty have ranged from almost 60 percent in the early 1930s to a mere 3.6 percent in 1982.

Why has the average tariff fluctuated so widely during U.S. history? Before turning to a specific analysis of changing conditions, one broad issue will help us answer the question. With the minor exception of tariffs imposed for revenues to fight wars or to apply foreign policy pressures, the average tariff has varied with business conditions, falling in periods of prosperity and rising during prolonged recessions or depressions. During periods of rising prosperity, manufacturers and workers displaced by free trade find little political support for the imposition of tariffs. When general business conditions turn downward, creating reduced demand for products and increased unemployment, the cry for protection grows louder in the political arena.

Early Tariff Policy

Tariffs were a major source of revenue for the Republic in its early years, as they are for some less-developed countries today. Post-1820 tariffs tended to vary with business conditions, rising during recessions and falling during

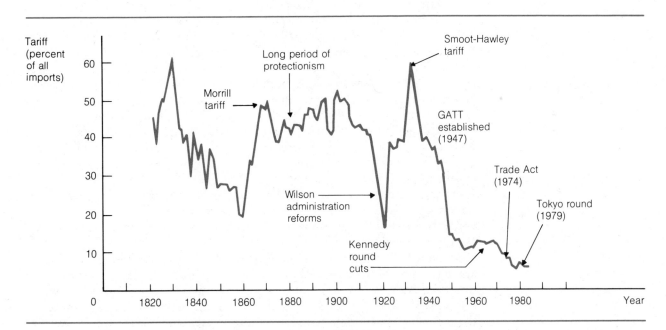

FIGURE 35–7   Level of Tariffs

The United States today has fewer tariffs than at any time in its history. Since the establishment of the General Agreement on Tariffs and Trade (GATT) in 1947, tariffs have followed a downward trend.

Sources: U.S. Bureau of the Census, *Historical Statistics of the U.S., 1976; Statistical Abstract of the United States,* 1984, p. 841.

more prosperous periods of business activity. On the eve of the Civil War, tariff revenues stood at about 19 percent of the value of imports, the lowest level in American history up to that time. In 1861, the Morrill tariff passed Congress under the pretext of generating much-needed revenue for the Union treasury. Whatever the initial intention, the Morrill tariff set off a wave of protectionism in U.S. policy that lasted until the second decade of the twentieth century. Tariffs declined dramatically during the Wilson administration, reaching an average of only 16 percent in 1920.

The dramatic downturn in the economy beginning in the late 1920s had a stark and lingering effect on protectionism in the United States. Manufacturing and commercial interests combined with a willing political climate to produce the Smoot-Hawley Act of 1930, which legislated the highest peacetime tariff in U.S. history. Average tariff levels were raised to almost 60 percent of the value of imports, setting off a frenzy of protectionist retaliations in other nations. The result of what was perceived as American self-interest was falling real incomes, sharply reduced consumer welfare, and a deepened and prolonged worldwide depression in the major world-trading countries.

Modern Tariff Policies

Contemporary tariff policy may be viewed as an attempt to overcome the disastrous effects of protectionism embodied in the Smoot-Hawley Act. The first Roosevelt administration, under the aegis of Secretary of State Cordell Hull, acted swiftly to counteract the protectionism of the Smoot-Hawley tariff. In 1934, the Reciprocal Trade Agreements Act was passed, giving the president power to negotiate bilateral tariff reductions of up to 50 percent with other countries. This act was the first important step in establishing the character of modern U.S. tariff policy, for two reasons: (1) It established a framework within which free trade is envisioned as a goal of policy, and (2) it removed from Congress, to a large extent, the tariff-making power (but not the power to set nontariff barriers). While Congress could legally remove the tariff-setting powers of the president, these powers have been renewed and strengthened a number of times since 1934.

In the post-World War II era, the powers of the president have been expanded to include multilateral negotiations for tariff reductions, that is, negotiations with all nations simultaneously. A major manifestation of America's multilateral policy was its support of the multinational organization called GATT (General Agreement on Tariffs and Trade) in 1947. Originally a twenty-two-member body, GATT now includes more than eighty world nations representing about 80 percent of world trade. GATT sets rules and conditions for tariff reductions and oversees bargaining with all participating nations simultaneously.

Modern U.S. trade policies evolve through GATT, and tremendous gains and expansion of world trade have occurred since the formation of this multinational body. Moreover, Congress has expressed some willingness to move toward free trade. In 1962, a Trade Expansion Act was passed permitting the president to negotiate tariff reductions on all commodities simultaneously rather than commodity by commodity. This act led to the so-called Kennedy round of tariff reductions (1964–1967), which produced huge concessions on manufactured and industrial products with special concessions for poor or less-developed countries.

Where Does the United States Stand in the Battle for Free Trade?

Protectionist interests in trading nations have not remained silent in the face of free trade. In fact, the era of lower tariffs coincided with the growth of nontariff trade restrictions such as quotas, outright prohibitions, and government subsidies to exporters. The requirement of import licensing is a feature of such restrictions. The expansion of these restrictions led to the Trade Act of 1974, giving the president expanded power over tariffs and new powers to deal with nontariff barriers. Negotiations under the auspices of GATT were conducted in Tokyo in 1979 (the Tokyo round) with promising results, especially on the issue of tariff reductions.

Economists are in virtually unanimous agreement that free trade is a critical goal, but some are doubtful about the viability of the move to free-trade policies. There are a number of reasons for skepticism, some revolving around the trade powers that Congress has retained and some relating to powers that Congress has delegated.

An escape clause is attached to the president's power to negotiate tariff reductions. The U.S. International Trade Commission is empowered by Congress through the Trade Act of 1974 to determine whether tariff reductions and imports would materially affect industries competing with the imports. The damage must be directly attributable to imports and not simply due to declining demand or inefficiency. This clause gives those opposed to free trade a wedge in the fight for protection, and a number of industries, such as color television manufacturers, have successfully invoked it.

The International Trade Commission is also charged with assessing damages done to workers in industries affected by tariff reductions. Trade adjustment—assistance payments to workers displaced by increased imports—is based on the fact that since government policies are responsible for the unemployment, government should address the problem with unemployment compensation and benefits. To date, workers in the automobile industry have been the major beneficiaries of trade adjustment, having received almost $3 billion in compensation. Some economists believe that such aid merely prolongs the inevitable and short-circuits the market adjustment process. In other words, this assistance may forestall movement of resources to industries in which the United States has a comparative advantage.

Most insidious of all to economists are the nontariff barriers and the hidden nontariff restrictions. Consent of Congress is still required of the president in negotiating reductions or elimination of nontariff barriers such as quotas, which Congress has the power to set. Protectionists therefore have simply changed the battlefield from tariffs to quotas. In periods of recession, protectionists lobby for trade restrictions other than tariffs, and they have been successful.

Conditions in the U.S. auto industry in the early 1980s are an example of this phenomenon. In the midst of a recession, reduced demand for U.S. automobiles, combined with competitive pressures from abroad, created pleas for protection: In the early 1980s, the U.S. automobile industry pressed for quota restrictions on Japanese imports. Fearing explicit and official congressionally imposed quotas and the retaliation they might bring, President Reagan in 1981 negotiated a three-year "self-imposed voluntary" lid on Japanese exports of automobiles in the amount of 1.76 million vehicles annually. In spite of Japan's adherence to the voluntary quota over the three-year period and notwithstanding a dramatic rise in U.S. automobile sales in 1982–1983, Senator Donald Reigle of Michigan demanded more protection. In 1983, when a Japanese minister indicated a possible unwillingness to renew the voluntary agreement, Reigle said, "The continuing Japanese attack on our basic industries is another Pearl Harbor. The time has come to close America's door to the flood of Japanese imported products."[4] The U.S. Congress threatened as much in early 1985.

As the senator's statement implied, we are a long way from free trade. Unfortunately, the U.S. auto industry is only one example of protectionist efforts. Any industry with enough power to affect political outcomes is in a position to gain some form of protection. In the battle, consumers are the inevitable losers.

Not only does free trade promote economic welfare; through the simultaneous creation of economic strength and interdependency, it also shores up the unity and strength of democratic countries. It is paradoxical that Western nations have long supported a goal of unity through multilateralism in foreign policy while simultaneously pursuing protectionist trade policies that limit economic growth and solidarity.

Summary

1. All individuals possess talents that give them advantages in trade. Nations, like individuals, benefit from specialization and trade. Nations trade not resources but rather the products and services created with resources. Differing resource endowments are the basis for trade in products because products are ordinarily more mobile than the resources that produce them.

2. Specialization and trade may take place between countries of vastly differing degrees of economic development. The reason is the law of comparative advantage, which states that trade is possible when the relative opportunity cost of producing two goods differs between two countries.

3. Specialization according to the law of comparative advantage permits the production possibilities and hence the rate of sustainable consumption of two nations to expand. That is, the two parties to trade may obtain more of both traded goods after specialization and trade.

4. There are both natural and artificial barriers to trade. Natural barriers are all exchange costs including transportation costs of moving goods from one country to another. Artificial barriers include taxes on imports of goods, called tariffs, and limitations or prohibitions on imported items, called quotas.

5. Tariffs on imports increase the profits of producers and the revenue of government, but consumers lose more than producers and governments gain. This net welfare loss is the reason most economists oppose protectionist policies.

6. Though tariffs carry a net loss to society, they are imposed whenever domestic producer-competitors are strong enough to supply gains to politicians. Consumer groups may oppose tariffs but are seldom well enough organized or vocal enough to do so successfully.

7. Two substantive arguments are made for protection: the national defense and infant industries arguments. Most economists, however, question the adequacy of these arguments in actual operation. Other arguments—such as "cheap foreign labor" or "buy American"—are regarded by most economists as only thinly veiled protectionist fallacies.

8. U.S. trade policies have historically waxed and waned with prosperity and depression, becoming more protectionist during economic downturns. Although import tariffs have been reduced dramatically in the past fifty years, protectionists' interests have turned to a new battleground—the imposition of various types of quotas.

Key Terms

absolute advantage	terms of trade	free trade	dumping
specialization	tariff	subsidy	terms of trade argument
comparative advantage	quota	infant industry	

[4]Quoted in "Uno's Surprise: Uncertainty About Auto Imports," *Time* (July 11, 1983), p. 19.

Questions for Review and Discussion

1. Evaluate and discuss the following statement: "Trade across international boundaries is essentially the same as trade across interstate boundaries."

2. Nations Alpha and Beta have the following production possibilities for goods X and Y:

Alpha	X	0	3	6	9
	Y	12	8	4	0
Beta	X	0	4	8	12
	Y	15	10	5	0

   a. Draw the production possibilities curves for both countries. What does 1 X cost in Alpha before trade? What does 1 Y cost in Beta? If these countries traded, which would export X and which would export Y?

   b. If Alpha can produce 100 units of X with all its resources or 60 units of Y, how much does 1 unit of Y cost? If Beta can produce 60 X or 40 Y, what does 1 unit of X cost in Beta? Which country has a comparative advantage in the production of X?

3. How can two countries simultaneously gain from trade? Under what circumstances can two countries not gain from trade?

4. What is the difference between natural barriers and artificial barriers to trade? Was the development of the Panama Canal an artificial encouragement of trade?

5. What does a tariff do to the terms of trade? Are consumers in both countries hurt by a tariff?

6. What are the differences and similarities between tariffs and quotas? Would consumers prefer one of these barriers to the other?

7. Who is hurt by a tariff? Who is helped? Who encourages government to impose tariffs?

8. What is the purpose of protective tariffs? Who or what is protected?

9. Evaluate the following statement: "Tariffs discourage the movement of goods between countries; therefore, they encourage the movement of resources such as capital and labor between countries."

ECONOMICS IN ACTION

When Is a Rose Not a Rose? The Flower Industry Seeks Protection

Changes in technology and in resource prices will often alter comparative advantage in surprising ways. Consider conditions in the American flower industry, particularly among flower and plant growers.

Imported plants and cut flowers were a mere trickle in 1973 compared to what they were in recent years. Colombia, Mexico, the Netherlands, and Israel have made big incursions into U.S. wholesale and retail markets for fresh cut flowers (especially roses, carnations, and chrysanthemums), and Belgium, Denmark, the Netherlands, Israel, and Costa Rica have done the same in the plant market. Colombia, America's largest foreign source of cut flowers, exported almost 73 million roses to the United States in 1982 (an increase of almost 25 percent over the previous year) and half a billion carnations. In fact, cut flowers were second only to coffee as Colombia's leading legal export item in 1982. Israel has also come on strong as a rose supplier to the United States with the aid of Israeli government subsidies.

A number of reasons can be offered for the lower prices and increased supplies of cut flowers and plants from foreign sources. The energy crisis of the 1970s raised the price of heating greenhouses, putting a cost squeeze on many domestic growers, esspecially those in the colder climates of North America. (Rose growing is presently centered in California and other southwest-

ern states for economic reasons.) Under such conditions, nations such as Colombia have a new source of comparative advantage in flower growing—no heating of greenhouses is necessary. A major Colombian flower grower said in 1983, "We work with what we have and only use varieties that will grow under our conditions. Even though OPEC prices are down, not heating is our only advantage."[a]

A second reason for the new strength of imports is the technical possibility of avoiding historical restrictions on the importation of plants potted in soil. Modern techniques permit the cultivation of plants in such nonsoil media as spagnum moss, unused peat, plastic particles, glass wool, and inorganic fibers. Dutch growers of such common plants as ferns and begonias have become especially adept at these new techniques for commercial propagation. Such plants (limited to seven varieties in mid-1983) are allowed into the United States under a U.S. Department of Argiculture (USDA) regulation called Quarantine 37. Under Quarantine 37, plants grown under very rigid specifications in foreign countries may be preinspected by USDA teams in the foreign country and admitted into U.S.

---

[a]Quoted in Barbara Bader, "How the Colombians Do It," *Florists' Review* 172 (April 14, 1983), p. 56.

markets. The plant-exporting nations (plus additional potential entrants) are asking for an expansion of permission to export more than forty additional varieties to the United States in nonsoil growing media.[b]

Predictably, the protectionists have drawn battle lines declaring "cheap foreign labor," "inferior quality products," "unemployment in the flower industry," "foreign government subsidies," and "health hazards through damaging insect pests." Leading the fight is the Washington-based lobby group of the industry called the Society of American Florists (SAF).

None of the reasons given by the flower industry, except the possible problem of the importation of damaging insects, holds a drop of water when compared to the benefits of free trade. If Colombian roses are of inferior quality, as U.S. growers allege, the market soon adjusts prices and quantities accordingly. (In fact, Colombia's initial stock of the Visa rose, the hearty mainstay of its export trade, had to be imported from France because of a refusal of American producers to trade.) If other nations wish to subsidize exports of flowers to the United States, American consumers gain.

How have the domestic flower growers fared? The

[b]Mike Branch, "Quarantine 37: What It Is, How It Works, What It Means," *Florida Foliage* (June 1983), pp. 9–12.

International Trade Administration of the U.S. Department of Commerce directed that a tariff of about 11½ percent of import value be imposed on Israeli roses to counteract the Israeli government's subsidy to promote flower exports. The lobbying efforts of SAF to impose further tariffs or quotas have concentrated on Congress and the USDA. The USDA is in a position to impose absolute and prohibitive quotas or restrictions on imported flowers and plants to prevent possible health hazards.

How much protection will be supplied in the flower industry? Politicians and government agencies have, in their own self-interest, been willing to grant protection to businesses before. But why should one small group of producers be subsidized at the larger costs of consumer welfare?

Question

Suppose that the elimination of all quotas and tariffs on the United States imports of Japanese automobiles would reduce employment in the American automobile industry by 200,000 jobs. Would you support the immediate elimination of all trade restrictions? Would you support a gradual elimination? Would you support the implementation of re-training programs and subsidies to auto workers?

36

The International Monetary System

When a car dealer in Virginia buys cars from a Detroit manufacturer to sell in Virginia, both use the same medium of exchange—U.S. dollars. Suppose, however, that the dealer is buying cars from Japanese or West German manufacturers to sell in Virginia. The Japanese firm will likely want payment in yen; the German firm, in marks. The car dealer thus confronts an international monetary problem. Are the prices stated in yen and marks fair? How are dollars converted into yen and marks? Is there a fixed rate for currency exchange, or does it vary from day to day or even from hour to hour? How would changes in exchange rates affect the price of the cars when they are sold in the United States? U.S. exporters face the same questions from a different perspective. They sell to citizens in foreign countries and seek payment for their goods and services in U.S. dollars. They must be careful, therefore, to pay attention to exchange rates because these rates can determine the prices and profits of their foreign sales.

In Chapter 35 we looked at the exchange of goods and services through foreign markets. The real, as opposed to monetary, side of the international economy concerns such topics as comparative advantage, resource specialization, the level of imports and exports between countries, and artificial barriers to free trade.

The monetary, or financial, side of international trade concerns the ways in which countries pay for the goods and services they exchange. In this chapter we examine many aspects of the relations among the currencies of the world, relations that reflect a kaleidoscope of changing conditions in each country, from interest rates and inflation rates to exports and imports of goods and capital. We look first at the surface—the relative values of national currencies—and then examine the many factors underlying these relative values.

The Foreign Exchange Market

Foreign exchange: The currencies of other countries that are demanded and supplied to conduct international transactions.

Foreign exchange markets: The institutions through which foreign exchange is bought and sold.

International sellers and buyers usually prefer to deal in the currency of their own country. American sellers prefer U.S. dollars for their products, and Japanese sellers prefer yen. The currency of another country that is required to make payment in an international transaction is called **foreign exchange.** Foreign exchange is bought and sold in **foreign exchange markets,** usually made up of large brokers and banks around the world.

Sometimes consumers demand foreign exchange; for example, when travelers land in a foreign airport, they often exchange dollars for the local currency. However, large firms dealing in international transactions simply keep bank deposits in foreign currencies to cover their foreign transactions. An American importer of French wine will likely pay for it by writing checks on a large American bank that holds an account in a French bank. The American bank will use the importer's dollars to purchase the francs needed for payment.

The demand for foreign exchange arises because a country's residents want to buy foreign goods. Conversely, the supply of foreign exchange arises because foreign customers want to buy U.S. goods. For example, we have a demand for francs by the U.S. importer who wishes to purchase French wine and a supply of francs by French customers who wish to purchase U.S. personal computers.

Rates of Exchange

Foreign exchange rate: The price of one country's currency stated in terms of another; for example, a dollar price of German marks of $0.37 means that an individual can buy one mark for $0.37.

The relation between dollars and francs is determined by the **foreign exchange rate:** the price of one currency in terms of another. These rates change daily and even hourly on the foreign exchange market. For example, on April 29, 1985, the value of the Philippine peso expressed in U.S. dollars was approximately $0.05, the Austrian schilling $0.05, the Canadian dollar $0.73, the British pound $1.23, the French franc $0.10, the Swedish krona $0.11, and the Mexican peso $0.004. (Figures have been rounded off.) On the same day, the exchange value of one U.S. dollar was approximately 252 Japanese yen, 1984 Italian lira, 952 Israeli shekels, 12 Indian rupees, and 4800 Brazilian cruzeiros.

For U.S. citizens, foreign exchange rates are used to convert foreign prices into U.S. prices and vice versa. Suppose that a bottle of French wine costs 1000 francs. Using the exchange rates just given, how expensive is the bottle of wine to the U.S. purchaser? The exchange rate of francs for dollars, expressed as the cost of 1 franc in U.S. dollars, is $0.10. The 1000-franc bottle of wine therefore costs the American connoisseur approximately $100.00.

Floating exchange rate system: An international monetary system in which exchange rates are set by the forces of demand and supply with minimal government intervention.

When the exchange rate is expressed in reverse, showing how much foreign currency can be purchased for a dollar, the foreign price is divided by the exchange rate to determine U.S. price. Suppose that a restaurant in Rome offers all the pasta you can eat for 10,000 lira. From the exchange rates given above, 1984 lira exchange for $1.00. The meal would thus cost you $10,000 \div 1984 = \$5.04$. As a final example of exchange rate conversion, suppose that a Swedish-made Volvo costs 100,000 krona if bought in the U.S. and 80,000 krona if bought at the plant in Sweden. How much do you save by buying the Volvo in Sweden? Since the price of the krona is $0.11, the Volvo would cost about $11,000 if purchased in the United States ($0.11 \times 100,000 = \$11,000$) and $8800 if purchased in Sweden ($0.11 \times 80,000 = \$8800$). You could save $2,200 by buying the Volvo in Sweden

Fixed exchange rate system:
An international monetary system in which each country's currency is set at a fixed level relative to other currencies, and this fixed level is defended by government intervention in the foreign exchange market.

Managed floating rate system: A system in which a country's currency is allowed to float freely in the foreign exchange market, within certain bounds; drastic changes in the value of the currency are mitigated by central bank intervention in the foreign exchange market.

(although when transportation costs and tariffs are added, the car may be cheaper in the United States).

How Are Exchange Rates Determined?

Exchange rates can be determined basically in two ways. The first is a **floating** or flexible **exchange rate system.** Under a floating exchange rate system, foreign exchange markets are allowed to operate without government intervention. A country's currency price is set by the forces of supply and demand in the foreign exchange market. The second case is a **fixed exchange rate system.** Under this system, a government intervenes in the foreign exchange market and tries to set, or fix, a price for its currency.

The present international monetary system is a blend of these two systems. Floating exchange rates are used in the United States and the other major industrial nations. Some smaller countries maintain fixed rates against such major currencies as the U.S. dollar and the English pound. No country, however, allows its exchange rate to float freely all the time. Wide swings in the exchange rate are controlled by government intervention in the currency market. For this reason the present international monetary system is called a **managed floating system.** In the next two sections we analyze how floating exchange rates and fixed exchange rates work.

Floating Exchange Rates

As we pointed out earlier, the U.S. demand for foreign currency arises because U.S. citizens wish to buy foreign products, travel in foreign countries, invest in foreign companies, and carry on other international activities. The U.S. supply of foreign currencies arises because foreigners want to buy U.S. goods, travel in the United States, send their children to school in the United States, invest in this country, and so on. All of these forces of supply and demand affect the exchange rate between the U.S. dollar and other currencies. In effect, they form the basis of the supply and demand schedules for U.S. dollars.

Exchange Rates Between Two Countries

We begin our analysis with a simple model of the determination of floating or flexible exchange rates. For simplicity we assume that foreign exchange takes place only between two countries, the United States and West Germany. The U.S. demand for foreign exchange is thus a demand for German marks; the supply of foreign exchange is a supply of German marks.

Figure 36–1 illustrates the U.S. demand and supply of marks. The vertical axis measures the exchange value of marks in U.S. dollars; the horizontal axis measures the quantity of marks supplied in the foreign exchange market. Like any other demand curve, the demand curve for marks is negatively sloped. Remember that Americans demand marks to buy West German goods. If the price that Americans pay for the mark falls, more marks can be purchased for a dollar. Suppose that the exchange value of a mark falls from $0.37 to $0.20 and American importers are buying German cameras that cost 100 marks. The old price of the cameras was $0.37 × 100 = $37.00; when the value of the mark falls, the new price is $0.20 × 100 = $20.00. Lower prices of the mark mean lower prices to U.S. purchasers of German goods. The lower the dollar price of marks, other things being equal, the greater the quantity demanded of German goods by U.S. consumers and the

FIGURE 36–1

A Two-Country Foreign Exchange Market

The vertical axis shows the exchange value of marks—how many dollars it takes to purchase one mark; the horizontal axis shows the quantity of marks available at various prices. The equilibrium price of marks is $0.37. At higher prices, such as $0.50, there is an excess supply of marks, which drives their price down. At lower prices, such as $0.20, there is an excess demand, which drives their prices up.

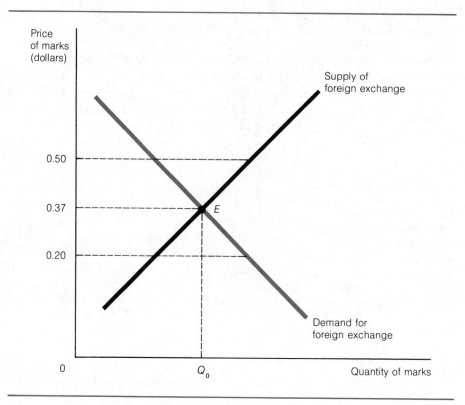

greater the quantity demanded of marks to make the additional purchases. The demand curve for foreign exchange is therefore negatively sloped, like the demand curve for any other economic good.

The supply curve of marks to the foreign exchange market is positively sloped, as drawn in Figure 36–1. As the price of the mark rises, the supply of marks on foreign exchange markets increases. Suppose the exchange value of marks rises from $0.37 to $0.50 and German consumers are buying American beef that costs $5.00 a pound. The old price of beef was $5.00 ÷ $0.37 = 13.51 marks; the new price is $5.00 ÷ $0.50 = 10 marks. As a result of the higher price of marks, German consumers face lower prices for U.S. goods such as beef. Germans will therefore increase their purchases of dollars to buy the now-cheaper U.S. goods. As they do so, the quantity supplied of marks is increased. The supply of foreign exchange is therefore positively sloped, like the supply curve of other economic goods.

The price of foreign exchange, or the exchange rate, reaches equilibrium where the demand and supply curves intersect. In Figure 36–1 the dollar price of marks is $0.37 at the point where the demand and supply curves of marks are equal, E, and a given quantity of marks are exchanged at this price. At prices below $0.37, such as $0.20, there is an excess demand for marks. Under these conditions pressure will be put on the dollar price of the mark to rise. There will be competition among U.S. importers of German goods over the relatively scarce supply of marks. As this competition causes the dollar price of marks to rise, the price that U.S. citizens pay for German goods will also rise, choking off U.S. imports, while the price Germans pay for U.S. goods will fall, stimulating U.S. exports and returning the exchange rate to equilibrium. Similarly, if the dollar price of marks is $0.50, there is an excess supply of marks, and the price of marks will be bid down to its equilibrium level of $0.37, just as it would be in any other competitive market.

Changes in Floating Exchange Rates

As the dollar price of marks changes, the quantity demanded and quantity supplied of marks also change. A change in the price of a good or service, in this case the price of foreign exchange, leads to movements along given demand and supply curves. This is the familiar other things being equal experiment by which all demand and supply curves are determined. In addition to movements along given demand and supply curves, such as those analyzed in Figure 36–1, we can investigate changes in demand or supply that lead to shifts in the curves.

Appreciation: The rise in the value of one currency relative to another.

Depreciation: The fall in the value of one currency relative to another.

Appreciation and Depreciation. An **appreciation** of a country's currency is an increase in its foreign exchange value. If it took 2 German marks yesterday and 3 German marks today to purchase 1 U.S. dollar, we say that the dollar has appreciated relative to the mark. This means that the price of U.S. goods abroad has increased. A Mack truck priced at $50,000 cost 100,000 marks yesterday. Today, it costs 150,000 marks. A **depreciation** of a country's currency means that its foreign exchange value has fallen. If yesterday it took 2 German marks to purchase 1 dollar and today it took only 1½ German marks, the U.S. dollar has depreciated. Yesterday's Mack truck cost 100,000 marks in Germany; today it costs 75,000 marks.

We now turn to the analysis of why exchange rates change, that is, why currencies appreciate and depreciate in value. In economic terms this means that we are going to investigate factors that shift the supply and demand curves for foreign exchange.

Inflation. Inflation is an important factor causing exchange rates to change. Suppose that the United States goes through an inflationary period during which prices rise by 30 percent while prices in other countries are stable and unchanged. In this situation, U.S. goods become more expensive relative to foreign goods. U.S. consumers will increase their purchases of foreign goods, and foreign consumers will reduce their purchases of U.S. goods. In other words, U.S. imports will increase and exports will decrease.

Figure 36–2 shows what happens in the foreign exchange market as a consequence of high U.S. inflation. To purchase more imports, U.S. demand for foreign exchange (marks) shifts right, from D to D_1. This shift causes the value of the dollar to depreciate and the value of the mark to appreciate. Moreover, these effects are reinforced as West Germans reduce their supply of foreign exchange from S to S_1, reflecting their lower demand for dollars with which to buy more expensive U.S. goods. On both counts, inflation causes the value of the dollar to depreciate. Generally, if a country has inflation rates higher than its trading partner, the former country's currency will depreciate. Conversely, lower inflation will cause currency to appreciate.

Purchasing power parity theory: A theory of exchange rate determination that states that differential inflation rates across countries affect the level of exchange rates.

According to the **purchasing power parity theory,** propounded by Gustav Cassel around 1917, the change in the exchange value of currency as a result of inflation allows international trade to go forward in the face of inflation. The depreciation of a currency is an adjustment in its purchasing power to account for the effect of inflation. Thus, a country with a high rate of inflation can still trade with other countries because its exchange rate depreciates to reflect the relative purchasing power of its currency. If the United States has an annual inflation rate of 100 percent and the value of the dollar in

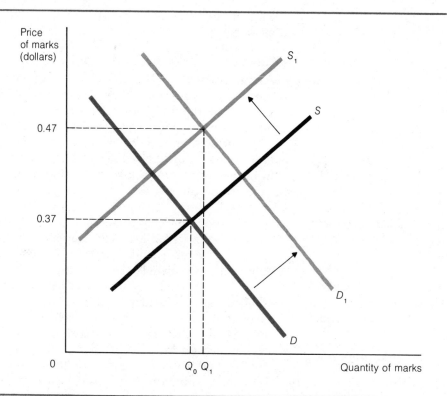

FIGURE 36–2   Inflation and a Floating Exchange Rate System

A higher inflation rate in the United States causes the demand for marks to shift from D to D_1 and the supply of marks to shift from S to S_1. These shifts drive the dollar price of marks up to $0.47. This change in the dollar price of marks reflects the desire of Americans to buy more German goods, which are now cheaper, and the desire of Germans to buy fewer American goods, which are now more expensive, in both cases because the United States has a higher inflation rate than West Germany.

exchange for marks falls by half, as the purchasing power parity theory predicts it will, then the prices of American goods will be unchanged to Germans. For example, as a result of inflation in the United States, the Mack truck that cost $50,000 rises to $100,000. The purchasing power parity theory predicts that the exchange value of the dollar will fall from 2 German marks to 1 German mark as a result of the same inflation. In this event the Mack truck will continue to cost Germans $50,000. The exchange rate adjusts to reflect the purchasing power of the dollar, leaving international trade undisturbed.

This theory works well in predicting the direction that exchange rates will take when there are large differences in inflation rates across countries. However, when inflation rates do not differ by much, movements of exchange rates are determined by other important factors such as the pattern of trade, capital movements, and interest rates.

Changes in Exports and Imports. Changes in the value of a country's exports or imports can lead to changes in its exchange rate. For example, the value of a nation's exports may increase relative to its imports because consumers' tastes shift to favor the former country's goods. For instance, U.S. tobacco products might become popular in Japan. Alternatively, a country might achieve comparative advantage in the production of certain products such as personal computers. For such reasons, the value of a country's exports can increase relative to the value of its imports.

Consequently, the country's exchange rate will appreciate, as Figure 36–3 illustrates. Suppose that West Germans increase their demand for personal computers produced in the United States. The supply of marks needed to get dollars with which to buy the computers will increase from S to S_1. The dollar therefore appreciates in value relative to the mark; that is, it now takes fewer dollars to purchase a mark ($0.20 versus $0.37). This appreciation of the dollar will raise U.S. prices, thereby stimulating U.S. imports, dampening U.S. exports, and restoring equilibrium in the foreign exchange market.

Obviously, this process can be reversed, and the value of a nation's imports can increase relative to its exports. Americans may choose to buy more foreign goods as their income rises. In this case, the U.S. demand for foreign exchange would increase, leading to a depreciation of the dollar. This depreciation would lead to a restoration of equilibrium in the foreign exchange market by stimulating U.S. exports and dampening U.S. imports.

Changes in Interest Rates. So far we have discussed foreign transactions as if only goods and services are exchanged among countries. In addition to these international transactions, financial capital flows between countries, and this flow can change. For example, short-run financial investments are sensitive to differences in interest rates across countries. Thus, financial capital will shift between countries as investors seek the highest rate of return on their money. How do these movements of money affect exchange rates?

Suppose that U.S. short-run interest rates rise above those in Western

FIGURE 36–3

Export and Import Changes

The German demand for U.S. computers increases, and the value of U.S. exports rises relative to its imports. The supply of marks therefore increases to S_1, and the value of the dollar appreciates. That is, a dollar is now worth more marks.

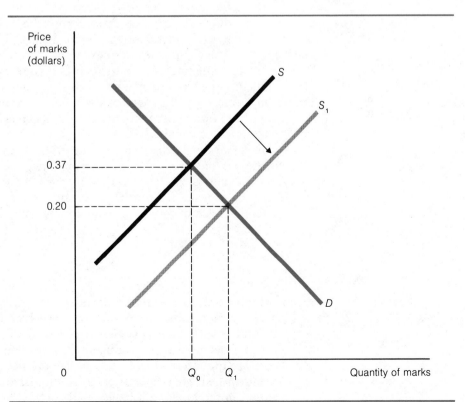

FIGURE 36-4

A Shift in Foreign Investment

As Americans invest their money abroad, the demand curve for foreign exchange increases, shifting from D to D_1. More marks are required to facilitate investments in West Germany. Foreign investment is thus associated with a depreciation of the U.S. dollar, which in turn will stimulate U.S. exports.

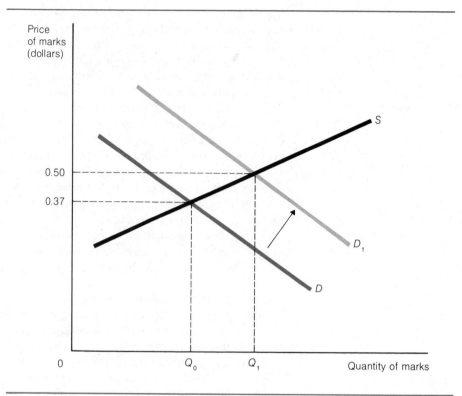

Europe. Then short-run financial investment will increase in the United States as Europeans seek to lend money in the United States where interest rates are higher. To make these loans, the Europeans will supply marks, francs, pounds, and so on to obtain dollars. This means that the dollar's exchange value will appreciate in response to higher U.S. interest rates.

The opposite analysis holds when U.S. interest rates are below those prevailing in other countries. Then short-run capital would flow out of the U.S. as investors sought higher returns in other countries. The demand for the currencies of these other countries would increase, and the dollar's exchange value would depreciate.

In Figure 36-4, the demand curve, D, represents the demand of U.S. citizens for foreign exchange to import foreign goods. When interest rates in foreign countries rise above those in the United States, demand for foreign exchange to make loans to those countries shifts from D to D_1. As a result, the dollar price of foreign exchange rises, and the dollar depreciates in exchange value.

Fixed Exchange Rates

As we have seen, a system of floating exchange rates allows the foreign exchange market to determine the prices at which currency will change hands. By contrast, a fixed exchange rate system does not allow exchange rates to float in a free market. Government intervenes in the foreign exchange market to fix the international price of its currency. Once fixed exchange rates are established, governments stand ready to protect the rates through intervention in the foreign exchange market.

The Operation of a Fixed Exchange Rate System

Although the United States and West Germany no longer trade on the basis of a fixed exchange rate, imagine what would happen if they did rather than letting the rate float to accommodate shifts in the market. Figure 36–5 illustrates what happens when a fixed exchange rate between marks and dollars does not happen to correspond to ever-shifting market conditions. If the current market equilibrium rate is $0.37 per mark but the U.S. government seeks to maintain an exchange rate of $0.50 per mark, the official or fixed rate is above the equilibrium rate. In this case, the dollar is undervalued relative to its current equilibrium value, and the mark is overvalued.

Without intervention, this situation would lead to an excess supply of marks in the foreign exchange markets, creating pressure on the dollar price of marks to fall to the equilibrium value of $0.37. Instead, to maintain the fixed rate, the U.S. government would act to purchase the excess supply of marks. To do so, the central bank (the Federal Reserve System) would build up its holdings of marks. In terms of imports and exports, the undervalued dollar and the overvalued mark make U.S. goods a good buy and German goods a poor buy. This situation basically leads to a surplus of exports over imports for the United States in its trade with West Germany.

The fixed exchange rate can also be set below the point where the de-

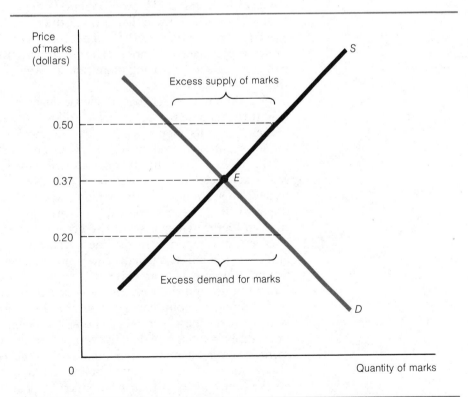

FIGURE 36–5   Fixed Exchange Rates Above and Below Equilibrium

Governments can interfere with floating foreign exchange markets and set official prices for their currencies. The equilibrium rate of dollars for marks is $0.37. An official fixed rate of $0.50 leads to an excess supply of marks, an undervalued dollar, and a U.S. surplus of exports over imports. An official rate of $0.20 leads to an excess demand for marks, an overvalued dollar, and a U.S. excess of imports over exports.

mand and supply of foreign exchange intersect. In this case the government would face the problem of keeping the value of the dollar from falling because there would be an excess demand for marks (an excess supply of dollars) in the foreign exchange market. The government would defend the fixed rate by increasing the supply of marks to the foreign exchange market from its holdings of official reserves, built up perhaps from its purchases when the official rate was above the equilibrium rate. This action would allow the U.S. dollars to be overvalued and the mark to be undervalued relative to their equilibrium values, and the imbalance would lead to an excess of imports over exports for the United States.

Changes in Fixed Exchange Rates

The fixed exchange rate set by government can stray significantly from the equilibrium exchange rate. A country will therefore experience persistent problems in international trade; namely, it will confront persistent surpluses or deficits in its foreign trade. In the balance of trade between exports and imports, a country whose exports have greater value than its imports is said to have a balance of trade surplus; one whose imports' value exceed its exports' value has a balance of trade deficit. What are the options under a fixed-rate system for alleviating such situations?

Changes in Domestic Macroeconomic Policy. Consider first the example of Figure 36–5, where the dollar price of the mark is maintained at $0.50 while the equilibrium price is $0.37. The dollar is undervalued, and the United States experiences a balance of trade surplus, while West Germany runs a persistent deficit, importing more than it exports. The Germans will demand that something be done.

One option is to change domestic macroeconomic policies in the two countries. Suppose the United States allows its money supply to rise and West Germany lets its money supply fall. In the United States the increase in money supply will lead to higher prices and costs and to a higher level of GNP. In West Germany the opposite effects will occur: There will be lower prices and costs and a lower level of GNP. Thus, U.S. exports will fall because they are now relatively more expensive on world markets, and U.S. imports will rise because foreign goods are now cheaper and income is higher. In West Germany exports will rise because German goods are now relatively cheaper, and imports will decline because U.S. goods are more expensive and German GNP has fallen. The basic tendency in both countries, then, is toward adjustments that erase the surplus of exports over imports in the United States and the corresponding deficit in Germany.

An important point emerges from this discussion. In a fixed exchange rate system, where the government chooses to defend the fixed rate at all costs, imbalances in foreign trade must be resolved through macroeconomic adjustment of entire domestic economies. To keep one price—the foreign exchange rate—fixed, the United States and Germany manipulate the level of all the other prices in their economies. A floating-rate system, by contrast, changes one price—the foreign exchange rate—to resolve balance of trade problems. (The relation between macroeconomic policy and the foreign exchange market is explored in more detail in Focus, "Macroeconomic Policy and the Foreign Exchange Market.")

Devaluation and Revaluation. The second option facing two fixed-rate countries with a trade imbalance is to change the official exchange rate. In

FOCUS   Macroeconomic Policy and the Foreign Exchange Market

Foreign trade can have dramatic effects on a government's macroeconomic policy. We consider briefly how monetary policy and fiscal policy are affected by foreign trade.

Monetary Policy

The link between the foreign exchange market and monetary policy depends on whether trade takes place under a system of fixed or floating exchange rates. In the former case, the central bank fixes the exchange rate and allows the money supply to be determined by the economic system; in the latter case, the central bank lets the exchange rate be determined in the market for foreign exchange and manipulates the domestic supply of money.

The Case of Fixed Exchange Rates. To support a fixed exchange rate, the central bank must be prepared to buy and sell whatever foreign exchange is necessary to keep the exchange rate at the fixed level. Suppose there is a surplus in the balance of trade: At the current exchange rate, the quantity of foreign exchange supplied by exporters exceeds the amount of foreign exchange demanded by importers. In this case the central bank must purchase foreign exchange by an amount equal to the excess supply to eliminate the tendency for the exchange rate to fall. The central bank purchases foreign exchange by printing more domestic currency, and this increase in the supply of money sets forces in motion that tend to restore equilibrium. Other things being equal, domestic prices rise with a larger money supply, and these higher prices will discourage exports and encourage imports, thus diminishing the trade surplus. In the case of a trade deficit, the central bank will be selling foreign exchange (in effect decreasing domestic currency) to keep the exchange rate from rising. This action will lower the domestic money supply, and the accompanying lower domestic prices will tend to discourage imports and encourage exports. The important point is that to fix one price—the exchange rate—the central bank must give up other money supply controls and allow all other prices to adjust.

The Case of Floating Exchange Rates. If the government instead is using a floating exchange rate system,

the central bank can use its domestic monetary policy tools, allowing the exchange rate to be determined by the market. Starting from equilibrium, an increase in the domestic money supply tends to raise prices, and these higher prices tend to encourage imports and discourage exports. Thus, there is a trade deficit (an excess demand for foreign exchange) at the old exchange rate. For individuals to be willing to supply the extra foreign exchange to finance these trade movements, the exchange rate must rise by whatever amount is necessary to eliminate the excess demand. The result of an increase in the domestic money supply is therefore a rise in the exchange rate by exactly the same proportion.

Fiscal Policy

The effects of a government's fiscal policy initiatives on exchange rates depend in large part on how the changes in government spending are financed. If we assume, for simplicity, that all increases in government expenditures are financed by borrowing, then expansionary fiscal policies stimulate aggregate demand, including the demand for imports on the one hand, and raise government borrowing on the other. The increased demand puts upward pressure on the exchange rate because of the additional foreign currencies required to finance the increased imports. But the increased government borrowing tends to raise interest rates, attracting foreign capital and placing downward pressure on the exchange rate. The net change in the exchange rate owing to the expansionary policy depends on which of the two effects dominates.

Similarly, contractionary fiscal policies may raise or lower the exchange rate. A lower level of government spending will reduce aggregate demand, reduce the demand for imports, and, at the same time, lower government borrowing. The lower import demand will tend to lower the exchange rate by fostering an excess supply of foreign currencies. But with less government borrowing, interest rates tend to fall. Capital outflows increase, putting upward pressure on exchange rates. Again, the net change of the exchange rate depends on which event—the fall-off in imports or the decline in interest rates—has greater effect.

Devaluation: A change in the level of a fixed exchange rate in a downward direction; a depreciation of a fixed rate.

this way, the two governments can avoid costly manipulation of their domestic economies to restore foreign trade equilibrium. There are two possible changes in this regard.

One is **devaluation** of one country's currency. When the dollar is overvalued ($0.20 per mark in Figure 36–5), the United States suffers a trade deficit and West Germany a trade surplus. With West Germany's agreement, the United States can lower the official exchange value of its currency. In this case we say that a devaluation has occured. In Figure 36–5 this means that the dollar price of marks is moved upward closer to is equilibrium value. This change in the exchange rate changes the relative prices of each country's imports and exports so as to help erase the U.S. deficit and the German surplus without each country's having to resort to costly deflation and inflation of its domestic economy.

Where the official rate is above the equilibrium rate (such as $0.50 per mark in Figure 36–5), the opposite process can be applied. In this case the United States has a trade surplus and West Germany a deficit. The two countries can support this exchange rate if the United States buys marks or if Germany sells dollars or some combination of both. However, to avoid inflation in the United States and deflation in Germany to correct the balance of trade situation, the two countries might agree to revalue the dollar. A **revaluation** occurs when the official price of a currency is raised. This means that the dollar price of marks would move downward in Figure 36–5, closer to its equilibrium value. The value of the dollar is thus raised, or revalued, and the corresponding value of the mark is lowered, or devalued.

Revaluation: A change in the level of a fixed exchange rate in an upward direction; an appreciation of a fixed rate.

Adjustments in official exchange rates seem less costly than the manipulation of whole economies. However, as we will see later, such adjustments are not always easy to accomplish. The Germans may reap advantages from a balance of trade surplus and resist efforts to devalue the dollar or to run their domestic economy so as to see this surplus dissipate. These are obviously complex matters of international negotiation that are inherent in a fixed-rate system.

Changes in Trade Policy. A third option for resolving balance of trade difficulties under a fixed-rate system is for a deficit country to erect barriers to international trade. Tariffs, quotas, and other barriers to the free movement of people and goods across international borders can be established to try to solve a trade problem. This is not an option that would find much favor among economists, however. As we learned in Chapter 35, such impediments to international trade cost the worldwide economy in terms of lost economic efficiency and specialization gains.

Both the floating and fixed exchange rate systems have been used in the past. The major countries presently operate essentially on a floating system and have done so since early 1973. Before that, starting shortly after World War II, a fixed exchange rate system governed international monetary relations. We will discuss the historical evolution of the international monetary system later in the chapter.

The Gold Standard

The **gold standard** is a type of fixed exchange rate system with a long history in international trade. David Hume (1711–1776), a Scottish philosopher and friend of Adam Smith, was the first person to explain how an international gold standard works. The world economy functioned on a gold stan-

Gold standard: An international monetary system in which currencies are redeemable at fixed rates in terms of gold.

dard most recently from the 1870s until World War I. Many observers today wistfully call for a return to the gold standard.

Under a gold standard, all trading nations are willing to redeem their currencies for gold. With each currency linked to gold, the precious metal becomes, in effect, a world currency. Suppose that the U.S. dollar is defined as being worth 1/20 ounce of gold. That is, the U.S. Treasury is prepared to redeem the dollars of both domestic and foreign citizens at this fixed rate in gold. The exchange value of an ounce of gold is set at $20.00. Other currencies are also denominated in gold. Let us say that the German mark is fixed at 1/5 ounce of gold. This means that the mark is backed by four times as much gold as the dollar. The U.S. converts dollars at $1 = 1/20 ounce of gold; Germany converts marks at 1 mark = 1/5 ounce of gold. The fixed exchange rate between dollars and marks in this case is $4 = 1 mark. No one would ever pay more than $4 for a German mark under this system. Why? You could buy 1/5 ounce of gold for $4 from the U.S. Treasury and use this gold to buy a mark from the German Treasury. The fixed exchange rate between dollars and marks of $4 = 1 mark is thus guaranteed by such behavior on the part of individuals.[1] Since each country defines its currency in terms of gold, national currencies basically become just gold certificates.

A second key element of the gold standard is that each country links its money supply to its supply or holdings of gold. This link is established as discussed above: Both foreign and domestic citizens are allowed to redeem gold in a country at a fixed rate. This action creates a natural equilibrating mechanism (Hume called it the **price-specie flow mechanism**), which keeps the imports and exports of countries in balance. Suppose that while operating on a gold standard, the United States imports more than it exports over a given period of time. This differential of purchases has to be paid in gold. Since money supplies are tied to gold, the gold payment causes the money supply of the United States to fall, and the money supplies of surplus countries to rise. Prices and costs in the United States would thus fall, and prices and costs in the trade surplus countries would rise. Deflation in the trade deficit countries such as the United States would stimulate exports and dampen imports, and inflation in trade surplus countries would dampen exports and stimulate imports. Balance of trade equilibrium is thus restored in both types of countries under a gold standard.

Price-specie flow mechanism: The mechanism by which the gold standard equilibrates the balance of trade between countries.

The gold standard is very much like the fixed exchange rate system discussed in the last section. However, a gold-based standard is not necessary to operate a fixed exchange rate system. What is required is that countries act as if they are on the gold standard. When a country runs a trade surplus, it increases its money supply, and its domestic prices and costs rise; ultimately, exports decrease and imports increase. When a country has a trade deficit, it decreases its money supply, prices and costs fall, and thus exports increase and imports decrease. In both cases a country is required to manipulate its whole economy to maintain equilibrium in its balance of trade with a fixed exchange rate.

The crucial difference between the gold standard and modern fixed-rate systems managed by government intervention in the foreign exchange mar-

---

[1] Obviously, shipping gold between the United States and Germany costs real resources. Thus, the dollar price of marks could vary between, for example, $3.98 and $4.01 if it costs $0.02 to ship 1/5 ounce of gold between the two countries. The values $3.98 and $4.01 are called gold points, or values of the exchange rate below or above which gold will be shipped between the two countries.

ket is that under a gold standard the domestic authorities do not have a choice about changing the money supply, thereby inflating or deflating its currency. No international negotiations are required. The system works automatically. A modern fixed-rate system requires that countries agree to follow the appropriate domestic monetary and fiscal policies or to devalue or revalue the official rates when disequilibrium exchange rates lead to persistent deficits and surpluses. To reach such agreements requires complex and costly international negotiations. Modern fixed-rate systems are thus at the mercy of individual action by each country. Countries may or may not follow the rules of the game. The gold standard left no such discretion to political authorities.

The Balance of Payments

Balance of payments: An official accounting record, following double-entry bookkeeping procedures, of all the foreign transactions of a country's citizens and firms. Exports are entered as credits and imports as debits.

As we saw in Chapter 22, countries calculate their gross national product to keep track of the level of aggregate production and national income over time. Countries also keep track of the flow of international trade and periodically publish a report of their transactions, called the **balance of payments.** The balance of payments is essentially an accounting record of a nation's foreign business. It contains valuable information about the level of exports, imports, foreign investment, and the transactions of government such as purchases and sales of foreign currencies.

The balance of payments, as an accounting statement, is kept according to the principles of double-entry bookkeeping. The concept of double-entry bookkeeping is simple: Each entry on the credit side of the ledger implies an equal entry on the debit side.[2] Thus, each international transaction creates both a debit ($-$) and a credit ($+$) item in the balance of payments ledger.

There is a simple rule to follow in classifying debits and credits. Any foreign transaction that leads to a demand for foreign currencies (or a supply of dollars) is treated as a debit, or a minus, item. The act of importing is a debit entry in the balance of payments. Any foreign transaction that leads to a demand for dollars (or a supply of foreign currency) is entered as a credit, or a plus, item. The act of exporting is a credit entry in the balance of payments.

Suppose a U.S. firm exports computers to France, and the French importer of the computers pays for the purchase with an IOU. The French IOU is entered on the credit side in U.S. balance of payments bookkeeping because the French importer must ultimately demand dollars to pay off its IOU to the U.S. firm. The sale of computers is entered on the debit side as the offsetting part of the transaction. The U.S. firm must give up a computer in return for the French importer's payment. Since each transaction implies an equal debit and credit, the balance of payments must always balance. In this respect, the balance of payment is like any other accounting balance sheet.

Of course, although the total balance of payments of a country must always balance (debits minus credits equal zero), its component parts need not balance. For example, the imports and exports of specific merchandise such as automobiles do not have to balance. But overall, surpluses in one part of the balance of payments must be canceled out by deficits in other parts. This is not to say, of course, that countries do not experience balance of payments

---

[2]Double-entry bookkeeping is discussed in detail in Chapter 7.

problems. Problems arise, as we will see, from persistent surpluses or deficits in the component parts of the balance of payments.

Exports and Imports: The Balance of Trade

Table 36–1 presents the data on the U.S. balance of payments for 1983. Items 1 and 2 represent exports and imports of goods and services. The difference between the sums of these aspects of the balance of payments—exports contrasted with imports—is the **balance of trade.** Note that trade in merchandise is only one facet of this balance. Merchandise trade figures are the result of a country's international trade in physical goods, such as books, cars, planes, and computers—the visible exports and imports. The data in Table 36–1 show that in 1983 merchandise exports (1a) were less than merchandise imports (2a) by $60.6 billion. This figure is sometimes called the merchandise trade balance and is equal to merchandise exports minus merchandise imports. If the figure is positive, there is a merchandise trade surplus. It was negative in 1983, so there was a merchandise trade deficit.

Many imports and exports are invisible. U.S. citizens and firms supply various services to foreigners, such as transportation, insurance, and loans.

Balance of trade: Roughly, the balance of a country's imports and exports of goods and services with other countries. If imports are greater than exports, the balance of trade is in deficit; if exports exceed imports, the balance is in surplus.

TABLE 36–1   U.S. Balance of Payments, 1983

U.S. exports enter the balance of payments with plus signs because buyers of these exports must pay U.S. firms with dollars. Dollars thus flow into the United States and are recorded as a plus, or credit, in the balance of payments. U.S. imports receive a minus sign, indicating that to pay for foreign goods U.S. buyers must supply dollars to foreign countries. Dollars thus flow out of the United States, and imports are then treated as a minus, or a debit, in the balance of payments. Similarly, outflows of U.S. capital and inflows of foreign capital are treated as debits and credits, respectively. To classify an item as a debit or a credit, think of whether dollars are leaving or entering the country. If they are leaving, the item is a debit; if they are entering, the item is a credit.

Item		Amount (billions of dollars)
1. Exports of goods and services		+332.2
a. Merchandise exports (including military sales)	+212.9	
b. Services	+ 42.5	
c. Income from U.S. assets abroad	+ 76.9	
2. Imports of goods and services		−365.1
a. Merchandise imports (including military purchases)	−273.5	
b. Services	− 37.7	
c. Income from foreign assets in U.S.	− 53.4	
3. Net unilateral transfers abroad		− 8.6
a. U.S. government grants and pensions	− 7.6	
b. Private remittances	− 1	
4. Current account balance		− 41.5
5. Net capital movements		+ 32.3
a. Outflow of U.S. capital	− 49.4	
b. Inflow of foreign capital	+ 81.7	
6. Statistical discrepancy		+ 9.3
7. Increase (−) in U.S. official reserve assets		− 1.2
8. Increase (+) in foreign official assets in U.S.		+ 5.1
9. Total		0

Source: U.S. Commerce Department, *Survey of Current Business* (December 1984), p. 47. Some categories may not add exactly due to rounding error.

Payments are received for these services just as in the case of visible exports. Items 1b and 1c and 2b and 2c represent invisible exports and imports in the balance of payments. When both visible and invisible exports and imports are considered in Table 36–1, U.S. exports were substantially less than U.S. imports, yielding a trade deficit of $32.9 billion.

Net Transfers Abroad

Net unilateral transfers abroad are one-way money payments from the United States to foreigners or United States citizens living abroad. Included in this category are such items as foreign aid and pension checks to retired U.S. citizens living abroad. Nothing tangible comes back to the United States for these transfers, but they do give rise to a demand for foreign exchange. They thus are entered as a debit item in the balance of payments. In 1983, the United States made $8.6 billion worth of such transfers.

Current Account Balance

The balance on current account is equal to exports of goods and services minus imports of goods and services minus net unilateral transfers abroad. The United States had a deficit on current account in 1983 of $41.5 billion.

Net Capital Movements

When U.S. citizens invest in foreign stocks and bonds, they hold a claim to foreign capital (buildings, factories, and so forth), and capital that would otherwise be invested in the United States is exported. When foreign citizens buy the stock and bonds of U.S. companies, capital is imported in the United States. In 1983, capital outflows were $49.4 billion.

Note what happens in bookkeeping terms when capital flows into or out of a country. Capital outflows are treated as a debit item in the balance of payments because they give rise to a demand for foreign exchange with which to make foreign investments. Capital inflows are a credit item because they increase the supply of foreign exchange traded for dollars with which to buy interests in U.S. firms.

Capital outflows exceeded capital inflows by $32.3 billion in 1983. This net outflow of capital helps explain why item 1c exceeds 2c in Table 36–1. The fact that the United States sends out more capital than it takes in is also reflected in the fact that the United States earns more on foreign investments than foreigners earn on investments in the United States.

Statistical Discrepancy

Item 6 in Table 36–1 is called statistical discrepancy, an accounting fudge factor. When all the data were collected and debits and credits computed, the debits outweighed the credits by $9.3 billion. The statistical discrepancy is added to make the balance of payments balance by compensating for imperfections in data gathering. Some credits have apparently gone unrecorded in computing the balance of payments. These could include hidden exports or unrecorded capital inflows. Most experts think that the latter is the source of most of the discrepancy, which is clearly quite large.

Transactions in Official Reserves

Items 7 and 8 reflect the activities of government in the foreign exchange market. The official reserve assets of a country are its holdings of gold, foreign exchange, and any other financial assets held by official agencies, such as the Federal Reserve System. Also included in reserve assets are special drawing rights (SDRs) with the International Monetary Fund, an international central bank located in Washington, D.C. SDRs are supplementary reserves of purchasing power that a country can draw on to pay international debts. SDRs are simply bookkeeping entries that member countries may draw against to settle an international payments deficit. They are not money in any physical sense, but SDRs may be used as a means of payment. Suppose that Mexico experiences a balance of payments deficit and is without dollars to repay Canada for its purchases of capital goods. Mexico, as a member of the IMF, may opt to settle the debt by transferring some of its special drawing rights credit to Canada. In this way international reserves and liquidity are increased without any actual transfer of funds. After the transfer, Canada may utilize the increased credit to settle some international obligations.

A government can finance an excess of imports over exports by drawing down its holdings of official reserves. Countries receiving surpluses may want to add to their official reserves. As we will see, during the 1947–1973 period of fixed exchange rates, these reserve accounts were quite important because countries used the reserves to stabilize their exchange rates. Deficit countries used reserves to defend the value of their currency, and surplus countries accumulated valuable international reserves.

Item 7 is the increase in U.S. holdings of official reserve assets. In 1983 these holdings increased by $1.2 billion. This item enters the balance of payments as a debit because an increase in official reserve holdings means an increase in the demand for foreign exchange. Item 7 is thus treated like a merchandise import.

Item 8 is the increase in investments of official agencies of foreign countries, such as the OPEC countries, in the United States. Since these investments represent demand for dollars, they are treated as a credit item in the balance of payments.

In 1983, the United States increased its net position in international reserves by $3.9 billion, the sum of items 7 and 8.

The Balance of Payments and the Value of the Dollar

The last entry in Table 36–1 is zero, indicating that the balance of payments must balance. This is the way any double-entry accounting system works out.

The balance of payments, however, is an aggregate record of a country's international payments for one year. Behind this record are the myriad transactions of U.S. citizens and firms with foreign citizens and firms. These transactions determine how the U.S. dollar fares in the foreign exchange market. If, over time, the value of the dollar rises against other currencies, we say that the dollar is strong. This is precisely what happened over the period 1982–1985. The dollar has appreciated in the range of 10 to 15 percent against most major foreign currencies. Most observers feel that this appreciation of the dollar reflects high U.S. interest rates that have attracted an inflow of foreign capital into the United States. (The important role of short-run capital flows and interest rates in the balance of payments is treated in Economics in Action at the end of the chapter.) This capital inflow leads to a higher demand for the dollar and thus to its appreciation. There are obviously costs and benefits associated with large inflows of foreign capital into a country. Foreigners thereby own more U.S. assets and firms, but the United States has an expanded supply of capital as a consequence. The important point is that the dollar, for such reasons, appreciated in currency markets during the early 1980s, and though this fact is not directly reflected in the balance of payments, it indicates that the U.S. foreign economic position might loosely be described as a surplus. That is, the impact of all our foreign transactions over time has been to cause the dollar to rise in value.

The Evolution of International Monetary Institutions

Major world trading nations today rely on floating exchange rates. These governments will occasionally intervene in the foreign exchange market when exchange rate changes are significant, but generally currency prices are free to adjust according to the forces of supply and demand. We briefly review the developments over the past hundred years that led the major trading nations to adopt floating exchange rates in 1973.

The Decline of the Gold Standard

In the period before World War I, roughly dating back to the 1870s, most currencies in the world economy were tied to gold. This means that each country was prepared to redeem its currency at a fixed price in gold to both its own citizens and foreigners. The United States, for example, stood ready to exchange an ounce of gold for $20.67 over this period. In addition, countries on the gold standard linked their money supplies to their holdings of gold bullion.

We have seen how the gold standard works. Basically, it is a fixed exchange rate system in which balance of trade deficits and surpluses are settled by shipments of gold. These shipments cause domestic money supplies and hence domestic national income and employment to fluctuate as a function of the state of the country's trade balance. Deficit countries experience gold outflows and domestic deflation, and surplus countries face gold inflows and domestic inflation. These adjustments in domestic price levels lead to a restoration of equilibrium in the balance of trade by changing the relative prices of a country's imports and exports.

The gold standard worked reasonably well in eliminating balance of trade surpluses and deficits over the period 1870–1914. Nonetheless, for a variety of reasons the gold standard broke down after World War I. There were attempts to return to the gold standard in the 1920s, but the shock to the international economy caused by the Great Depression in the 1930s brought the gold system down once and for all. Most historians point to three basic reasons for the failure of the gold standard.

First, some countries, notably the United States, stopped adhering to gold standard etiquette after World War I. In particular, these countries did not continue to allow their money supplies to be linked to their balance of trade. That is, when gold left the country to settle a trade deficit, a country did not necessarily contract its money supply in proportion to its loss of gold. The gold standard cannot work to restore balance of trade equilibrium under such circumstances.

Second, when Great Britain and France restored the gold standard in the 1920s, they set the "wrong" exchange rates. Britain overvalued the pound, and France undervalued the franc. These actions led to trade surpluses in France and unemployment in England. The pressures on domestic economies caused by inappropriately set exchange rates made it clear to policymakers that it was not worth sacrificing economic stability to the gold standard.

Third, and most important, the Great Depression (1929–1939) ended the movement to return to the gold standard. With worldwide unemployment of resources, countries could no longer accept the discipline of the gold standard, particularly when a balance of trade deficit contributed to unemployment. The gold standard still has a great bit of luster in some circles, however. Focus, "Should We Go Back to the Gold Standard?" discusses the current debate on this issue.

Bretton Woods and the Postwar System

It was not until after World War II that the problems of the international monetary system were addressed in a concerted way. Negotiators for the free world countries met in Bretton Woods, New Hampshire, in 1944 to develop a new international monetary order. The **Bretton Woods system,** which lasted almost thirty years, consisted of a system of fixed exchange rates and an international central bank, called the **International Monetary Fund** (IMF), to oversee the new system.

Bretton Woods system: The fixed-rate international monetary system established among Western countries after World War II.

International Monetary Fund: An international organization, established in the Bretton Woods system, designed to oversee the operations of the international monetary system.

We have already discussed the economics of fixed exchange rates. The basic idea of the Bretton Woods system was that over time countries would obtain or pay for their imports with their exports. However, there might be temporary periods over which a country might wish to run a trade deficit to obtain more imports than its current level of exports allowed it to obtain. Under the Bretton Woods system the country could do so by borrowing international reserves from the IMF. Over time, the country was expected to return to a trade surplus out of which the earlier loan of reserves could be repaid.

The IMF was the bank that held the reserves that allowed the system to operate in this way. When the IMF was formed, each member country was required to contribute reserves of its currency to the bank. The bank thus accumulated substantial holdings of dollars, marks, francs, pesos, and so on, and when member nations ran into balance of trade deficits, the bank would lend them reserves to resolve their difficulty. Each time a loan was made, the debtor nation was encouraged to reform its economic policies to avoid future deficits. Sound economic management might also lead to a trade surplus and a source of funds to repay the IMF loan.

The Bretton Woods system sounds fine in principle, and indeed it functioned tolerably well for a number of years. However, as we saw earlier, fixing the price of a currency is like fixing the price of any other good or service—it is very likely to cause surpluses and shortages in the currency market. As the conditions affecting imports and exports across countries change, the demand and supply of currencies will shift. Countries can find themselves with overvalued fixed exchange rates, which means that they will face persistent deficits in their balance of trade. Under the IMF system the deficit country could draw on its IMF reserves to settle the deficits, but it could not do this forever because it would exhaust its reserves. Chronic deficit countries with overvalued exchange rates were thus said to be in fundamental disequilibrium.

In our discussion of the theory of fixed exchange rates, we reviewed the various courses of action that a country could take in such a case. First, it could devalue its currency as a step toward restoring equilibrium in its balance of trade. Once its currency was devalued, the exchange rate would again be fixed and defended. Second, the country could attempt to improve its trade balance by imposing tariff and quota barriers to imports and perhaps subsidizing exports. In other words, the country could move away from free

FOCUS   Should We Return to a Gold Standard?

The relatively high inflation rates experienced by the United States during the late 1970s and early 1980s led to a variety of proposals for a return to a gold standard. Congress established a U.S. Gold Commission to study the question, and the members not only debated the desirability of again tying the dollar to gold but also considered how to implement a new gold standard.

Proponents of a return to the gold standard argue that the Federal Reserve Board has neither the willingness nor the ability to follow monetary policies that promote price stability and that some constraint on excessive monetary expansionism is necessary. What are the advantages and disadvantages of gold-backed dollars?

Advantages

Under a gold standard, currency is freely convertible into gold at some fixed rate, such as $35 per ounce, which was for many years the official U.S. price of gold. Thus, a nation's currency supply is directly related to its supply of gold. The supply of currency can be expanded relative to the supply of gold only by devaluing the dollar in terms of gold, that is, by raising gold's dollar price. A gold standard therefore enforces monetary discipline. In the absence of devaluation, the currency supply can be increased only at a rate equal to the increase in the output of gold, an expansion that has occurred historically at about 1.5 to 2 percent annually. Such a monetary policy would foster price stability and little or no inflation in the economy.

Disadvantages

There are several major drawbacks to a gold standard. First, real resources are tied up in money production—gold must be mined, stored, and transported. Second, the supply of gold is relatively inelastic. Expansions in the demand for money therefore place downward pressure on the general level of prices. Third, even though the dollar price of gold is fixed, a variety of demand-side factors can cause the market price of gold to diverge from its official price. Indeed, it would be the sheerest coincidence for the two prices to be equal. Cooperation among nations is required to maintain the official price in such circumstances. For instance, during the late 1960s, a two-tier system evolved in which a world price of $35 per ounce was supported through sales among Western central banks and by an agreement between governments not to buy or sell gold in the open market. Prior to that time a group of countries had formed a gold pool for the purpose of using their gold reserves to intervene periodically to stabilize the open market price at the official price.

Last, incentives to engage in monetary expansionism are not completely eliminated by a gold standard. Increases in the stock of currency can be purchased for a time by a willingness to suffer a drain on gold reserves. That is, domestic inflation tends to create a trade deficit, which under a gold standard is balanced by shipments of gold to trade surplus nations. However, the inflating country's currency becomes overvalued, and pressures to devalue gradually become irresistible. It was in fact persistent monetary expansion by the United States that ultimately led to the collapse of the earlier world gold standard.

---

trade to balance its imports and exports. Third, the country could behave as if it were on the gold standard. This would mean adopting restrictive monetary and fiscal policies designed to promote domestic deflation and high interest rates in the hope that such changes would restore the balance of trade to equilibrium. This, of course, is a problematic course of action for any country. For example, if the country's unemployment rate were already high, it is hard to believe that the nation's leaders would be sufficiently disciplined to undertake a deflationary course of action.

Speculation by buyers and sellers of currencies also undermines a fixed exchange rate system. Suppose Great Britain is running chronic trade deficits, and everyone, including speculators, expects that the pound will be devalued (even though British central bankers will deny such rumors vehemently). In effect, speculators are in a no-lose position. Will they continue to hold pounds? Clearly, they will not; pounds are about to become less

Speculation: The buying and selling of currencies with an eye to turning a profit on exchange rate changes, predicted devaluations, and so on.

valuable relative to other currencies. Speculators will sell their pounds for other currencies, increasing the supply of pounds to the foreign exchange market and putting additional downward pressure on the pound. By selling the weak currency and buying a strong currency, speculators make it more difficult for authorities to find a new fixed value for the pound. This is why speculation in a fixed exchange rate system is often termed destabilizing.

The Role of the Dollar in the Bretton Woods System

The U.S. dollar became the centerpiece of the Bretton Woods–IMF system. It assumed this role because it was the world's strongest currency after World Wall II and because the United States, which had a large stock of gold, stood ready to exchange the dollars of foreigners for gold at a rate of $35 an ounce.

The dollar thus became virtually an international currency in the Bretton Woods system. Many nations conducted their business in U.S. dollars and built up significant holdings of dollars. Indeed, a so-called Eurodollar emerged as firms, for various reasons, held dollar deposits in foreign banks and in foreign branches of U.S. banks. These dollars could be used to purchase goods and services not only in the United States but in many other nations as well. Moreover, they could be exchanged for gold. As international trade increased, the demand for dollars also increased. The use of the dollar as an international currency put the United States in a good position. Foreigners obtained dollars by supplying goods and services to U.S. citizens, so the United States was able to derive the benefits of imports such as cars and watches in exchange for dollars.

In the early period of the Bretton Woods system, the United States ran a large trade deficit, and this deficit provided a means of supplying dollar reserves to the rest of the world. It was thought that this deficit was a temporary problem that would soon go away to be replaced by a U.S. trade surplus. This was the premise of the Bretton Woods system. Yet the dollar was overvalued, and the U.S. deficits continued and grew larger into the 1950s and 1960s. The United States was in a difficult position. It was hard for it to devalue the dollar because the dollar was held by virtually every nation as an international reserve asset. A U.S. devaluation would have decreased the wealth of all those countries holding dollars. Several times, the United States tried to impose a restrictive macroeconomic policy to correct its balance of trade. However, when unemployment rose as a consequence, such policies were rapidly abandoned as pressures were brought to bear on policymakers. U.S. deficits continued to grow, and foreign holdings of dollars rose.

Speculators entered again. Many holders of dollars became concerned about what the United States was going to do. There were various runs on the U.S. gold stock as dollars were traded in by foreigners. In 1950, the United States had 509 million ounces of gold; by 1968, this stock had fallen to 296 million ounces. Confidence in the dollar fell further, and the stage was set for a drastic change in the international monetary system.

The Current International Monetary System

The present international monetary system was born in 1971. In the face of continuing large trade deficits and mounting speculation against the dollar, President Nixon broke the link between the dollar and gold in August 1971. No longer would the United States stand ready to exchange gold for dollars at $35 an ounce. In effect, the dollar was set free to fluctuate and seek its own level in the foreign exchange market. As a result, the overvalued dollar depreciated substantially against other major currencies. There were interim attempts to fix the price of the dollar again, but they failed. By early 1973, all the major currencies were floating.

The international monetary system that has been in effect since 1973 is a managed floating rate system. Exchange rates are allowed to seek their free-market values so long as fluctuations are within an acceptable range. If fluctuations fall outside this range, governments may intervene with their reserve holdings to dampen the fluctuations in their currencies. Thus, exchange rates are not completely free; government intervention in the foreign exchange market will be forthcoming if a country's currency falls dramatically in value. In 1978, for example, the dollar fell sharply, and the Carter administration intervened to restrict this decline with international reserves and loans from West Germany and Japan.

Although the new system abandoned fixed exchange rates, it did not abandon the IMF, which still exists as an international monetary organization. The role of the IMF under the managed floating rate system is still evolving, and to this point it has consisted of helping countries that have persistent balance of payments problems by giving them loans and policy advice about how to conduct their domestic macroeconomic policy to overcome these problems. The World Bank, which is part of the IMF system set up by the Bretton Woods agreement, has also played an increasing role in the new system. This sister institution of the IMF makes long-run development loans to poor countries.

Advantages of the Current System. The consensus view seems to be that the new system has worked very well under difficult circumstances. The following are the major advantages of the floating-rate system:

1. It allows countries to pursue independent monetary policies. If a country wants to inflate its economy, it can do so by letting its exchange rate depreciate, thereby maintaining its position in international markets without sacrificing its preferred monetary policy.
2. Under the fixed-rate system, a country sometimes had to deflate its economy to solve a balance of trade deficit. With floating rates it only has to let its exchange rate depreciate. Clearly, it is easier to change one price than to change all prices to resolve balance of trade difficulties.
3. When the OPEC countries dramatically raised the price of oil in 1973–1974, the world economy experienced a tremendous shock. The greatest achievement of the floating-rate system is the way it handled this shock. Oil-importing countries developed large trade deficits; huge trade surpluses built up in OPEC countries. These deficits and surpluses were accommodated by floating rates. Moreover, the huge oil revenues of the OPEC countries were recycled into Western investments. Though exchange rates changed significantly over the period, the new system weathered the storm and got the job done.

Disadvantages of the Current System. The floating-rate system is not, however, without its critics. Some of its problems are as follows:

1. Some observers point out that floating rates can be very volatile, and this volatility creates considerable uncertainty for international trade. Thus, exchange rate flexibility leads to conditions that can retard the amount of international trade and therefore the degree of specialization in the world economy. Table 36–2 lists the exchange rates for several major currencies in 1973–1984. It is clear that some rates have changed signif-

TABLE 36-2 Foreign Exchange Rates, 1973–1984 (U.S. cents per unit of foreign currency)

Year	French Franc	German Mark	Japanese Yen	British Pound	Swiss Franc
1973	22.536	37.758	0.36915	245.10	31.700
1974	20.805	38.723	0.34302	234.03	33.688
1975	23.354	40.729	0.33705	222.16	38.743
1976	20.942	39.737	0.33741	180.48	40.013
1977	20.344	43.079	0.37342	174.49	41.714
1978	22.218	49.867	0.47981	191.84	56.283
1979	23.504	54.561	0.45834	212.24	60.121
1980	23.694	55.089	0.44311	232.58	59.697
1981	18.489	44.362	0.45432	202.43	51.025
1982	15.293	41.236	0.40284	174.80	49.373
1983	13.183	39.235	0.42128	151.59	47.660
1984	11.474	35.230	0.42139	133.56	42.676

Source: Council of Economic Advisers, *Economic Report of the President* (Washington, D.C.: U.S. Government Printing Office, 1985), p. 351.

icantly—the Swiss franc and the yen have appreciated strongly, while the value of the pound dropped considerably. Moreover, the strength of the dollar against all these currencies is evident since 1982. Do these changes impede international trade? Careful studies suggest that they have not. They have found basically that the volume of a country's trade is not very sensitive to fluctuations in its exchange rate.[3]

2. A second criticism of floating rates is that they promote increased world inflation rates. Under a fixed-rate system, countries experience a balance of trade deficit if they inflate their economy. This link is severed under a floating-rate system. Hence, domestic political authorities can more easily give in to those interests who benefit from inflation. Some proponents of the gold standard argue for its return to guard against domestic inflation.

3. Some see the present system of various international monies as an anachronism. These critics argue that it would be more beneficial for the world to adopt a single currency and to become a unified currency area. Such a system would allow various benefits to the world economy and would do away with the need to convert one currency to another in international trade. Although monetary unification has been tried on a modest scale in some areas of the world, notably the recent monetary cooperation of the European countries in seeking to establish a European currency unit, a single global currency seems to be some distance in the future. After all, the forces of national autonomy are still strong, and virtually no government would lightly give up its power to control its money supply.

In sum, there are pros and cons with respect to the present international monetary system, but in general the system seems to have worked well over a difficult period in the international economy. Growing problems in the international payments system, however, call for the prospect of still further reform of the system. Focus, "World Debt Crisis," details the growing problem less-developed countries have in repaying debts to lenders such as the World Bank and large multinational banks.

[3]See Leland B. Yeager, *International Monetary Relations* (New York: Harper and Row, 1976), Chapter 13.

FOCUS   World Debt Crisis

In recent years, international debt in less-developed countries has become an increasing concern due to the apparent difficulties a number of governments have in meeting regular repayments. According to the International Monetary Fund, during 1983 about thirty developing countries completed or were engaging in debt refinancing—mostly consisting of postponing principal loan repayments due over the following year—involving a total debt of about $400 billion. Many Third World countries have become heavily in debt to international lending organizations such as the World Bank and to large private Western banks in recent years.

While most developed countries have a foreign debt of less than 20 percent of their GNP, the corresponding ratio in the case of many developing countries is much higher. Argentina, with a foreign debt of $45.3 billion, has a debt/GNP ratio of about 70.6 percent; Mexico's foreign debt is $89.8 billion, with a ratio of 60.5 percent; and Brazil's debt is $93.1 billion, with a ratio of 41.1 percent.

However, large indebtedness by itself does not indicate that a country is undergoing a debt crisis or that a country is likely to miss repayment. Indeed, some developed countries have high ratios of debt to GNP. Sweden's ratio is about 37 percent and Denmark's about 62 percent. For some developing countries, such as South Korea (with a debt/GNP ratio of about 53.5 percent), the comparatively large debt has represented foreign capital investment that has helped the countries develop rapidly and achieve impressive rates of economic growth while repaying their loans on schedule.

But some nations, such as Brazil and Mexico, have engaged in overambitious development projects based on projections of future revenues (oil in the case of Mexico) or costs (imported oil in the case of Brazil) that proved to be lower or higher, respectively. Other countries dependent on one major export (for example, Chile's dependence on copper) have experienced rapid declines in the prices of those exports that have further eroded their ability to meet their repayment schedules.

A large proportion of debt owed by less-developed nations represents loans from American banks, who are understandably concerned about the possibility of default on the part of debtor nations. For instance, Manufacturer's Hanover Trust in New York had $6.5 billion in loans outstanding to Latin American countries at the end of 1983. There are some hopeful signs, however. A long-awaited worldwide economic recovery may eventually help debtor nations meet regular loan payments. Also, the U.S. merchandise trade deficit represents good news for Third World debtor nations. In 1984, several developing countries, such as Mexico, Argentina, and Brazil, are expected to enjoy a trade surplus with the United States amounting to a total of over $9 billion, which will substantially assist those countries with especially heavy debt burdens to meet loan payments.

Various reforms have been proposed to provide short-run relief from current debt problems of Third World countries, including longer loan repayment periods and increased foreign private investment in developing countries. Over the longer run, however, successful economic development is the only viable solution to the problems Third World countries face.

Summary

1. Foreign exchange is the currency of another country that is needed to make payment in an international transaction.
2. Foreign exchange can be obtained in the worldwide foreign exchange market. The demand for foreign exchange arises from the desire to buy foreign goods. The supply of foreign exchanges arises from the desire of citizens of one country to buy the goods of another country.
3. A foreign exchange rate is the price of one currency in terms of another. A currency's value appreciates when its purchasing power in terms of other currencies rises. It depreciates when its purchasing power falls.
4. Under a floating or flexible exchange rate system, a country allows its exchange rate to be set by the forces of supply and demand in the foreign exchange market. Floating rates change when inflation rates are different across countries, when the value of a country's imports or exports changes, or when there are differences in interest rates across countries. In a floating exchange rate system, rates change to maintain equilibrium in a country's balance of trade.
5. Under a fixed exchange rate system, governments intervene in the foreign exchange market to set and defend the value of their currency. When a country's currency is overvalued relative to the equilibrium rate, the country will run a balance of trade deficit. It can seek to cure this deficit by deflating

the domestic economy, establishing trade barriers, or devaluing its currency to a new, lower fixed rate of exchange. When the fixed exchange rate is below its equilibrium value, a country will experience a balance of trade surplus and will sometimes revalue its currency upward.

6. Under a gold standard, a historical form of a fixed exchange rate system, each country makes its currency redeemable in gold and ties its money supply to its stock of gold. When a surplus of imports over exports appears, gold is shipped to foreigners to settle the balance of trade deficit, causing deflation in the domestic money supply of the deficit country and inflation in the surplus country, restoring equilibrium.

7. An important difference between floating- and fixed-rate systems is that the former changes one price to maintain equilibrium in the balance of trade while the latter changes all prices in the economy to maintain a fixed exchange rate.

8. The balance of payments is a record of the international transactions of a country's economy. It follows the principles of double-entry bookkeeping and must therefore always balance in an accounting sense. Transactions that give rise to a demand for foreign exchange are entered as debits; transactions that give rise to a supply of foreign exchange are treated as credits.

9. The modern history of the international monetary system began with the gold standard, followed by a system of fixed exchange rates managed by the International Monetary Fund. The present international monetary system is a managed floating rate system. Exchange rates are set by supply and demand, but governments can and have intervened if rate movements are too large.

Key Terms

foreign exchange
foreign exchange markets
foreign exchange rate
floating exchange rate system
fixed exchange rate system
managed floating rate system

appreciation
depreciation
purchasing power parity theory
devaluation
revaluation
gold standard

price-specie flow mechanism
balance of payments
balance of trade
Bretton Woods system
International Monetary Fund
speculation

Questions for Review and Discussion

1. In September 1983, the British pound was worth $1.50, and $1 was worth about 8 Swedish krona. How much would a BMW cost in U.S. dollars if the British price was 10,000 pounds? How much would a Volvo cost in U.S. dollars if the Swedish price were 84,000 krona?

2. The following exchange rates came from a newspaper report for September 7, 1982.

	U.S. Dollar Equivalent	
	Wednesday	Tuesday
Swiss franc	0.4596	0.4591
Italian lira	0.000624	0.0006255

	Currency per U.S. Dollar	
	Wednesday	Tuesday
Swiss franc	2.1755	2.1780
Italian lira	1602	1598.50

a. Did the Swiss franc rise or fall from Tuesday to Wednesday?
b. What about the Italian lira?
c. What happened to the dollar value of the franc?
d. What about the dollar value of the lira?

3. Suppose the United States and Sweden are on a floating exchange rate system. Explain whether the following events would cause the Swedish krona to appreciate or depreciate.
a. U.S. interest rates rise above Swedish interest rates.
b. American tourism to Sweden increases.
c. Americans fall in love with a newly designed Volvo, and Volvo sales in the United States skyrocket.
d. The U.S. government puts a quota on Volvo imports.
e. The Swedish inflation rate rises relative to the U.S. rate.
f. The United States closes a military installation in Sweden at the urging of the Swedish government.

4. What happens to the floating exchange rate of a country that has a higher inflation rate than other countries? Show your answer graphically.

5. How would you treat the following items in the bal-

ance of payments, that is, would you enter them as debits or credits?

a. An American travels to Canada.
b. A U.S. wine importer buys wine from Italy.
c. A California wine grower sells wine to England.
d. An Austrian corporation pays dividends to American stockholders.
e. General Motors pay dividends to French investors.
f. Several Japanese visit Hawaii on an American cruise ship.

g. A Swedish citizen invests in a U.S. company in Houston.

6. Explain why capital inflows are a credit entry in the balance of payments and capital outflows are a debit entry.

7. When we say that the gold standard required that all prices but one be changed to overcome a balance of trade deficit, what do we mean?

8. What are the primary advantages and disadvantages of the present international monetary system?

⌂ ECONOMICS IN ACTION

International Capital Movements and the Dollar

When reports in the media emphasize U.S. merchandise trade deficits, the impression is that the dollar is weak on the world market. Yet in the early and mid-1980s the dollar was quite strong internationally despite the fact that this country's merchandise, or current account, deficit continued to increase. The reason for the dollar's strength can be traced to the international capital market. During the same period when the United States was running trade deficits, foreigners invested more in the United States than U.S. residents invested overseas: The United States was a net capital importer.

With international capital movements, it is no longer true that a country's domestic saving must equal its domestic investment, for some of what is saved is invested abroad and some of what is invested domestically has been saved by residents of other countries. Recall that GNP is composed of consumption (C), investment (I), government expenditures (G), and net exports (exports, X, minus imports, M). In addition, income (Y = GNP) equals consumption plus saving (S). If we assume that government spending and taxes are zero, it follows that

$$C + S = C + I + X - M,$$

or

$$S - I = X - M.$$

This expression shows that international capital flows are reflected in the balance of trade. If, for example, there is an excess of saving over investment in a country, this excess will be reflected in an excess of exports over imports. The excess saving becomes a net export of capital to other countries. The export of capital will be transferred into physical goods through a current account surplus (exports exceed imports). On the other hand, net capital imports will be associated with balance of trade deficits.

Capital movements take place as investors seek the highest possible rate of return on their investments. If interest rates are high in the United States compared to the rest of the world, foreign capital will tend to be drawn to the United States. If the United States has a relatively low interest rate, however, domestic capital will tend to flow to foreign investment opportunities. Such capital movements serve to raise interest rates in countries with low interest rates and to lower them in countries with high interest rates. In the long run, with perfect capital mobility and in a simple world with no risks, interest rates would be the same in all countries.

The international capital market does not display such perfection, however. Investments in some countries are more risky than in others. For example, governments can confiscate private property, and revolutions can change economic systems. International interest rate differentials can therefore persist.

If we look at interest rate data for recent years, shown in Figure 36–6, we see that during 1981–1982 both long-run and short-run rates in the United States were generally higher than comparable rates in Western Europe. The interest rates displayed in the figure are of course nominal rates (the real rate of interest plus the expected rate of inflation). The long-run rate differentials in Figure 36–6a therefore suggest that during most of the period, the rate of inflation was expected to be higher in the United States than in Europe. In Figure 36–6b, short-run interest rates in the United States exceeded a weighted average of rates in eleven foreign countries. (The rates on ninety-day certificates of deposit in both the United States and Europe track each other quite closely, however.) Overall, the data suggest that, other things being equal, investing in the United States was relatively attractive, and one would therefore expect this country to have been a net capital importer during 1981–1982.

When one looks at the capital movements during the same period, however, a different pattern emerges. For example, capital outflows of $118.3 billion were recorded for 1982. Comparing this figure with capital inflows of $84.5 billion results in a net capital outflow of $33.8 billion; that is, the United States was apparently

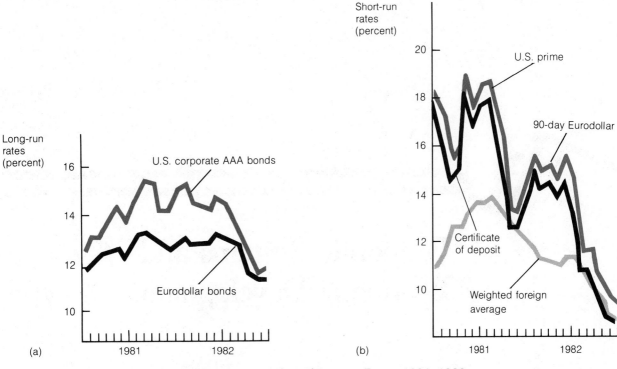

FIGURE 36–6   Selected Interest Rates, 1981–1982

Source: Federal Reserve Board, U.S. Department of Commerce, Bureau of Economic Analysis, *Survey of Current Business* (December 1982), p. 42.

a net capital exporter at a time when interest rates were relatively high here. The discrepancy can be resolved by noting unrecorded inflows of $41.9 billion during 1982—the statistical discrepancy shown in Table 36–1. These unrecorded inflows are due to errors and omissions in reporting transactions, and most of these errors are thought to occur in the capital account. Thus, the United States was probably a net capital importer during the early 1980s, resulting in an appreciation of the dollar on the world market. That appreciation continued well into 1984 and early 1985.

Question

In 1983 the United States trade deficit (balance on current account) reached $41 billion. In the first three quarters of 1984 the trade deficit had already doubled that amount. The American dollar, however, remained strong against other currencies. Should U.S. policy favor a strong dollar or a surplus in the balance of trade? Must we make a choice?

37

Economic Systems and Economic Development

*P*ast chapters have focused on the operation of a mixed capitalist economy, in which a private market sector plays the primary role in the production and distribution of goods and services. The study of **comparative economics,** which we introduce in this chapter, analyzes the differences between capitalist systems and other types of economic systems, particularly socialist economies. Under socialism, the state (the public sector) plays the primary economic role in the economy. The state controls all the resources and makes the decisions about the allocation and distribution of resources. Although there are various forms of socialism, we focus our attention on the Soviet economy.

This chapter also introduces the study of economic development and the economic challenges facing the poorer economies of the world. For example, per capita income in 1984 in India and Burma was less than $200. Why are these countries so poor? How can they be made richer? What is the proper role of the richer countries, such as the United States, in helping the poor countries reach higher standards of living? The incentives created by the economic systems in these countries are a key to understanding their low levels of economic development.

The theme of the chapter is that the choice of economic system, whether a country follows a command-type or a market-type structure, is often a crucial factor in a country's economic development and performance.

Comparative Economic Systems

What do we mean by an economic system? The economic system of a given country is based on institutions of ownership, incentives, and decision making, which underlie all economic activity.

The comparison of economic systems in different countries is made easier

Pure capitalism: An economic system in which all resources are owned and all relevant decisions are made by private individuals; the role of the state in such an idealized system is minimal or nonexistent.

Pure socialism: An economic system in which all basic means of production are owned by the state; the state operates the economy through a central plan.

by the classification of systems along a spectrum that ranges from pure capitalism to pure socialism. A **pure capitalist** economy is one in which all property is owned and all economic decisions are made by private individuals and in which government economic control is entirely absent. A pure capitalist economy would have no government at all. A **pure socialist** economy is one in which all property is owned by the state, which in turn makes all economic decisions and plans the economy's output in detail.

The real world is complex, however; there simply are no examples of either pure capitalist or pure socialist economic systems (although there are examples that come close to each). Moreover, it would be a mistake to draw a single line between pure capitalism and pure socialism and attempt to place various countries on either side of it. Some countries have economic systems in which most property is privately owned but in which government regulation and control is extensive; other countries have economic systems in which the opposite seems more nearly the case. Distinctions must therefore be made about the degree of capitalism and socialism in two dimensions—ownership and control of resources. Figure 37–1 illustrates the differences between existing economic systems in these two dimensions. The degree of

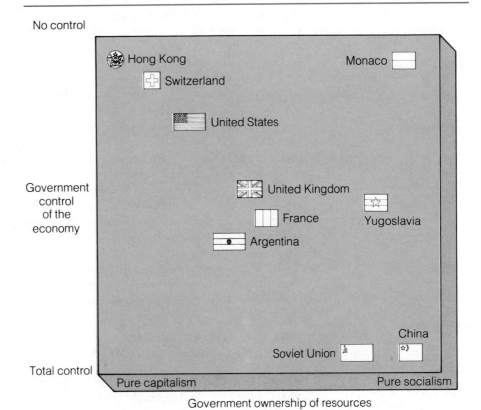

FIGURE 37–1   Classification of Economic Systems

The horizontal axis shows the degree of government ownership of productive property in an economy, ranging from pure capitalism on the left to pure socialism on the right. The vertical axis shows the degree of government involvement in economic decision making from no control at the top of the axis to total control at the bottom. Economies are classified by these two criteria. Thus, the economy of Hong Kong stands closest to the pure capitalism model; the Chinese economy is closest to the pure socialism model.

government ownership of property is along the horizontal axis, and the degree of government economic decision making is along the vertical axis.

Although there are many complex differences between any two economies, Figure 37–1 allows us to focus on two basic types of systems. To a varying degree, the economies of the United States, the United Kingdom, and France, for example, share the characteristics of **mixed capitalism.** In such a system the state is an important participant in the economy, but by and large the bulk of productive activity is undertaken by private firms and individuals. We need not dwell on the hallmarks of a capitalist economy here since this topic was covered in Chapter 2 and since this book so far has been based on the economic analysis of a mixed capitalist economy like that of the United States. (Note that one economy in Figure 37–1—that of Hong Kong—does come close to the pure capitalism model.)

In a large number of countries the dominant economic institutions are fundamentally different from those in capitalist societies. These are the socialist economies such as in China and the Soviet Union. Socialism is basically a system of economic organization in which resource allocation is determined by central planning rather than by market forces. Under socialism, at least in theory, resources are allocated and utilized according to a centrally determined and administered overall economic plan. The quantity and composition of output, the relative proportions of consumption and investment, the use of resources in production, and the allocation of final output are all decided by the central authorities. The central plan is also likely to include detailed decisions on quantities of raw materials and inputs, techniques of production, prices, wages, locations of plants and industries, and the pattern of employment of the labor force. The objectives of the central plan reflect the goals and preferences of the central planners. Consumers themselves have no direct input into the planning process; central planning effectively replaces the consumer sovereignty of capitalism.

Thus, capitalism and socialism are fundamentally different economic systems. Economists have been interested in how socialist economies function and how their performance compares to that of capitalist economies. (For a well-known debate among economists about socialism and how it works relative to capitalist markets, see Focus, "The Socialist Calculation Debate.") The Soviet economy is an example of socialism based on central planning. Soviet-style socialism is not the only type of socialism today, of course. Nonetheless, the Soviet economy can teach us a great deal about central planning and the performance of a socialist republic.

Mixed capitalism: An economic system in which most economic activities take place in a private sector but in which government also plays a substantial economic and regulatory role; an economy like that of the United States.

The Generality of Economic Analysis

Before we begin our analysis of the Soviet economy, a general point needs to be made: The basic economic problem of scarcity confronts all economic systems. Scarcity means that choices have to be made. To have more of one thing means having less of another. If a socialist economy wants to have a larger army, the resources for the army must come from alternative uses in the economy, such as agricultural labor. Changing an economy to a socialist basis does not make such trade-offs go away. Scarcity prevades economic organization; it is not something that confronts capitalism but not socialism.

Scarcity is not the only economic concept that applies generally to all economies. We consider the relevance of the following economic principles

FOCUS   The Socialist Calculation Debate

An important debate in this century among economists concerned whether socialism was possible at all, that is, whether a modern industrialized society could continue to exist if organized along socialist lines. This debate has enormous implications for the many economies in the world today that describe themselves as socialist.

The beginning of the controversy can be given as 1922, the year that Austrian economist Ludwig von Mises published a paper in German entitled "Economic Calculation in the Socialist Commonwealth." Mises attacked the position of contemporary socialist theorists that after a socialist state abolished money, the price system, and markets, it would be able to plan and direct all production. He argued that money prices determined across markets were necessary for rational economic calculation. The price system allowed resources to flow to their most highly valued uses in society. For example, although it would be technically feasible to build subway rails out of platinum, this would be an economically inefficient allocation if less-expensive substitutes were available for the rails. But only the price system, representing the competing bids of all potential users of platinum, allows for such judgments to be made. Without it, Mises argued, resources could not be allocated efficiently, and the economy could function only at a primitive level at best. A modern, technologically advanced, and complex economy would be impossible to achieve under socialism.

Socialists took the Mises challenge very seriously, with some of the more prominent writers (Oskar Lange and Abba Lerner in particular) acknowledging that Mises had identified an important weakness in the socialist position. Lange even half-seriously proposed that in the future socialist commonwealth a statue be erected in Mises' honor so that no one would forget that prices and markets would be essential under socialism too. In effect, the socialist counterattack took the form of a partial retreat from the original socialist position. Lange claimed that socialism might still work if central planning were abandoned in favor of a system wherein the state would set prices for goods and factors of production. Managers of state-owned firms would then produce until the marginal cost of their output equaled the assigned price of the good. These managers would simply requisition the inputs necessary to produce according to this rule; the state would adjust the prices it assigned in response to any shortages or surpluses of the factors of production.

While this plan seemed clever, Mises and his student, Friedrich von Hayek, countered that these "market socialist" schemes failed to solve the real problem of socialism. Although the socialist state could establish an arbitrary set of prices, it would not perform the function of a market price system unless these "prices" were able to convey an equal amount of information regarding the true opportunity costs associated with resource use. For the socialist state's prices to serve this purpose, they would have to reflect an enormous amount of information regarding the availability of resources, updated continually. If this task were possible at all, it would require high transaction costs. Further-

Ludwig von Mises (1881–1973)

Oskar Lange (1904–1965)

more, for such a system to approximate the efficiency of a market economy, incentives would have to be structured to ensure that individuals within the system would use information and resources efficiently. But this could be possible only where factors of production were privately owned, while the market socialists insisted that all resources be owned by the state.

In retrospect, was the socialist calculation debate relevant? Of existing socialist economies, only Yugoslavia resembles the market socialist proposals of Lange and Lerner. Although some of the other socialist countries—Hungary and Poland in particular—have introduced various reforms in recent years that involve a larger role for the private sector, most socialist economies are centrally planned in virtually all respects.

This observation would seem to imply that Mises and Hayek were wrong; socialism is indeed possible. However, two things are generally true about centrally planned economies: (1) Their economic performance is poor by comparison with capitalist market economies, and (2) the private sector in socialist economies, often in the form of illegal underground economies, is typically sizable and important. In these respects Mises and Hayek have been to some degree vindicated.

---

to the study of comparative economic systems: the law of demand, opportunity cost, diminishing marginal returns, and self-interest.

The Law of Demand
The law of demand is a general proposition about human behavior. It is not simply an economic law that applies to capitalism and private markets. The law of demand says that, other things being equal, price and quantity demanded will vary inversely. That is, the lower the price, the greater the quantity demanded, and vice versa. If socialist planners want to encourage the use of public transit and discourage the purchase of private cars, they will set relatively low prices on transit tickets and relatively high prices on cars. Consumers will respond according to the law of demand. If meat is scarce, its price will be set high to reflect its relative scarcity, and consumers will reduce their purchase of meat in state stores. If the socialist planners set zero prices on basic services such as medical care, demand will be so great that queues will develop, and forms of nonprice rationing to decide who gets to receive medical care will emerge.

Opportunity Cost
There is no such thing as a free lunch in any economic system. It may well be that a socialist republic offers "free" medical care, that is, medical care at zero price to recipients. But in no sense are the resources allocated to medical care free. They do not come out of thin air; they are directed away from alternative uses in the economy. These alternative uses of resources reflect their opportunity cost. If a socialist government wishes to increase its spending on space research, the resources must be reallocated from some other sector of the economy. The opportunity cost of a larger space program might be resources otherwise available for producing consumer goods.

Diminishing Marginal Returns
The law of diminishing marginal returns states that the increased application of a resource will ultimately lead to smaller and smaller increments of output, other things being equal. This phenomenon can be found in all economic systems. Worldwide, agriculture exhibits the law of diminishing returns. Generally, the stock of land in an economy is fixed, and the increased application of capital, such as tractors or fertilizer, to the fixed stock of land will lead to diminishing marginal returns. Soviet agriculture as well as American agriculture is subject to diminishing marginal returns. The only way that societies overcome the law of diminishing marginal returns over time is by

investing in capital goods production and technological change. That is, as societies expand their resource base and technical capabilities, they are able to produce more output.

Self-Interest

Economic analysis is based on the idea that individuals weigh the costs and benefits of economic choices and normally respond according to their individual self-interest. Self-interest does not mean that individuals are totally selfish. Individuals can demand or want anything, including a better life for others. Economics simply says that the amount of things individuals want is determined by the costs and benefits of having those things.

Some socialist scholars have argued that self-interested behavior will disappear under socialism and that individuals will come to work for the common good instead. This is perhaps an appealing thought, but it does not seem to be true. The structure of personal incentives helps determine the actions of individuals in both capitalist and socialist economies. For example, in the Soviet Union managers of oil prospecting teams were once rewarded in a piece-rate fashion according to the number of meters they drilled. But drilling goes harder and slower as one drills deeper, with a greater chance that the pipe and drill bits will crack or break. With meters drilled rather than oil discovered as the indicator of success, drillers quickly concluded that they should drill only shallow holes, and lots of them, significantly retarding the discovery of new oil reserves. As this one small example indicates, even socialist producers and consumers weigh costs and benefits in terms of self-interest. Ignoring such tendencies often results in poor economic performance.

The Soviet-Style Economy

With the understanding that the economic realities of the law of demand, opportunity cost, diminishing marginal returns, and self-interested behavior prevail in all economic systems, we turn now to an examination of how a socialist state—in particular, the Soviet Union—determines what gets produced, how, and for whom.

In 1917 the Bolsheviks, a group of revolutionary socialists, came into political power in Russia and formed a new state, the Soviet Union. The Bolsheviks, influenced by the writings of Karl Marx and Friedrich Engels and the leadership of Vladimir Ilich Lenin, sought to implement pure socialism— an economy in which all the means of production were owned and operated by the state and in which all economic activity would be centrally planned and controlled by the central government. Although Lenin, the revolutionary leader, did not live to see these goals accomplished, Joseph Stalin, his successor, was responsible for implementing them to a very great extent in the 1930s. Since that time, the Soviet economy has proven to be the model for most socialist economies founded in the twentieth century. In this section the basic organizational characteristics and performance of the Soviet economy are discussed.

A Command Economy

The Soviet economy is sometimes loosely described as a **command economy.** Under authoritarian control, its economy is directed by a central planning system that oversees the production and distribution of most of its resources. The managers of individual state-owned enterprises follow orders imposed by

Command economy: An economy that is centrally planned and controlled; all economic decisions are made by a state central planning agency.

the central plan that tell them what to produce, how much to produce, and how to organize production. In effect, socialist managers are more like floor supervisors than the chief executive officers of capitalist firms; they have little independent authority.

The Soviet government basically owns and operates all industry, including the capital goods sector, transportation, communication, banking and financial institutions, and even the entire wholesale and retail network. In one sense, the Soviet economy is like a giant vertically and horizontally integrated firm. However, the analogy should not be taken too far. As we will see, the socialist state is not like a giant corporation, and its citizens are not its shareholders. Whereas corporate decisions are implemented by voluntary contractual agreements, the decisions of the central planning board have the force of law.

In the case of the Soviet Union, the Communist party plays an important role in the management of the planned economy. This is true of many other planned economies based on the Soviet model, such as those of Poland, East Germany, and China. The Communist party is responsible for the selection of enterprise managers and monitors the performance of both labor and management for the central planning board. This party monitoring is carried out through the various ministries of industry, which administer the central plan in individual enterprises.

Central Planning

Although the details of the actual operation of a large economy with millions of participants are extremely complex, the basic pattern of operations of the centrally planned economy is briefly summarized in Figure 37–2.

The state central planning agency (called Gosplan in the Soviet Union) drafts an economic plan for the entire economy. This plan includes a long-range five-year plan as well as an annual plan. These plans have the force of law. The basic economic plan is directed to the more than 200,000 different Soviet enterprises, setting production targets and allocating inputs to each enterprise for all resources, including labor. Virtually all Soviet production is organized in the form of enterprises, which include everything from huge steel mills to small local baking plants. The manager of each enterprise is responsible for taking the resource inputs that have been allotted and using them to accomplish the goal assigned to the enterprise by the central plan. These targets are used to judge the performance of the firm over particular periods and to reward or penalize the responsible manager accordingly. For example, the plan might call for a steel mill to increase its production of rolled steel 5 percent over its previous year's output and reward the manager with a bonus for meeting the assigned quota (and a greater bonus for exceeding the quota).

Materials balancing: The method of central planning in the Soviet Union; the substitute for a price system in a capitalist economy; the attempt by Soviet planners to keep track of the availability of physical units of inputs and to program how these inputs are parceled out among state enterprises to produce final outputs.

Essentially, the central planning board attempts to ration raw materials throughout the entire economy without resort to a price system used in capitalist economies. It performs this task by the method of **materials balancing**—the maintaining of a balance sheet of all available supplies in the economy and of all sources of demand. In other words, the central planning board attempts to keep a comprehensive inventory of all raw materials and factor inputs—minerals, timber, labor, and so on—as well as a similar listing of the demand for these inputs by all the production units, given their assigned production goals. The method of materials balancing works out a consistent pattern of resource allocation throughout the economy that is compatible with the established planning priorities, that is, allocating necessary inputs

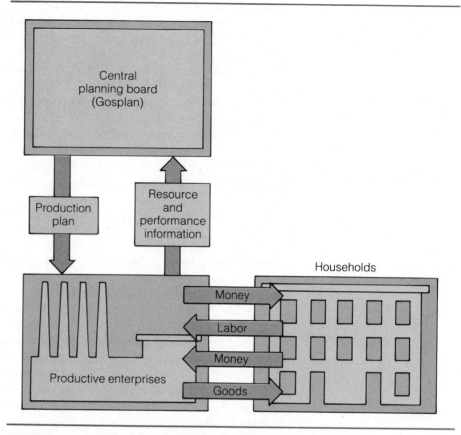

FIGURE 37–2   Basic Macroeconomic Sectors in a Soviet-Type Economy

This is a highly simplified model of the Soviet-style economy. Households supply labor inputs to enterprises in exchange for money. Labor inputs are assigned to firms by the planning process and not by a labor market. Money earned by labor is in turn exchanged for consumer goods and services produced by the enterprises. Exchange takes place in terms of money, although prices are fixed by the central planning board and remain fixed for long periods. In this simplified model, production takes place entirely within the state production establishment. Composition of output is determined not by consumers but by the central planning board.

perhaps first to military production, then to agricultural development, and so on.

By contrast with a capitalist economy, the socialist economy's central planners exercise essentially full control over production in the economy and can direct resources to whatever ends they choose. Production in the economic system is not directed by the final demand of consumers, who instead play a passive role. This is a fundamental difference in the two economies. In a capitalist economy the mix of goods provided reflects the demands of consumers. But in a centrally planned economy like the Soviet Union, the mix of goods provided reflects the decisions of the central planners. For example, if the Soviet government estimates that military spending should be 15 percent of GNP, it can reallocate resources accordingly to achieve this goal without having to convince consumers or voters to rearrange their priorities.

Central planning is not without its problems. The central planning board must coordinate millions of different inputs and potential outputs, an enormous task. Computer technology is available to make a coordinated overall plan feasible in principle. But the system must be constantly revised to accommodate unplanned-for circumstances. Droughts and other climatic events may cause crop failures, for example. When such events occur, agricultural plans must be revised. In recent years the planning system in agriculture has often failed to adjust to changed circumstances rapidly enough.

Another problem is production bottlenecks, which can develop if a supplier fails to deliver a critical input. The customer, who may in turn be some other enterprise's supplier, is not free to place an order with some other supplier of the necessary input. Enterprises have assigned suppliers, and securing official permission for shifting suppliers can be difficult and extremely time-consuming. Enterprise managers may be confronted with the choice of cutting corners where they can—perhaps radically reducing the quality of the goods they produce—or missing their assigned quota (or even failing to produce any output at all). An alternative is to make private, unofficial, and basically illegal private arrangements with other enterprise managers to overcome bottlenecks by switching suppliers. But such arrangements, however necessary, amount to cheating on the central planning process and to a partial introduction of the market process.

The Soviet planners must constantly monitor managerial performance to detect such cheating. To do so, the central planning board depends on information supplied by the managers themselves, whereas a capitalist economy would rely on a firm's success or failure in marketing its output. The incentives thus created can encourage managers to fabricate reports of available resources to avoid blame or to satisfy their output quota. While Soviet planners have long been aware of this problem and periodically undertake programs with great fanfare to overcome it, these reforms have often failed.

The Legal Private Sector in the Soviet Economy

Although it seems inconsistent with the rhetoric of socialist planning, a legal private sector does exist in the Soviet Union and in most other centrally planned economies. Private economic activity is not entirely prohibited by Soviet law, but it is restricted in two major respects. First, a private individual cannot act as a trade middleman; second, no employee can be hired for the purpose of making a profit. These two restrictions obviously greatly limit the extent to which legally sanctioned private enterprise may operate. Effectively, the only areas in which private enterprise is permitted are personal services and farming. Professionals such as doctors and teachers and craft laborers like carpenters and shoemakers are free to sell their services to consumers. And peasants on state-owned collective farms are permitted to grow crops on small assigned plots and are free to sell whatever they produce. In both cases, production of personal services or of goods for private sale can legally take place only after the completion of the day's labor in the individual's state-sector job. In effect, only moonlighting is legally permissible in the private sector.

Despite these restrictions, activities in the legal private sector realize a considerably higher rate of productivity than do comparable activities in the state sector. The primary evidence for this productivity differential is the output record of the private agricultural plots. These private plots, which constitute only slightly more than 1 percent of land under cultivation in the Soviet Union, account for approximately 25 percent of the nation's total agricultural output, in spite of the fact that the private plots are typically far below the optimum size for efficient production of the crops grown on them.

The productivity of private plots is strong evidence for the crucial role of incentives in determining rates of output. The private plot owners gain the full value of any increase in the efficiency of resource use on their own plots. In marked contrast, the collectivized state agricultural sector is plagued by poor efficiency and poor productivity. It is true that relatively more land-intensive agricultural products—such as cotton and wheat—are grown on collective farms, partially accounting for the disparity in productivity because the value per acre of these crops tends to be relatively low. But this fact only explains a small portion of the difference. Incentives clearly play a large role in the productivity of private Soviet agriculture.

Although evidence concerning the relative performance of the privately provided personal services industry is scarce, the productivity advantages there are similar. Quality seems higher, too. Although in theory medical treatment is provided free to all Soviet citizens, the quality of such care often tends to be low. It is reported that some surgeons are able to command private fees of over 1000 rubles (about $1350) for major operations from individual patients concerned with the quality of their care.

In addition to the legally sanctioned private sector in the Soviet Union, there is reported to be a booming underground or illegal private economy, somewhat like the underground economy in the United States. Focus, "The Soviet Underground Economy," describes these activities in the Soviet Union.

Comparative Performance

Despite the problems of centrally planned socialist economies, their performance relative to capitalist economies is worth noting. For example, a comparison is presented in Table 37–1 of the United States and the Soviet Union, the largest capitalist and socialist economies, with real GNPs of 2.9 and 1.4 trillion dollars, respectively, in 1980.

The total output of the U.S. economy is roughly twice that of the Soviet Union, although the Soviet Union has both a larger land area and a larger population. More food per person is produced in the United States by fewer agricultural workers; the United States also produces many more automobiles. However, the Soviet Union produces more oil and steel than the United States.

Looking at these two economies from a broader perspective, we must note that the investment rate in the Soviet Union is high. In 1980, for example, the Soviets allocated 29 percent of GNP to investment. The investment rate

TABLE 37–1   A Comparison of the U.S. and Soviet Production Characteristics, 1980

	United States	Soviet Union
Total output (per year)	$2.9 trillion	$1.4 trillion
Oil (barrels/day)	$8.57 million	$12.18 million
Steel (metric tons/year)	$100.8 million	$148.0 million
Automobiles (units/year)	$6.4 million	$1.3 million
Farm workers (percentage of population)	3 percent	24 percent
Meat (per person per year)	256 pounds	126 pounds
Grain (per person per year)	2552 pounds	1571 pounds
Population	230 million	248 million
Land area	3.6 million square miles	8.6 million square miles

Source: U.S. News and World Report (March 1, 1982), pp. 34–35.

FOCUS   The Soviet Underground Economy

The private sector in the Soviet economy is apparently much larger than what Soviet law permits. There is abundant evidence for the existence of a large Soviet underground economy, comprising private production activities that are officially prohibited or strongly discouraged by state authorities. In the Soviet Union the underground economy is called *na levo*—literally, "on the left."

Gur Ofer of Hebrew University, a specialist in the Soviet economy, calculates that up to 12 percent of the average Soviet citizen's income derives from the illegal private economy, and 18 percent of all consumer expenditures are made there. The size of the Soviet underground economy has been estimated to exceed 10 percent of the Soviet Union's GNP, or approximately $75 billion.[a] The underground economies in most Eastern European economies are of roughly comparable size, and significant underground sectors exist in other Communist countries like China and Vietnam.

Of course, some of this activity involves theft of resources owned by the state, but a large part of it seems to involve what might be termed capitalist acts between consenting adults. For example, records by the late John Lennon sell for up to 70 rubles (about $90) in Moscow. In Russia, buyers eagerly approach tourists

[a]*Time* (June 23, 1980), p. 50.

on the streets, offering them up to $200 for the blue jeans they are wearing. It has been reported that between 20 and 33 percent of liquor consumed in the Soviet Union is illicit samogen (a vodka-like moonshine often distilled to over 100 proof), produced by 250,000 to 300,000 people devoting the equivalent of full-time work to the activity.[b] And all of this trade is quite illegal.

Since the 1930s, the Soviet government has used its criminal code to try to suppress underground trading, making acting as a private entrepreneur or speculator punishable by severe penalties, including the death sentence. Elizaveta Tyntareva, a lawyer in Lithuania, opened an underground business a few years ago in which she bought consumer goods (watches, cameras, umbrellas, and other things) in areas where they were in oversupply and sold them in areas where they were in short supply. Early in 1980, she was arrested and convicted under the tough antispeculation laws; she was sentenced to twelve years in prison.[c] In May 1981, the newspaper *Bakinsky Rabochii* reported the execution of three men by firing squad. They were convicted of conspiring to turn the No. 3 Knitwear Shop in Baku into an underground factory to make private profits through an illegal two-shift work schedule that signifi-

[b]*New York Times* (March 8, 1981), p. 11.
[c]*Time* (Jan. 23, 1980), p. 50.

Black market activity in the Soviet Union.

cantly increased production. By selling off the additional production to satisfied customers, they had accumulated more than 2 million rubles in three years.[d] But despite draconian penalties, the underground economy continues to flourish.

It would be a mistake to regard the underground economy as functioning solely to shift consumer goods to the highest bidder. The illicit private sector helps ease shortages and reduce inefficiency in the planned sector of the economy as well. Managers of collective

[d]U.S. News and World Report (November 9, 1981), p. 42.

farms admit that often the only way to meet their production targets is to buy supplies on the black market. In the Soviet economy (and apparently Eastern Europe as well) planned production depends on an extensive system of bribery to function; Russian enterprise managers employ tolkachi, professional "expeditors" who basically bribe suppliers to provide necessary inputs.

In general, bribery (in the form of both actual payments and favors) performs a vital role in greasing the wheels of the planned economy. The underground private sector provides coordinating services that increase the efficiency of the aboveground planned sector.

in the United States is only 19 percent. In effect, the Soviet Union emphasizes investment over consumption to a much greater extent than the United States. Centralized planning allows planners to stress industrial development and capital accumulation. In the United States, both investment and capital accumulation are dependent on the savings and investment decisions of individuals and, in the long run, on the decisions of consumers.

While the investment rate in the Soviet Union has been relatively high, the efficiency of investment has remained generally low. Productivity of both labor and capital lags appreciably behind that of Western economies. For example, while Soviet investment in agriculture has increased fivefold since World War II, the growth in Soviet agricultural output has been low, and negative per capita growth rates in annual grain production are common. The data in Table 37–2 illustrate this point.

In the early 1970s, the growth rate of the Soviet economy appeared to be relatively high, however. Table 37–3 presents data on the growth of real gross domestic product during the period 1970–1980 for a variety of countries, including the United States and the Soviet Union. Over the entire

TABLE 37–2   Soviet Union per Capita Grain Production

	Grain Production (millions of metric tons)	Population, Midyear (millions)	Kilograms of Grain (per capita)
1970	186.8	242.8	769
1971	181.2	245.1	739
1972	168.2	247.5	680
1973	222.5	249.8	891
1974	195.7	252.1	776
1975	140.1	254.5	550
1976	223.8	256.8	871
1977	195.7	259.0	756
1978	237.4	261.3	909
1979	179.3	263.4	681
1980	189.1	265.5	712
1981	158.0[a]	267.7	590

[a]Unofficially reported by Soviet Union.

Source: Hearings Before the Joint Economic Committee of the Congress of the United States, Allocation of Resources in the Soviet Union and China, 1982, 97th Cong., 2nd sess. (Washington, D.C.: U.S. Government Printing Office, 1982), p. 269.

TABLE 37–3   Average Annual Rate of Growth in Real Gross Domestic Product For Various Economies, 1970–1980

Country	Rate of Growth, GDP	Rate per Capita
United States	3.7	2.4
Canada	4.9	3.5
West Germany	3.5	3.3
Sweden	3.2	2.4
United Kingdom	2.5	2.2
Australia	4.4	2.8
USSR[a]	5.1	4.1
Yugoslavia[a]	5.9	5.4
Japan	8.0	6.9
India	3.4	1.4
Brazil	7.1	4.8

[a]Data are for net material product; this figure does not include public administration, defense, and professional services.

Source: *World Development Report, 1982* (New York: Oxford University Press, 1982), Tables 1 and 2.

period, the Soviet growth rate was 5.1 percent. Only Japan's 8 percent exceeded this rate. The U.S. growth rate was 3.7 percent.

Soviet GNP growth has tended to be erratic, however. The rate of economic growth in 1961–1965 averaged 5 percent. This figure encompassed a high of 7.6 percent in 1964 and a low of 2.2 percent in 1963. Since 1975, Soviet growth rates have been generally falling, as illustrated in Figure 37–3. Analysts believe that Soviet GNP growth is gradually declining because of the increasing cost of raw materials as more accessible sources dry up. The Soviet Union is extraordinarily rich in raw materials, holding a large proportion of the world reserves of most economically important minerals, as well as oil and natural gas. But reserves of these resources are increasingly located in more remote areas, making exploitation more costly.

Finally, comparisons of Soviet and Western GNP do not take into ac-

FIGURE 37–3

Growth of Soviet GNP

Soviet economic growth in the late 1970s and early 1980s continued to be erratic and has generally declined over time. Data are constant-price data.

Source: Hearings Before the Joint Economic Committee of the Congress of the United States, *Allocation of Resources in the Soviet Union and China, 1982*, 97th Cong., 2nd sess. (Washington, D.C.; U.S. Government Printing Office, 1981), p. 4.

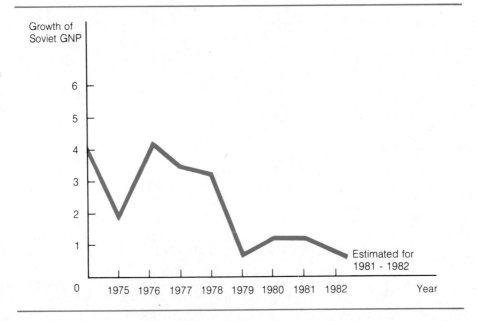

TABLE 37-4   Estimated Shares of Personal Income, by Quintile

Income Quintile	Share of Income (percent)		
	United States	Sweden	Soviet Union
Lowest	6.9	7.7	7.5
Second through fourth	52.5	56.5	55.0
Highest	40.6	35.9	37.5
Top 5 percent	15.9	12.9	14.0

Source: Lowell Galloway, "The Folklore of Unemployment and Poverty," in S. Pejovich, *Governmental Controls and the Free Market* (College Station: Texas A & M University Press, 1976), pp. 41–72.

count the qualitative aspects of the goods produced. The quality of Soviet consumer goods is generally recognized to be quite poor relative to similar Western goods. As the planned economy does not function on the basis of consumer sovereignty, a considerable proportion of production represents low-quality goods—indeed, goods that consumers do not want at all.

Income Distribution in the Soviet-Style Economy

One of the stated goals of socialist economies is to reduce income inequality among members of society. What can be said about the performance of socialist economies in this regard?

Data issued by the Soviet government are fragmentary, but they allow a rough comparison of income distribution in the Soviet Union with that in Western countries. One such comparison, involving the Soviet Union, the United States, and Sweden (a socialist democracy), is presented in Table 37–4. Interestingly, differences in the degree of inequality in the respective income distributions appear to be comparatively minor. The number of low-income individuals is about the same in the three economies, as is the number of individuals earning the highest incomes.

But this comparison may understate the degree of inequality in the Soviet economy. Special privileges are available to an enormous extent to Communist party members and bureaucrats, especially at the upper levels. For example, automobiles can usually be quickly purchased by members of the elite, while an ordinary buyer may have to place an order and wait a year or more. Soviet party officials and specially favored groups are permitted to shop in special stores stocked with foreign goods unavailable to ordinary citizens. Thus, the elite are often protected from the problems of shortages, waiting in long lines, and poor-quality goods that plague ordinary citizens. These special privileges mean that the effective degree of inequality in Soviet society is much higher than published figures suggest.

The Problem of Economic Development

Economic development: The processes through which a country attains long-run economic growth, such as by capital formulation and saving.

We turn now from comparisons of socialist and capitalist economic systems to a consideration of the disturbing differences between more-developed and less-developed countries.

In economic terms, development is not quite the same as growth. Economic growth usually refers to the increase in per capita GNP in an economy; **economic development** is the process of capital formation and the growth in economic productivity that causes rates of economic growth to rise. The history of the U.S. economy provides a model of successful economic development, leading to high and sustained rates of economic growth combined with a growing population. The U.S. economy in 1776 was in many respects an underdeveloped economy. The transition from a low to a high state of development occurred rapidly during the nineteenth century.

In the modern world, the developed nations are the relatively rich economies, and the less developed are the relatively poor economies. Many, though not all, less-developed countries are relatively stagnant economically. While their population is often growing rapidly, their per capita income grows slowly, if at all. As we will see, the form of economic institutions plays an important role in determining the extent to which economies develop.

The Characteristics of Less-Developed Countries

Less-developed countries: Countries with extremely low levels of real GNP per capita, dependence on subsistence agriculture, extremely low rates of savings, and high rates of population growth.

Most countries of the world, including China, the most populous in the world, are defined as **less-developed countries,** and three-quarters of the world's population live in less-developed countries. But there is an amazing amount of diversity among these countries; the only thing they have in common is their less-developed status. Economically, what characteristics are associated with the low development rates of less-developed countries?

In describing less-developed countries, there is general agreement that four features stand out: low per capita income, a dominant agricultural-household sector, relatively low savings rates, and rapid population growth.

A less-developed country is a poor country in the sense that its per capita GNP is low. The poverty of less-developed countries compared with that of developed countries is in some cases quite severe. Table 37–5 presents some relevant data. In 1980, 49.3 percent of the world's population (that is, about 2.16 billion people) lived in countries where the average per capita GNP was less than $420 (the low-income group in Table 37–5). At the same time, these countries generated only 4.9 percent of the world's total income. In sharp contrast, the nineteen countries with a per capita GNP of $4800 or more accounted for only 16.3 percent of the world's population but generated about 65 percent of the world's total income. In many of the countries in the low-income group, the present low level of development has not changed dramatically in hundreds of years.

The second defining characteristic of less-developed countries is a domi-

TABLE 37–5   Population, Income, and Growth, 1980

Country Group[a]	Population (millions)	Population (percentage of total population)	Per Capita Income (dollars)	Total Income (billions of dollars)	Total Income (percentage of total income)
19 industrial countries	714.4	16.3	10,320	7372.6	65.0
6 centrally planned economies	353.3	8.1	4640	1639.3	14.4
100 less developed countries, including:	3314.1	75.6	705.5	2338.1	20.6
4 high-income oil exporters	(14.4)	(0.3)	(12,630)	(181.9)	(1.6)
63 middle-income countries	(1138.8)	(26.0)	(1400)	(1594.3)	(14.0)
33 low-income countries	(2160.9)	(49.3)	(260)	(561.8)	(4.9)
Total (125 countries)	4381.8	100	2585	11,350	100

[a]Industrial countries have per capita incomes of at least $4,800; middle-income countries have per capita incomes of $420–4800; low-income countries have per capita incomes below $420.

Source: World Bank, *World Development Report, 1982* (New York: Oxford University Press, 1982), Annex Table 1, pp. 110–111.

nant agricultural-household sector. In a way this is merely another way of saying that less-developed countries are poor—a large proportion of their population is engaged in subsistence agriculture. Nearly two-thirds of the labor force of the low-income countries of Asia, Africa, and South America are employed in agriculture. In contrast, only 2 percent of the U.S. labor force is employed in agriculture. The size of the household (or nonmarket) sector in less-developed countries is generally much larger than in developed nations. Most households in less-developed countries are engaged in subsistence production in the sense that they raise their own food, make their own clothes, and construct their own homes. The degree of specialization and exchange in these economies is very limited.

The other two characteristic features found in most less-developed countries (although there are exceptions) are low savings rates and high rates of population growth.

Many poor economies have low savings rates because saving is very difficult (or even impossible) when income is at or near the subsistence level. Low savings rates mean that very little income is set aside each year for investment in capital goods for increased production. This inability to save in turn contributes to continuing low rates of development. The term sometimes used for this problem is the **vicious circle of poverty.**

Rapid population growth is another condition commonly associated with less-developed countries. The population of the poor countries of Asia, Africa, and South America has been expanding at an average rate of about 2.5 percent. At this rate, the populations of these nations double every twenty-five or thirty years. In contrast, the populations of developed nations grow less than 1 percent per year on average. This difference is often taken to mean that rapid population growth is a major contributor to low rates of development because such growth imposes an increasing burden on the limited resources of less-developed countries. The more people, the greater the need to provide education, health care, and other basic services—which require resources that would otherwise be available for capital investment. To the extent that rapid population growth represents more mouths to feed, per capita income will decline as population grows, other things being equal.

Vicious circle of poverty: The idea that countries are poor because they do not save and invest in capital goods and that they cannot save and invest because they are poor.

The Economic Gap Between Developed and Less-Developed Countries

Like wealth and poverty, development and underdevelopment are relative concepts. In 1750, England was probably the richest and most developed country in the world, but by modern standards mid-eighteenth-century England was relatively poor and undeveloped. Moreover, we cannot judge reliably whether any given country should be counted as a developed or less-developed economy on the basis of secondary characteristics—such things as the rate of population growth, degree of literacy, or degree of industrialization. There are, for example, some developed countries that have high rates of population growth and a lesser degree of industrialization than do some less-developed countries. A more reliable way to distinguish quantitatively between the developed and less-developed countries is by per capita income, that is, the level of a country's GNP divided by its population.

Comparing per capita income across countries involves some problems, however. Per capita income is a measure of the availability of goods and services to individuals, but it is obvious that we cannot draw meaningful comparisons between the incomes of Mexico, China, and Australia when GNP in each case is calculated in terms of a different national currency—

Exchange rate conversion method: A method of comparing economic well-being across countries by converting the currencies of different countries into a single currency, such as the U.S. dollar.

pesos, yuan, and pounds, respectively. The simplest means of overcoming this problem is to use the exchange rate between national currencies to convert the GNP of each nation into a common currency. Most international income comparisons follow this technique, with U.S. dollars usually constituting the common currency. The term for this technique is the **exchange rate conversion method.**

Unfortunately, even international income comparisons based on the exchange rate conversion method can be misleading. When international comparisons are made, the intention is to measure the differences in standards of living between different countries. Exchange rate conversion may reflect these differences poorly. While the exchange rate reflects differences in the purchasing power of currencies for goods that are traded across international markets, it may be a poor indicator of differences in the purchasing power of currencies with respect to goods and services that are not exchanged across international markets. For example, if 1 U.S. dollar will purchase 1½ Chinese yuan in the foreign exchange market, this exchange rate does not mean that the dollar will purchase exactly 1½ yuan of housing, dental care, or education in the United States (as 1½ yuan would in China). Differences in climate, culture, and tastes must be taken into account.

A technique for greatly improving international income comparisons involves expressing the conversion ratio between currencies in terms of their ability to purchase a typical bundle of goods and services in the countries where they are issued.[1] For example, the bundle of goods for the United Kingdom might include relatively expensive housing, heating, and food products, and the bundle for Thailand relatively cheap housing, heating, and food products. The United Nations International Comparison Project, a study begun in 1968 by the U.N. Statistical Office, has devised a workable purchasing power index for several currencies. In this **purchasing power parity method,** each category in the bundle of typical goods and services is weighted according to its contribution to GNP. The dollar cost of purchasing the typical bundle is then compared to the dollar cost of purchasing a similar bundle in the United States. After the purchasing power of the nation's currency is determined in terms of the typical bundle, this information is used to convert the GNP of the country in question to a common currency unit, the U.S. dollar.

The size of the gap in per capita income between the developed and the less-developed countries has been calculated by the World Bank using both methods. The results are shown in Table 37–6. Using the exchange rate conversion method, per capita income in the industrial market economies in

Purchasing power parity method: A method of comparing economic well-being across countries based on the idea of what it costs (in dollars, for example) to purchase a typical assortment of goods in each country.

TABLE 37–6   Measuring the Economic Gap Between Developed and Less-Developed Countries: Per Capita GNP, 1980

Country Group	Exchange Rate Conversion Method (dollars)	Purchasing Power Parity Method (dollars)
Less-developed countries	850	1790
Industrial market economies	10,660	8960

Source: World Bank, *World Bank Development Report 1981* (New York: Oxford University Press, 1981), p. 17.

[1]Dan Usher, *The Price Mechanism and the Meaning of National Income Statistics* (Oxford: Clarendon Press, 1968).

1980 was estimated to be $10,660, while that in less-developed countries was estimated to be $850. By this method of comparison the industrial market economies appeared to have a per capita income twelve times greater than that of the LDCs. However, the relative figures calculated in terms of the purchasing power parity method indicate that the more-developed countries' per capita GNP is only five times greater than that of the less-developed countries.

Purchasing power parity estimates of per capita income are a more accurate indicator of relative international performance. The problem is that they are more difficult to construct. However, even in purchasing power parity terms, there is still a very significant gap between the incomes of developed nations and less-developed nations.

Causes and Cures for Underdevelopment

It is clear that some countries are much worse off than others economically. It is not altogether clear, however, what causes the difference. Various factors have been blamed as causes of underdevelopment, including the vicious circle of poverty, rapid population growth, exploitation by more-developed nations, and the lack of economic incentives. It is important to determine which circumstances are significant in each country because attempts to narrow the gap between the haves and the have-nots cannot be successful unless they address the true causes of underdevelopment.

The Vicious Circle of Poverty

One idea that is often put forth to explain persistent underdevelopment is the vicious circle of poverty. Since incomes are low in underdeveloped countries, savings and investment rates tend to be low. A relatively large proportion of income is devoted to final consumption compared with that in developed countries. Low rates of investment retard capital formation, adoption of new technology, and the growth of future income. Thus, less-developed countries tend to remain less developed. Such countries may find themselves in a pit of underdevelopment from which they cannot escape.

Less-developed countries unquestionably could develop faster if they were able to allocate a higher proportion of their resources to economic growth. But the vicious circle theory contends that the relative poverty of poor countries actually prevents their economic development. Can this proposition be true? Clearly, all presently developed countries were at one time less developed. Yet these countries overcame their own dire conditions.

Whether the vicious circle of poverty is a cause of the lagging development of particular less-developed countries is basically an empirical question. We can determine a statistical answer to the question. Is much more income proportionately allocated to final consumption than to savings and investment in less-developed countries? The savings and investment rate in most countries of Africa and Southeast Asia is between 10 and 15 percent of GNP. In contrast, the savings and investment rate for industrial nations is generally between 20 and 25 percent of GNP. However, as Table 37–7 shows, some persistently underdeveloped countries have considerably lower rates of final consumption expenditure and therefore higher rates of savings and investment than do some highly developed economies. There evidently is no simple relation between the rate of savings and investment and the rate of economic development. While some less-developed countries, such as Bangladesh, have much higher rates of final consumption than developed

TABLE 37–7   **Final Consumption Rates and per Capita GNP**

	Final Consumption Expenditure, Government and Private (percentage of GNP)	Per Capita GNP (U.S. dollars)
Developed Countries		
United States	81	9869
Japan	68	8901
Spain	80	3968
Less-Developed Countries		
Bangladesh	96	103
Ethiopia	91	116
India	79	169
Indonesia	70	347
Mexico	73	1481
Sri Lanka	88	196

Source: U.S. Bureau of the Census, *Statistical Abstract of the United States,* 1982–1983, charts 1524, 1525.

countries in general do, other less-developed countries, such as Indonesia, have significantly lower rates of final consumption than most developed countries.

Population Growth and Economic Development

One of the characteristics shared by most developing countries is rapid population growth. The population of poor countries in Africa, Asia, and South America has been expanding in recent years at an annual rate of about 2.5 percent. By contrast, the population of the developed countries of Europe, Japan, and North America has been growing at a rate of around 1 percent per year. A cause-and-effect relation between rapid population growth and retarded economic growth of less-developed countries has often been claimed. Surely rapid population growth may impose substantial costs on less-developed countries, insofar as education and health care as well as elemental necessities like food and clothing must be provided from limited available resources. While this line of reasoning is plausible on the surface, however, there are reasons to believe that to blame underdevelopment on rapid population growth is an oversimplification that fails to consider that in economic terms both costs and benefits may be associated with population growth.

First, the high rate of population growth in the less-developed countries is fostered not by increasing birth rates but by stable birth rates combined with a sharp decline in mortality, or death, rates. Mortality rates in the less-developed world have fallen by 50 percent in the last thirty years because of advances in the control of diseases. Although infant mortality remains relatively high, life expectancy at birth in the less-developed world increased from about 35 years in 1950 to about 58 years in 1981. Surely, life expectancy at birth is an important measure of economic development. Hence, the population explosion in the less-developed countries is paradoxically evidence of economic advance because increased life expectancy is one result of economic progress.

High rates of fertility in less-developed countries—that is, a high demand for children—are also cited as a cause of underdevelopment. Both modern and traditional methods of birth control are generally available in less-developed countries, but this has not slackened demand for large families. High rates of fertility, other things being equal, do imply reduced per capita income. But the costs of having children that families in less-developed countries must face are lower than the costs faced by families in Western countries. In less-developed countries, children frequently represent valuable capital assets. It is common for quite young children—four or five years of age—to enter the labor force, especially in the agricultural sector. Only very young children are usually dependent in the sense that the marginal product of their labor does not substantially offset the cost to their families of clothing and feeding them. This usefulness of the young is reflected in the high labor force participation rates for many less-developed countries compared with developed countries. In short, children in less-developed countries apparently make a substantial contribution to aggregate output.

Thus, while population growth may explain underdevelopment, in whole or in part, in some cases it is clearly not a general explanation.

International Wealth Distribution and Economic Development

Is economic exploitation by richer countries a cause of underdevelopment? If so, will financial help from more-developed countries solve the problem? In 1974, the U.N. General Assembly adopted a Declaration on the Establishment of a New International Economic Order (NIEO), in which it asserted that only large-scale redistribution of wealth from rich to poor countries can accelerate the latter's agonizingly slow development process.

Two ideas lie behind the NIEO: the vicious circle of poverty and the responsibility of developed countries for the underdevelopment of poor countries. In the minds of NIEO advocates, presently developed countries are all colonial powers who developed their own economies largely through a redistribution of resources taken from less-developed countries. Today the less-developed countries' circle of poverty is unbreakable unless the redistribution of wealth is reversed through trade preferences for less-developed countries, commodity agreements (in effect, long-run contracts for various agricultural and other goods), and, most important, cancellation of some of the enormous foreign debts owed by less-developed countries to foreign banks. An important clause of the declaration states that discrimination (that is, any restrictions on the use of funds) should not be involved in grants of aid to governments of less-developed countries. What those governments choose to do with the aid should be entirely left to them.

The claims of the advocates of the NIEO are understandable. Many presently developed nations long colonized many of the presently less-developed nations. International trade has continued among developed and less-developed countries on a large scale, yet the less-developed countries have progressed slowly. Finally, common sense would seem to suggest that additional aid could only help matters.

The NIEO proposal, however, has many opponents, who refute the claim that past and present commercial contacts between the Western developed countries and the less-developed countries is a source of the less-developed countries' persistent poverty. Opponents point out that the list of the least-developed countries consists almost entirely of those that have had little or no involvement with international trade, such as Burundi, Chad, Rwanda, and Bhutan. Those with a record of the most extensive contacts with the

West, including Mexico, Singapore, and Brazil, are among the most advanced poorer countries.

Opponents of NIEO also challenge the notion that trade between the developed nations and the less-developed nations has been a one-way street. They stress that there were and are gains to the less-developed countries from this trade. Opponents also disagree with NIEO's proposal to abolish economic discrimination against the less-developed countries, which would effectively allow them to follow any sort of economic policy. Opponents cite economic policies of the less-developed countries that have severely retarded their own economic development, including persecution and expulsion of productive minority groups (Asians in East Africa, Chinese in Southeast Asia, and many others), enforced collectivization of farming, the establishment of state export monopolies, and the confiscation of the property of productive groups for political purposes.

Thus, the economic policies pursued by governments in less-developed countries seem to play a crucial role in their own economic development. Many government policies seem to inhibit economic growth and reduce the performance of the economy. In this context, critics question unrestricted aid to less-developed countries. The same governments that promulgate policies detrimental to economic development are unlikely to invest foreign aid in an economically efficient manner. To opponents of the NIEO proposals, the theory of exploitation from without underestimates the importance of internal institutions and economic policies in the development process. These critics argue that development is not determined by the greed or generosity of already developed nations but to a large degree by economic policies in the less-developed countries themselves.

Property Rights Arrangements and Economic Development

Can differences in economic institutions be pinpointed as a major cause of underdevelopment? It is often argued that economists are not able to bring societies into the laboratory and use the experimental method to test their theories. While this is certainly true, economists are able to make careful observations of how different economic institutions affect economic performance. There are contemporary examples of different economic systems that have a common basic cultural setting, a similar climate, and a common ethnic-religious origin but that have taken radically different routes to economic development. Comparison of such countries helps shed some light on the causes of underdevelopment. One relevant comparison can be made between North and South Korea.

North and South Korea. The two Koreas were one nation for hundreds of years, until 1950. When the two became separate nations, they adopted diametrically opposing paths to economic development. The People's Democratic Republic of (North) Korea has adhered strictly to the central planning model. In fact, by most accounts it has one of the world's most highly socialized and centrally planned economies. All industrial enterprises are either directly owned by the state or take the form of cooperatives, in which case they are owned indirectly by the state. Agriculture is carried out on either collective or state farms. Following the Soviet model, the central planning system has allocated priority to investment in heavy industry at the expense of the consumer and agricultural sectors. But despite this emphasis, the performance of the industrial sector has been disappointing. Productivity is low, and virtually all plant and equipment are obsolete, although North Korea has recently begun to purchase Western equipment and technology, including complete plants. Nominal per capita income in North Korea is only about $730 per year. There is little variety in consumer goods, and their quality is uniformly low. Shortages are common, and standing in long lines to purchase basic necessities is a way of life. Consumer durables like appliances are usually unavailable.

The land, climate, and people of South Korea are very similar to those of the North, but in other respects the differences between the two countries are dramatic. The Republic of (South) Korea has an economy based on private enterprise and a market economy. In little over twenty years South Korea transformed itself from an economy dominated by subsistence agriculture (characteristic of most poor nations) to a modern economy with an emphasis on light industry. A thriving export market has led South Korea's surge of development, with exports increasing at the impressive average annual rate of 40.4 percent between 1974 and 1977. This growth reflects in part the improving quality of its exports. The standard of living is among the highest in modern Asia; per capita GNP in 1978 was about $1200, with a diverse array of high-quality consumer goods available. Interestingly, South Korea has only 10 to 20 percent of the Korean peninsula's deposits of mineral resources, leaving the slower-developing north with the remainder.

Mainland China, Hong Kong, and Taiwan. Another example of the impact of different institutional settings on development is the comparative performance of mainland China and both Taiwan and Hong Kong. The same cultural and ethnic base has produced dramatically different results in terms of economic development.

The People's Republic of (Mainland) China is the world's most populous economy, with an estimated population in 1980 of about 1 billion (1,042,018,000). Unlike some other Communist countries, until recently China apparently did not have a legally sanctioned private sector in the

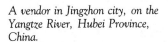

A vendor in Jingzhon city, on the Yangtze River, Hubei Province, China.

economy; economic activity was even more tightly controlled by the state than it is in the Soviet Union. Unlike the case of the Soviet economy, there has been little evidence in China of an underground economy, which in the Soviet case restores some degree of flexibility to the centrally planned system.

Since 1959, China has not issued economic statistics, so evaluation of the country's economic performance is based on outside estimates. But there is general agreement that despite promising reserves of oil, gas, coal, iron ore, and other natural resources, China is still one of the world's poorest countries. In 1979, the (estimated) GNP per capita was $516. However, even this low figure tends to overstate the standard of living in China, where quality consumer goods remain in scarce supply. Relative to average wages, goods tend in general to be more expensive than similar goods in the West. A worker can reasonably hope to save enough to buy a bicycle, but definitely not a car. Recently, the Chinese leadership, led by Communist Party Chairman Hua Guofeng and Deputy Chairman Deng Ziaoping, have begun relaxing the reliance on central planning by encouraging greater initiative on the part of enterprise management and encouraging private economic ventures to a limited extent.

In 1978, China introduced a contract responsibility system for some of its 800 million peasant farmers. The system allowed these farmers to sell their crops on the open market after they had turned over a certain percentage of their yield to the government. In 1984, China extended this free-market experiment to urban workers. In the past, urban workers had little incentive to produce; each job was guaranteed for a worker's lifetime, and both industrious and laggard workers were equally rewarded. Now, with the implementation of the new system, state-run plants are able to keep whatever profits they earn in excess of state taxes and distribute these profits to workers in the form of wage incentives. Managers are able to hire and fire workers and to set different wages for different jobs.

The most significant change in China's economy has been its adoption of a modified price system. The government has slowly relaxed its controls over prices, and the costs of many basic consumer items now fluctuate in response to supply and demand.

Although many Chinese are already enjoying greater prosperity because of these changes, the overall success of China's attempt to bring market principles into a command-type economy will depend on many factors, including the stability of the country's leadership. Eventually the world's most populous nation may become a powerful economic force in international trade.

Although the people of Taiwan and Hong Kong are ethnically identical to the Chinese people, Taiwan and Hong Kong are market economies, with relatively small public sectors. In fact, in both countries government regulation of the economy is considerably more limited than it is in the United States. Of the two, Taiwan has the greater endowment of natural resources. Hong Kong is very poor in natural resources. It has little level land for agriculture and must import both food and water from the mainland. Yet each of these economies has enjoyed successful economic development. Estimated 1979 GNP per capita for Taiwan was $1667; for Hong Kong, $2620. The standard of living in each economy is high and visibly improving. Largely unrestricted by government, individual entrepreneurship thrives. Whether Hong Kong's rapid economic growth will continue after it is returned to mainland China in 1999 is problematic, of course. The Chinese have assured

Hong Kong businessman of some freedom of action after the switch in regimes.

These real-world examples suggest that different systems of property rights and incentive structures play a vital role in determining a country's rate of economic development.

Conclusion: Explaining Economic Development

Why have economies developed at different rates? More specifically, why do less-developed countries seem to have such difficulty in matching the levels of development typical of Europe, Japan, and North America? This question is the object of intense popular controversy; we have already discussed some of the major issues in the debate, such as the vicious circle of poverty. It appears that neither the vicious circle of poverty nor large populations have prevented development of economies such as Hong Kong, Taiwan, and South Korea, which have developed rapidly by attracting foreign investment rather than relying primarily on foreign aid.

Likewise, the level of technology in most less-developed countries tends to be low. But this is an effect, not a cause, of low levels of development. Poor countries are unable to afford equipment and techniques common in rich countries. The technology itself—inventions, new processes and techniques—is widely available to less-developed countries. The knowledge necessary to improve productivity is not lacking, but the economic development making that technical knowledge worthwhile is. The lack of entrepreneurial ability in less-developed countries is also a dubious explanation. They often have untapped reserves of entrepreneurial ability, a problem that countries such as China are only beginning to address.

Of all the explanations for poor growth, economists are generally satisfied that economic incentives matter. Property rights arrangements and institutions that reward economic efficiency and permit consumers to choose freely the kinds and amounts of goods and services produced tend to promote development.

For an in-depth look at the economic underpinnings of development, Economics in Action at the end of this chapter describes the phenomenal success of the modern Japanese economy. Despite its dense population and sparse resource base, Japan has risen rapidly from a poor agricultural country to a model for what a capitalist market structure and incentives to savings and investment can accomplish.

Summary

1. The study of comparative economic systems focuses on the alternative institutional arrangements of different economies—such as who owns productive resources and who makes economic decisions—and the impact of these arrangements on economic performance.

2. Capitalism is an economic system in which a large proportion of economic activity is conducted by private individuals and firms. The private ownership of resources and the freedom to employ resources as the owner sees fit are a hallmark of capitalist economic systems. By contrast, socialism is an economic system in which the state owns and controls all the resources in the economy. Resource allocation under socialism is determined by central planning rather than by private initiative.

3. The basic economic concepts introduced in the earlier chapters of this book have general applicability to all economic systems. The law of demand, opportunity cost, diminishing marginal returns, and self-interest, for example, apply equally to capitalist and socialist economies.

4. The Soviet economy is a command economy in which virtually all economic decisions are made through a detailed system of central planning. The Soviet government owns and operates the whole economy.

5. There is a small legal private sector in the Soviet Union in agriculture and personal services. The private agricultural plots occupy a small percentage of cultivated land but are quite productive. There is also an extensive illegal private sector in the Soviet Union that eases the bottlenecks caused by central planning.

6. Comparisons of the performance of the Soviet economy and other economies are tricky. Data seem to indicate that the Soviet Union has achieved high but erratic growth rates over recent years. Moreover, when comparisons involving the quality of output are involved, Soviet performance in the consumer goods sector is generally poor.

7. Though an aim of Soviet policy is avowed to be greater equality in income distribution, the available evidence indicates that the effective distribution of income in the Soviet Union is very similar to that in the capitalist U.S. economy and the democratic socialist economy of Sweden.

8. Economic development is the study of how countries grow economically.

9. The less-developed countries of the world share several characteristics. They are characterized by low per capita income, large and growing populations, and low savings and investment rates.

10. Comparisons of per capita income betwen the industrial countries and the less-developed countries can be made by the exchange rate conversion method or the purchasing power parity method.

However the computation is handled, the economic gap between the developed and the less-developed countries is large.

11. The vicious circle of poverty refers to the fact that there is little savings in a poor economy and so little chance for future development. Economies are thus said to be poor tomorrow because they are poor today. This idea has some merit in explaining underdevelopment, but it must be applied carefully. Some poor economies, for example, appear to have high savings and investment rates.

12. Population growth is a serious obstacle to economic development. However, much of the recent population growth in less-developed countries is due to a declining death rate and to the need for young children to augment the labor supply in the agricultural sector, so the phenomenon of population growth in the less-developed countries must be interpreted cautiously.

13. There have been various responses to the New International Economic Order (NIEO) adopted by the United Nations in 1974. Under the NIEO, the rich countries of the world would transfer wealth to the poor countries. This and similar proposals are predicated on the idea that the poverty of the poor countries is due to the policies of the rich countries and not to policies and institutions that prevail in the poor countries.

14. Comparative economic development suggests that property rights, incentives, personal freedom, and other similar institutions are important in sparking economic growth. The examples of Hong Kong, Taiwan, and Japan support the validity of this approach to development.

Key Terms

comparative economics
pure capitalism
pure socialism
mixed capitalism
command economy

materials balancing
economic
    development
less-developed
    countries

vicious circle of poverty
exchange rate
    conversion method
purchasing power parity
    method

Questions for Review and Discussion

1. Would you expect the underground economy or the illegal private sector to be larger in the United States or in the Soviet Union? Why?

2. A socialist economy does not recognize private property rights. The state "owns" all resources. Name three consequences of this lack of private property rights.

3. Is income distribution in the Soviet Union fundamentally different from that in Western economies?

4. Give three examples not discussed in the text of how basic economic principles apply to the Soviet economy.

5. What are the primary characteristics of a less-developed economy?

6. The vicious circle of poverty asserts that the less-developed countries are poor today because they were poor yesterday. Do you agree? Explain.

7. Evaluate this statement: "To achieve economic

growth an economy requires lots of resources, including land, and a low rate of population growth."

8. Explain why you agree or disagree with the following statement: "A major difference between a planned economy and a free market economy is that the supply of any particular good in a planned economy is perfectly inelastic while in a free market it is elastic."

ECONOMICS IN ACTION   The Rising Japanese Economy

Industrial development in Japan began late by European standards. Japan's industrial revolution began in the early twentieth century, and by the time of World War II, Japan was a leading industrial power. But after the war, the Japanese economy was in shambles; its people were poor, and production processes were primitive. Japan was a less-developed country with a labor force employed predominantly in agriculture.

The growth and development of the Japanese economy since 1950 has been astonishing. Between 1950 and 1980 real GNP per capita in Japan rose elevenfold (from \$843 to \$9145). In 1980, Japanese GNP per capita was 80 percent of that of the United States; in 1950 it had been only 12.7 percent. Today the Japanese economy is the third largest in the world (after the United States and the Soviet Union).

To put the Japanese economy in perspective, it must be noted that Japan is 10 percent smaller than California, has only about half the population of the United States, and of the major industrial countries is the poorest in natural resources, importing virtually its entire supply of raw materials. Finally, Japan is one of the most densely populated countries on earth. Yet despite these obvious disadvantages, the Japanese development record is truly outstanding.

Several features of the Japanese economy are pertinent to an explanation of its success. The Japanese economy is primarily a capitalist market economy, but it is also characterized by extensive import restrictions. Virtually all productive resources are privately owned and operated. The labor market is different from that in the United States in that lifetime employment contracts are common in the case of larger firms (wherein workers can be dismissed only for misconduct prior to retirement at age fifty-five). There are no national labor unions. Local Japanese unions represent both blue- and white-collar employees and basically restrict their activities to establishing wage differentials between jobs. Assignment of workers to jobs is determined by individual firms without union involvement. The lifetime contract system seems to work well enough, and Japanese workers typically develop intense loyalty to

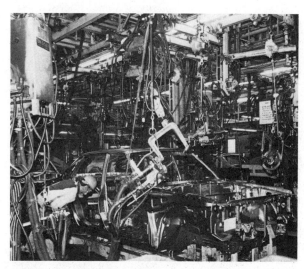

Machines do most of the work with a minimum of human attention at this automobile production line at Yokosuka, south of Tokyo.

their firms. The relative absence of labor market rigidity tends to increase economic efficiency.

Two features of Japanese economic affairs probably have played the most important role in Japan's rapid development: low marginal and absolute tax rates and high rates of savings and investment. Taxes consume a much smaller share of GNP in Japan than in Western economies. In 1978, taxes on productive effort (income, payroll, and profit taxes) consumed only 16.8 percent of GNP in Japan, compared to 21.4 percent in the United States, 20.5 percent in the United Kingdom, 23.8 percent in France, and more than 25 percent in both West Germany and Sweden. Moreover, tax rates on personal or business income have been steadily reduced by the Japanese government every year since 1950. In general, the tax rates are lower and the tax schedule is less progressive in Japan than in most Western countries.

The second significant influence on Japan's rapid development has been high rates of savings and investment. During the period from 1960 to 1979, almost one-third of the GNP of Japan was allocated to investment. This figure is considerably higher than the proportion of GNP invested in other developed countries.

Sources: The information in this feature was compiled from Steve Lohr, "The Japanese Challenge: Can They Achieve Technological Supremacy?" *New York Times Magazine* (July 8, 1984), pp. 18–41; and *Facts and Figures on Government Finance, 1981* (New York: Tax Foundation, 1981), Table 25.

The Japanese tax structure helps increase investment by taxing businesses and capital gains at much lower rates than in Western countries as well as exempting from taxation capital gains derived from sales of securities. Interest and dividends are taxed at a maximum rate of 25 percent, compared to marginal tax rates of up to 50 percent for similar income in the United States. Further, saving is encouraged by a system of tax credits. Unlike the case in centrally planned economies, where similarly high rates of investment by the state in certain areas of production are associated with shortages of consumer goods, Japan's market economy has induced high rates of voluntary investment to generate rapid economic growth and a high standard of living. In 1970, only 17 percent of Japanese families owned an automobile, compared with 62 percent in 1983; 26 percent owned TV sets in 1970, while 98.9 percent owned them in 1983.

Another feature of Japan's success is its openness to competition. At home, Japanese manufacturers must produce outstanding products at low prices in a competitive environment. Japan has nine auto manufacturers (the United States has only four), and its government has not intervened in the auto industry, allowing the extensive Japanese development of industrial robots. In exports, which constitute 13 percent of Japan's GNP, competition and price cutting are relentless. In 1977, Casio digital wristwatches cost $120; in 1984, they were selling for between $12 and $15. The Walkman portable stereos were introduced in 1979; by 1984, twelve firms were manufacturing them, with Sony offering eleven different models. Nine Japanese companies manufacture videocassette recorders (VCRs); in 1982, fifty-four new models were introduced.

The final ingredient of Japan's astonishing growth is its unabashed copying of technology developed in the West. Extremely successful in this area, it now is moving to achieve technological breakthroughs of its own by investing heavily in research and development. In 1984, 2.4 percent of its GNP was devoted to research—the same percentage as in the United States—and Japan plans to raise its expenditures in this area to 3 percent of its GNP. By 1990, the country plans to open the first of nineteen new centers of research and industrial development, each expected to cost at least $2.5 billion.

Japan is a model of market-directed growth and development. Private initiative and enlightened government policy have created perhaps the greatest success story of economic development in this century.

Question

What are some of the dangers in assessing other cultures, economies, and nations on the basis of our own value systems? In what ways might ancient Japanese culture affect the performance and evaluation of the contemporary Japanese economy?

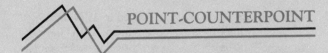

POINT-COUNTERPOINT   **David Ricardo and Gunnar Myrdal: Does Free Trade Always Benefit the Traders?**

David Ricardo

Gunnar Myrdal

DAVID RICARDO (1772–1823), contemporary and lifelong friend of Thomas Malthus, is widely regarded as the greatest of the early followers of Adam Smith. Ricardo initiated the use of economic models and developed a thorough and logical exposition of economic theory.

The study of economics began as an avocation for Ricardo. Extremely successful in securities investments and real estate, Ricardo accumulated a large fortune by his late twenties. He left school at the age of fourteen to work for his father, a Jewish immigrant from Holland, on the London Stock Exchange. Ricardo opened his own business at the age of twenty-two, became financially independent, and broke with his family and his religion in 1793 to marry a Quaker. After reading Smith's *Wealth of Nations* in 1799 while on vacation, Ricardo began to write on political economy. Despite his radicalism—he was a strong supporter of freedom of speech and an opponent of government corruption and religious persecution—his ideas gained rapid popularity. His articles were published in newspapers, and he retired from business in 1814 to concentrate on the study of economics.

It was through the publication of Ricardo's articles that Malthus became acquainted with Ricardo in 1809, and the two economists developed a strong and lasting friendship.

Ricardo agreed with Malthus's gloomy predictions about population and the economy and discussed this in his most famous work, *Principles of Political Economy,* published in 1817. In this book, Ricardo also set up a theoretical basis for the law of comparative advantage and developed the idea that free trade and exchange is beneficial to all parties involved. This book had a significant effect on the study of economics in England and has been influential to economists ever since its publication.

GUNNAR MYRDAL (b. 1898), sociologist and economist, was a joint recipient of the Nobel Memorial Prize in economics with Friedrich A. von Hayek in 1974. Myrdal is best known for his work on racial discrimination and social problems. His most famous work, *An American Dilemma: The Negro Problem and Modern Democracy,* which began as a study commissioned by the Carnegie Corporation in 1938, had an enormous impact on American attitudes toward integration. His later works explored trade and economic development: *The Political Element in the Development of Economic Theory* (1935) focused on the philosophical foundations of classical free-trade doctrine, and *Rich Lands and Poor: The Road to World Prosperity* (1957) discussed the application of the free-trade doctrine to the problems of economic development.

Myrdal was born in the Gustaf parish of Sweden. He studied law as an undergraduate at the University of Stockholm and went on to receive his Ph.D. in law from that university in 1927. In 1924, he married Alva Reimer, with whom he researched and wrote a study of Sweden's decreasing population, which was published in 1934. After graduation, Myrdal taught economics, traveled abroad, and served the government of Sweden both as a representative of the Social Democratic party and as cabinet minister. In 1947, he was appointed as executive secretary of the United Nations Economic Commission for Europe, a position he retained for the next ten years. In 1960, he became a professor of international economics at the University of Stockholm.

Is free trade always beneficial to the parties involved? Does the law of comparative advantage always apply? David Ricardo believed that the answer is yes. But Gunnar Myrdal, in his study of economic development among Third World nations, came to the conclusion that comparative advantage may not always work to the benefit of the poorer nation.

The Advantages of Free Trade

Ricardo, in *Principles of Political Economy and Taxation*, put forth the notion of comparative advantage in international trade. Briefly stated, comparative advantage means that nations should specialize in whatever good or service they can produce at lower opportunity costs than their trading partners. The example used by Ricardo was the exchange of wool and wine between England and Portugal. He showed that both countries gain when England specializes in wool production and Portugal specializes in wine production.

The key to explaining this seeming paradox lies in understanding what is meant by lower opportunity costs. Even though Portugal, in Ricardo's example, was absolutely more efficient at producing both wool and wine with its resources than England, it was relatively more efficient at producing wine. This means that there was a lower opportunity cost for Portugal to produce wine than wool because Portugal's resources were more suited to harvesting grapes. England, on the other hand, was relatively more efficient at wool production and had a lower relative opportunity cost in doing so. If each country specialized by producing only those goods for which it had lower opportunity costs, trade could take place that was beneficial to both parties.

For comparative advantage to work, trade between nations must be free from tariffs and other artificial barriers. When tariffs, quotas, or other import restrictions are set up to protect domestic industries, domestic consumers are robbed of the ability to acquire desired products at the lowest possible prices. This means that the benefits from specialization and comparative advantage are either reduced or, in the extreme, eliminated entirely. For example, Japanese automobile import restrictions in the United States artificially raise the import price of automobiles. Although voluntary, these restrictions have the effect of protecting higher-priced or lower-quality automobiles produced domestically. The impact of these regulations is to deny all or part of the advantages of specialization due to lower opportunity costs of producing automobiles in Japan.

Without such barriers, will the poorer nations of the world overspecialize in one or two goods and rob themselves of the opportunity to strengthen and diversify their economies? The answer, according to those who advocate free trade, is no.

Trade restrictions imposed for any reason diminish the domestic welfare of consumers and protect ineffi-

cient, higher-cost domestic producers. Tariffs or quotas to encourage import substitution and diversification in underdeveloped nations, although sometimes well intentioned, have seldom been beneficial to the general welfare of these countries or to their trading partners. Suppose that Indonesia, instead of purchasing communication satellites from the United States, decided to put prohibitive tariffs on such technology in order to develop a new domestic industry. The opportunity cost of doing so would be enormous in terms of the resources forgone from other productions in Indonesia. Total output in Indonesia would fall drastically to produce a good that could be acquired far more cheaply from abroad. Defenders of free trade argue that internal development of human and nonhuman resources will encourage true economic diversification and new comparative advantages in underdeveloped nations.

The Disadvantages of Free Trade

In Myrdal's view, the principle of comparative advantage and free trade works in favor of rich nations and keeps poor nations locked within a vicious cycle of poverty. In *Rich Lands and Poor: The Road to World Prosperity*, he writes, "The forces in the market tend to increase, rather than to decrease, the inequalities between regions."

To illustrate his argument, Myrdal points to the so-called banana republics, underdeveloped nations that virtually specialize in only one product. In Ricardo's view, such nations are protecting their self-interest because they can gain in trade what they could not gain by diversifying into other areas of production. But Myrdal believes that such overspecialization leaves a country extremely vulnerable to changes in demand. Should world demand for Costa Rica's bananas drop sharply, the country's economy would be in peril.

Myrdal characterizes the free-trade doctrine as outdated. "The English Classical economists did not set as their highest moral and political notion, directing their analysis, the welfare of mankind but rather the welfare of the British nation." In other words, Myrdal insists that free trade was relevant for nineteenth-century England and its self-interests, but not for the modern schism between developed nations and developing nations. He suggests that tariffs and subsidies would provide the impetus for developing nations to diversify production and to reduce dependence on one or a few exports. Ricardo and modern advocates of free trade argue that such restrictions are the certain path to welfare reductions and to lower economic growth and development.

Glossary

Ability-to-pay principle: The notion that tax bills should vary directly with income levels or the capacity of individuals to pay taxes.

Absolute advantage: The ability of a nation or a trading partner to produce a product with fewer resources than some other trading partner.

Absolute price: The price of a product measured in terms of money.

Absorption of growth: An increase in total spending that is great enough to purchase the extra output realized by an increase in productive capacity.

Accelerator principle: Changes in the level of investment expenditures (and thus GNP) take place because capital requirements change with changes in output rates.

Accommodation of cost-push inflation: An increase in the money supply by the Federal Reserve designed to offset the decrease in output that results from an increase in union and/or monopoly pressures.

Accounting costs: Direct costs of an activity measured in dollar terms; payments that a firm actually makes, in the form of bills or invoices; explicit costs.

Accounting profit: The amount by which total revenues exceed accounting costs; total revenue minus total explicit money expenditures.

Adaptive expectations theory: The proposition that present price expectations about the future are formed on the basis of actual experiences with prices in the recent past.

Adjusted income: The income of an individual after taxes are subtracted, transfer payments are added, and other items are accounted for.

Administrative lag: The period between recognition or passage of a proper policy and its implementation.

Advertising: Any communication that firms offer customers in an effort to increase demand for their product.

Age distributed income: The distribution of income over the various age brackets of the population, such as ten-year intervals.

Aggregate concentration: A measure of the percentage of total productive assets held by the largest one hundred or two hundred firms in the economy.

Aggregate demand: Spending, including consumption, investment, and government and foreign spending, over a specified period at a specific price level.

Aggregate demand curve: Graphical relation showing the level of demand at various price levels.

Aggregate supply: All the goods and services available over a specified period at specific price levels; the value of the total output of goods and services produced.

Aggregate supply curve: Curve that traces the relation between level of output and price levels.

Aggregate supply of land: The amount of land available to the entire economy at various rental rates.

Allocative efficiency: A situation in which the socially optimal amount of a good or service is produced in an industry, given the tastes and preferences of society and the opportunity cost of production.

Antitrust policy: The laws and agencies created by legislation in an effort to preserve competition.

Appreciation: The rise in the value of one currency relative to another.

Arc elasticity: A measure of average elasticity across all intermediate points between two points along a demand curve.

Artificial barriers to trade: Any restrictions created by government that inhibit trade, including quotas and tariffs.

Asset: Anything of value owned by the firm that adds to the firm's net worth.

Asset income: The income received from savings, capital investment, and land, all of which require forgone present consumption.

Austrian theory of the political business cycle: The argument that politically inspired monetary manipulation leads to costly distortions in investment decisions in the economy and ultimately to a business cycle.

Autonomous consumption: Consumption expenditures considered independent of the level of income.

Autonomous government expenditures: A level of government spending that does not change as income changes.

Autonomous investment: Investment expenditures considered independent of the level of income.

Average cost pricing: A form of price regulation that forces price equal to average cost and thus economic profits equal to zero.

Average fixed cost: Fixed cost divided by the level of output.

Average product: The output per unit of a variable input; total output divided by the amount of variable input.

Average propensity to consume (APC): The percentage of total income that is spent on consumption; total consumption expenditures divided by total income; C/Y; 1 − APS.

Average propensity to save (APS): The percentage of total income that is saved; total savings divided by total income; S/Y; 1 − APC.

Average tax rate: The percentage of income that is paid in taxes; total tax liability divided by total income.

Average total cost: Total costs divided by the level of output, or average fixed cost plus average variable cost; unit cost.

Average variable cost: Variable costs divided by the level of output.

Balance of payments: An official accounting record, following double-entry bookkeeping procedures, of all the foreign transactions of a country's citizens and firms. Exports are entered as credits and imports as debits.

Balance of trade: Roughly, the balance of a country's imports and exports of goods and services with other countries. If imports are greater than exports, the balance of trade is in deficit; if exports exceed imports, the balance is in surplus.

Balance sheet: An accounting representation of the assets and liabilities of a firm.

Barter: The trading of goods for goods with no medium of exchange such as money.

Barter economy: An economy in which all goods and services are traded for other goods and services without the use of a medium of exchange such as money.

Benefit principle: The notion that people who receive the benefits of publicly provided goods should pay for their production.

Bilateral monopoly: A market in which there is only one buyer and one seller of a resource or product.

Binding arbitration: An agreement between employers and labor to allow a third party to determine the conditions of a work contract.

Bretton Woods system: The fixed-rate international monetary system established among Western countries after World War II.

Budget constraint: A line that shows all possible combinations of two goods that an individual is able to purchase given a particular money income and price level for the two goods; budget line or consumption opportunity line.

Business cycle: Recurrent fluctuations in the level of business activity, as measured by real GNP.

Capital consumption: The loss of capital that occurs because the rate of depreciation is greater than the rate of capital formation.

Capital deepening: An increase in the total stock of capital relative to the amounts of other resources supplied.

Capitalist economy: An economic system in which the means of production are privately owned.

Capital formation: An increase in the stock of capital; the use of roundabout production to increase capital stock.

Capital/output ratio: The ratio of the amount of capital to the amount of output, or the ratio of investment expenditures to total expenditures.

Capital stock: Supply of items used in the production of goods and services; these items include tools, machinery, plant and equipment, and so on.

Cartel: A formal alliance of firms that reduces output and increases price in an industry in an effort to increase profits.

Cartel enforcement: An effort by the administrators of a cartel to prevent its members from secretly cutting price below the cartel price.

Cash reserves: Cash assets of commercial banks held in their vaults or at the Federal Reserve district bank.

Ceteris paribus: All other things held constant.

Change in demand: A shift of the demand curve or a situation in which different quantities are purchased at all previous prices.

Change in supply: A shift in the supply curve or a situation in which different quantities are offered at all of the previous prices.

Checkable deposits: Deposits at financial institutions such as credit unions or saving and loans against which checks may be written, often with restrictions, and funds withdrawn sometimes with interest or other penalties.

Choices at the margin: Decisions made by examining the benefits and costs of small, or one-unit, changes in a particular activity.

Circular flow of income: The flow of real goods and services, payments, and receipts between producers and suppliers.

Classical macroeconomic theory: The school of economic theory dominant during the late eighteenth, nineteenth, and early twentieth centuries. Classical theory viewed the macroeconomy as a self-adjusting mechanism capable of generating and sustaining maximum employment without government intervention.

Classical range: A perfectly inelastic aggregate supply curve that suggests that the price level can adjust with no change in output because of fully employed resources.

Coase theorem: Externalities will adjust to the same level when ownership rights are assigned and when the costs of negotiation are nonexistent or trivial, regardless of which party receives the rights.

Collective bargaining: The determination of a market

wage through a process in which sellers or buyers act as bargaining units rather than competing individually.

Collusion: An explicit or implicit agreement between firms in the same industry not to engage in competitive behavior.

Command economy: An economy that is centrally planned and controlled; all economic decisions are made by a state central planning agency.

Command society: An economic system in which the questions of "what," "how," and "for whom" are determined by a central authority.

Commercial banks: Chartered financial institutions that make commercial and consumer loans and accept various types of deposits, including demand deposits.

Commodity money: An item used as money that is also a salable commodity in its own right.

Common ownership: Lack of clearly specified ownership rights of a resource for whose use more than one person competes.

Comparative advantage: The ability of a nation or a trading partner to produce a product at a lower opportunity cost than some other trading partner.

Competition: A market situation satisfying two conditions—a large number of buyers and sellers and free entry and exit in the market—and resulting in prices equal to the costs of production plus a normal profit for the sellers.

Competitive labor market: A labor market in which the wage rate of a particular type of labor is determined by the forces of supply by a large number of sellers of labor and demand by a large number of buyers of labor.

Complementary inputs: Two or more inputs with a relation such that increased employment of one increases the marginal product of the other.

Complements: Products that have a relation such that an increase in the price of one will decrease the demand for the other or a decrease in the price of one will increase the demand for the other; two goods whose cross elasticity of demand is negative.

Constant-cost industry: An industry in which the minimum average cost of producing a good or service does not change as the number of firms in the industry changes; an industry for which the supply of resources is perfectly elastic, resulting in a perfectly elastic industry supply.

Constant returns to scale: The relation that suggests that as plant size changes, the long-run average total cost does not change.

Consumer equilibrium; A situation in which a consumer maximizes total utility within a budget constraint; equilibrium implies that the marginal utility obtained from the last dollar spent on each good is the same.

Consumer-initiated discrimination: A circumstance in which people prefer to purchase a good or service produced or sold by individuals of a particular sex, race, religion, or national origin; such consumers are willing to pay a premium to indulge their taste for discrimination.

Consumer price index (CPI): The device used to estimate the percent change in the prices of a particular bundle of consumption goods.

Consumers' surplus: The benefits that consumers receive from purchasing a particular quantity of a good at a particular price, measured by the area under the demand curve from the origin to the quantity purchased, minus price times quantity.

Consumption function: The curve that shows that the level of consumption is positively related to the level of income, holding all other important factors constant.

Consumption multiplier: A number that is multiplied by the change in autonomous consumption to reveal the change in income; 1/MPS.

Contractionary gap: The amount by which total expenditures fall short of full-employment income; recessionary gap.

Contrived scarcity: The action of a monopoly that reduces output and increases price and profits above the competitive level.

Corporation: A firm that is owned by one or more individuals who hold shares of stock that indicate ownership and rights to residuals but who have limited liability.

Cost-benefit analysis: A process used to estimate the net benefits of a good or project, particularly goods and services provided by government.

Cost-price spiral: A process in which increased production costs lead to increased product prices, leading to increased production costs, and so on. For the spiral to continue, the money supply must increase or real output must fall.

Cost-push inflation: Increases in the price level caused by monopoly and/or union pressure or by decreases in the supply of products or resources.

Costs: The value of opportunities forgone in making choices among scarce goods.

Costs of production: Payments made to the owners of resources to ensure a continued supply of resources for production.

Countercyclical policy: Macroeconomic policy to balance the business cycle.

Craft union: Workers with a common skill who unify to obtain market power and restrict the supply of labor in their trade; also called trade union.

Creative destruction theory: A business cycle theory that emphasizes the role of the entrepreneur in supplying new ideas and products and thereby creating profit opportunities that are emulated by competitors; such creativity and its emulation are hypothesized to occur in waves, leading to a business cycle.

Credit controls: The major methods used by the Federal Reserve, such as open market operations or changes in the reserve requirement, to control the monetary base.

Cross elasticity of demand: A measure of buyers' responsiveness to a change in the price of one good in terms of the change in quantity demanded of another good. The percent change in the quantity demanded of one good divided by the percent change in the price of another good.

Crowding-out effect. A fall in private investment as a result of an increase in government borrowing to finance expenditures.

Currency-deposit ratio: The desired ratio of currency holdings to demand deposit holdings, on average and for all money holders.

Curve: Any line, straight or curved, showing the correlation between two variables on a graph.

Decreasing-cost industry: An industry in which the minimum average cost of producing a good or service decreases as the number of firms in the industry increases; such an industry has a downward-sloping long-run supply curve.

Deflation: A sustained decrease in the overall price level.

Demand curve: A graphic representation of the quantities of a product that people are willing and able to purchase at all of the various prices.

Demand deficiency unemployment: Unemployment that results whenever aggregate demand is insufficient to purchase the quantity of goods and services produced under full-employment conditions.

Demand deposit: Any account against which funds may be transferred on demand and up to the value of the account.

Demand elasticity coefficient: The numerical representation of price elasticity of demand: $(\Delta Q/Q) \div (\Delta P/P)$.

Demand for loanable funds: A curve or schedule that shows the various amounts of money that people are willing and able to borrow at all interest rates.

Demand management: Government action that attempts to determine or change the level of total spending and thus aggregate demand.

Demand-pull inflation: Increases in the price level caused by increases in the demand for products and resources.

Demand-side policy: Fiscal or monetary policy that directs or changes the overall level of spending to create changes in macroeconomic variables.

Deposit expansion: The total value of bank deposits that an initial amount of reserves can support given some reserve requirement.

Depreciation: The wearing out of capital goods that occurs over a period of time; the fall in the value of one currency relative to another.

Depression: A severe, persistent contraction in business activity that may last for several years and result in 15 percent or more unemployment.

Deregulation: A situation in which government ceases to regulate a previously regulated industry in an effort to improve the performance of that industry.

Derived demand: The demand for factors of production that is a direct function of the demand for the product that the factors produce.

Devaluation: A change in the level of a fixed exchange rate in a downward direction; a depreciation of a fixed rate.

Differentiated products: A group of products that are close substitutes, but each one has a feature that makes it unique and distinct from the others.

Direct government purchases: Real goods and services such as equipment, buildings, and consulting services purchased by the government.

Discount rate: The interest rate that a firm uses to determine the present value of an investment in a capital good; the best interest rate that a firm can obtain on its savings; an interest rate charged by the Federal Reserve to depository institutions for loans backed up by some form of collateral.

Discretionary policy: An attempt to use temporary fiscal or monetary policy to reestablish economic stability when short-run disturbances have changed aggregate economic variables.

Diseconomies of scale: The relation between long-run average total cost and plant size that suggests that as plant size increases, the long-run average total cost curve increases.

Disposable personal income: Individuals' income less taxes, that is, spendable income.

Dissaving: The act of spending more than saving or borrowing against future income. Dissaving results from present consumption expenditures that are greater than present income.

Divestiture: A legal action that breaks a single firm into two or more smaller, independent firms.

Division of labor: Individual specialization in separate tasks involved in production of a good or service; increases overall productivity and economic efficiency.

Double coincidence of wants: A situation necessary for barter to be successful, in which the goods and services that an individual has to offer are the goods and services another individual desires.

Dumping: The selling of a product in a foreign nation at a price lower than the domestic market price.

Dynamic efficiency: A firm may at first glance impose welfare costs on the economy due to its monopoly power, but may on a closer look be a progressive, innovative firm. In other words, there may be a trade-off in analyzing real firms between static inefficiency and dynamic efficiency, between monopoly power and innovation, and so on.

Econometrician: One who utilizes economic theory, mathematics, and statistics to analyze and predict economic performance.

Economic development: The processes through which a country attains long-run economic growth, such as by capital formulation and saving.

Economic efficiency: Proper allocation of resources from the firm's perspective.

Economic goods: Scarce goods.

Economic growth: An increase in the sustainable productive capacity of society; increases in real GNP or real GNP per person through time.

Economic indicators: Key statistics, such as the net change in inventories, stock price index, and the money supply, that are used to forecast the pattern of business cycles. There are three varieties of indicators: leading, concurrent, and lagging.

Economic mobility: The ability or ease with which an individual may move from one income range to another.

Economic profit: The amount by which total revenues exceed total costs; total revenues minus total opportunity costs.

Economics: The study of how individuals and societies, experiencing limitless wants, choose to allocate scarce resources to satisfy their wants.

Economic stabilization: When aggregate variables such as the price level, the unemployment rate, and the economic growth rate are at acceptable levels, varying only slightly and temporarily from desired and achievable goals.

Economic system: A means of determining what, how, and for whom goods and services are produced.

Economic wage discrimination: A situation in which an employer pays individuals in the same occupation different wages, the wage difference based on race, sex, religion, or national origin rather than productivity differences.

Economic welfare: The situation in which products and services are offered to consumers at the minimum long-run average total cost of production.

Economies of scale: The relation between long-run average total cost and plant size that suggests that as plant size increases, the average cost of production decreases.

Elastic demand: A situation in which buyers are very responsive to price changes; the percent change in quantity demanded is greater than the percent change in price; $\epsilon_d > 1$.

Elasticity: A measure of the responsiveness of one variable caused by a change in another variable; the percent change in a dependent variable divided by the percent change in an independent variable.

Elasticity of demand for labor: A measure of the responsiveness of employment to changes in the wage rate; the percent change in the level of labor employed divided by the percent change in the wage rate.

Elasticity of supply: A measure of producers' or workers' responsiveness to price or wage changes; price elasticity of supply is the percent change in quantity supplied divided by the percent change in price.

Entrepreneur: An individual who organizes resources into productive ventures and assumes the uncertain status of a residual claimant in the resulting economic outcome.

Equalizing differences in wages: The differences in wages across all occupations that result in equality in total compensation.

Equation of exchange: An identity stating that the money supply multiplied by the income velocity of money is equal to the final output of all goods multiplied by the price level of all goods ($MV \equiv PQ$).

Equilibrium level of national income: The income level that exhibits no tendency to change. In the income-expenditures model, the level at which total expenditures equal total output.

Equilibrium price: The price at which quantity is equal to quantity supplied; when this price occurs there will be no tendency for it to change, other things being equal.

Escalator clause: Wage contracts that tie wage increases to a price index so that nominal wages will increase with increases in the price level.

Ex ante distribution: The distribution of income before government transfer payments and taxes are taken into account.

Excess capacity: A situation in which industry output is not produced at the lowest possible average total cost, the result of underutilized plant size.

Excess reserves: Cash reserves above required reserves.

Exchange costs: The value of resources used to make a trade; includes transportation costs, transaction costs, and artificial barriers to trade.

Exchange rate conversion method: A method of comparing economic well-being across countries by converting the currencies of different countries into a single currency, such as the U.S. dollar.

Expansionary gap: The amount by which total expenditures exceed the amount necessary to achieve full employment; inflationary gap.

Exploitation of labor: A situation in which the wage rate is less than the marginal revenue product of labor.

Exports (X): The goods, or the dollars received from goods, that are sold to people in foreign countries.

Ex post distribution: The distribution of income after the government influences the disposable income of individuals with taxes and transfer payments.

External costs: The cost of a firm's operation that it does not pay for.

Externality: Spillover costs or benefits; a situation in which the total costs or benefits to society of producing or consuming a good are greater than the costs or benefits to the individuals who produce or consume it.

Factor market: The market in which the prices of resources (factors of production, or inputs) are determined by the actions of businesses as the buyers of resources and households as the suppliers of resources; also called resource market.

Factors affecting demand: Anything other than price that determines the amount of a product that people are willing and able to purchase.

Factors affecting supply: Anything other than price that determines the amount of a product that producers are willing and able to offer.

Fallacy of composition: Generalization that what is true for a part is also true for the whole.

Family income: The sum of incomes earned by all members of a household.

Federal funds rate: The interest rate commercial banks can charge on overnight and short-run loans to other banks.

Federal Open Market Committee (FOMC): A committee of the Federal Reserve System, made up of the seven members of the Board of Governors of the Federal Reserve System and five presidents of Federal Reserve district banks, that directs the open market operations (buying and selling of securities) for the system.

Federal Reserve System: The central bank of the United States, established by Congress in 1913, which regulates the nation's banks and other financial institutions.

Fiat money: Money in the form of paper or some other inexpensive item that is not backed by a commodity such as gold but is legally certified by the government to function as the medium of exchange.

Final goods: Goods at the final stage of production that are not used to produce other goods.

Firm: An economic institution that purchases and organizes resources to produce desired goods and services.

Firm coordination: The process that directs the flow of resources into the production of a particular good or service through the forces of management organization within a firm.

Firm's long-run demand for labor: The various quantities of labor that a firm is willing to hire when all inputs are variable.

Fiscal policy: The use of taxation and government spending to effect changes in aggregate economic variables.

Fixed costs: Payments made to fixed inputs.

Fixed exchange rate system: An international monetary system in which each country's currency is set at a fixed level relative to other currencies, and this fixed level is defended by government intervention in the foreign exchange market.

Fixed input: Factors of production whose quantity cannot be changed as output changes in the short run.

Fixed plant: A situation in which the firm has a given size of plant and equipment to which it adds workers.

Flat-rate tax: A proportional income tax with no exemptions or deductions.

Floating exchange rate system: An international monetary system in which exchange rates are set by the forces of demand and supply with minimal government intervention.

Flow of earnings: The total amount of money earned by resource suppliers in a given period; as an approach to measuring GNP, also called income approach.

Flow of expenditures: The total amount of money that consumers, businesses, and government spend on final goods and services in a particular period.

Foreign exchange: The currencies of other countries that are demanded and supplied to conduct international transactions.

Foreign exchange markets: The institutions through which foreign exchange is bought and sold.

Foreign exchange rate: The price of one country's currency stated in terms of another's; for example, a dollar price of German marks of $0.37 means that an individual can buy one mark for $0.37.

Fractional reserve banking system: A monetary system in which a bank or other financial institution holds only a portion, or fraction, of its total assets in liquid reserves against its liabilities; also called partial reserve system.

Free enterprise: Economic freedom to produce and sell or purchase and consume goods without government intervention.

Free goods: Things that are available in sufficient quantity to fill all desires.

Free rider: An individual who is able to receive the benefits of a good or service without paying for it.

Free trade: The free exchange of goods between countries without artificial barriers such as tariffs or quotas.

Frictional unemployment: Unemployment that results from typical changes in the demand for goods and services.

Full employment: A situation in which the optimal level of employment is reached with no underemployment and no excessive unemployment; a situation in which all workers who desire employment at the current real wage are employed.

Full price: The total cost to an individual of obtaining a product, including money price and other costs such as transportation or waiting time.

Gini coefficient: A numerical estimation of the degree of inequality in the distribution of family income.

Gold standard: An international monetary system in which currencies are redeemable at fixed rates in terms of gold.

Goods: All tangible things that humans desire.

Government expenditures (G): The total amount of money that federal, state, and local governments spend on final goods and services.

Government license: A right granted by state or federal government to enter certain occupations or industries.

Government transfer payments: Money transferred by government through taxes from one group to another, either directly or indirectly, also called income security transfers.

Gross domestic product (GDP): The dollar value of all final goods and services produced in an economy in one year.

Gross national product (GNP): The market value of all final goods and services produced in an economy over a given period.

Gross private domestic investment (I): The total amount of money that private businesses spend on final goods, including capital goods.

Historical costs: Costs of production from the past.

Hoarding: Storing idle cash; accumulating money that is neither spent nor saved.

Homogeneous product: A good or service produced by many firms such that each firm's output is a perfect substitute for the other firms' output, with the result that buyers do not prefer one firm's product to another firm's.

Horizontal equity: A tax structure under which people with equal incomes pay equal amounts of taxes.

Human capital: Any nontransferable quality an individual acquires that enhances productivity, such as education, experience, and skills.

Human resources: All forms of labor used to produce goods and services.

Imperfect competition: A market model in which there is more than one firm but the necessary conditions for a purely competitive solution (homogeneous product, large number of firms, free entry) do not exist.

Implicit costs: The value of resources used in production for which no explicit payments are made; opportunity costs of resources owned by the firm.

Implicit price deflator: A price index that attempts to use the entire national output as the bundle of goods and thus to generate the most comprehensive price index and the best measure of inflation; also called GNP price deflator.

Imports (M): The goods, or the dollars spent on goods, sold by foreigners to domestic citizens.

In-kind transfer payments: Transfers of benefits other than money from government to citizens such as food stamps, public housing, and Medicare.

Income effect: The change in quantity demanded of a particular good that results from a change in real income, which has resulted in turn from a change in price.

Income elasticity of demand: A measure of buyers' response to a change in income in terms of the change in quantity demanded; the percent change in quantity demanded divided by the percent change in income.

Income-expenditures model: The theory of John Maynard Keynes showing that private expenditures are primarily determined by the level of national income and that such expenditures in turn determine the level of output and employment in the macroeconomy.

Income tax: A tax based on the level of income; an income-induced tax.

Income velocity: The average number of times that a dollar is used to purchase goods and services annually.

Increasing-cost industry: An industry in which the minimum average cost of producing a good or service increases as the number of firms in the industry increases; such an industry has an upward-sloping long-run supply curve.

Increasing returns to scale: The relation that suggests that the larger a firm becomes, the lower its costs are.

Indifference curve: A curve that shows all the possible combinations of two goods that yield the same total utility for a consumer.

Indifference map: A graph that shows two or more indifference curves for a consumer.

Indifference set: A group of combinations of two goods that yield the same total utility to a consumer.

Individual income: The sum of labor income, asset income, and government subsidies minus tax payments.

Industrial union: Workers within a single industry who organize regardless of skill in an effort to obtain market power.

Industry concentration ratios: An estimate of the degree to which assets, sales, or some other factor is controlled by the largest firms in an industry.

Industry regulation: Government rules to control the behavior or firms, particularly regarding prices and production techniques.

Inefficient tax: A tax that decreases the overall productive capacity of the country; a tax that decreases the supply of resources or decreases the efficiency of resource use.

Inelastic demand: A situation in which buyers are not very responsive to changes in price; the percent change in quantity demanded is less than the percent change in price: $\epsilon_d < 1$.

Infant industry: A new or developing domestic industry whose average costs of production are typically higher than those of established industries in other nations.

Inferior good: A product that an individual chooses to purchase in smaller amounts as income rises or larger amounts if income falls.

Inflation: A sustained increase in prices of goods and services; sustained increases in the overall price level; a reduction in the purchasing power of money.

Inframarginal rent: A type of rent that accrues to specialized factors of production.

Interest: The price a borrower pays for a loan or a lender receives for saving, measured as a percentage of the amount; the price of not consuming now but waiting to consume in the future.

Interest rate effect: The change in investment spending that results from a change in the interest rate.

Intermediate goods: Goods that are inputs in the production of final goods.

Intermediate range: An upward-sloping curve that suggests that the price level and output are directly and positively related.

International Monetary Fund: An international organization, established in the Bretton Woods system, designed to oversee the operations of the international monetary system.

Investment function: A curve that shows the relation between the level of investment and the level of some other variable such as income.

Investment multiplier: A number that is multiplied by the change in autonomous investment to reveal the change in income; 1/MPS.

Investment spending: The level of expenditures by business on capital goods and additions to inventories, by households on home construction, and by government on structures such as schools, roads, and housing.

Jobs creation bill: Legislation to increase the demand for labor either in the overall economy or in a particular sector or industry.

Joint unlimited liability: The unlimited liability condition in a partnership that is shared by all partners.

Keynesian range: An infinitely elastic aggregate supply curve that suggests that output can adjust with no change in the price level because of large-scale unemployment.

Kinked demand: A curve that has a discontinuous slope, the result of two distinct price reactions of competitors to changes in price.

Labor income: The payments an individual receives from supplying labor, equal to the individual's wage rate times the number of hours of labor supplied.

Labor-managed firm: A firm that is owned and thus managed by the employees of the firm, who have the right to claim residuals.

Labor union: A group of workers who organize to act as a unit in an attempt to affect labor market conditions.

Laissez-faire: A government policy of not interfering with market activities.

Laissez-faire economy: A market economy that is allowed to operate according to competitive forces with little or no government intervention.

Law of demand: The price of a product and the amount purchased are inversely related. If price rises, the quantity demanded falls; if price falls, the quantity demanded increases.

Law of diminishing marginal returns: A relation that suggests that as more and more of a variable input is added to a fixed input, the resulting extra output decreases, eventually to zero.

Law of increasing costs: As more scarce resources are devoted to producing one good, the opportunity costs per unit of the good tend to rise.

Law of one price: Exists in a perfect market. After the market forces of supply and demand reach equilibrium, a single price for a commodity prevails.

Law of supply: The price of a product and the amount that producers are willing and able to offer are directly related. If price rises, then quantity supplied rises; if price falls, then quantity supplied falls.

Legal barriers to entry: A legal franchise, license, or patent granted by government that prohibits other firms or individuals from producing particular products or entering particular occupations or industries.

Less-developed countries: Countries with extremely low levels of real GNP per capita, dependence on subsistence agriculture, extremely low rates of savings, and high rates of population growth.

Liability: Anything that is owed as a debt by a firm and therefore takes away from the net worth of the firm.

Limited liability: The legal term indicating that owners of corporations are not responsible for the debts of the firm except for the amount they have invested in shares of ownership.

Limit-entry pricing: A pricing policy by a firm that discourages entry of new firms by selling at a price below the short-run profit-maximizing price.

Limit pricing: The price behavior of an existing firm in which the firm charges a price lower than the current profit-maximizing price to discourage the entry of new firms and thus maximize its long-run profits.

Liquidity: The ease or speed with which any asset or commodity can be converted into money; the ability of a bank to meet current depositor liabilities in cash.

Liquidity preference theory: A proposition that the rate of interest is determined by overall money demand and supply and that economic activity flows from changes in the rate of interest.

Logrolling: The exchange of political favors, especially votes, to gain support for legislation.

Long run: An amount of time that is sufficient to allow all inputs to vary as the level of output varies.

Long-run aggregate supply curve: The relation between price and the output of goods and services after all short-run market adjustments of prices and wages have taken place.

Long-run average total cost: The lowest possible cost per unit of producing any level of output when all inputs can be varied.

Long-run competitive equilibrium: A situation in an industry in which economic profits are zero and each of the many firms is operating at minimum average total cost.

Long-run industry supply: The quantities of a product that all firms are willing and able to offer at all the various prices when the number of firms and scales of operation of each firm are allowed to adjust to the equilibrium level.

Long-run Phillips curve: A vertical line that shows that the rate of unemployment is independent of the rate of inflation when the expected rate of inflation adjusts to equal the actual rate of inflation.

Lorenz curve: A graph that shows the cumulative distribution of family income by comparing the actual distribution to the line of perfect equality.

Lump-sum tax: A fixed level of taxes or taxes that do not change as income changes; an autonomous tax.

M-1 money multiplier: The number by which the monetary base is multiplied to obtain the M-1 measure of the money supply.

Macroeconomics: Analysis of aspects of the economy as a whole.

Malthusian population principle: Population has the capability of increasing geometrically while food production can increase only arithmetically; the population can grow faster that the economy's ability to feed it.

Managed floating rate system: A system in which a country's currency is allowed to float freely in the foreign exchange market, within certain bounds; drastic changes in the value of the currency are mitigated by central bank intervention in the foreign exchange market.

Manager: An individual or group of individuals that organize and monitor resources within a firm to produce a good or service.

Margin: The difference in costs or benefits between the existing situation and a proposed change.

Marginal analysis: Study of the difference in costs and benefits between the status quo and the production or consumption of an additional unit of a specific good or service. This, not the average cost of all goods produced or consumed, is the actual basis for rational economic choices.

Marginal costs: The extra costs of producing one more unit of output; the change in total costs divided by the change in output.

Marginal cost pricing: A form of price regulation that forces price equal to marginal cost and results in optimal allocation of resources.

Marginal factor costs: The change in total cost that results from employing one more unit of a variable input.

Marginal opportunity costs: The extra costs associated with the production of an additional unit of a product; these costs are the lost amounts of an alternative product.

Marginal private costs: The increase in a firm's total costs resulting from producing one more unit.

Marginal product: The extra output that results from employing one more unit of a variable input.

Marginal product of labor: The change in total output that results from employing one more unit of labor.

Marginal propensity to consume (MPC): The percentage of an additional dollar of income that is spent on consumption; the change in consumption divided by the change in income; $1 - MPS$.

Marginal propensity to save (MPS): The percentage of an additional dollar of income that is saved; the change in savings divided by the change in income; $1 - MPC$.

Marginal rate of substitution: The amount of one good that an individual is willing to give up to obtain one more unit of another good.

Marginal revenue: The change in total revenue that results from selling one additional unit of output; the change in total revenue divided by the change in amount sold.

Marginal revenue product: The change in total revenue that results from employing one more unit of a variable input.

Marginal social costs: The increase in total costs to society (the firm plus everyone else) resulting from producing one more unit.

Marginal tax rate: The percentage of an increase that must be paid in taxes; the change in tax liability divided by the change in taxable income.

Marginal utility: The change in total utility that results from the consumption of one more unit of a good; the change in total utility divided by the change in quantity consumed.

Market: An arrangement that brings together buyers and sellers of products and resources; any area in which prices of products or services tend toward equality through the continuous negotiations of buyers and sellers.

Market coordination: The process that directs the flow of resources into the production of desired goods and services through the forces of the price mechanism.

Market demand: The total demand for a product at each of various prices, obtained by summing all of the quantities demanded at each price for all buyers.

Market demand for labor: The sum of, or overall, demand for a particular type of labor by all firms employing that labor; the total level of employment of a particular type of labor at all the various wage rates.

Market failure: A situation in which a private market does not provide the optimal level of production of a particular good.

Market or nominal interest rate: The current rate of interest paid at banks and lending institutions.

Market power: A situation characterized by barriers to entry of rival firms, giving an established firm control over price and, therefore, profit levels.

Market society: An economic system in which individuals acting in their self-interest determine what, how, and for whom goods and services are produced, with little or no government intervention.

Market supply: The total supply of a product, obtained by summing the amounts that firms offer at each of the various prices.

Market supply of labor: The total amount of labor that all individuals are willing to offer in a particular occupation at all the various wage rates.

Materials balancing: The method of central planning in the Soviet Union; the substitute for a price system in a capitalist economy; the attempt by Soviet planners to keep track of the availability of physical units of inputs and to program how these inputs are parceled out among state enterprises to produce final outputs.

Mean income: The income of the average income earner, found by dividing the total income by the number of income earners.

Measure of economic welfare: A measurement of social and economic well-being that accounts for the production of all goods, not just marketed goods. These include leisure, pollution, and certain government services.

Medium of exchange: Anything that is used as payment for goods and services, most commonly money. A medium of exchange eliminates the inefficiency and costs of barter.

Merger: The joining of two or more firms' assets that results in a single firm.

Microeconomics: Analysis of the behavior of individual decision-making units within an economic system, from specific households to specific business firms.

Mixed capitalism: An economy in which both market forces and government forces determine the allocation of resources.

Model: An abstraction from real-world phenomena that approximates reality and makes it easier to deal with; a theory.

Monetarism: A contemporary explanation of economic activity that takes into account both short-run and long-run adjustments of the economy and the different rates of change in the money supply. Monetarism combines elements of both the quantity theory of money and the liquidity preference theory plus a perspective on the inflation process that in-

cludes a new explanation of the demand for money, a distinction between the nominal interest rate and real interest rate, and the effect of people's expectations about future inflation rates on present behavior.

Monetary base: The sum of bank reserves plus currency held by the public.

Monetary policy: The act of changing the money supply or the rate of change in the money supply to create changes in aggregate economic variables.

Monetary rule: Monetary policy based on a prescribed rate of growth in the M-1 money supply.

Money: A generally accepted medium of exchange.

Money expansion: The increase of the money supply made possible by an initial amount of excess reserves in the banking system.

Money illusion: The assumption that nominal values are actually real values and that changes in nominal wages mean equal changes in real wages.

Money price: The dollar price that sellers charge buyers.

Monitor: An individual who coordinates team production and discourages shirking.

Monopolistic competition: A market model with freedom of entry and number of firms that produce similar but slightly differentiated products.

Monopoly rent: The payments to owners of capital in a monopolized industry that exceed the opportunity cost of capital.

Monopsony: A single buyer of a resource or product in a market.

Moral suasion: The Federal Reserve's ability to persuade banks that a particular course of action is desirable. Also called jawboning, moral suasion is one of the monetary tools of the Federal Reserve.

Mutual interdependence: A relation between firms in which the actions of one firm have significant effects on the actions and profits of other firms.

National income (NI): The income earned by suppliers of resources.

National income accounting: A statistical measuring of the nation's aggregate economic performance.

Natural monopoly: A monopoly that occurs because of a particular relation between industry demand and the firm's average total costs that makes it possible for only one firm to survive in the industry.

Natural rate hypothesis: Workers do not adapt instantly to changes in the demand for labor or in the wage rate but weigh their responses to changes in light of their expectations of future price and wage conditions.

Natural rate of unemployment: The rate of unemployment that occurs along with economic stability or equilibrium in labor markets; the frictional rate of unemployment plus all unemployment resulting from institutional changes that affect the decisions of workers to work and employers to demand labor.

Negative externality: A cost of producing or consuming a good that is not paid entirely by the sellers or buyers but is imposed on a larger segment of society; a situation in which the social costs of producing or consuming a good are greater than the private costs.

Negative income tax: A progressive income tax that allows for a negative tax rate (income subsidy) for income below a particular level. As income rises, the subsidy gradually diminishes to zero.

Negative, or inverse, relation: A relation between variables in which the variables change in opposite directions. A negative relation has a downward-sloping curve.

Net exports (X − M): The value of domestic goods and services purchased by foreigners minus the value of foreign goods purchased by domestic citizens.

Net foreign trade: Exports minus imports (X − M); the net amount of dollars spent on imports and exports.

Net national product (NNP): GNP minus depreciation; the value of the nation's output minus the capital used to obtain it.

Net worth: The value of a firm to the owners, determined by subtracting liabilities from assets; also called *equity*. For corporations, net worth is termed *capital stock*.

Neutrality of money: A proposition stating that the relative prices of all goods and services are independent of overall price level.

Neutral tax: A tax that has no effect on the production or consumption of goods.

Nominal GNP: Total output or total spending expressed in current dollars.

Nominal income: The amount of money or its equivalent received by income earners; money income.

Nominal rate of interest: The price of loanable funds measured as a percentage of the dollar or nominal amount of the loan.

Nonhuman resources: Inputs other than human labor involved in producing goods and services.

Nonmarket activities: Anything that an individual does while not earning income from working.

Nonprice competition: Any means that individual firms use to attract customers other than price cuts.

Nonprofit firm: A firm in which the costs of production and revenues must be equal and which does not have a residual claimant.

Nonproprietary: Relating to public ownership.

Normal good: A product that an individual chooses to purchase in larger amounts as income rises or smaller amounts as income falls.

Normative economics: Value judgments about how economics should operate, based on certain moral principles or preferences.

Normative public choice: The study of shortcomings and possible improvements of political arrangements, such as voting rules.

Oligopoly: A market model characterized by a few firms that produce either a homogeneous product or differentiated products and entry of new firms is very difficult or is blocked.

Open market operations: Purchases and sales of securities by the Federal Reserve that determine the size of the money supply by affecting the amounts of reserves in the banking system; the Federal Reserve's major method of directing the money supply and economic activity.

Opportunity cost: The value placed on opportunities forgone in choosing to produce or consume scarce goods.

Opportunity cost of capital: The value of the payments that could be received from the next-best alternative investment; the normal rate of return.

Overfull employment: A temporary situation in which the quantity of laborers employed is greater than the level of full employment.

Ownership claims: The legal titles that identify who owns the assets of a firm.

Partnership: A firm that has two or more owners who have unlimited liability for the firm's debts and who are residual claimants.

Patent: A monopoly granted by government to an inventor for a product or process, valid for seventeen years (in the United States).

Per capita income: The income per individual, found by dividing total income by the total number of people.

Perfect information: A condition in which information about prices and products is free to market participants; combined with conditions for pure competition, perfect information leads to perfect competition.

Perfect market: A market in which there are enough buyers and sellers so that no single buyer or seller can influence price.

Personal consumption expenditures (C): The total amount of money that households spend on final goods and services.

Personal income: The total amount of income actually received by individuals in a given period.

Personal saving: The portion of disposable personal income that consumers choose not to spend on final goods and services.

Phillips curve: A curve that shows the relation between the rate of inflation and rate of unemployment.

Political business cycle: The idea that incumbent politicians manipulate the economy, increasing employment before the election and fighting inflation after the election, to win reelection.

Positive economics: Observations or predictions of the facts of economic life.

Positive externality: A benefit of producing or consuming a good that does not accrue to the sellers or buyers but can be realized by a larger segment of society; a situation in which the social benefits of producing or consuming a good are greater than the private benefits.

Positive public choice: The analysis and explanation of political behavior.

Positive relation: A direct relation between variables in which the variables change in the same direction. A positive relation has an upward-sloping curve.

Post hoc fallacy: From *post hoc, ergo propter hoc*, "after this, therefore because of this." The inaccurate linking of unrelated events as causes and effects.

Poverty: A term describing family income below a defined level when other things such as size of family, location, and age are considered.

Precautionary demand: The amount of money that businesses and consumers want to hold to cover unforeseen expenses.

Present value: Today's value of a payment received in the future; future income discounted by the rate of interest.

Price ceiling: A maximum legal price established by government to protect buyers.

Price control: Government intervention in the natural functioning of supply and demand.

Price discrimination: The practice of charging one buyer or group a different price than another group for the same product. The difference in price is not the result of differences in the costs of supplying the two groups.

Price elasticity of demand: A measurement of buyers' responsiveness to a price change; the percent change in quantity demanded divided by the percent change in price.

Price expectations: Anticipations of future prices or inflation.

Price floor: A minimum legal price established by government.

Price index: A device used to estimate the average percent change in the prices of a particular bundle of goods.

Price leadership: A pricing behavior in which a single firm determines industry price.

Price level: The weighted average of the prices of all goods and services.

Prices: The opportunity costs established in markets for scarce goods, services, and resources.

Price searcher: A firm that must choose a price from a range of prices rather than have a single price imposed on it; such a firm has a downward-sloping demand curve for its product.

Price-specie flow mechanism: The mechanism by which the gold standard equilibrates the balance of trade between countries.

Price stability: A situation in which there is no change in the overall price level—the average price of all goods, services, and resources.

Price taker: A individual buyer or seller who faces a single market price and is able to buy or sell as much as desired at that price.

Principle of diminishing marginal utility: As more and more of a good is consumed, eventually its marginal utility to the consumer will fall, all things being equal.

Private costs: The total opportunity costs of production for which the owner of a firm is liable.

Private sector: All parts of the economy and activities that are not part of government.

Private-sector equilibrium: The state of equality between private expenditures and income or output; a condition of macroeconomic stability from which there is no tendency for change.

Producer price index (PPI): The device used to estimate the percent change in the prices of a particular bundle of inputs used by manufacturing business.

Production possibilities frontier: The situation represented by a curve that shows all of the possible combinations of two goods that a country or an economic entity can produce when all resources and technology are fully utilized and fixed in supply.

Productive efficiency: A situation in which the total output of an industry is obtained at the lowest possible cost for resources.

Productivity: The average output produced per worker in a given amount of time.

Productivity of labor: Output per worker.

Products market: The forces created by buyers and sellers that establish the prices and quantities of goods and services.

Profit-maximizing price: The price at which the difference between total revenue and total cost is greatest; the price at which marginal cost equals marginal revenue.

Profits: The amount by which total revenue exceeds total cost.

Progressive income tax: A tax based on a percentage of income that varies directly with the level of income.

Proportional income tax: A tax based on a fixed percentage of income for all levels of income.

Proprietary: Relating to private ownership and profit seeking by private owners.

Proprietorship: A firm that has a single owner who has unlimited liability for the firm's debts and who is the sole residual claimant.

Public choice: The economic analysis of politics; the branch of economics concerned with the application of economic principles to political decision making.

Public employees' union: Workers who are employed by the federal, state, or local government and who organize in an effort to obtain market power.

Public finance: The study of how governments at federal, state, and local levels tax and spend.

Public franchise: A right granted to a firm or industry allowing it to provide a good or service and excluding competitors from providing that good or service.

Public good: A good that no individual can be excluded from consuming, once it has been provided to another.

Publicly owned firm: A firm owned and operated by government.

Purchasing power parity method: A method of comparing economic well-being across countries based on the idea of what it costs (in dollars, for example) to purchase a typical assortment of goods in each country.

Purchasing power parity theory: A theory of exchange rate determination that states that differential inflation rates across countries affect the level of exchange rates.

Pure capitalism: An economic system in which all resources are owned and all relevant decisions are made by private individuals; the role of the state in such an idealized system is minimal or nonexistent.

Pure economic rent: The payment to a factor of production that is perfectly inelastic in supply.

Pure interest: The interest obtained from a risk-free loan.

Pure monopoly: An industry in which a single firm produces a product that has no close substitutes and in which entry of new firms cannot take place.

Pure socialism: An economic system in which all basic means of production are owned by the state; the state operates the economy through a central plan.

Purely competitive market: A coming together of a large number of buyers and sellers in a situation where entry is not restricted.

Quantity demanded: The amount of any good or service consumers are willing to purchase at some specific price.

Quantity supplied: The amount of any good or service that producers are willing to produce at some specific price.

Quantity theory of money: A hypothesis that there is a direct and predictable relation between the money supply and prices; as one increases so will the other.

Quasi-rent: The short-run payments to owners of capital in a competitive industry that exceed the opportunity cost of capital.

Quota: A limit on the quantity of an imported good.

Rate of inflation: The rate at which the overall price level of goods, services, and resources has increased over a specified time period; the percent increase in the average price of all items.

Rate of return on invested capital: Profits that are measured as a percentage of the costs of capital.

Rate of time preference: The percent increase in future consumption that is necessary to induce an individual to forgo some amount of present consumption.

Rational expectations hypothesis: A view of economic behavior that suggests that people completely and rationally anticipate changes in economic variables, including those created by government.

Rational expectations theory: A theory of macroeconomic behavior that assumes people will anticipate the effects of macroeconomic policy and will thereby be in a position to neutralize its intended effects.

Rational self-interest: The view of human behavior espoused by economists. People will act to maximize the difference between benefits and costs as determined by their circumstances and their personal preferences.

Rationing: Prices are rationing devices; the equilibrium price rations out the limited amount of a product produced by the most willing and able suppliers or sellers to the most willing and able demanders, or buyers.

Real balance effect: The change in purchasing power that results from a change in the price level and its effect on wealth.

Real GNP: Total output or total spending after adjustments for inflation or deflation.

Real GNP per capita: Output per person; real GNP divided by the population.

Real income: The purchasing power of income; the quantity of goods and services that an individual, a household, or a nation can consume with the money received from all sources of income; income adjusted for inflation or deflation.

Real rate of interest: The nominal interest rate minus the rate of inflation; the price of loanable funds measured as a percentage of the real buying power of the amount loaned.

Real wage rate: The nominal wage rate, W, divided by the price level, P.

Recession: A contraction in business activity, usually defined as a decline in real GNP over a period of six months or more.

Recognition lag: The period between a potentially destabilizing event in the economy and recognition of it by policymakers.

Relative price: The price of a product related in terms of other goods that could be purchased rather than in money terms.

Rent: A payment to a factor of production in excess of its opportunity cost.

Rent seeking: The activity of individuals who spend resources in the pursuit of monopoly rights granted by government; the process of spending resources in an effort to obtain an economic transfer.

Required reserves: A portion of a bank's cash assets that must be kept available at all times, according to the legal requirements of the Federal Reserve System.

Reserve bank credit: Loans and securities assets of the Federal Reserve System. Increases and decreases in these assets affect member institutions' reserves.

Reserve bank credit outstanding: The difference between the value of the factors increasing bank reserves and those decreasing reserves.

Reserve ratio: A percentage of the checkable or demand deposit reserves held by a bank.

Residual claimant: The individual or group of individuals that share in the excess of revenues over costs, that is, profits.

Resources: Inputs necessary to supply goods and services. Such inputs include land, minerals, machines, energy, and human labor and ingenuity (called the factors of production).

Resource specialization: The devotion of a resource to one particular occupation that is based on comparative advantage.

Resources market: The forces created by buyers and sellers that establish the prices and quantities of resources such as land, labor, and capital.

Revaluation: A change in the level of a fixed exchange rate in an upward direction; an appreciation of a fixed rate.

Right-to-work law: A law that prevents unions from forcing individuals to join a union as a prerequisite to employment in a particular firm.

Risk: The probability of a default or a failure of repayment of a loan.

Roundabout production: The production and use of capital goods to produce greater amounts of consumption goods in the future.

Saving: The act of forgoing present consumption in an effort to increase future consumption.

Saving function: The curve that shows that the level of savings is positively related to the level of income, holding all other factors, such as the interest rate, constant.

Say's law: An assumption of classical economists that the supply of goods and services in the economy creates a corresponding demand for those goods and services.

Scale of production: The relative size and rate of output of a physical plant that may be measured by the volume or value of firm capital.

Scarcity: Limitation of the amount of resources available to individuals and societies relative to their desires for the products that resources produce.

Service industry: An industry whose major product is customer service such as medical services, legal services, entertainment, food services and the like.

Services: All forms of work done for others—such as medical care and car washing—that do not result in production of tangible goods.

Share: The equal portions into which the ownership of a corporation is divided.

Share transferability: The power of an individual shareholder to sell his or her portion of ownership without the approval of other shareholders.

Shirking: A sometimes rational behavior of members of a team production process in which the individual exerts less than the normal productive effort.

Shortage: The amount by which quantity demanded exceeds quantity supplied when the price in a market is too low.

Short run: An amount of time that is not sufficient to allow all inputs to vary as the level of output varies.

Short-run firm supply: The portion of the marginal cost curve above the minimum average variable cost.

Short-run industry supply: A summation of all the existing firms' short-run supply curves.

Short-run Phillips curve: A downward-sloping curve that suggests that the rate of unemployment is inversely related to the rate of inflation when the expected rate of inflation is constant.

Shutdown: A loss-minimizing option of a firm in which it halts production in the short run to eliminate its variable costs, although it must still pay its fixed costs.

Simple money multiplier: The reciprocal of the reserve ratio.

Slope: The ratio of the change in (Δ) the x value to the change in (Δ) the y value; $\Delta y / \Delta x$.

Social costs: The total value of all resources used in the production of goods, including those used but not paid for by the firm.

Socialist economy: An economic system in which the means of production are owned and controlled by the government.

Socialization: Government ownership of a firm or industry.

Solvency: The ability of all assets to cover all liabilities.

Specialization: Performance of a single task in the production of a good or service to increase productivity.

Speculation: The buying and selling of currencies with an eye to turning a profit on exchange rate changes, predicted devaluations, and so on.

Speculative motive or demand: The amount of money desired by individuals and businesses to invest at various interest rates.

Standard of deferred payment: Money's acceptability in contractual arrangements, such as loans, involving future payments.

Standard of living: A measure of the value of real goods and services consumed by the average person.

Static inefficiency: A condition, related to the concept of welfare loss due to monopoly power, which is summarized as the production of too little output at too high a price.

Stock margin requirements: The percentage of the total price of a stock that must be paid in cash.

Store of value: A characteristic of money or any other commodity that allows it to maintain or increase its worth over time.

Strike: A refusal to work at the current wage or under current conditions.

Structural unemployment: Unemployment that occurs when there are major and permanent changes in the demand for goods or technology that affect the demand for labor.

Subsidy: A government cash grant to a favored industry.

Subsistence wage: A wage that is just enough to sustain the current population.

Substitute inputs: Two or more inputs with a relation such that increasing the employment of one decreases the marginal product of the other.

Substitutes: Products that have a relation such that an increase in the price of one will increase the demand for the other or a decrease in the price of one will decrease the demand for the other; two goods whose cross elasticity of demand is positive.

Substitution effect: The change in the quantity demanded of a particular good that results from a change of its price relative to other goods.

Sunk costs: Past payment for a presently owned resource.

Supply of loanable funds: A curve or schedule that shows the various amounts of money that people are willing and able to lend (save) at all interest rates.

Supply schedule: A schedule or curve that shows the quantities of a product that producers are willing and able to offer at all prices.

Supply shock inflation: An increase in the price level that results from a decrease in aggregate supply, usually caused by a sudden decrease in the availability of an essential resource.

Supply-side economics: The attempt to stimulate real production by providing incentives to workers and businesses.

Supply-side policy: Fiscal or monetary policy that directs or changes the overall productive capacity of the economy by affecting work effort, savings, or investment.

Surplus: The amount by which quantity supplied exceeds quantity demanded when the price in a market is too high.

Tangency solution: A long-run situation in which the firm's downward-sloping demand curve is just tangent to the average total cost curve, necessarily implying zero economic profits.

Tariff: A tax on imported goods designed to maintain or encourage domestic production.

Taxable income: The amount of income that is subject to income taxes; total income minus deductions and exemptions.

Tax incidence: The burden of a tax or fiscal resting place.

Tax rate: The percentage of income that is devoted to taxes.

Team production: An economic activity in which workers must cooperate, as team members, to accomplish a task.

Terms of trade: The ratio of exchange between two countries, based on the relative opportunity costs of production in each country; the price ratio or range of price ratios at which two entities are likely to trade.

Terms of trade argument: The use of export restrictions on goods in an effort to increase a country's monopoly or monopsony power in international markets.

Total compensation: The lifetime income that an individual receives from employment in a particular occupation, including all monetary and nonmonetary pay.

Total cost: All the costs of a firm's operations, including fixed and variable costs.

Total cost of production: The value of all resources used in production; explicit plus implicit costs.

Total product: The total amount of output that results from a specific amount of input.

Total revenue: The total amount of money received by a firm from selling its output in a given time period; price times quantity sold.

Total utility: The total amount of satisfaction obtained from the consumption of a particular quantity of a good; a summation of the marginal utility obtained from consuming each unit of a good.

Total wage bill: The total cost of labor to firms, equal to wage times total quantity of labor employed; the total income of all workers.

Trade-off: A situation in which the attainment of one desirable goal necessarily implies the loss of another desirable goal.

Traditional society: An economic system in which the "what," "how," and "for whom" questions are determined by customs and habits handed down from generation to generation.

Transaction costs: The value of resources used to make a purchase, including time, broker's fees, contract fees, and so on.

Transactions accounts: Any account, such as a demand or checkable deposit account, against which checks may be written or funds transferred.

Transactions demand: The amount of money desired by businesses and consumers to facilitate the purchase of goods and services.

Transfer payment: The transfer by government of income from one individual to another; it may take the form of cash or goods and services such as education, housing, health care, or transportation.

Transitivity of preferences: A rational characteristic of consumers that suggests that if A is preferred to B and B is preferred to C, then A is preferred to C.

Transportation costs: The value of resources used in the transportation of goods.

Trust: An institution that organizes firms in the same industry in an effort to increase profits by decreasing competition.

Unemployment of resources: A situation in which human or nonhuman resources that can be used in production are not so used.

Unemployment rate: The number of people that are unemployed divided by the number of people in the labor force; the percentage of the labor force that is unable to find jobs.

Unit elasticity of demand: A situation in which the percent change in quantity demanded is equal to the percent change in price: $\epsilon_d = 1$.

Unit of account: A common measure, such as dollars and cents, that expresses the relative values of goods and services.

Unlimited liability: A legal term that indicates that the owner or owners of a firm are personally responsible for the debts of a firm up to the total value of their wealth.

Utility: The ability of a good to satisfy wants; the satisfaction obtained from the consumption of goods.

Value added: The increase in the value or worth of a good at each stage of production.

Variable costs: Payments made to variable inputs that necessarily change as output changes.

Variable input: Factors of production whose quantity may be changed as output changes in the short run.

Vertical equity: A tax structure under which people with unequal incomes pay unequal taxes; people with higher incomes pay more taxes than people with lower incomes.

Vicious circle of poverty: The idea that countries are poor because they do not save and invest in capital goods and that they cannot save and invest because they are poor.

Wage-price spiral: A process in which increased product prices cause increased wages, increased wages cause increased prices, and so on. For the wage-price spiral to continue, the money supply must increase or real output must fall.

Wealth: The stock of monetary and nonmonetary assets.

Welfare loss: The consumer surplus that is lost to consumers but not transferred to the monopoly in profits when a competitive industry is monopolized.

Welfare loss due to monopoly: The lost consumers' surplus resulting from the restricted output of a monopoly firm.

x-axis, y-axis: Perpendicular lines in a coordinate grid system for mapping variables on a two-dimensional graph. The x-axis is the horizontal line; the y-axis is the vertical line. The intersection of the x- and y-axes is the origin.

X-inefficiency: The increase in costs of a monopoly resulting from the lack of competitive pressure to force costs to the minimum possible level.

Zero economic profits: The condition that faces the purely competitive firm in the long run; long-run equilibrium in a competitive industry leads to a condition where $P = MR = MC = LRATC$, which means that firms in the industry earn just a normal rate of return on their investment, or a zero economic profit.

Index

U.S. Employment, Wages, and Productivity

Year	Population (millions)	Civilian[a] Labor Force (millions)	Civilian Unemployment (millions)	Civilian Unemployment Rate (percent)	Index of Average Hourly Earnings, (1977 = 100)	Index of Output per Hour, (1977 = 100)	Percent Change in Output per Hour (1977 = 100)
1929	121.8	49.2	1.6	3.2			
1933	125.6	51.6	12.8	24.9			
1939	130.9	55.2	9.5	17.2			
1940	132.1	55.6	8.1	14.6			
1941	133.4	55.9	5.6	9.9			
1942	134.9	56.4	2.7	4.7			
1943	136.7	55.5	1.1	1.9			
1944	138.4	54.6	.7	1.2			
1945	139.2	53.9	1.0	1.9			
1946	141.4	57.5	2.3	3.9			
1947	144.1	60.2	2.4	3.9	58.5	43.7	
1948	146.6	60.6	2.3	3.8	58.9	46.1	
1949	149.2	61.3	3.6	5.9	62.3	46.7	
1950	152.3	62.2	3.3	5.3	64.0	50.4	
1951	154.9	62.0	2.1	3.3	63.6	51.8	2.8
1952	157.6	62.1	1.9	3.0	65.5	53.5	3.2
1953	160.2	63.0	1.8	2.9	68.7	55.2	3.2
1954	163.0	63.6	3.5	5.5	70.5	56.1	1.6
1955	165.9	65.0	2.9	4.4	73.3	58.3	4.0
1956	168.9	66.6	2.8	4.1	75.9	58.9	1.0
1957	172.0	66.9	2.9	4.3	76.9	60.4	2.5
1958	174.9	67.6	4.6	6.8	78.0	62.3	3.1
1959	177.4	68.4	3.7	5.5	80.0	64.3	3.2
1960	180.7	69.6	3.9	5.5	81.4	65.2	1.5
1961	183.7	70.5	4.7	6.7	83.0	67.4	3.3
1962	186.5	70.6	3.9	5.5	85.0	69.9	3.8
1963	189.2	71.8	4.1	5.7	86.3	72.5	3.7
1964	191.9	73.1	3.8	5.2	87.5	75.6	4.3
1965	194.3	74.5	3.4	4.5	89.0	78.3	3.5
1966	196.6	75.8	2.9	3.8	90.3	80.8	3.1
1967	198.7	77.3	3.0	3.8	92.2	82.6	2.3
1968	200.7	78.7	2.8	3.6	94.0	85.3	3.3
1969	202.7	80.7	2.8	3.5	95.0	85.5	.2
1970	205.1	82.8	4.1	4.9	95.7	86.2	.8
1971	207.7	84.4	5.0	5.9	98.3	89.3	3.6
1972	209.9	87.0	4.9	5.6	101.2	92.4	3.5
1973	211.9	89.4	4.4	4.9	101.1	94.8	2.6
1974	213.9	91.9	5.2	5.6	98.3	92.5	−2.4
1975	216.0	93.8	7.9	8.5	97.6	94.6	2.2
1976	218.0	96.2	7.4	7.7	99.0	97.6	3.3
1977	220.2	99.0	7.0	7.1	100.0	100.0	2.4
1978	222.6	102.3	6.2	6.1	100.5	100.5	.5
1979	225.1	105.0	6.1	5.8	97.4	99.3	−1.2
1980	227.8	106.9	7.6	7.1	93.5	98.8	−.5
1981	230.0	108.7	8.3	7.6	92.6	100.7	1.9
1982	232.3	110.2	10.7	9.7	93.4	100.9	.2
1983	234.5	111.6	10.7	9.6	94.8	103.7	2.7
1984[b]	236.7	113.5	8.5	7.5	94.7	107.4	3.6

[a]1929–1947 persons 14 and over
1948–1984 persons 16 and over
[b]preliminary estimate

Source: Economic Report of the President, 1985, Tables B–28, B–29, B–33, B–41, B–38.